Keys to the Cestode Parasites of Vertebrates

CABI is a trading name of CAB International

CABI Head Office
Nosworthy Way
Wallingford
Oxfordshire OX10 8DE
UK

Tel: +44 (0)1491 832111
Fax: +44 (0)1491 833508
Email: cabi@cabi.org
Web site: www.cabi.org

CABI North American Office
875 Massachusetts Avenue
7th Floor
Cambridge, MA 02139
USA

Tel: +1 617 395 4056
Fax: +1 617 354 6875
Email: cabi-nao@cabi.org

A catalogue record for this book is available from the British Library, London, UK

ISBN-13: 978-0-85198-879-5
ISBN-10: 0-85198-879-2

First published 1994
Transferred to print on demand 2006

Printed and bound in the UK by CPI Antony Rowe, Eastbourne.

Contents

Contributors

Dr Karin Andersen, *University of Oslo, Zoological Museum, Sars Gate 1, N-0562 Oslo 5, Norway.*

Dr Ian Beveridge, *Veterinary Clinical Centre, School of Veterinary Science, University of Melbourne, Princes Highway, Werribee, Victoria 3030, Australia.*

Professor Franco V. Bona, *Università di Torino, Dipartimento di Biologia Animale, Via Accademia Albertina 17, 10123 Torino, Italy.*

Dr Rodney A. Bray, *Department of Zoology, The Natural History Museum, Cromwell Road, London SW7 5BD, UK.*

Dr Ronald A. Campbell, *Department of Biology, University of Massachusetts Dartmouth, North Dartmouth, Massachusetts 02747–2300, USA.*

Professor Bogdan Czaplinski, *Medical School, Institute of Biostructure, Department of General Biology and Parasitology, Chalubinskiego 5, 02–004 Warsaw, Poland.*

Professor Louis Euzet, *Laboratoire de Parasitologie comparée, Université Montpellier II, Station Méditerranéenne de l'Environment Littoral, 1 Quai de la Daurade, 34200 Sète, France.*

Dr Boyko B. Georgiev, *Bulgarian Academy of Sciences, Institute of Parasitology, Acad. G. Bonchev Street, Bl.25, 1113 Sofia, Bulgaria.*

Dr David I. Gibson, *Department of Zoology, The Natural History Museum, Cromwell Road, London SW7 5BD, UK.*

the editors have had to maintain careful control over the length of chapters, which has led to necessary restrictions in some cases, for which we apologise.

Keys to the cestode orders and to the families of the Order Cyclophyllidea have been compiled by the editors with reference to the definitions provided by the authors.

The names of genera which do not appear in the Keys as valid genera or synonyms are presented in Chapter 31. While every effort has been made by the editors and contributors to ensure that all cestode generic names have been incorporated, some came to our attention too late for inclusion or have been inadvertently omitted. Others are larval or collective names, or names derived from the medical literature which have not been used in a Linnaean sense. Genera clearly defined in the text as *genera incertae sedis, genera inquirenda*, etc., are listed under the appropriate headings and text-references to them can be located by using the index.

A cumulative bibliography is presented at the end of the book. Only those references cited in the text are included; taxon authorities have been omitted.

A glossary based on definitions provided by the contributors is presented in order to explain and clarify, as far as possible, the terminology used. It should be noted, however, that the same term may have slightly different meanings when applied to different groups. Contributors have also provided some relevant explanations in the text pertaining to their groups such as left/right positions, dorsal/ventral and other information.

While work on the Keys was in progress, geopolitical changes in Europe and the ex-USSR have resulted in the adoption of different names by several states. Where country names appear in the text, they are those originally submitted and in some instances may differ from the name now in use.

The editors would like to offer their sincere thanks to all the authors for their contributions and to the many organizations and individuals who kindly provided specimens, references and other information.

A work of this nature which results in many new ideas and values is inevitably controversial and will provoke criticism from some cestodologists. We hope, however, that this will stimulate meaningful discussions and an improved understanding of cestode taxonomy.

L.F. KHALIL

A. JONES

R.A. BRAY

Acknowledgements

Many individuals and organizations have helped in the production of this volume.

The editors wish to extend their sincere thanks to all the authors for the time and effort they have devoted to this project. Their participation is warmly appreciated. At the International Institute of Parasitology, special thanks to Mrs Ann Hall who keyboarded the manuscript, to Mrs Olive Linton for assistance with the illustrations and to Mrs Christine Small for help with literature requisitions.

Many contributors wish to thank the following people for lending specimens for study: Mrs E.A. Harris, Dr D.I. Gibson and Dr R.A. Bray (The Natural History Museum, London); Dr J.R. Lichtenfels and Mrs P. Pillitt (US Department of Agriculture, Beltsville); Dr Cl. Vaucher (Muséum d'Historie Naturelle de Genève); Dr J. Mariaux (Université de Neuchâtel); Dr B.B. Georgiev (Institute of Parasitology, Sofia); Dr V.V. Kornyushin (Institute of Zoology, Kiev); Dr A. Jones and Dr L.F. Khalil (International Institute of Parasitology, St Albans); and the staff of the Museum d'Histoire Naturelle, Paris.

Mrs E. Gascoigne, Dr B.B. Georgiev and Mrs G. Bishop kindly translated Russian papers for Drs A. Jones and R.A. Bray.

Dr B.B. Georgiev acknowledges financial support from the National Scientific Research Foundation of the Republic of Bulgaria, Grant No. B-44/1991. Dr E.P. Hoberg's work was supported in part by the Natural Sciences and Engineering Research Council of Canada through an Operating Grant in Population Biology.

Editors and contributors also wish to acknowledge the use of the Host-Catalogue compiled by the Parasitic Worms Division, Department of Zoology, The Natural History Museum, London, from which lists of generic names were compliled.

1 Key to the Orders of the Cestoda

A. JONES[1], R.A. BRAY[2] & L.F. KHALIL[1]

[1]*International Institute of Parasitology, 395A Hatfield Road, St Albans, Herts AL4 0XU, UK:* [2]*Department of Zoology, The Natural History Museum, Cromwell Road, London SW7 5BD, UK.*

1a. External segmentation absent. Only one set of reproductive organs .... 2.
1b. External segmentation present or absent. More than one set of reproductive organs in each strobila .... 4.

2a. Genital pores near posterior end of body .... 3.
2b. Genital pores near anterior end of body. Cylindrical or funnel-like opening, often frilled, at posterior end of body .... Gyrocotylidea Poche, 1926. (Chapter 3.)

3a. Embryo with six hooks. Uterus coiled in posterior region of body. Scolex simple, with shallow grooves, loculi or frills .... Caryophyllidea van Beneden in Carus, 1863. (Chapter 5.)
3b. Embryo with ten hooks. Long uterus loops throughout length of body. Small proboscis or sucker-like apical organ present .... Amphilinidea Poche, 1922. (Chapter 2.)

4a. Scolex feebly developed or with funnel- or cup-shaped apical organ. No external segmentation; several sets of reproductive organs arranged serially .... Spathebothriidea Wardle & McLeod, 1952. (Chapter 4.)
4b. Scolex with suckers, bothria, bothridia or tentacles. External segmentation usually present, occasionally incomplete, exceptionally absent; usually one set of gonads per segment, occasionally two .... 5.

5a. True scolex with four tentacles .... 6.
5b. True scolex usually lacks tentacles, suckers usually present; exceptionally, scolex has more than four tentacles and has four suckers .... 7.

6a. Tentacles conspicuously armed with hooks. Scolex bears two or four bothridia. True scolex retained throughout life .... Trypanorhyncha Diesing, 1863. (Chapter 7.)

6b. Tentacles unarmed. True scolex retained only in primary strobila which fragments to form secondary strobila with pseudoscolex lacking tentacles .... Haplobothriidea Joyeux & Baer, 1961. (Chapter 11.)

7a. Scolex with single terminal sucker .... Nippotaeniidea Yamaguti, 1939. (Chapter 12.)

7b. Scolex with more than one sucker, with bothria or with bothridia .... 8.

8a. Scolex in two parts: anterior portion variable with muscular sucker, myzorhynchus or crown of unarmed tentacles; posterior portion has four suckers .... Lecanicephalidea Baylis, 1920. (Chapter 9.)

8b. Scolex not in two parts; with bothridia, bothria or suckers .... 9.

9a. Vitellarium compact .... 10.

9b. Vitellarium follicular .... 11.

10a. Vitellarium anteroventral to ovary. Bothridia usually rectangular to round, flat or sucker-like, with auriculate muscular appendages; bothridia rarely absent .... Tetrabothriidea Baer, 1924. (Chapter 14.)

10b. Vitellarium usually posterior to ovary. Scolex usually with four acetabulate suckers .... Cyclophyllidea van Beneden in Braun, 1900. (Chapters 15–30.)

11a. Scolex with four (occasionally five) suckers; suckers rarely absent .... Proteocephalidea Mola, 1928. (Chapter 13.)

11b. Scolex with bothria or bothridia .... 12.

12a. Scolex with two bothria; occasionally replaced by pseudoscolex .... Pseudophyllidea Carus, 1863. (Chapter 10.)

12b. Scolex with bothridia .... 13.

13a. Scolex with two or four bothridia; no armed rostellum. Genital pores lateral .... Tetraphyllidea Carus, 1863. (Chapter 8.)

13b. Scolex with two bothridia; with or without armed rostellum. Genital pores ventral .... Diphyllidea van Beneden in Carus, 1863. (Chapter 6.)

2 Order Amphilinidea Poche, 1922

D.I. Gibson

Department of Zoology, The Natural History Museum, Cromwell Road, London SW7 5BD, UK.

Introduction

The Amphilinidea is a small group of monozoic cestodes generally considered distinct from the eucestodes. They occur as adults in the body-cavity of chondrosteans, primitive freshwater teleosts, advanced marine teleosts and freshwater chelonians and use crustaceans as the intermediate hosts. Early revisions of the group include those of Poche (1922, 1926), and in recent years detailed studies have been carried out by Dubinina (1982a) and Bandoni & Brooks (1987a). Dubinina's work is a detailed monograph in which she recognizes six genera in four families, whereas that of Bandoni & Brooks is a rigorous application of the cladistic method, accepting only three genera in one family split into two subfamilies. In another, less critical, classification of the cestodes as a whole Schmidt (1986) has accepted eight genera in two families within this group. The phylogeny and other aspects of the biology of the amphilinideans have also been discussed by Xylander (1986).

The classification presented below falls somewhere in between those of Dubinina (1982a) and Bandoni & Brooks (1987a), although recognizing the same six genera as the former. The justification for the synonymy and acceptance of taxa in this work is based on the common and unique characters used in the keys. This classification differs from Dubinina's classification in that *Gephyrolina* is considered a member of the Schizochoeridae rather than the Amphilinidae, because they are parasites of teleosts rather than chondrosteans, the vagina does not cross the male duct and because Dubinina lays particular emphasis on the small size of the seminal receptacle which, according to Bandoni & Brooks, is a secondary reduction. Bandoni & Brooks considered *Gephyrolina* a synonym of *Schizochoerus* in spite of differences in the shape of the ovary, the number of vaginal pores and the size of the seminal receptacle. In Dubinina's classification, *Gigantolina* is recognized as a separate monotypic family on the basis of the presence of a common genital pore, while Bandoni & Brooks considered it the junior synonym of *Austramphilina*. In this work, it is accepted as a valid genus within the subfamily Austramphilininae. This classification also differs from that of Bandoni & Brooks in that, following Dubinina, *Nesolecithus* is recognized and not considered a synonym of *Schizochoerus*. Although the distinguishing characters, body-shape and the size of the seminal receptacle, may not be considered by some of generic significance, they are very distinctive.

Although it is widely accepted that the amphilinideans are a primitive group of cestodes, their origin has been the source of much speculation. Janicki (1930) suggested that adult amphilinideans are paedomorphic forms of polyzoic cestodes that have lost their plesiomorphic definitive hosts. Dubinina (1960) considered that the retention of larval hooks in the adult stage indicated a direct development from the Monogenea, and later (1982a) suggested that they were originally parasites of crustaceans. More recently, Gibson *et al.* (1987) have reviewed the theories on the origins of this group and consider the subject still problematical, although suggesting that the wide diversity of definitive hosts relative to other primitive members of the Cercomeromorphae and the site in the body-cavity might indicate derivation from paedomorphic eucestode plerocercoids. Nevertheless, in the same year Bandoni & Brooks (1987a), following Brooks *et al.* (1985), have considered the monozoic condition of the amphilinideans to be plesiomorphic and have suggested that they evolved in association with freshwater teleosts (these authors appear to include chondrosteans as teleosts). Preliminary DNA sequencing indicates that the Amphilinidea is a sister group of the Eucestoda (see Baverstock *et al.*, 1991).

The narrow host-specificity of extant amphilinideans indicates a long association with their hosts. Consequently, the presence of monotypic taxa within the group, caused by both host and parasite extinctions, is not surprising. As indicated by Gibson *et al.* (1987), the close association between the parasite and its host and the survival of the host in a little-changing environment has meant that selection pressure for change has been low. This is exemplified by the two species of *Nesolecithus* from osteoglossomorph fishes, one living in Brazil and the other in the zoogeographically adjacent region of west Africa: although these two species are morphologically very similar, it is 100 million years since these regions separated.

ORDER AMPHILINIDEA POCHE, 1922

Diagnosis: Body monozoic; foliiform to elongate; lacking distinct organs of attachment, although small sucker-like organ may be present at anterior extremity. Testes follicular in narrow, lateral, pre-ovarian bands. Cirrus-sac absent. Genital pores usually separate, occasionally male duct and vagina unite to form single pore; male and vaginal pores at or near posterior extremity. Ovary in posterior region of body; of various shapes. Vaginal pore(s) close to posterior extremity, ventral when single; vagina occasionally bifid distally opening through dorsal and ventral pores. Seminal receptacle minute to enormous; oval to elongate. Uterus long, tubular; N-shaped or looped; extending from ovary to anterior extremity; uterine pore at anterior extremity. Eggs unoperculate. Cercomer of larva with ten hooks; hooks may be retained as vestige in adult. Vitellarium follicular in lateral bands between testicular bands and body margins. Osmoregulatory system with two lateral vessels opening posteriorly through common pore. In body-cavity of chondrosteans, primitive freshwater teleosts, advanced marine teleosts and freshwater chelonians.

Key to families

1a. **Parasites of chondrosteans in the Holarctic region; vagina crosses male duct; terminal limb of uterus excessively convoluted relative to the other two limbs (Figs 2.1, 2.2) .... Amphilinidae Claus, 1879.**

1b. **Parasites of teleosts and chelonians mainly in or between southern continents derived from Gondwanaland; vagina not crossing male duct; terminal limb of uterus not excessively coiled relative to the other two limbs (Figs 2.3, 2.7) .... Schizochoeridae Poche, 1922.**
(Syns Austramphilinidae Johnston, 1931; Gigantolinidae Dubinina, 1982.)

Family Amphilinidae Claus, 1879

Diagnosis: Body fusiform, with blunt extremities. Male and vaginal pores separate; single vaginal pore; vagina crosses male duct. Ovary lobed. Seminal receptacle small. Uterus N-shaped, with terminal limb excessively coiled and descending limb adjacent to first ascending limb. In body-cavity of Holarctic chondrosteans. Type and only genus *Amphilina* Wagener, 1858.

Genus *Amphilina* Wagener, 1858 (Figs 2.1, 2.2)

(Syn. *Aridmostomum* Grimm, 1871)

Diagnosis: With characters of family. Type-species *A. foliacea* (Rudolphi, 1819).

Family Schizochoeridae Poche, 1922

(Syns Austramphilinidae Johnston, 1931; Gigantolinidae Dubinina, 1982)

Diagnosis: Body foliiform or elongate. Male and vaginal pores usually separate, occasionally united; vagina with single pore or bifid distally and opening via ventral and dorsal pores; vagina does not cross male duct. Ovary elongate, subglobular, reniform or bilobed. Seminal receptacle large or minute. Uterus N-shaped or looped; terminal limb not excessively coiled; descending limb usually (but not always) not adjacent to first ascending limb. In body-cavity of teleosts and chelonians in or between continents derived from Gondwanaland. Type-genus *Schizochoerus* Poche, 1922.

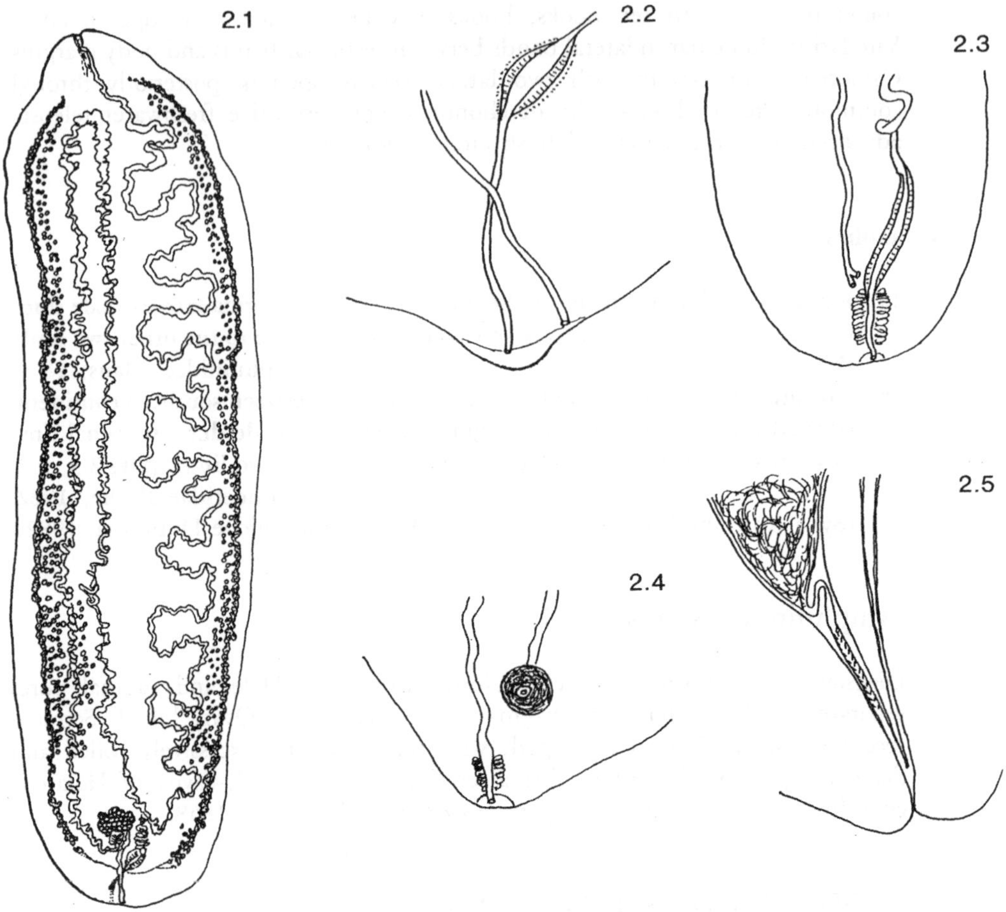

Fig. 2.1 *Amphilina japonica* Goto & Ishii, 1936, modified after Dubinina (1982a).
Figs 2.2–2.5 Posterior extremity (all redrawn after Dubinina, 1982a). 2.2. *Amphilina japonica* Goto & Ishii, 1936; 2.3. *Schizochoerus liguloideus* (Diesing, 1850); 2.4. *Gigantolina magna* (Southwell, 1915); 2.5. *Austramphilina elongata* Johnston, 1931.

Key to subfamilies

1a. Parasites of primitive freshwater teleosts in South America, Africa and India; uterus N-shaped (with terminal ascending limb lateral), narrow, tubular; seminal receptacle large, elongate, formed as anterior extension of vagina, or minute (Figs 2.6, 2.7, 2.8) .... Schizochoerinae Poche, 1922.
(Syn. Pseudogephyrolininae Gupta & Singh, 1992.)

1b. Parasites of advanced marine teleosts or chelonians in Indo-Australian region; uterus looped (with terminal limb median); uterus dilate; seminal receptacle oval to reniform, formed as median enlargement of vagina (Figs 2.9, 2.10) .... Austramphilininae Johnston, 1931.

Subfamily Schizochoerinae Poche, 1922

(Syn. Pseudogephyrolininae Gupta & Singh, 1992)

Diagnosis: Body foliiform or elongate. Male and vaginal pores separate; vagina with single pore or bifid distally and opening via dorsal and ventral pores. Ovary elongate or subglobular. Seminal receptacle large and elongate or minute. Uterus N-shaped with terminal ascending limb lateral; narrow, tubular. In body-cavity of primitive freshwater teleosts of South America, Africa and the Indian subcontinent.

Key to genera

1a. **Parasites of osteoglossomorph fishes of Brazil and West Africa; vagina bifurcates at distal extremity and opens on both dorsal and ventral surfaces; descending limb of uterus adjacent to terminal limb; ovary subspherical; seminal receptacle large, elongate, anterior to ovary (Figs 2.3, 2.7, 2.8) .... 2.**

1b. **Parasites of siluriform fishes of Indian subcontinent; with posterior invagination and caudal lobes; ovary elongate; vagina with single pore; descending limb of uterus not adjacent to terminal limb; seminal receptacle minute (Fig. 2.6) .... *Gephyrolina* Poche, 1926. (Fig. 2.6)**

(Syns *Hunteroides* Johri, 1959; *Pseudogephyrolina* Gupta & Singh, 1992.)

Diagnosis: Body very elongate; posterior extremity with distinct median depression enveloped by small caudal lobes. Vaginal pore single. Ovary elongate. Seminal receptacle minute. Uterus with descending limb adjacent to first ascending limb. In body-cavity of siluriform fishes of Indian subcontinent. Type-species *G. paragonopora* (Woodland, 1923).

2a. **Parasites of osteoglossomorph fishes in Brazil and West Africa; body fusiform; seminal receptacle about one-sixth of body-length (Fig. 2.7) .... *Nesolecithus* Poche, 1922. (Fig. 2.7)**

Diagnosis: Body foliiform; without terminal depression. Vagina bifid distally, opening through both dorsal and ventral pores. Ovary subglobular. Seminal receptacle large, elongate, about one-sixth of body-length. Uterus with descending limb adjacent to terminal ascending limb. In body-cavity of *Arápaima* and *Gymnarchus*. Type-species *N. janickii* Poche, 1922.

2b. **Parasites of osteoglossomorph fishes in Brazil; body elongate; seminal receptacle about one-third of body-length (Fig. 2.8) .... *Schizochoerus* Poche, 1922. (Figs 2.3, 2.8).**

Diagnosis: Body very elongate, without terminal depression. Vagina bifid distally, opening through dorsal and ventral pores. Ovary subglobular. Seminal receptacle large, extremely elongate, reaching about one-third of body length. Uterus with descending limb adjacent to terminal ascending limb. In body-cavity of *Arapaima*. Type-species *S. liguloideus* (Diesing, 1850).

2.6 2.7 2.8

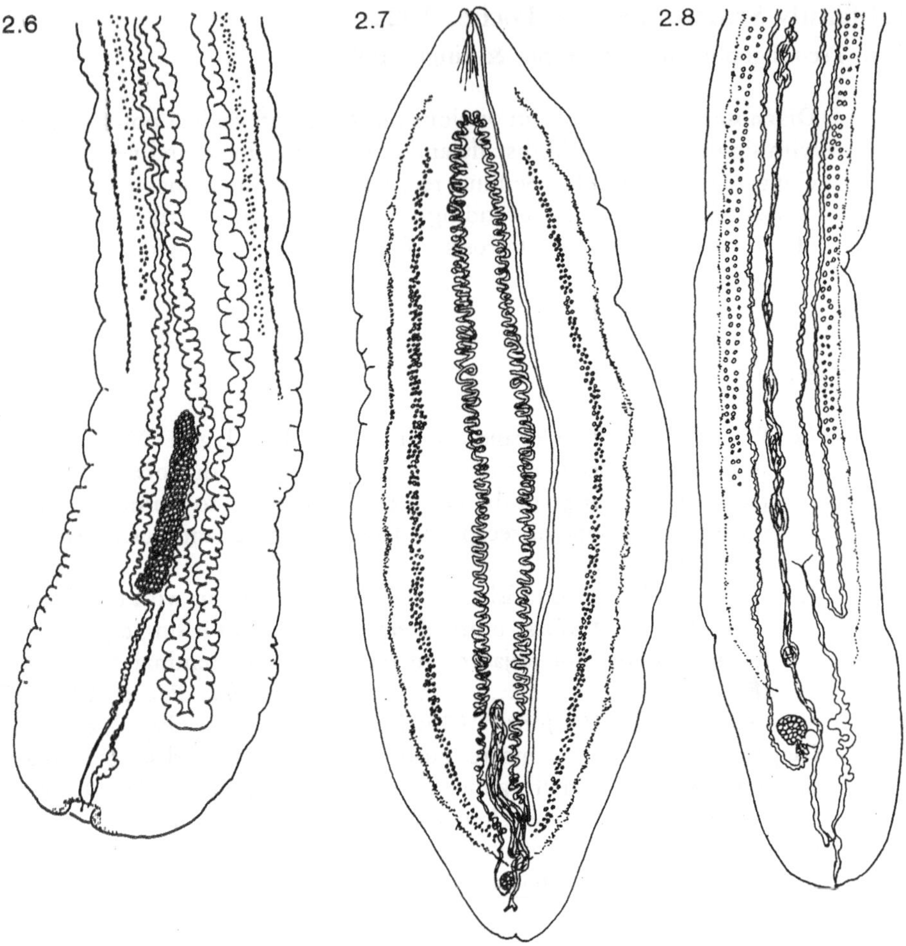

Fig. 2.6 *Gephyrolina paragonopora* (Woodland, 1923), modified after Bandoni & Brooks (1987a).
Fig. 2.7 *Nesolecithus janickii* Poche, 1922, modified after Dubinina (1982a).
Fig. 2.8 *Schizochoerus liguloideus* (Diesing, 1850), modified after Bandoni & Brooks (1987a).

Subfamily Austramphilininae Johnston, 1931

Diagnosis: Body elongate. Male and vaginal pores separate or united; vaginal pore single. Ovary subglobular to reniform or bilobed. Seminal receptacle large, saccular. Uterus looped, with terminal limb median. Parasitic in body-cavity of advanced marine (perciform) teleosts or freshwater chelonians of Indo-Australian region.

Key to genera

1a. **Parasites of perciform marine teleosts of the northern Indian Ocean and East Indies; male and female genital pores separate; posterior extremity with male papilla; vaginal pore muscular; ovary bilobed, lobes symmetrical; seminal receptacle anterior to ovary (Figs 2.4, 2.9) .... *Gigantolina* Poche, 1922. (Figs 2.4, 2.9)**
(Syn. *Gyrometra* Yamaguti, 1954.)

Diagnosis: Body large, very elongate; posterior margin bearing male papilla. Male and female pores separate. Vaginal pore muscular. Ovary bilobed; lobes symmetrical. Seminal receptacle anterior to ovary. Distinct duct (ductus yamagutii) of unknown function adjacent to terminal limb of uterus. In body-cavity of marine perciform teleosts from Indian Ocean and East Indies. Type-species *G. magna* (Southwell, 1915).

2.9

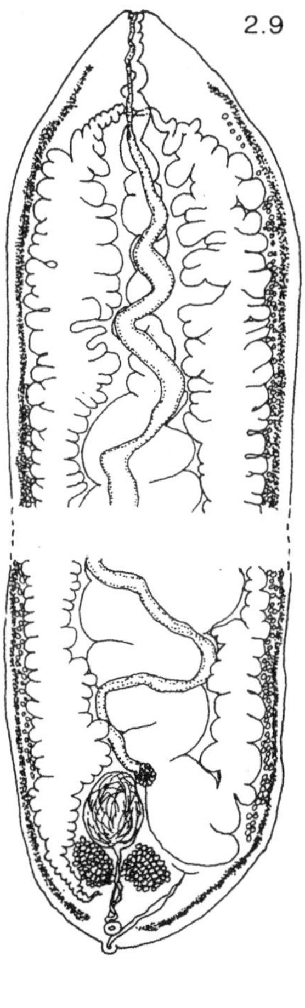

2.10

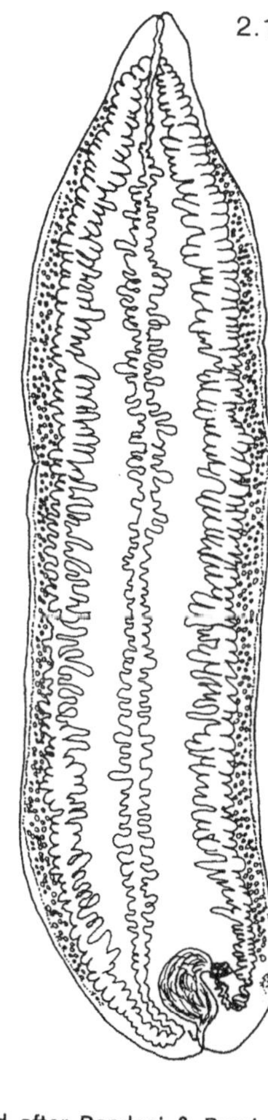

Fig. 2.9 *Gigantolina magna* (Southwell, 1915), modified after Bandoni & Brooks (1987a).
Fig. 2.10 *Austramphilina elongata* Johnston, 1931, modified after Bandoni & Brooks (1987a).

1b. Parasites of Australian freshwater chelonians; common genital pore present; vagina lined with spines over half its length; ovary irregularly reniform; seminal receptacle lateral to ovary (Figs 2.5, 2.10) .... *Austramphilina* Johnston, 1931. (Figs 2.5, 2.10)

(Syn. *Kosterina* Ihle & Ihle-Landenberg, 1932.)

Diagnosis: Body elongate. Posterior margin with slight median depression. Male duct and vagina united to open via common genital pore. Ovary irregularly subglobular. Seminal receptacle lateral to ovary. Ductus yamagutii absent. In body-cavity of Australian freshwater chelonians. Type-species *A. elongata* Johnston, 1931.

3 Order Gyrocotylidea Poche, 1926

D.I. GIBSON

Department of Zoology, The Natural History Museum, Cromwell Road. London SW7 5BD, UK.

Introduction

Gyrocotylideans are a primitive group of monozoic cestodes parasitic in the spiral valve of holocephalan fishes. In many phylogenetic classifications of the parasitic platyhelminths, including the recent cladistic analyses of Ehlers (1985), Brooks *et al.* (1985) and Brooks (1989), they are considered the most primitive group of cestodes and the most closely related to the Monogenea: indeed, they have been included in the Monogenea by some workers (e.g. Llewellyn, 1986) but this is not supported by preliminary DNA sequence analysis (Blair, 1993). According to Williams *et al.* (1987), the relationship between the gyrocotylideans and their holocephalan hosts has existed with little change for more than 350 million years and the origin of the group may extend back much further.

The nomenclatural and systematic history of the group is long and complicated. The most recent reviews are those of Williams *et al.* (1987) and Bandoni & Brooks (1987b). Problems have arisen because of the great morphological variability which occurs in terms of contractability and change of form in living material, artifactual changes caused by differences in fixation and the lack of meristic or other non-deformable characters. The group has 13 nominal species of which ten are generally considered valid as, for example, in Bandoni & Brooks' recent (1987b) revision of the group. A feature of the group is that only certain holocephalans serve as hosts and yet two or even three species often occur in the same host, although Colin *et al.* (1986) have cast doubts upon the validity of some of the sympatric forms (see below).

Poche (1926) considered the group to be composed of the single genus *Gyrocotyle* Diesing, 1850, but Joyeux & Baer (1951) recognized three genera: *Gyrocotyle* for large forms with a long, contractile posterior attachment organ bearing a small rosette, lacking lateral body folds and with body-spines restricted to two groups; *Amphiptyches* Grube & Wagener in Wagener, 1852 for small forms with the posterior attachment organ bearing a rosette, with fluted lateral body folds and with a wide distribution of body-spines; and *Gyrocotyloides* Fuhrmann, 1931 for large forms lacking a rosette on the posterior attachment organ, lacking lateral body folds and lacking body-spines. Although *Amphiptyches* was considered a subgenus of *Gyrocotyle* by van der Land & Dienske (1968), subsequent workers have tended to accept it as a synonym of the latter genus. The status of *Gyrocotyloides*, a monotypic genus, has been a matter of much debate over recent years, although recognized as valid as recently as 1986 by Schmidt. The controversy concerns the status of *Gyrocotyloides nybelini* Fuhrmann, 1931 and its relationship to *Gyrocotyle urna* (Grube & Wagener in Wagener, 1852), both parasites of *Chimaera monstrosa* off Norway. In a detailed study, Colin *et al.* (1986) came to the conclusion that both *Gyrocotyloides nybelini* and *Gyrocotyle confusa* van der Land & Dienske, 1968 are synonyms of *G. urna*, being within the limits of intra-specific variation of anomalous and/or damaged specimens of this species. This opinion has been disputed by Bandoni & Brooks (1987b) in their cladistic analysis of the group, by Bristow & Berland (1988) in a preliminary electrophoretic study, and by Berland *et al.* (1989) using fatty acid profiles. Although the validity of the species *Gyrocotyloides nybelini* is still in question, all of these authors are in agreement in considering *Gyrocotyloides* a synonym of *Gyrocotyle*.

ORDER GYROCOTYLIDEA POCHE, 1926

Diagnosis: Body monozoic, stout, fusiform to elongate. Anterior extremity with muscular sucker-like attachment organ. Posterior region of body attenuated and normally terminates in rosette-like adhesive organ (occasionally lost or ? lacking). Lateral margins fluted or not. Spines on body surface widely distributed, narrowly distributed or ? absent. Testes follicular, in two anterolateral fields. Cirrus-sac absent. Genital pores separate. Male pore ventromedial, between anterior extremity and uterine pore. Ovary follicular, in V- or U-shaped band around seminal receptacle; posterior to uterus. Vaginal pore dorsal, lateral to male pore. Vagina long, reaching back to open into seminal receptacle. Uterus coiled medially in middle of body between seminal receptacle and uterine pore; terminal part of uterus forms uterine sac; uterine pore ventromedial in anterior part of body. Eggs operculate; cercomer of larva bearing ten hooks. Vitellarium follicular, filling lateral regions of body. Osmoregulatory system reticulate, with two anterior pores. In spiral intestine of holocephalan fishes.

Family Gyrocotylidae Benham, 1901

Diagnosis: With characters of order. Type-genus *Gyrocotyle* Diesing, 1850.

3.1

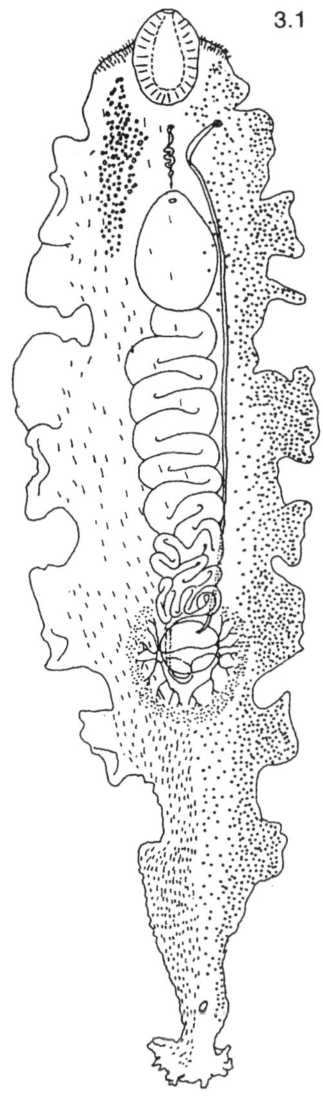

Fig. 3.1 Gyrocotylidean, dorsal view. Testes and body-spines shown only on left; vitelline follicles shown only on right. Modified after Williams *et al.* (1987).

Genus *Gyrocotyle* Diesing, 1850 (Fig. 3.1)
(Syns *Crobylophorus* Krøyer, 1852; *Amphiptyches* Grube & Wagener in Wagener, 1852; *Gyrocotyloides* Fuhrmann, 1931)

Diagnosis: With characters of family. Type-species *G. rugosa* Diesing, 1850.

4 Order Spathebothriidea Wardle & McLeod, 1952

D.I. GIBSON

Department of Zoology, The Natural History Museum, Cromwell Roaa, London SW7 5BD, UK.

Introduction

The spathebothriideans are a primitive group of cestodes parasitic in chondrosteans and teleosts which appear to be related to both the monozoic caryophyllideans and the polyzoic pseudophyllideans. This group is polyzoic in that serial multiplication of the reproductive system occurs down the strobila, but there is no distinct segmentation of the body.

Unfortunately, this group as a whole lacks a modern detailed study, the most recent being that of Nybelin (1922), whose figures are still copied in many works on the group. It is not known, for instance, whether the lack of segmentation is a primitive character or whether it is secondarily derived. The latter would seem more likely, since progenesis is widespread in the group, several forms being able to mature and become ovigerous in amphipod crustaceans. Mackiewicz (1981a), in his review of caryophyllidean systematics, indicated a position somewhere between the Caryophyllidea and the Pseudophyllidea, and Freeman (1973) indicated that the Spathebothriidea arose from 'protocestodes' independently of the Pseudophyllidea and gave rise to the Caryophyllidea.

The order was not recognized as being distinct from the Pseudophyllidea until the work of Wardle & McLeod (1952), who maintained that there was no evidence that the organs of attachment on the scolex were derived from the bothria typical of the pseudophyllideans. The group has not been universally accepted, still being considered as pseudophyllideans by Joyeux & Baer (1961) and Dubinina (1987). In his earlier revision of the group (as the family Cyathocephalidae), Nybelin (1922) included four genera, *Cyathocephalus* from freshwater and euryhaline teleosts (mainly salmonoids and percids), *Diplocotyle* from euryhaline and marine teleosts (mainly salmonoids and pleuronectids), *Bothrimonus* from chondrosteans and *Didymobothrium* from marine teleosts (soleids).

The justification for the synonymy and acceptance of taxa in this work is based on the common and unique characters used in the keys.

The appellation Acrobothriidae Olsson, 1872 (emended from 'Acrobothridea nova familia') is used as the oldest valid name for the family(ies) previously known as Diplocotylidae, Bothrimonidae and Cyathocephalidae. Although this family designation has been unused for 120 years, the next available name, Diplocotylidae, is not so well established as to warrant an application to the *ICZN* for establishment on the Official List.

The genus *Schyzocotyle* Akhmerov, 1960 was erected by Akhmerov (1960) as a member of the Cyathocephalidae and was accepted by Schmidt (1986), who placed it in the Bothrimonidae. Nevertheless, the genus is clearly pseudophyllidean in character and was considered a synonym of *Bothriocephalus* Rudolphi, 1808 by Dubinina (1982b).

ORDER SPATHEBOTHRIIDEA WARDLE & McLEOD, 1952

(Syn. Cyathocephalidea Renaud & Gabrion, 1988 ? *lapsus)**

Diagnosis: Body elongate-fusiform to taeniiform; dorso-ventrally flattened; polyzoic, with linear multiplication of reproductive organs throughout strobila, but without external evidence of segmentation. Distinct scolex differentiated or not; when distinguishable one or two sucker-like attachment organs present. Testes follicular; in continuous, irregular lateral bands throughout most of strobila. Cirrus-sac small to well developed; cirrus may be formed. Male and female pores close together, median, on same surface of body; pores irregularly alternate dorsal and ventral or just ventral. Vagina and uterus open posterior to male pore, together or through common atrium. Ovary rosette-shaped or bilobed; lobulate; median, posterior to genital pores. Vagina links vaginal pore and oviduct; may be enlarged proximally to form small seminal receptacle. Uterus tubular, convoluted; loops filling median field between ovary and genital pore, although they may extend anterior and/or posterior to these levels. Eggs operculate. Vitellarium follicular; in continuous lateral (cortical) fields between testes and body margins. Osmoregulatory system in form of lateral canals on each side of strobila; with irregular commissures and lateral pores. In intestine of freshwater, euryhaline and marine chondrosteans and teleosts from the Northern Hemisphere.

Key to families

1a. Worms lacking both a distinct scolex and any apparent organs of attachment (Fig. 4.4) .... Spathebothriidae Yamaguti, 1934.

1b. Worms with a differentiated scolex with one or two distinct sucker-like organs of attachment (Figs 4.1–4.3, 4.5) .... Acrobothriidae Olsson, 1872.
(Syns Bothrimonidae Perrier, 1897; Cyathocephalidae Lühe, 1899; Diplocotylidae Monticelli, 1892.)

* The name Cyathocephalidea was used by Renaud & Gabrion (1988) presumably in error for Cyathocephalidae.

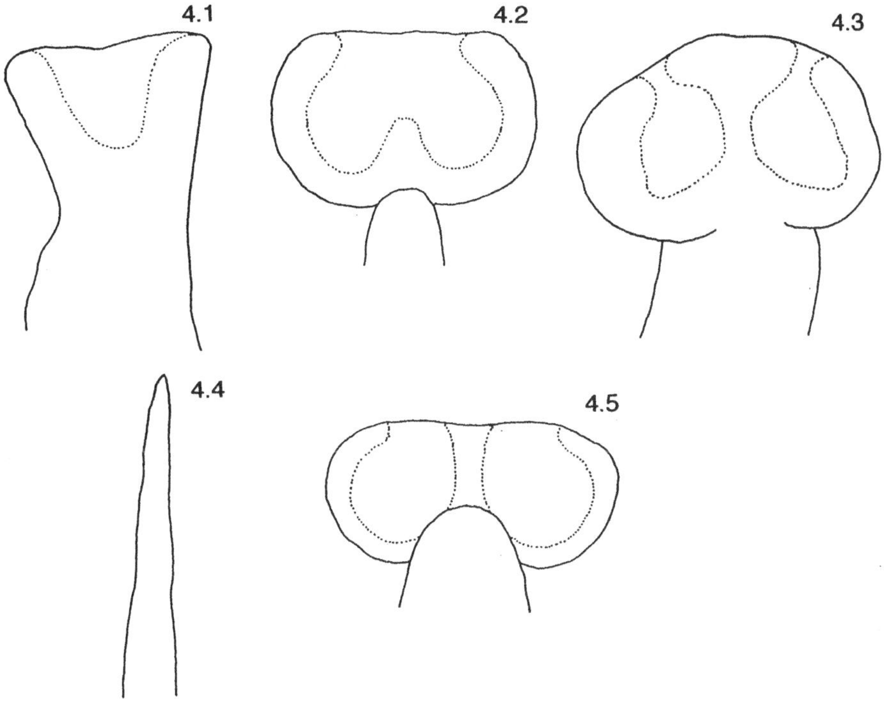

Figs 4.1-4.5 Scoleces. 4.1. *Cyathocephalus truncatus* (Pallas, 1781), original; 4.2 *Bothrimonus fallax* Lühe, 1900, modified after Nybelin (1922); 4.3. *Didymobothrium* sp., original; 4.4. *Spathebothrium simplex* Linton, 1922, modified after Hart & Guberlet (1936); 4.5. *Diplocotyle olrikii* Krabbe, 1874, modified after Nybelin (1922).

Family Spathebothriidae Yamaguti, 1934

Diagnosis: Body elongate-fusiform, but more attenuated posteriorly. No distinct scolex differentiated; organs of attachment lacking. Cirrus-sac small. Genital pores irregularly alternate. Uterine and vaginal pores adjacent. Ovary rosette-shaped. In intestine of marine teleosts. Type-genus *Spathebothrium* Linton, 1922.

Genus *Spathebothrium* Linton, 1922 (Fig. 4.4)

Diagnosis: With characters of family. Type-species *S. simplex* Linton, 1922.

Family Acrobothriidae Olsson, 1872

(Syns Bothrimonidae Perrier, 1897; Cyathocephalidae Lühe, 1899; Diplocotylidae Monticelli, 1892)

Diagnosis: Body elongate to taeniiform. Scolex differentiated; with one or two sucker-like organs of attachment. Cirrus-sac well developed; cirrus often formed. Genital pores irregularly alternate (but more often ventral than dorsal) or ventral. Vagina and uterus unite to form short, tubular genital atrium or open through common pore. Ovary bilobed; with smooth isthmus. In intestine of marine, euryhaline and freshwater teleosts and chondrosteans. Type-genus *Cyathocephalus* Kessler, 1868.

Key to genera

1a. With single, funnel-shaped organ of attachment. In freshwater and euryhaline teleosts, especially salmonoids (Fig. 4.1) .... *Cyathocephalus* Kessler, 1868. (Fig. 4.1)
(Syn. *Acrobothrium* Olsson, 1872.)

Diagnosis: Body elongate. Scolex separated from strobila by slight constriction; with funnel-shaped attachment organ. Genital pores irregularly alternate. Vagina and uterus open together through common pore. Ovarian lobes enveloping uterus. In Holarctic teleosts (especially salmonoids: there are also a few records from chondrosteans). Type-species *C. truncatus* (Pallas, 1781).

1b. With two forwardly directed, sucker-like attachment organs (Figs 4.2, 4.3, 4.5) .... 2.

2a. Lumina of attachment organs independent, separated by median septum; uterine fields wide, overlapping ovarian lobes. In teleosts (Figs 4.3, 4.5, 4.7) .... 3.

2b. Lumina of attachment organs united; uterine field narrow, enveloped laterally by ovarian lobes. In Holarctic chondrosteans (Figs 4.2, 4.6) .... *Bothrimonus* Duvernoy, 1842. (Figs 4.2, 4.6.)
(Syn. *Disymphytobothrium* Diesing, 1854.)

4.6

Fig. 4.6 *Bothrimonus fallax* Lühe, 1900, reproductive organs, modified after Nybelin (1922).

Fig. 4.7 *Diplocotyle olrikii* Krabbe, 1874, reproductive organs, modified after Nybelin (1922).

Diagnosis: Lumina of attachment organs completely fused, with only rudimentary septum at base. Genital pores irregularly alternate. Uterus and vagina unite to open through short, tubular atrium. Vitelline fields extend medially, partly covering ovarian lobes. In Holarctic chondrosteans. Type-species *B. sturionis* Duvernoy, 1842.

3a. **Genital pores ventral (Fig. 4.7)** .... ***Diplocotyle* Krabbe, 1874. (Figs 4.5, 4.7)**
Diagnosis: Lumina of attachment organs completely separated internally by median septum. Genital pores ventral. Vagina and uterus open adjacently through common pore. Uterine field broad, not enveloped by ovarian lobes. Vitelline follicles do not extend medially as far as ovarian lobes. In marine and euryhaline teleosts (especially salmonoids and pleuronectids). Type-species *D. olrikii* Krabbe, 1874.

3b. **Genital pores irregularly alternate on dorsal and ventral surfaces** .... ***Didymobothrium* Nybelin, 1922. (Fig. 4.3)**
Diagnosis: Lumina of attachment organs completely separated by median septum. Genital pores irregularly alternate. Vagina and uterus open adjacently through common pore. Uterine field broad, not enveloped by ovarian lobes. Vitelline follicles do not extend medially as far as ovarian lobes. In marine teleosts (soleids). Type-species *D. rudolphii* (Monticelli, 1890).

5 Order Caryophyllidea van Beneden in Carus, 1863

J.S. MACKIEWICZ

Department of Biological Sciences, State University of New York, Albany, New York 12222, USA.

Introduction

Caryophyllid cestodes are unique among the Cestoidea in having a monopleuroid body plan, i.e. without internal or external proglottidization and with a single set of reproductive organs. They thus superficially resemble cestodarians. In fact, Yamaguti (1959) considered them an order under the subclass Cestodaria Monticelli, 1892. Even *Hunteroides* Johri, 1959 (= *Gephyrolina* Poche, 1926), a true cestodarian, was mistakenly identified as a caryophyllidean cestode. Unlike the Cestodaria that have a 10-hooked lycophora, the caryophyllideans have a 6-hooked oncosphere, characteristic of the true cestodes or Cestoidea Rudolphi, 1808. However, the monopleuroid body plan has prompted some workers to view them as distinct from and of coequal systematic rank with the more dominant strobilate cestodes (Bazitov, 1976; Kulakovskaya & Demshin, 1978; and Ehlers, 1986). I support this view.

There are about 42 genera and 137 species in four families. They are widely distributed intestinal parasites of benthic-feeding siluriform and cypriniform freshwater fishes. Polyploidy is known for several species. One genus, *Archigetes* Leuckart, 1878, is commonly progenetic in tubificid annelids, the intermediate host for this well defined group of cestodes.

Classification at the family level has either had a single family with multiple subfamilies, or as followed here, the subfamilies have been raised to family rank. The primary family criterion is the arrangement of the longitudinal musculature (Figs 5.26–5.29) because it is a fundamental non-adaptive and conservative character, not significantly influenced by other anatomical features (Mackiewicz, 1982). There are many different scolex types (Figs 5.1–5.21).

Classification at the genus level reflects an ancient group where host and parasite isolation has played a major role in cestode evolution. As a consequence of this isolation, about 50% of genera are monotypic. Only four genera have 10 or more species and most of these, viz. *Lytocestus* Cohn, 1908, *Biacetabulum*

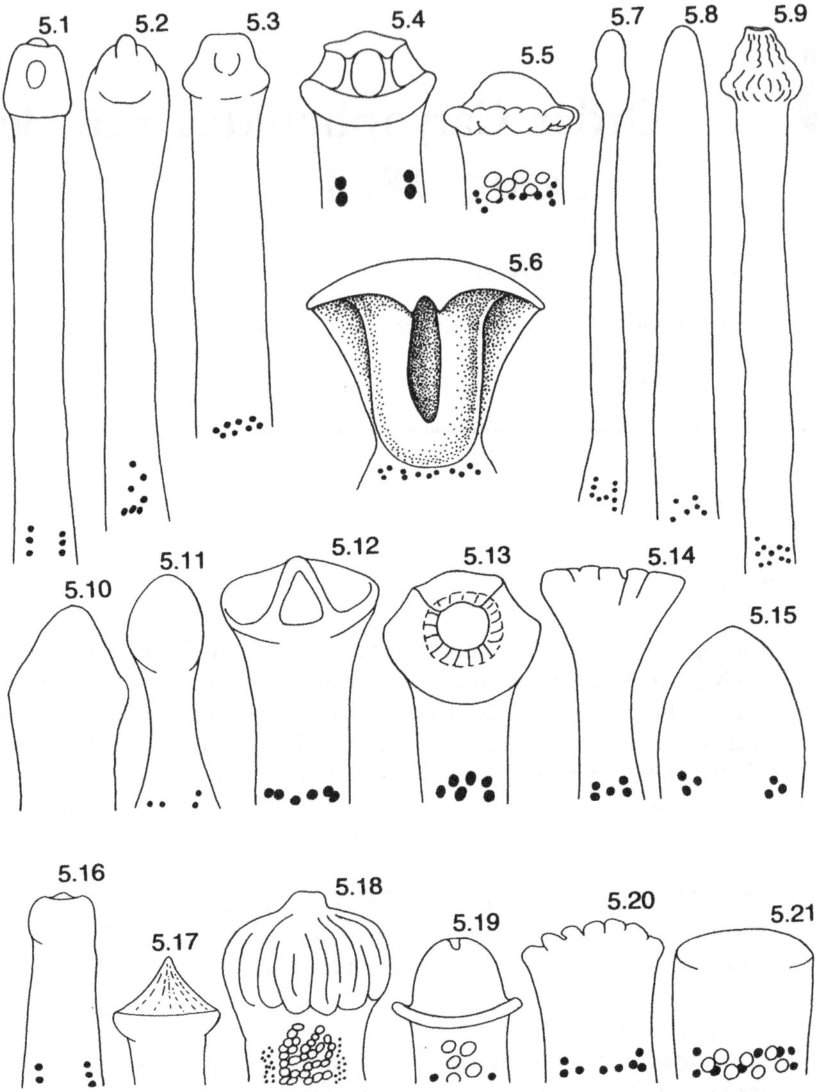

Figs 5.1-5.21 Scolex shapes and sample genera (original or after Mackiewicz, 1972, 1982, unless otherwise stated. 5.1. Loculomonobothriate, *Monobothrium*. 5.2. Loculopapillate, *Diefflluvium* (after Williams, 1978). 5.3. Loculotruncate, *Promonobothrium* (after Mackiewicz, 1968). 5.4. Bothrioloculodiscate, *Archigetes*. 5.5. Coronulate, *Balanotaenia*. 5.6. Fixomegabothriate, *Capingens*. 5.7. Spatulate, *Lytocestus*. 5.8. Digitiform, *Crescentovitus*. 5.9. Rugomonobothriate, *Monobothrioides*. 5.10. Hastate, (adapted from Hunter, 1930), *Pseudolytocestus*. 5.11. Bulbate, bulboacuminate, *Atractolytocestus*. 5.12. Cuneiloculate, *Glaridacris*. 5.13. Biacetabulate, *Biacetabulum*. 5.14. Cuneicrispitate, flabellate, *Caryophyllaeus*. 5.15. Tholate, *Hunterella*. 5.16. Monobothriate, *Monobothrium*. 5.17. Montanate, *Khawia* (= *Bothrioscolex*). 5.18. Rugomontanate, *Wenyonia*. 5.19. Choanocampanulate, *Caryoaustralus* (after Mackiewicz & Blair, 1980). 5.20. Cuneifimbriate, *Khawia*. 5.21. Cuneiform, *Caryophyllaeides*.

Hunter, 1927 and *Isoglaridacris* Mackiewicz, 1965 may reflect generic complexes. Our concept of the genus is far from stabilized. Perhaps introduction of the subgenus, not yet used in this group, may be helpful in partitioning some larger genera.

Many of the differences in systematic treatments of genera can be traced to various sources. Some are intrinsic to the organism and include variation associated with different states of maturity, different hosts and universal intraspecific variation. On the other hand, a factor such as small sample size may give a very limited view of a species but by far the two most important practices that have had a profound influence on and greatly hindered a genuine understanding of the systematics of caryophyllaeid cestodes have been (a) descriptions of new genera (or species) from specimens that have been dead in a host for a long period and may even be partially decomposed, and (b) the all too prevalent practice of compressing specimens between two glass slides prior to fixation. Consequently, natural characters are obliterated and, often, new fictitious ones created. While it is not always possible to collect adequate samples of specimens or even to observe them alive, new taxa should not be described from specimens that have been flattened or have come from hosts in which the gut has begun to decompose. Phantom genera born from such practices create subtle havoc and endless confusion whose legacy is a classification that is more illusion than reality.

Generic diagnoses in the present treatment contain the fewest number of pertinent characters (Mackiewicz, 1982) and thus should be considered as differential diagnoses. All species currently described in a genus may not fit within a diagnosis, hence reflecting systematic uncertainty. On the other hand, it is clear to me that some genera, as defined here, should be divided into one or more new genera. No new genera are proposed here; however, they will be erected elsewhere.

The following characters have been used in the keys to separate genera: scolex, vitelline follicle arrangement, ovarian morphology, gonopore condition and position, presence of external seminal vesicle, anterior extent of uterus and presence of postovarian vitelline follicles. The seminal receptacle is not used because its status in many genera is not known. Furthermore, it appears to be a transient structure visible only when full of sperm. The general topic of generic characters has been discussed by Mackiewicz (1969, 1982).

The following should be borne in mind when using these keys. Except for the massive, fixed scolex of *Capingens* (Fig. 5.6), the scolex of many other genera is highly mobile and capable of forming diverse shapes. Determination of gonopore condition can be difficult and is best done on well preserved, noncontracted specimens; unless the gonopore status is obvious, it is usually necessary to section specimens to clarify duct relationships. A single gonopore is formed when the cirrus (ejaculatory duct) joins the uterovaginal canal forming a hermaphroditic duct to the surface (Fig. 5.25) or when a deep, well defined genital atrium is formed. Gonopores are separate when (a) the cirrus and uterovaginal canal clearly open separately on the surface (Fig. 5.22), (b) when they open together into a shallow genital atrium as in some *Glaridacris* (Fig. 5.23) or when there is a distinct genital atrium (Fig. 5.24) as in *Khawia*. Although one can usually see both gonopores when a distinct genital atrium is present, the genital

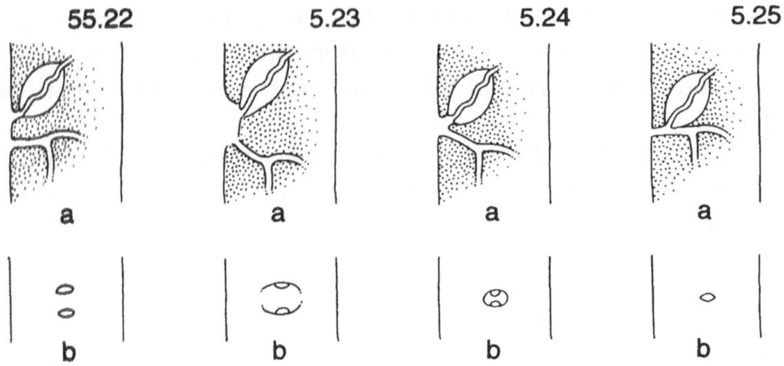

Figs 5.22–5.25 Gonopore status, (a) mid-sagittal and (b) surface view. 5.22. Separate. 5.23. Separate, shallow genital atrium. 5.24. Separate, distinct genital atrium. 5.25. Single, no genital atrium.

atrium itself may appear as a single, large gonopore if the specimen is even slightly contracted. If the maximum anterior extent of the uterus in mature worms is at the level of or only anterior to the cirrus-sac by the width of the uterus, then it is not considered to loop anterior to the cirrus-sac (Fig. 5.49). Preovarian vitelline follicles are designated as lateral and median when they are intermingled with or surround the testes (Figs 5.38, 5.47, 5.56, 5.60). In the diagnoses, 'lateral and median' means that the vitellarium is annular or the follicles are intermingled with testes because the two states can merge with each other (Figs 5.26–5.29). If the ovary is strongly follicular, the follicles of the posterior arms can easily be mistaken for postovarian vitelline follicles (Figs 5.40, 5.41). While the posterior ovarian follicles characteristically form a lateral cluster on each side above the excretory bladder (Fig. 5.40), postovarian vitelline follicles form a single median cluster (Figs 5.52, 5.65). If the ovary is strongly follicular and the posterior follicles form a single median cluster, special care must be taken to separate ovarian from vitelline follicles to determine if the ovary is shaped like an inverted letter A. In sections, or under high magnification, ovarian follicles lack the vacuolated nuclei of mature vitelline cells. Histochemically, ovarian follicles are negative for polyphenol oxidase, vitelline follicles are positive.

Terminology continues to cause confusion in this group of cestodes. Loculus and bothrium, for example, have been used interchangeably. To avoid further confusion in this key, a loculus is regarded as any shallow cavity that can appear or disappear, depending on scolex mobility. A species may have from one to three pairs of loculi (Figs 5.3, 5.4, 5.12). When a medial cavity is deeper, well developed and assumes a permanent form it becomes a bothrium (Figs 5.4, 5.6). As in the difossate scolex of the Pseudophyllidea, the bothria of caryophyllideans always occur as a single medial pair (Fig. 5.6) with or without accessory loculi (Fig. 5.4). It may be argued, and correctly so, that when there is only one pair of medial cavities or depressions they should be regarded as bothria, regardless of their state of development. While this may be technically correct, i.e. that medial loculi are just weakly developed bothria, it becomes difficult to apply this interpretation to *Isoglaridacris* or other genera with a wedge-shaped or cuneiform

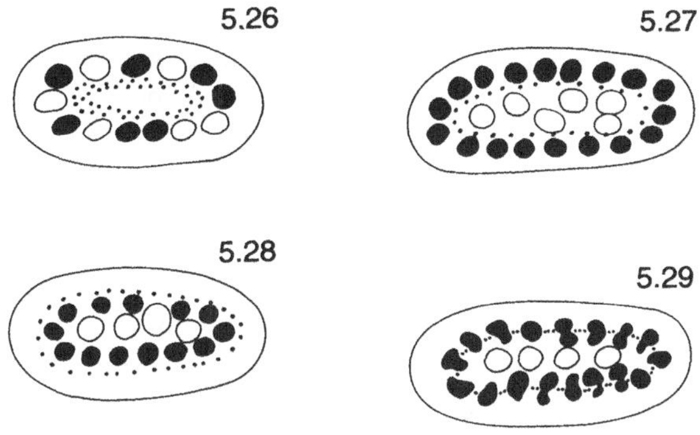

Figs 5.26–5.29 Placement of inner longitudinal muscles (vitelline follicles solid, testes open circles). 5.26. Balanotaeniidae. 5.27. Lytocestidae. 5.28. Caryophyllaeidae. 5.29. Capingentidae.

scolex bearing three pairs of shallow cavities or depressions (Fig. 5.12). All of these depressions appear similar in fixed material and thus are here regarded as loculi. To be consistent, all other shallow medial depressions on any other scolex are considered loculi.

Previously published regional keys include: USSR (Kulakovskaya, 1961), North America (Hoffman, 1967), Wisconsin, USA (Williams, 1977) and Iowa, USA (Williams, 1978). Keys to species of caryophyllidean genera include: *Khawia* by Calentine & Ulmer (1961); *Archigetes* by Kennedy (1965) and Calentine (1962); *Glaridacris* by Mackiewicz (1976); *Wenyonia* by Ukoli (1972); *Isoglaridacris* by Mackiewicz (1974); and *Monobothrium* by Calentine & Mackiewicz (1966).

Unless otherwise indicated, all figures are of the type of the genus or its character states. Partial figures of the posterior portion are of larger species (over 10 mm) and are diagrammatic, standardized, and not drawn to scale in order to emphasize generic characters. Smaller species, usually less than 10 mm, are pictured whole and with a scale. Since ovarian shape rather than structure is the primary focus, no attempt has been made to picture the compact ovarian follicles of many genera (Figs 5.51, 5.55). On the other hand, the loosely follicular ovary of other genera (Figs 5.39–5.41) is figured because it is so prominent as to be confused with vitelline follicles. Uterine glands have been omitted from all drawings in the interests of simplicity.

ORDER CARYOPHYLLIDEA VAN BENEDEN IN CARUS, 1863

Diagnosis: Elongate body without internal or external proglottization, with a single set of reproductive organs. Scolex unspecialized or with loculi, bothria, acetabula or a terminal introvert or disc; hooks absent. Genital pores ventral, median; uterovaginal pore opens posterior to or together with male gonopore. Testes preovarian. Vitelline follicles preovarian, with or without a postovarian cluster. Eggs operculate. Primarily intestinal parasites of cypriniform and siluriform freshwater fishes. Some genera with progenetic cycles and development in the coelom of intermediate host, aquatic Annelida.

Key to families

1a. **Testes and vitellarium in same plane in cortical parenchyma, neither internal to inner longitudinal muscles (Fig. 5.26) .... Balanotaeniidae Mackiewicz & Blair, 1978.**

Diagnosis: Scolex aloculate. Testes in medullary parenchyma. Vitellarium in cortical parenchyma, external to inner longitudinal muscles. Type-genus *Balanotaenia* Johnston, 1924.

1b. **Testes, vitellarium or both internal to inner longitudinal muscles .... 2.**

2a. **Vitellarium completely in cortical parenchyma. Inner longitudinal muscles separate medullary testes from cortical vitellarium (Fig. 5.27) .... Lytocestidae Hunter, 1927.**

Diagnosis: Scolex aloculate. Testes in medullary parenchyma. Vitellarium in cortical parenchyma, external to inner longitudinal muscles. Type-genus *Lytocestus* Cohn, 1908.

2b. **Vitellarium either completely in medullary parenchyma or partially in medullary and cortical parenchyma .... 3.**

3a. **Vitellarium and testes in medullary parenchyma. Inner longitudinal muscles external to vitellarium (Fig. 5.28) ... Caryophyllaeidae Leuckart, 1878.**

Diagnosis: Scolex variable. Testes and vitellarium in medullary parenchyma, internal to inner longitudinal muscles. Type-genus *Caryophyllaeus* Gmelin, 1790.

3b. **Vitellarium partially in medullary and cortical parenchyma. Inner longitudinal muscles between adjacent vitelline follicles (Fig. 5.29) .... Capingentidae Hunter, 1930.**

Diagnosis: Scolex variable. Testes and vitellarium partially in medullary and cortical parenchyma, neither completely external nor internal to inner longitudinal muscles. Type-genus *Capingens* Hunter, 1927.

Family Balanotaeniidae Mackiewicz & Blair, 1978

This small family has testes and vitellarium in the cortical parenchyma (Fig. 5.26), illustrating that testes need not be confined to the medullary parenchyma as in the other three families. Such an unusual arrangement allows for mingling of testes and vitelline follicles and theoretically would seem to preclude having vitelline follicles in two lateral rows, a common arrangement in the other families. Perhaps most interesting is that the cortical position allows for more testes per

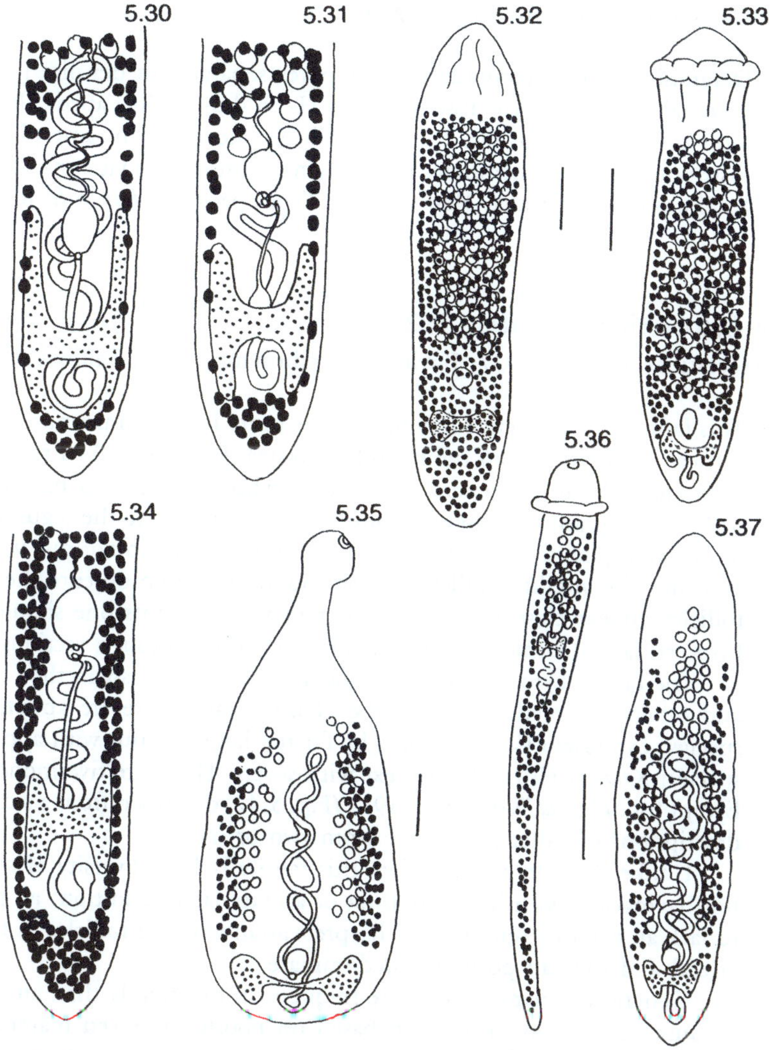

Figs 5.30–5.32, 5.34–5.37 Lytocestidae. **5.33.** Balanotaeniidae. 5.30. *Caryophyllaeides fennica.* 5.31. *Khawia sinensis.* 5.32. *Lytocestoides tanganyikae* (adapted from Baylis, 1928). 5.33. *Balanotaenia bancrofti.* 5.34. *Atractolytocestus.* 5.35. *Djombangia* (adapted from Satpute & Agarwal, 1974). 5.36. *Caryoaustralus sprenti.* 5.37. *Notolytocestus minor* Johnston & Muirhead, 1950. Figs 5.33, 5.36, 5.37 adapted from Mackiewicz & Blair (1980). Scales = 1 mm.

worm diameter, a fact that may be related to the very small size (less than 1 mm) of *Balanotaenia newguinensis* Mackiewicz & Blair, 1978, one of the smallest tapeworms known. This family contains the single genus *Balanotaenia* Johnston, 1924 with two species, both in the Australian zoogeographical region.

Genus *Balanotaenia* Johnston, 1924 (Figs 5.5, 5.33)

Diagnosis: Scolex coronulate. Gonopores separate, in definite genital atrium. External seminal vesicle absent. Ovary H- to dumbbell-shaped. Uterus does not loop anterior to cirrus-sac. Vitelline follicles and testes in same plane, annularly distributed around medullary parenchyma. Postovarian vitelline follicles absent. In Plotosidae. Australia, New Guinea. Type-species *B. bancrofti* Johnston, 1924.

Family Lytocestidae Hunter, 1927

(Syns Lytocestinae Hunter, 1927; Bovieninae Fuhrmann, 1931; Lallidae Johri, 1959)

This family is characterized by having the inner longitudinal muscles separate the cortical vitellarium from the medullary testes (Fig. 5.27). As a result of this configuration, the testes generally appear in wholemounts as a narrow central core with lateral or annular vitelline follicles arranged close to the tegument. By being in different planes, the testes and vitellarium usually do not mix with each other and thus the vitelline follicles can surround and often obscure the testes. If the follicles appear close to the tegument in wholemounts one should suspect the Lytocestidae, a suspicion that should be checked through sections.

The family Lytocestidae now consists of 16 genera, six in the Oriental region and others scattered in all but the Neotropical. There are about 47 described species; *Lytocestus* Cohn, 1908 is by far the largest genus with approximately 13 species; *Khawia* Hsu, 1935 is next with seven. There are five monotypic genera. Except for *Stocksia* Woodland, 1937 (Fig. 5.42) described from a single specimen, descriptions of other genera are from two or more.

Unlike the Caryophyllaeidae, this family lacks a diversity of scolex types by not having any with loculi, bothria or acetabula. In addition, all genera appear to lack an external seminal vesicle; preovarian vitelline follicles in the annular or lateral and median positions predominate.

Generic synonymies are more frequent in this family than any other, largely because descriptions have been based on poorly preserved material, have been incomplete or there has been misinterpretation of anatomical structure. *Bothrioscolex* Szidat, 1937 is an exception, however. Opinion has been divided as to whether the massive, eversible terminal cavity, described only by Szidat (1937), is alone sufficient to separate this genus from *Khawia* Hsu, 1935. General opinion is that it is not (Yamaguti, 1959; Kulakovskaya, 1961; Dubinina, 1962; Schmidt, 1986). Recent observation of this terminal cavity in live *Khawia* by Calentine & Ulmer (1961) appears to be convincing evidence that the two genera are the same. *Tsengia* Li, 1964, from *Cyprinus carpio* in Mongolia, differs from *Khawia* only in having an ovary in the shape of an inverted letter A. Williams & Sutherland (1981), unaware of the description by Li (1964), referred similar specimens from the same host in North America to *Khawia*, a conclusion accepted here because the fusion of ovarian arms is a common variation in other genera.

Re-examination of the type specimens reveals that ovarian follicles of *Lucknowia* Gupta, 1961 have been mistaken for postovarian vitelline follicles (Mackiewicz, 1981b). Consequently, this genus is synonymized with *Lytocestus*, here regarded as a complex because of the diverse forms, wide distribution and broad host spectrum, the widest of any 'genus' in the order. From Fig. 1C of Satpute & Agarwal (1980), it is apparent that ovarian follicles of *Introvertus* also have been mistaken for postovarian vitelline follicles. In addition, the swollen vas deferens has been misinterpreted as the external seminal vesicle and the apical elongation of the scolex as an introvert; *Introvertus* Satpute & Agarwal, 1980 is here synonymized with *Bovienia* Fuhrmann, 1931.

Comparison of the descriptions, illustration and examples of *Atractolytocestus* Anthony, 1958 and *Markevitschia* Kulakovskaya & Akhmerov, 1965 (one specimen) from *Cyprinus carpio* reveal that, except for testes number, there are no other major differences between the two (Jones & Mackiewicz, 1969). *Markevitschia* is, therefore, synonymized with *Atractolytocestus* (Fig. 5.34) even though the latter is a triploid.

The genera *Neolytocestus* and *Moravekia* reported briefly (1976a,b) by S.N. Sahay & U. Sahay in abstracts of the First National Convention of Indian Helminthologists, Bhubaneswar, India are lacking in sufficient detail to allow for critical comparison with other genera and thus are considered *genera inquirenda.*

Key to genera

1a. Uterus loops anterior to cirrus-sac .... 2.
1b. Uterus does not loop anterior to cirrus-sac .... 4.

2a. Postovarian vitelline follicles present. Ovary in shape of inverted letter A .... *Caryophyllaeides* Nybelin, 1922. (Figs 5.21, 5.30)
Diagnosis: Scolex cuneiform. Gonopore single. External seminal vesicle absent. Uterus loops anterior to cirrus-sac. Vitelline follicles lateral and median. In Cyprinidae. Scandinavia, Germany, Russia. Type-species *C. fennica* (Schneider, 1902) Nybelin, 1922.

2b. Postovarian vitelline follicles absent. Ovary H-shaped .... 3.

3a. Scolex tholate. Some testes and vitelline follicles anterior to uterus .... *Notolytocestus* Johnston & Muirhead, 1950. (Fig. 5.37)
Diagnosis: Gonopore single. External seminal vesicle absent. Ovary H-shaped. Uterus primarily anterior to cirrus-sac. Testes anterior and lateral to uterus. Vitelline follicles surround uterus and posterior testes. Postovarian vitelline follicles absent. In Plotosidae. Australia. Type-species *N. major* Johnston & Muirhead, 1950.

3b. Scolex bulbate with apical glandular structure. Testes and vitelline follicles lateral to uterus .... *Djombangia* Bovien, 1926. (Fig. 5.35)
Diagnosis: Gonopore single, genital atrium present. Cirrus-sac weakly developed. External seminal vesicle absent. Ovary bilobed. Uterus loops far anterior to cirrus-sac. Postovarian vitelline follicles absent. In Clariidae, Heteropneustidae. India, Java. Type-species *D. penetrans* Bovien, 1926.

4a. Postovarian vitelline follicles present .... 5.
4b. Postovarian vitelline follicles absent .... 8.

5a. Gonopore single in anterior half of body ....
.... ***Caryoaustralus*** **Mackiewicz & Blair, 1980. (Figs 5.19, 5.36)**

Diagnosis: Scolex choanocampanulate. External seminal vesicle absent. Ovary H-shaped. Uterus does not loop anterior to cirrus-sac, primarily postovarian. Vitelline follicles lateral and median, confluent with extensive postovarian follicles. In Plotosidae. Australia. Type-species *C. sprenti* Mackiewicz & Blair, 1980.

5b. Gonopore(s) single or double, in posterior quarter of body .... **6.**

6a. Scolex tholate. Neck absent. Single gonopore with deep genital atrium
.... ***Lytocestoides*** **Baylis, 1928. (Fig. 5.32)**

Diagnosis: (External seminal vesicle status unknown.) Ovary bilobed. Uterus does not loop anterior to cirrus-sac. Pre- and postovarian vitelline follicles form extensive layer covering all organs except for cirrus-sac. In Characidae. Africa. Type-species *L. tanganyikae* Baylis, 1928.

6b. Scolex not tholate. Neck present. Two gonopores in deep atrium ... **7.**

7a. Scolex cuneifimbriate or montanate, may form terminal eversible cavity. Neck short .... ***Khawia*** **Hsu, 1935. (Figs 5.17, 5.20, 5.31)**
(Syns *Bothrioscolex* Szidat, 1937; *Tsengia* Li, 1964.)

Diagnosis: Gonopores separate but close together in distinct genital atrium. External seminal vesicle absent. Ovary H-shaped, less often like an inverted letter A. Vitelline follicles lateral and median, postovarian follicles present. In Cyprinidae. North America, Europe, Russia, China. Type-species *K. sinensis* Hsu, 1935.

7b. Scolex bulbate to bulboacuminate. Distinct neck present ....
.... ***Atractolytocestus*** **Anthony, 1958. (Figs 5.11, 5.34)**
(Syn. *Markevitschia* Kulakovskaya & Akhmerov, 1965.)

Diagnosis: Gonopores separate in genital atrium. External seminal vesicle absent. Ovary H-shaped or bilobed. Uterus does not loop anterior to cirrus-sac. Vitelline follicles lateral and medial, confluent with postovarian follicles. In Cyprinidae. North America, Russia. Type-species *A. huronensis* Anthony, 1958.

8a. With one gonopore .... **9.**
8b. With two gonopores, may be in genital atrium .... **10.**

9a. Scolex tholate. Vitellarium annular ....
.... ***Tholophyllaeus*** **Mackiewicz & Blair, 1980. (Fig. 5.38)**

Diagnosis: Gonopore single. External seminal vesicle absent. Ovary H-shaped. Uterus does not loop anterior to cirrus-sac. Vitelline follicles lateral and median, postovarian follicles absent. In Plotosidae. Australia. Type-species *T. johnstoni* Mackiewicz & Blair, 1980.

9b. Scolex not tholate. Vitelline follicles predominantly in lateral rows ......
.... ***Stocksia*** **Woodland, 1937. (Fig. 5.42)**

Diagnosis: (Scolex shape undetermined; slightly expanded, possibly with grooves.) Gonopore single. Ovary H-shaped. Uterus does not loop anterior to cirrus-sac. Vitelline follicles crescentic to testes in cross section. Postovarian follicles absent. In Clariidae. Africa. Type-species *S. pujehuni* Woodland, 1937.

10a. Vitelline follicles in narrow lateral rows, no median follicles. Ovary generally H-shaped, rarely like inverted letter A .... *Bovienia* Fuhrmann, 1931. (Fig. 5.40)

(Syn. *Introvertus* Satpute & Agarwal, 1980.)

Diagnosis: Scolex digitiform to pyriform, may have protrusible apex. Gonopores separate. External seminal vesicle absent. Uterus does not loop anterior to cirrus-sac. Postovarian vitelline follicles absent. In Clariidae. Java, India. Type-species *B. serialis* (Bovien, 1926) Fuhrmann, 1931.

10b. Vitellarium annular or lateral and median .... 11.

11a. Scolex rugomonobothriate .... *Monobothrioides* Fuhrmann & Baer, 1925. (Fig. 5.9)

Diagnosis: Gonopores separate. External seminal vesicle absent. Ovary H-shaped. Uterus does not loop anterior to cirrus-sac. Vitelline follicles lateral and median, postovarian follicles absent. In Bagridae and Clariidae. Africa. Type-species *M. cunningtoni* Fuhrmann & Baer, 1925.

11b. Scolex digitiform unspecialized or spatulate .... 12.

12a. Ovary H-shaped, loosely follicular. Gonopores separate, far apart, genital atrium inconspicuous .... (Complex) *Lytocestus* Cohn, 1908. (Figs 5.7, 5.39)

(Syn. *Lucknowia* Gupta, 1961.)

Diagnosis: Scolex variable, from unspecialized to spatulate, pyriform; neck present. External seminal vesicle absent. Uterus does not loop anterior to cirrus-sac. Vitelline follicles lateral and median, postovarian follicles absent. In Mormyridae, Heteropneustidae, Clariidae, Characidae. Africa, India, Singapore, Hong Kong. Type-species *L. adhaerens* Cohn, 1908.

12b. Ovary loosely follicular shaped like inverted letter A, rarely H-shaped. Gonopores separate but in conspicuous genital atrium .... *Crescentovitus* Murhar, 1963. (Figs 5.8, 5.41)

Diagnosis: Scolex digitiform. Neck absent. Gonopores separate, close together in genital atrium. External seminal vesicle absent. Uterus does not loop anterior to cirrus-sac. Vitelline follicles lateral and median, postovarian follicles absent. In Heteropneustidae. India. Type-species *C. biloculus* Murhar, 1964.

Family Caryophyllaeidae Leuckart, 1878

(Syns Caryophyllaeinae Nybelin, 1922; Wenyoninae Hunter, 1927)

The distribution of both vitellarium and testes internal to the inner longitudinal muscles characterizes this family (Fig. 5.28). Two consequences follow from this

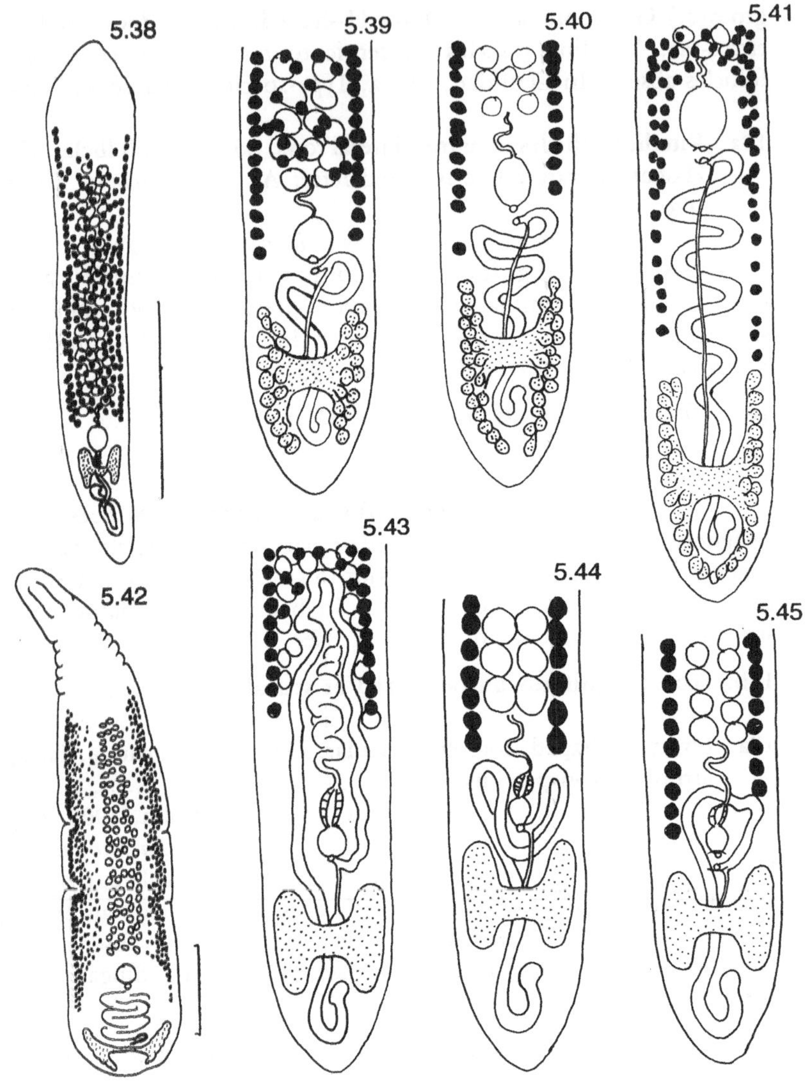

Figs 5.38–5.42 Lytocestidae. **5.43–5.45**. Caryophyllaeidae. 5.38. *Tholophyllaeus johnstoni* (adapted from Mackiewicz & Blair, 1980). 5.39. *Lytocestus birmanicus* Lynsdale, 1956. 5.40. *Bovienia serialis*. 5.41. *Crescentovitus biloculus*. 5.42. *Stocksia pujehuni* (adapted from Woodland, 1937). 5.43. *Diefluvium unipapillatum*. 5.44. *Rowardleus pennensis*. 5.45. *Hypocaryophyllaeus paratarius*. Scales = 1 mm.

organization that may aid in family allocation from wholemounts. One is that the testes and vitelline follicles can mingle, often imparting a mosaic effect; or, if the follicles are very numerous, they may surround and obscure the testes. A second is that the thickness of the cortical parenchyma is generally greater than in the Lytocestidae. As noted in the key, this family has an astonishing array of diverse scolex types, greater than in any other family.

The family Caryophyllaeidae consists of 20 genera with 17 in the Nearctic; nine are monotypic. *Isoglaridacris* Mackiewicz, 1968 is the largest genus with 12 species, *Biacetabulum* Hunter, 1927 is next with 10. Except for *Paracaryophyllaeus* Kulakovskaya, 1961, data on the disposition of the longitudinal muscles are known for all other genera. This genus is placed here on the basis of the provisional allocation of Kulakovskaya (1961). *Bialovarium* Fischthal, 1951 was described from a single individual, all other generic descriptions involved multiple specimens. The family includes about 73 described species, predominantly from catostomid fish with a few from cyprinid, cobitid and silurid freshwater fishes.

Generic synonymies are most evident with *Archigetes* Leuckart, 1878 which occurs progenetically in tubificid annelids and as an adult in fish. These synonymies reflect over 100 years of controversy revolving around whether *Archigetes* represents a valid genus with a number of species (Kennedy, 1965) or is the progenetic stage of other genera such as *Glaridacris* Cooper, 1920, *Biacetabulum* Hunter, 1927 and *Brachyurus* Szidat, 1938 (see Janiszweska, 1954; Kulakovskaya, 1962). The origin of this controversy is both the progenesis phenomenon involving tubificid annelids and the intraspecific variation or changing states with maturation that have led to a single species being placed into two separate genera. Unlike Kennedy (1965), I see no logical, biologically based reason to regard progenesis as a character restricted to only one genus, i.e., *Archigetes*; it is a universal, physiological and developmental phenomenon that can occur in various genera, as it does in the Trematoda. In support of this view is the unusually wide array of characters ascribed to *Archigetes* by Kennedy (1965). Based on a review of the literature and study of examples of most species involved in this controversy, I tend to concur with Janiszweska (1954) and Kulakovskaya (1962) in principle, i.e. that some *Archigetes* species are larval forms of other genera. On the other hand, I also concur with Calentine (1962) that *Archigetes* is a valid genus name. I, therefore, recognize *Archigetes* and *Brachyurus* as separate genera, a conclusion reinforced by my discovery (Mackiewicz, 1965) that the type specimen of *Paraglaridacris* has postovarian vitelline follicles and that the vitelline follicles of *A. iowensis* Calentine, 1962 (identified by the describer) are lateral and median, contrary to the original description. *Szidatinus* was proposed by McCrae (1961), in an abstract, for *Brachyurus* Szidat, 1938, that was preoccupied. Resolution of the *Archigetes* question, one of the most perplexing in all of cestodology, requires additional comparative studies and more extensive discussion of various genera and species, not appropriate in this publication. Because the name *Brachyurus* is preoccupied by *Brachyurus* Fischer-Waldheim, 1913 (rodent), *Paraglaridacris* Janiszewska, 1950 is here used in the key as the next available name.

Because of the great diversity of characters, several genera occur more than once in the key; these include: *Glaridacris* Cooper, 1920, *Biacetabulum* Hunter, 1927, *Monobothrium* Diesing, 1863 and *Isoglaridacris* Mackiewicz, 1965. Whether this phenomenon reflects hidden complexes of multiple genera or possibly subgenera remains to be determined after a thorough analysis of all genera in the family. They are each designated as a complex to reflect this systematic uncertainty.

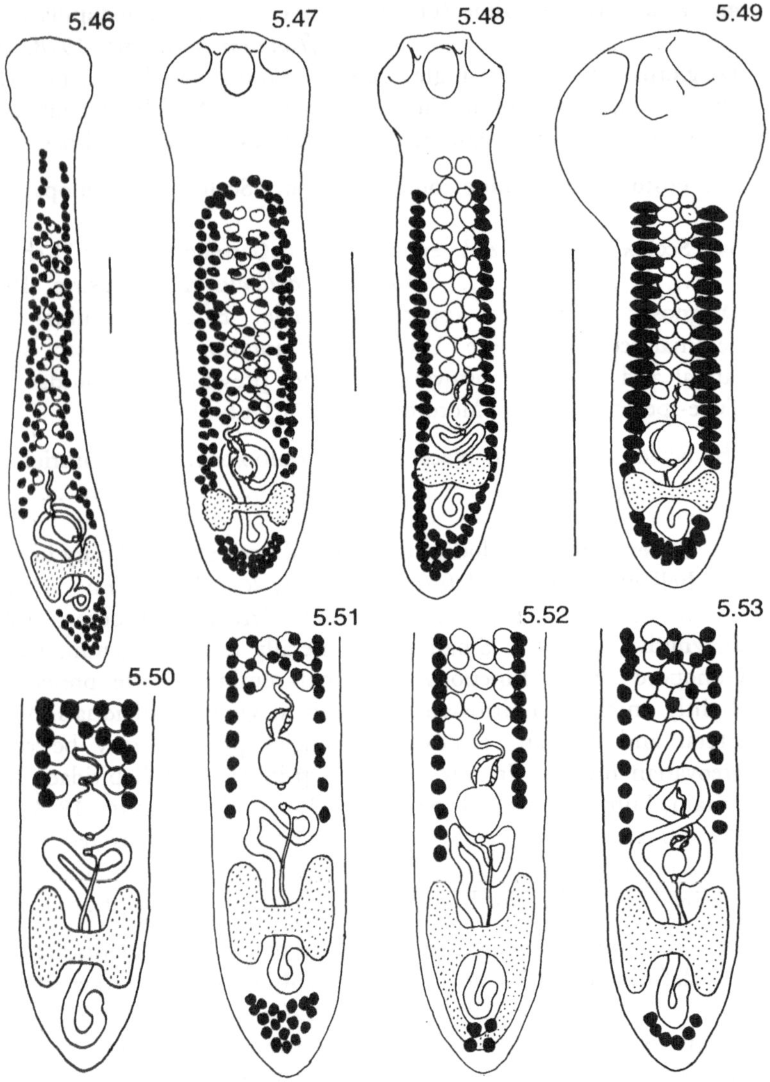

Figs 5.46–5.53 Caryophyllaeidae. 5.46. *Paracaryophyllaeus gotoi* (adapted from Motomura, 1927). 5.47. *Archigetes sieboldi* (adapted from Calentine & DeLong, 1966; original figure with cercomer on specimen from *Limnodrilus*). 5.48. *Paraglaridacris limnodrili* (original). 5.49. *Penarchigetes oklensis* (adapted from Mackiewicz, 1969). 5.50. *Rogersus rogersi*. 5.51. *Glaridacris catostomi*. 5.52. *Isoglaridacris bulbocirrus*. 5.53. *Biacetabulum infrequens*. Scales = 0.5 mm.

8a. Preovarian vitelline follicles distinctly in lateral rows .... 9.
8b. Preovarian vitelline follicles annular or lateral and median .... 15.

9a. With two gonopores .... 10.
9b. With one gonopore .... 11.

10a. Gonopores separate, in anterior half of body .... *Wenyonia* Woodland, 1923. (Figs 5.18, 5.61)
Diagnosis: Scolex rugomontanate. External seminal vesicle absent. Ovary H-shaped. Uterus does not loop anterior to cirrus-sac. Vitelline follicles in lateral rows. Postovarian vitelline follicles present. In Bagridae, Mochokidae. Africa. Type-species *W. virilis* Woodland, 1923.

10b. Gonopores separate, in posterior quarter of body .... (Complex) *Glaridacris* Cooper, 1920. (Fig. 5.51)
Diagnosis: Scolex cuneiloculate (Fig. 5.12) or bothrioloculodiscate (Fig. 5.4). Genital atrium may be present. External seminal vesicle present. Ovary H-shaped. Uterus does not loop anterior to cirrus-sac. Vitelline follicles in lateral rows or lateral and median, postovarian vitelline follicles present. In Catostomidae. North America. Type-species *G. catostomi* Cooper, 1920.

11a. Ovary V-shaped. Scolex loculotholate ... *Bialovarium* Fischthal, 1953. (Fig. 5.63)
Diagnosis: Gonopore single. External seminal vesicle present. Uterus does not loop anterior to cirrus-sac, all pre-ovarian. Vitelline follicles in lateral rows. Postovarian region and vitelline follicles absent. In Cyprinidae. North America. Type-species *B. nocomis* Fischthal, 1953.

11b. Ovary H- or dumbbell-shaped or like inverted letter A. Scolex not tholate .... 12.

12a. External seminal vesicle present .... 13.
12b. External seminal vesicle absent .... *Penarchigetes* Mackiewicz, 1969. (Fig. 5.49)
Diagnosis: Scolex bothrioloculodiscate, expanded or not. Gonopore single. Ovary dumbbell-shaped. Uterus does not loop anterior to cirrus-sac. Vitelline follicles in lateral rows, sometimes continuous with postovarian follicles. In Catostomidae. North America. Type-species *P. oklensis* Mackiewicz, 1969.

13a. Scolex cuneiloculate (Fig. 5.12). Ovary in shape of inverted letter A or nearly so .... (Complex) *Isoglaridacris* Mackiewicz, 1965. (Fig. 5.52)
Diagnosis: Gonopore single. External seminal vesicle present. Uterus does not loop anterior to cirrus-sac. Vitelline follicles in lateral rows, sometimes with a third median row or a few median follicles. Postovarian follicles present or absent. In Catostomidae. North America. Type-species *I. bulbocirrus* Mackiewicz, 1965.

13b. Scolex bothrioloculodiscate. Ovary H-shaped or dumbbell-like .... 14.

14a. Pre- and postovarian vitelline fields continuous. Ovary in shape of dumbbell or band. Defined neck absent. Small forms, generally less than 3 mm .... *Paraglaridacris* Janiszewska, 1950. (Fig. 5.48)
(Syns *Brachyurus* Szidat, 1938 *nec* Fischer-Waldheim, 1913; *Szidatinus* McCrae, 1961).
Diagnosis: Gonopore apparently single. External seminal vesicle present. Uterus does not loop anterior to cirrus-sac. Vitelline follicles in lateral rows, continuous

with postovarian follicles. Progenetic in tubificid annelids. In Cyprinidae, Cobitidae. Europe, Britain, Russia, Japan. Type-species *P. gobii* (Szidat, 1938), syn. of *P. limnodrili* n. comb. for *Glaridacris limnodrili* Yamaguti, 1934.

14b. Pre- and postovarian vitelline fields not continuous. Ovary H-shaped. Defined neck present. Larger forms, more than 4 mm .... *Janiszewskella* Mackiewicz & Deutsch, 1976. (Fig. 5.58)

Diagnosis: Gonopore single. External seminal vesicle present. Uterus does not loop anterior to cirrus-sac. Vitelline follicles in lateral rows, postovarian follicles present. In Catostomidae. North America. Type-species *J. fortobothria* Mackiewicz & Deutsch, 1976.

15a. With two gonopores .... 16.
15b. With one gonopore .... 21.

16a. Scolex loculate or monobothriate .... 17.
16b. Scolex aloculate .... 20.

17a. Postovarian vitelline follicles present .... 18.
17b. Postovarian vitelline follicles absent .... 19.

18a. Scolex bothrioloculodiscate (Fig. 5.4) or cuneiloculate. External seminal vesicle present (Figs 5.12, 5.51) .... (Complex) *Glaridacris* Cooper, 1920.
[See also 10b above]

18b. Scolex monobothriate. External seminal vesicle absent (Figs 5.16, 5.54) .... (Palaearctic) *Monobothrium* Diesing, 1863.

Diagnosis: Gonopores separate. External seminal vesicle present or absent. Ovary H-shaped. Uterus does not loop anterior to cirrus-sac. Vitelline follicles lateral and median, postovarian follicles present or absent. In Cyprinidae, Catostomidae. Europe, North America. Type-species *M. wageneri* Nybelin, 1922.

19a. Scolex monobothriate or loculomonobothriate. External seminal vesicle present .... (Nearctic) *Monobothrium* Diesing, 1863. (Figs 5.1, 5.16, 5.55)
[See also 18b above.]

19b. Scolex loculotruncate lacking terminal introvert. External seminal vesicle present .... *Promonobothrium* Mackiewicz, 1968. (Figs 5.3, 5.56)

Diagnosis: Gonopores separate. Ovary H-shaped. Uterus does not loop anterior to cirrus-sac. Vitelline follicles lateral and median, postovarian follicles absent. In Catostomidae. North America. Type-species *P. minytremi* Mackiewicz, 1968.

20a. Scolex flabellate or cuneicrispitate. External seminal vesicle absent .... *Caryophyllaeus* Gmelin, 1790. (Figs 5.14, 5.57)

Diagnosis: Gonopores separate, atrium present. Ovary H-shaped. Uterus does not loop anterior to cirrus-sac. Vitelline follicles lateral and median, postovarian follicles present. In Cyprinidae. Europe, Russia, (?Africa, Asia). Type-species *C. laticeps* (Pallas, 1781).

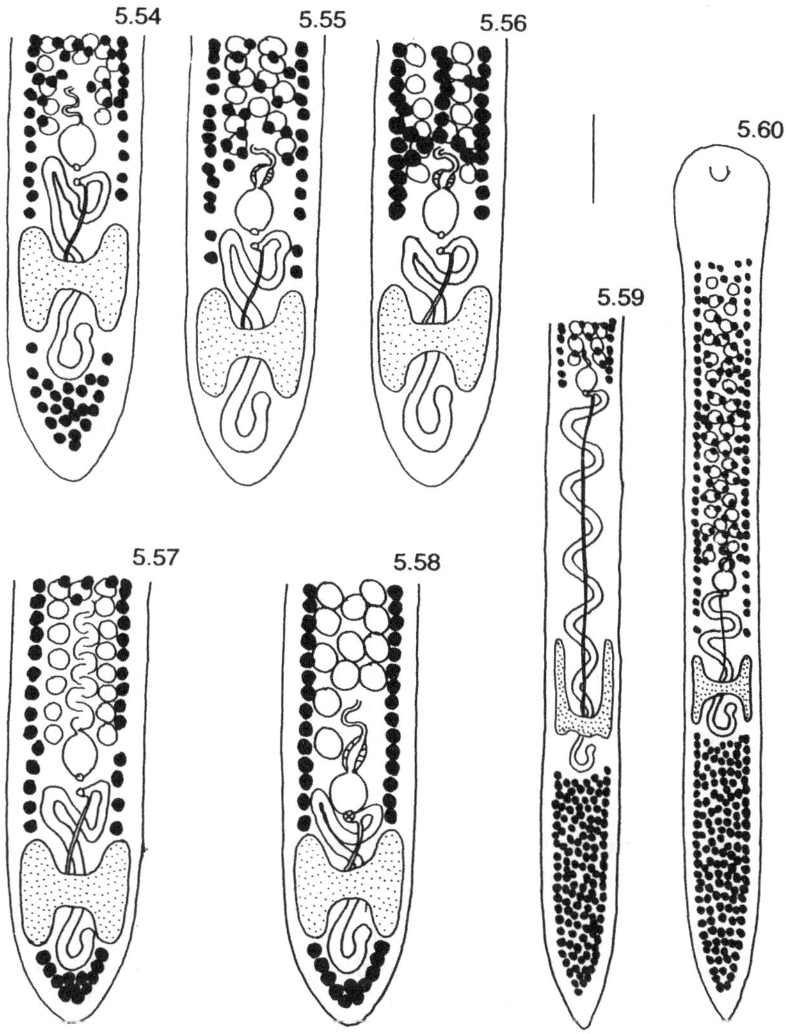

Figs 5.54–5.60 Caryophyllaeidae. 5.54. *Monobothrium wageneri* (Palaearctic). 5.55. *Monobothrium hunteri*. 5.56. *Promonobothrium minytremi*. 5.57. *Caryophyllaeus laticeps*. 5.58. *Janiszewskella fortobothria*. 5.59. *Calentinella etnieri*. 5.60. *Pliovitellaria wisconsinensis*. Scales = 1 mm.

20b. Scolex tholate. External seminal vesicle present .... ***Hunterella* Mackiewicz & McCrae, 1962. (Figs 5.15, 5.62)**

Diagnosis: Scolex tholate. Gonopores separate. External seminal vesicle present. Ovary bilobed or dumbbell-shaped. Uterus does not loop anterior to cirrus-sac. Vitelline follicles lateral and median, postovarian follicles present. In Catostomidae. North America. Type-species *H. nodulosa* Mackiewicz & McCrae, 1962.

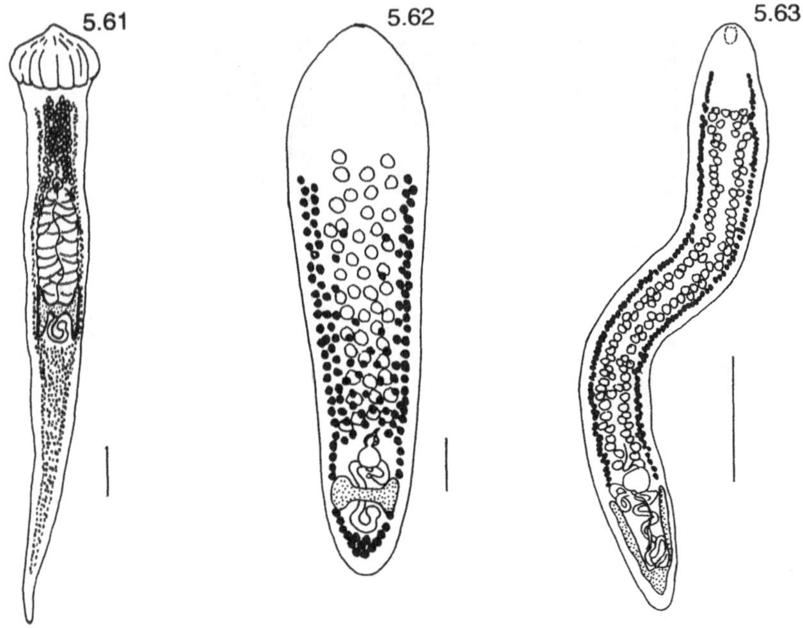

Figs 5.61–5.63 Caryophyllaeidae. 5.61. *Wenyonia virilis* (adapted from Mackiewicz, 1972). 5.62. *Hunterella nodulosa* (adapted from Mackiewicz & McCrae, 1962). 5.63. *Bialovarium nocomis* (adapted from Fischthal, 1954). Scales = 1 mm.

21a. External seminal vesicle present .... **22.**
21b. External seminal vesicle absent ....
.... ***Calentinella* Mackiewicz, 1974. (Fig. 5.59)**

Diagnosis: Scolex cuneiform (Fig. 5.21). Gonopore single. Ovary U-shaped. Uterus does not loop anterior to cirrus-sac. Vitelline follicles lateral and median, postovarian follicles present. In Catostomidae. North America. Type-species *C. etnieri* Mackiewicz, 1974.

22a. Gonopore in anterior half or two-thirds of body. Scolex loculotholate
.... ***Pliovitellaria* Fischthal, 1950. (Fig. 5.60)**

Diagnosis: Gonopore single. External seminal vesicle present. Ovary H-shaped. Uterus does not loop anterior to cirrus-sac. Vitelline follicles lateral and median, postovarian follicles extensive, more than one half preovarian vitelline zone. In Cyprinidae. North America. Type-species *P. wisconsinensis* Fischthal, 1951.

22b. Gonopore in posterior one-sixth of body. Scolex cuneiloculate ....
.... **(Complex) *Isoglaridacris* Mackiewicz, 1965.**
[See also 13a above.]

Family Capingentidae Hunter, 1930

(Syns Pseudolytocestinae Hunter, 1929; Capingentinae Hunter, 1930)

Because of the intermediate position of the inner longitudinal muscles that characterizes this family (Fig. 5.29) allocation to the Capingentidae can be a matter of judgement, more so than for any other family. Much depends on the extent of muscle development and on preovarian vitelline follicle size and shape. There is generally no problem when muscle fasciculi are numerous and arranged in a definite layer and where the vitelline follicles are large, lobed or irregular in shape so that most of them can be partially in both the cortical and medullary parenchyma. It is another matter, however, when muscles are small and irregularly spaced and vitelline follicles are small, round or large and not lobed. Unless the majority of vitelline follicles are lobed and protrude on either side of the muscle layer then allocation to the Capingentidae should be carefully reappraised.

Despite the fact that this is a small family of about 15 described species, it is the least well understood. Poor material, inadequate descriptions and misinterpretations of structures force one to recognize only six of the nine described genera with the remaining three relegated as *genera incertae sedis* or placed in synonymy. Of the six, five are monotypic and three are in the Nearctic. All original descriptions are based on multiple specimens and all except *Pseudocapingentoides* Verma, 1971 have data on the placement of longitudinal muscles.

Status of genera from India is uncertain (Mackiewicz, 1981b). Restudy of the type series of *Pseudocaryophyllaeus indica* Gupta, 1961 and *Capingentoides batrachii* Gupta, 1961 indicates that the musculature of both species appears to be that of the family Caryophyllaeidae and that the species may be conspecific. Furthermore, the *C. batrachii* series may consist of a mixed infection, a conclusion strengthened by the illustrations of Gupta (1961) that show annular vitelline follicles in cross sections but only lateral ones in whole mounts. Unfortunately, subsequent descriptions of *Pseudocaryophyllaeus mackiewiczi* Gupta & Parmar, 1982, *P. ritai* Gupta & Singh, 1983, *Capingentoides heteropneusti* Gupta & Sinha, 1969, *C. singhi* Verma, 1971, *C. moghei* Pandey, 1973 and *C. fotedari* Gupta & Parmar, 1985 do not include cross sections as well as consistently confusing the ovarian follicles as postovarian vitelline follicles. Until these two 'genera' can be critically restudied, *Capingentoides* is here considered a synonym of *Pseudocaryophyllaeus*. The latter is included in the key for this family pending clarification of the placement of the longitudinal musculature.

The descriptions and illustrations of *Pseudocapingentoides* Verma, 1971 lack sufficient detail to allow for family or generic allocation (Mackiewicz, 1981b). Schmidt (1986) placed this genus into synonymy with *Capingentoides* Gupta, 1961.

Key to genera

1a. Uterus loops anterior to cirrus-sac .... **2.**
1b. Uterus does not loop anterior to cirrus-sac .... **3.**

2a. Preovarian vitelline follicles in distinct lateral rows. Scolex cuneiloculate (Fig. 5.12) .... *Spartoides* Hunter, 1929. (Fig. 5.70)

Diagnosis: Gonopores separate. External seminal vesicle present. Ovary U-shaped. Postovarian vitelline follicles absent. In Catostomidae. North America. Type-species *S. wardi* Hunter, 1929.

2b. Preovarian vitelline follicles lateral and median or annular. Scolex massive, with pair of well developed bothria .... *Capingens* Hunter, 1927. (Figs 5.6, 5.69)

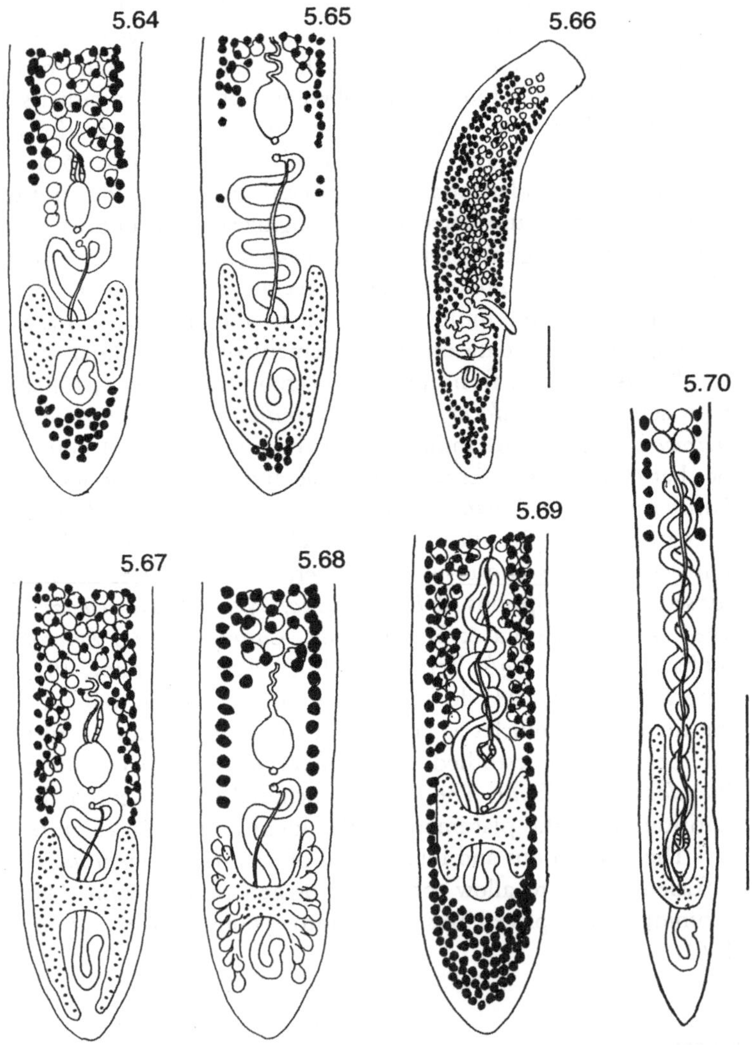

Figs 5.64-5.70 Capingentidae. 5.64. *Edlintonia ptychocheila.* 5.65. *Adenoscolex oreini.* 5.66. *Breviscolex orientalis* (adapted from Kulakovskaya, 1962). 5.67. *Pseudolytocestus differtus.* 5.68. *Pseudocaryophyllaeus indica* (adapted from Gupta, 1961). 5.69. *Capingens singularis.* 5.70. *Spartoides wardi.* Scales = 1 mm.

Diagnosis: Gonopores separate. External seminal vesicle present. Ovary H-shaped. Uterus loops anterior to cirrus-sac. Postovarian follicles present. In Catostomidae. North America. Type-species *C. singularis* Hunter, 1927.

3a. Postovarian vitelline follicles present .... 4.
3b. Postovarian vitelline follicles absent .... 6.

4a. Ovary in shape of inverted letter A .... ***Adenoscolex* Fotedar, 1958. (Fig. 5.65)**
Diagnosis: Scolex cuneiform (Fig. 5.21). Gonopores separate. External seminal vesicle absent. Uterus does not loop anterior to cirrus-sac. Vitelline follicles lateral and median, postovarian follicles present. In Cyprinidae. Kashmir. Type-species *A. oreini* Fotedar, 1958.

4b. Ovary not in shape of inverted letter A .... 5.

5a. External seminal vesicle present. Pre-and postovarian vitelline fields not continuous .... ***Edlintonia* Mackiewicz, 1970. (Fig. 5.64)**
Diagnosis: Scolex cuneiform (Fig. 5.21). Gonopores separate. Ovary H-shaped. Uterus does not loop anterior to cirrus-sac. Vitelline fields lateral and median, postovarian follicles present. In Cyprinidae. North America. Type-species *E. ptychocheila* Mackiewicz, 1970.

5b. External seminal vesicle absent. Pre-and postovarian vitelline fields continuous .... ***Breviscolex* Kulakovskaya, 1962. (Fig. 5.66)**
Diagnosis: Scolex short, truncate, no neck. (Gonopore status unknown.) External seminal vesicle present. Ovary bilobed. Uterus does not loop anterior to cirrus-sac. Vitelline fields lateral and median, postovarian follicles present. In Cyprinidae. Russia. Type-species *B. orientalis* Kulakovskaya, 1962.

6a. External seminal vesicle present .... ***Pseudolytocestus* Hunter, 1929. (Figs 5.10, 5.67)**
Diagnosis: Scolex weakly hastate, aloculate. Gonopores separate. Ovary H-shaped. Uterus does not loop anterior to cirrus-sac. Vitelline follicles lateral and median, postovarian follicles absent. In Catostomidae, Clariidae. North America; India? Type-species *P. differtus* Hunter, 1929.

6b. External seminal vesicle absent .... ***Pseudocaryophyllaeus* Gupta, 1961. (Fig. 5.68)**
(Syns *Capingentoides* Gupta, 1961; *Pseudocapingentoides* Verma, 1971.)
Diagnosis: Scolex bulbate, apex truncate. Gonopores separate. External seminal vesicle absent. Uterus does not loop anterior to cirrus-sac. Vitelline fields lateral and median, postovarian follicles absent. In Clariidae. India. Type-species *P. indica* Gupta, 1961.

macrocephalum from elasmobranchs caught off the west coast of Scotland and assigned it to the Diphyllidea. Schmidt (1970) erected a separate family, the Ditrachybothridiidae for this monotypic genus. Khalil & Abdul-Salam (1989) erected *Macrobothridium* for *M. rhynchobati* from elasmobranchs in the Arabian Gulf and assigned this monotypic genus to a new family Macrobothridiidae attached to the Diphyllidea. The validity of the Diphyllidea is now generally accepted and three families, each with a single genus, are assigned to it.

ORDER DIPHYLLIDEA VAN BENEDEN IN CARUS, 1863

Diagnosis: Scolex consisting of head proper and cephalic peduncle; two sessile bothridia present; armed rostellum present or absent, peduncle armed or unarmed. Proglottids acraspedote. Genital pore mid-ventral in posterior part of proglottid. Testes numerous, pre-ovarian. Cirrus-sac large, pyriform, opens ventrally immediately in front of vaginal pore. Ovary bilobed, posterior. Vitelline follicles lateral or surrounding proglottid. Uterine pore absent. Parasites of elasmobranchs.

Key to families and genera

1a. Cephalic peduncle long, armed with powerful straight hooks with roots .... Echinobothriidae Perrier, 1897.

Diagnosis: Scolex with well-developed rostellum armed on dorsal and ventral surfaces with hooks diminishing in size outwardly. Two oval bothridia present, covered with very small hair-like spines. Posterior proglottids longer than broad. Cirrus armed. Vitelline follicles in two lateral bands. Uterus median. Eggs round, may have short filament. In elasmobranchs. Cosmopolitan. Type and only genus *Echinobothrium* van Beneden, 1849.

Genus *Echinobothrium* van Beneden, 1849 (Figs 6.1, 6.2)

Diagnosis: With the characters of the family. Type-species *E. typus* van Beneden, 1849.

1b. Cephalic peduncle short, unarmed .... 2.

2a. Scolex with well developed rostellum, armed on dorsal and ventral surfaces with two alternating rows of large hooks .... Macrobothridiidae Khalil & Abdul-Salam, 1989.

Diagnosis: Rostellar hooks largest in middle row, decreasing in size towards extremities. Two unarmed bothridia present. Cephalic peduncle unarmed. Proglottids acraspedote, posterior proglottids longer than wide; last proglottid with long extension terminating in sucker-like structure, becomes detached from strobila and continues development in gut of host. Cirrus-sac large; cirrus armed

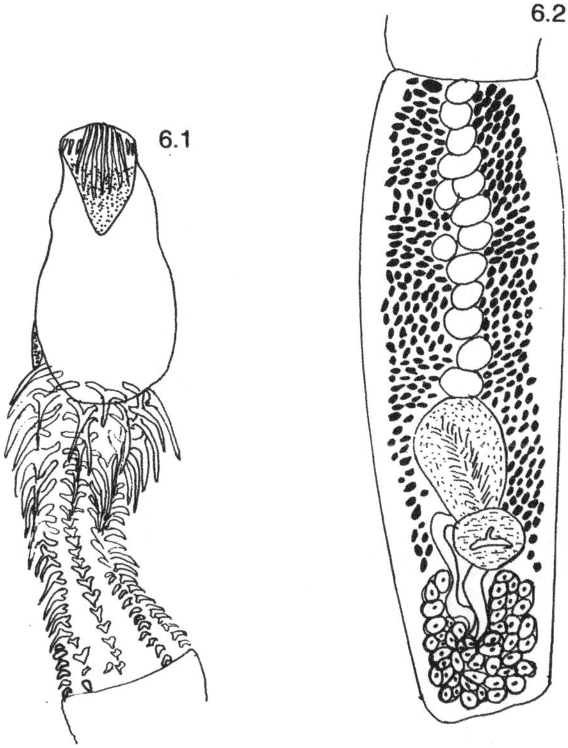

Figs 6.1, 6.2 *Echinobothrium bonasum* Williams & Campbell, 1980. 6.1. Scolex. 6.2. Mature proglottid. (After Williams & Campbell, 1980.)

with rose-thorn spines. Vitelline follicles in two lateral bands. Uterus initially saclike but later expands to fill anterior part of detached gravid proglottid, developing small simple lateral diverticula. Eggs small, spherical, without filaments. In elasmobranchs in Arabian Gulf. Type and only genus *Macrobothridium* Khalil & Abdul-Salam, 1989.

Genus *Macrobothridium* Khalil & Abdul-Salam, 1989 (Figs 6.3–6.6)

Diagnosis: With the characters of the family. Type and only species *M. rhynchobati* Khalil & Abdul-Salam, 1989.

2b. Scolex with bluntly pointed apex with feebly developed apical organ lacking any hooks .... Ditrachybothridiidae Schmidt, 1970.
Diagnosis: Scolex with two bothridia, one dorsal and one ventral; bothridia armed along border and outer convex surface with rows of small, backwardly directed spines; cephalic peduncle unarmed, terminated by velum. Strobila cylindrical, with poorly marked external segmentation anteriorly; posterior proglottids

6.3

6.4

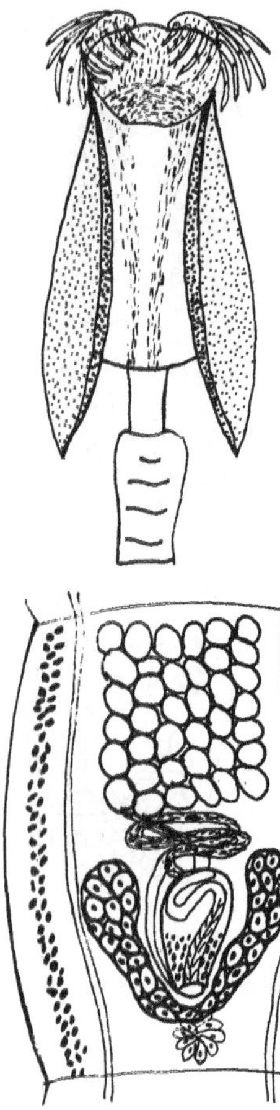

6.5

6.6

6.7

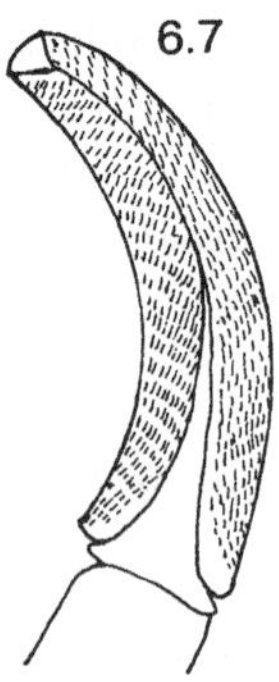

6.8

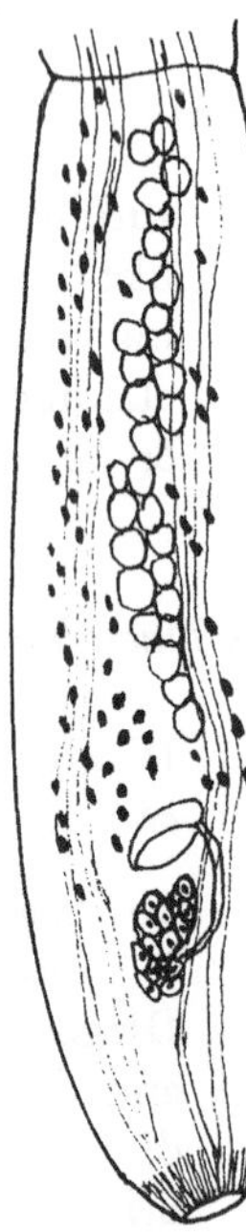

Figs 6.3–6.6 *Macrobothridium rhynchobati* Khalil & Abdul-Salam, 1989. 6.3. Scolex, lateral view. 6.4. Mature proglottid. 6.5. Last three proglottids of strobila. 6.6. Detached gravid proglottid. (After Khalil & Abdul-Salam, 1989.)

Figs 6.7, 6.8 *Ditrachybothridium macrocephalum* Rees, 1959. 6.7. Scolex, lateral view. 6.8. Last proglottid. (After Rees, 1959.)

four or five times longer than broad; last proglottid terminates in sucker-like structure. Cirrus-sac large, pyriform. Vitelline follicles inconspicuous, diffuse, surrounding most of the genitalia. Uterus and eggs not known. Parasite of elasmobranchs in Scottish waters. Type and only genus *Ditrachybothridium* Rees, 1959.

Genus *Ditrachybothridium* Rees, 1959 (Figs 6.7, 6.8)

Diagnosis: With the characters of the family. Type and only species D. *macrocephalum* Rees, 1959.

7 Order Trypanorhyncha Diesing, 1863

R.A. CAMPBELL[1] & I. BEVERIDGE[2]

[1]*Department of Biology, University of Massachusetts Dartmouth, North Dartmouth, Massachusetts 02747–2300, USA:* [2]*Veterinary Clinical Centre, School of Veterinary Science, University of Melbourne, Princes Highway, Werribee, Victoria 3030, Australia.*

Introduction

The order Trypanorhyncha consists of marine polyzoic cestode parasites, unique in the possession of a rhyncheal apparatus, whose adult stages inhabit the gastrointestinal system of elasmobranch fishes. Juvenile stages are known from teleosts, elasmobranchs and a variety of marine invertebrates, including zooplankton. The order ranks second only to the order Tetraphyllidea in prevalence among elasmobranchs.

Taxonomically, the Trypanorhyncha are complex and considered the most chaotic and confused of tapeworm groups (Wardle & McLeod, 1952; p. 287). The ordinal name Tetrarhynchidea Claus, preferred by Dollfus, has been superseded by Trypanorhyncha Diesing, 1863. It appears that confusion within the order has arisen mostly because of a multitude of incomplete descriptions, lost types, incorrect interpretations of tentacle armatures and lack of information on life histories. Guiart (1931), following the subtribal subdivisions of Diesing (1854; p. 589, 591), divided the trypanorhynchs into two suborders (Atheca and Thecaphora) on the basis of presence or absence of a larva with a blastocyst. Joyeux & Baer (1934) and Dollfus (1942; p. 434) considered the larval distinction a mistaken one. Actually the attempt to distinguish suborders is questionable for two reasons: (i) because of unknown uniformity regarding the absolute presence or absence of a larva enclosed by a blastocyst; (ii) homonymy. Guiart (1927) initially proposed the subordinal names Cystidea/Acystidea but discovered that those names had been used previously for echinoderms (Agassiz, 1846a; Guiart, 1931) and proposed the replacement names, Atheca/Thecaphora, from Diesing (1854). Fuhrmann (1931) accepted that classification and it is still in use (Schmidt, 1986). Exceptions to the subordinal definitions were discovered by Dollfus (1969b) for a postlarva in *Pseudogrillotia* (Thecaphora) and by Beveridge & Campbell (1988b) who described homeoacanthous armatures for the Tetrarhynchobothriidae (Thecaphora). Furthermore, we recently found that the names (Atheca/

Thecaphora) are also homonyms (Agassiz, 1846b; Thornely, 1904; Schulz *et al.*, 1928, 1936; Neave, 1939, 1940). Despite the lack of rules governing the use of suprafamilial names, *ICZN* Article 1(b)(4), we consider the use of suborders to be of little practical value and recognize families based upon armature type and general morphology.

The first workable organization of the Trypanorhyncha was by Dollfus (1942) and it is no enviable task to try to improve upon it. But it is important to note, firstly, he only included genera he could clearly recognize and, secondly, his disclaimer of creating a new 'Système des Tétrarhynques' (1942; p. 2). In spite of this disclaimer, Dollfus (1942) has become the most widely accepted basis for trypanorhynchan systematics.

The traditional systematics of the trypanorhynchs has been based upon scolex and proglottid anatomy, type of tentacular armature and structure of the plerocercus larva (Figs 7.1–7.9). The number and form of the scolex bothridia and characteristics of the rhyncheal apparatus within the scolex peduncle are essential diagnostic features (Figs 7.2, 7.4). The scolex peduncle (Fig. 7.1) may be subdivided into the region containing the tentacle sheaths (pars vaginalis), portion with the muscular bulbs (pars bulbosa) and space behind the bulbs (pars postbulbosa). The characteristics of the structures and the ratio of the lengths of the scolex regions have been taxonomically useful. Certain internal characteristics (Fig. 7.6a, b) have been presented (Woodland, 1927; p. 534) as distinctive of trypanorhynchs. These include: division of the parenchyma into cortex and medulla by a layer of longitudinal muscles; vitelline follicles forming a sleeve encircling the periphery of the medulla; testes extending posterior to the ovary; vagina ventral to the uterus and cirrus-sac; vaginal orifice beside the cirrus orifice, ventrally or a little posterior. In addition, the ovary is depicted as a multilobed structure, rather X-shaped in cross-section. Careful study of the numerous species more recently added to the literature show that there are exceptions to most of these characters. The ventral position of the vagina relative to the uterus is a consistent feature (Fig. 7.6b). The ovary may have two lobes, four lobes or multiple lobes joined by an isthmus or lack a central isthmus altogether. Thus far, the nongravid uterus appears to be either a median tube with or without preformed diverticula, or to be in the form of an inverted U (Fig. 7.7a–d). Uterine morphology appears particularly useful in separating the two suborders. The uterine pore may be preformed but is always ventral when present. The genital pore is usually on the lateral margin but may be ventrosubmarginal, usually irregularly alternate but may be unilateral, is often set into a notch-like depression of the proglottid margin and often demarcated by fleshy distensions. The terminal

Figs 7.1–7.5 Scolex and proglottid features. 7.1. Scolex regions. PBO, pars bothridialis; PBULB, pars bulbosa; PPB, pars postbulbosa; PS, pedunculus scolecis; PV, pars vaginalis; ST, strobila. 7.2. Rhyncheal apparatus. BA, basal armature; BO, bothridium; BULB, tentacle bulb; EPO, enigmatical prebulbar organ; MA, metabasal armature; RM, retractor muscle; TS, tentacle sheath. 7.3. Hook types. A, spiniform; B, bent spiniform; C, hastate; D, falciform; E, uncinate; F, uncinate; G, rose-thorn (upper extension base = 'heel'; lower extension of base = 'toe'); H, rebated; I, sagittate; J, serpe à bec; K, recurved tip; L, notched tip; M, single-winged; N, double-winged. 7.4. Enigmatical prebulbar organs. B, bulb; N, 'nucleus' (see Beveridge & Smith, 1988); RM, retractor muscle; TS, tentacle sheath. 7.5. Proglottid junctions. A, craspedote; B, acraspedote.

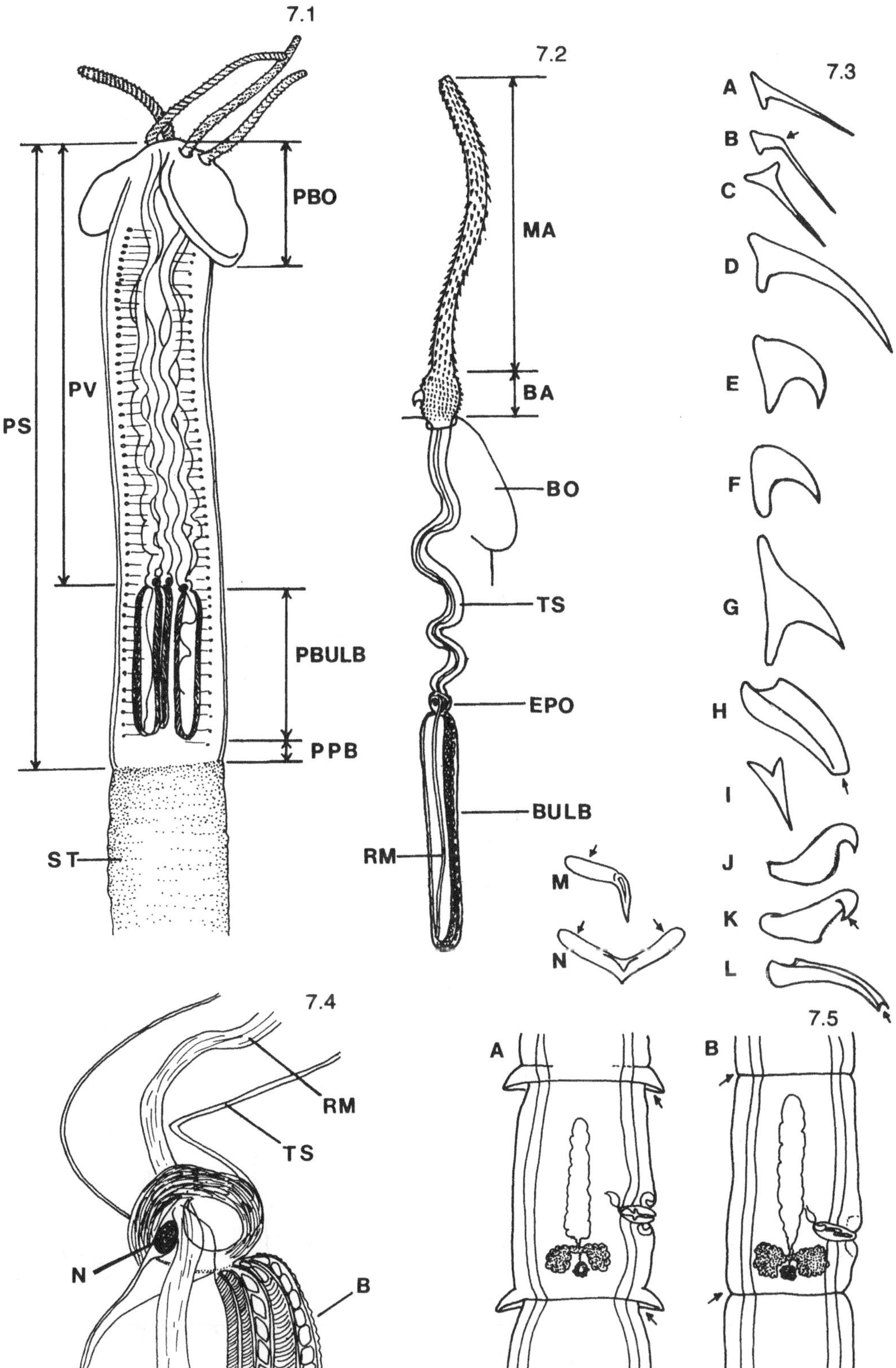
7.1
PBO
PV
PS
PBULB
PPB
ST
7.2
MA
BA
BO
TS
EPO
BULB
RM
7.3
A
B
C
D
E
F
G
H
I
J
K
L
M
N
7.4
RM
TS
N
B
7.5
A
B

7.6

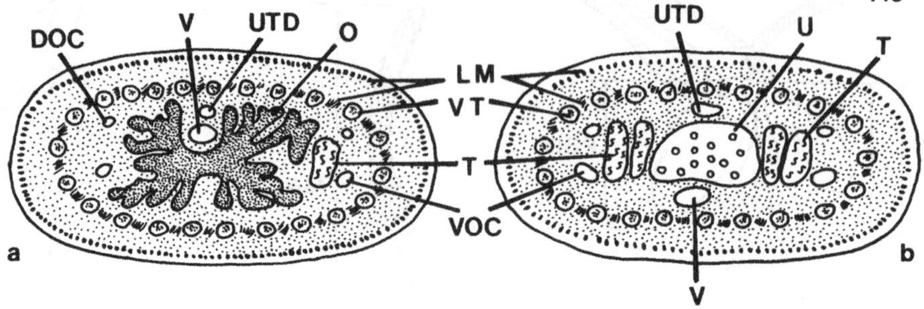

PROGLOTTIS ANATOMY

7.7

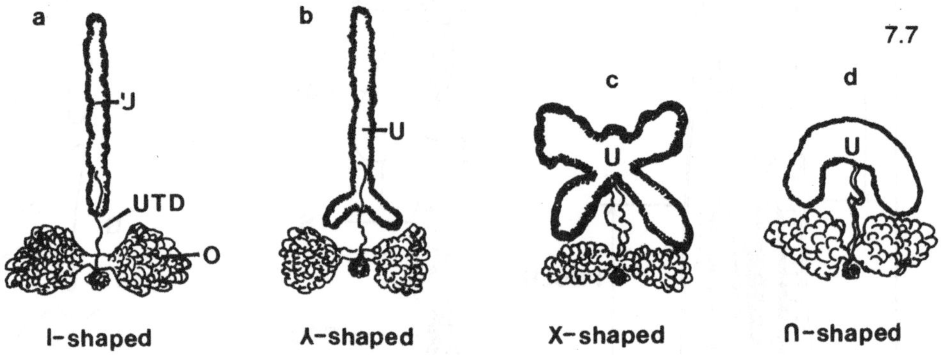

UTERINE MORPHOLOGY

7.8

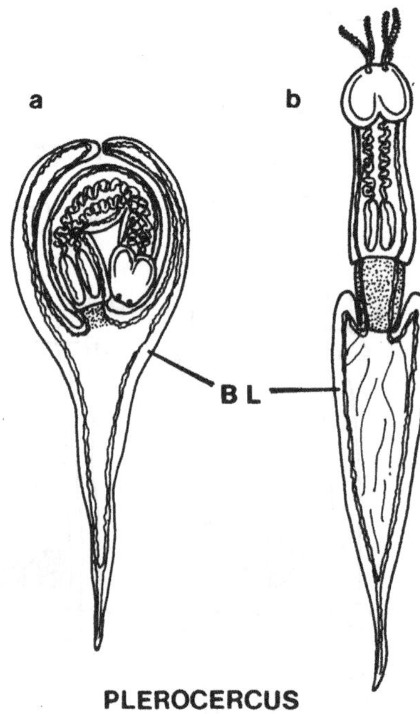

PLEROCERCUS

7.9

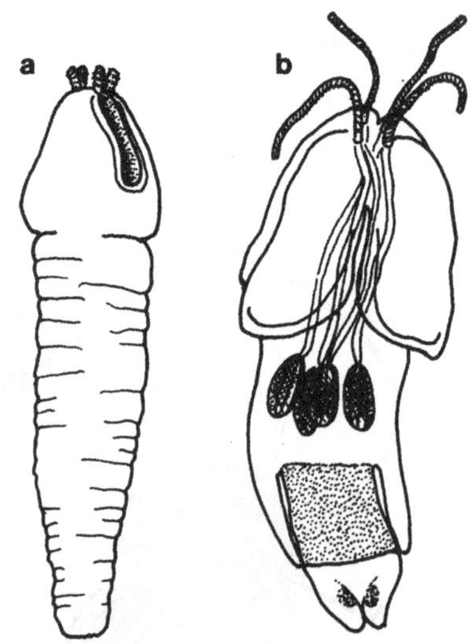

POSTLARVAE

genitalia have not received the attention they deserve despite the fact that four types were discussed by Dollfus (1942; 42–44). In addition to these, union of the vagina and male duct to form a hermaphroditic duct or hermaphroditic vesicle are now known (Figs 7.10–7.16) and future works should include examinations of sections of internal anatomy.

The pattern of the tentacle armature is an essential diagnostic character and can only be determined by studying the arrangement of hooks around the tentacle. To interpret correctly the armature, it is indispensable to recognize opposite surfaces of the tentacle. Opposing surfaces, or faces, are termed: 'external' opposite 'internal' and 'bothridial' opposite 'antibothridial' (Figs 7.17–7.24). The orientation of the armature pattern can only be determined from everted tentacles in wholemounts. However, drawing the entire pattern of the tentacle armature is most easily accomplished by cutting the tentacle free at its base and mounting it in a drop of glycerine beneath a coverglass on a microscope slide. By slight movements of the coverglass the tentacle can be rolled over to give a clear view of each surface.

Dollfus (1942) divided the tentacle armatures into three basic types: homeacantha, heteracantha (typica and atypica) and poecilacantha (Figs 7.25–7.35) and used them as superfamilies. As noted by Dollfus (1942; p. 294) the bilateral symmetry of the hook patterns typically runs between the internal/external faces when the hooks are arranged into half spiral rows. By drawing the hook pattern in a two-dimensional figure, the armature pattern and plane of bilateral symmetry becomes readily apparent (Fig. 7.24). Hook rows typically begin on the internal face (Fig. 7.18) and end on the external face (Fig. 7.19). Hooks in rows spiralling around the bothridial face are numbered in sequence as (1′2′3′, etc.), those spiralling to the antibothridial face are numbered (1, 2, 3, etc.) The bilateral nature of the pattern symmetry of the homeoacanths was not considered by Dollfus but, in fact, it does exist and is well exemplified by numerous species of the genus *Nybelinia*. The bilateral symmetry of the entire armature is evident in the changing hook form as the rows continue their spiral around to the opposite surface of the tentacle. Despite the fact that the hook files alternate so perfectly that they form quincunxes (Figs 7.25, 7.26, 7.36, 7.37), the heteromorphous hooks possessed by many species impart a clear plane of bilateral symmetry to the pattern between the internal and external surfaces of the tentacle (Figs 7.27, 7.28, 7.38, 7.39). In the genus *Kotorella* recently described by Euzet & Radujkovic (1989), an 'unwrapped' drawing of the armature in two dimensions clearly reveals the quincunxial pattern of the armature and heteromorphous nature of the hooks with the plane of bilateral symmetry extending through the bothridial to antibothridial faces (Figs 7.28, 7.39). Only in homeoacanths with

Figs 7.6–7.9 Proglottid anatomy and larvae. 7.6. Proglottid, internal anatomy. A, transverse section through ovary. B, Transverse section through preovarian region. DOC, dorsal osmoregulatory canal; LM, longitudinal muscle; T, testis; U, uterus; UTD, uterine duct; V, vagina; VOC, ventral osmoregulatory canal; VT, vitelline follicle. 7.7. Uterine morphology. A, I-shaped (*Eutetrarhynchus*). B, inverted Y-shaped (*Rhinoptericola*). C, X-shaped (*Didymorhynchus*). D, inverted U-shaped (*Nybelinia*). 7.8. Plerocercus. A, enclosed within blastocyst (BL). B, everted scolex. 7.9. Postlarvae. A, *Hepatoxylon*. B, *Nybelinia*.

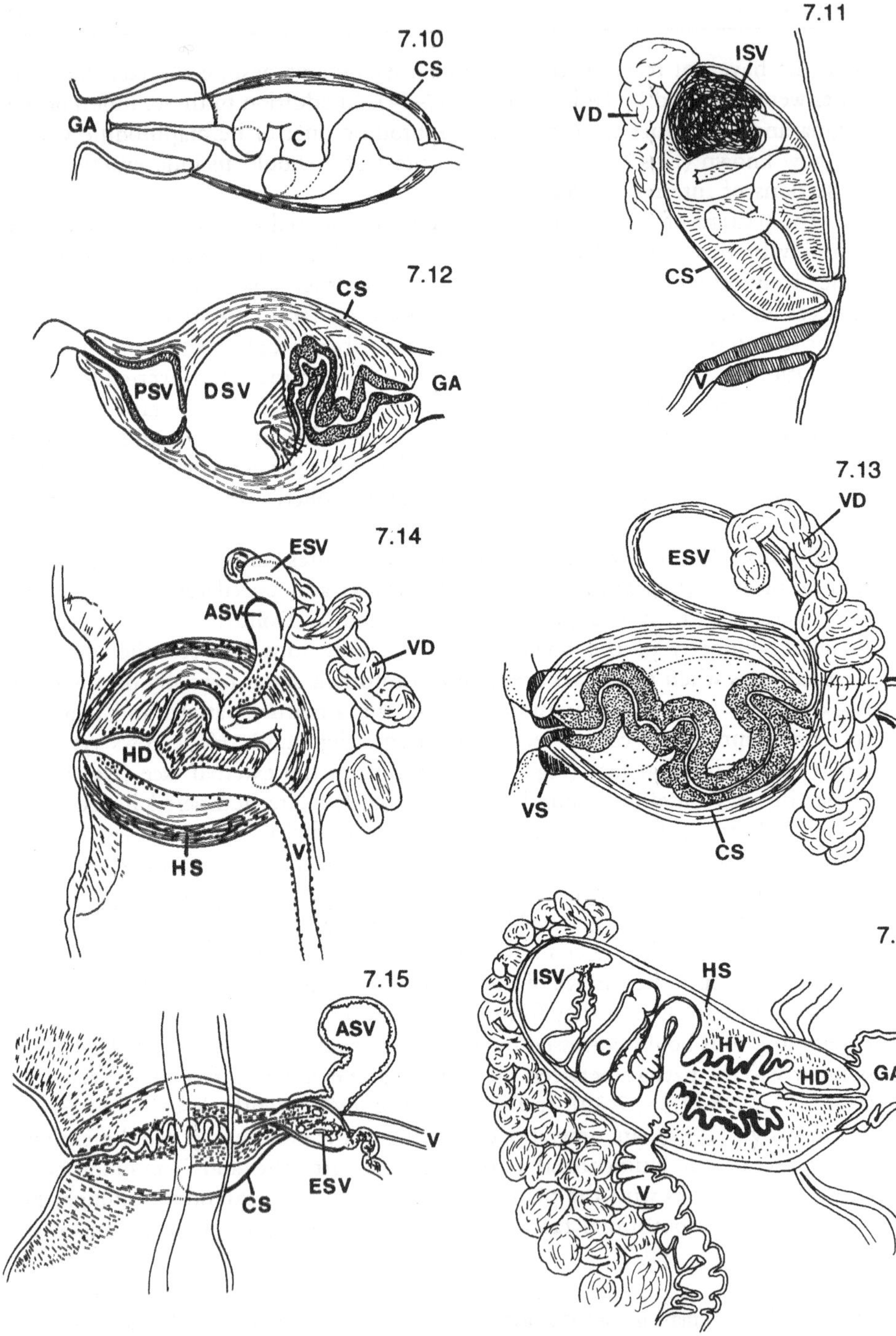

Figs 7.10–7.16 Terminal genitalia of the Trypanorhyncha. 7.10. *Shirleyrhynchus butlerae* Beveridge & Campbell, 1988. 7.11. *Eutetrarhynchus owensi* Beveridge, 1990. 7.12. *Tetrarhynchobothrium rossii* (Southwell, 1912) after Beveridge & Campbell, (1988b). 7.13. *Oncomegas wageneri* Dollfus, 1929 after Toth *et al.* (1992). 7.14. *Stragulorhynchus orectolobi* Beveridge & Campbell, 1989. 7.15. *Chimaerarhynchus rougetae* Beveridge & Campbell, 1989. 7.16. *Hornelliella macropora* (Shipley & Hornell, 1906). (Abbreviations: ASV, accessory seminal vesicle; C, cirrus; CS, cirrus-sac; DSV, distal seminal vesicle; ESV, external seminal vesicle; GA, genital atrium; HD, hermaphroditic duct; HS, hermaphroditic sac; HV, hermaphroditic vesicle; ISV, internal seminal vesicle; PSV, proximal seminal vesicle; V, vagina; VD vas deferens; VS, vaginal sphincter.)

homeomorphous hooks (of the same shape) arranged in quincunxes (Figs 7.25, 7.36), or in continuous spirals (Figs 7.26, 7.37), is the entire hook pattern so perfectly symmetrical that no particular plane of bilateral symmetry is apparent between the opposite faces of the tentacle. Not only are there numerous examples of homeoacanths with heteromorphous hooks and a predominant plane of bilateral pattern symmetry (Figs 7.27, 7.28, 7.38, 7.39) but the tendency for the hooks to form widely separate spiral rows by distortion of the quincunxial arrangement is clear (some species of *Nybelinia* Figs 7.26, 7.27; *Kotorella*; *Pseudonybelinia*; *Paranybelinia*) and further evidence of transition in the recognized hook patterns by alteration of hook arrangement and form along the rows between opposing faces of the tentacle.

Application of pattern isometry

The interpretation and differences between trypanorhynch armatures may be clarified by application of pattern isometry. Obviously the spiral distribution of hooks around a trypanorhynch tentacle represents a regularity of symmetry, though not balanced because of the perpetual rotation. Balanced symmetry is characteristic of patterns in a plane and is described by using the geometric tool of rigid motion, or **isometry**. In its most exact sense, an isometry or rigid motion, means that for plane figures all angles and distances, shapes and sizes, remain unchanged by a particular isometry. For plane figures there are only four possible isometries: **reflection, rotation, translation** and **glide reflection** (Albers *et al.*, 1988). Although it can be argued that the tentacle armatures are three-dimensional and do not occur in a flat plane, the transformation of complex three-dimensional figures into two-dimensional figures has been routinely employed by physicists in crystallography for ease of interpretation of symmetry. We have found that such transformations greatly facilitate the ease of distinguishing armature patterns of trypanorhynchs and their comparisons.

1. The simplest isometry is **reflection** in a line, a mirror image, commonly called bilateral symmetry. Certainly, the line of bilateral symmetry (reflection) for the three-dimensional human figure is recognized as the midsagittal plane through the dorsoventral body surfaces, leaving the lateral sides mirror images of each other. This is also true of tentacle armature patterns in trypanorhynchs where the rows begin on one surface of the tentacle and spiral around the tentacle to end on the opposite surface. For example, in heteroacanths (Figs 7.29–7.33) rows typically begin on the internal surface and end on the external tentacular surface. Lateral views of the hook pattern (bothridial/antibothridial) are as much mirror images as the right and left lateral views of the human body.
2. The second plane isometry is **rotation**. A figure has rotational symmetry if it has a centre about which it can be rotated by a certain angle but remains unchanged in appearance. A square is such a figure because rotation of 90 degrees (one-quarter turn) about its centre leaves it unchanged. A rectangle requires a 180 degree rotation (one-half turn) to leave its appearance unchanged – another example is rotation of the letter **S**. Rotational symmetry is, of course, easily

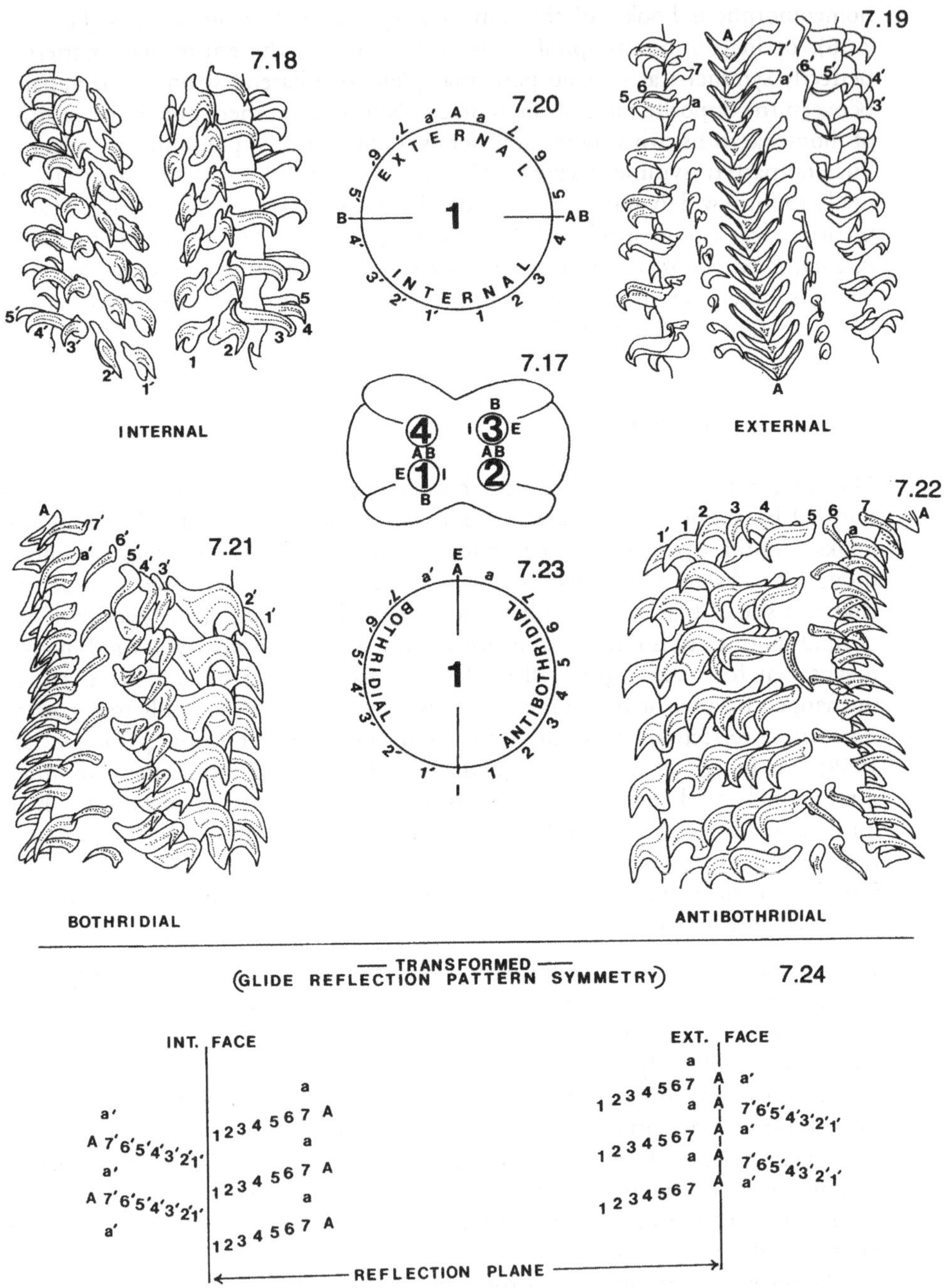

Figs 7.17–7.24 Tentacle armature of *Floriceps saccatus* Cuvier. 7.17. En face view of scolex showing position of four tentacles (diagrammatic). Numbering of tentacles and faces of tentacle labelled after Dollfus (1942). The numbering is arbitrary. Note that the orientation of the armatures on tentacles 1 and 3 are identical, as are those of tentacles 2 and 4. B, bothridial face; AB, antibothridial face; E, external face; I, internal face. 7.18. Internal face of tentacle number 1. The bothridial face is on the left (1′,2′,3′, etc). 7.19. External face of tentacle number 1. The

recognized in the quincunxes of the homeoacanthous armature patterns (Figs 7.25 & 7.35, 7.26 & 7.37).

3. The third plane isometry is **translation**, in which the figure has translation symmetry if it can be slid along itself (translated), without turning, and without appearing to have changed. An example is: VVVVVVVVVVVV.

4. The fourth plane isometry is called **glide reflection** which is a combination of **translation** (= glide) and **reflection** (bilateral symmetry) in a line parallel to the direction of translation. For example: bpbpbpbpbp (See Figs 7.28 & 7.39, 7.30 & 7.41, 7.34 & 7.45.)

Indeed, the fact that symmetries occur in combinations allows simple classification of all patterns, according to which combinations of the four fundamental isometries leave a pattern invariant (Albers *et al.*, 1988). Therefore, while the Fibonacci sequence used in phyllotaxis (Thompson, 1942), and digitalized computer image analysis have not been applied to trypanorhynch armatures, the pattern isometries of less complex two-dimensional figures can be readily recognized if the three-dimensional armature patterns are transformed ('unwrapped') into two-dimensional form. A correctly drawn figure must take into account the direction of the reflection plane (mirror image) whenever such a plane of bilateral symmetry exists.

Transformation of various hook armatures of trypanorhynchs into two-dimensional figures revealed the repetition of two plane isometries: rotational and glide reflection. **Rotational symmetry** is imparted by the **quincunxial** arrangement of hooks or the presence of continuous regularly spaced **spirals** in the homeoacanthous type armatures in which all hooks are of the same size and form (**homeomorphous hooks**). **Glide reflection symmetry** exists in armature patterns consisting of continuous **spirals** of hooks that undergo a regular change in form on opposing surfaces (**heteromorphous hooks**) or in armature patterns consisting of **half-spiral** rows of hooks, ascending to the right and left around the tentacle to end on the opposite surface. Glide reflection symmetry is characteristic of homeoacanths with hooks of different size and form (heteromorphous) on opposite tentacle surfaces. Glide reflection symmetry is also characteristic of all heteroacanthous and poeciloacanthous armatures because the hooks are arranged in identical righthand and lefthand half-spiral rows.

It should be noted that numerous variations in symmetries and combinations of symmetries are possible and do exist in trypanorhynch armature patterns (Figs 7.25–7.46). For example, close inspection of various homeoacanthous

bothridial face is on the right (3′,4′,5′,6′,7′) and satellite hook (a′). Chainette elements are marked (A). 7.20. Schematic diagram of the tentacle in cross-section. Note that the hooks are arranged in half-spiral rows beginning on the internal face 1(1′) and ending at the chainette (A) on the external face. 7.21. Bothridial face of tentacle number 1. 7.22. Antibothridial face of tentacle number 1. 7.23. Schematic diagram of the tentacle in cross-section. Note orientation of hook rows and compare with 7.20. 7.24. Armature of *F. saccatus* transformed into two-dimensional plane isometry. The pattern demonstrates glide reflection symmetry, a combination of bilateral (reflection) and translation (glide) symmetries. Note the repetition in the pattern and orientation of the bilateral (reflection) plane through the axis of the internal/external faces of the tentacle.

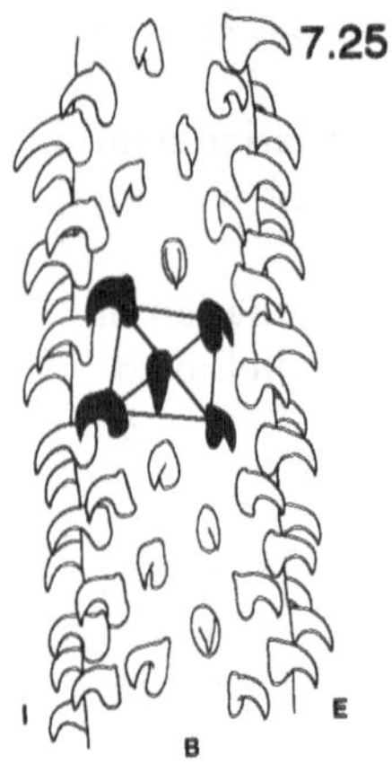

7.25

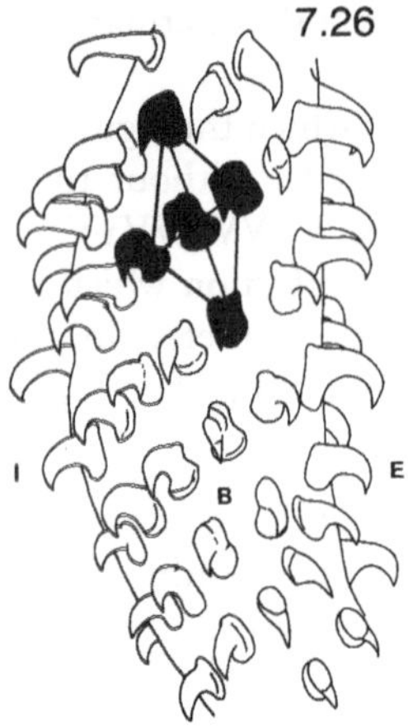

7.26

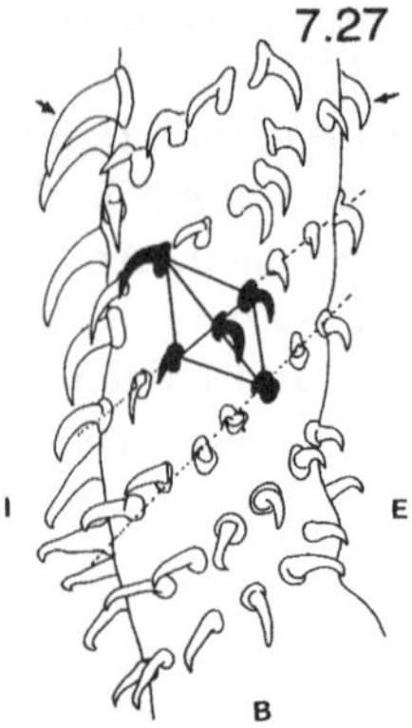

7.27

7.28

7.29

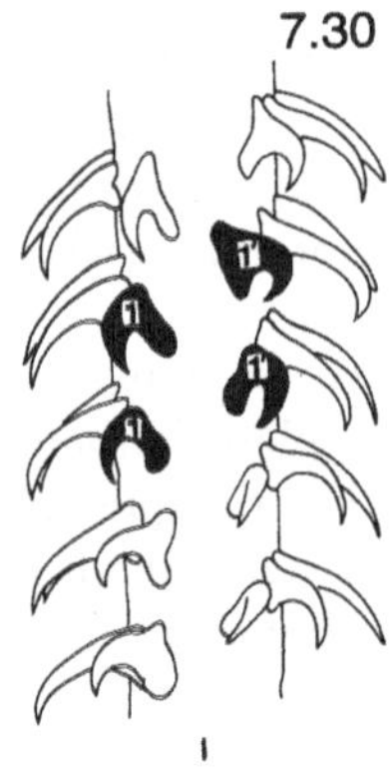

7.30

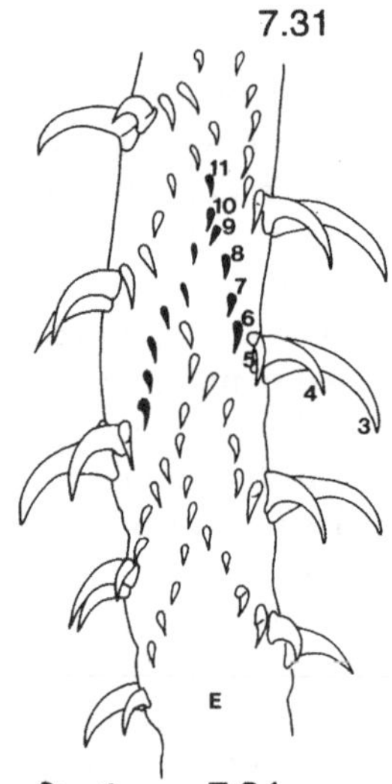

7.31

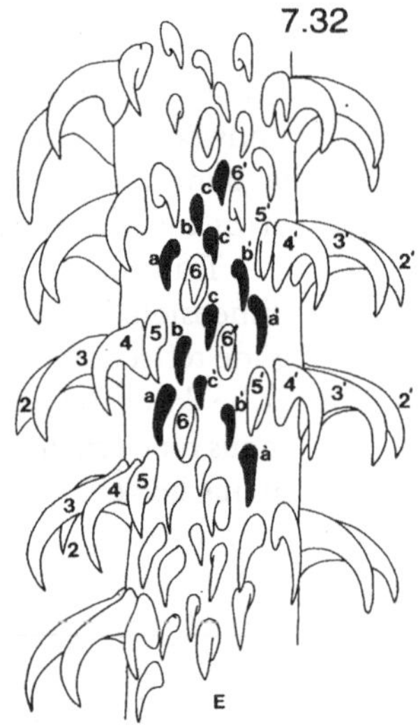

7.32

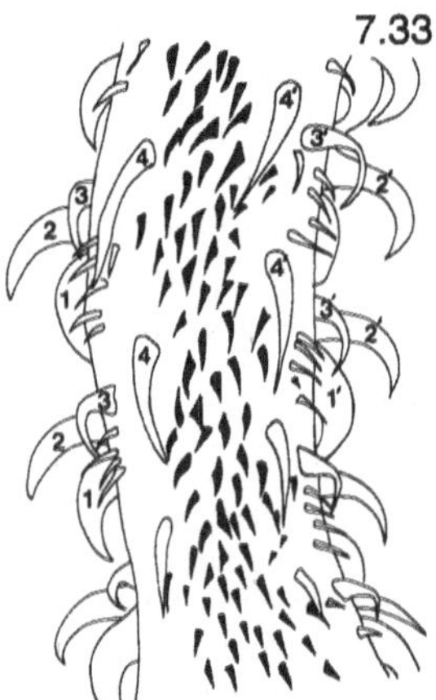

7.33

7.34

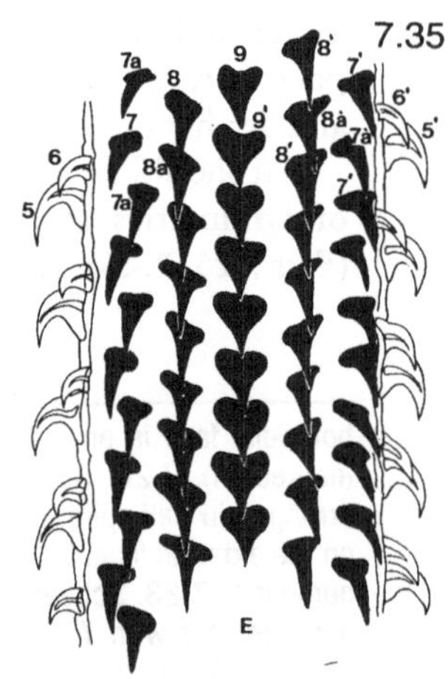

7.35

armatures reveals subtleties in the degree of rotational symmetry, especially in species of *Nybelinia*, where quincunxes show distortion from perfect 90 degree rotational symmetry. Examples of homeoacanths with homeomorphous hooks and near perfect 90 degree rotational symmetry are *N. oodes, N. riseri, N. edwinlintoni, N. goreensis, N. strongyla* (Fig. 7.25), and *Tentacularia coryphaenae*. Species showing distortion of the quincunxial arrangement into rectangles with 180 degree rotational symmetry are *N. lingualis*, and *N. rougetcampanae* (see Dollfus, 1960). The final result of distortion of the quincunxes is the formation of diagonally ascending rows as seen in *N. africana* (Fig. 7.26), *N. yamagutii* (Fig. 7.27), and the genera *Paranybelinia* and *Pseudonybelinia*. However, when homeoacanthous species have heteromorphous hooks, the change in hook form imparts a combination of bilateral (reflection) and translation (= glide) symmetry on the quincunxial pattern (Figs 7.27, 7.38; *N. yamagutii*). For most trypanorhynch armatures the plane of bilateral symmetry appears oriented through external/internal surfaces. This plane of bilateral symmetry is clearly seen in *N. yamagutii* and *Kotorella pronosoma* (Figs 7.27 & 7.38, 7.28 & 7.39). However, in the case of *N. estigmena* and *N. dakari* the heteromorphous hooks are opposite in a plane of bilateral symmetry through the bothridial and

Figs 7.25–7.35 Armatures of homeoacanths, heteroacanths and poeciloacanths. 7.25. *Nybelinia strongyla* Dollfus (after Dollfus, 1960). Homeoacanthous armature with homeomorphous hooks. Quincunxes (black) demonstrate almost perfect 90 degree rotational symmetry. 7.26. *Nybelinia africana* Dollfus (after Dollfus, 1960). Homeoacanthous armature with homeomorphous hooks. Hooks in files are more widely spaced than hooks in rows resulting in distortion of quincunxes (black) and 180 degree rotational symmetry. Distortion creates continuous spiral rows. 7.27. *Nybelinia yamagutii* Dollfus (after Dollfus, 1960). Homeoacanthous armature with heteromorphous hooks. Bothridial view. Note the distortion of quincunxes (black) to form spiral rows. Hooks of the internal face (left) are larger than those of the external face (right). 7.28. *Kotorella pronosoma* (Stossich). Drawn from types. Homeoacanthous armature with heteromorphous hooks. Hooks of bothridial face (B) are largest and form tight quincunxes (black) but quincunxes of smaller hooks form widely spaced quincunxes (black) on opposite face (AB). 7.29. *Oncomegas wageneri* Linton (after Toth *et al.*, 1992). Heteroacanthous armature. Internal face, metabasal armature. Note the close spacing of hooks 1(1′) at convergence of opposing rows. 7.30. *Parachristianella baverstocki* (after Beveridge, 1990a). Heteroacanthous armature. Internal face, metabasal armature. Note the distinct separation of hooks 1(1′) demonstrating divergence of the opposing rows. 7.31. *Parachristianella monomegacantha* Kruse (after Beveridge, 1990a). External face, metabasal armature demonstrating the typical heteroacanthous arrangement in which the principal rows form inverted V's (black) and no extra rows of hooks are interpolated between them. 7.32. *Otobothrium insigne* Linton (redrawn after Hildreth & Lumsden, 1985). External face, metabasal armature demonstrating the atypical heteroacanthous arrangement in which extra hook rows (a,b,c in black) are interpolated between the principal rows. 7.33. *Grillotia recurvispinis* Dollfus (after Dollfus, 1969a). External face, metabasal armature demonstrating the band of hooks (black). Note the rows of interpolated hooks between the principal hook rows and their proximity to the band of hooks. Considered an atypical heteroacanthous armature in this key. 7.34. *Halysiorhynchus macrocephalus* Shipley & Hornell (redrawn from types). External face of metabasal armature showing simple chainette of V-shaped elements (black) demonstrating poeciloacanthous armature. 7.35. *Patellobothrium quinquecatenatum* Beveridge & Campbell (after Beveridge & Campbell, 1989). External face of metabasal armature demonstrating poeciloacanthous armature consisting of five chainettes. Note the alternation of the nine hooks of the principal rows. A pair of interpolated hooks (7a, 8a) exists between successive principal rows and have become incorporated by form and position to form chainettes lateral to 9(′).

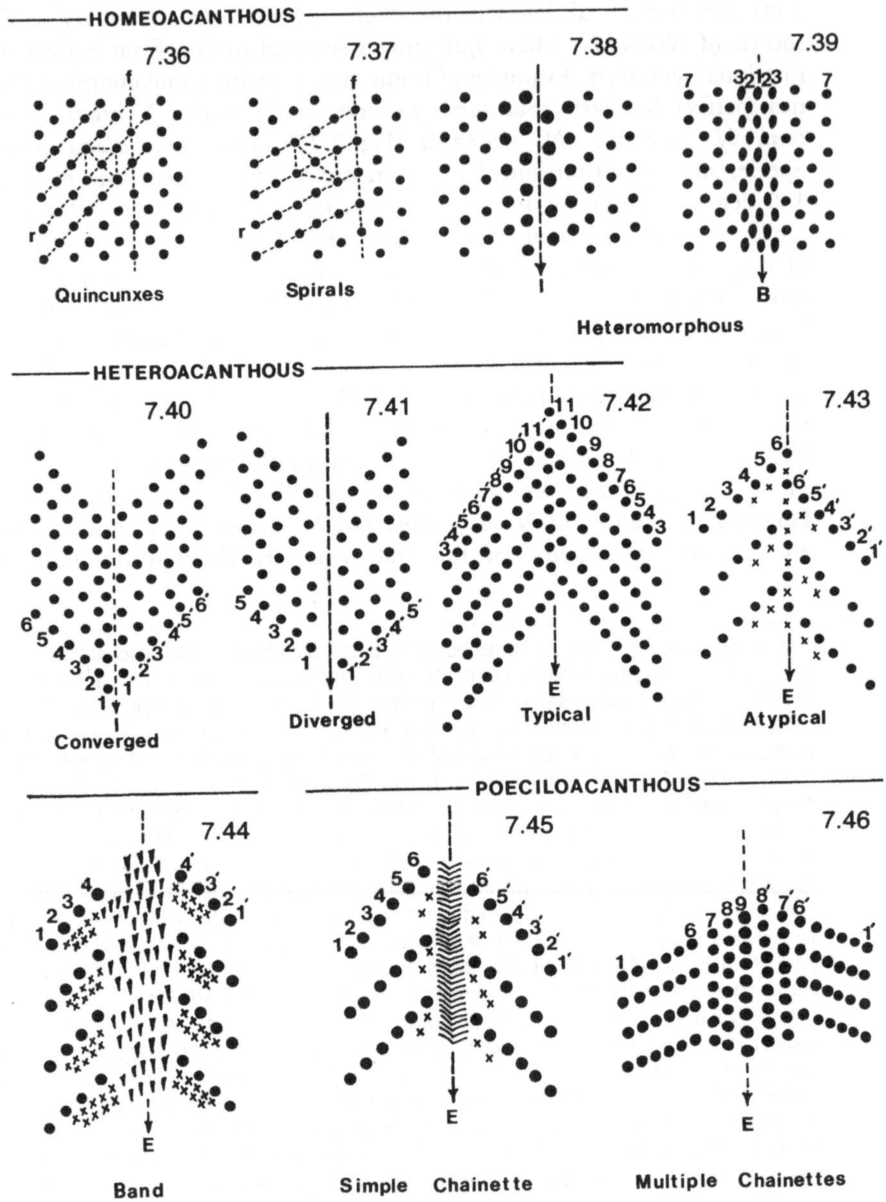

Figs 7.36–7.46 Transformed patterns of trypanorhynchs showing pattern symmetry. 7.36, 7.37. Homeoacanthous patterns with homeomorphous hooks. 7.36. Homeomorphous pattern showing 90 degree rotational symmetry of quincunxes, ex. *Nybelinia strongyla* Dollfus. 7.37. Homeomorphous pattern showing distortion of quincunxes into continuous spiral rows showing 180 degree rotational symmetry, ex. *Nybelinia africana* Dollfus. (abbreviations: r, principal row; f, longitudinal file.) 7.38, 7.39. Homeoacanthous patterns with heteromorphous hooks and glide reflection symmetry superimposed on rotational symmetry of the quincunxes. 7.38. *Nybelinia yamagutii* Dollfus showing largest hooks on the internal face. Arrow shows plane of bilateral symmetry viewed through the internal face (I). Bilateral symmetry is superimposed on 180 degree rotational symmetry of the quincunxes.

antibothridial faces (see Dollfus, 1960). In *Kotorella* both the spacing of the quincunxes and hooks change markedly in a plane of bilateral symmetry between the bothridial and antibothridial surfaces (Figs 7.28, 7.39).

The transformed two-dimensional pattern symmetry of the metabasal armatures of the heteroacanths, both typical and atypical, and the poeciloacanths are clearly of the glide reflection type. In these armature patterns the half-spiral rows are divided by a plane of bilateral symmetry through the external/internal surfaces and the direction of the translation of the pattern is along the long axis of the tentacle (Figs 7.29–7.35, 7.40–7.46). However, among the typical heteroacanthous armature patterns of the Eutetrarhynchidae, two distinctly different variations in pattern symmetry are seen. In the type-species, *Eutetrarhynchus ruficollis*, the half-spiral rows are almost exactly opposite one another and the pattern converges on a single longitudinal file of hooks on the plane of bilateral symmetry when viewed from either the external or internal face. This is also true of the genus *Oncomegas* (Figs 7.29, 7.40). In the eutetrarhynchid *Trigonolobium spinuliferum* there is a very slight separation but, in other eutetrarhynchid genera, distinctly separate half-spiral rows alternate to the right hand and left of the plane of symmetry stepwise up the tentacle (Figs 7.30, 7.41). Among all other known heteroacanths, typical or atypical, evidence of these two variations on the bilateral plane of pattern symmetry can be found (Figs 7.29 & 7.40, 7.30 & 7.41, 7.31 & 7.42, 7.32 & 7.43). For example, in the Otobothriidae, in *Otobothrium dipsacum* the opposing rows are convergent compared to the divergence of its congeners and in *Plesiorhynchus etmopteri* there is only a very slight separation where

7.39. *Kotorella pronosoma* (Stossich, 1901) showing large hooks in closely spaced quincunxes on bothridial face (B). Plane of bilateral symmetry oriented through the antibothridial and bothridial faces. Rotational symmetry is 180 degrees. Rotation of figures will demonstrate rotational symmetry. Translation is only possible in the long axis. 7.40–7.44. Heteroacanthous armature patterns. All types consist of hooks in half-spiral rows. Arrows indicate plane of bilateral symmetry through internal/external faces of the metabasal armature in all examples. All demonstrate glide reflection pattern symmetry. 7.40. Internal face of *Oncomegas wageneri* Linton showing convergence of half-spiral rows due to close proximity of hooks 1(1′). Note inverted V formation. Compare carefully with Figs 7.36 and 7.38 above. 7.41. Internal face of *Parachristianella baverstocki* Beveridge showing distinct space between hooks 1(1′) indicating divergence of principal rows characteristic of most heteroacanths and the poeciloacanths. 7.42. External face (E) of *Parachristianella monomegacantha* Kruse demonstrating inverted V formations at terminations of the alternating half-spirals characteristic of typical heteroacanths. 7.43. External face (E) of *Otobothrium insigne* Linton showing the interpolation of an extra row of three hooks (X) adjacent to principal hooks on the external face (E). The presence of additional hook rows on the external face is characteristic of atypical heteroacanths. 7.44. External face (E) of *Grillotia recurvispinis* Dollfus showing the band of small hooks occupying the space between the termination of the principal rows. Hooks of the band appear disorganized but follow a sinuous pattern along the reflection plane (arrow). Note the numerous interpolated hooks in rows adjacent to each principal row and how they merge with similar hooks forming the band (see 7.33). 7.45, 7.46. Poeciloacanthous patterns with chainettes. Both demonstrate glide reflection symmetry. The reflection plane is indicated by the arrow as viewed from the external face (E). 7.45. Simple chainette of a single file of V-shaped hooks. *Halysiorhynchus macrocephalus* (Shipley & Hornell). Note the interpolated hooks (X). 7.46. Multiple chainettes (five) of *Patellobothrium quinquecatenatum* Beveridge & Campbell. Note that all chainette hooks are of similar shape and form close files. Compare with 7.43.

other members of the Gilquiniidae are clearly separated. In all known poeciloacanths the rows are widely separate and arranged in alternate half-spirals (Figs 7.18, 7.19, 7.34, 7.35). Therefore, in the majority of heteroacanths and all poeciloacanths, the plane of bilateral symmetry lies in the distinct space separating two identical files of hooks which begin the principal rows, numbered 1(1′) by Dollfus (1942).

The isometries of the metabasal tentacle armatures of the trypanorhynchs currently recognized are as follows:

1. **Homeoacanths** – (hooks form quincunxes or continuous spiral rows; two symmetries)
   a) Type I (rotational symmetry; hooks homeomorphous, arranged in quincunxes)
   b) Type II (glide reflection symmetry; hooks heteromorphous imparting reflection plane)
      i. plane of reflection (bilateral) through internal/external tentacle surfaces.
      ii. plane of reflection (bilateral) through bothridial/antibothridial tentacle surfaces.
2. **Heteroacanths** – (hooks form half-spiral rows; glide reflection symmetry)
   a) Typical heteroacanths – number of hook rows constant on internal/external surfaces of tentacle.
      i. convergent files 1(1′) – a single file of hooks lies adjacent to the plane of bilateral symmetry at the origin where the V-pattern of half-spiral rows originate.
      ii. divergent files 1(1′) – a space exists between the first hooks of opposing half spiral rows in the plane of bilateral symmetry.
   b) Atypical heteroacanths – extra hooks or hook rows interpolated on external surface of tentacle.
      i. satellite hooks
      ii. extra rows
      iii. continuous band
3. **Poeciloacanths** – (a distinctive hook file or files forms one or more chainettes, typically on the reflection plane of the external tentacular surface; glide reflection symmetry)
   a) simple chainette (single or double files of hooks)
   b) compound chainette (several types and numbers of hooks)
   c) multiple chainettes

Classification

Historically the Order Trypanorhyncha has been divided into two suborders, Atheca (= Acystidea) and Thecaphora (= Cystidea). Species with a homeoacanthous armature and plerocercoid-like 'post-larva' presumed to lack a blastocyst are placed within the suborder Atheca. Species with a plerocercus enclosed by a blastocyst and heteroacanthous or poeciloacanthous armature have been placed in the Thecaphora. Within the Thecaphora, major groups were the 'typical' and 'atypical' heteroacanths, and the poeciloacanths with either a chainette or band of hooks, each group based upon its characteristic armature patterns (Figs 7.25–7.35).

The traditional subdivisions into the Atheca and Thecaphora on the basis of armature should not be considered fixed, immutable and inviolate. Beveridge & Campbell (1988b) showed that the thecaphoran family Tetrarhynchobothriidae, with three genera, has an armature pattern previously only associated with Atheca. This breaches the traditional division between the suborders but, more importantly, demonstrates that the variations in the armature patterns of thecaphoran trypanorhynchs are a gradual transition from homeoacanthous (= quincunxial) to heteroacanthous patterns rather than an abrupt division as implied by the subordinal classification. The armatures can be recognized as possessing pattern isometry in which there is a gradual transition in the pattern symmetry from the rotational symmetry of the homeoacanths to the glide reflection symmetry of the heteroacanths (Figs 7.36–7.46). Atypical heteroacanthous patterns are derived from typical heteroacanths. All the known poeciloacanth armatures can be derived from that of the typical heteroacanth simply by altering the spacings of hook files. Using *Lacistorhynchus* and *Callitetrarhynchus* as examples, increasing the space between hook files 8(8′) and 9(9′) leaves a central alternating chainette, with one element contributed from each principal hook row (Figs 7.340, 7.346). The chainette is more readily recognizable in these genera by a change in hook shape, by hooks 7(7′) and 8(8′) occupying a satellite position and by hooks 6(6′) being reduced in size and displaced anteriorly. Rather than forming a double chainette as described above, the elements of the chainette can merge to form a single file of hooks, as in the genus *Floriceps* (Fig. 7.325). In both single and double chainettes, the number of elements between principal rows can be increased. The chainette of *Halysiorhynchus* (Fig. 7.412) resembles that of *Floriceps*, but has an increased number of elements per principal row. In species of *Dasyrhynchus* (Fig. 7.353), there is an extra chainette element associated with each intercalary row of hooks, so that multiples of two and four times the numbers of chainette elements per principal hook row occur. A further modification in which the 'satellite' hooks 7(7′) and 8(8′) become chainettes gives rise to the three chainettes found in *Diesingium* (Fig. 7.382). A compound chainette of the type seen in *Hornelliella* (Fig. 7.363) is formed when hooks 7(7′) are separated from 8(8′) but hooks 8(8′) remain adjacent to hooks 9(9′).

Among the atypical heteroacanths, modifications are based on the interpolation of hooks or hook rows. Interpolations range from a single interpolated hook per principal row, a small row of three hooks in *Rhinoptericola* (Fig. 7.268), to a short row arising from each side as in *Otobothrium insigne* (Fig. 7.255). In *Poecilancistrum* (Fig. 7.263) there are four additional rows of hooks for each principal row, and these interweave to form a continuous band of hooks on the external surface of the tentacle. The more irregular bands of hooks seen in species of *Grillotia* (Fig. 7.298) may result from secondary loss, a process which is well demonstrated in *Pterobothrium* (Figs 7.276, 7.280). Thus, the term 'poeciloacanth', used in the past to include hook patterns with a chainette or with 'bands' of hooks is best restricted to families with chainettes and considered unrelated to those with 'bands' of hooks. The definition of the 'atypical heteroacanthous' armature pattern, in which 'there are more rows of hooks on the external face than on the internal face', suitably accommodates those families with 'bands' of hooks. We consider the Grillotiidae, Pterobothriidae and Molicolidae with their interpolated hooks and 'bands' of hooks to be 'atypical heteroacanths' and have separated these families from their former inclusion with the poeciloacanthous families. The proposed changes retain the concept of the superfamily classification of Dollfus (1942) but re-arrange the families in a more logical fashion based upon transitions in pattern symmetry as follows:

1. **Homeacanthoidea** (for Homeacantha Dollfus, 1942); homeoacanths (two symmetrical types) (Hepatoxylidae, Paranybeliniidae, Sphyriocephalidae, Tentaculariidae, Tetrarhynchobothriidae);
2. **Heteracanthoidea** (for Heteracantha Typica Dollfus, 1942); typical heteroacanths (two pattern types) (Eutetrarhynchidae, Gilquiniidae, Shirleyrhynchidae);
3. **Otobothrioidea**, status emend. (for Heteracantha Atypica Dollfus, 1942); atypical heteroacanths (with extra hooks, rows or a band) (Otobothriidae, Rhinoptericolidae, Grillotiidae, Pterobothriidae, Molicolidae);
4. **Poecilacanthoidea** (for Pecilacantha Dollfus, 1942); poeciloacanths (one or multiple chainettes) (Lacistorhynchidae, Dasyrhynchidae, Hornelliellidae, Mustelicolidae, Gymnorhynchidae, Mixodigmatidae).

It should be noted that Dollfus (1942) suggested three superfamilies (Homeacanthides, Heteracanthides, and Poecilacanthides) for the formal classification in a footnote (p. 85) but did not use the appropriate endings specified by recommendation 29A of the *International Code of Zoological Nomenclature*. In later publications Dollfus (e.g. 1969a) used four superfamily categories, citing his monograph (Dollfus, 1942), by splitting the Heteracantha into Typica and Atypica despite the fact that he did not suggest formal names. Spellings used by Dollfus were also inconsistent (Poecil-, Pecil-) and anglicized by other authors (Wardle & McLeod, 1952). We, like Joyeux & Baer (1961), prefer the use of superfamilies to the use of questionable suborders and have followed the *ICZN* recommendation (29A) in conserving and adapting Dollfus' concept to the formal classification. We suggest that the following organization not only facilitates identification but will allow application of pattern notation, morphometric analysis and may also reflect phylogeny.

Interpretation of armatures has been a source of confusion in classification of certain families. Dollfus attempted to subdivide the genus *Grillotia* by creating the subgenera *Grillotia* Dollfus, 1969, *Paragrillotia* Dollfus, 1969 and *Progrillotia* Dollfus, 1946. *Progrillotia* was elevated to genus (Dollfus, 1969a) but omitted by Schmidt (1986). We reject the subgenus *Paragrillotia* because the type-species, *G. (P.) simmonsi* (= *R. simile* Linton, 1900) does not meet the subgeneric definition and because its character, the interpretation of merging of the interpolated hooks with the external band, is easily confused. We concur with Schmidt (1986) in rejecting family status for *Pseudogrillotia* Dollfus, 1969 on the basis of a craspedote scolex and postlarval stage. Joyeux & Baer (1934) emended the subfamily diagnosis of the Eutetrarhynchinae to include provisionally *Tetrarhynchobothrium*. Dollfus (1969a) removed and elevated the genus to family rank, Tetrarhynchobothriidae, and placed it as a typical heteroacanth in yet another superfamily (Eutetrarhynchoidea). Beveridge & Campbell (1988b) showed the armature of *Tetrarhynchobothrium* was homeoacanthous (rotational pattern symmetry) unlike eutetrarhynchids (typical heteroacanths with glide reflection symmetry) and added two new genera to the family. The quincunxial armature, short bulbs and long pars postbulbosa of species of *Tetrarhynchobothrium* and *Didymorhynchus* show parallels with other homeoacanths (Acystidea/Atheca). The family Otobothriidae is a complex group of atypical heteroacanths in need of revision. *Otobothrium insigne* has only a single interpolated hook per principal row but in *Poecilancistrum* there are multiple interpolated rows. In *Otobothrium dipsacum* the hook files 1(1′) of the armature are unusual in that they are convergent instead of divergent as for congeners. *Bombicirhynchus* Pintner, 1931 was described as a homeoacanth but Dollfus (1942) believed the genus is heteroacanthous, belonging in the Otobothriidae. It is treated here as a *genus incertae sedis*. In some trypanorhynchs an obvious change in armature pattern was thought to be unique when it was not without parallel or change in isometry. For example, *Mixodigma* Dailey & Vogelbein, 1982 was interpreted as possessing two different kinds of armatures when it actually possesses a chainette of reduced length as occurs in the poeciloacanth *Dasyrhynchus talismani* (Dollfus, 1935). A comparable example among atypical heteroacanths is the reduced length of the band of hooks in *Pseudogrillotia peruviana*.

Dollfus (1969a, b) sought to emphasize the distinction between the armatures of *Grillotia* and *Lacistorhynchus* by erecting the new family Grillotiidae (1969b, p. 1046) because it separated the genera with chainettes (*Lacistorhynchus, Callitetrarhynchus, Floriceps*, see Beveridge & Campbell, 1989) from those with interpolated rows and a band of hooks on the external tentacular surface, i.e. *Grillotia*. Dollfus' (1969b) distinction stands in contradiction to the keys of Yamaguti (1959) and Schmidt (1986). Wardle & McLeod (1952) endorsed the classification scheme of Dollfus (1942) but in their key to families (pp. 298–299) erroneously included four families of the suborder Thecaphora (Cystidea) in the suborder Atheca (Acystidea). We agree with Dollfus (1969a) in separating genera with bands from those with chainettes and recognize a further distinction of the armature possessed by grillotiids as being suitably accommodated by the definition of 'atypical heteroacanthous'. Our interpretation is based upon transitions in pattern isometry explained above.

The classification of trypanorhynch genera is characterized by confusion for a variety of reasons. Dollfus (1946) noted that *Corynesoma* Leuckart in Diesing (1835) is a possible synonym of *Lacistorhynchus* but considered it a *nomen rejiciendum* and that *Cotylogenes* Lühe, 1902 was poorly defined, but is a probable synonym of *Lacistorhynchus*. *Christianella* Guiart, 1931 is a synonym of *Grillotia*. *C. minuta* van Beneden, the type-species, was described in detail by Nybelin (1940). This description was overlooked by Dollfus and more recent authors but it helps establish the synonymy of *Christianella* with *Grillotia*. *Armandia* Guiart, 1927 and *Christianella* are probably synonymous. The first and fourth species named to *Pterobothrium* Diesing, 1850, *P. macrourum* (Rudolphi, 1819; type-species) and *P. interruptum* (Rudolphi, 1819), have never been redescribed. Diesing (1850) based the generic characters of *Pterobothrium* upon the third named species, *P. heteracanthum* Diesing, 1850, which Dollfus (1942) accepted and designated the following synonyms: *Synbothrium fragile* Diesing, 1850; *Syndesmobothrium filicolle* Linton, 1890; *Synbothrium filicolle* Linton, 1897; *Gymnorhynchus gigas* (Cuvier, 1817) of Southwell, 1930; *Gymnorhynchus gigas* (Cuvier, 1817) of Chandler, 1935. Family Grillotiidae: Dollfus (1969a, b) used this family name several times without a formal diagnosis (p. 493 as Grillotiidae, Grillotiinae Dollfus, 1942; p. 1046 in 1969b). According to Dollfus (1942), *Diesingium* Pintner, 1929 is preoccupied but we have not been able to confirm this (e.g. Neave, etc.). Dollfus (1942) proposed that *Diesingella* Guiart, 1931 (= *Diesingium*) might belong in the Eutetrarhynchidae but later (Dollfus, 1969b) described it as an atypical heteroacanth, *Mustelicola* representing the family Mustelicolidae. Campbell & Beveridge (1988) showed that *Mustelicola* is a poeciloacanth with multiple chainettes and that *Diesingium* is its senior synonym (Beveridge & Campbell, 1994). As previously explained, the subordinal names (Atheca/Acystidea and Thecaphora/Cystidea) are preoccupied. Some families remain in question. The family Rhopalothylacidae Guiart, 1935 was proposed by Guiart for two genera, *Rhopalothylax* Guiart, 1935 and *Clujia* Guiart, 1935, both of which are genera *incertae sedis*. In the family Pseudogilquiniidae Bilqees & Khatoon, 1980 the type-species, *Pseudogilquinia karachiense*, is described as a heteroacanth with four bothridia but no description is available of the external face of the metabasal armature and it is therefore unrecognizable. Furthermore, *Pseudogilquinia magna* of Bilqees & Kurshid (1985) has been examined and found to have two bothridia and to be a poeciloacanth with a chainette very similar to *Dasyrhynchus*, differing only in bulb length and their projection into the strobila. Therefore, the genus *Pseudogilquinia* Bilqees & Khatoon, 1979 and family Pseudogilquiniidae are inadequately described and are considered as *inquirenda*. Genera *incertae sedis* are: *Acanthorhynchus* Diesing, 1850; *Balanophorus* Braun, 1894; *Bothriorhynchus* Jeude, 1829; *Eulacistorhynchus* Subhapradha, 1957; *Microbothriorhynchus* Yamaguti, 1952 (type specimens cannot be located); *Pintneriella* Yamaguti, 1934; *Rhynchobothrium* Rudolphi, 1819; *Rhynchobothrius* Rudolphi, 1819; *Tetrabothriorhynchus* Diesing, 1850; *Tetrarhynchus* Rudolphi, 1809; *Vaullegeardia* Guiart, 1927; *Wageneria* Monticelli, 1892.

In preparing this key we have attempted to examine all available type-species of recognized genera and as many representative congeners as possible. Numerous attempts were made to clarify 'problem' genera or species descriptions by working through various museum collections. Some of these efforts proved very worthwhile. For others, unfortunately, museum type specimens are no longer available, or their condition was too poor to be accurately included even with their published descriptions. The genera retained in this key are those that are based upon reliable and verified descriptions and satisfactory specimens. Even so, some genera and families could not be defined with the desired accuracy because proglottids are unknown. A formal diagnosis is provided for the family Grillotiidae and a new family, Shirleyrhynchidae n. fam., close to the Gilquiniidae, is proposed and a new genus erected in the Eutetrarhynchidae. The unique relationship of each is clearly shown within the keys.

Phylogeny

No attempt other than that of Dollfus (1942) has been made previously to determine the phylogenetic interrelationships of the various trypanorhynchan families and genera. It is almost certainly still premature, since the strobilar characters of many genera and even families remain unknown. A detailed cladistic analysis will only be possible when existing genera and families are better described and when the undescribed genera which already exist in collections are described. We have chosen simply to present the classification outlined above in a cladistic framework to show our intuitive assessment of possible phylogenetic relationships among major groups of families (Fig. 7.46a). We believe that too few characters are available to provide a reliable cladogram to the level of genus, but we hope that future work will remedy some of these deficiencies.

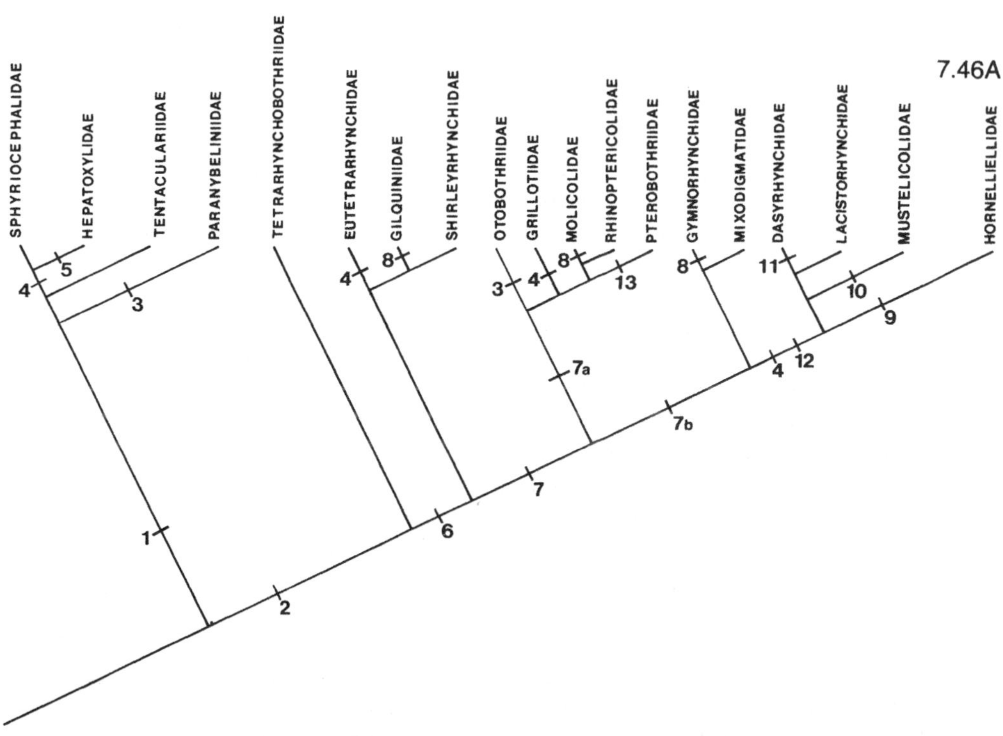

Fig. 7.46a Cladogram of possible phylogenetic relationships among the families of the Order Trypanorhyncha.

Plesiomorphic/Apomorphic Characters:

1. Uterus linear/uterus a transverse sac;
2. plerocercus without blastocyst/with blastocyst;
3. sensory fossettes absent/present;
4. four bothridia/two bothridia;
5. genitalia single/paired;
6. armature homeoacanthous/heteroacanthous;
7. typical/atypical (7a)/poeciloacanthous (7b);
8. accessory seminal vesicle absent/present;
9. chainette(s) single or double/compound;
10. one chainette/multiple chainettes;
11. intercalary hooks absent/present;
12. hermaphroditic duct absent/present;
13. bothridia non-pedicellate/pedicellate.

ORDER TRYPANORHYNCHA DIESING, 1863

Diagnosis: Class Cestoda. Polyzoic tapeworms of small to moderate size. Scolex with two or four very mobile, sessile or pedicellated bothridia and four retractile, hollow tentacles armed with spiral rows of hooks. Each tentacle everted or withdrawn by a 'rhyncheal apparatus' (lacking in *Aporhynchus*) consisting of tentacle sheath and retractor muscle originating within muscular bulb, all contained within elongated pedunculus scolecis. Tentacle armature patterns categorized as homeomorphous, heteroacanthous typical or atypical, or poeciloacanthous, and demonstrate characteristic pattern isometries. Segmentation of strobila usually distinct, proglottids craspedote or acraspedote, apolytic or anapolytic. Anterior of proglottid divided into cortical and medullary zones by longitudinal parenchymal muscle layer. Reproductive organs confined mostly to medullary zone. Testes numerous. Testicular fields may extend into postovarian space or over-reach lateral osmoregulatory canals. Vas deferens median. Cirrus enclosed within cirrus-sac or hermaphroditic sac. External, accessory and internal seminal vesicles present or absent. Vagina ventral to uterus and uterine duct. Genital atrium on lateral margin, often opening between lip-like swellings. Vaginal aperture separate from cirrus (lateral, ventral or posterior) or vagina penetrates hermaphroditic sac to form hermaphroditic vesicle or duct. Vaginal sphincters present or absent. Seminal receptacle present or absent. Ovary posterior, often well forward of posterior margin of proglottid; bilobed, tetra-lobed or multi-lobed in cross-section. Shell gland posterior to ovarian isthmus. Vitelline follicles cortical, forming sleeve around internal organs or separated into lateral bands, occasionally intruding into parenchymal muscle layer. Non-gravid uterus tubular, in midline or arching laterally, occasionally with diverticula in inverted U, Y or X-shape. Gravid uterus saccate, often with diverticula. Uterine pore often preformed. Intra-uterine eggs may contain oncospheres, extra-embryonic membranes with or without processes. Adults in elasmobranchs, rarely holocephalans. Larvae occur in marine plankton, pelagic and benthic invertebrates, fish and reptiles.

Key to superfamilies

1a. Bothridia typically narrow or not projecting beyond lateral margins of scolex, cavities sunken into muscular scolex, margins fused to scolex or only partly free, rims may be thick; tentacles short; pars bothridialis abutting or overlapping pars bulbosa; uterus begins as a transverse sac, uterine duct joining in centre, postlarva common (no blastocyst) .... 3.

1b. Bothridia occupying full width of scolex, usually projecting or even pedicellated, cavities shallow, margins often free, rims usually thin, sometimes notched posteriorly; tentacles relatively long; pars bothridialis rarely abutting or overlapping pars bulbosa; uterus a linear tube (rarely X-shaped) extending anteriorly from uterine duct; larva enclosed by blastocyst, postlarva uncommon .... 2.

2a. Armed tentacles and rhyncheal apparatus present .... 3.

2b. Armed tentacles and rhyncheal apparatus absent* .... 4b.

3a. Entire metabasal hook pattern comprised of quincunxes, providing rotational pattern symmetry or in continuous spiral rows; hooks homeomorphous or heteromorphous on opposite faces of tentacle; typically lacking special basal armature or tentacular swelling; homeoacanths .... Homeacanthoidea Dollfus, 1942.

3b. Metabasal hook pattern of half-spirals providing glide reflection pattern symmetry; principal rows form half-spirals extending between internal and external faces of tentacle; opposing rows form V-pattern at their origin but meet to form inverted V's or are interrupted at their termination; hook patterns on the external face include bands, chainettes or additional hook rows; special basal armature and basal enlargement of tentacle common .... 4.

4a. Interpolated hooks, external band of hooks or chainette(s) present in addition to principal rows .... 5.

4b. Interpolated hooks, external band of hooks, or chainette(s) absent; opposing principal rows form V-pattern at their origin, meet to form inverted V-pattern on opposite face; typical heteroacanths .... Heteracanthoidea Dollfus, 1942.

5a. Chainette absent, additional hook rows, interpolated hooks or band of hooks† present on external face; atypical heteroacanths .... Otobothrioidea Dollfus, 1942 status emend.

* *Aporhynchus*, family Gilquiniidae Dollfus, 1942 (Heteracanthoidea) is exceptional in lacking a rhyncheal apparatus.

† By definition atypical heteroacanths possess interpolated hooks on the external face that form more hook rows than are present on the internal face. Families with bands of hooks or chainettes were formerly placed together as poeciloacanthous types by Dollfus (1942). We consider a more accurate phylogenetic arrangement of families one that aligns those families with bands of hooks with the atypical heteroacanths, i.e. heteroacanths with interpolated hooks on the external face but no chainette, and restricts families with chainettes to the poeciloacanths. In doing so the original definition of the atypical heteroacanth remains unchanged but in fact combines families whose members show characteristics of once separate groups.

5b. Not as above; chainette(s) present on external face; poeciloacanths .... Poecilacanthoidea Dollfus, 1942.

Superfamily Homeacanthoidea

Key to families

1a. Two bothridia, hooks solid or hollow, scolex craspedote or acraspedote . ... 2.

1b. Four bothridia, hooks solid, scolex strongly craspedote with distinct velum .... Tentaculariidae Poche, 1926.
(Syns Kotorellidae Euzet & Radujkovic, 1989; Nybelinidae Poche, 1926; Rufferidae Guiart, 1927.)

Diagnosis: Bothridia sessile, bothridial margins partially free or entirely fused to scolex. Armature homeoacanthous. Hooks solid, usually homeomorphous. Pars vaginalis short. Tentacle sheaths never spiral. Bulbs short. Pars postbulbosa distinct. Proglottids rather wide, anapolytic. Genital pores pre-equatorial, marginal or submarginal. Testes medullary, in preovarian and postovarian space. Ovary separated from posterior proglottid margin by distinct space. Vitelline follicles form layer encircling internal organs. Uterus U-shaped. Adults in stomach or spiral valve of elasmobranchs. Postlarvae widespread in plankton; reported from teleosts in shallow, midwater and continental slope habitats, also from elasmobranchs and cephalopods, rarely in opisthobranchs and marine turtles. Type-genus *Tentacularia* Bosc, 1797.

2a. Bothridia with eversible sensory fossettes on posterior margin .... Paranybeliniidae Schmidt, 1970.

Diagnosis: Scolex short and stout, craspedote with velum. Tentacles short. Armature homeoacanthous, homeomorphous. Two fleshy sessile bothridia. Pars bothridialis longer than pars vaginalis. Tentacle sheath sinuous. Bulbs ellipsoidal, short. Proglottid anatomy and definitive hosts unknown. Postlarvae in plankton. Type-genus *Paranybelinia* Dollfus, 1966.

2b. Bothridia lack sensory fossettes .... 3.

3a. Scolex craspedote with velum, bothridium subdivided by median ridge, tentacle apertures enclosed by bothridial rim; single set of genitalia per proglottid .... Sphyriocephalidae Pintner, 1913.
(Syn. Bouchardidae Guiart, 1927.)

Diagnosis: Large muscular worms. Two deep, oval bothridia encircled by thickened rims. Pars bothridialis deeper than wide giving the scolex a mallet-like shape. Tentacles short. Armature homeoacanthous. Basal armature present. Hooks hollow, homeomorphous. Pars vaginalis and pars bulbosa short. Tentacle sheaths curved, not sinuous. Retractor muscle not entering bulbs. Bulbs short, oriented transversely. Proglottids craspedote, wide and quadrangular, anapolytic. Genital pores marginal. Testes medullary, in layers. Male terminalia bipartite consisting of muscular ductus ejaculatorius and cirrus enclosed within receptacle. Seminal vesicle present or absent. Ovary separated from posterior margin by space occupied by shell gland. Vitelline follicles circumcortical, encircling reproductive organs. Uterus displaced porally, pore preformed. Eggs with projection on each pole. Adults in sharks, postlarva in teleosts. Type-genus *Sphyriocephalus* Pintner, 1913 (syn. *Bouchardia* Guiart, 1927).

3b. **Scolex acraspedote, bothridium with or without median ridge, tentacle apertures not enclosed within bothridial rim .... 4.**

4a. **Two sets of genitalia per proglottid; bothridia with thickened rim and median ridge; tentacle apertures interrupt bothridial rim; pars bothridialis narrower than pedunculus scolecis .... Hepatoxylidae Dollfus, 1940.**

(Syns Dibothriorhynchidae Ariola, 1899; Diplogonimidae Guiart, 1931.)

Diagnosis: Large muscular worms. Bothridial pairs embedded in scolex, narrow, slit-like. Bothridial rims thickened, interrupted anteriorly by tentacle apertures. Tentacles short, stout. Armature homeoacanthous, hooks hollow. Bulbs short, ellipsoidal, sometimes oriented transversely. Tentacle sheaths arched, not sinuous. Retractor muscles originate in proximal ends of tentacle sheaths. Pars postbulbosa short. Proglottids anapolytic, wide and narrow. Genital pores marginal, pre-equatorial. Testes form dorsal and ventral layers in medulla. Male genitalia bipartite, consisting of ductus ejaculatorius expanded into muscular seminal vesicle partitioned off from cirrus enclosed in receptacle. Vagina with two sphincters and seminal receptacle. Ovaries small, near posterior extremity of proglottid. Vitelline follicles circumcortical. Uteri with large diverticula and preformed uterine pores. Eggs without projections. Adults in sharks, postlarva in teleosts and sharks. Type and only genus *Hepatoxylon* Bosc, 1811 (syns *Coenomorphus* Lönnberg, 1889; *Dibothriorhynchus* Blainville, 1828; *Diplogonimus* Guiart, 1931; *Tetrantaris* Templeton, 1836).

4b. **Single set of genitalia per proglottid; two patelliform bothridia without thickened rim, wider than pedunculus scolecis .... Tetrarhynchobothriidae Dollfus, 1969.**

Diagnosis: Scolex elongated, slender. Bothridia patelliform, protruding, ear-like. Metabasal armature quincunxial. Hooks hollow, homeomorphous or heteromorphous. Pars vaginalis long, tentacle sheaths sinuous or irregularly coiled. Gland cells present in pedunculus scolecis. Bulbs longer than bothridia, retractor muscles originating at bases of bulbs; prebulbar organs present; distinct extrinsic musculature attached to bulbs. Pars postbulbosa well developed, often longer than bulbs. Strobila craspedote, anapolytic. Genital pores marginal. Cirrus-sac contains one or two internal vesicles or bipartite cirrus. Testes medullary, arranged in layers, some lateral to ventral osmoregulatory canals. Ovary two or four-lobed in cross-section. Vitelline follicles encircling medulla or in lateral bands. Gravid uterus saccate or with numerous diverticula. Type-genus *Tetrarhynchobothrium* Diesing, 1854.

Superfamily Heteracanthoidea

Key to families

1a. **Two patelliform bothridia; bulbs elongate, prebulbar organs present .... Eutetrarhynchidae Guiart, 1927.**

Diagnosis: Scolex acraspedote. Metabasal armature heteroacanthous typical. Hooks homeomorphous or heteromorphous, solid or hollow. Tentacles long, with or without basal swelling. Distinctive basal armature present or absent. Pars vaginalis longer than pars bothridialis. Bulbs usually elongate, retractor muscle originates at base of bulb. Gland cells usually associated with retractor muscle within bulb. Prebulbar organs present. Pars postbulbosa negligible. Proglottids craspedote or acraspedote, long and narrow. Cirrus-sac and vagina open at marginal genital pore. External and internal seminal vesicles present or absent. Testes preovarian, may extend lateral to osmoregulatory canals. Ovary at posterior extremity of proglottid. Vitelline follicles circumcortical, anterior to ovary or form lateral bands. Gravid uterus saccate, often with numerous lateral diverticula. Type-genus *Eutetrarhynchus* Pintner, 1913.

1b. **Four bothridia; bulbs long or short, prebulbar organs present or absent .... 2.**

2a. **Bulbs long, prebulbar organs present, accessory seminal vesicle absent .... Shirleyrhynchidae n. fam.**

Diagnosis: Scolex elongate, craspedote or acraspedote. Four auriculate bothridia, velum present or absent. Metabasal armature heteroacanthous typical, basal armature and basal swelling present. Hooks heteromorphous. Tentacle sheaths sinuous. Pars vaginalis long. Retractor muscle originates at base of bulb. Pars postbulbosa short. Strobila acraspedote. Proglottids elongate. Genital pores marginal, irregularly alternate. Testes preovarian. Ovary posterior. Vitelline follicles circumcortical. Uterus median, tubular, extending to anterior extremity of proglottid. Type-genus *Shirleyrhynchus* Beveridge & Campbell, 1988.

2b. **Bulbs short, prebulbar organs absent, accessory seminal vesicle present .... Gilquiniidae Dollfus, 1942.**

Diagnosis: Scolex craspedote or acraspedote. Four sessile bothridia, rims incomplete medially or anteriorly. Rhyncheal system present or absent. Tentacles short, armature heteroacanthous typical. Alternating half-spiral rows of hooks do not overlap on external face. Hooks heteromorphous. Bulbs small and ellipsoidal. Pars postbulbosa long or absent. Proglottids acraspedote. Genital pores marginal, pre-equatorial. Accessory and external seminal vesicles present. Ovary separated from posterior margin by distinct space. Some testes postovarian. Vitelline follicles form sleeve around internal organs. Uterus tubular, distal portion deviated porally. Type-genus *Gilquinia* Guiart, 1927.

Superfamily Otobothrioidea*

Key to families

1a. Two bothridia .... **2.**
1b. Four bothridia .... **3.**

2a. Each bothridium with pair of eversible sensory fossettes .... **Otobothriidae Dollfus, 1942.**
Diagnosis: Scolex craspedote or much wider than strobila, velum well developed or absent. Two wide or elongated oval patelliform bothridia. Bothridia extend beyond lateral margins of scolex. Bothridial rims variously thickened, may be notched posteriorly. Tentacles short. Armature heteroacanthous atypical. Hooks heteromorphous, hollow. Tentacle sheaths coiled or sinuous. Bulbs relatively long or short. Retractor muscle originates in anterior region of bulb. Pars postbulbosa short or absent. Proglottids acraspedote, longer than wide. Genitalia single or double. Genital pores marginal, equatorial or postequatorial. Testes intervascular, some postovarian. Ovary set forward in proglottid. Vitelline follicles form sleeve around internal organs. Uterus linear forming lateral diverticula. No preformed uterine pore. Plerocerci in teleosts and sharks, adults in sharks. Type-genus *Otobothrium* Linton, 1890.

2b. Bothridia lack sensory fossettes .... **Grillotiidae† Dollfus, 1969.**
Diagnosis: Scolex acraspedote or craspedote. Two wide bothridia, cordiform or patelliform, with or without thickened rims, posterior and lateral margins free. Principal hook rows alternate, intercalary rows of hooks present, irregular band of small hooks on external face of tentacle. Basal armature present or absent. Hooks heteromorphous, hollow. Pars vaginalis long, tentacle sheaths sinuous. Bulbs long or short, prebulbar organs present or absent. Retractor muscle originates in posterior half of bulb. Proglottids acraspedote, apolytic. Mature proglottids elongated. Genital pores marginal, irregularly alternate. External seminal vesicle present in type-species. Testes medullary, some usually postovarian. Ovary not reaching posterior extremity of proglottid. Vitelline follicles form continuous sleeve around internal organs. Uterus saccate with lateral diverticula. Procercoid in copepods, plerocerci and postlarvae in teleosts, adults in elasmobranchs. Type-genus *Grillotia* Guiart, 1927.

3a. Four sessile bothridia .... **4.**
3b. Four pedicellate bothridia .... **Pterobothriidae Pintner, 1931.**
Diagnosis: Scolex slender, acraspedote; pars bothridialis cruciform. Bothridia with or without thickened rims. Tentacles long, sheaths spiral or not. Bulbs elongate, retractor originates in anterior or posterior regions. Prebulbar organs absent. Hooks heteromorphous, hollow; five hooks per principal row. Intercalary hook rows present. Band of hooks present on external face or not. Proglottids acraspedote, elongate, apolytic. Testes scattered, medullary, some postovarian.

*In this key **Poeciloacanthoidea** is restricted to families with chainettes. Families with true bands of hooks are included in the **Otobothrioidea**.

†Dollfus (1969a, p. 1046) referred to the family Grillotiidae, but never formally defined it.

Genital pores marginal, irregularly alternate. Ovary separated from posterior proglottid margin. Vitelline follicles circumcortical, continuous. Uterus simple, median; gravid uterus develops lateral diverticula. Plerocerci in teleosts, adults parasitic in elasmobranchs. Type-genus *Pterobothrium* Diesing, 1850 (syns *Myrmillorhynchus* Bilqees, 1980; *Neogymnorhynchus* Bilqees & Shah, 1982; *Synbothrium* Diesing, 1850; *Syndesmobothrium* Diesing, 1854).

4a. Five hooks per principal row, extra row of hooks between principal rows on external face, bulbs long .... Rhinoptericolidae Carvajal & Campbell, 1975.

Diagnosis: Scolex acraspedote; four sessile bothridia, well-separated, margins free. Tentacles long. Pars vaginalis long; tentacle sheaths sinuous. Metabasal armature heteroacanthous atypical; for each principal row on internal face there are two rows on external face. Hooks heteromorphous, hollow. Distinctive basal armature present. Bulbs long; retractor muscles originate at base of bulbs. Prebulbar organs absent. Proglottids acraspedote. Testes in longitudinal rows. External seminal vesicle present. Ovary posterior, tetra-lobed in cross-section. Vitelline follicles circumcortical, continuous. Linear uterus with two preformed diverticula. In spiral valve of rays. Type and only genus *Rhinoptericola* Carvajal & Campbell, 1975.

4b. More than five hooks per principal row, band of hooks present on external face, bulbs short .... Molicolidae Beveridge & Campbell, 1989.

Diagnosis: Scolex acraspedote. Four auriculate bothridia, posterior and lateral borders free, fused to scolex anteriorly. Intercalary row of hooks present. Hooks heteromorphous, hollow. Tentacle sheaths spiral. Prebulbar organs absent. Retractor muscles originate in anterior half of bulb. Proglottids acraspedote. Genital pores marginal. Accessory seminal vesicle present. Testes medullary, pre- and postovarian. Ovary separated from posterior margin by a space. Vitelline follicles circumcortical. Uterus median, tubular, not reaching anterior extremity; deviated toward genital pore in type-genus; uterine pore preformed or not. Plerocerci in teleosts, adults in sharks. Type-genus *Molicola* Dollfus, 1935.

Superfamily Poecilacanthoidea

Key to families

1a. Two bothridia .... 2.
1b. Four bothridia .... 5.

2a. Five hooks per principal row .... 3.
2b. More than five hooks per principal row .... 4.

3a. Single chainette, hooks 7 and 8 in satellite position .... Lacistorhynchidae Guiart, 1927.
(Syn. Floricipitidae Dollfus, 1929.)

Diagnosis: Scolex craspedote or acraspedote. Two bothridia, notched posteriorly. Armature poeciloacanthous, chainette simple; five large hooks per principal row, hook 6 small, displaced anteriorly. Intercalary rows of hooks absent. Basal armature present or absent. Hooks heteromorphous, hollow. Tentacle sheaths sinuous. Prebulbar organs absent. Retractor muscle anterior. Proglottids acraspedote; genital pores marginal. Some testes postovarian. Hermaphroditic duct present. Ovary separated from posterior margin by space. Vitelline follicles form sleeve around internal organs. Uterus simple, median, saccate without pore. Plerocerci in teleosts and copepods, adults in sharks. Type-genus *Lacistorhynchus* Pintner, 1913.

3b. Multiple chainettes present .... **Mustelicolidae Dollfus, 1969.**
Diagnosis: Scolex acraspedote. Two, well separated, sessile bothridia with free margins. Armature poeciloacanthous with multiple chainettes opposite principal rows. Hooks heteromorphous, hollow. Sheaths sinuous. Tentacles and bulbs relatively short. Retractor muscles originate in anterior half of bulb. Prebulbar organ absent. Hermaphroditic duct present. Seminal vesicles present. Testes numerous, intervascular, some postovarian. Ovary posterior, separated from posterior margin by space. Uterus simple, median, saccate with ventral pore. Vitelline follicles circumcortical. Adults in spiral valve of sharks. Type-genus *Diesingium* Guiart, 1931 (syn. *Mustelicola* Dollfus, 1969).

4a. Seven hooks per principal row, chainette composed of alternating pairs of large hooks similar to those of principal rows ....
.... **Hornelliellidae Yamaguti, 1954.**
Diagnosis: Scolex elongate, acraspedote. Two oval bothridia each with a median longitudinal ridge. Metabasal armature poeciloacanthous. Intercalary hooks absent. Distinctive basal armature present; no basal swelling. Tentacle sheaths coiled. Prebulbar organs absent. Retractor muscles originate in anterior region of bulb. Pars postbulbosa absent. Genital pores marginal, irregularly alternate. Testes scattered, anterior to genital pore. Internal seminal vesicle present. Male and female ducts unite to form hermaphroditic vesicle. Ovary not at posterior margin of proglottid. Vitelline follicles circumcortical, pre- and postovarian. Uterus simple, median; preformed uterine pore present. In elasmobranchs. Type-genus *Hornelliella* Yamaguti, 1954.

4b. Nine hooks per principal row, chainette hooks dissimilar to those of principal rows .... **Dasyrhynchidae Dollfus, 1935.**
(Syn. Dasyrhynchidae Guiart, 1935.)
Diagnosis: Scolex craspedote. Two bothridia. Chainette present, either single or double, bases of elements enlarged. Intercalary hooks present. Distinct basal armature present. Hooks heteromorphous, hollow. Retractor muscles insert at anterior end of bulb. Prebulbar organs absent. Internal and external seminal vesicles present. Testes numerous, intervascular, some postovarian. Hermaphroditic duct present. Ovary tetralobed. Uterus simple, median, saccate, without pore. Type and only genus *Dasyrhynchus* Pintner, 1928 (syn. *Sbesterium* Dollfus, 1929).

5a. **Paired chainette elements with single lateral wing, with or without double-winged (V-shaped) hooks, accessory seminal vesicle present .... Gymnorhynchidae Dollfus, 1935.**

Diagnosis: Four bothridia. Scolex acraspedote. Bulbs relatively short. Prebulbar organs absent. Genital pore on lateral margin. Uterus deviated, terminates adjacent to genital pore. Type-genus *Gymnorhynchus* Rudolphi, 1819.

5b. **Chainette of two-winged (V-shaped) elements, accessory seminal vesicle lacking .... Mixodigmatidae Dailey & Vogelbein, 1982.**

Diagnosis: Scolex acraspedote. Four bothridia. Bulbs elongate. Prebulbar organs present. Retractor muscle originates in base of bulb. Gland cells surround retractor muscle. Nine hooks per principal row of metabasal region. Hooks heteromorphous, solid. Uterus linear. Type-genus *Mixodigma* Dailey & Vogelbein, 1982.

Superfamily Homeacanthoidea

Family Tentaculariidae Poche, 1926

(Syns Kotorellidae Euzet & Radujkovic, 1989; Nybelinidae Guiart, 1931; Rufferidae Guiart, 1927.)

Key to genera

1a. **Four bothridia, scolex craspedote, armature in quincunxes, hooks heteromorphous, hooks of bothridial face dissimilar to those of antibothridial face orienting plane of bilateral (reflection) pattern symmetry through bothridial and antibothridial faces of tentacle .... *Kotorella* Euzet & Radujkovic, 1989. (Figs 7.47–7.50)**

Diagnosis: Scolex forms apical dome anterior to tentacle orifices. Four elongate bothridia, margins fused to scolex anteriorly. Pars bothridialis longer than pars bulbosa. Tentacles short, laterally swollen at base. Armature homeoacanthous, heteromorphous. Hooks solid. Quincunxial pattern tightly spaced on bothridial face, becoming widely spaced as rows proceed around tentacle to antibothridial face. Tentacle sheaths irregularly sinuous. Prebulbar organs absent. Bulbs small and ovoid. Retractor muscles originate at base of bulbs. Strobila acraspedote. Proglottid anatomy poorly known. Genital pore lateral, pre-equatorial, irregularly alternate. Cirrus unarmed. Ovary central. Testes surround ovary. Uterus median. In spiral valve of dasyatid rays. Adriatic Sea. Type-species *K. pronosoma* (Stossich, 1901) Euzet & Radujkovic, 1989.

1b. **Four bothridia, scolex craspedote, armature in quincunxes or continuous spirals, hooks homeomorphous or heteromorphous, but with plane of bilateral pattern symmetry between hook rows on internal and external faces of tentacle .... 2.**

2a. **Bothridia (four), sessile, long and narrow, posterior borders not free, pars bothridialis completely overlapping bulbs .... *Tentacularia* Bosc, 1797. (Figs 7.51–7.55)**

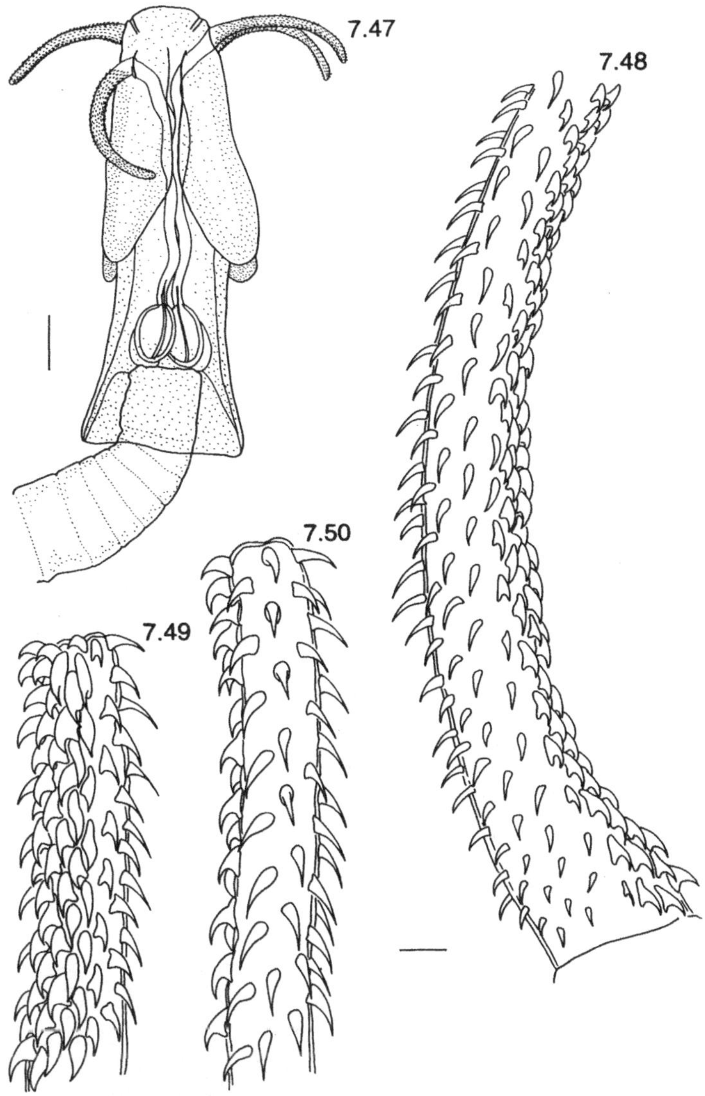

Figs 7.47–7.50 *Kotorella pronosoma* (Stossich, 1901). 7.47. Scolex. 7.48. Tentacle viewed from external face, metabasal region. 7.49. Tip of tentacle, bothridial face to left. 7.50. Tip of tentacle, antibothridial face. Scale lines: 7.47, 0.1 mm; 7.48–7.50, 0.01 mm.

(Syns *Abothros* Welch, 1876; *Pierretia* Guiart, 1927; *Stenobothrium* Diesing, 1850.)

Diagnosis: Scolex craspedote with distinct velum. Four elongate, sessile bothridia; elongate and narrow. Bothridial margins entirely fused to scolex. Tentacles short. Armature homeoacanthous. Hooks solid, usually homeomorphous. Sheaths sinuous or irregularly coiled, not helical. Prebulbar organs absent.

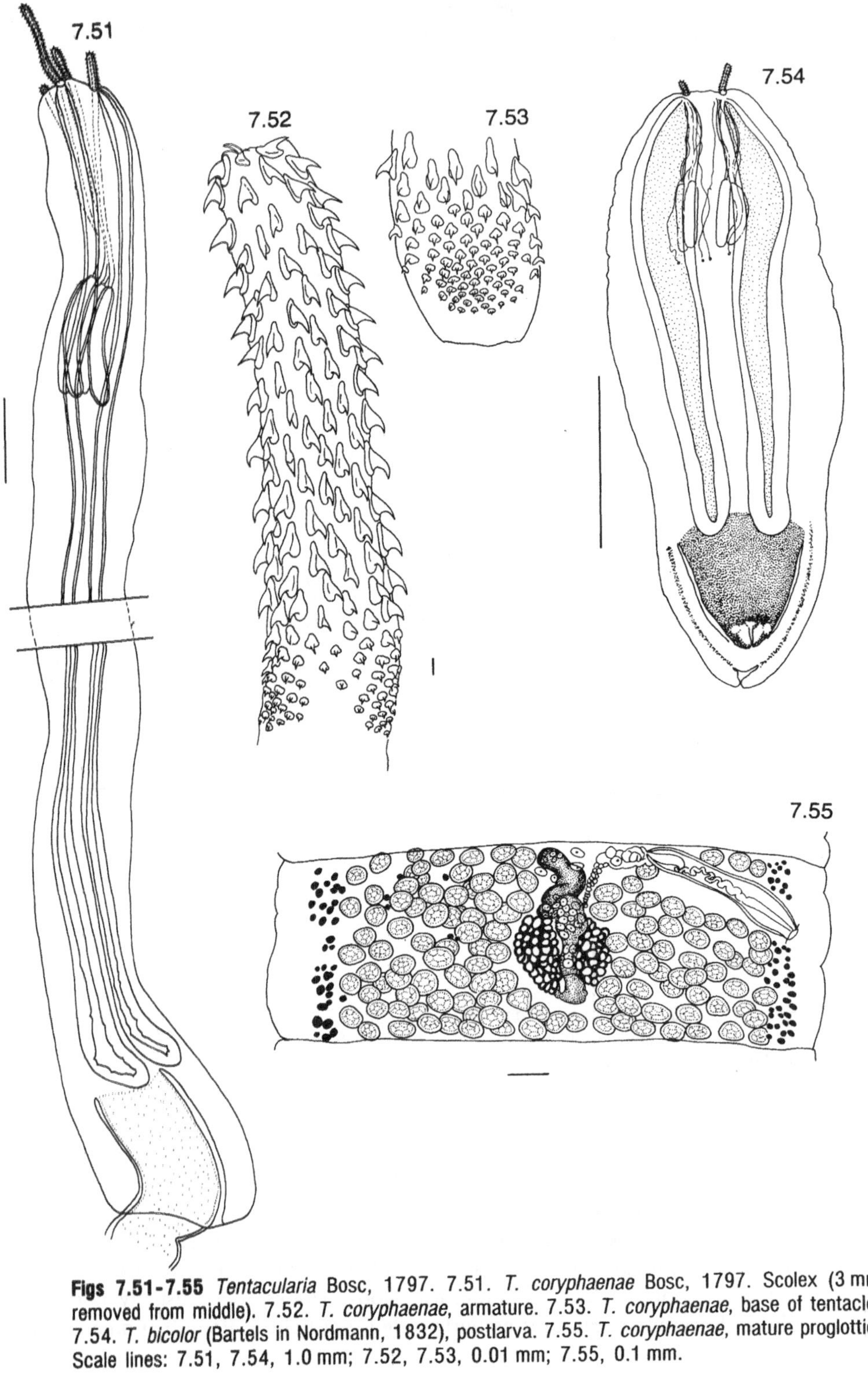

Figs 7.51–7.55 *Tentacularia* Bosc, 1797. 7.51. *T. coryphaenae* Bosc, 1797. Scolex (3 mm removed from middle). 7.52. *T. coryphaenae*, armature. 7.53. *T. coryphaenae*, base of tentacle. 7.54. *T. bicolor* (Bartels in Nordmann, 1832), postlarva. 7.55. *T. coryphaenae*, mature proglottid. Scale lines: 7.51, 7.54, 1.0 mm; 7.52, 7.53, 0.01 mm; 7.55, 0.1 mm.

Pars vaginalis short. Pars bulbosa in anterior region of pars bothridialis. Bulbs short, ellipsoidal. Retractor muscles originate in base of bulb. Proglottids craspedote or acraspedote, wider than long or never longer than twice their width. Genital pores pre-equatorial, marginal or submarginal. Testes medullary, in preovarian and postovarian space. Ovary separated from posterior proglottid margin by distinct space. Vitelline follicles medullary, encircling internal organs. Uterus U-shaped. Adults in elasmobranchs. Postlarvae widespread in plankton; from teleosts in shallow, midwater and continental slope habitats. Cosmopolitan. Type-species *T. coryphaenae* Bosc, 1797.

2b. Bothridia (four) sessile short and narrow, posterior borders free, pars bothridialis partially overlapping bulbs or not at all .... *Nybelinia* Poche, 1926. (Figs 7.56–7.63)
(Syns *Acoleorhynchus* Poche, 1926; *Aspidorhynchus* Molin, 1858 preocc.; *Congeria* Guiart, 1935; *Pleronybelinia* Sezen & Price, 1969; *Rufferia* Guiart, 1927.)

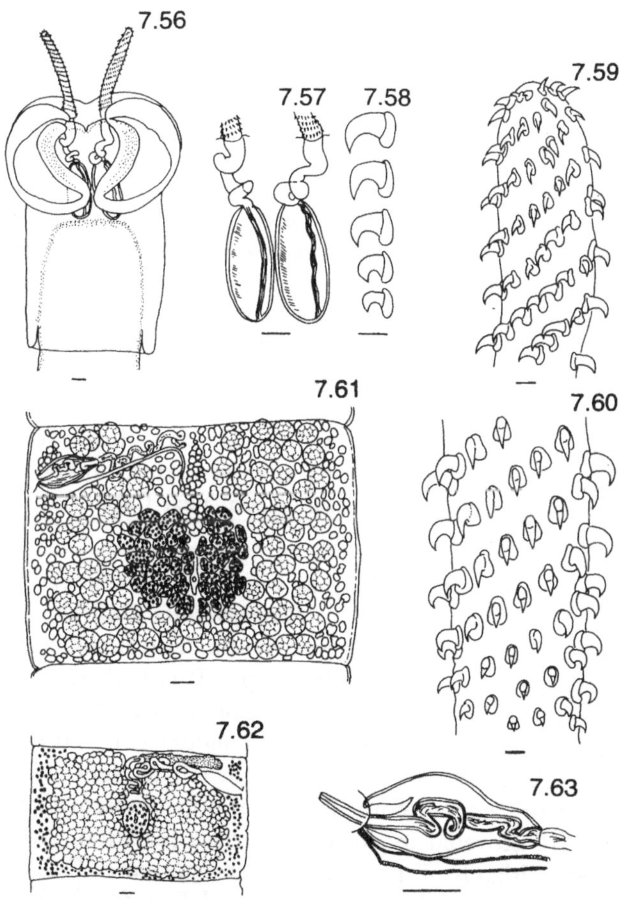

Figs 7.56-7.63 *Nybelinia thyrsites* Korotaeva, 1971. 7.56. Scolex. 7.57. Bulbs. 7.58. Hooks. 7.59. Metabasal armature. 7.60. Base of tentacle. 7.61. Mature proglottid. 7.62. Gravid proglottid. 7.63. Terminal genitalia. Scale lines: 7.56, 7.57, 7.61, 7.62, 7.63, 0.1 mm; 7.59, 7.60, 0.01 mm.

Diagnosis: Scolex craspedote, short and stout. Velum present. Bothridia paired on opposite surfaces, wide and reniform; posterior and lateral bothridial margins free. Bulbs usually short. Tentacles short. Armature homeoacanthous. Hooks solid, homeomorphous or heteromorphous on opposite sides of tentacle. Basal hooks reduced or enlarged. Tentacle sheaths sinuous or irregularly contorted. Prebulbar organs absent. Retractor muscles originate in base of bulb. Pars postbulbosa shorter than pars bothridialis. Proglottids wide and quadrangular, craspedote or acraspedote, anapolytic. Genital pores irregularly alternate, ventrosubmarginal. Cirrus-sac distinct, internal seminal vesicle present or absent. Testes medullary in layers in preovarian and postovarian space. Ovary small, X-shaped in cross-section, set forward of posterior extremity of proglottid. Uterus inverted U-shaped. Vitelline follicles circumcortical except around ovary. In elasmobranchs. Cosmopolitan. Postlarvae common in teleosts and elasmobranchs. Type-species *N. lingualis* (Cuvier, 1817) Poche, 1926.

(Dollfus (1942) proposed two subgenera: proglottids acraspedote (*Nybelinia*) and proglottids craspedote (*Syngenes*). However such a distinction is mostly unuseable because the strobilae of most species are unknown.)

Family Paranybeliniidae Schmidt, 1970

Key to genera

1a. **Bulbs about three times longer than wide, each hook with ventral tooth .... *Pseudonybelinia* Dollfus, 1966. (Figs 7.64, 7.65)**

Diagnosis: Scolex craspedote with velum. Two elongated separate sessile bothridia, each with pair of sensory fossettes on posterior margin. Pars bothridialis longer than pars vaginalis. Tentacles short. Armature homeoacanthous; hook files closely spaced distorting quincunxes into narrow rectangles, with 180° rotational pattern symmetry. Hooks aligned in spiral rows ascending from left to right. Hooks solid, each with ventral tooth and recurved tip. Tentacle sheaths sinuous. Prebulbar organs absent. Strobila unknown. Hosts unknown. Discovered in nocturnal surface plankton. Atlantic Ocean, near Cape Verde Islands. Type-species *P. odontacantha* Dollfus, 1966.

1b. **Bulbs about twice as long as wide, hooks without ventral tooth .... *Paranybelinia* Dollfus, 1966. (Figs 7.66, 7.67)**

Diagnosis: Scolex craspedote with velum. Two fleshy bothridia each with a pair of sensory fossettes on posterior margin. Pars bothridialis reaching bulbs. Tentacles short, sheaths sinuous. Armature homeoacanthous; hook files closely spaced distorting quincunxes into narrow rectangles, with 180° rotational pattern symmetry. Hooks form rows spiralling from left to right. Hooks solid, homeomorphous, lacking ventral tooth. Prebulbar organs absent. Strobila unknown. Hosts unknown. Discovered in nocturnal surface plankton. Atlantic Ocean taken off Cape Verde Islands. Type-species *P. otobothroides* Dollfus, 1966.

(The types of both *Paranybelinia* and *Pseudonybelinia* are faded, shrivelled

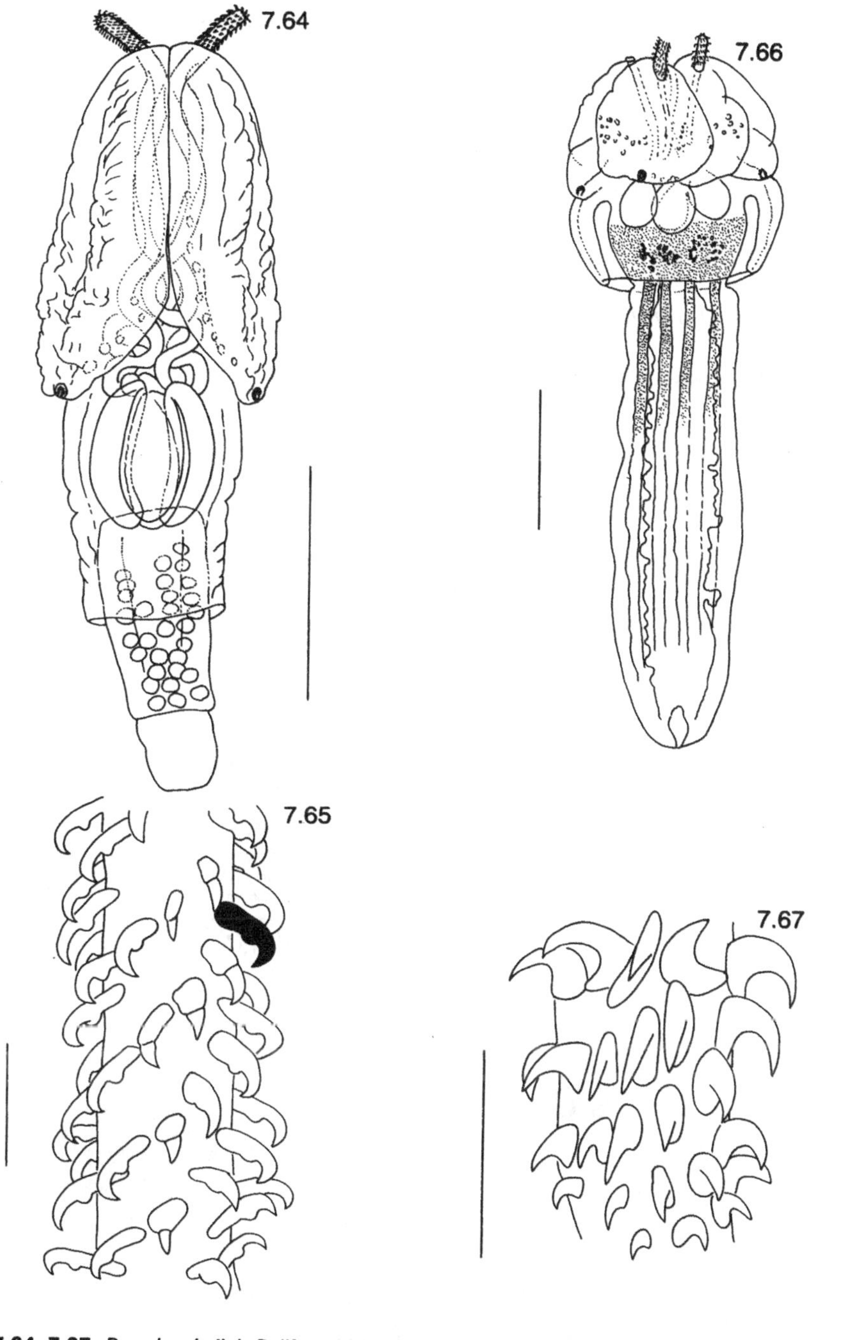

Figs 7.64–7.67 *Pseudonybelinia* Dollfus, 1966 and *Paranybelinia* Dollfus, 1966. 7.64. Postlarva, *Pseudonybelinia*. 7.65. *Pseudonybelinia*, armature, note rows of hooks bearing a spur (black). 7.66. *Paranybelinia* postlarva. 7.67. *Paranybelinia* armature, note rows of uncinate hooks. Scale lines: 7.64, 1.0 mm; 7.65, 0.05 mm; 7.66, 0.2 mm; 7.67, 0.025 mm.

and many details cannot be confirmed. The drawings and descriptions are based upon the observations of Dollfus.)

Family Hepatoxylidae Dollfus, 1940

(Syns Dibothriorhynchidae Ariola, 1899; Diplogonimidae Guiart, 1931)

Genus *Hepatoxylon* Bosc, 1811 (Figs 7.68–7.73)

(Syns *Coenomorphus* Lönnberg, 1889; *Dibothriorhynchus* Blainville, 1828; *Diplogonimus* Guiart, 1931; *Tetrantaris* Templeton, 1836)

Diagnosis: With the characters of the family. Type-species *H. trichiuri* (Holten, 1802).

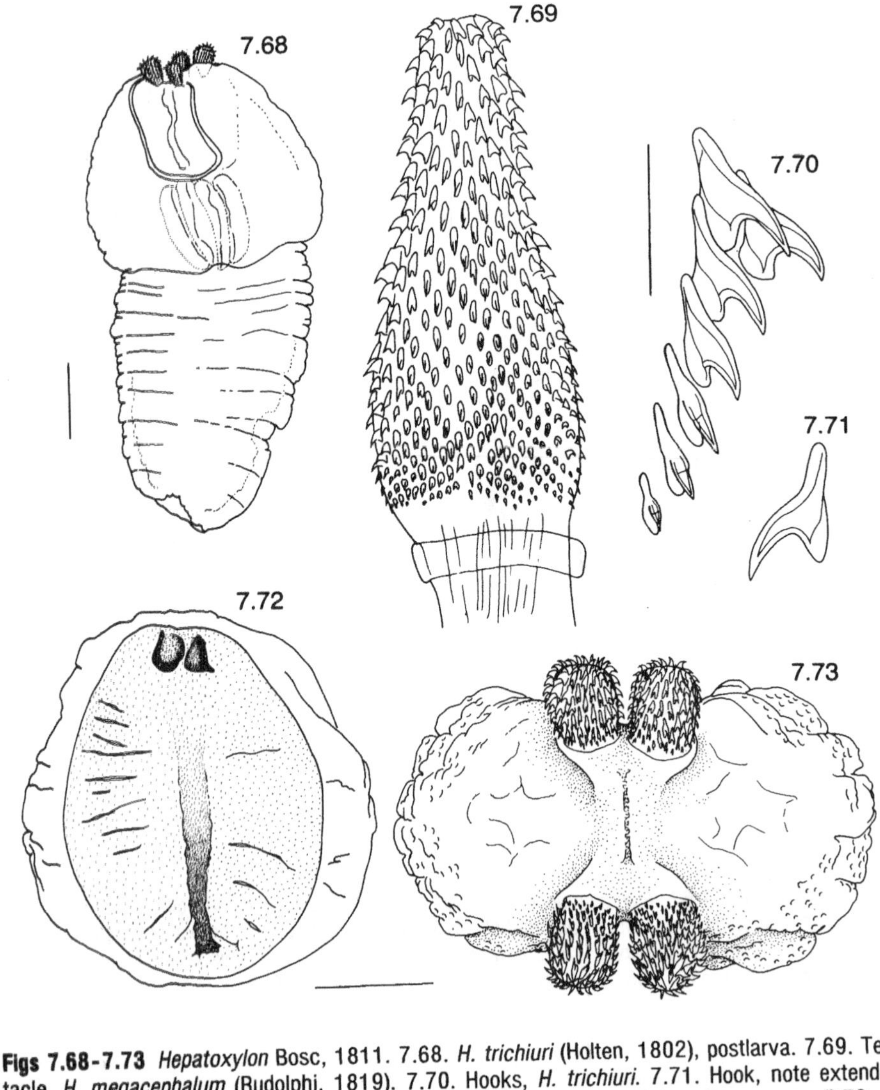

Figs 7.68–7.73 *Hepatoxylon* Bosc, 1811. 7.68. *H. trichiuri* (Holten, 1802), postlarva. 7.69. Tentacle, *H. megacephalum* (Rudolphi, 1819). 7.70. Hooks, *H. trichiuri*. 7.71. Hook, note extended heel. 7.72. Bothridium, *H. trichiuri*. Note tentacle apertures and partial median groove. 7.73. *H. trichiuri*, en face view of scolex. Note that bothridia are not joined across scolex. Scale lines: 7.68, 2.0 mm; 7.69, 7.70, 0.25 mm; 7.72, 7.73, 1.0 mm.

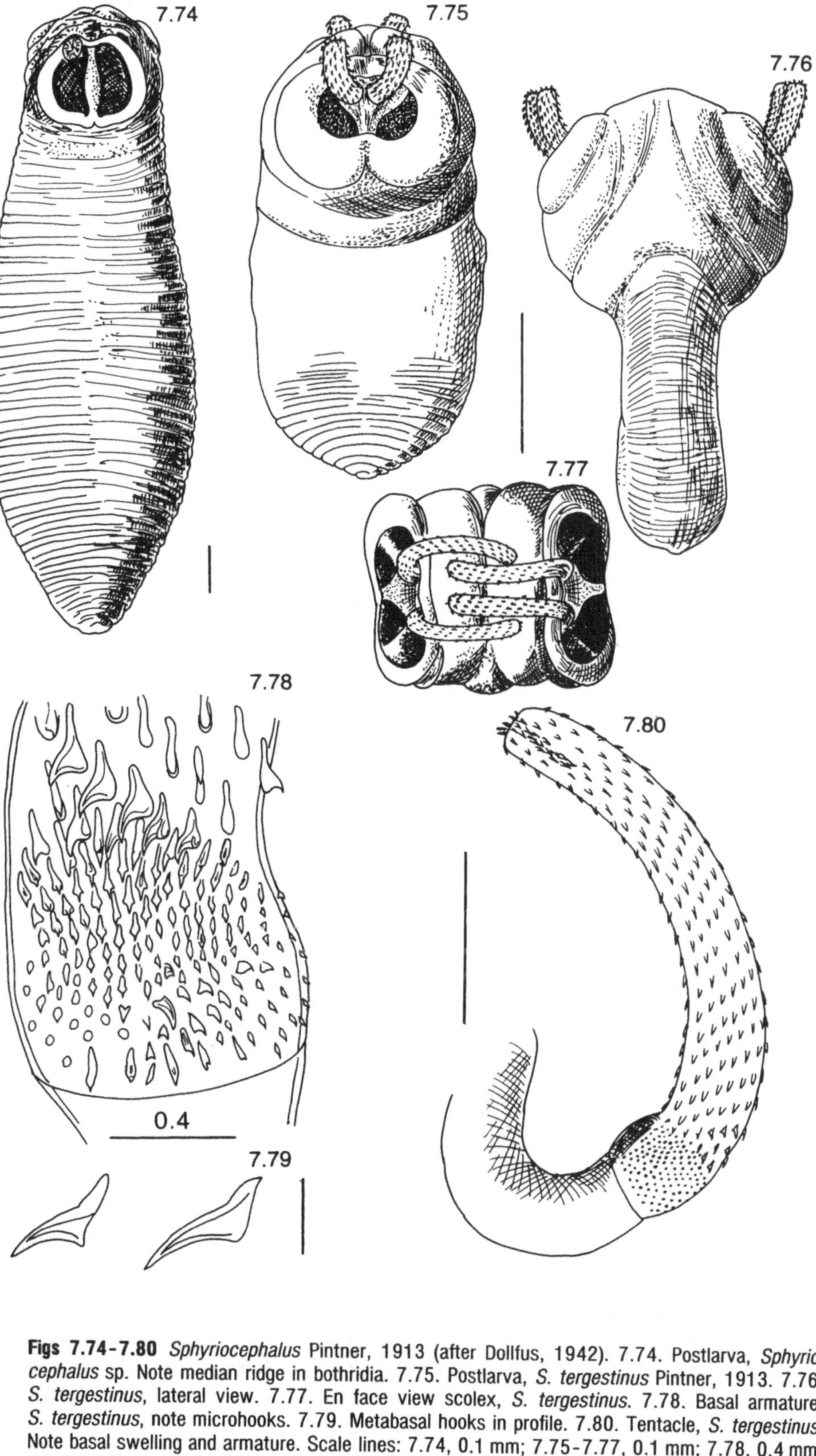

Figs 7.74–7.80 *Sphyriocephalus* Pintner, 1913 (after Dollfus, 1942). 7.74. Postlarva, *Sphyriocephalus* sp. Note median ridge in bothridia. 7.75. Postlarva, *S. tergestinus* Pintner, 1913. 7.76. *S. tergestinus*, lateral view. 7.77. En face view scolex, *S. tergestinus*. 7.78. Basal armature, *S. tergestinus*, note microhooks. 7.79. Metabasal hooks in profile. 7.80. Tentacle, *S. tergestinus*. Note basal swelling and armature. Scale lines: 7.74, 0.1 mm; 7.75–7.77, 0.1 mm; 7.78, 0.4 mm; 7.79, 0.05 mm; 7.80, 0.5 mm.

Family Sphyriocephalidae Pintner, 1913

(Syn. Bouchardidae Guiart, 1927)

Genus *Sphyriocephalus* Pintner, 1913. (Figs 7.74–7.80)
(Syn. *Bouchardia* Guiart, 1927)

Diagnosis: With the characters of the family. Type-species *S. viridis* (Wagener, 1854) Pintner, 1913.

Family Tetrarhynchobothriidae Dollfus, 1969

Key to genera

1a. Scolex sharply delineated, neck absent, proglottids craspedote, ovary bilobed, vitelline follicles in lateral bands, uterus with two preformed diverticula and pore .... *Didymorhynchus* Beveridge & Campbell, 1988. (Figs 7.81–7.89)

Diagnosis: Scolex acraspedote, spinose. Two oval bothridia with free margins. Hooks small, arranged in quincunxes, no basal armature or swelling. Tentacle sheaths irregularly coiled. Bulbs short, pars vaginalis and pars postbulbosa long; prebulbar organ present. Genital pores irregularly alternate. Cirrus-sac contains bipartite cirrus surrounded by muscle. Testes preovarian in two lateral fields. Vitelline follicles arching medially around testes. Dorsal osmoregulatory canals lateral to ventral osmoregulatory canals. In elasmobranchs. Sri Lanka. Type-species *D. southwelli* Beveridge & Campbell, 1988.

1b. Scolex delimited or not, neck present, proglottids acraspedote, ovary four-lobed, vitelline follicles encircle reproductive organs, uterus linear, lacking preformed diverticula and pore .... 2.

2a. Scolex acraspedote, well delineated, testes preovarian, internal seminal vesicle present, distal vagina surrounded by thick muscular wall .... *Zygorhynchus* Beveridge & Campbell, 1988. (Figs 7.90–7.101)

Diagnosis: Two oval bothridia with free margins. Hooks small, hollow, arranged in quincunxes; no basal swelling or armature. Tentacle sheaths sinuous. Pintner's cells present. Bulbs long, slender; prebulbar organ present; pars postbulbosa small or absent. Proglottids acraspedote. Genital pores alternate irregularly; genital ducts crossing osmoregulatory canals ventrally. Cirrus-sac globoid, muscular; internal seminal vesicle present. Testes medullary, scattered, crossing ventral osmoregulatory canals. Seminal receptacle present. Ovary four-lobed, posterior. Vitelline follicles encircling medulla. Uterus median, tubular, gravid uterus saccular, without diverticula; no preformed uterine pore. Dorsal osmoregulatory canals lateral to ventral canals. Type-species *Z. robertsoni* Beveridge & Campbell, 1988.

2b. Scolex not sharply delineated, postovarian testes present, cirrus-sac contains two internal seminal vesicles, vagina without thick muscle coat .... *Tetrarhynchobothrium* Diesing, 1854. (Figs 7.102–7.111)

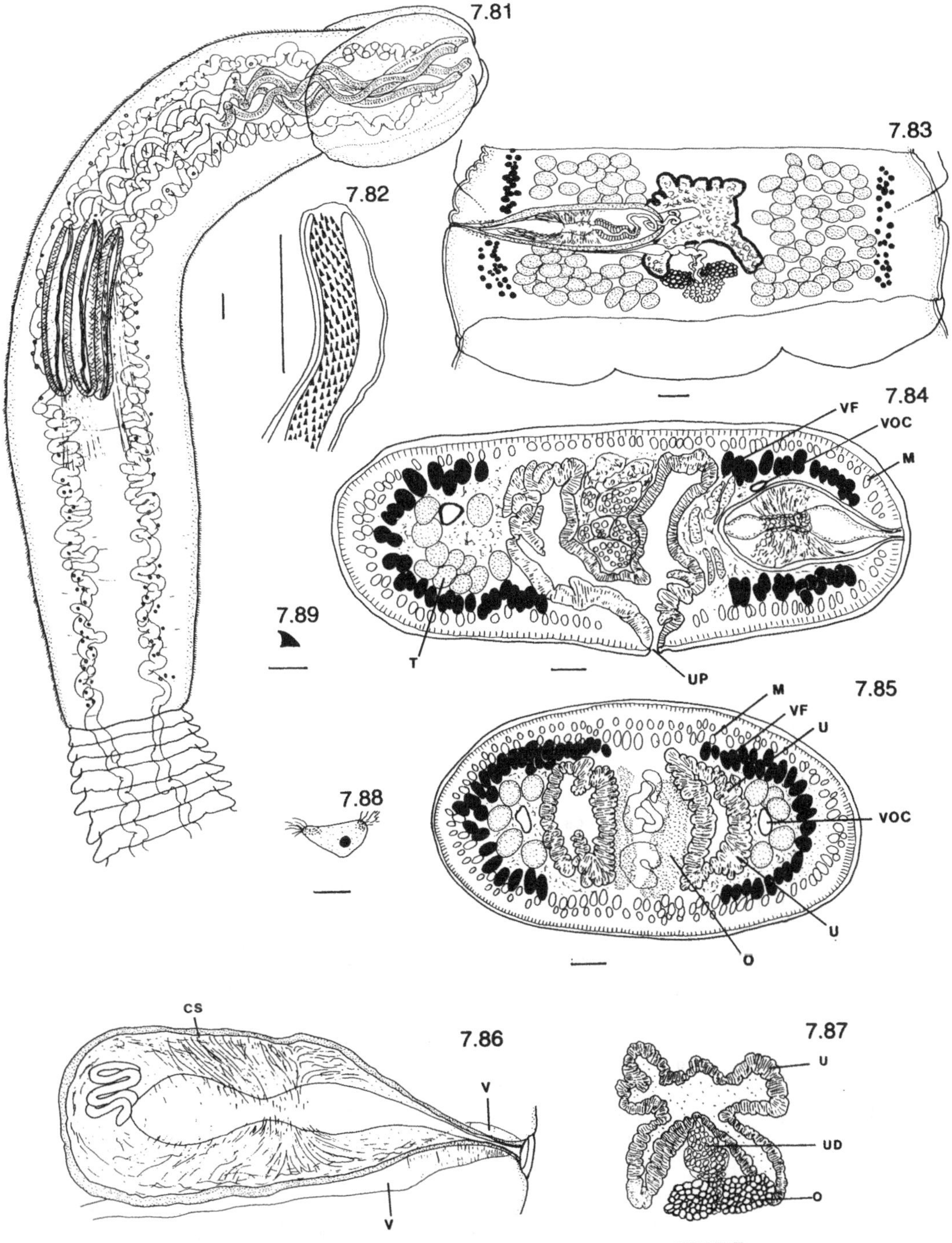

Figs 7.81–7.89 *Didymorhynchus southwelli* Beveridge & Campbell, 1988. 7.81. Scolex. 7.82. Tentacle armature. 7.83. Mature proglottid. 7.84. Transverse section of proglottid showing uterine pore. 7.85. Transverse section of proglottid showing uterine diverticula. 7.86. Terminal genitalia. 7.87. X-shaped uterus. 7.88. Egg. 7.89. Hook. Scale lines: 7.81–7.87, 0.1 mm; 7.88, 7.89, 0.01 mm. Abbreviations: CS, cirrus-sac; M, longitudinal muscle; O, ovary; T, testis; U, uterus; UD, uterine duct; UP, uterine pore; V, vagina; VOC, ventral osmoregulatory canal; VF, vitelline follicle.

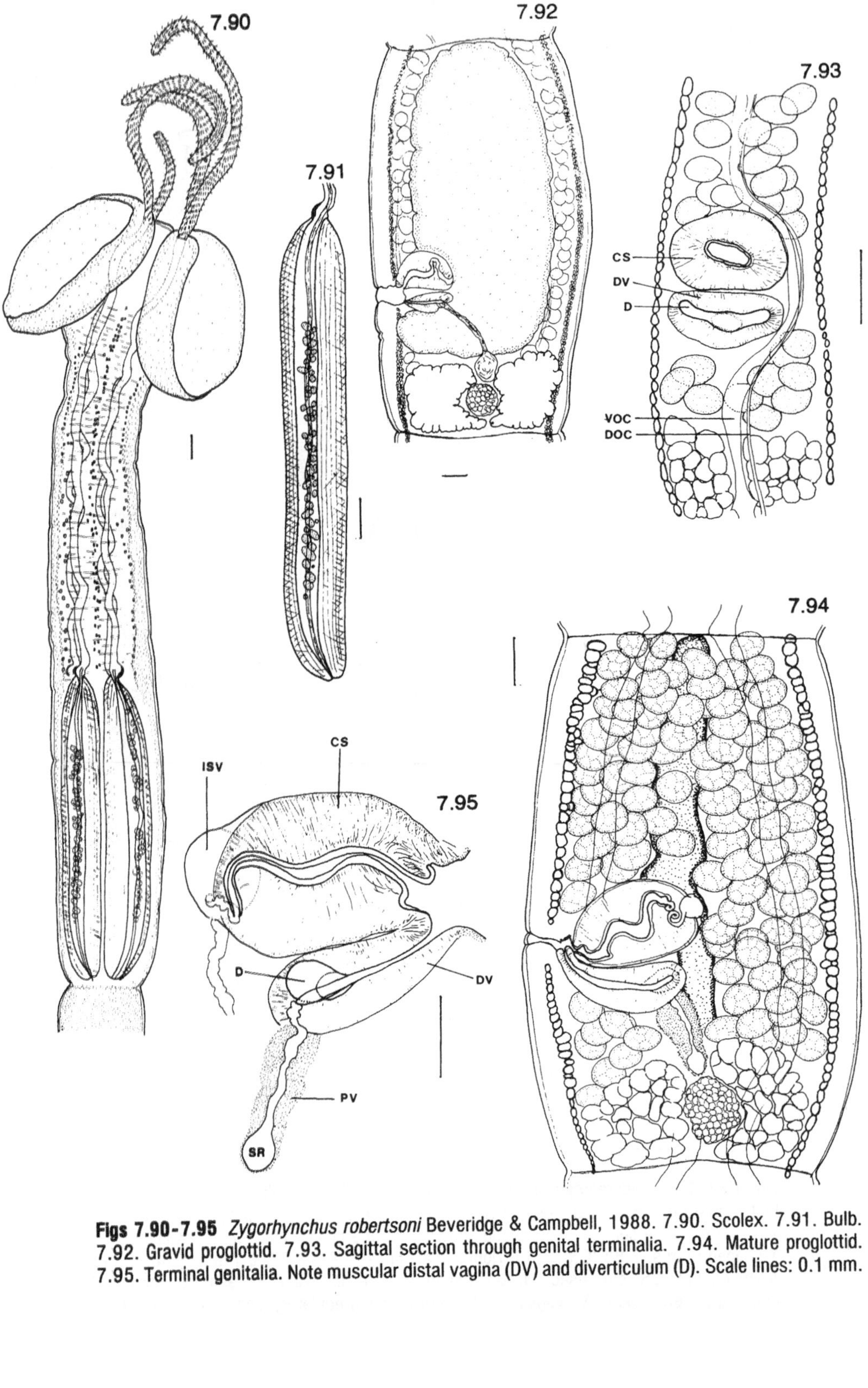

Figs 7.90–7.95 *Zygorhynchus robertsoni* Beveridge & Campbell, 1988. 7.90. Scolex. 7.91. Bulb. 7.92. Gravid proglottid. 7.93. Sagittal section through genital terminalia. 7.94. Mature proglottid. 7.95. Terminal genitalia. Note muscular distal vagina (DV) and diverticulum (D). Scale lines: 0.1 mm.

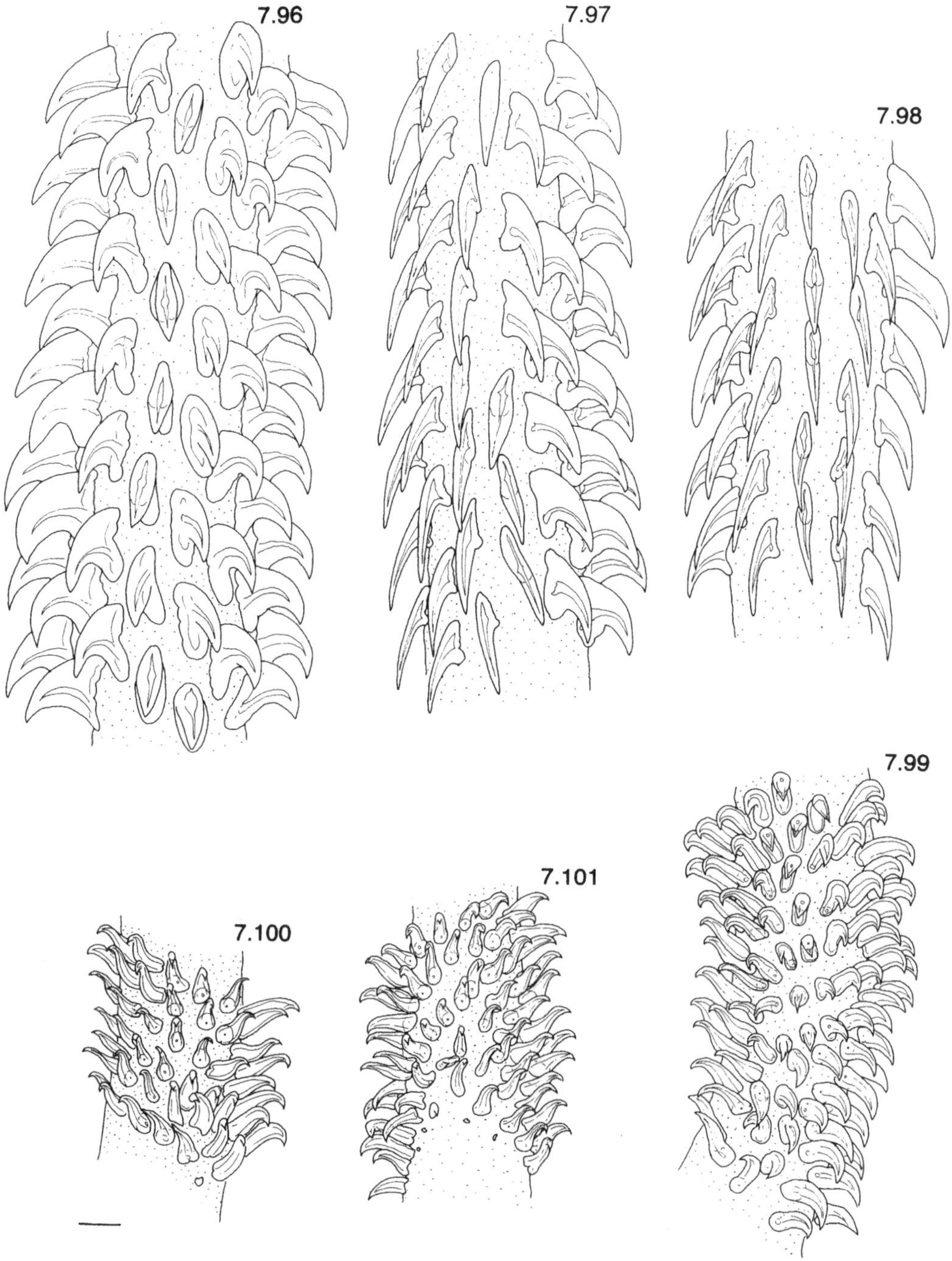

Figs 7.96–7.101 *Zygorhynchus robertsoni* Beveridge & Campbell, 1988. 7.96. Internal face, metabasal armature, 0.4 mm from base. 7.97. Bothridial face, metabasal armature, internal face on right side. 7.98. External face, distal region of metabasal armature. 7.99. Internal face, basal armature. 7.100. Bothridial face, basal armature, internal face on right. 7.101. External face, basal armature. Scale lines: 0.01 mm.

7.102

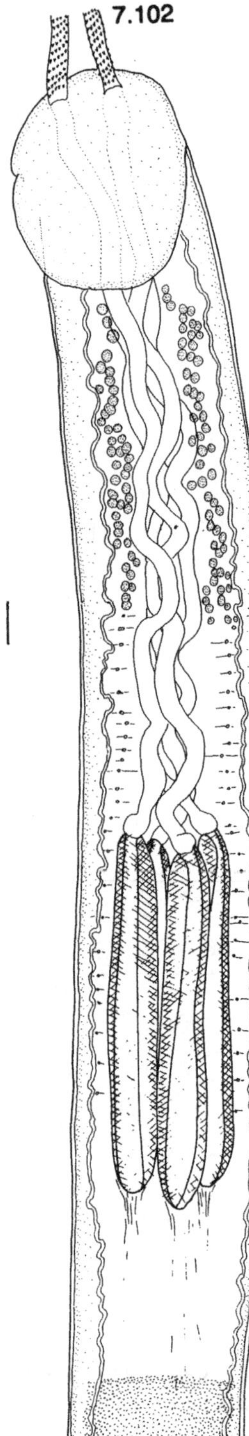

7.104

7.105

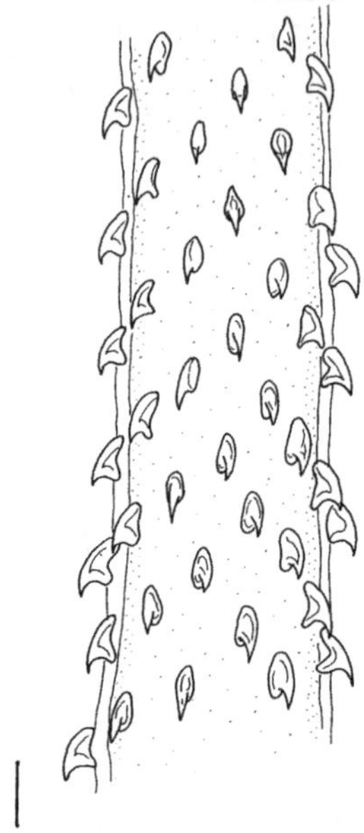

7.103

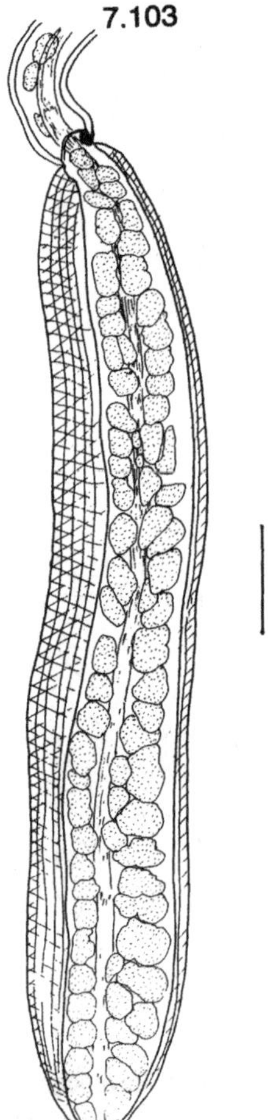

7.106

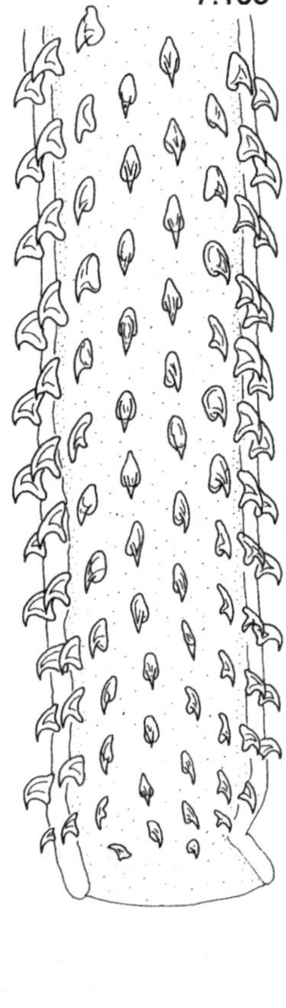

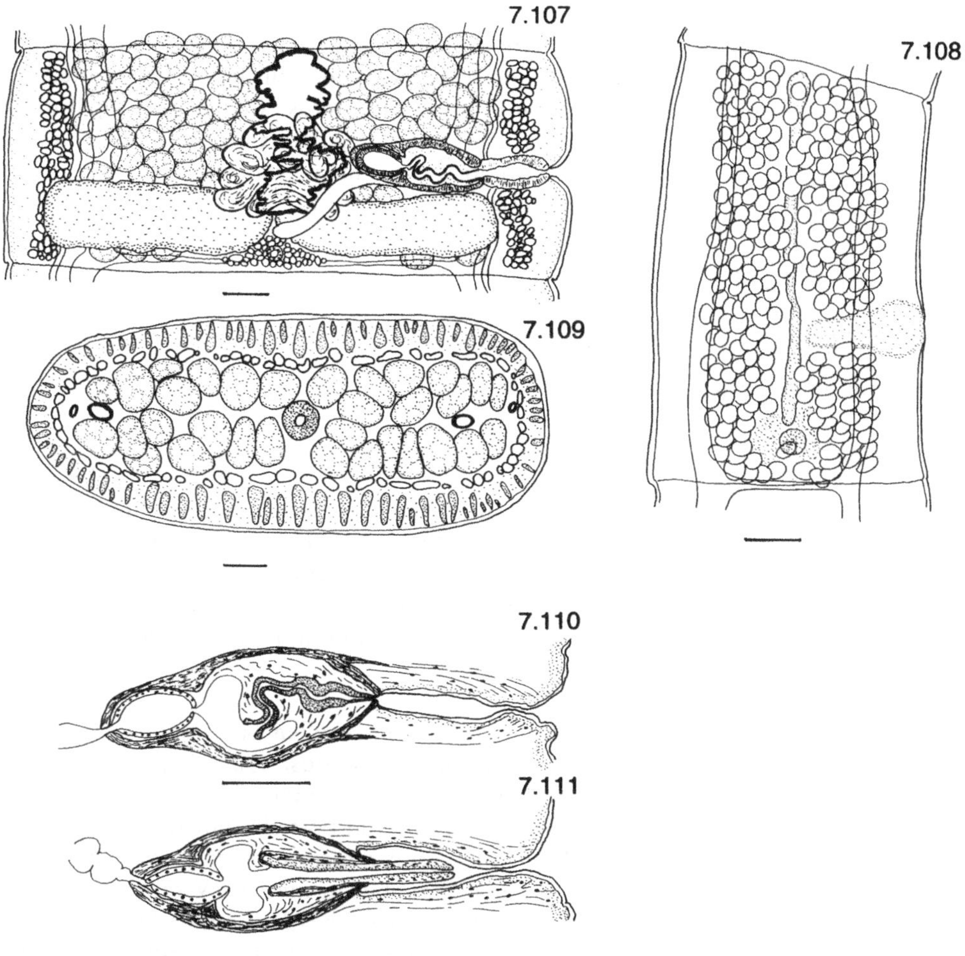

Figs 7.102–7.106 *Tetrarhynchobothrium australe* Beveridge & Campbell, 1988. 7.102. Scolex. 7.103. Bulb. 7.104. External face, base of tentacle. 7.105. Bothridial face, metabasal armature, internal face on right. 7.106. Bothridial face, base of tentacle, internal face on right. Scale lines: 7.102–7.103, 0.1 mm; 7.104–7.106, 0.01 mm.

Figs 7.107–7.111 *Tetrarhynchobothrium australe* Beveridge & Campbell, 1988. 7.107. Mature proglottid. 7.108. Immature proglottid showing testis distribution. 7.109. Transverse section, note layers of testes, median position of ventral osmoregulatory canals and uterus. 7.110. Cirrus-sac showing two internal seminal vesicles. 7.111. Cirrus-sac with partially everted cirrus. Scale lines: 0.1 mm.

Diagnosis: Two bothridia, not contiguous apically, margins free, not thickened. Hooks small, uncinate, arranged in quincunxes around tentacle. No basal armature or basal swelling. Tentacle sheaths sinuous; pars vaginalis longer than bulbs. Bulbs at least six times longer than wide. Pars postbulbosa well developed. Pintner's cells present; pseudosegmentation of pars vaginalis. Proglottids weakly craspedote. Genital pore irregularly alternate. Testes medullary, crossing ventral osmoregulatory canals, mostly pre-ovarian; some testes may be post-ovarian. Ovary tetralobed in cross-section, posterior. Vitelline follicles encircling medulla. Developing uterus simple median tube; gravid uterus saccular with small diverticula; no preformed uterine pore. Dorsal osmoregulatory canals lateral to ventral osmoregulatory canals. In elasmobranchs. Mediterranean, Indian Ocean, Australia. Type-species *T. tenuicolle* Diesing, 1854.

Superfamily Heteracanthoidea

Family Eutetrarhynchidae Guiart, 1927

(Syn. Renibulbidae Feigenbaum, 1975)

Key to genera

1a. Metabasal hooks homeomorphous or heteromorphous, arranged in ascending half-spiral rows, hook files 1(1′) abutting with hook file (1′) at apex of uniform V-shaped pattern of opposing hook rows .... 2.

1b. Metabasal hooks distinctly heteromorphous, arranged in ascending half-spiral rows, hook files 1(1′) widely separated and forming distinct alternating alignment of left and right hook rows .... 4.

2a. Basal armature with single macrohook on swollen base of tentacle .... *Oncomegas* Dollfus, 1929. (Figs 7.112–7.117)

Diagnosis: Small cestodes. Scolex slender, acraspedote. Two oval, patelliform bothridia with free margins, thickened rims, and posterior notch. Tentacles long with asymmetrical basal swelling. Metabasal armature heteroacanthous typical, files 1(1′) abut forming V's, terminal files abut in inverted V formation; hooks homeomorphous or heteromorphous. Tentacle sheaths irregularly coiled proximally, sinuous in region of bothridia. Prebulbar organs present. Bulbs long, retractor muscles originate in base of bulb. Pars postbulbosa short or absent. Strobila apolytic, proglottids few, acraspedote. Genital pores irregularly alternate or unilateral. External seminal vesicle present, internal seminal vesicle absent. Testes medullary, in two parallel longitudinal rows or scattered in lateral fields; in one or more dorsoventral layers. Vagina with one or two sphincters. Seminal receptacle present. Ovary at posterior extremity, tetralobed in cross-section. Non-gravid uterus median, tubular; gravid uterus with lateral diverticula. Uterine pore present or absent. Vitellarium follicular, circumcortical. Osmoregulatory canals helical or straight, pass dorsal to cirrus-sac. In spiral valve of batoids. Northwest Atlantic coast, USA, Pacific coast of Mexico, Australia; plerocerci in Atlantic and Pacific teleosts, plankton and sea cucumber. Type-species *O. wageneri* (Linton, 1890) Dollfus 1929.

2b. Basal armature lacks macrohook, basal region swollen or not .... 3.

3a. Basal armature and swelling present; testes tandem, internal seminal vesicles absent .... *Dollfusiella* n. g. (Figs 7.118–7.128, 7.137)

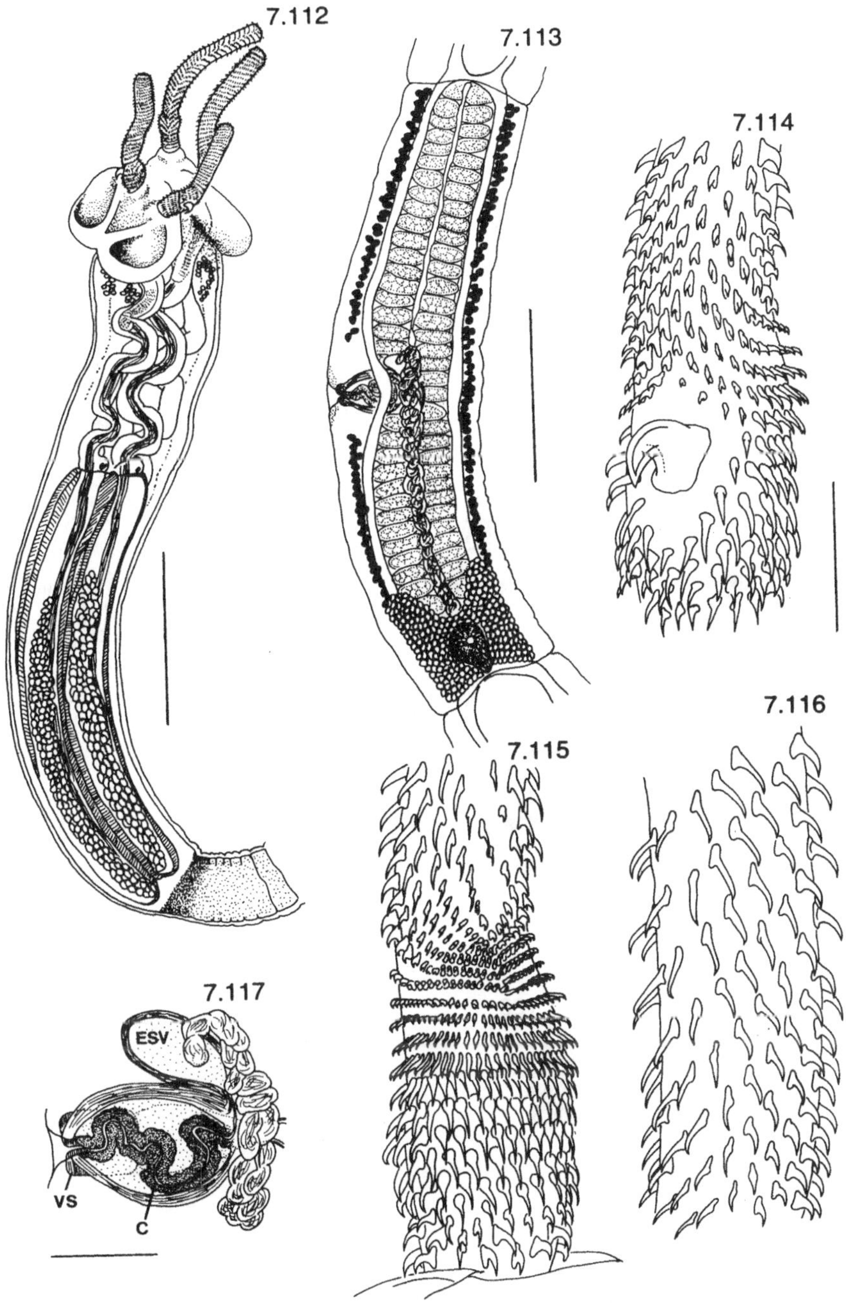

Figs 7.112-7.117 *Oncomegas wageneri* (Linton, 1890) (after Toth *et al.*, 1992.) 7.112. Scolex, note internal ridge in bothridium. 7.113. Mature proglottid. 7.114. External face in basal region. Note macrohook. 7.115. Internal face, basal region of tentacle. 7.116. Bothridial face, metabasal armature. 7.117. Terminal genitalia showing external seminal vesicle and vaginal sphincter. Scale lines: 7.112, 0.5 mm; 7.113, 0.3 mm; 7.114-7.116, 0.05 mm; 7.117, 0.1 mm.

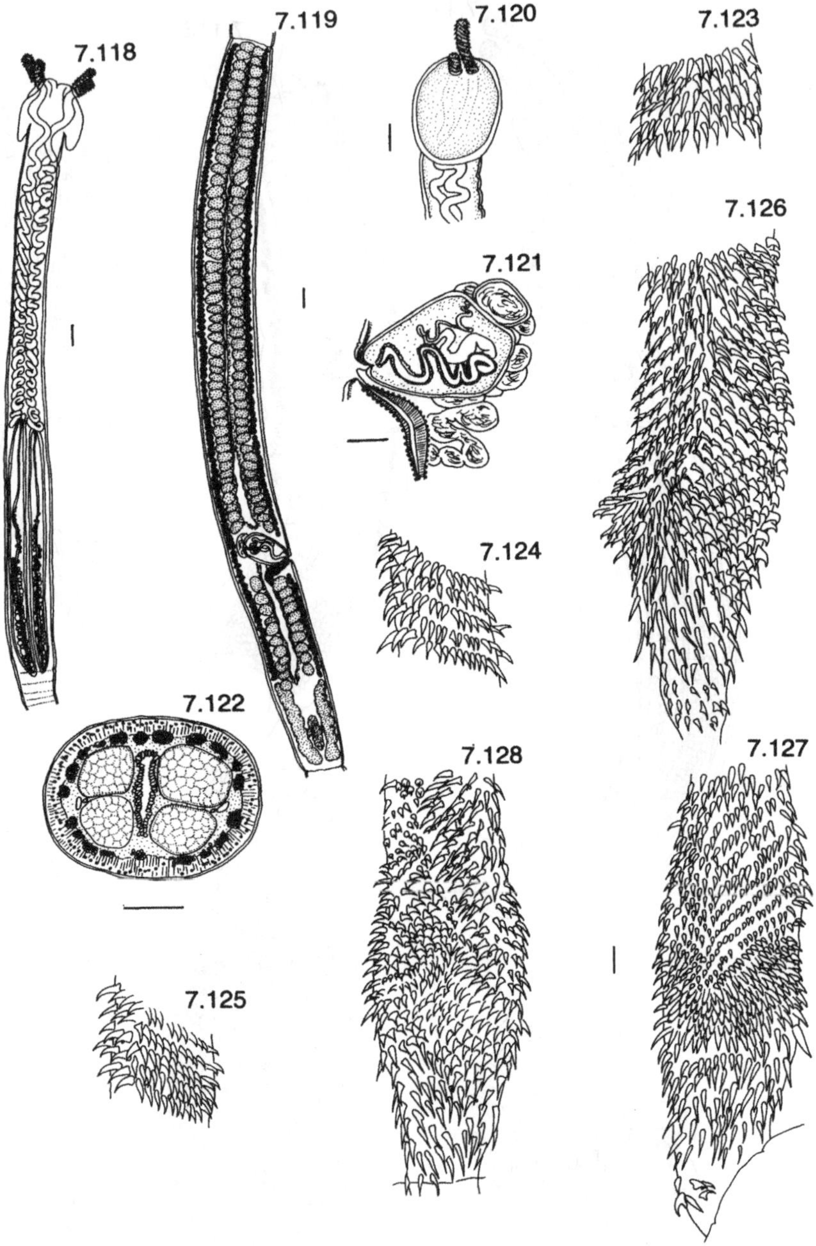

Figs 7.118-7.128 *Dollfusiella australis* (Prudhoe, 1969) gen. et comb. n. 7.118. Scolex. 7.119. Mature proglottid. 7.120. Bothridium. 7.121. Terminal genitalia. 7.122. Transverse section showing layering of testes. 7.123. Bothridial face, metabasal armature. 7.124. External face, metabasal armature. 7.125. Internal face, metabasal region. 7.126. Bothridial face, basal armature. 7.127. Internal face, basal armature. 7.128. External face, basal armature. Scale lines: 7.118-7.122, 0.1 mm; 7.123-7.128, 0.01 mm.

Diagnosis: Scolex acraspedote. Two oval patelliform bothridia with free margins. Armature heteroacanthous typical, hook files 1(1′) abut forming V-shaped origin of principal rows; rows terminate in inverted V formation on opposite face. Hooks homeomorphous or heteromorphous. Basal swelling and special basal armature present. Tentacle sheaths spiral or sinuous. Bulbs long, retractor originates in base of bulb. Prebulbar organs present. Proglottids acraspedote, elongated. Testes in two intervascular parallel bands, layered dorsoventrally in two or four parallel rows. Genital pore marginal. Ovary at posterior extremity. Vitelline follicles form sleeve around internal organs. Uterus median, tubular. In elasmobranchs. Caribbean, Pacific, Mediterranean, Amazon. Type-species *D. australis* (Prudhoe, 1969) n. comb. (syn. *Eutetrarhynchus australis* Prudhoe, 1969).

3b. Basal armature and swelling absent; testes scattered in lateral fields, one or two internal seminal vesicles present ....
.... *Eutetrarhynchus* Pintner, 1913. (Figs 7.129–7.136)

Diagnosis: Scolex acraspedote, occasionally spinose. Two oval bothridia, margins free. Armature heteroacanthous typical, principal rows abut single hook file at origin of half-spiral rows forming V's; at termination of rows on opposite face principal rows abut forming inverted V's. Hooks homeomorphous or heteromorphous, hollow. Tentacle sheaths sinuous or spiral. Bulbs long, retractor muscles originate in base of bulb. Prebulbar organs present. Proglottids craspedote or acraspedote, slender. Testes may overreach lateral osmoregulatory canals. Genital pore marginal. External seminal vesicle present or absent. Ovary in posterior extremity. Vitelline follicles form sleeve around internal organs. Uterus median, tubular; saccate in gravid proglottids. Plerocercus and postlarva known. Adults in elasmobranchs, primarily batoids. Atlantic, Pacific, Mediterranean. Type-species *E. ruficollis* (Eysenhardt, 1829).

4a. Bulbs shorter than bothridia ....
.... *Mecistobothrium* Heinz & Dailey, 1974. (Figs 7.138–7.142)
(Syn. *Renibulbus* Feigenbaum, 1975.)

Diagnosis: Small worms with numerous proglottids. Scolex acraspedote. Two oval, sessile bothridia, joined by velum in type-species. Armature heteroacanthous typical. Basal swelling and basal armature present. Hooks heteromorphous. Metabasal hook files 1(1′) distinctly separate, hooks arranged in alternating ascending half-spirals meeting to form inverted V's on external face of tentacle. Principal rows begin with large rosethorn-shaped hooks on internal face, remaining hooks small and spiniform. Tentacle sheaths only slightly sinuous or spiral. Prebulbar organs present. Retractor muscles originate in base of bulbs. Pars postbulbosa present or absent. Proglottids craspedote. Immature proglottids wider than long; terminal proglottids longer than wide. Ovary large, occupying posterior extremity of proglottid. Testes preovarian in multiple layers. Genital pore marginal, slightly postequatorial, irregularly alternate. Cirrus-sac distinct. Uterus median, tubular, terminating near anterior extremity. Vitelline follicles form sleeve encircling internal organs. Adults in spiral valve of batoids. Pacific

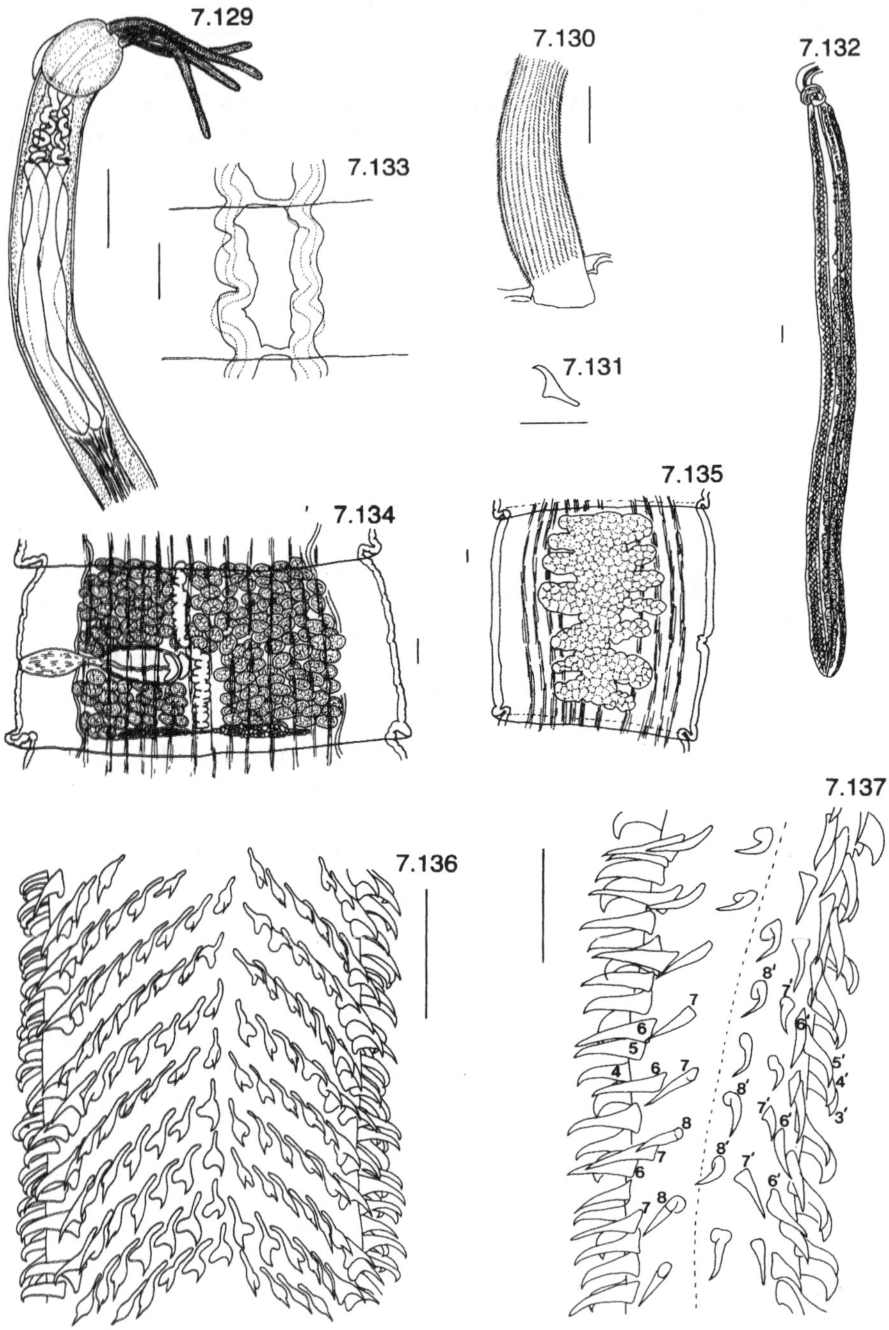

Figs 7.129–7.136 *Eutetrarhynchus* Pintner, 1913. 7.129. Scolex, *E. leucomelanus* (Shipley & Hornell, 1906) (after Beveridge, 1990a). 7.130. Basal armature, *E. leucomelanus*. 7.131. Hook from 7.130. 7.132. Tentacular bulb. 7.133. Osmoregulatory canals. 7.134. Mature proglottid. 7.135. Gravid proglottid. 7.136. External face, metabasal armature, *E. ruficollis* (Eysenhardt, 1829) (after Dollfus, 1942). **Fig. 7.137**. External face, metabasal armature, *Dollfusiella carayoni* (Dollfus, 1942) comb. n. Scale lines: 7.129, 1.0 mm; 7.130, 7.132, 7.133, 7.134, 7.135, 0.1 mm; 7.131, 0.01 mm; 7.136, 0.05 mm; 7.137, 0.025 mm.

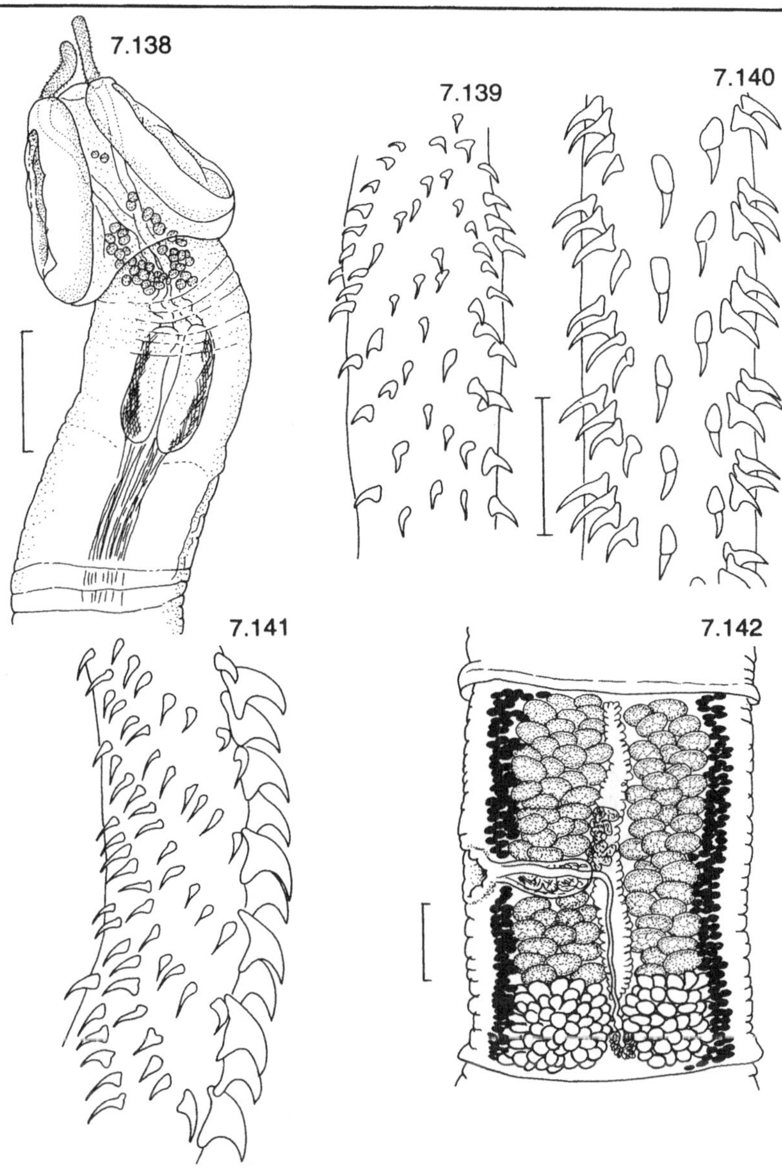

Figs 7.138–7.142 *Mecistobothrium myliobati* Heinz & Dailey, 1974. 7.138. Scolex. 7.139. External face, metabasal armature. 7.140. Internal face, metabasal armature. 7.141. Antibothridial face, metabasal region, internal face on the right. 7.142. Mature proglottid. Scale lines: 7.138–7.141, 0.25 mm; 7.142, 0.2 mm.

and Atlantic coasts of North America. Type-species *M. myliobati* Heinz & Dailey, 1974.

4b. Bulbs longer than bothridia .... **5.**

5a. Proglottids craspedote with lobed velum ....
.... ***Trigonolobium* Dollfus, 1929. (Figs 7.143–7.150)**

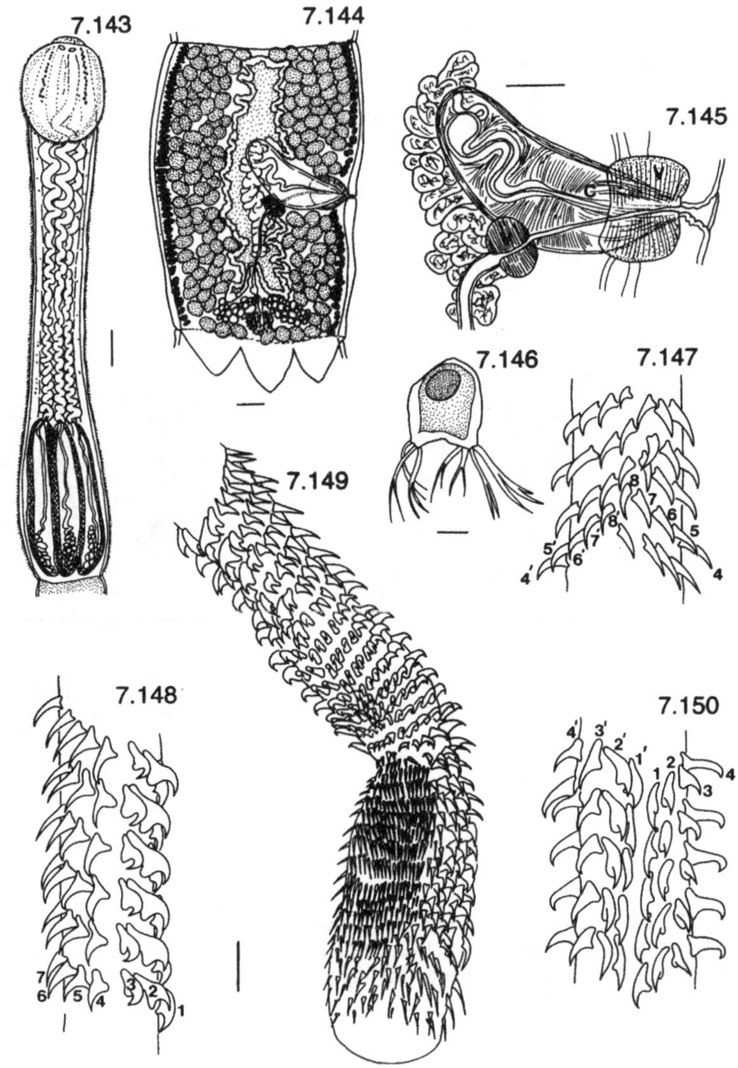

Figs 7.143-7.150 *Trigonolobium spinuliferum* (Southwell, 1911) (after Beveridge, 1990a). 7.143. Scolex. 7.144. Mature proglottid. 7.145. Terminal genitalia showing vaginal sphincters. 7.146. Egg. 7.147. External face, metabasal region. 7.148. Bothridial face, metabasal region. 7.149. Basal armature, internal face to the left. 7.150. Internal face, metabasal armature. Scale lines: 7.143-7.145, 0.1 mm; 7.146-7.150, 0.01 mm.

Diagnosis: Scolex acraspedote, spinose. Two bothridia widely oval. Pars vaginalis long, Pintner's cells present. Tentacle sheaths partly spiral or sinuous. Prebulbar organs present. Retractor muscles originate in bases of bulbs. Metabasal armature heteroacanthous typical; distinct basal armature present on basal swelling. Hooks heteromorphous. Principal rows alternate; hook files 1(1″) separated by distinct space, terminal files abut on external face in inverted V pattern. Genital pore marginal. Internal and external seminal vesicles absent. Testes occupy available space in lateral medullary fields. Vagina with sphincters. Ovary tetralobed, near posterior extremity. Vitelline follicles form lateral bands arching around testes. Uterus linear, bifurcated proximally, pore absent; gravid uterus saccate. Eggs with branching filaments. In elasmobranchs. Sri Lanka, Australia. Type-species *T. spinuliferum* (Southwell, 1911).

5b. Proglottids acraspedote .... **6.**

6a. Basal swelling and armature absent .... ***Parachristianella*** **Dollfus, 1946. (Figs 7.151–7.159)**
Diagnosis: Small cestodes with few proglottids. Scolex acraspedote. Two wide, oval bothridia with thin rims. Armature heteroacanthous, heteromorphous, hooks solid. Hooks 1(1′) well separated, larger than other hooks in principal row. Tentacle sheaths irregularly spiral or sinuous. Bulbs elongated, retractor muscles originate in base of bulb. Prebulbar organs present. Proglottids acraspedote, elongated, maturing rapidly. Genital pores marginal. Testes intervascular in two parallel rows anterior to ovary. Cirrus-sac prominent, internal and external seminal vesicles absent. Vagina enters genital atrium anterior to cirrus-sac. Ovary prominent, at posterior extremity. Vitelline follicles encircle organs anterior to ovary. Uterus simple, median; gravid uterus saccate. Plerocerci in Crustacea and teleosts, adults in elasmobranchs. Atlantic, Mediterranean, Pacific. Type-species *P. trygonis* Dollfus, 1946.

6b. Basal swelling and armature present .... **7.**

7a. Hooks 1(1′) small, increasing in size towards middle of row then diminishing in size .... ***Prochristianella*** **Dollfus, 1946. (Figs 7.160–7.168)**
Diagnosis: Small cestodes with few proglottids. Scolex craspedote or acraspedote. Two bothridia, widely oval, rims thin or moderately developed. Metabasal armature heteroacanthous typical. Hooks heteromorphous, hollow. Hook files 1(1′) well separated. Largest hooks at middle of principal rows. Basal swelling and distinctive basal armature present. Tentacle sheaths irregularly sinuous or spiral. Bulbs elongate, retractor muscles originate in base. Prebulbar organs present. Proglottids acraspedote, elongate. Genital pores marginal. Testes intervascular, in two preovarian parallel rows. Cirrus-sac prominent. Internal and external seminal vesicles absent. Ovary prominent, at posterior extremity. Vitelline follicles encircle internal organs anterior to ovary. Uterus simple, median; gravid uterus saccate, lacking pore. Plerocerci in Crustacea and teleosts, adults in elasmobranchs and holocephalans. Atlantic, Mediterranean, Pacific. Type-species *P. papillifer* (Polyarkoff, 1909).

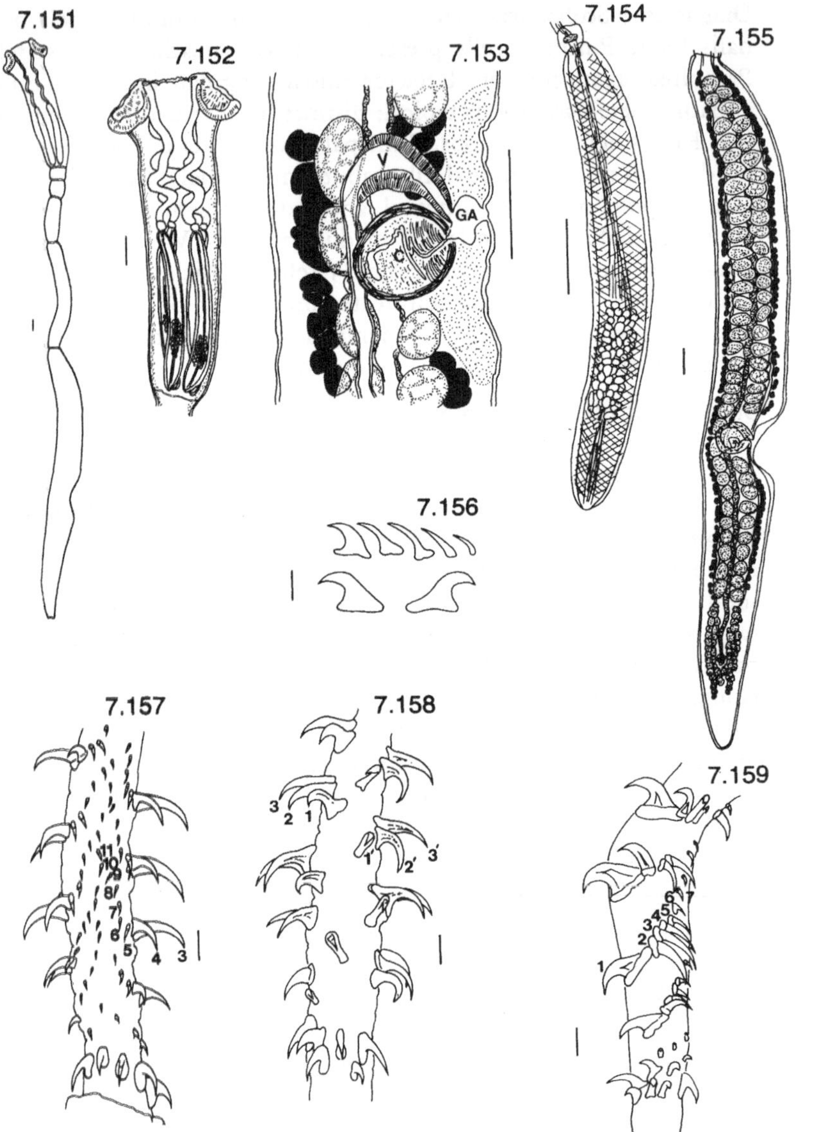

Figs 7.151-7.159 *Parachristianella monomegacantha* Kruse, 1959 (after Beveridge, 1990a). 7.151. Entire worm. 7.152. Scolex. 7.153. Terminal genitalia. 7.154. Bulb. 7.155. Mature proglottid. 7.156. Profiles of hooks 1-6. 7.157. Internal face, basal region. 7.158. External face, basal region. 7.159. Bothridial face, basal region. Scale lines: 7.151-7.155, 0.1 mm; 7.156-7.159, 0.01 mm.

7b. **Hooks 1(1′) largest, remainder of hooks decreasing along row, basal swelling and armature present** .... **8.**

8a. **Triad of large hooks present on asymmetrical basal swelling** .... ***Trimacracanthus*** **Beveridge & Campbell, 1987. (Figs 7.169–7.182)**

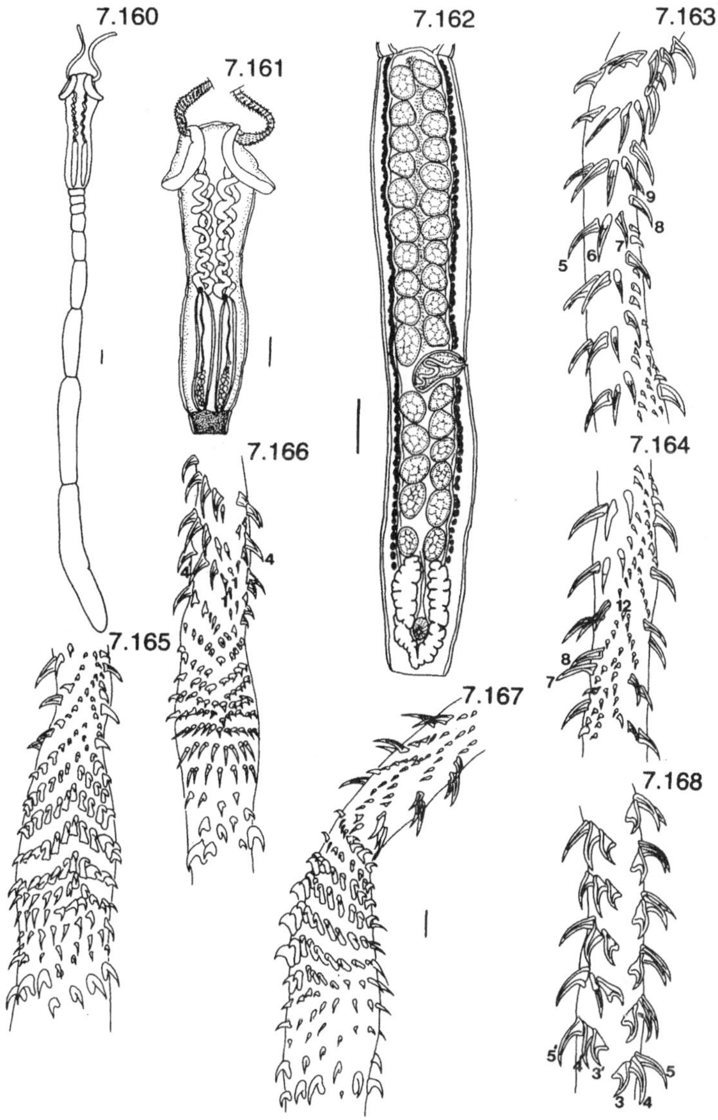

Figs 7.160-7.168 *Prochristianella mooreae* Beveridge, 1990 (after Beveridge, 1990a). 7.160. Entire worm. 7.161. Scolex. 7.162. Mature proglottid. 7.163. Bothridial face, metabasal region. 7.164. External face, metabasal region. 7.165. External face, basal armature. 7.166. Internal face, basal armature. 7.167. Bothridial face, basal armature. 7.168. Internal face, metabasal armature. Scale lines: 7.160-7.162, 0.1 mm; 7.163-7.168, 0.01 mm.

Diagnosis: Scolex slender, elongate, acraspedote. Two bothridia, margins thick, free, not contiguous apically, may be notched posteriorly suggesting incipient subdivision. Pars bulbosa and pars vaginalis much longer than pars bothridialis. Bulbs long and slender. Tentacle sheaths sinuous but not spiral. Prebulbar organs present. Retractor muscles originate in posterior extremity of bulb. Metabasal

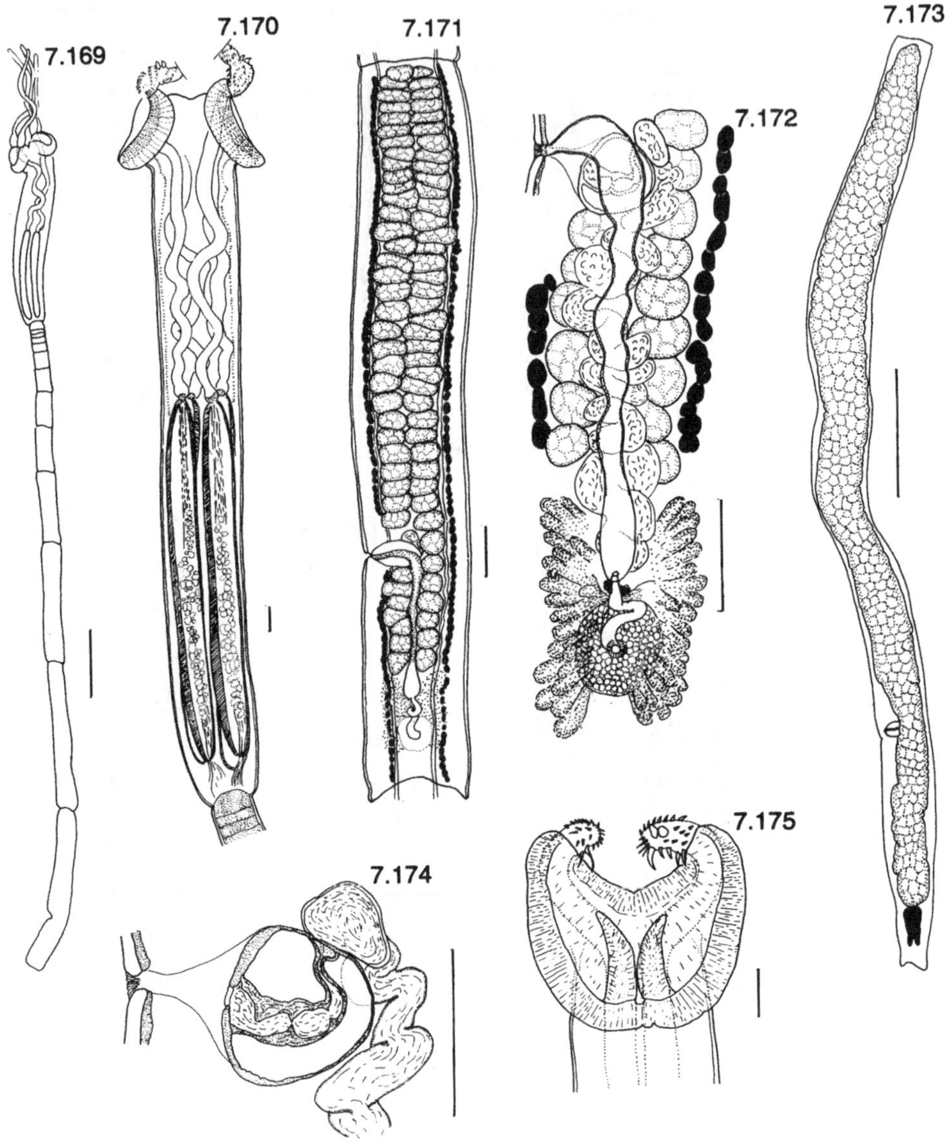

Figs 7.169-7.175 *Trimacracanthus aetobatidis* (Robinson, 1959) (after Beveridge & Campbell, 1987). 7.169. Entire worm. 7.170. Scolex. 7.171. Mature proglottid. 7.172. Reproductive system. 7.173. Gravid proglottid. 7.174. Terminal genitalia. 7.175. Bothridium. Scale lines: 7.169, 7.173, 1.0 mm; 7.170-7.172, 7.174-7.175, 0.1 mm.

Figs 7.176-7.182 *Trimacracanthus aetobatidis* (Robinson, 1959) (after Beveridge & Campbell, 1987). 7.176. Bothridial face, near tip of tentacle. 7.177. External face, metabasal region. 7.178. Antibothridial face, metabasal region. 7.179. External face, metabasal region. 7.180. Bothridial face, basal region. 7.181. External face, basal region. 7.182. Internal face, basal region. Scale lines: 0.1 mm.

7.176

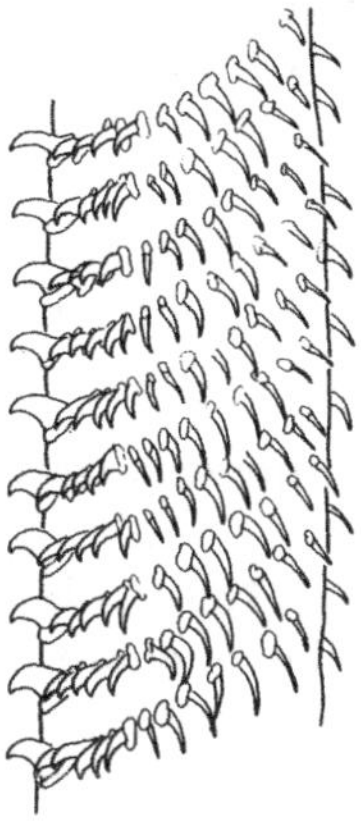

7.177

7.178

7.179

7.180

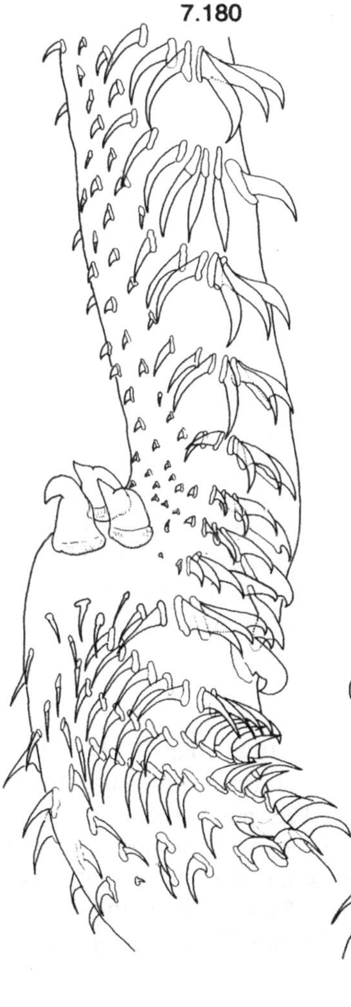

7.181

7.182

armature heteroacanthous, heteromorphous, typical. Hook solid. Metabasal armature of alternating half-spiral rows beginning with large hooks 1(1′), hooks 2(2′) only slightly smaller; remaining hooks of row gradually decrease in size. Strobila acraspedote, hyperapolytic. Genital pore marginal. Testes medullary, preovarian. External seminal vesicle present. Ovary posterior, tetralobed in cross section. Seminal receptacle present. Uterus median. Vitelline follicles circumcortical. In spiral valve of elasmobranchs. Northwest Atlantic, New Zealand, Australia. Type-species *T. aetobatidis* (Robinson, 1959) Beveridge & Campbell, 1987.

8b. Basal armature lacks triad of large hooks, distinctive basal armature restricted to external face ....
.... ***Pseudochristianella* Campbell & Beveridge, 1990. (Figs 7.183–7.189)**

Diagnosis: Small worms with two bothridia and elongate bulbs. Prebulbar organs present. Pars postbulbosa and velum absent. Metabasal armature heteroacanthous, heteromorphous typical. Metabasal hooks arranged in alternating half-spiral rows. Hooks 1(1′) large, separated by prominent space, remaining hooks of principal row decreasing in size. Basal swelling on tentacle. Proglottids acraspedote; testes in two longitudinal rows. Internal and external seminal vesicles absent. In selachians. Indian Ocean. Type-species *P. southwelli* Campbell & Beveridge, 1990.

Family Gilquiniidae Dollfus, 1942

Key to genera

1a. Rhyncheal system present .... **2.**

1b. Rhyncheal system absent ....
.... ***Aporhynchus* Nybelin, 1918. (Figs 7.190–7.194)**

Diagnosis: Scolex acraspedote. Four auriculate bothridia; rims incomplete anteriorly. Strobila acraspedote. Genital atrium near anterior extremity of proglottid, enclosed by sphincter. Accessory, external and internal seminal vesicles present. Testes scattered, layered, in two lateral fields, anterior and posterior to ovary. Ovary small, separated from posterior proglottid margin by distinct space. Uterus median, deviated porally; uterine pore present; gravid uterus saccate. Vitelline follicles circum-medullary. In sharks. Northeast Atlantic, Tasmania. Type-species *A. norvegicus* (Olsson, 1868).

2a. Craspedote scolex with prominent velum ....
.... ***Deanicola* Beveridge, 1990. (Figs 7.195–7.201)**

Diagnosis: Four auriculate bothridia with thickened rims; rims incomplete anteriorly. Pars vaginalis longer than pars bulbosa. Pars postbulbosa present or absent. Bulbs ovoid or elongate; retractor muscles originate in base. Tentacles short, sheaths sinuous. Metabasal armature heteroacanthous, heteromorphous; principal rows abut on external face, do not overlap. Basal swelling and distinct basal armature present. Strobila acraspedote. Proglottids quadrangular to

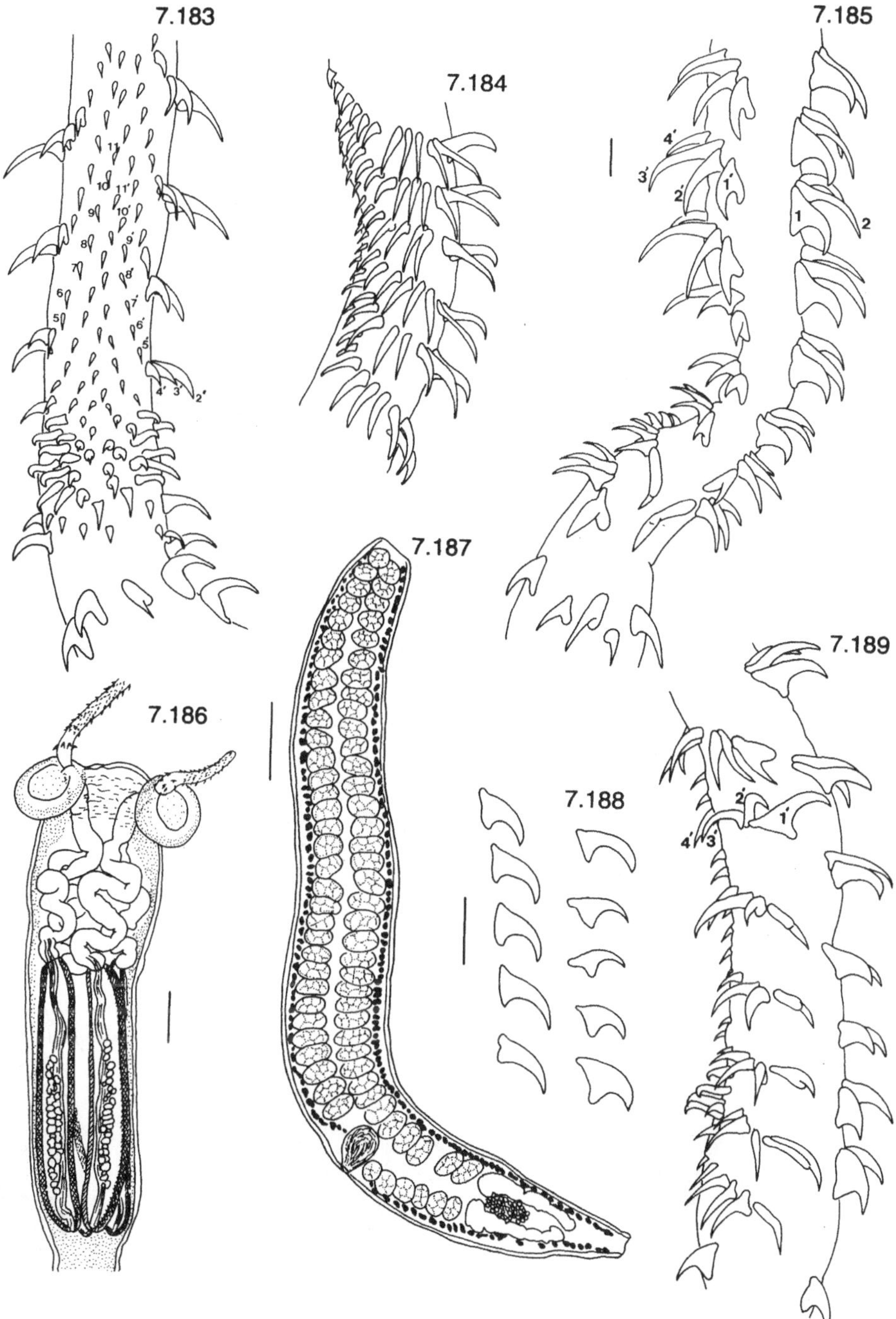

Figs 7.183-7.189 *Pseudochristianella southwelli* Campbell & Beveridge, 1990. 7.183. External face, basal region. 7.184. Bothridial face, metabasal region. 7.185. Internal face, basal region. 7.186. Scolex. 7.187. Mature proglottid. 7.188. Profiles of hooks 1(1′). 7.189. Bothridial face, basal region. Scale lines: 7.183-7.185, 7.189, 0.01 mm; 7.188, 0.01 mm.

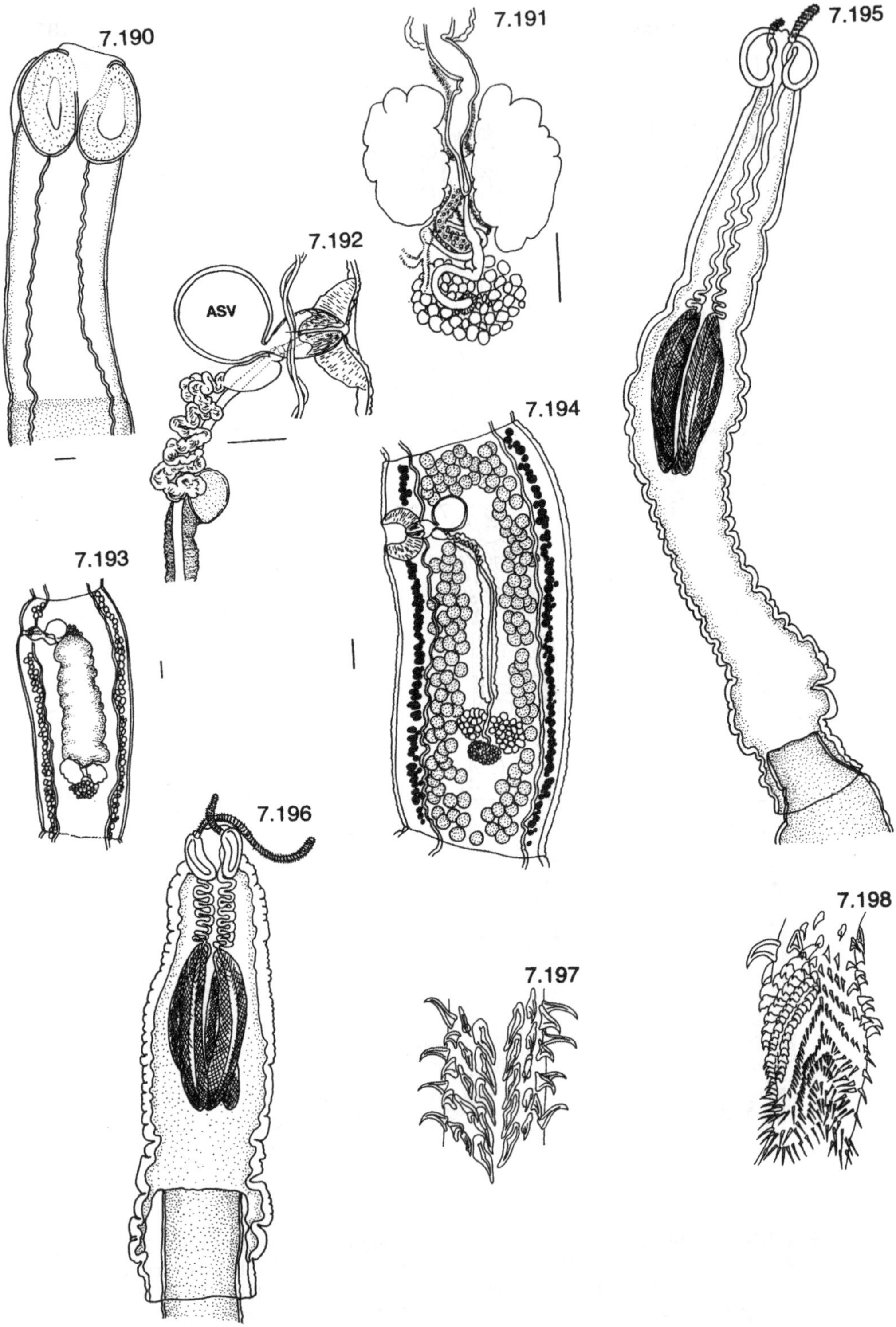
7.190
7.191
7.192
ASV
7.193
7.194
7.195
7.196
7.197
7.198

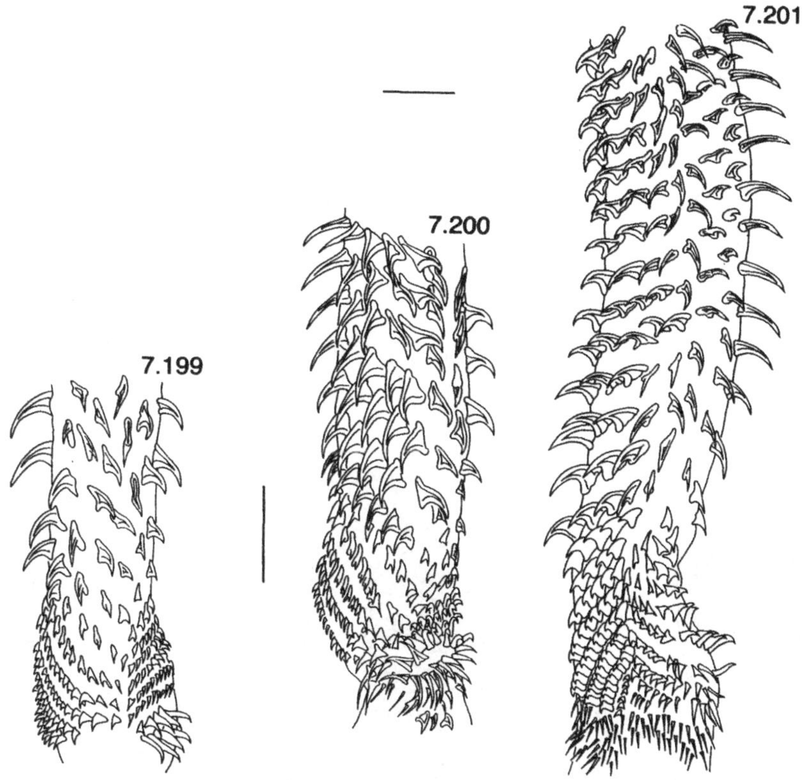

Figs 7.190–7.194 *Aporhynchus norvegicus* (Olsson, 1868) (after Beveridge, 1990b). 7.190. Scolex. 7.191. Female genital complex. 7.192. Male genitalia. 7.193. Gravid proglottid. 7.194. Mature proglottid. Scale lines: 0.1 mm.

Figs 7.195–7.201 *Deanicola protentus* Beveridge, 1990. 7.195. Scolex, relaxed specimen. 7.196. Scolex, contracted specimen. 7.197. Internal face, metabasal region. 7.198. External face, transition from basal to metabasal armature. 7.199. Internal face, transition from basal to metabasal armature. 7.200. Bothridial face, internal face on right side. 7.201. Antibothridial face of tentacle, external face on right side; base twisted showing internal face. Scale lines: 7.195–7.196, 1.0 mm; 7.197–7.201, 0.1 mm.

elongate. Genital pore on lateral margin near anterior extremity of proglottid, enclosed by sphincter. Accessory seminal vesicle present. Testes scattered in two lateral bands contiguous anterior to uterus and posterior to ovary. Ovary small, set well forward of posterior proglottid margin. Uterus median, tubular, deviated toward genital pore; saccate when gravid; preformed uterine pore present or absent. Vitelline follicles circum-medullary. In sharks (*Deania*). Tasmania and North Atlantic (Campbell, unpublished). Type-species *D. protentus* Beveridge, 1990.

2b. Acraspedote scolex, velum absent .... **3.**

3a. Pars postbulbosa absent, hooks 1(1′) divergent ....
.... ***Gilquinia* Guiart, 1927. (Figs 7.202–7.211)**

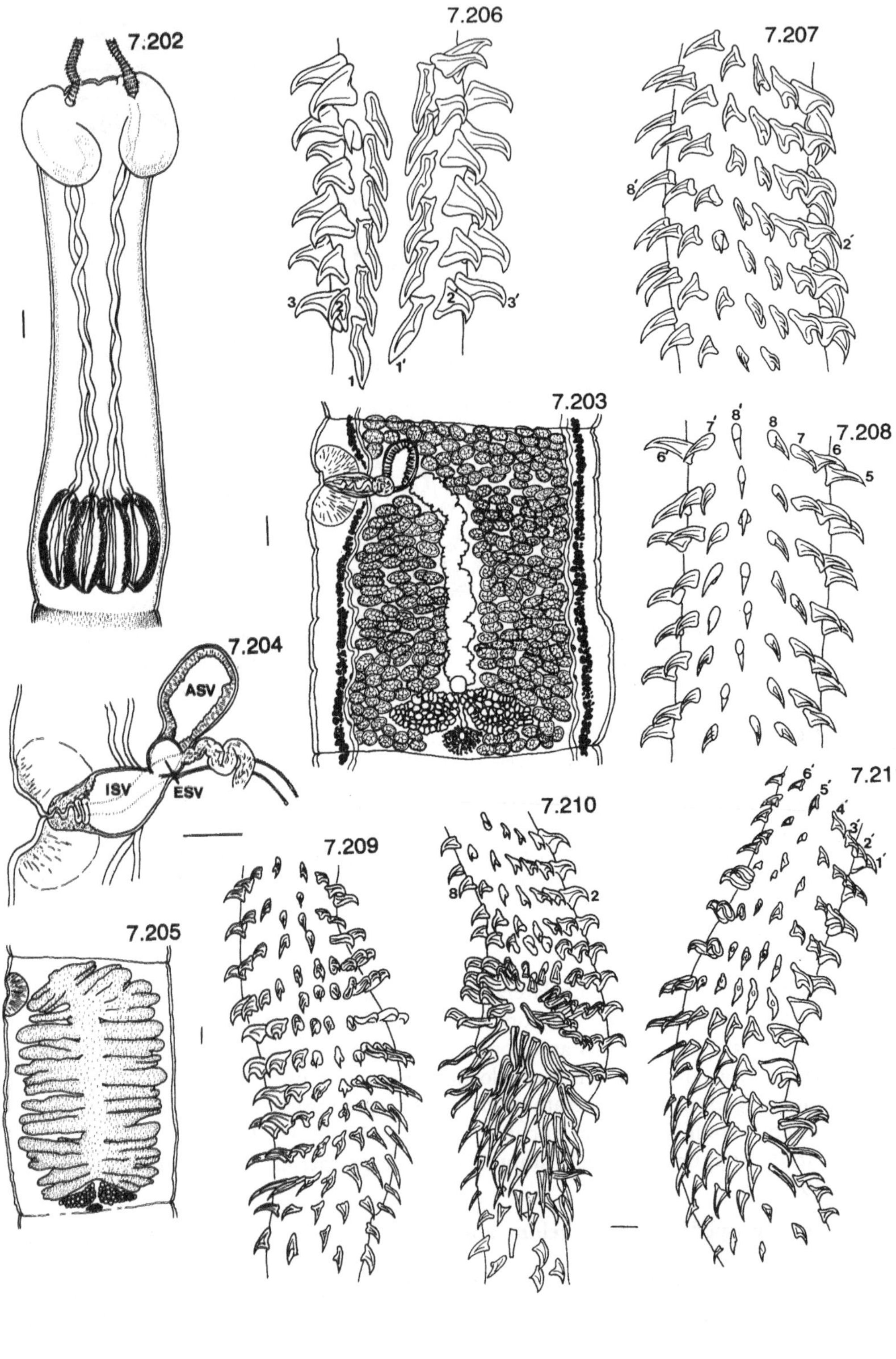
7.202
7.206
7.207
7.203
7.208
7.204
ASV
ISV
ESV
7.205
7.209
7.210
7.211

Diagnosis: Four ear-like bothridia. Medial bothridial margins incomplete, fused with scolex. Rhyncheal system well developed. Pars vaginalis longer than pars bulbosa. Tentacles short. Metabasal armature heteroacanthous, typical. Hooks heteromorphous, arranged in alternating half-spirals. Hook files 1(1′) divergent; half-spiral rows abut on external face but do not overlap. Basal swelling and distinctive basal armature present. Tentacle sheaths sinuous. Bulbs relatively short, ovoid. Retractor muscles originate in base of bulb. Prebulbar organs lacking. Strobila acraspedote, proglottids elongate. Genital pore marginal, near anterior extremity. Testes numerous, scattered in available intervascular space, anterior and posterior to ovary. Accessory seminal vesicle present in addition to external and internal seminal vesicles. Ovary separated from posterior extremity of proglottid by space. Uterus median, simple, tubular, distal portion deviated toward genital pore; uterine pore absent. Vitelline follicles circum-medullary. In squalid sharks. Atlantic, Pacific. Type-species *G. squali* (Fabricius, 1794).

3b. Pars postbulbosa present, hook files 1(1′) not divergent .... *Plesiorhynchus* Beveridge, 1990. (Figs 7.212–7.219)
Diagnosis: Four oval bothridia, rims thick, incomplete anteriorly. Tentacles short, sheaths slightly sinuous. Pars vaginalis only slightly longer than pars bulbosa. Bulbs short, retractor muscles originate in base of bulbs. Metabasal armature heteroacanthous, heteromorphous typical, rows do not overlap on external face. Distinctive basal armature and swelling absent. Strobila acraspedote. Genital pores marginal, enclosed by sphincter, near anterior extremity of proglottid; irregularly alternate. Accessory, external and internal seminal vesicles present. Testes scattered, in two bands lateral to ovary, may overreach lateral osmoregulatory canals, contiguous at extremities or not. Ovary small, set forward of posterior proglottid margin. Uterus tubular, deviated toward genital pore; gravid uterus saccate; uterine pore present or absent. Vitelline follicles circum-medullary. In sharks. Tasmania. Type-species *P. etmopteri* Beveridge, 1990.

Family Shirleyrhynchidae n. fam.

Key to genera

1a. Basal armature on slight swelling; hooks 1(1′) divergent, rows abut on external face .... *Shirleyrhynchus* Beveridge & Campbell, 1988. (Figs 7.220–7.234)
Diagnosis: Scolex long, acraspedote. Four narrow, cupped bothridia, aperture resembling a longitudinal slit. Bulbs elongate, retractor muscles originate at base

Figs 7.202–7.211 *Gilquinia squali* (Fabricius, 1794) (after Beveridge, 1990b). 7.202. Scolex. 7.203. Mature proglottid. 7.204. Terminal genitalia. 7.205. Gravid proglottid. 7.206. Internal face, metabasal armature. 7.207. Bothridial face, metabasal region, internal face on right. 7.208. External face, metabasal region. 7.209. External face, basal armature. 7.210. Internal face, basal armature. 7.211. Bothridial face, basal armature. Scale lines: 7.202–7.205, 0.1 mm; 7.206–7.211, 0.01 mm.

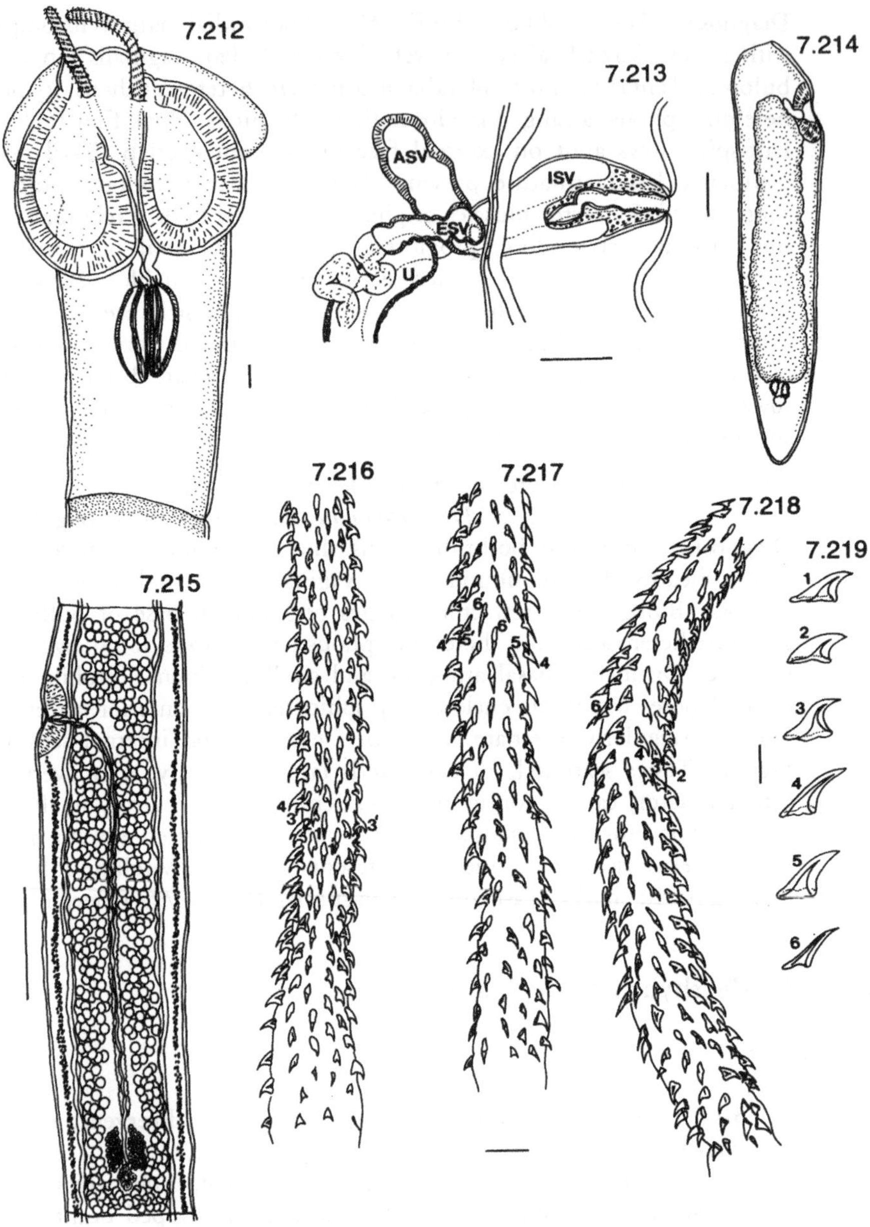

Figs 7.212-7.219 *Plesiorhynchus etmopteri* Beveridge, 1990 (after Beveridge, 1990b). 7.212. Scolex. 7.213. Male terminal genitalia, dorsal view. 7.214. Gravid proglottid. 7.215. Mature proglottid. 7.216. Internal face, basal and metabasal regions. 7.217. External face, basal and metabasal regions. 7.218. Bothridial face, basal and metabasal regions. 7.219. Profiles of hooks 1-6. Scale lines: 7.212, 7.213, 0.1 mm; 7.214, 7.215, 1.0 mm; 7.216-7.218, 0.025 mm; 7.219, 0.01 mm.

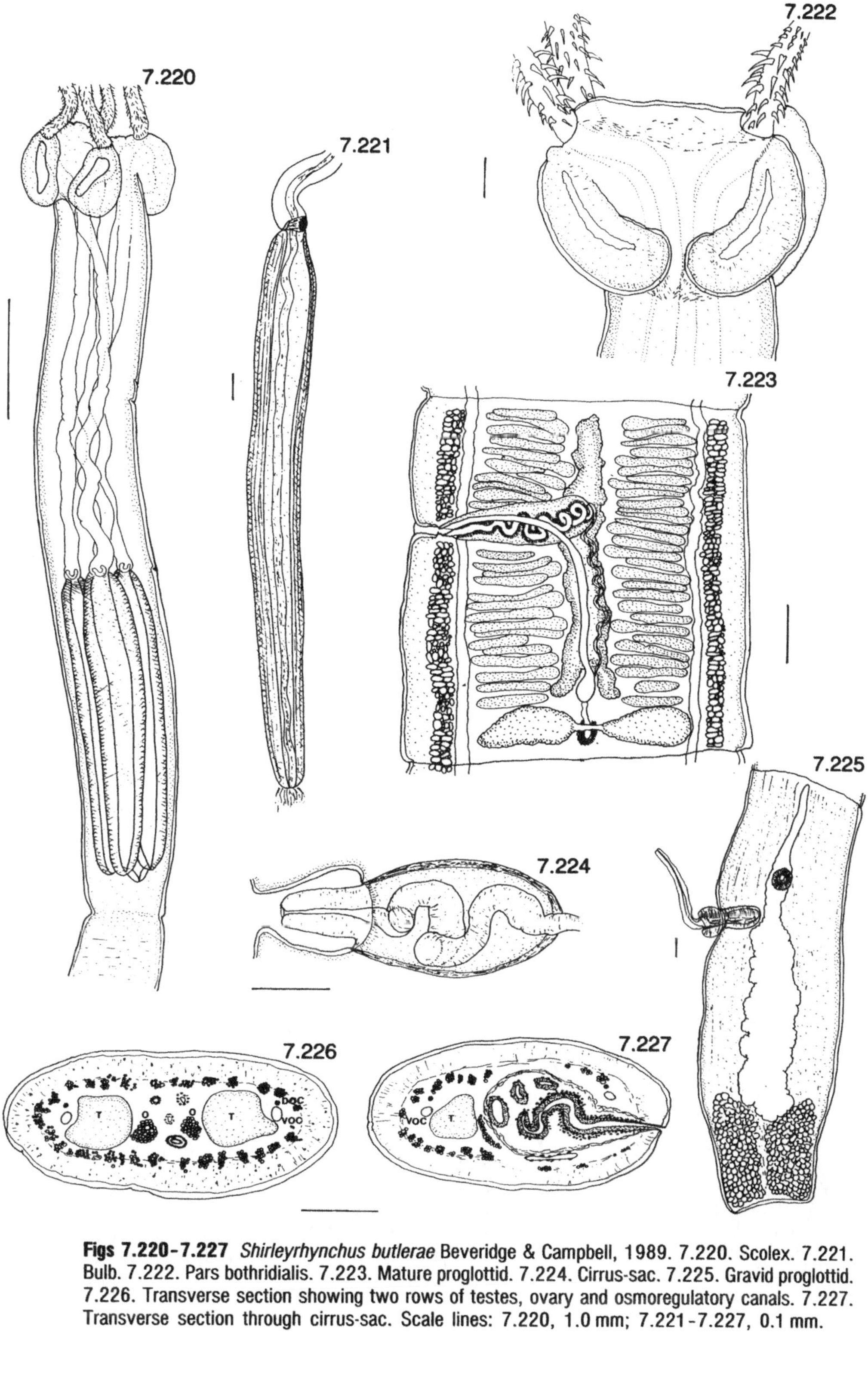

Figs 7.220–7.227 *Shirleyrhynchus butlerae* Beveridge & Campbell, 1989. 7.220. Scolex. 7.221. Bulb. 7.222. Pars bothridialis. 7.223. Mature proglottid. 7.224. Cirrus-sac. 7.225. Gravid proglottid. 7.226. Transverse section showing two rows of testes, ovary and osmoregulatory canals. 7.227. Transverse section through cirrus-sac. Scale lines: 7.220, 1.0 mm; 7.221–7.227, 0.1 mm.

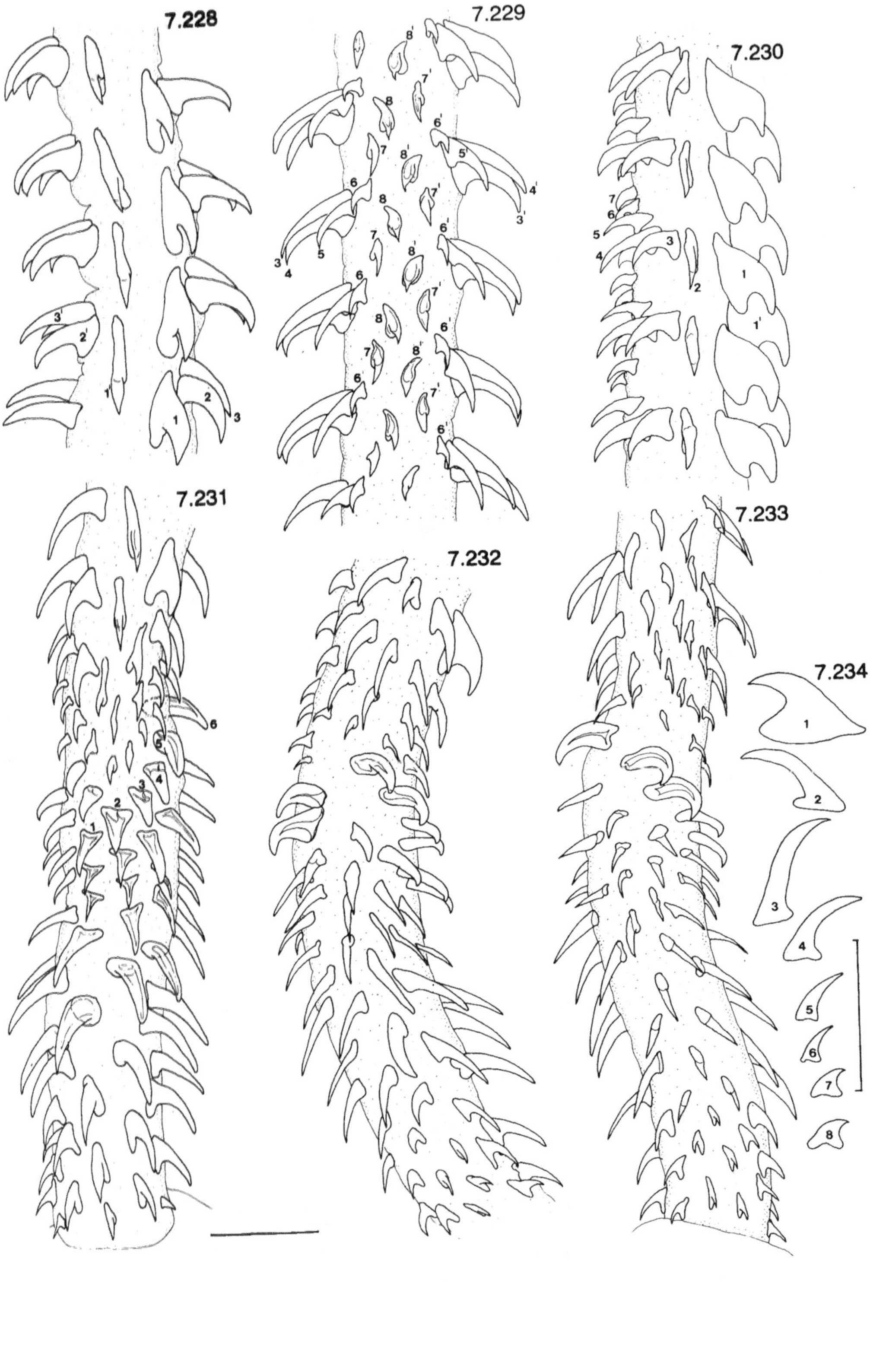

7.228
7.229
7.230
7.231
7.232
7.233
7.234

of bulb. Prebulbar organs present. Pars vaginalis long; tentacle sheaths sinuous; pars postbulbosa distinct. Metabasal armature heteroacanthous, heteromorphous, typical. Basal armature distinctive. Basal hooks hollow; metabasal hooks solid. Strobila acraspedote; proglottids elongate, numerous, apolytic. Testes preovarian, in parallel longitudinal rows. Genital pores marginal, irregularly alternate. Accessory seminal vesicle absent. Seminal vesicles absent. Vitelline follicles circumcortical. Uterus tubular, median, terminates near anterior extremity of proglottid; uterine pore present. Eggs tanned, spherical. In Batoidea (Dasyatidae). Australia. Type-species *S. butlerae* Beveridge & Campbell, 1988.

1b. **Basal armature present; hook files 1(1′) widely separate on internal face; rows do not abut on external face** .... *Cetorhinicola* **Beveridge & Campbell, 1988. (Figs 7.235–7.240)**
Diagnosis: Scolex elongate, acraspedote. Four bothridia, joined on posterior margin by velum. Tentacles long, slender. Basal armature present. Metabasal armature heteroacanthous, heteromorphous, typical. Hooks solid, arranged in half-spiral rows ascending from internal to external face. Pars vaginalis long; tentacle sheaths sinuous. Bulbs elongate, retractor muscles originate at base of bulb. Prebulbar organs present. Pars postbulbosa short. Strobila unknown. In basking sharks. South Australia. Type-species *C. acanthocapax* Beveridge & Campbell, 1988.

Superfamily Otobothrioidea

Family Otobothriidae Dollfus, 1942

Key to genera

1a. **Two sets of reproductive organs per proglottid; scolex craspedote with velum** .... *Diplootobothrium* **Chandler, 1942. (Figs 7.241–7.249)**
Diagnosis: Large trypanorhynchs. Two subspherical bothridia with lateral sensory fossettes. Tentacle sheaths spiral. Bulbs slender. Pars bulbosa only slightly longer than pars bothridialis. Pars vaginalis more than four times length of bothridia. Retractor muscles originate in base of bulb. Armature heteroacanthous atypical. Hooks heteromorphous, hollow. Rows begin on bothridial face and end on antibothridial face. Alternating principal rows of large hooks separated by interpolated row of small hooks on antibothridial face. Strobila wide, flat, proglottids acraspedote. Genital pores marginal, postequatorial. Cirrus-sac globose, cirrus armed. External (accessory ?) seminal vesicle present. Testes numerous, scattered throughout available medullary space. Vagina posterior to cirrus-sac. Ovaries bilobed, situated forward of posterior margin to either side of median line. Uteri elongate, tubular, parallel, reaching to anterior proglottid border. Uterine duct

Figs 7.228–7.234 *Shirleyrhynchus butlerae* Beveridge & Campbell, 1989. 7.228. Internal face, metabasal armature. 7.229. External face, metabasal armature. 7.230. Bothridial face, metabasal armature. 7.231. Basal region, internal face. 7.232. Basal region, antibothridial face. 7.233. Basal region, external face. 7.234. Metabasal hooks 1-8 in profile. Scale lines: 0.1 mm.

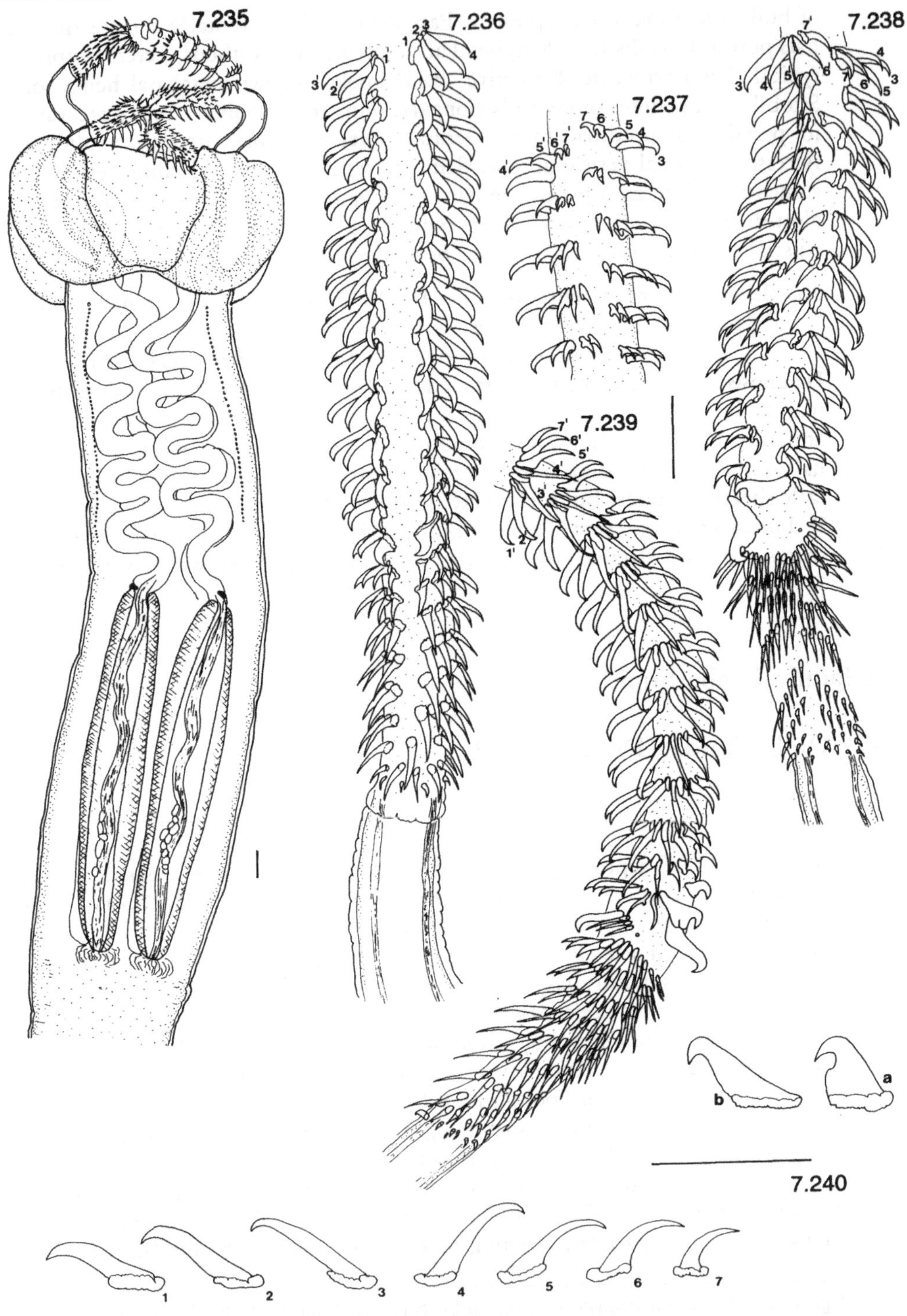

Figs 7.235-7.240 *Cetorhinicola acanthocapax* Beveridge & Campbell, 1988. 7.235. Scolex. 7.236. Tentacle, basal and metabasal regions, internal face. 7.237. External face, metabasal region. 7.238. Tentacle, basal and metabasal regions, external face. 7.239. Tentacle, basal and metabasal regions, antibothridial face. 7.240. Principal hooks 1-7 in profile; enlarged hooks of antibothridial (a) and bothridial (b) faces. Scale lines: 0.1 mm.

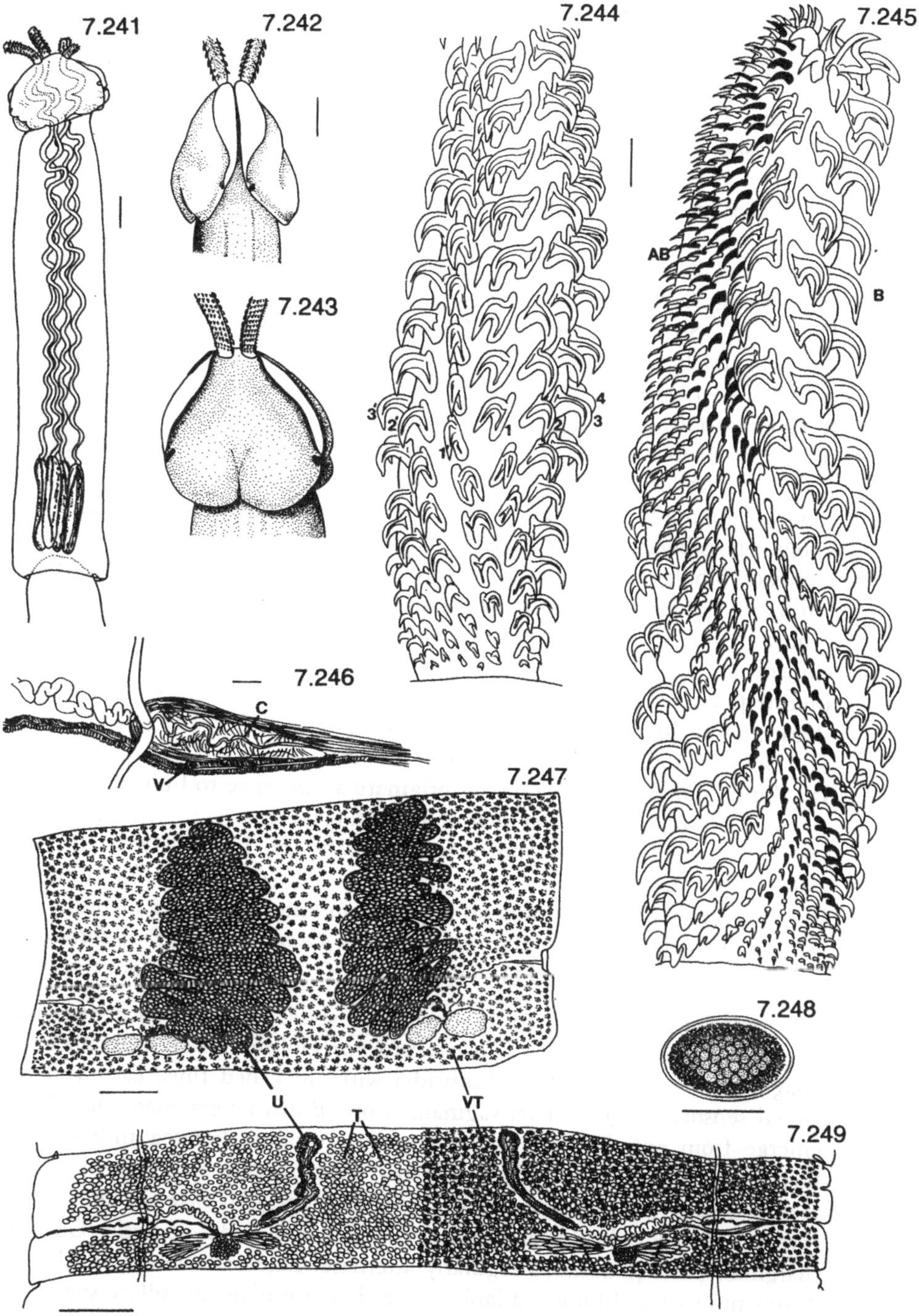

Figs 7.241-7.249 *Diplootobothrium springeri* Chandler, 1942. 7.241. Scolex (after Chandler, 1942a). Retractor originates in base of bulb. 7.242. Bothridia, lateral view. 7.243. Bothridia, facial view. 7.244. Tentacle, bothridial face. 7.245. Tentacle, antibothridial face, note torsion. Intercalary row in black. 7.246. Terminal genitalia, note hermaphroditic duct. 7.247. Gravid proglottid. 7.248. Egg. 7.249. Mature proglottid. Scale lines: 7.241-7.243, 7.247, 7.249, 1.0 mm; 7.244, 7.245, 7.246, 0.1 mm; 7.248, 0.05 mm.

joins proximal end of uterus. Gravid uterus with numerous diverticula. Vitelline follicles scattered throughout cortical zone. In sharks. Western North Atlantic, Gulf of Mexico, Indian Ocean. Type-species *D. springeri* Chandler, 1942.

1b. One set of reproductive organs per proglottid; scolex craspedote or acraspedote .... 2.

2a. Scolex craspedote, tentacle sheaths spiral or strongly sinuous .... *Otobothrium* Linton, 1890. (Figs 7.250–7.255)
(Syns *Paramecistobothrium* Bilqees & Kurshid, 1987; *Symbothriorhynchus* Yamaguti, 1952.)

Diagnosis: Scolex craspedote. Two patelliform bothridia with or without posterior notches, apically inclined; rims more or less thickened, pair of eversible sensory fossettes on posterolateral margins. Tentacles relatively short. Armature heteroacanthous atypical; one or more rows or groups of small hooks on external face for each row of principal hooks originating on internal face. Hooks heteromorphous, hollow. Bulbs short with retractor originating at entrance or long with retractor originating in anterior quarter of bulb. Pars postbulbosa absent. Strobila apolytic possibly euapolytic, slender, narrow at its origin. Proglottids acraspedote, elongate. Single set of genitalia per proglottid. Genital pores marginal, irregularly alternate, equatorial or postequatorial. Testes in medullary intervascular field; some postovarian. Ovary well forward of posterior extremity. Vitelline follicles circumcortical. Uterus median, tubular; saccate in gravid proglottids; no preformed uterine pore. Blastocyst oval without caudal extension. Plerocerci in teleosts and sharks; adults in sharks. Cosmopolitan. Type-species *O. crenacolle* Linton, 1890.

Dollfus (1942) designated two subgenera:

Ia. Bulbs short; retractor muscles originate at entrance to bulb; more than one row of small hooks per row of large hooks .... *Otobothrium (Otobothrium)* Linton, 1890. Type-species *O. (O.) crenacolle* Linton, 1890.

Ib. Bulbs long; retractor muscles originate in anterior quarter of bulb; one row small hooks per row large hooks .... *Otobothrium (Pseudotobothrium)* Dollfus, 1942. Type-species *O. (P.) magnum* Southwell, 1924.

2b. Scolex acraspedote, tentacle sheaths not spiral but abruptly bent at junction with bulbs .... ***Poecilancistrum* Dollfus, 1929. (Figs 7.256–7.264)**

Diagnosis: Two widely oval bothridia with thickened rims, each with pair of lateral sensory fossettes. Pars vaginalis short. Bulbs longer than wide. Tentacles emerge from anterior bothridial margins. Armature heteroacanthous atypical, hooks heteromorphous, hollow. Principal rows of large hooks begin on internal face, separated by more numerous, somewhat irregular, rows of small hooks on external face. Proglottids acraspedote, longer than wide, apolytic. Genital pore marginal, postequatorial, irregularly alternate. External seminal vesicle present. Testes numerous, filling available pre-and postovarian medullary space. Vagina posterior to cirrus-sac. Ovary bilobed, X-shaped in cross-section, separated from posterior margin by a space. Vitelline follicles encircle internal organs except ovary and genitalia. Uterus narrow, median, reaching near anterior extremity; no preformed uterine pore. Blastocyst with long caudal extension. Plerocerci in

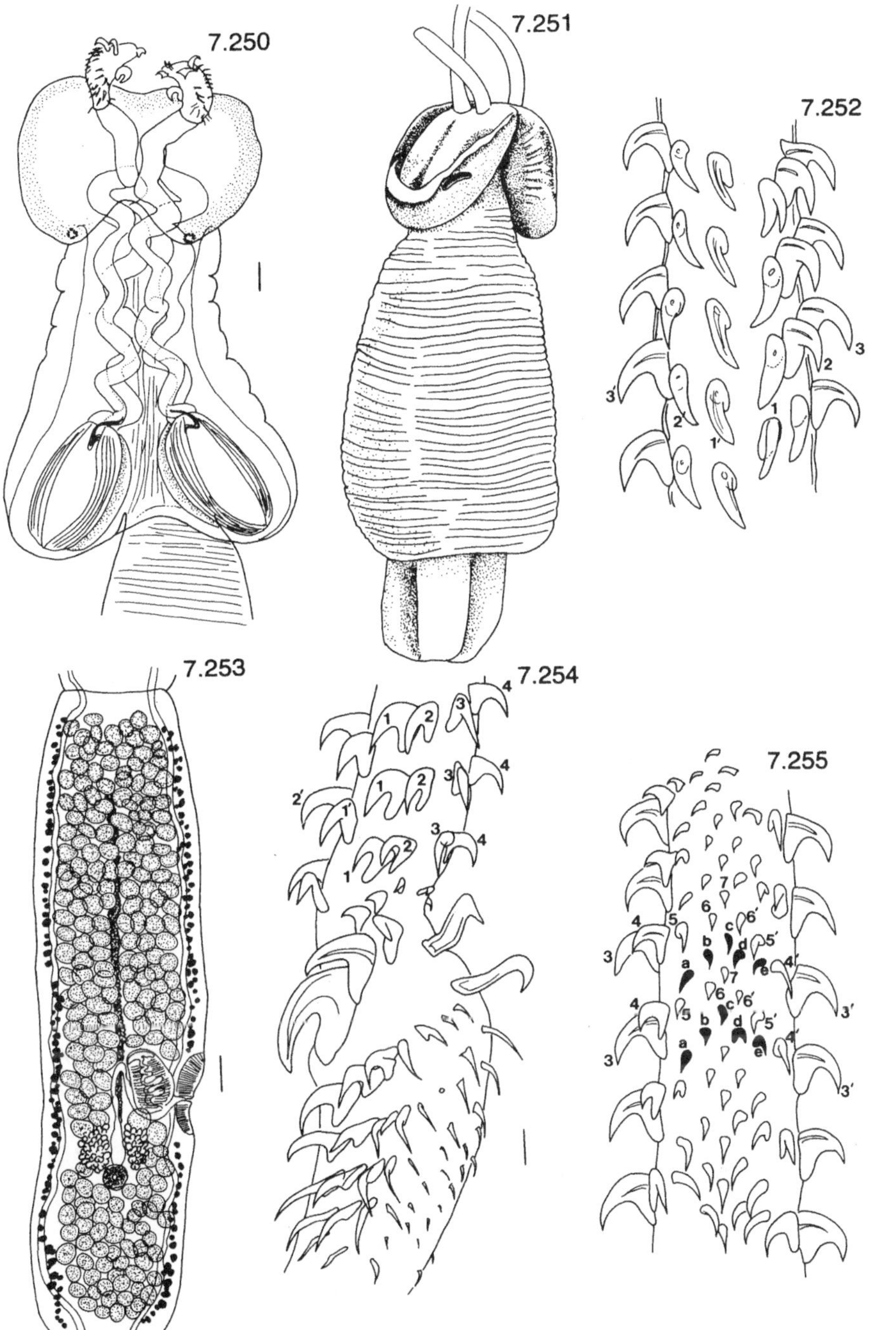

Figs 7.250–7.255 *Otobothrium* Linton, 1890. 7.250. *Otobothrium* sp., scolex (from Australia). 7.251. *O. insigne* Linton, 1905, postlarva (drawn from Hildreth & Lumsden, 1985). 7.252. Internal face, *Otobothrium* sp. 7.253. Mature proglottid, *Otobothrium* sp. 7.254. Basal region, *Otobothrium* sp., internal face to left. 7.255. External face, *O. insigne* (drawn and renumbered from Hildreth & Lumsden, 1985). Scale lines: 7.250, 7.253, 0.1 mm.

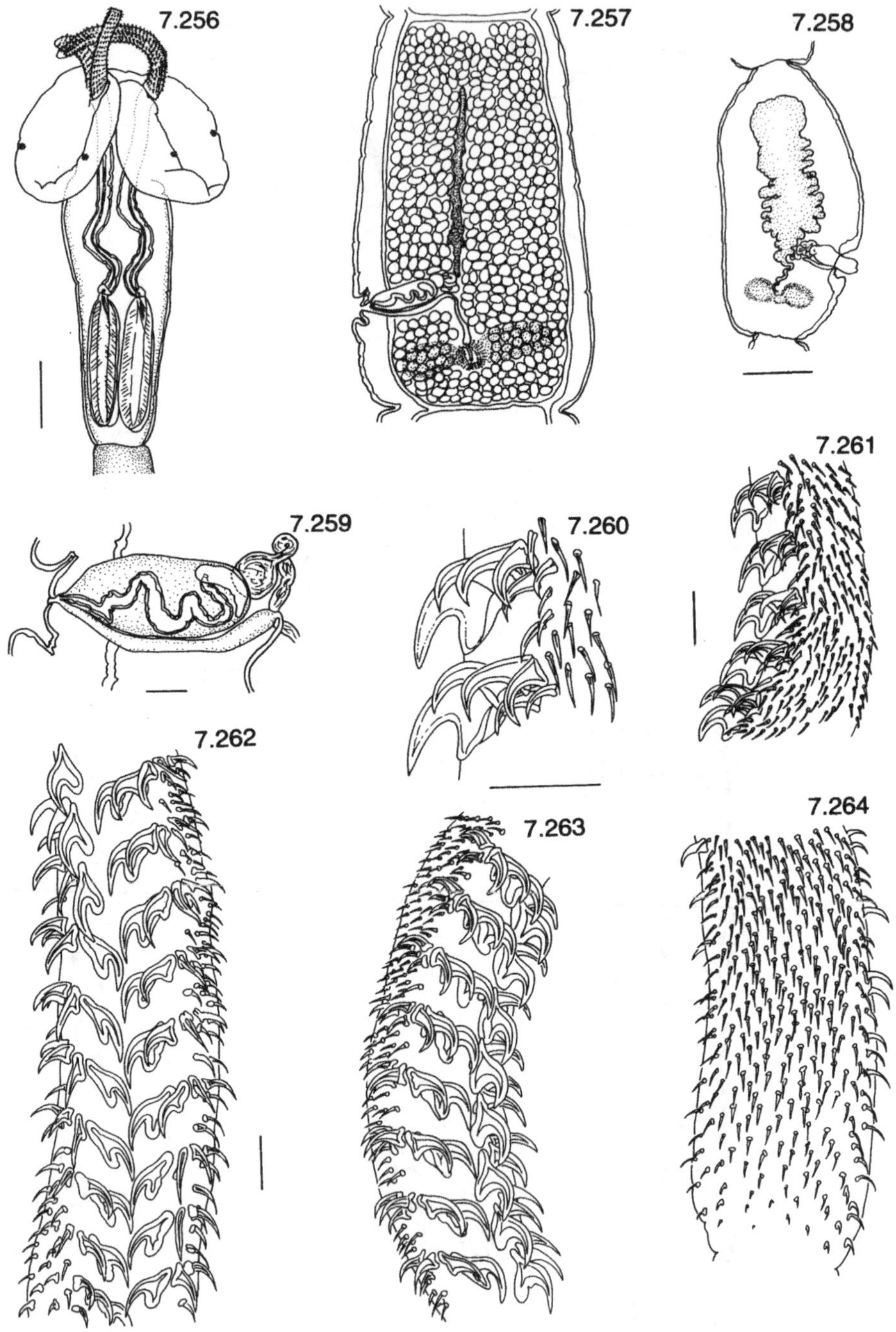

Figs 7.256–7.264 *Poecilancistrum caryophyllum* (Diesing, 1850). 7.256. Scolex. 7.257. Mature proglottid. 7.258. Gravid proglottid. 7.259. Terminal genitalia. 7.260. Detail of principal hooks, antibothridial face. 7.261. Bothridial face. 7.262. Internal face. 7.263. Antibothridial face. 7.264. External face. Scale lines: 7.256–7.258, 1.0 mm; 7.259–7.264, 0.1 mm.

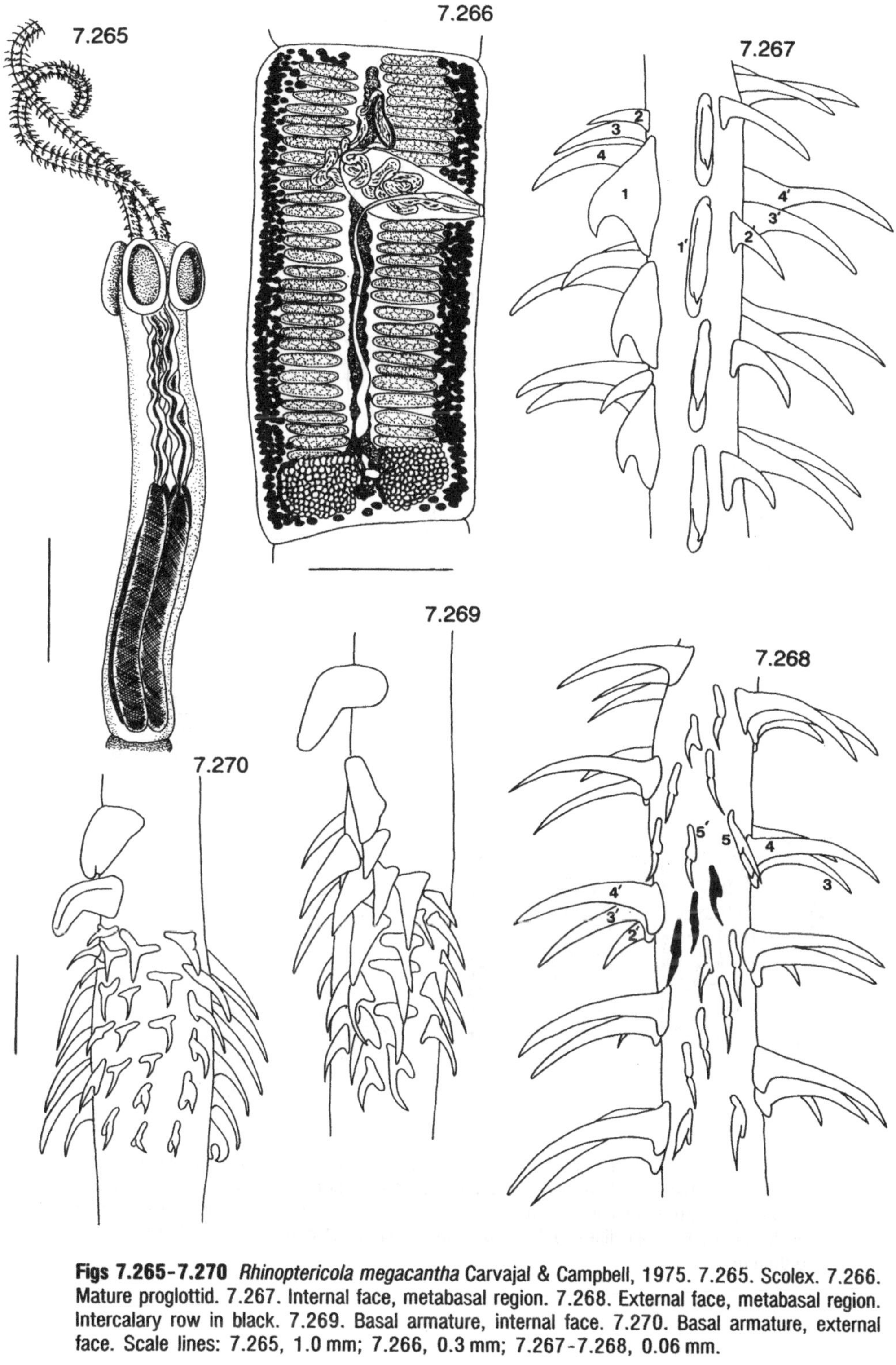

Figs 7.265–7.270 *Rhinoptericola megacantha* Carvajal & Campbell, 1975. 7.265. Scolex. 7.266. Mature proglottid. 7.267. Internal face, metabasal region. 7.268. External face, metabasal region. Intercalary row in black. 7.269. Basal armature, internal face. 7.270. Basal armature, external face. Scale lines: 7.265, 1.0 mm; 7.266, 0.3 mm; 7.267–7.268, 0.06 mm.

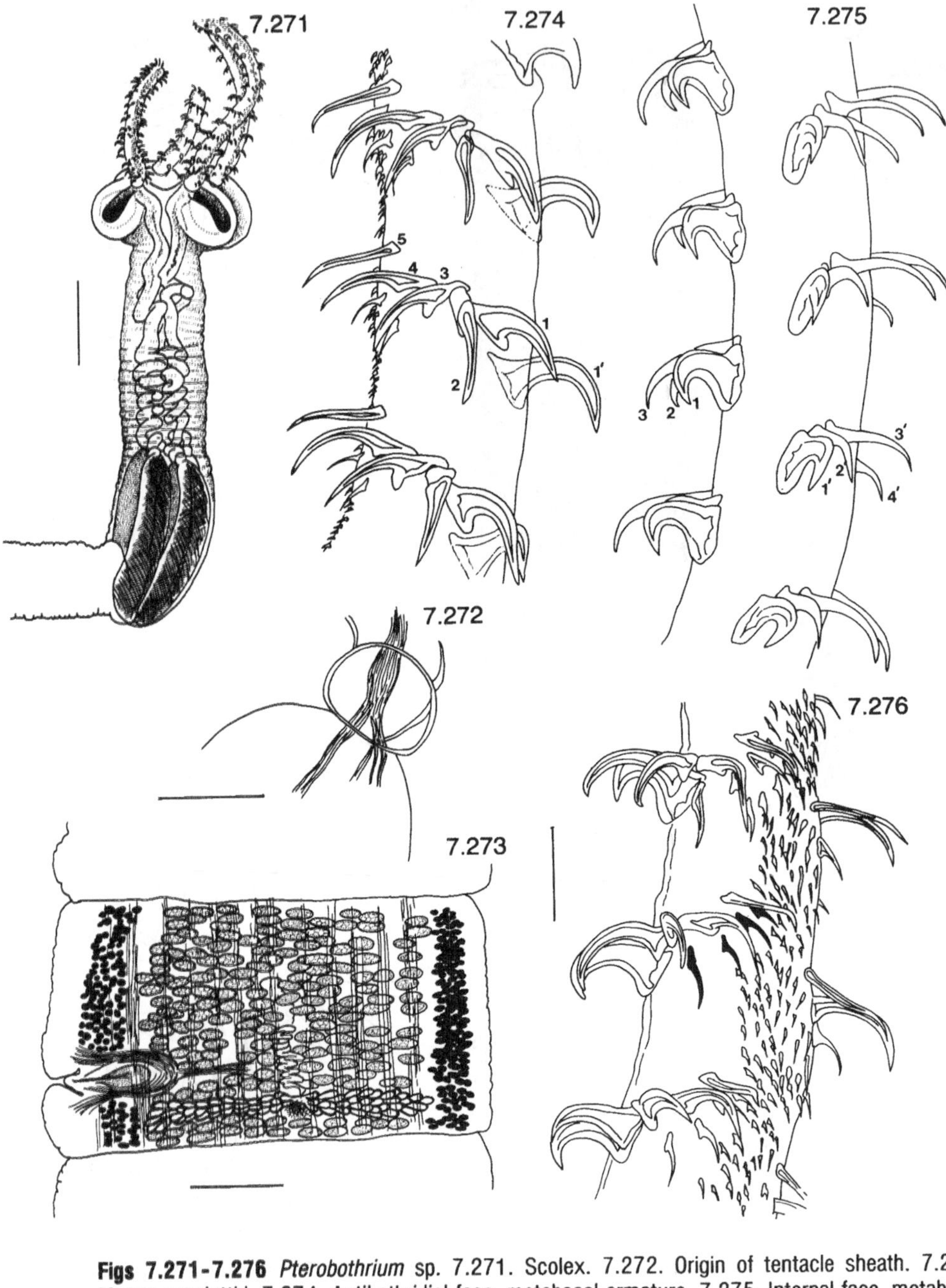

Figs 7.271-7.276 *Pterobothrium* sp. 7.271. Scolex. 7.272. Origin of tentacle sheath. 7.273. Mature proglottid. 7.274. Antibothridial face, metabasal armature. 7.275. Internal face, metabasal armature. 7.276. Bothridial face, metabasal armature, external face on right (note band). Intercalary hooks in black. Scale lines: 7.271, 0.8 mm; 7.272, 0.1 mm; 7.273, 0.25 mm; 7.274-7.276, 0.08 mm.

teleosts, adults in sharks. Gulf of Mexico, Central Atlantic, Indian Ocean, Australia. Type-species *P. caryophyllum* (Diesing, 1850) Dollfus, 1929.

Family Rhinoptericolidae Carvajal & Campbell, 1975

Genus *Rhinoptericola* Carvajal & Campbell, 1975 (Figs 7.265–7.270)

Diagnosis: With the characters of the family. Type-species *R. megacantha* Carvajal & Campbell, 1975.

Family Pterobothriidae Pintner, 1931

Genus *Pterobothrium* Diesing, 1850 (Figs 7.271–7.281)

(Syns *Myrmillorhynchus* Bilqees, 1980; *Neogymnorhynchus* Bilqees & Shah, 1982; *Synbothrium* Diesing, 1850; *Syndesmobothrium* Diesing, 1854)

Diagnosis: With characters of the family. Type-species *P. macrourum* (Rudolphi, 1819)*.

Family Grillotiidae Dollfus, 1969

(Syn. Pseudogrillotiidae Dollfus, 1969)

Key to genera

1a. **Craspedote scolex, bulbs short; vitelline follicles in lateral bands; postovarian testes present; postlarva in life-cycle .... *Pseudogrillotia* Dollfus, 1969. (Figs 7.286–7.291)**

Diagnosis: Scolex long, slender, weakly or strongly craspedote with velum. Strobila craspedote, euapolytic, proglottids numerous. Two bothridia, notched on posterior margin. Pars vaginalis long. Tentacle sheaths spiral or sinuous. Prebulbar organs present. Bulbs much shorter than pars vaginalis. Pars postbulbosa absent. Metabasal armature with band of small hooks on external face, band may or may not extend full length of tentacle; basal armature with or without special hooks. Basal swelling present or absent. Proglottids elongate. Genital pores marginal, irregularly alternate. Testes numerous, occupying preovarian and postovarian medullary region. Ovary in posterior one-third of proglottid separated from posterior extremity by distinct space. Uterus simple, median, tubular, not reaching anterior extremity. Adults in spiral valve of sharks, postlarvae in teleosts. Gulf of Mexico, Chile, Australia. Type-species *P. pleistacantha* Dollfus, 1969.

1b. **Acraspedote scolex, bulbs long or short, vitellaria circumcortical; prebulbar organs and postovarian testes present or absent; no postlarva in life cycle .... 2.**

* Southwell's (1929b, p. 278) claim that Diesing (1855, p. 177) named *P. heteracanthum* type-species after creation of *Pterobothrium* in Diesing (1850) is an incorrect translation.

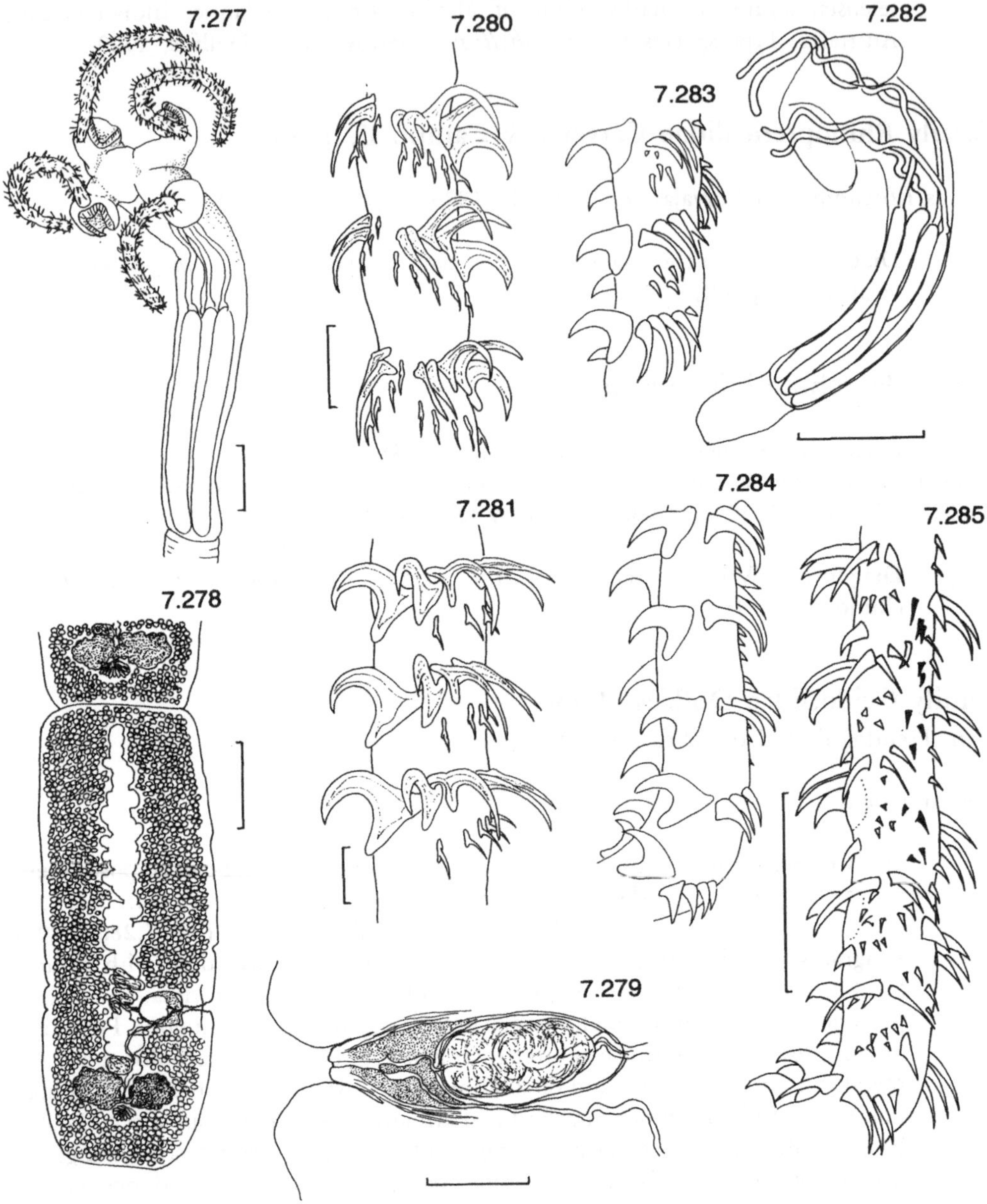

Figs 7.277–7.285. Figs 7.277–7.281. *Pterobothrium hawaiiense* Carvajal, Campbell & Cornford, 1976. 7.277. Scolex. 7.278. Mature proglottid. 7.279. Terminal genitalia. 7.280. External face, metabasal armature. 7.281. Bothridial face, metabasal armature. **Figs 7.282–7.285.** *Progrillotia louiseuzeti* Dollfus, 1969. 7.282. Scolex. 7.283. Metabasal armature, internal face on left. 7.284. Metabasal armature, view of lateral face. 7.285. Metabasal armature, external face slightly to right. Scale lines: 7.279, 7.280, 0.1 mm; 7.281, 0.05 mm, 7.277, 7.278, 7.282–7.285, 0.5 mm.

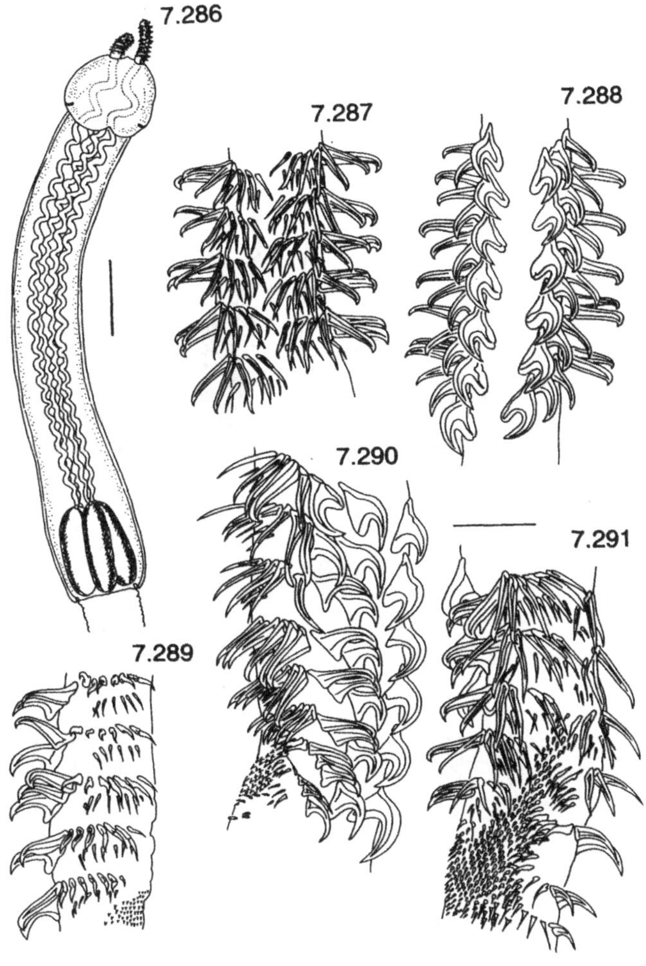

Figs 7.286–7.291 *Pseudogrillotia spratti* Campbell & Beveridge, 1993. 7.286. Scolex. 7.287. External face, metabasal region. 7.288. Internal face, metabasal region. 7.289. Antibothridial face just anterior to base, note intercalary rows. 7.290. Bothridial face just anterior to base, internal face on right. 7.291. External face, basal region. Scale lines: 7.286, 1.0 mm; 7.287–7.291, 0.1 mm.

2a. Bulbs typically longer than pars bothridialis, prebulbar organs and post-ovarian testes present .... *Grillotia* Guiart, 1927. (Figs 7.292–7.300)
(Syns *Christianella* Guiart, 1931; *Heterotetrarhynchus* Pintner, 1929; *Paragrillotia* Dollfus, 1969.)

Diagnosis: Scolex elongate. Two bothridia, inclined apically, sessile, patelliform or cordiform, margins free, rims thickened or not. Tentacle sheaths more or less sinuous. Bulbs slender, typically longer than pars bothridialis, rarely shorter. Retractor muscles originate in posterior half of bulb. Metabasal armature with alternating half-spiral rows of four to six principal hooks separated by intercalary rows; longitudinal band of small hooks separates principal rows on external face.

7.292

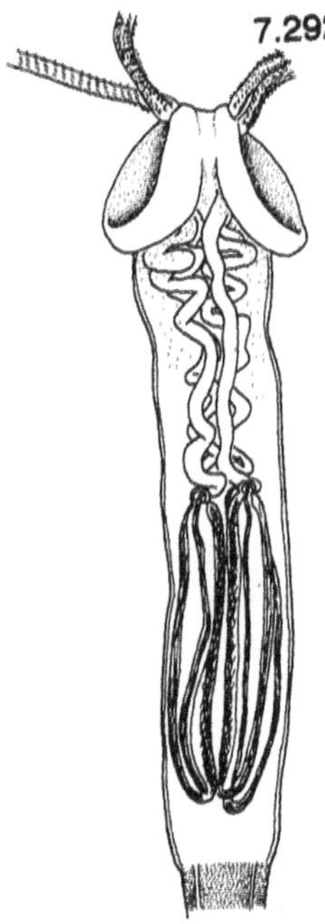

7.293

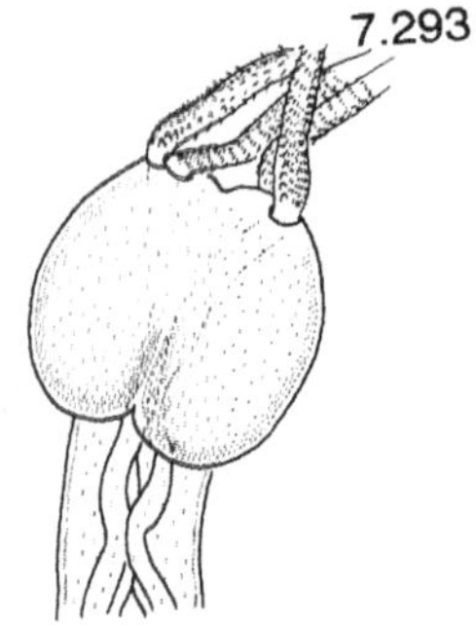

7.294

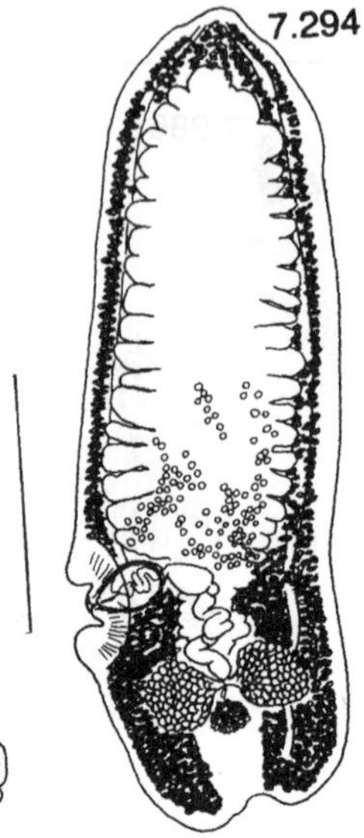

7.295

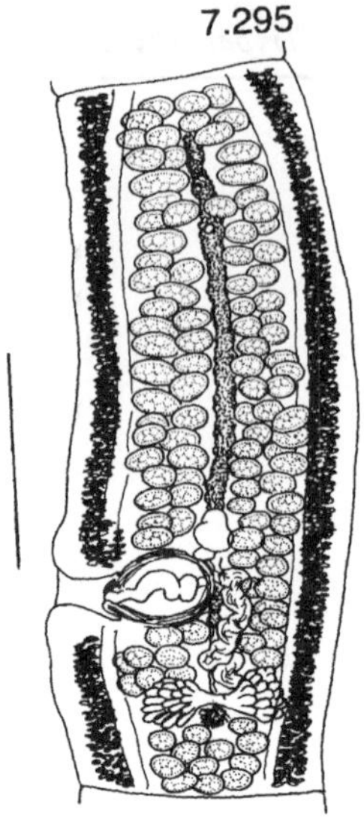

7.296

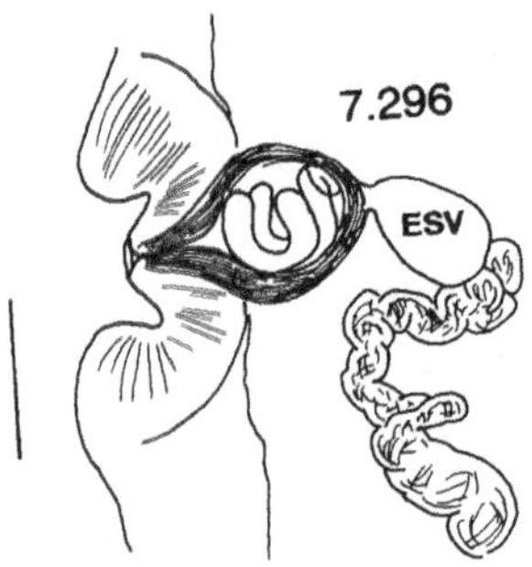

7.297

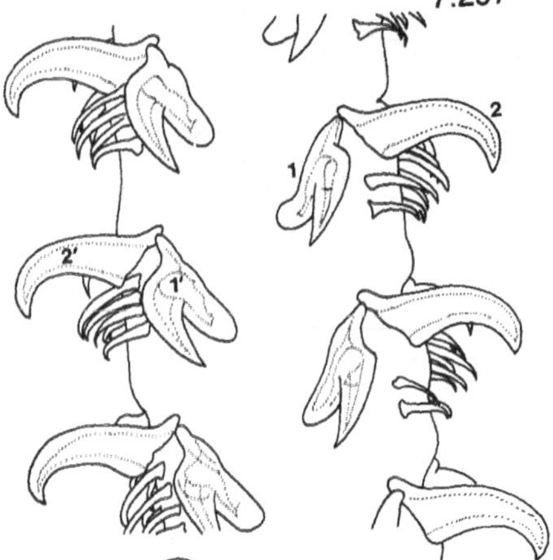

7.298

7.299

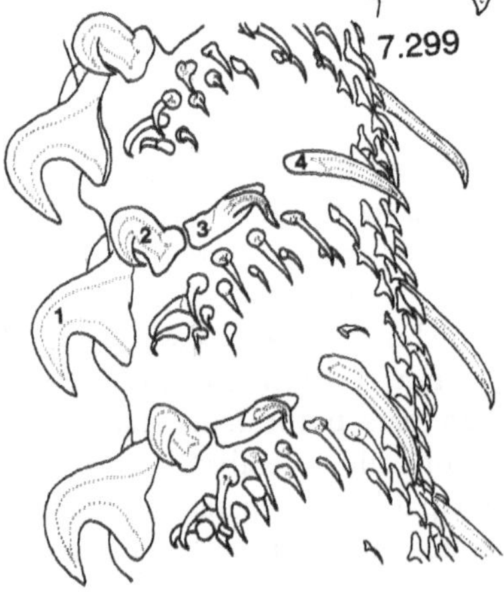

7.300

Basal armature absent. Basal swelling present or absent. Hooks heteromorphous, hollow. Proglottids elongate, acraspedote, euapolytic. Genital pores postequatorial, irregularly alternate. External and internal seminal vesicles present. Accessory seminal vesicle present or absent. Hermaphroditic sac present. Testes medullary, often numerous, more or less in parallel rows, some postovarian. Ovary tetralobed, small, separated from posterior extremity or not. Vitelline follicles circumcortical, interspersed among longitudinal muscles. Uterus median, tubular, with small diverticula; gravid uterus saccate with ventral pore. Plerocerci common in teleosts; adults in elasmobranchs, particularly batoids. Atlantic, Mediterranean, Pacific, Australia. Type-species *G. erinaceus* (van Beneden, 1858).

2b. Bulbs very long, prebulbar organs lacking, postovarian testes absent .... *Progrillotia* Dollfus, 1946. (Figs 7.282–7.285)
Diagnosis: Bothridial margins not indented posteriorly. Longitudinal band of hooks interrupted opposite each principal row. Special basal armature may be present. Bulbs very long, prebulbar organs lacking. Testes in longitudinal rows. Accessory seminal vesicle absent. Ovary at posterior extremity of proglottid. In elasmobranchs. Type-species *P. pastinacae* Dollfus, 1946.

Family Molicolidae Beveridge & Campbell, 1989

Key to genera

1a. Interpolated rows of hooks present, hermaphroditic duct present; uterus straight, lacking pore .... *Stragulorhynchus* Beveridge & Campbell, 1988. (Figs 7.301–7.313)
Diagnosis: Scolex elongate, acraspedote. Four auriculate, 'spinose' bothridia. Tentacles emerge from apices of bothridia; basal swelling present. Metabasal armature with broad band of hooks. Distinctive basal armature present. Tentacle sheaths spiral. Prebulbar organs absent. Retractor muscles originate in anterior half of bulb. Pars postbulbosa absent. Neck absent. Proglottids acraspedote, hyperapolytic. Testes numerous, pre- and postovarian. Genital pore marginal, irregularly alternate. External and accessory seminal vesicles present. Male and female terminalia unite to form hermaphroditic duct. Ovary well separated from posterior extremity. Vitelline follicles circumcortical. Uterus simple, median, tubular, not reaching anterior extremity; uterine pore absent. Parasites of sharks (Orectolobidae). Spencer Gulf, South Australia. Type-species *S. orectolobi* Beveridge & Campbell, 1988.

Figs 7.292-7.300 *Grillotia erinaceus* (van Beneden, 1858). 7.292. Scolex. 7.293. Facial view of bothridium. 7.294. Gravid proglottid. 7.295. Mature proglottid. 7.296. Male terminal genitalia. 7.297. Internal face, metabasal region. 7.298. External face, metabasal region. 7.299. Antibothridial face, metabasal region. 7.300. Antibothridial face, basal region. 7.297-7.300 after Dollfus (1942). Scale lines: 7.292, 1.0 mm; 7.294, 1.5 mm; 7.295, 0.5 mm; 7.296, 0.32 mm; 7.297-7.300, 0.05 mm.

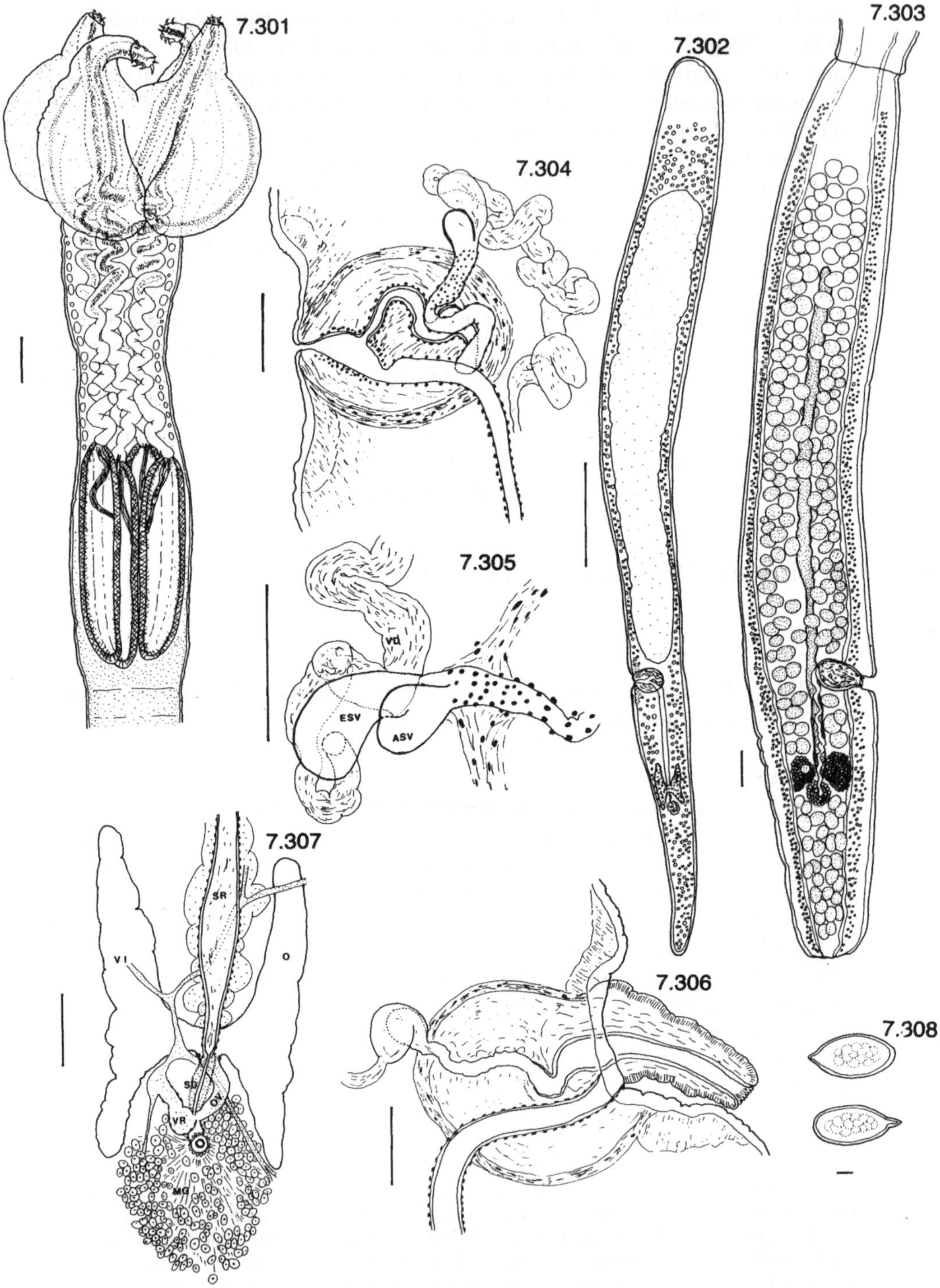

Figs 7.301-7.308 *Stragulorhynchus orectolobi* Beveridge & Campbell, 1988. 7.301. Scolex. 7.302. Gravid proglottid. 7.303. Mature proglottid. 7.304. Terminal genitalia and hermaphroditic duct. 7.305. Male vesicles and vas deferens. 7.306. Hermaphroditic sac with 'cirrus' everted. 7.307. Female genital complex. 7.308. Eggs. Scale lines: 0.1 mm.

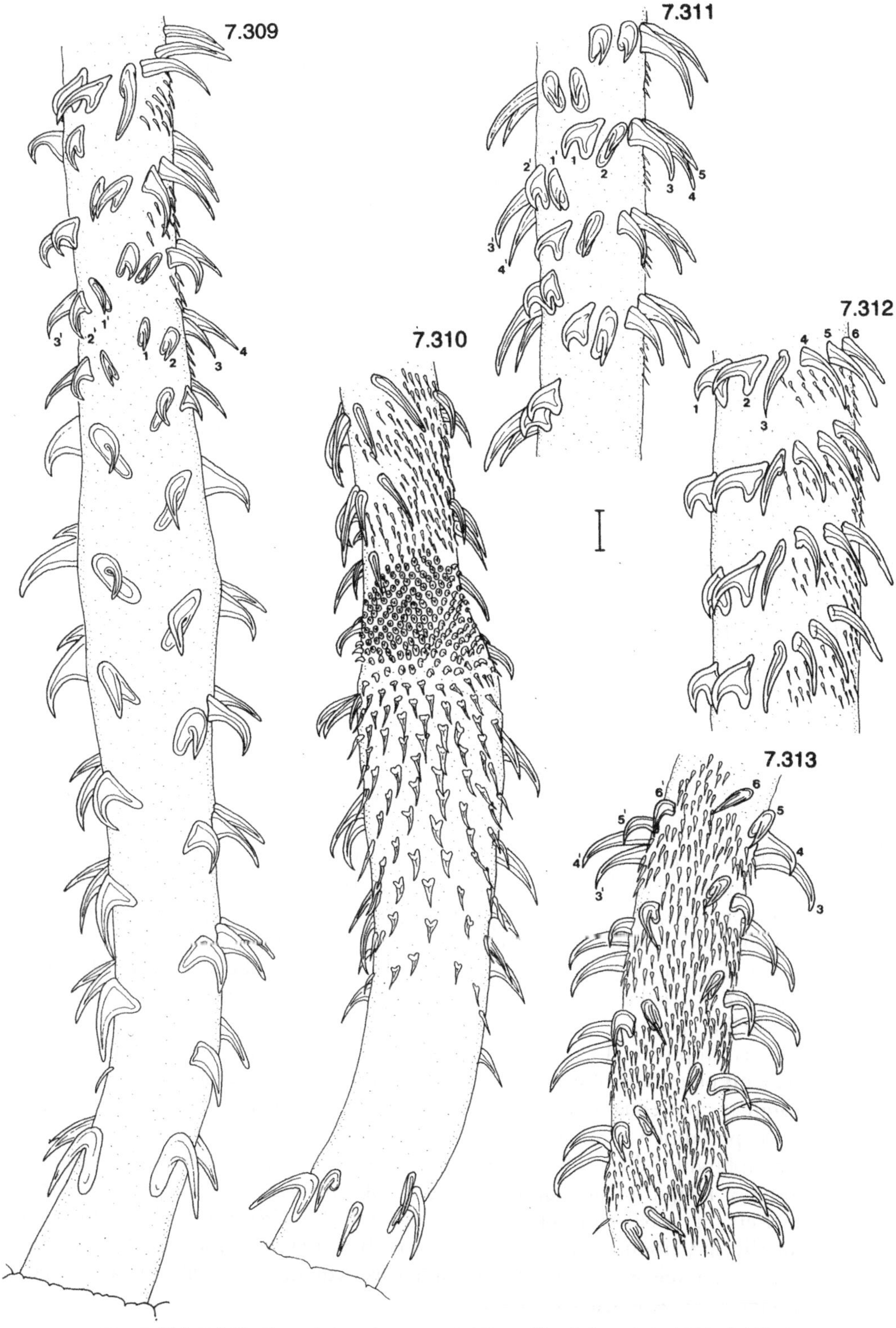

Figs 7.309–7.313 *Stragulorhynchus orectolobi* Beveridge & Campbell, 1988. 7.309. Internal face, basal region of tentacle. 7.310. External face, basal region. 7.311. Internal face, metabasal region. 7.312. Bothridial face, metabasal region. 7.313. External face, metabasal region. Scale lines: 0.01 mm.

1b. Interpolated rows of hooks absent, hermaphroditic duct absent; uterus deviated, uterine pore adjacent to genital atrium .... *Molicola* Dollfus, 1935. (Figs 7.314–7.319)

Diagnosis: Scolex elongate, acraspedote. Four auriculate bothridia, curved and inclined apically, merged with scolex anteriorly, rims thickened. Tentacle orifices on anterior surface of bothridial rims. Tentacle sheaths sinuous. Prebulbar organs absent. Bulbs elongate; retractor muscles originate in anterior half of bulb. Hooks heteromorphous, hollow. Principal rows separated by band of small hooks on middle of external face. Intercalary rows of hooks absent. Distinctive basal armature present, base of tentacle circumscribed with ring of large hooks. Genital pores marginal, irregularly alternate. Vagina ventral to cirrus-sac. Accessory and external seminal vesicles present. Testes medullary, intervascular, in pre-and postovarian fields. Ovary well forward of posterior extremity. Vitelline follicles circumcortical. Uterus simple, tubular, deviated toward genital pore; uterine pore lies adjacent to genital atrium. Plerocerci in teleosts; adults in sharks. Atlantic, Pacific. Type-species *M. horridus* (Goodsir, 1841).

Superfamily Poecilacanthoidea

Family Lacistorhynchidae Guiart, 1927

(Syn. Floricipitidae Dollfus, 1929)

Key to genera

1a. Chainette hooks two-winged .... *Floriceps* Cuvier, 1817. (Figs 7.320–7.327)

Diagnosis: Scolex large, elongate, slightly craspedote. Two reverse cordiform bothridia, indented on posterior margins. Tentacles emerge near apex of scolex. Pars vaginalis much longer than pars bulbosa. Tentacle sheaths spiral, surrounded by numerous gland cells. Bulbs relatively short. Retractor muscles originate in anterior half of bulb. Pars postbulbosa absent. Armature poeciloacanthous, consisting of alternating half-spiral rows. Hooks heteromorphous, hollow. No basal armature or swelling. Intercalary hooks absent. Hooks 7 and 8 of principal rows in satellite position adjacent to chainette. Hooks 9(9′) two-winged, much enlarged in proximal half of tentacle, forming simple chainette of closely spaced elements on middle of external face. Strobila acraspedote, apolytic. Genital pores marginal, postequatorial. Testes numerous, intervascular, some postovarian. Hermaphroditic sac contains internal seminal vesicle. Vagina enters sac to form hermaphroditic duct. Ovary small, tetralobed in cross-section, set forward of posterior margin of proglottid. Uterus simple, median, extending about three-quarters of proglottid length. Vitelline follicles circumcortical, interrupted dorsally and around ovary and shell gland. Blastocyst without caudal extension. Plerocerci in teleosts; adults in elasmobranchs. Atlantic and Pacific Oceans. Type-species *F. saccatus* Cuvier, 1817.

1b. Chainette hooks not winged .... 2.

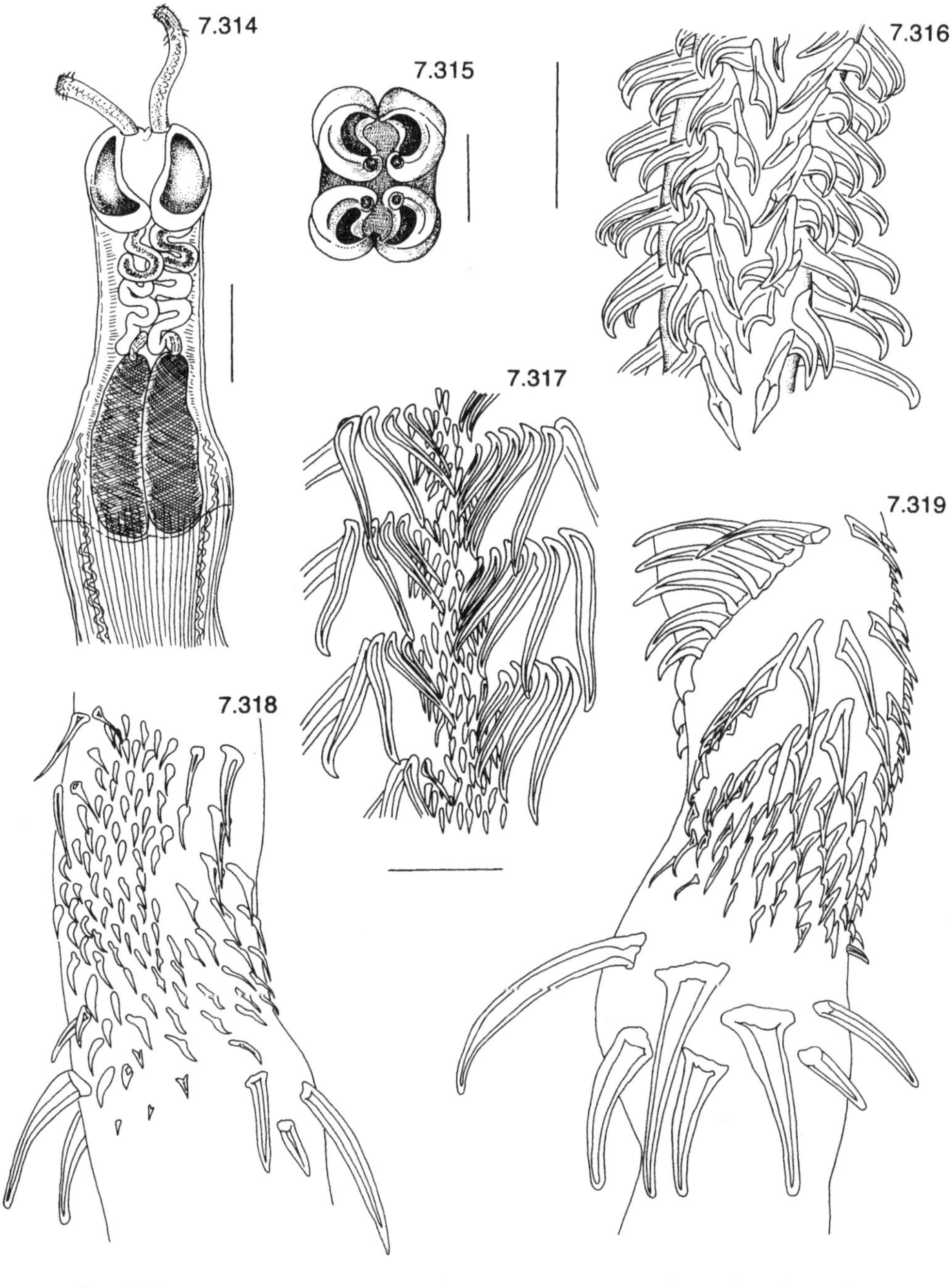

Figs 7.314-7.319 *Molicola horridus* (Goodsir, 1841). 7.314. Scolex (after Dollfus, 1942). 7.315. En face view of scolex (after Dollfus, 1942). 7.316. Internal face, metabasal region. 7.317. External face, metabasal region. 7.318. Basal armature, external face. 7.319. Basal armature, antibothridial face, external face on right. Scale lines: 7.314, 1.0 mm.

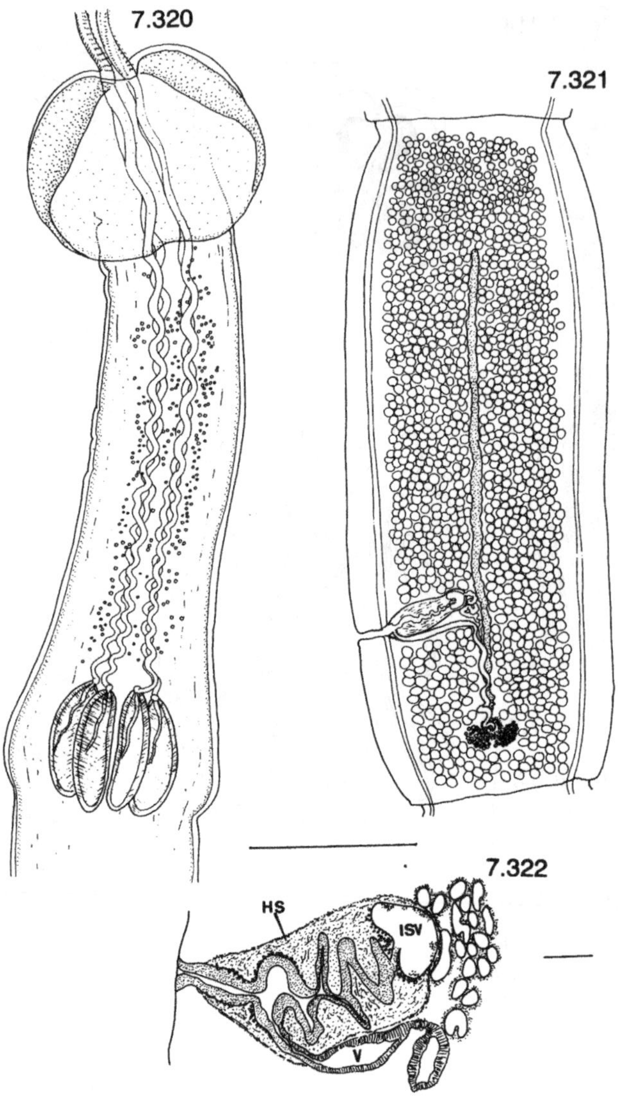

Figs 7.320–7.322 *Floriceps minacanthus* Campbell & Beveridge, 1987. 7.320. Scolex. 7.321. Mature proglottid (vitelline follicles not shown). 7.322. Section of terminal genitalia showing hermaphroditic sac (after Richmond & Caira, 1991). Scale lines: 7.320, 7.321, 1.0 mm; 7.322, 0.1 mm.

2a. **Basal swelling and armature present** .... ***Lacistorhynchus*** **Pintner, 1913. (Figs 7.328–7.343)**

Diagnosis: Small, weak worms with acraspedote scolex bearing two bothridia. Bothridia notched posteriorly. Pars bothridialis shorter than or equal to pars bulbosa. Pars vaginalis relatively long. Pars postbulbosa present. Tentacles emerge from anterior edge of bothridia, slender with slight basal swelling and basal

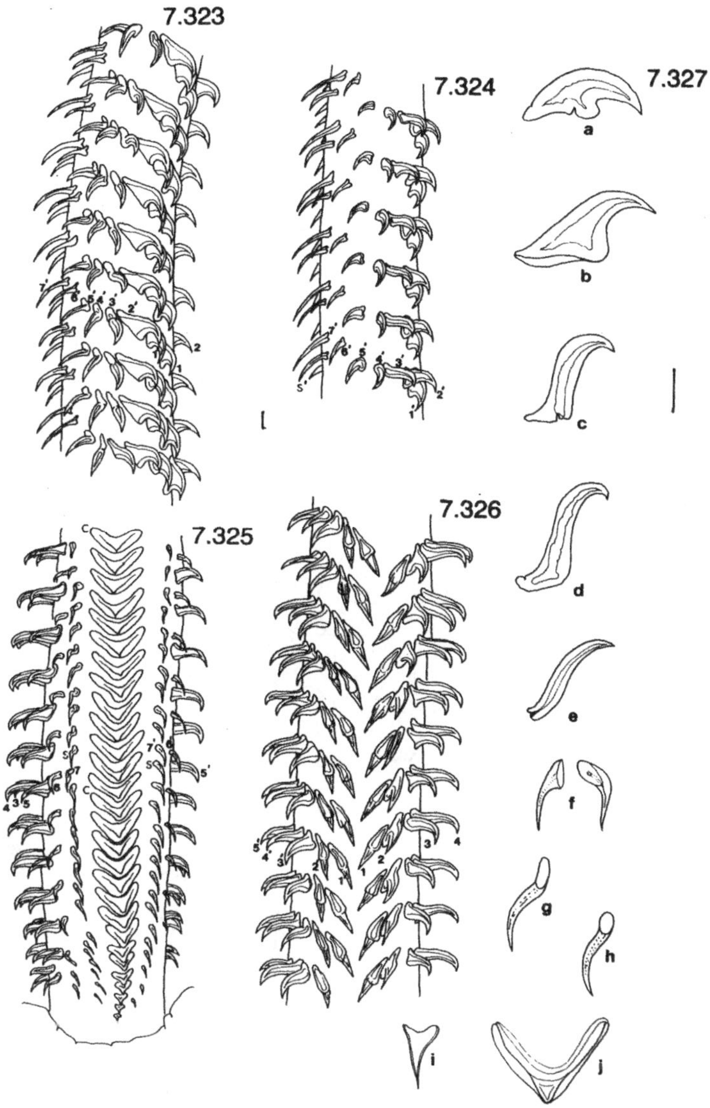

Figs 7.323-7.327 *Floriceps minacanthus* Campbell & Beveridge, 1987. 7.323–7.324. Bothridial face, metabasal armature. 7.325. External face, basal and metabasal regions. 7.326. Internal face, metabasal region. 7.327. Principal hooks 1-7 (a-g); h, satellite hook 8; i, chainette element from mid-tentacle; j, chainette element near basal region of tentacle. Scale lines: 7.323-7.326, 0.1 mm; 7.327, 0.01 mm.

armature with characteristic serpe à bec hooks. Metabasal armature poeciloacanthous with alternating half-spiral rows of eight hooks each. Hooks 7 and 8 displaced as satellites of the chainette; hooks 9(9′) of opposing rows form simple chainette on middle of external face. Hooks heteromorphous, hollow. Tentacle sheaths spiral. Prebulbar organs absent. Bulbs relatively short. Retractor muscles

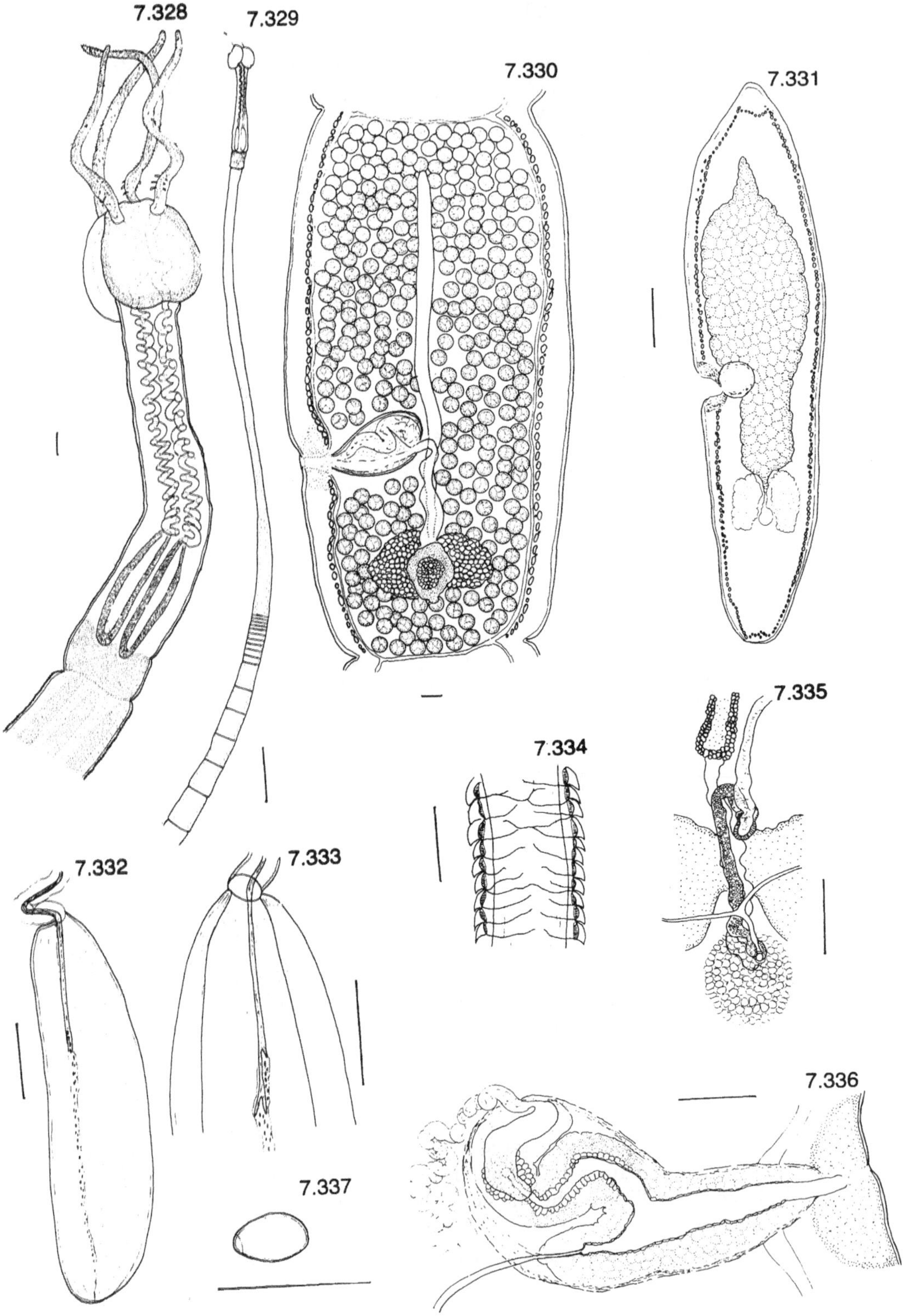
7.328
7.329
7.330
7.331
7.332
7.333
7.334
7.335
7.336
7.337

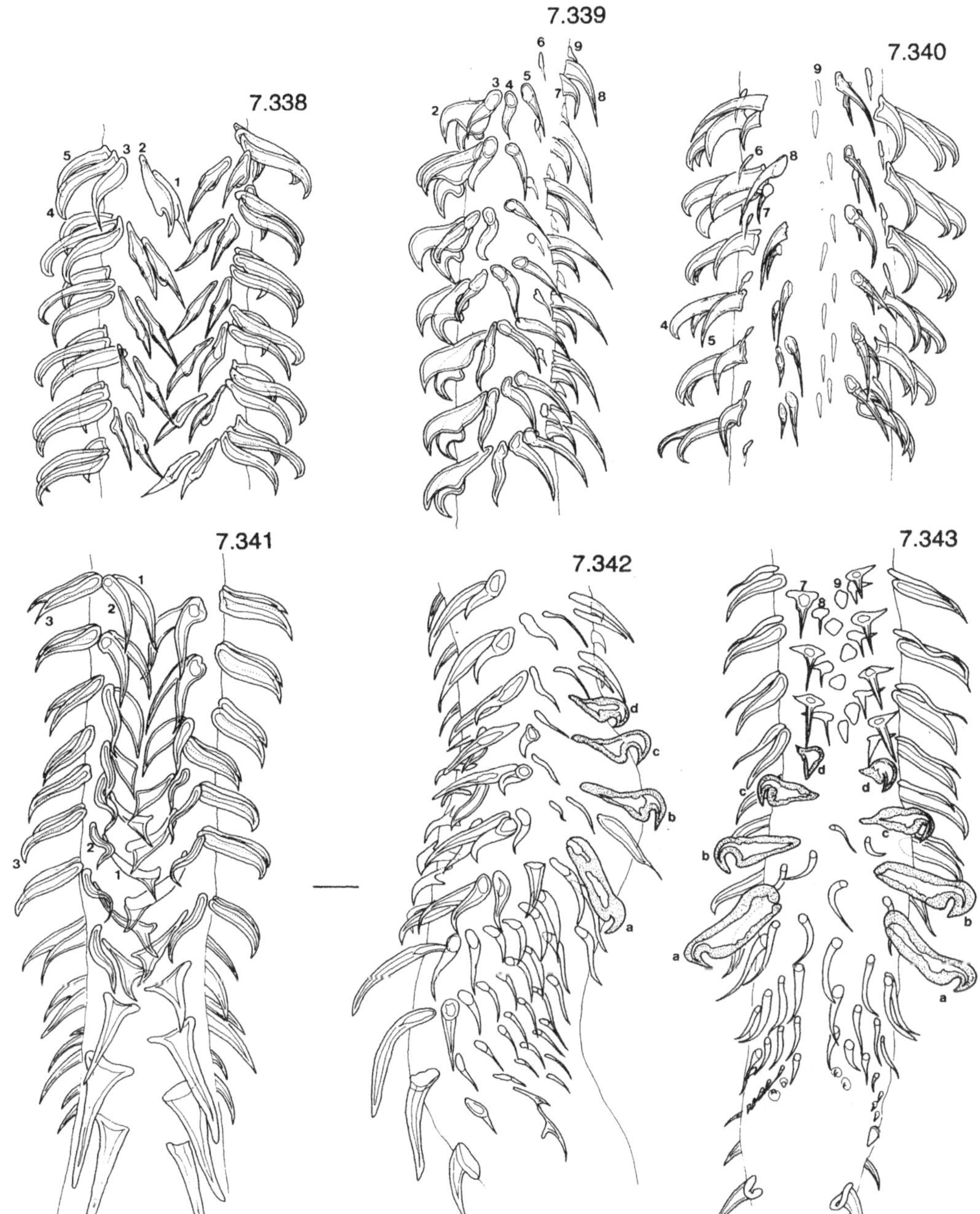

Figs 7.328–7.337 *Lacistorhynchus dollfusi* Beveridge & Sakanari, 1987. 7.328. Scolex. 7.329. Scolex and strobila, note elongated neck-like region. 7.330. Mature proglottid. 7.331. Gravid proglottid. 7.332. Bulb. 7.333. Detail of retractor muscle origin. 7.334. Detail of newly formed proglottids. 7.335. Female genital complex. 7.336. Terminal genitalia and hermaphroditic sac. 7.337. Egg. Scale lines: 0.1 mm.

Figs 7.338–7.343 *Lacistorhynchus dollfusi* Beveridge & Sakanari, 1987. 7.338. Internal face, metabasal region. 7.339. Antibothridial face, metabasal region. 7.340. External face showing simple chainette, metabasal region. 7.341. Basal armature, internal face. 7.342. Basal armature, antibothridial face, external face to right. 7.343. Basal armature, internal face, serpe à bec hooks (a–d). Scale lines: 0.01 mm.

originate in anterior half of bulb. Strobila acraspedote, slender, proglottids numerous. Testes numerous, intervascular, some postovarian. Genital pores lateral, irregularly alternate. Vagina enters hermaphroditic sac to form hermaphroditic duct. Ovary tetralobed in cross section, set slightly forward of posterior extremity. Uterus median, tubular, terminating near anterior quarter of proglottid; saccate when gravid. Vitelline follicles encircle internal organs, interrupted at hermaphroditic sac, ovary and shell gland. Plerocerci in copepods and teleosts; adults in sharks. North Atlantic, Mediterranean, Pacific Coast USA, Australia. Type-species *L. tenuis* (van Beneden, 1858) Pintner, 1913.

2b. Basal swelling with distinctive armature absent .... ***Callitetrarhynchus* Pintner, 1931. (Figs 7.344–7.348)**
(Syns *Anthocephalus* Rudolphi, 1819; *Callotetrarhynchus* Pintner, 1931; *Lintoniella* Yamaguti, 1934 *nec* Woodland, 1927.)

Diagnosis: Scolex long, slender, feebly craspedote. Two patelliform bothridia, notched on posterior margin. Pars vaginalis long, tentacle sheaths regularly sinuous, enlarged anteriorly. Numerous gland cells surround tentacle sheaths from bulbs through most of pars vaginalis. Prebulbar organ absent. Bulbs relatively short. Retractor muscles originate in anterior third of bulb. Pars postbulbosa absent. Armature poeciloacanthous, without basal armature or enlargement. Principal rows half-spiral, alternate, hooks 6,7,8 form distinct triad separate from remainder of principal row, hook 7 set forward of hook 8 as satellites of the chainette. Intercalary rows absent. Chainette simple, bases of elements without wings, consisting of hooks 9(9′) well separated and forming single file on middle of external face. Strobila long, thin, acraspedote. Genital pores lateral, irregularly alternate. Vagina pierces hermaphroditic sac and joins male duct to form hermaphroditic duct. Hermaphroditic sac pyriform. Testes numerous, small, some postovarian. Ovary small, set forward of posterior proglottid border. Uterus median, tubular. Vitelline follicles in lateral cortex. Plerocercus with caudal extension, in teleosts; adults in sharks. Cosmopolitan. Type-species *C. gracilis* (Rudolphi, 1819) Pintner, 1931.

Family Dasyrhynchidae Dollfus, 1935

Genus *Dasyrhynchus* Pintner, 1928. (Figs 7.349–7.355)
(Syn. *Sbesterium* Dollfus, 1929)

Diagnosis: With the characters of the family. Type-species *D. variouncinatus* (Pintner, 1913) Pintner, 1928.

Family Hornelliellidae Yamaguti, 1954

Genus *Hornelliella* Yamaguti, 1954 (Figs 7.356–7.366)

Diagnosis: With characters of the family. Type-species *H. macropora* (Shipley & Hornell, 1906).

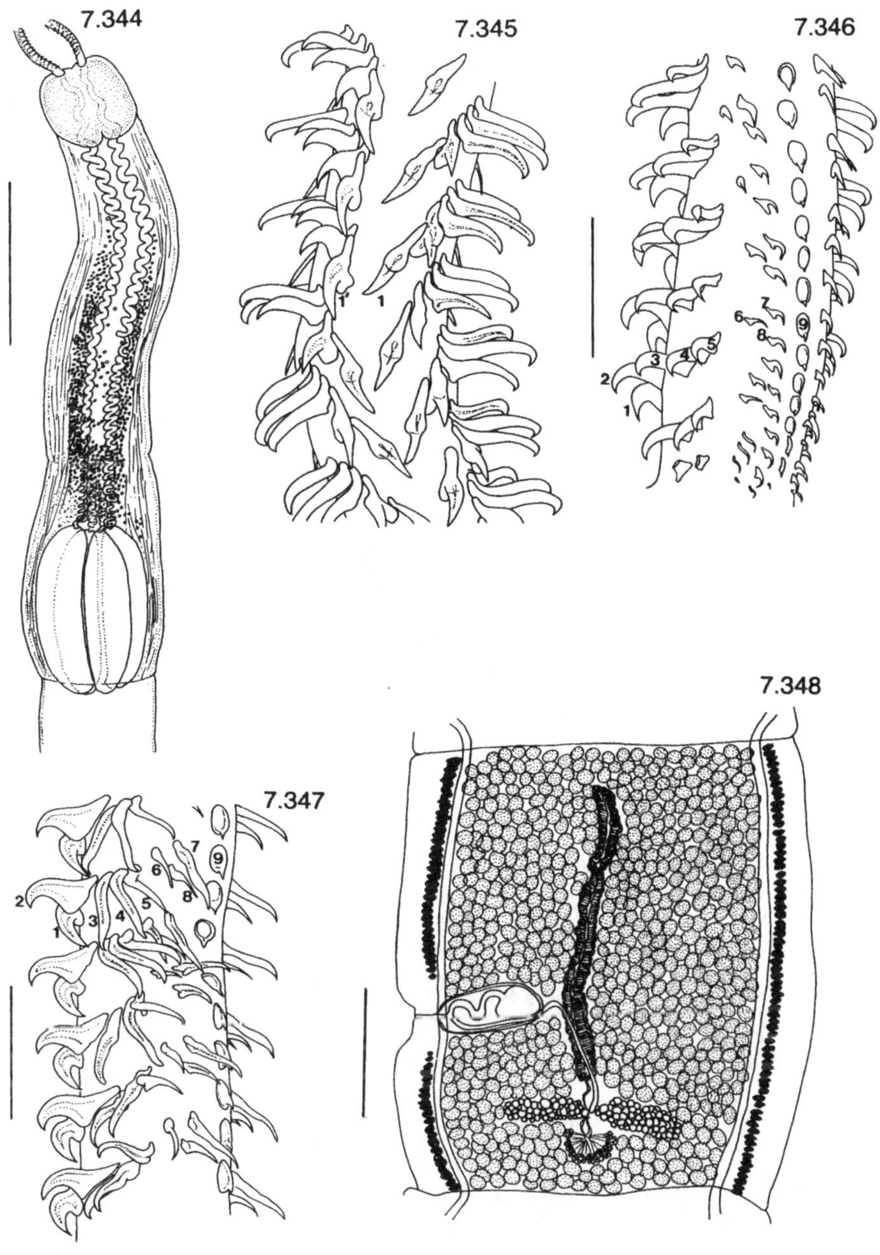

Figs 7.344-7.348 *Callitetrarhynchus gracilis* (Rudolphi, 1819). 7.344. Scolex. 7.345. Internal face, metabasal region. 7.346. External face, near base. 7.347. Antibothridial face, external face on right. 7.348. Mature proglottid, note hermaphroditic duct and internal seminal vesicle. Scale lines: 7.344, 1.0 mm; 7.345-7.347, 0.05 mm; 7.348, 0.5 mm. 7.344-7.347 after Dollfus, 1942.

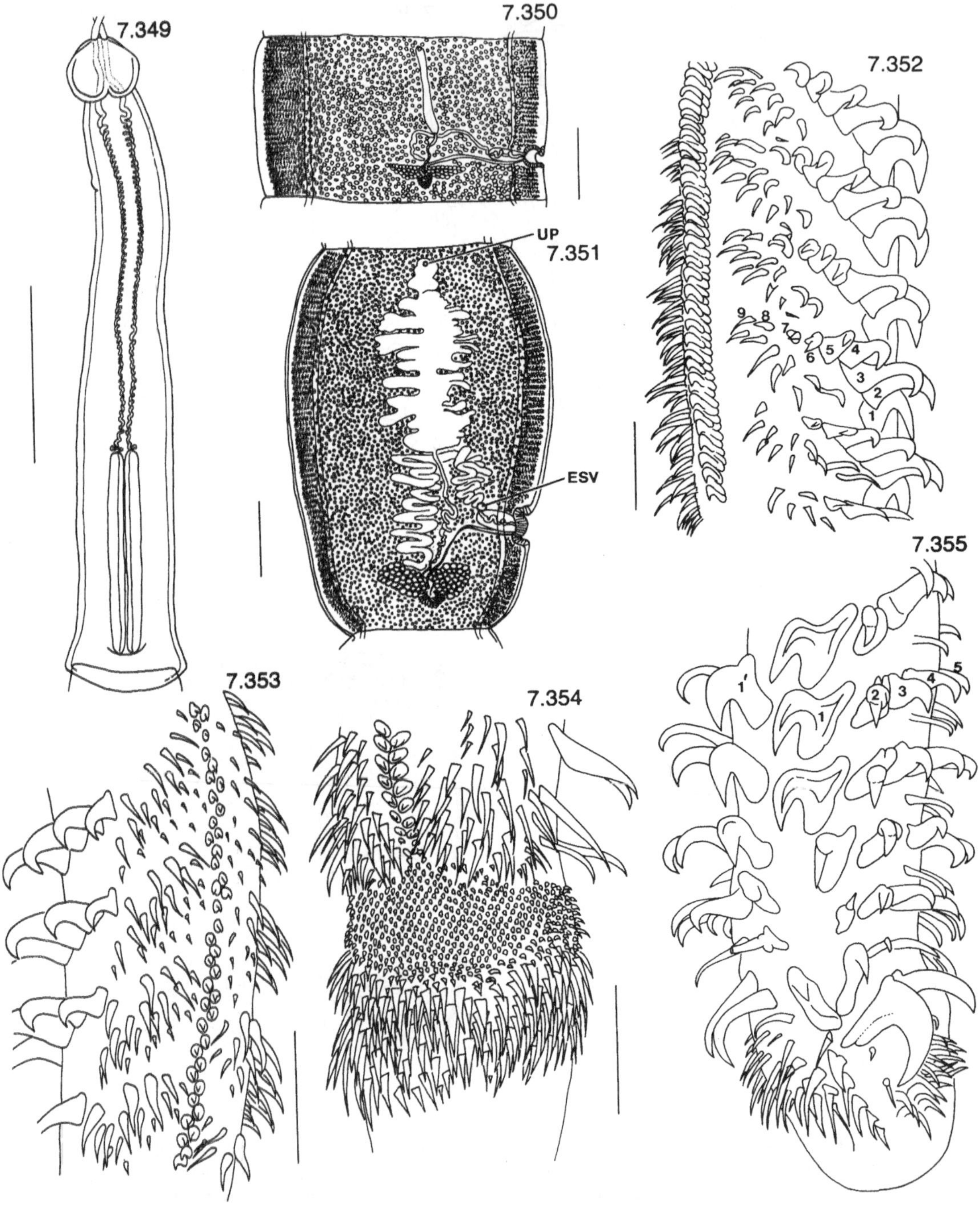

Figs 7.349–7.355 *Dasyrhynchus giganteus* (Diesing, 1850) Pintner, 1929. 7.349. Scolex. 7.350. Mature proglottid. 7.351. Gravid proglottid, note external seminal vesicle. 7.352. Antibothridial face, chainette on left. 7.353. External face, metabasal region. 7.354. External face, basal armature. 7.355. Internal face, basal armature. 7.353, 7.354 after Carvajal *et al.* (1976), all other figures after Dollfus (1942). Scale lines: 7.349, 5.0 mm; 7.350, 7.351, 1.0 mm; 7.352–7.355, 0.1 mm.

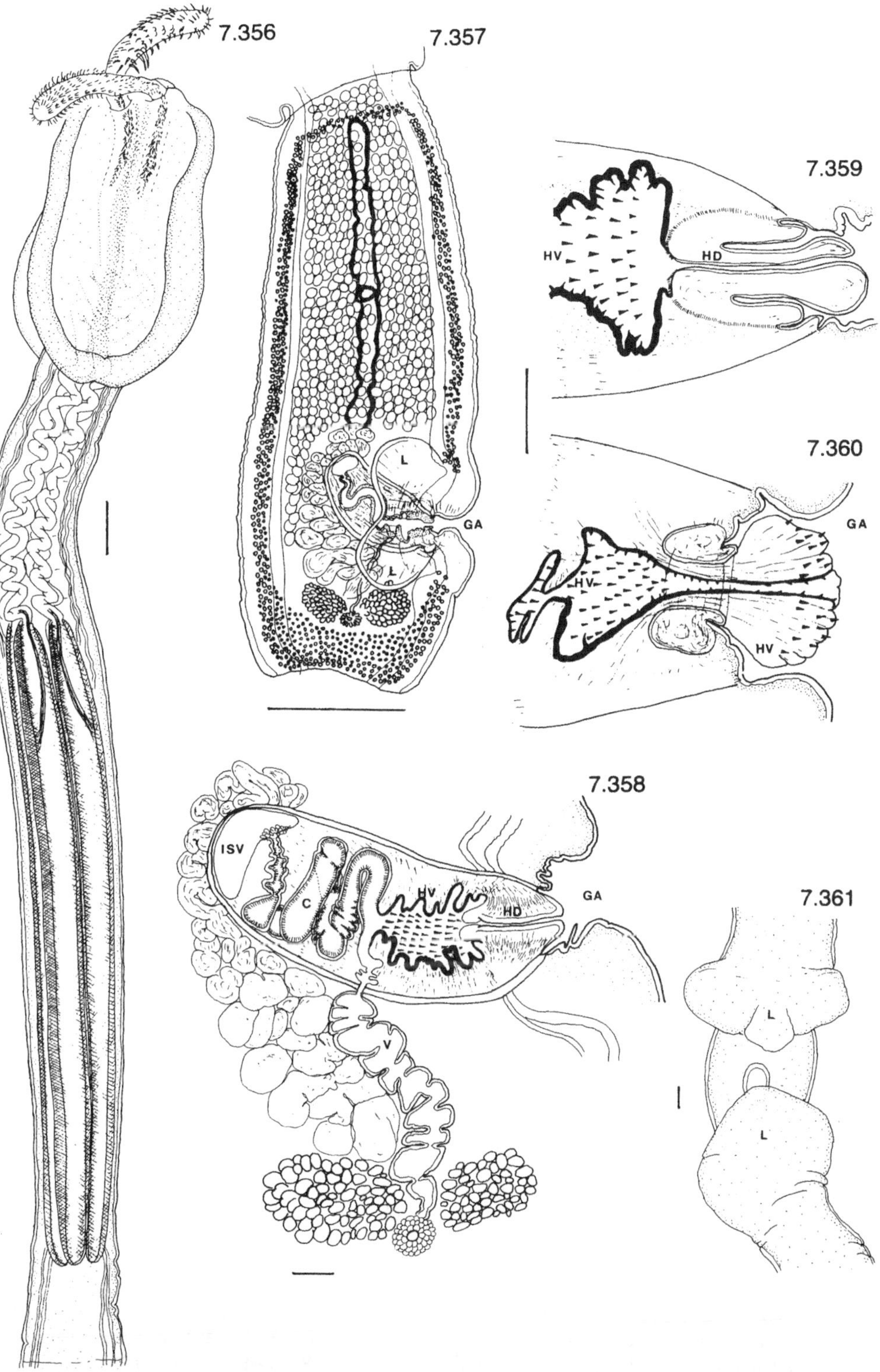

Figs 7.356–7.361 *Hornelliella macropora* (Shipley & Hornell, 1912). 7.356. Scolex. 7.357. Mature proglottid. 7.358. Genital system, note internal seminal vesicle, cirrus, hermaphroditic vesicle, hermaphroditic duct. 7.359. Genital atrium with partially everted hermaphroditic vesicle (HD). 7.360. Genital atrium (GA) with partially everted hermaphroditic vesicle (HV). 7.361. Genital pore, sagittal view showing lappets (L). Scale lines: 7.356, 7.357, 1.0 mm; 7.358–7.361, 0.1 mm.

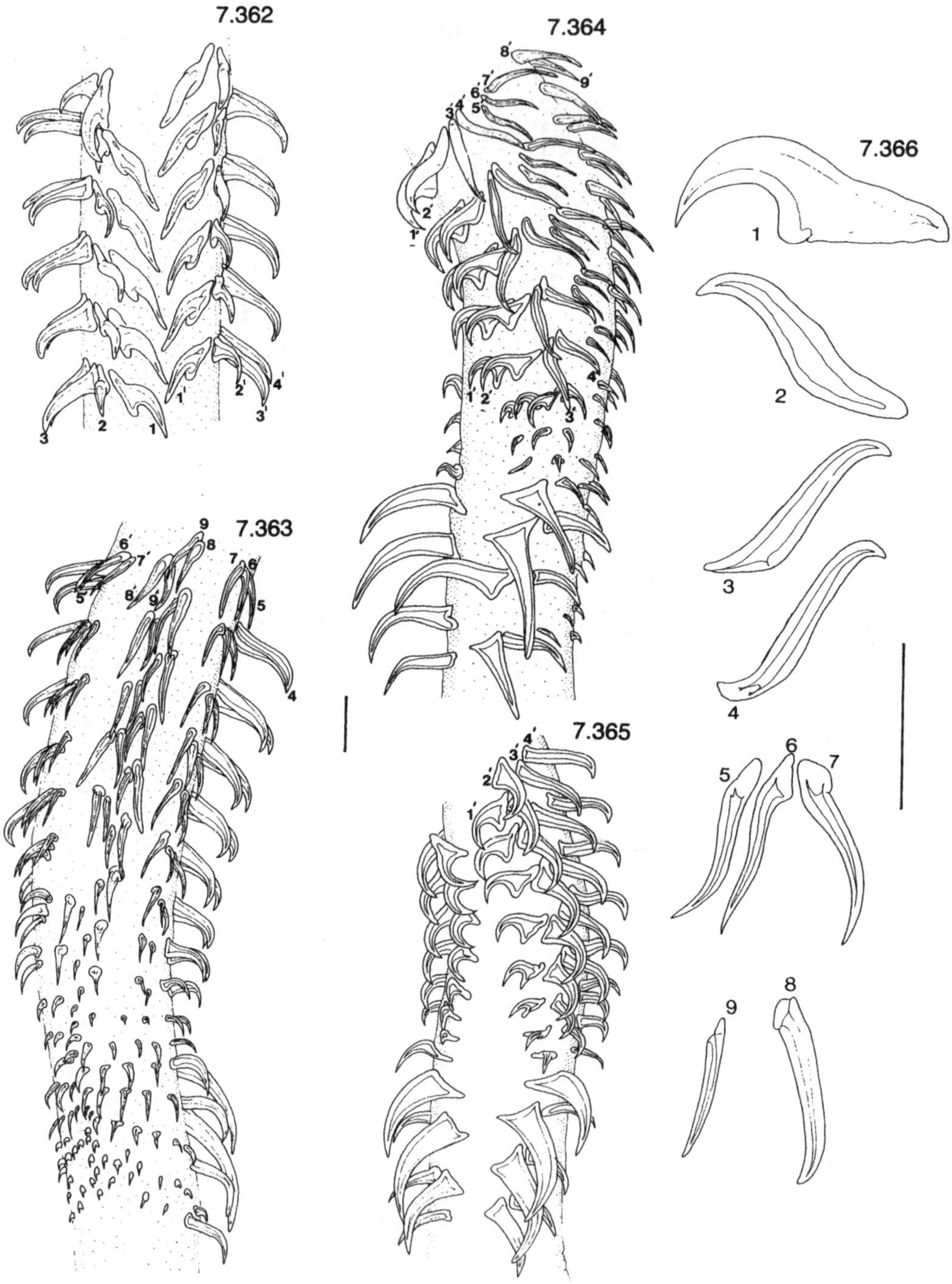

Figs 7.362–7.366 *Hornelliella macropora* (Shipley & Hornell, 1912). 7.362. Internal face, metabasal armature. 7.363. External face, basal region. 7.364. Bothridial face, basal region. 7.365. Internal face, basal region. 7.366. Profiles of hooks 1–9. Scale lines: 0.1 mm.

Family Mustelicolidae Dollfus, 1969

Key to genera

1a. Pars bothridialis overlapping pars bulbosa, five chainettes present .... ***Patellobothrium*** **Beveridge & Campbell, 1989. (Figs 7.367–7.376)**

Diagnosis: Scolex short, acraspedote. Pars vaginalis shorter than pars bothridialis. Bulbs short, sheaths sinuous. Prebulbar organ absent. Retractor muscles originate at anterior end of bulb. Two large, patelliform bothridia. Armature

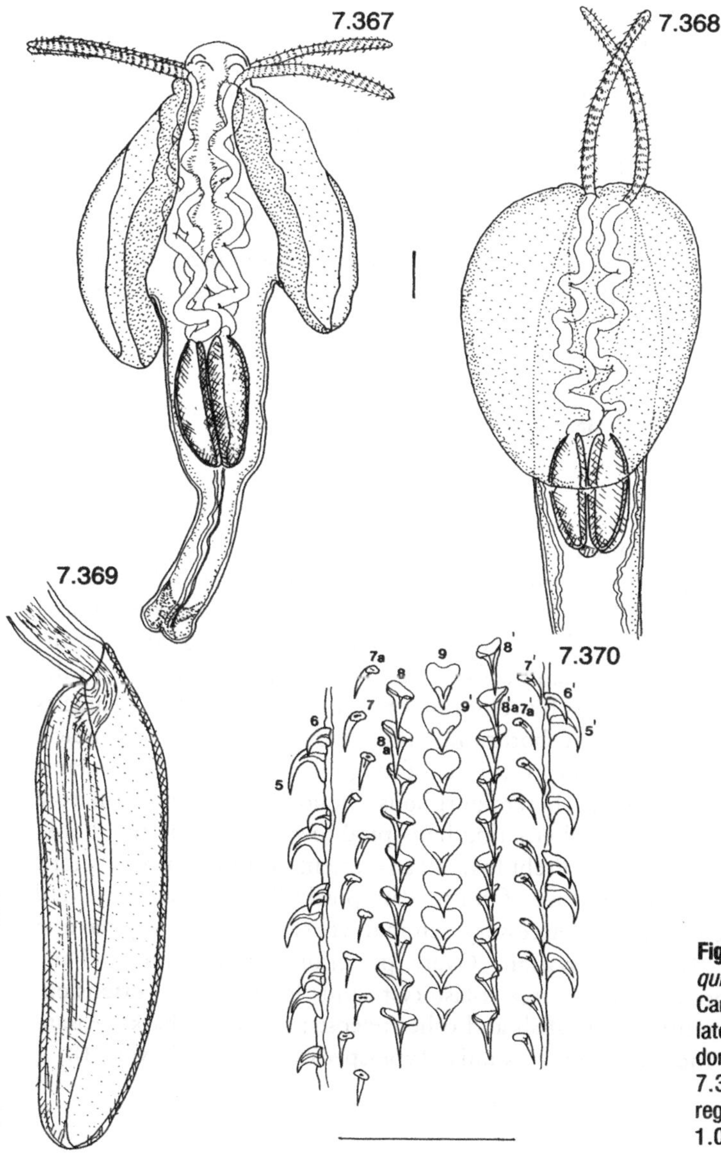

Figs 7.367-7.370 *Patellobothrium quinquecatenatum* Beveridge & Campbell, 1989. 7.367. Scolex, lateral view. 7.368. Scolex, dorsoventral view. 7.369. Bulb. 7.370. External face, metabasal region. Scale lines: 7.367-7.368, 1.0 mm; 7.369-7.370, 0.1 mm.

poeciloacanthous, heteromorphous, hooks hollow. Basal swelling and distinctive basal armature absent. External face of tentacle with five chainettes. Strobila unknown. Adults in elasmobranchs; plerocercus in teleosts. Australia. Type and only species *P. quinquecatenatum* Beveridge & Campbell, 1989.

1b. Pars bothridialis not overlapping pars bulbosa, three chainettes present .... *Diesingium* Pintner, 1929. (Figs 7.377–7.382)
(Syns *Mustelicola* Dollfus, 1969; *Diesingella* Guiart, 1931.)
Diagnosis: Scolex acraspedote. Two patelliform bothridia, rimmed; not contiguous apically. Pars vaginalis longer than pars bothridialis. Bulbs short. Pars postbulbosa very short. Tentacle sheaths irregularly sinuous. Prebulbar organs absent. Gland cells numerous in prebulbar region of pedunculus scolecis. Retractor muscles originate in anterior portion of bulb. Tentacles lack basal swelling. Armature consists of three double chainettes opposite hooks 1(1′) of principal rows. Principal rows alternate. Proglottids acraspedote. Genital pores irregularly alternate. Hermaphroditic sac present, cirrus present. Seminal vesicles present. Testes medullary, intervascular, occupying available pre-and postovarian space. Ovary bilobed in dorsoventral view, near posterior extremity. Uterus median, saccate, with anterior pore. Vitelline follicles encircle internal organs. Eggs anoperculate. Osmoregulatory system complex, with anastomoses. In sharks. Mediterranean, Atlantic and Australia. Type-species *D. lomentaceum* (Diesing, 1850).

Family Gymnorhynchidae Dollfus, 1935

Key to genera

1a. Double chainette of paired elements, all single-winged .... *Gymnorhynchus* Rudolphi, 1819. (Figs 7.383–7.391)
(Syn. *Acanthocephalus* Cobbold, 1879, preoccupied and *lapsus*.)
Diagnosis: Large trypanorhynchs. Scolex acraspedote. Four long and narrow bothridia, posterior and lateral margins free, fused to scolex anteriorly. Bothridia well separated but joined in pairs by velum on opposite faces of scolex. Tentacles emerge from anterior border of bothridial rims. Proximal region of tentacle unarmed. Basal armature begins with corona of long falciform hooks. Distinctive basal armature present. Hooks heteromorphous, hollow. Metabasal armature poeciloacanthous, consisting of alternating half-spiral rows of eight or nine hooks each. Tentacle sheaths sinuous. Prebulbar organs absent. Bulbs relatively short. Retractor muscles originate in base of bulb. Pars postbulbosa distinct. Genital pore marginal, pre-equatorial. Accessory seminal vesicle present. Vagina and cirrus-sac separate. Testes preovarian. Ovary posterior. Uterus and vitellaria undescribed. Plerocercus with large blastocyst and long caudal extension. Plerocerci in ocean sunfish, swordfish and other teleosts; adults in sharks. Atlantic, Mediterranean, Indian Ocean, Australia. Type-species *G. gigas* (Cuvier, 1817) Rudolphi, 1819.

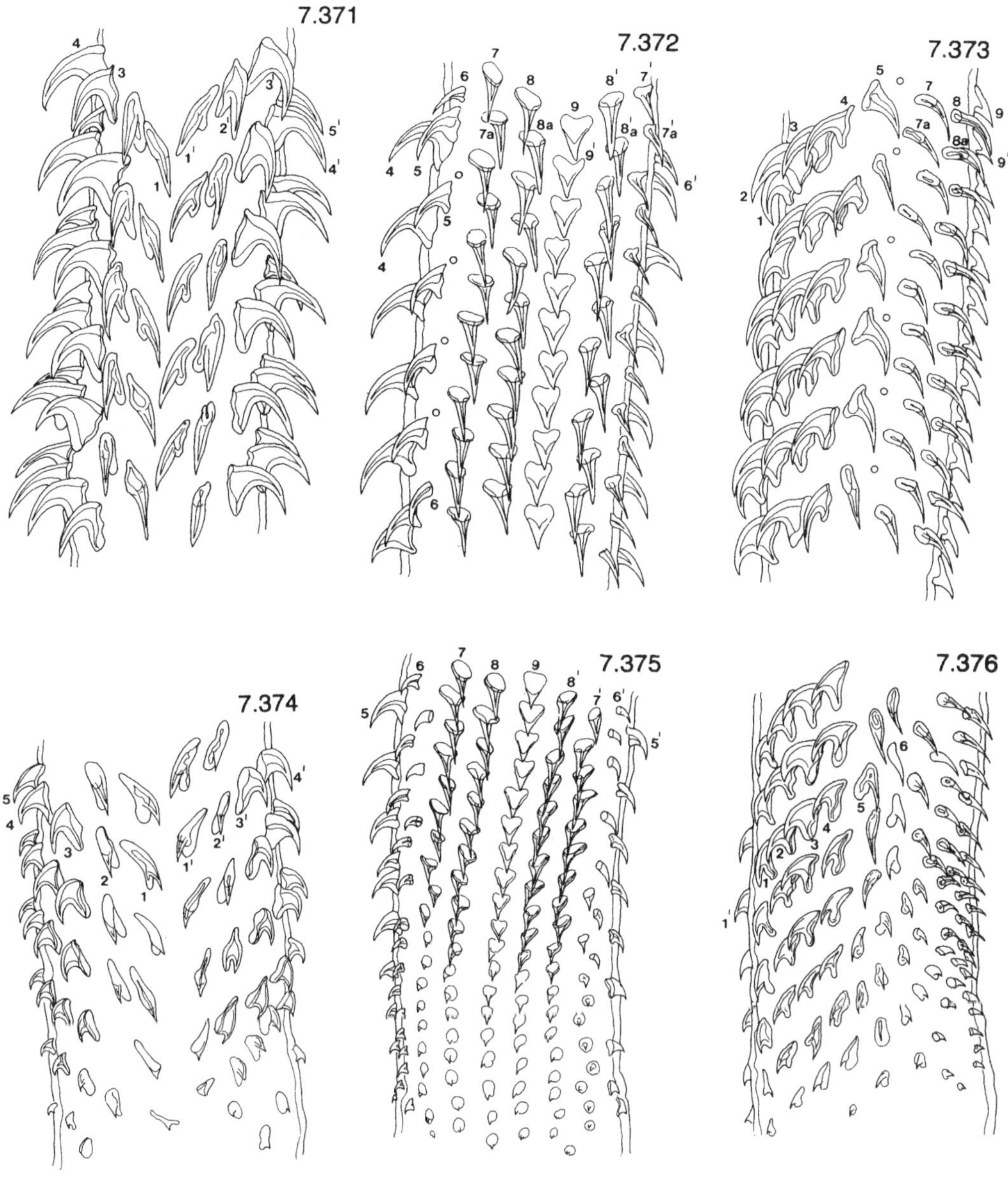

Figs 7.371–7.376 *Patellobothrium quinquecatenatum* Beveridge & Campbell, 1989. 7.371. Internal face, metabasal region. 7.372. External face, metabasal region. 7.373. Bothridial face, metabasal region. 7.374. Internal face, basal region. 7.375. External face, basal region. 7.376. Bothridial face, basal region. Circles indicate absence of hook 6.

1b. **Compound chainette composed of single-winged pairs of hooks alternating with single two-winged hooks** ....
..... ***Chimaerarhynchus*** **Beveridge & Campbell, 1989. (Figs 7.392–7.402)**

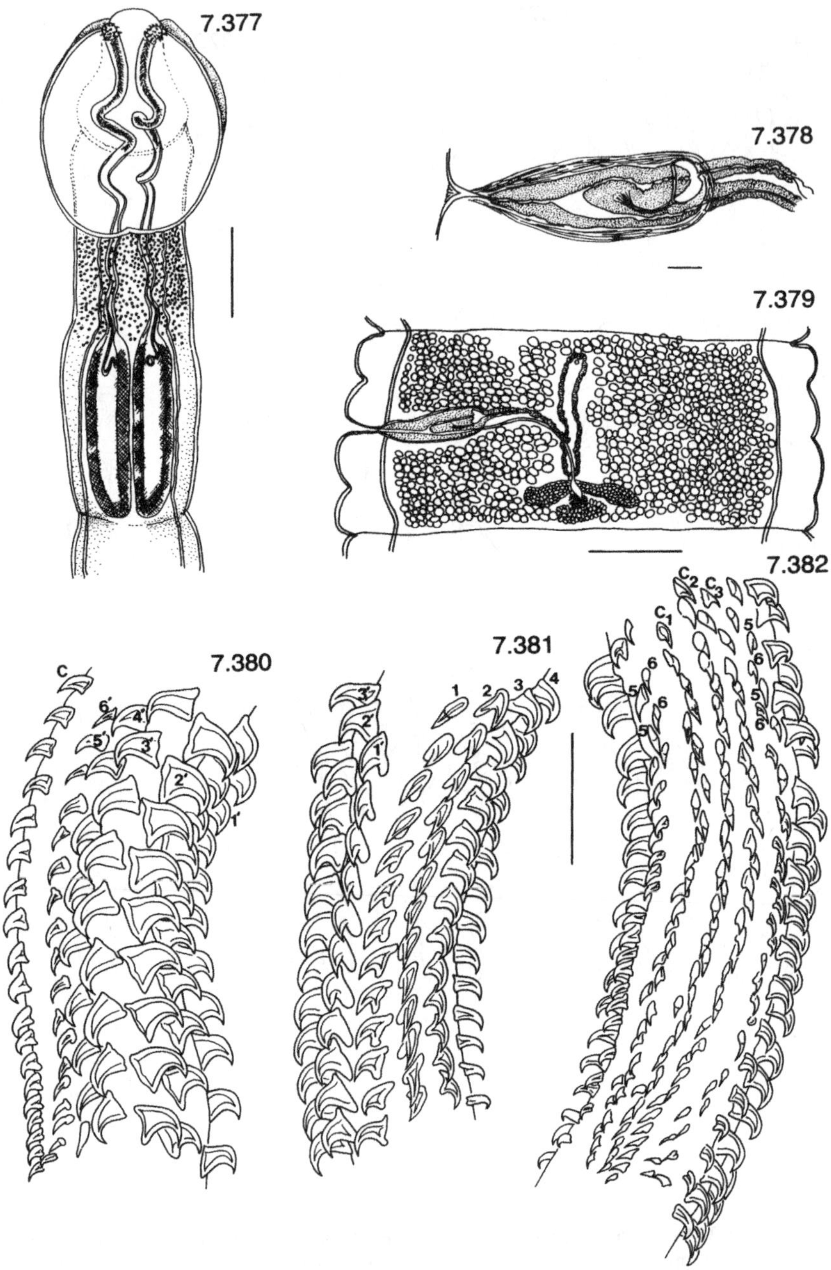

Figs 7.377-7.382 *Diesingium antarcticum* (Campbell & Beveridge, 1988). 7.377. Scolex. 7.378. Hermaphroditic sac. 7.379. Mature proglottid. 7.380. Antibothridial face, metabasal region. 7.381. Internal face, metabasal region. 7.382. External face, metabasal region showing three double chainettes (C). Scale lines: 7.377, 7.379, 0.5 mm; 7.378, 7.380-7.382, 0.1 mm.

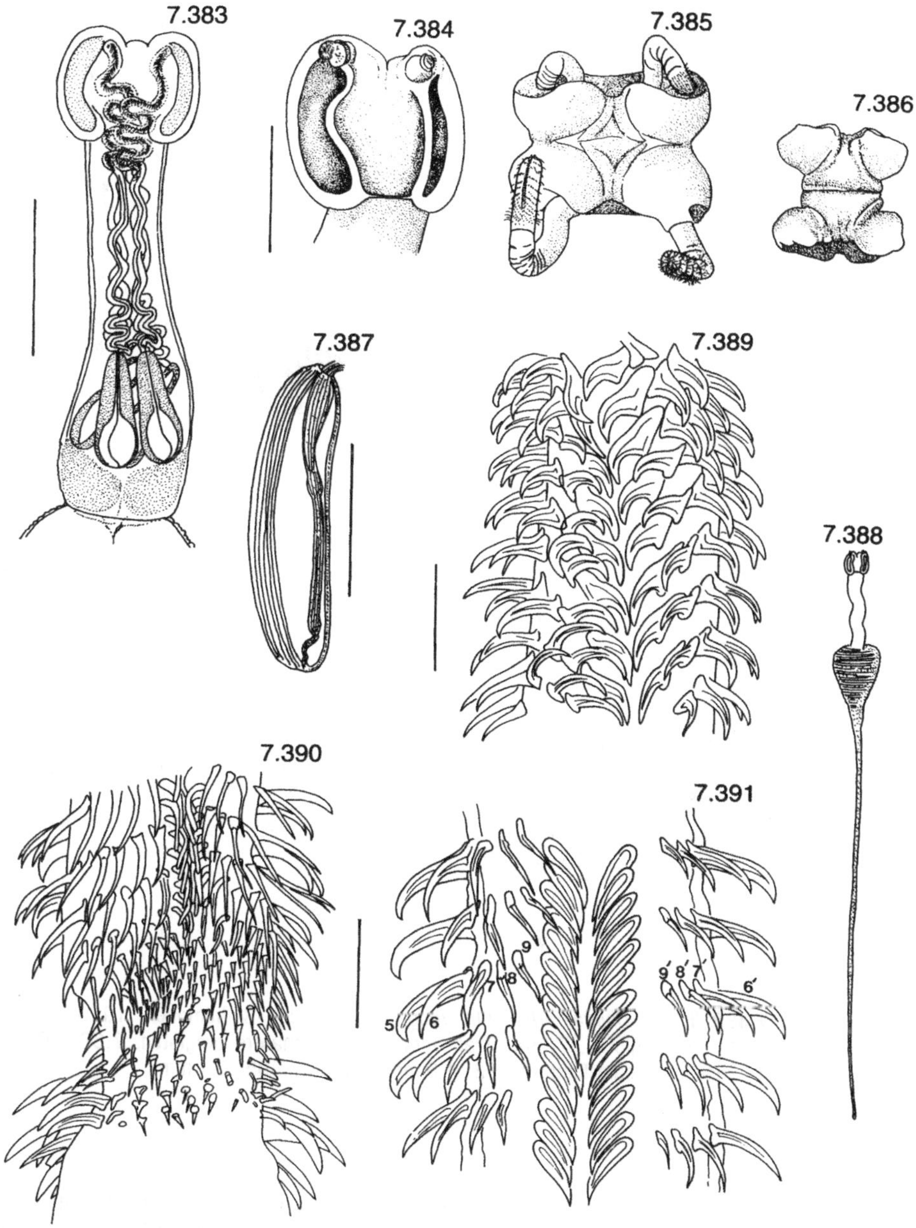

Figs 7.383–7.391 *Gymnorhynchus gigas* (Cuvier, 1817) (after Dollfus, 1942). 7.383. Scolex, plerocercus. 7.384. Pars bothridialis. 7.385. En face view scolex showing tentacles. 7.386. En face view scolex showing bothridia. 7.387. Bulb. 7.388. Plerocercus. 7.389. Internal face, metabasal armature. 7.390. External face, basal region. 7.391. External face, metabasal region. Scale lines: 7.383, 1.0 mm; 7.384, 7.387, 2.0 mm; 7.389–7.391, 0.2 mm.

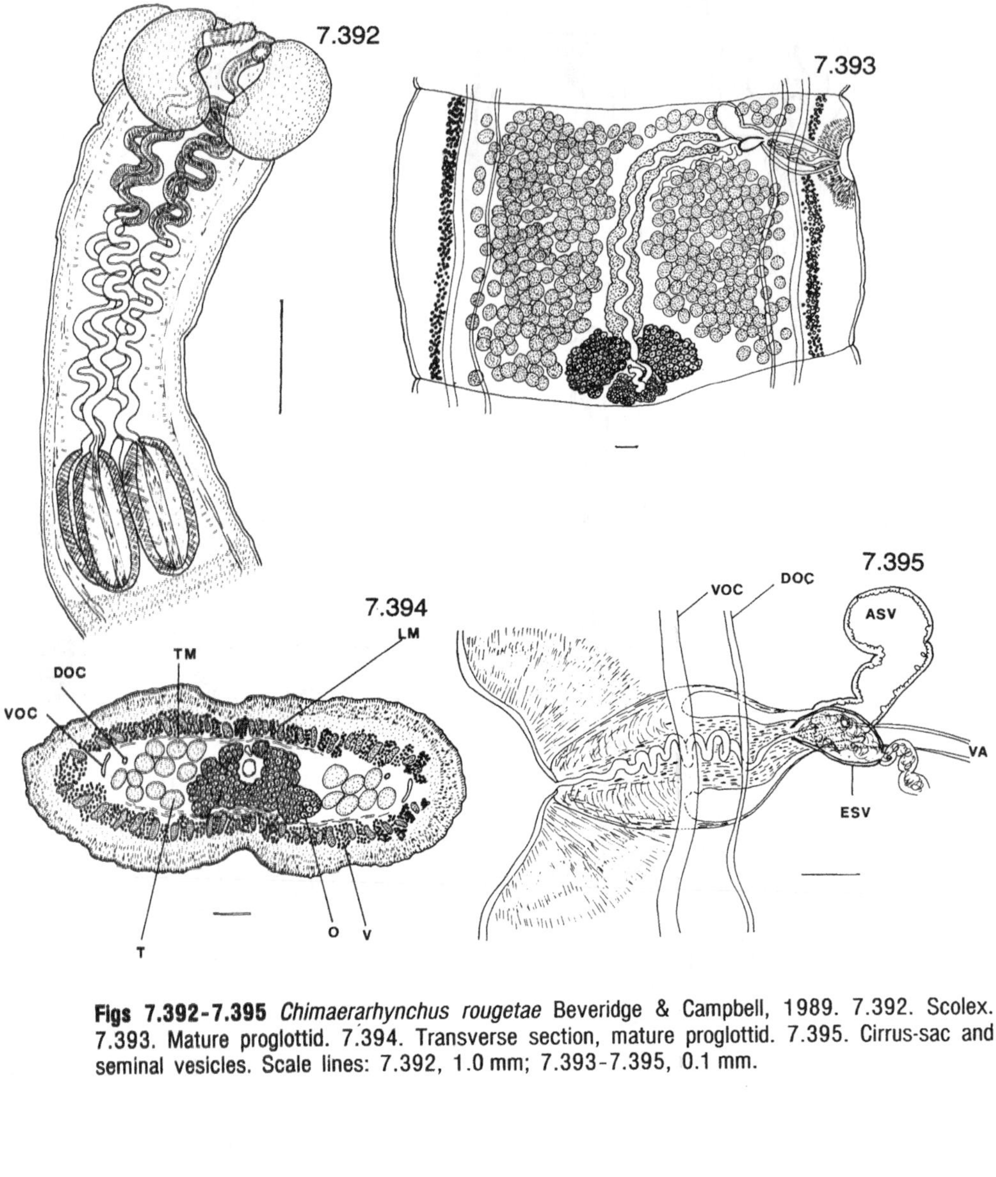

Figs 7.392-7.395 *Chimaerarhynchus rougetae* Beveridge & Campbell, 1989. 7.392. Scolex. 7.393. Mature proglottid. 7.394. Transverse section, mature proglottid. 7.395. Cirrus-sac and seminal vesicles. Scale lines: 7.392, 1.0 mm; 7.393-7.395, 0.1 mm.

Diagnosis: Scolex elongate, acraspedote. Pars vaginalis and pars bulbosa longer than pars bothridialis. Retractor muscles originate at base of bulb. Four auriculate bothridia with free margins. Basal swelling and basal armature absent. Metabasal armature poeciloacanthous, heteromorphous; hooks hollow. Strobila acraspedote. Proglottids wider than long. Testes numerous, intervascular, scattered in poral and aporal fields, none postovarian. Genital pores pre-equatorial, irregularly alternate. External, internal and accessory seminal vesicles present. Vagina ventral to cirrus-sac. Ovary posterior. Uterus arches toward genital pore; preformed uterine pore absent. Vitelline follicles circum-medullary. In sharks. Southeast

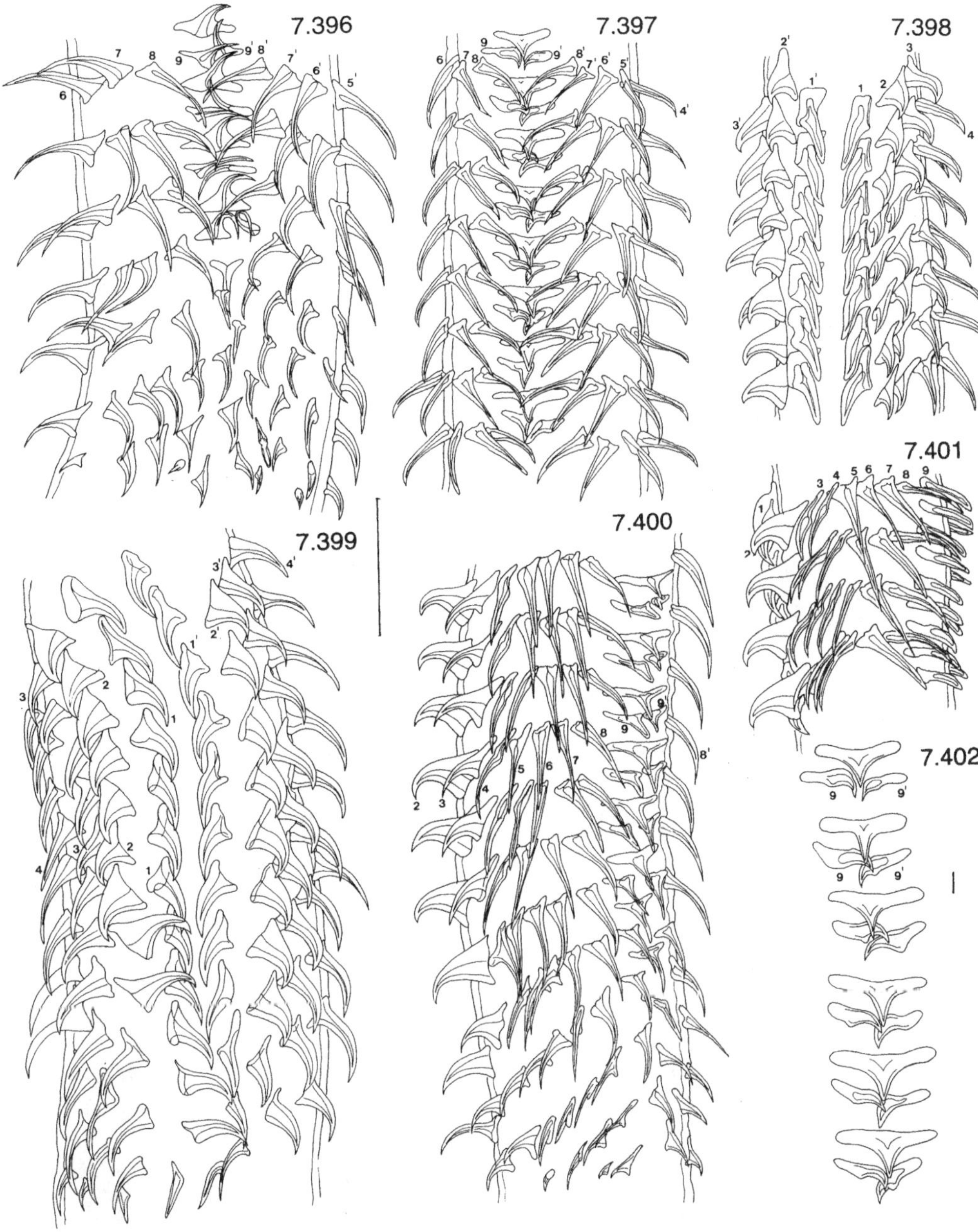

Figs 7.396-7.402 *Chimaerarhynchus rougetae* Beveridge & Campbell 1989. 7.396. External face, basal region. 7.397. External face, metabasal region. 7.398. Internal face, metabasal region. 7.399. Internal face, basal region. 7.400. Bothridial face, basal region, chainette to right. 7.401. Bothridial face, metabasal region. 7.402. Detail of chainette (compound type). Scale lines: 7.396-7.401, 0.1 mm; 7.402, 0.01 mm.

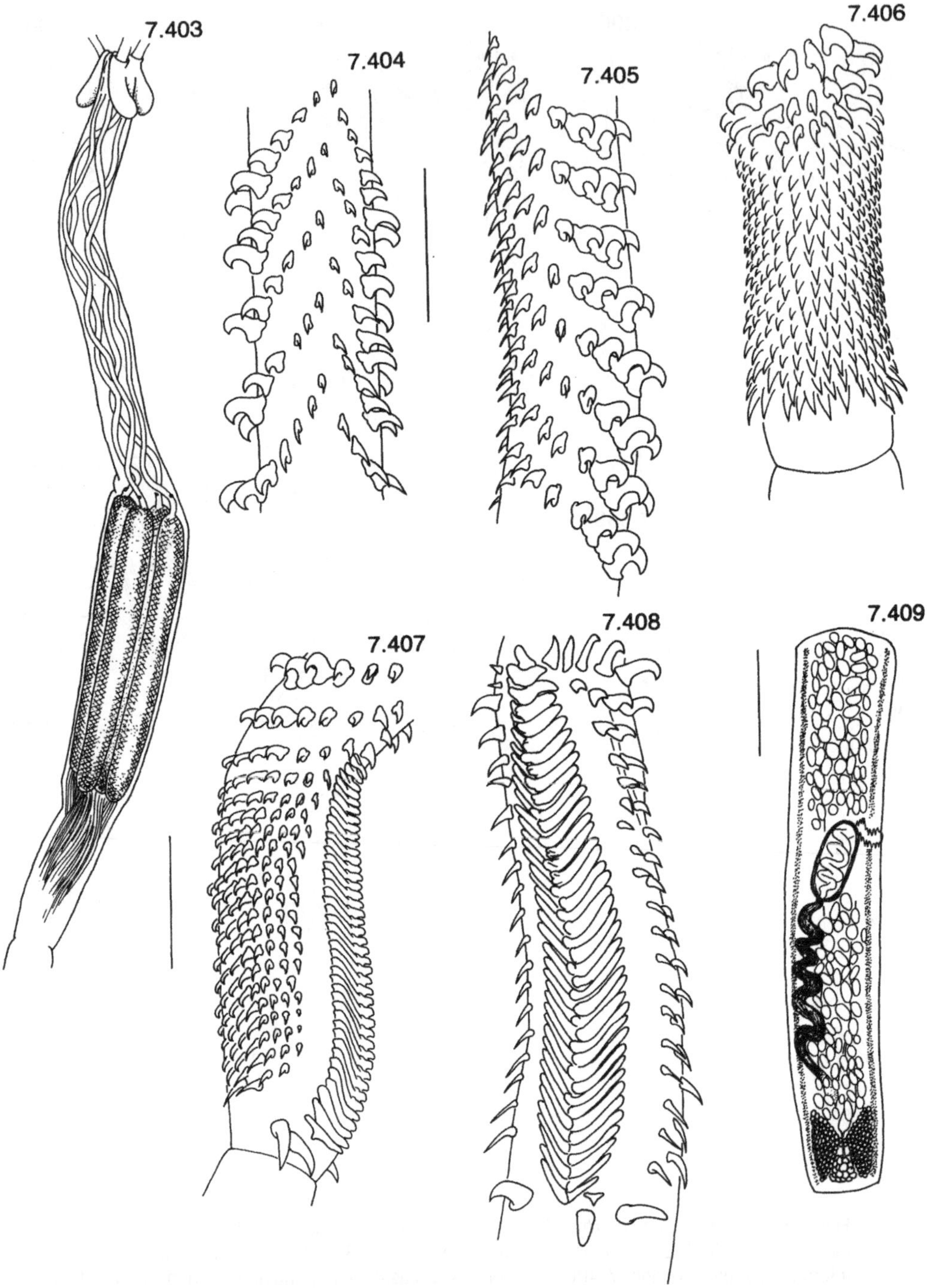

Figs 7.403-7.409 *Mixodigma leptaleum* Dailey & Vogelbein, 1982. 7.403. Scolex. 7.404. External face, metabasal region. 7.405. Bothridial face, metabasal region. 7.406. Internal face, basal armature. 7.407. Antibothridial face, basal armature. 7.408. External face, basal armature. 7.409. Terminal proglottid. All figures except 7.408 after Dailey & Vogelbein (1982). Scale lines: 7.403, 1.0 mm; 7.404-7.408, 0.2 mm; 7.409, 0.5 mm.

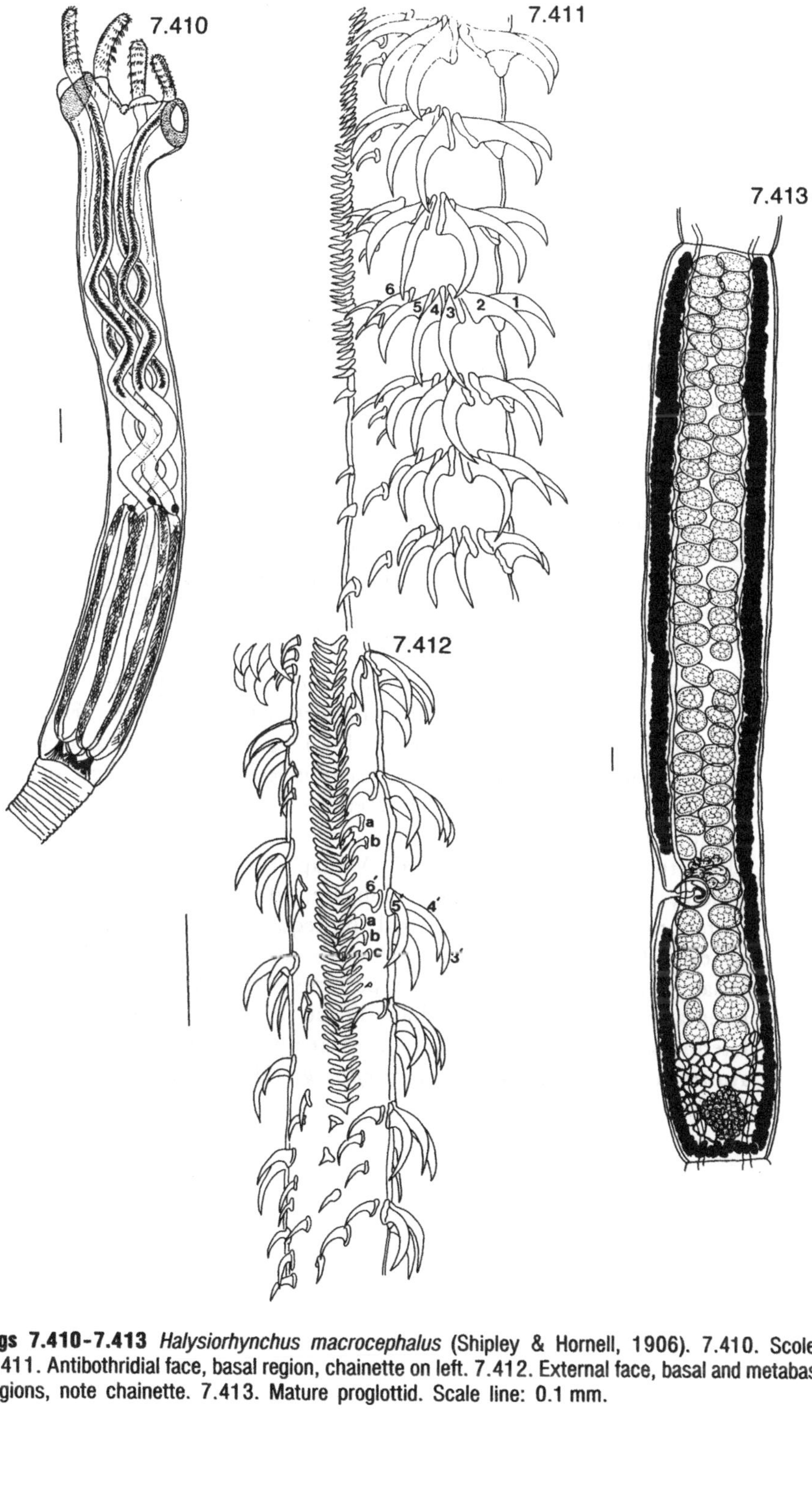

Figs 7.410–7.413 *Halysiorhynchus macrocephalus* (Shipley & Hornell, 1906). 7.410. Scolex. 7.411. Antibothridial face, basal region, chainette on left. 7.412. External face, basal and metabasal regions, note chainette. 7.413. Mature proglottid. Scale line: 0.1 mm.

Atlantic Ocean off Senegal, Africa. Type-species *C. rougetae* Beveridge & Campbell, 1989.

Family Mixodigmatidae Dailey & Vogelbein, 1982

Key to genera

1a. Bothridia sessile, paired; scolex acraspedote ....
.... ***Mixodigma* Dailey & Vogelbein, 1982. (Figs 7.403–7.409)**
Diagnosis: Scolex long, bothridia merged into dorsoventral pairs. Pars postbulbosa present. Pars vaginalis long, tentacle sheaths sinuous. Prebulbar organs present. Bulbs long, retractor muscles originate at base of bulbs. Gland cells associated with retractor muscles. Armature poeciloacanthous; hooks heteromorphous, solid. Chainette of V-shaped hooks at base of tentacle in type species. Metabasal region with alternating half-spiral rows of nine hooks each, meeting on external face in typical heteroacanthous pattern. Basal region of tentacle lacks basal swelling. Strobila acraspedote, anapolytic. Genital pores marginal. Testes medullary, preovarian. Cirrus-sac present, seminal vesicles lacking. Vagina with seminal receptacle. Vitelline follicles encircle medullary region. Uterus median, linear, terminating at anterior extremity. In deep-sea sharks. Pacific. Type-species *M. leptaleum* Dailey & Vogelbein, 1982.

1b. Bothridia on short pedicels, separated, scolex craspedote ....
.... ***Halysiorhynchus* Pintner, 1913. (Figs 7.410–7.413)**
Diagnosis: Scolex elongate. Tentacles long; basal swelling lacking; tentacle sheaths sinuous. Bulbs long, retractor muscles originate in base of bulb. Gland cells surround retractor muscle. Metabasal armature poeciloacanthous; hooks heteromorphous, solid. Nine hooks per half-spiral row in metabasal region. Chainette hooks 9(9′) biwinged (V-shaped) on external face. Strobila acraspedote, apolytic. Proglottids elongate, flattened anteriorly, subcylindrical posteriorly. Genital pores marginal, postequatorial, irregularly alternate. Cirrus-sac extends near midline. External and internal seminal vesicles present. Testes in lateral preovarian fields. Ovary posterior, tetralobed in cross-section. Seminal receptacle absent. Uterus median, linear, terminating near anterior extremity. Uterine pore present. Vitelline follicles circum-medullary. In Batoidea. Sri Lanka, Australia. Type-species *H. macrocephalus* (Shipley & Hornell, 1906) Pintner, 1913.

8 Order Tetraphyllidea Carus, 1863

L. EUZET

Laboratoire de Parasitologie comparée, Université Montpellier II, Station Méditerranéenne de l'Environment Littoral, 1 Quai de la Daurade, 34200 Sète, France.

Introduction

A historical account of the systematics of the cestodes of selachians would be too long to give here. It may be found in the monograph by Southwell (1925) or in the text by Wardle & McLeod (1952). I remark simply that the individuality of these parasites was recognized as early as 1850 by van Beneden. At this date, this author grouped together under the name 'Tetraphyllides' the 'phyllobothriens' and 'phyllacanthiens' (currently in the Tetraphyllidea) and the 'phyllorhynchiens' (currently Trypanorhyncha). The last were separated by Diesing (1863) into a distinct order.

The initial subdivision by van Beneden into the 'phyllobothriens' (now the Phyllobothriidae) and the 'phyllacanthiens' (now the Onchobothriidae) is to be found in almost all the subsequent works on the systematics of the order Tetraphyllidea.

A third family, the Lecanicephalidae Braun, 1900 (Gamobothriidae Linton, 1901) was added to the first two. It is roughly this subdivision of the order into three families which is found in the work of Southwell (1925), although the Lecanicephalidae (in which the scolex has four round suckers) was placed in the Cyclophyllidea. Since 1930, new families and orders have been proposed to define the systematics of the Tetraphyllidea.

The family Disculicipiditae Joyeux & Baer, 1936 was erected to replace the Discocephalidae Pintner, 1928, preoccupied. Joyeux & Baer (1936a) proposed the generic name *Disculiceps* in place of *Discocephalum* with, as type-species, *Disculiceps pileatum* (Linton, 1890), synonym *Discocephalum pileatum* Linton, 1890. This amendment appears to be in conformity with Article 55(b) of the *International Code of Zoological Nomenclature*, but has apparently not been submitted for a decision by the appropriate Commission. Pending this, we will conserve *Disculiceps* and the family Disculicipitidae Joyeux & Baer, 1936.

In 1952, Wardle & McLeod elevated some families to ordinal rank. The

classification of the Tetraphyllidea at that time was as follows:
Tetraphyllidea Carus, 1863: Phyllobothriidae Braun, 1900; Onchobothriidae Braun, 1900.
Disculicipitidea Wardle & McLeod, 1952: Disculicipitidae Joyeux & Baer, 1936.
Lecanicephala Wardle & McLeod, 1952: Lecanicephalidae Braun, 1900; Cephalobothriidae Pintner, 1928.

In 1959, Euzet completed a classification of the Tetraphyllidea proposed as a draft in 1953. Subsequent work has shown that the criteria utilized did not have the significance previously attributed to them. In particular, assembling in the superfamily Prosobothrioidea the families Prosobothriidae, Platybothriinae, Phoreiobothriidae, Reesiinae and Gastrolecithidae is artificial. This classification should be abandoned.

Yamaguti (1959) erected the family Triloculariidae with *Trilocularia* Olsson, 1867 as type-genus. This parasite of *Squalus acanthias* (L., 1758) does in fact present a very singular arrangement of the genital system. In my opinion, this anatomy is a direct consequence of the fact that *Trilocularia acanthiaevulgaris* (Olsson, 1867) is a hyperapolytic species. This anatomy, which is close to that found in the proglottid of *Potamotrygonocestus magdalenensis* Brooks & Thorson, 1976, is not fundamentally different from that of the phyllobothriid tetraphyllideans.

Schmidt (1986) added to the Triloculariidae the genera *Pentaloculum* Alexander, 1963, a parasite of *Typhlonarke aysoni* (Hamilton, 1902) (Torpedinidae) and *Zyxibothrium* Hayden & Campbell, 1981, a parasite of *Raja senta* Garman, 1885 (Rajidae). These three genera do not appear to have much in common except the scolex, in which the four bothridia have a surface divided by muscular septa and a small number of loculi. Although in my opinion these three genera do not constitute a natural group, I have retained them together but in the subfamily Triloculariinae Yamaguti, 1959, which I place in the Phyllobothriidae.

After 1960, some new genera have been classified in new families of the Tetraphyllidea, or the same taxa have been placed in distinct orders. These are, in chronological order: *Litobothrium* Dailey, 1969, family Litobothriidae Dailey, 1969, order Litobothriidae Schmidt, 1986; *Dioecotaenia* Schmidt, 1969, family Dioecotaeniidae Schmidt, 1969, order Dioecotaeniidea Schmidt, 1986; *Cathetocephalus* Dailey & Overstreet, 1973, family Cathetocephalidae Dailey & Overstreet, 1973; and *Chimaerocestos* Williams & Bray, 1984, family Chimaerocestidae Williams & Bray, 1984.

In all these cestodes, one finds the characters of the Tetraphyllidea, whether in morphology of the scolex or in the anatomy of the genital system. I propose to maintain them within the Tetraphyllidea but, in recognition of their uniqueness, to distinguish them at family level.

Comments on controversial genera are given below.

Family Onchobothriidae Braun, 1900

The systematic position of the genera *Ceratobothrium* Monticelli, 1892 and *Thysanocephalum* Linton, 1890 which many authors have placed in the

Onchobothriidae will be discussed below. I have placed them in the Phyllobothriidae.

Balanobothrium Hornell, 1912. Fuhrmann (1931) gave an excellent illustration of the scolex of *B. parvum*, which he placed in the Onchobothriidae. The anterior part of the scolex, sunken into the intestinal wall of the host, forms a globular mass within which one can, however, distinguish four bothridia each marked by an anterior sucker and a pair of minute forked hooks. The posterior part of the scolex forms a sort of collar representing the free part of the four bothridia. After Fuhrmann (1931), authors were divided in their opinions on the systematic position of *Balanobothrium*. Yamaguti (1959) and Schmidt (1986) maintained it in the Lecanicephalidae; Perrenoud (1931), Wardle & McLeod (1952) and Joyeux & Baer (1961) placed it in the Onchobothriidae. Baer & Euzet (1962), having examined Southwell's material, were of the second opinion.

Cylindrophorus Diesing, 1863. Euzet (1952) discussed the validity of this genus erected by Diesing (1863) from the illustration given by Wagener (1854) of *Tetrabothrium aus Charcharias rondeletii*. It has been recognized (Cohn, 1902a; Mola, 1903) that these diagrams represent two different tetraphyllideans. One, *Prosobothrium* Cohn, 1902 has an unarmed scolex; the other, *Platybothrium* Linton, 1890 has a scolex with bothridia armed with characteristic hooks. The genus *Cylindrophorus*, not having a type-species, represents a *nomen nudum* and should disappear from the nomenclature.

Phyllobothroides Southwell, 1911. This generic name was used by Southwell (1911) for *P. kerkhami*, a parasite of *Chiloscyllium indicum*. Southwell (1925) synonymized this species with *Pedibothrium longispine* Linton, 1909. *Phyllobothroides* Southwell, 1911 falls into synonymy with *Pedibothrium* Linton, 1909.

Pinguicollum Riser, 1955. Riser (1955) created this genus for *Onchobothrium pinguicollum* Sleggs, 1927. The host is a *Raja* sp. caught on the Pacific coast of California. Hart (1936) described under the name *Pedibothrium pinguicollum* an onchobothriid collected from *Raja binoculata* Girard, 1854 in Puget Sound (Pacific coast of the USA). He assigned this parasite to the species described by Sleggs but placed it in the genus *Pedibothrium* Linton, 1909. This generic attribution was seriously doubted by Wardle & McLeod (1952). Riser (1955) found, on the Pacific coast of the USA in several species of *Raja* (including *R. binoculata*), a common onchobothriid which he compared with the species described by Sleggs and recovered by Hart. Riser made this species the type of the new genus *Pinguicollum*. It differs from *Acanthobothrium* in having 'the hooks fused in a common matrix, immobile phyllidia and a pseudoscolex'. The definition of this pseudoscolex has not been determined and the presence of a special structure at the level of the scolex is not evident in the illustration of *Pinguicollum pinguicollum* by Riser. The special morphology of the scolex of this cestode probably results from its being embedded in the wall of the spiral intestine. The fact that the bothridial hooks are united into one structure (matrix) does not seem to be a genuine character. In species of *Acanthobothrium*, the base of the hooks is always close and sometimes united (for example, in *A. confusum* Baer & Euzet, 1962). It does not seem useful to create a genus for this single character. I consider *Pinguicollum* Riser, 1955 synonymous with *Acanthobothrium* van Beneden, 1849.

Spiniloculus Southwell, 1925. The scolex of the type-species, *S. mavensis* Southwell, 1925, a parasite of *Mustelus* sp., was described as being formed by four bothridia. Each bothridium, divided into two by a transverse septum, is separated from a broad flat muscular anterior region by a cushion bearing a C-shaped hook on its lateral extremities. In an addendum, Southwell (1930a) amended the description of *S. mavensis* from material collected from *Chiloscyllium indicum*. He emphasized that the bothridia are not divided by a transverse septum and that this morphology was found in the bothridia of the type-specimens of *S. mavensis* from *Mustelus* species. Southwell then gave a new diagnosis of the genus. As a result of this, the onchobothriid tetraphyllidean in *Ginglymostoma cirratum* Muller & Henle (Ginglymostomatidae) found at Ratnagiri (India) and named *Spinibiloculus* by Deshmukh & Shinde (1980) comes within the genus *Spiniloculus*. *Spinibiloculus* was differentiated from *Spiniloculus* by having four simple bothridia without a transverse septum. On the other hand, the hooks, situated at the same site in the bothridium, have a similar morphology. The anatomy is very close but the position of the vagina of *Spinibiloculus ratnagiriensis*, given as posterior to the cirrus-sac, should be re-examined. It should be added that Subhapradha (1957) described *Spiniloculus mavensis*, from *Chiloscyllium griseum* (Muller & Henle, 1841), with bothridia not subdivided by a transverse septum. Williams (1964) and Caira (1987) have already noted the difference between the description by Subhapradha and that by Southwell. Caira (1990) returned to the problem of the scolex morphology of *Spiniloculus*. She gave a diagnosis detailing it as 'each bothridium divided into 3 loculi with 2 transverse costae, anterior costa conspicuous, posterior costa inconspicuous externally. Each bothridium with a pair of simple, hollow, U-shaped hooks; bases of hooks widely separated; 1 hook on each posterolateral corner of anteriormost loculus'. This diagnosis has not been followed here. The anterior loculus should be regarded as part of the muscular plateau found in front of the hooks in the bothridium of the majority of the Onchobothriidae. The second fold, which is very inconsistent in *Spiniloculus*, corresponds without doubt to the shape of the bothridium when the scolex is attached to the intestinal wall. Its presence is perhaps linked to the area of insertion of the longitudinal musculature on the internal face of the bothridium.

There remains the question of the systematic position of the hosts. *Spiniloculus mavensis* was described from a *Mustelus* sp. from the eastern coast of Australia. It seems that this cestode has not since been seen in a host of this genus. In contrast, *S. mavensis* was reported from *Chiloscyllium indicum* by Southwell (1925), from *C. griseum* by Subhapradha (1957) and from *Hemiscyllium punctatum* (= *C. punctatum*) by Williams (1964) and Caira (1990).

S. ratnagiriensis (Deshmukh & Shinde, 1980) is an intestinal parasite of *Ginglymostoma concolor* (= *Nebrius ferrugineus*). *Chiloscyllium* and *Ginglymostoma* are classified in the Orectolobiformes. Given this, the presence of *Spiniloculus* in *Mustelus* (Carchariniformes, Triakidae) is doubtful.

I should add that the genera *Chiloscyllium* and *Nebrius* are parasitized by another onchobothriid genus, *Yorkeria* Southwell, 1927, which is an argument in favour of the individuality of the tetraphyllideans of Orectolobiformes.

Acrobothrium Baer, 1948 (*nec* Olsson, 1872). The genus has been synonymized with *Acanthobothrium* van Beneden, 1849 by Yamaguti (1959) and Euzet (1959). The only character separating the two genera, the morphology of the hooks, is not sufficient, this being quite variable in *Acanthobothrium*. Williams (1966b) further specified that the name *Acrobothrium* should not have been used because it had already been used by Olsson, 1872 for another cestode, *Acrobothrium typicum*, a parasite of the teleost *Lota vulgaris* in fresh water. [See Chapter 4.]

Circulobothrium Shinde & Chincholikar, 1975. The type-species, *C. singhi*, was described from a single individual collected from a *Trygon* sp. at Ratnagiri (India). The material appears to be in bad condition. The illustration of the scolex shows only a single hook of peculiar morphology although the authors state in their description 'one pair of hooks per bothridium'. The anatomy of the proglottids appears to be of the tetraphyllidean type although the ovary is not shown. This cestode is not recognizable and cannot be found again because the host species is unknown. A type-specimen does not seem to have been deposited. *Circulobothrium* is regarded as a *genus inquirendum*.

Family Phyllobothriidae Braun, 1900

Subfamily Thysanocephalinae Euzet, 1953

Thysanocephalum Linton, 1890. Many authors have placed *Thysanocephalum* in the Onchobothriidae but, in *T. thysanocephalum*, the bothridia have, on the flat anterior region, two small lateral projections. These muscular structures are very different from the sclerified hooks which characterize the Onchobothriidae. For this reason, I retain *Thysanocephalum* in the Phyllobothriidae. The description and illustration of the anterior region of young stages of *T. thysanocephalum* by Subhapradha (1957) show that the metascolex is, in the beginning, composed of four small simple lobes which form, by folding, the metascolex of the adult. Including all tetraphyllideans with a well-developed metascolex in the subfamily Thysanocephalinae is practical but probably artificial. *Thysanocephalum* is a parasite of large sharks of the family Carcharinidae while *Myzocephalus* and *Myzophyllobothrium* are parasites of stingrays (Myliobatidae and Dasyatidae). Of these genera, *Thysanocephalum* is the best known.

Three other genera of tetraphyllideans parasitic in Myliobatiformes have been described as having a metascolex.

Rhoptrobothrium Shipley & Hornell, 1906. This genus has not been retained because it is based on a single immature individual, the anatomy of which is unknown. As Southwell (1925) noted, this parasite of Myliobatidae is close to *Myzocephalus* Shipley & Hornell, 1906, as the bothridia have an anterior sucker. It is treated as a *genus inquirendum*.

Myxophyllobothrium Shinde & Chincholikar, 1981. The genus is characterized by a scolex with four bothridia, each with an anterior sucker and a metascolex with four distinct lobes, highly folded. Apart from the nature of the host, *Trygon sephen* = *Hypolophus sephen* (Forsskal, 1775) which is a dasyatid, nothing separates this genus from *Myzocephalus*, of which I consider it a synonym.

Shindeobothrium Shinde & Chincholikar, 1975. The description of the type-species, *S. indica*, is based on a single individual found in the spiral valve of *Trygon* sp. at Ratnagiri on the west coast of India. This parasite appears to have a small scolex and a metascolex. The resemblance to *Myzophyllobothrium* was noted by Shinde & Chincholikar (1975). However, the description is incomplete and the illustration too poor to allow recognition of this cestode, which has not been found again, the host species being unknown. *Shindeobothrium* should be suppressed and is treated here as a *genus inquirendum*.

Subfamily Echeneibothriinae de Beauchamp, 1905

Discobothrium van Beneden, 1870. The validity of this genus is controversial. For some authorities, *Discobothrium*, type-species *D. fallax* van Beneden, 1870, characterized by a scolex with a large apical myzorhynchus and four round suckers of the acetabulate type, should be placed in the Lecanicephalidae. However, the anatomy (Euzet, 1959; Williams, 1966a), with the vagina anterior to the cirrus-sac and crossing the vas deferens, is of the phyllobothriid type. According to Euzet (1959), who studied live specimens of *D. fallax*, the scolex has a voluminous apical myzorhynchus and the four small bothridia have a loculate surface. The bothridia can be withdrawn into the scolex leaving, on the surface, a round depression which could be taken for an acetabulum. With loculate bothridia and proglottids lacking postvaginal testes, this parasite belongs to *Echeneibothrium* and should be designated *E. fallax* (van Beneden, 1870). This proposal will be retained here. *Discobothrium* is a synonym of *Echeneibothrium*. The remaining species placed in *Discobothrium* should be renamed and their systematic position determined.

Subfamily Phyllobothriinae de Beauchamp, 1905

Many phyllobothriids have been grouped together in this subfamily. If their anatomy is of the tetraphyllidean type, the scolex is extremely varied. It is, moreover, on the morphology of the scolex that the genera are distinguished. It follows that, in certain cases, the state of contraction of this organ is what impedes the generic attribution of a parasite. Among the genera classified in this family, a certain number should no longer be recognized either because they are synonyms of an older genus or because they are inadequately described. Amongst these, the following should be mentioned.

Anthocephalum Linton, 1890. The type-species, *A. gracile* Linton, 1890, a parasite of *Dasyatis centroura* (Mitchill, 1815) has been renamed by Linton (1890) as *Anthobothrium gracile* (Linton, 1890). The scolex consists of four bothridia, each with an anterior accessory sucker and a border marked by a regular row of loculi. It should be placed in *Phyllobothrium*. However, in *Phyllobothrium*, the species name *P. gracile* Wedl, 1855 has been used for a parasite of *Torpedo marmorata*. Southwell (1925) renamed the parasite of *Dasyatis centroura* as *P. centrurum* Southwell, 1925. *Anthocephalum* could be resurrected by grouping together the species currently placed in *Phyllobothrium* which have bothridia with an accessory sucker and a loculated margin, and an anatomy without postvaginal testes on the poral side.

Aocobothrium Mola, 1907. The genus contains one species parasitic in an unknown teleost living in fresh water. Based on the anatomy, particularly the route of the vagina anterior to the cirrus-sac, this cestode would be placed in the Tetraphyllidea. The scolex with four sacciform bothridia opening anteriorly is similar to that of *Scyphophyllidium* Woodland, 1927. I propose to suppress this genus, since the host is not a selachian but an unknown freshwater teleost; it is treated here as a *genus inquirendum*.

Bilocularia Obersteiner, 1913. *B. hyperapolytica* Obersteiner, 1913, the only species of the genus, was considered synonymous with *Calyptrobothrium rigii* Monticelli, 1893 by Euzet (1959). In these two hyperapolytic tetraphyllideans, the morphology of the scolex with a large globular accessory sucker and an anatomy in which the postvaginal testicular field is absent are similar. However, taking into account the generally oioxenous specificity of the Tetraphyllidea, this synonymy is doubtful. *Bilocularia* is said to be a parasite of *Centrophorus granulosus* (Schneider, 1801), while *Calyptrobothrium* has been recovered from *Torpedo nobiliana* Bonaparte, 1835 on both sides of the North Atlantic. Since its discovery, *B. hyperapolytica* has not been found again in the type-host. The genus is considered a *genus inquirendum*.

Biporophyllaeus Subramanian, 1939. The genus was erected with the type *B. madrasensis* Subramanian, 1939, a parasite of *Chiloscyllium indicum* (Gmelin, 1789). Obviously, it is a case of a detached proglottid from a tetraphyllidean parasite, which cannot be linked with a known species in this selachian. The genus *Biporophyllaeus* should be abandoned; it is treated here as a *genus inquirendum*. The family Biporophyllaeidae Subramanian, 1939 and the order Biporophyllidea [*sic*] Wardle & McLeod, 1952 should also disappear.

Carpobothrium Shipley & Hornell, 1906. The very brief diagnosis given by Shipley & Hornell (1906) deals only with the morphology of the scolex. My proposed diagnosis is based on the description of *C. megaphallum* Subhapradha, 1957, a parasite of *Chiloscyllium griseum* Muller & Henle, 1838. *C. indicum* (Gmelin, 1789), the host of *Carpobothrium chiloscylli* Shipley & Hornell, 1906, is closely related to *Chiloscyllium griseum*. It transpires that the parasites of *Rhyn-*

chobatus djeddensis (Forsskal, 1775) (Rhynchobatidae) and of *Urogymnus asperimus* Bloch & Schneider, 1801 (Dasyatidae), described by Southwell (1925) under the name of *Carpobothrium chiloscylli*, do not belong to *Carpobothrium* as defined here. These species, which should be collected again from the type-hosts, correspond to a taxonomic unit which has yet to be defined.

Cyatocotyle Mola, 1908. Mola (1908) created this genus for *Tetrabothrius aus charcharias rondeletti* Wagener, 1854 which is said to be a composite species. Wagener (1854) did not describe the anatomy. That given by Mola for *C. marchesetti* is notable for a vagina opening in front of the cirrus-sac. The scolex is reminiscent of that of *Prosobothrium* Cohn, 1902 and partly of *Platybothrium auriculatum* Yamaguti, 1952. *Cyatocotyle* should be suppressed; it is treated here as a *genus inquirendum*.

Hoaleshwaria Shinde & Chincholikar, 1975. This genus with the type-species *H. marathwadensis* Shinde & Chincholikar (1975) was described from a single individual collected from *Trygon* sp. The species cannot be recognized either from its description or illustration. The host species being unknown, it cannot be obtained again. *Hoaleshwaria* should be suppressed; it is treated here as a *genus inquirendum*.

Pithophorus Southwell, 1925. Southwell (1925) described this cestode with four tubular bothridia open at each end from *Rhynchobatus djeddensis* (Forsskal, 1775) (Rajiformes, Rhynchobatidae). *P. tetraglobus* does not seem to have been recovered since its original description. The other species placed in *Pithophorus* are parasites of sharks (Orectolobiformes, Lamniformes, Carcharhiniformes). Their systematic position should be reviewed, particularly their relationship with the genera *Scyphophyllidium* Woodland, 1927 and *Marsupiobothrium* Yamaguti, 1952, also reported from sharks.

Polipobothrium Mola, 1908. The scolex of the type-species *P. vaccarii* Mola, 1908, has four bothridia each with a regular row of nine small suckers on the margin. From the illustration, the scolex is borne on a long cephalic peduncle. The anatomy conforms to the tetraphyllidean type. The type-host is a large shark, *Selache maxima* = *Cetorhinus maximus* (Günner, 1765). Since the work of Mola, no phyllobothriid corresponding to *Polipobothrium* has been found in *C. maximus*. The genus and species should be considered doubtful; it is treated as a *genus inquirendum*.

Reesium Euzet, 1955. Euzet erected the genus for *Dinobothrium paciferum* Sproston, 1948 from *Cetorhinus maximus* (Günner, 1765). The characters on which this separation was based (presence of strong spines on the strobila, structure of the ovary) are not of major significance at generic level. *Reesium* is considered to be a synonym of *Dinobothrium* van Beneden, 1889.

Spongiobothrium Linton, 1889. The scolex of the type-species, *S. variabile* Linton, 1889 from *Dasyatis centroura*, has four highly folded bothridia with a surface divided by septa into loculi but without an accessory sucker. The last character led Southwell (1925) to synonymize the genus with *Anthobothrium* but the synonymy is doubtful. *S. variabile* should be obtained again from the type-host to verify whether the genus is valid. The discovery in *D. pastinaca* of *Phyllobothrium pastinacae* Mokhtar-Maamouri & Zamali, 1981 with highly folded loculated bothridia, but with an accessory sucker, calls into question the exact structure of the bothridia of *Spongiobothrium*. It is treated here as a *genus inquirendum*.

Myliobatibothrium Shinde & Mohekar, 1983. The type-species, *M. alii* Shinde & Mohekar, 1983 was found in *Myliobatis nieuhofii* (Muller & Henle) at Ratnagiri on the west coast of India. The illustration given of the scolex makes this species very difficult to place. From the description, the bothridia are simple except for a row of eight loculi on the posterior border, but the diagram shows, in the centre of each bothridium, a pouch-like structure broadly open anteriorly. The anatomy is of the phyllobothriid type although the vagina has not been noted as anterior to the cirrus-sac. Shinde & Mohekar [*sic*] (1983) placed this genus close to *Duplicibothrium* Williams & Campbell, 1978. The description and illustration being inadequate, *Myliobatibothrium* is considered a *genus incertae sedis* pending the recovery and redescription of this parasite from material from the type-host. On the subject of this tetraphyllidean, there is a question of what is represented by the host name *Myliobatis nieuhofii* (Muller & Henle) quoted by Shinde & Mohekar (1983). Researches of the date of the description by Muller & Henle have failed to reveal any trace of this fish in the ichthyological literature, but *Aetomyleus nichofii* Schneider, 1801 has been found. What is the relationship between these two species?

Pelichnibothrium Monticelli, 1889. The three individuals described in this genus (Monticelli, 1889; Yamaguti, 1952; Euzet, 1959) correspond to a postlarval stage of the phyllobothriid tetraphyllideans. The scolex has four bothridia with a simple border and an accessory sucker, and an apical sucker as in all the plerocercoids of the *Scolex pleuronectis* type. Behind the scolex, a strobilate region is followed by a long appendix corresponding to the posterior vesicle of the plerocercoid. This posterior part is comparable with that which characterizes the large plerocercoids found in the intestine of cephalopods. The anatomy of proglottids of the strobilate region is characteristic of the Tetraphyllidea. It is not possible to say to which adult this postlarva corresponds.

Pillersium Southwell, 1927. The genus is known only from a single species, *P. oweni* Southwell, 1927, parasite of *Urogymnus asperrimus* (Bloch & Schneider, 1901). According to Southwell (1927), the scolex has only two bothridia with highly folded margins. The anatomy is not known. The systematic position of this strange cestode is difficult to determine. I believe it to be affiliated to the Phyllobothriidae because, from Southwell's illustration, it is possible that the Y-shaped scolex is composed of four bothridia fused into dorsoventral pairs.

Subfamily Triloculariinae Yamaguti, 1959

Kowsalyabothrium Muralidhar, Shinde & Jadhav, 1989. I place this genus, with some doubt, close to the Triloculariinae but it should be treated as a *genus inquirendum*. In this parasite of *Trygon centrura*, the bothridia with a distinct (?) accessory sucker resemble those of species of *Crossobothrium*. The same applies to the anatomy of the proglottids. The presence of a longitudinal septum dividing the bothridium into two, the main character of the genus, is to be confirmed. This will be difficult because the host cited is doubtful. Is *Trygon centrura* from India a synonym of *Dasyatis centroura* (Mitchill, 1815)? *D. centroura* is a species whose geographical distribution is limited to the North Atlantic. This stingray has never been reported from the Indian Ocean.

Subfamily Rhinebothriinae Euzet, 1953

Rhinebothrium Linton, 1890 currently contains about forty species (see Butler, 1987b) which do not correspond in all things, particularly in the structure of the bothridium, to the type-species. In *R. flexile* Linton, 1890, the surface of the bothridium is divided by muscular septa into two symmetrical rows of loculi plus an unpaired loculus at the extremity. From the structure of the bothridia, divisions will be proposed in the genus but only when the Rhinebothriinae described by Shipley & Hornell and by Southwell from Myliobatiformes in the Indian Ocean have been reviewed.

It would probably be necessary to include in the Rhinebothriinae the parasites of selachians initially described as *Echeneibothrium* although lacking a myzorhynchus. Several species currently placed in the genus *Anthobothrium*, which have a scolex with four bothridia with a surface partly or entirely divided by muscular septa, should be placed in the Rhinebothriinae. Examples are *A. taeniurae* Saoud, 1963 and *A. sexorchidum* Williams, 1964, both parasites of *Taeniura lymna* (Forsskal, 1875) (Dasyatidae) and *A. amuletum* Butler, 1987, a parasite of *Rhinobatus armatus* (Rhinobatidae).

Shindeiobothrium Jadhav, Shinde & Deshmukh, 1981. The scolex of this cestode has four bothridia said to be sessile, which is difficult to verify given the apparent state of compression of the scolex. These bothridia may correspond to the *Rhinebothrium* type but the presence of a large central space renders this doubtful. Regarding the distribution of the testes in the proglottid, there is a contradiction between the description (preovarian testes) and the illustration in which the testes descend on each side of the ovary almost as far as the posterior extremity of the proglottid. Pending a redescription of the type-species, *S. karbharae*, a parasite of *Trygon sephen* = *Hypolophus sephen* (Forsskal, 1775), *Shindeiobothrium* is considered a *genus inquirendum*. The type-species is close to *Echeneibothrium ceylonicum* Shipley & Hornell, 1906, a parasite of *T. kuhlii* (= *Dasyatis kuhlii*) (Muller & Henle, 1841).

Two species of tetraphyllideans described under *Echeneibothrium* may in fact represent distinct genera.

Echeneibothrium smitii Shinde, Deshmukh & Jadhav, 1981. This parasite from the spiral valve of *Trygon sephen* = *Hypolophus sephen* (Forsskal, 1775) on the west coast of India, should be recovered again for a more detailed study. The original description, based on two individuals, states that the centre of the scolex has a myzorhynchus, the presence of which is not evident in the illustration (Shinde *et al.*, 1981). If this observation is accurate, this parasite would represent the first *Echeneibothrium* found in Myliobatiformes. In *E. smitii*, the bothridia, which have a single row of seven loculi, are carried at the extremity of a very long pedicel (four times the length of the bothridium) which forms an enlargement at the base of the bothridium. The anatomy is of the *Rhinebothrium* or *Echeneibothrium* type without postvaginal testes on the poral side. If this species is found again, the unique morphology of the scolex would characterize a new genus. This parasite, which is a phyllobothriid, was erroneously classified by its authors in the Onchobothriidae.

Echeneibothrium javanicum Shipley & Hornell, 1906. This species, parasite of *Rhinoptera javanica*, considered by Southwell as a synonym of *Rhinebothrium cancellatum* Linton, 1890, was placed in the genus *Echeneibothrium* although the scolex has no myzorhynchus. The synonymy being incorrect, *E. javanicum* appears to be a species which should be placed in the Rhinebothriinae, since the bothridia are subdivided into numerous loculi by muscular septa. However, the distribution of the loculi is unique. From the description and illustration, two regions can be distinguished in the bothridia: the anterior region divided into three longitudinal rows of loculi and the posterior into one transverse row of seven loculi. Unfortunately, the anatomy is poorly known. This cestode appears close to *Duplicibothrium* Williams & Campbell, 1978. In this parasite of *Rhinoptera bonasus*, the bothridium is also divided into two parts but the anterior region has only a single longitudinal row of loculi (Figs 8.1, 8.2). The anatomy of *Duplicibothrium* is unusual, with testes throughout the length of the proglottid (anterior and posterior to the ovary) and circum-medullary vitelline follicles. I retain *Echeneibothrium javanicum* as a *species incertae sedis* pending its recovery from the type-host, *Rhinoptera javanica* Muller & Henle, 1841, in the type locality.

Tiarabothrium Shipley & Hornell, 1906. The type-species, *T. javanicum* Shipley & Hornell, 1906, is a parasite of *Rhinoptera javanica*. The scolex has four sessile bothridia divided into twelve loculi by transverse septa. The bothridia are on the surface of a voluminous central mass of dense connective tissue. From the illustration by Shipley & Hornell (1906), this mass does not have the structure of a myzorhynchus. Southwell (1925) synonymized *Tiarabothrium* with *Echeneibothrium* but, given the systematic position of the host (Dasyatidae), I think it should be placed with Rhinebothriinae. Known only from two individuals, this parasite should be obtained again from the type-host and revised to determine its exact status and position. It is treated here as a *genus inquirendum*.

Maccallumiella Yamaguti, 1959. Yamaguti (1959) created for *Taenia patini* MacCallum, 1921, from *Arius* sp. (Teleostei, Siluriformes), the genus *Maccallumiella* and the family Maccallumiellidae which he tentatively appended to the Tetraphyllidea. The nature of the host and the route of the vagina posterior to the cirrus-sac suggest that this may be a proteocephalidean. It is treated here as a *genus incertae sedis*.

Mastacembellophyllaeus Shinde & Chincholikar, 1977 should be suppressed because it is based on an isolated tetraphyllidean proglottid. It is treated here as a *genus inquirendum*.

Platybothrium Hornell, 1923 is represented by a larva. It cannot be assimilated to *Platybothrium* Linton, 1890 (Onchobothriidae) and the characters of the larva are not sufficiently distinctive for the creation of a new genus. It is considered to be a *genus inquirendum*.

Dittocephalus Parona, 1887. This cestode has an extraordinary scolex with two suckers (bothria, bothridia?) each borne on a trisegmented peduncle. What little is known of the anatomy indicates that it is of the pseudophyllidean type because the genital pore is mid-ventral. The operculated egg also resembles that of the Pseudophyllidea. The selachian host is doubtful. The genotype is unknown and the taxon is not a tetraphyllidean. It is regarded as a *genus inquirendum*.

Phanobothrium Mola, 1907. The host is unknown but was presumed to be a selachian. The anatomy is tetraphyllidean with the vagina anterior to the cirrus-sac. The spinous scolex has four oblong suckers and a single large apical organ like that of some proteocephalideans. The single vitelline gland described by Mola (1907) could be Mehlis's gland. The genus is treated here as a *genus inquirendum*.

Neolitobothrium Bilqees, 1983 is represented by an individual from the spiral valve of *Myrmillo manazo* (Mustelidae). The route of the vagina was not described but the anatomy indicates that this is a tetraphyllidean. *N. hexalaciniatum* Bilqees, 1983 may be only a fragment of a *Calliobothrium* sp. which has lost its scolex and the anterior region of the strobila. This has been observed in many individuals of *C. verticillatum* collected *post mortem* from *Mustelus mustelus*. *Neolitobothrium* should be considered a *genus inquirendum*.

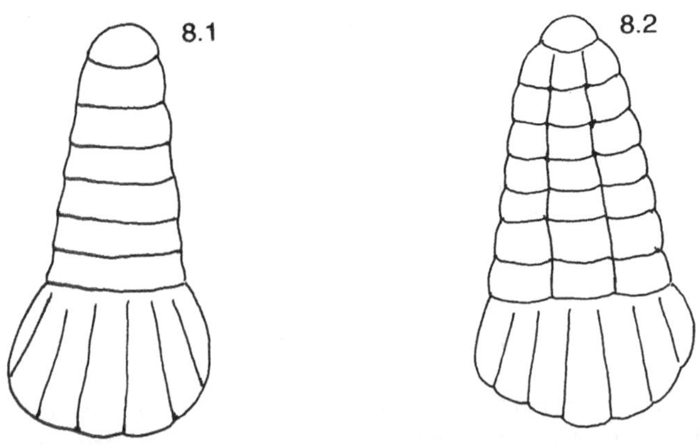

Fig. 8.1 Septa and loculi in bothridium of *Duplicibothrium*. (After Williams & Campbell, 1978.)
Fig. 8.2 Septa and loculi in bothridium of *Echeneibothrium javanicum*. (After Shipley & Hornell, 1906.)

ORDER TETRAPHYLLIDEA CARUS, 1863

Diagnosis: Scolex with four sessile or pedunculated bothridia of various shapes; hooks present or absent. Rarely, scolex lacks bothridia, suckers and armature; or has four sessile glandular discs; or may be cushion-like with a complex velum; or consists of a single apical sucker. Metascolex sometimes present. Strobila usually hermaphroditic, sexes rarely separate; usually single, rarely multiple; craspedote or not; anapolytic, euapolytic or rarely hyperapolytic. Genital pores lateral or sublateral, irregularly alternating. Testes numerous; postvaginal testes present or not on poral side. Ovary posterior, bilobed or tetralobed in cross-section. Vagina anterior to cirrus-sac crossing the vas deferens. Vitelline follicles medullary, usually lateral, sometimes encircling proglottid, rarely condensed around the ovary. Uterus medioventral; uterine pores sometimes present, median. Eggs may be grouped in cocoons. Adults in spiral valve of elasmobranchs and holocephalids.

Key to families

1a. Scolex with one attachment organ .... 2.
1b. Scolex with four attachment organs .... 3.

2a. Scolex a large organ lacking bothridia. One to 24 strobilae per scolex .... Cathetocephalidae Dailey & Overstreet, 1973.
2b. Scolex a single sucker, followed by pseudoscolex modified from anterior proglottids .... Litobothriidae Dailey, 1969.

3a. Scolex with four organs forming a crumpled collar and one globular apical organ embedded in gut wall of host .... Disculicipitidae Joyeux & Baer, 1936.
3b. Scolex with only four organs .... 4.

4a. Scolex with four cup-shaped glandular organs .... Prosobothriidae Baer & Euzet, 1955.
4b. Scolex with four muscular organs (bothridia) .... 5.

5a. Strobila with sexes completely separate .... Dioecotaeniidae Schmidt, 1969.
5b. Strobila with hermaphroditic proglottids .... 6.

6a. Bothridia with hooks .... Onchobothriidae Braun, 1900.
6b. Bothridia unarmed .... 7.

7a. Strobila with vitelline follicles in two lateral fields throughout proglottid .... Phyllobothriidae Braun, 1900.
7b. Strobila with vitelline follicles confined to two lateral fields in posterior part of proglottid .... Chimaerocestidae Williams & Bray, 1984.

Family Cathetocephalidae Dailey & Overstreet, 1973

Diagnosis: Scolex a single transverse fleshy organ lacking bothridia, suckers or armature. Apical surface highly rugose with fleshy papilliform projections on leading and trailing edges. Posterior surface of scolex smooth with central transverse fold extending entire width. One to 24 strobila per scolex. Strobila slightly craspedote, apolytic. Genital pores lateral, irregularly alternating. Testes numerous. Large ovary, median, posterior, bilobed in cross-section. Vagina anterior to cirrus-sac. Vitelline follicles circum-medullary. Type and only genus *Cathetocephalus* Dailey & Overstreet, 1973.

Genus *Cathetocephalus* Dailey & Overstreet, 1973 (Figs 8.3, 8.4)

Diagnosis: With the characters of the family. In Carcharinidae. Cosmopolitan. Type-species *C. thatcheri* Dailey & Overstreet, 1973.

Family Litobothriidae Dailey, 1969

Diagnosis: Scolex (pseudoscolex?) a single apical sucker followed by anterior modified proglottids. Neck present or absent. Strobila craspedote and laciniated. Euapolytic. Genital pores lateral, irregularly alternating. Testes numerous; postvaginal testes present on poral side. Ovary median, posterior, bilobed or tetralobed in cross-section. Vagina anterior to cirrus-sac. Vitelline follicles circum-

8.3

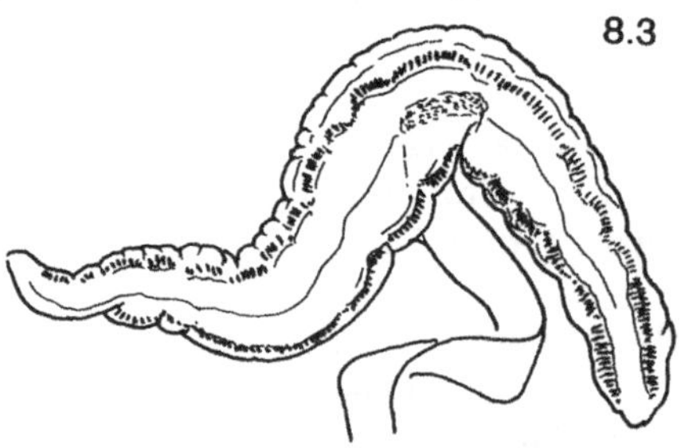

8.4

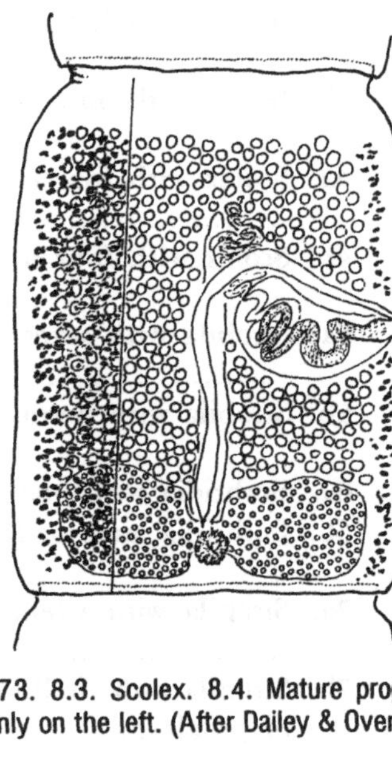

Figs 8.3, 8.4 *Cathetocephalus* Dailey & Overstreet, 1973. 8.3. Scolex. 8.4. Mature proglottid, dorsal view. Vitelline follicles, circum-medullary, drawn only on the left. (After Dailey & Overstreet, 1973.)

medullary, extending throughout length of proglottid including posterior to ovary. Uterus medioventral, from ovary to level of cirrus-sac. Type and only genus *Litobothrium* Dailey, 1969.

Genus *Litobothrium* Dailey, 1969 (Figs 8.5–8.7)

(Syn. *Renyxa* Kurochkin & Slankis, 1973)

Diagnosis: With the characters of the family. In Lamniformes (Alopiidae, Odontaspididae, Mitsukurinidae). Pacific Ocean. Type-species *L. alopias* Dailey, 1959.

Renyxa Kurochkin & Slankis, 1973 is synonymized with *Litobothrium*. Although the number and morphology of the proglottids following the apical sucker may be good criteria for separating species, they are insufficient to distinguish two genera among these tetraphyllidean parasites of *Alopias*. I note

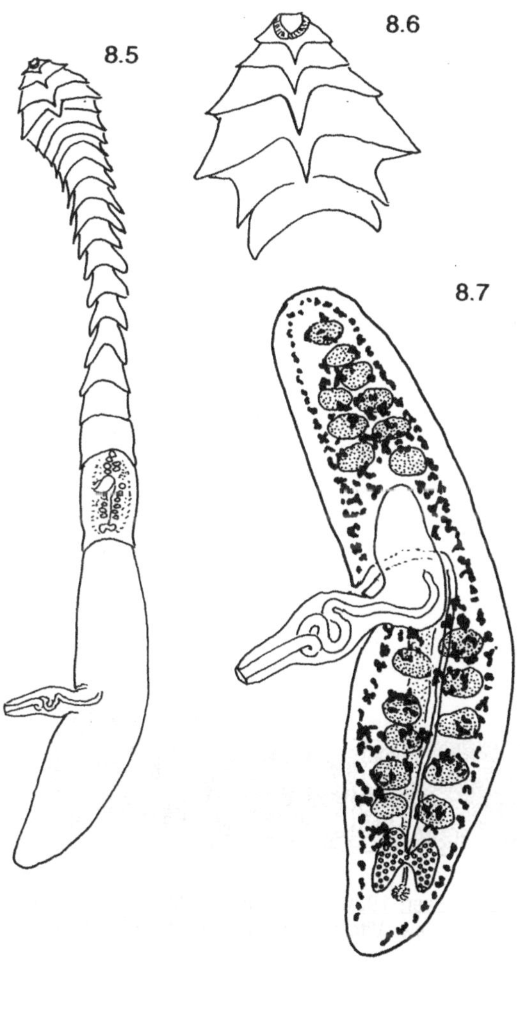

Figs 8.5–8.7 *Litobothrium* Dailey, 1969. 8.5. Strobila *in toto*. 8.6. Anterior part of strobila showing apical sucker and modified proglottid. 8.7. Mature proglottid, dorsal view. (After Dailey, 1969.)

the resemblance of the strobila to that of *Eniochobothrium* Shipley & Hornell, 1906 (see Chapter 9).

Family Disculicipitidae Joyeux & Baer, 1936

(Syn. Discocephalidae Pintner, 1928)

Diagnosis: Scolex divided into large cushion-like region variable in size, embedded within intestinal wall and followed by complex, frilled velum, probably in four parts (bothridia?). Strobila acraspedote, anapolytic. Genital pores sublateral ventral, irregularly alternating. Testes very numerous; postvaginal testes present on poral side. Ovary medioposterior, bilobed in cross-section. Vagina anterior to cirrus-sac. Vitelline follicles circum-medullary. Uterus median with lateral outgrowths. Type and only genus *Disculiceps* Joyeux & Baer, 1936.

Genus *Disculiceps* Joyeux & Baer, 1936 (Figs 8.8–8.10)

(Syn. *Discocephalum* Linton, 1890, preoccupied)

Diagnosis: With the characters of the family. In Carcharinidae. Cosmopolitan. Type-species *D. pileatum* (Linton, 1890).

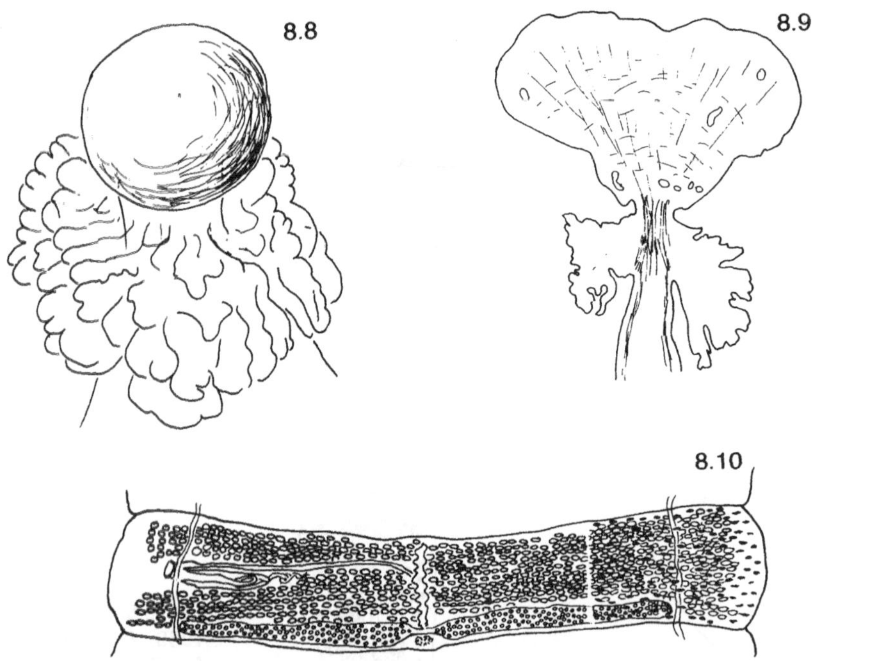

Figs 8.8–8.10 *Disculiceps* Joyeux & Baer, 1936. 8.8. Scolex (original). 8.9. Longitudinal section of the scolex. 8.10. Anatomy, ventral view. Vitelline follicles circum-medullary, drawn only on the right. (After Nock & Caira, 1988.)

Family Prosobothriidae Baer & Euzet, 1955

Diagnosis: Scolex with four sessile glandular discs. Neck densely spinose. Strobila acraspedote, euapolytic. Genital pores lateral, irregularly alternating. Testes numerous; postvaginal testes present on poral side. Ovary posterior, bilobed in cross-section. Vagina anterior to cirrus-sac. Vaginal sphincter present. Vitelline follicles circum-medullary (some postovarian). Eggs (24) clustered in cocoon. Type and only genus *Prosobothrium* Cohn, 1902.

Genus *Prosobothrium* Cohn, 1902 (Figs 8.11–8.13)
(Syn. *Lintoniella* Woodland, 1927)

Diagnosis: With the characters of the family. In Carcharinidae. Cosmopolitan. Type-species *P. armigerum* Cohn, 1902.

Family Dioecotaeniidae Schmidt, 1969

Diagnosis: Sexes separated. Strobilae dimorphic in size and shape. Scolex with four bothridia, each subdivided into loculi by transverse and longitudinal septa. Myzorhynchus absent. **Male strobila**. Proglottids with genital pores lateral irregularly alternating. Testes in circle in two layers on both sides of cirrus-sac. External seminal vesicle absent, internal seminal vesicle present. Cirrus long, spinose. **Female strobila**. Proglottid with posterior bilobed ovary. Vaginal pore absent. Vagina median, convoluted. Seminal receptacle present in lobe of ovary, irregularly alternating. Two vitelline glands surrounding lobes of ovary laterally.

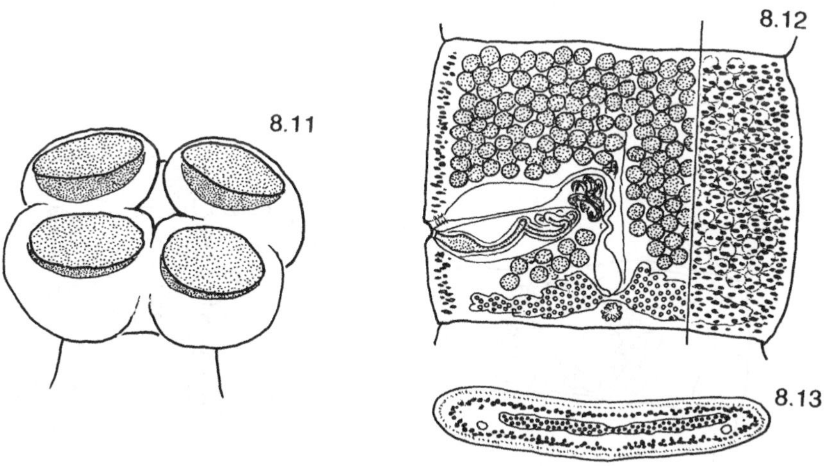

Figs 8.11–8.13 *Prosobothrium* Cohn, 1902. 8.11. Scolex. 8.12. Mature proglottid, dorsal view, vitelline follicles circum-medullary, drawn only on the right. 8.13. Transverse section through the ovarian region. (After Euzet, 1959.)

Uterus a bilobed transverse sac anterior to ovary. Sperm transfer by hypodermic impregnation. Type and only genus *Dioecotaenia* Schmidt, 1969.

Genus *Dioecotaenia* Schmidt, 1969. (Figs 8.14–8.17)

Diagnosis: With the characters of the family. In Rhinopteridae. American North Atlantic. Type-species *D. cancellata* (Linton, 1890).

Family Onchobothriidae Braun, 1900

Diagnosis: Scolex with four sessile muscular bothridia each armed with simple or forked hooks; bothridial surface simple or divided by transverse septa, anterior muscular pad with or without sucker. Strobila acraspedote or craspedote; apolytic or euapolytic. Genital pores lateral, irregularly alternating. Testes numerous; postvaginal testes present or absent on poral side. Ovary posterior, bilobed or tetralobed in cross-section. Uterus medioventral; uterine pores sometimes present, median. Eggs may be grouped in cocoons. Adults in spiral valve of elasmobranchs. Type-genus *Onchobothrium* De Blainville, 1828.

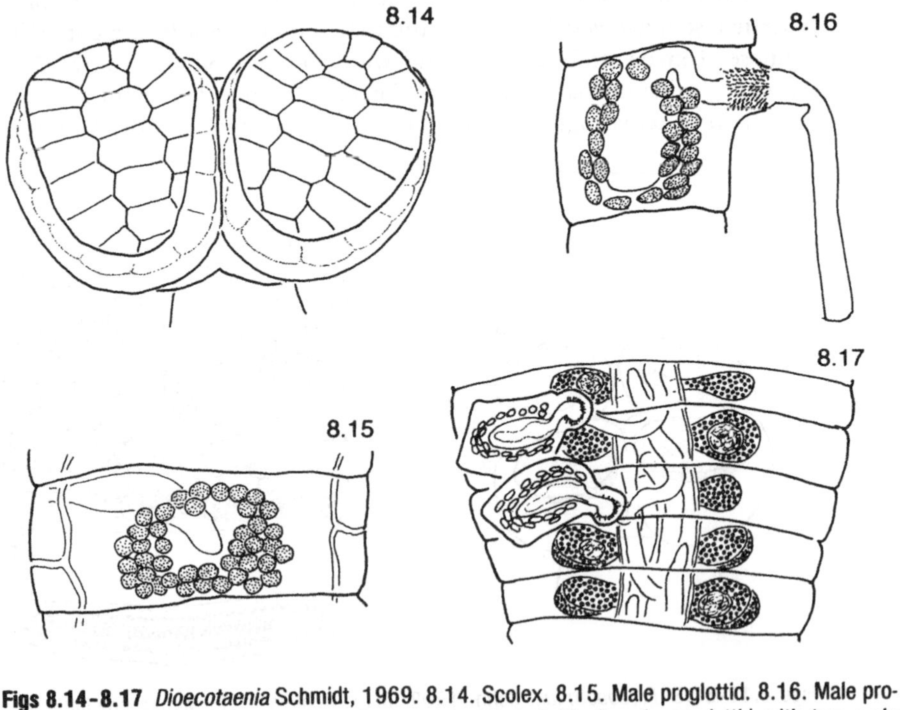

Figs 8.14–8.17 *Dioecotaenia* Schmidt, 1969. 8.14. Scolex. 8.15. Male proglottid. 8.16. Male proglottid with cirrus evaginated and seminal vesicle filled. 8.17. Female proglottid with two male proglottids attached by hypodermic impregnation. (After Schmidt, 1969.)

Key to genera

1a. **Bothridia divided by one or two transverse septa** .... **2.**
1b. **Bothridia undivided** .... **10.**

2a. **Bothridia divided by two transverse septa** .... **3.**
2b. **Bothridia divided by one transverse septum** .... **8.**

3a. **Bothridia with two pairs of hooks** ....
.... ***Calliobothrium* van Beneden, 1850. (Figs 8.18, 8.19)**
Diagnosis: Scolex with four bothridia each divided into three loculi by two transverse septa. In front of the bothridium a pad with one or three suckers and two pairs of simple articulated hooks. Strobila craspedote, laciniated or not, euapolytic. Genital pore lateral, irregularly alternating. Testes numerous, in two lateral fields; postvaginal testes present on poral side. Ovary posterior. Vagina anterior to cirrus-sac. Vitelline follicles lateral. Uterus sac-shaped, medioventral. Uterine pores medioventral. Eggs grouped in cocoon. In Mustelidae. Cosmopolitan. Type-species *C. verticillatum* (Rudolphi, 1819).

3b. **Bothridia with one pair of hooks** .... **4.**

4a. **Paired hooks similar** .... **5.**
4b. **Paired hooks dissimilar** .... **6.**

5a. **Hooks with one prong, base joined** ....
.... ***Onchobothrium* De Blainville, 1828. (Figs 8.20, 8.21)**

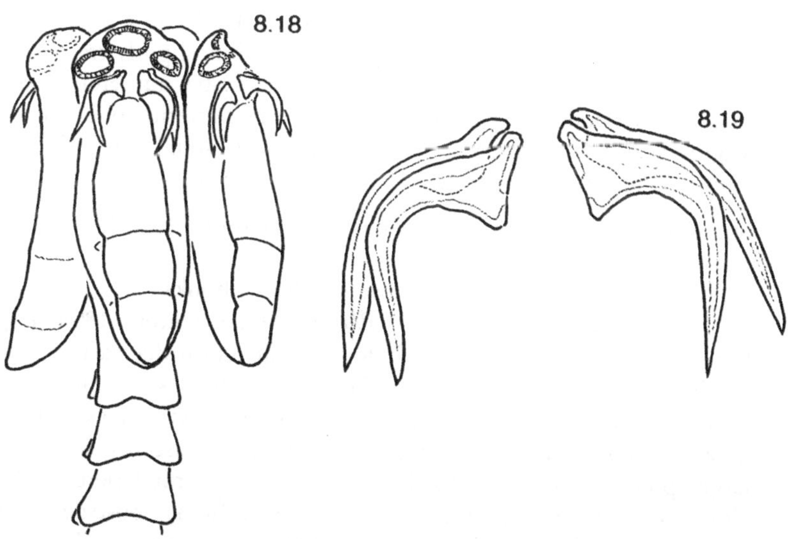

Figs 8.18, 8.19 *Calliobothrium* van Beneden, 1850. 8.18. Scolex. 8.19. Hooks. (After Euzet, 1959.)

Diagnosis: Scolex with four bothridia, each divided into three loculi by two transverse septa. In front of each bothridium a muscular pad with one accessory sucker and two thorn-shaped hooks. Pair of hooks joined at their base by a horseshoe-like plate. Bases with tubercles or hair-like process. Strobila acraspedote, apolytic. Genital pore lateral, irregularly alternating. Testes numerous in two lateral fields; postvaginal testes present on poral side. Vagina anterior to cirrus-sac. Ovary posterior, tetralobed in cross-section. Vitelline follicles lateral. Uterus medioventral. Eggs spherical. In Rajidae. Cosmopolitan. Type-species *O. uncinatum* (Rudolphi, 1819).

5b. Hooks forked with two prongs .... ***Acanthobothrium* van Beneden, 1849. (Figs 8.22–8.24)** (Syns *Acrobothrium* Baer, 1948; *Euonchobothrium* Diesing, 1854; *Petalocephalus* Jeude, 1829; *Petalostoma* Jeude, 1829; *Prosthecobothrium* Diesing, 1863.)

Diagnosis: Scolex with four bothridia, each divided into three loculi by two transverse septa; in front of each bothridium a muscular pad with accessory sucker and pair of symmetrically forked hooks. Cephalic peduncle sometimes present. Strobila acraspedote, euapolytic. Genital pore lateral, irregularly alternating. Testes numerous; postvaginal testes present on poral side. Cirrus spinose. Ovary posterior, tetralobed in cross-section. Vagina anterior to cirrus-sac. Vitelline follicles lateral. Uterus sac-shaped, medioventral. Eggs spherical. In elasmobranchs. Cosmopolitan. Type-species *A. coronatum* (Rudolphi, 1819).

6a. Hooks dissimilar, lateral hook with one prong, median hook forked with two prongs .... ***Acanthobothroides* Brooks, 1977. (Figs 8.25–8.27)**

Diagnosis: Scolex with four sessile bothridia, each divided into three loculi by two transverse septa. In front of each bothridium, a muscular pad with accessory sucker and pair of dissimilar hooks. Medial hook with base and single prong, lateral hook with handle and two prongs. Strobila slightly craspedote, euapolytic. Genital pores lateral, irregularly alternating. Testes numerous; postvaginal testes present on poral side. Cirrus spinose. Ovary posterior bilobed in cross-section. Vagina anterior to cirrus-sac. Vitelline follicles lateral. In freshwater Dasyatidae. South America. Type-species *A. thorsoni* Brooks, 1977.

6b. Hooks forked with two prongs but dissimilar handles .... 7.

7a. One sclerotized plate between handle of hooks .... ***Platybothrium* Linton, 1890. (Figs 8.28, 8.29)**

Diagnosis: Scolex spinose with four sessile bothridia each divided into three loculi by two transverse septa. Anterior loculus with one lateral lobule. In front of each bothridium a muscular pad with accessory sucker and two dissimilar hooks. Hooks with two prongs and unequal base. Neck and strobila with large spines. Strobila acraspedote, euapolytic. Genital pores lateral, irregularly alternating. Testes numerous in two lateral fields; postvaginal testes present or not on poral side. Cirrus spinose. Ovary posterior, bilobed in cross-section. Vagina anterior to cirrus-sac. Vitelline follicles circum-medullary. Uterus

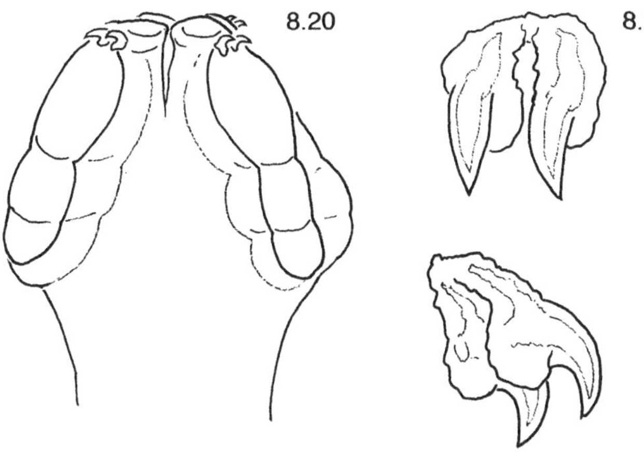

Figs 8.20, 8.21 *Onchobothrium* De Blainville, 1828. 8.20. Scolex. 8.21. Hooks, ventral and lateral view. (After Euzet, 1959.)

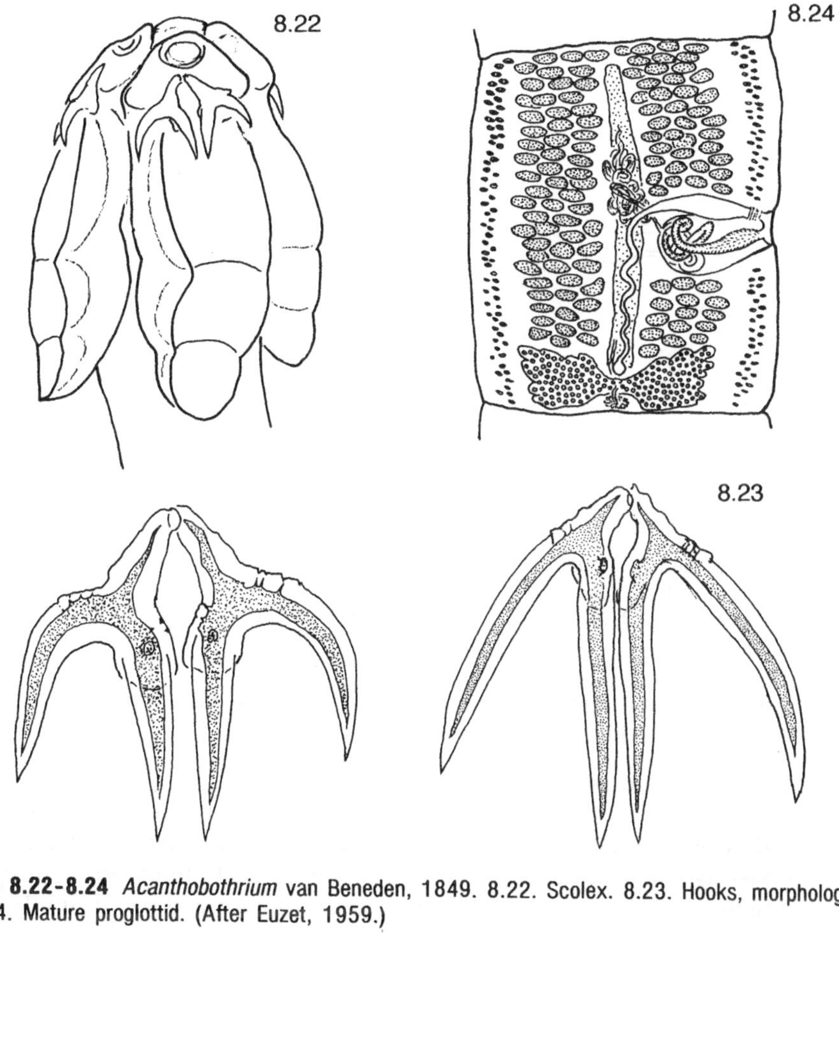

Figs 8.22–8.24 *Acanthobothrium* van Beneden, 1849. 8.22. Scolex. 8.23. Hooks, morphology. 8.24. Mature proglottid. (After Euzet, 1959.)

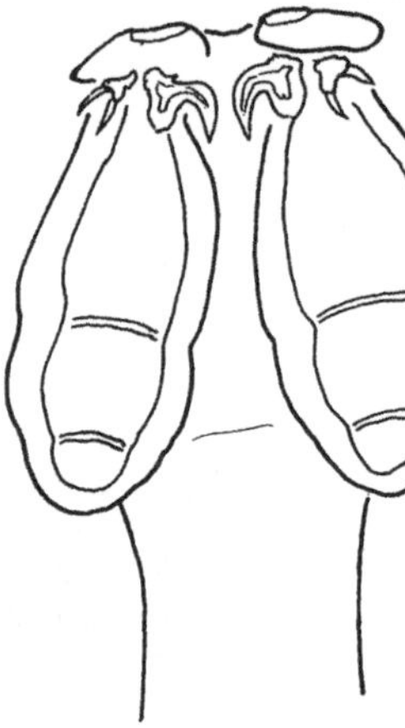

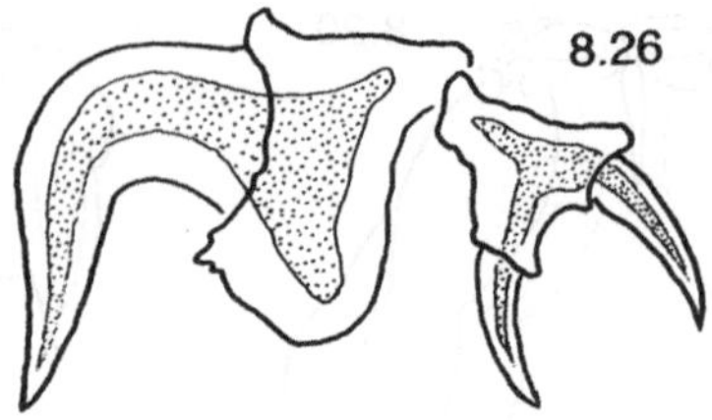

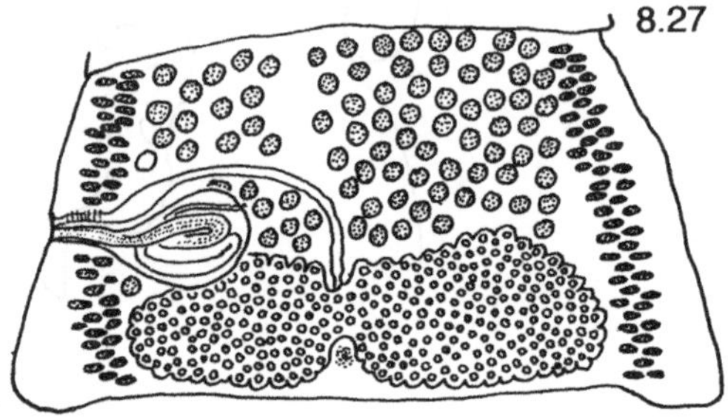

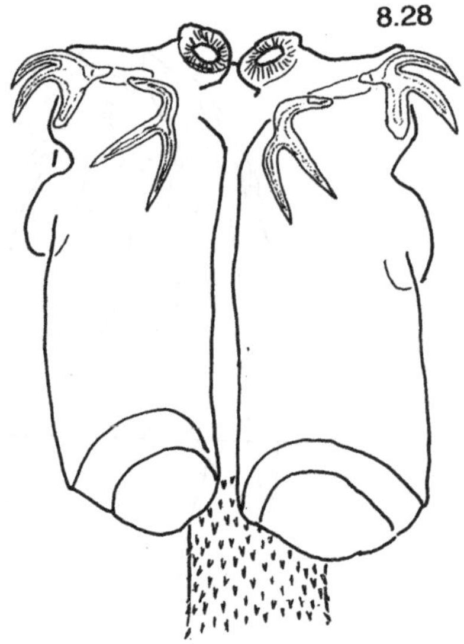

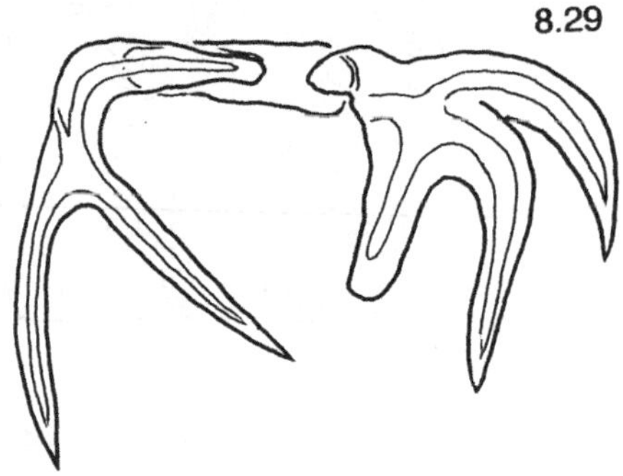

Figs 8.25–8.27 *Acanthobothroides* Brooks, 1977. 8.25. Scolex. 8.26. Hooks. 8.27. Mature proglottid. (After Brooks, 1977.)
Figs 8.28, 8.29 *Platybothrium* Linton, 1890. 8.28. Scolex. 8.29. Hooks. (After Euzet, 1959.)

medioventral. Eggs subspherical. In Carcharinidae. Cosmopolitan. Type-species *P. cervinum* Linton, 1890.

7b. No plate between handle of hooks ....
.... ***Dicranobothrium* Euzet, 1953. (Figs 8.30, 8.31)**

Diagnosis: Scolex with four sessile bothridia, each divided into three loculi by two transverse septa. Anterior loculus with lateral minute lobe. In front of each bothridium, a muscular pad with accessory sucker and two dissimilar hooks with

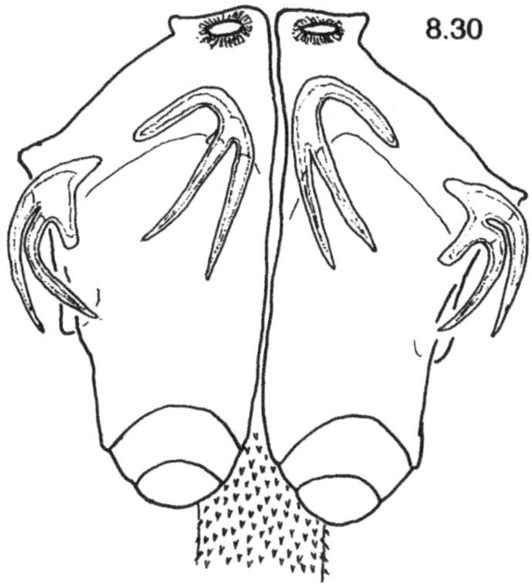

8.30

8.31

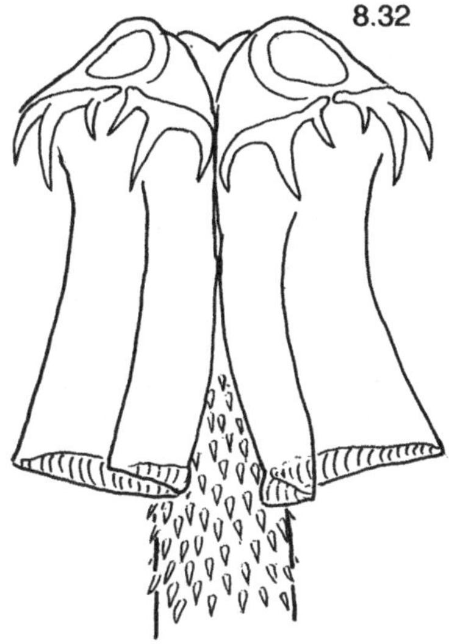

8.32

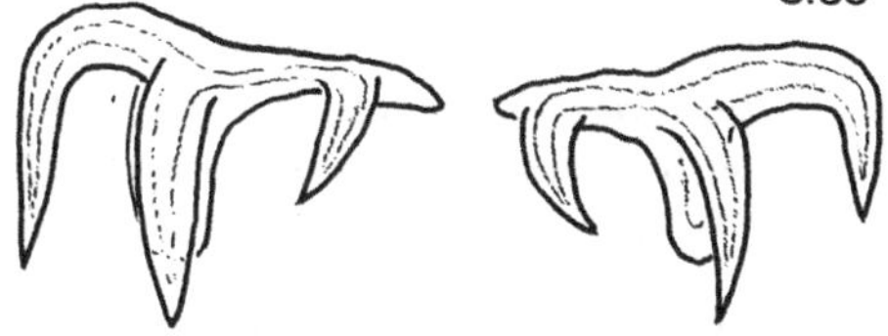

8.33

Figs 8.30, 8.31 *Dicranobothrium* Euzet, 1953. 8.30. Scolex. 8.31. Hooks (no plate between hooks). (After Euzet, 1959.)
Figs 8.32, 8.33 *Phoreiobothrium* Linton, 1889. 8.32. Scolex. 8.33. Hooks. (After Euzet, 1959.)

two prongs. Neck and strobila spinose. Strobila acraspedote, euapolytic. Genital pore lateral, irregularly alternating. Testes numerous, in two lateral fields; postvaginal testes present on poral side. Ovary posterior, bilobed in cross-section. Vagina anterior to cirrus-sac. Vitelline follicles circum-medullary. Uterus medio-ventral. In Carcharinidae. Cosmopolitan. Type-species *D. spinulifera* (Southwell, 1912).

8a. Bothridia with loculated posterior compartment ....
.... *Phoreiobothrium* Linton, 1889. (Figs 8.32, 8.33)

Diagnosis: Scolex with four bothridia. In front of each a muscular pad with accessory sucker and one pair of hooks with three (or two) unequal prongs. Bothridia simple with posterior margin divided into single row of loculi. Neck present, with large spines. Strobila acraspedote, euapolytic. Genital pore lateral, irregularly alternating. Testes numerous, in two lateral fields; postvaginal testes present on poral side. Ovary posterior, bilobed in cross-section. Vagina muscular, anterior to cirrus-sac. Vitelline follicles lateral. Uterus sac-shaped, medioventral. In Carchariniformes. Cosmopolitan. Type-species *P. lasium* Linton, 1889.

8b. Bothridia with one posterior compartment, not loculated .... 9.

9a. Bothridium with two forked hooks each with small handle .... *Uncibilocularis* Southwell, 1925. (Figs 8.34, 8.35)

Diagnosis: Scolex with four sessile bothridia each divided into two loculi by one transverse septum. In front of each bothridium one small muscular pad (accessory sucker?) and one pair of symmetrical forked hooks with unequal prongs. Strobila acraspedote, apolytic. Genital pores lateral, irregularly alternating. Testes few, in two lateral fields; postvaginal testes present on poral side. Cirrus spinose. Ovary posterior. Vagina anterior to cirrus-sac. Vitelline follicles lateral. Uterus ventral with lateral projections. Eggs spherical. In Myliobatiformes. Indian and Pacific Oceans. Type-species *U. trygonis* Shipley & Hornell, 1927.

9b. Bothridia with one pair of hooks, each hook with two prongs and large handle .... *Megalonchos* Baer & Euzet, 1962. (Figs 8.36, 8.37)

Diagnosis: Scolex with four bothridia, each divided into two unequal loculi by transverse septum. In front of each bothridium, one pair of forked hooks dissimilar in size with very long handle. Strobila acraspedote, apolytic. Genital pores lateral, irregularly alternating. Testes numerous; postvaginal testes absent

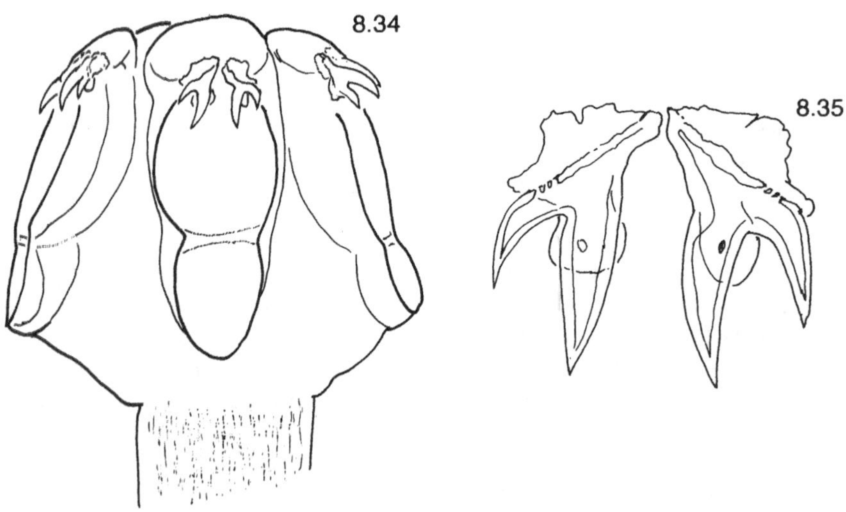

Figs 8.34, 8.35 *Uncibilocularis* Southwell, 1925. 8.34. Scolex. 8.35. Hooks. (Original.)

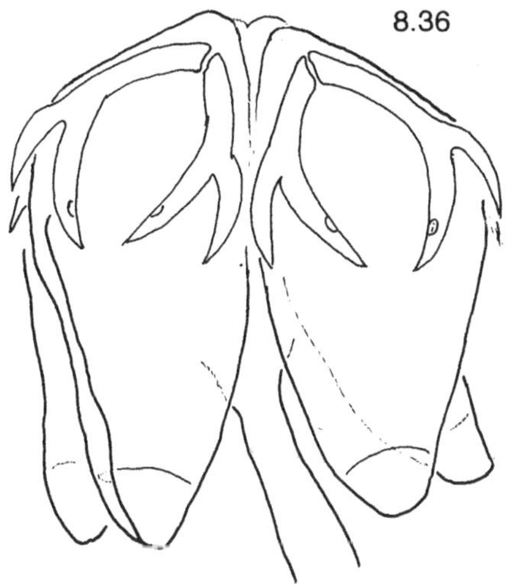

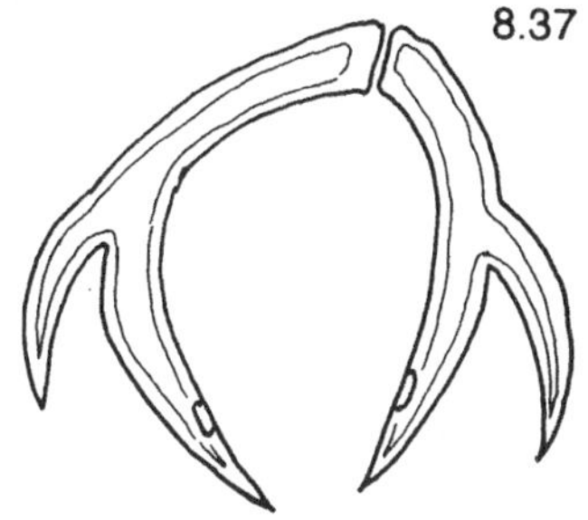

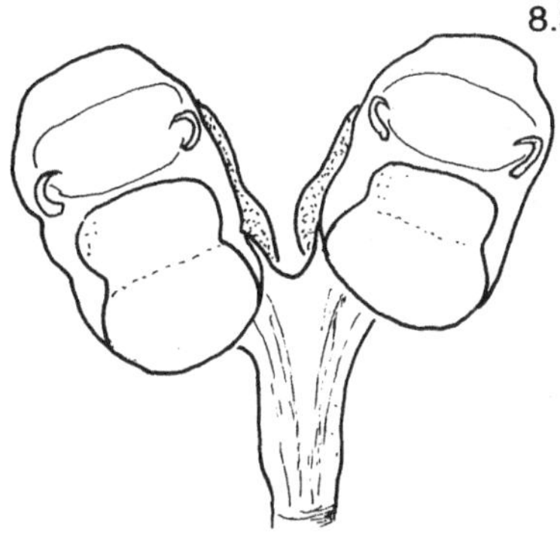

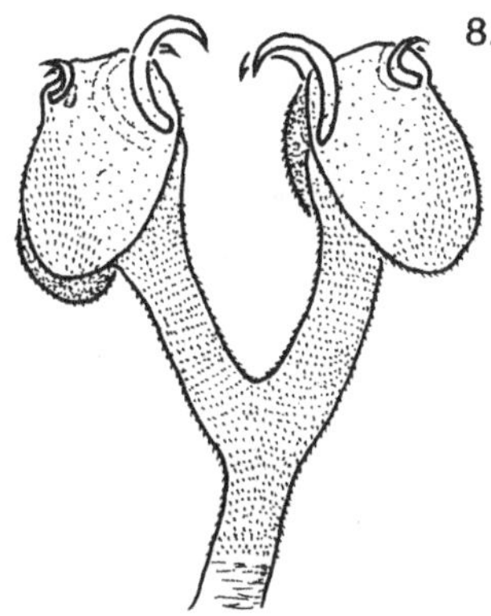

Figs 8.36, 8.37 *Megalonchos* Baer & Euzet, 1962. 8.36. Scolex. 8.37. Hooks. (After Southwell, 1930a.)
Fig. 8.38 *Spiniloculus* Southwell, 1925. Scolex. (After Williams, 1964.)
Fig. 8.39 *Yorkeria* Southwell, 1930. Scolex. (After Williams, 1964.)

on poral side. Ovary posterior. Vagina anterior to cirrus-sac. Vitelline follicles? In *Hemigaleus balfouri* (Carchariniformes). Sri Lanka. Type-species *M. mandleyi* (Southwell, 1927).

10a. Scolex forked, bothridia paired, united dorsoventrally .... 11.
10b. Scolex not forked, with anterior muscular pad .... 12.

11a. Bothridia with two C-shaped hooks of equal size. One large accessory sucker .... *Spiniloculus* Southwell, 1925. (Fig. 8.38)
(Syn. *Spinibiloculus* Deshmukh & Shinde, 1980)

Diagnosis: Scolex forked, spinose, with four simple bothridia each with one very large accessory sucker. One pair of C-shaped hooks situated at the posterolateral corner of accessory sucker. Strobila acraspedote, euapolytic. Genital pores lateral, irregularly alternating. Testes numerous; postvaginal testes present on poral side. Ovary posterior. Vagina anterior to cirrus-sac. Vitelline follicles lateral. Uterus medioventral, extending from ovary to level of cirrus-sac. In Orectolobiformes. Indian and Pacific Oceans. Type-species *S. mavensis* Southwell, 1925.

11b. Bothridia with two C-shaped hooks of different size. One small accessory sucker .... *Yorkeria* Southwell, 1927. (Fig. 8.39)

Diagnosis: Scolex forked, spinose, with four bothridia in two pairs united dorsoventrally. Bothridium subcircular with one anterior muscular pad; two C-shaped hooks unequal in size, the larger medial. Strobila acraspedote, euapolytic. Genital pores lateral, irregularly alternating. Testes numerous, postvaginal testes absent on poral side. Cirrus spinose. Ovary about one-quarter from posterior end of proglottid, four-lobed in cross-section. Vagina anterior to cirrus-sac. Vitelline follicles lateral. Uterus medioventral, extending only to level of cirrus-sac. In Orectolobiformes. Indian and Pacific Oceans. Type-species *Y. parva* Southwell, 1927.

12a. Bothridium with one pair of simple hooks .... **13.**
12b. Bothridium with one pair of forked hooks .... **14.**

13a. Anterior muscular pad globular. One pair of hooks, each hook rosethorn-shaped, embedded in anterior pad .... *Pachybothrium* Baer & Euzet, 1962. (Figs 8.40, 8.41)

Diagnosis: Scolex with four large and thick sessile bothridia, uniloculate, armed with one pair of simple hooks in centre of anterior muscular pad. Neck present. Strobila acraspedote, euapolytic. Genital pores lateral, irregularly alternating. Testes numerous, postvaginal testes absent on poral side. Ovary posterior, bilobed in cross-section. Vagina anterior to cirrus-sac. Vitelline follicles lateral. Uterus medioventral between ovary and cirrus-sac. In Orectolobiformes. Indian and Pacific Oceans. Type-species *P. hutsoni* (Southwell, 1911).

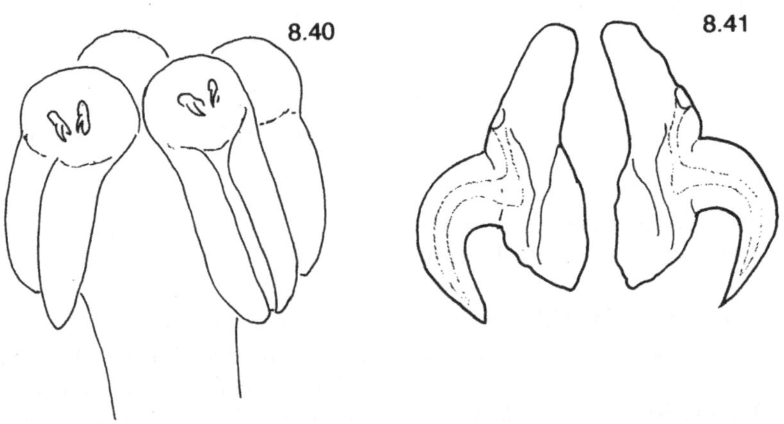

Figs 8.40, 8.41 *Pachybothrium* Baer & Euzet, 1962. 8.40. Scolex. 8.41. Hooks. (After Baer & Euzet, 1962.)

13b. Anterior accessory sucker. Bothridium with one pair of simple hooks .... *Potamotrygonocestus* Brooks & Thorson, 1976. (Figs 8.42–8.44)

Diagnosis: Scolex with four sessile non-septate bothridia. Each with accessory sucker and pair of symmetrical hooks, each hook with one prong. Scolex and neck spinose. Strobila acraspedote, euapolytic. Genital pores lateral, irregularly alternating. Testes numerous, in two lateral fields; postvaginal testes absent on poral side. Ovary posterior, bilobed in cross-section. Vagina anterior to cirrus-sac. Vitelline follicles lateral. Uterus sac-shaped medioventrally. Eggs spherical. In freshwater Dasyatidae. South America. Type-species *P. magdalenensis* Brooks & Thorson, 1976.

14a. Postvaginal testes absent. Scolex with four sessile non-septate bothridia .... *Pedibothrium* Linton, 1909. (Figs 8.45, 8.46)

(Syn. *Phyllobothroides* Southwell, 1911.)

Diagnosis: Scolex with four large undivided bothridia each armed with one pair of asymmetrical two-pronged hooks; anterior muscular pad present. Strobila acraspedote, euapolytic. Genital pores lateral, irregularly alternating. Testes

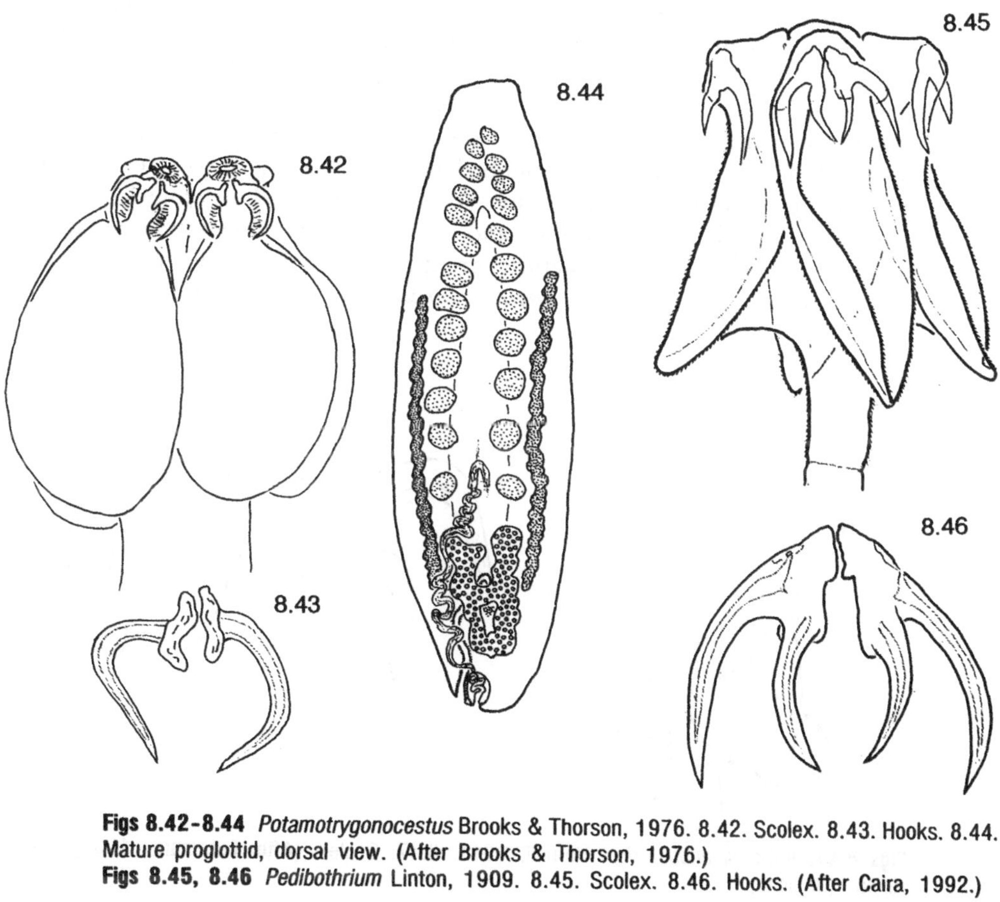

Figs 8.42–8.44 *Potamotrygonocestus* Brooks & Thorson, 1976. 8.42. Scolex. 8.43. Hooks. 8.44. Mature proglottid, dorsal view. (After Brooks & Thorson, 1976.)
Figs 8.45, 8.46 *Pedibothrium* Linton, 1909. 8.45. Scolex. 8.46. Hooks. (After Caira, 1992.)

numerous; postvaginal testes absent on poral side. Ovary posterior. Vagina anterior to cirrus-sac. Vitelline follicles lateral (some postovarian). Uterus medioventral between ovary and cirrus-sac. In Orectolobiformes. Cosmopolitan. Type-species *P. globicephalum* Linton, 1909.

14b. Postvaginal testes present. Scolex bulbous, embedded in gut wall .... *Balanobothrium* Hornell, 1911. (Figs 8.47, 8.48)

Diagnosis: Scolex with four flattened sessile undivided bothridia, each with apical accessory sucker and pair of minute hooks, each hook with two prongs. Strobila acraspedote, apolytic. Genital pores lateral, irregularly alternating. Testes numerous; postvaginal testes present on poral side. Cirrus spinose. Ovary posterior, bilobed in cross-section. Vagina anterior to cirrus-sac. Vitelline follicles lateral. Uterus ventral with lateral diverticulum. Entire scolex enlarged, bulbous, embedded in gut wall of host. In elasmobranchs. Pacific and Indian Oceans. Type-species *B. tenax* Hornell, 1912.

Family Phyllobothriidae Braun, 1900

Diagnosis: Scolex with four sessile or pedunculated muscular bothridia, simple or divided by septa, with or without anterior accessory sucker; apical myzorhynchus may be present; metascolex sometimes present. Strobila acraspedote or craspedote; anapolytic, apolytic, euapolytic or hyperapolytic. Genital pores lateral, irregularly alternating. Testes numerous; postvaginal testes present or not on poral side. Ovary posterior, bilobed or tetralobed in cross-section. Vagina anterior to cirrus-sac, crossing the vas deferens. Vitelline follicles lateral, sometimes well developed dorsally and ventrally. Uterus medioventral, uterine pore may be present. Eggs sometimes grouped in cocoons. Adults in spiral valve of elasmobranchs. Type-genus *Phyllobothrium* van Beneden, 1850.

8.47

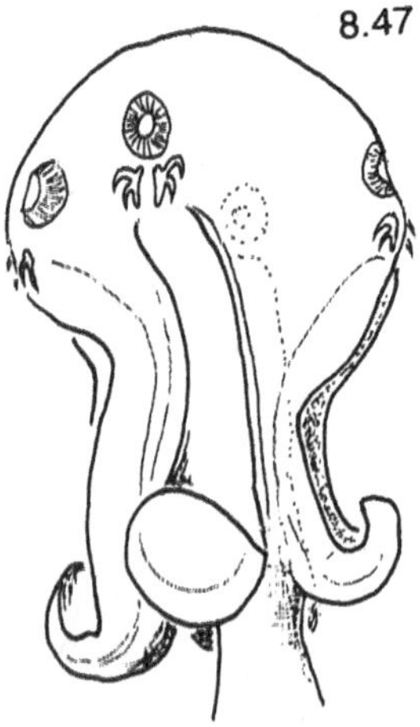

8.48

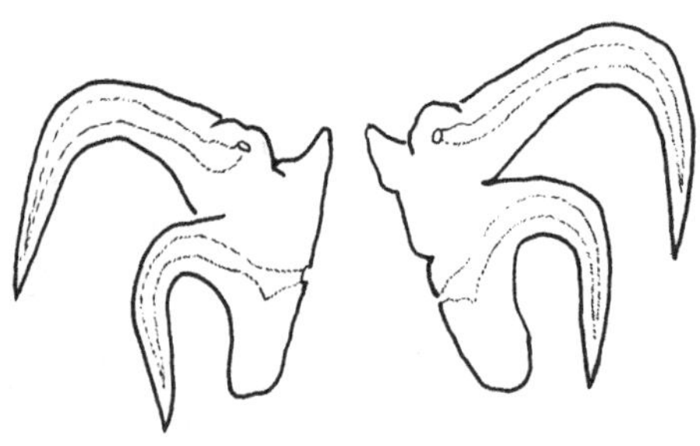

Figs 8.47, 8.48 *Balanobothrium* Hornell, 1911. 8.47. Scolex. (After Fuhrmann, 1931). 8.48. Hooks. (Original.)

Key to subfamilies

1a. Scolex with four small bothridia followed by strong metascolex .... Thysanocephalinae Euzet, 1953.
1b. Scolex with four bothridia without metascolex .... 2.

2a. Scolex with apical glandulomuscular myzorhynchus .... Echeneibothriinae de Beauchamp, 1905.
2b. Scolex without myzorhynchus .... 3.

3a. Scolex with four undivided bothridia with simple, crumpled or loculated margin .... Phyllobothriinae de Beauchamp, 1905.
3b. Scolex with four bothridia totally divided into loculi by muscular septa .... 4.

4a. Number of loculi reduced .... Triloculariinae Yamaguti, 1959.
4b. Loculi numerous .... Rhinebothriinae Euzet, 1953.

Subfamily Thysanocephalinae Euzet, 1953

Diagnosis: Phyllobothriidae. Scolex with four minute bothridia with or without anterior muscular pad or sucker and one large metascolex crumpled or tetralobed. Postvaginal testes present. Vitelline follicles lateral.

Key to genera

1a. Metascolex with four indistinct parts, very frilled and crumpled. Scolex with four minute bothridia each with anterior muscular pad .... *Thysanocephalum* Linton, 1889. (Figs 8.49–8.51)
Diagnosis: Scolex very small with four sessile bothridia each with two minute muscular prongs and one anterior muscular pad forming lateral auriculum. Strobila acraspedote, apolytic. Genital pores irregularly alternating. Testes numerous, postvaginal testes present on poral side. Ovary posterior. Vagina anterior to cirrus-sac. Vitelline follicles forming marginal strips along entire length of proglottid. In Carcharinidae. Cosmopolitan. Type-species *T. thysanocephalum* (Linton, 1889).

1b. Metascolex with four distinct folded lobes .... 2.

2a. Scolex with four bothridia each with anterior sucker .... *Myzocephalus* Shipley & Hornell, 1906. (Fig. 8.52)
(Syn. *Myxophyllobothrium* Shinde & Chincholikar, 1981.)
Diagnosis: Scolex small with metascolex in four crumpled folds. Genital pores irregularly alternating. Strobila anatomy unknown. In Myliobatidae. Indian Ocean. Type-species *M. narinari* Shipley & Hornell, 1906.

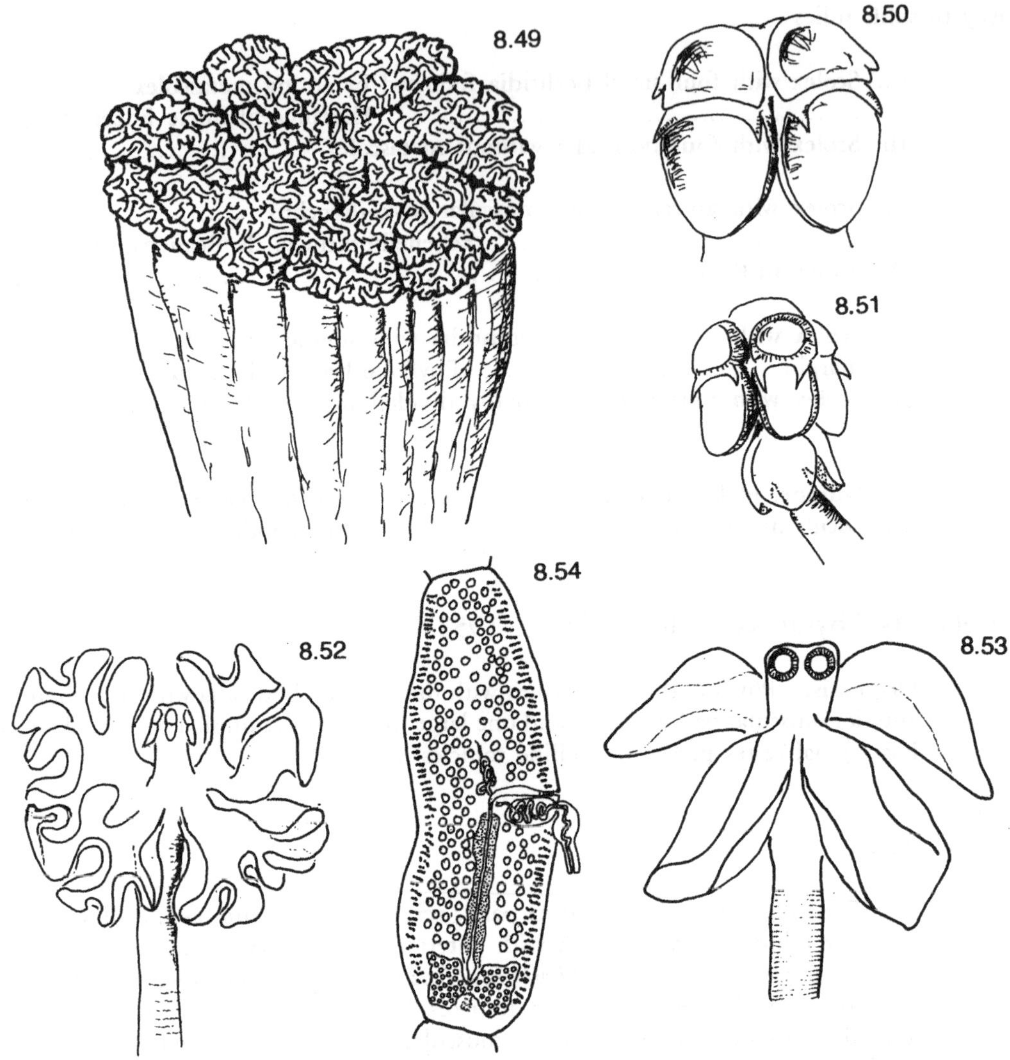

Figs 8.49–8.51 *Thysanocephalum* Linton, 1889. 8.49. Adult scolex and metascolex. 8.50. Scolex. (After Euzet, 1959.) 8.51. Young scolex and metascolex. (After Subhapradha, 1957.) **Fig. 8.52** *Myzocephalus* Shipley & Hornell, 1906. Scolex. (After Shipley & Hornell, 1906.) **Figs 8.53, 8.54** *Myzophyllobothrium* Shipley & Hornell, 1906. 8.53. Scolex and metascolex. 8.54. Mature proglottid. (After Shipley & Hornell, 1906.)

2b. Scolex with four bothridia without anterior suckers .... *Myzophyllobothrium* Shipley & Hornell, 1906. (Figs 8.53, 8.54)

Diagnosis: Scolex small with four large circular bothridia. Metascolex divided into four distinct parts with smooth edges. Red pigment at base of metascolex. Strobila acraspedote, apolytic. Genital pores lateral, irregularly alternating. Testes numerous; postvaginal testes present on poral side. Ovary posterior. Vagina anterior to cirrus-sac. Vitelline follicles lateral. Uterus median, ventral, extending from ovary to level of cirrus-sac. In Myliobatiformes. Indian Ocean. Type-species *M. rubrum* Shipley & Hornell, 1906.

Subfamily Echeneibothriinae de Beauchamp, 1905

Diagnosis: Phyllobothriidae. Scolex with apical myzorhynchus. Four bothridia, smooth or divided by septa. Postvaginal testes absent. Parasites of Rajidae.

Key to genera

1a. Scolex with four smooth bothridia each with simple or loculated margin .... 2.
1b. Scolex with four bothridia partially or totally divided by septa ...... 3.

2a. Bothridia cup-shaped with simple margin .... *Pseudoanthobothrium* Baer, 1956. (Fig. 8.55)
Diagnosis: Scolex with four pedunculated bothridia. Myzorhynchus present. Strobila acraspedote, euapolytic. Genital pores irregularly alternating. Testes few, anterior; postvaginal testes absent on poral side. Ovary posterior, X-shaped in cross-section. Vagina anterior to cirrus-sac. Vitelline follicles lateral. In Rajidae. North Atlantic. Type-species *P. hanseni* Baer, 1956.

2b. Bothridia with one marginal row of loculi .... *Clydonobothrium* Euzet, 1959. (Fig. 8.56)
Diagnosis: Scolex with small apical myzorhynchus and four large bothridia with crumpled and loculated margin. Accessory sucker absent. Strobila acraspedote, anapolytic. Genital pores lateral, irregularly alternating. Testes anterior to cirrus-sac; postvaginal testes absent on poral side. Ovary posterior. Vagina anterior to cirrus-sac. Vitelline follicles lateral. In Rajidae. North Atlantic. Type-species *C. elegantissimum* (Lönnberg, 1889).

3a. Bothridia with incomplete septa .... *Phormobothrium* Alexander, 1963. (Fig. 8.57)
Diagnosis: Scolex apical with myzorhynchus and four pedunculate bothridia. Each bothridium with few incomplete longitudinal ridges. Strobila acraspedote, euapolytic (?) Genital pores lateral, irregularly alternating. Testes few; postvaginal testes absent on poral side. Ovary posterior. Vagina anterior to cirrus-sac. Vitelline follicles lateral. In Rajidae. New Zealand. Type-species *P. affine* (Olsson, 1867).

3b. Bothridia totally septate .... 4.

4a. Bothridia divided into three loculi .... *Tritaphros* Lönnberg, 1889. (Fig. 8.58)
Diagnosis: Scolex with small myzorhynchus and four sessile bothridia. Each bothridium divided into three large loculi by two longitudinal septa. Strobila slightly craspedote, anapolytic. Genital pores lateral, irregularly alternating.

8.55

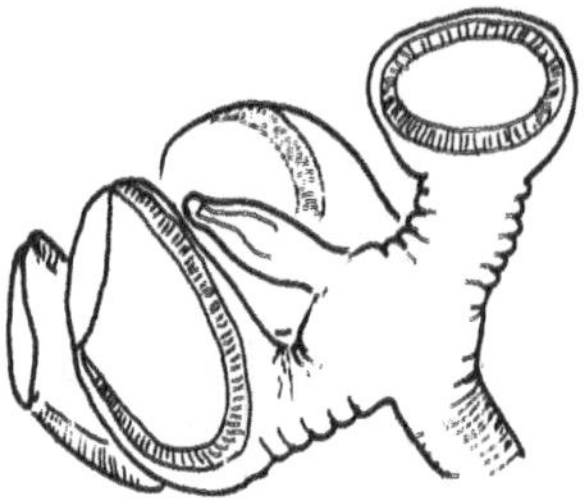

8.56

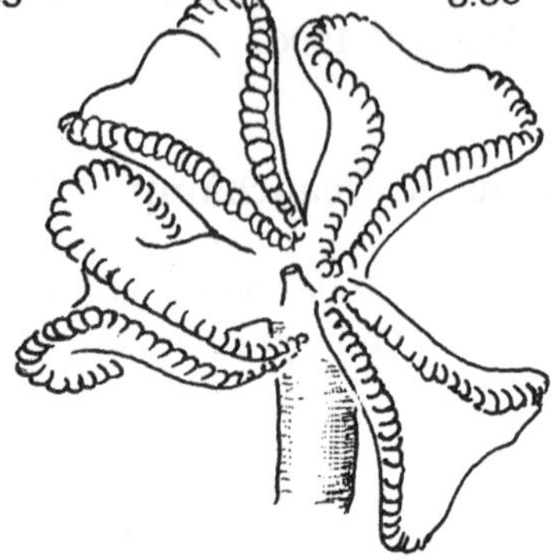

8.58

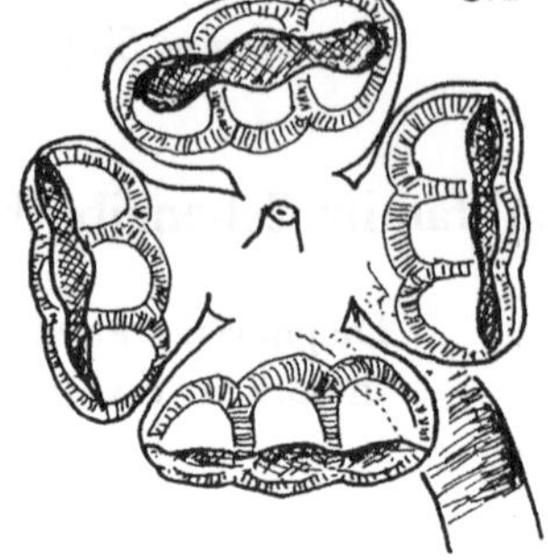

8.57

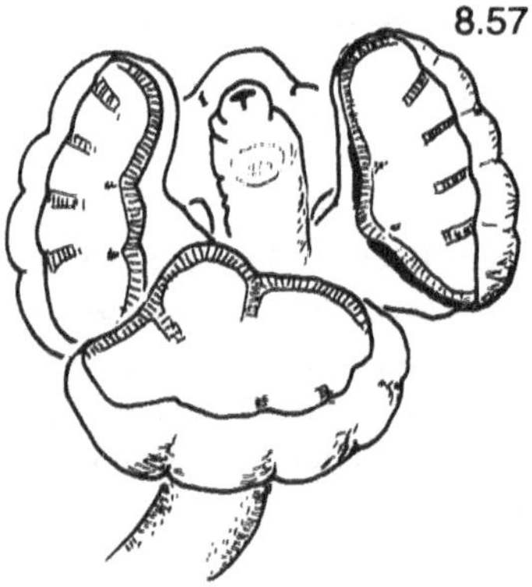

Fig. 8.55 *Pseudanthobothrium* Baer, 1956. Scolex. (After Baer, 1956.)
Fig. 8.56 *Clydonobothrium* Euzet, 1959. Scolex. (After Euzet, 1959.)
Fig. 8.57 *Phormobothrium* Alexander, 1963. Scolex. (After Alexander, 1963.)
Fig. 8.58 *Tritaphros* Lönnberg, 1889. Scolex. (After Euzet, 1959.)

Testes anterior; postvaginal testes absent on poral side. Ovary posterior, X-shaped in cross-section. Vagina anterior to cirrus-sac. Vitelline follicles in lateral fields. In Rajidae. North Atlantic, western Mediterranean. Type-species *T. retzii* Lönnberg, 1889.

4b. Bothridia divided into numerous loculi ....
.... ***Echeneibothrium* van Beneden, 1850. (Figs 8.59–8.61)**
(Syn. *Discobothrium* van Beneden, 1870.)

Diagnosis: Scolex with apical retractile myzorhynchus of variable size and four pedunculated or sessile bothridia each divided by transverse and longitudinal septa. Strobila acraspedote or slightly craspedote, euapolytic. Genital pores lateral, irregularly alternating. Testes few; postvaginal testes absent on poral side. Ovary posterior, X-shaped in cross-section. Vagina anterior to cirrus-sac. Vitelline follicles lateral. Uterus median tubular. In Rajidae. Cosmopolitan. Type-species *E. variabile* van Beneden, 1850.

Subfamily Phyllobothriinae de Beauchamp, 1905

Diagnosis: Phyllobothriidae. Scolex unarmed, with four bothridia without loculi, with simple, crumpled or loculated margin. No myzorhynchus.

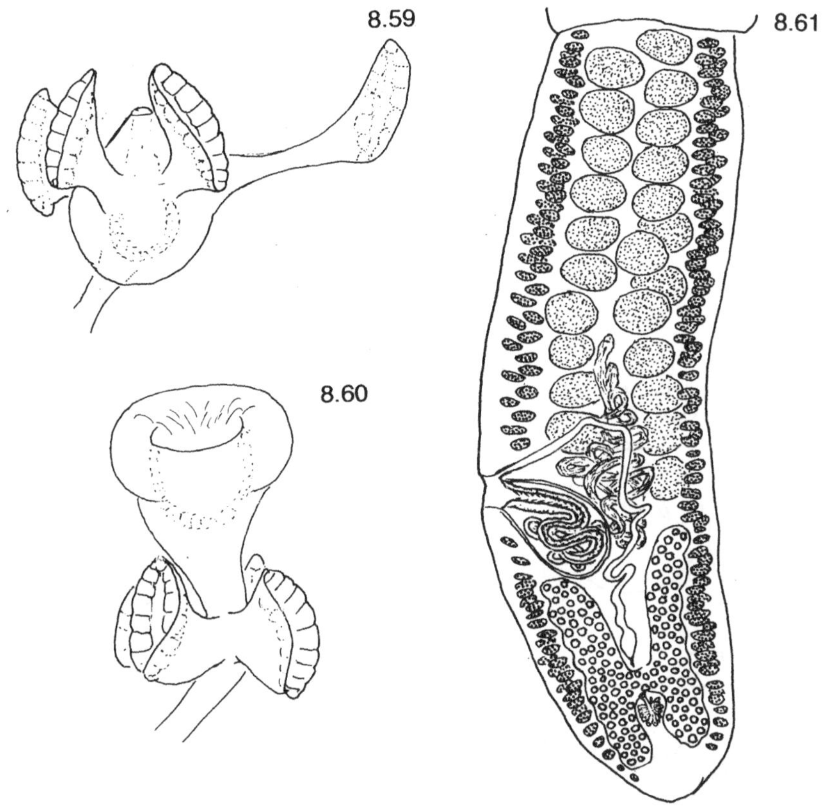

Figs 8.59–8.61 *Echeneibothrium* van Beneden, 1850. 8.59. Scolex with invaginated myzorhynchus. 8.60. Scolex with everted myzorhynchus. 8.61. Mature proglottid. (After Euzet, 1959.)

Key to genera

1a. Bothridia without accessory sucker .... **2.**
1b. Bothridia with accessory sucker .... **7.**

2a. Bothridia sessile, tubular, with anterior and posterior opening .... ***Pithophorus* Southwell, 1925. (Figs 8.62, 8.63)**
Diagnosis: Scolex with four globular or cylindrical hollow bothridia, each open at both anterior and posterior ends. Strobila acraspedote. Genital pores lateral, irregularly alternating. Testes numerous; postvaginal testes present on poral side. Ovary bilobed in cross-section. Vagina anterior to cirrus-sac. Vitelline follicles lateral. In elasmobranchs. Pacific Ocean, Indian Ocean, Sea of Japan. Type-species *P. tetraglobus* (Southwell, 1911).

2b. Bothridia non-tubular .... **3.**

3a. Bothridia globular, pouch-shaped .... **4.**
3b. Bothridia non-globular with simple or crimped margin .... **5.**

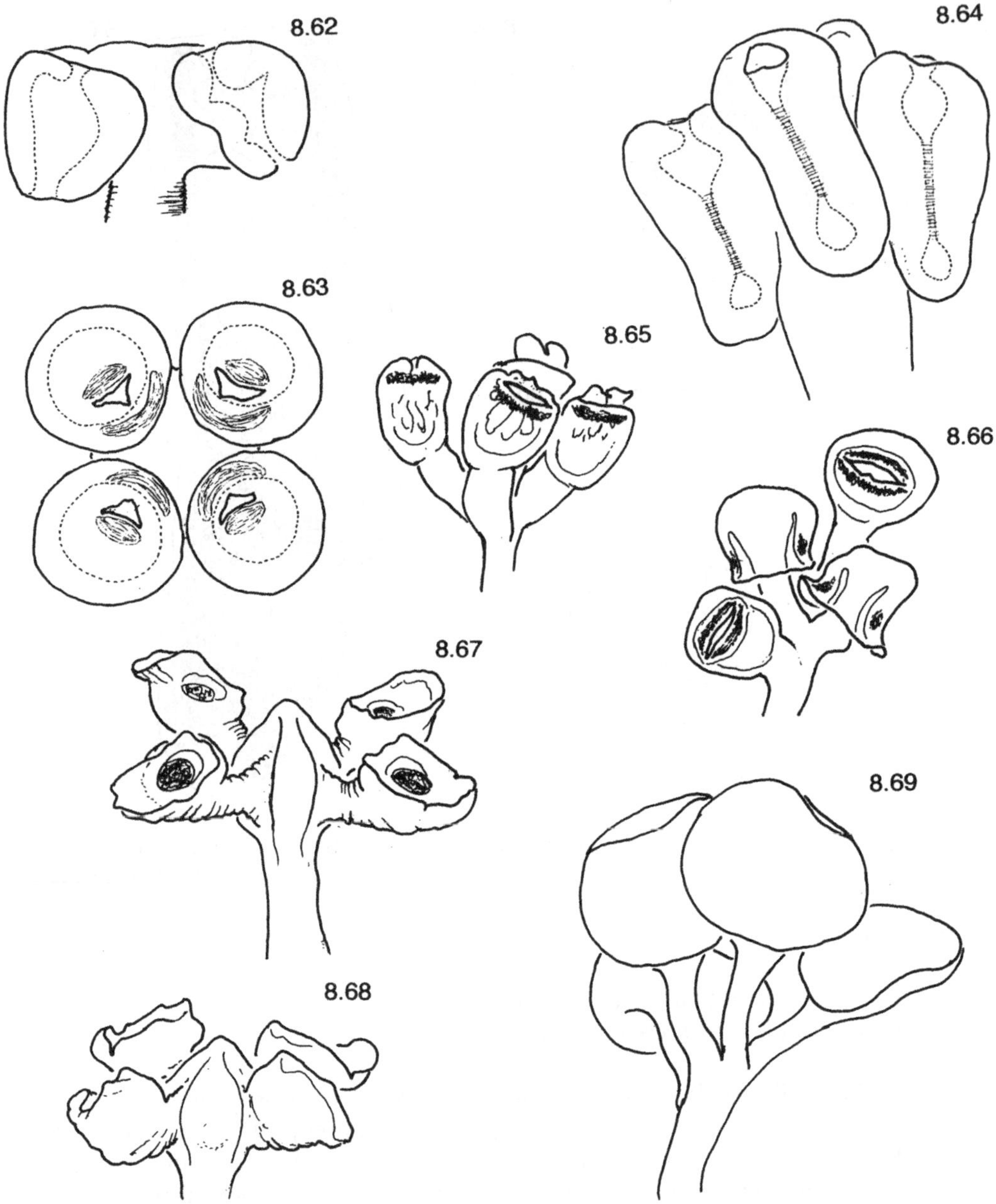

Figs 8.62, 8.63 *Pithophorus* Southwell, 1925. 8.62. Scolex, lateral view. 8.63. Scolex, apical view. (After Subhapradha, 1957.)
Fig. 8.64 *Scyphophyllidium* Woodland, 1927. Scolex. (After Euzet, 1959.)
Figs 8.65, 8.66 *Carpobothrium* Shipley & Hornell, 1906. 8.65. Scolex contracted. 8.66. Scolex with everted bothridia. (After Subhapradha, 1957.)
Figs 8.67, 8.68 *Clistobothrium* Dailey & Vogelbein, 1990. 8.67. Scolex, with bothridia extended. 8.68. Scolex contracted. (After Dailey & Vogelbein, 1990.)
Fig. 8.69 *Anthobothrium* van Beneden, 1850. Scolex. (After Euzet, 1959.)

4a. Bothridial opening without muscular pad ....
.... ***Scyphophyllidium* Woodland, 1927. (Fig. 8.64)**
Diagnosis: Scolex with four globular, sessile bothridia with irregular anterior opening. Accessory sucker absent. Strobila acraspedote, anapolytic. Genital pores lateral, irregularly alternating. Testes numerous; postvaginal testes present on poral side. Ovary posterior, bilobed in cross-section. Vitelline follicles in wide lateral bands. In Carcharinidae. North Atlantic. Type-species *S. giganteum* (van Beneden, 1858).

4b. Bothridial opening with muscular pad ....
.... ***Carpobothrium* Shipley & Hornell, 1906. (Figs 8.65, 8.66)**
Diagnosis: Scolex with four pedunculated bothridia each with a single flap surrounding a slit-like opening. No marginal loculi on the flap. A pair of muscular pads at the base of the flap. Strobila acraspedote, apolytic. Genital pores lateral, irregularly alternating. Testes numerous; no postvaginal testes on poral side. Cirrus-sac U-shaped. Ovary posterior. Vagina large, anterior to cirrus-sac. Vitelline follicles lateral. Uterus median, extending from ovary to cirrus-sac. In Orectolobiformes. Indian Ocean. Type-species *C. chiloscyllii* Shipley & Hornell, 1906.

5a. Bothridia with one central sucker ....
.... ***Clistobothrium* Dailey & Vogelbein, 1990. (Figs 8.67, 8.68)**
Diagnosis: Scolex with four pedunculated bothridia separated by a large cruciform apex. Each bothridium with large bowl-shaped sucker and folding lappet that projects over sucker opening when extended. Strobila acraspedote, anapolytic. Genital pore lateral, irregularly alternating. Testes numerous; postvaginal testes present on poral side. Ovary posterior. Vagina anterior to cirrus-sac. Vitelline follicles in wide lateral bands. Uterus median, reaching only to posterior margin of cirrus-sac. In Carcharinidae. American coast of Pacific. Type-species *C. carcharodoni* Dailey & Vogelbein, 1990.

5b. Bothridia without central sucker .... **6.**

6a. Bothridia simple, disc-shaped ....
.... ***Anthobothrium* van Beneden, 1850. (Fig. 8.69)**
Diagnosis: Scolex with four bothridia, pedunculate with a simple edge. Accessory sucker absent. Sometimes one central muscular sucker. Strobila craspedote, laciniated, euapolytic. Genital pores irregularly alternating. Testes numerous; postvaginal testes present on poral side. Ovary posterior. Vitelline follicles lateral. In Carchariniformes. Cosmopolitan. Type-species *A. cornucopia* van Beneden, 1850.

6b. Bothridia crimped, sometimes retractile ....
.... ***Rhodobothrium* Linton, 1889. (Figs 8.70, 8.71)**
(Syns *Proboscidosaccus* Gallien, 1949; *Inermiphyllidium* Riser, 1955; *Sphaerobothrium* Euzet, 1959.)
Diagnosis: Scolex with four large subspherical bothridia supported by pedicel. Bothridia trumpet-shaped when relaxed with convex adherent surface traversed by numerous convolutions forming an irregular pattern. Accessory sucker absent.

Strobila acraspedote, apolytic, proglottid triangular. Testes numerous; postvaginal testes present on poral side. Ovary posterior, tetralobed in cross-section. Vagina anterior to cirrus-sac. Vitelline follicles lateral. Uterus median, sac-like with lateral diverticula. In Myliobatiformes. Cosmopolitan. Type-species *R. pulvinatum* Linton, 1889.

7a. Bothridia with central glandular disc .... ***Orygmatobothrium* Diesing, 1863. (Fig. 8.72)**

Diagnosis: Scolex with four bothridia each with simple or crumpled margin. Accessory sucker present. One glandular circular organ near centre of each bothridium. Strobila acraspedote, euapolytic. Genital pores irregularly alternating. Testes numerous; postvaginal testes present on poral side. Ovary posterior, tetralobed in cross-section. Vitelline follicles lateral. In Mustelidae. Cosmopolitan. Type-species *O. musteli* (van Beneden, 1850).

7b. Bothridia without central glandular disc .... **8.**

8a. Bothridia globular, pouch-shaped .... ***Marsupiobothrium* Yamaguti, 1952. (Figs 8.73, 8.74)**

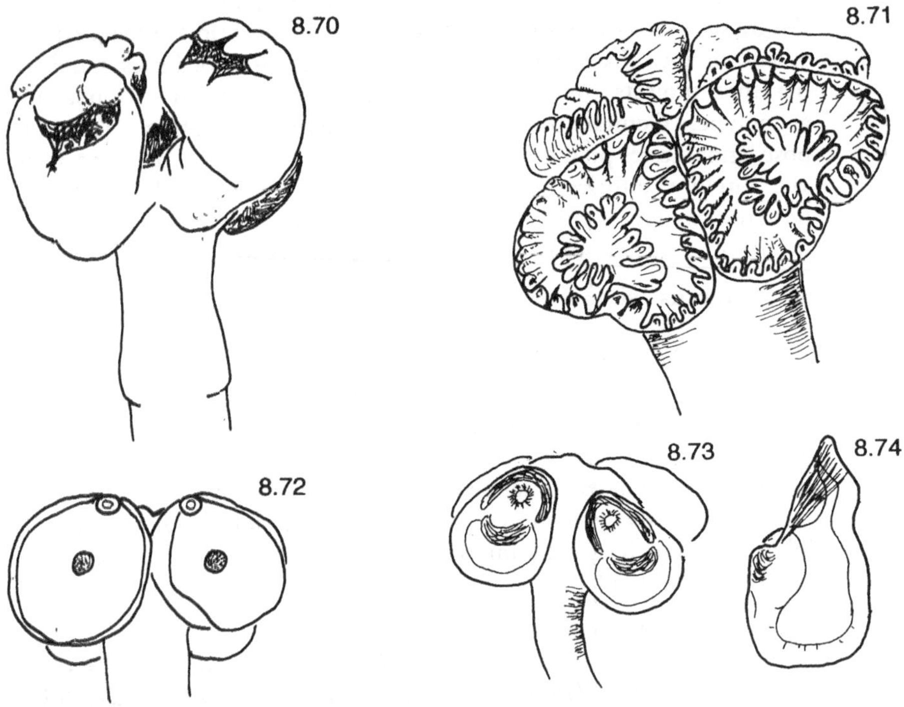

Figs 8.70, 8.71 *Rhodobothrium* Linton, 1889. 8.70. Scolex contracted. 8.71. Scolex extended. (After Euzet, 1959.)
Fig. 8.72 *Orygmatobothrium* Diesing, 1863. Scolex. (After Euzet, 1959.)
Figs 8.73, 8.74 *Marsupiobothrium* Yamaguti, 1952. 8.73. Scolex. 8.74. Bothridium. (After Yamaguti, 1952.)

Diagnosis: Scolex with four sessile bothridia each shaped like pyriform sac, opening of which is provided with sphincter-like muscle especially in posterior region. Submarginal accessory sucker present at anterior end of each bothridium. Strobila acraspedote, euapolytic. Genital pores lateral, irregularly alternating. Testes numerous; postvaginal testes present on poral side. Ovary posterior with two long slender wings. Vitelline follicles in two extensive lateral fields. In Alopiidae (Carchariniformes). Pacific. Type-species *M. alopias* Yamaguti, 1952.

8b. Bothridia not pouch-shaped .... **9.**

9a. Bothridia with crumpled or loculated margin ....
.... ***Phyllobothrium* van Beneden, 1850. (Figs 8.75–8.79)**
(Syn. *Anthocephalum* Linton, 1890.)

Diagnosis: Scolex with four bothridia each sessile or pedunculated. One accessory sucker present. Margin of bothridia folded, curled, with or without loculi. Apical glandular organ sometimes present. Strobila craspedote or acraspedote, apolytic or anapolytic. Genital pores lateral, irregularly alternating, sometimes unilateral along part of strobila. Testes numerous; postvaginal testes present or not on poral side. Ovary posterior, four-lobed or bilobed in cross-section. Vitelline follicles lateral. In elasmobranchs. Cosmopolitan. Type-species *P. lactuca* van Beneden, 1850.

9b. Bothridia with simple margin .... **10.**

10a. Bothridia with anterolateral bifid lobe .... **11.**
10b. Bothridia without bifid lobe .... **12.**

11a. Proglottid with lateral vitelline follicles ....
.... ***Dinobothrium* van Beneden, 1889. (Figs 8.80, 8.81)**
(Syns *Diplobothrium* van Beneden, 1889; *Reesium* Euzet, 1955.)

Diagnosis: Scolex with four sessile bothridia, large, not folded, on margin each surmounted by muscular pad with two little prongs, extremity of muscular pad terminated by bifid lobe. One accessory sucker present on medial surface of muscular pad. Strobila acraspedote, anapolytic. Genital pores lateral, irregularly alternating. Testes numerous; postvaginal testes present on poral side. Ovary posterior, tetralobed in cross-section. Vitelline follicles lateral. In Lamniformes. Cosmopolitan. Type-species *D. septaria* van Beneden, 1889.

11b. Proglottid with ventral vitelline follicles ....
.... ***Gastrolecithus* Yamaguti, 1952. (Figs 8.82, 8.83)**

Diagnosis: Scolex large with four sessile concave bothridia, paired dorsoventrally. Each bothridium with one anterior muscular pad, one accessory sucker and two little lateral prongs and bifid crest-like appendage at its anterodorsal corner. Strobila craspedote, anapolytic. Genital pores lateral, irregularly alternating. Testes numerous; postvaginal testes present on poral side. Ovary bilobed in cross-section. Vagina anterior to cirrus-sac. Vitelline follicles ventral. In Cetorhinidae. Cosmopolitan. Type-species *G. planus* (Linton, 1922).

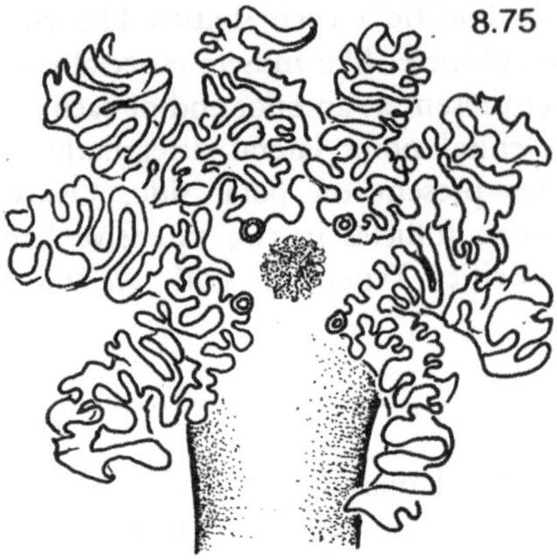
8.75

8.76

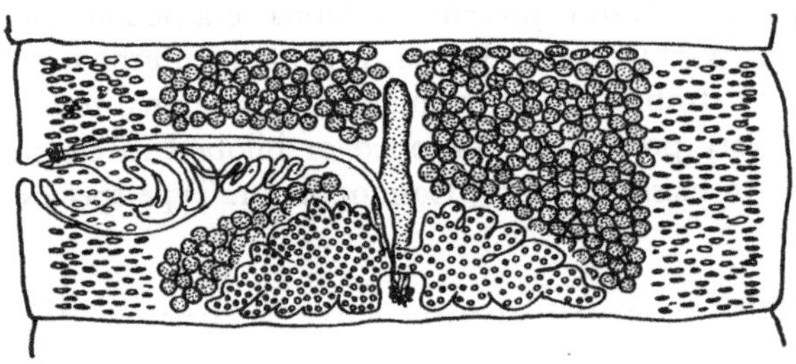
8.77

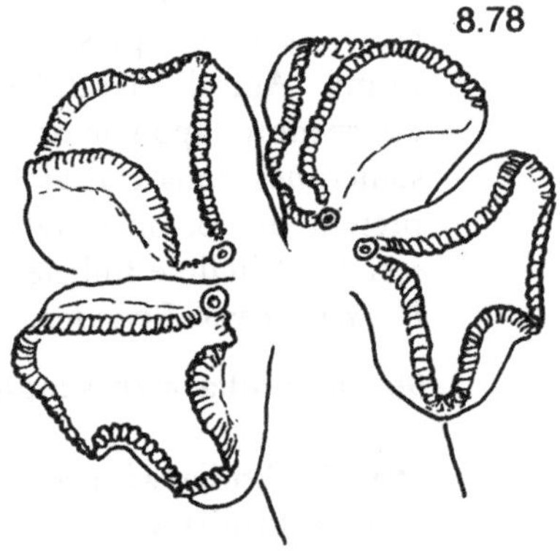
8.78

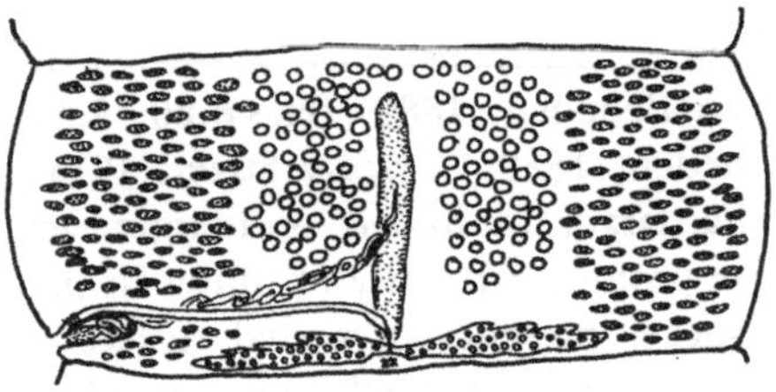
8.79

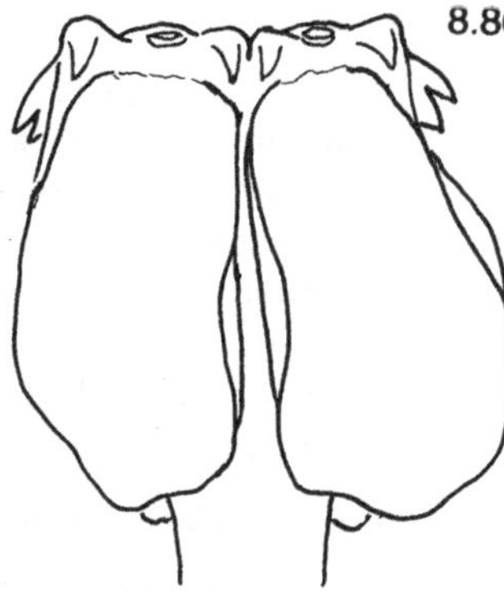
8.80

8.82

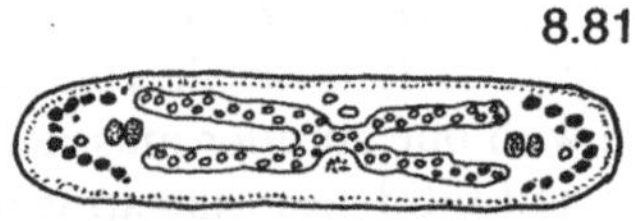
8.81

8.83

12a. Accessory sucker circular .... **13.**
12b. Accessory sucker triangular with two posterolateral muscular prongs ....
.... **14.**

13a. Accessory sucker cup-shaped. Postvaginal testes present ....
.... ***Crossobothrium* Linton, 1889. (Fig. 8.84)**
Diagnosis: Scolex with four bothridia, each sessile with accessory sucker and simple edge. Strobila acraspedote, euapolytic. Genital pore lateral, irregularly alternating. Testes numerous; postvaginal testes present on poral side. Ovary posterior, tetralobed in cross-section. Vagina anterior to cirrus-sac. Vitelline follicles lateral. In Pleurotremata. Cosmopolitan. Type-species *C. angustum* (Linton, 1889).

13b. Accessory sucker globular. Postvaginal testes absent ....
.... ***Calyptrobothrium* Monticelli, 1893. (Fig. 8.85)**
Diagnosis: Scolex with four sessile bothridia with simple edge. Accessory sucker large, subspherical. Strobila acraspedote, hyperapolytic. Genital pores lateral, irregularly alternating. Testes numerous; postvaginal testes absent on poral side. Ovary posterior. Vagina anterior to cirrus-sac. Vitelline follicles lateral. In Torpedinidae. North Atlantic and Mediterranean. Type-species *C. riggii* Monticelli, 1893.

14a. Vitelline follicles lateral ....
.... ***Ceratobothrium* Monticelli, 1892. (Fig. 8.86)**
Diagnosis: Scolex with four sessile bothridia each directly below large muscular pad bearing two projecting horn-like appendages consisting mainly of circular muscle fibres. Yellowish scolex (scolex deformatus) often embedded in gut wall. Strobila acraspedote, apolytic. Genital pores lateral, irregularly alternating. Testes numerous; postvaginal testes present on poral side. Cirrus-sac long, oblique anteriorly. Ovary posterior, X-shaped in cross-section. Vagina anterior to cirrus-sac. In Lamniformes. Cosmopolitan. Type-species *C. xanthocephalum* Monticelli, 1892.

14b. Vitelline follicles circum-medullary ....
.... ***Monorygma* Diesing, 1863. (Fig. 8.87)**
Diagnosis: Scolex with apical glandular mass and four large sessile bothridia each with anterior muscular pad, horseshoe-shaped, bearing two small lateral appendages. Strobila acraspedote, anapolytic. Genital pores lateral, irregularly alternating. Genital atrium present. Testes numerous; postvaginal testes present on poral side. Ovary posterior. Vagina anterior to cirrus-sac. Uterus saccate, between ovary and cirrus-sac. In Squaliformes. North Atlantic and Mediterranean. Type-species *M. perfectum* (van Beneden, 1853).

Figs 8.75–8.79 *Phyllobothrium* van Beneden, 1850. 8.75. Scolex of *P. lactuca* van Beneden, 1850. 8.76. Bothridial complication in *P. lactuca.* 8.77. Mature proglottid of *P. lactuca* with postvaginal testes. 8.78. Scolex of *P. gracile* Wedl, 1855. 8.79. Mature proglottid of *P. gracile* without postvaginal testes. (After Euzet, 1959.)

Figs 8.80, 8.81 *Dinobothrium* van Beneden, 1889. 8.80. Scolex. 8.81. Cross-section showing lateral vitelline follicles. (After Euzet, 1959.)

Figs 8.82, 8.83 *Gastrolecithus* Yamaguti, 1952. 8.82. Scolex, lateral view. 8.83. Cross-section showing ventral vitelline follicles. (After Euzet, 1959.)

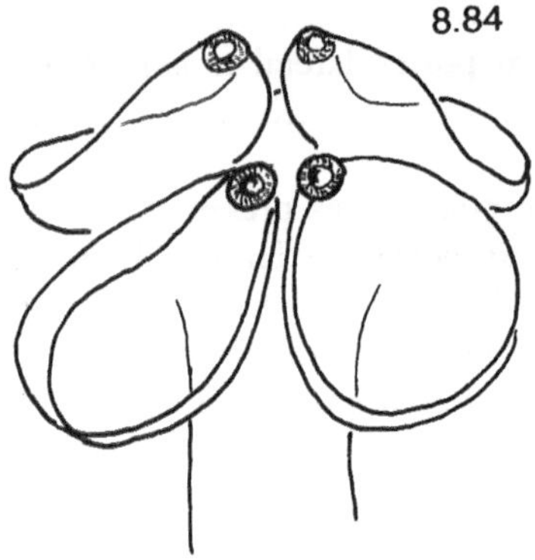
8.84

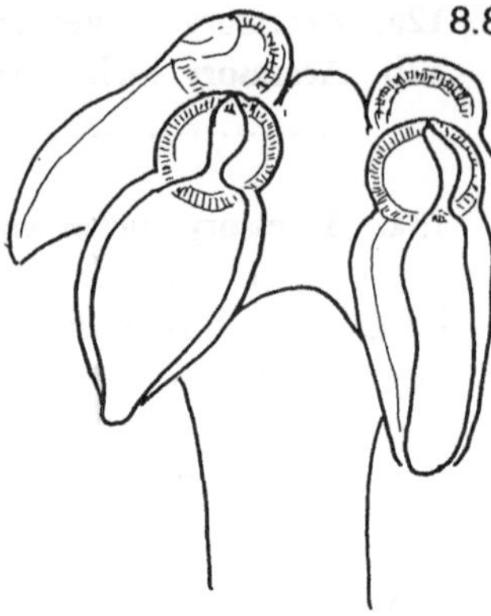
8.85

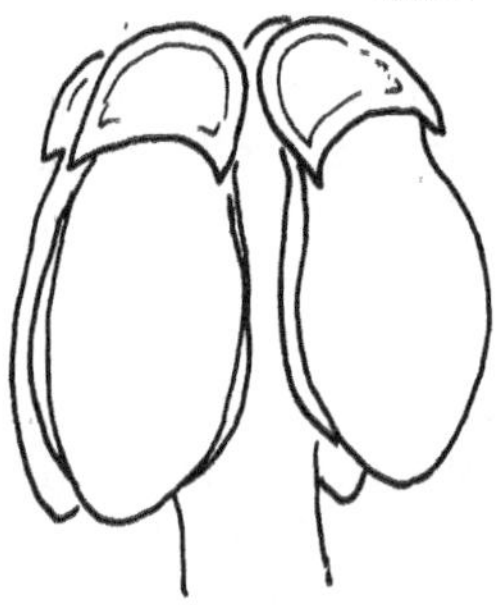
8.86

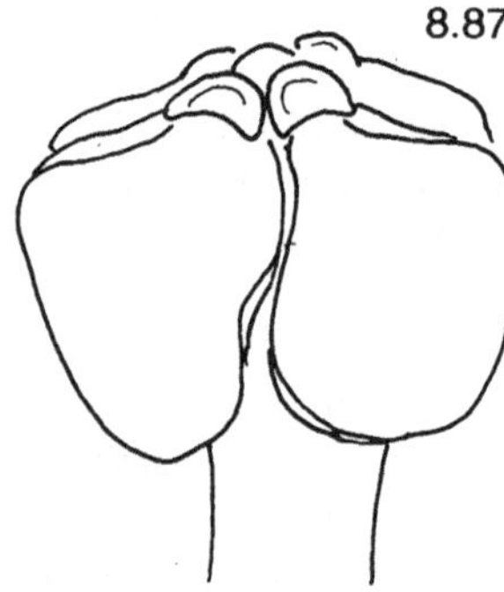
8.87

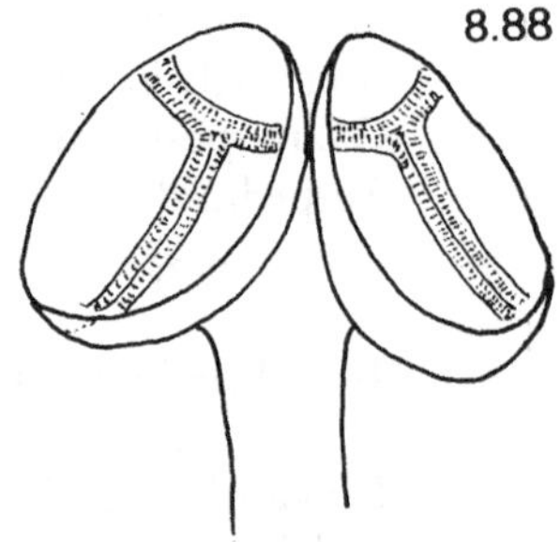
8.88

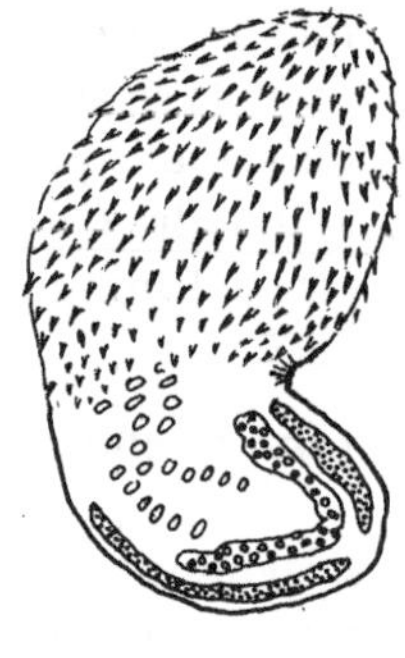
8.89

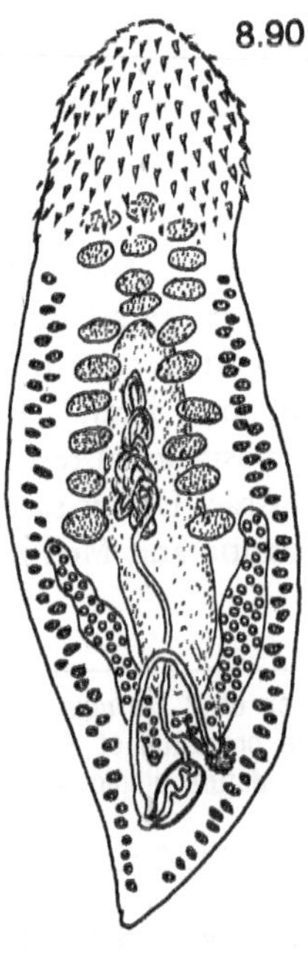
8.90

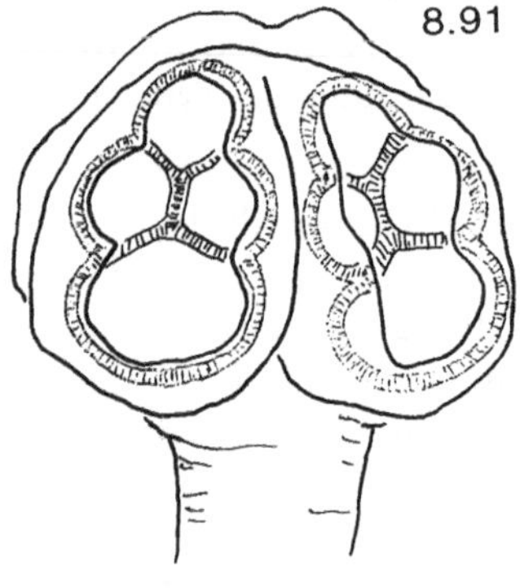
8.91

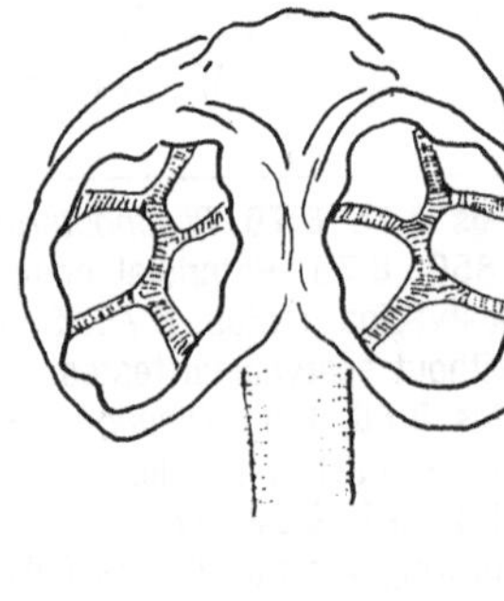
8.92

Subfamily Trilocularinae Yamaguti, 1959

Diagnosis: Scolex with four bothridia divided by septa into few loculi. No metascolex or myzorhynchus.

Key to genera

1a. Scolex with four bothridia each divided into three loculi. Hyperapolytic. Genital pore posterior ...... *Trilocularia* Olsson, 1867. (Figs 8.88–8.90)
(Syns *Phyllobothrideum* Olsson, 1867; *Urogonoporus* Lühe, 1901.)
Diagnosis: Scolex with four sessile bothridia each divided into three loculi (one apical, two posterior) by muscular septa. No myzorhynchus. Strobila filiform, hyperapolytic. Free proglottids with large anterior spines. Genital pore near posterior end of proglottid. Testes few, anterior; postvaginal testes absent on poral side. Ovary posterior, U-shaped. Vitelline follicles lateral. Uterus median with lateral diverticulum. In Squalidae. North Atlantic and Mediterranean. Type-species *T. acanthiaevulgaris* Olsson, 1867.

1b. Scolex with four bothridia each, with more than three loculi. Apolytic. Genital pore lateral .... 2.

2a. Bothridia divided into four symmetrical loculi .... *Zyxibothrium* Hayden & Campbell, 1981. (Fig. 8.91)
Diagnosis: Scolex with four sessile bothridia each divided into four loculi (one anterior, two median, one posterior) by muscular septa. No myzorhynchus. Strobila slightly craspedote, apolytic. Genital pore marginal, irregularly alternating. Testes anterior; postvaginal testes absent on poral side. Ovary posterior, U-shaped, bilobed in cross-section. Vagina anterior to cirrus-sac. Vitelline follicles lateral. Uterus sac-shaped. In *Raja*. North Atlantic. Type-species *Z. kamienae* Hayden & Campbell, 1981.

2b. Bothridia divided into five asymmetrical loculi .... *Pentaloculum* Alexander, 1963. (Fig. 8.92)
Diagnosis: Scolex with four sessile bothridia each divided by muscular septa into five radial loculi more or less equal in size. No myzorhynchus. Strobila small, slightly craspedote, apolytic. Genital pore lateral, irregularly alternating. Testes numerous; postvaginal testes absent on poral side. Ovary posterior. Vagina anterior to cirrus-sac. Vitelline follicles lateral. In *Torpedo aysoni*. New Zealand. Type-species *P. macrocephalum* Alexander, 1963.

Fig. 8.84 *Crossobothrium* Linton, 1889. Scolex. (After Euzet, 1959.)
Fig. 8.85 *Calyptrobothrium* Monticelli, 1893. Scolex. (After Euzet, 1959.)
Fig. 8.86 *Ceratobothrium* Monticelli, 1892. Scolex. (After Euzet, 1959.)
Fig. 8.87 *Monorygma* Diesing, 1863. Scolex. (After Euzet, 1959.)
Figs 8.88–8.90 *Trilocularia* Olsson, 1867. 8.88. Scolex. 8.89. Young proglottid. 8.90. Mature proglottid. (After Euzet, 1959.)
Fig. 8.91 *Zyxibothrium* Hayden & Campbell, 1981. Scolex. (After Hayden & Campbell, 1981.)
Fig. 8.92 *Pentaloculum* Alexander, 1963. Scolex. (After Alexander, 1963.)

Subfamily Rhinebothriinae Euzet, 1953

Diagnosis: Scolex with four bothridia divided by septa into numerous loculi. No myzorhynchus.

Key to genera

1a. Postvaginal testes absent on poral side .... **2.**
1b. Postvaginal testes present on poral side .... **3.**

2a. Ovarian lobes unequal, poral lobes shorter ....
...... ***Rhinebothroides* Mayes, Brooks & Thorson, 1981. (Figs 8.93, 8.94)**
Diagnosis: Scolex with four pedunculated bothridia, subdivided into loculi by several transverse septa, and one longitudinal, muscular septum. No myzorhynchus. Strobila acraspedote, apolytic. Genital pores lateral, irregularly alternating. Testes numerous, anterior to cirrus-sac; postvaginal testes absent on poral side. Ovary posterior with greatly reduced poral lobe in dorsal view, X-shaped in cross-section. Vagina anterior to cirrus-sac. Vitelline follicles lateral. Only in freshwater Dasyatidae (*Potamotrygon*). South America. Type-species *R. moralarai* (Brooks & Thorson, 1976).

2b. Ovarian lobes equal .... ***Rhinebothrium* Linton, 1889. (Figs 8.95, 8.96)**
Diagnosis: Scolex with four pedunculated bothridia, subdivided into loculi by several transverse septa and one longitudinal septum. Bothridial margin entire or loculated. No myzorhynchus. Cephalic peduncle short or absent. Strobila acraspedote or slightly craspedote. Euapolytic. Genital pore lateral. Testes few to numerous, anterior to cirrus-sac; postvaginal testes absent on poral side. Ovary posterior, X-shaped in cross-section. Vagina anterior to cirrus-sac. Vitelline follicles lateral. Uterus simple, median. Eggs isolated or in cocoon. In Myliobatiformes, Rajiformes. Cosmopolitan. Type-species *R. flexile* Linton, 1890.

3a. Bothridium in two parts, posterior part with longitudinal septa only ....
.... ***Duplicibothrium* Williams & Campbell, 1978. (Figs 8.97, 8.98)**
Diagnosis: Scolex forked with four sessile bothridia, dorsal and ventral fused into two pairs. Bothridia subdivided into loculi by transverse septa except at base where they form cup with few radial loculi. No myzorhynchus. Cephalic peduncle present. Strobila acraspedote, apolytic. Genital pores submarginal, irregularly alternating. Testes medullary; postvaginal testes present on poral side. Cirrus-sac anterior. Ovary posterior with radiating lobules. Vagina anterior to cirrus-sac. Vitelline follicles circum-medullary. In Rhinobatidae. North Atlantic. Type-species *D. minutum* Williams & Campbell, 1978.

3b. Bothridium with transverse and longitudinal septa .... **4.**

4a. Bothridium sessile with three longitudinal rows of loculi ....
.... ***Glyphobothrium* Williams & Campbell, 1977. (Figs 8.99, 8.100)**
Diagnosis: Scolex globular with four sessile bothridia fused to scolex. Each bothridium subdivided into three longitudinal rows of loculi by several transverse and two longitudinal septa. No myzorhynchus. Strobila craspedote, apolytic.

8.93

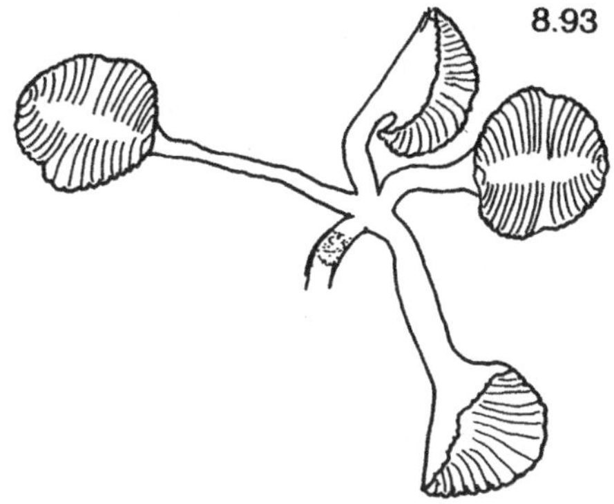

8.94

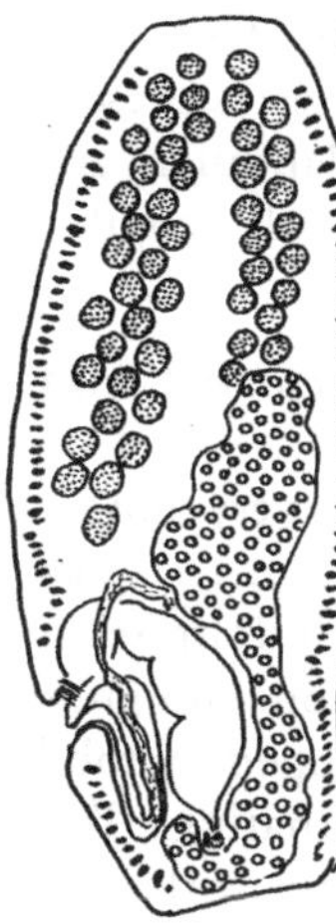

8.96

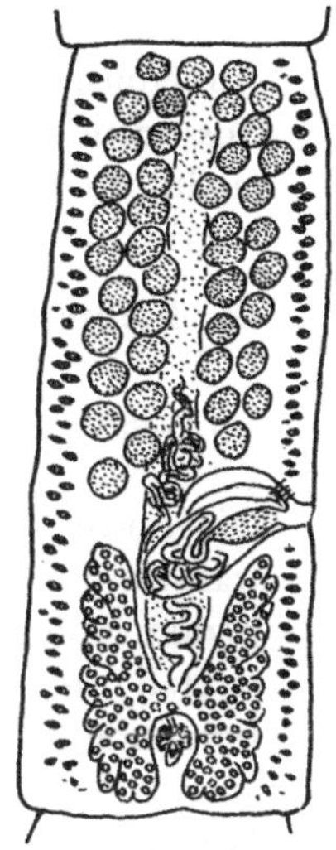

8.95

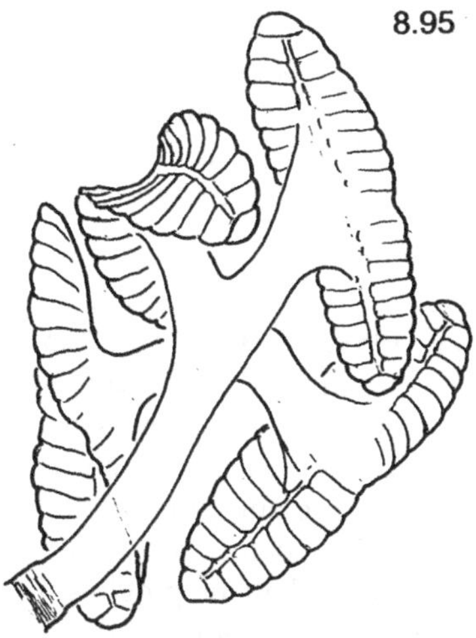

8.97

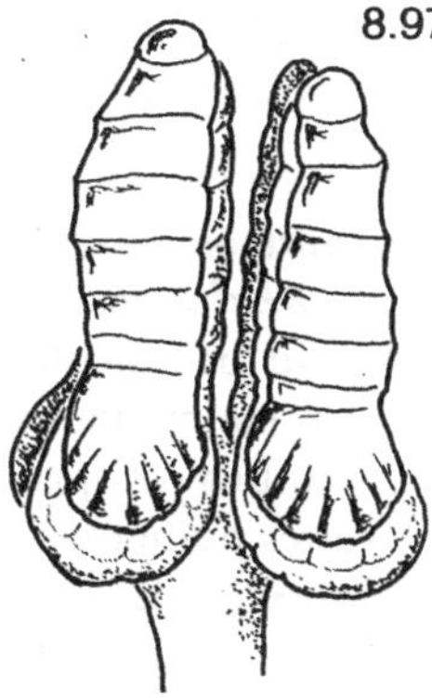

8.98

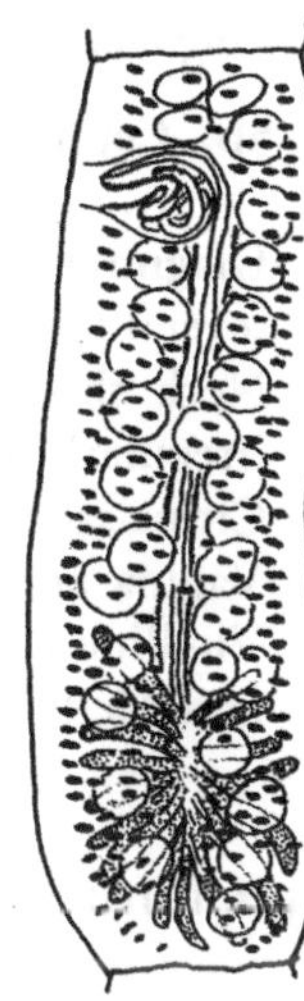

8.99

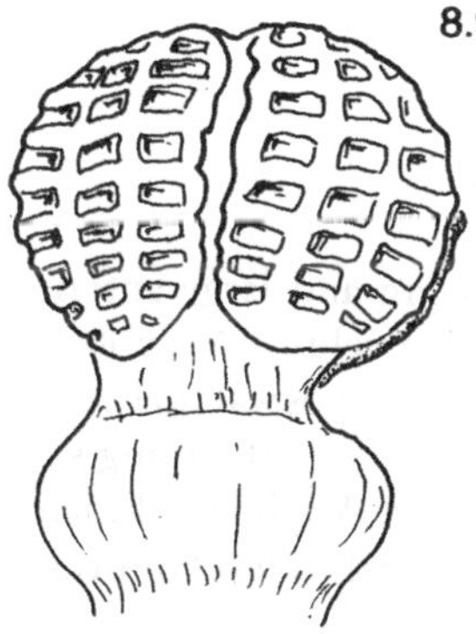

8.100

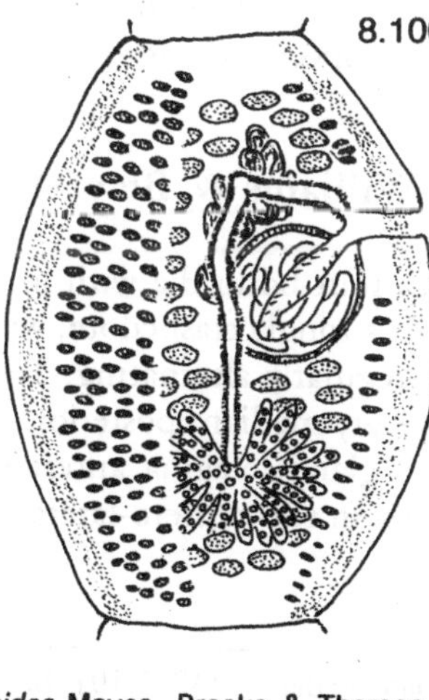

Figs 8.93, 8.94 *Rhinebothroides* Mayes, Brooks & Thorson, 1981. 8.93. Scolex. 8.94. Mature proglottid. (After Mayes *et al.*, 1981.)

Figs 8.95, 8.96 *Rhinebothrium* Linton, 1889. 8.95. Scolex. 8.96. Mature proglottid. (After Euzet, 1959.)

Figs 8.97, 8.98 *Duplicibothrium* Williams & Campbell, 1978. 8.97. Scolex. 8.98. Mature proglottid. (After Williams & Campbell, 1978.)

Figs 8.99, 8.100 *Glyphobothrium* Williams & Campbell, 1977. 8.99. Scolex. 8.100. Mature proglottid, vitelline follicles only on the left side. (After Williams & Campbell, 1978.)

Genital atrium present. Genital pore lateral, irregularly alternating. Testes numerous; postvaginal testes present on poral side. Ovary posterior, X-shaped in cross-section. Vagina glandular, anterior to cirrus-sac. Vitelline follicles circum-medullary, extending full length of proglottid. In Rhinobatidae. American North-Atlantic. Type-species *G. zwerneri* Williams & Campbell, 1974.

4b. Bothridium pedunculated with two longitudinal rows of loculi ...... 5.

5a. Cephalic peduncle very long .... *Caulobothrium* **Baer, 1948. (Figs 8.101–8.103)**

Diagnosis: Scolex with four sessile bothridia; divided into loculi by several transverse septa, and one longitudinal septum. No myzorhynchus. Cephalic peduncle greatly developed. Strobila craspedote or not, apolytic. Genital pores lateral, irregularly alternating. Testes numerous; postvaginal testes present on poral side. Cirrus-sac large with spinose cirrus. Ovary posterior, X-shaped in cross-section. Vagina anterior to cirrus-sac. Vaginal sphincter present. Vitelline follicles lateral, well-developed dorsally and ventrally. Uterus median with lateral diverticula. In Myliobatidae. North Atlantic and Mediterranean. Type-species *C. longicolle* (Linton, 1890).

5b. Cephalic peduncle short or absent .... *Rhabdotobothrium* **Euzet, 1953. (Figs 8.104, 8.105)**

Diagnosis: Scolex with four pedunculated bothridia, subdivided into loculi by several transverse septa and one median longitudinal septum. No myzorhynchus. Neck short or absent. Strobila craspedote, euapolytic. Genital pores lateral, irregularly alternating. Testes numerous; postvaginal testes present on poral side. Cirrus-sac large with spinose cirrus. Ovary posterior, X-shaped in cross-section. Vagina anterior to cirrus-sac, forming large loop in anterior half of proglottid. Vitelline follicles lateral. Uterus median, sac-like. In Dasyatidae. North Atlantic and Mediterranean. Type-species *R. dollfusi* Euzet, 1953.

Family Chimaerocestidae Williams & Bray, 1984

Diagnosis: Scolex with four bothridia with loculated margin and large anterior accessory sucker. Strobila strongly craspedote, with four-lobed velum. Anapolytic. Genital pores lateral, irregularly alternating. Testes numerous; postvaginal testes present on poral side. Ovary median, posterior, in form of swollen lobed ring surrounding large Mehlis's gland. Vagina anterior to cirrus-sac. Vaginal sphincter present. Vitelline follicles in two postero-lateral fields, restricted to level of ovary,

---

Figs 8.101, 8.103 *Caulobothrium* Baer, 1948. 8.101. Scolex with large cephalic peduncle. 8.102. Scolex. 8.103. Mature proglottid. (After Euzet, 1959.)

Figs 8.104, 8.105 *Rhabdotobothrium* Euzet, 1953. 8.104. Scolex. 8.105. Mature proglottid. (After Euzet, 1953.)

Figs 8.106, 8.107 *Chimaerocestos* Williams & Bray, 1984. 8.106. Scolex. 8.107. Mature proglottid. (After Williams & Bray, 1984.)

8.101

8.102

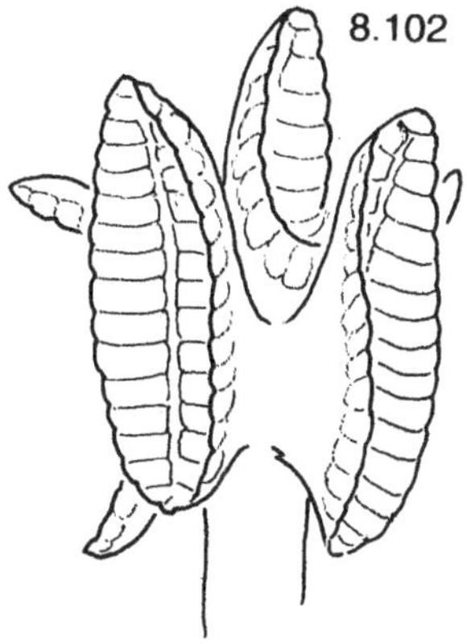

8.103

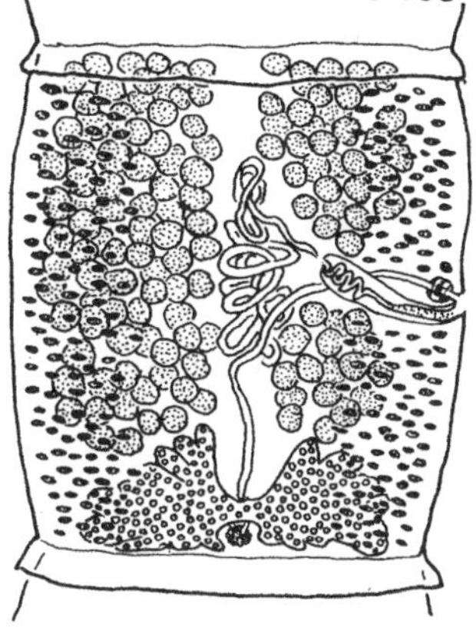

8.104

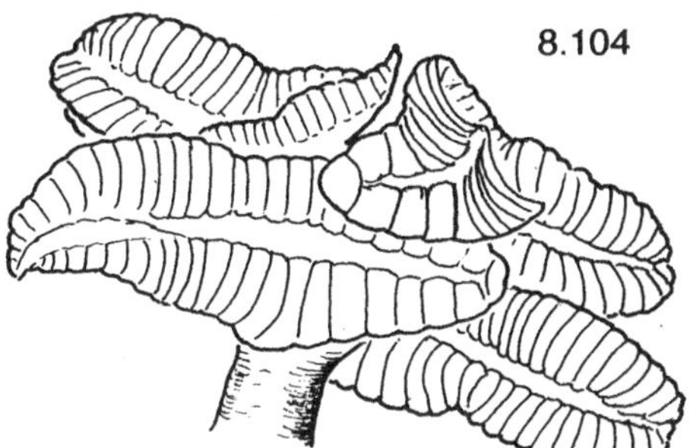

8.105

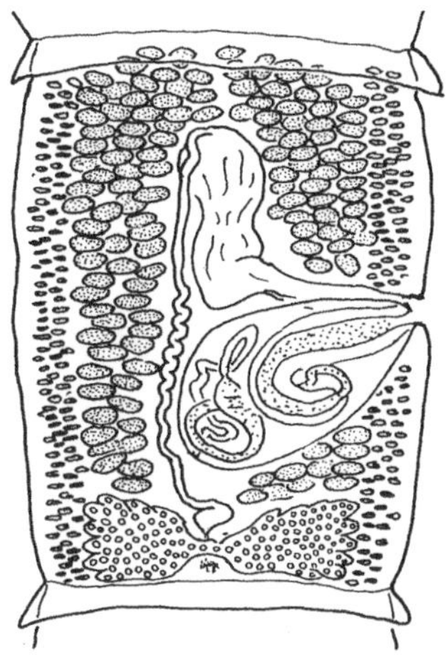

8.106

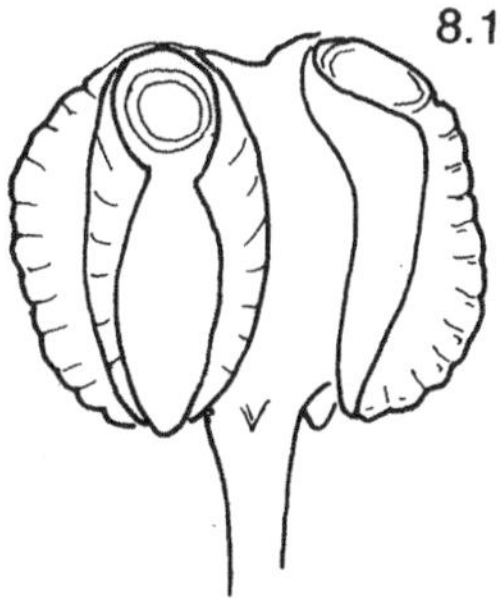

8.107

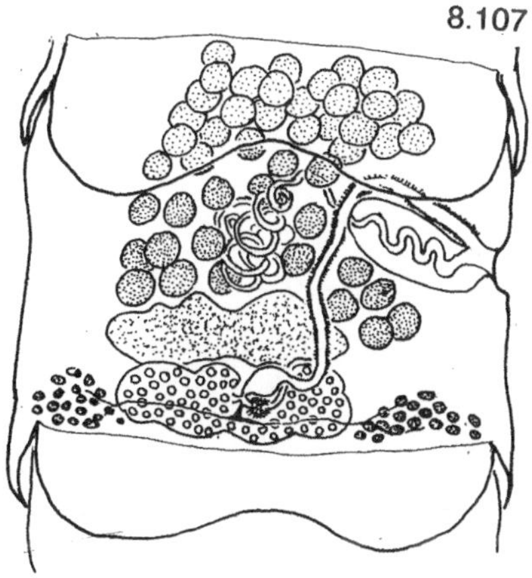

some follicles in four lobes of velum. Uterus sac-shaped. Eggs (15) clustered in cocoon. In Holocephala. Type and only genus *Chimaerocestos* Williams & Bray, 1984.

Genus *Chimaerocestos* Williams & Bray, 1984 (Figs 8.106, 8.107)

Diagnosis: With the characters of the family. In Holocephala (*Rhinochimaera atlantica*). North Atlantic. Type-species *C. prudhoei* Williams & Bray, 1984.

9 Order Lecanicephalidea Wardle & McLeod, 1952

L. EUZET

Laboratoire de Parasitologie comparée, Université Montpellier II, Station Méditerranéene de l'Environment Littoral, 1 Quai de la Daurade, 34200 Sète, France.

Introduction

Butler (1987a) gave an excellent historical review of the cestodes of selachians placed in the families Lecanicephalidae Braun, 1900 and Cephalobothriidae Pintner, 1928. The question of their systematic position remains outstanding. Some authors, such as Wardle & McLeod (1952), Yamaguti (1959) and Schmidt (1986) grouped them in a special order; others such as Southwell (1930a) and Euzet (1959) kept them in the Tetraphyllidea but in a separate superfamily. Butler (1987a), on the other hand, was inclined to classify them as a single family within the Tetraphyllidea.

When one lists the unique characters of these selachian parasites, one finds only the following features. The scolex is bipartite, the posterior part (pars basalis) bearing four circular suckers of the acetabular type and the anterior part (pars apicalis) very variable: a muscular sucker, a massive metoporhynchus or a crown of tentacles. Furthermore, there is the anatomy of the proglottid, in which the vagina opens posterior (sometimes to the side of) the cirrus-sac and does not cross the male duct.

Apart from the nature of the host, there are as many differences between these cestodes and the Tetraphyllidea as between the Proteocephalidea and the Tetraphyllidea. Therefore I have followed the scheme of Wardle & McLeod, accepted by Yamaguti (1959) and Schmidt (1986), and classified these cestodes in the order Lecanicephalidea Wardle & McLeod, 1952. However, the systematics within this order remain very confused. The old descriptions and some recent ones are too imprecise to be used. Many poorly characterized and poorly illustrated genera must be rejected. Some others can be retained but as *genera incertae sedis*.

In agreement with Campbell & Williams (1984) and Butler (1987a), I think that the majority of these parasites should be collected again from the type host, in the type locality (if possible) and redescribed to determine their morphology and anatomy.

To attempt a systematic scheme of the Lecanicephalidea, one can use only a few works. Apart from Butler (1987a), one may cite Baer (1948) for *Lecanicephalum* Linton, 1890 and *Parataenia* Linton, 1889; Subhapradha (1957) for *Polypocephalus* Braun, 1878; Euzet & Combes (1965) for *Tetragonocephalum* Shipley & Hornell, 1905; Campbell & Williams (1984) for *Tylocephalum* Linton, 1890; and Butler (1987a) for *Polypocephalus* Braun, 1878.

The question of the Balanobothriidae Pintner, 1928, often placed close to the Lecanicephalidae, has already been discussed (see Chapter 8 on the Tetraphyllidea). These cestodes are typical onchobothriid tetraphyllideans with a pair of double-pointed hooks per bothridium but the anterior region of the scolex, embedded in the intestinal wall, assumes a globular shape which has masked the presence of four bothridia. While recognizing the difficulties, I propose to subdivide the Lecanicephalidea, according to the morphology of the pars apicalis of the scolex, into four families.

ORDER LECANICEPHALIDEA WARDLE & McLEOD, 1952

(Syn. Lecanicephala Wardle & McLeod, 1952)

Diagnosis: Scolex bipartite; posterior part (pars basalis) with four acetabulate suckers; anterior part (pars apicalis) variable, with muscular sucker, massive metoporhynchus or crown of tentacles. Strobila craspedote or acraspedote, apolytic. Genital pores lateral, irregularly alternating. Testes few or numerous. Vagina opens posteriorly, sometimes laterally, to cirrus-sac, does not cross male duct. Ovary posterior, lobed or bilobed in cross-section. Vitelline follicles in lateral fields or encircling proglottid. Adults in spiral valve of elasmobranchs.

Key to families

1a. Pars apicalis divided into numerous unarmed tentacles .... Polypocephalidae Meggitt, 1924.
1b. Pars apicalis undivided .... 2.

2a. Pars apicalis a pedunculated disc, pars basalis with four large acetabula .... Anteroporidae n. fam.
2b. Pars apicalis non-pedunculated .... 3.

3a. Pars apicalis a muscular fungiform metoporhynchus .... Tetragonocephalidae Yamaguti, 1959.
3b. Pars apicalis a large muscular evaginable sucker .... Lecanicephalidae Braun, 1900.

Comments

Polypocephalidae Meggitt, 1924. Subhapradha (1951) described two species of *Polycephalus* (*P. coronatus* from *Rhynchobatus djeddensis* Forsskal, 1871 and *P.*

affinis from *Rhinobatos granulatus* Cuvier, 1824) with a scolex in which the tentacles are united in pairs but bifid at their extremity. This special character could represent a criterion for a new genus, but the apical division of the tentacles is very difficult to observe while they are retracted into the interior of the pars basalis.

Genera incertae sedis

Two doubtful genera are close to *Polypocephalus*.

Anthemobothrium Shipley & Hornell, 1906. The type-species is *A. pulchrum* Shipley & Hornell, 1906, a small cestode parasitic in *Trygon sephen* =*Hypolophus sephen*. The difference is in the tentacles which according to Shipley & Hornell (1906) have a plumed appearance. The anatomy is unknown. This parasite does not seem to have been reported since the original description.

Calycobothrium Southwell, 1911 (syn. *Cyclobothrium* of Southwell, 1911) is based on a single individual collected from *Aetobatis narinari*. The scolex does not appear to have a pars basalis with four acetabula. The short, bulky tentacles are arranged in a circle around a disc in the centre of which a pore opens. This arrangement of the tentacles is close to that illustrated by Butler (1987b) in *Polypocephalus* sp. from *Himantura* sp. However, in this cestode, the scolex has a pars basalis with four acetabula. As described by Southwell (1911), the anatomy of *Calycobothrium typicum* is of the tetraphyllidean type. This parasite should be redescribed from the type host.

Anteroporidae n. fam.

The genus *Anteropora* was created with a single species, *A. indica* Subhapradha, 1957. This parasite from the spiral valve of *Narcine timlei* represents, from the evidence, a detached proglottid. Schmidt (1986) considered the genus to be doubtful. When examining the anatomy of this proglottid, one observes a marked similarity to the anatomy of *Discobothrium japonicum* found by Yamaguti (1934) in the spiral valve of *Narke japonica*. In both cestodes, the vagina opens behind the cirrus-sac, there is a small number of testes in a single median row, and the lateral vitelline follicles do not extend in front of the genital pore. These two cestodes of Torpediniformes belong to the same genus, which I place in the Lecanicephalidea, as did Wardle & McLeod (1952). I have discussed, in connection with *Discobothrium fallax* van Beneden, 1870, the validity of *Discobothrium* van Beneden, 1870 and concluded that it was a synonym of *Echeneibothrium* van Beneden, 1870 (see Chapter 8). The species placed in *Discobothrium* (such as *D. japonicum*) should be renamed at the generic level. In accordance with Article 23 (f) of the *International Code of Zoological Nomenclature*, I propose to retain *Anteropora* Subhapradha, 1957 and to place the parasite found by Yamaguti in *Narke japonica* in this genus. We therefore have *Anteropora japonica* (Yamaguti, 1934) n. comb. (syn. *Discobothrium japonicum* Yamaguti, 1934). This new concept of *Anteropora* entails the suppression of the Monoporophylleidae

Subhapradha, 1957, since the family name is to be rejected as it does not correspond to that of a genus included in the family. This leads to the suppression of *Monoporophylleus* Shinde & Chincholikar, 1977, created to replace *Anteropora* in order to put the generic name into agreement with that of the family. It becomes a synonym of *Anteropora.*

Tetragonocephalidae Yamaguti, 1959

The description by Euzet & Combes (1965) of a lecanicephalidean from *Dasyatis violacea* in the Mediterranean, attributed to *Tetragonocephalum uarnak* Shipley & Hornell, 1905, poses the problem of the type of host specificity exhibited by the Lecanicephalidea. Given the confusion in the group, no conclusion can be reached at present.

Given the inadequacy of old descriptions, several questions remain to be answered.

Adelobothrium Shipley, 1900. Shipley's (1900) description of *A. aetobatidis*, parasite of *Aetobatis narinari*, shows that the scolex has a structure of the metoporhynchus type. The description of the anatomy suggest that the vitelline follicles are circumcortical. If this is so, *Adelobothrium* will be synonymous with *Tylocephalum*. The species described by Southwell (1925) as *Adelobothrium aetobatidis* Shipley, 1900 but from *Rhynchobatus djeddensis* is perhaps a different species but should be also placed in *Tylocephalum* because the vitelline follicles are circum-medullary.

Lecanicephalidae Braun, 1900

The type-species was redescribed by Baer (1948). It is probable that several cestodes placed in the genera *Cephalobothrium* Shipley & Hornell, 1906, *Hexacanalis* Perrenoud, 1931 or *Tylocephalum* belong to this family.

Cephalobothrium Shipley & Hornell, 1906. The three species described in this genus do not provide adequate information for a definition of the genus. The anatomy of *C. aetobatidis* Shipley & Hornell, 1906, the type-species, is virtually unknown. The scolex and anatomy of *C. abruptum* Southwell, 1911 from *Pteroplatea micrura* are consistent with the genus *Lecanicephalum*, to which it is assigned here. *C. variabile* Southwell, 1911 from *Pristis cuspidatus* and *Trygon (Dasyatis) kuhli* has a scolex like that of *Tylocephalum* but the vagina is anterior to the cirrus-sac and the anatomy is, therefore, of the tetraphyllidean type. *Cephalobothrium* is treated here as a *genus inquirendum*.

Hexacanalis Perrenoud, 1931. This genus is based mainly on the osmoregulatory system of *Cephalobothrium abruptum*. As this is the type-species, which has now been transferred to *Lecanicephalum, Hexacanalis* becomes a synonym of this genus.

Staurobothrium Shipley & Hornell, 1905. *S. aetobatidis* Shipley & Hornell, 1905 from *Aetobatis narinari* has an anatomy of the lecanicephalidean type with the vagina posterior to the cirrus-sac. The scolex was described as having four large round pedunculated bothridia and a tiny apical papilla but Southwell (1925), who examined specimens from *A. narinari*, did not see an apical papilla. Because of the anatomy of the proglottid, I place *Staurobothrium* Shipley & Hornell, 1905 in the Lecanicephalidea but as a *genus incertae sedis* because the peculiar morphology of the scolex may justify a new family.

Comments on problematic genera

Spinocephalum Deshmukh, 1980. The type-species *S. rhinobatii* Deshmukh, 1980 was collected from *Rhinobatus granulatus* = *Rhinobatos granulatus* at Veraval (west coast of India). This species belongs to *Tetragonocephalum*, having lateral vitelline follicles. The only character given to distinguish the genus was the presence of small spines on the metoporhynchus (Deshmukh, 1980). However, as already remarked by Campbell & Williams (1984), these small spines had been noted by Euzet & Combes (1965) on the scolex of *T. uarnak* Shipley & Hornell, 1906. *Spinocephalum* is a synonym of *Tetragonocephalum*.

A certain number of genera classified as Lecanicephalidea do not belong to this order and are so badly defined that they cannot be classified. They are, in chronological order:

Aphanobothrium Seurat, 1906. Southwell (1925) gave this genus as a synonym of *Tylocephalum* Linton, 1890. *Aphanobothrium* was created by Seurat (1906) for larvae encapsulated in the flesh of *Pintada margaritifera* in the Gambier Islands (French Polynesia). The larvae were attributed to adult cestodes parasitic in *Aetobatis narinari* but without a description. As *A. narinari* has been cited as host of about ten lecanicephalideans, apparently belonging to several of the families defined above, one cannot decide on the synonymy of *Aphanobothrium*. It should be considered a *genus inquirendum*.

Kystocephalus Shipley & Hornell, 1906. The scolex of the type-species from *Aetobatis narinari* is described as having a myzorhynchus covered by a membrane. This could correspond to a scolex of *Tetragonocephalum* detached from the intestinal wall with a fragment of epithelium or a *Lecanicephalum* with the apical sucker more or less everted. Since its anatomy is poorly known, the precise synonymy of *Kystocephalus* cannot be decided and it should be considered a *genus inquirendum*.

Yogeshwaria Chincholikar & Shinde, 1976. The type-species of the genus, *Y. nagabushani* Chincholikar & Shinde, 1976, was collected from a *Trygon* sp. at Ratnagiri (west coast of India). This species, based on a single immature specimen, was characterized by a scolex with two bothria (*sic*). This species is not recognizable and, the host species being unknown, cannot be recovered again. *Yogeshwaria* should be suppressed. It is treated here as a *genus inquirendum*. Chincholikar & Shinde (1976) compared their species with *Diagonobothrium asymmetricum* Shipley & Hornell, 1906 from *Myliobatis maculata* = *Aetomyleus maculatus* which, described from a single specimen, probably represents a diphyllidean scolex which has lost the rostellar spines and those of the cephalic peduncle.

Flapocephalus Deshmukh, 1979. The type-species is *F. trygonis* Deshmukh, 1977 from *Trygon sephen* = *Hypolophus sephen* collected at Ratnagiri (west coast of India). From the morphology of the scolex, this cestode belongs to the Lecanicephalidea. Its very compressed scolex appears to be of the *Lecanicephalum* type but its anatomy, with the vagina opening anterior to the cirrus-sac, is of the tetraphyllidean type. The morphological and anatomical features given by Deshmukh (1979) to characterize *Flapocephalus* are not found in *F. saurashtri* Shinde & Deshmukh, 1979. In this species, based on a single individual from *A. sephen* = *Hypolophus sephen*, one cannot distinguish the two lobes of the pars apicalis given as the principal character of the genus. However, the anatomy with the vagina posterior to the cirrus-sac is of the lecanicephalid type. *Flapocephalus* is not recognizable and should be suppressed. It is treated here as a *genus inquirendum*.

Trygonicephalum Shinde & Jadhav, 1984. The type-species *T. ratnagiriensis* Shinde & Jadhav, 1984, was collected from *Trygon sephen* = *Hypolophus sephen* at Ratnagiri on the west coast of India. According to Shinde & Jadhav (1984), the scolex has a large oval apical disc and two bothridia. The anatomy is of the tetraphyllidean type with the vagina anterior to the cirrus-sac. This description does not correspond to any known genus. *Trygonicephalum* cannot be classified among the tetraphyllideans or the lecanicephalideans and should be suppressed. It is treated here as a *genus inquirendum*.

Guptaia Malhotra, 1985. The type-species, placed in the Lecanicephalidae, is *G. garhwalensis* Malhotra, 1985. It was collected in fresh water in a cyprinid teleost, *Schizothorax richardsoni*. This cestode probably belongs to another order and should be excluded from the Lecanicephalidea. It is to be treated as a *genus incertae sedis*.

Lecanicephalidea incertae sedis

Eniochobothrium Shipley & Hornell, 1906 has only one species, *E. gracile* Shipley & Hornell, 1906, collected once in Sri Lanka from *Rhinoptera javanica*. Since Southwell (1930a), it has been considered to belong to the Lecanicephalidae or as *incertae sedis* in this family (Yamaguti, 1959) or as *genus inquirendum* among the Lecanicephala (Wardle & McLeod, 1952). From the general description, the scolex has ‘four suckers and a conspicuous rostellum’. The morphology of the strobila is unique. There are three regions: an anterior with very enlarged proglottids and a posterior region of four to five broad sexually mature proglottids, separated by a mid-region of about ten narrow proglottids. The genital pores alternate irregularly and there is a large cirrus-sac containing an evaginable cirrus. The vitelline follicles are lateral. The remainder of the anatomy is unknown. Despite a strobila which renders this cestode easily recognizable, this parasite has not been found again in the type-host (*R. javanica*) since the original description, although this stingray had been examined several times by Southwell and by Indian researchers.

When studying cestodes of elasmobranchs, one is struck by the strange resemblance between the strobila of *Eniochobothrium* and that of *Litobothrium* Dailey, 1969 (syn. *Renyxa* Kurochin & Slankis, 1973). However, in *Litobothrium* the scolex is represented by a simple apical sucker. The anatomy is of the tetraphyllidean type and it is parasitic in Lamniformes (Alopiidae and Odontaspididae).

Several hypotheses could be advanced to explain these similarities and dissimilarities: *Eniochobothrium* and *Litobothrium* are different cestodes and the morphological resemblance of the strobila is the result of convergence; or *Eniochobothrium* is identical with *Litobothrium* and *Rhinoptera javanica* is an accidental host in which the parasite does not develop fully and has retained the true scolex; or the host has been confused and its presence in *Rhinoptera* is an error. Only further research will permit a decision between these different hypotheses.

Hornellobothrium Shipley & Hornell, 1906. Type-species *H. cobraeformis* Shipley & Hornell, 1906. This minute species (2 mm) from *Aetobatis narinari* was recently recovered by Butler (1987b) who described its morphology precisely. The scolex is composed of four sessile masses (bothridia?) covered with spines and an apical organ which appears to be erectile. The species seems to be euapolytic. The anatomy is very poorly known. We have provisionally classified this genus in the Lecanicephalidea.

Family Polypocephalidae Meggitt, 1924

Diagnosis: Scolex with pars apicalis divided into evaginable tentacles (sometimes paired) able to project suddenly to the exterior; pars basalis cushion-like, bearing four acetabula. Strobila craspedote or acraspedote, apolytic. Genital pores lateral, irregularly alternating. Testes few (four or six), large, aligned along median axis of proglottid. External and internal seminal vesicles present. Ovary bilobed, posterior. Vagina posterior to cirrus-sac. Vitelline follicles lateral. Type and only genus *Polypocephalus* Braun, 1878.

Genus *Polypocephalus* Braun, 1878 (Figs 9.1, 9.2)

(Syns *Parataenia* Linton, 1889; *Thysanobothrium* Shipley & Hornell, 1906)

Diagnosis: With the characters of the family. In Rajiformes and Myliobatiformes. Cosmopolitan. Type-species *P. radiatus* Braun, 1878.

Family Anteroporidae n. fam.

Diagnosis: Scolex with pars apicalis a pedunculated muscular disc, pars basalis with four large bothridia-like acetabula. Strobila acraspedote, apolytic. Genital

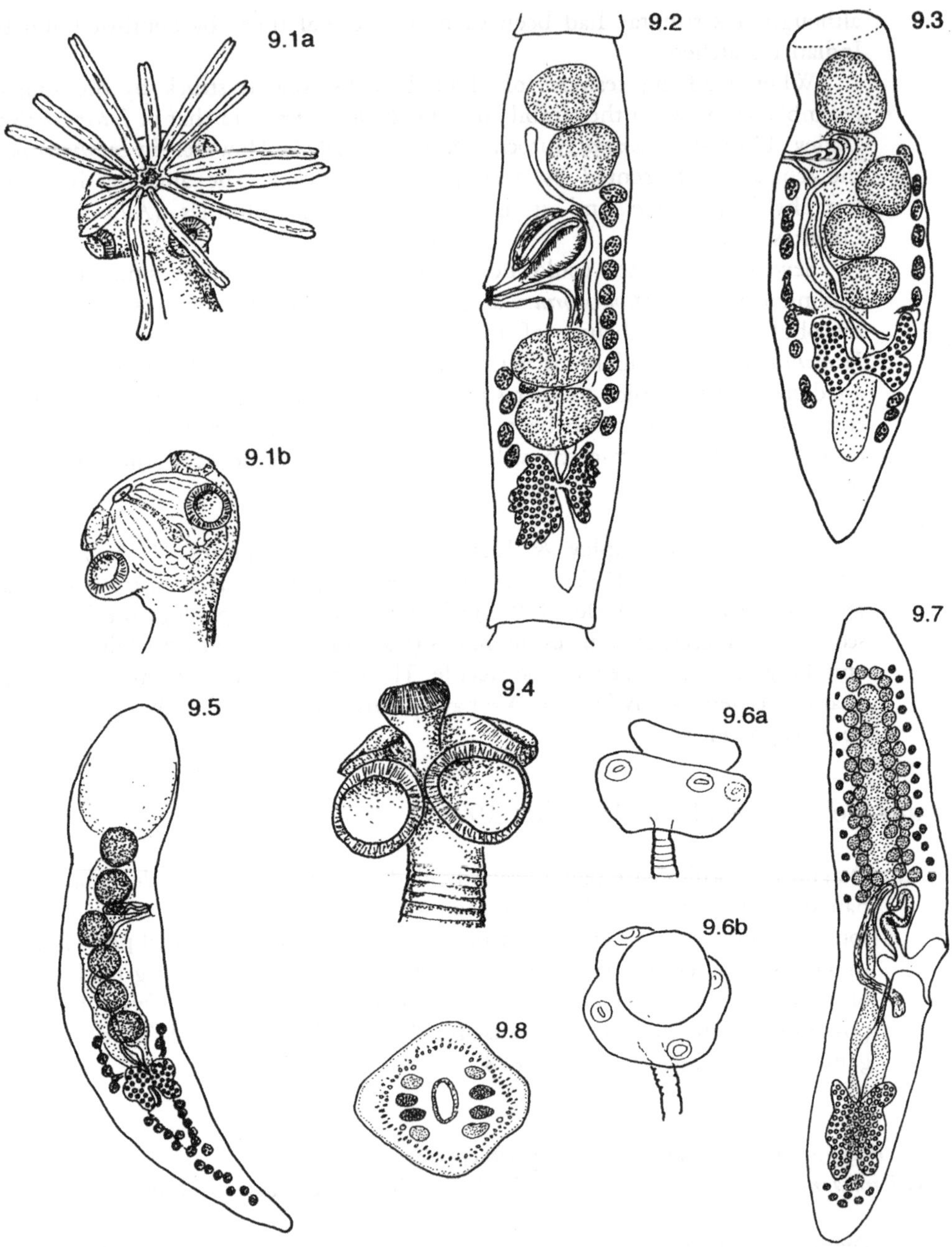

Figs 9.1, 9.2 *Polypocephalus* Braun, 1978. 9.1a, b. Scolex showing two different shapes. 9.2. Mature proglottid, dorsal view. (After Baer, 1948.)
Figs 9.3–9.5 *Anteropora* Subhapradha, 1957. 9.3. Mature proglottid, dorsal view. (After Subhapradha, 1957.) 9.4, 9.5. *A. japonica* (Yamaguti, 1934). 9.4. Scolex. 9.5. Mature proglottid, dorsal view. (After Yamaguti, 1934.)
Figs 9.6–9.8 *Tetragonocephalum* Shipley & Hornell, 1905. 9.6a, b. Scolex, showing two shapes. 9.7. Mature proglottid, dorsal view. 9.8. Transverse section showing lateral vitelline follicles. (After Euzet & Combes, 1965.)

pores lateral, irregularly alternating. Testes few (four or six) aligned along median axis of proglottid. Ovary lobate, posterior. Vagina posterior to cirrus-sac. Vitelline follicles lateral, posterior to genital pore. Type and only genus *Anteropora* Subhapradha, 1957.

Genus *Anteropora* Subhapradha, 1957 (Figs 9.3–9.5)
(Syn. *Monoporophylleus* Shinde & Chincholikar, 1977)

Diagnosis: With the characters of the family. In Torpedinidae. Japan, India. Type-species *A. indica* Subhapradha, 1957.

Family Tetragonocephalidae Yamaguti, 1959

Diagnosis: Scolex with pars apicalis a fungiform metoporhynchus, not invaginable. Type-genus *Tetragonocephalum* Shipley & Hornell, 1905.

Key to genera

1a. Vitelline follicles lateral ....
.... ***Tetragonocephalum*** **Shipley & Hornell, 1905. (Figs 9.6–9.8)**
(Syn. *Spinocephalum* Deshmukh, 1980.)

Diagnosis: Scolex with pars apicalis a fungiform metoporhynchus covered with minute spines; pars basalis cushion-like bearing four acetabula. Strobila acraspedote, apolytic. Proterogynous. Genital pores lateral, enlarged, irregularly alternating. Testes anterior, numerous, no postvaginal testes on poral side. External and internal seminal vesicles present. Ovary posterior, bilobed, circular in cross-section. Vagina posterior to cirrus-sac. Vitelline follicles anterior and posterior to ovary. Uterus bisacculate. In Myliobatiformes. Cosmopolitan. Type-species *T. trygonis* Shipley & Hornell, 1905.

1b. Vitelline follicles circum-medullary ....
.... ***Tylocephalum*** **Linton, 1890. (Figs 9.9–9.11)**

Diagnosis: Scolex with pars apicalis a fungiform metoporhynchus, not invaginable; pars basalis cushion-like, bearing four acetabula. Strobila craspedote. Genital pores lateral, irregularly alternating. Testes numerous, anterior. Postvaginal testes present on poral side. External seminal vesicle only. Ovary posterior, bilobed in cross-section. Uterus medioventral. In Rhinopteridae. North Atlantic. Type-species *T. pingue* Linton, 1890.

Family Lecanicephalidae Braun, 1900

Diagnosis: Scolex with pars apicalis a large circular sucker, highly polymorphic, invaginable or evaginating entirely to resemble a metoporhynchus; pars basalis cushion-like, bearing four acetabula. Strobila acraspedote, apolytic. Genital pores

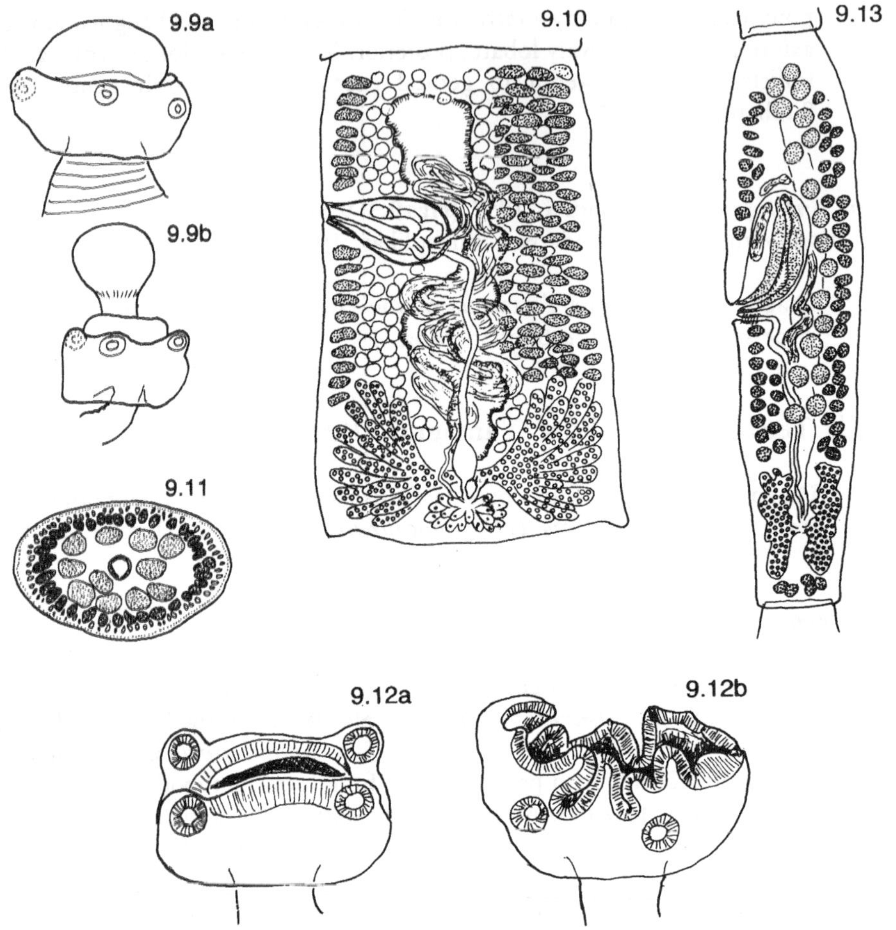

Figs 9.9–9.11 *Tylocephalum* Linton, 1890. 9.9a, b. Scolex showing two shapes. 9.10. Mature proglottid, dorsal view. 9.11. Transverse section showing circum-medullary vitelline follicles. (After Campbell & Williams, 1984.)
Figs 9.12, 9.13 *Lecanicephalum* Linton, 1890. 9.12a, b. Scolex showing two shapes. 9.13. Mature proglottid, dorsal view. (After Baer, 1948.)

lateral, irregularly alternating. Testes numerous; postvaginal testes present on poral side. External and internal seminal vesicles present. Ovary posterior, bilobed in cross-section. Vitelline follicles lateral, anterior and posterior to ovary. Type and only genus *Lecanicephalum* Linton, 1890.

Genus *Lecanicephalum* Linton, 1890 (Figs 9.12, 9.13)
(Syn. *Hexacanalis* Perrenoud, 1931)

Diagnosis: With the characters of the family. In Myliobatiformes. Cosmopolitan. Type-species *L. peltatum* Linton, 1890.

10 Order Pseudophyllidea Carus, 1863

R.A. BRAY[1], A. JONES[2] & K.I. ANDERSEN[3]

[1]*Department of Zoology, The National History Museum, Cromwell Road, London SW7 5BD, UK:* [2]*International Institute of Parasitology, 395A Hatfield Road, St Albans, Herts AL4 0XU, UK:* [3]*University of Oslo, Zoological Museum, Sars Gate 1, N–0562 Oslo 5, Norway.*

Introduction

During this revision of the Order Pseudophyllidea Carus, 1863, characters regarded as significant at the family and generic levels have been re-evaluated, supported by re-examination of type material whenever possible. For reasons given in more detail in the text, the use of such characters as eggs operculate or not, or embryonated or not when laid, is not considered to be useful at the family level. A number of synonymies at the family level have therefore been made.

ORDER PSEUDOPHYLLIDEA CARUS, 1863

Diagnosis: Scolex usually present; with two bothria, one dorsal, one ventral; armed or unarmed. Scolex rarely absent or replaced by pseudoscolex or scolex deformatus. Bothria rarely absent. External segmentation usually present, complete or incomplete; rarely absent. Internal segmentation complete or not. Proglottids craspedote or acraspedote, usually wider than long, anapolytic. Reproductive organs usually single, rarely double or multiple. Genital pores single or double, dorsal, ventral or marginal. Testes numerous, medullary. Cirrus-sac usually present, rarely absent. Ovary medullary, posterior. Vitellarium follicular, rarely compact; follicles usually cortical, sometimes in medulla, sometimes intruding between longitudinal muscle bundles. Uterus variable in shape; uterine sac present or absent; uterine pore present, ventral. First larval stage in crustaceans, second usually in fish but rarely in other vertebrate groups. Adults predominantly in fish, also in other vertebrate groups.

Key to families:

1a. **Genital pore and uterine pore both on ventral surface** .... **5.**
1b. **Genital pore mediodorsal to marginal, uterine pore ventral** .... **2.**

2a. **Genital pore dorsal, median** .... **Bothriocephalidae Blanchard, 1849.**
(Syns Ptychobothriidae Lühe, 1902; Acompsocephalidae Rees, 1969.)
2b. **Genital pore marginal or submarginal** .... **3.**

3a. **Genital pore submarginal** .... **Echinophallidae Schumacher, 1914.**
(Syn. Parabothriocephalidae Yamaguti, 1959.)
3b. **Genital pore marginal** .... **4.**

4a. **Vitellarium compact** .... **Philobythiidae Campbell, 1977.**
4b. **Vitellarium follicular** .... **Triaenophoridae Lönnberg, 1889.**
(Syn. Amphicotylidae Ariola, 1899.)

5a. **Cirrus-sac present** .... **Diphyllobothriidae Lühe, 1910.**
(Syn. Dibothriocephalidae Lühe, 1902.)
5b. **Cirrus-sac absent** .... **Cephalochlamydidae Yamaguti, 1959.**

Family Bothriocephalidae Blanchard, 1849

(Syns Ptychobothriidae Lühe, 1902; Acompsocephalidae Rees, 1969)

The families Bothriocephalidae and Ptychobothriidae have been distinguished traditionally by the presence of operculated, unembryonated eggs in the first family and of unoperculated, embryonated eggs in the second. Both characters are difficult to apply accurately and consistently. The presence or absence of an operculum is frequently difficult to establish and the information is often missing from descriptions and diagnoses. The degree of embryonation *in utero* is readily established in some cases but not in others and is known to vary in some pseudophyllideans according to temperature, e.g. in species of *Triaenophorus* (see Kuperman, 1973, transl. 1981). While the family and generic diagnoses of the Bothriocephalidae and Ptychobothriidae given by Yamaguti (1959) and Schmidt (1986) are consistent within each author's system in respect of the operculate or non-operculate condition of the eggs, there are inconsistencies with regard to embryonation. Schmidt's diagnosis of the Ptychobothriidae states that they have embryonated eggs but in his diagnoses of *Senga* Dollfus, 1934 and *Polyonchobothrium* Diesing, 1854 the eggs are described as not embryonated. The information is not given for the eight other genera in the family. Yamaguti's (1959) treatment of *Senga* is similar to that of Schmidt. Personal observations on *Polyonchobothrium polypteri* have shown that some embryos are developed to the point of having oncospheral hooks, although the majority in the terminal part of the uterus are not developed to this stage. These inconsistencies highlight the absence of reliable information on these points. The use of operculation and of embryonation as family characters is considered unreliable and consequently the Ptychobothriidae has been synonymized with the Bothriocephalidae. The status of individual genera allocated to both families and to the Acompsocephalidae Rees, 1969 is discussed below.

Acompsocephalum Rees, 1969 is synonymous with *Anantrum* Overstreet, 1968, which has priority. Rees (1969) erected the family Acompsocephalidae because her genus lacked external segmentation and had no scolex or specialized attachment organs. *Anantrum* is the only pseudophyllidean genus completely devoid of external segmentation although in many genera it is incomplete. Overstreet (1968) placed his genus in the Bothriocephalidae. Jensen & Heckmann (1977), also placed *Anantrum* in the Bothriocephalidae because of the variation in scolex shape and in external segmentation found in this family. Overstreet (1968) and Jensen & Heckmann (1977) did not specify their view of the composition of the family. As treated by Yamaguti (1959), it contained three genera all with normal scoleces. In the Bothriocephalidae of Schmidt (1986), *Anantrum* is the only genus to lack attachment organs. Its eggs, according to Jensen & Heckmann (1977) and Overstreet (1968), but not Rees (1969), are operculate. We consider it a member of the Bothriocephalidae.

Penetrocephalus Rao, 1960 was erected for *Bothriocephalus ganapatii*, a name first used by Rao (1954) without a description and used again in 1960. The same cestode was described by Subhapradha (1955) as *B. penetratus*. Rao (1960) distinguished it by its initially dibothriate scolex, lacking an apical disc, which penetrated the gut wall of the host to enter the liver, subsequently becoming encapsulated and degenerate. Also, the posterolateral margins of the proglottids were markedly craspedote. Protasova (1977) declared *B. ganapatii* Rao, 1954 a *nomen nudum* and retained *B. penetratus* in *Bothriocephalus*. *Penetrocephalus* is treated here as a synonym of *Bothriocephalus* since, following Protasova (1977), the location of the scolex and the craspedote proglottids are not considered sufficient to justify the genus.

Alloptychobothrium Yamaguti, 1968 is provisionally retained. Erected for *A. spilonotopteri* from flying fish, *Cypselurus spilonotopterus* from Hawaii by Yamaguti (1968), it closely resembles *Plicatobothrium* Cable & Michaelis, 1967 from *C. bahiensis* from Curaçao. The major difference is that *Plicatobothrium* has a conspicuous Y-shaped uterine sac (Cable & Michaelis, 1967); a uterine sac is said to be lacking in *Alloptychobothrium* and the eggs are said to be embryonated (Yamaguti, 1968).

Circumoncobothrium Shinde, 1968 was distinguished by an uninterrupted circle of hooks on the apical disc. The type and subsequent species are all based on single specimens (Shinde, 1968a, 1977; Shinde & Chincholikar, 1977; Shinde & Jadhav, 1976; Chincholikar & Shinde, 1977) and the continuity of the hook circle is probably a fixation artefact resulting from pressure. It is synonymous with *Senga* Dollfus, 1934, species of which occur in the same hosts.

Oncobothriocephalus Yamaguti, 1959 was established by Yamaguti (1959) for *Ptychobothrium armatum* Fuhrmann, 1902 from *Turdus parochus* in Egypt. The species had been listed as a species of *Polyonchobothrium* Diesing, 1854 by Meggitt (1930). According to Tadros (1968), it is morphologically consistent with *P. polypteri* (Leydig, 1853) and may have been a fish parasite ingested by a bird, or the collection data attributing it to a bird host may have been in error. Following Tadros (1968), Protasova (1977) and Schmidt (1986), it is regarded as a synonym of *Polyonchobothrium*.

Ptychobothrioides Yamaguti, 1959 was erected for *Bothriocephalus spiraliceps* Volz, 1900 from *Falco concolor* in Abyssinia. As pointed out by Tadros (1966), its characters resemble those of *Ptychobothrium* Lönnberg, 1889, with which he synonymized it. It is regarded as a *genus inquirendum*, following Protasova (1977).

Family diagnosis

Scolex variable in shape, usually with two bothria; rarely lacking attachment organs. Apical disc present or absent; if present, armed or unarmed. Bothria weakly to strongly developed; margins inrolled or crenulated or not. External segmentation present or absent, complete or not. Strobila rarely spirally twisted. Proglottids craspedote or not, usually wider than long. Genital pores dorsal, median or submedian. Testes in dorsal medulla. Ovary posterior. Vitelline follicles usually cortical, may be intermingled with longitudinal muscle fibres. Uterus opens ventrally through median or submedian uterine pore. Eggs operculate or not, embryonated or not. In teleosts. Type-genus *Bothriocephalus* Rudolphi, 1808.

Key to genera

1a. External segmentation completely absent. Scolex unarmed. Bothria and apical disc absent .... *Anantrum* Overstreet, 1968. (Figs 10.1–10.3)
(Syn. *Acompsocephalum* Rees, 1969.)

Diagnosis: Scolex variable in shape (conical to papilliform; mushroom-shaped). Strobila spirally twisted or not. Genital pores dorsomedian. Testes medullary in two lateral bands. Ovary bilobed, posterior, median. Vitelline follicles cortical, mainly in two lateral fields dorsally and ventrally. Uterus with uterine sac; uterine pore ventral, slightly submedian, irregularly alternating. Eggs operculate, unembryonated. In marine teleosts (*Synodus* spp.). Western Atlantic, eastern Pacific. Type-species *A. tortum* (Linton, 1905).

1b. External segmentation present. Bothria present; apical disc present or absent .... 2.

2a. Scolex armed .... 3.
2b. Scolex unarmed .... 5.

3a. Scolex strongly compressed laterally; apical disc armed marginally with numerous tiny hooks. In marine teleosts .... *Oncodiscus* Yamaguti, 1934. (Figs 10.4–10.6)

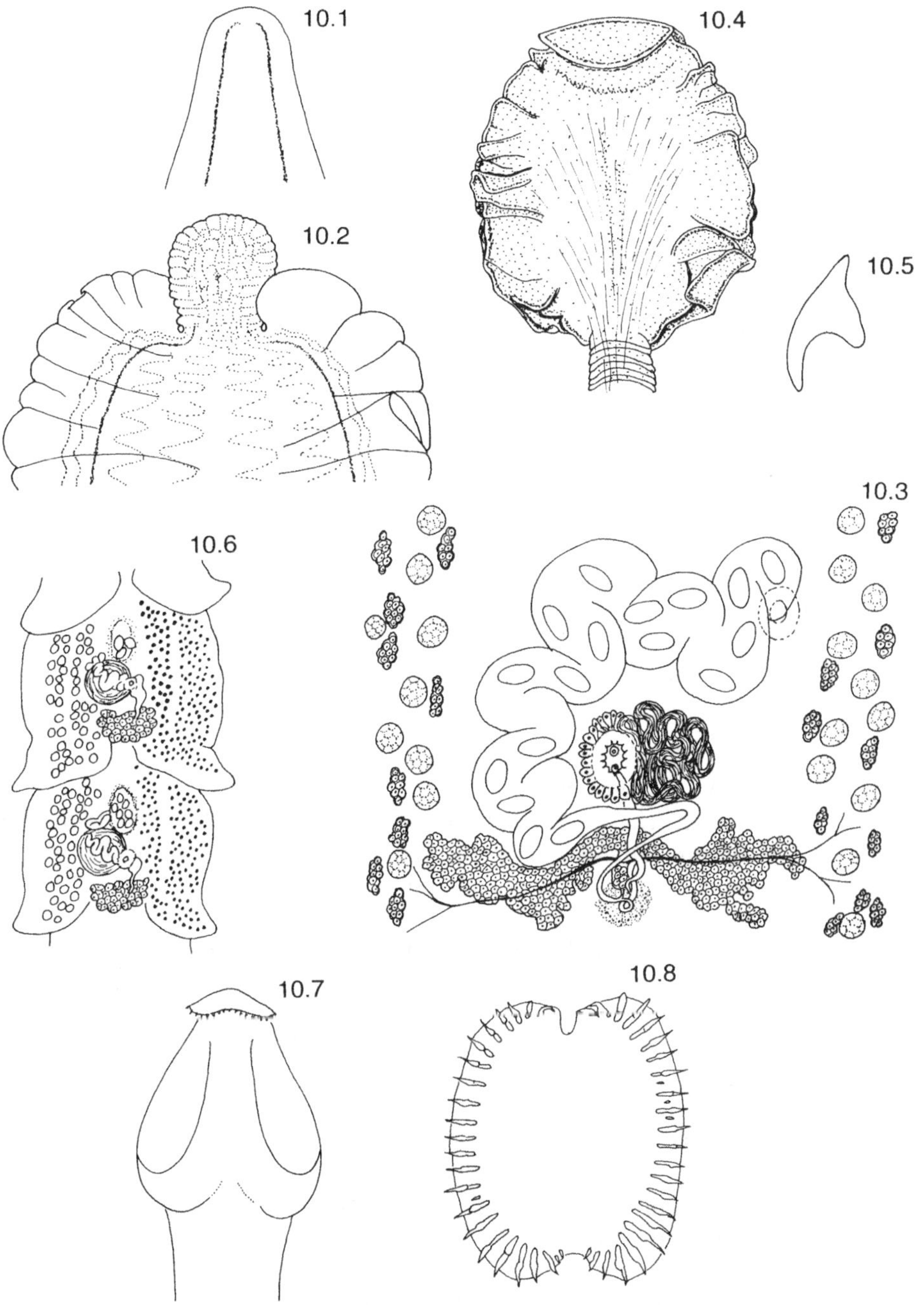

Figs 10.1-10.3 *Anantrum tortum* (Linton, 1905). 10.1. Scolex, relaxed. 10.2. Scolex, contracted. 10.3. Reproductive organs. After Rees (1969).
Figs 10.4-10.6 *Oncodiscus sauridae* Yamaguti, 1934. 10.4. Scolex. 10.5. Hook. 10.6. Gravid proglottids, testes shown on left, vitelline follicles on right. After Yamaguti (1934).
Figs 10.7, 10.8 *Senga besnardi* Dollfus, 1934. 10.7. Scolex. 10.8. Apical disc. After Dollfus (1934a).

Diagnosis: Bothria well-developed with strongly crenulated margins. Apical disc bilobed. Neck absent. External segmentation complete. Proglottids craspedote with a median posterior notch. Genital pore dorsomedian, slightly post-equatorial. Cirrus-sac large, thin-walled. Cirrus protrusible, unarmed. Ovary transversely elongate, lobed, posterior. Vagina distended and muscular distally, opens anteriorly to cirrus-sac. Vitelline follicles circumcortical. Uterine duct coils forward to large uterine sac; uterine pore ventral, median, equatorial. Eggs operculate, unembryonated. In marine teleosts. Japan, India. Type-species *O. sauridae* Yamaguti, 1934.

3b. Scolex rectangular; apical disc armed with hooks arranged in two semicircles or four quadrants. In freshwater teleosts .... 4.

4a. Hooks on scolex arranged in two semicircles .... *Senga* Dollfus, 1934. (Figs 10.7, 10.8)
(Syn. *Circumoncobothrium* Shinde, 1968.)

Diagnosis: Bothria shallow with well-developed margins. Apical disc indented dorsally and ventrally. Neck absent. External metamerism present but incomplete. Proglottids acraspedote, wider than long except when gravid. Genital pores dorsal, virtually median. Testes medullary, in two lateral fields. Ovary compact, median, posterior. Vitelline follicles cortical, continuous around margins of proglottid. Uterus loops forward, uterine sac opens ventrally by median pore near anterior margin of proglottid. Eggs anoperculate, unembryonated. In freshwater teleosts. India, Pakistan, China. Type-species *S. besnardi* Dollfus, 1934.

4b. Hooks arranged in four quadrants .... *Polyonchobothrium* Diesing, 1854. (Figs 10.9–10.11)
(Syns *Tetracampos* Wedl, 1861; *Oncobothriocephalus* Yamaguti, 1959.)

Diagnosis: Scolex rectangular. Bothria shallow; apical disc present, armed. External segmentation present but feebly demarcated. Genital pore median, dorsal, pre-equatorial. Testes medullary, in two lateral fields. Ovary posterior, bilobed, transversely elongated. Vitelline follicles cortical, in lateral bands dorsally and ventrally, occasionally continuous around lateral margins of proglottid. Uterus loops forward, forms small uterine sac which opens midventrally near anterior margin of proglottid. Eggs anoperculate, not fully embryonated when laid. In freshwater teleosts. Africa, Asia, India, Australia, Russia. Type-species *P. polypteri* (Leydig, 1853).

5a. Scolex fan-shaped or sagittate; bothrial margins crenulated .... 6.
5b. Scolex otherwise .... 8.

6a. Uterine pore distinctly submedian .... *Ptychobothrium* Lönnberg, 1889. (Figs 10.12, 10.13)

Diagnosis: Scolex unarmed, laterally compressed, heart- or arrowhead-shaped in lateral view. Bothria well-developed with smooth or slightly crenulated margins. Apical disc absent. Neck absent. External segmentation incomplete. Proglottids acraspedote. Genital pore dorsal, median, pre-equatorial. Cirrus-sac thin-walled. Testes in dorsal medulla, in two lateral fields. Ovary compact, posterior. Vagina

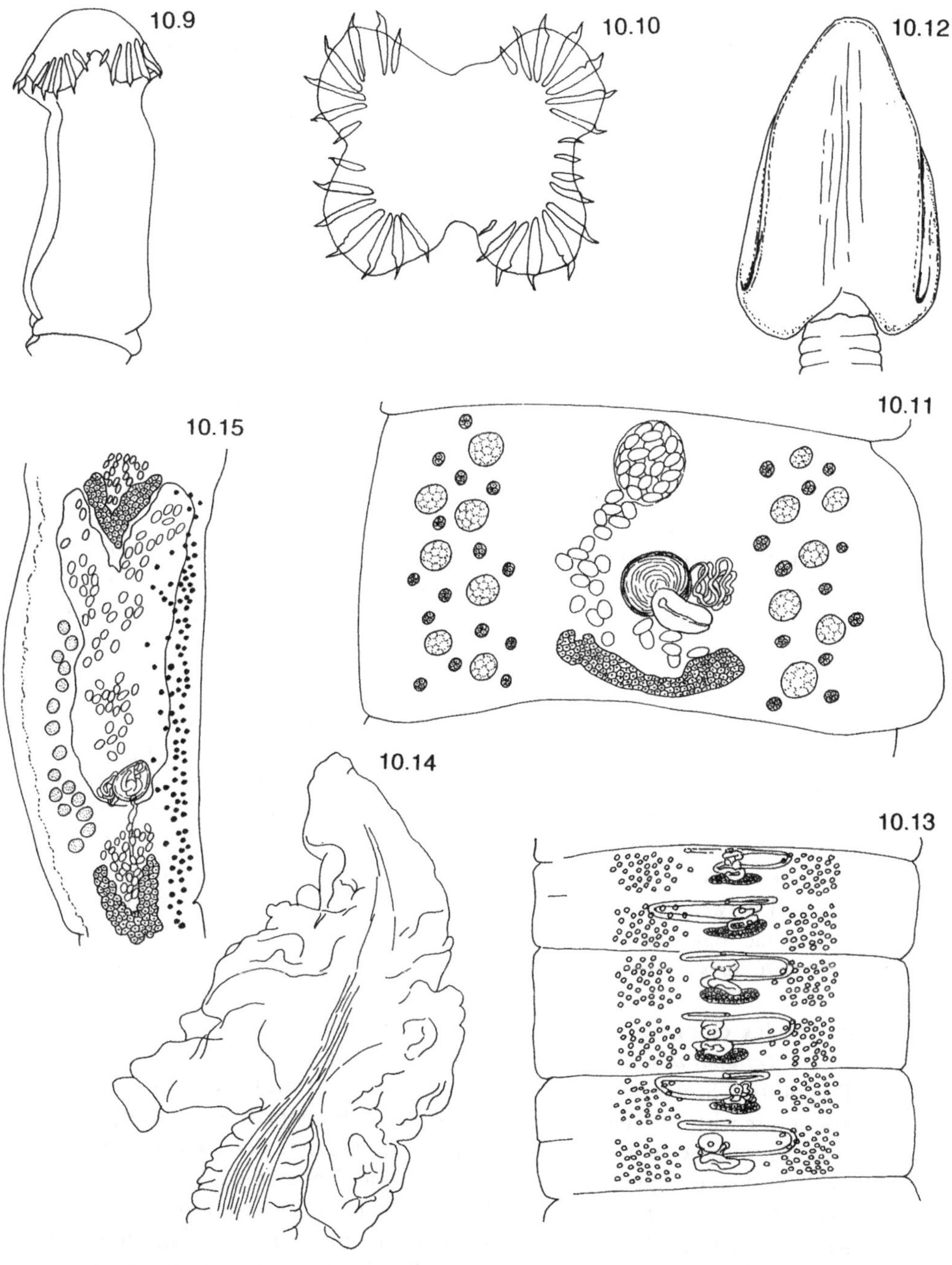

Figs 10.9–10.11 *Polyonchobothrium polypteri* (Leydig, 1853). 10.9. Scolex. 10.10. Apical disc. Both after Jones (1980). 10.11. Mature proglottid.
Figs 10.12, 10.13 *Ptychobothrium belones* (Dujardin, 1845). 10.12. Scolex. 10.13. Maturing proglottids. After Yamaguti (1934).
Figs 10.14, 10.15 *Plicatobothrium cypseluri* Cable & Michaelis, 1967. 10.14. Scolex. 10.15. Gravid proglottid. Original.

recurved, opens anterior to cirrus-sac. Vitelline follicles intermingled with the longitudinal muscle bundles. Uterus sigmoid, alternating irregularly. Uterine sac present; uterine pore ventral, opening near anterior margin of proglottid. Eggs thin-shelled, anoperculate, embryonated. In marine teleosts (Belonidae). Pacific, Atlantic. Type-species *P. belones* (Dujardin, 1845).

6b. Uterine pore midventral .... 7.

7a. Uterine sac Y-shaped ....
.... ***Plicatobothrium*** **Cable & Michaelis, 1967. (Figs 10.14, 10.15)**

Diagnosis: Scolex triangular to fan-shaped, unarmed, no apical disc. Bothria deep with crenulated margins, not connected by apical groove. Neck absent. Strobila acraspedote, external segmentation incomplete. Genital pore dorsomedian, post-equatorial. Testes medullary, lateral, extend from level of ovary to behind arms of uterine sac. Ovary posterior, median, enclosed by arms of uterine sac of succeeding proglottid. Vitelline follicles in lateral fields, intermingled with longitudinal muscle fibres, which are not grouped into bundles. Uterine sac with short anterior arms extending into the preceding proglottid; opens ventrally by median pore at posterior end of sac. Eggs anoperculate, not developed to point of having oncospheral hooks when laid. In marine teleosts (flying fish, *Cypselurus*). Caribbean, Indian Ocean. Type-species *P. cypseluri* Cable & Michaelis, 1967.

7b. Uterine sac inconspicuous or absent ....
.... ***Alloptychobothrium*** **Yamaguti, 1968. (Figs 10.16, 10.17)**

Diagnosis: Scolex arrowhead-shaped in lateral view; bothrial margins strongly crenulated. Neck absent. External segmentation complete or not. Proglottids wider than long, craspedote or not. Genital pore mid-dorsal. Testes in one or two layers in submedian medulla. Ovary posterior, median, arcuate and transversely elongated. Vagina well-developed. Vitelline follicles lateral, continuous in midline and between proglottids. Uterus in median field, opening midventrally; no uterine sac. Eggs not operculated, contain oncospheres. In marine teleosts (flying fish, *Cypselurus*). Pacific Ocean. Type-species *A. spilonotopteri* Yamaguti, 1968.

8a. Scolex elongated; strong, with powerful plicated apical disc ....
Tetrapapillocephalus **Protasova & Mordvinova, 1986. (Figs 10.18, 10.19)**

Diagnosis: Scolex strongly developed with powerful, unarmed, four-lobed apical disc with plicated edges; bothria poorly developed. External segmentation present, with secondary divisions anteriorly. Genital pore dorsal, more or less median. Vagina opens anteriorly to cirrus-sac. Testes medullary, in two lateral fields. Ovary posterior, lobed, bialate. Vitelline follicles cortical, continuous between proglottids. Uterine sac median, opens midventrally? Eggs anoperculate, partly embryonated. In marine teleosts (*Saurida undosquamis, Beryx splendens*). Indian Ocean. Type-species *T. magnus* Protasova & Mordvinova, 1986.

8b. Scolex elongated or not; apical disc present or absent, not strongly developed or plicated .... 9.

9a. Eggs operculate when laid .... 10.
9b. Eggs not operculate .... 11.

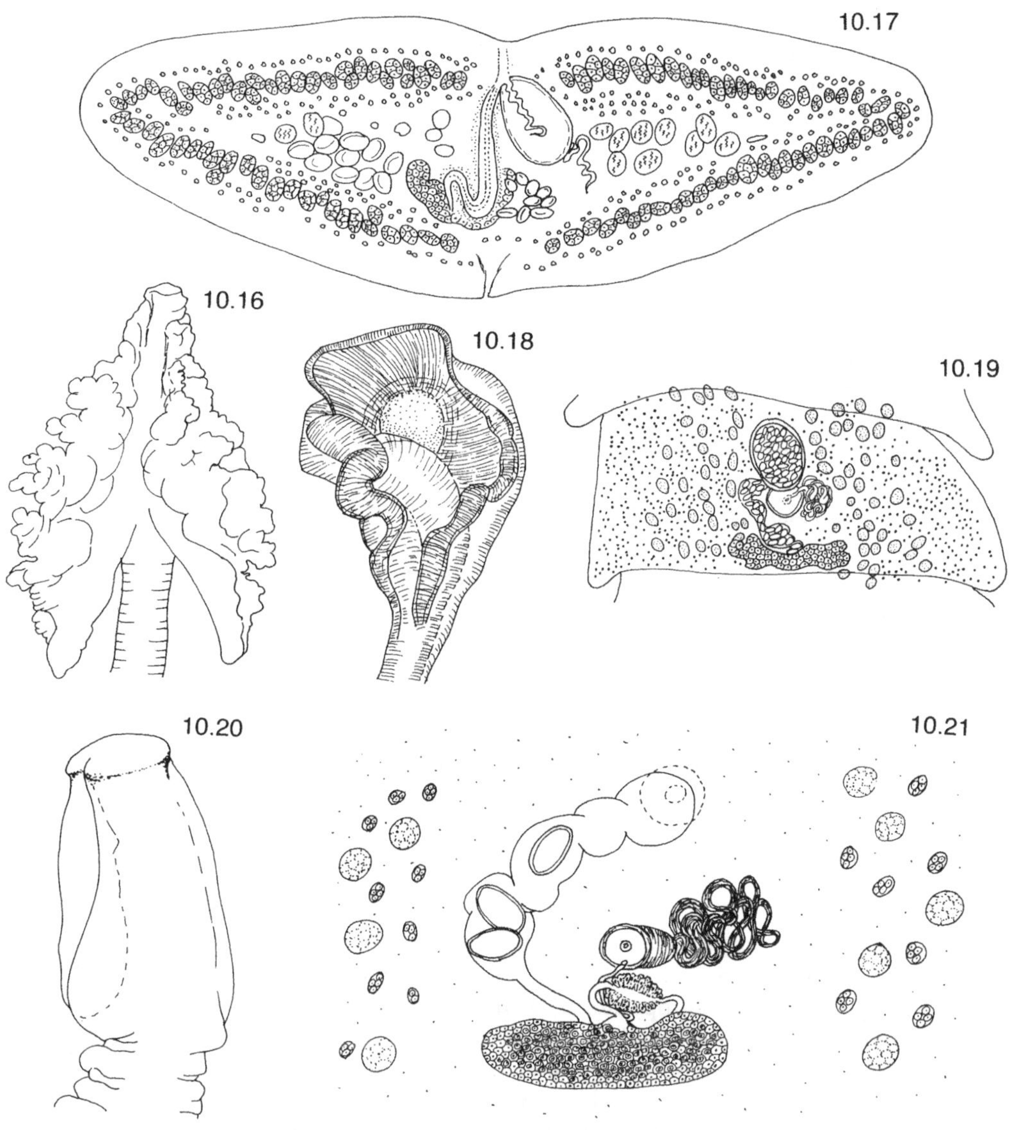

Figs 10.16, 10.17 *Alloptychobothrium spilonotopteri* Yamaguti, 1968. 10.16. Scolex, original. 10.17. Mature proglottid in transverse section, after Yamaguti (1968).
Figs 10.18, 10.19 *Tetrapapillocephalus magnus* Protasova & Mordvinova, 1986. 10.18. Scolex. 10.19. Gravid proglottid. After Protasova & Mordvinova (1986).
Figs 10.20, 10.21 *Bothriocephalus scorpii* (Müller, 1776). 10.20. Scolex. 10.21. Reproductive organs.

10a. Vitelline follicles cortical. Eggs unembryonated when laid .... ***Bothriocephalus*** **Rudolphi, 1808. (Figs 10.20, 10.21)** (Syns *Dibothrium* Diesing, 1850; *Penetrocephalus* Rao, 1960; *Schyzocotyle* Akhmerov, 1960.)

Diagnosis: Scolex usually elongated, occasionally pyriform or rounded; unarmed; apical disc present; bothria shallow or deep, with non-crenulated margins, may be connected by apical groove. External segmentation present. Proglottids wider than long. Genital pores dorsomedian. Testes medullary in two lateral fields. Cirrus unarmed. Ovary median, posterior, transversely elongated, bialate or not. Uterine sac present. Uterine pore ventral, median or slightly submedian, anterior to genital pore. Eggs operculate. In marine and freshwater teleosts. Cosmopolitan. Type-species *B. scorpii* (Müller, 1776).

10b. Vitelline follicles medullary, intermingled with testes. Eggs embryonated when laid .... *Taphrobothrium* Lühe, 1899. (Figs 10.22, 10.23)

Diagnosis: Scolex elongate, apical disc present but unarmed, bothria surficial. External segmentation present but incomplete. Proglottids wider than long. Genital pores dorsal, median. Cirrus-sac muscular. Ovary posterior, transversely elongated. Uterus sigmoid. Uterine sac present; uterine pore ventral submedian. Eggs thin-shelled, operculate. In marine teleosts. Japan. Type-species *T. japonense* Lühe, 1899.

11a. Bothrial margins fused except for anterior aperture surrounded by powerful sphincter. In marine teleosts .... *Clestobothrium* Lühe, 1899. (Fig. 10.24)

Diagnosis: Scolex broadly oval, robust with deep bothria connected by apical groove. Neck absent. Strobila muscular. External segmentation distinct. Genital pores mid-dorsal. Testes in two lateral fields. Ovary median, transversely elongated. Vitelline follicles cortical, continuous around lateral margins of proglottid. Uterine coils median, terminating in large uterine sac; uterine pore midventral. Eggs anoperculate, unembryonated. In marine fish. Mediterranean, Atlantic. Type-species *C. crassiceps* (Rudolphi, 1819).

11b. Bothrial margins not fused; sphincter absent. In freshwater teleosts .... 12.

12a. Scolex subpyriform; bothrial cavities deep. In freshwater teleosts of the Middle East .... *Coelobothrium* Dollfus, 1970. (Fig. 10.25)

Diagnosis: Scolex subpyriform, unarmed, apex rounded, no apical disc or apical groove. Bothrial cavities large, very deep with narrow apertures. Neck absent. External segmentation incomplete. Genital pore dorsal, almost median. Testes medullary, in two lateral fields. Ovary posterior, transversely elongated. Seminal receptacle well-developed. Vitelline follicles present throughout cortex. Uterine sac large, occupying all the proglottid; uterine pore ventral, slightly submedian. Eggs anoperculate. In freshwater fish (*Varicorhinus damascinus umbla*). Iran. Type-species *C. monodi* Dollfus, 1970.

12b. Scolex thimble-shaped; bothrial cavities shallow. In African freshwater teleosts .... *Ichthybothrium* Khalil, 1971. (Figs 10.26, 10.27)

Diagnosis: Scolex slightly dorsoventrally flattened, unarmed. Bothria shallow, margins not inrolled. Strobila weakly muscled; external segmentation incomplete; proglottids acraspedote. Genital pore dorsal, median. Testes medullary in lateral fields. Cirrus-sac thick-walled; cirrus unarmed. Vagina opens behind cirrus-sac. Ovary posterior, bialate. Uterine duct sinuous, uterine sac opens at ventral submedian uterine pore. Vitelline follicles cortical, in two lateral fields dorsally and

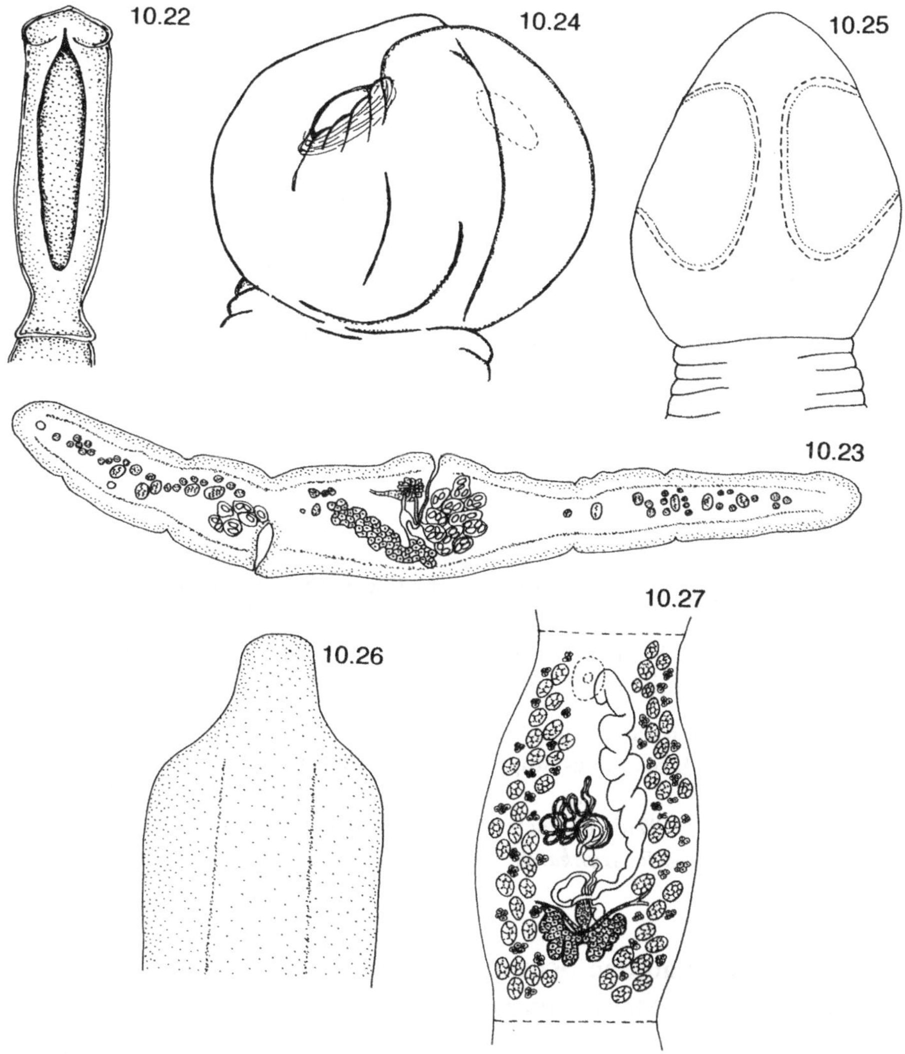

Figs 10.22, 10.23 *Taphrobothrium japonense* Lühe, 1899. 10.22. Scolex. 10.23. Gravid proglottid, transverse section. After Yamaguti (1934).
Fig. 10.24 *Clestobothrium crassiceps* (Rudolphi, 1819). Scolex.
Fig. 10.25 *Coelobothrium monodi* Dollfus, 1970. Scolex. After Dollfus (1970).
Figs 10.26, 10.27 *Ichthybothrium ichthybori* Khalil, 1971. 10.26. Scolex. 10.27. Mature proglottid. After Khalil (1971).

ventrally. Eggs anoperculate, embryonated. In African freshwater teleosts (*Ichthyborus*). Type-species *I. ichthybori* Khalil, 1971.

Family Philobythiidae Campbell, 1977

Our knowledge of the morphology of members of this small family is based on the descriptions of the two known species by Campbell (1977, 1979). The family is unusual in that its members have a single, transversely elongate, lobed vitellarium.

Family diagnosis

Scolex with bothria; apical disc weakly developed or absent. Strobila distinctly segmented, craspedote. One set of genitalia per proglottid. Testes medullary, median. Cirrus-sac with unarmed cirrus. Genital pores marginal; irregularly alternating. Vagina enters genital atrium anterior to cirrus-sac. Ovary in posterior medulla. Vitellarium single, lobed, transversely elongate, medullary, postovarian. Uterus diverticulate. Uterine pore ventromedian. In deep-sea teleosts. Type-genus *Philobythos* Campbell, 1977.

Key to genera

1a. Scolex with apical disc. Testes in field surrounding female system. Oncospheres in clusters of three to five within membranous capsule .... *Philobythos* Campbell, 1977. (Figs 10.28–10.30)

Diagnosis: With characters of family. In Stephanoberycidae and Macrouridae. North Atlantic Ocean. Type-species *P. atlanticus* Campbell, 1977.

1b. Scolex without apical disc. Testes restricted to anterior extremity of proglottid. Oncospheres single within membranous capsule .... *Philobythoides* Campbell, 1979. (Figs 10.31, 10.32)

Diagnosis: With characters of family. In Alepocephalidae. Northern Atlantic. Type-species *P. stunkardi* Campbell, 1979.

Family Echinophallidae Schumacher, 1914

(Syn. Parabothriocephalidae Yamaguti, 1959)

The Parabothriocephalidae cannot be distinguished from the Echinophallidae, which has priority. The differences of double genitalia and a spiny cirrus are not valid at family level and, in addition, *Bothriocotyle* and *Paratelemerus* with single sets of genitalia and conspicuously armed cirri represent intermediate forms.

The type-species of *Parabothriocephaloides, P. segmentatus* Yamaguti, 1934, and that of *Parabothriocephalus, P. gracilis* Yamaguti, 1934, occur together, in the same host species and individuals of *Psenopsis anomala*, the type host of both species. Yamaguti (1934) distinguished *Parabothriocephaloides* by its pseudoscolex, prominent posterolateral proglottid margins in the anterior part of the strobila, vaginal sphincter and cortical vitelline follicles. He stated, however, that the scolex of *Parabothriocephalus gracilis* was easily detached. Possibly, detachment and replacement of the scolex by a pseudoscolex may occur in *Parabothriocephaloides segmentatus*. This occurs in the triaenophorid *Fistulicola plicatus* (Rudolphi, 1819) Lühe, 1899, indicating that the presence of a pseudoscolex is

10.28

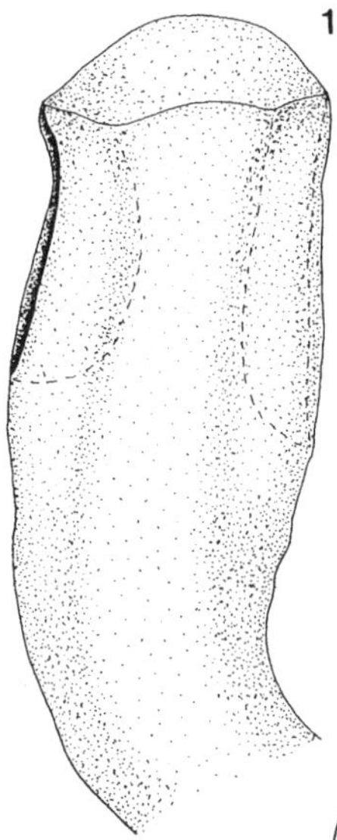

10.29

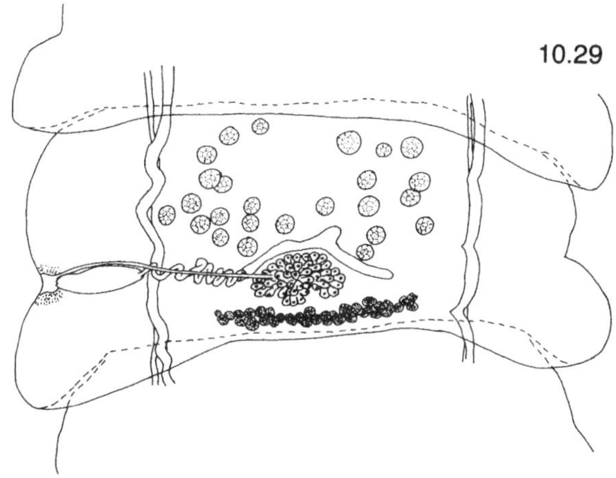

10.30

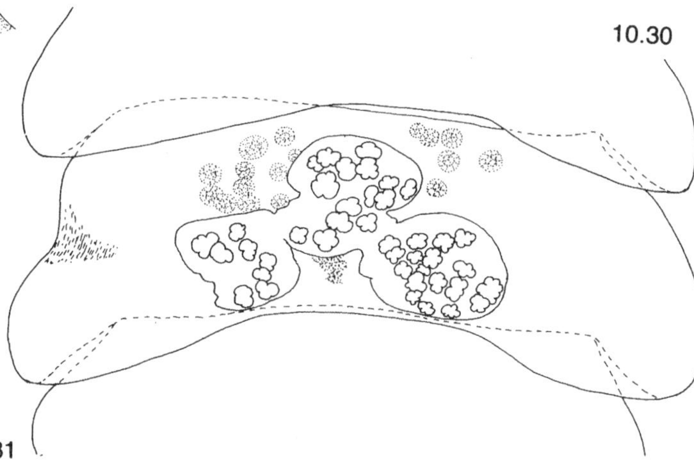

10.31

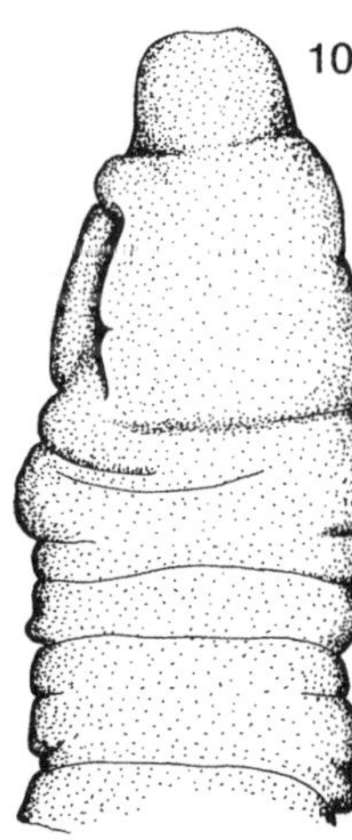

10.32

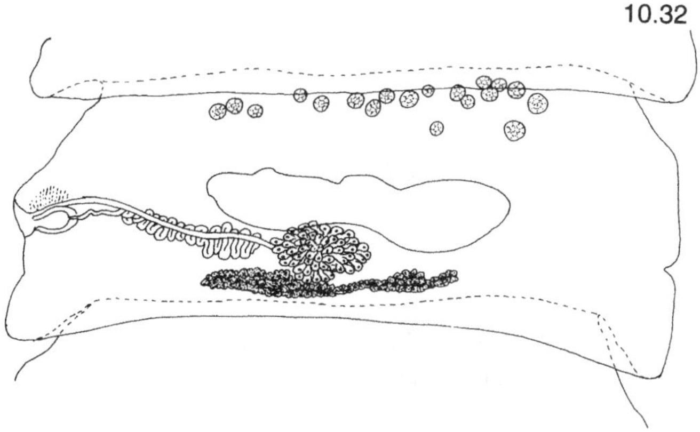

Figs 10.28–10.30 *Philobythos atlanticus* Campbell, 1977. 10.28. Scolex. 10.29. Mature proglottid. 10.30. Gravid proglottid.

Figs 10.31, 10.32 *Philobythoides stunkardi* Campbell, 1979. 10.31. Scolex, modified after Campbell (1979). 10.32. Early gravid proglottid.

not necessarily a generic feature but that both states may co-exist. Furthermore, Campbell, Correia & Haedrich (1982) have shown that a vaginal sphincter may be present or absent in the genus *Parabothriocephalus* and that in this genus the distribution of the vitelline follicles is variable. It is medullary in the type-species, circumcortical with intrusions into the medulla in *P. macruri* Campbell *et al.*, 1982 and circumcortical in *P. sagitticeps* (Sleggs, 1927) (see Yamaguti, 1934; Campbell *et al.*, 1982; Jensen, 1976). The posterolateral expansion of the anterior proglottids is not a generic character. *Parabothriocephaloides*, therefore, falls within the range of variation now known to occur in *Parabothriocephalus*, with which it is synonymized.

Atelemerus and *Paratelemerus* have in the past been differentiated from closely related forms by the presence of a field of spines on the posterior border of each proglottid. Our observations on worms identified as *Echinophallus* indicate that this is not a good generic criterion, as it is easily lost at fixation. This feature apart, the type-species of *Atelemerus* and *Echinophallus* are indistinguishable (and from the same host) and we consider them synonymous.

We have retained *Paratelemerus* as distinct, but without much conviction. We believe that careful critical study of new material of a number of genera from centrolophid fishes, i.e. *Paratelemerus, Bothriocotyle, Parabothriocephalus, Neobothriocephalus* and *Glossobothrium*, will lead to a reduction in the number of recognized valid genera. *Metabothriocephalus* Yamaguti, 1968 has also been provisionally retained pending the availability of new material.

Family diagnosis

Scolex persistent or replaced by pseudoscolex. Apical disc present or absent; if present, may be inconspicuous; unarmed. Bothria usually shallow, weak; rarely with posterior appendages. Strobila usually flattened, rectilinear, rarely spiral and concave ventrally. Segmentation complete or not. Proglottids usually wider than long. One or two sets of genitalia per proglottid. Genital pores single or double, dorsosubmarginal. Cirrus-sac small to large, may cross midline. Cirrus armed or not. Testes medullary. Ovary posterior. Vitelline follicles either cortical with or without intrusions into medulla, or medullary. Uterine pore ventral, median or submedian. Eggs operculate or not, unembryonated. In marine teleosts. Type-genus *Echinophallus* Schumacher, 1914.

Key to genera

1a. Genitalia single .... 2.
1b. Genitalia double .... 7.

2a. Bothria with prominent posterior appendages ....
.... ***Glossobothrium* Yamaguti, 1952. (Figs 10.33, 10.34)**

Diagnosis: Scolex elongated with unarmed apical disc; each bothrium with a linguiform fluted appendage projecting from its base. Neck absent. External

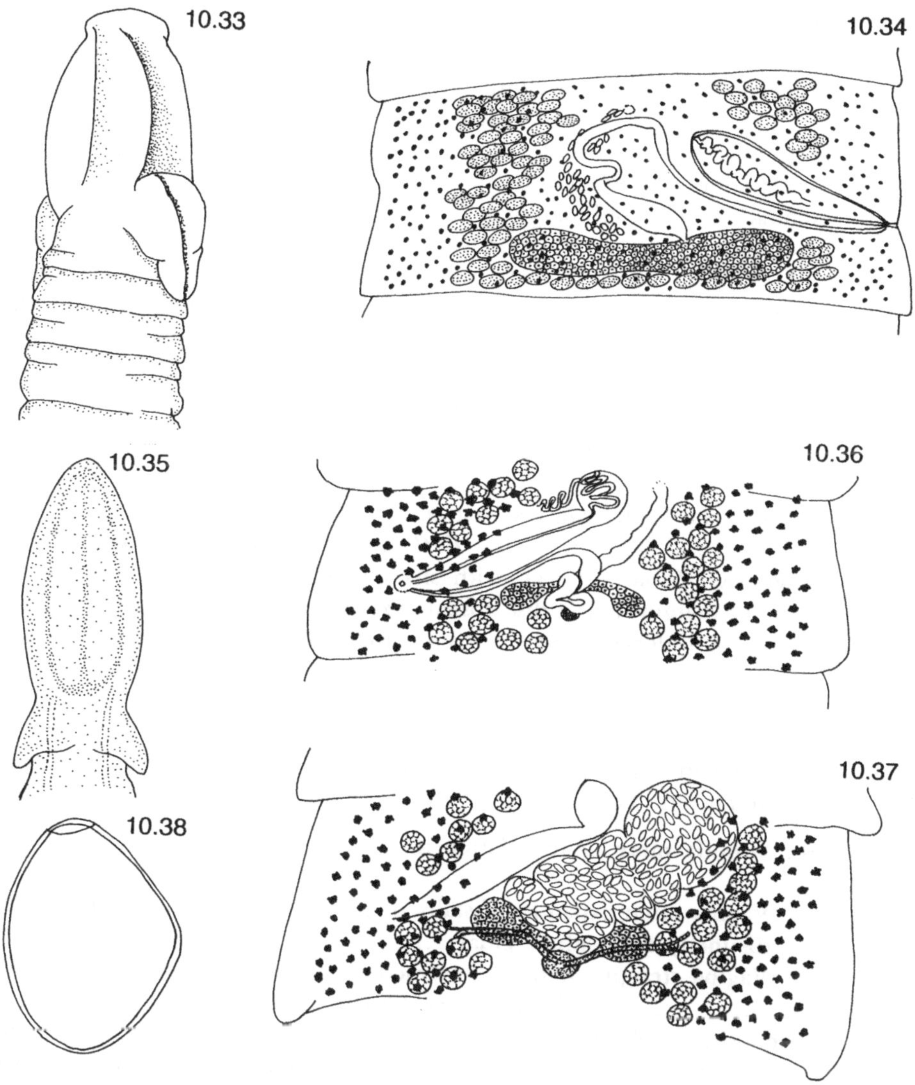

Figs 10.33, 10.34 *Glossobothrium nipponicum* Yamaguti, 1952. 10.33. Scolex. 10.34. Gravid proglottid, after Yamaguti (1952).
Figs 10.35–10.38 *Neobothriocephalus aspinosus* Mateo & Bullock, 1966. 10.35. Scolex. 10.36. Mature proglottid. 10.37. Gravid proglottid. 10.38. Egg. After Mateo & Bullock (1966).

metamerism present. Proglottids short, imbricated anteriorly, with salient posterior margins. Genital pore marginal, postequatorial. Cirrus-sac claviform, invested with gland cells. Cirrus armed with tiny spines. Ovary bialate, posterior, slightly poral. Vagina with proximal muscular dilation, opens behind cirrus-sac. Vitelline follicles cortical, diffuse. Uterine duct sigmoid; uterine sac opens midventrally near anterior margin of proglottid. Eggs operculate, unembryonated. In marine teleosts. Japan. Type-species *G. nipponicum* Yamaguti, 1952.

2b. Bothria lacking posterior appendages .... 3.

3a. Eggs with lateral swelling ....
.... ***Neobothriocephalus* Mateo & Bullock, 1966. (Figs 10.35–10.38)**

Diagnosis: Scolex ovoid, unarmed; bothria shallow. External segmentation almost complete. Proglottids cuneate, wider than long except at posterior end. Genital pore dorsosubmarginal. Cirrus-sac oblique, large. Cirrus unarmed. Testes medullary in two lateral bands. Vaginal sphincter present; vagina aspinose, without distal dilation. Ovary posterior, slightly poral, bialate. Vitelline follicles mainly cortical in dorsal and ventral fields, not continuous around margin of proglottid; few medullary. Uterus sigmoid or sinuous; uterine sac opens ventromedially. Eggs operculate, unembryonated. In marine fish. Peru, Chile. Type-species *N. aspinosus* Mateo & Bullock, 1966.

3b. Eggs without lateral swelling .... 4.

4a. Strobila spiral, concave ventrally ....
.... ***Bothriocotyle* Ariola, 1900. (Figs 10.39, 10.40)**

Diagnosis: Scolex conical, with shallow bothria and distinct small sucker at posterior margin; apical disc absent. External segmentation distinct, craspedote. Genital pore dorsal; closer to margin than median line; irregularly alternating. Cirrus-sac large. Cirrus armed with large spines. Testes numerous; in single medullary layer. Ovary bilobed, medullary, near cirrus-sac. Vitelline follicles medullary, ventral. Uterus with long narrow duct and small sac. Uterine pore ventral; median. Eggs anoperculate. Parasite of marine teleosts (*Centrolophus*). Mediterranean. Type-species *B. solinosomum* Ariola, 1900.

4b. Strobila more or less flattened .... 5.

5a. Vagina often expanded distally or proximally, spinose or not; genital pore distinctly dorsosubmarginal .... 6.

5b. Vagina straight, narrow, unarmed; genital pore almost marginal ....
.... ***Metabothriocephalus* Yamaguti, 1968. (Figs 10.41–10.43)**

Diagnosis: Scolex not clearly demarcated from neck; apical disc indistinct; small, oval, subapical bothrium, surrounded by condensed tissue, on each flat surface. Strobila muscular, fleshy; proglottids wider than long with salient posterior margins. Genital pore pre-equatorial. Cirrus-sac small. Testes medullary, in two lateral fields, relatively few. Ovary slightly poral, transversely elongated. Vagina opens anterior to cirrus-sac. Vitelline follicles entirely cortical. Uterine sac median, opens midventrally. Eggs thick-shelled, operculate, unembryonated. In marine teleosts (*Myripristis* spp.). Hawaii. Type-species *M. menpachi* Yamaguti, 1968.

Figs 10.39, 10.40 *Bothriocotyle solinosomum* Ariola, 1900. 10.39. Two views of scolex. 10.40. Three proglottids showing positions of terminal genitalia. Modified after Ariola (1902).

Figs 10.41–10.43 *Metabothriocephalus menpachi* Yamaguti, 1968. 10.41. Scolex. 10.42. Mature proglottid. 10.43. Gravid proglottid.

Figs 10.44, 10.45 *Paratelemerus seriolella* Gulyaev, Korotaeva & Kurochkin, 1989. 10.44. Pseudoscolex. 10.45. Mature proglottid. Modified after Gulyaev *et al.* (1989).

10.39

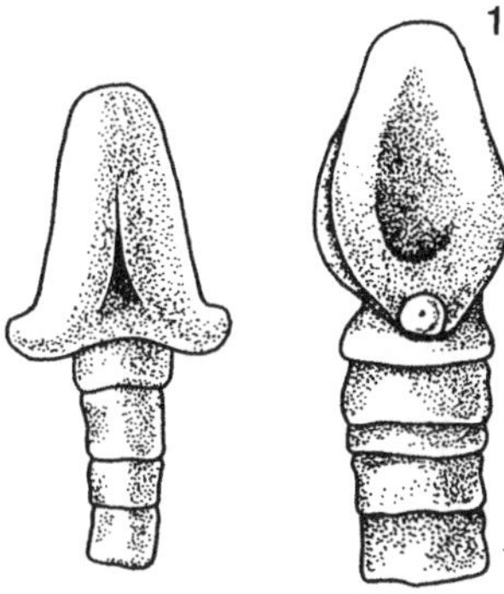

10.40

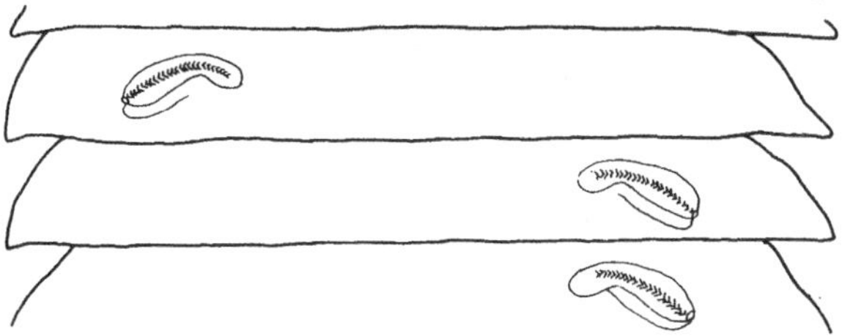

10.41

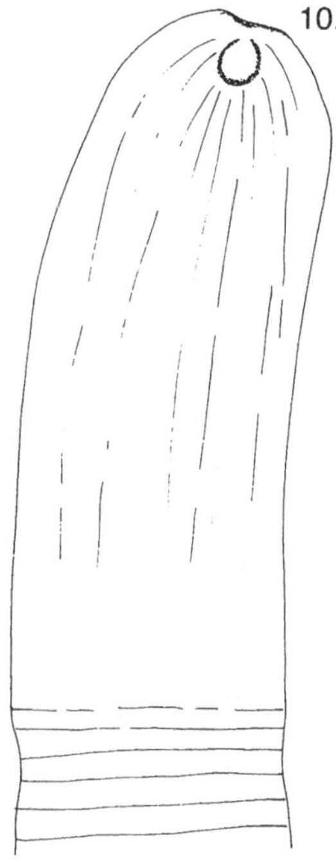

10.42

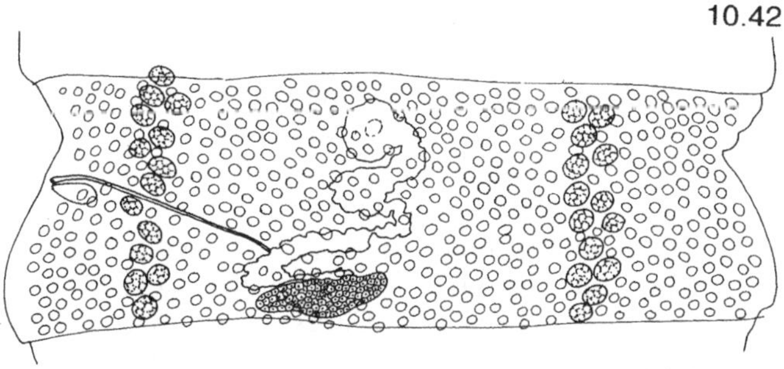

10.43

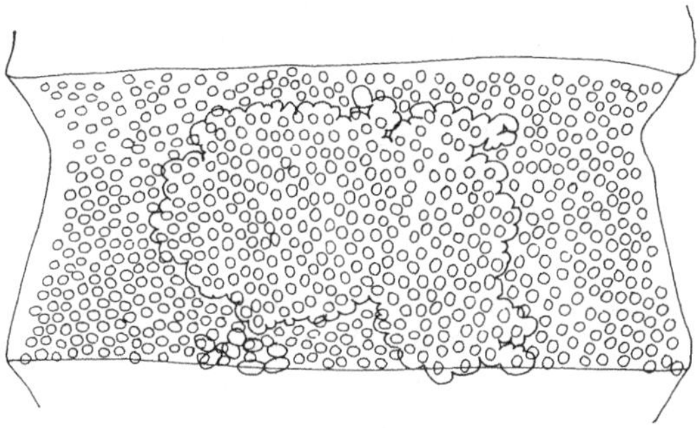

10.44

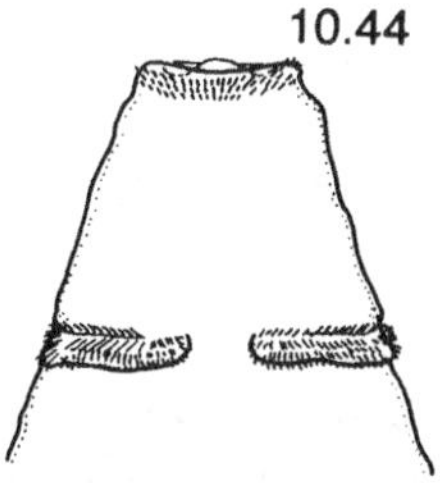

10.45

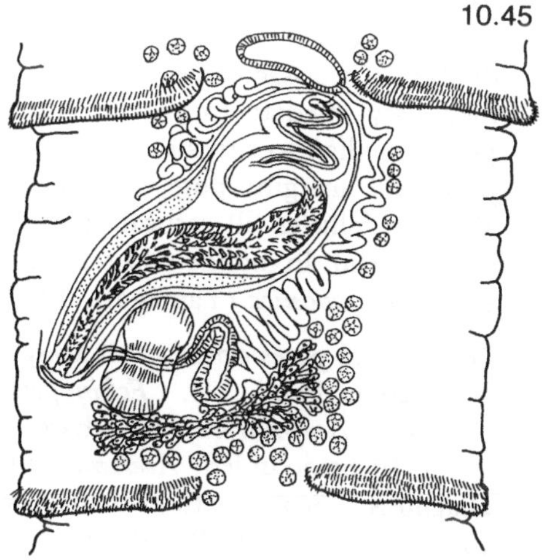

6a. Cirrus-sac large, reaching into aporal part of proglottid and to anterior edge of proglottid .... ***Paratelemerus* Gulyaev, Korotaeva & Kurochkin, 1989. (Figs 10.44, 10.45)**

Diagnosis: Scolex replaced by pseudoscolex. External segmentation distinct. 18–21 proglottids, only the last one to three have fully-formed genital apparatus. Posterolateral margins of proglottid spined. Genital pore dorsosubmarginal, postequatorial. Testes lateral, confluent behind ovary. Cirrus conspicuously armed with large spines. Ovary posterior, median or slightly poral, bialate. Vagina strongly recurved, muscular, dilated proximally. Vaginal sphincter present or absent. Uterine duct winding on aporal side of cirrus-sac. Uterine sac small, median, opens ventromedially? Eggs? In marine fish (Centrolophidae, Stromateidae). Off north-western Australia. Type-species *P. seriolella* Gulyaev, Korotaeva & Kurochkin, 1989.

6b. Cirrus-sac smaller, not reaching midline, not reaching anterior edge of proglottid .... ***Parabothriocephalus* Yamaguti, 1934. (Figs 10.46, 10.47)** (Syn. *Parabothriocephaloides* Yamaguti, 1934.)

Diagnosis: Scolex with two surficial bothria but no apical disc; easily detached. Neck absent. External segmentation almost complete. Proglottids wedge-shaped, longer than wide. Genital pore dorsosubmarginal. Testes medullary in two lateral fields, or confluent in mid-line. Cirrus-sac muscular. Cirrus spinose. Vagina narrow proximally, often expanded and sometimes muscular distally; vaginal armature and/or sphincter present or absent. Ovary poral, lobed, bialate. Vitelline follicles may be circumcortical, perhaps intruding into medulla, or medullary. Uterine duct sigmoid or not; uterine sac invades preceding proglottid. Uterine pore midventral or slightly antiporal. Eggs operculate, unembryonated; anopercular boss present or absent. In marine teleosts. Japan, Atlantic and Pacific Oceans. Type-species *P. gracilis* Yamaguti, 1934.

7a. Genital pores dorsosubmarginal, very close to lateral margin .... ***Paraechinophallus* Protasova, 1975. (Fig. 10.48)**

Diagnosis: Scolex with two shallow bothria and inconspicuous apical disc, lost in older worms. Pseudoscolex trapezoidal. External segmentation distinct; posterior border of proglottid lobed, overlapping following proglottid. Genitalia double. Cirrus-sàc large, muscular. Cirrus with large spines at base. Testes medullary, median to ovaries, but with gap in median region. Ovary lobed, posteromedian to cirrus-sac. Uterus forms sac just closer to median line than to proglottid margin; pore ventral, anterior. Eggs operculate, unembryonated. Vitelline follicles fill cortical parenchyma, intrude into medulla in lateral parts of proglottid. In marine teleosts (Centrolophidae). Pacific (Japan, Hawaii). Type-species *P. japonica* (Yamaguti, 1934).

7b. Genital pores dorsal, well away from margin of proglottid up to halfway to median line ...... ***Echinophallus* Schumacher, 1914. (Figs 10.49–10.51)** (Syns *Amphitretus* Blanchard, 1894 preoccupied; *Acanthophallus* Lühe, 1910 preoccupied; *Atelemerus* Guiart, 1935.)

Diagnosis: Scolex with two shallow bothria and inconspicuous apical disc; in young worms only. Pseudoscolex trapezoidal. Proglottids weakly to strongly

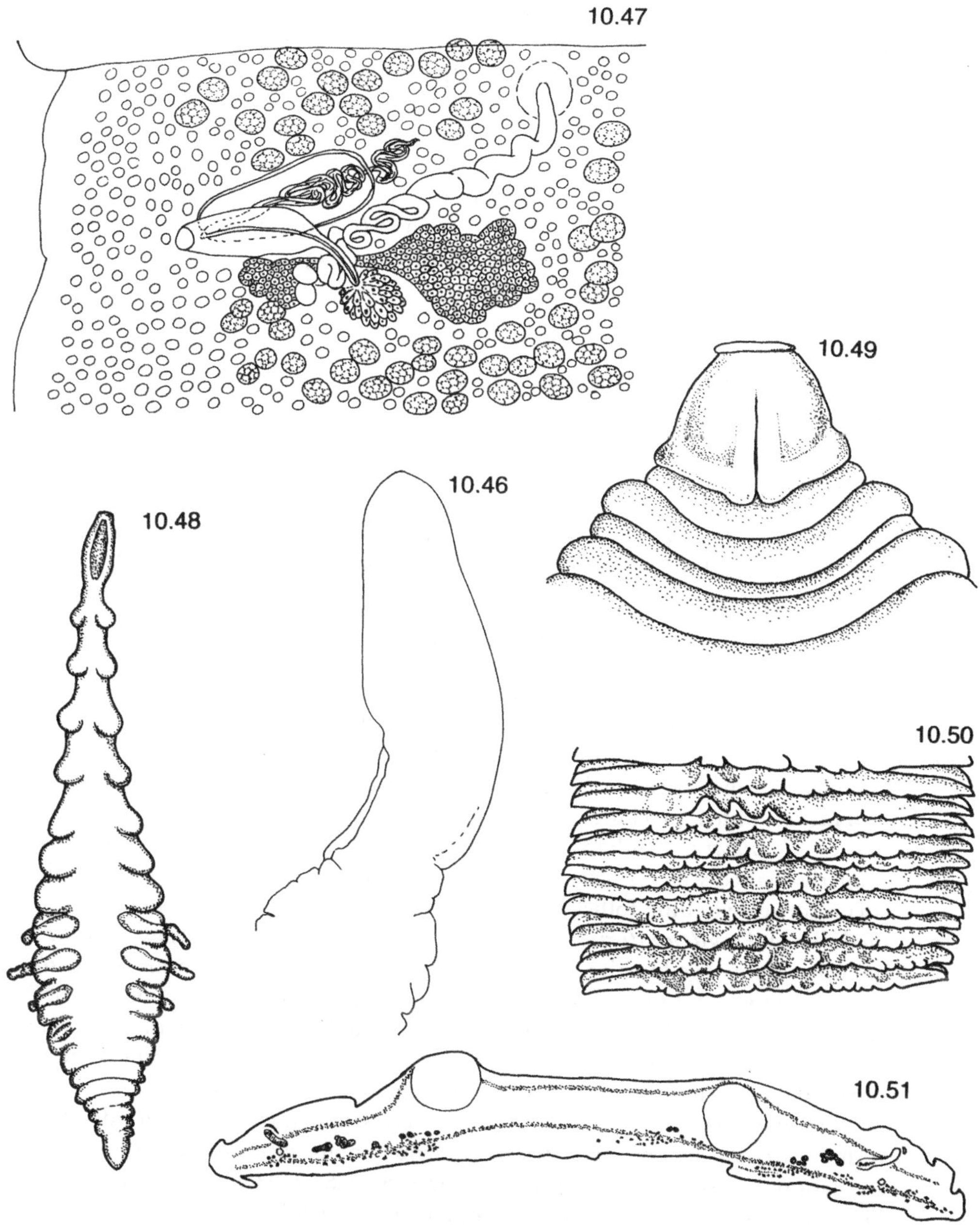

Fig. 10.46. *Parabothriocephalus johnstoni* Prudhoe, 1969. Scolex, original. **Fig. 10.47**. *P. macruri* Campbell, Correia & Haedrich, 1982, mature proglottid, adapted from Campbell *et al.* (1982). **Fig. 10.48** *Paraechinophallus japonicus* (Yamaguti, 1934.) Whole worm. Modified after Yamaguti (1934).

Figs 10.49-10.51 *Echinophallus wageneri* (Monticelli, 1890). 10.49. Scolex. 10.50. Surface view of proglottids. 10.49 and 10.50 modified after Ariola (1902). 10.51. Transverse section of gravid proglottid, modified after Schumacher (1914).

craspedote with armed or unarmed thickened posterior borders. Genitalia double per proglottid. Cirrus-sac large, muscular. Cirrus with long spines. Testes numerous, medullary, in two fields median to genital pores, either side of and a few posterior to ovary, continuous between proglottids. Ovary lobate, posterior to cirrus-sac. Vitelline follicles mostly cortical, abundant laterally, continuous between proglottids. Uterus tubular, convoluted; later forming globular sac, median to genital pore. Uterine pore submedian, near anterior margin of proglottid. Eggs operculate, unembryonated. In marine teleosts (Centrolophidae and Sparidae). Mediterranean, North Atlantic, Great Australian Bight, Pacific (Japan). Type-species *E. wageneri* (Monticelli, 1890).

Family Triaenophoridae Lönnberg, 1889

(Syn. Amphicotylidae Ariola, 1899)

The families Triaenophoridae and Amphicotylidae have been distinguished by the presence of operculated eggs in the former but not the latter. Amphicotylid eggs are embryonated when laid but triaenophorid eggs may be embryonated or not. As discussed above for the Bothriocephalidae and Ptychobothriidae, these characters are not suitable for use at family level. The Amphicotylidae is, therefore, synonymized with the Triaenophoridae.

Fistulicola Lühe, 1899 and *Pseudeubothrium* Yamaguti, 1958, both with type-species which parasitize *Xiphias gladius*, ostensibly differ in the (occasional) presence of a pseudoscolex, a slightly poral ovary in *Fistulicola* and in the presence of a neck and a median ovary in *Pseudeubothrium*. It is not known whether a pseudoscolex may occur in *Pseudeubothrium* but it may be present or absent in *Fistulicola*, which precludes its use for separating the genera. The two type-species, *F. plicatus* (Rudolphi, 1819) and *P. xiphiados* Yamaguti, 1968, have similarly-shaped scoleces, very wide proglottids with markedly expanded margins, circumcortical vitelline follicles, a large cirrus-sac which may just extend into the medulla in *Pseudeubothrium*, and thick-shelled embryonated eggs. *Fistulicola* has operculate eggs but the presence or absence of an operculum in *Pseudeubothrium* eggs was not mentioned by Yamaguti (1968). The type specimens (USNM Helm. Coll. No. 63871) do not include gravid proglottids and the internal organs cannot be clearly distinguished. The differences between the two genera are not sufficient to separate them and they are regarded as synonymous.

Pseudeubothrioides Yamaguti, 1968 was synonymized by Schmidt (1986) with *Pseudeubothrium*, here treated as synonymous with *Fistulicola*. Following an examination of the type specimens, however, (USNM Helm. Coll. No. 63872), *Pseudeubothrioides* is provisionally recognized as valid.

Pachybothrium Pozdniakov, 1983 is preoccupied, but is treated here as a synonym of *Amphigonoporus*. *A. carvalhoi* is reported from *Mola mola* off Brazil (Mendes, 1944) and *P. dogieli* from *M. ramsayi* in the Tasman Sea (Pozdniakov, 1983). The only feature which may be significantly different is the quoted distribution of the vitelline follicles. In *A. carvalhoi* the follicles are essentially cortical with a few in the muscular region and the medulla, the follicles are only in the 'ventral' plane (dorsal in figure). In *P. dogieli* the follicles are said to be mainly medullary and partly cortical, and to surround the testes. These distinctions may be of value at the specific level, or may be interpretational differences. In either case, we do not consider them enough to distinguish the genera, or to warrant the erection of a new name.

Probothriocephalus Campbell, 1979 is transferred to the Triaenophoridae because it has a marginal genital pore and we can find no feature of generic value distinguishing *Probothriocephalus, Flexiphallus* Protasova & Parukhin, 1986, *Heterovitellus* Protasova & Parukhin, 1986 and *Partitiotestis* Protasova & Parukhin, 1986. In erecting their three genera, Protasova & Parukhin (1986) did not refer to *Probothriocephalus*. *Partitiotestis* is said to be distinguished by the testes being discontinuous between proglottids but Protasova & Parukhin's (1986) figure 2b shows the testicular fields of adjacent proglottids coming very close to each other. Protasova & Parukhin (1986) distinguished *Flexiphallus* from *Heterovitellus* by the cirrus-sac of the former lying at an angle to the longitudinal axis of the proglottid, rather than perpendicular, and by the vitellarium being cortical rather than cortical and intermuscular. We do not consider these features adequate to distinguish genera and, therefore, consider *Flexiphallus, Heterovitellus* and *Partitiotestis* synonyms of *Probothriocephalus*.

Blanchardella Moniez, 1895 (new name for *Leuckartia* Moniez, 1880 *nec* Agassiz, 1862 (coelenterate) or Claus, 1862 (crustacean)) may be a synonym of *Eubothrium* as suggested by Schmidt (1986), but if so it takes precedence. It seems unwise, however, to do away with the well-known genus *Eubothrium* and, therefore, we consider *Blanchardella* a *genus inquirendum*.

Pseudobothrium Guiart, 1935 is not recognizable from Guiart's description. Dollfus (1940) was unable to pronounce on its validity. It is regarded here as a *genus inquirendum*.

Family diagnosis

Scolex usually present but may be replaced by pseudoscolex or scolex deformatus. Apical disc present or absent; if present, armed or unarmed. Bothria present or absent; weak or shallow to well-developed, occasionally with projecting posterior margins. Segmentation present or absent. Proglottids usually wider than long. Reproductive organs single or double. Genital pores marginal. Testes medullary. Cirrus-sac small to large. Cirrus armed or not. Ovary posterior. Vitelline follicles cortical and/or medullary; may intrude between longitudinal muscles. Uterine pore ventral, usually median; rarely double. Eggs operculate or not, embryonated or not. In freshwater and marine teleosts. Type-genus *Triaenophorus* Rudolphi, 1793.

Key to genera

1a. Scolex armed .... **2.**
1b. Scolex unarmed .... **3.**

2a. Scolex armed with two dorsal and two ventral trident-shaped hooks .... *Triaenophorus* Rudolphi, 1793. (Figs 10.52, 10.53)
(Syns *Rhytelminthus* Zeder, 1800; *Rhytis* Zeder, 1800; *Tricuspidaria* Rudolphi, 1793.)

Diagnosis: Scolex with apical disc bearing hooks. Bothria shallow. Neck absent. Strobila transversely folded but external metamerism absent. Genital pore marginal. Testes medullary, numerous. Cirrus-sac transverse; cirrus unarmed. Ovary posterior, slightly poral, multilobed, bialate. Vitelline follicles cortical, diffuse. Uterus weakly coiled; uterine sac present; uterine pore opens ventrally, slightly submedian on poral side. Eggs operculate, unembryonated or wholly or partly embryonated, depending on temperature. In carnivorous freshwater teleosts (mainly *Esox*). Circumboreal. Type-species *T. nodulosus* (Pallas, 1781) (type by monotypy).

2b. Scolex armed with several alternating rows of small hooks on margin of apical disc .... *Anchistrocephalus* Monticelli, 1890. (Figs 10.54–10.56)
(Syn. *Ancistrocephalus* Lühe, 1899.)

Diagnosis: Apical disc hooks most numerous adjacent to the bothrial surfaces. Neck absent. External metamerism distinct; proglottids wider than long; longitudinal musculature well-developed. Testes medullary in two lateral fields, confluent at posterior end of proglottid. Cirrus protrusible. Genital pore marginal. Ovary posterior, lobed. Vitelline follicles mainly medullary but with cortical layer just external to longitudinal muscles. Uterus alternating irregularly without reference to genital pores. Eggs thick-shelled, operculate, unembryonated. In marine teleosts (sunfish). Mediterranean Sea, Atlantic and Pacific Oceans. Type-species *A. microcephalus* (Rudolphi, 1819).

3a. Reproductive organs double per proglottid .... *Amphigonoporus* Mendes, 1944. (Figs 10.57–10.59)
(Syn. *Pachybothrium* Pozdniakov, 1983 *nec* Baer & Euzet, 1962.)

Diagnosis: Scolex with distinct apical disc and deep bothria, scolex deformatus occasionally present. External segmentation clearly visible. Genital pores lateral. Cirrus-sac thick-walled. Cirrus protrusible, unarmed. Testes numerous; fields continuous across medulla; median to cirrus-sacs. Ovary small, lobate; just median to ventral osmoregulatory canals. Uterus double, sacciform. Uterine pores double. Vitelline follicles mainly cortical, but some reach into medulla, in dorsal field only or surrounding testicular region. Osmoregulatory canals not known. In *Mola* spp. Type-species *A. carvalhoi* Mendes, 1944.

3b. Reproductive organs single per proglottid .... 4.

Figs 10.52, 10.53 *Triaenophorus nodulosus* (Pallas, 1781). 10.52. Scolex. 10.53. Hooks.
Figs 10.54-10.56 *Anchistrocephalus microcephalus* (Rudolphi, 1819). 10.54. Scolex. 10.55. Mature proglottid. 10.56. Gravid proglottid.
Figs 10.57-10.59 *Amphigonoporus carvalhoi* Mendes, 1944. 10.57. Scolex. 10.58. Mature proglottid. 10.59. Transverse section of mature proglottid. In 10.58 and 10.59, testes omitted on left, vitelline follicles omitted on right. All modified after Mendes (1944).

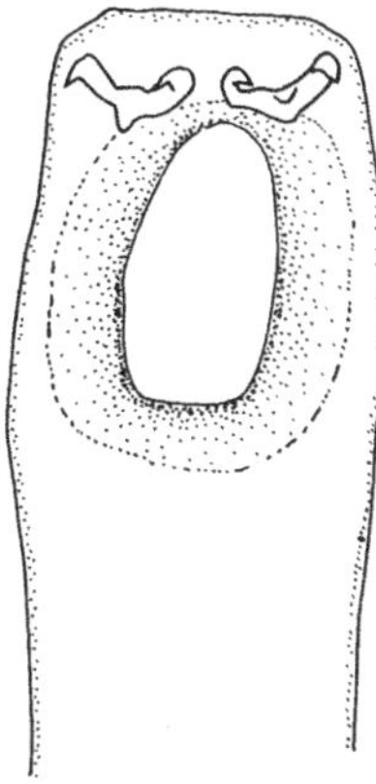

10.52

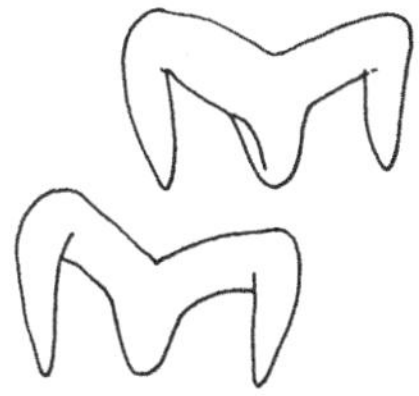

10.53

10.54

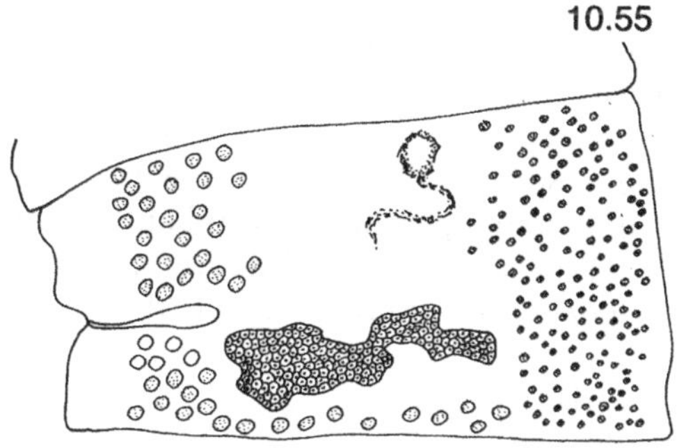

10.55

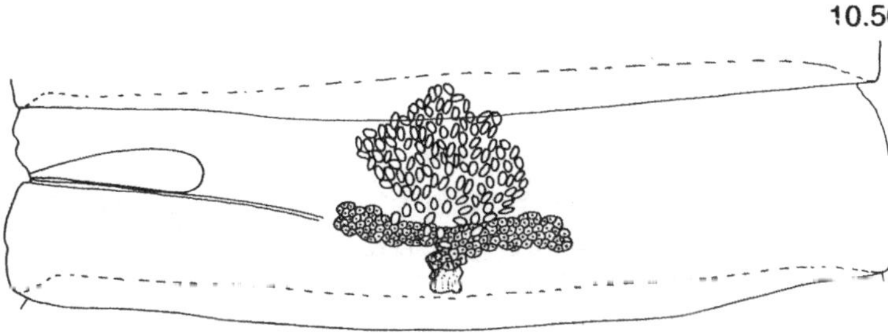

10.56

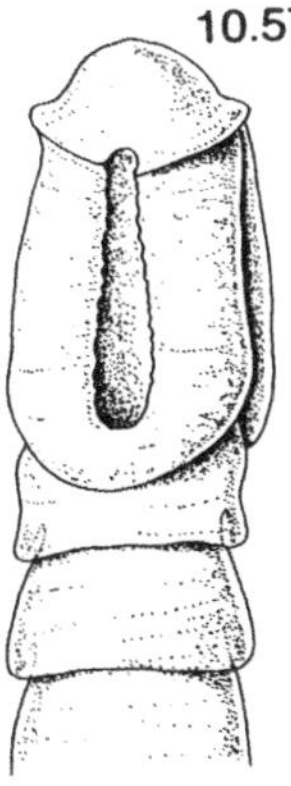

10.57

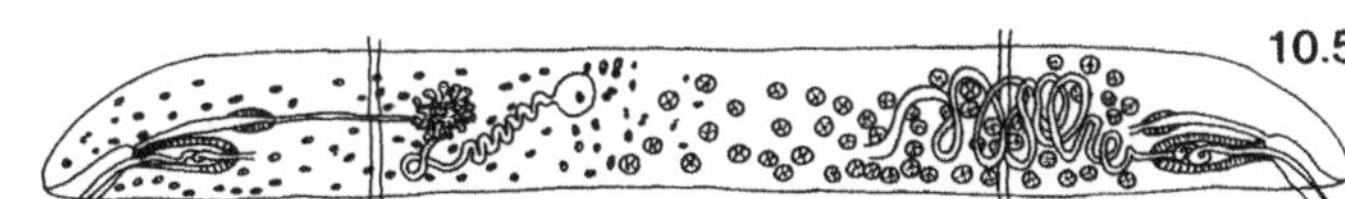

10.58

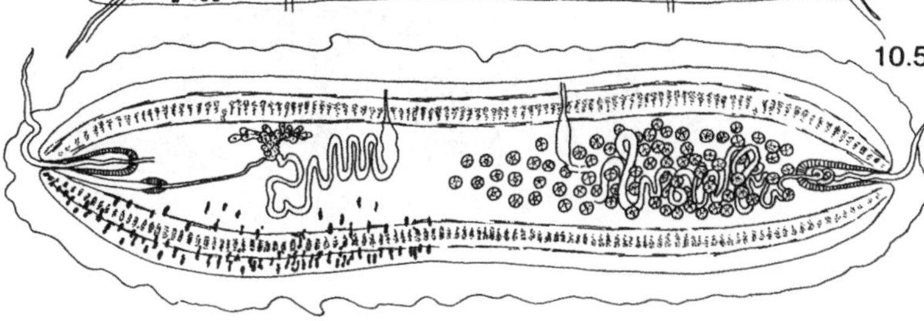

10.59

4a. Scolex replaced by scolex deformatus .... 5.
4b. Scolex [or pseudoscolex?] usually present .... 6.

5a. Vitelline follicles intermingled with testes, continuous throughout length of proglottid .... ***Abothrium*** **van Beneden, 1871. (Figs 10.60, 10.61)**

Diagnosis: Scolex deformatus present, unarmed, apical disc and bothria absent. Strobila robust. External segmentation distinct. Genital pores lateral, alternating irregularly. Cirrus-sac small. Cirrus unarmed. Testes numerous, in lateral fields. Ovary symmetrical, median. Uterus with median coils. Uterine pore median. Vitelline follicles medullary, in two lateral fields. Osmoregulatory canals 10–25. In teleosts, mainly Gadiformes. Type-species *A. gadi* van Beneden, 1871.

5b. Vitelline follicles partly cortical and partly medullary, not intermingled with testes .... ***Parabothrium*** **Nybelin, 1920. (Figs 10.62–10.64)**

Diagnosis: Scolex deformatus present, unarmed, apical disc and bothria absent. Strobila robust. External segmentation distinct. Genital pores lateral, alternating irregularly. Cirrus-sac large. Testes numerous, in lateral fields, confluent in anterior part of proglottid. Ovary lobate, median. Uterus saccular, becoming lobulate to stellate. Uterine pore median. Vitelline follicles in two ventrolateral fields; in posterior part of proglottid only. Osmoregulatory canals 18–50. In gadid teleosts. Type-species *P. bulbiferum* Nybelin, 1920 [now considered a synonym of *P. gadipollachii* (Rudolphi, 1810)].

6a. Bothria divided .... 7.
6b. Bothria undivided .... 8.

10.60

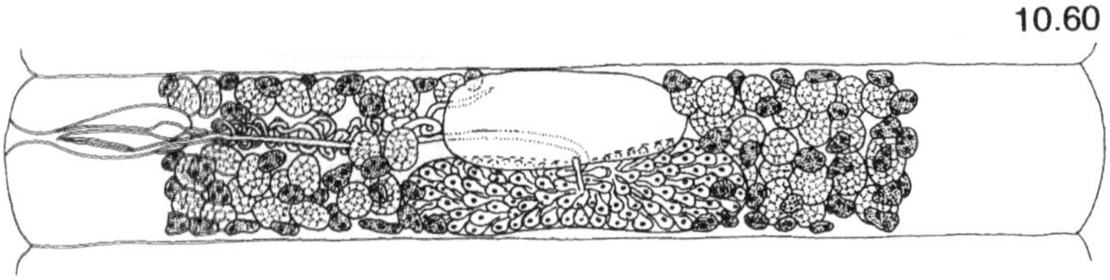

10.61

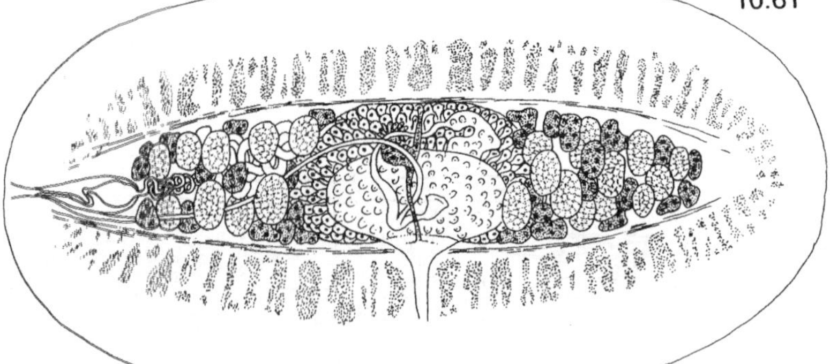

10.62

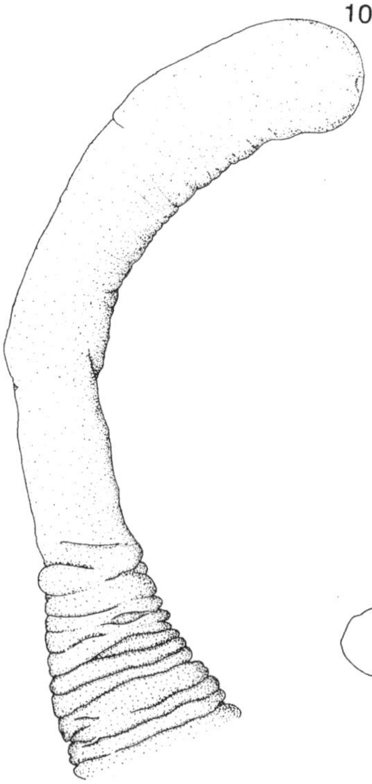

10.63

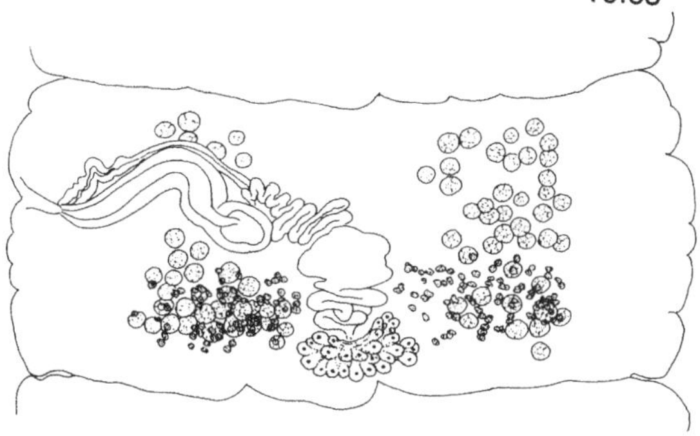

10.64

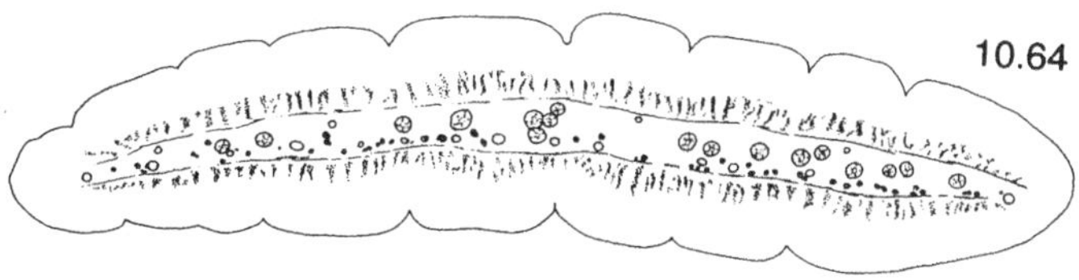

10.65

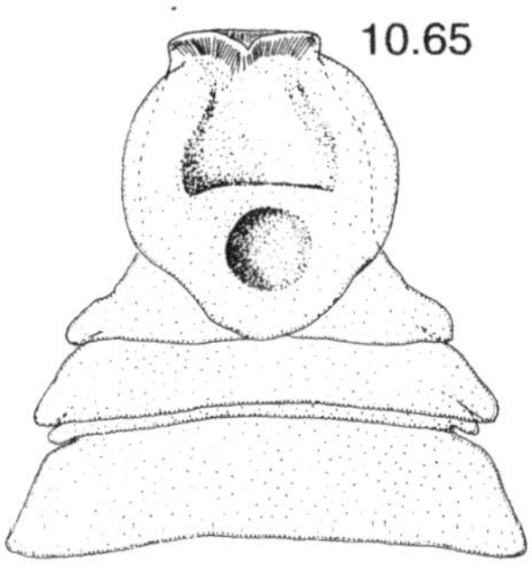

10.66

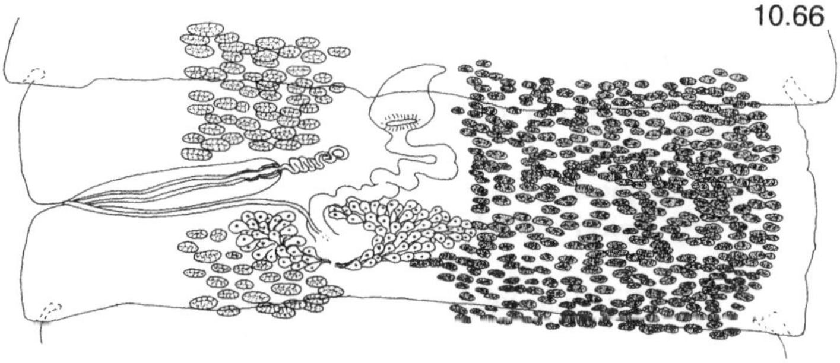

Figs 10.60, 10.61 *Abothrium gadi* van Beneden, 1871. 10.60. Mature proglottid. 10.61. Mature proglottid in transverse section. Both modified after Nybelin (1922).
Figs 10.62–10.64 *Parabothrium bulbiferum* Nybelin, 1920. 10.62. Scolex deformatus. 10.63. Mature proglottid. 10.64. Transverse section of mature proglottid.
Figs 10.65, 10.66 *Amphicotyle heteropleura* (Diesing, 1850). 10.65. Scolex. 10.66. Mature proglottid; vitelline follicles omitted porally, testes omitted aporally.

7a. **Bothria each with one division forming posterior sucker. Vitelline follicles cortical, in lateral fields. Osmoregulatory canals four .... *Amphicotyle* Diesing, 1863. (Figs 10.65, 10.66)**

Diagnosis: Scolex trapezoid, with apical disc. Strobila robust. External segmentation distinct anteriorly, obscured by secondary wrinkles posteriorly. Genital pores lateral, alternating irregularly. Cirrus-sac large. Cirrus protrusible, armed.

Testes numerous in lateral fields or single field. Ovary symmetrical, median to poral. Uterus with median coils. Vitelline follicles in two lateral dorsal and ventral fields, continuous longitudinally. In teleosts. Type-species *A. heteropleura* (Diesing, 1850).

7b. Bothria divided into several loculi. Vitelline follicles medullary and cortical, surrounding proglottid. Osmoregulatory canal system reticulate, about 10–14 .... *Pseudamphicotyla* Yamaguti, 1959. (Figs 10.67, 10.68)

Diagnosis: Scolex slightly elongate, with apical disc. Bothria two. External segmentation distinct, with or without secondary wrinkles. Genital pores, irregularly alternating. Cirrus-sac present. Cirrus spined. Testes numerous, in one continuous layer. Ovary slightly poral. Uterus saccular, globular to oval. Uterine pore median. Vitelline follicles continuous longitudinally. In teleosts (Pentacerotidae). Type-species *P. quinquarii* (Yamaguti, 1952).

8a. Ovary dendritic. In abyssal fish .... *Pistana* Campbell & Gartner, 1982. (Figs 10.69–10.71)

Diagnosis: Scolex unarmed, arrowhead-shaped; bothria deep with freely-projecting posterior margins; apical disc absent. Neck absent. External metamerism distinct. Genital pore marginal, postequatorial. Cirrus-sac elongated, extending into medulla. Cirrus unarmed. Testes in two lateral fields. Ovary posterior, slightly poral, bialate. Vagina expanded distally, opens posterior to cirrus-sac. Vitelline follicles circumcortical. Uterus median. Uterine sac pyriform, opens midventrally. Eggs operculate, embryonated. In abyssal marine teleosts (*Eurypharynx pelecanoides*). Western North Atlantic. Type-species *P. eurypharyngis* Campbell & Gartner, 1982.

8b. Ovary not dendritic .... 9.

9a. Apical disc present .... 10.
9b. Apical disc absent .... 13.

10a. Vitelline follicles mainly medullary. Uterus with long lateral pouches .... *Marsipometra* Cooper, 1917. (Figs 10.72–10.74)

Diagnosis: Scolex pyramidal or sagittate. Bothria oval, deep or shallow. External segmentation present. Genital pores lateral, alternating irregularly. Cirrus-sac large. Cirrus protrusible. Testes numerous, lateral fields confluent anterior to uterus, and usually posterior to ovary. Ovary reniform, with lobulate wings, median. Uterus saccate, forms long lateral pouches. Uterine pore median.

Figs 10.67, 10.68 *Pseudamphicotyla quinquarii* (Yamaguti, 1952). 10.67. Scolex. 10.68. Mature proglottid, vitelline follicles omitted aporally. Both modified after Yamaguti (1952).
Figs 10.69–10.71 *Pistana eurypharyngis* Campbell & Gartner, 1982. 10.69. Scolex. 10.70. Early mature proglottid. 10.71. Gravid proglottid.
Figs 10.72–10.74 *Marsipometra hastata* (Linton, 1898). 10.72. Scolex. 10.73. Mature proglottid. Vitelline follicles omitted porally, testes omitted aporally. 10.74. Gravid proglottid. All modified after Beaver & Simer (1940).

10.67

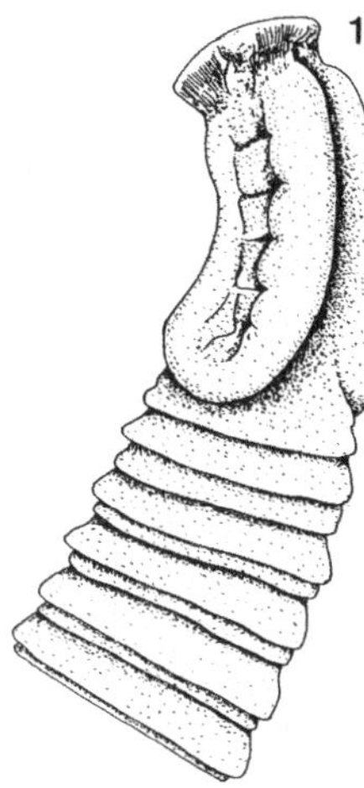

10.68

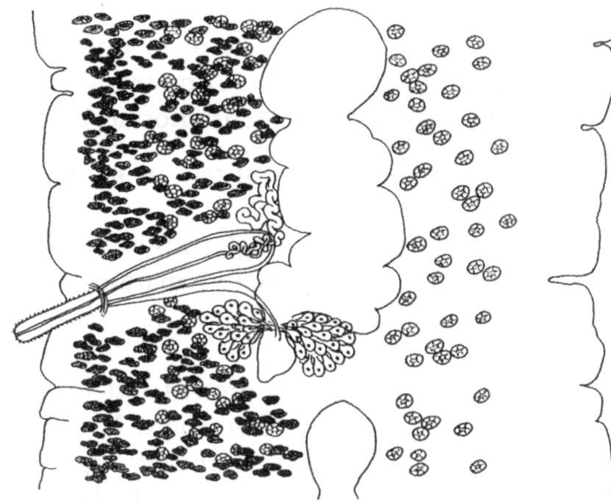

10.69

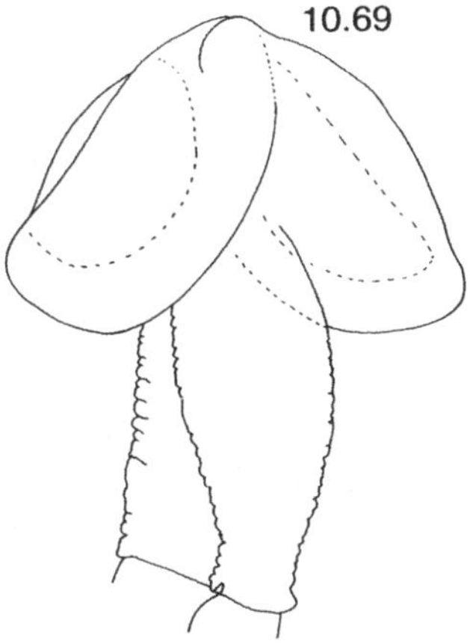

10.70

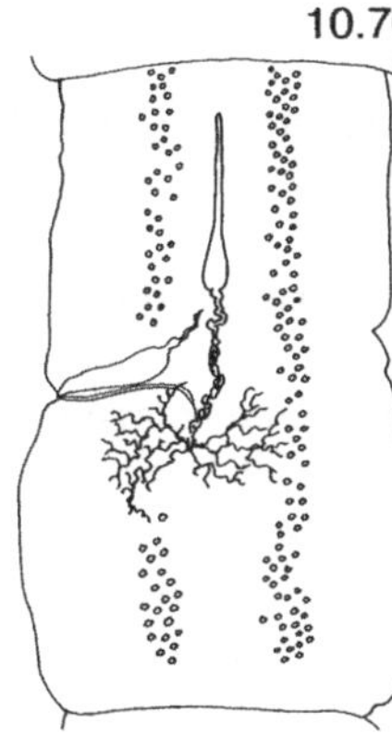

10.71

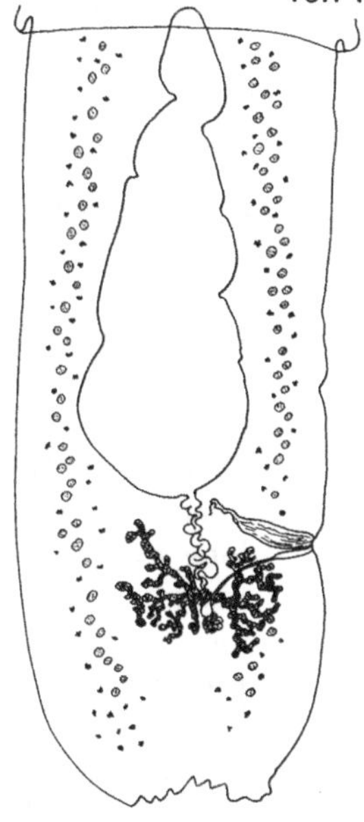

10.72

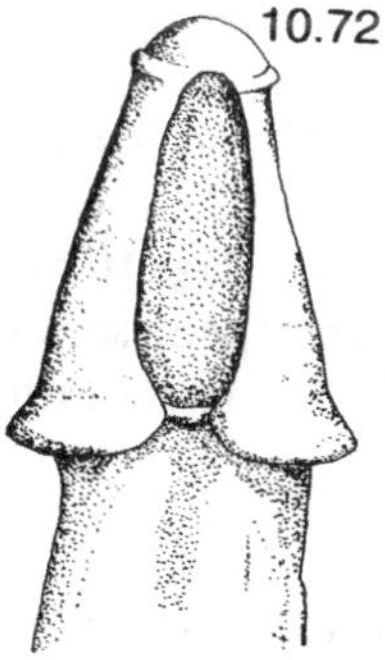

10.73

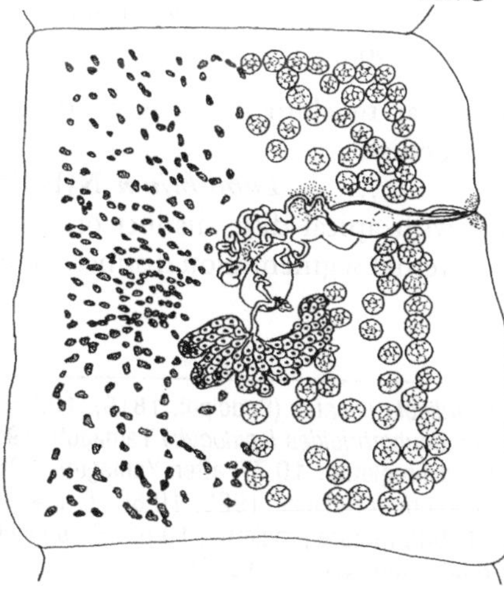

10.74

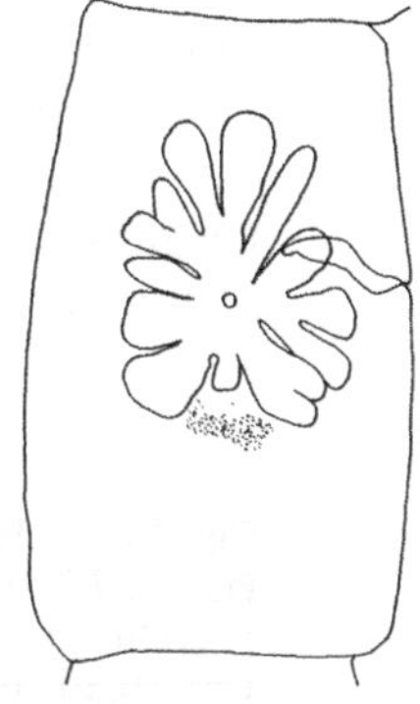

Vitelline follicles medullary or in muscle zone, in right and left ventral fields, continuous longitudinally. Osmoregulatory canals 4–6 or 4 convoluted and branched. In *Polyodon* (Acipenseriformes). Type-species *M. hastata* (Linton, 1898).

10b. Vitelline follicles entirely or mainly cortical .... 11.

11a. Scolex sagittate. Proglottids much wider than long, strongly craspedote with extended posterolateral margins .... ***Fistulicola* Lühe, 1899. (Figs 10.75, 10.76)**
(Syn. *Pseudeubothrium* Yamaguti, 1968.)

Diagnosis: Apical disc present but unarmed; bothrial margins smooth. Scolex may be replaced by pseudoscolex. Neck absent. External metamerism complete. Genital pores marginal. Cirrus-sac large. Cirrus protrusible. Testes medullary, in single field. Ovary slightly poral, bialate. Vitelline follicles circumcortical. Uterus strongly coiled, muscular distally. Uterine pore anterior beneath velum of preceding proglottid, ventrally submedian on poral side. Eggs operculate, embryonated. In marine teleosts (*Xiphias gladius, Zeus faber*). Atlantic, Pacific, Mediterranean. Type-species *F. plicatus* (Rudolphi, 1819).

11b. Scolex not sagittate. Proglottids lack extended posterolateral margins ... . **12.**

12a. Scolex elongate. Uterus strongly coiled, forms half-loop; uterine pore submedian, poral .... ***Pseudeubothrioides* Yamaguti, 1968. (Figs 10.77–10.79)**

Diagnosis: Scolex with simple bothria with smooth margins. Apical disc present, unarmed. Neck present. External metamerism present. Proglottids wider than long except when fully gravid; not strongly craspedote, with salient posterior margins and almost parallel sides. Genital pores marginal, pre-equatorial. Cirrus-sac cortical, oblique, muscular. Testes medullary, confluent posteriorly but few present in preovarian region. Ovary median, transversely elongate, bialate. Vagina opens behind cirrus-sac. Vitelline follicles cortical. Uterine sac muscular. Uterine pore at or just behind level of genital pore. Eggs thick-shelled, not embryonated. In marine teleosts (*Lepidocybium flavobrunneum*). Hawaii. Type-species *P. lepidocybii* Yamaguti, 1968.

12b. Scolex subglobular to elongate, may be deformed. Uterus saccate, oval; uterine pore median .... ***Eubothrium* Nybelin, 1920. (Figs 10.80–10.82)**

Diagnosis: Scolex with apical disc, distinct or poorly developed; scolex deformatus may occur. External segmentation distinct. Genital pores lateral, irregularly

---

Figs 10.75, 10.76 *Fistulicola plicatus* (Rudolphi, 1819). 10.75. Scolex. 10.76. Gravid proglottid.
Figs 10.77–10.79 *Pseudeubothrioides lepidocybii* Yamaguti, 1968. 10.77. Scolex. 10.78. Mature proglottid. 10.79. Gravid proglottid. 10.77 after Yamaguti (1968).
Figs 10.80–10.82 *Eubothrium* Nybelin, 1920. 10.80. *E. crassum* (Bloch, 1779), scolex. 10.81. *E. salvelini* (Schrank, 1790), mature proglottid. 10.82. *E. rugosum* (Batsch, 1786), transverse section of mature proglottid, after Nybelin (1922).

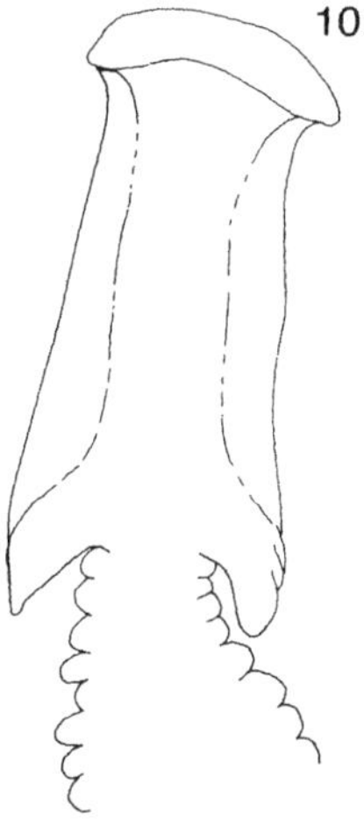
10.75

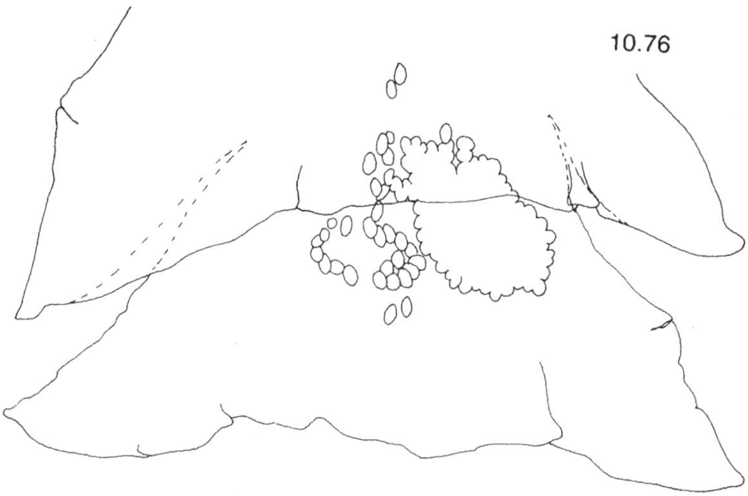
10.76

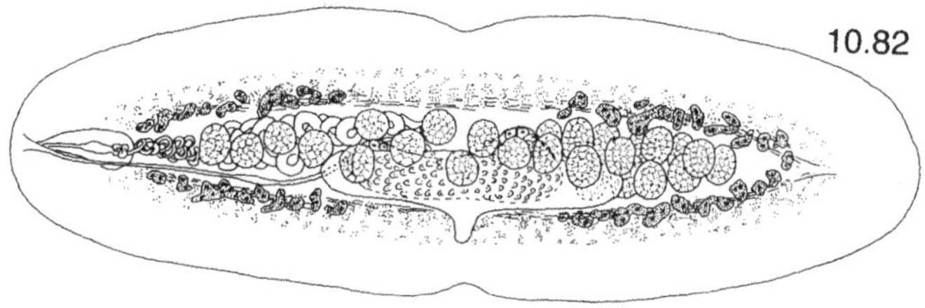
10.82

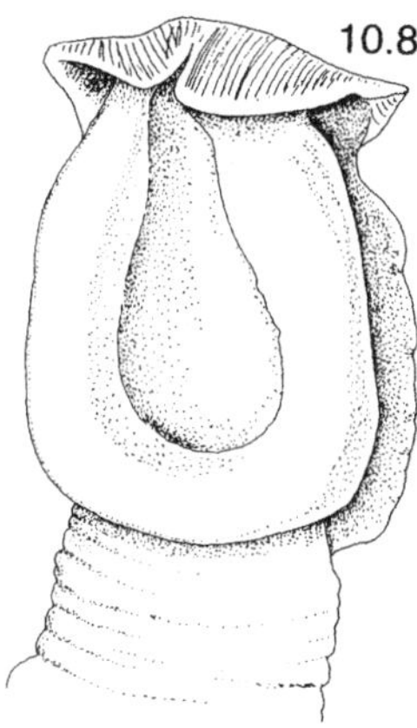
10.80

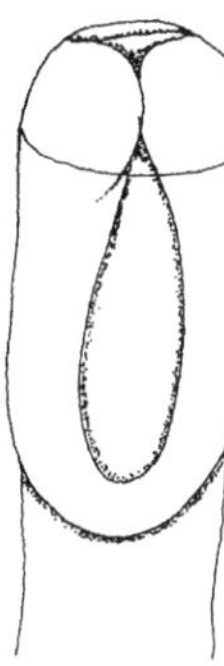
10.77

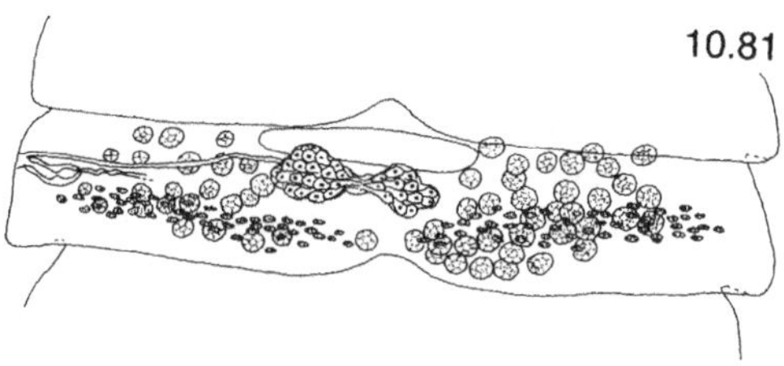
10.81

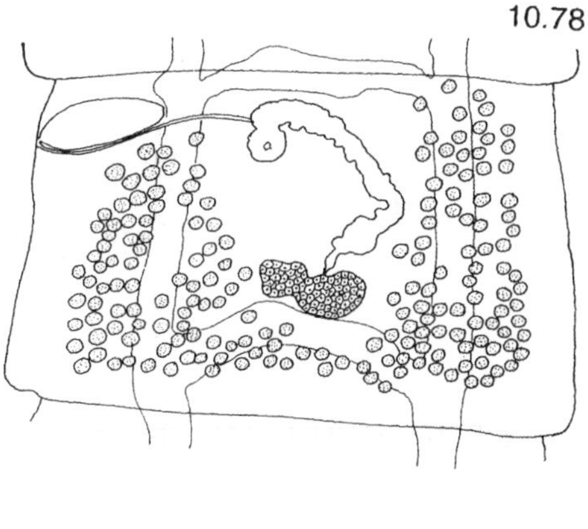
10.78

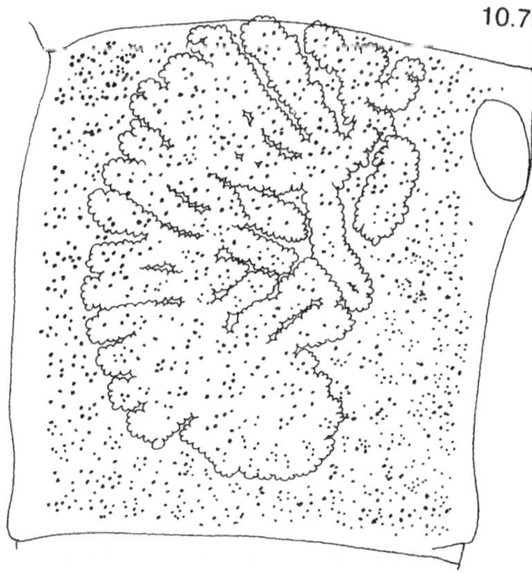
10.79

alternating. Cirrus-sac small. Cirrus unarmed. Testes numerous, in two lateral fields which may be almost confluent. Ovary reniform, median. Vitelline follicles cortical, or among longitudinal muscles or occasionally in medulla; in lateral fields, not confluent ventrally or dorsally, not continuous longitudinally. In freshwater and marine teleosts. Type-species *E. rugosum* (Batsch, 1786).

13a. Vitelline follicles medullary .... **14.**
13b. Vitelline follicles cortical .... **16.**

14a. Scolex sagittate with freely projecting posterior margins ....
.... ***Anonchocephalus* Lühe, 1902. (Figs 10.83, 10.84)**

Diagnosis: Scolex unarmed. Apical disc absent. Neck absent. External segmentation present. Proglottids wider than long. Genital pore marginal, postequatorial. Cirrus-sac elongated, muscular. Cirrus armed with spines. Testes medullary, in two lateral fields united at the posterior margin of the proglottid. Ovary posterior, poral, compact. Vagina opens posterior to cirrus-sac. Vitelline follicles ventral to testes. Uterus coiled, narrow, with large, muscular uterine sac opening midventrally. Eggs thin-shelled, operculate, embryonated. In marine teleosts. Chile. Type-species *A. chilensis* (Riggenbach, 1896).

14b. Scolex otherwise .... **15.**

15a. Vitelline follicles in lateral bands ....
.... ***Bathybothrium* Lühe, 1902. (Figs 10.85–10.87)**

Diagnosis: Scolex oval to heart-shaped, no apical disc. External segmentation distinct. Reproductive organs single. Genital pores lateral, alternating irregularly. Testes numerous, lateral fields unite anterior to ovary. Cirrus-sac large. Cirrus unarmed. Ovary irregular, median. Uterus lobulate, median, pre-ovarian. Uterine pore median or slightly poral. Vitelline follicles few, in medulla. Osmoregulatory canals? In cyprinids. Type-species *B. rectangulum* (Bloch, 1782).

15b. Vitelline follicles surrounding proglottid ....
.... ***Fissurobothrium* Roitman, 1965. (Figs 10.88–10.90)**

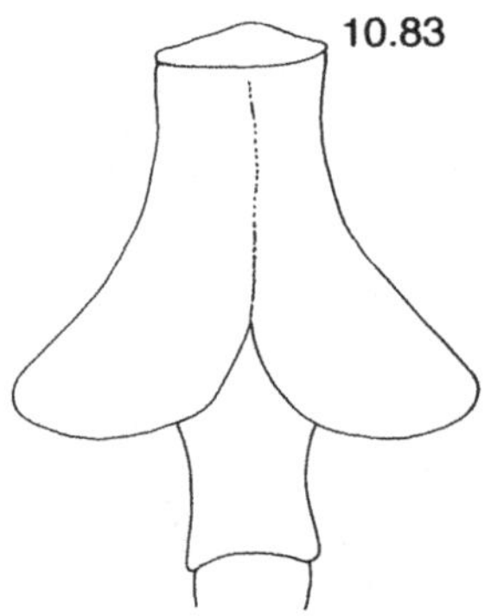

Figs 10.83, 10.84 *Anonchocephalus chilensis* (Riggenbach, 1896). 10.83. Scolex (after Riggenbach, 1895). 10.84. Mature proglottid.

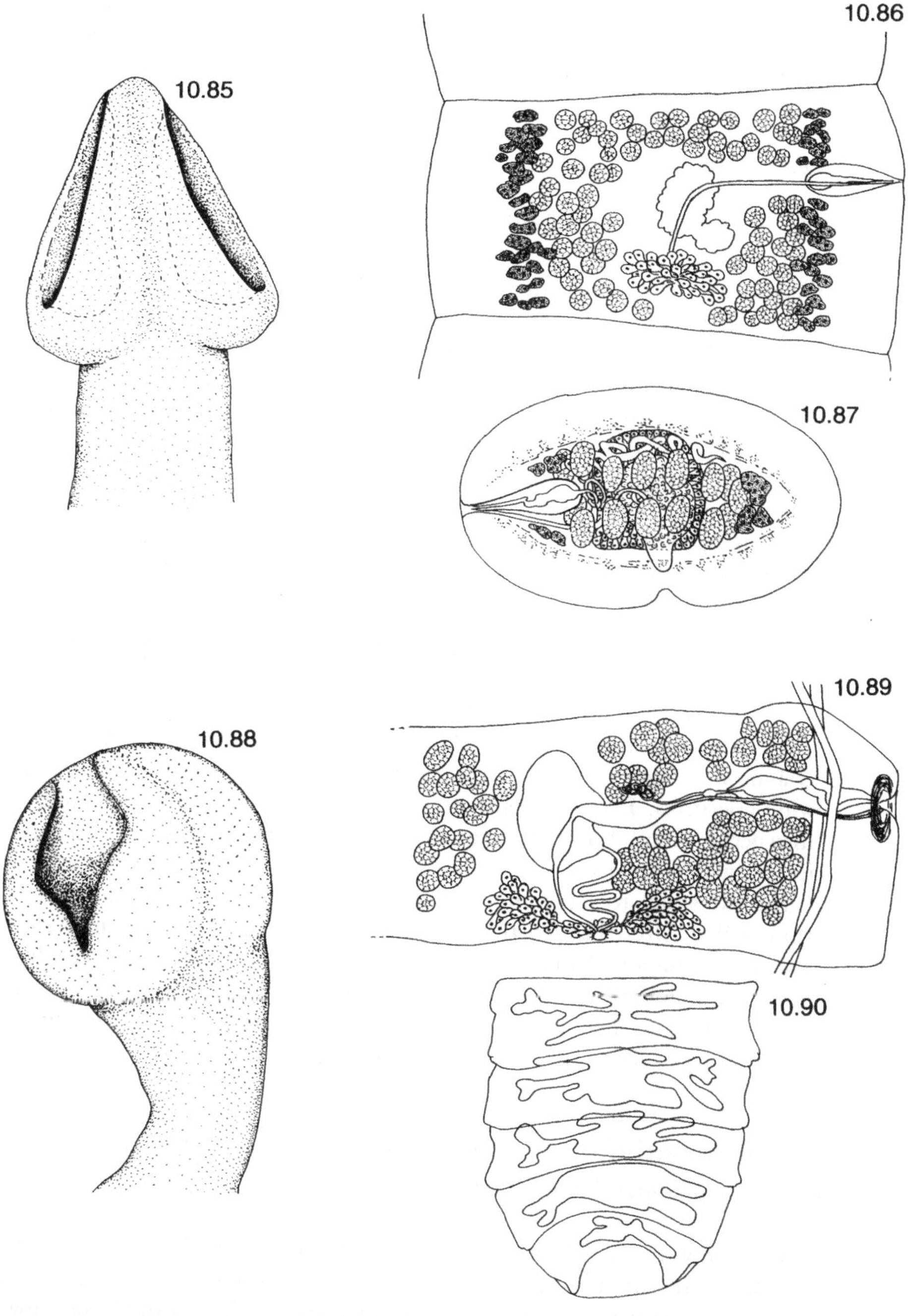

Figs 10.85–10.87 *Bathybothrium rectangulum* (Bloch, 1782). 10.85. Scolex. 10.86. Mature proglottid. 10.87. Transverse section of mature proglottid, modified after Nybelin (1922).
Figs 10.88–10.90 *Fissurobothrium unicum* Roitman, 1965. 10.88. Scolex. 10.89. Mature proglottid, aporal extremity and vitelline follicles omitted. 10.90. Terminal proglottids, showing uterine configuration. All modified after Roitman (1965).

Diagnosis: Scolex rounded, no apical disc. Bothria deep, sucker-like. External segmentation present. Genital pores lateral, [alternating?]. Cirrus-sac present. Cirrus unarmed. Testes numerous, in two lateral fields. Ovary symmetrical, lobed, median. Uterus with large lateral branches. Uterine pore slit-like. Vitelline follicles medullary. Osmoregulatory canals four. In cyprinids. Type-species *F. unicum* Roitman, 1965.

16a. Scolex sagittate. Vitelline follicles confined to posterior half of proglottid .... *Eubothrioides* Yamaguti, 1952. (Figs 10.91, 10.92)

Diagnosis: Scolex unarmed. Bothria elongated; margins may be slightly crenulated. Neck absent. External metamerism distinct. Strobila fleshy. Proglottids imbricated, trapezoidal, much wider than long. Genital pore marginal, slightly postequatorial. Cirrus-sac small, not extending into medulla. Cirrus unarmed. Testes in two narrow lateral fields medial to nerve cords. Ovary posterior, slightly poral, transversely elongated, not distinctly bialate. Vagina dorsal or ventral to cirrus-sac. Vitelline follicles cortical, diffuse. Uterus in mid-field; uterine sac small, opens midventrally near anterior margin of proglottid. Eggs operculate, elongated pyriform with attenuated end inclined to one side; partly embryonated. In marine teleosts (*Zenopsis nebulosa*). Japan. Type-species *E. lamellatus* Yamaguti, 1952.

16b. Scolex not sagittate. Vitelline follicles continuous longitudinally .... *Probothriocephalus* Campbell, 1979. (Figs 10.93, 10.94)
(Syns *Heterovitellus* Protasova & Parukhin, 1986; *Partitiotestis* Protasova & Parukhin, 1986; *Flexiphallus* Protasova & Parukhin, 1986.)

Diagnosis: Scolex conical to elongate, no apical disc. Bothria shallow, weak with wide apertures. External segmentation distinct or indistinct, secondary wrinkles may occur. Genital pores postequatorial, alternating irregularly. Testes numerous, in two widely-separated medullary fields, confluent medially posterior to ovary. Cirrus-sac small to large, not reaching mid-line. Ovary bilobed, poral. Vagina thin-walled. Uterus coiled tubular proximally, saccular distally, may intrude into preceding proglottid when gravid. Uterine pore median. Eggs operculate or not, unembryonated, may have tubercle at poles. Vitelline follicles circumcortical, occasionally also between muscle bundles, continuous longitudinally. In deep-water teleosts. Type-species *P. muelleri* Campbell, 1979.

Family Diphyllobothriidae Lühe, 1910

(Syn. Dibothriocephalidae Lühe, 1902)

The family name Diphyllobothriidae was established by Lühe (1910) as a substitute for his family Dibothriocephalidae erected in 1902. The family was originally meant to contain a number of forms similar to the long-known broad tapeworm of man – first described as *Taenia lata* by Linnaeus (1758) which was also the type-species of the family. However, due to priority rules when changing the family name in 1910, the relatively unknown and unsatisfactorily described *Diphyllobothrium stemmacephalum* Cobbold, 1858 from porpoise became the type-species of the family.

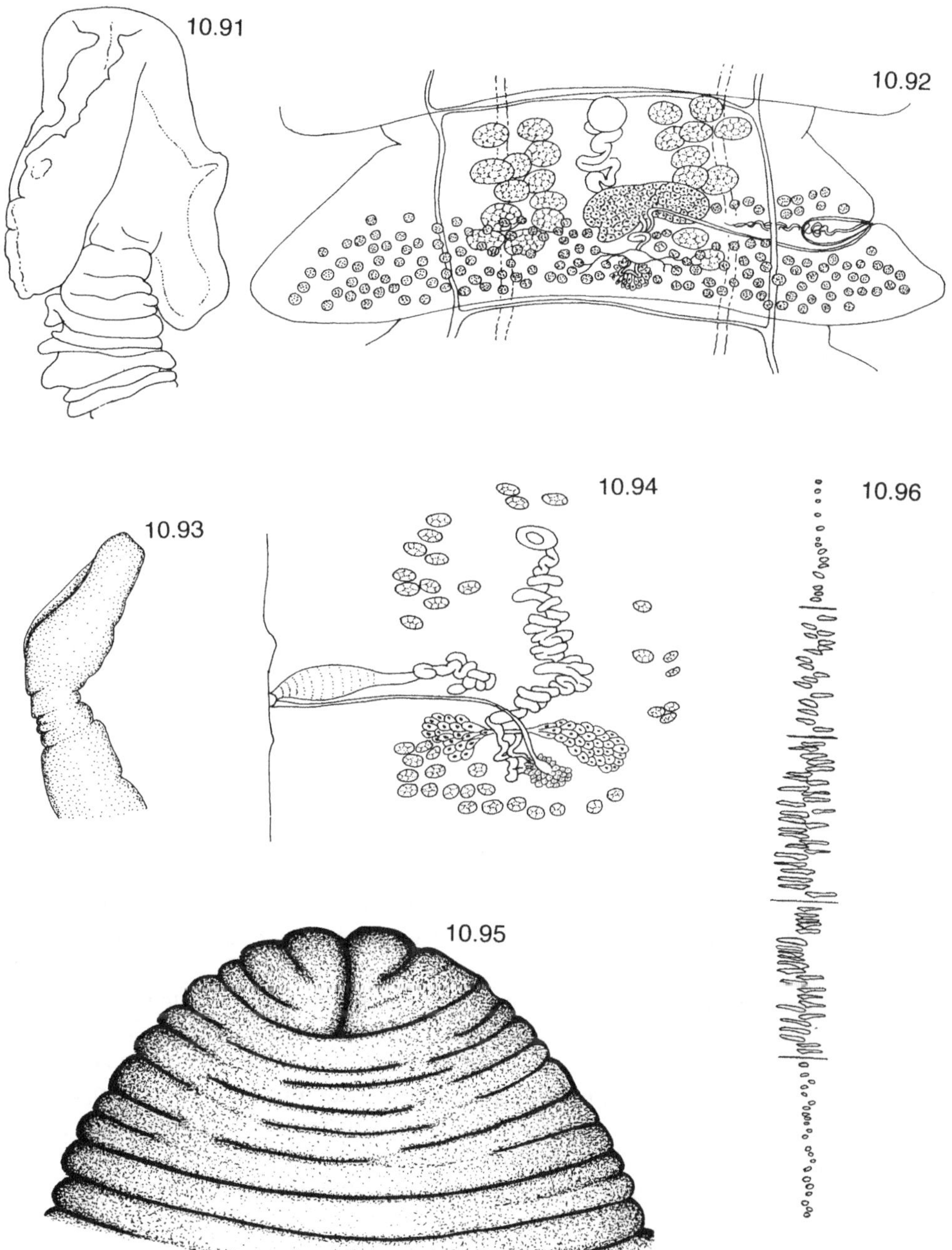

Figs 10.91, 10.92 *Eubothrioides lamellatus* Yamaguti, 1952. 10.91. Scolex. 10.92. Mature proglottid. After Yamaguti (1952).

Figs 10.93, 10.94 *Probothriocephalus muelleri* Campbell, 1979. 10.93. Scolex. 10.94. Mature proglottid. Both after Campbell (1979).

Figs 10.95, 10.96 *Ligula intestinalis* (Linnaeus, 1758). 10.95. Anterior end of plerocercoid, after Cooper (1918). 10.96. Schematic representation of genital primordia, after Dubinina (1966).

The descriptions of *Tetragonoporus* Skryabin, 1961 and *Polygonoporus* Skryabin, 1967 show many characters similar to *Diplogonoporus* Lönnberg, 1872. They are synonymized with *Diplogonoporus* until better descriptions may prove otherwise. They are all parasites of cetaceans.

Adenocephalus Nybelin, 1931, *Cordicephalus* Wardle, McLeod & Stewart, 1947 and *Flexobothrium* Yurakhno, 1988 are all regarded as synonyms of *Diphyllobothrium*. Presently, no characters can be found which justify generic separation between them.

Capooria Malhotra, 1985 superficially resembles members of the family Bothriocephalidae. The characters employed to allocate it to the Diphyllobothriidae require confirmation. Pending a redescription, it is treated here as a *genus inquirendum*.

Pyramicocephalus Monticelli, 1890 is also synonymized with *Diphyllobothrium*. Monticelli founded the genus on small wrinkled plerocercoids with scoleces with deep marginal bothrial grooves, found in some marine fish species but the adult worm, so far, lacks a distinct description. A few *Diphyllobothrium* species exist (e.g. *D. hians* from the same host) which possess scoleces with deep marginal bothrial grooves and crenulated bothrial edges. Due to the lack of a good description of the adult worm, *Pyramicocephalus* is synonymized with *Diphyllobothrium*.

Spirometra Faust, Campbell & Kellogg, 1929 is recognized as a valid genus, although some, e.g. Schmidt (1986), synonymized it with *Diphyllobothrium*. There are, however, enough morphological and life-cycle differences between them to justify separation at generic level.

Family diagnosis

Pseudophyllidea. Strobila medium to large, usually acraspedote, anapolytic. External segmentation complete or not. Scolex of various shapes, occasionally poorly developed. Bothria usually well developed, of various shapes; occasionally represented only by shallow slits; rarely hypertrophied. Osmoregulatory canals cortical or medullary, numerous or not. Reproductive organs usually single, may be double or occasionally multiple. Testes numerous, medullary. Cirrus-sac present. Ovary posterior. Male and female ducts open together or separately on midventral surface. Vitelline follicles cortical, may intrude between longitudinal muscle bundles. Uterus of various shapes, opens ventrally, posteriorly to genital pore(s). Larval stages in fish, reptiles and occasionally other groups. Adults in reptiles, birds and mammals. Type-genus *Diphyllobothrium* Cobbold, 1858.

Key to genera

1a. External segmentation confined to anterior portion of strobila .... 2.
1b. External segmentation complete throughout strobila .... 3.

2a. One set of reproductive organs per proglottid ....
.. *Ligula* Bloch, 1782. (Figs 10.95, 10.96)
(Syn. *Braunia* Leon, 1908.)

Diagnosis: Fleshy worms up to 400 mm long by 7–8 mm wide. Anterior portion of strobila segmented, rest unsegmented but transversely wrinkled. Anterior end of strobila bluntly pointed with poorly developed scolex; bothria represented only by two shallow tiny slits. One set of reproductive organs per proglottid. Adults in fish-eating birds. Large fleshy plerocercoids in body-cavity of cyprinids. Circumboreal. Type-species *L. intestinalis* (L., 1758).

2b. Two sets of reproductive organs per proglottid ....
.... *Digramma* Cholodkowsky, 1915. (Figs 10.97, 10.98)

Diagnosis: As *Ligula* except two sets of genitalia per proglottid. Type-species *D. alternans* (Rudolphi, 1810).

3a. Scolex poorly developed, bothria represented only by two tiny slits in anterior-most part of strobila .... *Schistocephalus* Creplin, 1829. (Fig. 10.99)
(Syn. *Schistorhynchus* Zschokke, 1896.)

Diagnosis: Fairly small and lanceolate worms, maximum length around 200 mm. Proglottids wider than long and distinctly craspedote. Well-developed

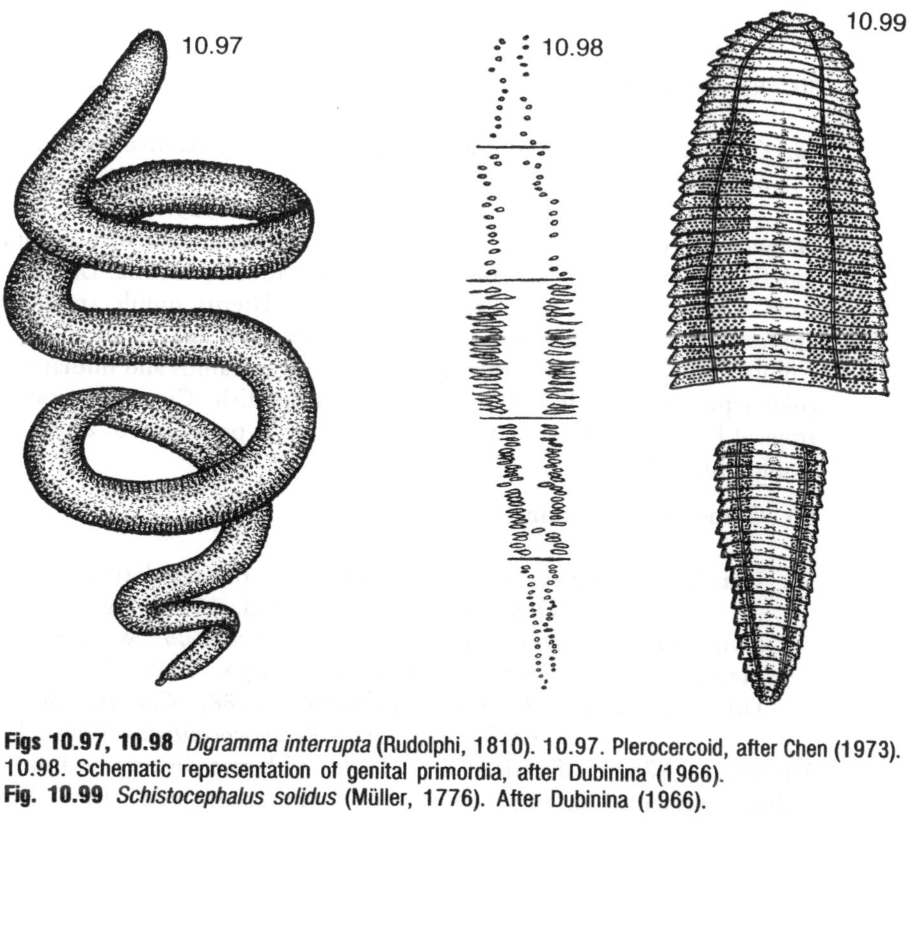

Figs 10.97, 10.98 *Digramma interrupta* (Rudolphi, 1810). 10.97. Plerocercoid, after Chen (1973). 10.98. Schematic representation of genital primordia, after Dubinina (1966).
Fig. 10.99 *Schistocephalus solidus* (Müller, 1776). After Dubinina (1966).

parenchymal longitudinal muscles. Adults in fish-eating birds. Large segmented plerocercoids in body-cavity of sticklebacks and other small freshwater fish. Type-species *S. solidus* (Müller, 1776).

3b. Scolex distinct and well developed .... **4.**

4a. Scolex broad and fan-like, bothrial margins frilled or crenulated .... ***Duthiersia*** **Perrier, 1873. (Figs 10.100, 10.101)**

Diagnosis: Small worms not exceeding 200 mm in length. Posterior bothrial apertures present in Asiatic forms, absent in African forms. Vaginal sphincter present; vagina opens behind cirrus-sac. Uterus tubular, laterally looped and rosette-like. In varanid lizards in south eastern Asia and Africa. Type-species *D. expansa* Perrier, 1873.

4b. Scolex more or less spoon-shaped with two distinct bothria, bothrial edges usually plain, no inrolling of bothrial edges .... **5.**

4c. Scolex modified .... **10.**

5a. Proglottids very craspedote. Vitelline follicles cortical as single layer in wing-like expansions .... ***Glandicephalus*** **Fuhrmann, 1921. (Fig. 10.102)**

Diagnosis: Fairly small worms around 200 mm long. Strobila thick, external metamerism well marked, proglottids with wing-like expansions. Testes in single layer. Uterus strongly serpentine. In Antarctic seals. Type-species *G. antarcticus* (Blair, 1853).

5b. Normal segmentation .... **6.**

6a. Cirrus-sac composite, dorsal portion representing external seminal vesicle of other genera .... ***Spirometra*** **Faust, Campbell & Kellogg, 1929. (Figs 10.103, 10.104)**

Diagnosis: Scolex spoon- or finger-like in marginal outline with bothria fading indefinitely into mid-dorsal and mid-ventral furrows of strobila. Usually medium sized worms, weakly muscular, distinct neck. Uterus simple spiral of closely appressed coils, widening into sac before uterine pore. Eggs pointed at both ends. Adults mainly in cat-like carnivores, occasionally in canids and humans. Plerocercoids (spargana) in all vertebrate groups except fish. Circumtropical and subtropical but also in USA and southern Europe. Type-species *S. erinaceieuropaei* (Rudolphi, 1819).

6b. Cirrus-sac and seminal vesicle separate .... **7.**

7a. Usually only one set of reproductive organs per proglottid .... ***Diphyllobothrium*** **Cobbold, 1858. (Figs 10.105–10.107)**
(Syns *Adenocephalus* Nybelin, 1931; *Cordicephalus* Wardle, McLeod & Stewart, 1947; *Diancyrobothrium* Bacigalupo, 1945; *Dibothriocephalus* Lühe, 1899; *Flexobothrium* Yurakhno, 1988; *Gatesius* Stiles, 1908; *Lueheella* Baer, 1924; *Pyramicocephalus* Monticelli, 1890.)

Diagnosis: Worms of various sizes. Scolex with two distinct bothria. Vitelline follicles numerous in single layer continuous across proglottids. Uterus with

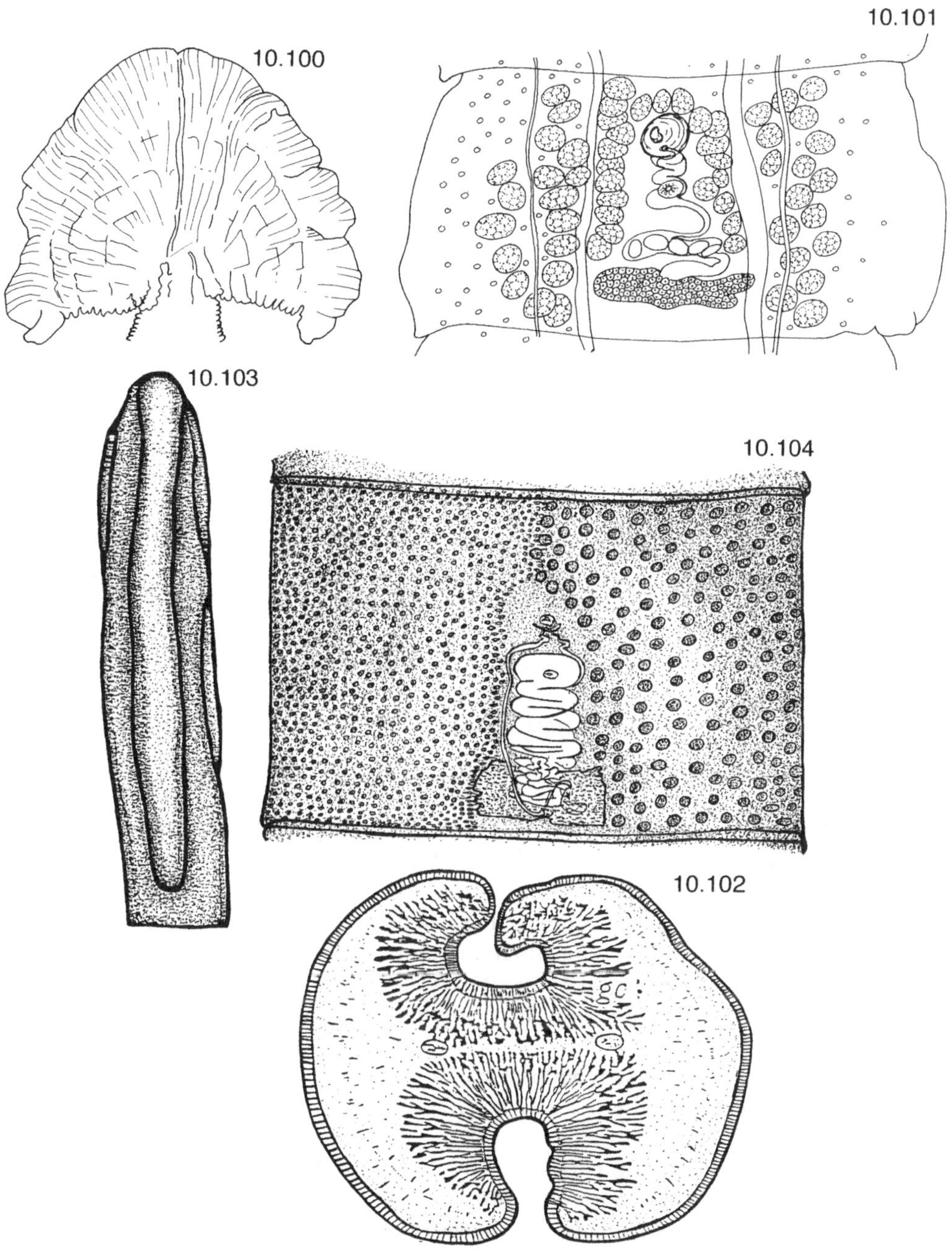

Figs 10.100, 10.101 *Duthiersia fimbriata* (Diesing, 1850). 10.100. Scolex. 10.101. Mature proglottid.

Fig. 10.102 *Glandicephalus antarcticus* (Blair, 1853). Scolex in transverse section, showing glands. After Fuhrmann (1920b).

Figs 10.103, 10.104 *Spirometra erinaceieuropaei* (Rudolphi, 1819). 10.103. Scolex. After Meggitt (1924b). 10.104. Mature proglottid. After Faust, Campbell & Kellogg (1929).

several loops which may be parallel or rosette-shaped. Testes usually in single layer. Parenchymal muscles fairly well developed. Adults in fish-eating birds and mammals (terrestrial and marine). Plerocercoids in fish. Type-species *D. stemmacephalum* Cobbold, 1858.

7b. Two or more sets of reproductive organs per proglottid .... 8.

8a. Not more than two sets of reproductive organs per proglottid .... 9.
8b. More than two sets of reproductive organs per proglottid .... *Hexagonoporus* Gubanov in Delyamure, 1955. (Fig. 10.108)
(Syns *Polygonoporus* Skryabin, 1967; *Tetragonoporus* Skryabin, 1961.)
Diagnosis: Strobila up to 30 m long. Apical disc fleshy; unlobed. Bothria short, fleshy. Strobila expands rapidly behind scolex, then narrows. Proglottids much wider than long. Four to 14 pairs of reproductive organs per proglottid, arranged side by side. External seminal vesicle present. Ovary bilobed, transversely elongated. Vitelline follicles cortical. Uterus with five to seven loops on each side. In cetaceans. Antarctica. Type-species *H. physeteris* Gubanov in Delyamure, 1955.

9a. Osmoregulatory canals cortical, not numerous .... *Diplogonoporus* Lönnberg, 1892. (Figs 10.109, 10.110)
(Syn. *Krabbea* Blanchard, 1894.)
Diagnosis: Very large worms with weak musculature. Proglottids short and very wide. Large seminal vesicle. Uterine coils parallel or rosette-shaped. In cetaceans. Type-species *D. balaenopterae* Lönnberg, 1899.

9b. Osmoregulatory canals medullary, numerous (35 to 70) .... *Multiductus* Clarke, 1962. (Figs 10.111–10.113)
Diagnosis: Strobila long (up to 18 m), proglottids numerous, craspedote. Scolex globular, bothria deep. Two layers of longitudinal muscle fibres, inner layer always arranged in bundles, outer layer in bundles in some regions. Osmoregulatory canals connected by anastomoses. Two sets of genitalia, symmetrically placed. Testes medullary, in more than one layer. Vitelline follicles cortical, between longitudinal muscle layers. Uterine sac and uterine duct present. In sperm whale. Antarctica. Type-species *M. physeteris* Clarke, 1962.

10a. Scolex with fused bothrial edges .... 11.
10b. Scolex otherwise .... 12.

11a. Bothria forming cups .... *Baylisia* Markowski, 1952. (Fig. 10.114)
Diagnosis: Scolex short with two cup-shaped bothria. Inner longitudinal and transverse muscles strongly developed. Two sets of reproductive organs per proglottid. In crabeater seal. Antarctic. Type-species *B. baylisi* Markowski, 1952.

11b. Bothria forming tubes .... *Bothridium* Blainville, 1824. (Figs 10.115, 10.116)
(Syns *Prodicoelia* Leblond, 1836; *Solenophorus* Creplin, 1839.)
Diagnosis: Moderate-sized worms up to 500 mm long by 6 mm wide. Strobila with numerous short proglottids. Uterus with thick duct and dorsal sac. Testes

10.105

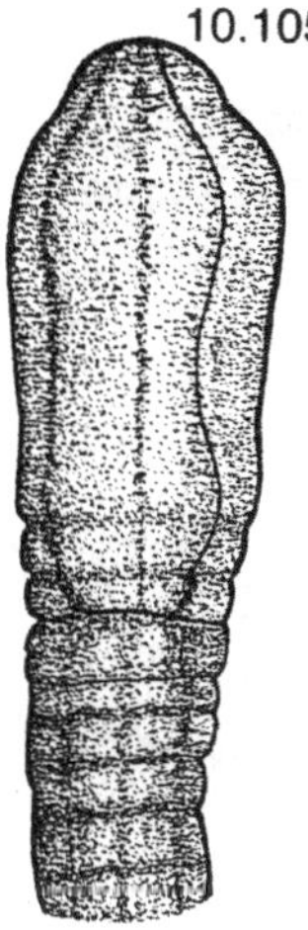

10.106

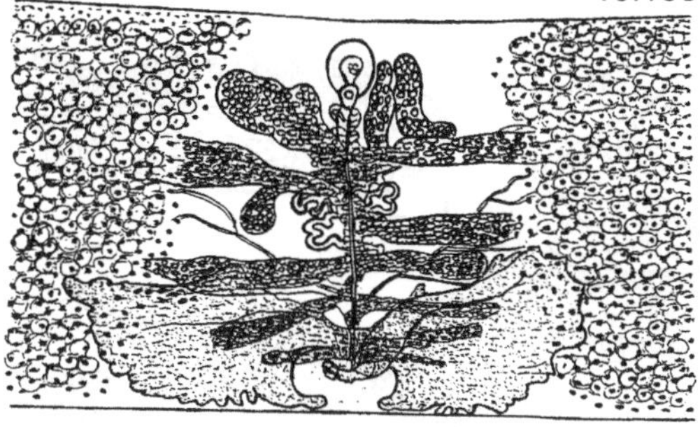

10.107

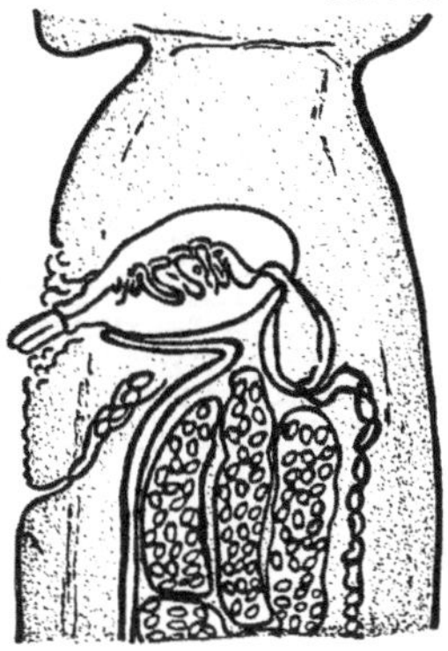

10.108

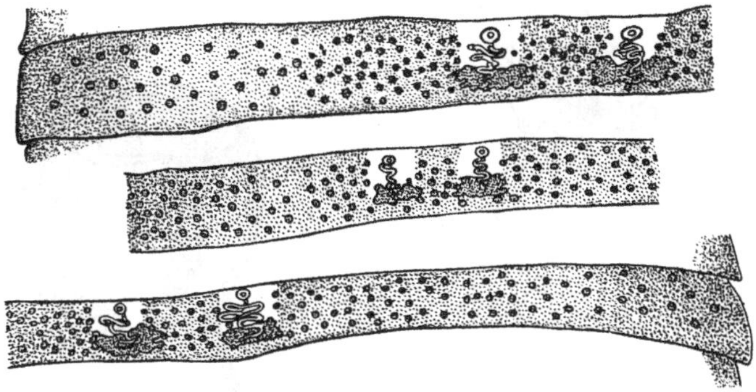

10.109

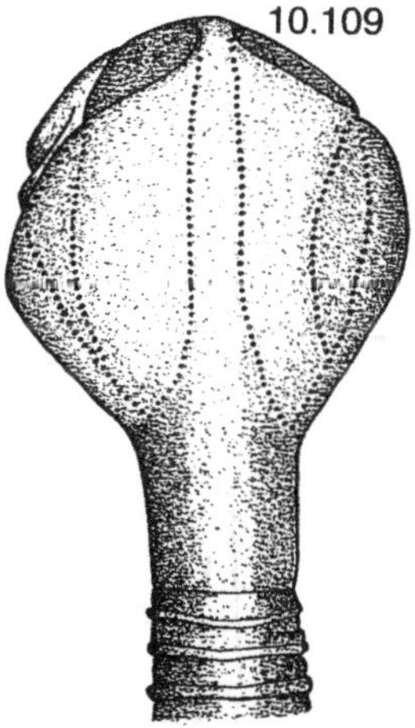

10.110

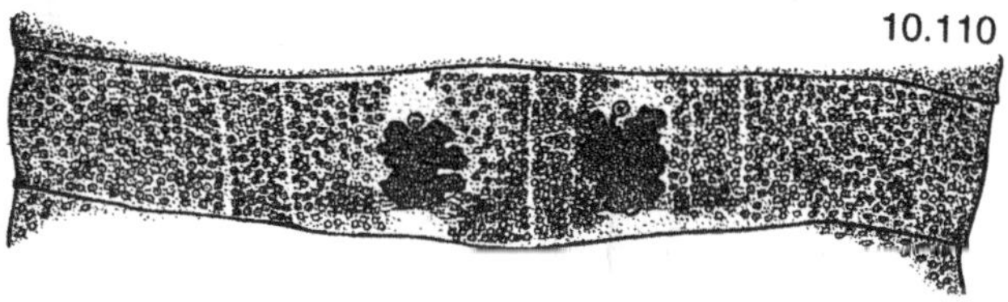

Figs 10.105–10.107 *Diphyllobothrium latum* (Linnaeus, 1758). 10.105. Scolex. 10.106. Mature proglottid, central region. 10.107. Relationships of terminal genital ducts in sagittal section. After Rausch & Hilliard (1970).

Fig. 10.108 *Hexagonoporus physeteris* Gubanov in Delyamure, 1955. Mature proglottid. After Gubanov in Delyamure (1955).

Figs 10.109, 10.110 *Diplogonoporus balaenopterae* (Lönnberg, 1891). 10.109. Scolex. After Delyamure (1955). 10.110. Mature proglottid. After Yamaguti (1942).

10.112

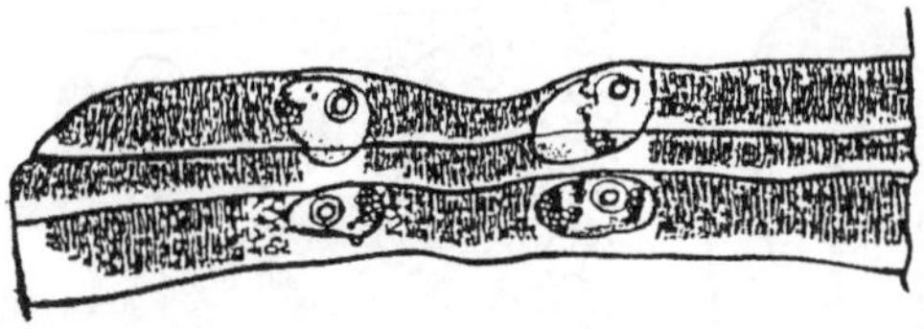

10.111

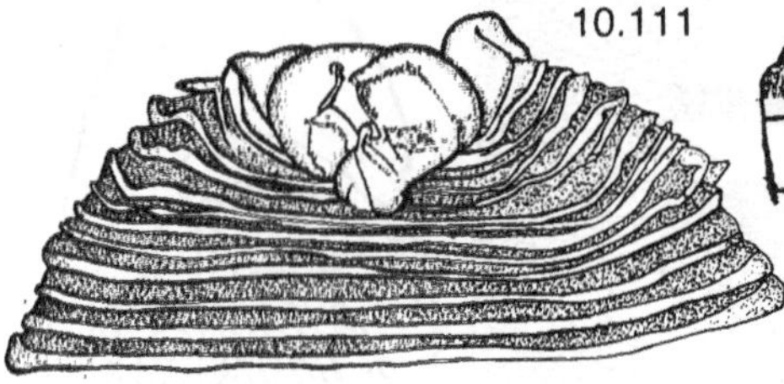

10.113

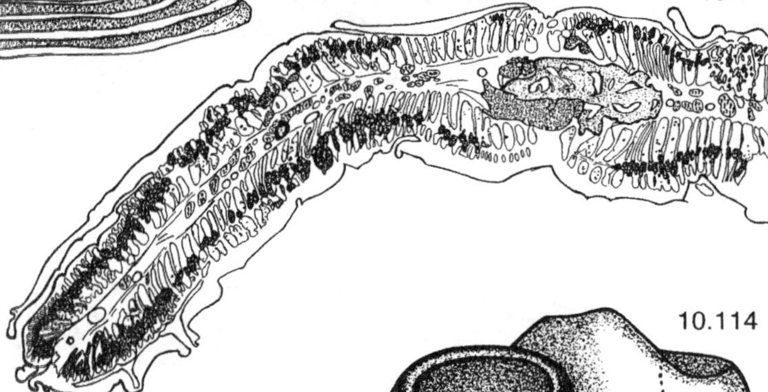

10.114

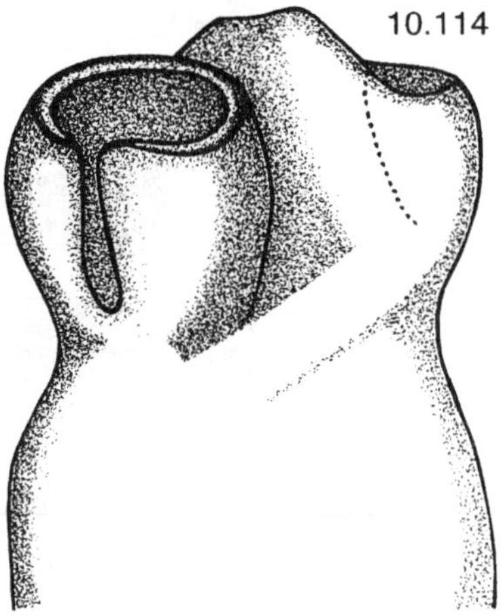

10.116

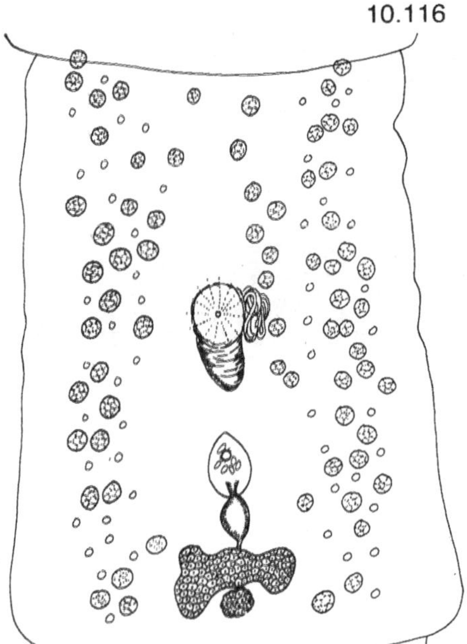

10.115

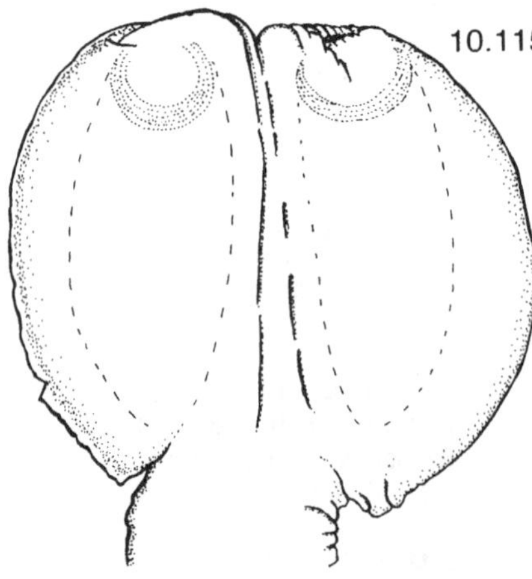

10.117

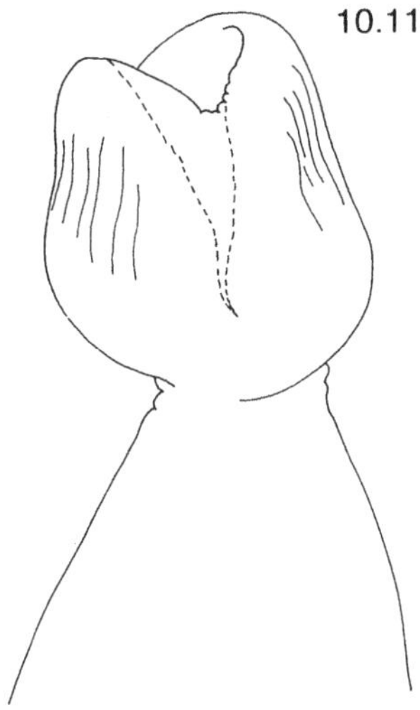

10.118

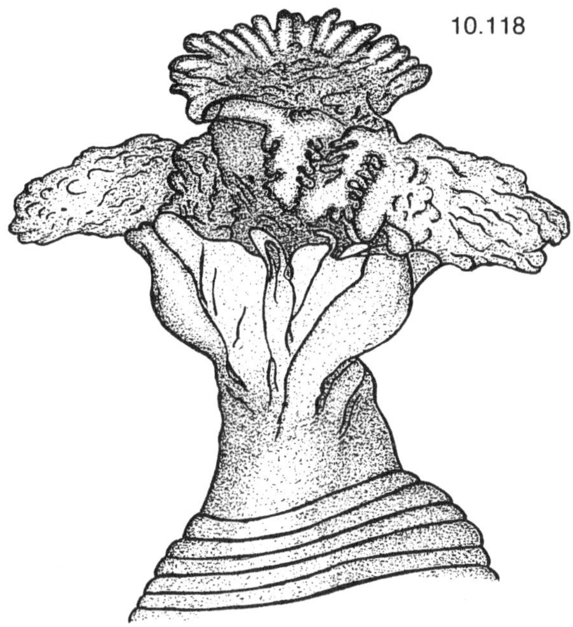

Figs 10.111-10.113 *Multiductus physeteris* Clarke, 1962. 10.111. Scolex. 10.112. Mature proglottids. 10.113. Transverse section of half of a mature proglottid. After Clarke (1962).
Fig. 10.114 *Baylisia baylisi* Markowski, 1952. Scolex. After Markowski (1952).
Figs 10.115, 10.116 *Bothridium pithonis* Blainville, 1824. 10.115. Scolex. 10.116. Mature proglottid.
Fig. 10.117 *Scyphocephalus secundus* Tubangui, 1938. Scolex. After Tubangui (1938).
Fig. 10.118 *Baylisiella tecta* (Linstow, 1892). Scolex. After Markowski (1952).

in two lateral bands. In boid snakes and varanid lizards. Africa, Sri Lanka, India and Philippines. Type-species *B. pithonis* Blainville, 1824.

12a. Bothria hypertrophied to form many complex folds .... 13.
12b. Anterior end of scolex invaginated to form sucking organ; bothria rudimentary .... *Scyphocephalus* Riggenbach, 1898. (Fig. 10.117)

Diagnosis: Proglottids wider than long. Genital pore median, pre-equatorial. Testes numerous, confluent medially in anterior and posterior regions of proglottid. Cirrus-sac occupies entire depth of medulla. Ovary bilobed, posterior. Seminal receptacle present. Vaginal opening posterior to that of cirrus-sac. Vitelline follicles circumcortical. Uterus with few coils. In varanid lizards. South-east Asia. Type-species *S. bisulcatus* Riggenbach, 1898.

13a. Anterior bothrial margins expanded to form cauliflower-like apical organ .... *Baylisiella* Markowski, 1952. (Fig. 10.118)

Diagnosis: Scolex usually embedded deep in intestinal wall of host. Strobila thick, tapering towards end. Proglottids short and wide. Testes in two or three layers. In elephant seals. Antarctic. Type-species *B. tecta* (Linstow, 1892).

13b. All of bothria hypertrophied to form many complex folds .... *Plicobothrium* Rausch & Margolis, 1969. (Fig. 10.119)

Diagnosis: Strobila muscular with many proglottids (more than 6000) all much wider than long. Arrangement of proglottids in sequential groups throughout strobila indicates multiplication by transverse division of primary segment. Two sets of genital organs in tandem position, sometimes present in segments not divided externally. Testes and vitelline follicles in separate lateral fields. Large cirrus-sac. In cetaceans. Atlantic. Type-species *P. globicephalae* Rausch & Margolis, 1969.

10.119

10.120

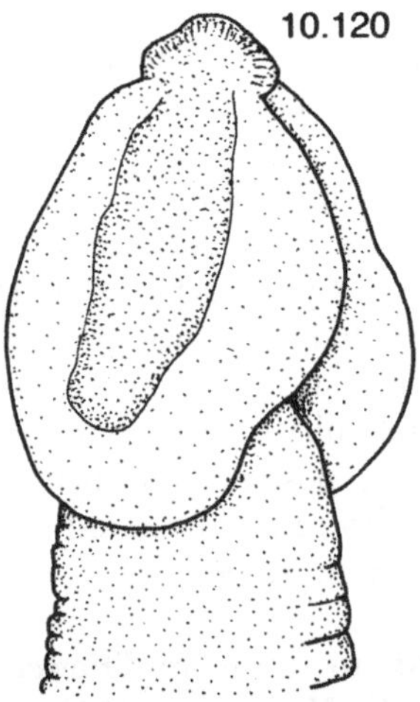

10.121

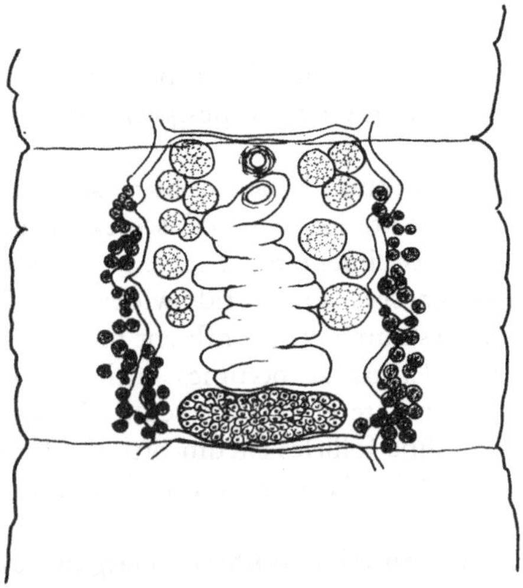

Fig. 10.119 *Plicobothrium globicephalae* Rausch & Margolis, 1969. Scolex. After Rausch & Margolis (1969).

Figs 10.120–10.121 *Cephalochlamys namaquensis* (Cohn, 1906). 10.120. Scolex. 10.121. Mature proglottid.

Family Cephalochlamydidae Yamaguti, 1959

Family diagnosis

Pseudophyllidea. Scolex like an arrowhead, with two (dorsal and ventral) elongated bothria united at apex. Neck distinct. Segmentation apparent after appearance of genitalia. Inner longitudinal musculature well developed. Longitudinal osmoregulatory canals connected by transverse canal in posterior region of each proglottid. Few testes, in medulla, arranged in two submedian groups anterior to ovary. No cirrus-sac. Male aperture joins vagina at base of genital atrium, which opens ventrally in median line near anterior border of proglottid. Ovary bilobed or entire, situated posteriorly in proglottid. Vitelline follicles extend length of proglottid in two lateral fields of cortex, and may converge posteriorly. Uterus in transverse slings between longitudinal osmoregulatory vessels anterior to ovary; ventral uterine pore posterior to genital atrium. Eggs thin-shelled, embryonated, not operculate. Parasites of the intestine of anurans and caudates. Type and only genus *Cephalochlamys* Blanchard, 1908.

Genus *Cephalochlamys* Blanchard, 1908 (Figs 10.120, 10.121)
(Syns *Chlamydocephalus* Cohn, 1906, *nec* Diesing, 1850, *nec* Schmarda, 1859; *Pseudocephalochlamys* Yamaguti, 1959)

Diagnosis: With characters of family. Type-species *C. namaquensis* (Cohn, 1906).

11 Order Haplobothriidea Joyeux & Baer, 1961

A. JONES

International Institute of Parasitology, 395A Hatfield Road, St Albans, Herts AL4 0XU, UK.

Introduction

The affinities and systematic position of *Haplobothrium* Cooper, 1914 are controversial. The genus, the only one in the order, is known only from a North American freshwater fish, *Amia calva* L., an archaic holostean. Two species have been described: *H. globuliforme* Cooper, 1914 and *H. bistrobilae* Premvati, 1969. This genus has a primary scolex with four tentacles reminiscent of those of the Trypanorhyncha but the anatomy of the proglottids (Cooper, 1914a, b; 1917) and the life-cycle pattern are characteristic of the Pseudophyllidea (Essex, 1929; Thomas, 1930; Meinkoth, 1947). Although Nybelin (1922), Poche (1924) (who placed it in its own suborder) and Fuhrmann (1931) aligned *Haplobothrium* with the trypanorhynchs, Woodland (1927) and Dollfus (1942) believed that its affinities were pseudophyllidean and that its tentacles and those of the trypanorhynchs represented a case of parallel evolution. Dollfus (1942) pointed out that a digenean, *Rhopalias* Stiles & Hassall, 1898 has a pair of armed retractile proboscides. Thomas (1983) concluded that the tentacle muscle system of *H. globuliforme* differed from that of the trypanorhynchs and that the genus was closer to the pseudophyllideans.

Cooper's (1914a, b) original descriptions of *H. globuliforme* included only the secondary scolex, which lacks tentacles. Consequently, he placed the genus in the family Diphyllobothriidae (= Dibothriocephalidae). Later, having discovered the tentacle-bearing primary scolex, he (1917) erected the subfamily Haplobothriinae in the same family; subsequently, Meggitt (1924a) elevated it to family rank. According to Article 50(c) of the *International Code of Zoological Nomenclature*, the authority for the family should be Cooper, 1917, as attributed by Meggitt (1924a). The family was accepted in the Pseudophyllidea by Wardle and McLeod (1952), Yamaguti (1959) and Schmidt (1986).

Joyeux & Baer (1961) erected a separate order, Haplobothrioidea [*sic*], for *Haplobothrium* on the grounds that it occupied an important place in cestode

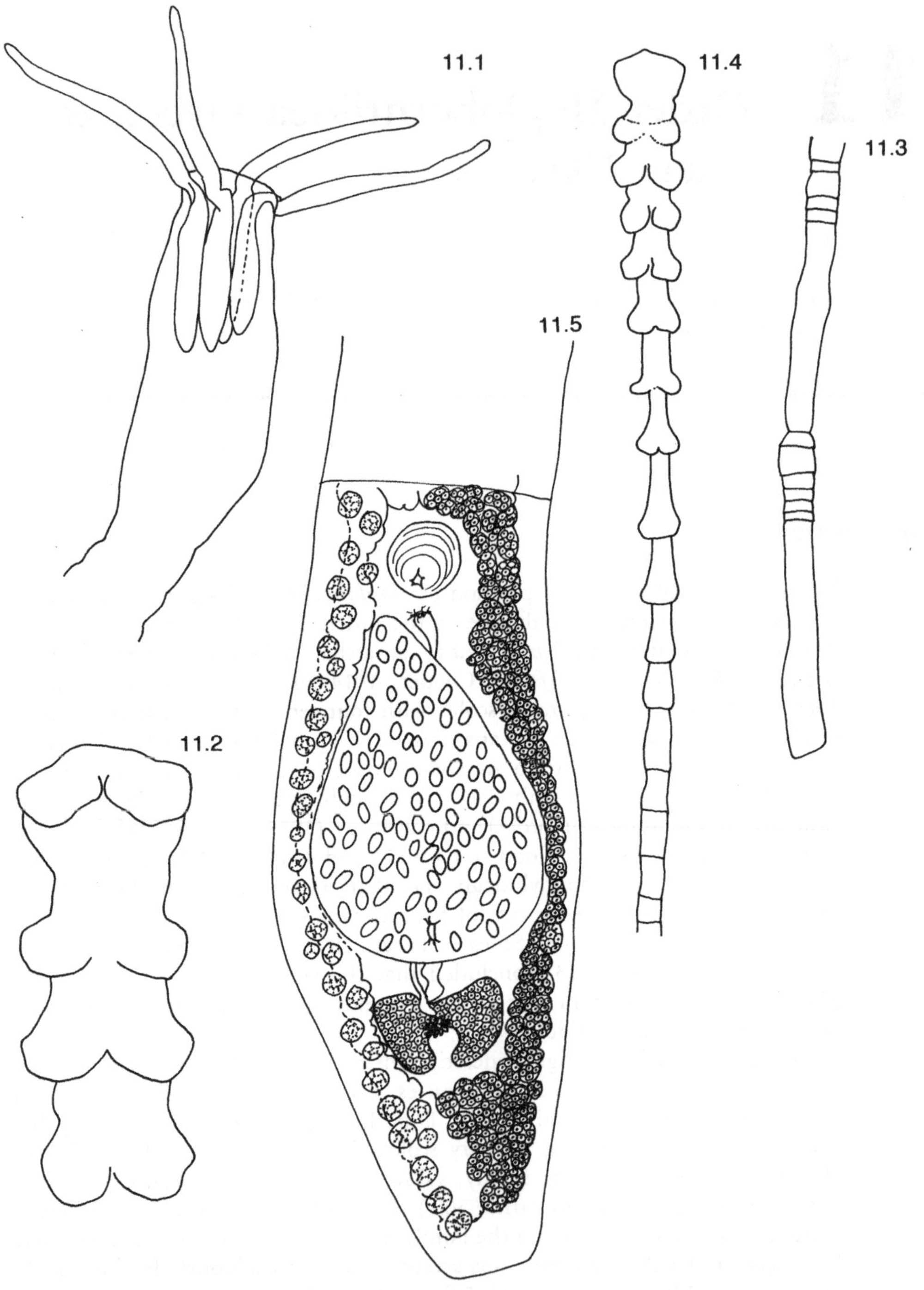

Figs 11.1–11.5 *Haplobothrium globuliforme* Cooper, 1914. 11.1. Primary scolex. 11.2. Secondary scolex. 11.3. Primary strobila. 11.4. Secondary strobila. 11.5. Gravid proglottid.

phylogeny and was morphologically distinct from both the trypanorhynchs and pseudophyllideans. This has been adopted by MacKinnon, Jarecka & Burt (1985) and MacKinnon & Burt (1985a, b, c). Significance was also attached to its restriction to a relict fish species. MacKinnon & Burt (1985a) found that sperm structure in *H. globuliforme* indicated a close relationship with the Pseudophyllidea rather than the Trypanorhyncha. According to MacKinnon, Jarecka & Burt (1985), *H. globuliforme* exhibits a form of asexual reproduction unique in the cestodes.

ORDER HAPLOBOTHRIIDEA JOYEUX & BAER, 1961

Diagnosis: Primary scolex club-shaped with four tentacles, with spine-like microtriches basally, which can be withdrawn into muscular sacs. Bothria absent. Primary strobila does not itself develop proglottids but has segmented regions, at intervals behind the primary scolex, each of which separates off to become a secondary strobila. Anterior segment of secondary strobila modified as pseudoscolex (or secondary scolex). Pseudoscolex flattened anteriorly, with four shallow indentations around raised central dome. Posterior margin of pseudoscolex and succeeding proglottids form four flat projecting appendages. One set of genitalia per proglottid. Genital pore midventral, anterior. Cirrus-sac median, anterior. Cirrus armed with minute spines. Testes medullary in two lateral fields. Ovary posterior, median, horseshoe-shaped with anterior isthmus. Vagina opens behind cirrus-sac. Seminal receptacle present. Vitelline follicles medullary, confluent in midline anteriorly and posteriorly. Uterus with coiled uterine duct and dilated uterine sac. Uterine pore persistent or transient. Eggs operculate, embryonated. In North American freshwater teleosts (bowfin, *Amia calva* L.). Type-family Haplobothriidae Cooper, 1917.

Family Haplobothriidae Cooper, 1917

Diagnosis: With the characters of the order. Type and only genus *Haplobothrium* Cooper, 1914.

Genus *Haplobothrium* Cooper, 1914 (Figs 11.1–11.5)

Diagnosis: With the characters of the family. Type-species *H. globuliforme* Cooper, 1914.

12 Order Nippotaeniidea Yamaguti, 1939

R.A. BRAY

Department of Zoology, The Natural History Museum, Cromwell Road, London SW7 5BD, UK.

Introduction

This small order of cestodes includes about six species from freshwater fishes in Japan, China, Russia and New Zealand. Hine (1977) studied three species in New Zealand and sustained two genera *Nippotaenia* Yamaguti, 1939 and *Amurotaenia* Akhmerov, 1941 (see Yamaguti, 1939 and Akhmerov, 1941). Several authors before and since, including Akhmerov (1960) and, most recently, Demshin (1985), Dubinina (1987) and Brooks & McLennan (1993) have not considered *Amurotaenia* valid. Schmidt (1986) followed Hine's findings and Wang (1984) used the name *Amurotaenia* for the species found in China. I am retaining the genera as distinct, following Hine, while pointing out that this is not the view of many workers. The group is considered by Yamaguti (1951) to be intermediate between the Pseudophyllidea and the Cyclophyllidea, but Freeman (1973) thought that this group, along with the Spathebothriidea and the Caryophyllidea, was the nearest to the protocestode 'stalk'. In Brooks, Hoberg & Weekes' (1991) analysis of cestode phylogeny, the Nippotaeniidae is considered a 'sister-group of all other non-pseudophyllidean cestodes'. Life-cycle studies show that the single apical sucker alone is present from the earliest stages and that the metacestodes are of a 'primitive' type with no lacuna visible (Yamaguti, 1951; Mamaev, 1971; Demshin, 1985; Weekes, 1986).

ORDER NIPPOTAENIIDEA YAMAGUTI, 1939

Diagnosis: Single apical sucker present, no other attachment organs found on scolex. Strobila relatively short, anapolytic to hyperapolytic. Inner longitudinal muscles not bundled, sheath-like, attached to sphincter-like anterior end of apical sucker. Longitudinal osmoregulatory vessels numerous in cortex, few in medulla. Nerve tract ventral to cirrus-sac and vagina. Testes in pre-vitelline medulla, pre-

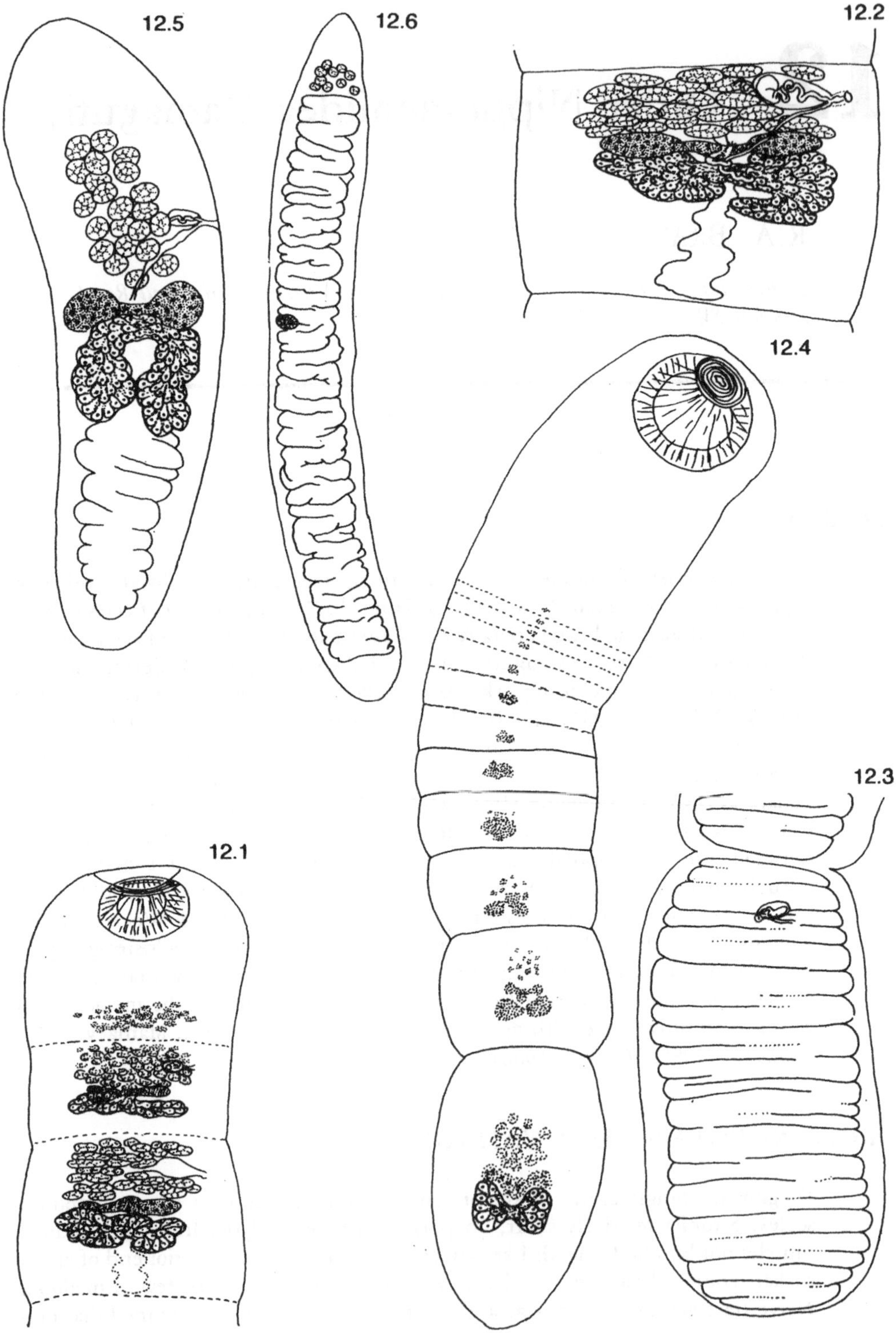

12.5 12.6 12.2 12.4 12.3 12.1

sent or absent in gravid proglottids. Cirrus-sac thin-walled, mainly in medulla, with convoluted ejaculatory duct; proximal part may form internal seminal vesicle. Cirrus simple. Genital pore opens near lateral margin, in anterior half of proglottid. Ovary symmetrical with two lateral lobes; in mid- to posterior part of proglottid. Vitellarium two symmetrical lobes, anterior to ovary. Vagina opens into genital atrium posterior to cirrus-sac. Uterus in transverse coils in medulla, may fill almost entire medulla. Eggs with three shells. In freshwater teleosts in Japan, Russia, China and New Zealand. Type and only family Nippotaeniidae Yamaguti, 1939.

Family Nippotaeniidae Yamaguti, 1939

Diagnosis: With characters of order. Type-genus *Nippotaenia* Yamaguti, 1939.

Key to genera

1a. Anapolytic or hypoanapolytic. Testes lost in gravid proglottids. In freshwater teleosts of Japan, Russia and New Zealand .... *Nippotaenia* Yamaguti, 1939. (Figs 12.1–12.3)

Diagnosis: Characters as above. Type-species *N. chaenogobii* Yamaguti, 1939.

1b. Hyperapolytic. Testes partly or completely retained in gravid proglottids. In freshwater teleosts of Japan, Russia, China and New Zealand .... *Amurotaenia* Akhmerov, 1941. (Figs 12.4–12.6)

Diagnosis: Characters as above. Type-species *A. perccotti* Akhmerov, 1941.

Figs 12.1–12.3 *Nippotaenia chaenogobii* Yamaguti, 1939. 12.1. Scolex and early proglottids. 12.2. Mature proglottid. 12.3. Gravid proglottid. (After Yamaguti, 1939.)

Figs 12.4, 12.5 *Amurotaenia mogurndae* (Yamaguti & Miyata, 1940). 12.4. Strobila. 12.5. Detached gravid proglottid. (After Dubinina, 1971.)

Fig. 12.6 *Amurotaenia decidua* Hine, 1977, detached gravid proglottid. (After Hine, 1977.)

13 Order Proteocephalidea Mola, 1928

A.A. Rego

Departamento de Helmintologia, Instituto Oswaldo Cruz, Av. Brasil 4365, C.P. 926, 21045–900 Rio de Janeiro, Brazil.

Introduction

The proteocephalids originated in primitive fishes in Gondwana, probably in South American rivers. Most existing species are actually found in Gondwana (South America, Africa, India, Australia). Some species, however, have dispersed to temperate areas and into hosts other than fishes (e.g. into Europe, Asia, North America) as has occurred with the cosmopolitan genus *Proteocephalus* (family Proteocephalidae), whose species are found in freshwater fishes, amphibians and reptiles. In contrast, representatives of the family Monticelliidae are restricted to tropical areas of South America and their hosts are freshwater fishes, in particular siluriforms. Groups other than the siluroids have recently been found to harbour both monticelliids and proteocephalids, i.e. non-siluriform fishes and other vertebrates. Species of *Nomimoscolex* (Monticelliidae: Zygobothriinae) were recently found in Amphibia and Reptilia (de Chambrier, 1987; de Chambrier & Vaucher, 1992). This may be evidence of the monticelliids commencing a colonization of hosts other than siluriforms. Proteocephalids of fishes are found only in hosts living in freshwater environments, none in marine waters. The subfamily Prosobothriinae Yamaguti, 1959, from marine elasmobranchs, listed by some authors in the order Proteocephalidea, is in my opinion more closely related to the Tetraphyllidea. However, Wardle & McLeod (1952) referred these species to the order Lecanicephalidea. [See also Euzet, this volume.]

Some important contributions to the knowledge of the order Proteocephalidea are the works of La Rue (1914), Woodland (1925, 1933a, b, 1934a, b, c), Freze (1965), Brooks (1978), Brooks & Rasmussen (1984) and Rego (1987). The classification of the order is mostly based on the work of Woodland who supported his studies with abundant material collected in South America (the Amazon region) and Africa. The taxonomy at the family level has its foundations in the relationship of the reproductive organs to the longitudinal muscle bundles, as seen in cross-section. The relationship of vitelline follicles to the longitudinal muscles is fundamental in the subdivision into families. The Proteocephalidae has medullary vitelline follicles and the Monticelliidae has cortical vitelline follicles.

Woodland's scheme of classification has been useful for decades but is now becoming inadequate due to the discovery of forms with intermediate characters. For example, in the recently described subfamily Nupeliinae (see Pavanelli & Rego, 1991) the reproductive organs and vitelline follicles are placed partly in the medulla and partly in the cortex.

It is possible that, in future, modification to the scheme of classification will emphasize the characters of the scolex and suckers more than the present system. Our knowledge of the group so far is incomplete and any tentative attempt to change the current taxonomy will not succeed. For these reasons I shall adopt here a conservative scheme of classification similar to that utilized in reference books and papers such as those of Wardle & McLeod (1952), Yamaguti (1959), Brooks (1978), Brooks & Rasmussen (1984), Schmidt (1986) and principally the excellent treatise on proteocephalids by Freze (1965). Some minor modifications and corrections of some errors of previous authors have been included. For example, *Choanoscolex* La Rue, 1911 was originally placed in Corallobothriinae of the Proteocephalidae but after Brooks & Rasmussen (1984) and Rego (1990) noted the cortical position of the vitelline follicles and reproductive organs, the genus was transferred to Monticellinae of the Monticellidae. I also accept three genera in Peltidocotylinae (= Othinoscolecinae): *Peltidocotyle, Othinoscolex* and *Woodlandiella*, differing from Schmidt (1986) who included only one genus, *Peltidocotyle*. Conservative authors accepted only one or two genera in Monticellinae while I am of the opinion that it is advantageous to split *Monticellia* into several genera, as suggested by Brooks & Rasmussen (1984) and Freze (1965). Consequently, I accept *Spatulifer, Goezeella, Monticellia* and *Spasskyellina*, and the recently proposed genus *Paramonticellia* Pavanelli & Rego, 1991. *Jauella*, a genus previously placed in Jauelliinae Rego & Pavanelli, 1985, is now moved to Peltidocotylidae, resulting in the abolition of the subfamily.

For this key, only two families are accepted. The Proteocephalidae is primarily characterized by medullary vitelline follicles, i.e. internal to the longitudinal muscle bundles (when these muscles are weakly developed or absent it is difficult, or not possible, to define the disposition of vitelline follicles and gonads). The Monticelliidae contains genera with mainly cortical vitelline follicles, i.e. external to the longitudinal muscle bundles.

In the Proteocephalidae, the gonads are usually medullary, but in the Monticelliidae they are cortical and medullary, in varied combinations, permitting division into subfamilies.

The relative importance of vitelline follicles and reproductive organs remains to be ascertained. If the position of the vitellarium is fundamental, *Marsypocephalus* (Marsypocephalinae) must remain in the Proteocephalidae, due to the medullary position of the vitellarium in this family but, in contrast to the other Proteocephalidae, the Marsypocephalinae have their testes in the dorsal cortex. This is apparently a monticelliid character. Nevertheless, I prefer to treat the Marsypocephalinae as a member of the Proteocephalidae pending the elucidation of this problem.

The newly established Nupeliinae (Pavanelli & Rego, 1991) was defined as having its vitellarium and reproductive organs partly medullary and partly cortical. This group is clearly intermediate between the Proteocephalidae and Monticelliidae. The Nupeliinae is provisionally placed here in Monticelliidae.

The identification of proteocephalid and monticelliid species is a difficult task. A major problem is the inadequacy of the descriptions of some species; for instance, in *Endorchis* Woodland, 1934 and *Myzophorus* Woodland, 1934, both of the subfamily Endorchiinae, the position of the uterus, whether cortical or medullary, was not clearly stated which led some workers to doubt the validity of these genera. Rego (1991) invalidated the genus and species of *Endorchis* (synonymizing it with *Nomimoscolex*) and also the subfamily Endorchiinae and provisionally transferred *Myzophorus* to Zygobothriinae. However, I now think that *Myzophorus* is a controversial genus that needs to be re-evaluated.

The subfamilies of the Proteocephalidae are separated generally by the characteristics of the scolex, i.e. presence or absence of a metascolex, armed rostellum, piercing organ and spines on the scolex.

The monticelliid subfamilies are differentiated solely by the position of the reproductive organs in the cortex and/or in the medulla. In the subfamily Monticelliinae, the testes, ovary and uterus are situated in the cortex. In the Zygobothriinae, gonads are entirely medullary but the vitelline follicles are cortical. The other subfamilies exhibit intermediate forms of these characters.

The examination of more specimens is crucial to a precise definition of taxa and phylogenetic affinities. Of particular interest would be phylogeny and affinities of South American proteocephalids and those described from Africa and other continents.

ORDER PROTEOCEPHALIDEA MOLA, 1928

Diagnosis: Strobila small to medium in size. Scolex variable; with four simple acetabula, or suckers may be biloculate, triloculate or tetraloculate; apical sucker or armed rostellum occasionally present. Metascolex present or absent. Segmentation usually distinct. Strobila usually anapolytic. Mature proglottids wider than long or square. Gravid proglottids usually longer than wide. Parenchyma divided into cortical and medullary regions, generally separated by internal longitudinal musculature; longitudinal musculature may be reduced or inconspicuous, especially in Monticelliidae. Reproductive organs medullary or, in various combinations, in medulla and cortex. Vitellarium follicular, generally in lateral fields. Genital pores lateral. Uterus generally with diverticula and with longitudinal apertures in gravid proglottids. Adults in freshwater fish, amphibians and reptiles.

Key to families

1a. **Vitelline follicles cortical** .... **Monticelliidae La Rue, 1911.**
Diagnosis: Scolex of varied forms. Testes, ovary and uterus, in various combinations, situated in cortex or medulla. Internal longitudinal musculature sometimes

weakly developed or inconspicuous. In freshwater fishes, rarely amphibians and reptiles. Tropical and subtropical areas of South America. Type-genus *Monticellia* La Rue, 1911.

1b. Vitelline follicles medullary .... Proteocephalidae La Rue, 1911.
Diagnosis: Scolex of varied forms. Ovary and uterus medullary. Testes medullary except in *Marsypocephalus*. Parenchyma usually divided into cortical and medullary regions by distinct layer of longitudinal muscle fibres. Parasites of fishes, amphibians and reptiles. North and South America, Africa, Europe, Asia, Australia. Type-genus *Proteocephalus* Weinland, 1858.

Family Proteocephalidae La Rue, 1911

Key to subfamilies

1a. Scolex without metascolex .... 2.
1b. Scolex with metascolex .... Corallobothriinae Freze, 1965.
Diagnosis: Suckers covered by folds of metascolex tissue, with or without sphincter. Vitellarium and genitalia medullary. In siluriforms. South America, North America, Africa, Russia. Type-genus *Corallobothrium* Fritsch, 1866.

2a. Testes medullary .... 3.
2b. Testes cortical .... Marsypocephalinae Woodland, 1933.
Diagnosis: Ovary, vitellarium and uterus medullary. Scolex round, with four lobes, each lobe bearing large sucker with muscular sphincter. Longitudinal musculature conspicuous. In siluriform fishes from Africa. Type and only genus *Marsypocephalus* Wedl, 1861.

3a. Vitellarium follicular, lateral .... 4.
3b. Vitellarium not follicular, in form of two massive bodies, posterior to ovary .... Sandonellinae Khalil, 1960.
Diagnosis: Scolex without metascolex, but with highly modified apical organ formed of two lappets. Genitalia medullary. Uterus saccular. Vitellarium near posterior margin of segment. Uterus forms round sac that invades cortex. Strobila craspedote. In African osteoglossid fishes. Type and only genus *Sandonella* Khalil, 1960.

4a. Scolex lacking rostellum or spines. Vitellarium and reproductive organs medullary. Scolex globular. Suckers of various forms .... Proteocephalinae Mola, 1929.
Diagnosis: Rostellum and metascolex absent. Suckers acetabular uniloculate, biloculate or tetraloculate. Fifth or apical sucker may be present. Vitellarium lateral, medullary. Reproductive organs also medullary (Fig. 13.1). In fishes, amphibians and reptiles. Cosmopolitan. Type-genus *Proteocephalus* Weinland, 1858.

4b. Scolex with armed rostellum or 'piercing organ' .... 5.

5a. Scolex with armed rostellum .... **Gangesiinae Mola, 1929.**
Diagnosis: Rostellum armed, non-retractile. One, two or several circles of hooks present. Scolex, suckers and neck spined or not. Vitellarium lateral. Ovary bilobed. In siluroid fishes. Africa, Europe, Asia. Type-genus *Gangesia* Woodland, 1924.

5b. Scolex with specialized rostellum (piercing organ) of muscular-glandular structure .... **Acanthotaeniinae Freze, 1963.**
Diagnosis: Scolex and anterior part of strobila covered with dense network of spines. Rostellum terminal. Reproductive organs medullary. Mehlis's gland highly developed. Longitudinal musculature weakly developed. In reptiles of genus *Varanus*. Asia, Australia. Type-genus *Acanthotaenia* Linstow, 1903.

Subfamily Proteocephalinae Mola, 1929

Key to genera

1a. Suckers of normal acetabular type .... 2.
1b. Suckers modified in various ways .... 4.

2a. Vitellarium apparently absent ..... ***Travassiella* Rego & Pavanelli, 1987.**
Diagnosis: Small worms, filiform. Suckers salient, directed laterally. Vitellarium not observed (Fig. 13.2). Gonads and uterus medullary. Longitudinal musculature inconspicuous, distinction between medulla and cortex difficult to discern (Fig. 13.3). In South American siluroid fishes. Type-species *T. avitellina* Rego & Pavanelli, 1987.

2b. Vitellarium present .... 3.

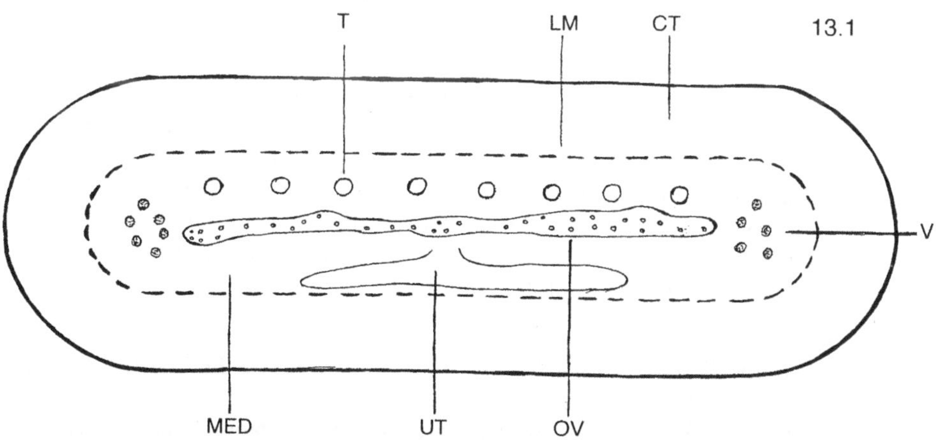

Fig. 13.1 Proteocephalinae, schematic section. T = testes, LM = longitudinal muscle bundles, CT = cortex, V = vitellaria, MED = medulla, UT = uterus, OV = ovary. (Original.)

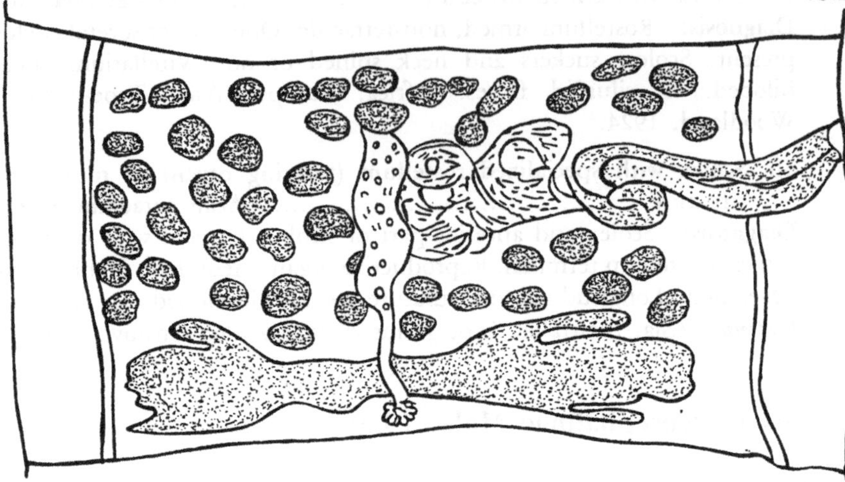

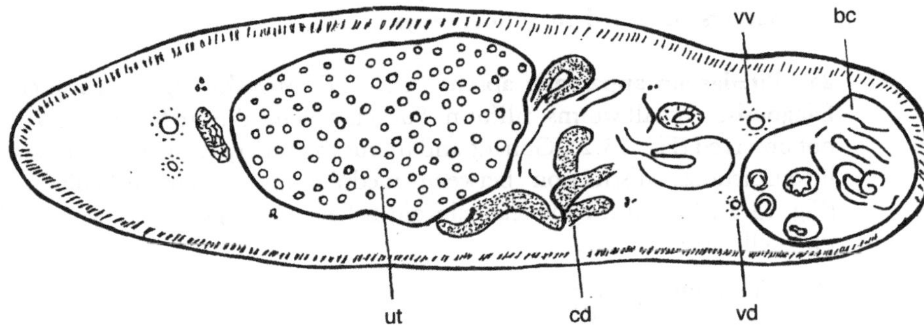

Figs 13.2, 13.3 *Travassiella avitellina* Rego & Pavanelli, 1987. 13.2. Mature proglottid. 13.3. Transverse section of gravid proglottid. (After Rego & Pavanelli, 1987.)

3a. **Testes in continuous field throughout proglottids (Fig. 13.5)** .... ***Proteocephalus*** **Weinland, 1858.**
(Syn. *Ichthyotaenia* Lönnberg, 1894.)

Diagnosis: Scolex with four suckers of normal type, fifth or apical sucker may be present (Fig. 13.4). Reproductive organs and vitellarium medullary. Vitellarium lateral. In fishes, amphibians and reptiles. Cosmopolitan. Type-species *P. filicollis* (Rudolphi, 1802).

3b. **Testes in two lateral fields** .... ***Ophiotaenia*** **La Rue, 1911.**
(Syns *Batrachotaenia* Rudin, 1917; *Solenotaenia* Beddard, 1913; *Testudotaenia* Freze, 1965.)

Diagnosis: Scolex and internal features as in *Proteocephalus*, except for testes distributed in two lateral fields (Fig. 13.6). In fishes, amphibians and reptiles. Cosmopolitan. Type-species *O. perspicua* La Rue, 1911.

13.4

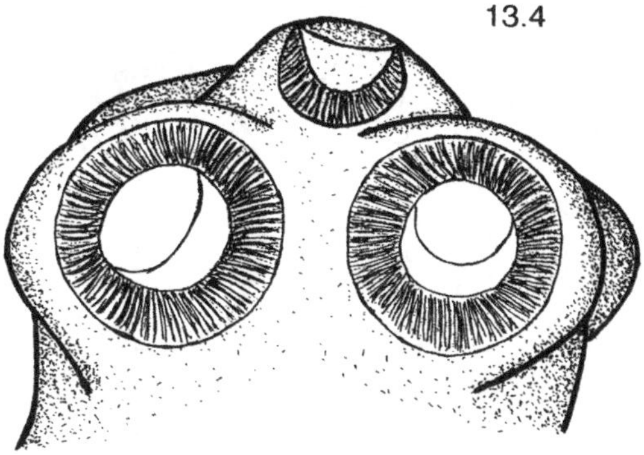

13.5

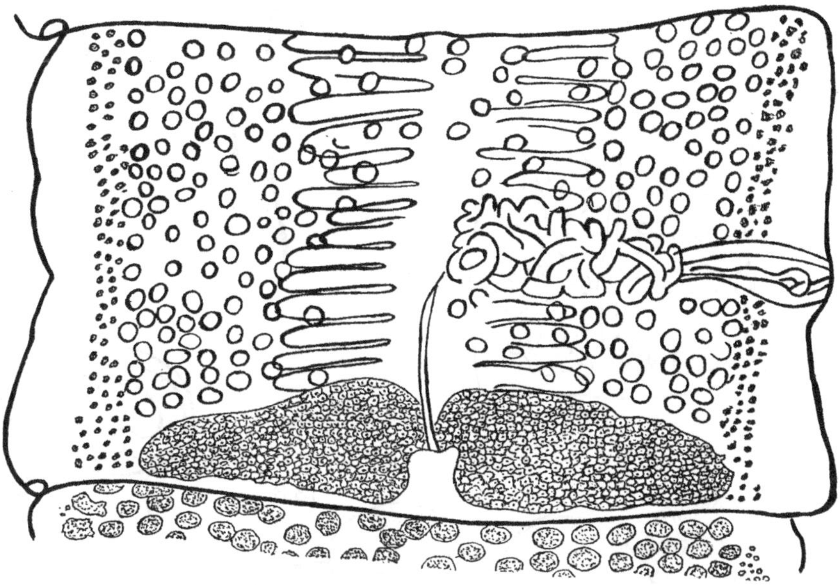

Figs 13.4, 13.5 *Proteocephalus osculatus* (Goeze, 1782). 13.4. Scolex. 13.5. Mature proglottid. (After Nybelin, 1942.)

4a. Suckers not highly modified, large, posterior margin notched, sucker heart-shaped (Fig. 13.7) .... ***Crepidobothrium* Monticelli, 1900.**

(Syn. *Ophidotaenia* Beddard, 1913.)

Diagnosis: Scolex large. Testes in two lateral fields. Ovary bilobed, near posterior margin of proglottid. Uterus median, with lateral branches. In South American snakes. Type-species *C. gerrardi* (Baird, 1860).

4b. Suckers highly modified .... **5.**

5a. Suckers pedunculate, located in protruding part of scolex (Fig. 13.8) .... ***Macrobothriotaenia* Freze, 1965.**

Diagnosis: Scolex wider than neck. Genital pore slightly postequatorial. Proglot-

13.6

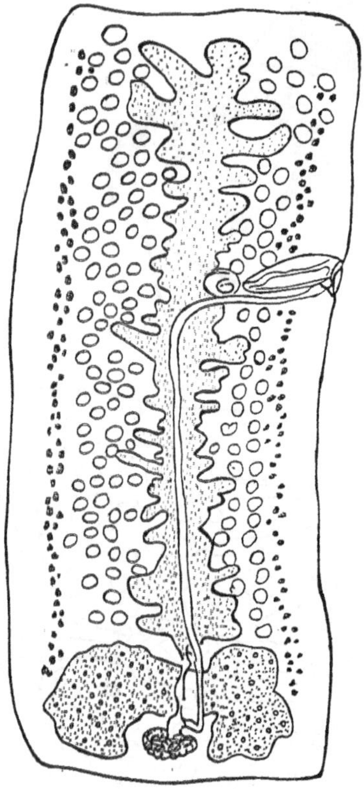

13.9

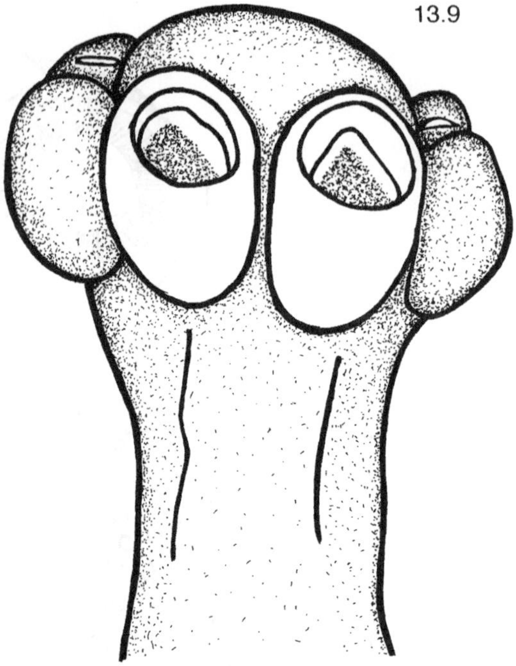

13.7

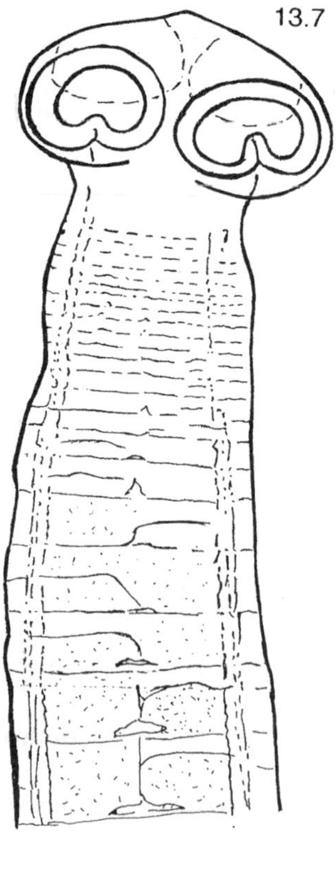

13.8

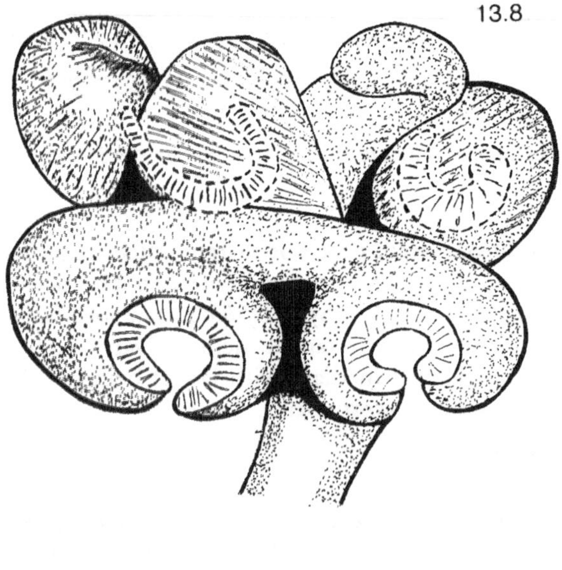

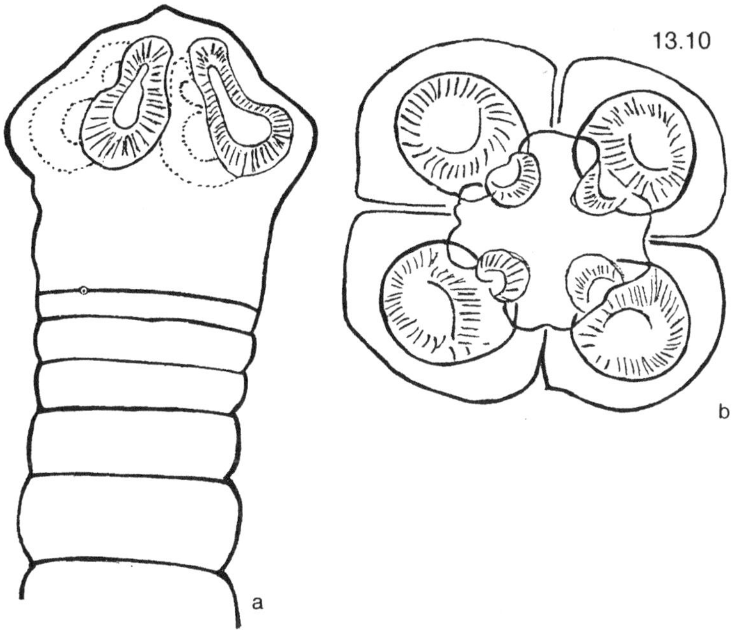

Fig. 13.6 *Ophiotaenia cohospes* (Cordero, 1946). Mature proglottid. (After Cordero, 1946.)
Fig. 13.7 *Crepidobothrium gerrardi* (Baird, 1860). Scolex and immature proglottids. (After Dollfus, 1932.)
Fig. 13.8 *Macrobothriotaenia ficta* (Meggitt, 1927). Scolex. (After Meggitt, 1927d.)
Fig. 13.9 *Brayela karuatayi* (Woodland, 1934). Scolex. (After Woodland, 1934a.)
Fig. 13.10 *Tejidotaenia appendiculata* (Baylis, 1947). a. Anterior region. b. Scolex, apical view. (After Baylis, 1947.)

tids longer than wide. Testes in two fields. Cirrus-sac large. Ovary transversely elongated. In Indian snakes. Type-species *M. ficta* (Meggitt, 1927).

5b. Suckers not pedunculate .... 6.

6a. Each sucker divided into two equal compartments (biloculate), apparently eight suckers, but only four have apertures (Fig. 13.9) .... *Brayela* Rego, 1984.

Diagnosis: Scolex dome-shaped. Testes in two lateral fields. Uterus longitudinal tube without diverticula. In South American siluroid fish. Type-species *B. karuatayi* (Woodland, 1934).

6b. Suckers with two unequal cavities, or four cavities .... 7.

7a. Suckers divided into two unequal cavities (Fig. 13.10) .... *Tejidotaenia* Freze, 1965.

Diagnosis: Small worms. Strobila consists of few proglottids. Testes in two lateral fields. Ovary large, alate. Vitellarium lateral. Uterus with diverticula. In Surinam monitors of South America. Type-species *T. appendiculata* (Baylis, 1947).

13.11a 13.11b

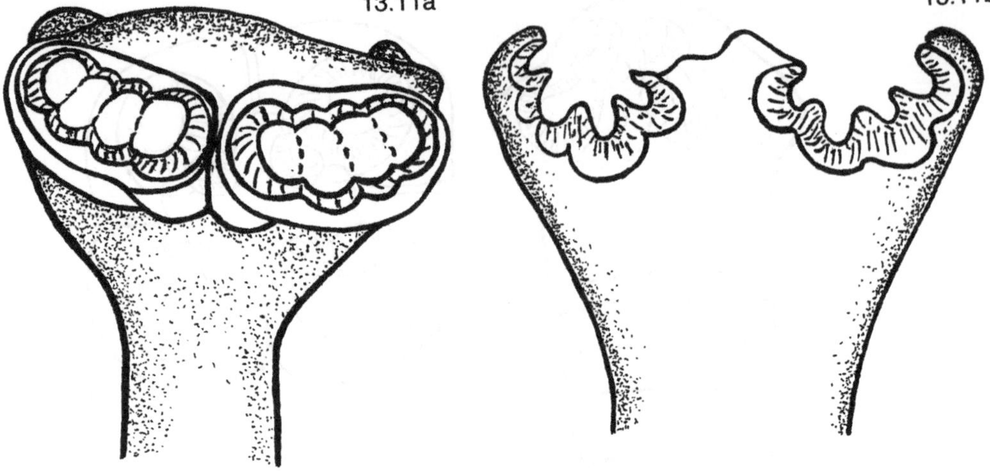

13.12

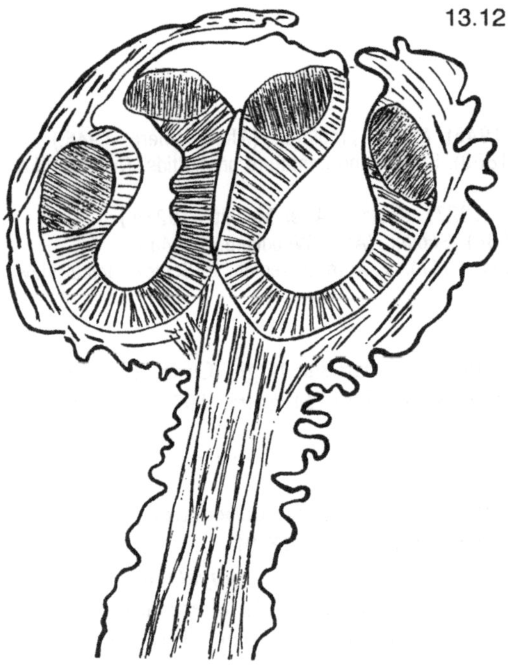

Fig. 13.11 *Deblocktaenia ventosaloculata* (Deblock, Rosé & Broussart, 1962). a. Scolex. b. Section of scolex. (After Deblock *et al.*, 1962.)
Fig. 13.12 *Megathylacus jandia* Woodland, 1934. Section of scolex. (After Woodland, 1934a.)

7b. Cavity of each sucker divided into four muscular chambers by septa (Fig. 13.11) .... ***Deblocktaenia*** **Odening, 1963.**

Diagnosis: Suckers transverse, apertures directed anteriorly. Scolex much wider than neck. Proglottids very elongated. Genital pores pre-equatorial. Testes in two lateral fields. Vagina with sphincter. Uterus with diverticula. In colubrid snakes from Madagascar. Type-species *D. ventosaloculata* (Deblock, Rosé & Broussart, 1962).

Subfamily Corallobothriinae Freze, 1965

Key to genera

1a. Suckers with sphincter .... 2.
1b. Suckers without sphincter .... 3.

2a. Suckers large, saccular, hidden under folds of tissue; each sucker has strong muscular sphincter that completely encircles aperture of sucker (Fig. 13.12) .... *Megathylacus* Woodland, 1934.

Diagnosis: Scolex large, quadrilobate. Metascolex developed. Suckers with sphincter (Fig. 13.12). Longitudinal musculature well developed. Genital pore near middle of proglottid margin. Reproductive organs medullary. Vitellarium lateral. In South American siluriform fishes. Type-species *M. jandia* Woodland, 1934.

2b. Scolex as in *Megathylacus* but differs in having sphincter muscles concentrated at only one side of sucker orifice .... *Megathylacoides* Jones, Kerly & Sneed, 1956.

Diagnosis: Suckers with incomplete sphincter (Fig. 13.13). Mature proglottids

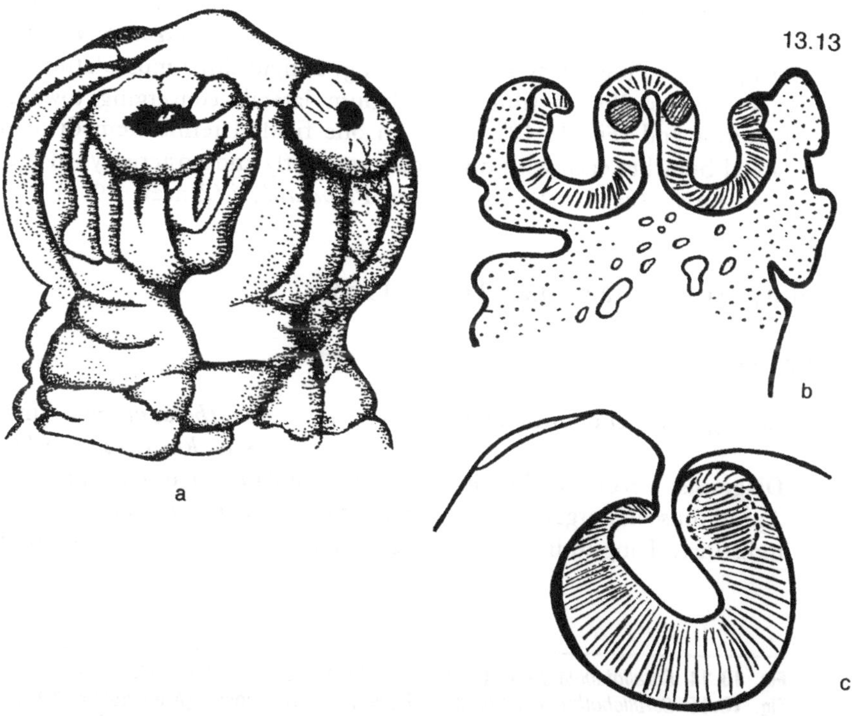

Fig. 13.13 *Megathylacoides giganteum* (Essex, 1928). a. Scolex. b. Section of scolex. c. Detail of sucker. (After Essex, 1928.)

longer than wide. Testes in single layer. Ovary bilobed, posterior. Vitelline follicles lateral. Uterus with anterior pore. In siluroid fishes of North America. Type-species *M. giganteum* (Essex, 1928).

3a. Metascolex well developed .... **4.**
3b. Metascolex relatively weakly developed ....
.... ***Corallotaenia* Freze, 1965.**

Diagnosis: Small worms. Suckers of common type, surrounded by some folds of tissue (Fig. 13.14). Vitelline follicles lateral. Mature proglottids longer than broad. Testes in one layer. In fishes from North and South America. Type-species *C. parva* (Larsch, 1941).

4a. Scolex with corolla-like metascolex. Suckers hidden by folds of metascolex (Figs 13.15, 13.16). Uterine diverticula mainly lateral ....
.... ***Corallobothrium* Fritsch, 1886.**

Diagnosis: Large worms. Proglottids numerous, broader than long. Genital pore in anterior half of proglottid margin. Longitudinal musculature massive, forming bundles. Uterus with lateral diverticula. Vitellarium lateral, converging at posterior margin. Testes in several layers. In siluroid fishes from Africa and North America. Type-species *C. solidum* Fritsch, 1886.

4b. Scolex with folded metascolex, four typical suckers and small apical sucker (Fig. 13.17). Uterus transverse, with anterior and posterior diverticula ...
.... ***Paraproteocephalus* Chen, 1962.**

Diagnosis: Proglottids wider than long. Longitudinal musculature consisting of powerful bundles of fibres. Vitellarium lateral, converging towards ovary at posterior end (Fig. 13.18). Testes in one field. Uterine stem divided into two lateral branches each with diverticula anteriorly and posteriorly. In siluroid fishes from Russia (Far-East). Type-species *P. parasiluri* (Zmeev, 1936).

Subfamily Acanthotaeniinae Freze, 1963

Key to genera

1a. Eggs not contained in capsules ...... ***Acanthotaenia* von Linstow, 1903.**
(Syn. *Rostellotaenia* Freze, 1963.)

Diagnosis: Scolex and anterior part of strobila covered with spines (Fig. 13.19). Apical muscular organ (piercing organ) present. External segmentation sometimes indistinct. Longitudinal musculature weakly developed. Testes in two fields.

Fig. 13.14 *Corallotaenia parva* (Larsch, 1941). Scolex. (After Larsch, 1941.)
Fig. 13.15 *Corallobothrium fimbriatum* (Essex, 1928). Scolex. (After Essex, 1928.)
Fig. 13.16 *C. solidum* Fritsch, 1886. Longitudinal section of scolex. (After Fritsch, 1886.)
Figs 13.17, 13.18 *Paraproteocephalus parasiluri* (Zmeev, 1936). 13.17. Scolex. (After Zmeev, 1936.) 13.18. Gravid proglottid. (After Freze, 1965.)

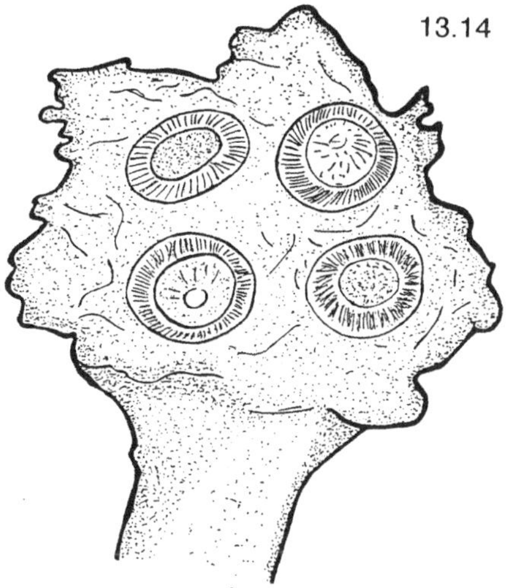

13.14

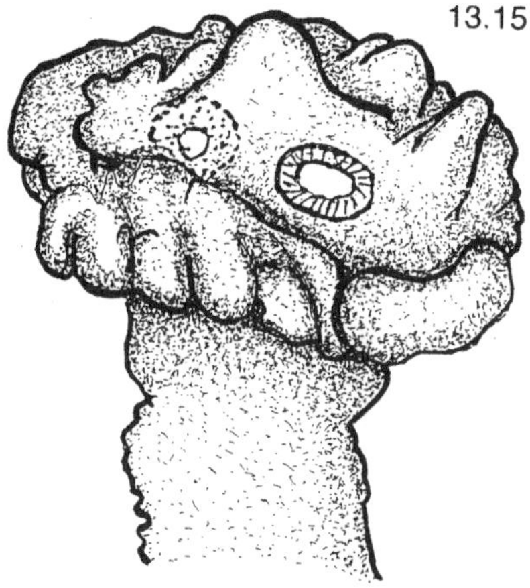

13.15

13.16

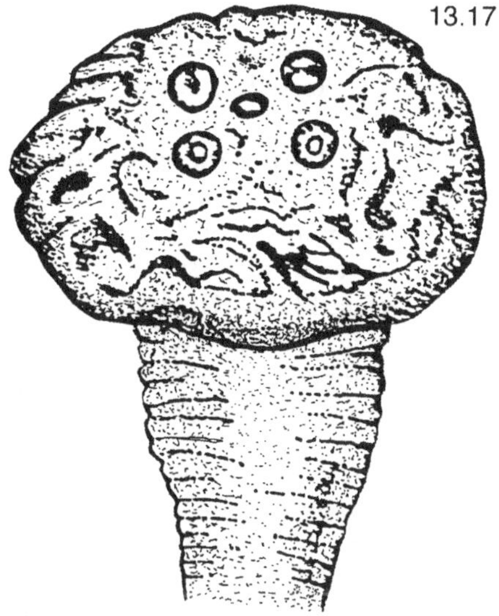

13.17

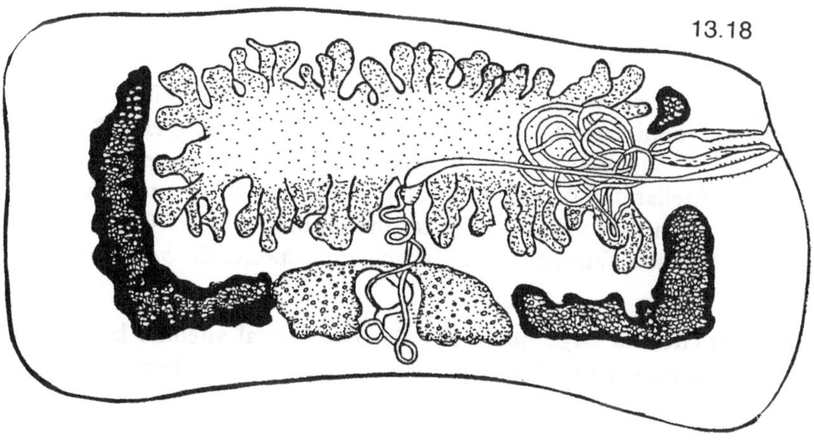

13.18

Uterus with numerous, irregular diverticula (Fig. 13.20). In varanid reptiles from Africa, Australia, India, Malaysia. Type-species *A. shipleyi* von Linstow, 1903.

1b. Eggs contained in membranous capsules (Fig. 13.23) .... *Kapsulotaenia* Freze, 1963.
(Syn. *Capsulotaenia* Freze, 1963.)

Diagnosis: Scolex and anterior part of strobila covered with spines. Apical region of scolex developed as 'piercing organ' (Fig. 13.21). Proglottids usually longer than wide (Fig. 13.22). Uterus with developed diverticula. Mehlis's gland strongly developed. Musculature weakly developed. In monitors (Reptilia) in Indonesia, Australia. Type-species *K. sandgroundi* (Carter, 1943).

Subfamily Gangesiinae Mola, 1929

Key to genera

1a. Rostellum armed with one or two circles of hooks .... 3.
1b. Rostellum armed with several circles of minute hooks .... 2.

2a. Testes lacking in anterior median region of proglottids .... *Electrotaenia* Nybelin, 1942.

Diagnosis: Scolex quadrilobulate, each lobe bearing sucker. Scolex and neck covered with small spines. Rostellum with several circles of minute hooks (Fig. 13.24). Genital atrium deep, muscular. Cirrus-sac small. Testes in two fields (Fig. 13.25). Vitellarium lateral, fields curved in posterior part of segment. Ovary bilobed, at posterior margin of proglottid. In siluroid fishes from Africa. Type-species *E. malopteruri* (Fritsch, 1886).

2b. Testes fill entire field anterior to ovary .... *Silurotaenia* Nybelin, 1942.
Diagnosis: Large worms. Scolex, suckers and neck covered with small spines (Fig. 13.26). Rostellum armed with several circles of minute hooks. Genital atrium shallow, not muscular. Cirrus-sac small, round. Testes in single field (Fig. 13.27). Ovary bilobed, large, at posterior margin of proglottid. Vitelline follicles lateral to osmoregulatory canals. In siluroid fishes from Europe. Type-species *S. siluri* (Batsch, 1786).

3a. One or two circles of small hooks (Fig. 13.28). Vitelline bands extend throughout lateral edge of proglottid ...... *Gangesia* Woodland, 1924.
Diagnosis: Suctorial surface of sucker covered with spines. Mature proglottid square. Vitellarium lateral, extending entire length of proglottid (Fig. 13.29), medullary, partly cortical (?). Testes in continuous field. Cirrus-sac variable. In siluroid fishes of Europe, India, Japan. Type-species *G. bengalensis* (Southwell, 1913).

3b. One circle of large hooks (Fig. 13.30). Lateral vitelline bands concentrated in posterior part of proglottid .... *Vermaia* Nybelin, 1942.

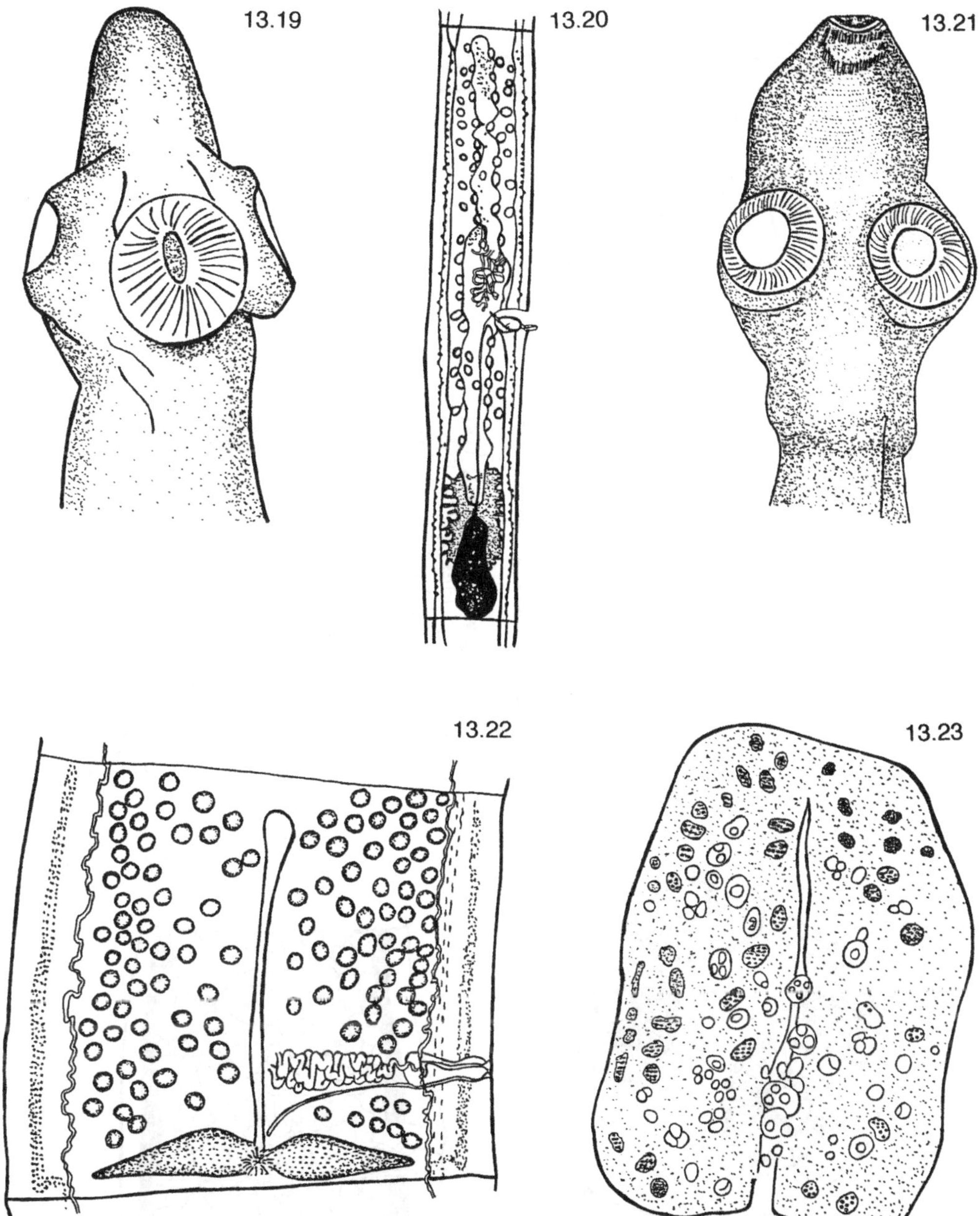

Figs 13.19, 13.20 *Acanthotaenia shipleyi* von Linstow, 1903. 13.19. Scolex. 13.20. Mature proglottid. (After Yamaguti, 1954b.)
Figs 13.21, 13.22 *Kapsulotaenia sandgroundi* (Carter, 1943). 13.21. Scolex. (After Freze, 1965.) 13.22. Mature proglottid. (After Carter, 1943.)
Fig. 13.23 *K. varia* (Beddard, 1913). Gravid proglottid. (After Beddard, 1913.)

13.24

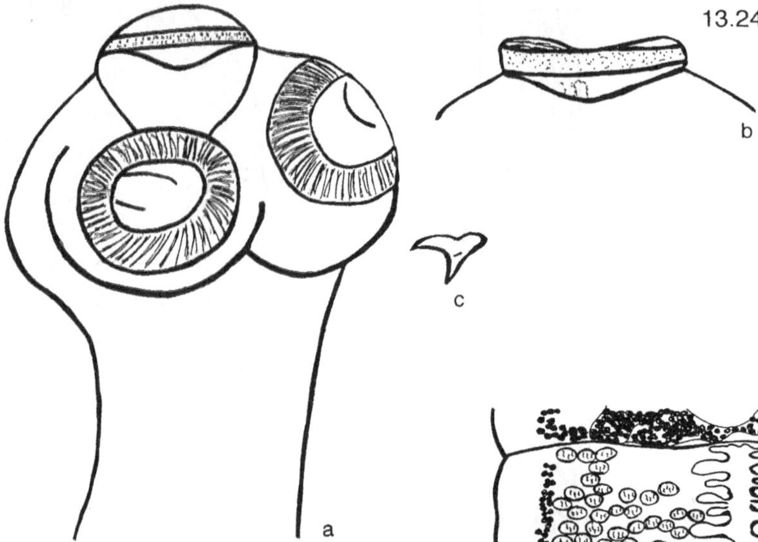

13.25

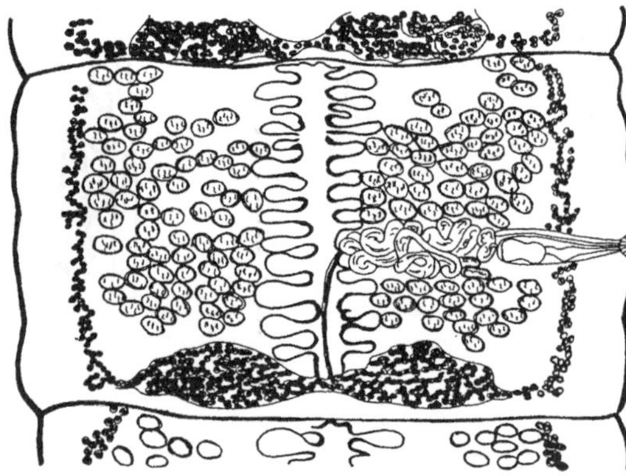

13.26

13.27

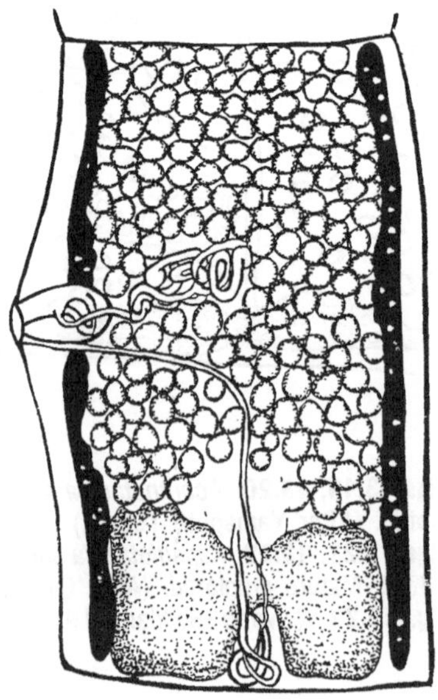

13.28

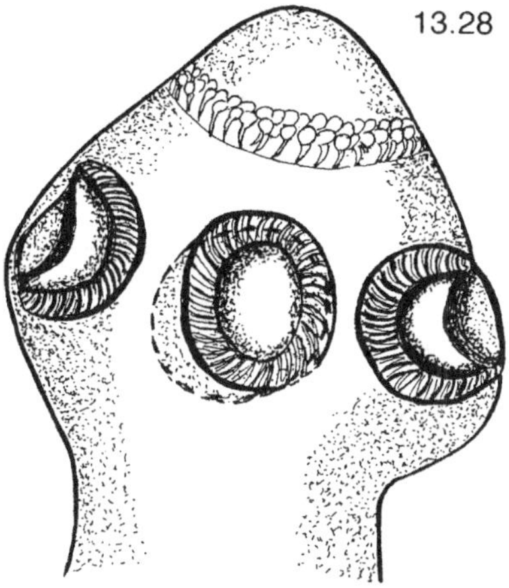

13.29

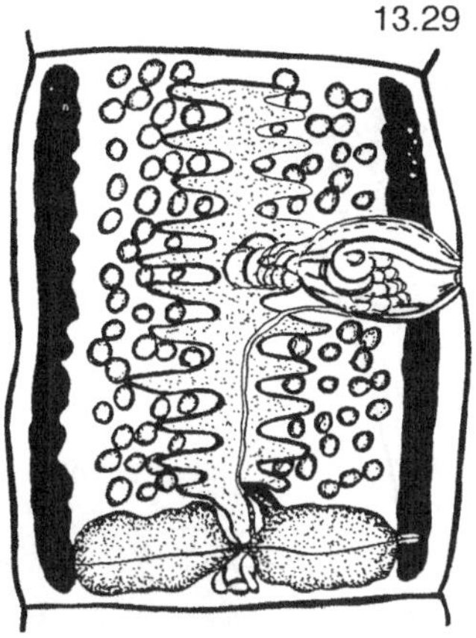

Figs 13.24, 13.25 *Electrotaenia malopteruri* (Fritsch, 1886); 13.24. a. Scolex; b. Detail of rostellum; c. Hook. (After Nybelin, 1942.) 13.25. Gravid proglottid. (After Janicki, 1928.)
Figs 13.26, 13.27 *Silurotaenia siluri* (Batsch, 1786). 13.26. Scolex. 13.27. Gravid proglottid. (After Freze, 1965.)
Figs 13.28, 13.29 *Gangesia parasiluri* Yamaguti, 1934. 13.28. Scolex. 13.29. Mature proglottid. (After Yamaguti, 1934.)

Diagnosis: Scolex, neck and proglottids covered with spines. Posterior proglottids longer than wide. Longitudinal musculature weakly developed. Vitellarium in short lateral fields (Fig. 13.31), but completely developed only in posterior proglottids. Genital pore in anterior half of lateral margin. Cirrus-sac small. Testes in two fields, anterior to ovary. Uterus with numerous diverticula. Eggs embryonated in uterus. In Indian siluroid fishes. Type-species *V. pseudotropii* (Verma, 1928).

Subfamily Marsypocephalinae Woodland, 1933

The subfamily contains one genus only.

Genus *Marsypocephalus* Wedl, 1861 (Figs 13.32–13.34)
(Syn. *Loennbergia* Fuhrmann & Baer, 1925)

Diagnosis: With the characters of the subfamily. Type-species *M. rectangulus* Wedl, 1861.

Subfamily Sandonellinae Khalil, 1960

The subfamily contains one genus only.

13.30

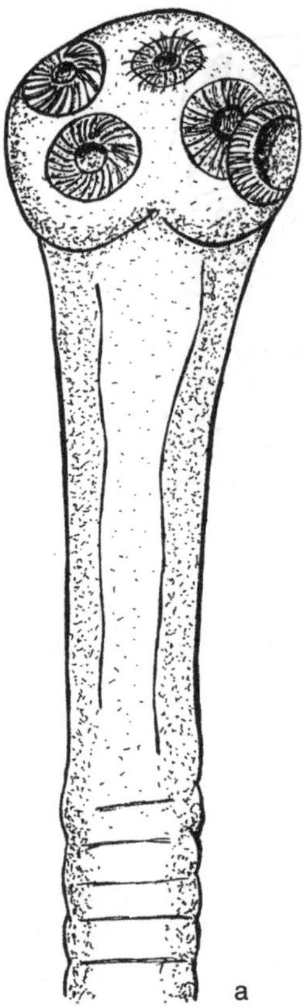

13.31

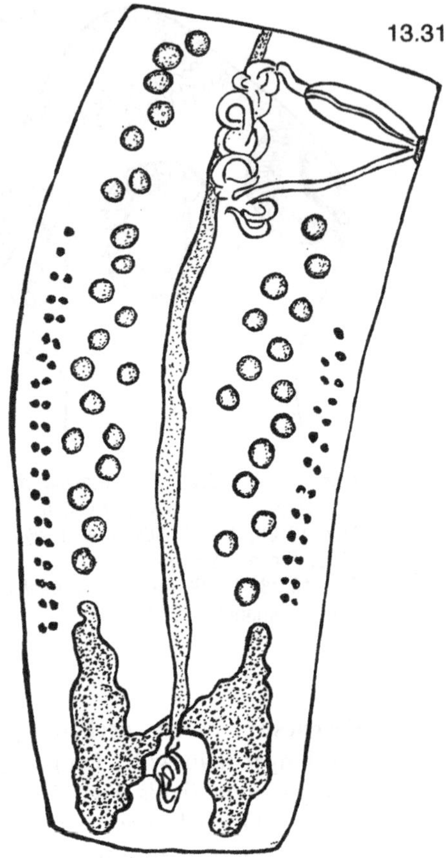

13.32

13.33

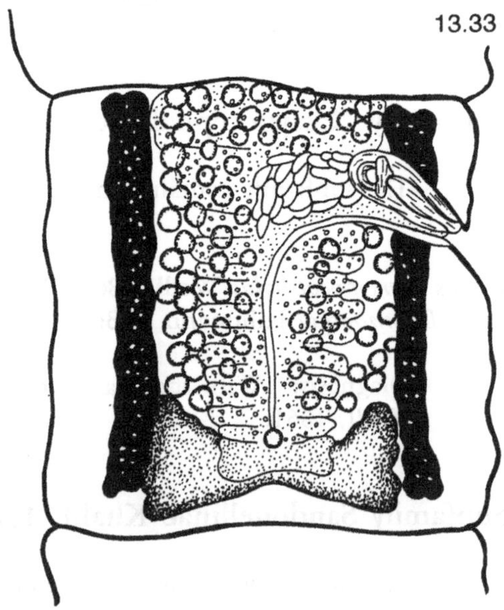

13.34

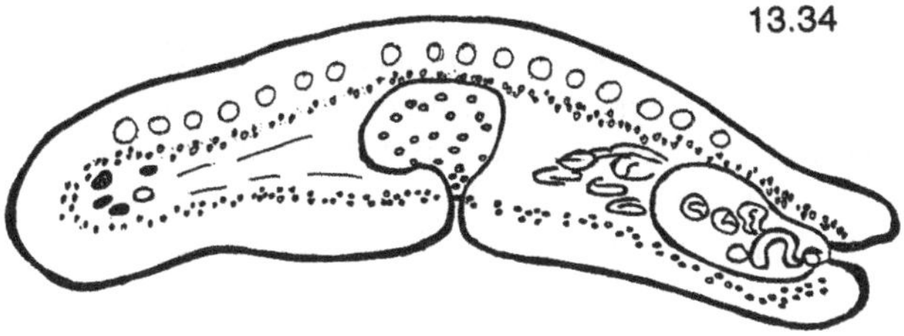

13.35

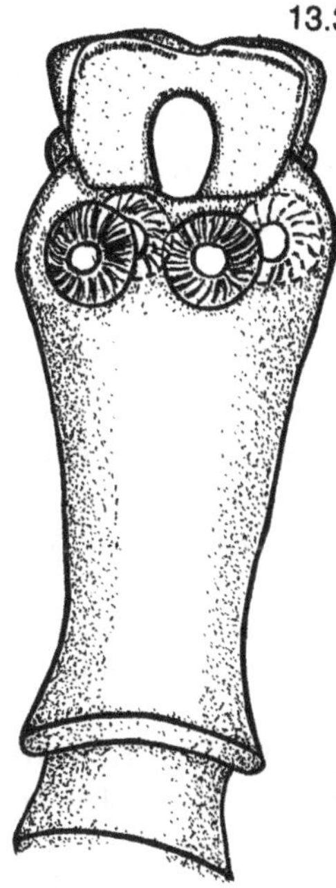

13.36

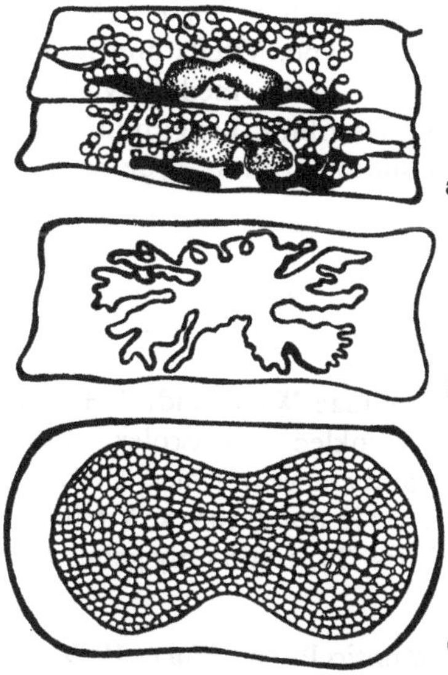

Figs 13.30, 13.31 *Vermaia pseudotropii* (Verma, 1928); 13.30. a. Scolex; b. Rostellum and hooks. 13.31. Mature proglottid. (After Verma, 1928.)
Figs 13.32–13.34 *Marsypocephalus* Wedl, 1861. *M. rectangulus* Wedl, 1861. 13.32. Scoleces. 13.33. Mature proglottid. (After Janicki, 1928.) 13.34. *M. heterobranchus* Woodland, 1925. Transverse section of proglottid.
Figs 13.35, 13.36 *Sandonella sandoni* (Lynsdale, 1960). 13.35. Scolex. 13.36. a. Gravid proglottids; b. Uterus (without eggs); c. Uterus (with eggs). (After Khalil, 1960.)

Genus *Sandonella* Khalil, 1960 (Figs 13.35, 13.36)

Diagnosis: With the characters of the subfamily. Type-species *S. sandoni* (Lynsdale, 1960).

Family Monticelliidae La Rue, 1911

Key to subfamilies

1a.	Testes medullary	2.
1b.	Testes cortical	3.

2a. Ovary and uterus entirely medullary ....
.... **Zygobothriinae Woodland, 1933. (Fig. 13.37)**
(Syn. Postgangesiinae Akhmerov, 1969.)

Diagnosis: Scolex varied, with metascolex present or not. Suckers uniloculate or biloculate. Gonads medullary. Vitellarium cortical, lateral or dorsoventral. In freshwater fishes from South America and Russia. Type-genus *Zygobothrium* Diesing, 1850.

2b. Ovary and uterus partly cortical, partly medullary ....
.... **Nupeliinae Pavanelli & Rego, 1991. (Fig. 13.38)**

Diagnosis: Scolex simple. Suckers of normal type. Vitellarium mostly cortical, some follicles in medulla. Uterus and ovary partly medullary. Testes mostly medullary, some between longitudinal muscle fibres. Longitudinal musculature scattered, irregular. In South American siluroid fishes. Type and only genus *Nupelia* Pavanelli & Rego, 1991.

3a. Ovary cortical or partly cortical .... 4.
3b. Ovary medullary .... 5.

4a. Ovary partly cortical .... **Rudolphiellinae Woodland, 1935. (Fig. 13.39)**

Diagnosis: Scolex truncate. Massive wrinkled metascolex present, with longitudinal furrows. Suckers in central protuberance of scolex. Uterus medullary. Ovary partly medullary, partly cortical. Testes in dorsal cortex. Vitellarium in ventral cortex. In South American siluroid fishes. Type and only genus *Rudolphiella* Fuhrmann, 1916.

4b. Ovary entirely cortical .... **Monticelliinae Mola, 1929. (Fig. 13.40)**

Diagnosis: Scolex bears four suckers, uniloculate or biloculate. Metascolex present or absent. Uterus and gonads cortical. Vitelline fields lateral, ventral or crescent-shaped. Uterus with lateral diverticula. Longitudinal musculature variable. In South American freshwater fishes. Type-genus *Monticellia* La Rue, 1911.

5a. Uterus medullary .... **Ephedrocephalinae Mola, 1929. (Fig. 13.41)**

Diagnosis: Scolex with folded metascolex, collar-like, behind suckers. Four stalked suckers of common type. Strobila with median furrows. Uterus and ovary entirely medullary. Testes in dorsal cortex. Vitellarium in ventral cortex. Longitudinal musculature very strong. In South American siluroid fishes. Type and only genus *Ephedrocephalus* Diesing, 1850.

5b. Uterus cortical .... **Peltidocotylinae Woodland, 1934. (Fig. 13.42)**
(Syn. Othinoscolecinae Woodland, 1933.)

Diagnosis: Scolex with four suckers or no suckers. Suckers of normal type (uniloculate) or biloculate, forming two separate cavities. Metascolex present. Ovary medullary. Uterus cortical. Testes in dorsal cortex. Vitelline follicles cortical, in two bands, dorsal and ventral. In South American siluroid fishes. Type-genus *Peltidocotyle* Diesing, 1850.

13.37

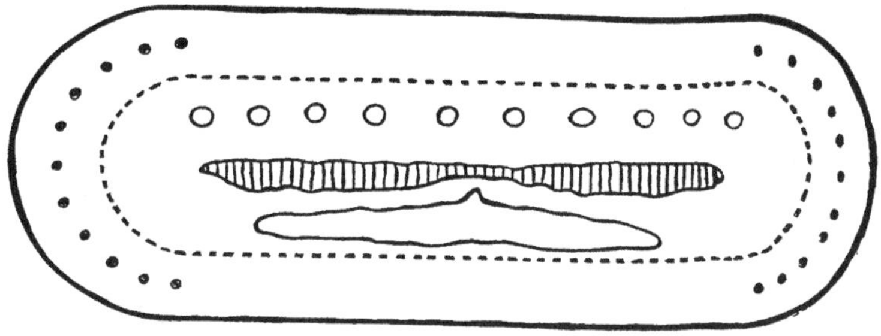

13.38

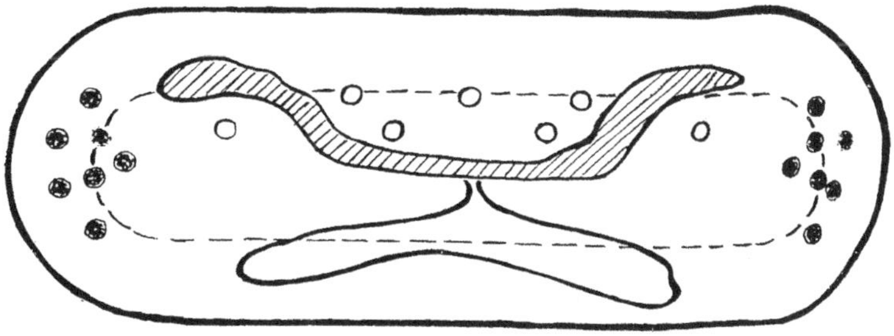

13.39

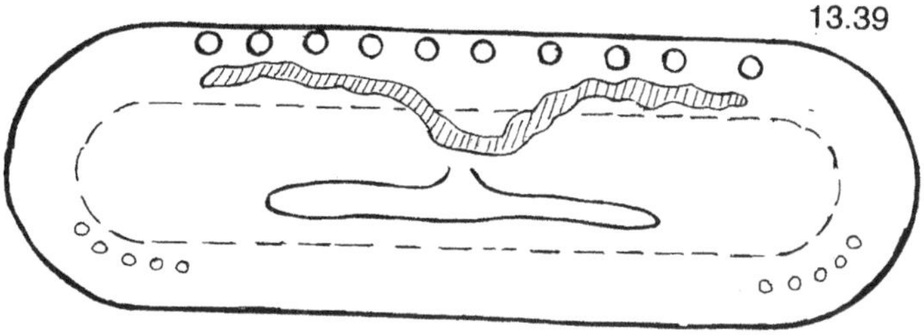

13.40

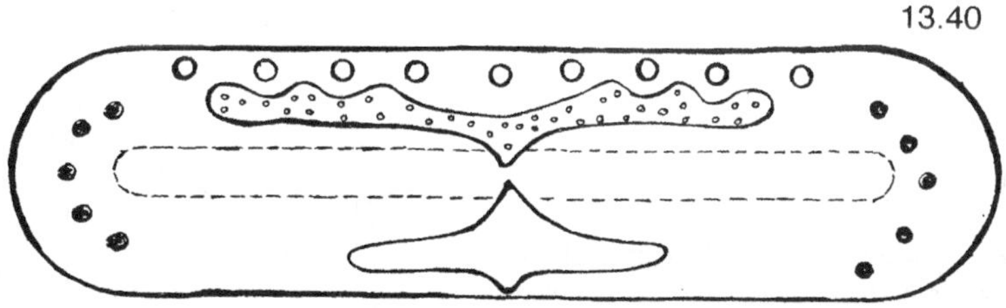

Fig. 13.37 Zygobothriinae. Schematic section. Vitelline follicles cortical, genitalia medullary. (Original.)

Fig. 13.38 Nupeliinae Pavanelli & Rego, 1990. Transverse section of proglottid. Schematic. (Original.)

Fig. 13.39 Rudolphiellinae. Schematic section. Vitelline follicles and testes cortical; uterus medullary; ovary partially medullary, partially cortical. (Original.)

Fig. 13.40 Monticelliinae. Schematic section. Vitelline follicles and gonads in the cortex. (Original.)

13.41

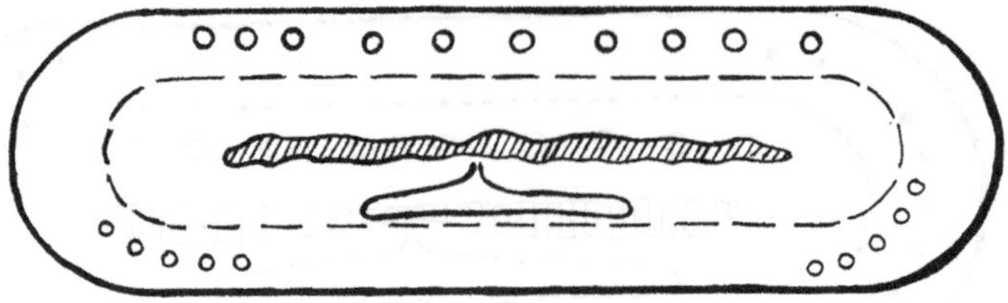

13.42

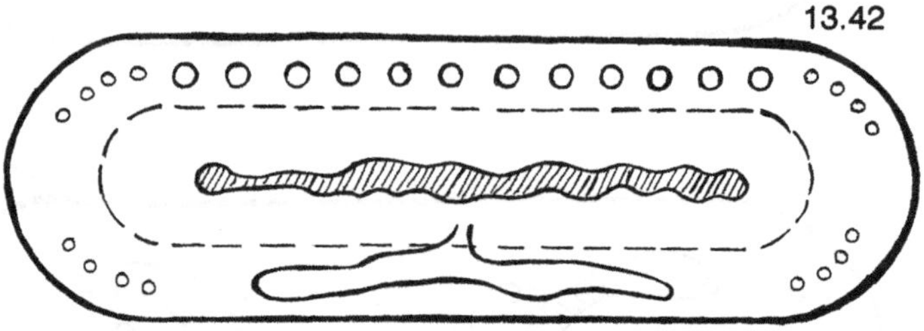

13.44

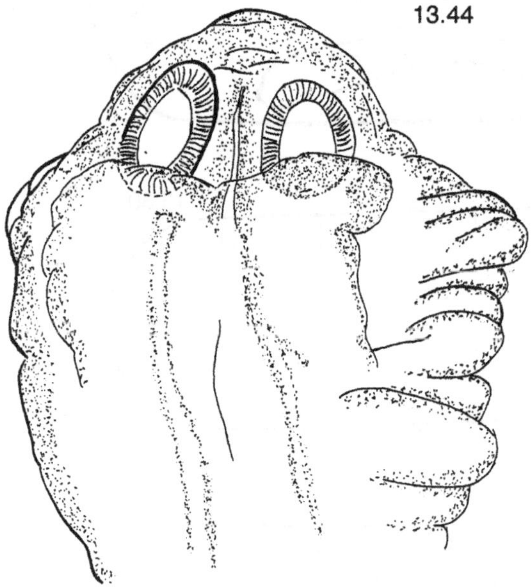

13.43

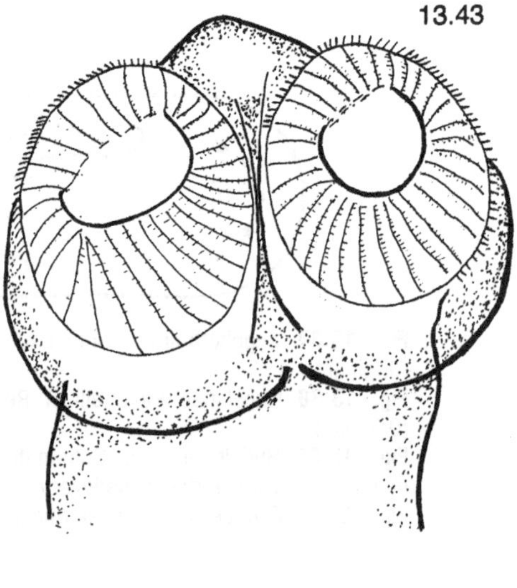

Fig. 13.41 Ephedrocephalinae. Vitelline follicles cortical; testes cortical; ovary and uterus medullary. (Original.)
Fig. 13.42 Peltidocotylinae. Schematic section. Ovary medullary; vitelline follicles, testes and uterus cortical. (Original.)
Fig. 13.43 *Spasskyellina spinulifera* (Woodland, 1934). Scolex. (After Rego, 1990.)
Fig. 13.44 *Spatulifer rugosa* (Woodland, 1935). Scolex. (After Rego, 1989.)

Subfamily Monticelliinae Mola, 1929

Key to genera

1a. **Suckers armed with several rows of minute spines. Uterine diverticula weakly developed .... *Spasskyellina* Freze, 1965.**

Diagnosis: Scolex globose (Fig. 13.43). No metascolex. Small worms. Ovary, testes and uterus entirely cortical. Uterine slit formed early. Uterine diverticula narrow. Longitudinal musculature weakly developed. In South American siluroid fishes. Type-species *S. lenha* (Woodland, 1933).

1b. **Suckers without spines. Uterus with well developed diverticula ...... 2.**

2a. **Scolex with well developed metascolex. Suckers hidden by folds of tissue .... 3.**

2b. **Scolex with weakly developed metascolex or without metascolex ..... 4.**

13.45

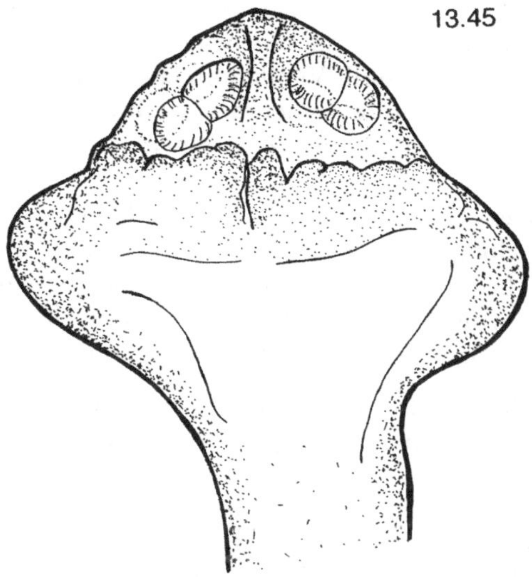

13.46

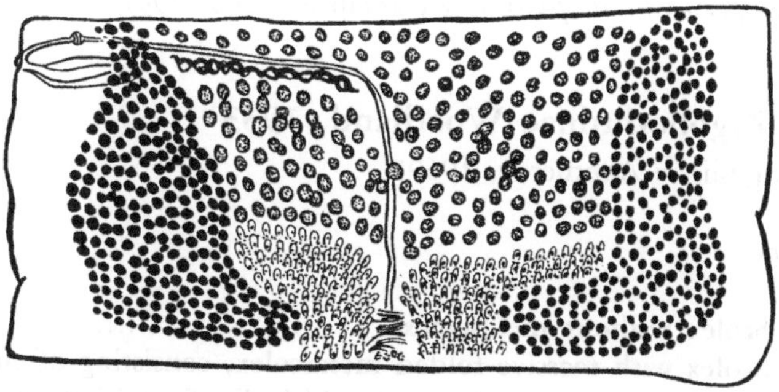

Figs 13.45, 13.46 *Goezeella siluri* Fuhrmann, 1916. 13.45. Scolex. 13.46. Mature proglottid. (After Brooks & Rasmussen, 1984.)

3a. Uniloculate suckers, in apical part of scolex .... *Spatulifer* **Woodland, 1934.**

Diagnosis: Developed metascolex (Fig. 13.44). Vitelline fields in semi-crescent pattern. Longitudinal musculature generally weakly developed. In South American siluroid fishes. Type-species *S. surubim* Woodland, 1934.

3b. Biloculate suckers (Fig. 13.45), hidden by folds of tissue .... *Goezeella* **Fuhrmann, 1916.**

Diagnosis: Metascolex developed, collar-like. Ovary mostly cortical, but central parts of ovary remain in medulla. Vitelline follicles distributed in several rows in ventral cortex, curved towards ovary (Fig. 13.46). Uterus with developed diverticula. Secondary transverse folds in strobila. In South American siluroid fishes. Type-species *G. siluri* Fuhrmann, 1916.

4a. Metascolex weakly developed, represented by some folds of tissue in base of suckers .... **5.**

4b. No metascolex, no wrinkles, no folds of tissue in base of suckers .... *Monticellia* **La Rue, 1911.**

Diagnosis: Scolex round. Suckers of normal type (Fig. 13.47). Vitellarium and reproductive organs cortical. Vitellarium lateral or semi-crescentic in cross-section. Longitudinal musculature sometimes weakly developed. In South American freshwater fishes. Type-species *M. coryphicephala* (Monticelli, 1892).

5a. Metascolex, represented by some folds of tissue in base of sucker (Fig. 13.48), suckers elongate .... *Choanoscolex* **La Rue, 1911.**

Diagnosis: Longitudinal musculature weakly developed. Uterus with many lateral outgrowths. Vitellarium lateral or crescent-shaped in section. In freshwater fishes. South America. Type-species *C. abscisus* (Riggenbach, 1895).

5b. Metascolex weakly developed, represented by some wrinkles in tissue of scolex; suckers of normal type, covered by layers of tissue (Fig. 13.49) .... *Paramonticellia* **Pavanelli & Rego, 1991.**

Diagnosis: Ovary mainly cortical but partly medullary (Fig. 13.50). Uterus cortical. Vitellarium crescent-shaped in cross-section. Genital pore anterior. Longitudinal musculature well developed. In siluroid fishes of South America. Type-species *P. itaipuensis* Pavanelli & Rego, 1991.

Subfamily Zygobothriinae Woodland, 1933

(Syn. Postgangesiinae Akhmerov, 1969)

Key to genera

1a. Scolex without metascolex .... **2.**

1b. Scolex with massive folded metascolex, consisting of circular wrinkled wall enclosing cavity at base of which lie four bilobate uniloculate suckers (Figs 13.51, 13.52) .... *Amphoteromorphus* **Diesing, 1850.**

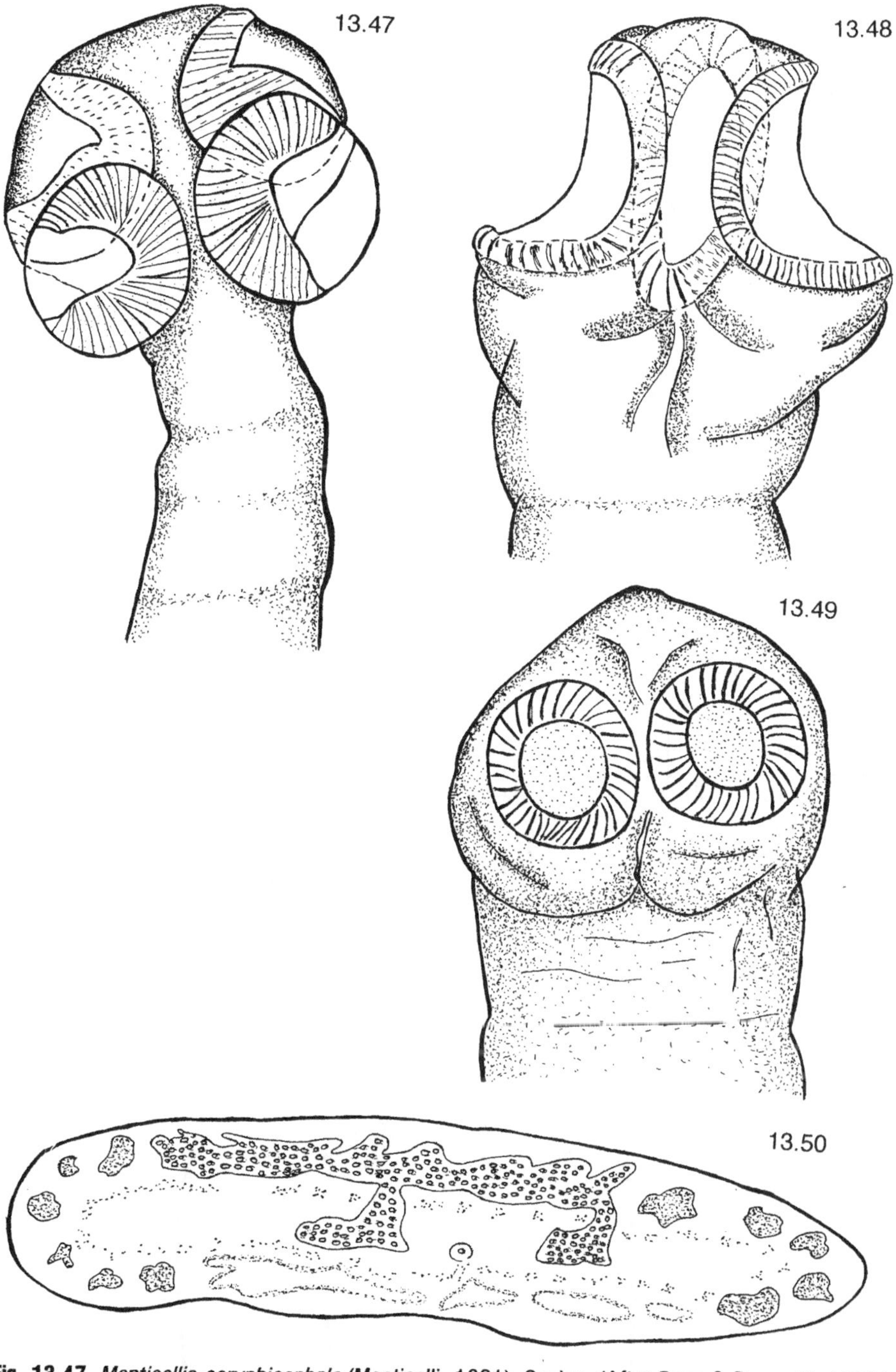

Fig. 13.47 *Monticellia coryphicephala* (Monticelli, 1891). Scolex. (After Rego & Pavanelli, 1990.)
Fig. 13.48 *Choanoscolex abscisus* (Riggenbach, 1895). Scolex. (After Rego, 1990.)
Figs 13.49, 13.50 *Paramonticellia itaipuensis* Pavanelli & Rego, 1991. 13.49. Scolex. 13.50. Section of gravid proglottid. (After Pavanelli & Rego, 1991.)

13.51

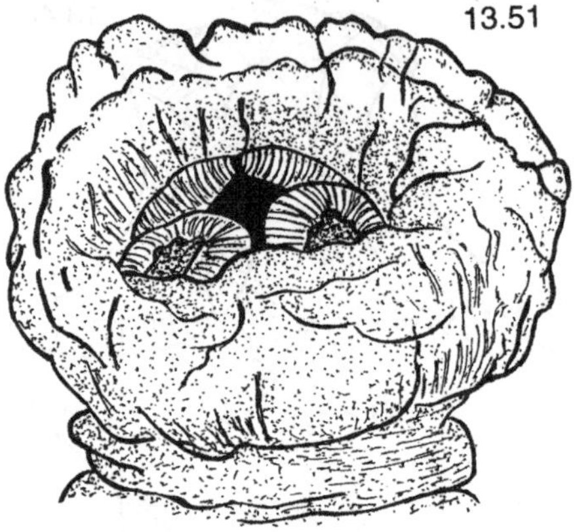

13.52

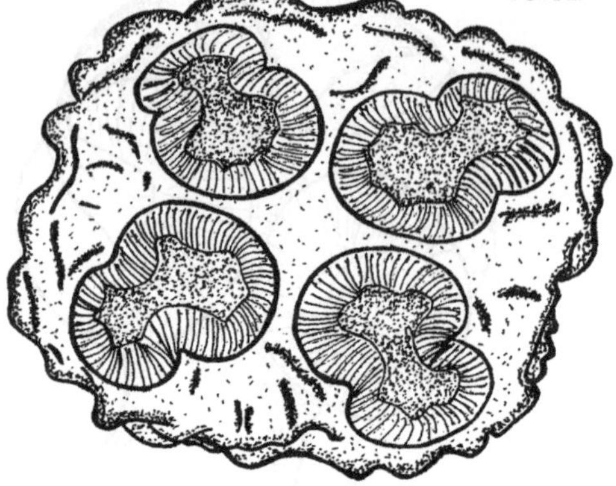

13.54

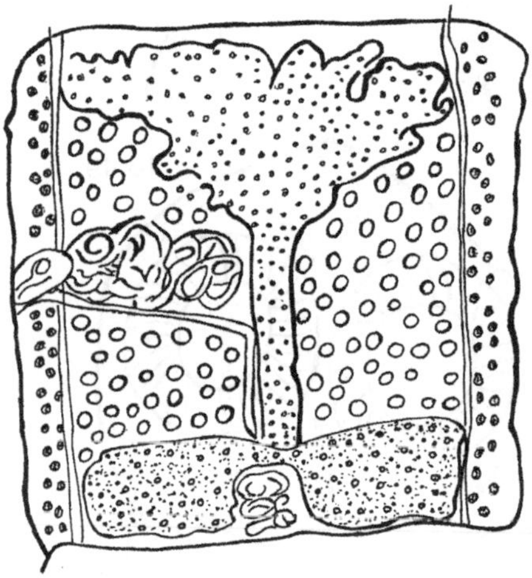

13.53

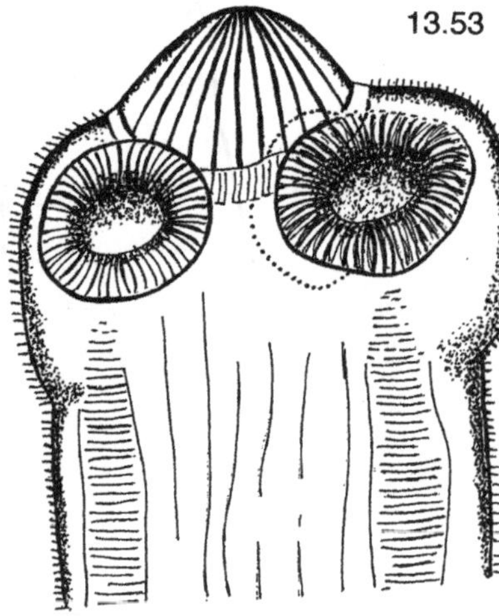

13.55

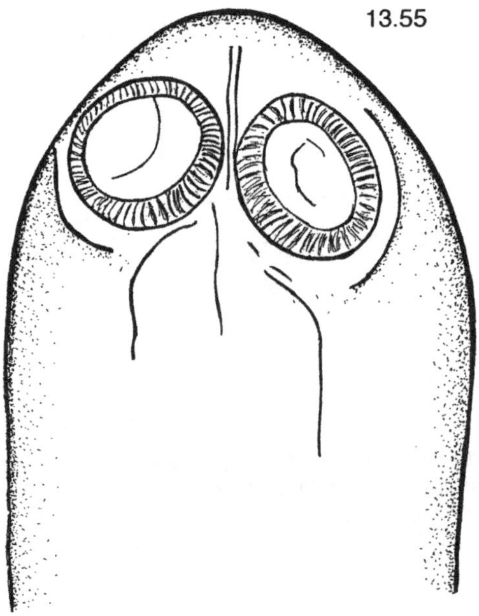

13.56

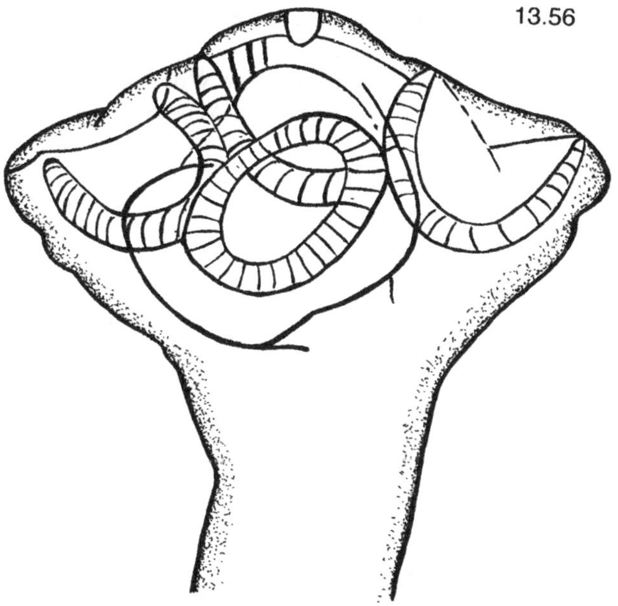

13.57

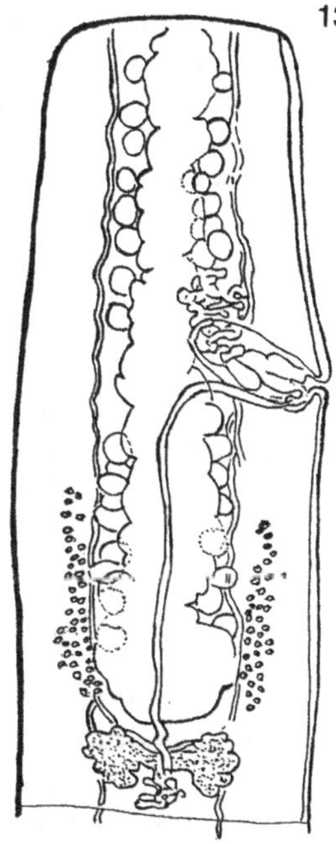

Figs 13.51, 13.52 *Amphoteromorphus peniculus* Diesing, 1850. 13.51. Scolex. 13.52. Scolex, section. (After Woodland, 1933.)
Figs 13.53, 13.54 *Postgangesia orientale* Akhmerov, 1969. 13.53. Scolex. 13.54. Proglottid. (After Akhmerov, 1969.)
Fig. 13.55 *Nomimoscolex lopesi* Rego, 1990. Scolex. (After Rego, 1990.)
Figs 13.56, 13.57 *Vaucheriella bicheti* de Chambrier, 1987. 13.56. Scolex. 13.57. Proglottid. (After de Chambrier, 1987.)

Diagnosis: Strobila with numerous proglottids, broader than long. Thin layer of longitudinal muscle fibres. Genital pores usually unilateral. Vitellarium distributed in more or less triangular pattern, in dorsal and ventral cortex. Gonads and uterus medullary. In South American siluroid fishes. Type-species *A. peniculus* Diesing, 1850.

2a. Apical organ present. Scolex and neck with spines. (Fig. 13.53) .... *Postgangesia* Akhmerov, 1969.
Diagnosis: Ovary bilobed, massive. Uterus with lateral diverticula, outgrowths begin anteriorly (Fig. 13.54). Vitellarium lateral, in cortex. Testes, ovary and uterus medullary. Genital pore median. In siluroid fishes from Russia. Type-species *P. orientale* Akhmerov, 1969.

2b. No apical organ, no spines in scolex .... **3.**

3a. Suckers modified, bothridia-like, with two openings, or with projections .... **6.**
3b. Suckers not modified .... **4.**

4a. Vitellarium occupies entire length of proglottids. Suckers of normal type (Fig. 13.55). No metascolex present .... *Nomimoscolex* Woodland, 1934.
(Syn. *Endorchis* Woodland, 1934.)

Diagnosis: Proglottids usually longer than wide. Gonads and uterus medullary. Vitellarium cortical, lateral. Uterus with developed diverticula. Longitudinal musculature developed or inconspicuous. In freshwater fishes from South America. Type-species *N. piraeeba* Woodland, 1934.

4b. Vitellarium lateral but restricted to posterior part of proglottid, behind genital pore .... *Vaucheriella* de Chambrier, 1987.

Diagnosis: Suckers of normal type (Fig. 13.56). Genital pore lateral, median. Vagina with sphincter. Vitellarium in ventral cortex, lateral, limited to posterior part of proglottid (Fig. 13.57). Gonads and uterus medullary. Longitudinal musculature weakly developed. In snakes from South America. Type-species *V. bicheti* de Chambrier, 1987.

5a. Longitudinal musculature in one layer, conspicuous .... 6.

5b. Longitudinal musculature in two layers, inconspicuous .... *Myzophorus* Woodland, 1934.

Diagnosis: Scolex swollen, sometimes with slight folds (Fig. 13.58). Proglottids laciniated or not (Fig. 13.59). Ovary medullary, but some portions intrude into dorsal cortex. Uterus mostly medullary. Testes medullary. Vitellarium lateral, cortical. In siluroid fishes. South America. Type-species *M. admonticellia* Woodland, 1934.

6a. Suckers triloculate, bothridia-like (Fig. 13.60) .... *Gibsoniela* Rego, 1984.

Diagnosis: Proglottids wider than long. Longitudinal musculature in regular bundles. Vitellarium cortical, disposed in semi-crescent pattern, some follicles invade medulla close to muscles. Ovary medullary, but some follicles cross longitudinal muscles into cortex (Fig. 13.61). Uterus with few diverticula. In siluroid fishes. South America. Type-species *G. mandube* (Woodland, 1935).

6b. Suckers not triloculate .... 7.

7a. Suckers large, globose, each has two openings (Fig. 13.62) .... *Zygobothrium* Diesing, 1850.

Diagnosis: Surface of scolex is very wrinkled. Proglottids craspedote (Fig. 13.63), with median notch on each surface. Proglottids very numerous, broader than long. Vitellarium in dorsal and ventral cortex. In South American siluroid fishes. Type-species *Z. megacephalum* Diesing, 1850.

Figs 13.58, 13.59 *Myzophorus pirara* Woodland, 1935. 13.58. Scolex and immature proglottids. 13.59. Mature proglottid. (After Rego, 1984a.)
Figs 13.60, 13.61 *Gibsoniela mandube* (Woodland, 1935). 13.60. Scolex. (After Rego, 1984b.) 13.61. Section at level of ovary; vitelline follicles partially cortical. (Original.)

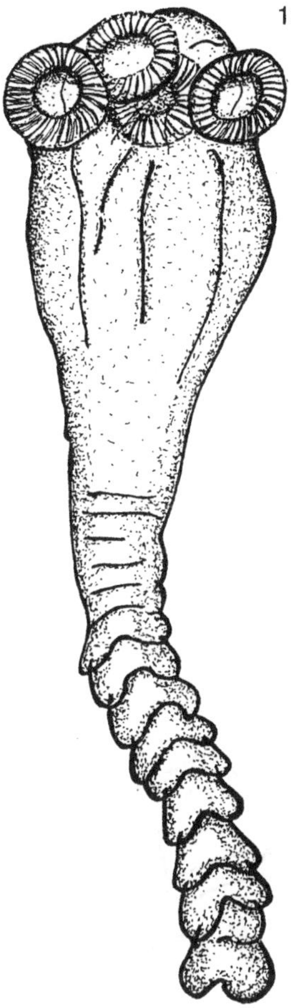

13.58

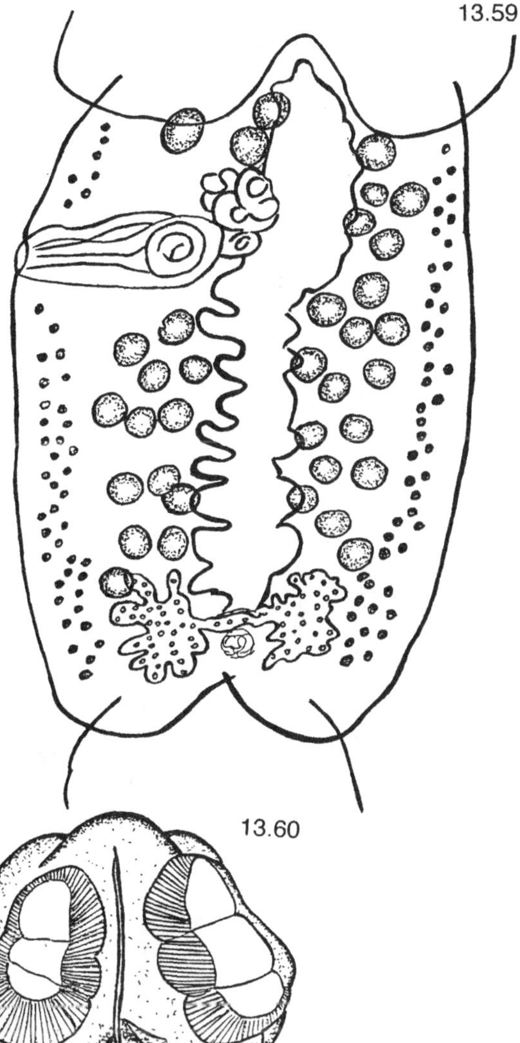

13.59

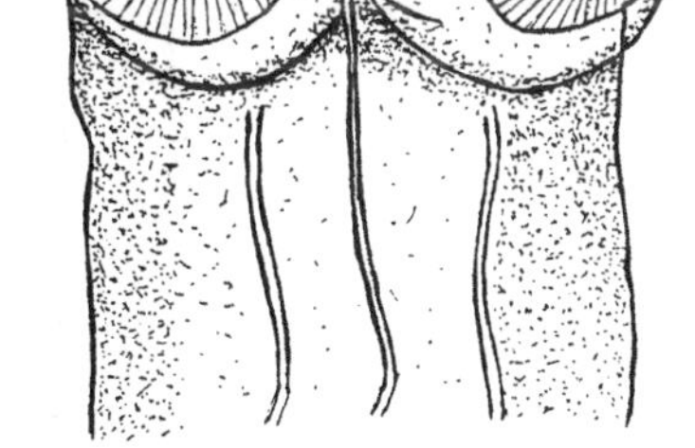

13.60

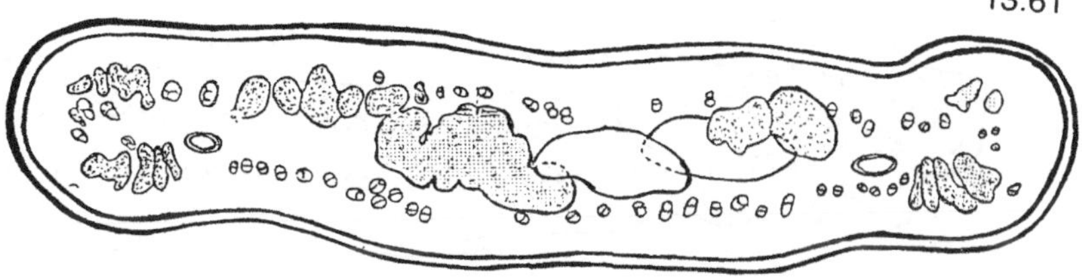

13.61

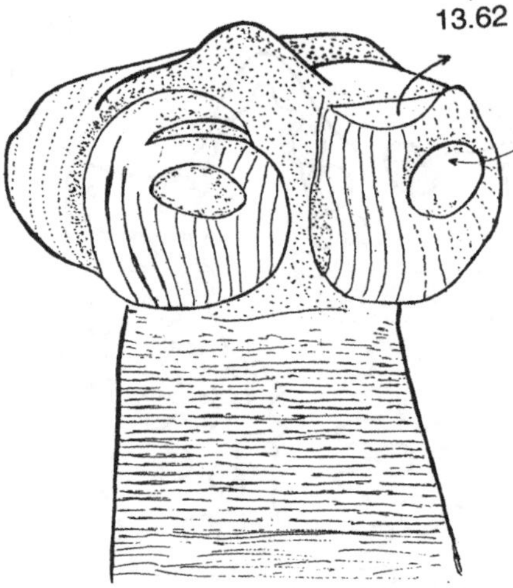

13.62

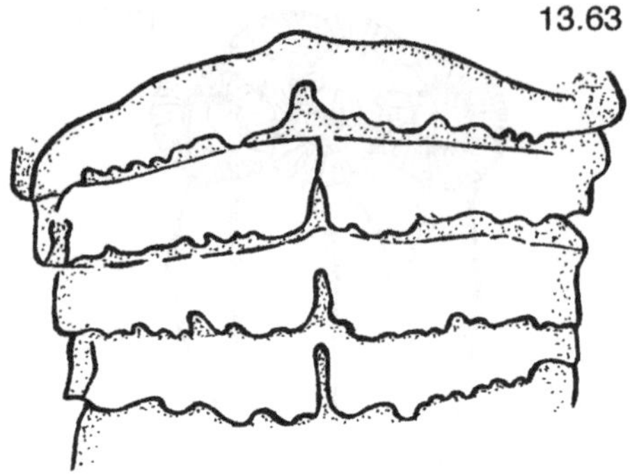

13.63

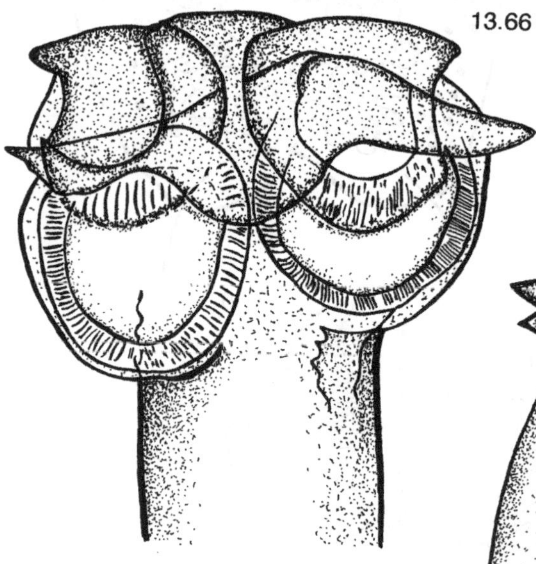

13.66

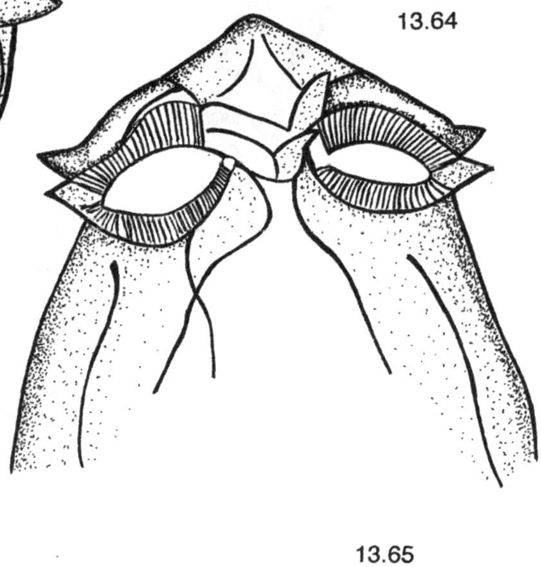

13.64

13.65

13.67

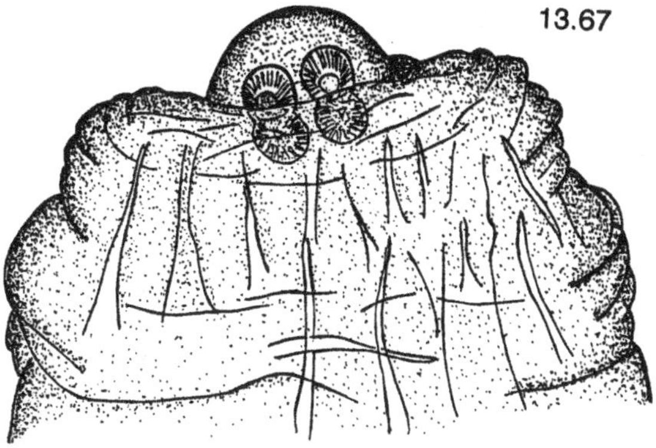

13.68

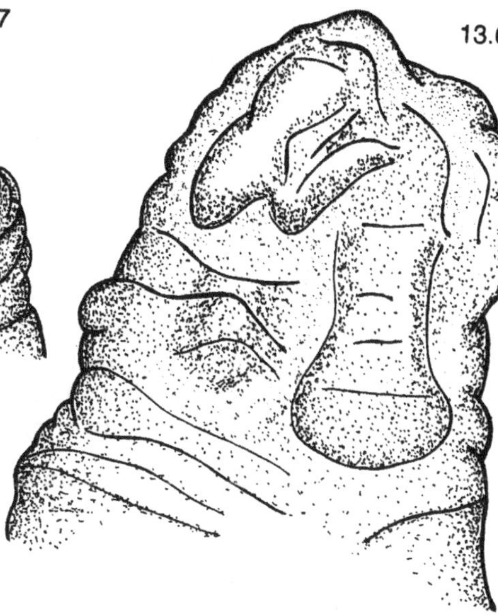

Figs 13.62, 13.63 *Zygobothrium megacephalum* Diesing, 1850. 13.62. Scolex. 13.63. Proglottids with lappets. (After Woodland, 1933b.)
Figs 13.64, 13.65 *Houssayela sudobim* (Woodland, 1935). 13.64. Scolex. 13.65. Transverse section of proglottid. (After Woodland, 1935a.)
Fig. 13.66 *Harriscolex kaparari* (Woodland, 1935). Scolex. (After Rego, 1990.)
Fig. 13.67 *Peltidocotyle rugosa* Diesing, 1850. Scolex. (After Rego, 1990.)
Fig. 13.68 *Othinoscolex lenha* Woodland, 1933. Scolex. (After Woodland, 1933b.)

7b. Suckers without two openings .... **8.**

8a. Suckers small, each with four horny projections (Fig. 13.64) .... ***Houssayela* Rego, 1987.**
Diagnosis: Scolex small. Strobila with numerous proglottids. Most proglottids broader than long. Longitudinal musculature well developed. Uterus with numerous diverticula. Vitellarium cortical, in semi-crescent pattern, some follicles invade medulla (Fig. 13.65). In South American siluroid fishes. Type-species *H. sudobim* (Woodland, 1935).

8b. Suckers not small, with two triangular projections (Fig. 13.66) .... ***Harriscolex* Rego, 1987.**
Diagnosis: Scolex with dome-shaped anterior end. Proglottids broader than long. Vitellarium cortical, in dorsal and ventral bands. Testes in single field. Longitudinal musculature well developed. In siluroid fishes of South America. Type-species *H. kaparari* (Woodland, 1935).

Subfamily Peltidocotylinae Woodland, 1934

(Syn. Othinoscolecinae Woodland, 1933)

Key to genera

1a. Suckers divided into two loculi by transverse septum (Fig. 13.67) .... *Peltidocotyle* **Diesing, 1850.**

Diagnosis: Scolex large, with well developed metascolex provided with longitudinal furrows. Strobila wide behind scolex, tapering toward posterior end. Secondary transverse grooves present in strobila. Longitudinal musculature strongly developed. Vitellarium cortical in dorsal and ventral bands. Ovary medullary. Uterus mostly cortical. Testes in dorsal cortex. In South American siluroid fishes. Type-species *P. rugosa* Diesing, 1850.

1b. Suckers absent or of common type .... 2.

2a. Suckers absent .... *Othinoscolex* **Woodland, 1933.**

Diagnosis: Surface of scolex covered by irregular mass of thick folds (Fig. 13.68). No suckers or attachment areas observed. Secondary transverse creases of strobila highly developed. Proglottids many times broader than long. Uterus with small diverticula. Vitellarium cortical, in dorsal and ventral bands. Longitudinal musculature layer thin. In South American siluroid fishes. Type-species *O. lenha* Woodland, 1933.

2b. Suckers present, uniloculate .... 3.

3a. Uterus medullary, but with diverticula in cortex (Fig. 13.70) .... *Jauella* **Rego & Pavanelli, 1985.**

Diagnosis: Scolex small, retractile. Suckers small, apical. Metascolex cone-shaped, with few folds of tissue (Fig. 13.69). Vitellarium in dorsal and ventral bands. Testes in dorsal cortex. In South American siluroid fishes. Type-species *J. glandicephalus* Rego & Pavanelli, 1985.

3b. Uterus entirely cortical .... *Woodlandiella* **Freze, 1965.**

Diagnosis: Scolex with very wrinkled metascolex. Four suckers on apical part of scolex (Fig. 13.71). Proglottids much broader than long. Secondary folds of strobila highly developed. Longitudinal musculature well marked. Vitellarium in dorsal and ventral bands. In South American siluroid fishes. Type-species *W. myzofera* (Woodland, 1933).

Subfamily Ephedrocephalinae Mola, 1929

The subfamily contains one genus only.

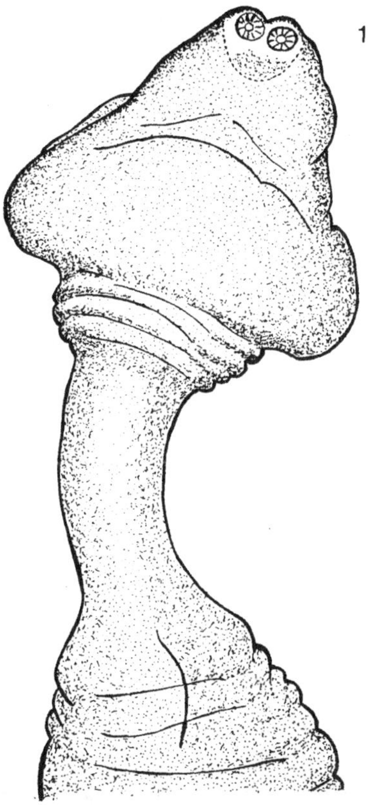

13.69

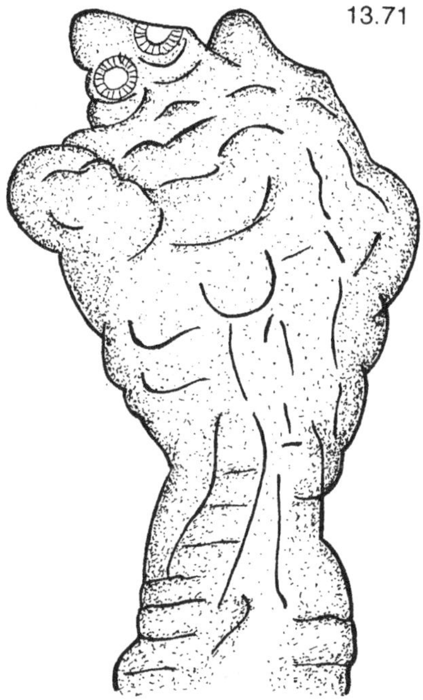

13.71

13.70

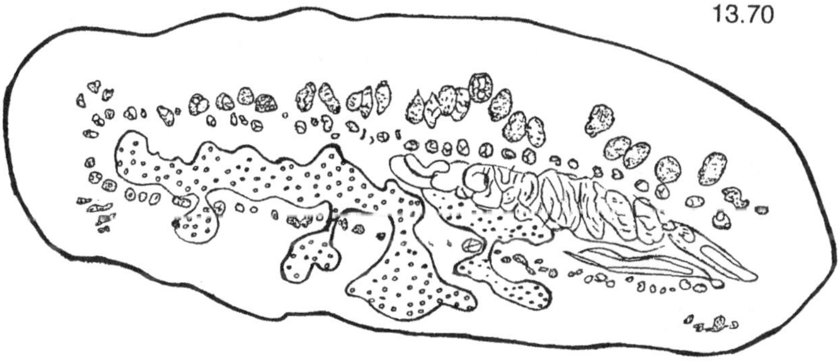

Figs 13.69, 13.70 *Jauella glandicephalus* Rego & Pavanelli, 1985. 13.69. Scolex. 13.70. Section of gravid proglottid; uterus partially cortical. (After Rego & Pavanelli, 1985.)
Fig. 13.71 *Woodlandiella myzofera* (Woodland, 1933). Scolex. (After Woodland, 1933b.)

Genus *Ephedrocephalus* Diesing, 1850 (Fig. 13.72)

Diagnosis: With the characters of the subfamily. Type-species *E. microcephalus* Diesing, 1850.

13.72

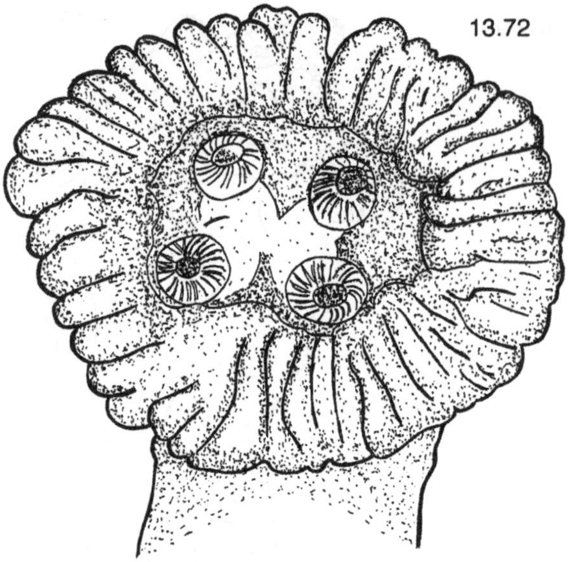

13.73

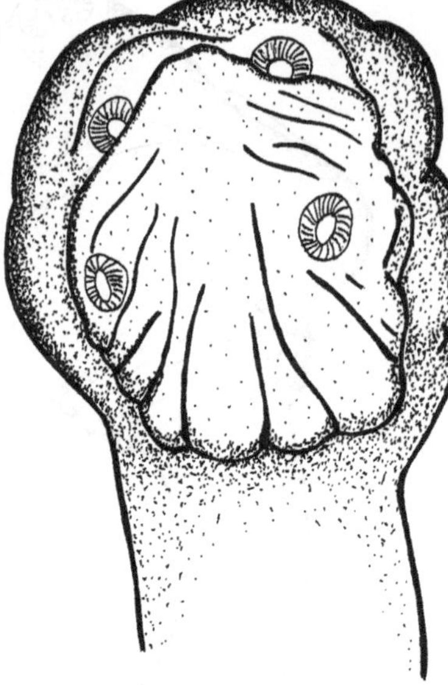

13.74

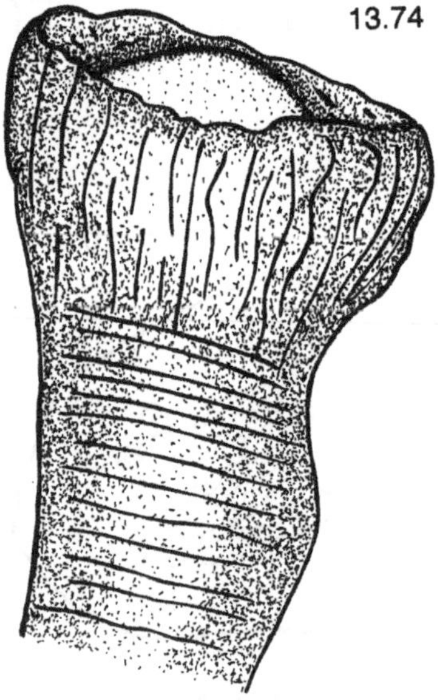

13.75

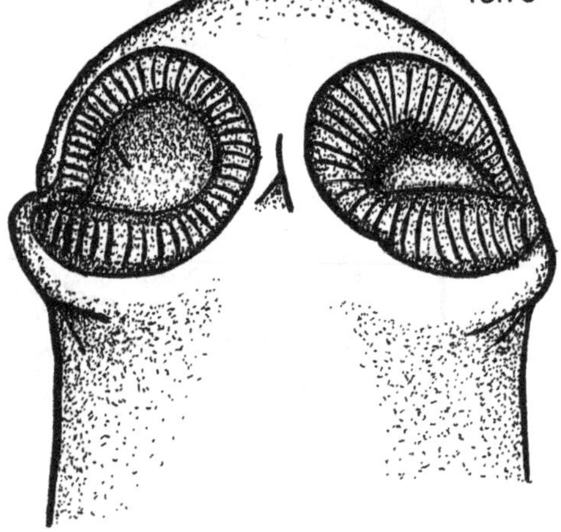

13.76

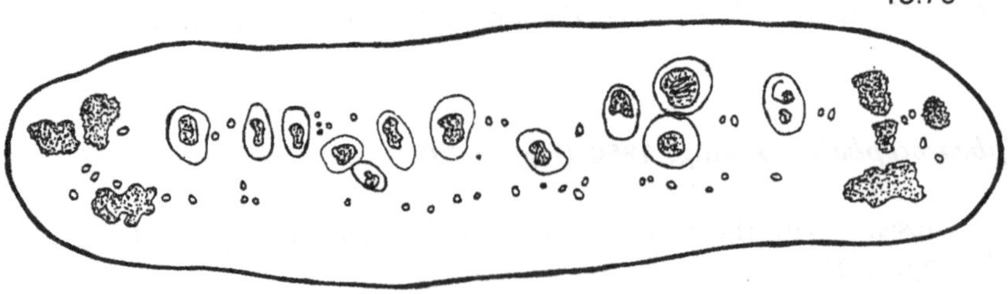

Subfamily Rudolphielliinae Woodland, 1935

(Syn. Amphilaphorchidiinae Woodland, 1934)

The subfamily contains one genus only.

Genus *Rudolphiella* Fuhrmann, 1916 (Figs 13.73, 13.74)

(Syn. *Amphilaphorchis* Woodland, 1934)

Diagnosis: With the characters of the subfamily. Type-species *R. lobosa* (Riggenbach, 1896).

Subfamily Nupeliinae Pavanelli & Rego, 1991

The subfamily contains one genus only.

Genus *Nupelia* Pavanelli & Rego, 1991 (Figs 13.75, 13.76)

Diagnosis: With the characters of the subfamily. Type-species *N. portoriquensis* Pavanelli & Rego, 1991.

Genera inquirenda

Genus *Sciadocephalus* Diesing, 1850

Diagnosis: Scolex umbrella-shaped (Fig. 13.77). Four central suckers and small apical sucker. Internal longitudinal muscle layer absent. No differentiation between cortex and medulla. Type-species *S. megalodiscus* Diesing, 1850 *species inquirenda*.

The position of the gonads and vitellarium was not established by Woodland (1933a), who redescribed the species. The scolex is similar to Corallobothriinae.

Genus *Manaosia* Woodland, 1935

Diagnosis: Scolex large, with large suckers submerged in scolex tissue and opening as small apertures on its surface (Fig. 13.78). Longitudinal muscle bundles well

---

Fig. 13.72 *Ephedrocephalus microcephalus* Diesing, 1850. Scolex. (After Mola, 1906.)
Fig. 13.73 *Rudolphiella myoides* (Woodland, 1934). 13.73. Scolex. (After Woodland, 1934b.)
Fig. 13.74 *R. piranabu* (Woodland, 1934). Scolex. (After Woodland, 1934b.)
Figs 13.75, 13.76 *Nupelia portoriquensis* Pavanelli & Rego, 1991. 13.75. Scolex. 13.76. Section of proglottid. (After Pavanelli & Rego, 1991.)

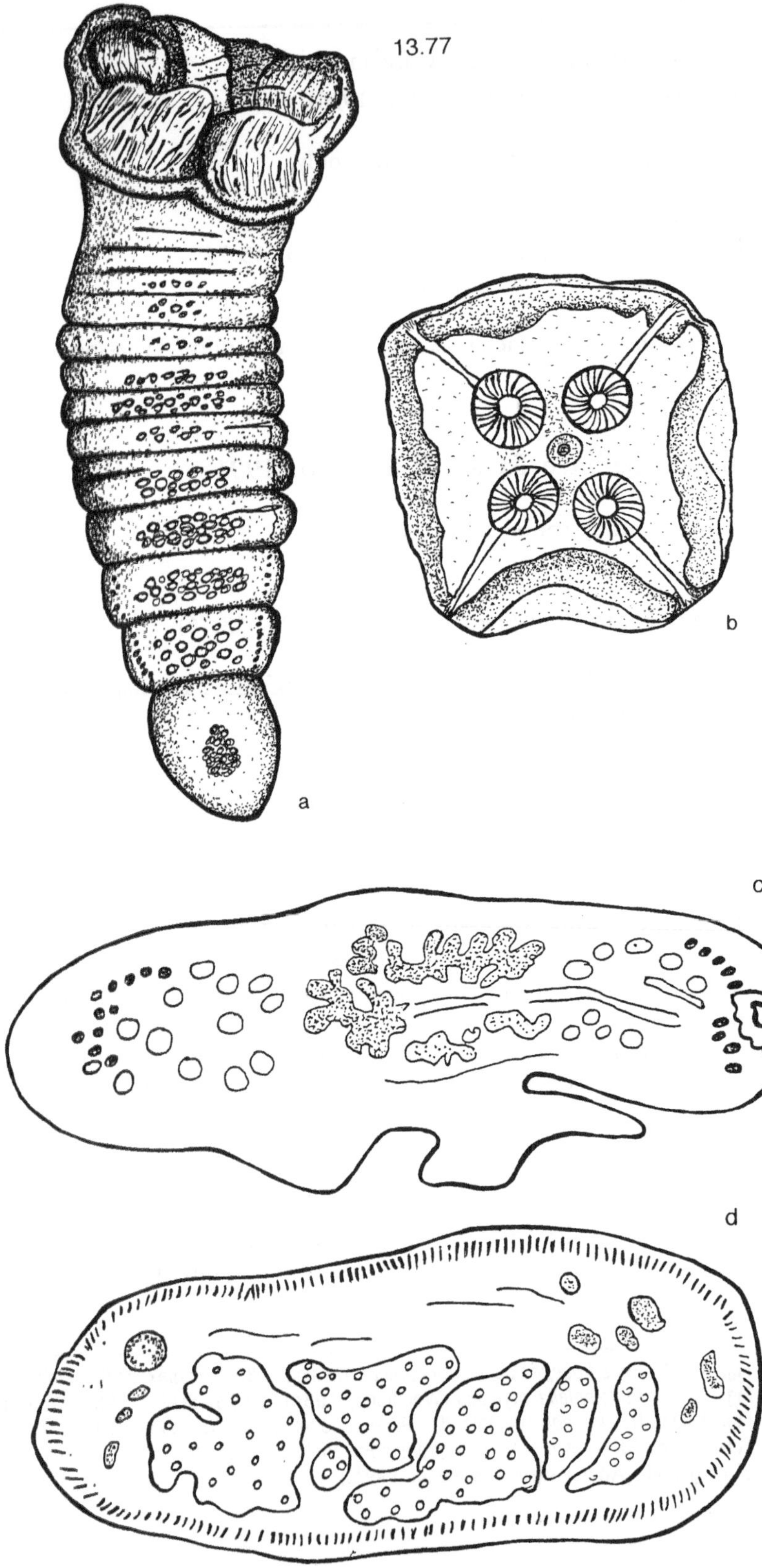
13.77
a
b
c
d

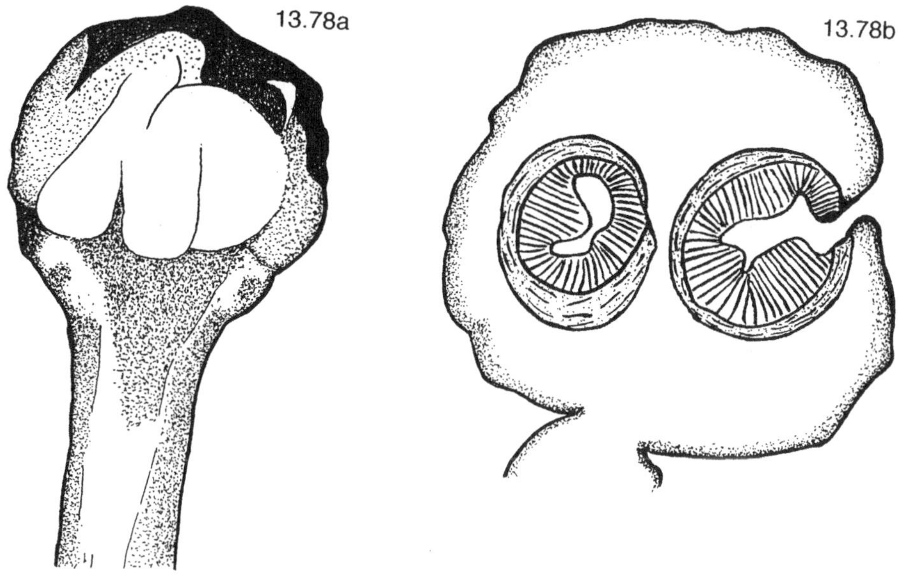

Fig. 13.77 *Sciadocephalus megalodiscus* Diesing, 1850. a. Entire specimen. b. Apical view of scolex. c. Transverse section at level of ovary. d. Transverse section at level of uterus. (After Woodland, 1933b.)

Fig. 13.78 *Manaosia bracodemoca* Woodland, 1935. a. Scolex. b. Scolex in section. (After Woodland, 1935b.)

developed. Testes probably cortical. Ovary partly cortical and partly medullary. Type-species *M. bracodemoca* Woodland, 1935 *species inquirenda*.

Woodland (1935b) does not refer to the position of the vitellarium and uterus.

The genus *Cangatiella* Pavanelli & Machado dos Santos, 1991 was inadvertently omitted from this key. It belongs in the subfamily Proteocephalinae. For a full discussion, see Pavanelli & Machado dos Santos (1991).

Genus *Cangatiella* Pavanelli & Machado dos Santos, 1991

The important generic characters are: Longitudinal osmoregulatory canals medullary. Vitelline follicles do not extend beyond osmoregulatory canals; follicles intrude into the longitudinal muscle layer. Egg provided with polar filaments. In siluriform fish. Brazil. Type-species *C. arandasi* Pavanelli & Machado dos Santos, 1991.

14 Order Tetrabothriidea Baer, 1954

E.P. HOBERG

USDA, Agricultural Research Service, Biosystematic Parasitology Laboratory, Beltsville Agricultural Research Center, Beltsville, Maryland 20705, USA.

Introduction

The family Tetrabothriidae Linton, 1891 is a monophyletic assemblage of six genera that occur as parasites of marine homoeotherms (Hoberg, 1989). Although cestodes of this family are a predominant component of the parasite faunas of seabirds, pinnipeds and cetaceans, aspects of their higher taxonomy and biology have remained enigmatic. Historically, tetrabothriids have been classified among the Pseudophyllidea (Nybelin, 1922), Cyclophyllidea (Fuhrmann, 1932; Wardle & McLeod, 1952; Yamaguti, 1959; Schmidt, 1986), as the suborder Tetrabothriata of the Tetraphyllidea (Spasskii, 1958, 1992; Temirova & Skryabin, 1978; Ryzhikov *et al.*, 1985), or in the separate order Tetrabothriidea (Baer, 1954a; Galkin, 1987a). Divergent opinions concerning the ordinal classification of this taxon have centred upon interpretations of the structural homology of the scolex (Baylis, 1926; Rees, 1956; Andersen & Lysfjord, 1982) and the significance of a typically compact and anteroventral vitelline gland that defines the family (Wardle & McLeod, 1952; Hoberg, 1989). Unambiguous support of a tetraphyllidean ancestry for the Tetrabothriidae resides in recognition of an homologous pattern of morphogenesis for the holdfast in metacestodes representing these taxa (Hoberg, 1987a; Brooks *et al.*, 1991). Whether considered to represent a separate order with tetraphyllidean affinities (as presented herein) or a suborder of the Tetraphyllidea, it is clear that the placement of the Tetrabothriidae as a family of the Cyclophyllidea (as presented in most traditional classifications) cannot be supported based on morphological, phylogenetic and biological criteria (Hoberg, 1989).

Cycles for tetrabothriids have yet to be elucidated but are considered to involve crustaceans, cephalopods, and (or) teleosts as intermediate and paratenic hosts, and homoeotherms as definitive hosts (Baer, 1954a; Temirova & Skryabin, 1978; Jones, 1988). Although patterns of transmission are unknown, the infective larval stage has been identified as a uniacetabulate plerocercoid (Hoberg, 1987a).

A unique heterochronic sequence in postlarval ontogeny of the adult scolex, initiated within the definitive host, has been defined and postulated as a uniform pattern of development within the family (Hoberg, 1989; Hoberg *et al.*, 1991). The apparent morphological similarities of undifferentiated larval tetrabothriids and other tetraphyllideans will complicate recognition of the former in the intermediate host (Hoberg, 1987a).

Identification at the generic level relies primarily on the structure of the scolex, bothridia and auricular appendages, and the configuration of the genital atrium. The genus *Chaetophallus* Nybelin, 1916, recognized here, is considered a synonym of *Tetrabothrius* Rudolphi, 1819 by some authorities (Baer, 1954a; Temirova & Skryabin, 1978). The validity of other genera is not currently disputed. However, differentiation among congeners, particularly in the genera *Tetrabothrius, Anophryocephalus* Baylis, 1922 and *Trigonocotyle* Baer, 1932 may be problematic due to interspecific homogeneity and broad intraspecific variation of some otherwise diagnostic attributes (Hoberg, 1987b, 1990; Hoberg *et al.*, 1991). Incomplete descriptions in previous taxonomic treatments have also hindered the separation of some species. Re-evaluation of many species of *Tetrabothrius* appears to be necessary due to an inadequate documentation of intraspecific variation and a lack of consistency for some morphological traits (Hoberg, 1989). Particularly significant is the progressive increase in dimensions (with increasing maturity and changes in physiological condition) of the genital atrium (and associated male and female ducts), cirrus-sac and vitelline gland (Murav'eva & Popov, 1976; Hoberg, 1987b). To allow for accurate comparisons among taxa, intraspecific variation must be documented and measurements must be presented with reference to the degree of development of the proglottids and strobila. Thus, to characterize the full extent of intraspecific variation for specific characters, it is necessary to define the stage of ontogeny of the strobila where maximum dimensions are observed. The following terminology is recommended (see Hoberg, 1987a): immature (genital anlagen visible to presence of undifferentiated male and female organs); mature (male and female systems completely functional, tubular uterus narrow); postmature (male and female systems well defined; tubular uterus expanded but not containing developing oncospheres); pregravid (expanded uterus contains undifferentiated oncospheres); gravid (oncospheres fully developed; uterine pore patent).

Taxonomic characters at the species level have included unique features of the scolex (when apparent), composition of the longitudinal musculature, structure of the genital atrium (incorporating length of male canal and aspects of the atrial vagina) and cirrus-sac, and number of testes. Apparent overlap in some diagnostic characters in species of *Tetrabothrius, Anophryocephalus* and *Trigonocotyle* is extensive and mensural attributes may be dramatically influenced by individual variation and physiological condition (age) of the proglottid (Murav'eva & Popov, 1976; Hoberg, 1987b; Hoberg *et al.*, 1991). Consequently, qualitative characters, particularly the form of the genital atrium and genital papillae, relative positions of the apertures of the male canal and vagina, presence and location of genital sphincters and occasionally the structure of the bothridia and auricles (especially in *Anophryocephalus* and *Trigonocotyle*) are of critical importance. The most significant characters, those of the genital atrium, cannot be completely evaluated from whole-mounted specimens. The longitudinal musculature and genital atrium with associated ducts must be viewed either in histological or hand-cut thick sections (transverse), thus augmenting the difficulty of accurate identification of genera and species.

The speciose genus *Tetrabothrius* with 42 nominal species in avian and eight species in cetacean hosts has been partitioned into four subgenera (Baer, 1954a; Murav'eva, 1975; Temirova & Skryabin, 1978; Odening, 1982). Recognition of *Tetrabothrius* (*Tetrabothrius*) spp., *T.* (*Neotetrabothrius*) spp., *T.* (*Oriana*) spp. and *T.* (*Culmenamniculus*) spp. was based on apparent structural differences in the genital atrium, presence or absence of genital papillae within the atrium, and the relationships of the male canal and atrial region of the vagina. A phylogenetic basis for these subgenera has not yet been established and inadequate descriptions of the genital atrium for many species continue to obscure their accurate placement. Thus designation at the subgeneric level necessitates careful consideration of the variation inherent in some diagnostic characters and may be limited by the quality of material available for study.

Phylogenetic analyses currently support monophyly for the Tetrabothriidae and provide foundation for the six genera (*Tetrabothrius, Chaetophallus, Trigonocotyle, Anophryocephalus, Strobilocephalus* Baer, 1932 and *Priapocephalus* Nybelin, 1922) presented in the key (Hoberg, 1989). Preliminary comments on taxonomic problems within the genus *Trigonocotyle* have been presented (Hoberg, 1990). The genus *Anophryocephalus* has been revised (Hoberg *et al.*, 1991) and results of phylogenetic reconstruction, historical biogeography and host-parasite co-evolution summarized (Hoberg & Adams, 1992; Hoberg, 1992).

ORDER TETRABOTHRIIDEA BAER, 1954

Diagnosis: Eucestoda. Scolex usually with four muscular bothridia. Vitelline gland compact, anteroventral to ovary. Uterus a transverse tube dorsal to ovary; with single or multiple pores. Parasites of marine homoeotherms. Type-family Tetrabothriidae Linton, 1891.

Family Tetrabothriidae Linton, 1891

Diagnosis: Strobila diminutive to large, proglottids numerous, usually craspedote, wider than long. Scolex lacking rostellum. Bothridia rectangular to round, flat or sucker-like, possessing auriculate muscular appendages; or bothridia and appendages lacking (*Priapocephalus* Nybelin, 1922). Neck present. Genital atrium unilateral, muscular, complex. Cirrus-sac ovoid or elongate. Testes few to numerous. Ovary highly lobate, transversely elongate. Vitelline gland compact, anteroventral to ovary. Uterus a transverse tube; dorsal pore(s) present. Strobilar stage in seabirds and marine mammals. Type-genus *Tetrabothrius* Rudolphi, 1819.

Key to genera

1a. **Scolex with four auriculate bothridia. Genital atrium muscular, well developed, with distinct male atrial canal. Vitelline gland compact, anteroventral to ovary. Genital ducts usually between osmoregulatory canals. Uterine pore single, dorsal, median .... 2a.**

1b. **Scolex massive, globular to conical with collar at base, without bothridia or appendages. Genital ducts ventral to osmoregulatory canals. Cirrus-sac elongate, pyriform. Genital atrium without muscular modification. Testes primarily lateral to female organs. Vitelline gland with tendency to be follicular. Uterine pores dorsal, multiple .... *Priapocephalus* Nybelin, 1922. (Figs 14.1–14.3)**

Diagnosis: Scolex conical to globular, massive, with collar-like basal region; bothridia and auricles absent. Genital pores ventrolateral. Male atrial canal absent. Cirrus-sac elongate to pyriform; cirrus spinose. Testes numerous (> 200) situated laterally. Vas deferens voluminous. Vagina armed with hair-like spines, aperture ventral to cirrus. Vaginal and inner seminal receptacles present. Ovary slightly poral, highly lobate. Vitelline gland branched, anteroventral to ovary. Uterus saccate. In Cetacea. Cosmopolitan. Type-species *P. grandis* Nybelin, 1922.

2a. **Bothridia each with single, laterally directed auricular appendage. Auricles fused anteriorly to form apical complex or organ .... 3.**

2b. **Auricles independent, not fused to form apical organ. Genital atrium muscular or reduced .... 4.**

3a. **Bothridia rectangular. Cirrus-sac ovoid. Testes surround female organs .... *Tetrabothrius* Rudolphi, 1819. (Figs 14.4–14.9)**

(Syns *Tetrabothrium (Eutetrabothrium)* Diesing, 1854; *Amphoterocotyle* Diesing, 1863; *Diplobothrium* Lönnberg, 1891 *nec* van Beneden, 1889; *Prosthecocotyle* Monticelli, 1892; *Bothridiotaenia* Lönnberg, 1896; *Oriana* Leiper & Atkinson, 1914; *Porotaenia* Szpotanska, 1917; *Neotetrabothrius* Nybelin, 1929; *Trigonocotyle* Guiart, 1935 *nec* Baer, 1932; *Paratetrabothrius* Yamaguti, 1940).

Diagnosis: Scolex rectangular with four flat or cup-shaped rectangular bothridia, each with a laterally directed muscular auricular appendage. Auricles fused to form apical organ. Genital pores unilateral; ducts between poral osmoregulatory canals, rarely ventral. Genital atrium deep, muscular. Cirrus-sac not extending beyond osmoregulatory canals. Cirrus enters genital atrium through well defined male atrial canal, dorsal to vaginal aperture. Vaginal and inner seminal receptacles present. Ovary multilobate, extending to osmoregulatory canals. Vitelline gland compact, preovarian, ventral. Uterus saccate, slightly lobate, situated dorsally and extending beyond osmoregulatory canals when gravid. Median dorsal uterine pore present. In marine birds (Podicipediformes, Gaviiformes, Sphenisciformes, Procellariiformes, Pelecaniformes and Charadriiformes; rarely Anseriformes) and mammals (Odontoceti and Mysticeti). Cosmopolitan. Type-species *T. macrocephalus* (Rudolphi, 1810).

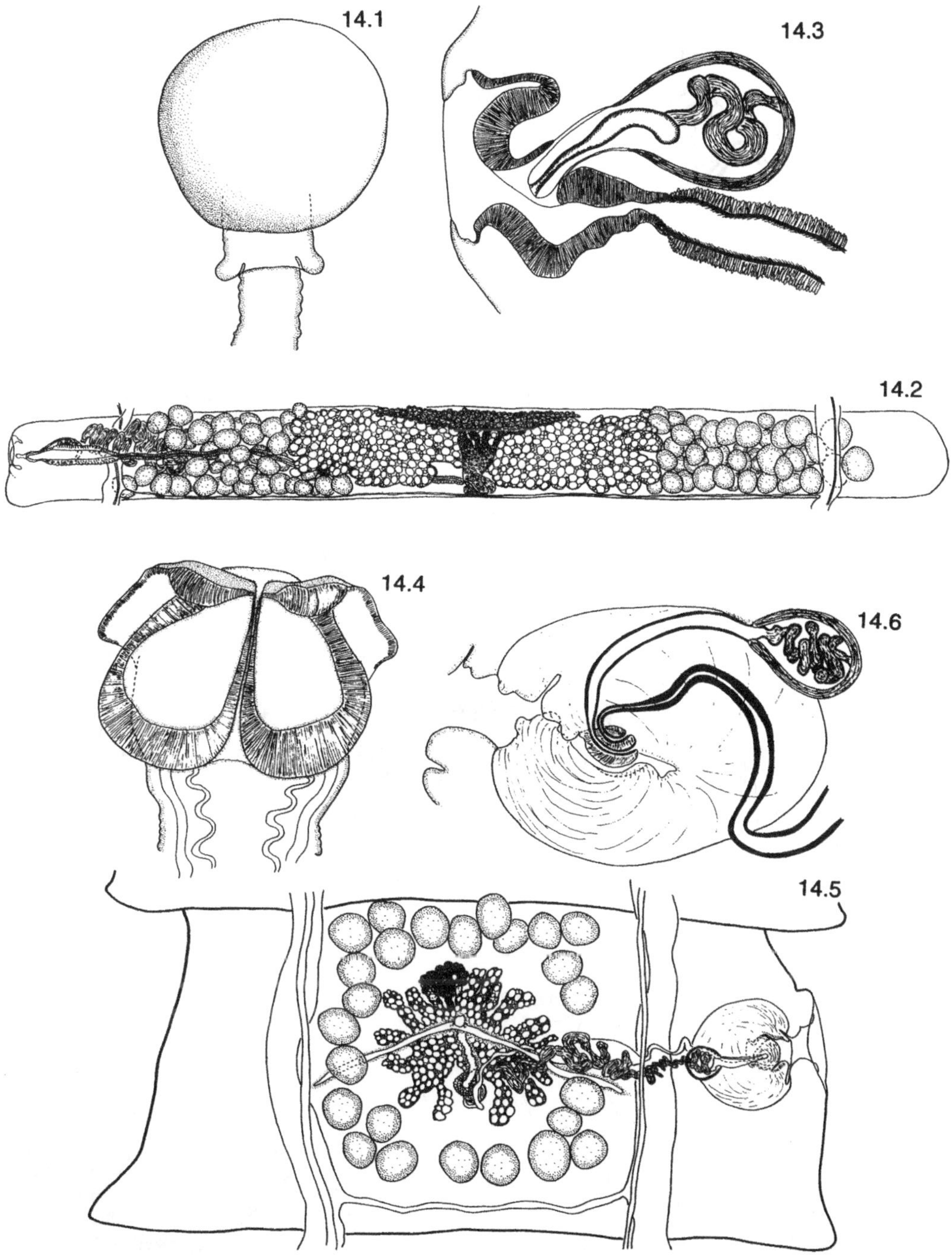

Figs 14.1-14.3 *Priapocephalus eschrichtii* Murav'eva & Treshchev, 1970. 14.1. Scolex. 14.2. Mature proglottid, ventral view. 14.3. Genital atrium, transverse section, view from anterior. **Figs 14.4-14.6** *Tetrabothrius (Culmenamniculus) laccocephalus* Spätlich, 1909. 14.4. Scolex. 14.5. Mature proglottid, dorsal view. 14.6. Genital atrium, transverse section, view from anterior.

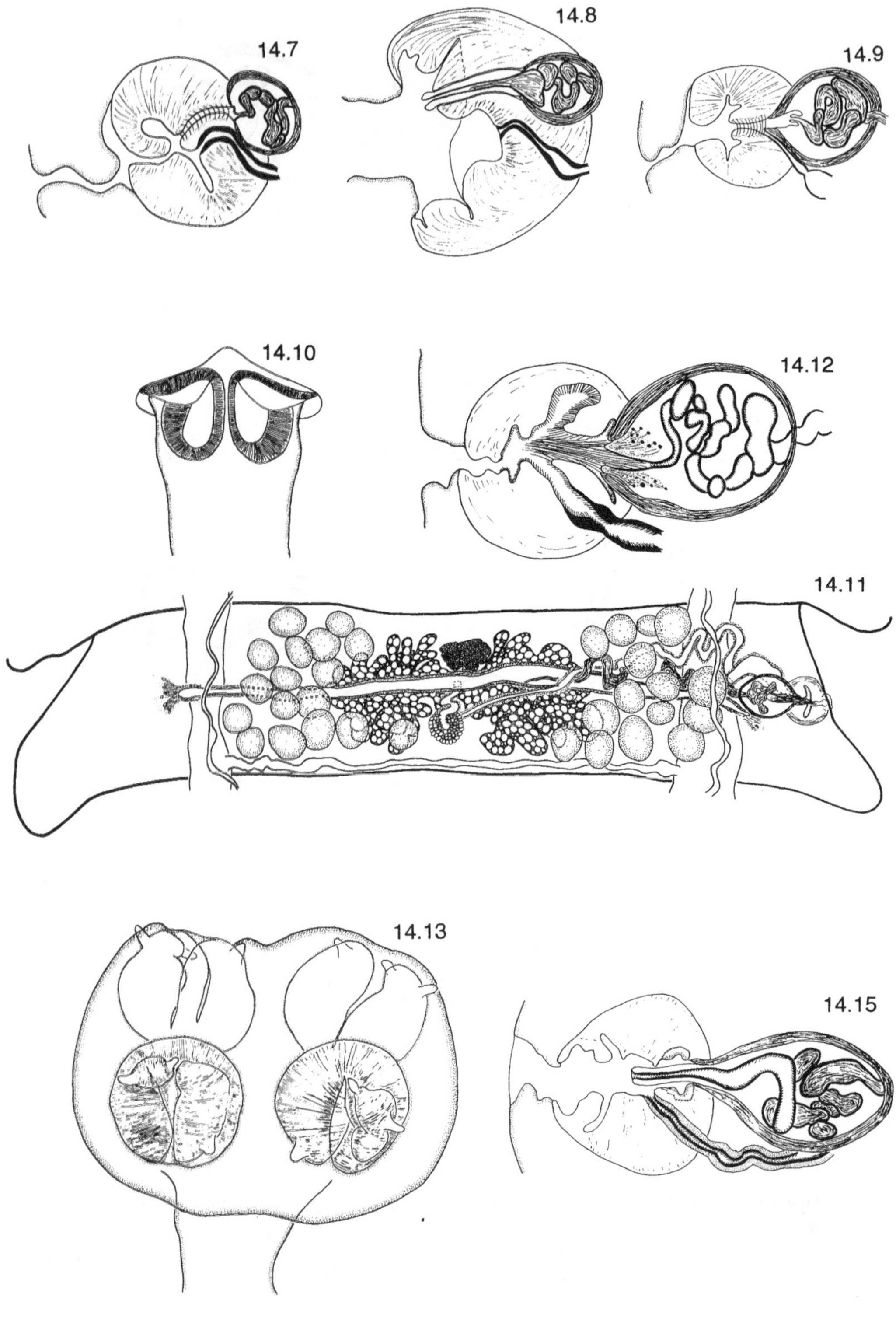
14.7
14.8
14.9
14.10
14.12
14.11
14.13
14.15

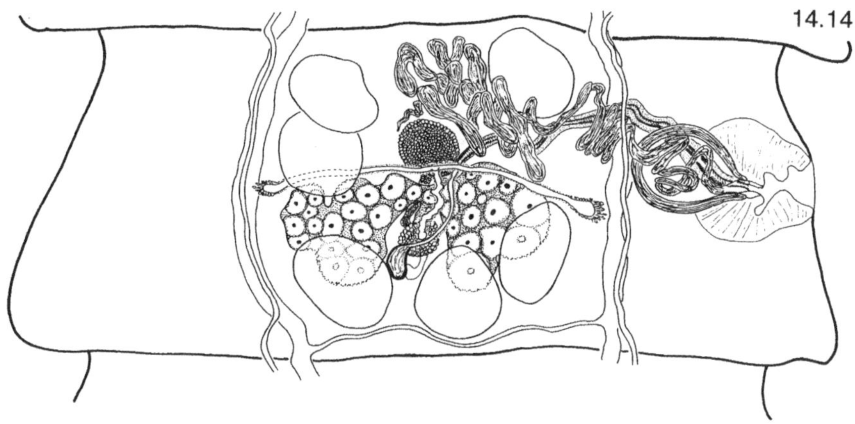

Figs 14.7–14.9 Genital atria among subgenera of *Tetrabothrius* spp. shown in transverse section, viewed from the anterior. 14.7. *Tetrabothrius (Tetrabothrius) macrocephalus* Rudolphi, 1819. Modified from Baer (1954a). 14.8. *Tetrabothrius (Oriana) erostris* Lönnberg, 1896. 14.9. *Tetrabothrius (Neotetrabothrius) eudyptidis* (Lönnberg, 1896). Modified from Baer (1954a).
Figs 14.10–14.12 *Chaetophallus umbrella* (Fuhrmann, 1899). 14.10. Scolex. Modified from Fuhrmann (1921). 14.11. Mature proglottid, dorsal view. 14.12. Genital atrium, transverse section, view from anterior.
Figs 14.13–14.15 *Trigonocotyle sexitesticulae* Hoberg, 1990. 14.13. Scolex. 14.14. Mature proglottid, dorsal view. 14.15. Genital atrium, transverse section, view from anterior. All modified from Hoberg (1990).

Key to subgenera

Ia. Genital atrium lacking distinct genital papilla II
Ib. Genital atrium with prominent genital papilla III
IIa. Male canal and atrial region of vagina independent, apertures separate near centre of atrium, vagina ventral ....
.... ***Tetrabothrius (Tetrabothrius)*** **Rudolphi, 1819. (Fig. 14.7)**
IIb. Male canal and atrial vagina fused distally to form common duct with single aperture opening into genital atrium ....
.... ***Tetrabothrius (Neotetrabothrius)*** **Nybelin, 1929. (Fig. 14.9)**
(Syn. *Tetrabothrius (Uniamniculus)* Murav'eva, 1975.)
IIIa. Apertures of male canal and vagina located on apex of ventrally decurved papilla ....
.... ***Tetrabothrius (Culmenamniculus)*** **Murav'eva, 1975. (Fig. 14.6)**
IIIb. Aperture of male canal on apex of papilla, vaginal aperture separate, ventral at base of papilla ....
.... ***Tetrabothrius (Oriana)*** **Leiper & Atkinson, 1914. (Fig. 14.8)**
(Syn. *Tetrabothrius (Biamniculus)* Murav'eva, 1975.)

3b. Bothridia rounded. Cirrus-sac elongate. Genital atrium spinose. Testes lateral and postovarian ...... ***Chaetophallus*** **Nybelin, 1916. (Figs 14.10–14.12)**
Diagnosis: Scolex rounded to rectangular with four rounded bothridia, each with a laterally directed muscular auricle. Genital pores unilateral, slightly ventral; ducts between osmoregulatory canals. Genital atrium deep, muscular, with

short bristle-like spines lining luminal surface; long hair-like spines extend from near base of cirrus-sac for length of male canal. Cirrus-sac not extending beyond osmoregulatory canals. Vagina entering genital atrium ventral to male canal; dilated proximally to form vaginal seminal receptacle; inner seminal receptacle present. Ovary multilobate, not attaining osmoregulatory canals. Vitelline gland compact, preovarian. Uterus saccate; median dorsal uterine pore present. In Procellariiformes. Southern Hemisphere. Type-species *C. robustus* Nybelin, 1916.

4a. Auricular appendages independent, often weakly muscular. Three auricles on each bothridium. Cirrus-sac elongate. Genital atrium muscular, well developed .... *Trigonocotyle* Baer, 1932. (Figs 14.13–14.15)

Diagnosis: Scolex rectangular to cuboidal, with four robust and highly muscular bothridia, each with three fleshy to muscular auricular appendages. Auricles independent; apical development slight. Genital pores unilateral; ducts between osmoregulatory canals. Genital atrium deep, genital papillae absent; aperture of atrial vagina ventral and adjacent to orifice of male atrial canal. Cirrus-sac elongate to pyriform; cirrus spinose. Testes relatively few, surrounding female gonads. Vagina with spinose atrial region; inner seminal receptacle present. Ovary strongly bilobed, slightly lobate. Vitelline gland compact, preovarian. Uterus saccate, extending beyond osmoregulatory canals when gravid; median pore present. In Odontoceti. Cosmopolitan. Type-species *T. globicephalae* Baer, 1954.

4b. Auricular appendages paired or single. Genital atrium reduced dorsally. Prominent muscular chamber ventral to vaginal aperture. Genital pores ventrolateral .... 5.

5a. Scolex rounded to rectangular. Bothridia sucker-like. Auricular appendages paired, muscular, located along anterior margin of each bothridium. Neck relatively long. Dorsal osmoregulatory canals atrophied. Ovary in posterior region of proglottid, surrounded and overlapped by testes .... *Anophryocephalus* Baylis 1922. (Figs 14.16–14.18)

Diagnosis: Scolex intricate, with four ornate, rounded sucker-like bothridia each with two fused or independent auricular appendages at anterior margin; apertures of bothridia complex, with membranous opercula or enveloped in parenchyma. Dorsal osmoregulatory canals present, atrophied posteriorly. Genital pores unilateral, often ventrolateral; ducts between osmoregulatory canals. Genital atrium with prominent muscular chamber ventral to vaginal aperture and muscular pad slightly dorsal and adjacent to orifice of male canal. Cirrus-sac ovoid to elongate. Testes large, relatively few; dorsal and overlapping ovary. Ovary massive. Vitelline gland compact, ovoid to elongate; anteroventral to ovary. Vaginal and inner seminal receptacles present. Atrial vagina spinose in some species, entering genital atrium ventral to male canal. Uterus a transverse sac, often contained within osmoregulatory canals when gravid; median dorsal pore present. In Phocidae and Otariidae. Holarctic. Type-species *A. anophrys* Baylis, 1922.

5b. Scolex globular, massive (may exceed 5 mm diameter). Bothridia sucker-like, slightly triangular, located at base of hypertrophied apex of scolex. Auricular

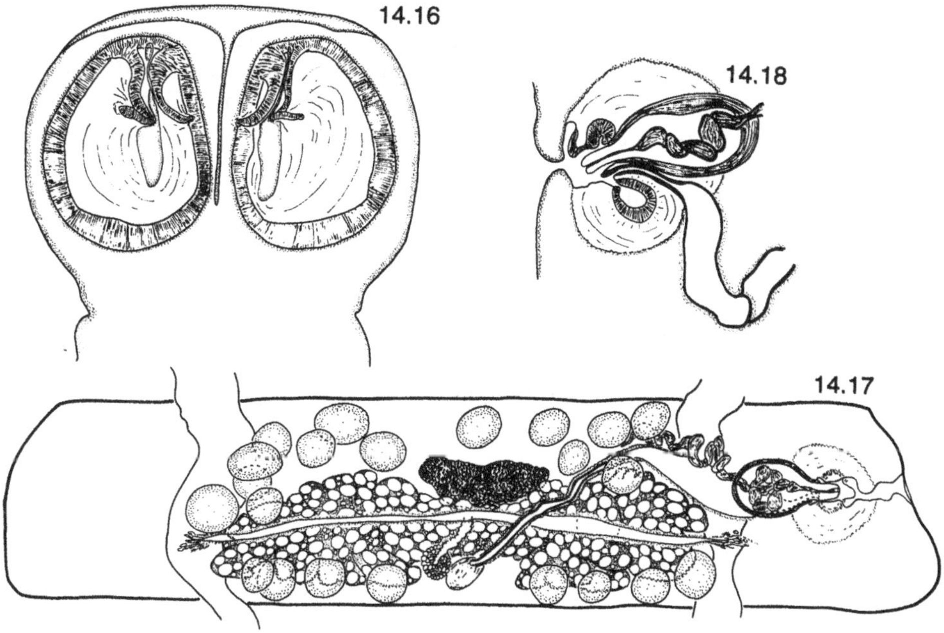

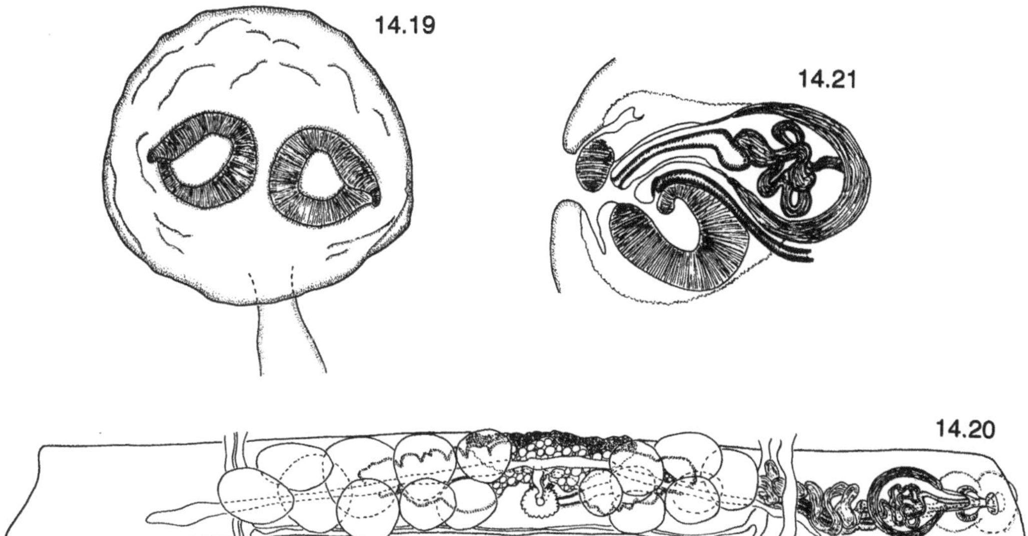

Figs 14.16–14.18 *Anophryocephalus anophrys* Baylis, 1922. 14.16. Scolex. 14.17. Mature proglottid, dorsal view. 14.18. Genital atrium, transverse section, view from anterior. All modified from Hoberg *et al.*, 1991.

Figs 14.19–14.21 *Strobilocephalus triangularis* (Diesing, 1850). 14.19. Scolex. 14.20. Mature proglottid, dorsal view. 14.21. Genital atrium, transverse section, view from anterior.

apppendages single, laterally directed. Cirrus-sac massive, ovoid, thick-walled .... *Strobilocephalus* Baer, 1932. (Figs 14.19–14.21)

Diagnosis: Scolex with hypertrophied apical region and four triangular, muscular, sucker-like, auriculate bothridia located near base. Genital pores ventrolateral; ducts between osmoregulatory canals. Genital atrium with prominent muscular chamber ventral to aperture of vagina; muscular pad situated adjacent to orifice of male canal. Male canal opens into atrium on papilla dorsal to vagina. Cirrus spinose. Testes few, surrounding female gonads. Vagina armed with hair-like spines. Inner seminal receptacle present. Ovary bilobed, transversely elongate, extending to osmoregulatory canals. Vitelline gland compact, anteroventral to ovary. Uterus a transverse sac extending between osmoregulatory canals when gravid; median dorsal pore present. In Odontoceti. Cosmopolitan. Type-species *S. triangularis* (Diesing, 1850).

15 Order Cyclophyllidea van Beneden in Braun, 1900

A. JONES[1], R.A. BRAY[2] & L.F. KHALIL[1]

[1]*International Institute of Parasitology, 395A Hatfield Road, St Albans, Herts AL4 OXU, UK:* [2]*Department of Zoology, The Natural History Museum, Cromwell Road, London SW7 5BD, UK.*

ORDER CYCLOPHYLLIDEA VAN BENEDEN IN BRAUN, 1900

Diagnosis: Scolex usually with four suckers. Rostellum usually present but may be absent, armed or not. Strobila usually with distinct metamerism, usually hermaphroditic. Genital pores usually lateral (ventral in Mesocestoididae). Vitellarium compact, usually posterior to ovary. Uterus variable; may be persistent or replaced by paruterine organ(s) or other derivatives. Parasitic as adults in amphibians, reptiles, birds and mammals.

Key to families

1a. Genital pores median ... Mesocestoididae Perrier, 1897. (Chapter 16.)
1b. Genital pores marginal or submarginal .... 2.

2a. Strobila cylindrical; external segmentation evident only posteriorly. Two (rarely three) testes per proglottid. Uterus replaced by two or more paruterine organs. Adults in reptiles and amphibians .... Nematotaeniidae Lühe, 1910. (Chapter 19.)
2b. Strobila dorsoventrally flattened; external segmentation usually present, rarely absent. Testes number variable. Paruterine organs present or absent .... 3.

3a. Dioecious. Male and female strobila entirely separate or female strobila retains male copulatory apparatus .... Dioecocestidae Southwell, 1930. (Chapter 22.)
3b. Strobila hermaphroditic .... 4.

4a. Mature uterus persistent, a median stem with lateral branches ..... 5.

4b. Mature uterus otherwise if persistent, or replaced by uterine capsules or paruterine organs .... 6.

5a. Armed rostellum present (rarely absent) with two rows of hooks. Eggs with thick-walled embryophore composed of polygonal blocks. Adults in carnivores .... Taeniidae Ludwig, 1886. (Chapter 30.)
5b. Armed rostellum absent. Eggs otherwise. Adults in rodents .... Catenotaeniidae Spasskii, 1950. (Chapter 18.)

6a. Rostellum usually present, rarely absent, typically armed with very numerous minute hammer-shaped hooks. Hooks rarely otherwise. Rostellar spines present or absent. Suckers often spinose. Uterus persistent or replaced by uterine capsules or paruterine organs .... Davaineidae Braun, 1900. (Chapter 24.)
6b. Rostellum present or absent. Rostellar hooks not hammer-shaped. Uterus as above .... 7.

7a. Rostellum absent; suckers unarmed. Proglottids wider than long, including gravid proglottids. Uterus persistent or replaced by uterine capsules or paruterine organs. Eggs often with pyriform apparatus .... Anoplocephalidae Cholodkowsky, 1902. (Chapter 17.)
7b. Rostellum usually present, rarely absent. Uterus as above. Pyriform apparatus absent .... 8.

8a. Typical vagina normally present, rarely absent .... 9.
8b. Typical vagina absent or vestigial, may be functionally replaced by other ducts .... 13.

9a. Uterus replaced by single paruterine organ. Rostellum without saccular sheath. In non-aquatic birds .... Paruterinidae Fuhrmann, 1907. (Chapter 27.)
9b. Uterus persistent or replaced by uterine capsules .... 10.

10a. Rostellum usually present (rarely absent or vestigial); rostellar pouch present .... 11.
10b. Rostellum usually present; rostellar pouch absent .... 12.

11a. Rostellar hooks in one row, exceptionally two. Genitalia single, rarely double. Genital pores usually unilateral, rarely alternating. Testes usually three, rarely more or less. Uterus persistent, sacciform, occasionally reticular .... Hymenolepididae Ariola, 1899. (Chapter 29.)
11b. Rostellar hooks in one or more rows. Genitalia single, rarely double. Genital pores unilateral, alternating or double (rarely). Testes numerous. Uterus persistent or replaced by egg capsules .... Dilepididae Railliet & Henry, 1909. (Chapter 25.)

12a. Genitalia single. Uterus persistent, sacciform. In non-aquatic birds .... Metadilepididae Spasskii, 1959. (Chapter 28.)
12b. Genitalia double. Uterus replaced by uterine capsules containing one or several eggs. In mammals ... Dipylidiidae Stiles, 1896. (Chapter 26.)

13a. Proterogynous. Proglottids hermaphroditic or male and female proglottids alternate regularly .... Progynotaeniidae Fuhrmann, 1936. (Chapter 20.)
13b. Protandrous .... 14.

14a. Secondary supplementary ducts present in female reproductive system .... Amabiliidae Braun, 1900. (Chapter 23.)
14b. Supplementary ducts absent .... Acoleidae Fuhrmann, 1899. (Chapter 21.)

16 Family Mesocestoididae Fuhrmann, 1907

R.L. RAUSCH

University of Washington, Department of Comparative Medicine, T-142 Health Sciences, SB-42, Seattle, Washington 98195, USA.

Introduction

Cestodes of the family Mesocestoididae, placed in two genera, differ from other taxa in the Cyclophyllidea in some important characteristics. The cycle of *Mesocestoides* spp. evidently requires three hosts and is thus unique among those of the cyclophyllideans. The cycle of the type- and only species of *Mesogyna*, the second genus, has not been determined. Among anatomical features, the median ventral position of the genital atrium in both genera and the bipartite vitelline gland in *Mesocestoides* spp. appear also to be unique among the cyclophyllideans.

Chandler (1946) provided a detailed description of the reproductive system of a cestode identified as *Mesocestoides latus* Mueller, 1927, confirmed that a true paruterine organ is present and concluded that the genital atrium is situated on the ventral surface of the proglottid. Byrd & Ward (1943), in a report overlooked by Chandler, interpreted the location of the genital atrium as dorsal, on the grounds that the ovary was more closely apposed to the opposite surface, which therefore would be ventral. They considered also that the longitudinal osmoregulatory canals lay in the same plane and thus were not useful in establishing dorsoventral relationships. The interpretation of Byrd & Ward was later accepted by Chandler (1947), but most subsequent descriptions of *Mesocestoides* spp. state that the genital atrium is ventral. Chertkova & Kosupko (1978, p. 129) correctly described the atrium as being ventral, with the ovary situated more closely to the dorsal surface of the proglottid. In *M. kirbyi* Chandler, 1944, and in cestodes identified as *M. lineatus* (Goeze, 1782), I observed also that the ovary and testes lie in approximately the same dorsal plane; in *M. kirbyi*, the numerous testes clearly had a distribution dorsal to the ventral longitudinal osmoregulatory canals. In both, the genital atrium opened ventrally.

The anatomy of *Mesogyna hepatica* Voge, 1952 has not been fully described, and specimens appropriate for detailed study have not been available.

The systematic arrangement of cestodes now placed in the suborder Mesocestoidata (see Chertkova & Kosupko, 1978) remains controversial. Recognizing that the absence of a paruterine organ in *Mesogyna hepatica* was not compatible with its assignment to the Mesocestoididae as then defined, Voge (1952) emended the family diagnosis to accommodate it. The presence or absence of a paruterine organ usually has been ranked as an important character in discriminating higher categories in the classification of cyclophyllideans. All other taxa possessing that structure have been placed in the superfamily Paruterinoidea Matevosyan, 1962, in the suborder Hymenolepidata Skryabin, 1940 (see Matevosyan, 1969; Matevosyan & Movsesyan, 1970). Consequently, the inclusion of *Mesocestoides* and *Mesogyna* in a single family seems artificial. Anatomically, the mature proglottid of *Mesogyna hepatica* is similar to that of *Mesocestoides* spp., although the vitelline gland consists of a single, transversely elongate organ posterior to the ovary rather than of two separate organs ventral to the ovarian lobes. The sacculate uterus of *Mesogyna hepatica* contains only a few (ca. 25) large (approximately 50 μm in diameter) eggs, which have a thick outer membrane. The affinities of these cestodes will be better understood after more adequate anatomical comparison and elucidation of their cycles.

Chertkova & Kosupko (1977) retained *Mesogyna* in the suborder Mesocestoidata, but established for it the family Mesogynidae Chertkova & Kosupko, 1977. Schmidt (1986) reduced the Mesogynidae to the rank of subfamily in the Mesocestoididae. The systematic arrangement accepted here is as follows: order Cyclophyllidea; suborder Mesocestoidata Skryabin, 1940 (the subordinal category is not universally accepted in the classification of cyclophyllideans); family Mesocestoididae Fuhrmann, 1907; subfamily Mesocestoidinae Lühe, 1894, with one genus, *Mesocestoides* Vaillant, 1863; and the subfamily Mesogyninae Chertkova & Kosupko, 1977, with one genus, *Mesogyna* Voge, 1952.

Attempts to define the cycles of *Mesocestoides* spp. have been unsuccessful, and the identity of the first intermediate host remains unknown. Soldatova (1944) conducted an investigation on a fur-farm where foxes commonly harboured a strobilar stage identified as *M. lineatus*. Oncospheres and developing first-stage larvae (as indicated by form of embryonic hooks) were found in naturally infected free-living mites (Oribatei). Soldatova exposed 3000 mites representing four families to the eggs of the cestode. Early-stage cysticercoids were found in mites of three genera 122–125 days post-exposure. Since the cysticercoids were not fully developed, experimental infection of rodents known to serve as second intermediate host could not be attempted. Tetrathyridia (second larval stage) of *Mesocestoides* spp. have been reported from vertebrates representing about 200 species, including amphibians, reptiles, birds and mammals (Chertkova & Kosupko, 1978). The organism(s) that serve as first intermediate host of *Mesocestoides* spp. must be widespread and abundant.

Nothing is known concerning the intermediate hosts of *Mesogyna hepatica*. Bjotvedt & Hendricks (1982) reported the presence of 'larval stages' of *M. hepatica*, apparently resembling tetrathyridia, in the bile-ducts of kit foxes, *Vulpes macrotis*, probably captured in the State of Arizona. Cestodes of the genus *Mesocestoides* occur commonly in the small intestine of fissiped carnivores, particularly those of the families Canidae and Mustelidae, but records exist from mammals of various other families. Two species have been described from birds.

Specimens usually designated *M. lineatus* have been obtained from man from widely separated localities. Such infections in Japan have been attributed to the consumption (as food) of uncooked tissues of snakes (Kamegai *et al.*, 1967). The wide range of hosts reported for *M. lineatus* suggests that the strobilar stage of that, and perhaps other, species is capable of development in any mammal that ingests tetrathyridia. Such a lack of host-specificity is not typical of cyclophyllideans, but is similar to that of some cestodes of the order Pseudophyllidea, e.g. *Diphyllobothrium dendriticum* (Nitzsch, 1824). So far as is known, *Mesogyna hepatica* is a host-specific cestode of the kit fox, in which it inhabits the bile-ducts, gall-bladder, and possibly also the upper duodenum (Bjotvedt & Hendricks, 1982). That apparent host-specificity is possibly ecological (*sensu* Baer); otherwise, *M. hepatica* would seem in that respect to contrast strongly with *Mesocestoides* spp.

The taxonomy of *Mesocestoides* is complicated by a high degree of non-significant morphological variation, apparently host-induced, which has been particularly evident in cestodes identified by us as *M. lineatus* from carnivores of various species in Alaska. Chertkova & Kosupko (1978) recognized 12 species of *Mesocestoides*, and listed 11 other nominal species as *species inquirendae*. Loos-Frank (1980) concluded that the taxon *M. lineatus* (Goeze, 1782) is not identifiable, and proposed a new name, *M. leptothylacus* Loos-Frank, 1980, for the cestode common in red foxes, *Vulpes vulpes*, in southwestern Germany. The taxon designated *M. corti* Hoeppli, 1925 has been the object of numerous studies, because its designated tetrathyridium is capable of asexual multiplication by longitudinal fission (Specht & Voge, 1965; Eckert *et al.*, 1969; Hess, 1972). Hoeppli (1925) described *M. corti* from a house mouse, *Mus musculus*, which, for various reasons, cannot be considered a natural host. Beaver (1989) pointed out that the type-host must have been abnormal, '. . . presumably lacking natural immune status.' Since cestodes reared from the tetrathyridia that multiply by fission have not been convincingly related to the taxon studied by Hoeppli, Etges (1991) has proposed a new name, *Mesocestoides vogae* Etges, 1991, for the tetrathyridium. *M. corti* was listed as a *species inquirenda* by Chertkova & Kosupko (1978).

At the present time, few species of *Mesocestoides* can be reliably distinguished. Since tetrathyridia often occur in large numbers in small mammals such as shrews and voles, the use of single-source larvae for the experimental infection of a range of carnivores (and perhaps rodents, which are not natural final hosts of these cestodes) would do much toward definition of the extent of host-induced morphological variation. In some regions, rates of infection by tetrathyridia in small mammals range from about 1% to 4% (Gubanov & Fedorov, 1970). Possibly also, morphological differences in tetrathyridia could be correlated with their respective strobilar stages. The following diagnosis of the Mesocestoididae is accurate so far as *Mesocestoides* is concerned. The lack of specimens of *Mesogyna hepatica* of adequate quality prevents an understanding of all morphological details.

Family Mesocestoididae Fuhrmann, 1907

(Syns Mesocestoidae Perrier, 1897; Mesocestoidae Ariola, 1899)

Diagnosis: Strobila thin, numerous proglottids. Scolex with four suckers; rostellum lacking. Mature proglottids wider than long, with relative length increasing posteriad; gravid proglottids longer than wide, or not. Single set of reproductive organs in each proglottid. Cirrus-sac and vaginal duct entering genital atrium on ventral surface of proglottid at midline. Testes usually numerous, arranged in one or two layers dorsally, and forming two lateral groups typically extending throughout length of proglottid. Ovary consisting of two separate or contiguous lobes near posterior margin of proglottid, situated dorsally, in same plane as testes. Two vitelline glands present, ventral to respective ovarian lobes, or vitelline gland single, transversely elongate, on midline posterior to ovary. Uterus ventral, developing as sinuous, median longitudinal tube. Eggs passing posteriad into paruterine organ, or into sac-like uterus lacking paruterine organ. Eggs small and abundant, or large and relatively few. First larval stages unknown; second larval stage a tetrathyridium, occurring in amphibians, reptiles, birds and mammals. Strobilar stage in mammals; rarely in birds. Type-genus *Mesocestoides* Vaillant, 1863.

Key to subfamilies

1a. **With characteristics of the family. Paruterine organ present. Inhabiting small intestine of carnivorous mammals; rarely in birds .... Mesocestoidinae Lühe, 1894.**

(Syns Mesocestoidinae Stiles, 1896; Mesocestoidina Lühe, 1899.)

Diagnosis: Strobila thin, relatively narrow, up to 1.5 m in length. Scolex with four suckers; rostellum lacking. Mature proglottids wider than long, with relative length increasing posteriad; gravid proglottids longer than wide. Reproductive organs single. Cirrus-sac and vaginal duct opening into genital atrium at midline on ventral surface. Testes numerous, in one or two layers, dorsal, usually forming two lateral groups typically extending throughout length of proglottid. Ovary bilobed, lobes separated or contiguous, dorsal at midline near posterior margin of proglottid. Two vitelline glands, ventral to respective ovarian lobes. Uterus ventral, sinuous longitudinal tube at midline. Eggs abundant, passing posteriad into paruterine organ. First larval stage unknown; second larval stage a tetrathyridium, in amphibians, reptiles, birds and mammals. Strobilar stage in carnivorous mammals; rarely in birds. Type and only genus *Mesocestoides* Vaillant, 1863.

Genus *Mesocestoides* Vaillant, 1863 (Figs 16.1–16.3)

(Syns *Monodoridium* Walter, 1866; *Ptychophysa* Hamann, 1885)

Diagnosis: With the characteristics of the subfamily Mesocestoidinae. Type-species *M. ambiguus* Vaillant, 1863.

16.2

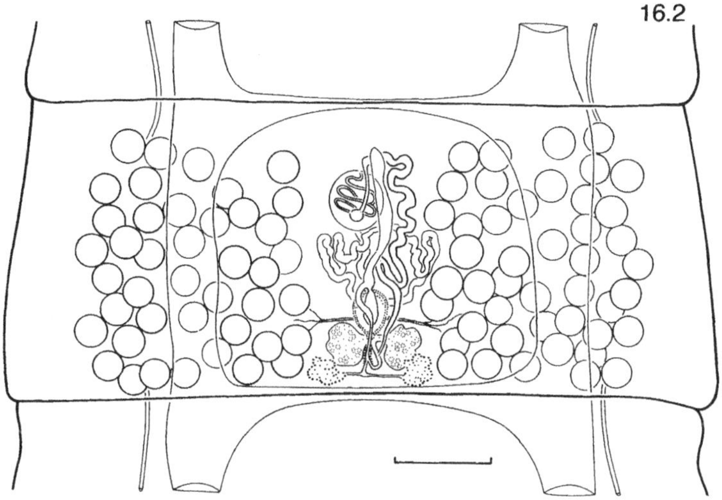

16.3

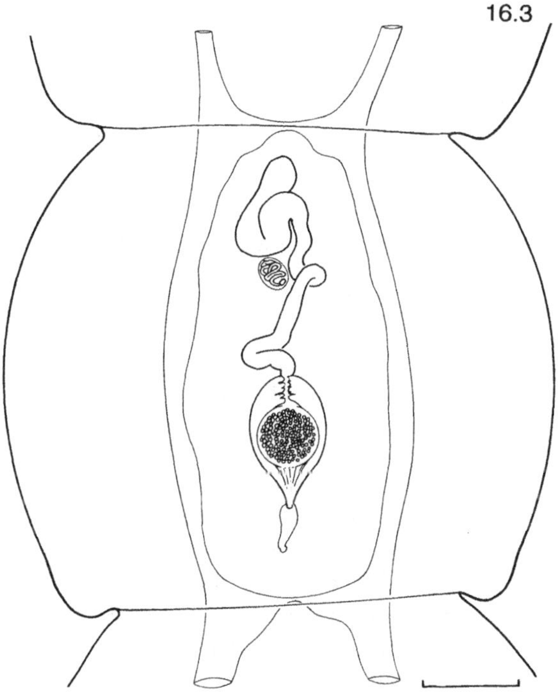

16.1

Fig. 16.1 Scolex of *Mesocestoides kirbyi* Chandler, 1944. Scale 250 μm.
Fig. 16.2 Mature proglottid of *Mesocestoides kirbyi*, ventral view. Scale 250 μm.
Fig. 16.3 Gravid proglottid of *Mesocestoides kirbyi*. Scale 500 μm.

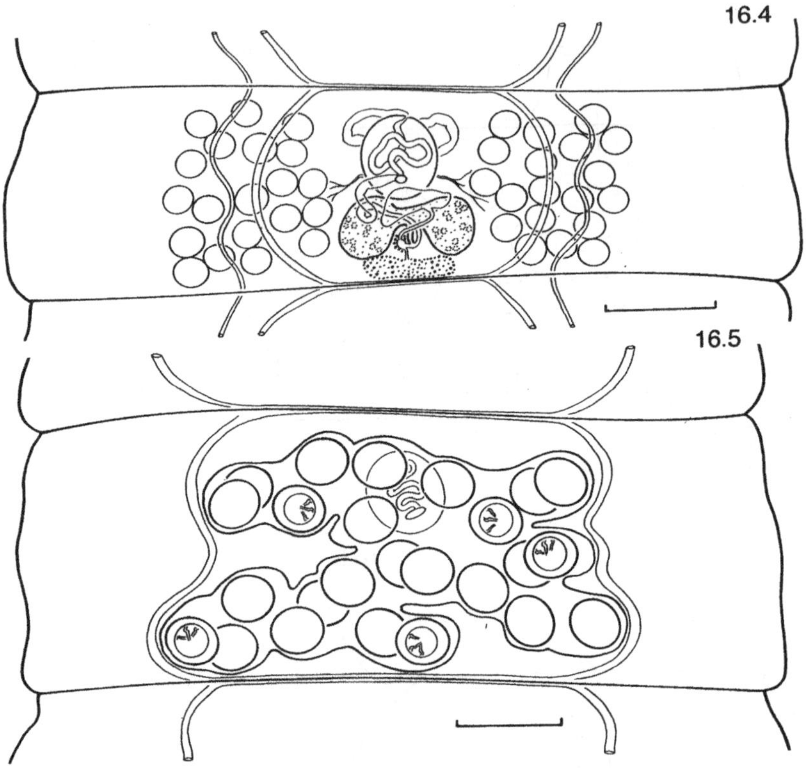

Fig. 16.4 Mature proglottid of *Mesogyna hepatica* Voge, 1952, ventral view (paratype). Scale 100 μm.
Fig. 16.5 Gravid proglottid of *Mesogyna hepatica* (paratype). Scale 100 μm.

1b. **With characteristics of the family. Paruterine organ lacking; uterus sacculate. Known only from gall-bladder and bile-ducts of the kit fox, *Vulpes macrotis* .... Mesogyninae Chertkova & Kosupko, 1977.**

Diagnosis: Strobila thin, small, with ca. 60 proglottids. Scolex with four suckers; rostellum lacking. All proglottids wider than long. Reproductive organs single. Cirrus-sac and vagina opening into genital atrium situated on ventral surface of proglottid at midline. Testes not numerous, dispersed dorsally in parenchyma, tending to form two lateral groups. Ovary bilobed, on midline near posterior margin of proglottid. Vitelline gland transversely elongate, situated posterior to ovary. Uterus sacculate; first visible as small, vesicular structure posterior to genital atrium, enlarging to occupy field between longitudinal osmoregulatory canals. Eggs large, few. First larval stage unknown; second larval stage apparently tetrathyridium-like. Strobilar stage in gall-bladder and bile-ducts of kit fox, *Vulpes macrotis*. Type and only genus *Mesogyna* Voge, 1952.

Genus *Mesogyna* Voge, 1952 (Figs 16.4, 16.5)

Diagnosis: With the characteristics of the subfamily Mesogyninae. Type-species *M. hepatica* Voge, 1952.

17 Family Anoplocephalidae Cholodkovsky, 1902

I. BEVERIDGE

Veterinary Clinical Centre, School of Veterinary Science, University of Melbourne, Princes Highway, Werribee, Victoria 3030, Australia.

Introduction

Traditionally, the cyclophyllidean family Anoplocephalidae has been characterized by the lack of a rostellum and any form of scolex armature. The implication is that this condition is plesiomorphic or primitive when compared with related families in which an armed rostellum is present, such as the Davaineidae, Dilepididae and Hymenolepididae. However, the rostellum and/or its armature can be lost as a secondary phenomenon and there are numerous species or genera within the latter three families which have secondarily lost their rostellum and/or their rostellar armature. Thus, because of the obvious limitations of the principal character defining the family, the four subfamilies of the Anoplocephalidae could conceivably be a polyphyletic assemblage of taxa, and there is considerable morphological evidence to indicate that this is indeed the case (see Voge, 1969). Four types of uterine development have been described within the family, and these form the basis of the division into the subfamilies Anoplocephalinae, Linstowiinae, Inermicapsiferinae and Thysanosomatinae. In the Anoplocephalinae, the uterus is tubular or reticular but persists in gravid proglottids; in the Linstowiinae, eggs are embedded singly in the parenchyma in so-called 'uterine' capsules; in the Inermicapsiferinae eggs are enclosed in groups in 'fibrous' parenchymatous capsules, while in the Thysanosomatinae, eggs are enclosed in paruterine organs which develop as appendages of the uterus.

Lopez-Neyra (1954, 1955) and Joyeux & Baer (1961) drew attention to the remarkable similarity in uterine structure between the various subfamilies of the Anoplocephalidae, Dilepididae and Davaineidae. In each family, there are subfamilies with tubular uteri, with parenchymatous egg capsules, with paruterine organs, or with eggs embedded singly in the parenchyma. Lopez-Neyra (1954, 1955) attempted to translate this observation into a taxonomic system based on uterine structure rather than rostellar armature by transferring many davaineid genera to the Anoplocephalidae, while Joyeux & Baer (1961) transferred a number of linstowiid genera to the Davaineinae. The latter authors hesitated to follow this course to its conclusion because of the immensity of the taxonomic upheavals involved.

Considerable confusion surrounds the relationships of the Inermicapsiferinae. The Inermicapsiferinae resembles very closely those genera of the Davaineinae with several eggs per egg capsule, not only in uterine structure but also in features of the mature proglottid, differing solely in lacking rostellar and sucker armature. Species of the davaineid genus *Baerfainia* Yamaguti, 1955 are known which have a vestigial rostellum (Baer & Fain, 1955), providing an obvious link between the Inermicapsiferinae and the Davaineidae. So close is the morphological resemblance between these two taxa, that Baer & Fain (1955) transferred most of the genera of the Linstowiinae (which at that time included the Inermicapsiferinae) to the Davaineidae.

There are also important biological differences between the anoplocephalid subfamilies supporting the morphological differences, and these have been summarized by Stunkard (1961, 1969). The anoplocephaline members have oribatid mites as intermediate hosts, the linstowiine members have coleopterans as intermediate hosts and the thysanosomatine members have psocopterous insect intermediate hosts. The life-cycles of the inermicapsiferines have not been adequately documented.

More recently, Freeman (1973) proposed a phylogeny of the Cyclophyllidea based on the characters of the metacestode, placing the Anoplocephalinae close to the Hymenolepididae, Davaineidae and Dipylidiinae in possessing a cysticercoid and cercomer, while the Linstowiinae was aligned with the Taeniidae, Amabiliidae and some Dilepidinae, all of which possess an acaudate metacestode. In his view, the Anoplocephalidae is clearly a polyphyletic assemblage, with the Linstowiinae quite distinct from the Anoplocephalinae. He did not consider the two other subfamilies.

In spite of these recent advances in our understanding of cestode phylogeny, it seems unlikely that the time is yet ripe for the taxonomic upheaval foreseen by Joyeux & Baer (1961) or for a classification based on phylogeny. While the phylogenetic relationships of several representative genera from the various anoplocephalid subfamilies are now much clearer, the positions of many others still remain obscure, and it seems reasonable to insist on the clarification of the relationships of these genera before major dislocations occur. In particular (discussed below), *Taufikia* Woodland, 1928, traditionally placed in the Anoplocephalinae, is probably a genus of unarmed dilepidids which has been included in the Anoplocephalidae for convenience. Similarly, the 'rostellar sac' of *Sudarikovina* Spasskii, 1951 (see Hunkeler, 1974) could well represent a vestigial rostellum, and this genus may properly belong in a taxon other than the Anoplocephalinae, probably the Dilepididae or Hymenolepididae. Because of these constraints, the traditional structure of the Anoplocephalidae has been utilized in this key, since the presence of an unarmed rostellum is a useful character in identification, if not in elucidating the phylogeny of cestodes. Hopefully, in the not too distant future, it will be possible to construct a key which more closely reflects the phylogeny of the anoplocephalid cestodes.

The following key to genera of the family Anoplocephalidae has been based, wherever possible on an examination of specimens, preferably types, rather than on the literature. In several instances, the whereabouts of types are unknown and the extant descriptions are too poor to allow the validity of a genus to be assessed. In such cases, there has been no option other than to exclude the genera from consideration. Genera omitted and regarded as *genera inquirenda* are: *Coelodela*, described by Shipley (1900), *Neoaporina* erected by Saxena & Baugh (1973) for *Aporina borealis* Linstow, 1905 and *Spasskofuhrmina* erected by Palladwar & Kalyankar (1989). Specimens were not available for examination of the following genera, but they have been well or adequately described in the literature: *Diuterinotaenia* Gvozdev, 1961, *Witenbergitaenia* Wertheim, Schmidt & Greenberg, 1986, *Thysanotaenia* Beddard, 1911 and *Eranuides* Semenova, 1972.

Generally, a conservative approach has been taken to the genera of each subfamily. Where the generic characters appeared poor or in any way dubious, the genus has not been allowed to stand, although several exceptions were made. Additional work is required in a number of areas to establish the validity or otherwise of particular genera, and these are indicated below in the text.

Illustrations provided are diagrammatic only and are intended to show the principal features of the mature and gravid proglottids. The dorsoventral relationships of various organs have frequently been sacrificed for the sake of clarity, particularly with respect to the position of the uterus, which provides a series of useful taxonomic characters.

Anoplocephalinae

The subfamily Anoplocephalinae is characterized by a persistent uterus which is initially either tubular or reticular, but which finally becomes saccate. Three principal modes of uterine development have been recognized (Spasskii, 1951; Rausch, 1976). In the commonest form of development, the uterine anlage is first visible as a cord-like aggregation of cells extending transversely across the proglottid. With further development, the uterus becomes a well-defined tube of uniform diameter with a clearly detectable lumen. The diameter of the tube increases as eggs appear in its lumen (Figs 17.23, 17.24), and diverticula may appear along its anterior and posterior margins (Fig. 17.26). As the eggs develop, the uterus fills the entire proglottid (Rausch, 1976). A modification of this form of development occurs in species such as *Anoplocephaloides wimerosa*. In this species, the uterine anlage extends to the poral excretory vessels, but only two-thirds of the way across the proglottid to the aporal canals (Fig. 17.15). As the uterus fills, it extends aporally to reach the aporal canals. A more extreme situation occurs in *Crossotaenia baeri* in which the uterine anlage commences as a circular sac anterior to the ovary, and as it fills, extends laterally across the proglottid medulla (Mahon, 1954) (Figs 17.4, 17.5).

In the second form of development, the uterine anlage is first visible as a cord of cells crossing the proglottid. The lateral extremities of the uterus subsequently increase in width and become fenestrated, assuming a reticulate structure, followed by the central part of the uterus. The central part of the uterus subsequently develops anterior and posterior diverticula (Fig. 17.53), and the entire uterus eventually fills the proglottid. This type of development occurs in *Paranoplocephala omphalodes* (see Rausch, 1976).

In a third form of development, the uterine anlage appears as a broad cellular band anterior to the ovary and spreads to cover the ventral surface of the medulla. The uterus subsequently becomes reticulate, forming a uniform, complex reticulum across the entire ventral aspect of the proglottid. As eggs fill the uterus, the reticulate appearance is lost and the uterus appears saccate. Such development occurs in the genera *Diandrya* (Fig. 17.58) and *Moniezia* (Fig. 17.59).

These types of uterine development form a continuum from a simple tubular uterus to a complex reticulated type, but in each case the uterus is persistent in gravid proglottids, immediately distinguishing them from the Linstowiinae, Thysanosomatinae and Inermicapsiferinae.

While the various types of uterine development appear to be distinctive and the subdivision of the Anoplocephalidae (*sensu* Spasskii, 1951) into the Anoplocephalinae with a tubular uterus and the Monieziinae with a reticular uterus appears as a perfectly logical consequence, the distinction of the two modes of uterine development in practice remains difficult. Rausch (1976) showed that the uterine development in the genera *Andrya* and *Paranoplocephala* from rodents had been completely misunderstood by numerous earlier workers, and that while some species of *Paranoplocephala* had a slightly reticulated uterus, other species included in the genus had a simple tubular uterus. He resurrected the genus *Anoplocephaloides* Baer, 1923 for the species with a simple tubular uterus, but (Rausch, 1976) remained uncertain as to whether there were any major differences between the genera *Andrya* and *Paranoplocephala* in terms of pattern of uterine development. However, Tenora, Murai & Vaucher (1984, 1986) have subsequently reported that early uterine development differs in the two genera. Further detailed study is required to resolve these and other uncertainties, but for the present it has been decided not to utilize the Monieziinae at any level, following Schmidt (1986), until detailed studies on the uterine development of a number of other genera have been made. Particular problems remain with genera such as *Lentiella* and *Perutaenia* which have been described from extremely poor specimens, and in which the detail of the development of the uterus cannot be determined from the only existing specimens.

Uterine development in *Taufikia* Woodland, 1928 (syns *Gidhaia* Johri, 1934, *Neophronia* Saxena, 1967) is not typical of the Anoplocephalinae. In this genus, the uterus develops as an outgrowth of the uterine duct, with three main branches (Fig. 17.6), gradually branching laterally and anteriorly throughout the proglottid (Fig. 17.7). The mature uterus is branched but not reticulated and resembles representatives of the Dilepididae (see Bona, 1975) rather than the Anoplocephalinae. This genus may in fact belong to the Dilepididae, but without life-cycle studies, no definitive decision can be made.

The pyriform apparatus surrounding the embryophore figures prominently in definitions of the Anoplocephalinae. Typically characterized by two elongate horns (Spasskii, 1951), more recent studies have revealed a much greater diversity in its structure (Freeman, 1952; Rausch & Maser, 1977; Beveridge, 1976, 1985). From a taxonomic point of view, its presence may be difficult to establish, since in certain species, the pyriform apparatus only develops after eggs have been voided (Beveridge, 1976). Nevertheless, the pyriform apparatus is present in virtually all genera of Anoplocephalinae occurring in mammals, but is absent from most of the genera present in birds except for *Stringopotaenia* and possibly *Paronia* in which it is described as 'vestigial' by Burt (1939b). A detailed study of unfixed material would be needed to see if these differences, apparent from the literature, are real or not.

A major problem associated with the taxonomy of anoplocephalid cestodes is genera of which the scolex is unknown and, therefore, the lack of armature is not certain. *Aporina* Fuhrmann, 1902 has been included in spite of the lack of description of a scolex, primarily because of the unique structure of the uterus. However, the genus has been restricted to the type species. *Crossotaenia* Mahon, 1954 was initially described in the absence of a scolex, but new specimens examined in the Natural History Museum collections (BMNH 1984.6.4.4) indicate that the scolex is indeed unarmed and the genus belongs to the Anoplocephalinae. Other genera in which the scolex has been described subsequent to the erection of the taxon are *Pulluterina* Smithers, 1954 (see Weekes, 1981) and *Triuterina* Fuhrmann, 1921 (see Joyeux, Gendre & Baer, 1928; Jones, 1982). *Killigrewia* Meggitt, 1927, described without a scolex and from an unknown host is also retained for reasons discussed below. Other genera such as *Francolina* Capoor, Sawada, Bhalya & Rastogi, 1987 have been erroneously assigned to the Anoplocephalidae in spite of possessing an unarmed rostellum and obviously belong to another family, probably the Paruterinidae. By contrast, *Himalaya* Malhotra, Sawada & Capoor, 1983, described as having an unarmed rostellum, is clearly *Mosgovoyia pectinata*.

Several genera admitted as valid by Schmidt (1986) have been reduced to synonymy following examination of type or other specimens. *Perutaenia* Parra, 1953 is synonymized with *Monoecocestus* Beddard, 1914 since the vagina opens to the genital atrium anterior to the cirrus-sac. The type and only specimens are extremely macerated and few internal features can be seen clearly, particularly the uterus. Thus, new specimens are required to confirm the relationships of this species, but the position of the vagina anterior to the cirrus-sac (Parra, 1953) and the posterior distribution of the testes suggest that this species belongs within *Monoecocestus*.

Lentiella is also a synonym of *Monoecocestus*. The testes lie in the posterior part of the medulla and the vagina, which is only clearly visible in two proglottids, opens to the genital atrium anterior to the cirrus-sac, in contradistinction to Rego's (1964) description. The uterus is prominently lobed and appears to be slightly reticulated; however, in the only specimen available, from the collections of the Instituto Oswaldo Cruz (IOC 29.779), the proglottids in which the critical stages of uterine development occur are damaged. More material is required for a detailed study of uterine development, and pending this, the genus is made a synonym of *Monoecocestus*.

Eranuides Semenova, 1972 was described without a scolex and placed in the Anoplocephalinae because it had paired 'sac-like' uteri. The drawings associated with the description (Semenova, 1972, Fig. 5) in fact show a reticulated uterus and the genus is, therefore, a synonym of *Moniezia*.

Fuhrmannella Baer, 1925 was synonymized with *Moniezia* by Baylis (1935) and the decision is upheld here following a study of relevant specimens. *Flabelloskrjabinia* was erected by Spasskii (1951) for *Anoplocephala tapirus* Chin, 1938 found in a tapir from the Philippines (Chin, 1938). Chin (1938) stated that the testes were mainly in the aporal region of the proglottid and this is shown in his Fig. 2. His Fig. 3, however, shows a large number of testes in the poral region. The whereabouts of Chin's specimens are unknown; however, examination of new material from the same host species in the collections of the International Institute of Parasitology, St Albans, UK (IIP 1700) indicate that Chin's description rather than his illustration is correct and that virtually all the testes are aporal. Thus, *A. tapirus* is allied to *Anoplocephaloides* rather than *Anoplocephala* as suggested by Spasskii (1951). In addition, Sawada & Tongchai (1966) described a new cestode, *Paranoplocephala indicata*, from the Malaysian tapir, *Tapirus indicus* which, because the testes were aporal, they assigned to *Paranoplocephala*. Following Rausch's (1976) resurrection of *Anoplocephaloides*, *P. indicus* would appear to belong here. Rausch (1976) considered *Flabelloskrjabinia* distinct from *Anoplocephaloides* because of the distribution of testes shown in Chin's Fig. 3. If Chin's Fig. 3 is incorrect, *Flabelloskrjabinia* would fall as a synonym of *Anoplocephaloides*. Through the kindness of Dr Sawada, a paratype of *P. indicata* was also examined and conforms closely with the IIP material. There is probably only one species of cestode in tapirs at present and its name is *Anoplocephaloides tapirus* (syn. *P. indicata*). Thus, *Anoplocephaloides* would contain two species from perissodactyls, *A. mamillana* and *A. tapirus*. As indicated by Rausch (1976) and Genov & Georgiev (1988), *Anoplocephaloides* currently contains several groups of species, each of which may eventually warrant generic rank. If so, the generic name *Flabelloskrjabinia* would have priority as a name for the taxa parasitic in perissodactyls. Pending further study, *Flabelloskrjabinia* is here retained, provisionally, as a synonym of *Anoplocephaloides*.

Species of *Anoplocephaloides* parasitizing lagomorphs have recently been removed to a new genus *Leporidotaenia* Genov, Murai, Georgiev & Harris, 1990 based on the presence of a large, armed cirrus with a retractor muscle, a 'spinose' tegument, a protrusible genital atrium and the vagina posterior to the cirrus-sac (Genov *et al.*, 1990). Their arrangement of species differs from that utilized by Rausch (1976) and can be criticized from the point of view that most of the characters selected are relative and gradations are possible. Baer (1927) utilized the degree of development of the cirrus-sac and its associated structures in separating *Prototaenia* from *Bertiella*, a separation which has not stood the test of time. *Leporidotaenia* may, therefore, prove to be a valid taxon along with the other species groups defined by Genov & Georgiev (1988), but pending corroborating evidence it is maintained provisionally as a synonym of *Anoplocephaloides*.

In contrast to Schmidt (1986), *Gallegoides* is retained as a valid genus based on the descriptions by Tenora & Mas-Coma (1978) and Feliu *et al.* (1984) as well as examination of type specimens. *Gallegoides* differs significantly from *Anoplocephaloides* in that the uterus terminates anterior to the cirrus sac, the testes lying posterior to the uterus. A similar difference exists between *Mosgovoyia* and *Cittotaenia* when compared with *Pseudocittotaenia*, all the latter genera having paired genitalia. Beveridge (1978) hypothesized that *Cittotaenia* was derived from species of *Anoplocephaloides* by doubling of the genitalia; *Gallegoides* would likewise be a suitable ancestor for *Pseudocittotaenia*.

A number of anoplocephaline genera occur in birds rather than mammals. Several genera occur in Psittaciformes, namely *Triuterina*, recently redescribed by Jones (1982), *Biporouterina*, erected by Burt (1973), *Hemiparonia*, recently redefined by Schmidt (1972), *Paronia*, studied extensively by Burt (1939b), *Pulluterina*, recently redescribed by Weekes (1981) and *Stringopotaenia*, described by Beveridge (1978). The taxonomy of the remaining anoplocephaline genera present in psittaciform and non-psittaciform birds has been the source of considerable controversy, particularly the genera *Aporina*, *Taufikia* and *Killigrewia*. *Aporina* was erected by Fuhrmann (1902a) for a cestode from a parrot from Brazil. The scolex was not described, but the uterus is characteristic, being single, arched anterior to the female genital complex, with two lateral branches extending anteriorly. Genital pores are absent. Although several species have been added to the genus subsequently (see Spasskii, 1951, Yamaguti, 1959, Schmidt, 1986), none has the characteristic uterus or lacks genital pores, and the genus is here considered to be restricted to the type species. Validation of its position must await a redescription based on specimens with scoleces.

The genus *Killigrewia* was established by Meggitt (1927b) for a cestode, *K. frivola*, without a scolex, taken from an unknown host in Egypt. Fuhrmann (1932) reduced the genus to synonymy with *Aporina*, but Johri (1934) reinstated *Killigrewia*, redescribing what he believed to be *K. frivola* from a pigeon in India, while Saxena & Baugh (1973) have also provided evidence that the two are not synonymous, since the structure of the uterus differs and a genital pore is absent in *Aporina* but present in *Killigrewia*. They transferred *Aporina delafondi* (Railliet, 1892) to *Killigrewia*, thus conserving an otherwise dubious genus. The type specimens of *K. frivola* (BMNH 1934.5.11.3–4) were re-examined and are re-illustrated below (Fig. 17.41). The label associated with the types gives the host as 'probably *Neophron percnopterus*'. *Killigrewia* appears to be a valid genus and Saxena & Baugh (1973) appear to have been correct in their decision to transfer *A. delafondi* to *Killigrewia*. *Nepalesia* Sharma, 1943, is treated as a synonym of *Killigrewia* following Schmidt (1986). *Columbia* Srivastava & Capoor, 1966 is also a synonym of *Killigrewia* according to Capoor & Srivastava (1967) who had mistaken a seminal receptacle for a paruterine organ in their original description of *Columbia*.

The second species of *Killigrewia* described by Meggitt (1927b), *K. pamelae*, possesses a scolex and was incompletely described by this author. Examination of the description indicated that the uterus is similar to *Taufikia* and it is, therefore, transferred to the latter genus as *T. pamelae* comb. nov.

Taufikia was described from a vulture by Woodland (1928). Spasskii (1951) implied (p. 197) that the genus belonged to the Dilepididae, and as mentioned above, examination of the development of the uterus strongly supports this view. However, lacking life-cycle data, it is impossible to make a definitive decision. It is here retained only provisionally within the Anoplocephalinae.

Gidhaia Johri, 1934 resembles *Taufikia* in uterine development, but the testes in *Taufikia* are mainly anterior to the female genitalia, while in *Gidhaia* many extend laterally to the female genitalia as well. *Neophronia* Saxena, 1967 is a synonym of *Gidhaia*. Dollfus (1963) employed a definition of *Taufikia*, with *Gidhaia* as its synonym, which clearly allowed for a lateral distribution of the testes. The two genera differ slightly in testis distribution, but the patterns merge, as they do with *Neophronia*. The differences in testis distribution noted by Saxena (1969) are valid at the species level, but their generic value has yet to be determined. Consequently, *Gidhaia* and *Neophronia* are synonymous with *Taufikia*.

Biporouterina Burt, 1973, with its single species, *B. psittaculae*, was distinguished from the related genera *Hemiparonia* and *Pulluterina* by the spur-like anterior extension of the uterus in mature proglottids, a character it shared with *Triuterina*. The presence of so-called 'uterine pores', from which the generic name is derived, is not considered significant. Baylis (1927) thought that uterine pores were a significant diagnostic feature in *Bertiella anapolytica* from rodents but more recent studies (Beveridge, 1985) have not supported this notion. Burt (1973) distinguished *Biporouterina* from *Triuterina* by the presence of polar processes on the eggs of *Biporouterina* and unspecified differences in the osmoregulatory systems. Neither character seems to be of generic significance and the two taxa have, therefore, been merged.

Burt (1939b) also described a species of *Paronia*, *P. calcaruterina*, from the red-vented bulbul, *Molpastes haemorrhous*, which differed from congeners in the presence of a spur on the anteromedian aspect of the uterus, from whence the specific name was derived. A new genus, *Bulbultaenia*, is, therefore, proposed to accommodate *P. calcaruterina*. *P. calcaruterina* and *B. psittaculae* differ only in having paired rather than single genitalia, thus resembling the morphological similarities seen between the genera *Hemiparonia* – *Paronia* and *Pulluterina* – *Stringopotaenia*, all from Psittaciformes, the first of each pair having single genitalia, and the second having paired genitalia.

Moniezoides rouxi Fuhrmann, 1918 was placed in an independent genus because of morphological peculiarities of the suckers and the presence of four glands on the scolex. The uterus has been described and illustrated as being single. However, examination of the type specimens, which are in poor condition, showed apparently paired uteri. Until new material is available, it is impossible to resolve this apparent dichotomy and reliance has been placed on the characters of the scolex in the interim.

The genus *Progamotaenia* Nybelin, 1917 *sensu* Beveridge, 1976 was subdivided into four genera by Schmidt (1986). *Progamotaenia* and *Adelataenia* Beveridge, 1976 had testes in two lateral groups, with either paired uteri or a single uterus while *Fuhrmannodes* Strand, 1942 and *Wallabicestus* Schmidt, 1975, each had a single band of testes but with two uteri and a single uterus, respectively. The division of testes into one or two groups is unreliable as a specific character in several species of *Progamotaenia* (see Beveridge, 1976) and a generic division on this basis is, therefore, not advisable. *Adelataenia* Beveridge, 1976 is a *lapsus* in Schmidt (1986), and should have been *Adelataenia* gen. nov. The division of the genus into those species with a single uterus and those with paired uteri is straightforward at first sight, but genetic evidence on speciation within this taxon (Baverstock *et al.*, 1985b) suggests that the presence of a single uterus in *P. effigia* Beveridge, 1976 is a secondary reversion to a plesiomorphic character state. This being the case, a more conservative approach has been adopted here and the various genera mentioned have been included as synonyms of *Progamotaenia*. Schmidt's (1986) analysis may prove to be correct, but more evidence is needed.

Neoctenotaenia Tenora, 1976 differs from *Mosgovoyia* Spasskii, 1951 in the structure of the osmoregulatory system, which Tenora (1976) believed was a conservative taxonomic character warranting generic rank. By contrast, the species of *Mosgovoyia* and *Neoctenotaenia* share a number of characters in the distribution of the testes and the anatomy of the genital ducts (Beveridge, 1978) and are here considered congeneric.

As defined here, the genus *Moniezia* consists of a diverse collection of cestodes. Apart from the well-known representatives from ruminants and suids, there is a single species (*M. bequarti*) in rodents, as well as one (*M. rhea*) in ratite birds and one (*M. rugosa*) in primates. *M. rhea* differs from congeners in that Mehlis' gland is poorly developed (Beveridge, 1978), though this does not appear to be a useful generic character. The genus clearly warrants a detailed revision, but its subdivision at present would be premature.

Palaia Shipley, 1900 is unrecognizable and is treated as a *nomen nudum*.

Shindeia Gaikwad & Shinde, 1985 is unrecognizable and is almost certainly not an anoplocephalid. It is treated here as a *genus incertae sedis*.

Linstowiinae

In members of the Linstowiinae, the uterus is ephemeral, either persisting briefly or not being detectable at any stage in developing proglottids. The uterine epithelium forms a thin outer envelope around individual eggs (Conn, 1985) which are embedded singly in the parenchyma of the proglottid.

The genera within the Linstowiinae have been in a constant state of flux since Spasskii (1951) attempted to subdivide the large genus *Oochoristica* Lühe, 1898. His approach was rejected entirely by Della Santa (1956), while both Dollfus (1957) and Stunkard (1961) have pointed out that some genera erected by Spasskii (1951) could have been usefully retained.

In this key, where reliable morphological criteria for distinguishing genera, such as the distribution of testes, presence of lappets on the suckers and the presence of a velum on proglottids exist, then the genera have been maintained as valid. Genera or subgenera based on dubious criteria have been eliminated.

Spasskii (1987a,b, 1988a,b) erected a series of new linstowiid genera (*Peramelinia, Pseudolinstowia, Sharpilia, Timorenia, Hickmania, Priodontia, Linstoparonia, Mangustella, Vasoramia*) which again appear to be based on host distribution rather than reliable morphological characteristics. None is accepted in this review.

One major source of disagreement has been over the validity of the genus *Mathevotaenia* Akhumyan, 1946 which was described as having craspedote proglottids, rather than acraspedote ones as in *Oochoristica* and being parasitic in mammals rather than reptiles. Spasskii (1951) was criticized for basing his taxonomy on the hosts rather than on the morphology of the cestodes themselves

(Dollfus, 1954). Some species of *Oochoristica* have been described which are acraspedote but occur in mammals (see Beveridge, 1977). As well, the monotypic genus *Sinaiotaenia*, parasitic in gerbillid rodents, has acraspedote proglottids. In this key, the presence of craspedote or acraspedote proglottids has been accepted as a valid morphological criterion, and the genus *Mathevotaenia* has been allowed to stand. However, using this criterion, *Oochoristica* contains cestodes from both mammals and reptiles. The presence or absence of a seminal receptacle is not considered an important generic character, in contradistinction to Spasskii (1951), and has not been used as a diagnostic character.

Oschmarenia Spasskii, 1951 was erected to contain *Thysanotaenia incognita* Meggitt, 1927 described from a wallaby in a zoo in Rangoon, Burma. The type specimens were re-examined by Beveridge (1976) who noted that they appeared to be specimens of the davaineid cestode genus *Calostaurus* which had lost their rostellar hooks, a common phenomenon with poorly preserved specimens of this genus. Consequently, *Oschmarenia* and its nominal subgenus, *Oschmarenia*, are excluded from the Anoplocephalidae and are considered *genera inquirenda*. Spasskii (1951) erected two additional subgenera within *Oschmarenia*, namely *Inversia* and *Morosovella*. The type specimens of the relevant species in the United States National Museum Helminth Collection (USNMHC) have been examined and both *O. (I.) oklahomensis* (Peery, 1939) (USNMHC 93306) and *O. (M.) mephitis* (Skinker, 1935) (USNMHC 39861) have slightly craspedote proglottids, although this character is not mentioned in any published descriptions. Consequently, both subgenera become synonyms of *Mathevotaenia*.

The genus *Atriotaenia* Sandground, 1926 has been retained. It is based on the presence of extremely well-developed radial musculature surrounding a prominent genital atrium and the entry of the genital ducts through the posterior wall of the atrium. Joyeux (1945) believed that the presence of muscles around the genital atrium was not a reliable character. However, when taken in conjunction with the site at which the ducts enter an enlarged atrium (posterior), it appears to be a useful secondary character. Members of the subgenus *Ershovia* Spasskii, 1951 also possess a well developed genital atrium and associated musculature (see Chandler, 1942c, Joyeux & Baer, 1936b). Consequently the subgenus, with its two species *A. procyonis* (Chandler, 1942) and *A. incisa* (Railliet, 1899), becomes a synonym of *Atriotaenia*.

Atriotaenia baltazardi, described more recently by Quentin (1967) has the genital ducts opening directly into the genital atrium and is made a synonym of *Quentinotaenia mesovitellinica* Rego, 1967 (see Quentin, this volume). Spasskii (1969b), in a conference proceedings, erected a new genus, *Quentinia*, within the Anoplocephalidae for *A. baltazardi*. This genus now becomes a member of the Catenotaeniidae.

Opossumia Spasskii, 1951 was removed as a subgenus of *Linstowia* by Beveridge (1983) and is here included as a synonym of *Mathevotaenia*.

Schizorchodes Bienek & Grundmann, 1973, from rodents, was established as a new genus within the Anoplocephalinae. Examination of a paratype specimen indicates, however, that the genus belongs to the Linstowiinae since the uterus is ephemeral and the eggs are embedded singly within the parenchyma. The genus was illustrated as having the testes distributed in two separate groups posterior to the female genitalia. In the material studied, the testes were arranged either in two groups or a single group, both arrangements occurring within a single strobila. Because of this, *Schizorchodes* is treated as a synonym of *Mathevotaenia*. *Witenbergitaenia* Wertheim, Schmidt & Greenberg, 1986 was also erected for a cestode from gerbillid rodents and was characterized by having the testes distributed in two lateral groups posterior to the female genitalia. Since the new genus was established for a single specimen, the reliability of the generic characteristic needs to be established from a greater range of material given the variability seen in *Schizorchodes*. *Witenbergitaenia* is here treated provisionally as a valid genus pending the availability of additional material.

There is some confusion in the literature over the relative positions of the genital ducts in *Panceriella*. Southwell (1926) showed the vagina entering the genital atrium anterior to the cirrus-sac, while Baer (1927) showed the reverse orientation. The material upon which both workers based their descriptions has been re-examined. Baer's specimens are poorly stained, while those of Southwell clearly show the vagina anterior to the cirrus-sac. This character thus becomes an important means of distinguishing *Panceriella*.

Cleberia was described as an anoplocephalid, in spite of the fact that the scolex was lacking (Rego, 1967a), and was placed in the Linstowiinae. Through the kindness of Dr Rego, a specimen was made available for examination. The eggs develop in capsules, but no fully gravid proglottids are present and it is not clear whether the numerous cells in each capsule will form a single egg or several eggs. If they form several eggs, then the genus belongs to the Dilepididae rather than the Anoplocephalidae. As a scolex is lacking, the genus is considered *inquirendum*.

Doublesetina Srivastava & Srivastav, 1989 is treated here as a *genus inquirendum*. It is poorly described (see Srivastava & Srivastav, 1989) and cannot be confidently allocated to a family.

Multicapsiferina Fuhrmann, 1922 is a genus of uncertain family and subfamily status, treated here as a *genus incertae sedis*. *Zschokkea* Fuhrmann, 1902 and *Zschokkeella* Ransom, 1909, both preoccupied, are synonyms of *Multicapsiferina*.

Thysanosomatinae

Genera of the subfamily are characterized by the development of paruterine organs adjacent to or attached to the uterus and the movement of eggs from the uterus into the paruterine organs. The development of paruterine organs in the subfamily has been described in considerable detail by Spasskii (1951). Eggs first enter a tubular uterus. In *Thysaniezia*, *Wyominia* and *Thysanosoma*, numerous paruterine organs with fibrous walls develop as exvaginations of the uterus. Several eggs pass into each paruterine organ and the paruterine organs remain attached by one pole to the uterus. In *Stilesia*, the paruterine bodies commence as solid structures adjacent to the uterus. They then become hollow. The uterus disappears as eggs pass into the two paruterine organs present in each proglottid.

In *Avitellina*, a single paruterine organ is present, but it contains paruterine capsules, each with several eggs.

The family Avitellinidae was erected by Skryabin (1926) for all anoplocephalid genera possessing paruterine organs but Stunkard (1961) pointed out that the rules of nomenclature had not been followed in the selection of the name and that Thysanosomatinae was the correct name. Stunkard's (1961) usage is followed here.

Most genera of the family are well characterized. *Wyominia* is an exception in that it was poorly described by Scott (1941). Baer (1954b) re-examined Scott's specimens and provided a more detailed description, which was utilized by Schmidt (1986) in his key to genera. However, the type and only species *W. tetoni* still warrants redescription as the redescription by Baer (1954b) is inadequate by contemporary standards.

Inermicapsiferinae

In the Inermicapsiferinae, the uterus is again ephemeral, but the eggs are finally enclosed within fibrous or parenchymatous capsules, each containing several eggs, indicating a close similarity with the Davaineinae. The structure differs clearly from that seen in the Linstowiinae, within which subfamily these genera at times have been included. Their placement in the separate subfamily, the Inermicapsiferinae, is fully justified on the grounds of uterine development. Spasskii (1951) recognized four genera, *Inermicapsifer, Pericapsifer, Metacapsifer* and *Thysanotaenia*. However, Deblock & Diaouré (1962) suggested that *Thysanotaenia* may be a synonym of *Inermicapsifer*. It has been allowed to stand for the time being.

Metacapsifer was erected by Spasskii (1951) for *I. aberratus* Baer, 1925 which Baer (1925c) considered to be an aberrant species of *Inermicapsifer*. It is the only species of the subfamily with the testes exclusively aporal. The type and only specimens consist of poorly oriented serial sections. As a consequence, a detailed redescription is impossible, and Spasskii's genus is accepted provisionally, pending a redescription of the species. Baer's (1925c) figure is reproduced.

'Cotype' specimens of *Megacapsula leiperi* Wahid, 1961 (in IIP, C1225–1226) were examined and the genus is transferred to synonymy with *Oochoristica*.

Phylogeny

The phylogeny of the anoplocephalid cestodes has been considered in detail by a number of authors (Baer, 1927; Spasskii, 1951; Tenora, 1976) leading to the proposal of several different schemes for their evolution. Genera in the Australasian region have been investigated by Beveridge (1976, 1985) and in North America and Europe by Rausch (1976, 1982) and Beveridge (1978). The origin of the Anoplocephalinae from the Linstowiinae (Baer, 1927) is no longer tenable following Freeman's (1973) analysis based on morphogenesis of the metacestodes; however, the derivation suggested by Freeman (1973) of the Linstowiinae from the Proteocephalidae in aquatic reptiles, amphibians and fish is consistent with host distribution, since most Linstowiinae occur in reptiles, with relatively few genera in insectivorous mammals. Thus a transition is apparent in the hosts, from aquatic vertebrates to terrestrial reptiles where a major radiation has occurred, and finally a few genera have invaded mammals. The Inermicapsiferinae are evidently derived from the Davaineidae by the loss of a rostellum and sucker armature. The Anoplocephalinae, as suggested by the present morphological analysis, consists of two major groups, one parasitic in birds and the other parasitic in mammals. The evolution of the genera in birds apparently centres on the Psittaciformes, with a few genera only occurring in Passeriformes. The genus *Taufikia* probably does not belong within the Anoplocephalidae and has not been considered from a phylogenetic point of view. *Moniezia rhea* presents a phylogenetic anomaly which cannot at present be resolved, as does the occurrence of one species of *Monoecocestus* in a rheiform bird. Whether or not the principal avian lineage of anoplocephalids predates the mammalian lineage is not clear. The hypothesis of an avian origin of the subfamily is attractive in view of Freeman's (1973) demonstration of a close association between the Anoplocephalinae, Davaineidae and Dilepidinae, all of which occur principally in birds, an idea also suggested by Joyeux & Baer (1961). However, the genus *Anoplocephala*, the origin of the mammalian anoplocephalids, occurs in a wide range of mammals, namely perissodactyls, primates, elephants and hyraxes, possibly suggesting a long association between hosts and parasites. Lacking a detailed analysis of the genus, the hypothesis can only be considered tentative, since the antiquity of the hosts is not necessarily related to the length of association with their parasites. Tenora (1976) considered that the most primitive genera of anoplocephalids were those in marsupials, since the hosts demonstrate a range of primitive characteristics. By contrast, Beveridge (1985) has provided evidence to suggest that the anoplocephalids in fact are relatively recent parasites of Australasian diprotodont marsupials and are derived from precursors in rodents which invaded the Australian continent 5–10 million years ago. Whatever their origin, the most important radiation of the Anoplocephalinae has been with the rodents and lagomorphs, based on the number of genera present in these hosts. Transfer from rodents has apparently occurred to primates, dermopterans and marsupials involving the genera *Bertiella* and *Progamotaenia* (see Beveridge, 1985) while *Andrya* and *Diandrya* in rodents have given rise to *Moniezia* in rodents and in ruminants (Beveridge, 1978; Rausch, 1982). *Anoplocephala* appears to have a number of ancestral characteristics, but which, if any, of the current hosts of the genus were the original ones is not clear. Other examples of transfer to perissodactyls and primates within the Anoplocephalinae are known, and a detailed analysis of species of *Anoplocephala* is required to determine the origins of the genus. The importance of the duplication of genitalia as an evolutionary mechanism in the Anoplocephalidae was emphasized by Baer (1955b). Currently two pairs of such genera are recognized in the Linstowiinae (*Linstowia–Echidnotaenia*, *Oochoristica–Panceriella*), while in the Anoplocephalinae, a series of such pairs exist, *Paronia–Hemiparonia*, *Pulluterina–Stringopotaenia*, *Biporouterina–Bulbultaenia* among the avian genera (Baer, 1955b; Beveridge, 1978) and *Bertiella–Progamotaenia* in marsupials, *Schizorchis–Mosgovoyia* in lagomorphs and *Anoplocephaloides–Ctenotaenia*, *Andrya–Diandrya* and *Gallegoides–*

Pseudocittotaenia in rodents. Duplication of genitalia thus occurs invariably within the same host group and is clearly a common phenomenon within the Anoplocephalidae.

Paramoniezia represents an anomaly in any attempt to provide a phylogenetic analysis of the Anoplocephalinae (Beveridge, 1976). The principal characteristics of the genus (e.g. distribution of testes) are plesiomorphic, and this provides little or no information about its phylogenetic position. Consequently, it has been omitted from consideration. Although a common parasite of vombatid marsupials, its possible occurrence in feral ungulates in Australia (Beveridge, 1976) and the presence of a species in suids from Africa (Baylis, 1927; Graber *et al.*, 1980) have to be resolved before definitive comments can be made about its relationships.

The position of the Thysanosomatinae is also uncertain. All constituent genera are characterized by a paruterine organ and are parasitic in ruminants. Mahon (1954) suggested that *Crossotaenia* was intermediate between the Anoplocephalinae and the Thysanosomatinae since it and several genera of the Thysanosomatinae possess a poorly developed Mehlis' gland. Her suggestion is adopted here. Paruterine organs also occur in genera of the Nematotaeniidae, Paruterinidae and Idiogeninae; however, Jones (1988) concluded that the extent of morphological differences between the so-called paruterine organs in the different families indicated parallel rather than related evolutionary developments.

Thus, the evolution of the anoplocephalids is poorly understood. Clearly some general hypotheses can be advanced, but a rigorous, detailed examination of anoplocephalid phylogeny is warranted. A speculative phylogeny based on the schemes of various authors and on currently available information is presented in Fig. 17.89. Much of the phylogeny has clearly been derived intuitively but is presented here in a cladistic framework. A cladistic approach does present some difficulties, such as with *Triplotaenia*, a genus apparently derived from *Progamotaenia*, but so modified that many of the characteristics used in breaking up the subfamily are inappropriate. It and *Phascolotaenia* have been inserted following the relationships suggested by Beveridge (1976). *Triplotaenia* is so bizarre morphologically that it was placed in a separate family by Yamaguti (1959) and a separate subfamily by Beveridge (1976). Viewed from a cladistic view point, these taxa are inappropriate and neither has been used in the preparation of the key.

Family Anoplocephalidae Cholodkovsky, 1902

Diagnosis: Cestodes of small to large size; scolex without rostellum; suckers unarmed; proglottids craspedote or acraspedote; reproductive organs single or double; testes numerous; genital pores marginal, alternating regularly or irregularly. Uterus either saccate or reticulate, persisting in gravid proglottids, or ephemeral and breaking down into egg capsules or developing paruterine organs; egg frequently with pyriform apparatus. In mammals, birds and reptiles. Cosmopolitan. Type-genus *Anoplocephala* Blanchard, 1848.

Key to subfamilies

1a. **Uterus tubular (Figs 17.23, 17.24), reticular (Figs 17.47, 17.53) or saccate (Fig. 17.5) in mature proglottids, persistent in gravid proglottids .... Anoplocephalinae Blanchard, 1891.**

Diagnosis: Anoplocephalidae; uterus saccate or reticular, persisting in gravid proglottids. In mammals and birds. Cosmopolitan.

1b. **Uterus ephemeral, breaking down into capsules. Eggs scattered through parenchyma singly, with a thin outer membrane termed the uterine capsule .... Linstowiinae Fuhrmann, 1907.**

Diagnosis: Anoplocephalidae; uterus ephemeral, breaking down into simple egg capsules. Eggs lodged singly in parenchyma. In mammals and reptiles. Cosmopolitan.

1c. **Uterus ephemeral, breaking down into 'fibrous', parenchymatous capsules. Eggs in groups in parenchymatous capsules (Fig. 17.87) .... Inermicapsiferinae López-Neyra, 1943.**

Diagnosis: Anoplocephalidae; uterus ephemeral, breaking down into thick, fibrous capsules enclosing eggs; several eggs per capsule. In mammals. Africa, Asia.

1d. **Uterus developing paruterine organs into which eggs pass (Figs 17.74–17.77, 17.80). Distinct ovary and vitellarium often absent, replaced by germovitellarium .... Thysanosomatinae Skryabin, 1933.**

Diagnosis: Anoplocephalidae; uterus with one or several paruterine organs. In ruminants. Cosmopolitan.

Key to genera of the Anoplocephalinae

1a. **Two strobilae per scolex (Figs 17.1, 17.2) .... *Triplotaenia* Boas, 1902. (Figs 17.1–17.3)**

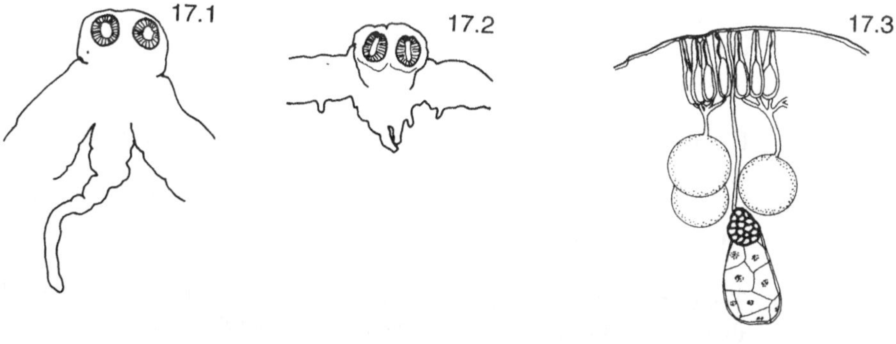

Figs 17.1, 17.2 *Triplotaenia undosa* Beveridge, 1976. Scoleces with paired strobilae originating from them. Redrawn from Beveridge (1976).
Fig. 17.3 *Triplotaenia undosa* Beveridge, 1976. Genital system. Modified from Beveridge (1976).

Diagnosis: Strobilae elongate, undulate or spirally coiled; external segmentation lacking; genital pores on external margin; internal margin irregularly fringed. Each testis supplies several cirrus-sacs (Fig. 17.3). Uterus saccate, persistent. Pyriform apparatus present. In macropodid marsupials. Australia. Type-species *T. mirabilis* Boas, 1902.

1b. Single strobila per scolex .... 2.

2a. Uterus initially spherical or trilobed, anterior to ovary, extending to fill proglottid .... 3.
2b. Uterus initially tubular or reticulate .... 4.

3a. Uterus initially spherical (Fig. 17.4), extending laterally (Fig. 17.5) to fill proglottid. Testes posterior to ovary .... *Crossotaenia* **Mahon, 1954. (Figs 17.4, 17.5.)**

Diagnosis: Strobila small; proglottids prominently craspedote, wider than long. Genitalia single. Genital pores regularly alternating. Genital ducts pass between canals. Seminal vesicles absent. Testes in single band across medulla posterior to female genitalia. Ovary poral. Vagina posterior to cirrus-sac. Seminal receptacle present. Uterus anterior to female genitalia. Pyriform apparatus absent. In bile-ducts of ruminants. Africa. Type-species *C. baeri* Mahon, 1954.

3b. Uterus initially trilobed (Fig. 17.6), elongating and branching laterally and anteriorly to fill proglottid (Fig. 17.7) .... *Taufikia* **Woodland, 1928. (Figs 17.6, 17.7)**
(Syns *Gidhaia* Johri, 1934; *Neophronia* Saxena, 1967.)

Diagnosis: Strobila slender. Proglottids ascraspedote, wider than long. Genitalia single. Genital pores alternate irregularly. Genital ducts cross ventral osmoregulatory canals dorsally. Internal seminal vesicle present. Testes pre-ovarian or scattered. Ovary central. Vagina posterior to cirrus-sac. Seminal receptacle present. Pyriform apparatus absent. In accipitriform birds (vultures). Africa, India, Asia. Type-species *T. edmondi* Woodland, 1928.

4a. Uterus tubular initially, becoming saccate, developing diverticula, never reticulate .... 5.
4b. Uterus reticulate during part of its development (Figs 17.46–17.48) .... 27.

5a. Uterus a straight transverse or longitudinal tube (Figs 17.8, 17.12) .... 6.
5b. Uterus arched anterior to female genitalia, or with paired posterior diverticula on either side of vitellarium (Figs 17.36, 17.41, 17.43) .... 19.

6a. Testes originate anterior to uterus (Figs 17.8, 17.10); (occasionally becoming displaced posteriorly in *Progamotaenia ewersi*) .... 7.
6b. Testes not entirely anterior to uterus .... 8.

17.4

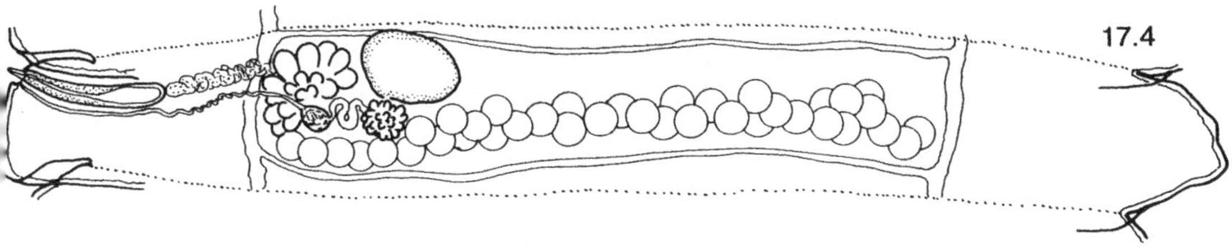

17.5

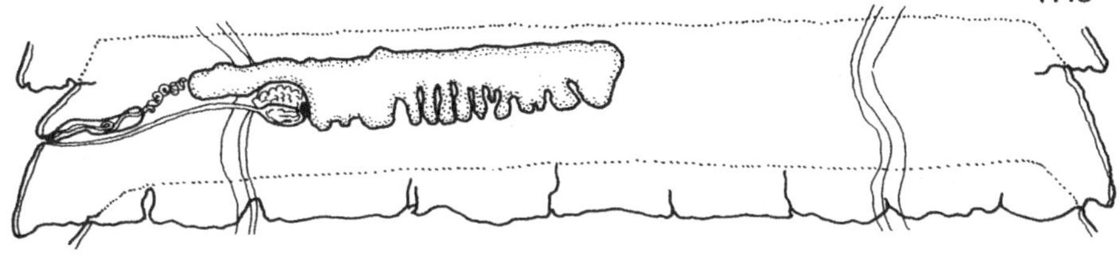

17.6

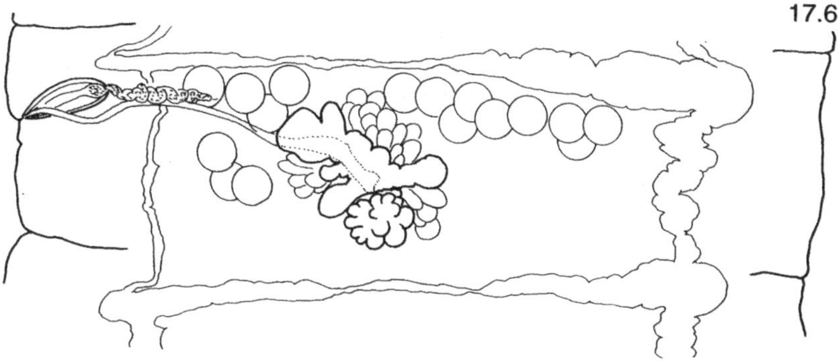

17.7

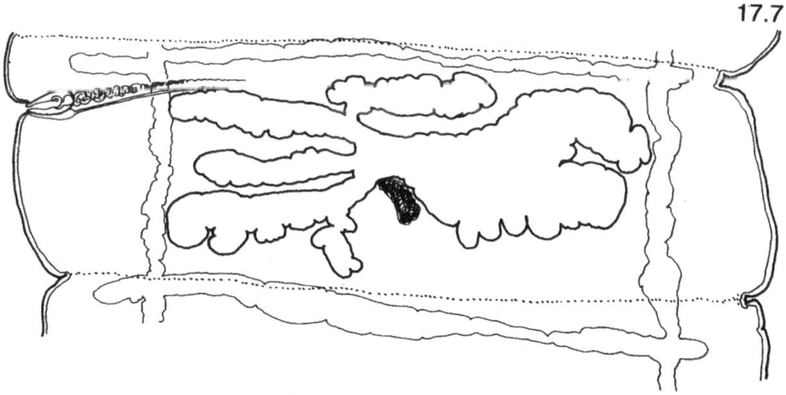

Figs 17.4, 17.5 *Crossotaenia baeri* Mahon, 1954. 17.4. Mature proglottid. Original. 17.5. Developing uterus. Redrawn from Mahon (1954).
Fig. 17.6 *Taufikia edmondi* Woodland, 1928. 'Cotype', mature proglottid. Original.
Fig. 17.7 *Taufikia edmondi* Woodland, 1928. 'Cotype', developing uterus. Original.

7a. **Genitalia single** ...... ***Bertiella*** **Stiles & Hassall, 1902. (Figs 17.8, 17.9)**
(Syns *Bertia* Blanchard, 1891; *Parabertiella* Nybelin, 1917; *Indotaenia* Singh, 1962; *Prototaenia* Baer, 1927; *Beveridgia* Spasskii, 1988.)

Diagnosis: Strobilae of various sizes. Proglottids craspedote, wider than long. Genital pores unilateral or irregularly alternate. Genital ducts cross canals dorsally. Internal seminal vesicle present. Testes in single band or two groups, entirely anterior to uterus. Ovary poral, central or aporal. Vagina posterior to cirrus-sac. Seminal receptacle present or absent. Uterus single, transverse, tubular (Fig. 17.9). Pyriform apparatus present. In primates (including man), rodents, dermopterans and Australasian marsupials. Cosmopolitan. Type-species *B. studeri* (Blanchard, 1891).

7b. **Genitalia paired** .... ***Progamotaenia*** **Nybelin, 1917. (Figs 17.10, 17.11)**
(Syns *Hepatotaenia* Nybelin, 1917; *Baeriella* Fuhrmann, 1932; *Fuhrmannodes* Strand, 1942; *Adelataenia* Beveridge, 1976 *lapsus calami* in Schmidt, 1986; *Wallabicestus* Schmidt, 1975.)

Diagnosis: Strobilae of variable size. Proglottids craspedote often fringed, wider than long. Genital ducts cross canals dorsally; internal and external seminal vesicles present. Testes originate anterior to uterus in single band or two groups, sometimes become displaced posteriorly in *P. ewersi*. Ovaries poral. Vagina posterior to cirrus-sac. Seminal receptacle present. Uterus single or paired, transverse, tubular (Fig. 17.11). Pyriform apparatus present. In bile-ducts or intestines of macropodid marsupials. Australia. Type-species *P. bancrofti* (Johnston, 1912).

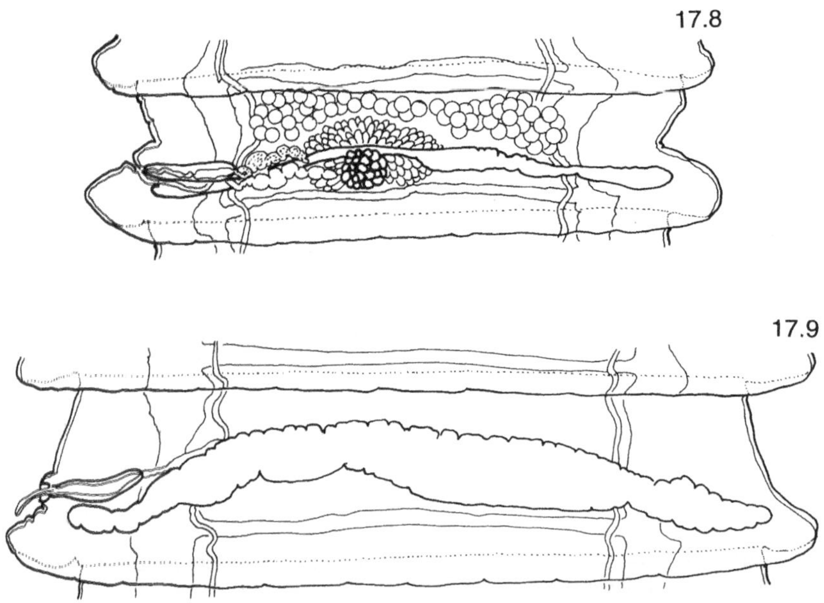

Figs 17.8, 17.9 *Bertiella phalangeris* Beveridge, 1985. 17.8. Mature proglottid. 17.9. Gravid proglottid. Redrawn from Beveridge (1985).

17.10

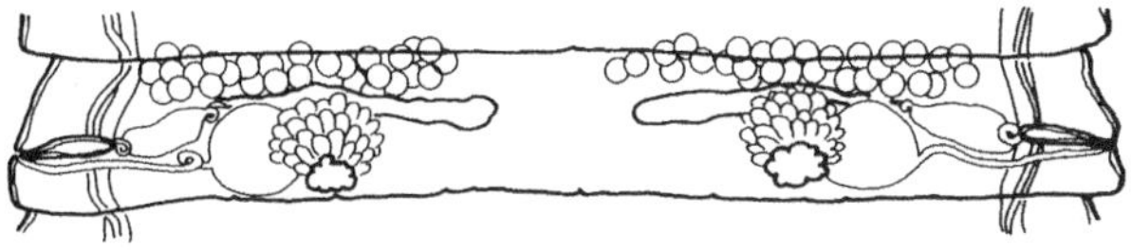

17.11

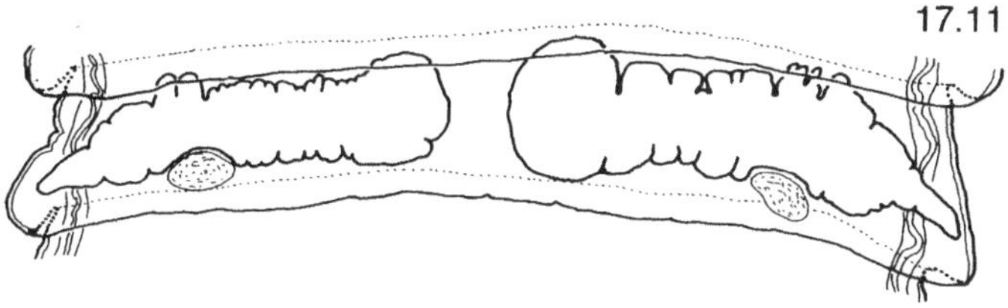

17.12

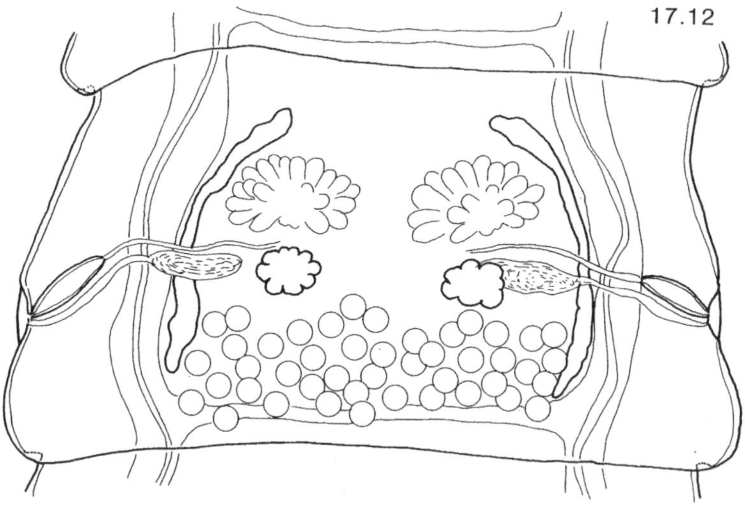

17.13

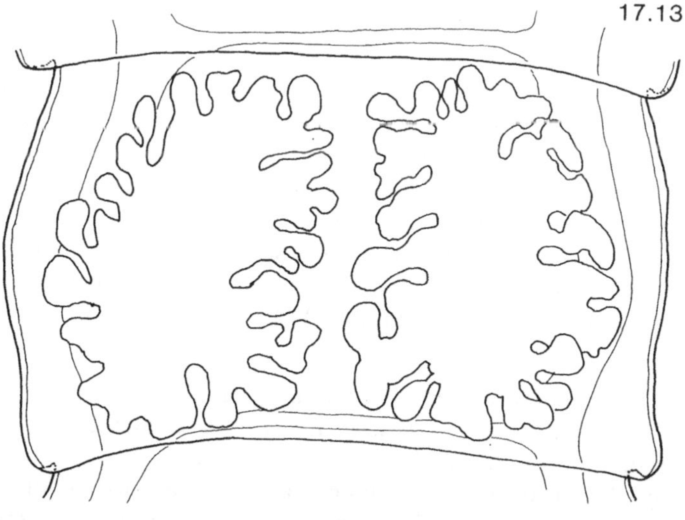

Figs 17.10, 17.11 *Progamotaenia festiva* (Rudolphi, 1819). 17.10. Mature proglottid. 17.11. Gravid proglottid. Redrawn from Beveridge (1976).
Figs 17.12, 17.13 *Diuterinotaenia spasskii* Gvozdev, 1961. 17.12. Mature proglottid. 17.13. Gravid proglottid. Redrawn from Gvozdev (1961a).

8a. **Testes exclusively posterior to uterus or in posterior half of proglottid (Figs 17.12, 17.14, 17.15)** .... **9.**
8b. **Testes not exclusively posterior to uterus** .... **15.**

9a. **Uteri paired, longitudinal (Figs 17.12, 17.13)** ....
.... ***Diuterinotaenia*** **Gvozdev, 1961. (Figs 17.12, 17.13)**
Diagnosis: Strobila small, slender. Proglottids craspedote; wider than long. Genitalia paired. Seminal vesicles absent. Genital ducts cross osmoregulatory canals dorsally. Testes in single band posterior to female genitalia. Ovaries poral. Vagina posterior to cirrus-sac. Seminal receptacle present. Uteri tubular. Pyriform apparatus absent. In ochotonid lagomorphs. Asia. Type-species *D. spasskii* Gvozdev, 1961.

9b. **Uterus single or double, transverse** .... **10.**

10a. **Lateral terminations of uteri anterior to cirrus-sac and vagina (Figs 17.14, 17.17)** .... **11.**
10b. **Lateral termination of uteri posterior to cirrus-sac and vagina (Figs 17.19, 17.21)** .... **13.**

11a. **Genitalia single** .... **12.**
11b. **Genitalia paired** .... ***Pseudocittotaenia*** **Tenora, 1976. (Fig. 17.14)**
Diagnosis: Strobilae small. Proglottids craspedote, wider than long. Genital ducts cross osmoregulatory canals dorsally. Internal seminal vesicle present. Testes posterior to uterus (Fig. 17.14). Ovaries poral. Vagina posterior to cirrus-sac; seminal receptacle present. Uterus single, transverse, tubular, extending ventrally beyond osmoregulatory canals, terminating anterior to cirrus-sac. Pyriform apparatus present. In geomyid rodents. North America. Type-species *P. praecoquis* (Stiles, 1895).

12a. **Testes aporal to ovary (Fig. 17.15)** ....
.... ***Anoplocephaloides*** **Baer, 1923 (in part). (Figs 17.15, 17.16)**
(Syn. *Leporidotaenia* Genov, Murai, Georgiev & Harris, 1990.)
Diagnosis: Strobila small to medium sized. Proglottids craspedote, wider than long. Genitalia single. Genital pores usually unilateral. Genital ducts cross osmoregulatory canals dorsally; internal and external seminal vesicles present. Ovary poral. Vagina posterior to cirrus-sac. Uterus transverse, tubular (Fig. 17.16). Pyriform apparatus present. In rodents, lagomorphs, perissodactyls. North and Central America, Africa, Europe, Asia. Type-species *A. infrequens* (Douthitt, 1915).

12b. **Testes poral and aporal to ovary (Fig. 17.17)** ....
.... ***Gallegoides*** **Tenora & Mas-Coma, 1978. (Fig. 17.17)**
Diagnosis: Strobila small. Proglottids craspedote, wider than long. Genitalia single. Genital ducts cross osmoregulatory canals dorsally. Internal and external seminal vesicles present. Testes in band posterior to uterus. Ovary slightly poral. Vagina posterior to cirrus-sac. Seminal receptacle present. Uterus single, tubular, transverse, extending beyond osmoregulatory canals, terminating anterior to

17.14

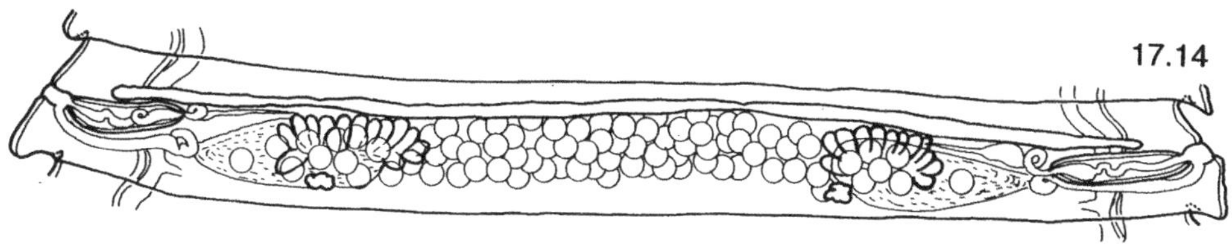

17.15

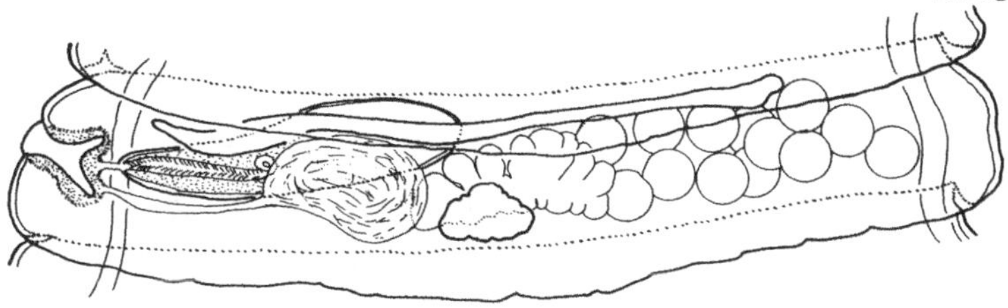

17.16

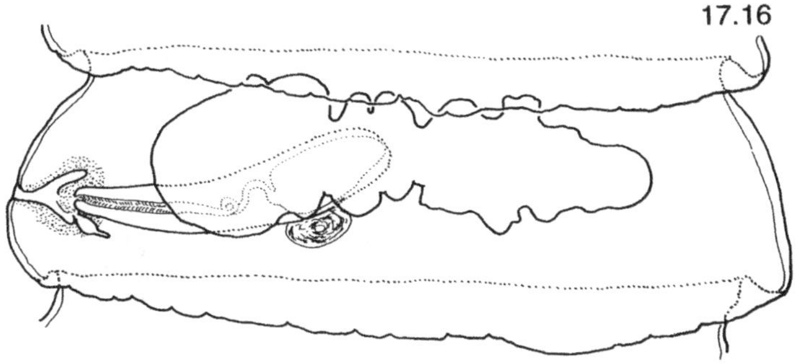

Fig. 17.14 *Pseudocittotaenia praecoquis* (Stiles, 1895). Mature proglottid. Redrawn from Beveridge (1978).
Figs 17.15, 17.16 *Anoplocephaloides wimerosa* (Moniez, 1880). 17.15. Mature proglottid. 17.16. Gravid proglottid. Original.

cirrus-sac. Pyriform apparatus present. In arvicoline rodents. Asia, Europe. Type-species *G. arfaai* (Mobedi & Ghadirian, 1977).

13a. Neck greatly elongated, scolex covered with glands (Fig. 17.18) .... *Ectopocephalium* Rausch & Ohbayashi, 1974. (Figs 17.18, 17,19)
Diagnosis: Strobila small. Scolex with prominent apical glands. Genitalia paired. Genital ducts cross osmoregulatory canals dorsally; internal seminal vesicle present. Testes in band posterior to uterus (Fig. 17.19). Ovaries poral. Vagina posterior to cirrus-sac; distal vagina with gland cells. Seminal receptacle present. Uterus single, transverse, tubular, crossing osmoregulatory canals, terminating posterior to cirrus-sac. Pyriform apparatus present. In ochotonid lagomorphs. Nepal. Type-species *E. abei* Rausch & Ohbayashi, 1974.

13b. Neck not elongated, scolex without glands .... 14.

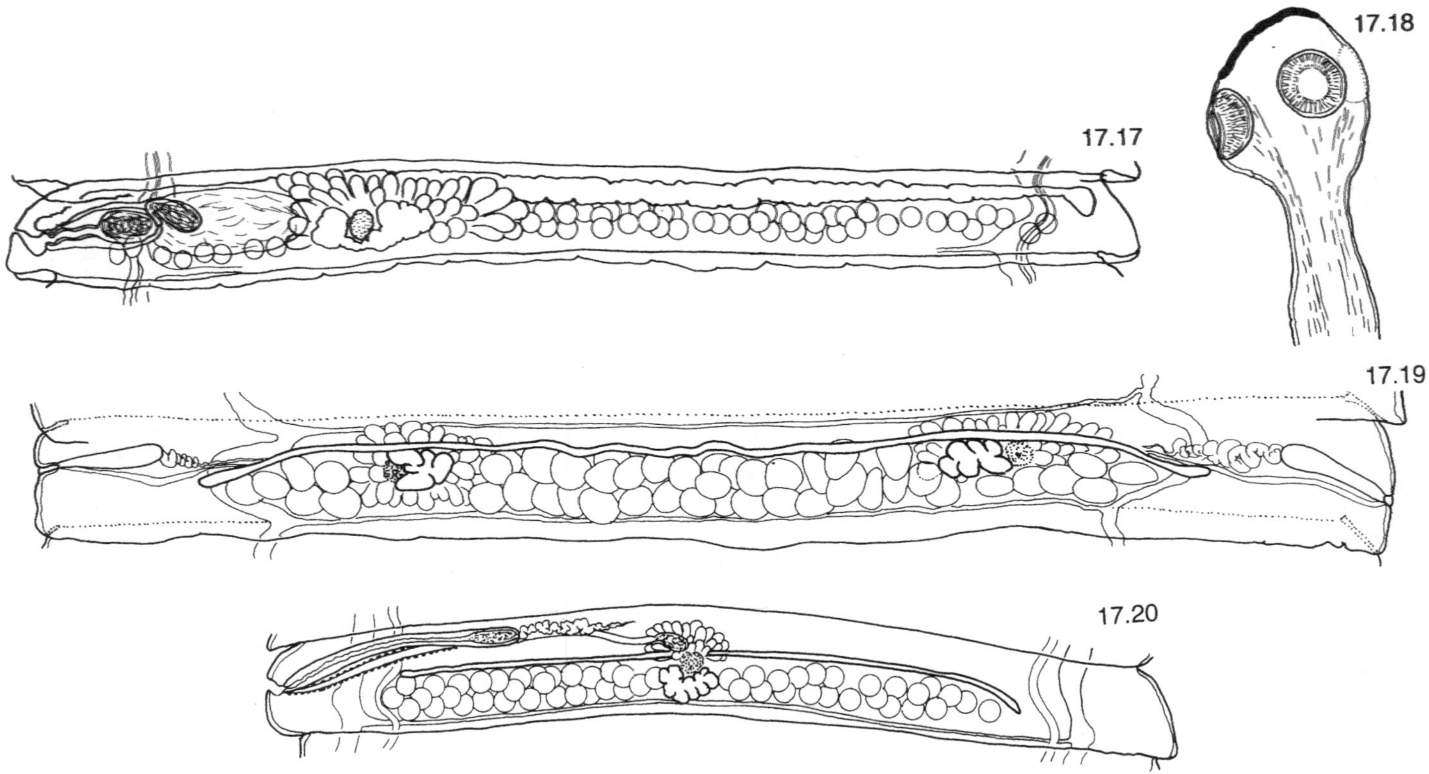

Fig. 17.17 *Gallegoides arfaai* (Mobedi & Ghadirian, 1977). Mature proglottid. Original. **Figs 17.18, 17.19.** *Ectopocephalium abei* Rausch & Ohbayashi, 1974. 17.18. Holotype, scolex; note scolex glands. 17.19. Mature proglottid. Original.
Fig. 17.20 *Schizorchis ryzhikovi* Rausch & Smirnova, 1984. Mature proglottid. Original.

14a. Genitalia single (Fig. 17.20) ... ***Schizorchis* Hansen, 1948. (Fig. 17.20)**
Diagnosis: Strobila small. Proglottids craspedote, wider than long. Genital pores alternate irregularly. Genital ducts cross osmoregulatory canals dorsally; internal and external seminal vesicles present. Testes posterior to uterus. Ovary central. Vagina posterior to cirrus-sac. Distal vagina with gland cells. Seminal receptacle present. Uterus single, transverse, tubular, restricted to medulla. Pyriform apparatus present. In ochotonid lagomorphs. North America, Asia. Type-species *S. ochotonae* Hansen, 1948.

14b. Genitalia paired (Fig. 17.21) ....
.... ***Mosgovoyia* Spasskii, 1951. (Fig. 17.21)**
(Syns *Neoctenotaenia* Tenora, 1976; *Himalaya* Malhotra, Sawada & Capoor, 1983.)
Diagnosis: Strobila large. Proglottids craspedote, wider than long. Genital ducts cross osmoregulatory canals dorsally; internal seminal vesicle present. Testes in single band or two groups posterior to uterus. Ovaries poral. Vagina posterior to cirrus-sac. Distal vagina with gland cells. Seminal receptacle present. Uterus single, transverse, tubular, restricted to medulla or crossing canals dorsally and terminating posterior to cirrus-sac. Pyriform apparatus present. In leporid lagomorphs, rarely in rodents. North and South America, Europe, Africa, Asia. Type-species *M. pectinata* (Goeze, 1782).

15a. Testes aporal to ovary, anterior and posterior to uterus (Fig. 17.22) ....
.... **16.**
15b. Testes poral and aporal to ovary, anterior and posterior to uterus (Figs 17.23, 17.25) .... **17.**

16a. Genitalia single ....
.... ***Anoplocephaloides* Baer, 1923 (in part). (See 12a above.)**
(Syn. *Flabelloskrjabinia* Spasskii, 1951.)
16b. Genitalia paired .... ***Ctenotaenia* Railliet, 1893. (Fig. 17.22)**
Diagnosis: Strobila large. Proglottids craspedote, wider than long. Genital ducts cross osmoregulatory canals dorsally; internal and external seminal vesicles present. Testes in single band between ovaries, anterior and posterior to uterus. Vagina posterior to cirrus-sac. Seminal receptacle present. Uterus single, transverse, tubular, restricted to medulla (Fig. 17.22). Pyriform apparatus present. In sciurid rodents. Europe, Asia. Type-species *C. marmotae* (Froehlich, 1802).

17a. Genitalia single. Genital pores unilateral ....
.... ***Anoplocephala* Blanchard, 1848. (Figs 17.23, 17.24)**
(Syn. *Plagiotaenia* Peters, 1871.)
Diagnosis: Strobila moderate to large. Proglottids craspedote, wider than long. Genital pores unilateral. Genital ducts cross canals dorsally. Internal and external seminal vesicles present. Testes scattered throughout medulla. Ovary poral. Vagina posterior to cirrus-sac. Seminal receptacle present. Uterus single, transverse, tubular, restricted to medulla (Figs 17.23, 17.24). Pyriform apparatus

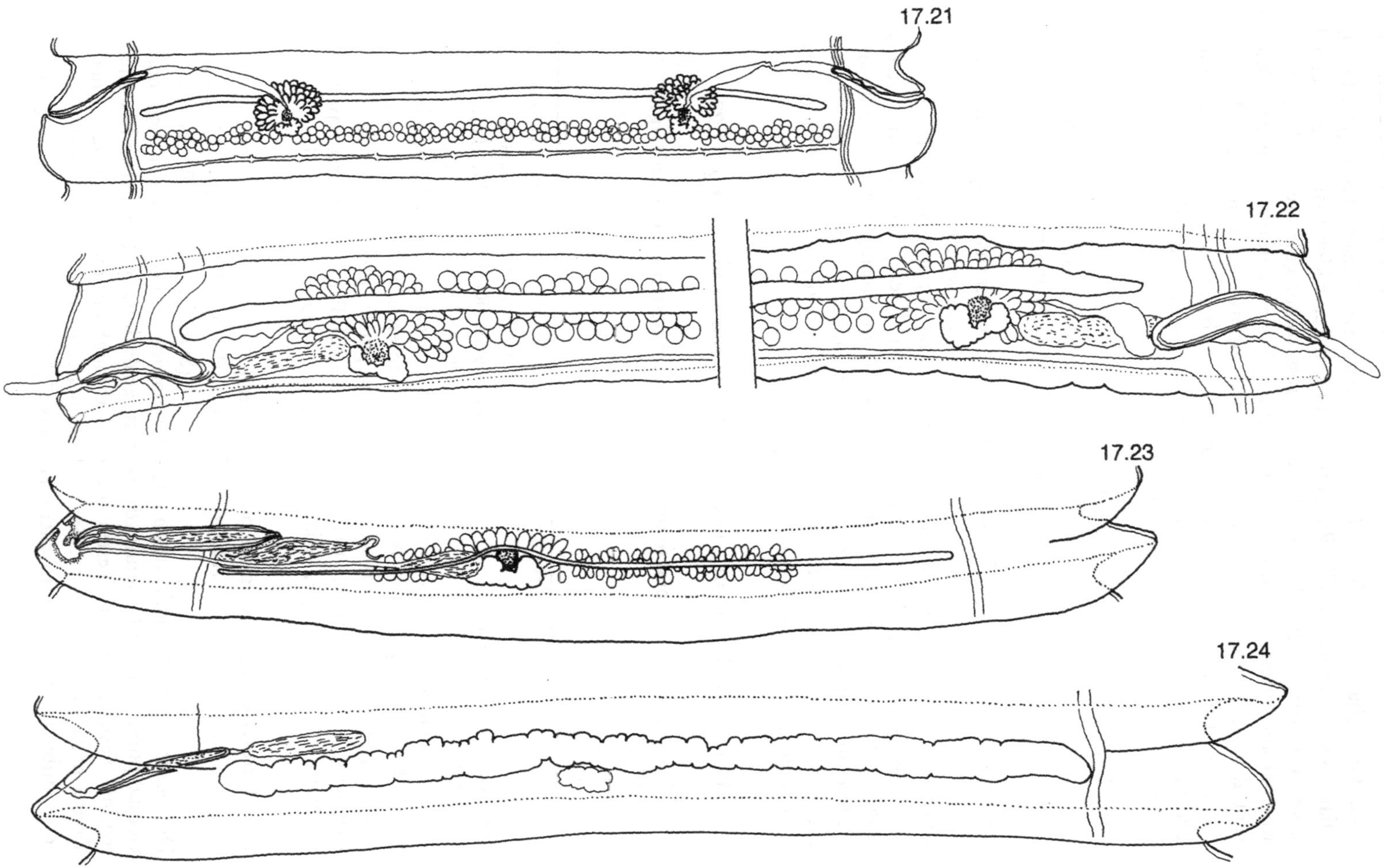

Fig. 17.21 *Mosgovoyia pectinata* (Goeze, 1782). Mature proglottid. Redrawn from Beveridge (1978). **Fig. 17.22.** *Ctenotaenia marmotae* (Froehlich, 1802). Mature proglottid. Redrawn from Beveridge (1978). **Figs 17.23, 17.24.** *Anoplocephala perfoliata* (Goeze, 1782). 17.23. Mature proglottid. 17.24. Gravid proglottid. Original.

present. In hyracoids, perissodactyls, proboscoids, primates. Cosmopolitan. Type-species *A. perfoliata* (Goeze, 1782).

17b. Genitalia paired .... **18.**

18a. Testes scattered through medulla (Fig. 17.25). Seminal receptacle elongate, merging into vagina (Fig. 17.26) ....
.... ***Paramoniezia* Maplestone & Southwell, 1923. (Figs 17.25, 17.26)**

Diagnosis: Strobila moderate. Proglottids craspedote, wider than long. Genitalia paired. Genital ducts cross osmoregulatory canals dorsally. Internal seminal vesicle present. Ovaries poral. Vagina posterior to cirrus-sac. Seminal receptacle present. Uterus single, transverse, tubular. Pyriform apparatus present. In vombatid marsupials and suids. Australia, Africa. Type-species *P. suis* Maplestone & Southwell, 1923.

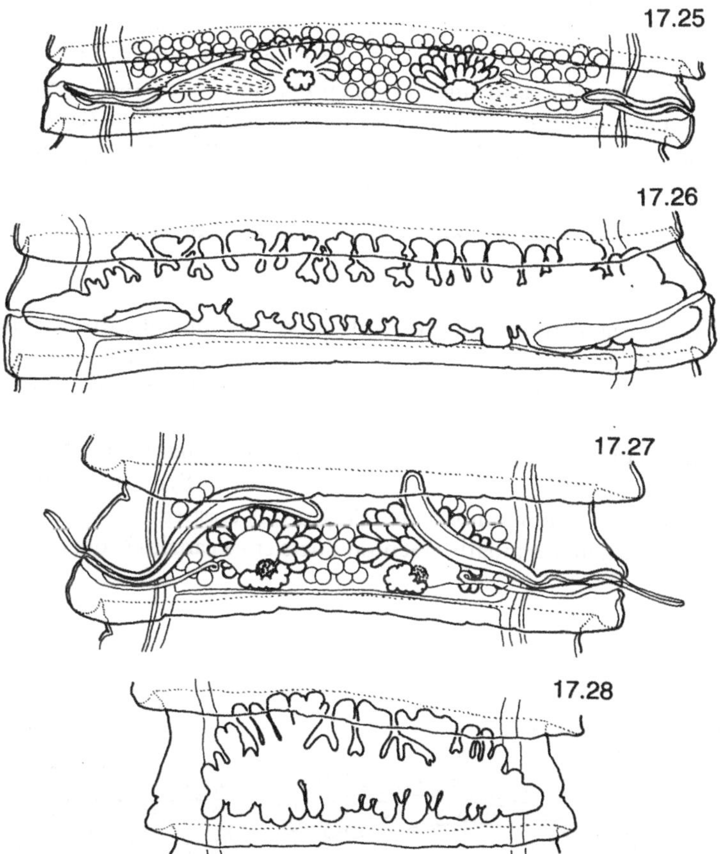

Figs 17.25, 17.26 *Paramoniezia johnstoni* Beveridge, 1976. 17.25. Mature proglottid. 17.26. Gravid proglottid. Redrawn from Beveridge (1976).
Figs 17.27, 17.28 *Phascolotaenia comani* Beveridge, 1976. 17.27. Mature proglottid. 17.28. Gravid proglottid. Redrawn from Beveridge (1976).

18b. Testes with central group posterior to uterus, two lateral groups anterior to uterus. Seminal receptacle circular, distinct from vagina (Fig. 17.27) .... *Phascolotaenia* Beveridge, 1976. (Figs 17.27, 17.28)

Diagnosis: Strobila small. Proglottids craspedote, wider than long. Genitalia paired. Genital ducts cross canals dorsally. Cirrus-sacs cross canals, reaching mid-region of proglottid. Internal and external seminal vesicles present. Testes posterior to uterus between ovaries, anterior and posterior to uterus porally. Vagina posterior to cirrus-sac. Seminal receptacle present. Uterus single, transverse, tubular. Pyriform apparatus present. In vombatid marsupials. Australia. Type-species *P. comani* Beveridge, 1976.

19a. Genital pores absent. Uterus single, extends beyond osmoregulatory canals, with two anterior diverticula on each side of proglottid (Figs 17.29, 17.30) .... *Aporina* Fuhrmann, 1902. (Figs 17.29, 17.30)

Diagnosis: Strobila moderate. Proglottids craspedote, wider than long. Genitalia single. Cirrus-sac and vagina alternate irregularly. Genital ducts cross canals dorsally. Cirrus-sac vestigial, non-functional. Seminal vesicles absent. Testes in two broad lateral bands, few extending posterior to ovary. Ovary central or slightly poral to mid-line. Vagina posterior to cirrus-sac. Seminal receptacle present. Uterus tubular, loops anteriorly around seminal receptacle with paired, extra-vascular, anteriorly-directed diverticula. Pyriform apparatus absent. In parrakeet. Brazil. Type-species *A. alba* Fuhrmann, 1902.

19b. Genital pores present. Uterus without paired, anterior diverticula .... 20.

20a. Uterus with central anteriorly directed, spur-like diverticulum (Figs 17.31–17.33) .... 21.
20b. Uterus without central anteriorly directed, spur-like diverticulum 22.

21a. Genitalia single ...... *Triuterina* Fuhrmann, 1922. (Figs 17.31, 17.32)
(Syn. *Biporouterina* Burt, 1973.)

Diagnosis: Strobila moderate. Proglottids craspedote, wider than long. Genital pores alternate irregularly. Genital ducts cross canals (?) dorsally. Internal seminal

17.29

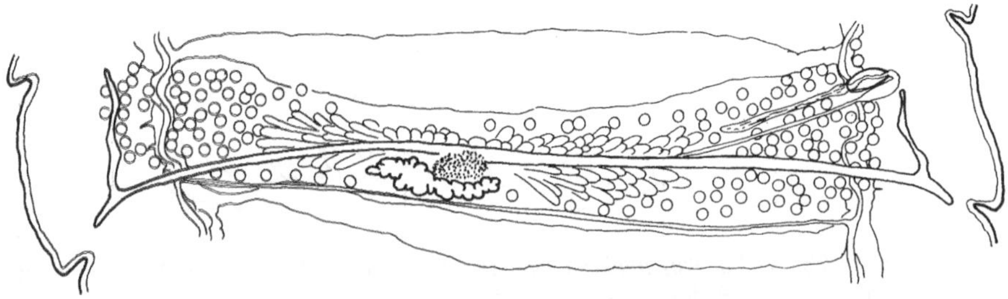

Fig. 17.29 *Aporina alba* Fuhrmann, 1902. Holotype; mature proglottid reconstructed from sections. Original.

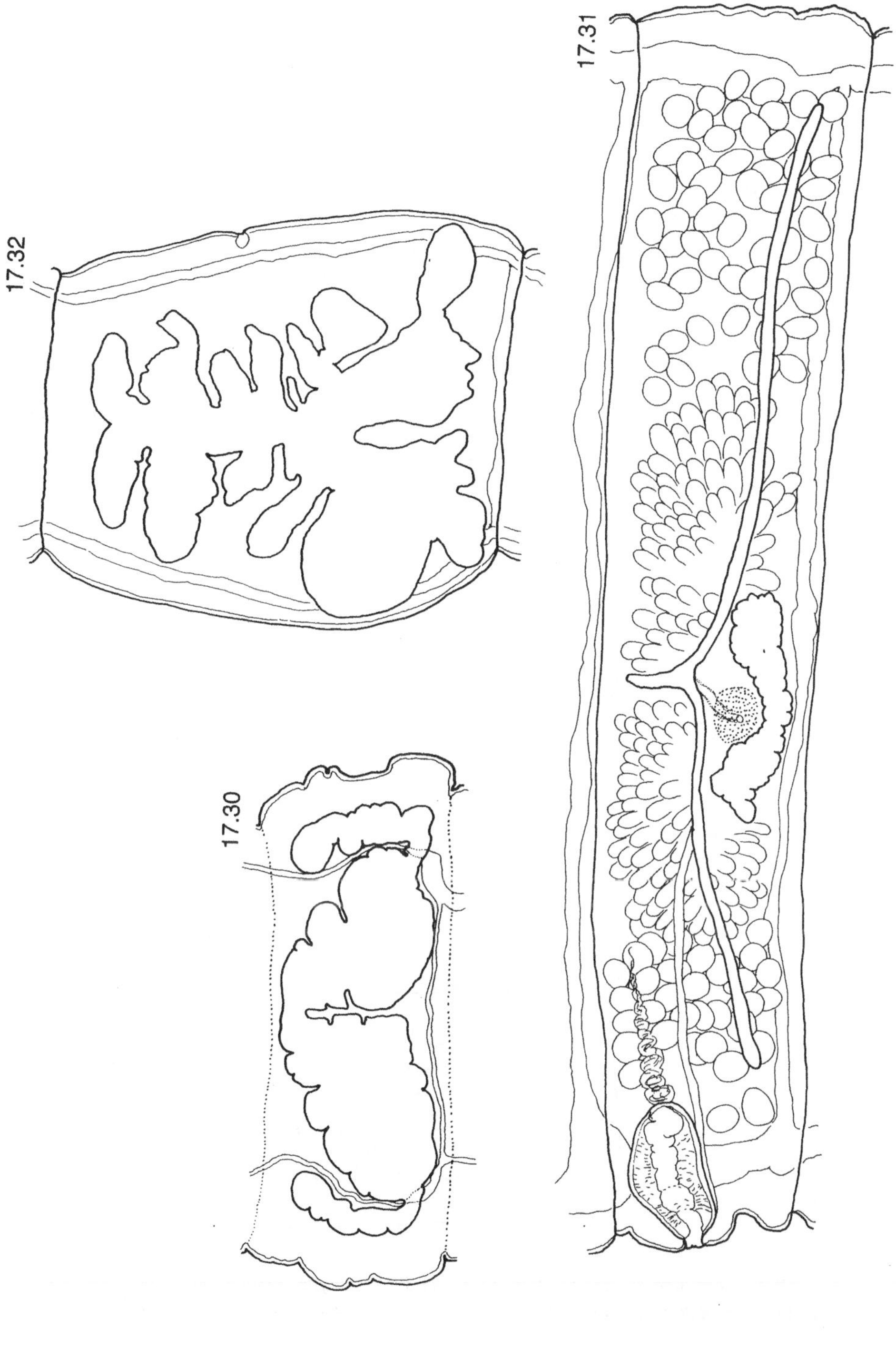

Fig. 17.30 *Aporina alba* Fuhrmann, 1902. Holotype; section of gravid proglottid. Original. **Figs 17.31, 17.32.** *Triuterina anoplocephaloides* (Fuhrmann, 1902). 17.31. Mature proglottid. 17.32. Gravid proglottid. Original.

vesicle present. Testes scattered throughout medulla. Ovary central. Vagina posterior to cirrus-sac. Seminal receptacle present. Uterus tubular, looped anteriorly around seminal receptacle, with anteriorly directed, median spur (Figs 17.31, 17.32). Pyriform apparatus absent. In psittaciform birds. Africa, Sri Lanka. Type-species *T. anoplocephaloides* Fuhrmann, 1902.

21b. Genitalia paired .... ***Bulbultaenia* n. g. (Figs 17.33, 17.34)**
Diagnosis: Strobila moderate. Proglottids craspedote, wider than long. Genital ducts cross canals dorsally. Internal seminal vesicles present. Testes scattered through medulla. Ovaries poral. Vagina posterior to cirrus-sac. Seminal receptacle present. Uteri paired, tubular, looped anteriorly around seminal receptacles, each with single, median, anteriorly directed spur. Pyriform apparatus absent. In red-vented bulbul, *Molpastes haemorrhous*. Sri Lanka. Type-species *B. calcaruterina* (Burt, 1939) n. comb. (Syn. *Paronia calcaruterina* Burt, 1939.)

22a. Uterus simple, U-shaped, surrounding vitellarium anteriorly (Figs 17.35–17.38) .... **23.**
22b. Uterus with prominent horizontal branches, small anterior U-shaped deviation anterior to vitellarium and paired posterior diverticula on either side of vitellarium (Figs 17.43, 17.45) .... **25.**

17.33

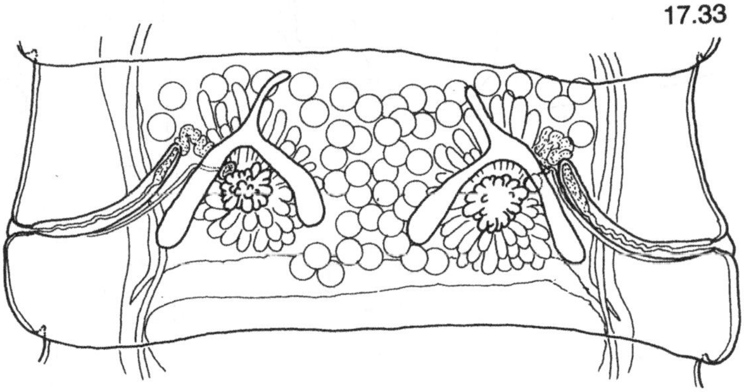

17.34

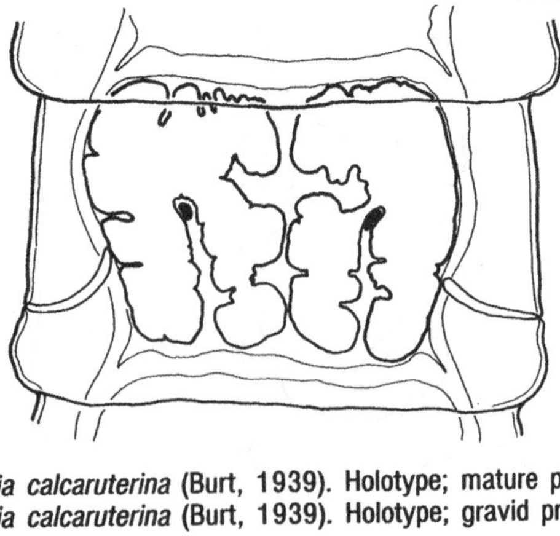

Fig. 17.33 *Bulbultaenia calcaruterina* (Burt, 1939). Holotype; mature proglottid. Original.
Fig. 17.34 *Bulbultaenia calcaruterina* (Burt, 1939). Holotype; gravid proglottid. Original.

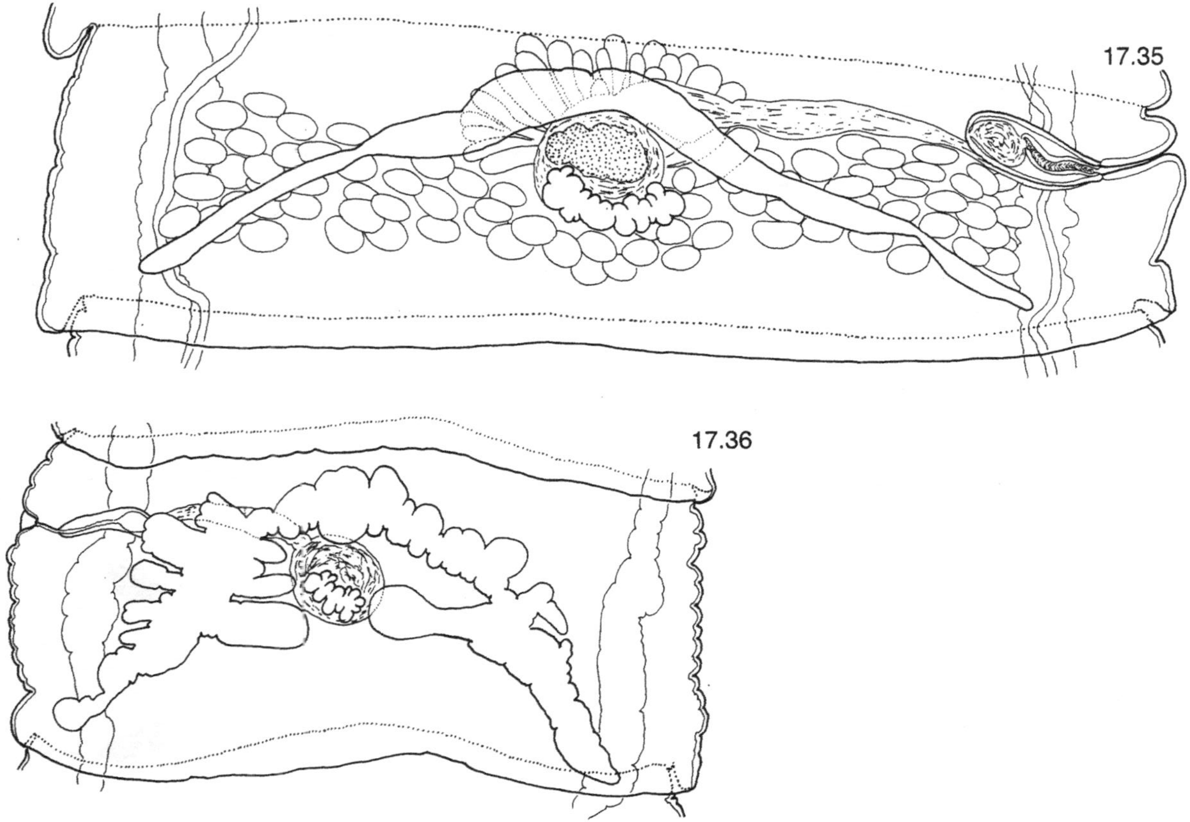

Figs 17.35, 17.36 *Hemiparonia bancrofti* (Johnston, 1912). 17.35. Mature proglottid. 17.36. Gravid proglottid. Original.

23a. Genitalia single .... ***Hemiparonia*** **Baer, 1925. (Figs 17.35, 17.36)**
Diagnosis: Strobila moderate. Proglottids craspedote, wider than long. Genital ducts cross canals dorsally. Seminal vesicles present. Testes in band posterior and lateral to ovary. Ovary central. Vagina posterior to cirrus-sac. Seminal receptacle present. Uterus single, tubular, looped anteriorly around seminal receptacle. Pyriform apparatus absent. In psittaciform birds. Australia. Type-species *H. cacatuae* (Maplestone, 1922).

23b. Genitalia paired .... **24.**

24a. Uterus double, sometimes merging, scolex lacking glands anterior to suckers .... ***Paronia*** **Diamare, 1900. (Figs 17.37, 17.38)**
Diagnosis: Strobila moderate. Proglottids craspedote, wider than long. Genitalia paired. Genital ducts cross canals dorsally. Internal seminal vesicles present or absent; external vesicle absent. Testes scattered throughout medulla. Ovaries poral. Vagina posterior to cirrus-sac. Seminal receptacle present. Uteri tubular, looped anteriorly around seminal receptacles. Pyriform apparatus rudimentary or absent. In birds. Asia, Africa, South America, Australia. Type-species *P. trichoglossi* (Linstow, 1888).

17.37

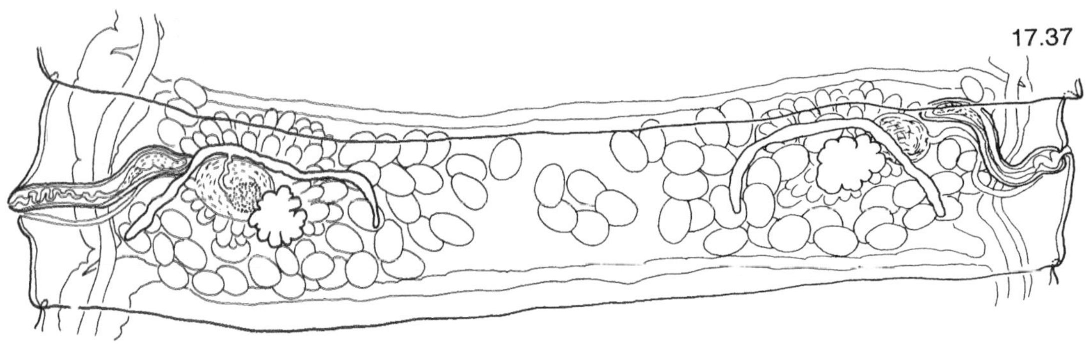

17.38

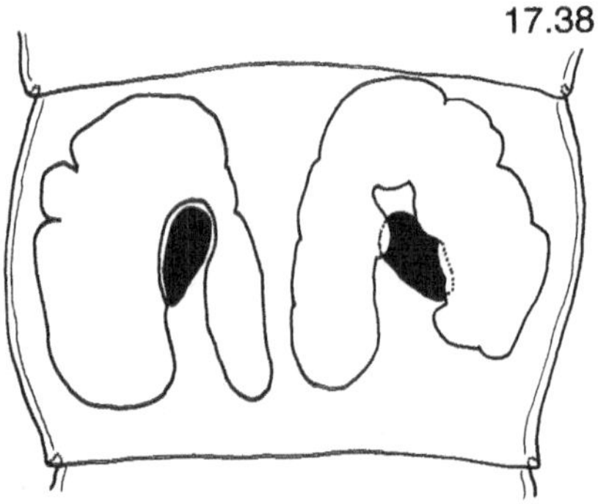

17.39

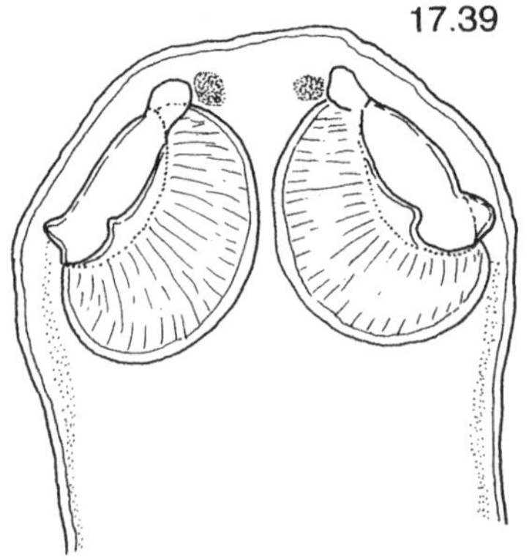

Fig. 17.37 *Paronia trichoglossi* (Linstow, 1888). Type; mature proglottid. Original.
Fig. 17.38 *Paronia trichoglossi* (Linstow, 1888). Gravid proglottid. Original.
Fig. 17.39 *Moniezoides rouxi* Fuhrmann, 1918. Type; scolex; note lappets on suckers and glands anterior to suckers. Original.

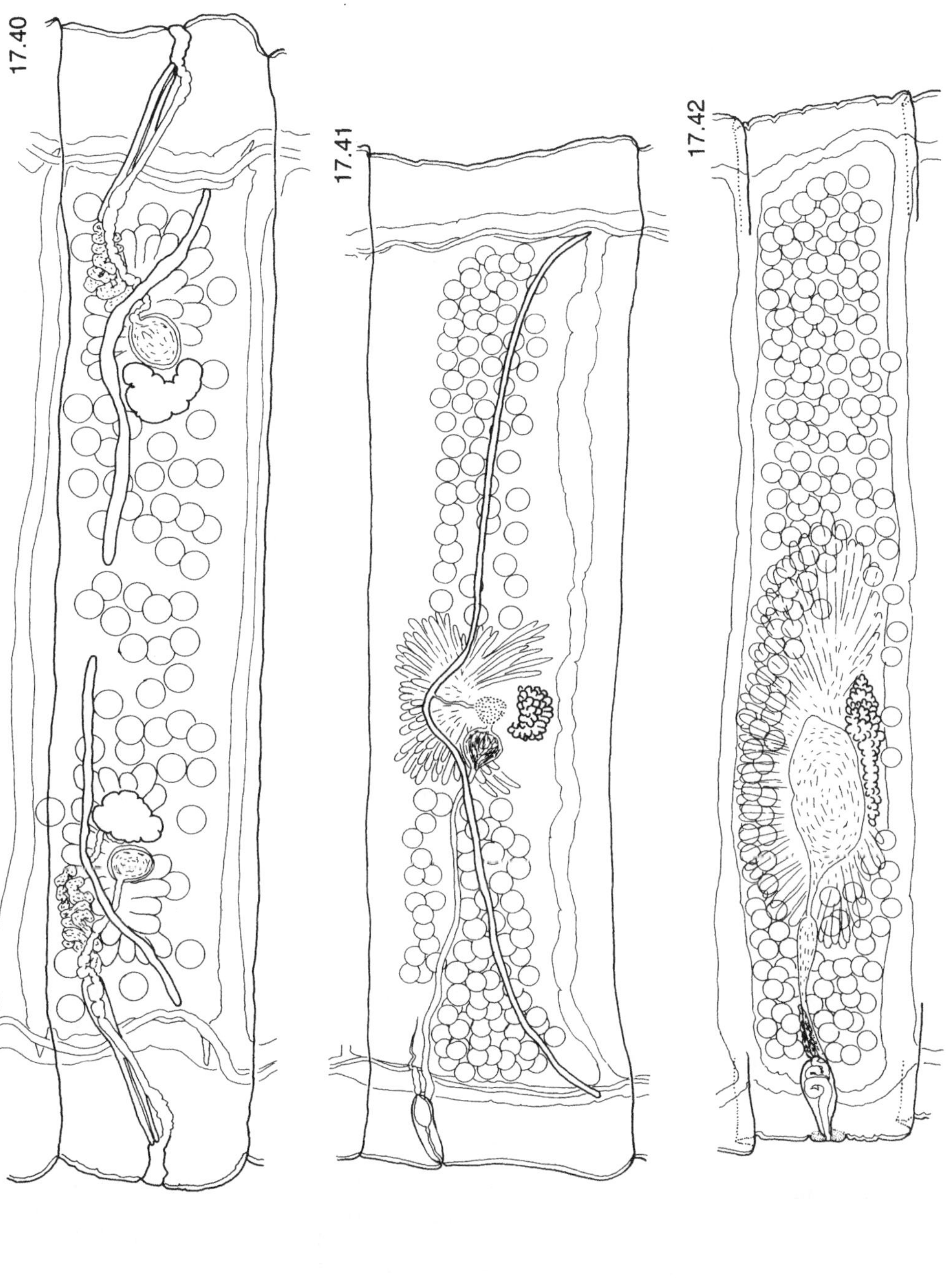

Fig. 17.40 *Moniezoides rouxi* Fuhrmann, 1918. Type; mature proglottid. Original. **Fig. 17.41.** *Killigrewia frivola* Meggitt, 1927. 'Cotype'; mature proglottid. Original. **Fig. 17.42.** *Pulluterina nestoris* Smithers, 1954. Mature proglottid. Original.

24b. Uterus (?) single (Fig. 17.40); glands present anterior to each sucker. Suckers with prominent lappets (Fig. 17.39) .... *Moniezoides* Fuhrmann, 1918. (Figs 17.39, 17.40)

Diagnosis: Strobila moderate. Scolex with four groups of glands anterior to suckers. Proglottids craspedote, wider than long. Genitalia double. Genital ducts cross canals dorsally. Internal and external seminal vesicles absent. Testes scattered throughout medulla. Vagina posterior to cirrus-sac. Seminal receptacle present. Uterus (?) single, tubular, looped anteriorly around both ovaries. Pyriform apparatus (?) absent. In psittaciform birds. South Pacific. Type-species *M. rouxi* Fuhrmann, 1918.

25a. Testes in two distinct, lateral groups (Fig. 17.41) .... *Killigrewia* Meggitt, 1927. (Fig. 17.41)
(Syns *Pseudoaporina* Saxena & Baugh, 1973; *Nepalesia* Sharma, 1943; *Columbia* Srivastava & Capoor, 1966.)

Diagnosis: Strobila small. Proglottids craspedote, wider than long. Genitalia single. Genital pores alternate irregularly. Genital ducts cross canals dorsally. Internal seminal vesicles present. Testes in two groups either side of ovary. Ovary slightly poral. Vagina posterior to cirrus-sac. Seminal receptacle present. Uterus single, tubular, looped broadly around seminal receptacle, terminating in posterolateral corners of medulla. Pyriform apparatus (?) absent. In birds. Africa, Asia, Europe, North and South America, Australia. Type-species *K. frivola* Meggitt, 1927.

25b. Testes distributed across medulla .... 26.

26a. Genitalia single .... *Pulluterina* Smithers, 1954. (Figs 17.42, 17.43)

Diagnosis: Strobila large. Proglottids craspedote, wider than long. Genital pores alternate irregularly. Genital ducts cross canals dorsally. Internal seminal vesicle present. Testes scattered throughout medulla. Ovary poral. Vagina posterior to cirrus-sac. Seminal receptacle present. Uterus single, tubular, looped anteriorly around seminal receptacle with pair of posterior diverticula on either side. Pyriform apparatus absent. In psittaciform birds. New Zealand. Type-species *P. nestoris* Smithers, 1954.

26b. Genitalia double .... *Stringopotaenia* Beveridge, 1978. (Figs 17.44, 17.45)

Diagnosis: Strobila moderate. Proglottids craspedote, wider than long. Genital ducts cross canals dorsally. Internal seminal vesicle present. Testes scattered throughout medulla. Ovaries poral. Vagina posterior to cirrus-sac. Seminal recep-

Fig. 17.43 *Pulluterina nestoris* Smithers, 1954. Gravid proglottid. Original.
Figs 17.44, 17.45 *Stringopotaenia psittacea* (Fuhrmann, 1904). 17.44. Holotype; part of mature proglottid. 17.45. Holotype; section through gravid proglottid. Redrawn from Beveridge (1978).
Figs 17.46–17.48 *Cittotaenia denticulata* (Rudolphi, 1804). 17.46. Mature proglottid. 17.47. Developing uterus showing reticulations. 17.48. Gravid uterus; note lack of reticulation. Redrawn from Beveridge (1978).

17.43

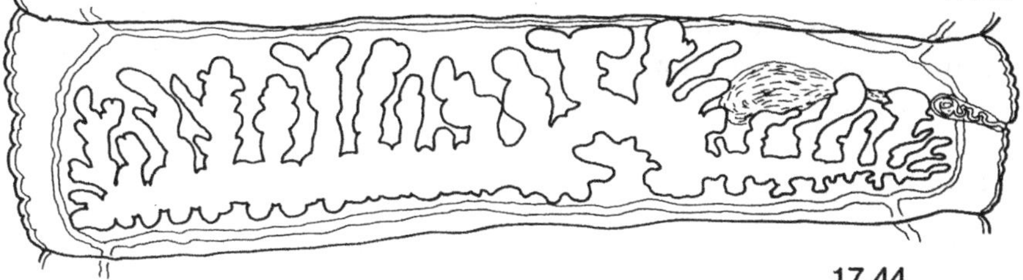

17.44

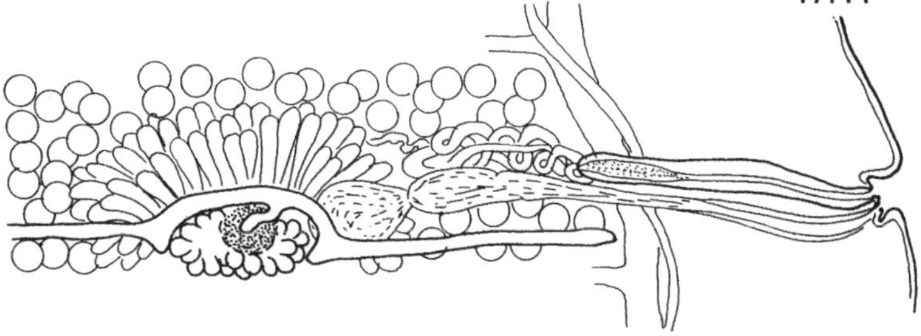

17.45

17.46

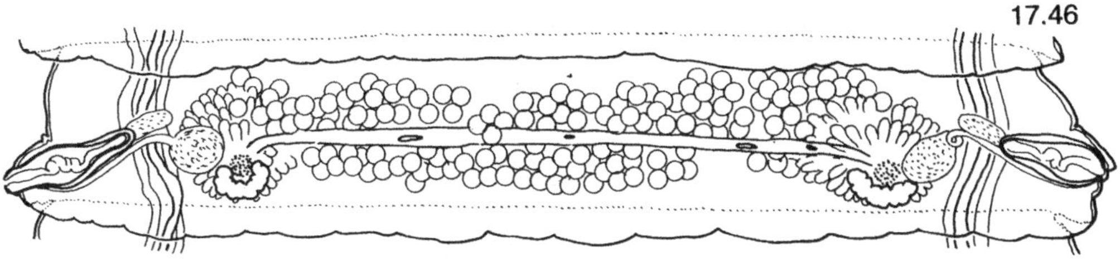

17.47

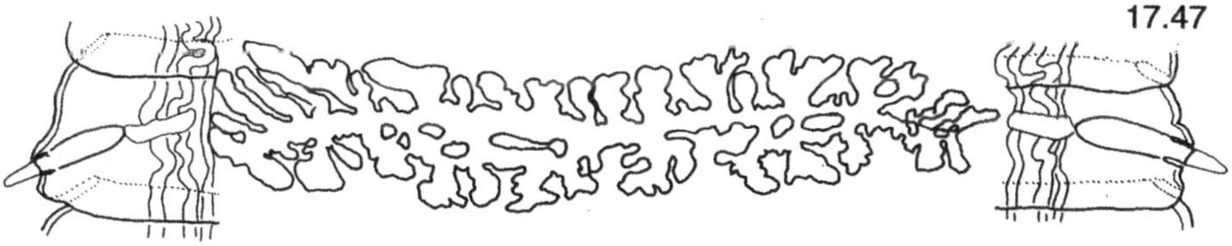

17.48

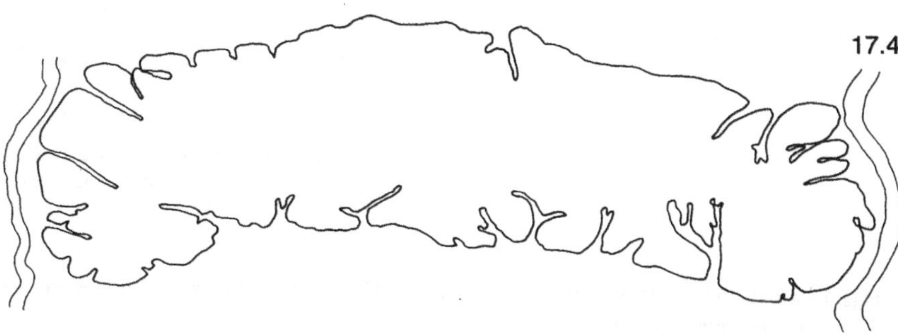

tacle present. Uterus single, tubular, transverse, looping anteriorly around each seminal receptacle, with paired, posteriorly directed diverticula on either side. Pyriform apparatus present. In psittaciform birds. New Zealand. Type-species *S. psittacea* (Fuhrmann, 1904).

27a. Uterus transverse, tubular, with few, simple reticulations (Figs 17.46, 17.47, 17.53) .... **28.**
27b. Uterus develops as complex, finely reticulated net, finally becoming saccate when gravid (Figs 17.55, 17.57–17.59) .... **30.**

28a. Genitalia single .... **29.**
28b. Genitalia paired .... ***Cittotaenia* Riehm, 1881. (Figs 17.46–17.48)**

Diagnosis: Strobila large. Proglottids craspedote, wider than long. Genital ducts cross canals dorsally. Cirrus-sac very powerfully developed. Internal and external seminal vesicle absent. Testes scattered throughout medulla or in two groups. Ovaries poral. Vagina posterior to cirrus-sac. Seminal receptacle present. Uterus single, transverse, slightly reticulate during development. Pyriform apparatus present. In leporid lagomorphs and chinchillid rodents. Europe, South America. Type-species *C. denticulata* (Rudolphi, 1804).

29a. Cavity present in apex of scolex (Fig. 17.49). Testes posterior and lateral to ovary (Fig. 17.50) ..... ***Sudarikovina* Spasskii, 1951. (Figs 17.49–17.51)**

Diagnosis: Strobila small. Proglottids craspedote, wider than long. Genitalia single. Genital pores unilateral. Genital ducts cross canals dorsally. Internal and

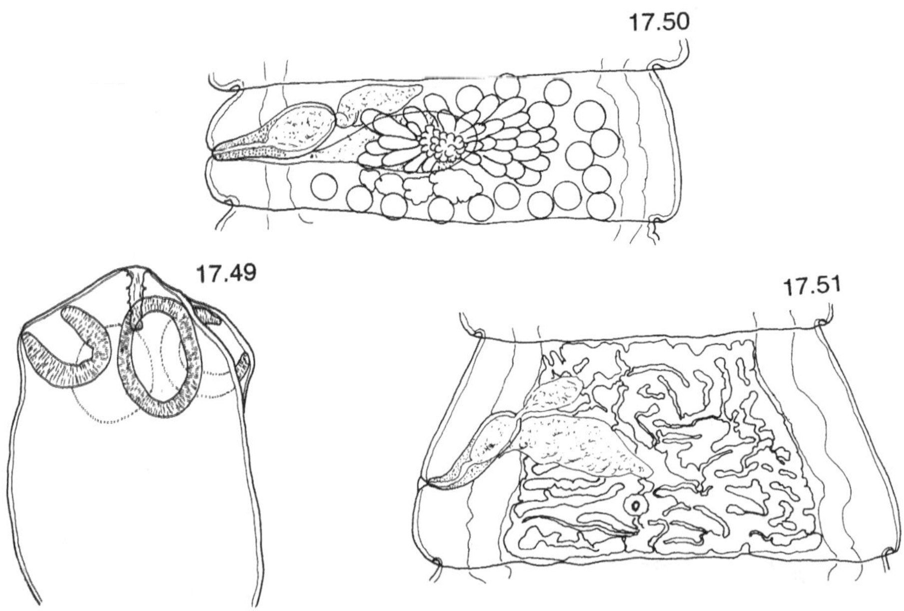

Figs 17.49–17.51 *Sudarikovina taterae* Hunkeler, 1972. 17.49. Scolex. 17.50. Mature proglottid 17.51. Gravid proglottid. Original.

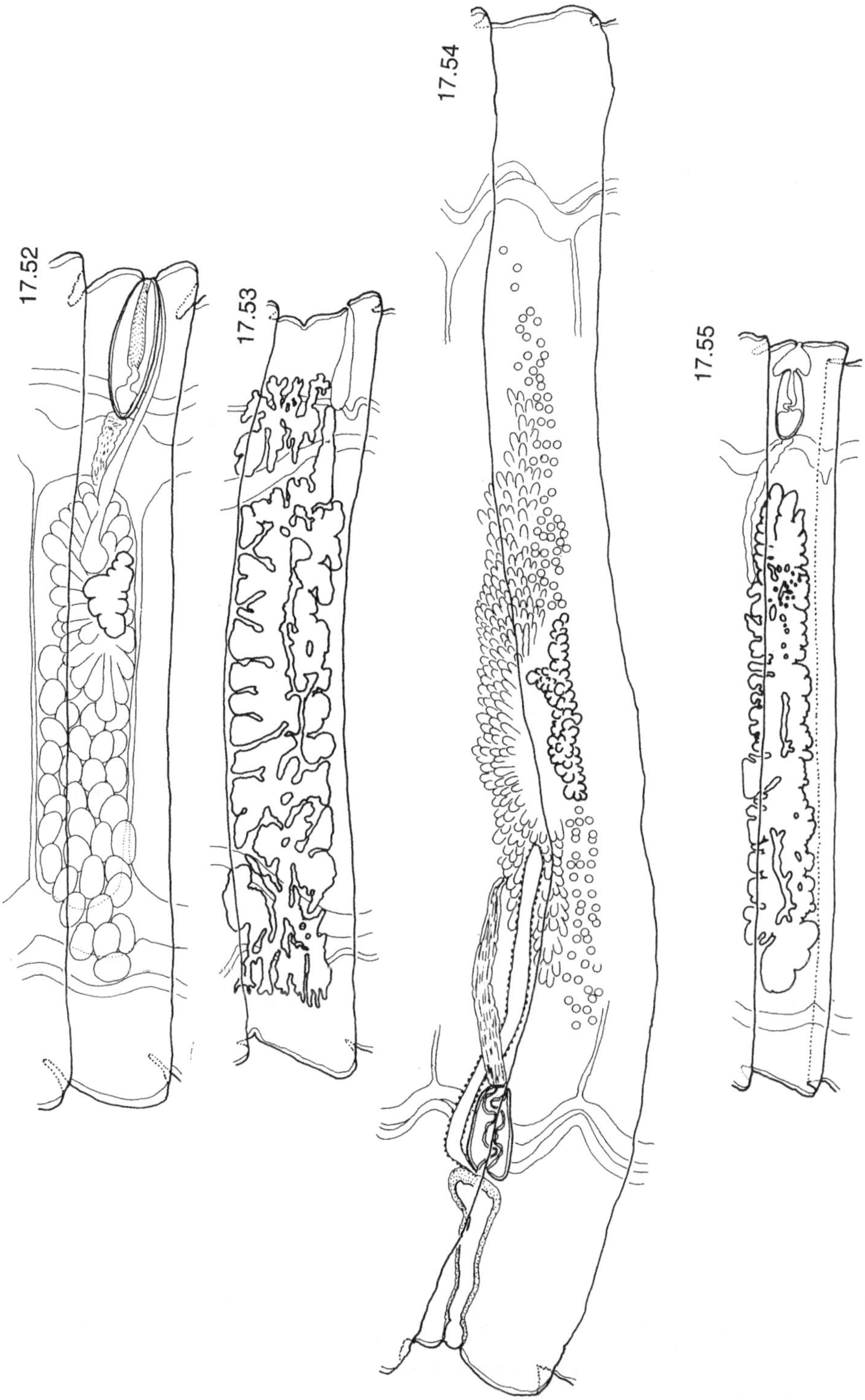

Figs 17.52–17.53 *Paranoplocephala omphalodes* (Hermann, 1783). 17.52. Mature proglottid. 17.53. Gravid proglottid. Original. **Figs 17.54, 17.55.** *Monoecocestus hagmani* (Janicki, 1904). 17.54. Mature proglottid. 17.55. Gravid proglottid. Original.

external seminal vesicles present. Ovary central. Vagina posterior to cirrus-sac. Seminal receptacle present. Uterus reticulate. Pyriform apparatus present or absent. In rodents. Africa. Type-species *S. monodi* (Joyeux & Baer, 1930).

29b. Cavity in scolex absent. Testes aporal to ovary .... ***Paranoplocephala* Lühe, 1910. (Figs 17.52, 17.53)** (Syn. *Aprostatandrya* Kirschenblatt, 1932.)

Diagnosis: Strobila small to medium-sized. Proglottids craspedote, wider than long. Genitalia single. Genital pores unilateral or irregularly alternating. Genital ducts cross canals dorsally. Internal and external seminal vesicles present. Uterus single, at first rod-like, transverse, becoming reticulate with diverticula. Pyriform apparatus present. In rodents. Europe, Asia, North America. Type-species *P. omphalodes* (Hermann, 1783).

30a. Vagina opens to genital atrium anterior to cirrus-sac. Testes posterior to ovary (Fig. 17.54) .... ***Monoecocestus* Beddard, 1914. (Figs 17.54, 17.55)** (Syns *Schizotaenia* Janicki, 1904 preoccupied; *Perutaenia* Parra, 1953; *Lentiella* Rego, 1964.)

Diagnosis: Strobila small to large. Proglottids craspedote, wider than long. Genitalia single. Genital pores alternate regularly. Genital ducts cross canals dorsally. Internal and external seminal vesicles present. Testes in band in posterior half of medulla. Ovary central or poral. Vagina anterior to cirrus-sac. Seminal receptacle present. Uterus reticulate. Pyriform apparatus present. In rodents, artiodactyls, rheiform birds. North and South America. Type-species *M. decrescens* (Diesing, 1856).

30b. Vagina opens to genital atrium posterior to cirrus-sac. Testes anterior and posterior to ovary .... **31.**

31a. Genitalia single .... ***Andrya* Railliet, 1893. (Figs 17.56, 17.57)**

Diagnosis: Strobila moderate. Proglottids craspedote, wider than long. Genital pores alternate regularly. Genital ducts cross canals dorsally. Internal and external seminal vesicles present; external vesicle with prominent covering of 'glandular' cells. Testes scattered, mainly aporal. Ovary poral. Vagina posterior to cirrus-sac. Seminal receptacle present. Uterus finely reticulate when first develops. Pyriform apparatus present. In leporid lagomorphs. Europe, Africa, Asia. Type-species *A. rhopalocephala* (Riehm, 1881).

31b. Genitalia paired .... **32.**

Figs 17.56, 17.57 *Andrya rhopalocephala* (Riehm, 1881). 17.56. Mature proglottid. Note thickened glandular covering of external seminal vesicle. 17.57. Developing reticulate uterus. Original.
Fig. 17.58 *Diandrya composita* Darrah, 1930. Lateral region of mature proglottid. Note thickened glandular covering of external seminal vesicle. Original.
Fig. 17.59 *Moniezia expansa* (Rudolphi, 1810). Mature proglottid. Original.

17.56

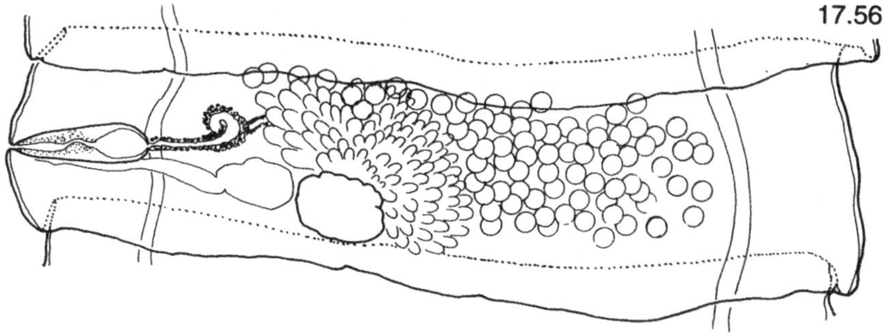

17.57

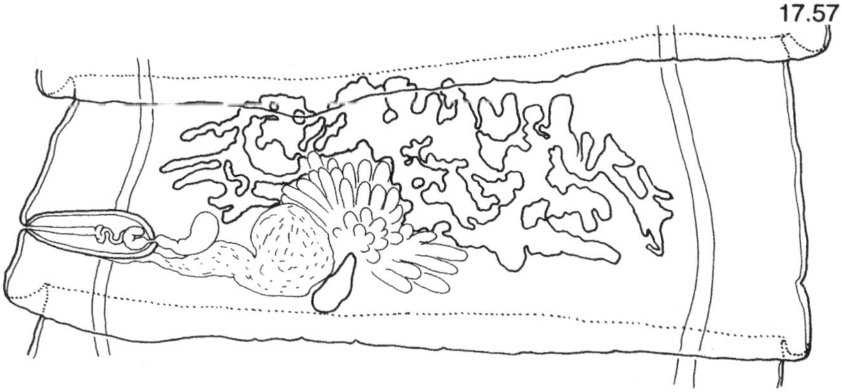

17.58

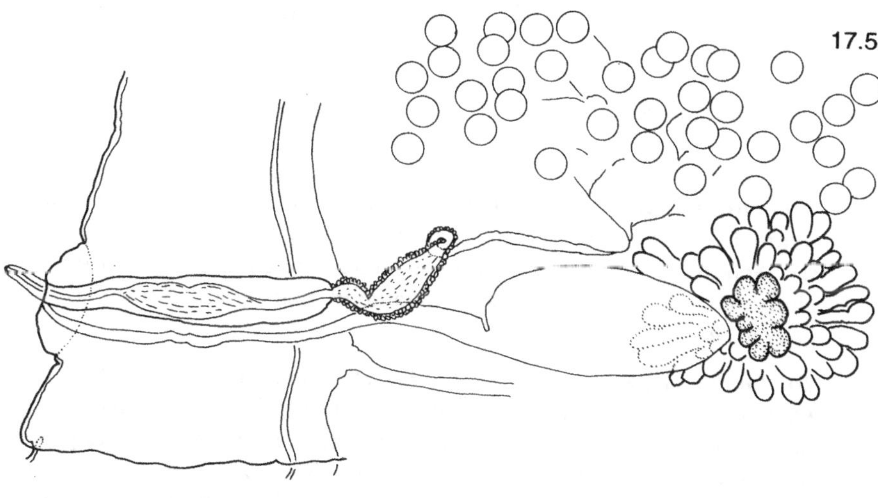

17.59

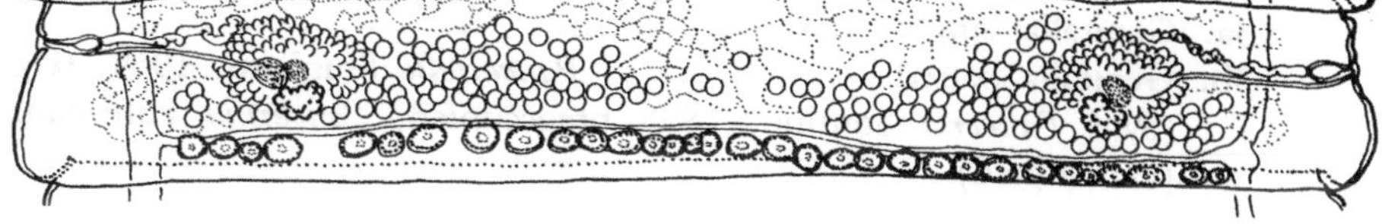

32a. External seminal vesicle covered with gland cells (Fig. 17.58) .... ***Diandrya* Darrah, 1930. (Fig. 17.58)**

Diagnosis: Strobila large. Proglottids craspedote, wider than long. Genitalia paired. Genital ducts cross canals dorsally. Internal and external seminal vesicles present. Testes scattered throughout medulla. Ovaries poral. Vagina posterior to cirrus-sac. Seminal receptacle present. Uterus finely reticulate when first develops. Pyriform apparatus present. In sciurid rodents. North America. Type-species *D. composita* Darrah, 1930.

32b. External seminal vesicle absent .... ***Moniezia* Blanchard, 1891. (Fig. 17.59)**
(Syns *Baeriezia* Skryabin & Schulz, 1937; *Blanchardiezia* Skryabin & Schulz, 1937; *Eranuides* Semenova, 1972; *Fuhrmannella* Baer, 1925.)

Diagnosis: Strobila large. Proglottids craspedote, wider than long. Genitalia paired. Genital ducts cross canals dorsally. Testes scattered throughout medulla. Ovaries poral. Vagina posterior to cirrus-sac. Seminal receptacle present. Uterus finely reticulate when first formed. Interproglottidal glands present or absent. Pyriform apparatus present. In ruminants, suids, rodents, primates, ratite birds. Cosmopolitan. Type-species *M. expansa* (Rudolphi, 1810).

Key to genera of the Linstowiinae

1a. Vitellarium compact, post-ovarian (Figs 17.65, 17.66, 17.68) .... 3.
1b. Vitellarium transversely elongated, extending beyond ovary (Figs 17.60, 17.61) .... 2.

2a. Genitalia single .... ***Linstowia* Zschokke, 1899. (Fig. 17.60)**
(Syn. *Peramelinia* Spasskii, 1987.)

Diagnosis: Strobila large. Proglottids craspedote, wider than long. Genital pores alternate irregularly. Genital ducts cross canals ventrally. Internal and external seminal vesicles present. Testes scattered. Ovary poral. Vagina posterior to cirrus-sac. Seminal receptacle diminutive. Uterus ephemeral. Vitellarium transversely elongated. In monotremes and peramelid marsupials. Australia. Type-species *L. echidnae* (Thompson, 1893).

2b. Genitalia paired .... ***Echidnotaenia* Beveridge, 1980. (Fig. 17.61)**
Diagnosis: Strobila small. Proglottids craspedote, wider than long. Genital ducts cross canals ventrally. Internal and external seminal vesicles present. Testes scattered. Vagina posterior to cirrus-sac. Seminal receptacle present. Uterus ephemeral. Vitellarium transversely elongated. In monotremes. Australia. Type-species *E. tachyglossi* (Johnston, 1913).

3a. Genitalia single .... 4.
3b. Genitalia paired .... ***Panceriella* Stunkard, 1969. (Fig. 17.62)**
(Syns *Panceria* Sonsino, 1895 preoccupied; *Pancerina* Fuhrmann, 1899 preoccupied)

17.60

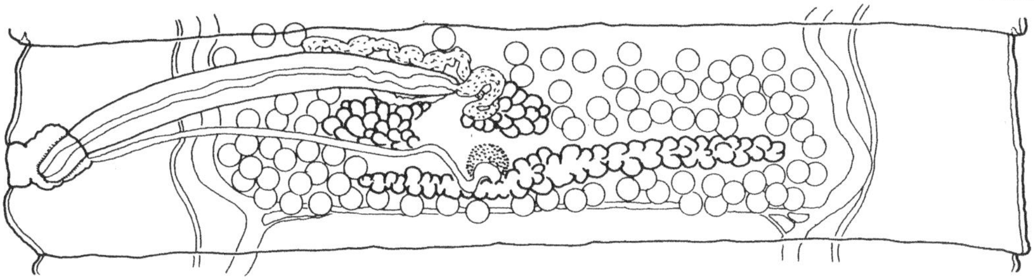

17.61

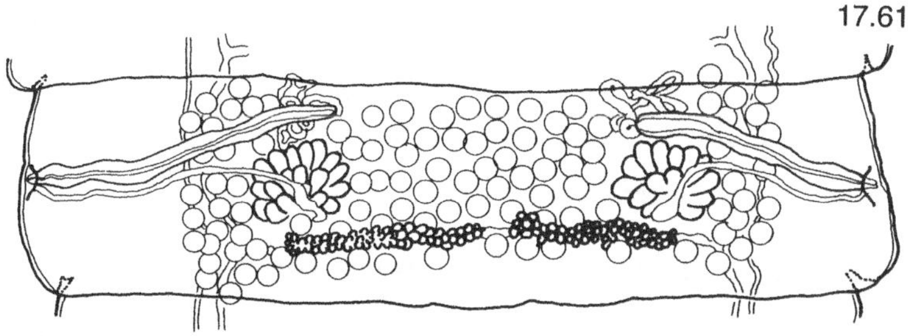

17.62

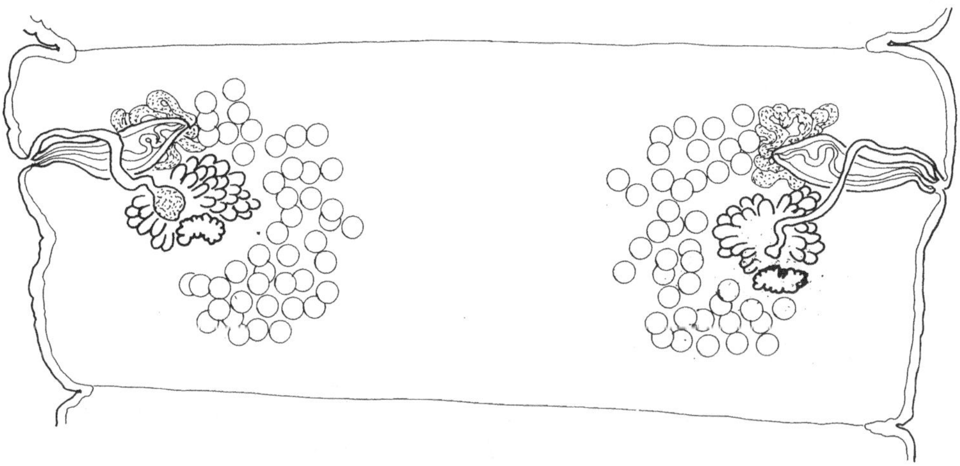

Fig. 17.60 *Linstowia echidnae* (Thompson, 1893). Mature proglottid. Note transversely extended vitellarium. Redrawn from Beveridge (1983).
Fig. 17.61 *Echidnotaenia tachyglossi* (Johnston, 1913). Mature proglottid. Note transversely extended vitellaria. Redrawn from Beveridge (1980).
Fig. 17.62 *Panceriella varani* (Stossich, 1895). Mature proglottid. Original.

Diagnosis: Strobila small. Proglottids acraspedote; mature proglottids wider than long. Genital ducts lie between canals. Seminal vesicles absent. Testes in two groups anterior, medial and posterior to female genitalia. Vagina opens to genital atrium anterior to cirrus-sac. Seminal receptacle present. Uterus ephemeral. Vitellarium compact. In varanid lizards. Africa. Middle East. Type-species *P. varani* (Stossich, 1895).

4a. Testes present anterior to ovary (Figs 17.63–17.66) .... **5.**
4b. Testes posterior to ovary and/or on either side, not anterior (Figs 17.68–17.71) .... **8.**

5a. Testes entirely anterior to ovary (Fig. 17.63) .... ***Sinaiotaenia* Wertheim & Greenberg, 1971. (Fig. 17.63)**

Diagnosis: Strobila moderate. Proglottids acraspedote, longer than wide. Genitalia single. Genital pores alternate irregularly. Genital ducts lie between canals. Seminal vesicles absent. Testes scattered anterior and lateral to female genitalia. Ovary at posterior end of proglottid. Vagina posterior to cirrus-sac. Seminal receptacle present. Uterus ephemeral. In gerbillid rodents. Sinai. Type-species *S. witenbergi* Wertheim & Greenberg, 1971.

5b. Some testes posterior to ovary .... **6.**

6a. Testes anterior to ovary in two distinct lateral groups (Fig. 17.64) .... ***Cycloskrjabinia* Spasskii, 1951. (Fig. 17.64)**

Diagnosis: Strobila small. Proglottids acraspedote, longer than wide. Genitalia single. Genital pores alternate irregularly. Genital ducts cross canals ventrally. Internal and external seminal vesicles absent. Testes posterior to ovary and in two lateral bands extending to anterior end of proglottid. Ovary central. Vagina posterior to cirrus-sac. Seminal receptacle absent. Uterus ephemeral. Vitellarium compact. In Chiroptera. North America, Europe. Type-species *C. taborensis* (Loewen, 1934).

6b. Testes anterior to ovary not in two distinct lateral groups .... **7.**

7a. Proglottids acraspedote (Fig. 17.65) .... ***Semenoviella* Spasskii, 1951. (Fig. 17.65)**

Diagnosis: Strobila small. Genitalia single. Genital pores alternate irregularly. Genital ducts pass canals ventrally. Internal and external seminal vesicles absent. Testes surround female genital complex. Cirrus-sac large, crosses medulla. Ovary central. Vagina posterior to cirrus-sac. Uterus ephemeral. Vitellarium compact. In lizards. South America. Type-species *S. amphisbaenae* (Rudolphi, 1819).

7b. Proglottids craspedote .... ***Paralinstowia* Baer, 1927. (Fig. 17.66)**
(Syn. *Pseudolinstowia* Spasskii, 1987.)

Diagnosis: Strobila small. Genitalia single. Genital pores unilateral or irregularly alternate. Genital ducts cross canals ventrally. Internal and external seminal vesicles absent. Testes scattered throughout medulla. Ovary central. Vagina posterior to cirrus-sac. Seminal receptacle present. Uterus ephemeral. Vitellarium compact. In didelphid, dasyurid and peramelid marsupials. South America, Australia. Type-species *P. iheringi* (Zschokke, 1904).

8a. Suckers with prominent, paired lappets (Fig. 17.67) .... ***Tupaiataenia* Schmidt & File, 1977. (Fig. 17.67)**

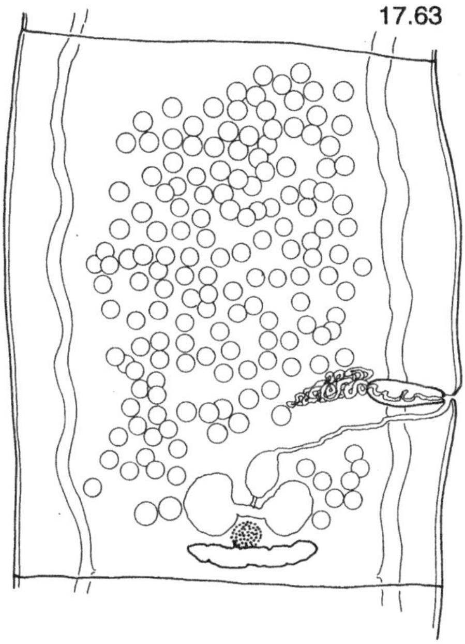

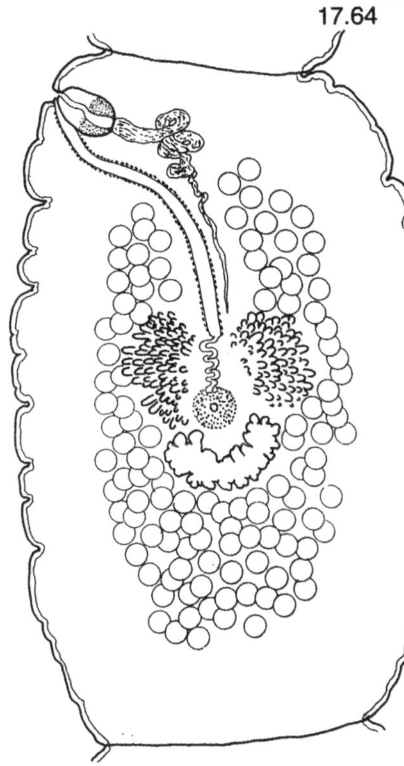

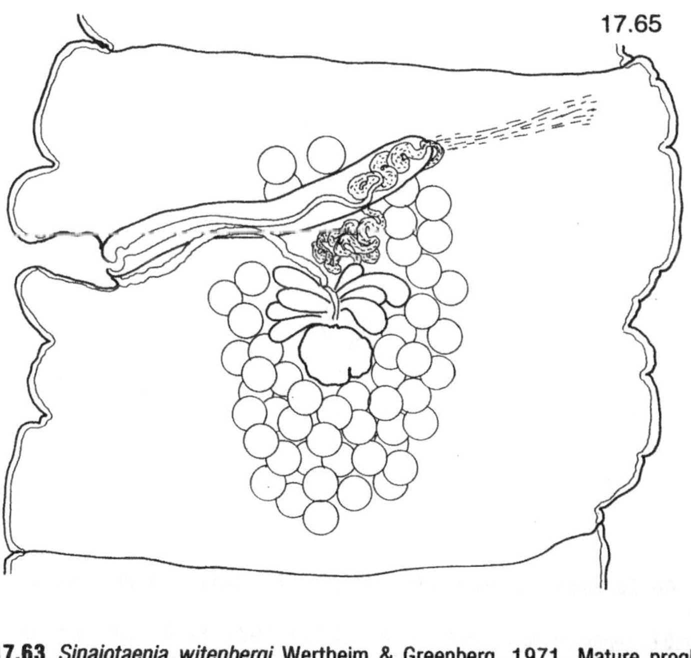

Fig. 17.63 *Sinaiotaenia witenbergi* Wertheim & Greenberg, 1971. Mature proglottid. Original.
Fig. 17.64 *Cycloskrjabinia taborensis* (Loewèn, 1934). Mature proglottid. Original.
Fig. 17.65 *Semenoviella amphisbaenae* (Rudolphi, 1819). Mature proglottid. Original.

17.66

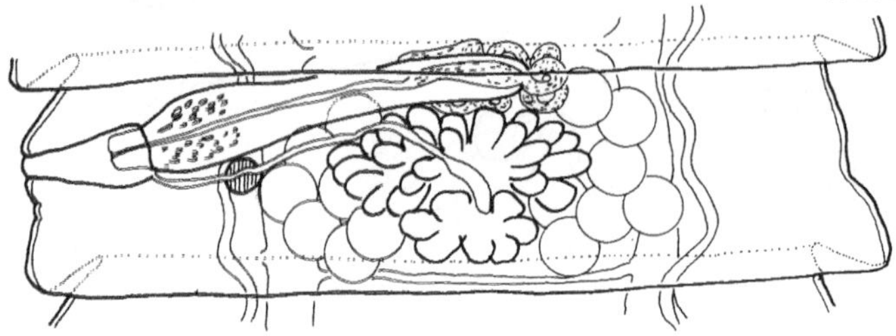

17.67

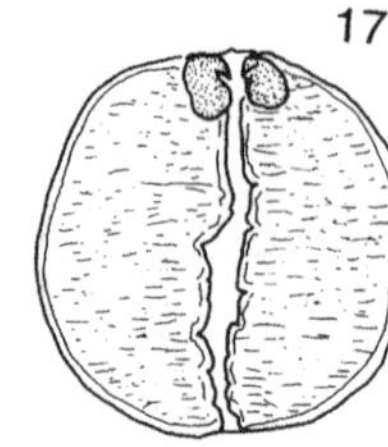

17.68

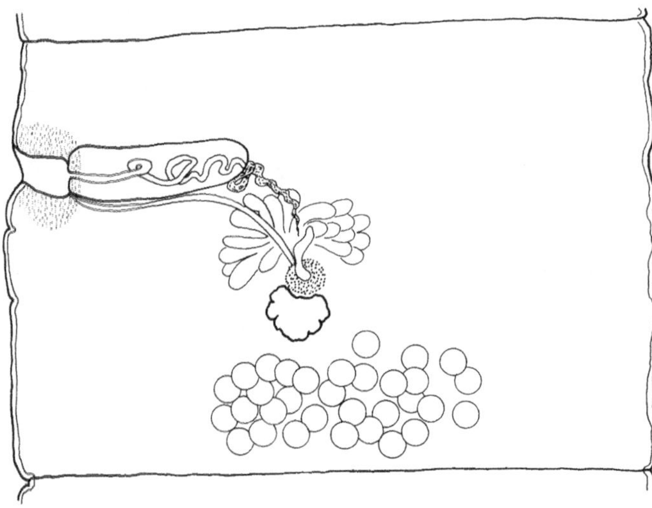

17.69

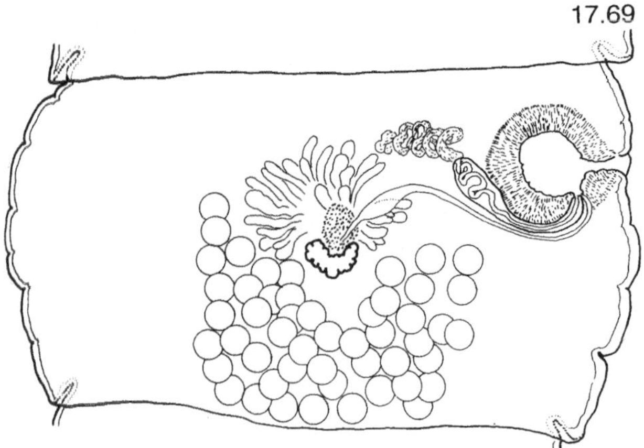

Fig. 17.66 *Paralinstowia semoni* (Zschokke, 1896). Mature proglottid. Redrawn from Beveridge (1983).
Fig. 17.67 *Tupaiataenia quentini* Schmidt & File, 1977. Sucker with anterior lappets. Original.
Fig. 17.68 *Oochoristica tuberculata* (Rudolphi, 1819). Mature proglottid. Original.
Fig. 17.69 *Atriotaenia sandgroundi* (Baer, 1935). Paratype; mature proglottid. Original.

Diagnosis: Strobila small. Proglottids craspedote, wider than long. Genitalia single. Genital pores alternate irregularly. Genital ducts cross canals dorsally. Internal seminal vesicle present; external vesicle absent. Testes posterior to vitellarium. Ovary central. Vagina posterior to cirrus-sac. Seminal receptacle present. Uterus ephemeral. Vitellarium compact. In tree-shrews. South-east Asia. Type-species *T. quentini* Schmidt & File, 1977.

8b. Suckers without lappets .... **9.**

9a. Proglottids acraspedote .... ***Oochoristica* Lühe, 1898. (Fig. 17.68)**
(Syns *Diochetos* Harwood, 1932; *Skrjabinochora* Spasskii, 1948; *Megacapsula* Wahid, 1961; *Sharpilia* Spasskii, 1988; *Semenochetos* Palladwar & Kalyankar, 1989).

Diagnosis: Strobila small. Proglottids longer than wide. Genitalia single. Genital pores alternate irregularly. Genital ducts lie between canals. Internal and external seminal vesicles absent. Testes posterior to vitellarium. Ovary central. Vagina posterior to cirrus-sac. Seminal receptacle present or absent. Uterus ephemeral. Vitellarium compact. In reptiles, mammals. Cosmopolitan. Type-species *O. tuberculata* (Rudolphi, 1819).

9b. Proglottids craspedote .... **10.**

10a. Genital atrium prominent with well-developed radial musculature, genital ducts open into posterior part of atrium (Fig. 17.69) ....
.... ***Atriotaenia* Sandground, 1926. (Fig. 17.69)**
(Syn. *Ershovia* Spasskii, 1951.)

Diagnosis: Strobila small. Proglottids craspedote, wider than long. Genitalia single. Genital pores alternate irregularly. Genital ducts cross canals dorsally or lie between them. Cirrus-sac enters genital atrium posteriorly. Internal and external seminal vesicles absent. Testes posterior and lateral to female genital complex. Ovary central. Vagina posterior to cirrus-sac. Seminal receptacle present. Uterus ephemeral. Vitellarium compact. In mustelid and procyonid mammals. Europe, North and South America. Type-species *A. sandgroundi* (Baer, 1935).

10b. Genital atrium without prominent muscular atrium. Genital ducts open into medial part of atrium .... **11.**

11a. Testes invariably divided into two lateral groups (Fig. 17.70) ....
***Witenbergitaenia* Wertheim, Schmidt & Greenberg, 1986. (Fig. 17.70)**

Diagnosis: Strobila small. Proglottids craspedote, as wide as long. Genitalia single. Genital pores alternate irregularly. Genital ducts cross canals dorsally. Internal and external seminal vesicles absent. Testes posterior to female genitalia. Ovary central. Vagina posterior to cirrus-sac. Seminal receptacle present. Uterus ephemeral. Vitellarium compact. In murid rodents. Israel. Type-species *W. sinaica* Wertheim, Schmidt & Greenberg, 1986.

11b. Testes in single group or some proglottids in strobila with testes in single group .... ***Mathevotaenia* Akhumyan, 1946. (Fig. 17.71)**
(Syns *Opossumia* Spasskii, 1951; *Inversia* Spasskii, 1951; *Morosovella*

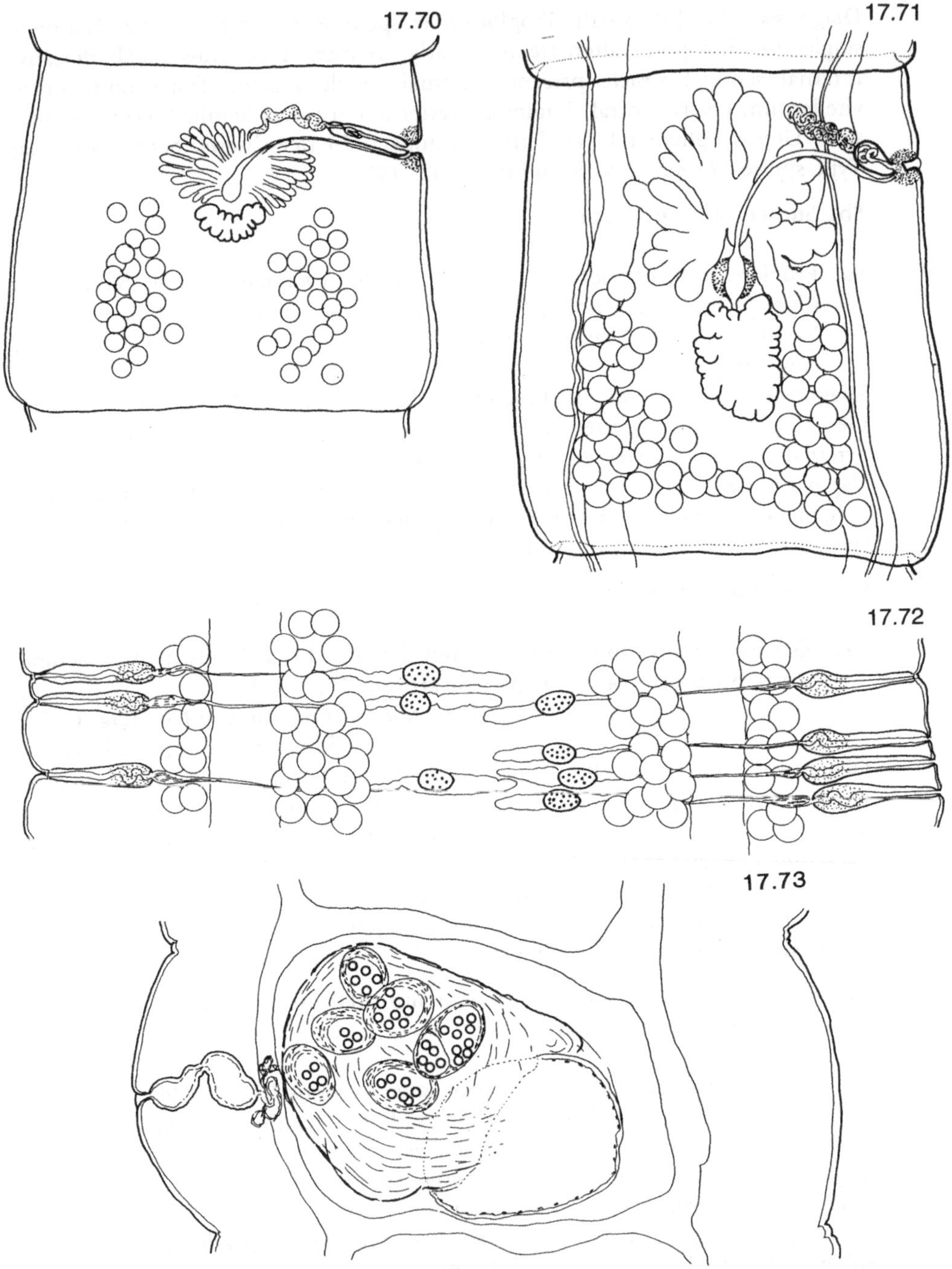

Fig. 17.70 *Witenbergitaenia sinaica* Wertheim, Schmidt & Greenberg, 1986. Mature proglottid. Redrawn from Wertheim *et al.* (1986).
Fig. 17.71 *Mathevotaenia symmetrica* (Baylis, 1927). Type; mature proglottid. Original.
Figs 17.72, 17.73 *Avitellina centripunctata* (Rivolta, 1874). 17.72. Mature proglottids; note lack of segmentation. 17.73. Gravid proglottid with single paruterine organ and eggs and paruterine capsules. Original.

Spasskii, 1951; *Paratriotaenia* Stunkard, 1965; *Markewitschitaenia* Sharpilo & Kornyushin, 1975; *Timorenia* Spasskii, 1987; *Hickmania* Spasskii, 1987, as *Hickmawia*; *Schizorchodes* Bienek & Grundman, 1973; *Priodontia* Spasskii, 1987; *Linstoparonia* Spasskii, 1987; *Mangustella* Spasskii, 1987; *Vasoramia* Spasskii, 1987.)

Diagnosis: Strobila small. Proglottids craspedote. Genitalia single. Genital pores alternate irregularly. Genital ducts cross canals dorsally or lie between them. Internal and external seminal vesicles absent. Testes posterior and lateral to female genitalia. Ovary central. Vagina posterior to cirrus-sac. Seminal receptacle present. Uterus ephemeral. Vitellarium compact. In mammals. Cosmopolitan. Type-species *M. symmetrica* (Baylis, 1927).

Key to genera of the Thysanosomatinae

1a. One or two paruterine organs per proglottid (Figs 17.73, 17.77) .... 2.
1b. Numerous small paruterine organs develop from each uterus (Figs 17.79, 17.80, 17.82) .... 3.

2a. Single paruterine organ per proglottid (Fig. 17.73) .... *Avitellina* Gough, 1911. (Figs 17.72, 17.73)
(Syns *Hexastichorchis* Blei, 1921; *Anootypus* Woodland, 1928; *Ascotaenia* Baer, 1927.)

Diagnosis: Strobila large. Proglottids poorly demarcated particularly in anterior region; wider than long. Genitalia single. Genital pores alternate irregularly. Genital ducts cross canals dorsally. Internal and external seminal vesicles absent. Testes in two lateral groups, each subdivided by osmoregulatory canal. Germovitellarium (combined ovary and vitellarium) present. Single paruterine organ with fibrous capsules each containing several eggs. In ruminants. Europe, Asia, Africa, North America. Type-species *A. centripunctata* (Rivolta, 1874).

2b. Two paruterine organs per proglottid (Fig. 17.77) .... *Stilesia* Railliet, 1893. (Figs 17.74–17.77)
(Syn. *Aliezia* Shinde, 1969.)

Diagnosis: Strobila large. Proglottids craspedote, wider than long. Genitalia single. Genital pores alternate irregularly. Genital ducts lie between canals. Internal and external seminal vesicles absent. Testes in two lateral groups external to dorsal canals. Germovitellarium (combined ovary and vitellarium) present. Vagina posterior to cirrus-sac. Seminal receptacle absent. Uterus initially transverse tubular with enlarged ends; each extremity develops paruterine organ. In bile-ducts of ruminants. Asia, Africa. Europe. Type-species *S. globipunctata* (Rivolta, 1874).

3a. Testes in two lateral, extravascular groups (Figs 17.78, 17.79), genitalia single .... *Thysaniezia* Skryabin, 1926. (Figs 17.78, 17.79)
(Syn. *Helictometra* Baer, 1927.)

Diagnosis: Strobila large. Proglottids craspedote, wider than long. Genital pores

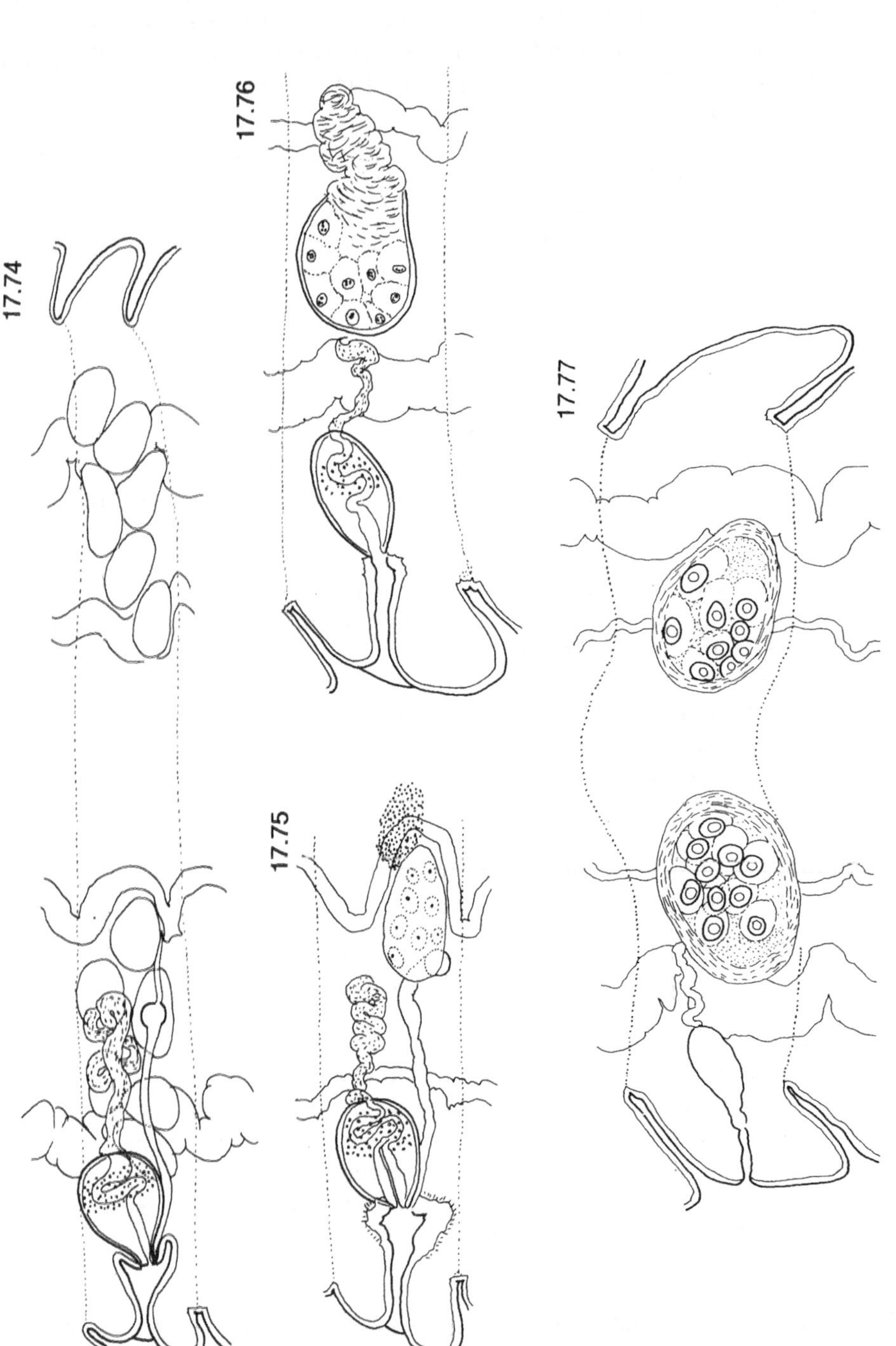

Figs 17.74–17.77 *Stilesia globipunctata* (Rivolta, 1874). 17.74. Mature proglottid. 17.75. Gravid proglottid, 17.76. Developing paruterine organ. 17.77. Gravid proglottid with two paruterine organs. Original.

17.78

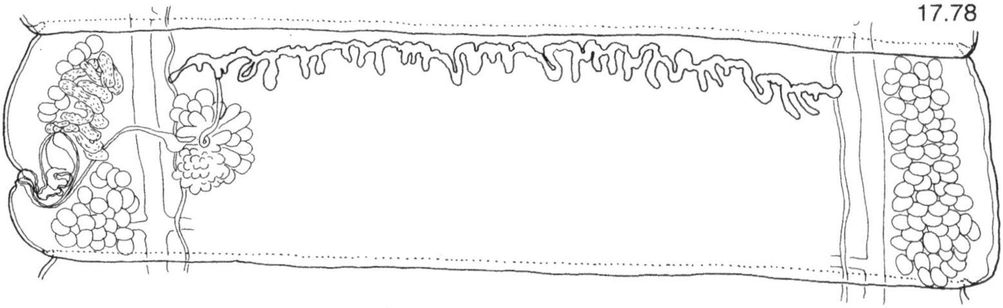

17.79

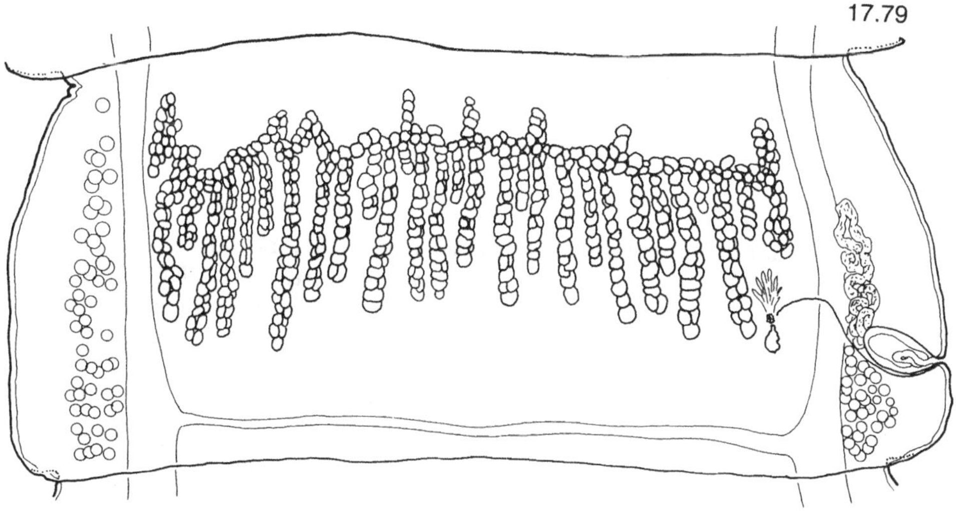

17.80

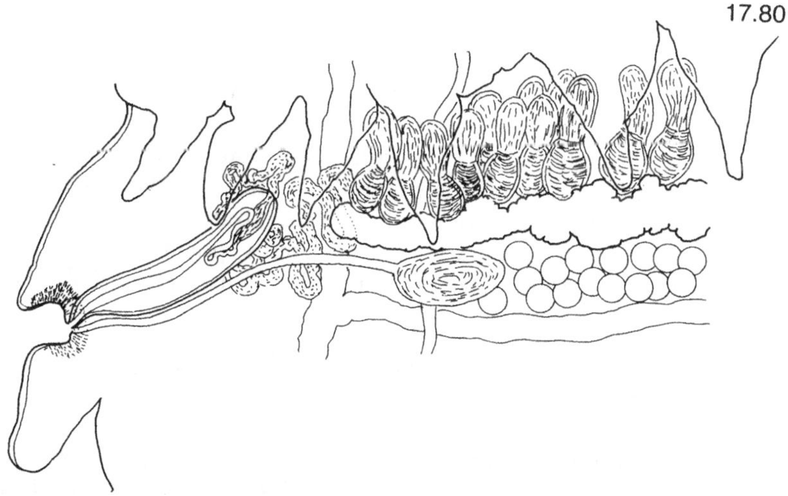

Figs 17.78, 17.79 *Thysaniezia ovilla* (Rivolta, 1878). 17.78. Mature proglottid. 17.79. Gravid proglottid. Original.

Fig. 17.80 *Thysanosoma actinioides* Diesing, 1835. Lateral region of mature proglottid showing fringed velum, uterus and paruterine organs. Original.

alternate irregularly. Genital ducts lie between canals. Internal seminal vesicles present. Ovary poral. Vitellarium post-ovarian. Vagina posterior to cirrus-sac. Seminal receptacle present. Uterus initially tubular, transverse, replaced by >300 paruterine capsules. In ruminants. Cosmopolitan. Type-species *T. ovilla* (Rivolta, 1878).

3b. **Testes in single continuous band posterior to uterus (Figs 17.80, 17.81). Genitalia paired** .... 4.

4a. **Proglottid prominently fringed (Fig. 17.80)** .... ***Thysanosoma* Diesing, 1835. (Fig. 17.80)**

Diagnosis: Strobila large. Proglottids craspedote, wider than long. Genitalia paired. Genital ducts lie between canals. Seminal vesicles absent. Testes numerous, in band posterior to uterus. Germovitellarium present; posterior to level of cirrus-sac. Seminal receptacle present. Uterus single, transverse. Numerous (>250) small paruterine organs develop anteriorly from uterus. In bile-ducts of ruminants. North and South America. Type-species *T. actinioides* Diesing, 1835.

4b. **Proglottid not prominently fringed (Figs 17.81, 17.82)** .... ***Wyominia* Scott, 1941. (Figs 17.81, 17.82)**

Diagnosis: Strobila large. Proglottids craspedote, wider than long. Velum folded. Genitalia paired. Genital ducts lie between canals. Seminal vesicles absent. Testes numerous, in band posterior to uterus. Ovaries poral. Vitellarium post-ovarian. Vagina posterior to cirrus-sac. Seminal receptacle present. Uterus single, transverse, replaced by about 20 paruterine organs. In bile-ducts of ruminants. North America. Type-species *W. tetoni* Scott, 1941.

Key to genera of the Inermicapsiferinae

1a. **Testes and egg capsules intervascular. Ovary central (Fig. 17.83)** .... ***Thysanotaenia* Beddard, 1911. (Fig. 17.83)**

Diagnosis: Strobila large. Proglottids craspedote, wider than long. Genitalia single. Genital ducts cross canals dorsally. Seminal vesicles absent. Testes absent only from anterior, poral region of proglottid. Ovary subcentral. Vagina posterior to cirrus-sac. Seminal receptacle present. Uterus ephemeral; fibrous egg capsules intervascular. In lemurs. Malagasy Republic. Type-species *T. lemuris* Beddard, 1911.

1b. **Testes and egg capsules extending beyond osmoregulatory canals. Ovary poral (Figs 17.85, 17.86)** .... 2.

2a. **Testes scattered uniformly throughout medulla (Fig. 17.84)** .... ***Pericapsifer* Spasskii, 1951. (Fig. 17.84)**

Diagnosis: Strobila small. Proglottids craspedote, wider than long. Genitalia single. Genital pores unilateral. Genital ducts cross canals dorsally. Seminal vesicles absent. Testes extending beyond canals. Ovary poral. Vagina posterior to

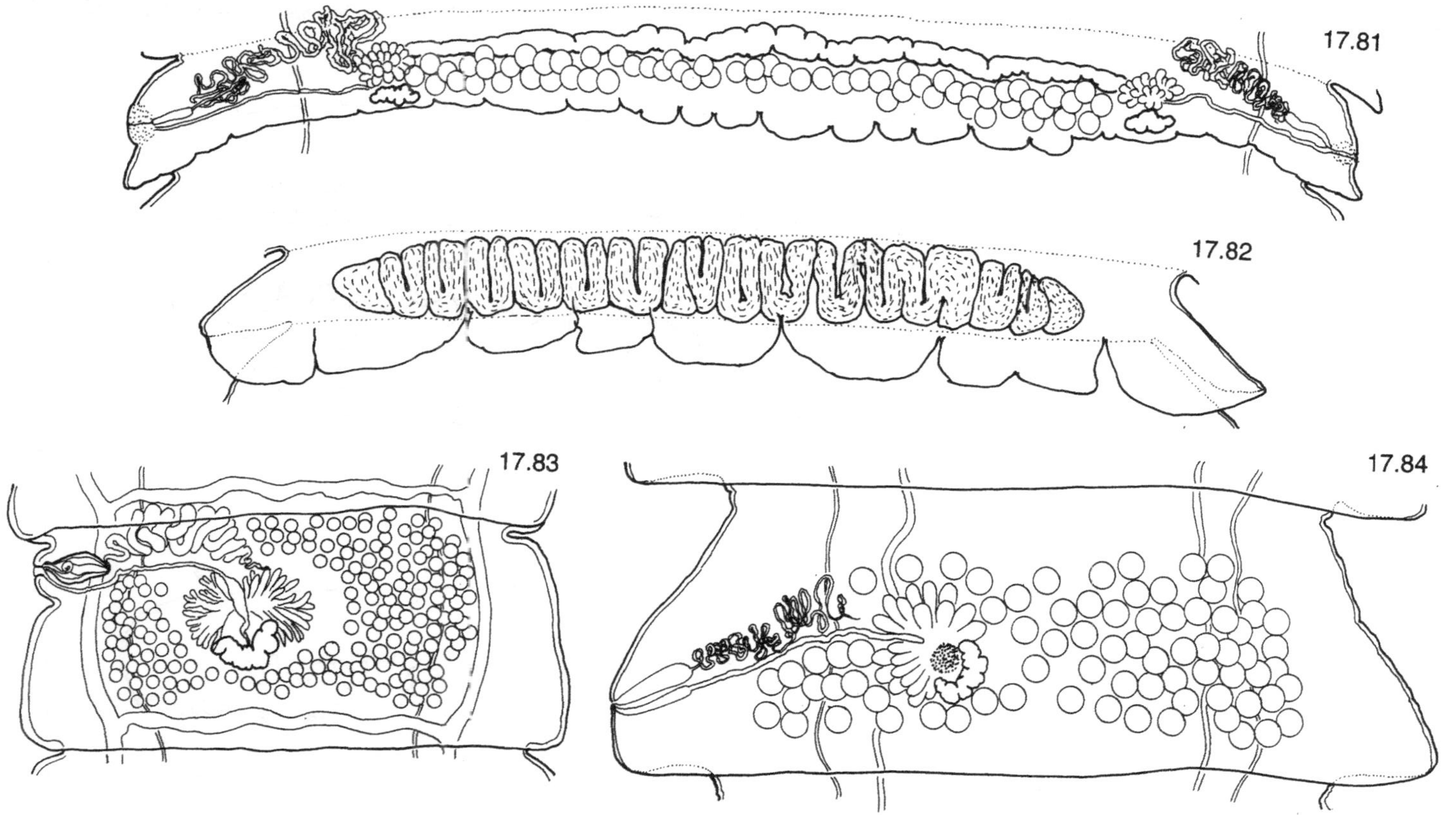

Figs 17.81, 17.82 *Wyominia tetoni* Scott, 1941. 17.81 Mature proglottid. 17.82. Gravid proglottid. Original.
Fig. 17.83 *Thysanotaenia lemuris* Beddard, 1911. Mature proglottid. Redrawn from Deblock & Diaouré (1962).
Fig. 17.84 *Pericapsifer pagenstecheri* (Setti, 1897). Mature proglottid. Original.

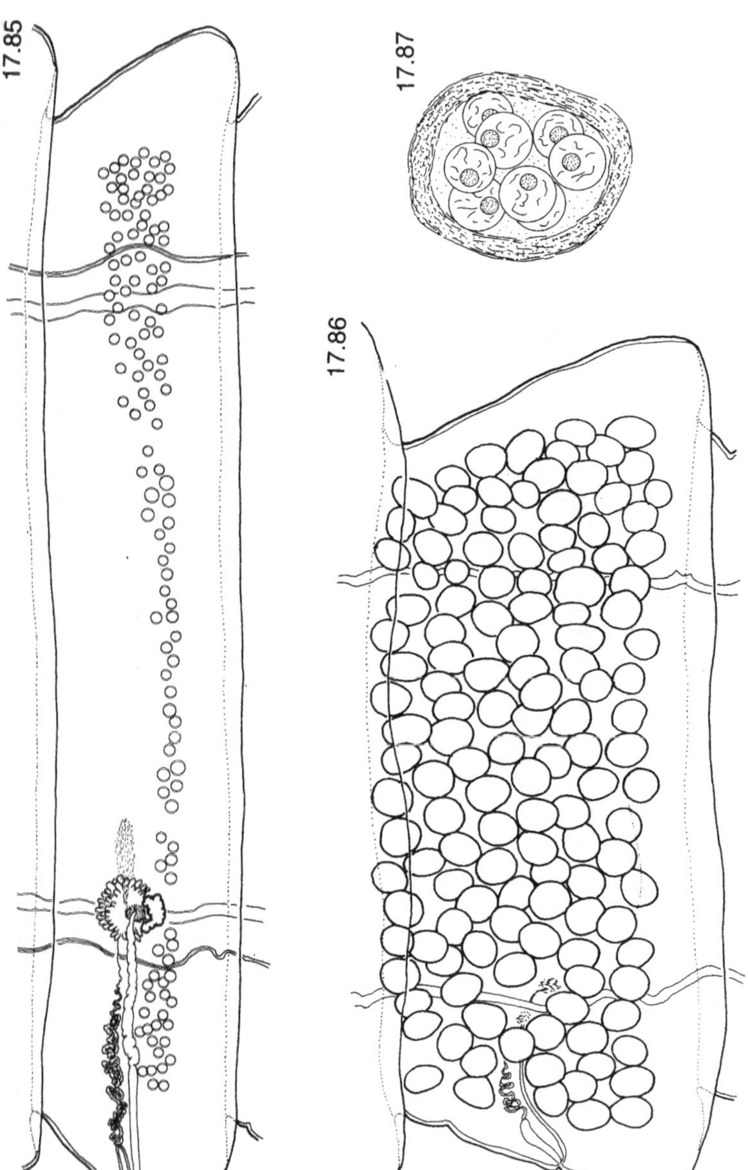

Figs 17.85–17.87 *Inermicapsifer hyracis* (Rudolphi, 1808). 17.85. Mature proglottid; note extent of testes beyond canals. 17.86. Gravid proglottid; note extravascular egg capsules. 17.87. Egg capsule. Original.

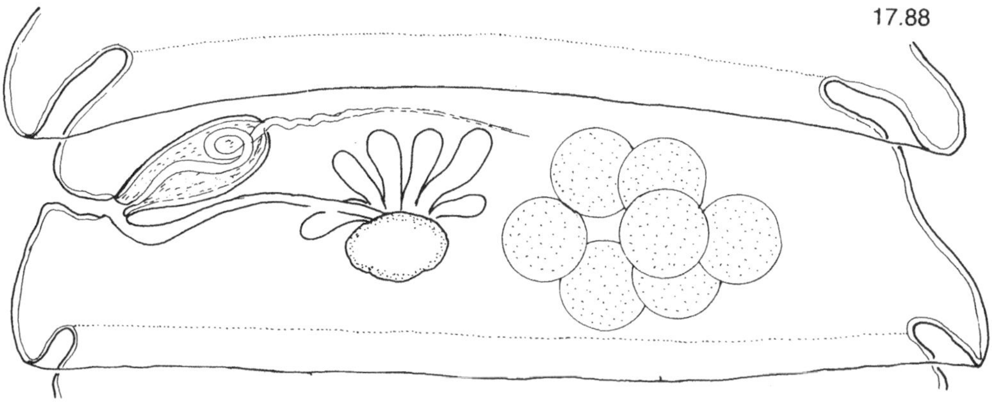

Fig. 17.88 *Metacapsifer aberratus* (Baer, 1925). Mature proglottid. Redrawn from Baer (1925c).

cirrus-sac. Seminal receptacle present. Uterus ephemeral; fibrous egg capsules extend laterally beyond canals. In hyracoids. Africa. Type-species *P. pagenstecheri* (Setti, 1897).

2b. Testes restricted in distribution .... **3.**

3a. Testes posterior to ovary or posterior and lateral or in two lateral groups (Fig. 17.85) .... ***Inermicapsifer*** **Janicki, 1910. (Figs 17.85–17.87)** (Syns *Arhynchotaenia* Pagenstecher, 1877 preoccupied; *Hyracotaenia* Beddard, 1912.)

Diagnosis: Strobila small to large. Proglottids craspedote, wider than long. Genitalia single. Genital pores unilateral. Genital ducts cross canals dorsally. Seminal vesicles absent. Testes in single band at posterior margin of proglottid, or posterior and lateral to ovary, or two lateral groups, extending beyond canals. Ovary poral. Vagina posterior to cirrus-sac. Seminal receptacle present. Uterus ephemeral; fibrous egg capsules extend laterally beyond canals. In hyracoids, rodents, lagomorphs. Africa, Asia. Type-species *I. hyracis* (Rudolphi, 1808).

3b. Testes restricted to aporal side of proglottid (Fig. 17.88) .... ***Metacapsifer*** **Spasskii, 1951. (Fig. 17.88)**

Diagnosis: Strobila small. Proglottids craspedote. Genitalia single. Genital pores unilateral. Seminal vesicles absent. Testes few. Ovary poral. Vagina posterior to cirrus-sac. Seminal receptacle present. Uterus ephemeral; fibrous egg capsules replace uterus. In rodents. Africa. Type-species *M. aberratus* (Baer, 1925).

17.89

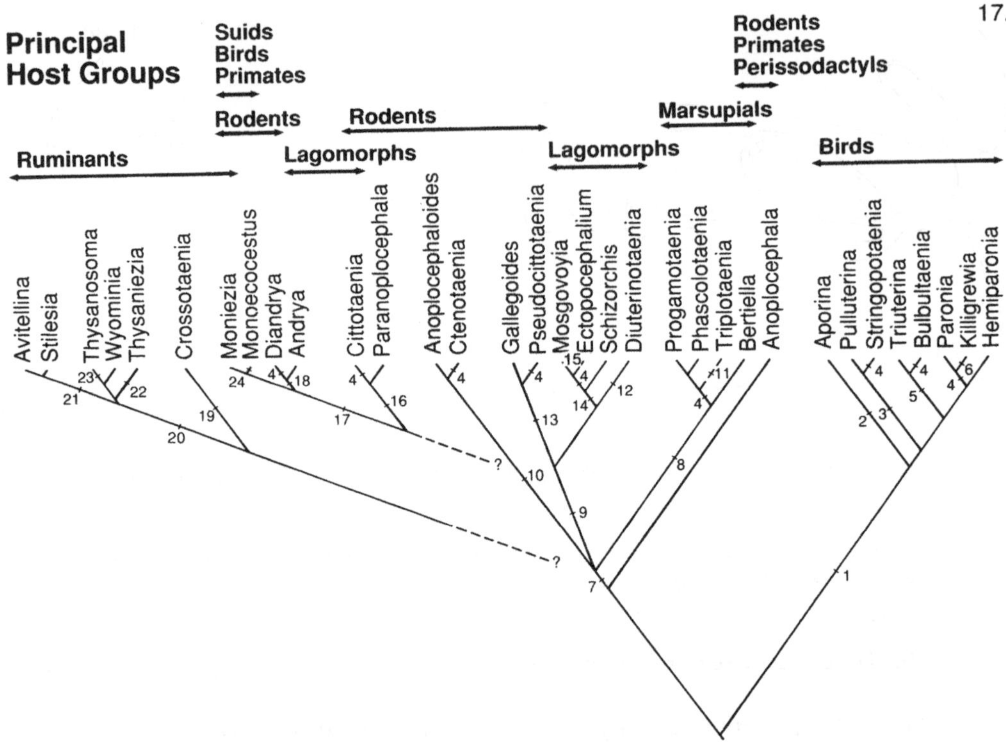

Fig. 17.89 Speculative view of phylogenetic relationships within Anoplocephalinae and Thysanosomatinae. Figures on diagram represent the following character states which are considered to be derived or apomorphic; where relevant, the primitive or plesiomorphic state follows.
1. Uterus looped anteriorly around female genital complex with ends terminating at posterolateral corners of medulla (or beyond) - uterus a straight tube. 2. Uterus with paired anterolateral diverticula (Fig. 17.30) - without diverticula. 3. Uterus with pair of posterior diverticula on either side of seminal receptacle (Fig. 17.43) - diverticula absent. 4. Genitalia paired - single. 5. Uterus with anteriorly-directed spur (Fig. 17.33) - spur absent. 6. Testes in two lateral groups (Fig. 17.41) - testes uniformly distributed. 7. Testes distributed other than throughout medulla - distributed throughout medulla. 8. Testes entirely anterior to uterus (or seminal receptacle circular). 9. Testes entirely posterior to uterus. 10. Testes aporal to female genitalia. 11. Strobila subdivided (Figs 17.1, 17.2) - strobila entire. 12. Uteri longitudinal (Fig. 17.12) - uteri transverse. 13. Uterus crosses vessels ventrally, terminates anterior to cirrus-sac (Fig. 17.14) - uterus medullary. 14. Uterus crosses vessels dorsally, terminates posterior to cirrus-sac (Fig. 17.19) - uterus medullary. 15. Scolex elongate with glands (Fig. 17.18) - scolex not elongate, without glands. 16. Uterus slightly reticular (Fig. 17.47) - uterus tubular. 17. Uterus highly reticular - slightly reticular. 18. External seminal vesicle with glandular coat (Fig. 17.58) - gland cells absent. 19. Mehlis' gland reduced or paruterine organs present - gland not reduced, organs absent. 20. Paruterine organs present - absent. 21. Germovitellarium present - absent. 22. Testes extravascular - testes intravascular. 23. Velum fringed - not fringed. 24. Vagina anterior to cirrus-sac - vagina posterior to cirrus-sac.

18 Family Catenotaeniidae Spasskii, 1950

J.C. QUENTIN

Université d'Orléans, UFR Faculté des Sciences, Biologie Animale, BP 6759–45067, Orléans Cedex 2, France.

Introduction

Hypothesis of the phylogeny of cestodes of the family Catenotaeniidae

According to Freeman (1973) the Catenotaeniidae have common ancestors with the Proteocephalidae and with the more recently derived Taeniidae.

A comparative analysis of the genital apparatus, uterine structures and geographical distribution of the hosts suggests that the most primitive catenotaeniids are cestodes of the genus *Hemicatenotaenia* Tenora, 1977, the type-species of which is *H. geosciuri*, a parasite of an African sciurid. This genus is fundamentally a parasite of sciuromorphic rodents which appeared in Holarctic, Asiatic and African Regions during the Oligocene and it was also at this period a parasite of the old rodent families Geomyidae and Heteromyidae. It is characterized by a long uterus with numerous lateral branches (30–60 pairs) and a great number of testes.

Secondarily, a new line appeared during the Miocene: the genus *Catenotaenia* Janicki, 1904, the type-species of which is *C. pusilla*, a parasite of mice. This genus is a parasite of North African Xerini, which appeared in the Miocene (Thenius, 1979), and diversified also in the Cricetidae, Muridae and Microtidae in the Holarctic Region. Also at this time, the genus *Catenotaenia* parasitized the old families Geomyidae and Heteromyidae. Consequently, these cestodes have fewer testes and fewer main lateral uterine branches. Penetration of the Catenotaeniinae into the Neotropical Region occurred only recently.

Quentinotaenia mesovitellinica (Rego, 1967) Tenora, Mas-Coma, Murai & Feliu, 1980, a parasite of Caviidae in Brazil, originated probably through capture from Nearctic Sciuridae or Cricetidae to Caviidae, these sciurid and cricetid rodents being introduced into South America only in the Quaternary (Thenius, 1972). This cestode has few testes and a horseshoe-shaped ovary.

The second phase of the evolutionary line of the Catenotaeniidae began in the Palaearctic Region with the genus *Pseudocatenotaenia* Tenora, Mas-Coma, Murai & Feliu, 1980, a parasite of murid rodents of the genus *Apodemus*. *Pseudocatenotaenia* still belongs to the subfamily Catenotaeniinae because of its primitive anatomical characters: an osmoregulatory system with two pairs of longitudinal canals, and testes situated in the posterior part of the proglottid behind the ovary. This genus is also linked with the evolutionary line of the subfamily Skrjabinotaeniinae in so far as the gross morphology of *Pseudocatenotaenia* already resembles some *Skrjabinotaenia* species (e.g. *S. lobata*), with wide mature proglottids and with the testes and the ovary overlapping the lateral osmoregulatory canals.

The subfamily Skrjabinotaeniinae is, according to Tenora *et al.* (1980), of recent appearance. Its origin will undoubtedly be found in the Palaearctic or the Asiatic Region, probably in myomorphic rodents of the Muroidea group. The hosts and geographical distribution of the Skrjabinotaeniinae suggest that they arose in the Late Tertiary in connection with the migration of rodents of the families Muridae and Gerbillidae to Africa. The ancestral form of the Skrjabinotaeniinae is represented by *S. rhombomydis*, a species parasitic in Gerbillidae from the Caspian Coast. This species is morphologically similar to *Skrjabinotaenia psammomi*, parasitic in Gerbillidae of the genus *Psammomys* in Egypt. This new evolutionary line of the Skrjabinotaeniinae differs from the Catenotaeniinae in having a network of numerous osmoregulatory canals and by testes surrounding the ovary in different ways.

According to Quentin (1971) and Tenora *et al.* (1980), two main evolutionary lines can be detected in the Skrjabinotaeniinae. One line (genus *Skrjabinotaenia*) includes the historically oldest forms, in all of which the last gravid proglottid of the strobila is always longer than wide and never has a posterior longitudinal fissure. In this group, it is possible to observe a gradual shortening in the length of the strobila and a decrease in the number of proglottids through the different species. *S. lobata*, the longest species and the only species distributed in both Europe and Africa, represents one of the most primitive forms. All other species of the genus are exclusively Ethiopian. The extreme case in this evolutionary line is *S. pauciproglottis*, the shortest and youngest living form with a strobila constituted of only two or three proglottids. The differentiation of *Skrjabinotaenia* occurred in Africa, mainly in Muridae. These cestodes exhibit stenoxenous specificity and are found also in Gerbillidae, Dendromuridae and Malagasy Cricetidae. The presence of Skrjabinotaeniinae in Madagascar is probably the result of a fortuitous and recent introduction of these helminths from the African continent into Madagascar with the genus *Rattus*.

Further differentiation within the subfamily Skrjabinotaeniinae occurred in the Ethiopian Region. Another line, the genus *Meggittina* Lynsdale, 1953, arose from *Skrjabinotaenia* parasitic in North African Gerbillidae, and in this line it is possible to detect in the last gravid proglottid some signs of a longitudinal fissure. *Meggittina* is fundamentally parasitic in Gerbillidae and expanded recently to the rest of the African continent and to Madagascar, adapting secondarily to other families of rodents, the Muridae and Cricetidae.

These hypotheses on the phylogeny of cestodes of the family Catenotaeniidae preserve the systematics and taxonomy of the Catenotaeniidae given by Tenora *et al.* (1980). The family Catenotaeniidae is divided into two subfamilies: Catenotaeniinae and Skrjabinotaeniinae, with the six genera *Hemicatenotaenia, Catenotaenia, Quentinotaenia, Pseudocatenotaenia, Skrjabinotaenia* and *Meggittina*.

However, my concept of the evolutionary line of the Catenotaeniidae is slightly different:

1. The genus *Hemicatenotaenia* with the type-species *H. geosciuri* (Ortlepp, 1938) is regarded as the more primitive.
2. The genus *Catenotaenia* with the type-species *C. pusilla* (Goeze, 1782) is derived from *Hemicatenotaenia*.
3. In the phylogeny of the Catenotaeniidae, the genus *Pseudocatenotaenia* is placed between Catenotaeniinae and *Skrjabinotaenia* and not between *Hemicatenotaenia* and *Catenotaenia*.
4. The species *Atriotaenia (Ershovia) baltazardi* Quentin, 1967 is placed in synonymy with *Quentinotaenia mesovitellinica* (Rego, 1967).
5. The genus *Neoskrjabinotaenia* Bilqees, 1982 is placed in synonymy with the genus *Catenotaenia*. The number of uterine branches, the asymmetrical ovary, the number and the position of the testes correspond to the characters of *Catenotaenia*.

Family Catenotaeniidae Spasskii, 1950

Diagnosis: Scolex rounded or semi-oval, provided with four suckers and sometimes with small apical sucker. Osmoregulatory system composed of two main pairs of longitudinal canals or network of numerous canals. Mature and gravid proglottids elongated either transversely or longitudinally. Genital apparatus single. Testes in posterior part of proglottid or around ovary in various ways. Ovary asymmetrical (except *Quentinotaenia*). Genital atrium in anterior third of proglottid, irregularly alternating. Uterus of taenioid type with central stem and lateral branches subdivided into secondary branches. Larval stage of merocercoid type developing in tyroglyphid acarians. Larva has apical sucker. Oncosphere possesses oval membrane or membrane with processes. Adults parasitic in rodents. Type-genus *Catenotaenia* Janicki, 1904.

Key to subfamilies

1a. **Two pairs of osmoregulatory longitudinal canals (Fig. 18.1); testes always posterior to the ovary in the posterior part of the proglottid (Fig. 18.2) .... Catenotaeniinae Spasskii, 1949.**

Diagnosis: Catenotaeniidae. Testes confined to postovarian field between lateral osmoregulatory canals or overlapping them. Lateral uterine branches at right angles to median stem. Gravid proglottids longer than wide. Type-genus *Catenotaenia* Janicki, 1904.

1b. **Osmoregulatory system with numerous canals (Fig. 18.3); testes surrounding ovary in various ways but never confined to the posterior part of the proglottid (Fig. 18.4) .... Skrjabinotaeniinae Genov & Tenora, 1979.**

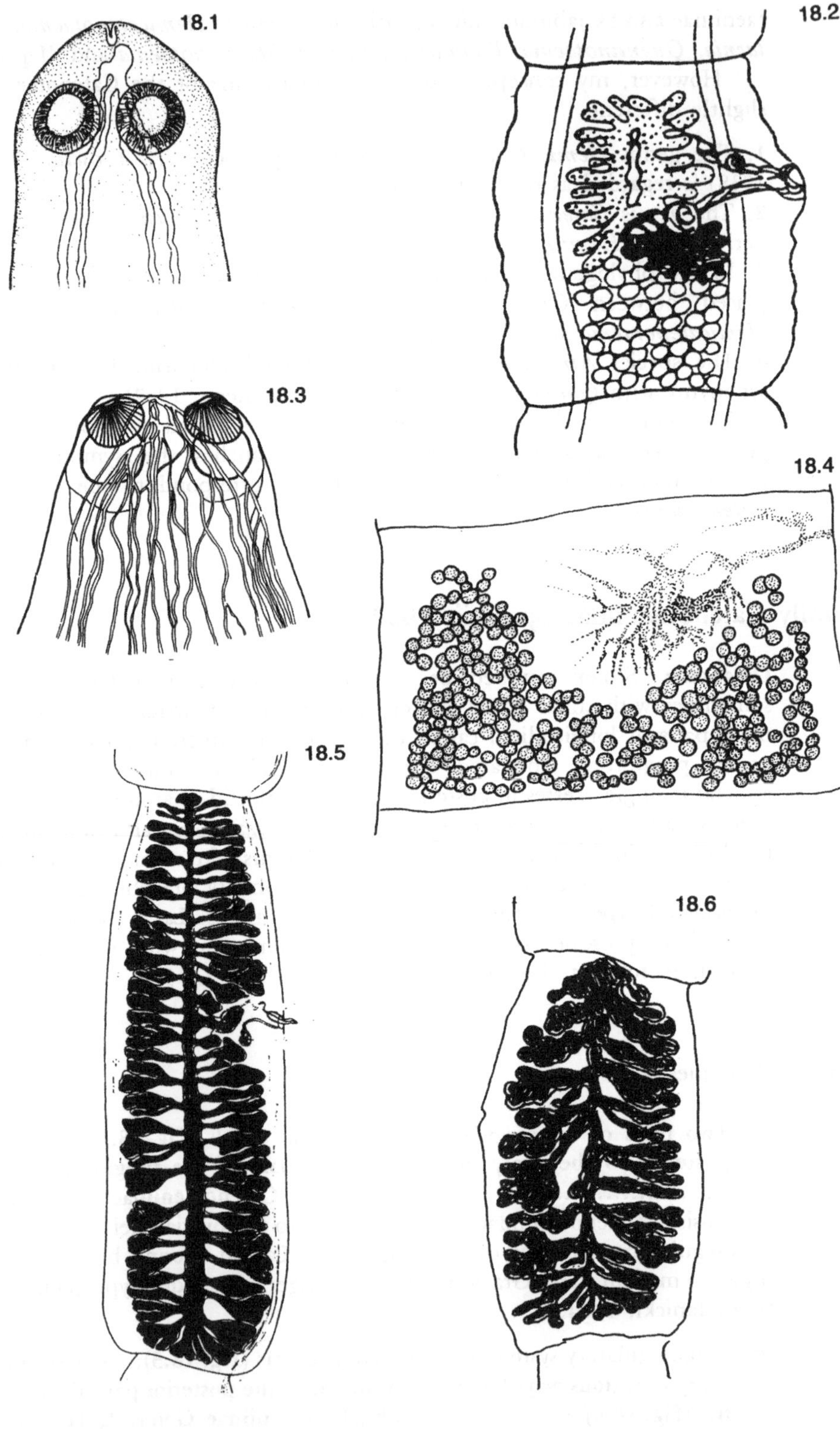
18.1
18.2
18.3
18.4
18.5
18.6

Diagnosis: Catenotaeniidae. Osmoregulatory system a network of numerous longitudinal and transverse canals. Testes surrounding ovary in various ways, including as two separate lateral groups. Lateral uterine branches at various angles to the median stem. Gravid proglottids of various sizes. Type-genus *Skrjabinotaenia* Akhumyan, 1946.

Key to genera of the subfamily Catenotaeniinae Spasskii, 1949

1a. More than 30 pairs of lateral uterine branches in gravid proglottid (30–60 pairs) (Fig. 18.5); parasites of Sciuridae (fundamental hosts) and Microtidae .... *Hemicatenotaenia* Tenora, 1977.
Diagnosis: Catenotaeniinae. Ovary asymmetrical. 150–300 testes situated posteriorly to female organs, do not cross the longitudinal osmoregulatory canals. Parasites of Holarctic, Ethiopian and Asiatic Sciuridae (fundamental hosts) and Microtidae. Type-species *H. geosciuri* (Ortlepp, 1938) Tenora, 1977.

1b. Fewer than 30 pairs of lateral uterine branches in gravid proglottid (Fig. 18.6) .... 2.

2a. Testes not overlapping osmoregulatory canals bilaterally (Fig. 18.8), less than 150 .... 3.
2b. Testes overlapping lateral osmoregulatory canals bilaterally; mean number of testes greater than 150 (Fig. 18.9); parasites of murid rodents of genus *Apodemus* ....
.... *Pseudocatenotaenia* Tenora, Mas-Coma, Murai & Feliu, 1980.
Diagnosis: Catenotaeniinae. Strobila acraspedote. Mature proglottids wider than long. Ovary asymmetrical, highly lobate, extensive, in anterior half of proglottid. Lobes of ovary reach lateral osmoregulatory canals or overlap them. Vitelline gland in poral part of proglottid. 140–180 testes in posterior half of proglottid. Uterus with lateral branches overlapping excretory canals bilaterally. Eggs oval with two processes. Parasites of Palaearctic Muridae of the genus *Apodemus*. Type-species *P. matovi* (Genov, 1971) Tenora, Mas-Coma, Murai & Feliu, 1980.

3a. 70 to 150 testes (Fig. 18.7); parasites of Muroidea (fundamental hosts) and also Holarctic Heteromyidae and Sciuridae ....
.. *Catenotaenia* Janicki, 1904.
(Syns *Cattotaenia* Shipley, 1908, *lapsus*; *Neoskrjabinotaenia* Bilqees, 1982.)
Diagnosis: Catenotaeniinae. Ovary asymmetrical. Fewer than 150 testes situated

---

Fig. 18.1 Scolex of *Pseudocatenotaenia matovi* (Genov, 1971). After Tenora *et al.* (1980).
Fig. 18.2 Mature proglottid of *Catenotaenia pusilla* (Goeze, 1782). After Meszaros & Murai (1979).
Fig. 18.3 Scolex of *Skrjabinotaenia lobata* (Baer, 1925). After Quentin (1971).
Fig. 18.4 Proglottid of *Skrjabinotaenia lobata* (Baer, 1925). After Quentin (1971).
Fig. 18.5 Gravid proglottid of *Hemicatenotaenia ris* (Yamaguti, 1942). After Yamaguti (1942).
Fig. 18.6 Gravid proglottid of *Catenotaenia pusilla* (Goeze, 1782). Original.

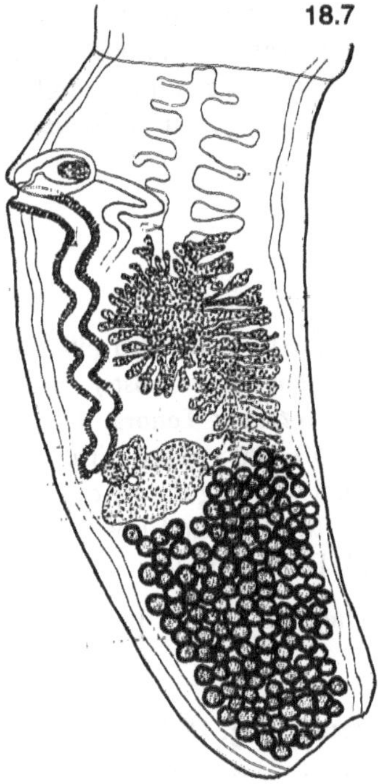

18.7

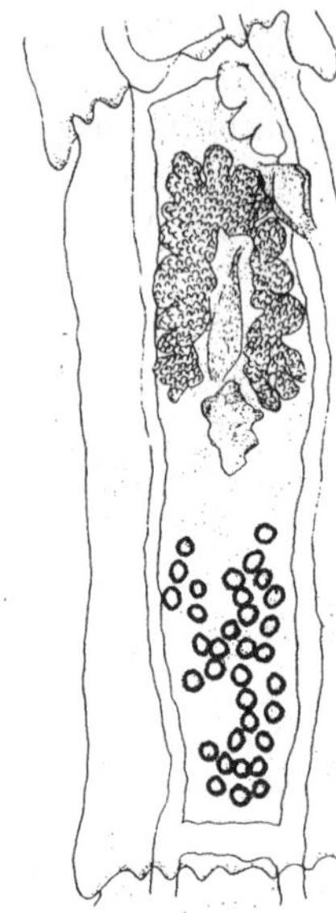

18.8

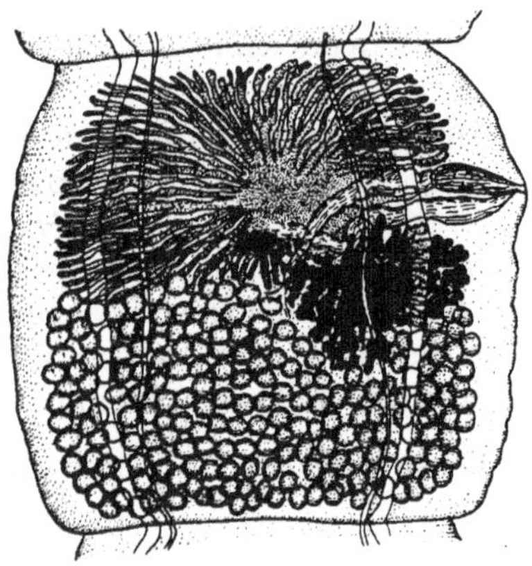

18.9

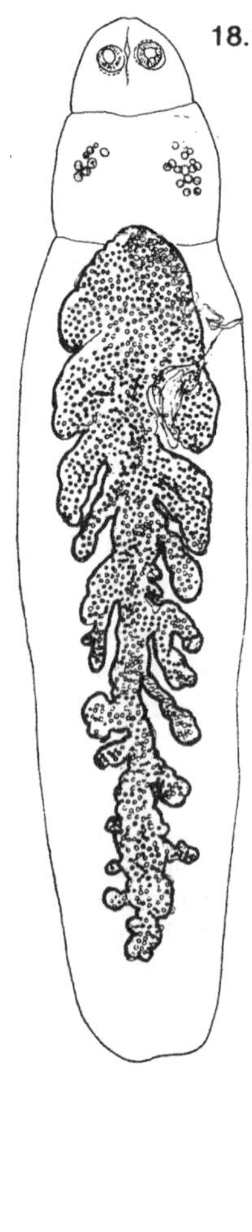

18.10

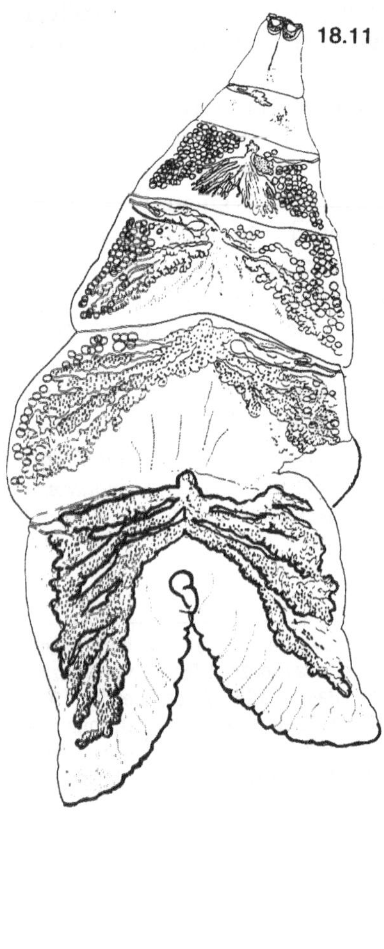

18.11

posteriorly to female organs, do not cross longitudinal osmoregulatory canals. Uterus with fewer than 30 pairs of lateral uterine branches. Parasites of Holarctic and Asiatic Sciuridae, Cricetidae, Muridae, Microtidae, Heteromyidae, Geomyidae and ? Erinaceidae. Type-species *C. pusilla* (Goeze, 1782) Janicki, 1904.

3b. 30 to 40 testes (Fig. 18.8); parasites of Neotropical Caviidae .... *Quentinotaenia* Tenora, Mas-Coma, Murai & Feliu, 1980.
(Syn. *Quentinia* Spasskii, 1969.)

Diagnosis: Catenotaeniinae. Ovary in anterior half of proglottid, almost symmetrical, horseshoe-shaped. Vitelline gland compact, enclosed by ovarian lobes or postovarian. Testes few, posterior to vitelline gland between lateral osmoregulatory canals. Uterus with 15–20 compact lateral branches. Parasites of Brazilian Caviidae. Type-species *Q. mesovitellinica* (Rego, 1967) Tenora, Mas-Coma, Murai & Feliu, 1980.

Key to genera of the subfamily Skrjabinotaeniinae Genov & Tenora, 1979

1a. Gravid proglottid undivided, longer than wide (Fig. 18.10); parasites of Palaearctic and Ethiopian Muridae (fundamental hosts) and also Palaearctic and Ethiopian Gerbillidae, Ethiopian and Malagasy Cricetidae and Dendromuridae .... *Skrjabinotaenia* Akhumyan, 1946.

Diagnosis: Skrjabinotaeniinae. Strobila acraspedote with three to sixty proglottids. Mature proglottids longer than wide. Ovary asymmetrical, highly lobate, extensive. Testes numerous, distribution relative to ovary variable (posterior, lateral, sometimes surrounding ovary or in two lateral groups). Uterus taenioid with median stem and numerous lateral branches shorter than median stem. Parasites of Palaearctic and Ethiopian Muridae (fundamental hosts) and also Palaearctic and Ethiopian Gerbillidae, Ethiopian and Malagasy Cricetidae and Dendromuridae. Type-species *S. oranensis* (Joyeux & Foley, 1930) Akhumyan, 1946.

1b. Gravid proglottid deeply cut posteriorly by a longitudinal fissure and transversely elongated (Fig. 18.11); parasites of Ethiopian Gerbillidae (fundamental hosts) and also of Ethiopian and Malagasy Muridae and Cricetidae .... *Meggittina* Lynsdale, 1953.
(Syn. *Rajotaenia* Wertheim, 1954.)

Diagnosis: Skrjabinotaeniinae. Strobila acraspedote, strongly reduced, consisting of small scolex, wide neck and only few proglottids. Gravid proglottid widening to form two lateral wings. Ovary and vitelline gland poral. Testes numerous,

---

Fig. 18.7 Mature proglottid of *Catenotaenia chabaudi* Dollfus, 1953. After Dollfus (1953).
Fig. 18.8 Proglottid of *Quentinotaenia mesovitellinica* (Rego, 1967). After Rego (1967b).
Fig. 18.9 Mature proglottid of *Pseudocatenotaenia matovi* (Genov, 1971). After Tenora *et al.* (1980).
Fig. 18.10 *Skrjabinotaenia pauciproglottis* Quentin, 1965. After Quentin (1971).
Fig. 18.11 *Meggittina gerbilli* (Wertheim, 1954). After Quentin (1971).

in two lateral groups or in two groups anterior to female organs. Median longitudinal stem of uterus very short, lateral uterine branches each with secondary branches from their inner sides. Parasites of African and Malagasy Gerbillidae (fundamental hosts), Muridae and Cricetidae. Type-species *M. baeri* Lynsdale, 1953.

Note added in proof

The genus *Quentinia* Spasskii, 1969 proposed for the species *Q. baltazardi* (Quentin, 1967) predates *Quentinotaenia* Tenora, Mas Coma, Murai & Feliu, 1980. Consequently, *Quentinotaenia* falls into synonymy with *Quentinia*. The correct name of the type-species of *Quentinia* is *Q. baltazardi* (Quentin, 1967), syn. of *Q. mesovitellinica* (Rego, 1967).

19 Family Nematotaeniidae Lühe 1910

M.K. JONES

Centre for Microscopy and Microanalysis, The University of Queensland, Brisbane, Queensland 4072, Australia.

Introduction

The Nematotaeniidae Lühe, 1910 was revised recently by Jones (1987b). These keys follow the taxonomic system proposed by Jones (1987b) and recognize four genera. Three other genera, *Baerietta* Hsü, 1935, *Hexaparuterina* Palacios & Barroeta, 1967 and *Nematotaenoides* Ulmer & James, 1976 were included previously in the Nematotaeniidae, but are not recognized here.

Prior to Jones' (1987b) revision, *Baerietta* was the largest genus of the Nematotaeniidae. *Baerietta*, with two testes per mature proglottid, was distinguished from *Cylindrotaenia*, which was said to have only one testis per proglottid. Jones (1987b) re-examined types of *Cylindrotaenia* and *Baerietta* species and noted that this sole distinguishing character was invalid, as members of both genera had two testes per proglottid. In the absence of other distinguishing characters, *Baerietta* became a junior synonym of *Cylindrotaenia*. *Nematotaenoides* shared some characters with nematotaeniids, such as parasitism in an anuran and possession of paruterine organs (Ulmer & James, 1976). Jones (1987a) argued, however, that *Nematotaenoides* had more in common with members of the paruterinine genus *Anonchotaenia*; in particular, the structure and formation of the paruterine organs in the two genera were almost indistinguishable. Therefore, Jones (1987a) proposed that *Nematotaenoides* become a synonym of *Anonchotaenia*. When originally proposed by Palacios & Barroeta (1967), *Hexaparuterina* was considered to be a member of the Paruterininae. This genus shares few characters with nematotaeniids except the possession of paruterine organs and parasitism in anuran amphibians, yet Schmidt (1986) included *Hexaparuterina* within the Nematotaeniidae in his keys. In the present key, the original designation of *Hexaparuterina*, within the Paruterininae, is favoured. [Georgiev & Kornyushin, this volume, consider it a synonym of *Metroliasthes*. Editors.]

Four diagnostic characters of the Nematotaeniidae, originally listed by Douglas (1958), are worthy of consideration as aids to confirming identifications made in these keys. These characters, modified slightly from Douglas (1958), are as follows: (i) parasitism as adults in amphibians or reptiles; (ii) possession of two testes (rarely three) per mature proglottid; (iii) a cylindrical form with acraspedote wider-than-long immature proglottids in which proglottidization is not evident externally; and (iv) a distinctive pattern of paruterine capsule formation, characterized by the possession of conical paruterine organs, which lie immediately adjacent to the uterus, and paruterine capsules, which are thin-walled, egg-bearing capsules that develop from specialized regions at the apices of the paruterine organs (Jones 1987b; 1988). A typical nematotaeniid scolex and the morphology of a mature proglottid as reconstructed in transverse section are illustrated in Figs 19.1 and 19.2, respectively.

Family Nematotaeniidae Lühe, 1910

Diagnosis: Small elongate cestodes, apolytic or hyperapolytic. Scolex with four simple suckers, without rostellum or apical organ. Strobila cylindrical; proglottids acraspedote. Muscular system poorly developed; thin layer of inner longitudinal muscle divides body into cortes and medulla. Osmoregulatory canals form transverse loop in scolex; dorsal canals unjoined elsewhere; ventral canals joined by transverse commissure near posterior border of each proglottid. Genital pores lateral, alternating irregularly. Testes two, medullary. Cirrus-sac present, extending into medulla. Vagina posterior or ventral to cirrus. Ovary and vitellarium compact, medullary. Uterus saccate, forming thin envelope around each embryo. Paruterine capsules two or more per proglottid, developing from precursors at apex of conical paruterine organs. Adults in amphibians and reptiles. Type-genus *Nematotaenia* Lühe, 1899.

Key to genera

1a. **Two paruterine capsules or organs per proglottid** .... 2.
1b. **More than two paruterine capsules or organs per proglottid** .... 3.

2a. **Paruterine capsules surrounded by second envelope** .... ***Bitegmen*** **Jones, 1987. (Figs 19.3, 19.4)**

Diagnosis: Strobila apolytic. Proglottids few in number, acraspedote. Suckers simple, unarmed. Ventral osmoregulatory canals large, joined posteriorly in pro-

---

Fig. 19.1 Scolex, *Nematotaenia*. After Jones (1987b). Scale 100 μm.
Fig. 19.2 Reconstruction of transverse sections, *Nematotaenia*. After Jones (1987b). T testis; O ovary; V vitellarium. Scale 100 μm.
Figs 19.3, 19.4 *Bitegmen*. 19.3. Pre-gravid proglottid, showing paruterine organs. 19.4. Gravid proglottid. After Jones (1987b). Scale 100 μm.
Figs 19.5, 19.6 *Cylindrotaenia*. 19.5. Pre-gravid proglottid. 19.6. Gravid proglottid. After Jones (1987b). PO paruterine organ. PC paruterine capsule. Scale 100 μm.

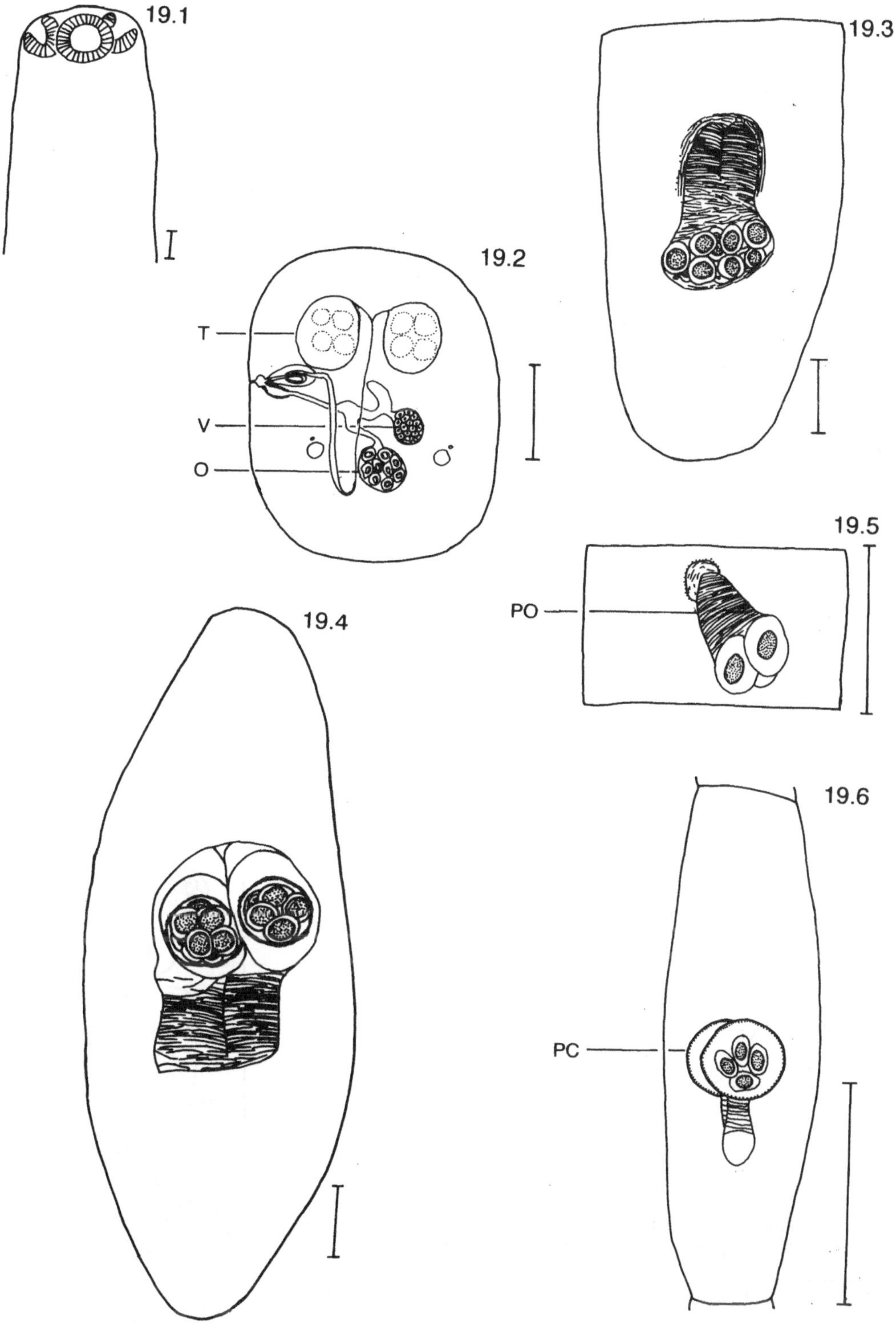
19.1
19.3
19.2
T
V
O
19.5
PO
19.4
19.6
PC

19.7

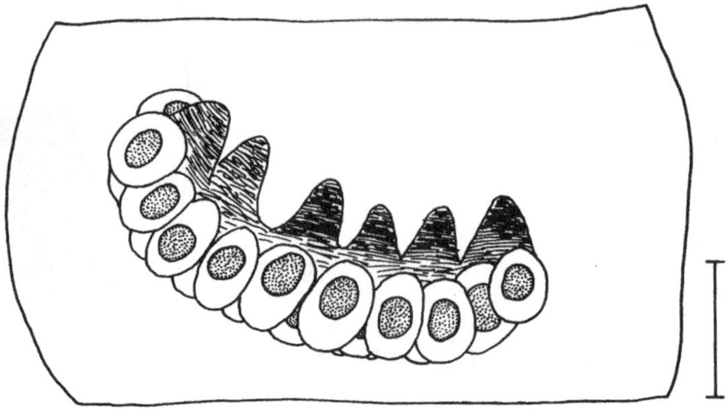

19.8

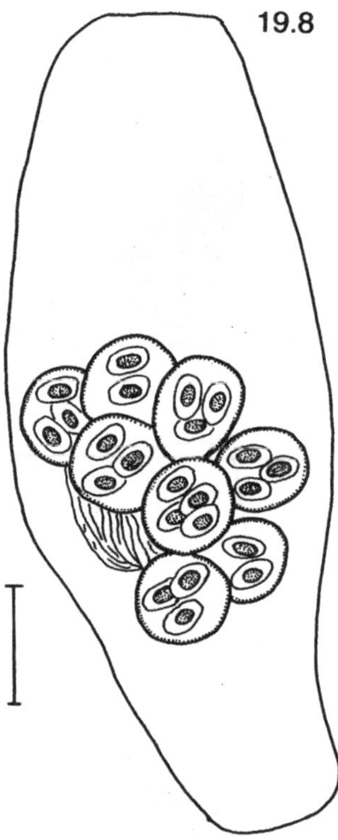

19.9

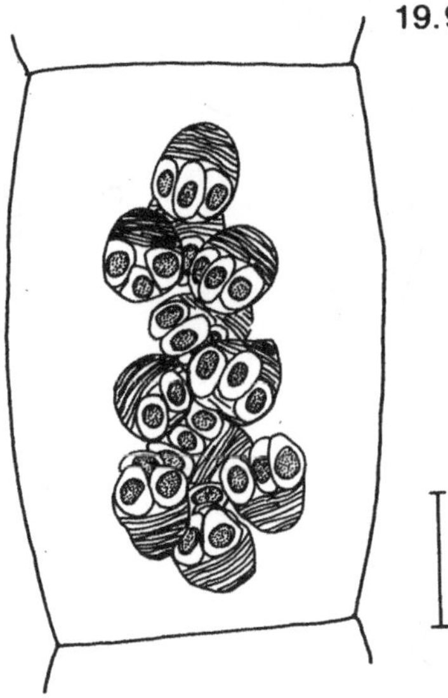

19.10

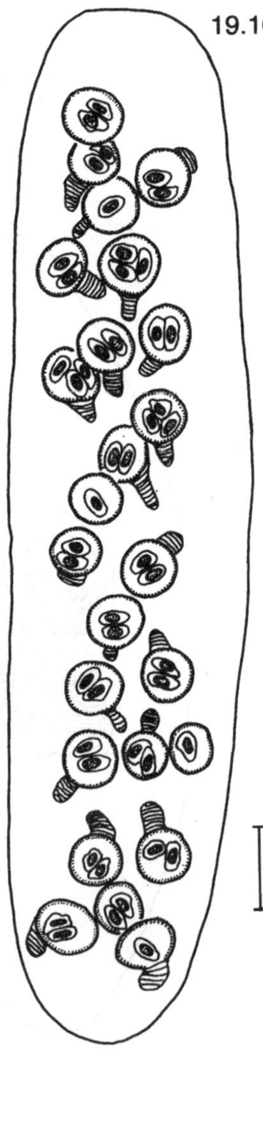

Figs 19.7, 19.8 *Distoichometra*. 19.7. Pre-gravid proglottid. After Jones (1987b). 19.8. Gravid proglottid, specimen courtesy of Dr D.B. Conn. Scale 100 μm.

Figs 19.9, 19.10 *Nematotaenia*. 19.9. Pre-gravid proglottid. 19.10. Gravid proglottid. After Jones (1987b). Scale 100 μm.

glottid by transverse commissure; dorsal canals evident only at scolex. Genital pores alternating irregularly. Genital atrium heavily muscular. Cirrus-sac thick-walled, separated into proximal and distal regions by septum. Vas deferens not looping in cirrus-sac. External seminal vesicle present. Testes two, dorsal. Ovary ventral, spherical. Vitellarium lateral to ovary. Vagina not heavily muscular distally. Paruterine organs paired. Paruterine capsules two, opposite. In Sauria (Anguidae). North America. Type-species *B. gerrhonoti* (Telford, 1965).

2b. Paruterine capsules not surrounded by outer enveloping structure .... *Cylindrotaenia* Jewell, 1916. (Figs 19.5, 19.6)
(Syn. *Baerietta* Hsü, 1935.)

Diagnosis: Strobila apolytic or hyperapolytic. Proglottids few or numerous, acraspedote. Suckers simple, unarmed. Ventral osmoregulatory canals large, joined posteriorly in proglottid by transverse commissure; dorsal canals unjoined. Genital pores alternating irregularly. Genital atrium lightly muscular. Cirrus-sac thin or thick-walled, not divided by septum. Vas deferens looping once in cirrus-sac. External seminal vesicle absent. Testes two (rarely three), dorsal. Vagina not heavily muscular distally (except *C. japonica*). Paruterine organs paired. Paruterine capsules two, opposite or tandem. In Anura, Caudata, Sauria (Scincidae, Gekkonidae). North America, South America, Africa, Asia, Australia, New Zealand. Type-species *C. americana* Jewell, 1916.

3a. Paruterine organs in row of two to six pairs of organs; four to twelve paruterine capsules clustered in posterior half of proglottids .... *Distoichometra* Dickey, 1921. (Figs 19.7, 19.8)

Diagnosis: Strobila apolytic. Proglottids numerous, acraspedote. Suckers simple, unarmed. Ventral osmoregulatory canals large, joined posteriorly in proglottid by transverse commissure; dorsal canals unjoined. Genital pores alternating irregularly. Genital atrium lightly muscular. Cirrus-sac thin-walled, lacking septum. Vas deferens looping once in cirrus-sac. External seminal vesicle absent. Ovary ventral, spherical. Vitellarium dorsolateral to ovary. Vagina not heavily muscular distally. Partuterine organs in row of two to six pairs of organs. Paruterine capsules 4–12 per proglottid, clustered in posterior half of proglottid. In Anura and Caudata. North America. Type-species *D. bufonis* Dickey, 1921.

3b. Paruterine organs not paired; five to 150 paruterine capsules scattered throughout proglottid .... *Nematotaenia* Lühe, 1899. (Figs 19.1, 19.2, 19.9, 19.10)

Diagnosis: Strobila apolytic. Proglottids few or numerous, acraspedote; suckers simple, unarmed. Ventral osmoregulatory canals large; joined posteriorly in proglottid by transverse commissure; dorsal canals unjoined. Genital pores alternating irregularly. Genital atrium lightly muscular. Cirrus-sac thin-walled, lacking septum. Vas deferens looping once or more in cirrus-sac. External seminal vesicle present or absent. Testes two, dorsal. Ovary ventral, spherical. Vitellarium dorsolateral to ovary. Paruterine capsules 5–150 per proglottid, scattered throughout proglottid. In Anura, Caudata, Sauria (Gekkonidae). Africa, Europe, Asia, Australia. Type-species *N. dispar* (Goeze, 1782).

20 Family Progynotaeniidae Fuhrmann, 1936

L.F. KHALIL

International Institute of Parasitology, 395A Hatfield Road, St Albans, Herts AL4 0XU, UK.

Introduction

The family Progynotaenidae (later emended to Progynotaeniidae by Burt, 1939a) was proposed by Fuhrmann (1936a) and subdivided into two subfamilies: Progynotaeniinae for the genera *Progynotaenia* Fuhrmann, 1909, *Proterogynotaenia* Fuhrmann, 1922 and *Leptotaenia* Cohn, 1901 and the subfamily Gynandrotaeniinae for the genus *Gynandrotaenia* Fuhrmann, 1936. Some of these genera were previously assigned to the Acoleidae because they lack a vaginal pore but Fuhrmann felt that as they all show distinct protogyny, where the female genital system develops and matures before the male system, they should be separated from the Acoleidae into a distinct family. Three more genera were later added to the family, namely *Andrepigynotaenia* Davies & Rees, 1947; *Thomasitaenia* Ukoli, 1965 and *Paraprogynotaenia* Rysavy, 1966, all parasitizing birds. *Andrepigynotaenia* was originally described without rostellar hooks, but Williams (1960) later found two rows of rostellar hooks in newly collected material.

Baer (1940) suggested that further research is needed to justify the establishment of this family. Wardle & McLeod (1952) did not accept it and argued that the establishment of a family on physiological grounds rather than upon morphology is justifiable only if the physiological phenomenon is restricted to members of this family. They regarded the genera assigned to the family as cyclophyllideans of uncertain status. Yamaguti (1959) accepted the family and the two subfamilies proposed by Fuhrmann. Schmidt (1970, 1986) accepted the genera in the family but without separate subfamilies.

Ryzhikov & Tolkacheva (1981) pointed out that *Andrepigynotaenia* resembles *Proterogynotaenia* very closely but differs from it only in that the cirrus-sac is dorsal to the osmoregulatory canals and not between them. They felt that this did not justify the separation of the two genera and regarded *Andrepigynotaenia* as a synonym of *Proterogynotaenia*. This view was supported by Schmidt & Canaris

(1992) who found this character to be variable in *P. marcusae*. In this review, the family is accepted and only six genera are assigned to it.

Family Progynotaeniidae Fuhrmann, 1936

Diagnosis: Small to medium-sized cestodes with weak musculature. Scolex may or may not be divided into pro- and metascolex. Rostellum present or absent, armed or unarmed. Protogynous genitalia; each proglottid hermaphroditic or one proglottid may contain only female organs and the next only male organs with regular alternation. Male genitalia completely developed in posterior proglottids only. Male pore alternating regularly or irregularly. Female genitalia develop very early. Vagina absent. Seminal receptacle large, median. Uterus sac-like, lobed. Parasites of birds. Type-genus *Progynotaenia* Fuhrmann, 1909.

Key to genera

1a. Separate male and female proglottids alternating regularly .... **2.**
1b. Hermaphroditic proglottids .... **3.**

2a. Scolex divided into proscolex with armed rostellum and metascolex with armed suckers .... ***Gynandrotaenia* Fuhrmann, 1936. (Fig. 20.1)**
Diagnosis: Small cestodes with up to 19 proglottids, first proglottid behind scolex always female, last proglottid in strobila often male. Scolex comparatively large; proscolex dome-shaped, retractable within metascolex. Rostellum retractable with single crown of six hooks. Male genital pores open laterally near posterior margin of proglottid, alternate irregularly in successive male proglottids. Cirrus-sac large, containing prostate cells, cirrus armed at base. Testes numerous. Ovary lobed, occupying most of medulla. Vitellarium compact, transversely elongate, near posterior margin of proglottid. Uterus appears first as median longitudinal duct, becoming sac-like with lateral branches when gravid. Seminal receptacle develops between ovary and vitellarium, becomes large, quickly degenerates. Eggs oval, with recurved polar processes on innermost membrane. In flamingos. Cosmopolitan. Type-species *G. stammeri* Fuhrmann, 1936.

2b. Scolex not divided, rostellum absent, suckers not armed ....
.... ***Thomasitaenia* Ukoli, 1965. (Figs 20.2–20.4)**
Diagnosis: Medium-sized cestodes with up to 80 proglottids, no neck. Comparatively large, spherical, suckers small. Male genital pores open laterally, irregularly alternating in successive male proglottids. Cirrus-sac well developed; cirrus long, armed. Testes numerous, closely packed. Ovary large, fills most of developing proglottid. Vitellarium compact, transversely elongate, postovarian. Uterus first appears as small transverse sac becoming horseshoe-shaped and finally large transverse sac in gravid female proglottids. Seminal receptacle large, posterior to uterus. In Charadriiformes. Africa. Type-species *T. nunguae* Ukoli, 1965.

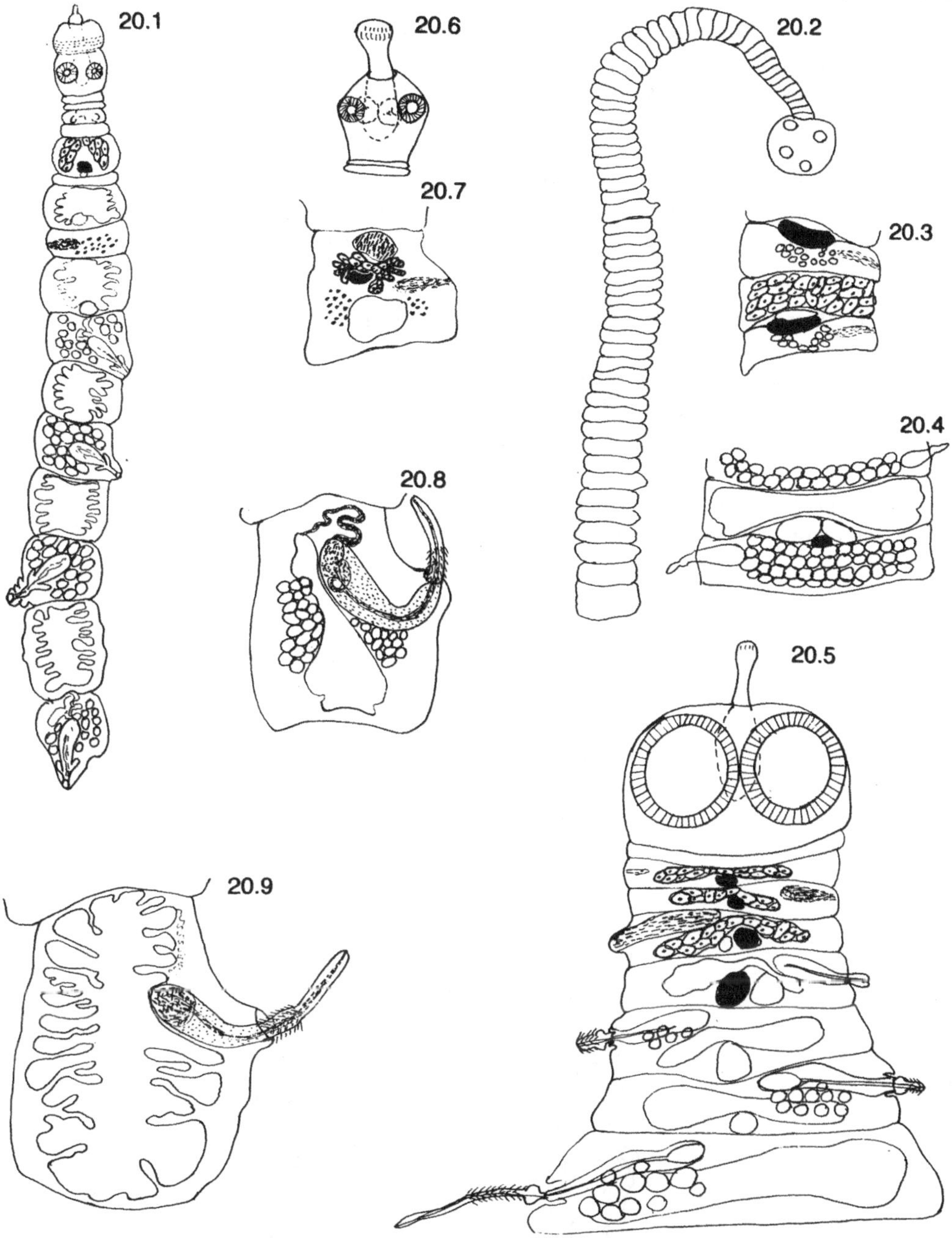

Fig. 20.1. *Gynandrotaenia stammeri* Fuhrmann, 1936 entire worm. Original.
Figs. 20.2-20.4. *Thomasitaenia nunguae* Ukoli, 1965. 20.2. Entire worm. 20.3. Female proglottid. 20.4. Mature male proglottid. After Ukoli (1965).
Fig. 20.5. *Leptotaenia ischnorhyncha* (Lühe, 1898), entire worm. Original.
Figs 20.6-20.9. *Proterogynotaenia haematopodis* (Davies & Rees, 1947). 20.6. Scolex. 20.7. Proglottid showing functional female genitalia. 20.8. Proglottid showing functional male organs. 20.9. Proglottid showing gravid uterus and cirrus-sac. Original.

3a. Testes in single group on poral side only .... *Leptotaenia* **Cohn, 1901. (Fig. 20.5)**

Diagnosis: Small cestodes, strobila wedge-shaped, up to 15 proglottids. Scolex comparatively large. Rostellum retractable with single crown of hooks. Cirrus-sac large, oblique, regularly alternating. Cirrus long, armed. Ovary bilobed, large. Vitellarium compact, longitudinally median, postovarian. Seminal receptacle large, longitudinally elongate. Uterus saccular filling whole of medulla. Eggs small, round. In flamingos. Cosmopolitan. Type-species *L. ischnorhyncha* (Lühe, 1898).

3b. Testes in two lateral groups, poral and aporal .... **4.**

4a. Single circle of rostellar hooks .... **5.**

4b. Two circles of rostellar hooks .... *Proterogynotaenia* **Fuhrmann, 1911. (Figs 20.6–20.9)**

(Syn. *Andrepigynotaenia* Davies & Rees, 1947.)

Diagnosis: Small to medium-sized cestodes, proglottids craspedote. Scolex elongate with prominent rostellum and large, unarmed suckers. Male genital

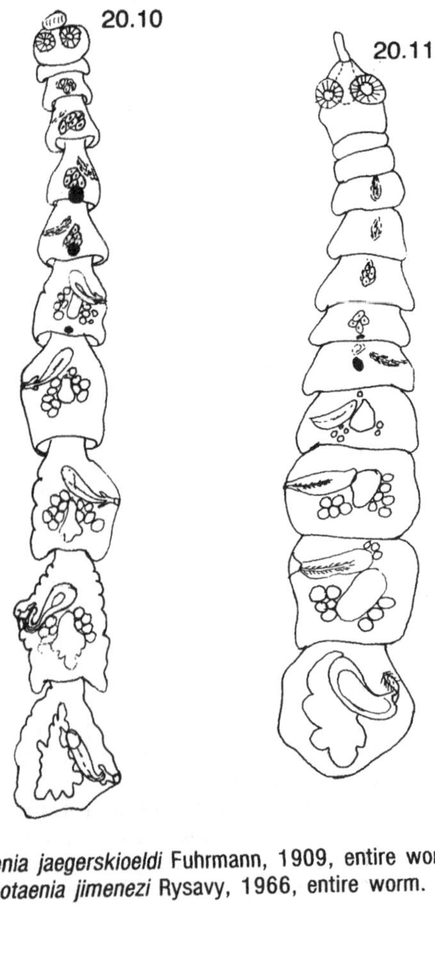

Fig. 20.10. *Progynotaenia jaegerskioeldi* Fuhrmann, 1909, entire worm. Original.
Fig. 20.11. *Paraprogynotaenia jimenezi* Rysavy, 1966, entire worm. After Rysavy (1966).

pores regularly or irregularly alternate. Testes numerous, in two lateral groups (except *P. marcusae* Schmidt & Canaris, 1992). Cirrus-sac large; cirrus armed. Ovary in anterior part of proglottid, lobate. Vitellarium compact, postovarian. Seminal receptacle large, anterior to ovary. Uterus sac-like, may protrude into preceding proglottid. Eggs round. In Charadriiformes. Cosmopolitan. Type-species *P. rouxi* Fuhrmann, 1911.

5a. Male genital pores alternate regularly .... ***Progynotaenia*** **Fuhrmann, 1909. (Fig. 20.10)**
Diagnosis: Small cestodes with few proglottids. Scolex with well developed rostellum with single circle or zigzag row of hooks. Neck absent. Testes numerous in two groups, one on each side of uterus. Cirrus-sac large, oblique. Cirrus armed. Ovary central, bilobed. Vitellarium small, compact, postovarian. Seminal receptacle transversely elongate, between ovary and vitellarium. Uterus sac-like, lobed. Type-species *P. jaegerskioeldi* Fuhrmann, 1909.

5b. Male genital pores alternate irregularly .... ***Paraprogynotaenia*** **Rysavy, 1966. (Fig. 20.11)**
Diagnosis: Small cestodes, proglottids craspedote, wider than long except last few proglottids. Scolex conical. Rostellum armed with single circle of hooks. Testes few, in two lateral groups separated by uterus. Cirrus-sac well developed, large. Cirrus armed. Ovary lobed, antiporal. Vitellarium compact, postovarian. Seminal receptacle present. Uterus elongate, lobed, median. In Charadriiformes. Cuba, Taiwan. Type-species *P. jimenezi* Rysavy, 1966.

21 Family Acoleidae Fuhrmann, 1899

L.F. KHALIL

International Institute of Parasitology, 395A Hatfield Road, St Albans, Herts AL4 0XU, UK.

Introduction

The family Acoleidae is based on the genus *Acoleus* Fuhrmann, 1899. Fuhrmann (1907b) elevated the subfamily Acoleininae Fuhrmann, 1899 to family rank [see *ICZN* Art. 36(a)] under the name Acoleinidae, with the main character being the absence of a vaginal pore. Ransom (1909) pointed out that the name is grammatically incorrect and amended it to Acoleidae, accepting in it the genera *Acoleus, Gyrocoelia* Fuhrmann, 1899, *Diplophallus* Fuhrmann, 1900, *Dioecocestus* Fuhrmann, 1900 and *Shipleya* Fuhrmann, 1908. Fuhrmann later (1909a, 1911) added *Progynotaenia* Fuhrmann, 1909 and *Proterogynotaenia* Fuhrmann, 1911.

The validity of the family and the genera attached to it was disputed by several workers. Meggitt (1924a) accepted the family and added *Monoecocestus* Beddard, 1914; *Urocystidium* Beddard, 1912 and *Diplposthe* Jacobi, 1896 to it. Fuhrmann (1932) pointed out that *Monoecocestus* is an anoplocephalid, *Urocystidium* is probably a taeniid larva and *Diplposthe* is a hymenolepidid because there are usually three testes in each proglottid. Poche (1926) established the family Diploposthidae for *Diplposthe* while Southwell & Hilmy (1929) moved *Diplophallus* and *Dioecocestus* to the Diploposthidae. Southwell (1930a) established the Dioecocestidae for cestodes with entirely separate male and female strobilae. Fuhrmann (1932) retained seven genera within the Acoleidae but subsequently (1936a) established the Progynotaeniidae for *Progynotaenia* and *Proterogynotaenia*. He distributed the remaining genera of the Acoleidae between the subfamily Acoleinae for forms without a vaginal pore and Dioecocestinae for separately sexed forms with a vaginal pore.

Burt (1939a) accepted the Dioecocestidae and retained only *Acoleus* and *Diplophallus* in the Acoleidae. More recently, Wardle & McLeod (1952), Yamaguti (1959) and Schmidt (1970) took the view that the Acoleidae contains only *Acoleus*. Spasskii & Spasskaya (1968) considered the Diploposthidae to be synonymous with the Hymenolepididae.

Ukoli (1965) established *Himantocestus* and assigned it to the Diploposthidae because of the partial duplication of the male genital organs and the absence of vaginal pores. The rostellum in his specimens was unarmed. His specimens are very near to *Diplophallus* and he was probably unaware of *D. andinus* Voge & Reed, 1953 which has an unarmed rostellum. This led Ryzhikov & Tolkacheva (1981) justifiably to synonymize *Himantocestus* and its only species *H. blanksoni* with *Diplophallus* and *D. andinus*, respectively. Olsen (1966) and Ahern & Schmidt (1976) pointed out that many scoleces of *Diplophallus* lose some or all their rostellar hooks during collection and suggested that careful collection and examination of *D. andinus* would probably demonstrate the presence of rostellar hooks. Olsen (1966) reported *D. taglei* from a mammalian host but, as all previous records of the genus came from charadriiform birds, Ryzhikov & Tolkacheva (1981) excluded this species from the genus. Schmidt (1986) accepted *Acoleus, Diplophallus, Diploposthe* and *Jardugia* in the Acoleidae. In this revision, *Diploposthe* and *Jardugia* are assigned to the Hymenolepididae while *Acoleus* and *Diplophallus* are the only genera retained in the Acoleidae.

Family Acoleidae Fuhrmann 1899

Diagnosis: Relatively large, thick cestodes with short proglottids and strong musculature. Scolex with armed or unarmed rostellum and simple unarmed suckers. Single or double male genitalia in each proglottid. Testes numerous. Cirrus-sac large, strongly muscular. Cirrus robust, strongly armed. External vaginal pores absent, vagina represented by transverse seminal receptacle. Ovary transversely elongate. Vitellarium compact, postovarian. Uterus sac-like. Parasites of birds. Type-genus *Acoleus* Fuhrmann, 1899.

Key to genera

1a. Male and female genitalia single .... *Acoleus* **Fuhrmann, 1899. (Figs 21.1–21.3)**

Diagnosis: Scolex with armed rostellum and simple unarmed suckers. Male genital pores alternate regularly. Cirrus-sac large; cirrus armed. Male duct ventral to poral osmoregulatory canals. Testes numerous, in band across proglottid. Ovary median, transversely elongate, lobed. Young uterus a transverse tube, becoming sac-like, occupying most of proglottid. Eggs round or oval with small polar thickenings. In Charadriiformes and Ralliformes. Cosmopolitan. Type-species *A. vaginatus* (Rudolphi, 1819) Fuhrmann, 1899.

1b. Male genitalia double, female genitalia single .... *Diplophallus* **Fuhrmann, 1900. (Figs 21.4–21.6)** (Syn. *Himantocestus* Ukoli, 1965).

Diagnosis: Rostellum armed or unarmed. If armed, with single circle of hooks. Suckers well developed, unarmed. Proglottids wider than long. Male genital pores

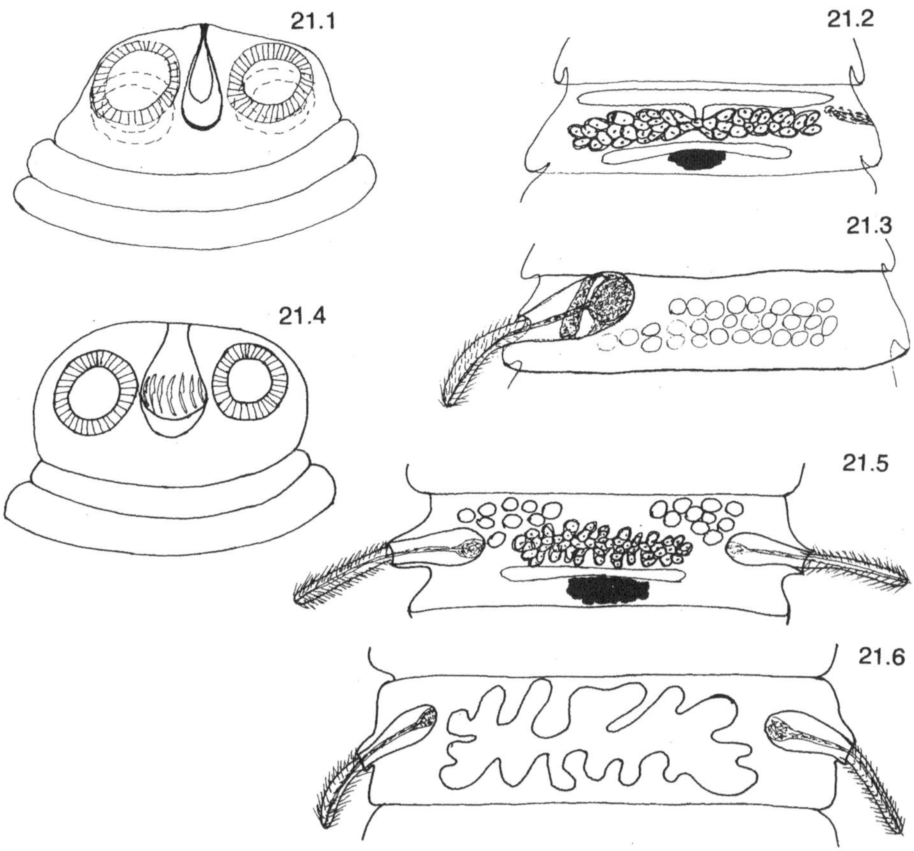

Figs 21.1-21.3 *Acoleus vaginatus* (Rudolphi, 1819). 21.1. Scolex. 21.2. Functional female proglottid. 21.3. Functional male proglottid.
Figs 21.4-21.6 *Diplophallus coili* Ahern & Schmidt, 1976. 21.4. Scolex. 21.5. Mature proglottid. 21.6. Gravid proglottid.

bilateral. Cirrus-sac well developed; cirrus long, heavily armed. Male ducts ventral to osmoregulatory canals. Testes numerous, in band across proglottid or in two lateral fields. Ovary transversely elongate, lobed. Young uterus a transverse tube becoming sac-like, occupying most of proglottid. Eggs numerous, oval. In Charadriiformes. Cosmopolitan. Type-species *D. polymorphus* (Rudolphi, 1819) Fuhrmann, 1899.

22 Family Dioecocestidae Southwell, 1930

A. JONES

International Institute of Parasitology, 395A Hatfield Road. St Albans, Herts AL4 0XU, UK.

Introduction

The family Dioecocestidae was first proposed by Southwell (1930b) with *Dioecocestus* Fuhrmann, 1900 as the type and only genus, distinguished by possession of entirely separate male and female strobilae. The composition of the family has varied as other dioecious cyclophyllideans from birds were added or removed. The absence of a vagina with its own aperture led to its being grouped close to or combined with other families which share this character. A number of species are imperfectly known, which has added to the confusion. Fuhrmann (1936b) suppressed it, placing *Dioecocestus* in the Acoleidae Fuhrmann, 1899* but in a distinct subfamily, the Dioecocestinae Southwell, 1930, with a separate subfamily Acoleinae for *Shipleya* Fuhrmann, 1907, *Gyrocoelia* Fuhrmann, 1899 and others. Most subsequent authorities, however, have accepted family status. Burt (1939a) restored the Dioecocestinae to family rank, including in it *Dioecocestus*, his new genus *Infula* and both *Shipleya* and *Gyrocoelia* on the grounds that the dioecious state was more significant than the absence of a vaginal pore. Burt's scheme was adopted by Wardle & McLeod (1952) and by Yamaguti (1959) who restored the Dioecocestinae for *Dioecocestus* and established the new subfamily Gyrocoeliinae for *Gyrocoelia, Shipleya, Infula* and his new genus *Pseudoshipleya*. Subsequent additions are *Neodioecocestus* Siddiqi, 1960 and *Echinoshipleya* Tolkacheva, 1979. On the basis of embryonic development and egg structure, Coil (1963, 1966, 1968, 1970a,b) retained Yamaguti's family and subfamilies, with *Dioecocestus* the sole genus in the Dioecocestinae and *Gyrocoelia* and *Shipleya* in the Gyrocoeliinae. He believed the latter two genera to be more closely related to each other than either *Infula* or *Dioecocestus* were to them and that *Infula* and *Shipleya* were sufficiently different in egg structure to be placed in separate families or at least subfamilies. Coil (1970a)

* Fuhrmann (1899a) erected the subfamily Acoleinae, elevating it to family rank (1907b) as the Acoleinidae, amended by Ransom (1909) to Acoleidae. According to *ICZN* Art. 36(a) the date for the family should be 1899.

recommended a separate subfamily for *Dioecocestus* on the grounds that the male and female organs were completely separated into different strobilae, but was unable to reach any conclusions on relationships from a study of the relatively unspecialized egg of *D. acotylus*.

More recently, Ryzhikov & Tolkacheva (1981) recognized the Dioecocestidae as a monotypic family and amended the status of the Gyrocoeliinae to family rank for the remaining genera, the main distinction being the retention of male copulatory organs in the female strobila in the Gyrocoeliidae. Schmidt (1986) adopted only the Dioecocestidae, without subfamilies, for all the dioecious genera of cyclophyllideans.

Traditionally, the characters used to distinguish taxa at the suprageneric level have been the completely or functionally dioecious nature of the strobila, the retention of male copulatory organs in the female strobila (of all genera except *Dioecocestus*), duplication of the male organs (in *Dioecocestus* only) and the absence of a typical vagina with an independent external aperture. Genera have been separated by the shape of the uterus, and the presence or absence of a rostellum and of rostellar armature.

There has been a great deal of debate on the dioecious nature of these cestodes, particularly in regard to *Gyrocoelia* and *Shipleya*. Reproductive processes in dioecious cestodes are not clearly understood. There are reports of both monoecious and dioecious species in the genus *Gyrocoelia* (see Burt, 1939a, Cable & Myers, 1956) but, while both conditions may exist, descriptions of some species are inadequate or contradictory and Burt (1939a) concluded that the evidence for the presence of testes in female strobilae was unsatisfactory. More recently described species have been established as dioecious (e.g. Mettrick, 1962; Rego, 1968; Cable & Myers, 1956). *G. albaredai* Lopez-Neyra, 1952, described as hermaphroditic, was based on a single specimen (Lopez-Neyra, 1952), attempts to trace which have been unsuccessful. The species was regarded as invalid by Baer (1959) and Ryzhikov & Tolkacheva (1981). While *Gyrocoelia* has been generally accepted as functionally dioecious, Rausch & Rausch (1990) have shown that specimens of the related genus *Shipleya* can be hermaphroditic. *S. inermis* is normally functionally dioecious but some members of either sex were found with rudimentary organs of the other and two predominantly male strobilae had fully functional female organs capable of egg production. It appears, however, that testes and female organs are generally not found in the same strobila. In female *S. inermis* with testes, the testes were underdeveloped, lacking a connection to the genital duct, and the worms were functionally female (Rausch & Rausch, 1990).

The absence of a typical vagina, although not restricted to the dioecious cyclophyllideans, has been one of their main characters. *Dioecocestus* females have a vagina with an occluded distal region which does not open externally but a separate vagina with an exterior aperture has generally been regarded as absent in the other genera. Various authors have, however, described other ducts which fulfil the same function. Burt (1939a) described a narrow duct connecting the cirrus-sac in female *Infula burhini* to the seminal receptacle and believed that insemination by a male strobila took place via the cirrus-sac. Coil (1955) reported a similar duct in *I. macrophallus* (synonymized with *I. burhini* by Ryzhikov &

Tolkacheva, 1981) and (1963) in *Gyrocoelia pagollae*. It has also been found in *Shipleya inermis* by Rausch & Rausch (1990). They suggested that *S. inermis* had evolved from a hermaphroditic ancestor and that, during evolution to the dioecious state, the cirrus-sac was retained in the female where it functioned as a vagina, a separate vagina was lost and a genital duct was formed by fusion of the vas deferens with the proximal region of the original vagina. It seems clear that at least in *Shipleya* and *Infula*, and possibly in *Gyrocoelia*, there is a modified duct which functions as a vagina but a typical separate vagina with an external aperture is lacking in all genera.

The key presented here is intended to facilitate identification and does not attempt to reflect phylogenetic relationships which will become clearer when more information is available on life-cycles and embryonic development. The family Dioecocestidae is considered to include cyclophyllideans which are anatomically or functionally dioecious, in which the male copulatory apparatus (but not the testes) is retained in the female strobila of all but one genus and may be modified for insemination, and in which a typical vagina with an external aperture is lacking. The Dioecocestinae has been retained for *Dioecocestus*, in which the sexes are entirely separate, and the Gyrocoeliinae for the remaining genera which are normally functionally dioecious and in which the female strobila retains a cirrus-sac like that of the male in appearance.

Neodioecocestus Siddiqi, 1960, described from a single male specimen lacking a rostellum or hooks by Siddiqi (1960) has not been accepted as valid. *Dioecocestus* as presently constituted includes species with or without a rostellum, rostellar armature or suckers and *Neodioecocestus* falls within the current concept of the genus, in which these differences have been treated as significant at the species level only. Several species are poorly known and attempts to obtain specimens of these and the type of *Neodioecocestus cablei* have not been successful. Outside *Dioecocestus*, the presence of a rostellum and rostellar armature has been awarded greater significance. The presence of a rostellum and the presence and nature of its armature are the characters distinguishing *Shipleya* and *Echinoshipleya* Tolkacheva, 1979, which are both accepted here, following Ryzhikov & Tolkacheva (1981) and Schmidt (1986). However, with regard to *Dioecocestus*, it is considered preferable to continue the traditional conservative approach rather than to proliferate genera based on species which are still, as yet, inadequately known. *Hamulocestus* Spasskii, 1992 is considered synonymous with *Dioecocestus*.

Following Ryzhikov & Tolkacheva (1981), *Pseudoshipleya* Yamaguti, 1959 is synonymized with *Infula*. The type-species, *Shipleya farrani*, described by Inamdar (1942), can be excluded from *Shipleya* by the presence of an unarmed rostellum and a ring-shaped uterus. The original description differs from *Infula* in that a small number of testes were said to be present 'only in a few mature segments'. The type specimens are not available for examination but it seems possible that ovarian lobes have been misinterpreted as testes.

Most illustrations are original; others are redrawn from the sources acknowledged in the figure legends.

Family Dioecocestidae Southwell, 1930

(Syn. Gyrocoeliidae Yamaguti, 1959)

Diagnosis: Large cestodes with numerous proglottids. Entirely or partly dioecious. Male and female strobilae entirely separate or female strobila retains male copulatory apparatus. Rostellum present or absent; if present, armed or unarmed. Suckers present or absent. Longitudinal musculature in two layers, transverse musculature in three. Male organs single or double. If single, genital pores alternate regularly in male strobila. Testes numerous, in one median group or two lateral groups. Internal seminal vesicle well-developed or not, external seminal vesicle absent. Cirrus-sac muscular. Cirrus robust, armed with spines. Female organs single. Genital pores absent in *Dioecocestus* female strobila, alternate regularly or irregularly in other genera. Ovary lobed, bialate to fan-shaped. Vitellarium postovarian. Seminal receptacle poral to vitellarium or between ovary and vitellarium. Typical vagina with external opening absent, may be replaced functionally in female strobila by the modified male duct. Young uterus transverse, ring-shaped or horseshoe-shaped. Uterine pore present or absent. Eggs oval. Parasites of birds (Charadriiformes and Podicipediformes). Type-genus *Dioecocestus* Fuhrmann, 1900.

Key to subfamilies

1a. Male and female strobilae entirely separate. Male genitalia duplicated. Vestigial vagina present, lacking external opening. Uterus a transverse tube .... Subfamily Dioecocestinae Southwell, 1930.

Diagnosis: Strobila large. Rostellum present or absent; if present, armed or unarmed. Suckers present or absent. Genital ducts pass between osmoregulatory canals. Male genitalia duplicated, male pores bilateral. Cirrus-sac small, muscular; cirrus spinous. Testes numerous, in two lateral groups. Female organs single. Ovary multilobed, usually bialate. Vitellarium lobed, postovarian. Seminal receptacle slightly poral to vitellarium, connects with vagina. Vagina alternates irregularly. Early uterus a transverse tube, becoming lobate. In grebes.

1b. Female strobila retains male copulatory apparatus. Male genitalia single. Typical vagina absent, may be functionally replaced by modified duct. Early uterus ring- or horseshoe-shaped .... Subfamily Gyrocoeliinae Yamaguti, 1959.

Diagnosis: Functionally dioecious. Rostellum present or absent; if present, armed or unarmed. Suckers present. Reproductive organs single. Genital ducts pass between osmoregulatory canals. Male copulatory apparatus present in female strobila. Cirrus-sac large, muscular. Cirrus robust, armed with numerous spines. Testes numerous, in single central group. Genital pores alternate regularly in male strobila, regularly or irregularly in female strobila. Ovary submedian, multilobed, bialate to fan-shaped. Vitellarium lobed, postovarian. Seminal receptacle between ovary and vitellarium. Uterus lobate when mature. Uterine pore present or absent. In charadriiform birds.

Subfamily Dioecocestinae Southwell, 1930

This subfamily contains only one genus.

Genus *Dioecocestus* Fuhrmann, 1900 (Figs 22.1, 22.2)
(Syns *Hamulocestus* Spasskii, 1992; *Neodioecocestus* Siddiqi, 1960)

Diagnosis: With the characters of the subfamily. In Ciconiiformes and Podicipediformes. Cosmopolitan. Type-species *D. paronai* Fuhrmann, 1900.

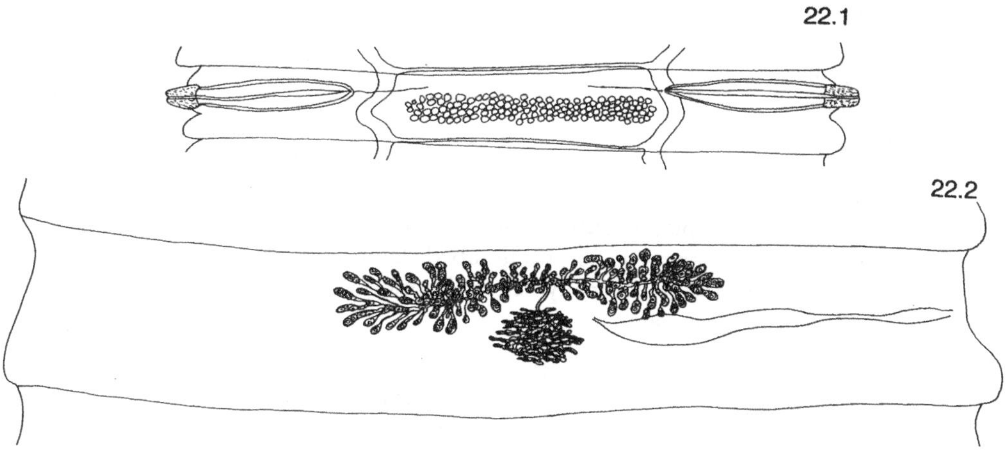

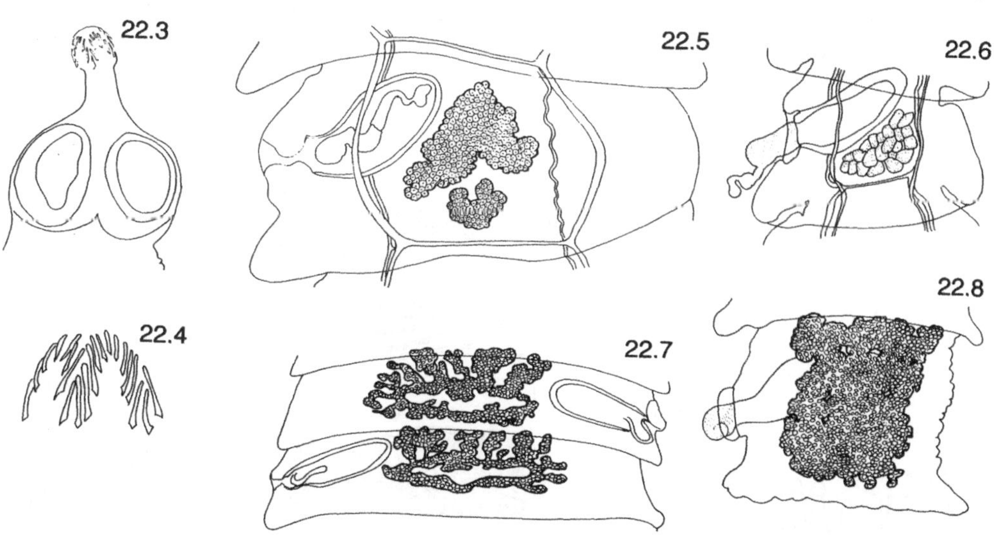

Figs 22.1, 22.2 *Dioecocestus acotylus* Fuhrmann, 1904. 22.1. Male proglottid. 22.2. Female proglottid. After Rysavy & Macko (1973).
Figs 22.3–22.8 *Gyrocoelia perversa* Fuhrmann, 1899. 22.3. Scolex. 22.4. Rostellar hooks. 22.5. Female proglottid. 22.6. Male proglottid. 22.7. Early gravid proglottid. 22.8. Gravid proglottid.

Subfamily Gyrocoeliinae Yamaguti, 1959

Key to genera

1a. Rostellum present. Uterus ring-shaped, uterine pores present .... **2.**
1b. Rostellum present or absent. Uterus horseshoe-shaped, uterine pores absent .... **3.**

2a. Rostellum armed with a single zig-zag row of hooks .... ***Gyrocoelia* Fuhrmann, 1899. (Figs 22.3–22.8)**
(Syn. *Brochocephalus* Linstow, 1906.)

Diagnosis: Strobila functionally dioecious. Rostellum capitate. One set of reproductive organs per proglottid. Genital pores alternate regularly in males, regularly or irregularly in females. Cirrus-sac large, muscular, retained in female strobila; cirrus robust, conspicuously armed. Genital ducts pass between osmoregulatory canals. Testes numerous, in median group. External seminal vesicle absent, internal seminal vesicle well-developed or not. Ovary median, bialate to multilobed, fan-shaped. Seminal receptacle between ovary and vitellarium. Typical vagina absent, may be replaced functionally by cirrus-sac and modified male duct. Young uterus ring-shaped, becoming lobed. Uterine pores present. Eggs oval. In Charadriiformes and occasionally Pelecaniformes. Cosmopolitan. Type-species *G. perversa* Fuhrmann, 1899.

2b. Rostellum present but unarmed .... ***Infula* Burt, 1939. (Figs 22.9–22.12)**
(Syn. *Pseudoshipleya* Yamaguti, 1959.)

Diagnosis: Strobila functionally dioecious. One set of reproductive organs per proglottid. Genital pores alternate regularly in males, regularly or irregularly in

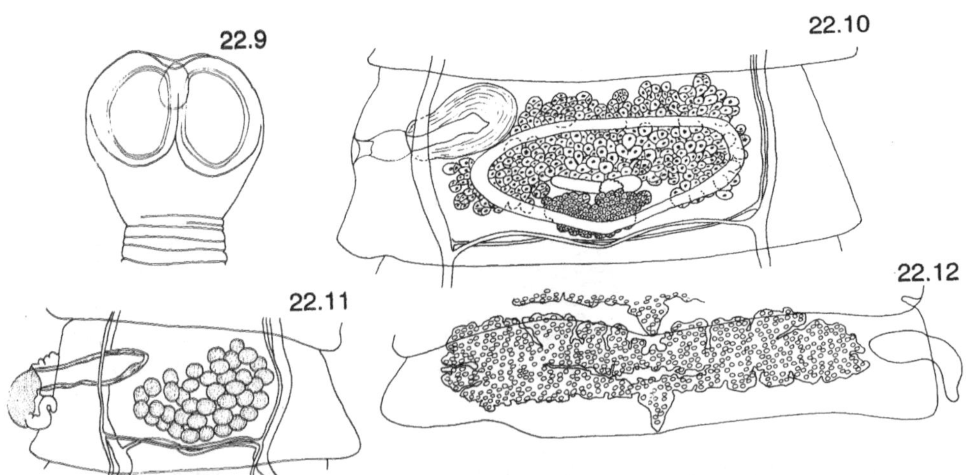

Figs 22.9–2.12 *Infula burhini* Burt, 1939. 22.9. Scolex. 22.10. Female proglottid. 22.11. Male proglottid. 22.12. Gravid proglottid.

females. Cirrus-sac large, muscular, retained in female strobila. Cirrus robust, armed. Genital ducts pass between osmoregulatory canals. External seminal vesicle absent, internal seminal vesicle present. Testes numerous, in median group. Ovary median, bialate, multilobed. Seminal receptacle between ovary and vitellarium. Typical vagina absent but replaced functionally by cirrus-sac and modified male duct. Uterus ring-shaped, becoming lobate. Uterine pore present at posterior margin of proglottid. Eggs oval; middle envelope thick, with polar plugs. In Charadriiformes. Cosmopolitan. Type-species *I. burhini* Burt, 1939.

3a. **Rostellum absent** .... ***Shipleya* Fuhrmann, 1908. (Figs 22.13–22.16)**
Diagnosis: Normally functionally dioecious but may be incompletely gonochoristic or hermaphroditic. Reproductive organs single. Genital pores alternate regularly. Genital ducts pass between osmoregulatory canals. Cirrus-sac large, muscular. Cirrus armed. Testes numerous, in median group. Ovary median, multilobed, fan-shaped. Cirrus-sac retained in female strobila, functions as vagina. Typical vagina absent. Seminal receptacle between ovary and vitellarium, linked to cirrus-sac by narrow duct. Early uterus horseshoe-shaped with anterior arm. Uterine pores absent. In Charadriiformes. North and South America. Type-species *S. inermis* Fuhrmann, 1908.

3b. **Rostellum present, armed with rows of numerous small aploparaksoid hooks** .... ***Echinoshipleya* Tolkacheva, 1979. (Figs 22.17, 22.18)**

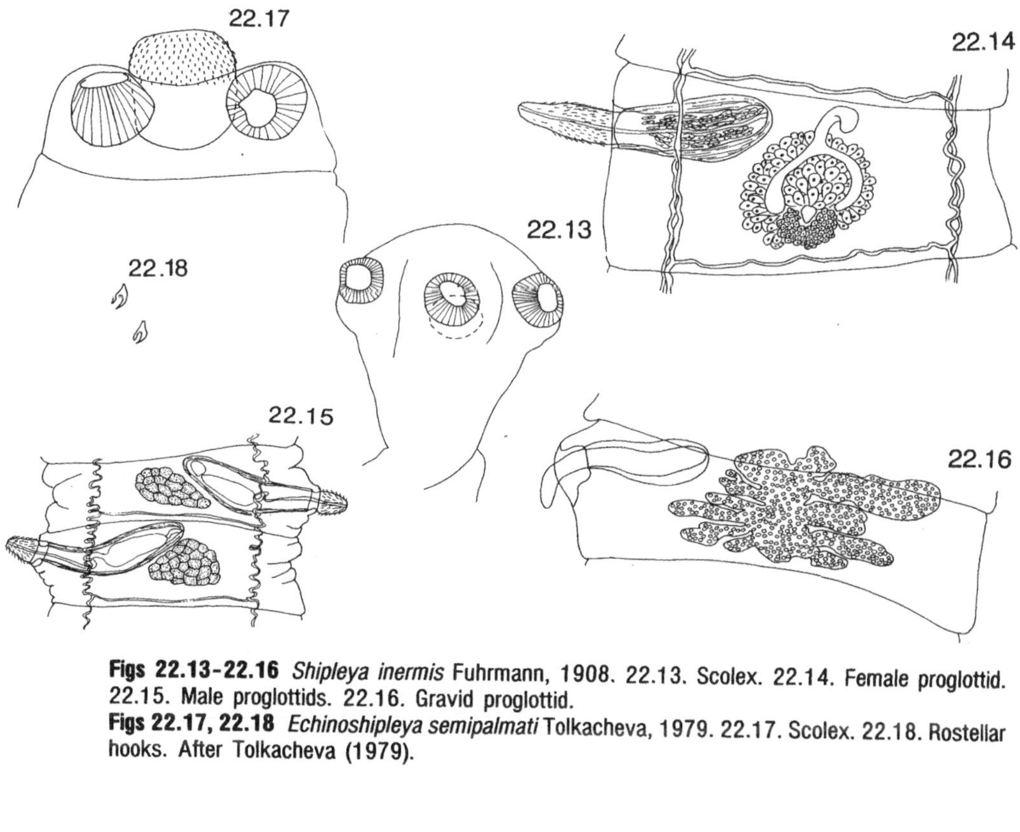

Figs 22.13–22.16 *Shipleya inermis* Fuhrmann, 1908. 22.13. Scolex. 22.14. Female proglottid. 22.15. Male proglottids. 22.16. Gravid proglottid.
Figs 22.17, 22.18 *Echinoshipleya semipalmati* Tolkacheva, 1979. 22.17. Scolex. 22.18. Rostellar hooks. After Tolkacheva (1979).

Diagnosis: Strobila functionally dioecious. Reproductive organs single. Genital pores alternate regularly in both sexes. Cirrus-sac large, muscular, retained in female strobila. Genital ducts pass between osmoregulatory canals. Testes in single median group. Ovary median, multilobed, fan-shaped. Typical vagina absent. Early uterus horseshoe-shaped with anterior arm. Uterine pores absent. Eggs oval. In Charadriiformes. USSR. Type-species *E. semipalmati* Tolkacheva, 1979.

23 Family Amabiliidae Braun, 1900

A. JONES

International Institute of Parasitology, 395A Hatfield Road, St Albans, Herts AL4 0XU, UK.

Introduction

This family is characterized by possession of accessory or supplementary ducts associated with the female reproductive organs. It was originally established as the subfamily Amabiliinae in the Taeniidae by Braun (1900) and raised to family rank by Fuhrmann (1907a) as the Amabilinidae, amended to Amabiliidae by Ransom (1909). According to the *ICZN*, Article 36(a), Braun, 1900 is the authority for the family.

The family originally contained *Amabilia* Diamare, 1893 (syn. *Aphanobothrium* Linstow, 1906), *Schistotaenia* Cohn, 1900 and *Tatria* Kowalewski, 1904 and was accepted with this composition by Ransom (1909), Wardle & McLeod (1952) and Yamaguti (1959). Subsequent additions are *Diporotaenia* Spasskaya, Spasskii & Borgarenko, 1971, *Laterorchites* Fuhrmann, 1932, *Pseudoschistotaenia* Fotedar & Chishti, 1976, *Joyeuxilepis* Spasskii, 1947 and *Ryjikovilepis* Gulyaev & Tolkacheva, 1987.

Subfamilies were introduced when Johri (1959) revived the Amabiliinae for *Amabilia* and proposed Schistotiinae, later amended to Schistotaeniinae, for *Schistotaenia* and *Tatria*. Ryzhikov & Tolkacheva (1975) erected the Diporotaeniinae for *Diporotaenia*. These subfamilies were recognized by Ryzhikov & Tolkacheva (1981) but not by Schmidt (1986). The Amabiliinae was distinguished chiefly by paired male organs and a dendritic ovary, the Diporotaeniinae by a scolex with petaloid lobes and the Schistotaeniinae mainly by single male gonads in combination with a normal scolex. Variations in the route of the accessory ducts were also used as characters. These features are regarded here as significant at the generic rather than subfamilial level and subfamilies have not been adopted.

Pseudoschistotaenia Fotedar & Chishti, 1976 was originally named in a conference abstract and later described in more detail (Fotedar & Chishti, 1976, 1980). It was distinguished from *Schistotaenia* by regularly alternating male pores, seminal receptacles linked in successive proglottids, a small rather than wide ovary and absence of a blindly-ending, aporally-directed accessory duct. These characters occur in species of *Schistotaenia* (see Baer, 1940; Chandler, 1948; Schell, 1955; Johri, 1959; Rausch, 1970), with which *Pseudoschistotaenia* is considered synonymous. The type-specimens of *P. indica* Fotedar & Chishti, 1976, the type-species, were not available for examination.

Joyeuxilepis Spasskii, 1947 was first proposed for *Echinorhynchotaenia biuncinata* Joyeux & Baer, 1943, based on a cysticercoid, the adult of which was unknown, with distinctive rostellar armature (Joyeux & Baer, 1943; Spasskii, 1947). Originally in the Hymenolepididae, it was transferred to the Amabiliidae as a subgenus of *Tatria* by Spasskii & Spasskaya (1976) who considered the species *biuncinata* to be a synonym of *T. (J.) decacantha* Fuhrmann, 1913. *Joyeuxilepis* was raised to generic rank by Gulyaev & Tolkacheva (1987), distinguished from *Tatria* by the shape of the rostellar hooks ('aploparaksoid-like') and protandric maturation of the reproductive organs. They also erected *Ryjikovilepis* for *Tatria* (sensu lato) *dubininae* Ryzhikov & Tolkacheva, 1981, with hooks similar to *Joyeuxilepis* but different in characters of the osmoregulatory system, male terminal ducts and accessory ducts.

Borgarenko & Gulyaev (1990) described the adult of *J. biuncinata* for the first time and considered its synonymy with *J. decacantha* to be doubtful. They found that important points, which would be taxonomically significant, relevant to species of *Tatria* s.l. remain to be clarified and considered it premature to present a generic diagnosis for *Joyeuxilepis*. In view of this, *Joyeuxilepis* is provisionally maintained here as a synonym of *Tatria*, pending further investigation.

Laterorchites Fuhrmann, 1908 was transferred from the Dilepididae to the Amabiliidae by Spasskii (1969a) and again by Rysavy & Macko (1973). This was accepted by Ryzhikov & Tolkacheva (1981). The generic diagnosis given below is based on the redescription of *L. bilateralis* (Fuhrmann, 1908) by Rysavy & Macko (1973).

The illustrations are original or modified and redrawn from the sources acknowledged in the captions.

Family Amabiliidae Braun, 1900

Diagnosis: Small to medium cestodes. Proglottids usually craspedote, wider than long with prominent lateral processes (rarely absent). Scolex usually normal, rarely hypertrophied to form petaloid lobes. Armed rostellum usually present, with single crown of rostellar hooks. Spines present or absent on rostellum, scolex or suckers. Testes posterior or lateral, may surround female organs. Male ducts single or double. Male pores alternate regularly or irregularly. Cirrus-sac reaches osmoregulatory canals or not. External seminal vesicle present, internal present or absent. Prostatic vesicle present or absent. Cirrus armed. Female organs single. Ovary and vitellarium median, lobed, rarely dendritic. True vagina absent, may be functionally replaced by accessory ducts. Accessory or supplementary ducts may connect seminal receptacles between successive proglottids, with apertures on dorsal and ventral surfaces, or with osmoregulatory anastomosis; they may pass from seminal receptacle to open or end blindly near lateral margin aporal to male

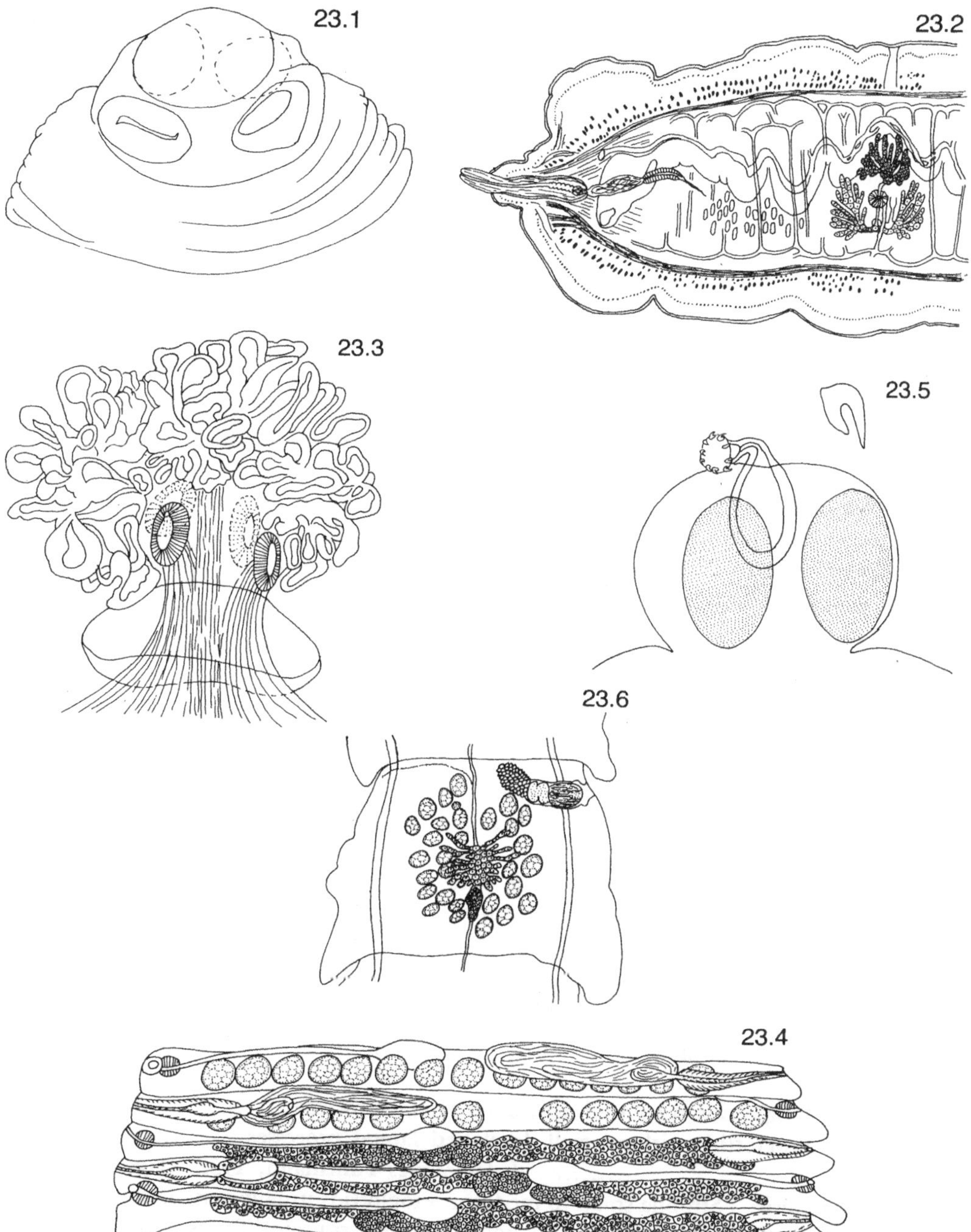

Figs 23.1, 23.2 *Amabilia lamelligera* (Owen, 1832). 23.1. Scolex. 23.2. Transverse section of part of mature proglottid, redrawn after Clausen (1915).
Figs 23.3, 23.4 *Diporotaenia colymbi* Spasskaya, Spasskii & Borgarenko, 1971. 23.3. Scolex. 23.4. Mature proglottids. After Spasskaya *et al.* (1971).
Figs 23.5, 23.6 *Laterorchites bilateralis* (Fuhrmann, 1908). 23.5. Scolex and rostellar hook. 23.6. Mature proglottid. After Rysavy & Macko (1973).

pore. Uterus usually saccate, rarely reticulate or initially horseshoe-shaped. Eggs round or oval, or fusiform with polar filaments. In birds. Type-genus *Amabilia* Diamare, 1893.

Key to genera

1a. Male genital ducts double .... *Amabilia* **Diamare, 1893. (Figs 23.1, 23.2)**
(Syn. *Aphanobothrium* Linstow, 1906.)

Diagnosis: Robust cestodes. Proglottids imbricated with conspicuous lateral processes. Armed rostellum apparently absent. Suckers unarmed. Testes in two lateral fields, occasionally confluent. Cirrus-sac extravascular, muscular. Cirrus robust, armed. Small internal seminal vesicle present. Male ducts pass between osmoregulatory canals. Ovary and vitellarium median, dendritic. Vagina absent. Seminal receptacle communicates with dorsoventral accessory canal which opens on both surfaces in mid-line of strobila and also opens into ventral transverse osmoregulatory anastomosis. Uterus finely reticulate. Eggs fusiform with polar filaments. In Phoenicopteriformes. Africa, India, Asia. Type-species *A. lamelligera* (Owen, 1832).

1b. Male genital ducts single .... **2.**

2a. Scolex hypertrophied, forming petaloid lobes ....
Diporotaenia **Spasskaya, Spasskii & Borgarenko, 1971. (Figs 23.3, 23.4)**

Diagnosis: Strobila medium-sized. Proglottids short, very wide. Anterior part of scolex modified, obscuring armed rostellum. Suckers unarmed. Fleshy collar present behind suckers. Testes in single transverse row. Cirrus-sac small, reaches osmoregulatory canals. Cirrus armed. Internal seminal vesicle small, external seminal vesicle conspicuous. Male pores alternate regularly. Ovary wide, slightly lobed. Vitellarium lobed, slightly aporal to male pore. Accessory duct with distal sphincter passes from seminal receptacle to open on lateral margin aporally to male pore. Uterus saccate, lobed. In Colymbiformes. Tadzhikistan, Turkmenia. Type-species *D. colymbi* Spasskaya, Spasskii & Borgarenko, 1971.

2b. Scolex normal .... **3.**

3a. Proglottids with well-developed lateral processes .... **4.**

3b. Proglottids without well-developed lateral processes ....
.... *Laterorchites* **Fuhrmann, 1932. (Figs 23.5, 23.6)**

Diagnosis: Strobila small to medium. Proglottids usually craspedote. Rostellum armed with single row of ten hooks. Suckers spinose. Testes surround female organs. Prostatic vesicle present between cirrus-sac and external seminal vesicle. Cirrus armed. Male pores alternate irregularly. Ovary median, lobed. Vitellarium postovarian. Accessory duct from seminal receptacle ends blindly near aporal lateral margin. Uterus initially horseshoe-shaped, becoming lobed. In Podicipediformes. Central America. Type-species *L. bilateralis* (Fuhrmann, 1908).

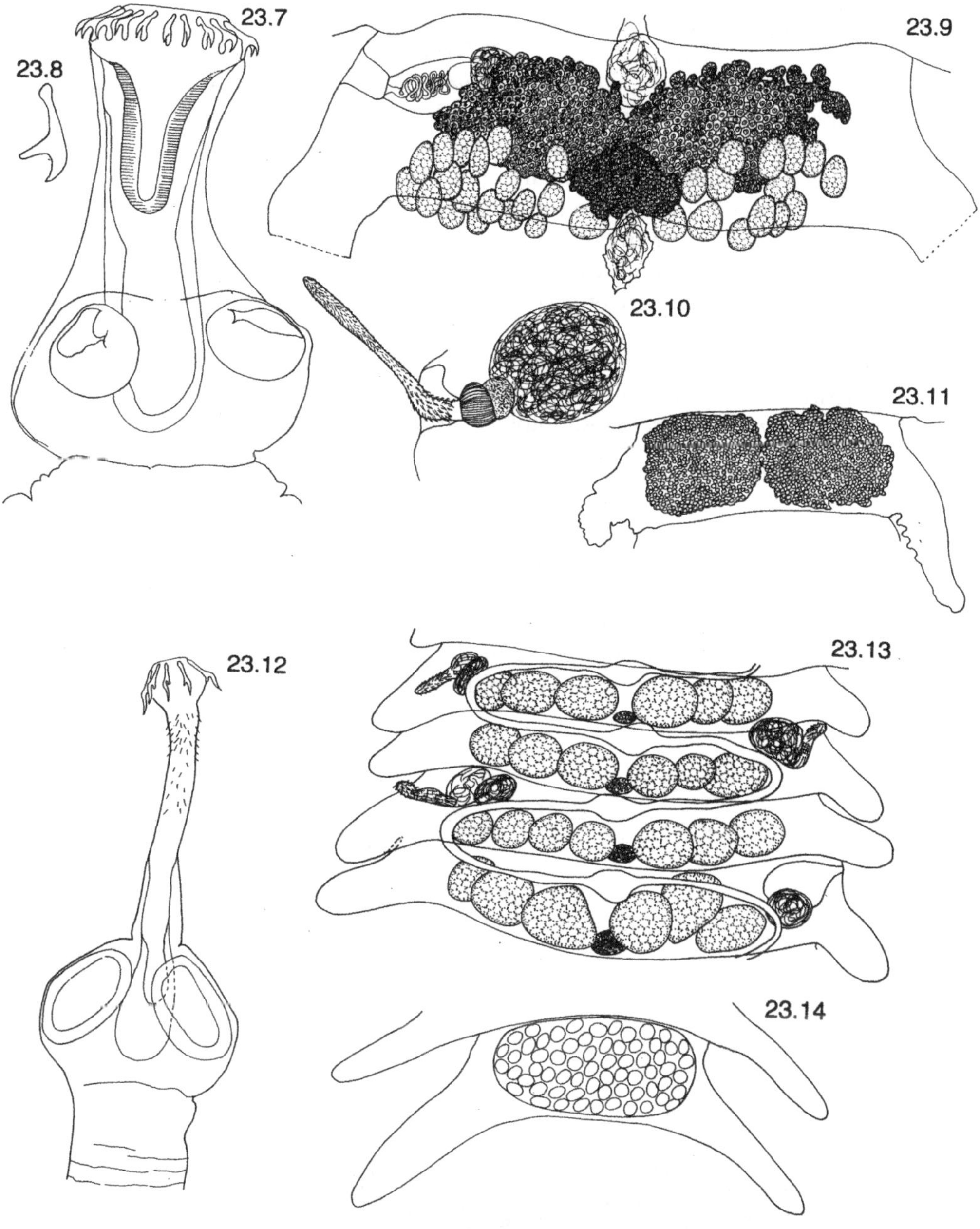

Figs 23.7–23.11 *Schistotaenia colymba* Schell, 1955. 23.7. Scolex. 23.8. Rostellar hook. 23.9. Mature proglottid. 23.10. Male terminal ducts. 23.11. Gravid proglottid.
Figs 23.12–23.14 *Tatria biremis* Kowalewski, 1904. 23.12. Scolex. 23.13. Mature proglottids, male organs. 23.14. Gravid proglottid.

4a. Strobila medium. Scolex robust. Rostellar hooks 16–26 .... *Schistotaenia* Cohn, 1900. (Figs 23.7–23.11)

(Syn. *Pseudoschistotaenia* Fotedar & Chishti, 1976)

Diagnosis: Rostellum robust, armed with single row of strong hooks. Spines present or not on rostellum or suckers. Proglottids wide, with lateral processes. Testes usually in two groups, occasionally in one field. Cirrus-sac small. Prostatic vesicle present between cirrus-sac and external seminal vesicle. Cirrus armed. Male pores alternate regularly or irregularly. Ovary median, lobed, bi-alate or transversely elongated. Vitellarium median, postovarian, lobed. Accessory ducts present. Seminal receptacles interconnected by median longitudinal accessory duct; rarely, sigmoid duct also present. Additional accessory ducts from seminal receptacle open on dorsal and ventral surfaces or only ventrally. Uterus saccate. Eggs spherical. In Podicipediformes. Africa, Asia, Europe, North America. Type-species *S. macrorhyncha* (Rudolphi, 1810) Cohn, 1900.

4b. Strobila small, delicate. Scolex delicate. Rostellar hooks 8–14 .... 5.

5a. Two pairs of longitudinal osmoregulatory canals .... *Tatria* Kowalewski, 1904. (Figs 23.12–23.14)

(Syn. *Joyeuxilepis* Spasskii, 1947.)

Diagnosis: Strobila small. Proglottids wide with digitiform lateral processes. Rostellum delicate, elongated, armed with single row of small hooks. Hooks stylet-like or aploparaksoid-like. Spines present or not on rostellum, scolex and

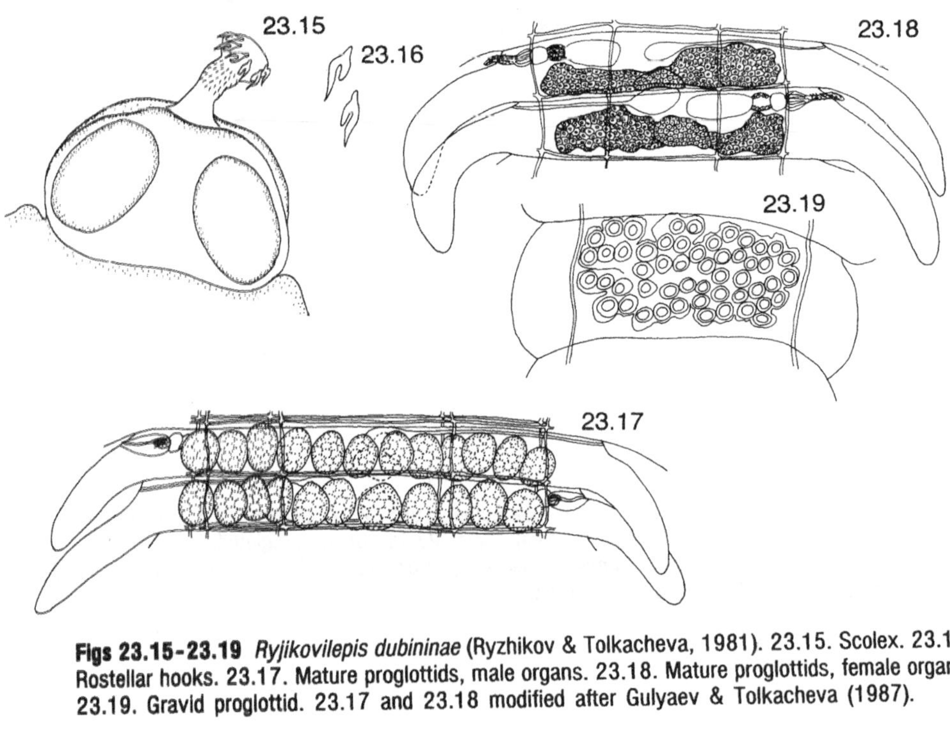

Figs 23.15–23.19 *Ryjikovilepis dubininae* (Ryzhikov & Tolkacheva, 1981). 23.15. Scolex. 23.16. Rostellar hooks. 23.17. Mature proglottids, male organs. 23.18. Mature proglottids, female organs. 23.19. Gravid proglottid. 23.17 and 23.18 modified after Gulyaev & Tolkacheva (1987).

suckers. Testes relatively few, in two lateral groups or one transverse row. Internal and external seminal vesicles present. Cirrus-sac small. Cirrus armed. Male pores usually alternate regularly. Ovary median, lobed, bialate or very wide. Vitellarium postovarian, median or slightly poral. Seminal receptacles connected between successive proglottids by accessory ducts which are median longitudinal or sigmoid or, exceptionally, both. Uterus saccate. Eggs round or oval. In Podicipediformes. Africa, Asia, Europe, Russia, North and South America. Type-species *T. biremis* Kowalewski, 1904.

5b. Four pairs of longitudinal osmoregulatory canals ....
.... ***Ryjikovilepis* Gulyaev & Tolkacheva, 1987. (Figs 23.15–23.19)**
Diagnosis: Strobila small. Proglottids with well-developed lateral processes. Rostellum with single crown of aploparaksoid-like hooks. Spines on rostellum and suckers. Testes in single row. Internal seminal vesicle absent; external seminal vesicle present, separated by prostatic vesicle into prebursal and postprostatic parts. Cirrus armed. Male pore alternates regularly. Vagina absent. Ovary bilobed, transversely elongate. Accessory duct passes from seminal receptacle to aporal lateral margin. Uterus saccular, not crossing osmoregulatory canals. Adult in Podicipediformes. Russia, Europe. Type-species *R. dubininae* (Ryzhikov & Tolkacheva, 1981).

24 Family Davaineidae Braun, 1900

A. JONES[1] & R.A. BRAY[2]

[1]*International Institute of Parasitology, 395A Hatfield Road, St Albans, Herts AL4 0XU, UK:* [2]*Department of Zoology, The Natural History Museum, Cromwell Road, London SW7 5BD, UK.*

Introduction

The family Davaineidae Braun, 1900* has often been divided into subfamilies based on the nature of the uterus: Davaineinae with parenchymatous uterine capsules each containing one or several eggs; Ophryocotylinae with a persistent saccular uterus and the eggs not in capsules (except when small portions of the periphery of the uterus are pinched off by parenchyma); and Idiogeninae with a large paruterine organ surrounding the uterus with its main bulk placed anteriorly to the uterus, which is saccular or an inverted U-shape with no egg-capsules.

The relationship between the davaineids and the anoplocephalids is clearly close, as indicated by the parallel development of the uterus and related structures. We have adopted a pragmatic approach in our treatment of the Davaineidae, with the intention of producing a key for identification rather than to reflect phylogenetic relationships which are currently not well understood. As treated here, the davaineids remain a polyphyletic assemblage. This key does not take account of the likelihood that some davaineid genera are probably more closely related to the anoplocephalids than to most davaineids, or of the premise that anoplocephalids are essentially unarmed davaineids. It is also probable that some idiogenine genera are very closely related to some unarmed paruterinids. These relationships are suggested by the nature of the uterus and its derivatives, which clearly needs much more attention in cestode systematics. It is becoming clear that the paruterine organ in the Cyclophyllidea has polyphyletic origins (Georgiev & Kornyushin, this volume).

The nature and origin of the egg-capsule is not clear in such genera as *Davainea, Skrjabinia, Cotugnia* and *Ophryocotyle*. The situation may parallel that of the anoplocephalid subfamily Linstowiinae (see Conn, 1985; Beveridge, this volume) where in some cases the eggs are said to be embedded in the parenchyma in uterine capsules. Conn (1985) described the eggs of *Oochoristica* from

*According to *ICZN* Art. 50(c), the author of this taxon should be Braun, 1900.

transmission electron microscope observations but, as far as we are aware, there are no equivalent observations on davaineids. Light microscope observations of *Skrjabinia* (by Reid, Ackert & Case, 1938) and Cotugnia (by Singh & Baugh, 1984) give conflicting information. Reid *et al.* (1938) described the two outer membranes around the oncosphere as being connected by smaller membranes which are probably the remains of the uterine wall, but they did not state whether they considered the outer membranes as remains of the uterus and did not discuss a 'uterine capsule'. Singh & Baugh (1984) described various concentric layers around the oncosphere as three membranes and three envelopes and contested the presence of parenchymatous capsules in *Cotugnia*. Their observations were based only on the use of light microscopy and are, therefore, limited in value. We are not convinced that the situation in the subfamilies Davaineinae and Ophryocotylinae is well enough understood to separate them confidently. As this key is for ease of identification, we have departed from the traditional recognition of three subfamilies in the Davaineidae and have not recognized the Davaineinae and Ophryocotylinae as distinct (cf. Spasskii & Spasskaya, 1976).

The family characteristically has small, numerous hammer-shaped rostellar hooks. This is the one feature totally diagnostic of the Davaineidae, and the position of taxa which lack it has often been, and still remains, controversial.

Some characters previously used to separate genera are not recognized at that level here. The scolex characters significant at generic level are: presence or absence of the rostellum; shape and continuity or discontinuity of the rostellar crown; hook shape and size; number of rows of rostellar hooks and the presence (or rarely absence) and shape of the suckers. Accessory rostellar spines and sucker armature are of only specific significance. Significant strobilar characters include: the number of proglottids and of osmoregulatory canals; single or paired genitalia; and modifications of the genital atrium and terminal genital ducts. The number of testes is significant only when consistently very low, e.g. two. Their distribution is significant only when distinctly and consistently poral and is generally used in conjunction with other characters. The position of the ovary when distinctly poral is regarded as generically significant but its shape is not. Unilaterality or alternation of the genital pores is significant at generic level. In gravid proglottids, the presence of a paruterine organ is significant at subfamily level. Where egg capsules are present, they either have thick fibrous walls and enclose several eggs or have thin delicate walls and enclose a single egg. The number of eggs per capsule (i.e. one or several) is of generic significance. Only in one instance does a species have one or several eggs per capsule: *Calostaurus mundayi* Beveridge, 1975 has one to three eggs per capsule according to Beveridge's (1975) original description, confirmed by examination of the type specimens (BMNH 1976.1.6. 29–30). The capsule walls are, however, thin and delicate like those of taxa with monovular capsules and are quite unlike the fibrous-walled capsules in most taxa with several eggs per capsule. Other species of *Calostaurus* have a single egg enclosed in a thin-walled capsule. *C. mundayi* is regarded here as exceptional. The number of capsules per proglottid is generically significant only if consistently very low. Capsules containing a single egg are invariably numerous.

Davaineinae

The davaineids with paired genitalia form an easily recognizable group which has been elevated to subfamily rank by Movsesyan (1969) but the Cotugniinae was not recognized by Schmidt (1986) and is not recognized here because the character is of generic rather than subfamily significance. It has not been used to distinguish subfamilies in the anoplocephalids or the dilepidids (see this volume). Six nominal genera have consistently paired genitalia: *Cotugnia* Diamare, 1893, *Abuladzugnia* Spasskii, 1973, *Erschovitugnia* Spasskii, 1973, *Rostelugnia* Spasskii, 1984, *Pavugnia* Spasskii, 1984 and *Multicotugnia* López-Neyra, 1943. Of these, Movsesyan (1987, 1989) recognized only *Cotugnia* and *Multicotugnia*. Schmidt (1986) accepted *Abuladzugnia* but synonymized *Erschovitugnia* and *Multicotugnia* with *Cotugnia*; the remaining two genera were published too late to be available for his key.

Abuladzugnia is recognized here as valid.

Multicotugnia was erected by López-Neyra (1943a) for *Cotugnia brotogerys* Meggitt, 1915 because each egg capsule contained several eggs, a character regarded here as significant at the generic level. Meggitt (1915) stated clearly twice in his description of the species that the capsules contain several eggs but his illustration suggests that each capsule contains only one. The genus is provisionally retained. Descriptions of *C. brotogerys* by Southwell (1930b) and Mudaliar (1939) confirm the presence of several eggs per capsule, but our attempts to obtain specimens for examination proved unsuccessful.

The major feature of *Erschovitugnia* is the size of its rostellar hooks (70–90 μm). This is large for the group but is regarded here as of specific rather than generic significance; the synonymy of *Erschovitugnia* with *Cotugnia* is upheld, following Schmidt (1986) and Movsesyan (1987). Although large, the hooks are of the typical davaineid hammer shape.

Manitaurus Spasskaya & Spasskii, 1971 is considered valid following examination of the type specimens of *M. rahmi* (Baer & Fain, 1955) (BMNH 1960.7.12. 8–14), which confirmed that the gravid proglottids each have two to three large egg capsules each containing several eggs. This precludes synonymy with *Diorchiraillietina contorta* (Zschokke, 1895) Yamaguti, 1959 which has 30–40 egg capsules per proglottid, each with one egg (confirmed from specimens at the BMNH). Schmidt's (1986) synonymy of *M. rahmi* with *D. contorta* and hence of *Manitaurus* with *Diorchiraillietina* cannot be supported. Movsesyan (1987) regarded *Manitaurus* as an inermicapsiferine anoplocephalid on the grounds that the scolex was unarmed. This view is not shared here. Probably, as stated in Baer & Fain's (1955) original description, the type specimens have a rostellum from which the hooks have been lost. Specimens examined by us lacked scoleces.

Movsesyan (1987, 1989) transferred *Baerfainia* from the Davaineidae to the Hymenolepididae but the type species has a few large fibrous egg capsules each containing several eggs rather than a persistent uterus, which seems to preclude this. We have retained it in the Davaineidae because of its otherwise close resemblance to other genera in this family. Movsesyan (1987) quoted Spasskaya & Spasskii's (1971) opinion that *Baerfainia*, *Manitaurus* and *Metadavainea* Baer & Fain, 1955 were inermicapsiferines and that this subfamily belonged in the

Anoplocephalidae. The position of the two former genera will remain rather equivocal until the nature of the scolex in both can be elucidated by the examination of new specimens. A transfer to the Inermicapsiferinae might become appropriate for *Baerfainia* if the absence of a rostellum in *B. anoplocephaloides* were confirmed, but would be inappropriate for *Metadavainea* which has a rostellum armed with large, robust hooks. These hooks are not typically davaineid in shape or number but other morphological characters are consistent with the Davaineidae and the genus has been retained in this family, a course also taken by Movsesyan (1987, 1989). Schmidt (1986) appears inadvertently to have omitted the genus from his keys.

Delamuretta Spasskii, 1977 (syn. *Delamurella* Spasskii & Spasskaya, 1976 preoccupied, also spelled *Dalamurella*) was erected for *Davainea polycalceola* Janicki, 1902 and characterized by possession of six longitudinal osmoregulatory canals. Schmidt (1986) listed *D. polycalceola* as a species of *Davaineoides* Fuhrmann, 1920, the type-species of which, *D. vigintivasus* (Skryabin, 1914) Fuhrmann, 1920, has 20 osmoregulatory canals. He followed Fuhrmann (1920a) and Yamaguti (1959) in defining *Davaineoides* as possessing 6–20 osmoregulatory canals but inappropriately listed *Delamurella* [sic] and *Delmuretta* [sic] as synonyms of *Raillietina (Skrjabinia)*, not of *Davaineoides*. The *Raillietina* group *sensu lato* has a maximum of four osmoregulatory canals. *Delamurella* was listed as a synonym of *Davaineoides* by Movsesyan (1987, 1989) without mention of the later name *Delamuretta*. The genus has been retained pending further evaluation.

The large *Raillietina* group has traditionally been treated as one genus, *Raillietina* Fuhrmann, 1920, with four subgenera *R. (Raillietina)* Fuhrmann, 1920, *R. (Fuhrmannetta)* Stiles & Orleman, 1926, *R. (Skrjabinia)* Fuhrmann, 1920 and *R. (Paroniella)* Fuhrmann, 1920. Although many authors have followed this system (e.g. Fuhrmann, 1920a; Stiles & Orleman, 1926; López-Neyra, 1931; Artyukh, 1966; Schmidt, 1986), the subgenera were elevated to generic rank by Movsesyan (1966). Following recognition of the generic significance of the number of eggs per capsule plus genital pores unilateral or irregularly alternating, the generic status of *Raillietina, Fuhrmannetta, Skrjabinia* and *Paroniella* is recognized here. Their further subdivision into subgenera by Movsesyan (1966) based on the presence or absence of sucker armature is not accepted because this character is of specific significance only. Accordingly, the subgenus *Armacetabulum* Movsesyan, 1966 is a synonym of *Skrjabinia*. The subgenus *Nonarmiella* erected by Movsesyan (1966) in the genus *Paroniella* was based on a species described as new by Purvis (1932), *Raillietina (Paroniella) acanthovagina*. Purvis later (1940) accepted an opinion by Baylis that his specimens were misidentified specimens of *R. (R.) volzi*. There is some confusion in Purvis's text, because at one point he stated that *R. (P.) southwelli*, also described as new in his first paper, was synonymous with *R. (R.) volzi* but this is presumably an error because he subsequently accepted the synonymy, also proposed by Baylis, of *R. (P.) southwelli* with *R. (P.) rangoonica*. Examination of Purvis's specimens (BMNH 1977.11.11.79) suggests that the eggs are encapsulated in groups, whereas in *Paroniella* the eggs are encapsulated singly. *Nonarmiella* is therefore based on a misidentified taxon and must be synonymous with *Raillietina* as its type-species is a synonym of a *Raillietina* species.

Corvinella Spasskaya & Spasskii, 1971 was characterized primarily by the arrangement of the testes in two lateral groups, a character of only specific significance; it is synonymized with *Paroniella*. The name is preoccupied by *Corvinella* Lesson, 1830 (Aves).

Tetraonetta Spasskaya & Spasskii, 1971 has, according to its original generic diagnosis, genital pores which alternate irregularly or are unilateral (Spasskaya & Spasskii, 1971) but a later key by Spasskii & Spasskaya (1976) states that the pores are unilateral. López-Neyra (1931) established that specimens described by various authors under the name of the type-species, *Davainea urogalli* (Modeer, 1790) Blanchard, 1891, which he discussed under the name *Meggittia urogalli*, belonged to more than one species. He stated that the genital pores are unilateral. Consequently, *Tetraonetta* has nothing to distinguish it from *Paroniella*.

Metaparonia Spasskii & Spasskaya, 1976 was erected in a key, without a formal generic diagnosis, with *R. (P.) bovieni* Baer & Fain, 1955 as type-species. From Spasskii & Spasskaya's (1976) key and from the original description of the type-species by Baer & Fain (1955), its distinguishing character is a long cirrus-sac figured as intersecting the osmoregulatory canals but not approaching the mid-line. We do not consider this sufficient to separate it from *Paroniella*.

Numidella Spasskaya & Spasskii, 1971 has been retained *pro tem* because it has an unusually large cirrus-sac which crosses the mid-line of the proglottid, a cirrus armed with large spines, and a distal inflation of the vagina conspicuously armed in several species allocated to the genus. Examination of specimens of *N. numida* (Fuhrmann, 1912) has confirmed these characters. The terminal genital ducts of *Davainea* are similar but the strobila in this genus consists of only a few proglottids.

Gvosdevinia Spasskii, 1973 is recognized as valid because of its combination of a distinctly poral ovary and the restriction of the testes to a poral, postovarian field.

Daovantienia Spasskii & Spasskaya, 1976, erected for *Raillietina (Skrjabinia) centropi* Southwell, 1922, was distinguished by the possession of a bilobed ovary and of testes surrounding the female organs and extending laterally beyond the osmoregulatory canals. Examination of the type (BMNH 1977.11.1.27) and other specimens (BMNH 1983.6.15.13–15; 1983.6.20.15; 1983.7.12.7–10; 1983.6.16.11) has confirmed these characters but they are of specific rather than generic significance and the genus is synonymized with *Skrjabinia*, following Movsesyan (1987, 1989) and Schmidt (1986) (as *R. (Skrjabinia)*).

Markewitchella Spasski & Spasskaya, 1976, based on *Raillietina (Skrjabinia) bonini* (Megnin, 1899) Fuhrmann, 1932, does not differ significantly from *Skrjabinia*, and their synonymy is recognized here, following Schmidt (1986) and Movsesyan (1987, 1989).

Soninotaurus Spasskii, 1973 is accepted as valid. The type-species, *S. rhynchota* (Ransom, 1909), is distinguished by an undulating crown of rostellar hooks which forms eight lobes. The undulating crown separates it from all other davaineine genera (*sensu stricto*) except *Calostaurus* and *Abuladzugnia*. *Calostaurus* species parasitize Australasian marsupials and the rostellar crown forms four or six lobes or a Maltese cross. *Abuladzugnia* has a laterally interrupted rostellar crown and paired genitalia. *S. rhynchota*, a parasite of North American

Piciformes, has a continuous crown and single genitalia. Its characters have been confirmed by examination of type and other specimens (USNM 7235).

Paspalia is recognized as the only *Raillietina*-like form with hooks of this large size or relatively small number. Schmidt (1986) and Movsesyan (1987) consider it a synonym of *Raillietina*, but we have examined type specimens and found that, contrary to the original description by Paspalewa & Woidowa (1969), egg capsules containing several eggs are not present. Hook number and shape are not typical of the davaineids and the genus probably has affinities with the dilepidids. The uterus is not clear in the available material, nor is the presence of single-egg capsules. The material is not in good condition and some features of the mature proglottid, e.g. the testes distribution and number, can be interpreted differently from Paspalewa & Woidowa (1969).

Idiogenoides is retained as valid as it shows an unusual combination of an *Idiogenes*-like mature proglottid and parenchymatous egg-capsules enclosing several eggs (Kotlan, 1921). The genus was erected by López-Neyra (1929) and, as well as the type-species, included *I. oligorchida* (Fuhrmann, 1911) and *I. birmanica* (Meggitt, 1926). Spasskaya & Spasskii (1971) added a further five species, all from parrots, to the genus. It is not entirely clear what features unite these species; Spasskii & Spasskaya (1976) in their key stress the two groups of testes, the few testes (<20), well-developed cirrus-sac and the thick and powerfully armed cirrus. On the other hand, we provisionally recognize the genus based on the large size of the cirrus-sac, which in the type-species, *I. allagea*, reaches close to the mid-line of the proglottid and on this basis we recognize only the type-species as belonging to the genus. Unfortunately the type-species has only been described once (Kotlan, 1921).

The following synonyms of *Raillietina* were differentiated from *Raillietina* on minor features, not considered significant by us. The only relevant distinguishing feature of *Kotlanotaurus* as keyed by Spasskii & Spasskaya (1976) is the penetration of the egg-capsules into the cortical parenchyma. We do not consider that this feature, which may be ontogenetic, has generic importance. The subgenus *Nonarmina* Movsesyan, 1966 has unarmed suckers, a feature which we do not consider of generic importance as it is so often a fixation artifact. *Roytmania* was distinguished in Spasskii & Spasskaya's (1976) key by minor features of the testicular distribution which are, in fact, not exhibited by the type-species *R. weissi* (Joyeux, 1923). *Skrjabinotaurus* was distinguished by its incomplete ring of sucker spines, the size of its rostellum relative to the suckers and the relatively few (*ca.* 4–9) testes. The first two characters are not of generic importance. The number of testes is significant when consistently low, e.g. *Diorchiraillietina*, but not when variable.

Spasskii (1973) stated that *Vadifresia* and *Raillietina* are the only genera and subgenera of davaineids with a single row of rostellar hooks but the type of *Raillietina, R. tetragona*, often has two rows of rostellar hooks, as do most species in the genus. We examined specimens of the type-species of *Vadifresia, V. baeri* (Meggitt & Subramanian, 1927), and confirm that only a single row of rostellar hooks is present. The hooks in the row are not all entirely aligned, but show no regular alternation of rows as found in related genera. This suggests that the single row of rostellar hooks is a consistent feature, but other authors have different views. Baer (1959) described *Raillietina baeri* with two rows of hooks. Quentin (1964) described 'les crochets sont disposés en une seule couronne, ou bien en une double couronne' and illustrated a single row. Hunkeler (1974) described a single row, but thought more study was needed to evaluate the validity of this character, pointing out that Chandler (1942b) described *R. bakeri* with 'double crown . . . appearing as single row, alternate hooks being set only about 2 μm out of line'. The hooks in *V. baeri* cannot be said to alternate in this way. The genus is provisionally retained, consistent with our recognition elsewhere of the number of rows of rostellar hooks as a generic character.

Demidovella is provisionally recognized here. It was erected by Spasskii & Spasskaya (1976) based on the description of *Davainea leptotrachela* by Hungerbühler (1910) from *Pterocles namaqua*. The genus *Gvosdevinia* Spasskii, 1973 is from similar hosts and has a similar mature proglottid morphology, but unlike *Demidovella* does not have an egg-capsule enclosing several eggs. Schmidt (1986) considered *Demidovella* a synonym of *Raillietina (Fuhrmannetta)* and Movsesyan (1987) of *Fuhrmannetta*.

Mathevossianetta Movsesyan, 1966 and *Fuhrmannetta* Stiles & Orleman, 1926 differ in the reported absence or presence, respectively, of hooks on the suckers. This is not a generic criterion and *Mathevossianetta* is synonymized here with *Fuhrmannetta*. Movsesyan (1977) considered these taxa to be congeneric subgenera.

Dollfusoquenta Spasskii, 1973, erected for *Dilepis dollfusi* Quentin, 1964, a parasite of African rodents, has been retained in the Davaineidae as a pragmatic measure. It was originally placed in the Dilepididae by Quentin (1964) who described the uterus as persistent, a feature recognized in Spasskii's (1973) generic diagnosis. Despite this, Schmidt (1986) placed it in the Davaineinae although he recognized the uterus as a persistent irregular sac and inclusion in the Ophryocotylinae would have been more appropriate. Movsesyan (1987, 1989) did not include it in his lists of davaineid genera.

Daveneolepis Spasskii, 1979 is a genus of uncertain affinities which Spasskii (1979) appeared to regard as a dilepidid, although based on *Raillietina campanulata* (Fuhrmann, 1909) (syn. *Davainea campanulata*), previously in the Davaineidae. It was excluded from the davaineids by Movsesyan (1987, 1989) but synonymized with *R. (Skrjabinia)* by Schmidt (1986). The hooks of the type-species are atypical of the Davaineidae and the nature of the gravid proglottids is uncertain (Fuhrmann, 1909c; López-Neyra, 1931; Artyukh, 1966). The position and validity of the genus are unclear, as it is based on an inadequately described species. It is treated here as a *genus incertae sedis*.

Spasskii & Kornyushin (1977) considered the Ophryocotylidae to contain only *Ophryocotyle* and *Burtiella*. *Burtiella*, erected by Spasskii & Kornyushin (1977) for *Ophryocotyle zeylanica* (Linstow, 1906), apparently differs from *Ophryocotyle* in having the genital ducts passing between the osmoregulatory canals rather than dorsal to them (Burt, 1962). We have reservations as to the value of this character. We have examined several *Ophryocotyle* species and can confirm that usually the genital ducts pass dorsally to the osmoregulatory vessels, but in transverse sections of *O. zeylanica* we have not confirmed Burt's (1962) observation. In fact, we can detect only the large ventral osmoregulatory vessel and no sign of the dorsal vessel.

The cirrus-sac of *O. zeylanica* has an internal seminal vesicle, not detected by us in other *Ophryocotyle* species but the figure of *O. proteus* in Spasskii & Kornyushin (1977) suggests that it may occur in this species and the character is not, therefore, considered by us to validate *Burtiella*.

Ophryocotylus was described, but not figured, by Srivastava & Capoor (1977) with the type-species *O. dinopilium*. Although no illustrations were given, the description was detailed enough and the status of the taxon was discussed in enough detail to make the names available. The form was described again, as new, by Srivastav [*sic*] & Capoor (1982) with the type-species spelled *O. dinopii*. The genus is distinguished from related forms by its single row of rostellar hooks. Despite the somewhat perfunctory nature of the description and figures, the genus is provisionally recognized here as the authors emphasize the number of hook rows seen. We stress, however, that we are not totally convinced of its validity.

Ophryocotyloides is poorly known. The type-species, *O. uniuterina*, is known only from its original description by Fuhrmann (1909c). The type-material in the Naturhistorisches Museum, Vienna (No. V.2799), consists of two fragments, a piece of pre-mature strobila and a piece of gravid strobila. No scolex or mature proglottids are present. The unilaterality of the genital pores is confirmed. The uterus develops as a distended sac, becomes very lobate as the eggs develop and then in the later gravid proglottids the septa of the lobations disappear and the uterus fills the whole proglottid up to the margins of the body, overlapping the longitudinal osmoregulatory canals. No indication of egg-capsules was seen. The characters of the genus are taken from this material and Fuhrmann's (1909c) description, and not from later descriptions of species supposedly in the genus (e.g. Singh, 1960) as Spasskii & Kornyushin (1977) stated that the systematic position of the genus is unclear and it contains representatives of several (at least four) other genera. We follow Schmidt in recognizing *Ophryocotyloides*, but only provisionally and pending a careful redescription of the type-species.

Fernandezia is, according to Spasskii & Kornyushin (1977), close to the davaineine genus *Skrjabinia*, which is said to have single eggs in uterine capsules. Binder (1971) described the uterus of *F. spinosissima* (von Linstow, 1894) as saccular, with deeply lobate margins which can simulate capsules.

Dasyurotaenia Beddard, 1912 has been included by Schmidt (1986) in the Davaineidae, subfamily Ophryocotylinae. Other authors (e.g. Baer, 1925b; Sandars, 1957; Yamaguti, 1959; Abuladze, 1964) placed it in the Taeniidae but Rausch (1981, this volume) rejected this on phylogenetic and morphological grounds. However, the morphology of the scolex and proglottids, but not the eggs, of *Dasyurotaenia*, as described by Beveridge (1984), is more similar to the taeniids than the davaineids. Beveridge (1984) speculated that the genus might be derived from the linstowiids of marsupials, but the absence of information on the life-cycle renders allocation to a family difficult. At present, we see no reason to include this genus in the Davaineidae and regard it as a *genus incertae sedis*.

The genus *Barbusa* Capoor & Srivastava, 1975, and the tribe Barbuseini, erected in the Davaineinae by Capoor & Srivastava (1975) are based on poorly-described specimens. It is not clear from the description that these are definitely davaineids. *Barbusa* is treated here as a *genus inquirendum*. The shape and arrangement of the rostellar hooks are unknown for the Davaineidae and require confirmation.

Idiogeninae

The type-species of the type-genus, *Idiogenes otidis*, was described by Krabbe (1868) from the bustard, *Otis tarda*, from a zoological collection in Jutland, Denmark. The bustard is not, and probably never has been, a regular breeding bird in Denmark (Dr Niels Otto Preuss, pers. comm.). Krabbe did not describe a scolex and we can confirm that no scoleces were present in the dried-out type material in the Zoological Museum, Copenhagen. Some authors have considered that the strongly craspedote proglottids can attach as pseudoscoleces. Several authors have subsequently described scoleces for species they considered *I. otidis*. Yeh (1957) and Gómez (1981) have reported on worms from *O. tarda* in Spain: both described four rows of rostellar hooks, and Gómez described a patch of small accessory spines proximally to the rostellar hooks. Our observations of specimens from *O. tarda* from London Zoo (IIP 1246) confirm the presence of four rows of hooks (could be interpreted as three or five in parts of the circle) and the accessory spines (Fig. 24.84). Joyeux, Gendre & Baer (1928) described two rows of rostellar hooks for *I. otidis* from *Eupodotis* [as *Trachelotis*] *senegalensis* from Guinea, West Africa. Mahon (1954) renamed this form *I. pseudotidis*. Borgarenko (1981) figured a scolex of *I. otidis* with, apparently, one row of rostellar hooks and listed the hosts *O. tarda* and, mainly, *O. undulata* in Tadzhikistan. This evidence suggests to us that *I. otidis* is characterized by four rows of rostellar hooks and the patch of accessory spines, and this calls into question the validity of *Ersinogenes* Spasskaya, 1961 and *Paraidiogenes* Movsesyan, 1971. *Ersinogenes* is said to differ from *Idiogenes* in having five rows of rostellar hooks, and also possesses accessory spines. The type-species, *E. spinatum* Spasskaya, 1961 was described from *Otis tarda* in Tuva, USSR. As it is possible to interpret the number of rows in the specimens of *I. otidis* we have seen, as from three to five, we do not consider this a useful differentiating character. In every other way this genus resembles *Idiogenes*. Because of this, and because of its host, we consider *Ersinogenes* a synonym of *Idiogenes*. *Paraidiogenes* is supposedly differentiated from *Idiogenes* by having three rows of rostellar hooks and accessory spines are described. The type-species of *Paraidiogenes, P. mongolica* (Danzan in Spasskaya & Spasskii, 1971) was originally described as *Idiogenes kori* Ortlepp, 1938 by Spasskaya (1961b) from *Otis tarda* in Mongolia and renamed *Idiogenes mongolica* by Danzan in a thesis in 1964. The first published use of the name was by Spasskaya & Spasskii (1971). Movsesyan (1970) mentioned the name *Paraidiogenes* in a list and key, but no constituent species were mentioned and he formally erected the genus in 1971. We believe there are no grounds for considering this genus distinct from *Idiogenes*. Our own observations on several species make it clear, however, that many of the species currently considered to belong to *Idiogenes* have only two rows of rostellar hooks. We have, therefore, defined *Idiogenes* sensu stricto as having three to five rows of rostellar hooks and with accessory spines present, based on the type-species *I. otidis*.

Movsesyan (1970) created *Pseudidiogenes* as a subgenus of *Idiogenes*, but no constituent species were mentioned and, therefore, this name is not valid from this date (*ICZN* Art. 13(b)). The name was erected legally by the same author in 1971, with the type-species *I. (P.) flagellum*, and in both papers the new subgenus was differentiated from *Idiogenes (I.)* by having armed suckers. Kornyushin (1989) raised *Pseudidiogenes* to generic rank, based on the same criteria as Movsesyan, i.e. whether the suckers were armed or not. The *Idiogenes* forms with two rows of rostellar hooks should, we believe, be considered generically distinct from those with three to five rostellar hook rows, and we consider that the genus *Pseudidiogenes* can be redefined to accommodate them.

The names *Otiditaenia* and *Schistometra* were both coined in publications in 1912. They are generally considered synonymous, but different authors recognize different priority. Beddard (1912) was 'published, March 1912' whereas Cholodkowsky (1912) has no date other than the year and, therefore, according to the *ICZN* should be dated as 31 December 1912. We believe that Baer (1955a) is correct in arguing for the priority of *Otiditaenia*.

We have examined 'co-types' of the type-species of *Paraschistometra*, i.e. *P. macqueeni* Woodland, 1930 (BM(NH) 1965.2.19.1–23) and agree with Baer (1955a) that it is synonymous with *Otiditaenia*.

Capsodavainea is an objective synonym of *Chapmania* as it is based on the same type-species. Our observations on *Chapmania* are based mainly on Kornyushin (1989).

We have recognized *Satyanarayana* Khan, 1984 although it is similar to *Chapmania* Monticelli, 1893. The type-species of the former genus, *S. satyanarayani* was described by Khan (1984) from the lungs of the heron *Ardea cinerea* in India. In many respects the species appears close to *Chapmania*, but the ovary is poral with all the testes aporal to it. In *Chapmania*, the ovary is median or just porally submedian (see Baer, 1955a; Hörchner, 1962; Movsesyan, 1977; Kornyushin, 1989) with the testes passing poral to the ovary. We consider this character to be of generic importance, but it should be noted that the site of *S. satyanarayani* is very unusual and that Khan (1984) had to tease the worms from the lung tissue with the help of needles.

Most text illustrations are original; others have been modified and redrawn from the sources named in the figure legends.

Family Davaineidae Braun, 1900

Diagnosis: Rostellum usually present, rarely rudimentary, varies from small to enormous, armed with hooks in usually two rows but occasionally in one, three, five or 10–12 rows. Hook crown round, oval or undulating, interrupted or not. Hooks characteristically numerous, small and hammer-shaped, rarely otherwise. Rostellar spines present or absent. Suckers present, rarely absent, unarmed or armed with rows of minute spines. Strobila small to large, rarely tiny. Proglottids usually numerous, rarely few. Osmoregulatory canals four (dorsal and ventral pairs), two (ventral pair only), or rarely six or twenty. Genitalia single or double; when single, genital pores unilateral or alternate irregularly. Cirrus-sac small to large, variable in extent relative to osmoregulatory canals. Testes few to numerous, distribution variable. Ovary commonly median, exceptionally poral, lobed, occasionally distinctly bilobed. Vitellarium postovarian. Uterus persistent or replaced by paruterine organ or by egg capsules containing one or several eggs. In birds and mammals. Type-genus *Davainea* Blanchard, 1891.

Key to subfamilies

1a. Paruterine organ absent .... Davaineinae Braun, 1900.
(Syn. Ophryocotylinae Fuhrmann, 1907.)
Diagnosis: Davaineidae. Uterus persistent or replaced by egg capsules, either numerous thin-walled capsules containing a single egg or few to numerous fibrous capsules containing several eggs.

1b. Paruterine organ present .... Idiogeninae Fuhrmann, 1907.
Diagnosis: Davaineidae. Paruterine organ present.

Subfamily Davaineinae Braun, 1900

Key to genera

1a. Egg capsules present, each containing several eggs .... 2.
1b. Not as above .... 11.

2a. Genital organs double per proglottid ....
... *Multicotugnia* López-Neyra, 1943.
(Not illustrated; this genus resembles *Cotugnia* (Figs 24.30–24.32) except for having several eggs per capsule.)
Diagnosis: Rostellum armed with double row of small hammer-shaped hooks. Suckers unarmed. Proglottids numerous, craspedote. Paired dorsal and ventral osmoregulatory canals present. Genital pores bilateral. Cirrus-sac small, extravascular. Testes numerous, confluent behind female organs, extending laterally beyond osmoregulatory canals. Ovaries lateral, strongly lobed. Vitelline glands postovarian. Egg capsules numerous, each containing several eggs. In Psittaciformes. Brazil. Type-species *M. brotogerys* (Meggitt, 1915).

2b. Genital organs single per proglottid .... 3.

3a. Egg capsules few (<5) per proglottid. Testes few (<5). Parasites of Pholidota .... 4.
3b. Egg capsules more than 5 per proglottid. Testes few or numerous . 5.

4a. Rostellum rudimentary, unarmed ....
.... *Baerfainia* Yamaguti, 1959. (Figs 24.1–24.3)
Diagnosis: Suckers armed. Proglottids numerous, craspedote. Only ventral osmoregulatory canals present. Genitalia single. Genital pores unilateral. Cirrus-sac small, extravascular. Testes few (three to four), one poral, remainder aporal.

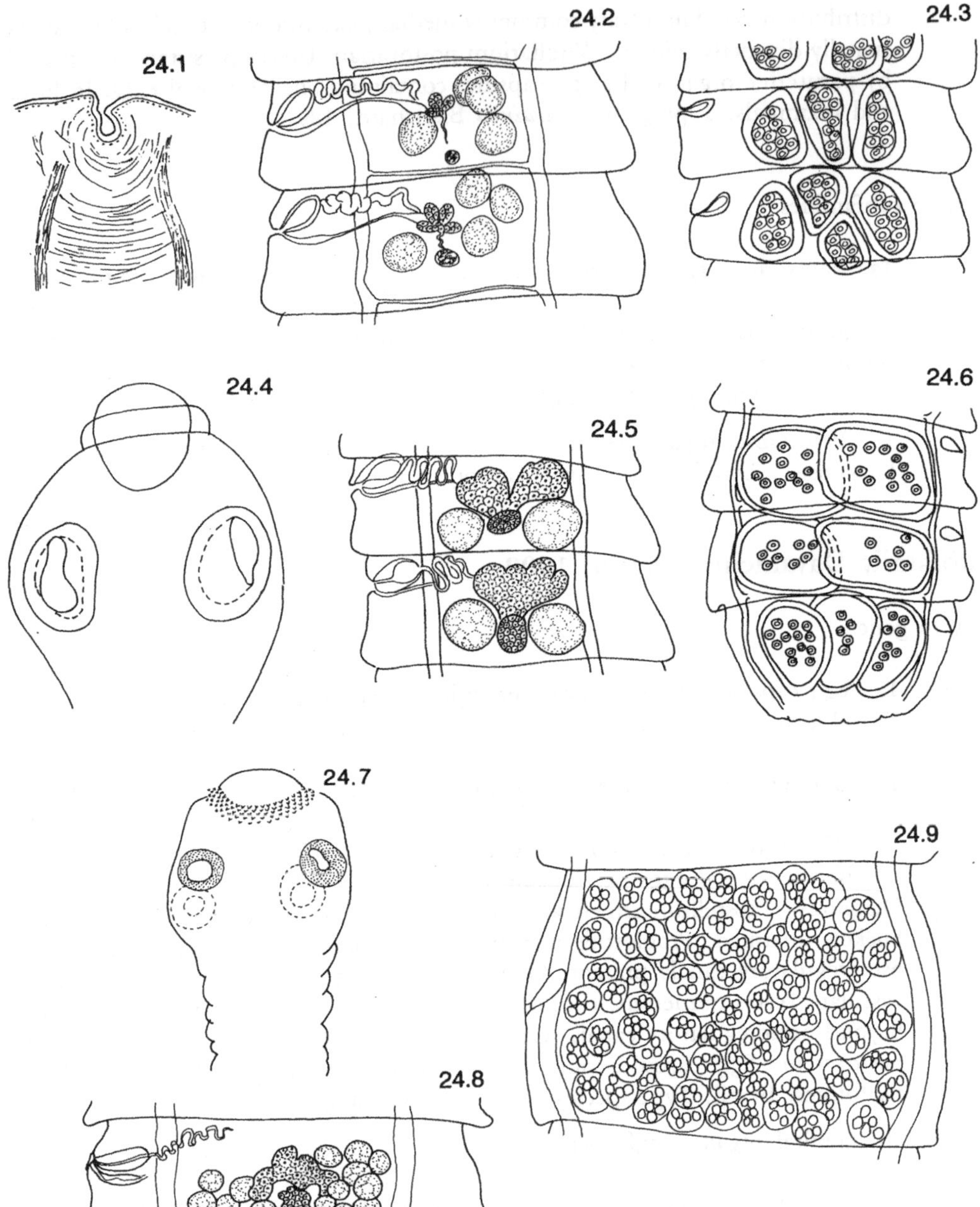

Figs 24.1-24.3 *Baerfainia anoplocephaloides* (Baer & Fain, 1955). (After Baer & Fain, 1955.) 24.1. Scolex. 24.2. Mature proglottids. 24.3. Gravid proglottids.
Figs 24.4-24.6 *Manitaurus rahmi* (Baer & Fain, 1955). (Figs 24.4, 24.5 after Baer & Fain, 1955.) 24.4. Scolex. 24.5. Mature proglottids. 24.6. Gravid proglottids.
Figs 24.7-24.9 *Pentocoronaria rusannae* Matevosyan & Movsesyan, 1966. (After Matevosyan & Movsesyan, 1966.) 24.7. Scolex. 24.8. Mature proglottid. 24.9. Gravid proglottid.

Ovary median, lobed. Vitellarium postovarian. Egg capsules very large, few (three to four), each containing numerous eggs. In Pholidota. Africa. Type-species *B. anoplocephaloides* (Baer & Fain, 1955).

4b. Rostellum present, well-developed ....
.... ***Manitaurus*** **Spasskaya & Spasskii, 1971. (Figs 24.4– 24.6)**
Diagnosis: Hooks unknown. Suckers unarmed. Proglottids numerous, craspedote. Genitalia single. Genital pores unilateral. Cirrus-sac reaching osmoregulatory canals. Testes two, one either side of ovary. Ovary median, lobed. Vitellarium postovarian. Egg capsules few (two to three), large, containing numerous eggs. In Pholidota. Africa. Type-species *M. rahmi* (Baer & Fain, 1955).

5a. Rostellar hooks in five circles ....
.... ***Pentocoronaria*** **Matevosyan & Movsesyan, 1966. (Figs 24.7–24.9)**
Diagnosis: Rostellum armed with small hammer-shaped hooks. Suckers armed. Proglottids numerous, craspedote. Only ventral paired osmoregulatory canals present. Genitalia single. Genital pores unilateral. Cirrus-sac pyriform, muscular, extravascular in mature proglottids. Testes relatively few, lateral and posterior to ovary. Ovary median, slightly lobed. Vitellarium postovarian. Egg capsules numerous, each containing several eggs. In Columbiformes. USSR. Type-species *P. rusannae* Matevosyan & Movsesyan, 1966.

5b. Rostellar hooks in two rows .... **6.**
5c. Rostellar hooks in one row ....
.... ***Vadifresia*** **Spasskii, 1973. (Figs 24.10, 24.11)**
Diagnosis: Suckers armed. Proglottids craspedote. Reproductive organs single. Genital pores unilateral. Cirrus-sac small. Testes numerous. Ovary median. In rodents and lagomorphs; Ethiopian, Nearctic. Type-species *V. baeri* (Meggitt & Subramanian, 1927).

6a. Rostellar hooks 24 or less ....
.... ***Metadavainea*** **Baer & Fain, 1955. (Figs 24.12–24.15)**
Diagnosis: Rostellum armed with double row of relatively few large robust hooks with well-developed blades, thick guards and reduced handles. Suckers armed. Accessory rostellar spines present. Proglottids numerous, craspedote. Pair of dorsal and of ventral osmoregulatory canals present. Genitalia single. Genital pores unilateral. Cirrus-sac extravascular. Testes numerous, lateral and anterior to ovary. Ovary median, slightly lobed. Vitellarium postovarian. Egg capsules relatively few (8–15), each containing several eggs. In Pholidota. Africa, Asia. Type-species *M. aelleni* Baer & Fain, 1955.

6b. Rostellar hooks numerous .... **7.**

7a. Ovary poral .... **8.**
7b. Ovary median .... **9.**

8a. Genital pores unilateral. Testes in poral and aporal fields. In Struthioniformes and Rheiformes ....

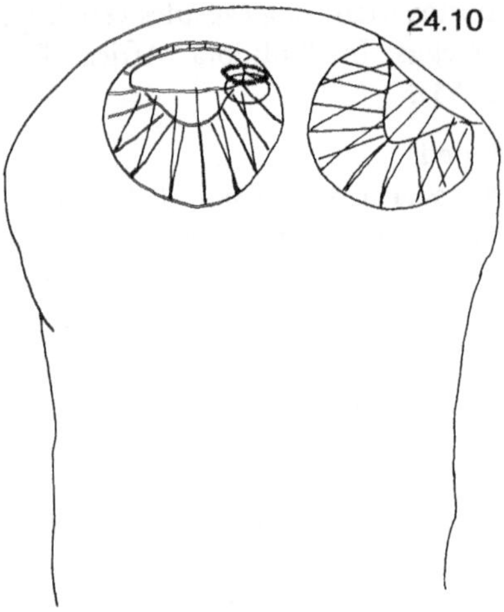

24.10

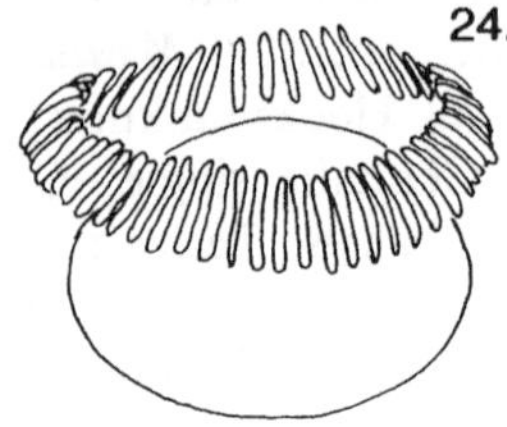

24.11

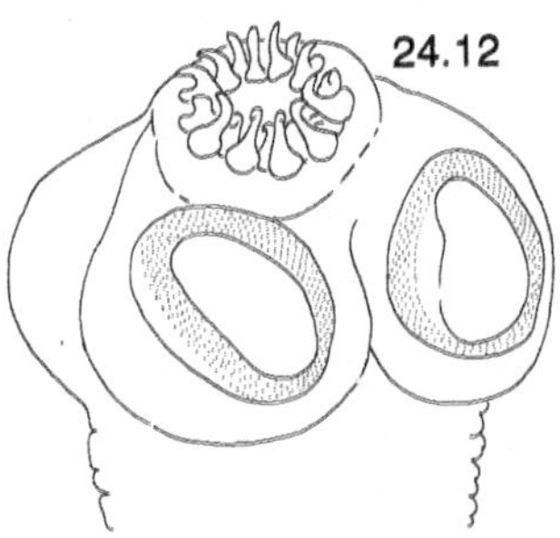

24.12

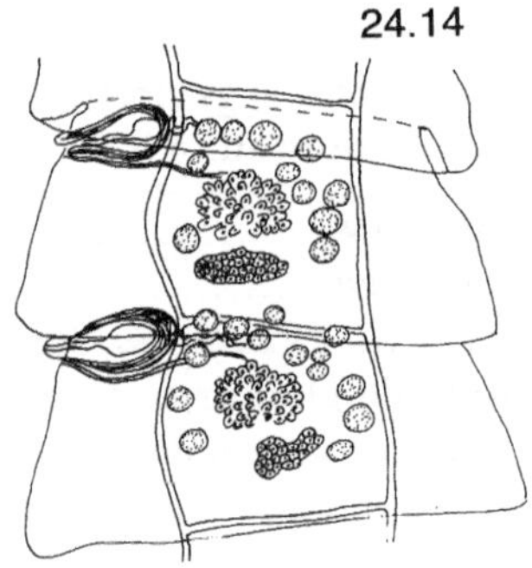

24.14

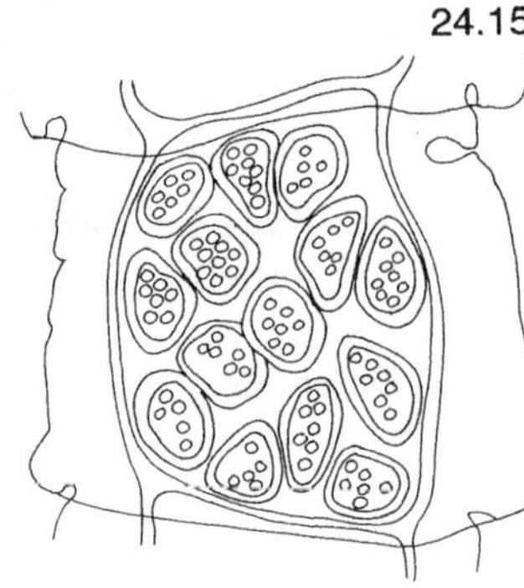

24.15

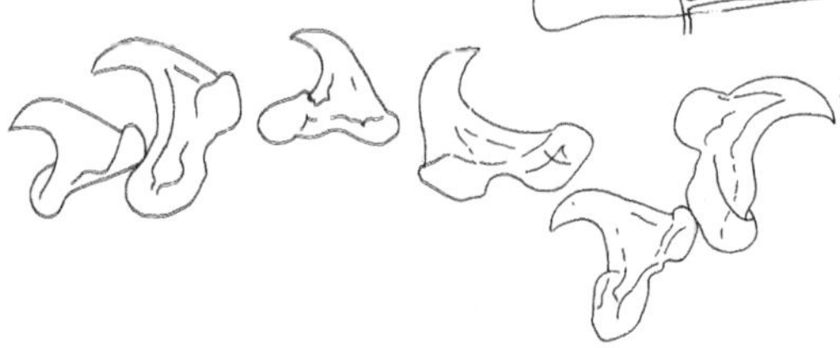

24.13

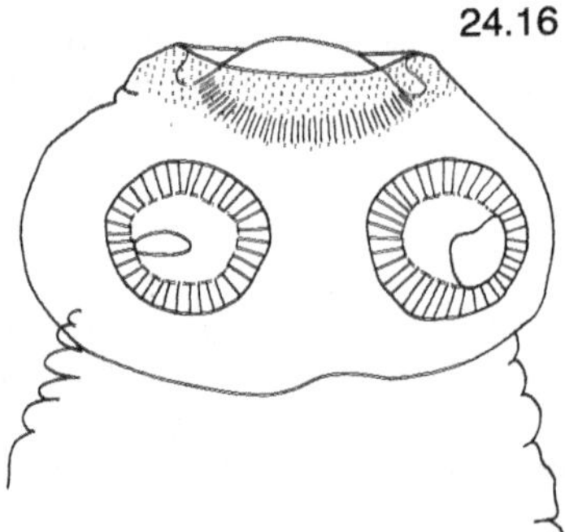

24.16

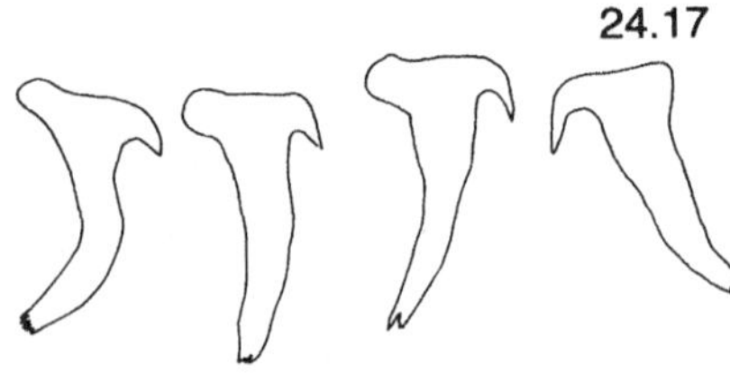

24.17

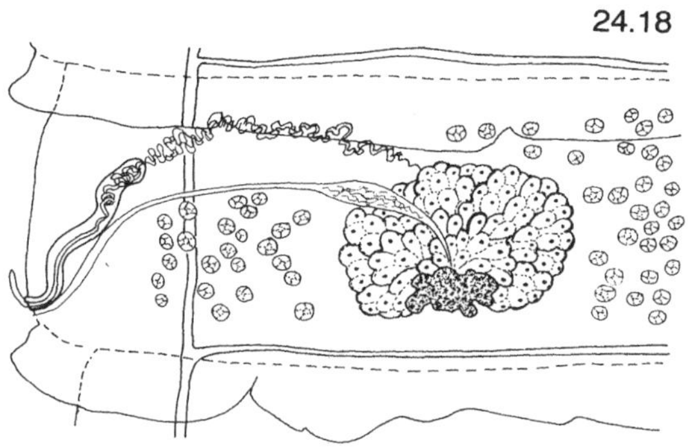

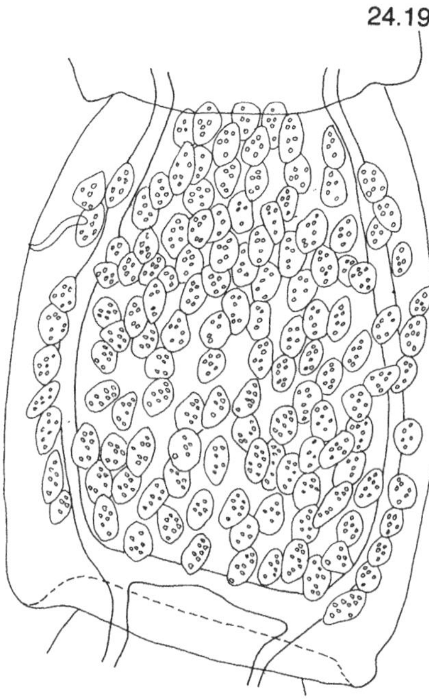

Figs 24.10, 24.11 *Vadifresia baeri* (Meggitt & Subramanian, 1927). 24.10. Hook. ring. 24.11. Hook ring and rostellum (detail).
Figs 24.12–24.15 *Metadavainea aelleni* Baer & Fain, 1955. 24.12. Scolex. 24.13. Rostellar hooks. 24.14. Mature proglottids. 24.15. Gravid proglottids.
Figs 24.16–24.19 *Houttuynia struthionis* (Houttuyn, 1772). 24.16. Scolex. 24.17. Rostellar hooks. 24.18. Mature proglottid, poral region. 24.19. Gravid proglottid.

.... ***Houttuynia* Fuhrmann, 1920. (Figs 24.16–24.19)**
Diagnosis: Rostellum wide, armed with double row of large hammer-shaped hooks. Rostellar spines present. Suckers unarmed. Strobila large. Proglottids numerous, craspedote. Paired ventral osmoregulatory canals present. Genitalia single. Genital pores unilateral. Cirrus-sac elongate, not crossing osmoregulatory canal. Testes numerous, in poral and aporal fields lateral to ovary, extend lateral to osmoregulatory canals. Ovary slightly poral, lobed, fan-shaped. Vitellarium postovarian. Egg capsules numerous, contain several eggs. In Struthioniformes and Rheiformes. Africa, South America. Type-species *H. struthionis* (Houttuyn, 1772).

8b. Genital pores alternate irregularly. Testes restricted to poral field. In Pteroclidae .... *Demidovella* Spasskii & Spasskaya, 1976. (Fig. 24.20)
Diagnosis: Rostellar hooks in two circular rows. Suckers armed. Proglottids craspedote. Reproductive organs single. Three to four eggs per capsule. In sandgrouse. Ethiopia. Type-species *D. leptotrachela* (Hungerbühler, 1910).

9a. Genital pores unilateral .... **10.**
9b. Genital pores alternate irregularly ....

24.21

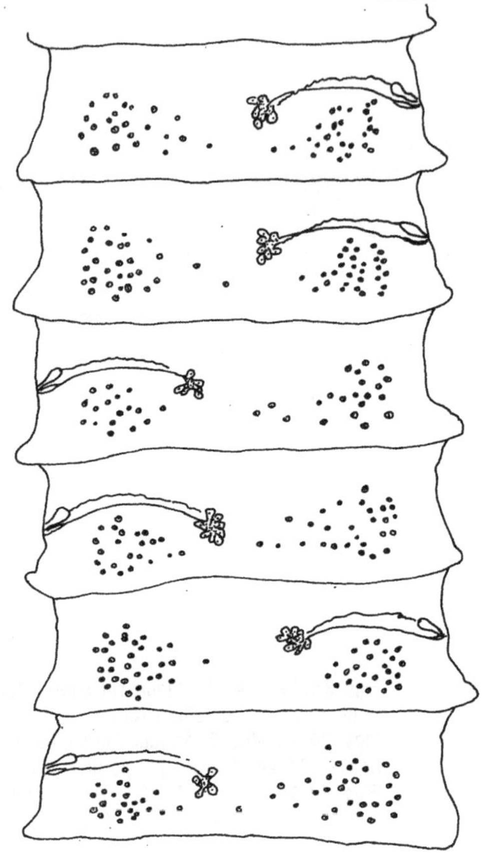

24.20

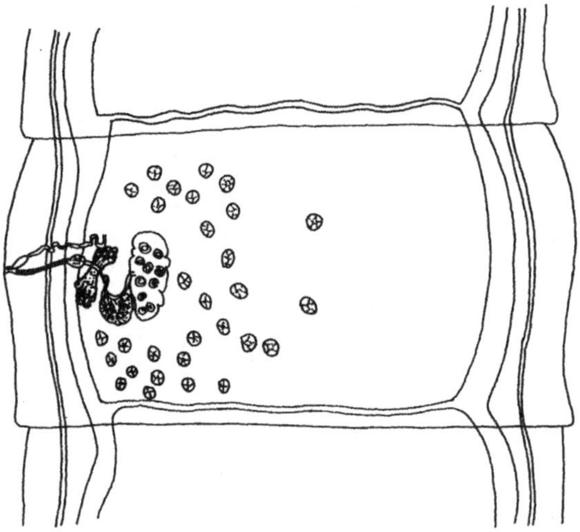

24.22

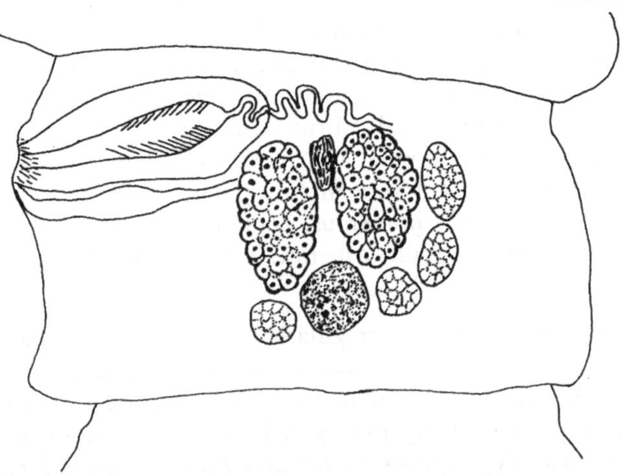

Fig. 24.20 *Demidovella leptotrachela* (Hungerbühler, 1910). (After Hungerbühler, 1910.)
Fig. 24.21 *Fuhrmannetta crassula* (Rudolphi, 1819). Mature proglottids.
Fig. 24.22 *Idiogenoides allagea* (Kotlan, 1921). Mature proglottid. (After Kotlan, 1921.)

.... ***Fuhrmannetta*** **Stiles & Orleman, 1926. (Fig. 24.21)**
(Syns *Johnstonia* Fuhrmann, 1921 preoccupied; *Mathevossianetta* Movsesyan, 1966.)

Diagnosis: Rostellar hooks in two circular rows. Suckers armed or not. Proglottids wider than long. Reproductive organs single. Testes numerous. Ovary median. Three to five eggs per capsule. In birds and mammals. Cosmopolitan. Type-species *F. crassula* (Rudolphi, 1819).

10a. Cirrus-sac large, closely approaching mid-line ....
.... ***Idiogenoides*** **López-Neyra, 1929. (Fig. 24.22)**

Diagnosis: Rostellar hooks in two circular rows. Suckers armed. Proglottids craspedote. Reproductive organs single. Genital pores unilateral. Testes few (four

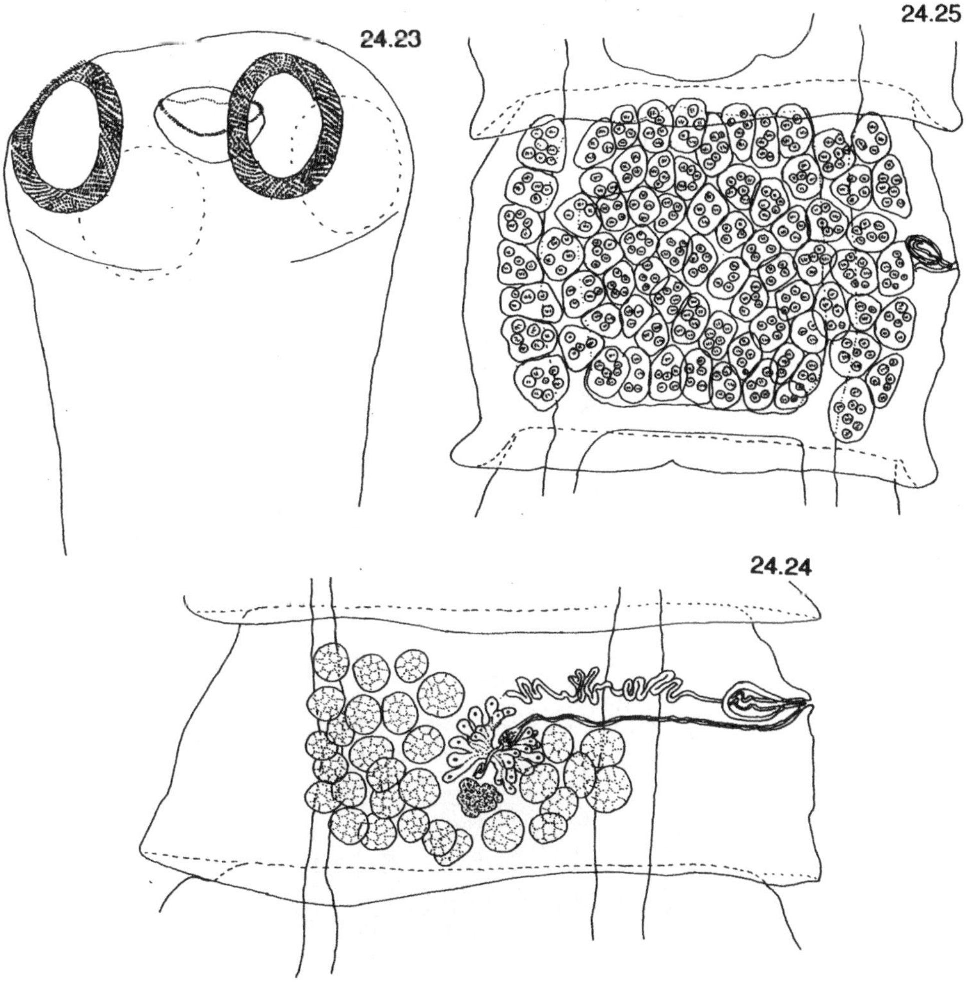

Figs 24.23–24.25 *Raillietina* Fuhrmann, 1920. 24.23. *R. echinobothrida* (Megnin, 1880). Scolex. 24.24. *R. tetragona* (Molin, 1858). Mature proglottid. 24.25. *R. tetragona*. Gravid proglottid.

to six). Ovary median. Six to seven eggs per capsule. In parrot. New Guinea. Type-species *I. allagea* (Kotlan, 1921).

10b. Cirrus-sac small, not reaching or just crossing the osmoregulatory canals .... *Raillietina* Fuhrmann, 1920. (Figs 24.23–24.25)
(Syns *Kotlania* López-Neyra, 1929; *Kotlanotaurus* Spasskii, 1973; *Nonarmiella* Movsesyan, 1966; *Nonarmina* Movsesyan, 1966; *Oschmarinetta* Spasskii, 1984; *Ransomia* Fuhrmann, 1921; *Roytmania* Spasskii, 1973; *Skrjabinotaurus* Spasskii & Yurpalova, 1973.)

Diagnosis: Rostellar hooks in circular row. Suckers armed, partly armed or unarmed. Proglottids craspedote. Reproductive organs single. Genital pores unilateral. Testes numerous. Ovary median. Two to eight eggs per capsule. In birds and mammals. Cosmopolitan. Type-species *R. tetragona* (Molin, 1858).

11a. Suckers absent .... *Pluviantaenia* Jones, Khalil & Bray, 1992. (Figs 24.26–24.29)

Diagnosis: Rostellum inflated, enormous, armed proximally with double row of many small hammer-shaped hooks. Proglottids numerous, craspedote. Two pairs of osmoregulatory canals present. Genital pores generally alternate regularly but

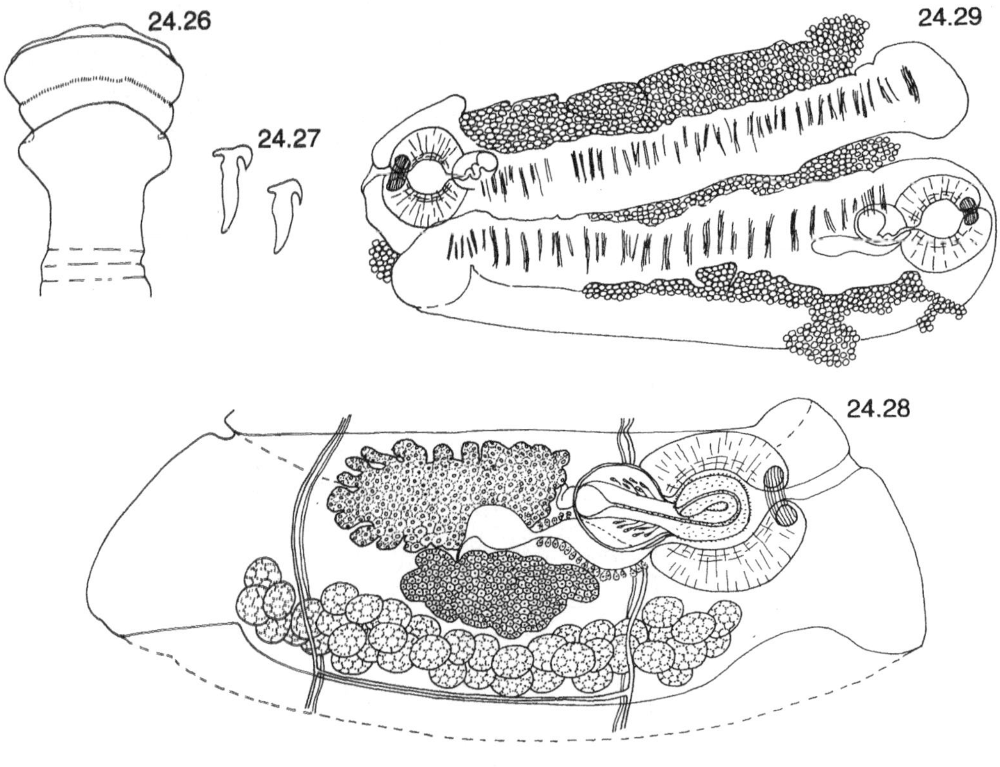

Figs 24.26–24.29 *Pluviantaenia kassalensis* Jones, Khalil & Bray, 1992. (After Jones *et al.*, 1992.) 24.26. Scolex. 24.27. Rostellar hooks. 24.28. Mature proglottid. 24.29. Gravid proglottid.

may alternate irregularly or be bilateral. Genital atrium enormous, sucker-like with massive sphincter distally; atrial region protrusible. Duplication of genital atria and cirrus-sacs may occur. Cirrus-sac thin-walled. Cirrus large, armed. Testes numerous, along posterior margin of proglottid. Vagina sigmoid, inflated and glandular distally. Ovary median. Vitellarium postovarian. Uterus persistent. In Glareolidae. Sudan. Type-species *P. kassalensis* Jones, Khalil & Bray, 1992.

11b. Suckers present .... **12.**

12a. Rostellar hook crown round or oval .... **13.**
12b. Rostellar hook crown undulating or lobed .... **14.**

13a. Genital organs consistently double per proglottid ....
.... ***Cotugnia*** **Diamare, 1893. (Figs 24.30–24.32)**
(Syns *Erschovitugnia* Spasskii, 1973; *Pavugnia* Spasskii, 1984; *Rostelugnia* Spasskii, 1984.)

Diagnosis: Rostellum broad, armed with a double row of small (rarely large) hammer-shaped hooks. Suckers usually unarmed, rarely armed. Proglottids numerous, craspedote. Dorsal pair of osmoregulatory canals present or absent, ventral pair present. Genital pores bilateral. Cirrus-sac small, extravascular. Testes numerous, in one or two fields. Ovaries lateral, lobed. Vitelline glands postovarian. Egg capsules numerous, each containing single egg. In birds. Cosmopolitan. Type-species *C. digonopora* (Pasquale, 1890).

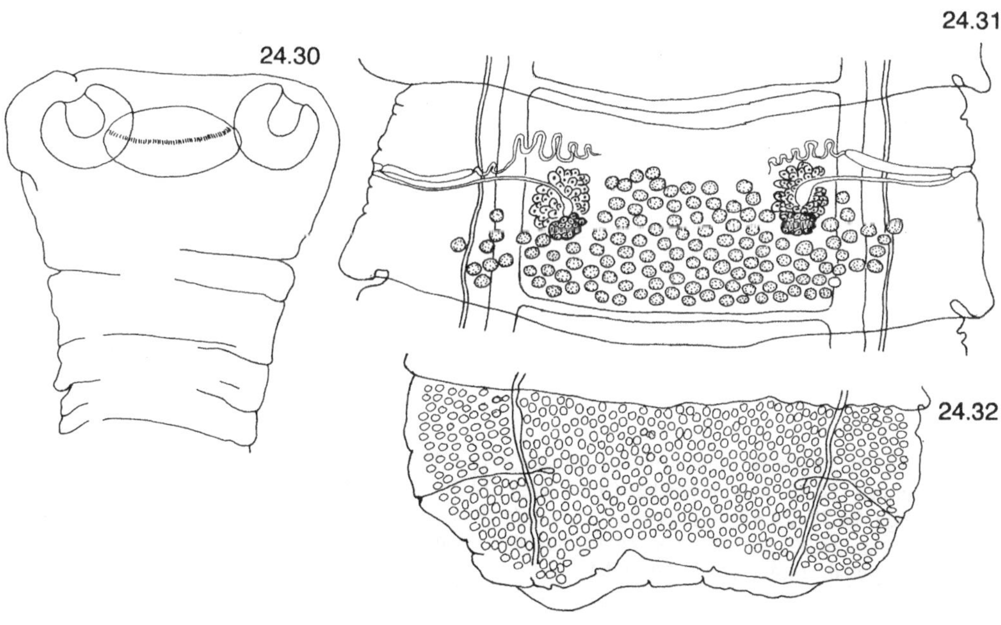

Figs 24.30–24.32 *Cotugnia digonopora* (Pasquale, 1890). 24.30. Scolex. 24.31. Mature proglottid. 24.32. Gravid proglottid.

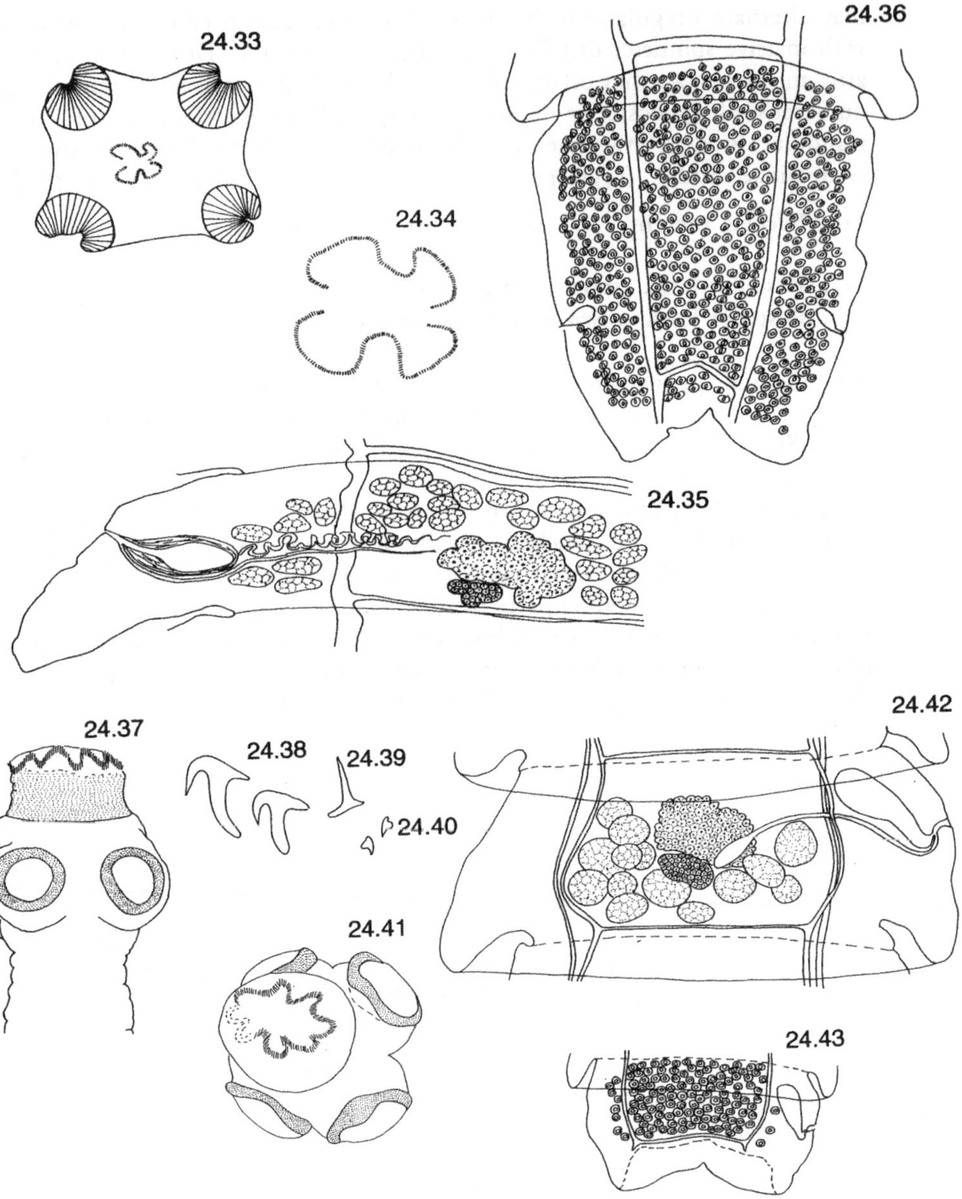

Figs 24.33-24.36 *Abuladzugnia gutterae* (Ortlepp, 1938). 24.33. Apex of scolex. 24.34. Rostellar crown. 24.35. Mature proglottid (one side). 24.36. Gravid proglottid.

Figs 24.37-24.43 *Soninotaurus rhynchota* (Ransom, 1909). 24.37. Scolex. 24.38. Rostellar hooks. 24.39. Sucker spine. 24.40. Rostellar spines. 24.41. Scolex, apical view. 24.42. Mature proglottid. 24.43. Gravid proglottid.

13b. Genital organs single per proglottid .... **18.**

14a. Genital organs double per proglottid. Rostellar crown four-lobed, discontinuous laterally ..... ***Abuladzugnia* Spasskii, 1973. (Figs 24.33–24.36)**
Diagnosis: Rostellum large, armed with a double row of small hammer-shaped hooks. Suckers unarmed. Proglottids numerous, craspedote. Only ventral osmoregulatory canals present. Genital pores bilateral. Cirrus-sac small, extravascular. Testes numerous, internal and external to osmoregulatory canals. Ovaries lateral, lobed. Vitelline glands postovarian. Numerous egg capsules, each containing single egg. In Galliformes. Africa. Type-species *A. gutterae* (Ortlepp, 1938).

14b. Genital organs single. Rostellar crown uninterrupted .... **15.**

15a. Rostellar crown lobed. Uterus replaced by egg capsules containing a single egg .... **16.**
15b. Rostellar crown undulating or in flattened rows. Uterus persistent .... **17.**

16a. Rostellar crown forms eight lobes. Parasites of North American Piciformes .... ***Soninotaurus* Spasskii, 1973. (Figs 24.37–24.43)**
Diagnosis: Rostellum armed with double row of small, hammer-shaped hooks. Accessory spines present. Suckers armed. Proglottids numerous, craspedote. Dorsal and ventral pairs of osmoregulatory canals present. Genitalia single. Genital

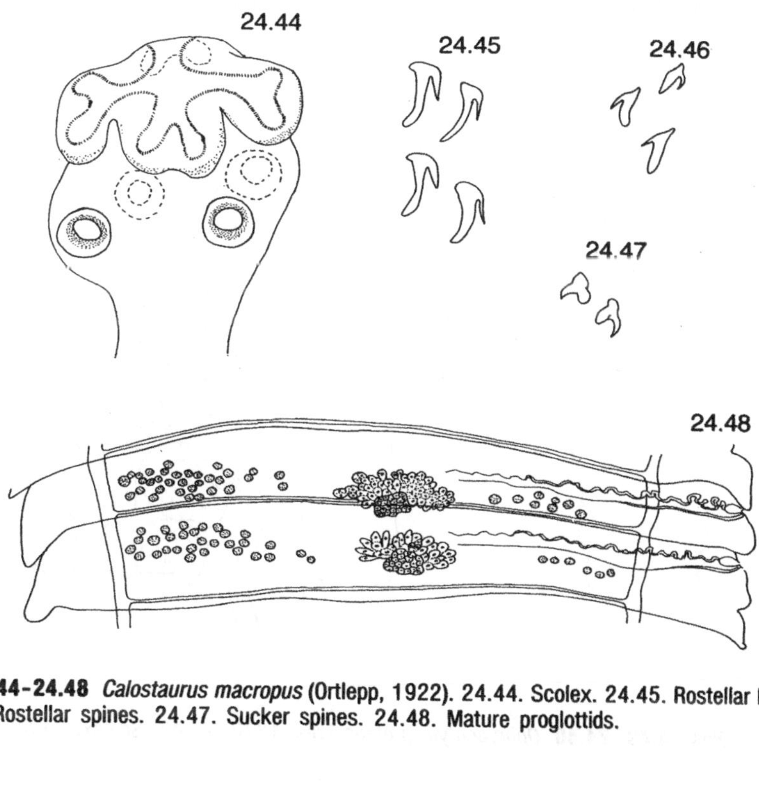

Figs 24.44–24.48 *Calostaurus macropus* (Ortlepp, 1922). 24.44. Scolex. 24.45. Rostellar hooks. 24.46. Rostellar spines. 24.47. Sucker spines. 24.48. Mature proglottids.

pores unilateral. Cirrus-sac small, extravascular. Testes numerous, lateral and posterior to female organs. Ovary median, lobed. Vitellarium postovarian. Egg capsules numerous, containing single egg. In Piciformes. North America. Type-species *S. rhynchota* (Ransom, 1909).

16b. Rostellar crown forms four or six lobes or a Maltese cross. In Australian marsupials .... *Calostaurus* Sandars, 1957. (Figs 24.44–24.48)

Diagnosis: Rostellum enormous, armed with double row of small hammer-shaped hooks. Accessory spines present. Suckers armed. Proglottids numerous, craspedote or not. Paired dorsal and ventral osmoregulatory canals present. Genitalia single. Genital pores unilateral. Cirrus-sac small, extravascular. Testes numerous, usually in two lateral fields but occasionally confluent behind female organs. Ovary median, lobed, often distinctly bilobed. Vitellarium compact, postovarian. Egg capsules numerous, containing one egg (exceptionally up to three). In marsupials. Australia, Papua New Guinea. Type-species *C. macropus* (Ortlepp, 1922).

17a. Rostellar hooks in undulating rows .... *Ophryocotyle* Friis, 1870. (Figs 24.49, 24.50)

(Syn. *Burtiella* Spasskii & Kornyushin, 1977.)

Diagnosis: Rostellar hooks small, in two rows. Minute spines may be present on base of rostellum. Suckers armed. Proglottids craspedote. Reproductive organs single. Genital pores alternating regularly or irregularly. Cirrus-sac small. Ovary median to aporal. Uterus sac-like, lobed. In birds. Holarctic, Ethiopian, Neotropical, Oriental. Type-species *O. proteus* Friis, 1870.

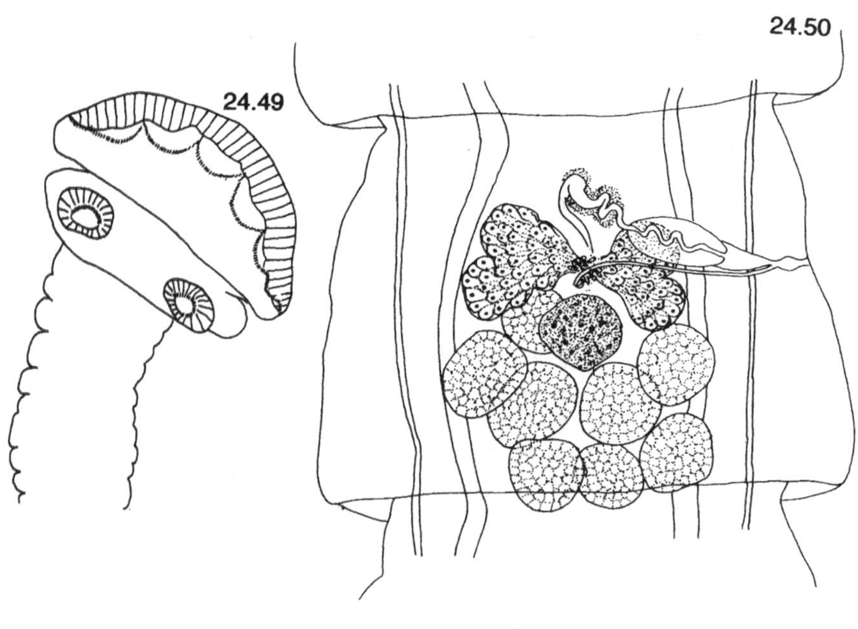

Figs 24.49, 24.50 *Ophryocotyle proteus* Friis, 1870. 24.49. Scolex. 24.50. Mature proglottid.

24.51

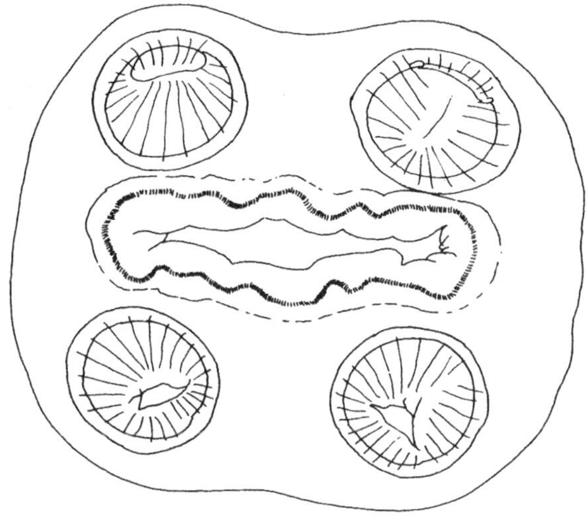

24.52

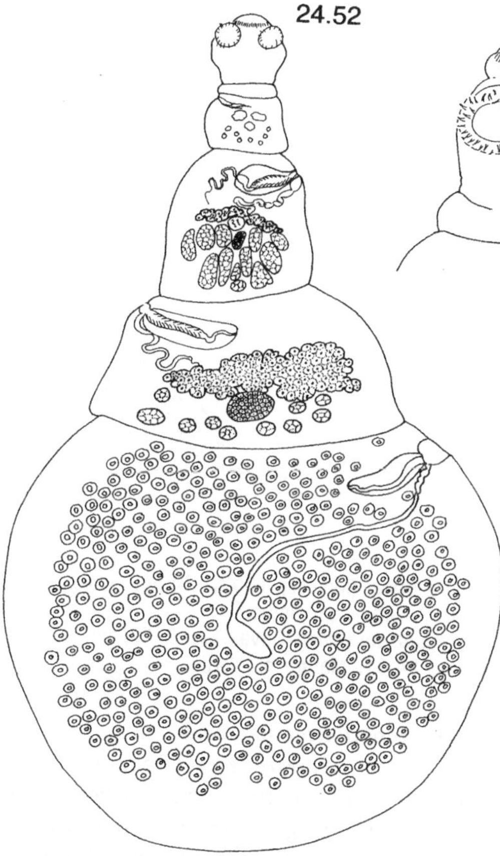

24.53

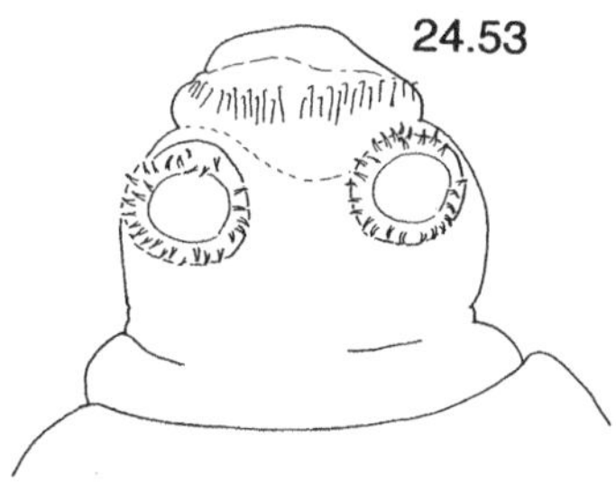

24.54

24.55

Fig. 24.51 *Fernandezia goizuetai* López-Neyra, 1936. Scolex, apical view.
Figs 24.52–24.55 *Davainea proglottina* (Davaine, 1860). 24.52. Entire worm. 24.53. Scolex. 24.54. Rostellar hooks. 24.55. Sucker spines.

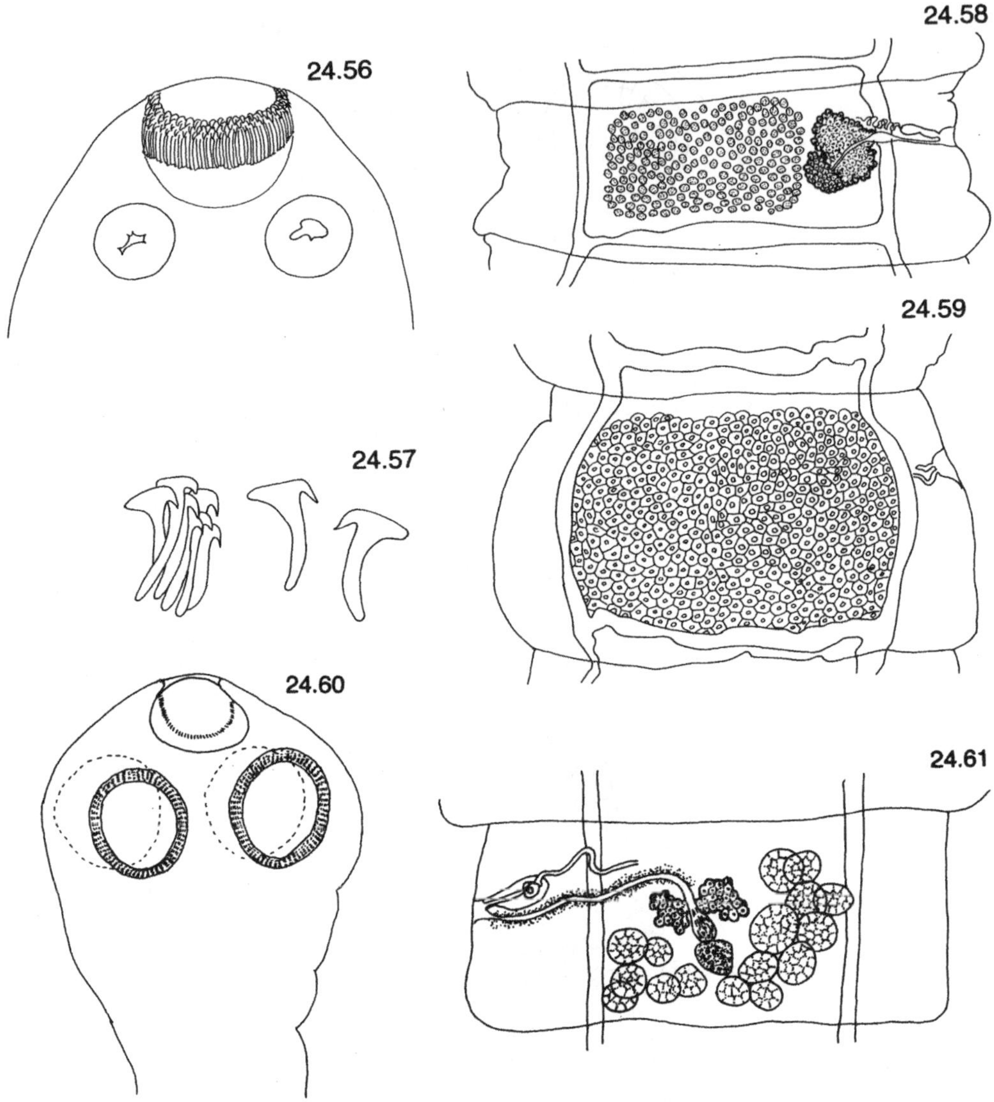

Figs 24.56–24.59 *Porogynia paronai* (Moniez, 1892). 24.56. Scolex. 24.57. Rostellar hooks. 24.58. Mature proglottid. 24.59. Gravid proglottid.
Figs 24.60, 24.61 *Ophryocotylus dinopilium* Srivastava & Capoor, 1977. (After Srivastava & Capoor, 1982.) 24.60. Scolex. 24.61. Mature proglottid.

17b. Rostellar crown a flattened irregular oval .... ***Fernandezia* López-Neyra, 1936. (Fig. 24.51)**

Diagnosis: Rostellar hooks small, in two rows. Suckers unarmed. Proglottids craspedote. Reproductive organs single. Genital pores irregularly alternating. Cirrus-sac small. Testes numerous. Ovary median. Uterus persistent, deeply lobate. In Passeriformes. Palaearctic, Ethiopian, Oriental. Type-species *F. goizuetai* López-Neyra, 1936.

18a. **Proglottids few** .... ***Davainea* Blanchard, 1891. (Figs 24.52–24.55)**
Diagnosis: Rostellum armed with a double row of small hammer-shaped hooks. Suckers armed or not. Strobila small, proglottids very few. Genitalia single. Genital pores unilateral or alternating irregularly. Cirrus-sac large, crosses osmoregulatory canals and often crosses mid-line of proglottid. Cirrus conspicuously armed. Testes few, mainly postovarian. Vagina armed. Ovary median or slightly poral, lobed, often strongly bilobed. Vitellarium postovarian. Egg capsules numerous, each containing one egg. In birds. Cosmopolitan. Type-species *D. proglottina* (Davaine, 1860).

18b. **Proglottids numerous** .... **19.**

19a. **Rostellar hooks in three circles. Testes in a single aporal field. Ovary and vitellarium poral** ....
.... ***Porogynia* Railliet & Henry, 1909. (Figs 24.56–24.59)**
(Syn. *Polycoelia* Fuhrmann, 1907, preoccupied.)
Diagnosis: Rostellum armed with large hammer-shaped hooks. Suckers unarmed. Proglottids numerous, craspedote. Only ventral pair of osmoregulatory

24.62

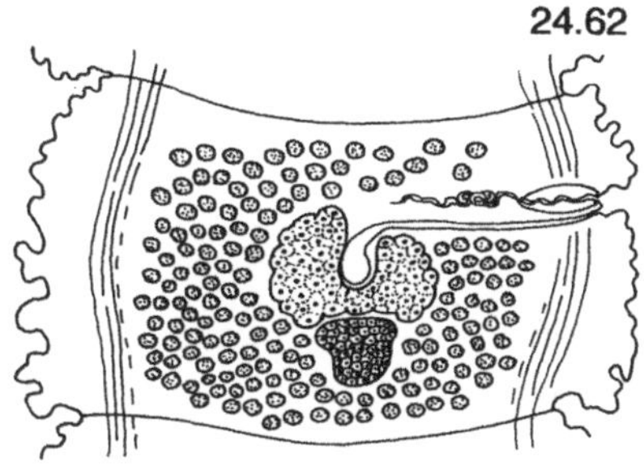

24.63

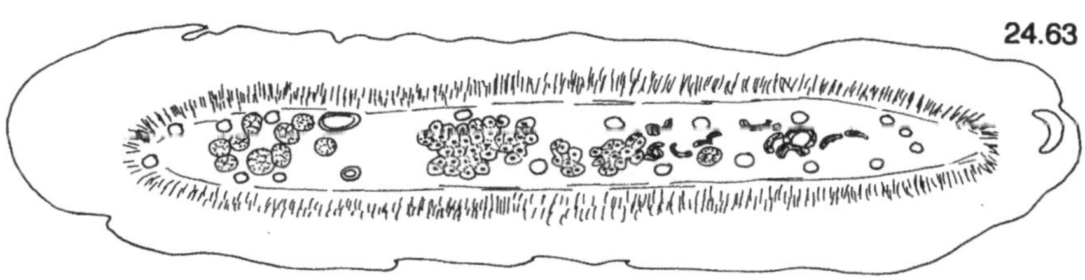

24.64

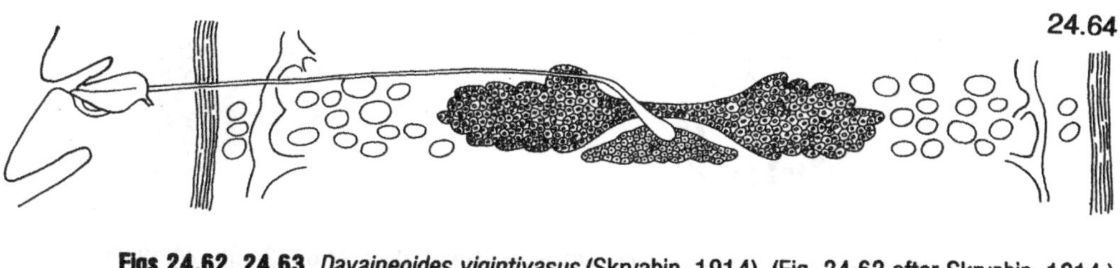

Figs 24.62, 24.63 *Davaineoides vigintivasus* (Skryabin, 1914). (Fig. 24.62 after Skryabin, 1914.) 24.62. Mature proglottid. 24.63. Same in transverse section.
Fig. 24.64 *Delamuretta polycalceola* (Janicki, 1902). (After Janicki, 1902.) Mature proglottid.

canals present. Genital pores unilateral. Cirrus-sac small, extravascular. Testes numerous. Ovary poral, fan-shaped. Vitellarium behind and slightly aporal to ovary. Egg capsules numerous, each containing single egg. In Galliformes and Hyracoidea. Europe, Africa. Type-species *P. paronai* (Moniez, 1892).

19b. Rostellar hooks in two circles .... **20.**

19c. Rostellar hooks in one circle ....
.... ***Ophryocotylus* Srivastava & Capoor, 1977. (Figs 24.60, 24.61)**

Diagnosis: Rostellar hooks numerous, small. Suckers armed. Proglottids craspedote. Reproductive organs single. Genital pores unilateral. Cirrus-sac small. Testes numerous. Ovary median. Uterus sac-like, persistent. In Piciformes. India. Type-species *O. dinopilium* Srivastava & Capoor, 1977.

20a. Longitudinal osmoregulatory canals 20 ....
.... ***Davaineoides* Fuhrmann, 1920. (Figs 24.62, 24.63)**

Diagnosis: Scolex unknown. Proglottids numerous, wider than long. Osmoregulatory canals numerous, ten dorsal and ten ventral. Genitalia single. Genital pores alternate irregularly. Cirrus-sac small, reaching but not crossing poral osmoregulatory canals. Testes numerous, surrounding ovary. Ovary median,

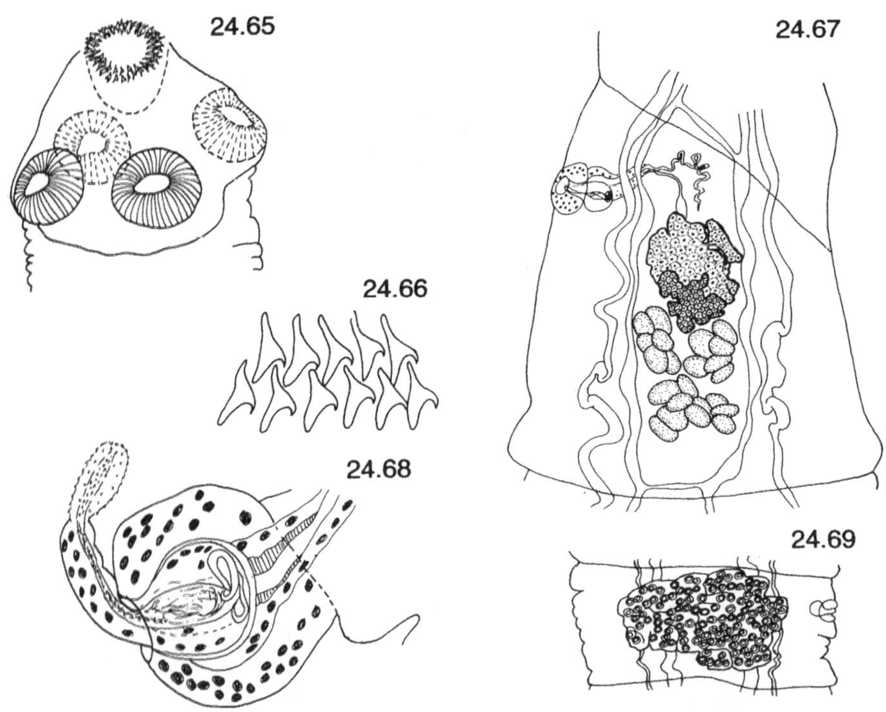

Figs 24.65–24.69 *Dollfusoquenta dollfusi* (Quentin, 1964). (Figs 24.65, 24.68, 24.69 after Quentin, 1964.) 24.65. Scolex. 24.66. Rostellar hooks. 24.67. Mature proglottid. 24.68. Protruded atrial region. 24.69. Gravid proglottid.

bilobed. Vitellarium postovarian, compact. Egg capsules numerous, containing single egg. In Galliformes. Brazil. Type-species *D. vigintivasus* (Skryabin, 1914).

20b. Longitudinal osmoregulatory canals six ....
.... *Delamuretta* Spasskii, 1977. (Fig. 24.64)
(Syn. *Delamurella* Spasskii & Spasskaya, 1976, preoccupied.)
Diagnosis: Rostellum armed with a double row of hammer-shaped hooks. Rostellar spines present. Suckers armed. Proglottids numerous. Genitalia single. Genital pores unilateral or alternate irregularly. Cirrus-sac small, extravascular. Testes numerous, in two fields lateral to ovary. Ovary median, bilobed. Vitellarium postovarian, compact. Egg capsules numerous, containing single egg. In murid rodents. Africa, India. Type-species *D. polycalceola* (Janicki, 1902).

20c. Longitudinal osmoregulatory canals two to four .... 21.

21a. Uterus persistent .... 22.
21b. Uterus replaced by egg capsules with a single egg .... 23.

22a. Rostellar hooks not typically davaineid. Genital atrium protrusible, encloses cirrus-sac and distal vagina. In African rodents ....
.... *Dollfusoquenta* Spasskii, 1973. (Figs 24.65–24.69)
Diagnosis: Scolex large in relation to strobila. Rostellum armed with double row of about 100 hooks with short blades and massive guards. Proglottids numerous. Paired dorsal and ventral osmoregulatory canals present. Genitalia single. Genital pores unilateral (occasionally alternating irregularly). Testes numerous, in single postovarian field. Ovary median, lobed. Vitellarium postero-dorsal to ovary, lobed. Uterus persistent, saccular. In rodents. Africa. Type-species *D. dollfusi* (Quentin, 1964).

22b. Rostellar hooks typically davaineid. Genital atrium not as above. In birds, Ethiopian, Oriental, Neotropical ....
.... *Ophryocotyloides* Fuhrmann, 1920. (Not illustrated.)
Diagnosis: Rostellar hooks numerous, small. Suckers armed. Proglottids craspedote. Reproductive organs single. Genital pores unilateral. Cirrus-sac small. Testes numerous. Ovary median. Uterus persistent, lobate. In birds. Ethiopian, Oriental, Neotropical. Type-species *O. uniuterina* (Fuhrmann, 1908).

23a. Testes two. In Pholidota ....
.... *Diorchiraillietina* Yamaguti, 1959. (Figs 24.70, 24.71)
Diagnosis: Rostellum armed with a double row of small hammer-shaped hooks. Suckers armed. Proglottids numerous. Paired dorsal and ventral osmoregulatory canals present. Genitalia single. Genital pores unilateral. Cirrus-sac crosses osmoregulatory canals. Testes aporal to ovary. Ovary median, bilobed. Vitellarium postovarian. Egg capsules numerous, each containing single egg. In Pholidota. Sri Lanka, Java. Type-species *D. contorta* (Zschokke, 1895).

23b. Testes more than two .... 24.

24.70

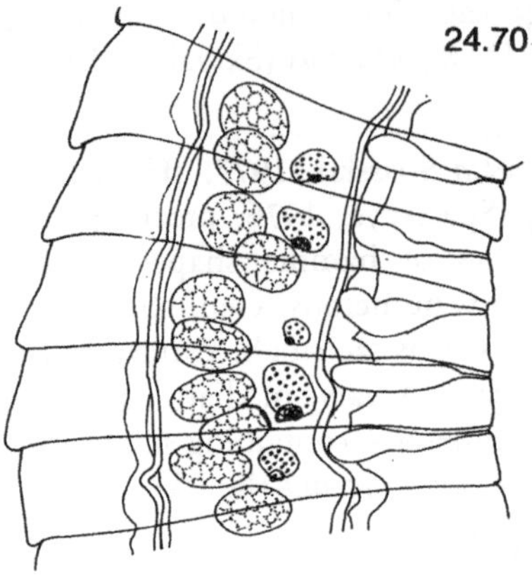

24.71

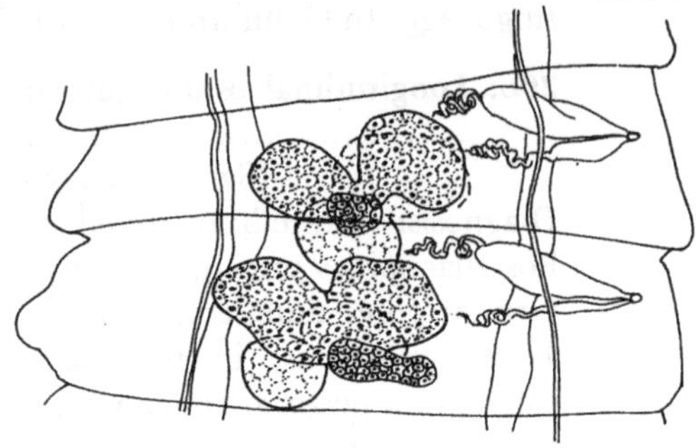

24.72

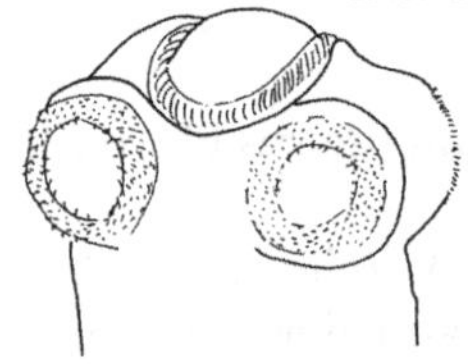

24.73

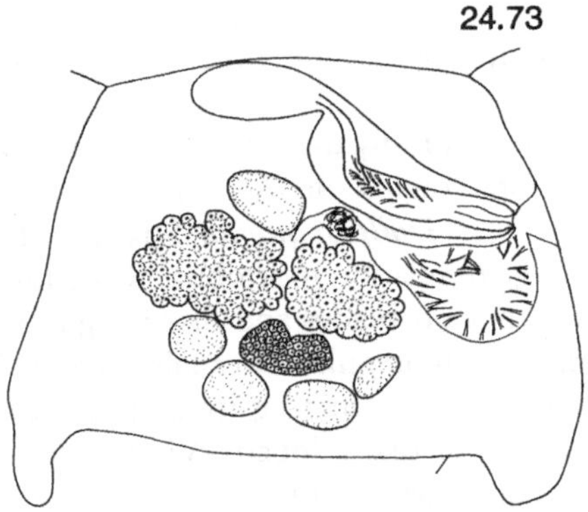

24.75

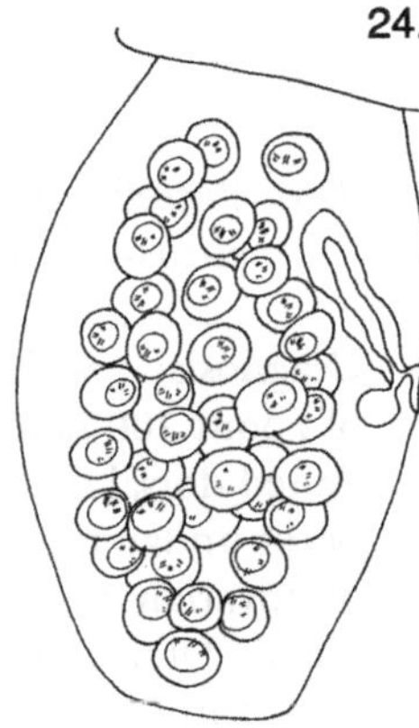

24.74

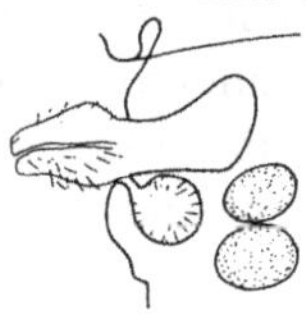

24.76

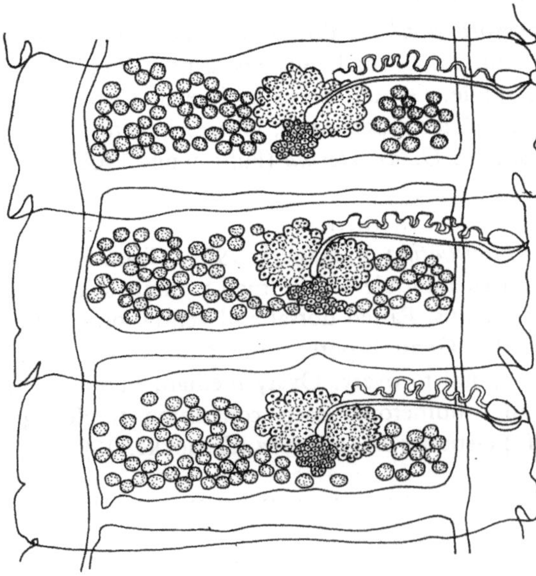

24.77

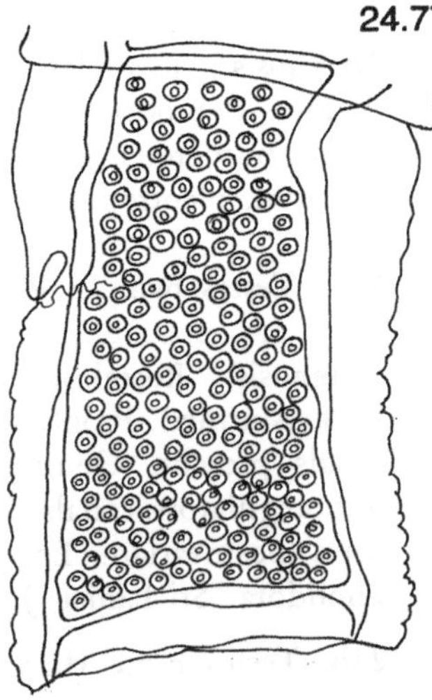

24a. Cirrus-sac crosses mid-line of proglottid. Cirrus armed with large spines. Vagina distally inflated, armed or not .... *Numidella* Spasskaya & Spasskii, 1971. (Figs 24.72–24.75)

Diagnosis: Rostellum armed with double row of small hammer-shaped hooks. Suckers armed. Proglottids numerous, craspedote. Genitalia single. Genital pores unilateral. Vagina usually armed distally. Testes relatively few, in semicircle around female organs. Ovary median to submedian, usually bilobed. Vitellarium postovarian. Egg capsules numerous, contain single egg. In Galliformes. Cosmopolitan. Type-species *N. numida* (Fuhrmann, 1912).

24b. Genital ducts not as above .... 25.

25a. Genital pores unilateral .... 26.
25b. Genital pores alternate irregularly .... 27.

26a. Rostellar hooks small, numerous .... *Paroniella* Fuhrmann, 1920. (Figs 24.76, 24.77)
(Syns *Corvinella* Spasskaya & Spasskii, 1971, preoccupied; *Meggittia* López-Neyra, 1929*; *Metaparonia* Spasskii & Spasskaya, 1976; *Raillietina* (*Paroniella*) Fuhrmann, 1920; *Tetraonetta* Spasskaya & Spasskii, 1971.)

Diagnosis: Rostellum armed with double row of small hammer-shaped hooks. Suckers armed or not. Proglottids numerous, usually craspedote. Genitalia single. Genital pores unilateral. Cirrus-sac usually small and extravascular but may be larger, crossing osmoregulatory canals. Testes usually numerous, variable in distribution. Ovary median or submedian. Vitellarium postovarian. Egg capsules numerous, each containing single egg. In birds and mammals. Cosmopolitan. Type-species *P. longispina* (Fuhrmann, 1909).

26b. Rostellar hooks few (26–32), large .... *Paspalia* Spasskaya & Spasskii, 1971. (Figs 24.78, 24.79)

Diagnosis: Large rostellar hooks in two circular rows. Suckers armed. Proglottids craspedote. Reproductive organs single. Genital pores unilateral. Cirrus-sac small, not crossing osmoregulatory canals. Testes numerous. Ovary median. Eggs single in capsules (?). In woodpecker. Palaearctic. Type-species *P. macracanthos* (Paspalewa & Woidowa, 1969).

27a. Ovary poral. Testes confined to poral postovarian field. In Pterocletiformes .... *Gvosdevinia* Spasskii, 1973. (Fig. 24.80)

Diagnosis: Rostellum broad, armed with double row of small hammer-shaped hooks. Suckers unarmed. Proglottids numerous, wider than long. Pair of dorsal

---

Figs 24.70, 24.71 *Diorchiraillietina contorta* (Zschokke, 1895). 24.70. Early mature proglottid. 24.71. Mature proglottid.

Figs 24.72–24.75 *Numidella numida* (Fuhrmann, 1912). 24.72. Scolex. 24.73. Mature proglottid. 24.74. Protruded cirrus. 24.75. Gravid proglottid.

Figs 24.76, 24.77 *Paroniella urogalli* (Modeer, 1790). 24.76. Mature proglottid. 24.77. Gravid proglottid.

* See footnote p. 437.

24.78

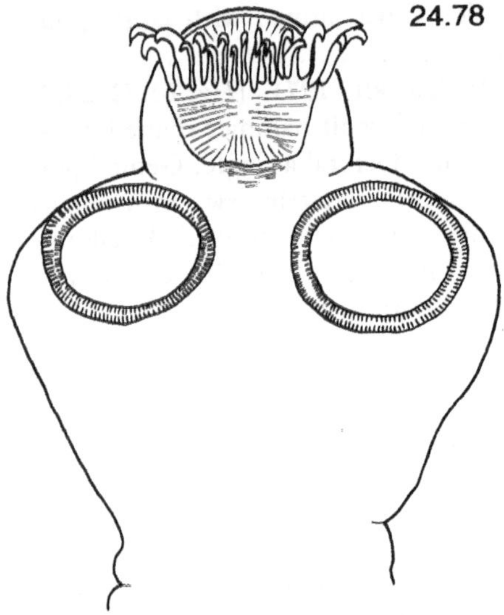

24.79

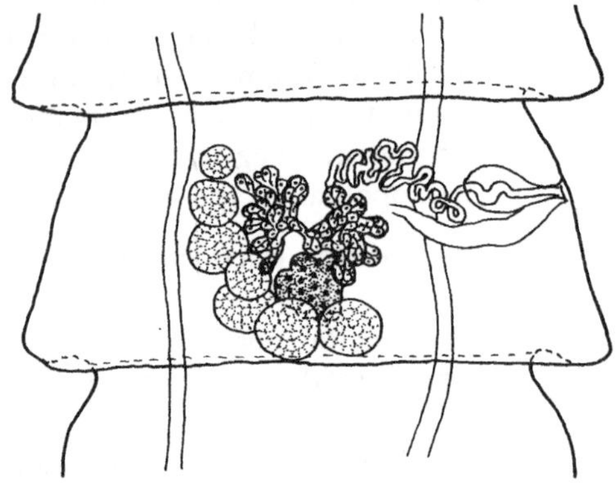

24.80

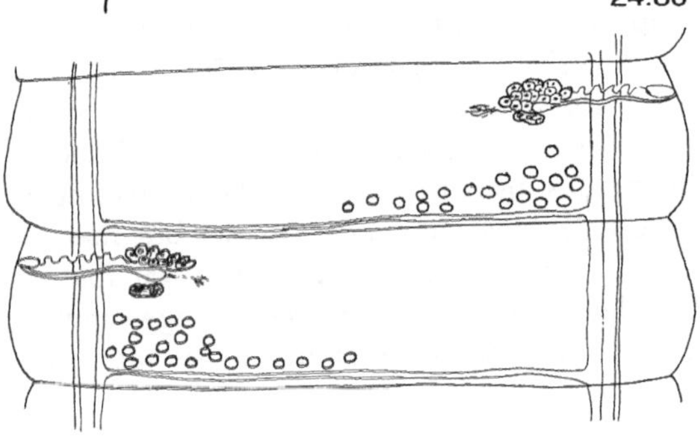

24.83

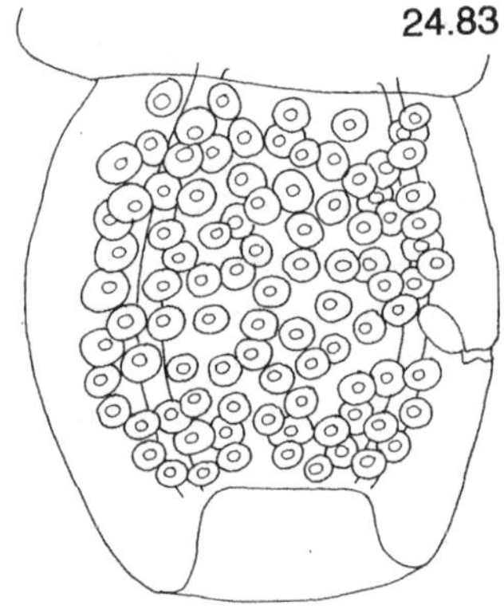

24.82

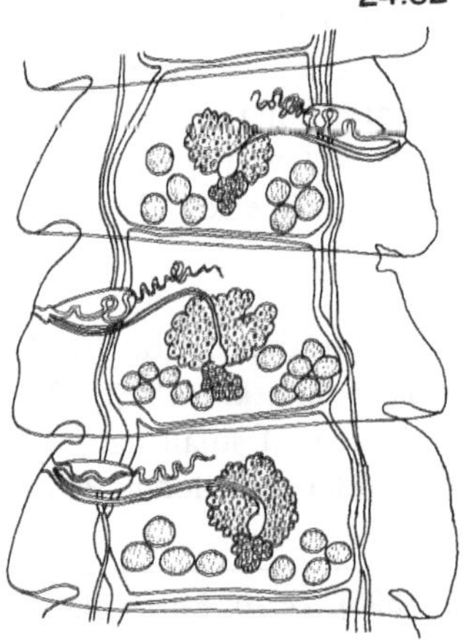

24.81

Figs 24.78, 24.79 *Paspalia macracanthos* (Paspalewa & Woidowa, 1969). 24.78. Scolex. 24.79. Mature proglottid.

Fig. 24.80 *Gvosdevinia pterocleti* (Gvozdev, 1961). (After Gvozdev, 1961b.) Early mature proglottids.

Figs 24.81–24.83 *Skrjabinia cesticillus* (Molin, 1858). 24.81. Scolex. 24.82. Mature proglottids. 24.83. Gravid proglottid.

and of ventral osmoregulatory canals present, dorsal canals poral. Genitalia single. Genital pores alternate irregularly. Cirrus-sac extravascular. Testes numerous. Ovary fan-shaped. Vitellarium postovarian. Egg capsules numerous, containing single egg. In Pterocletiformes. Palaearctic. Type-species *G. pterocleti* (Gvozdev, 1961).

27b. Ovary median. Testes poral and aporal .... ***Skrjabinia* Fuhrmann, 1920. (Figs 24.81–24.83)**
(Syns *Armacetabulum* Movsesyan, 1966; *Brumptiella* López-Neyra, 1929*; *Daovantienia* Spasskii & Spasskaya, 1976; *Markewitchella* Spasskii & Spasskaya, 1972; *Raillietina (Skrjabinia)* Fuhrmann, 1920.)

Diagnosis: Rostellum armed with double row of small hammer-shaped hooks. Suckers armed or not. Proglottids numerous, craspedote. Genitalia single. Genital pores alternate irregularly. Cirrus-sac usually small and extravascular but may cross osmoregulatory canals. Testes usually numerous. Ovary median to submedian, lobed. Egg capsules numerous, each containing single egg. In birds and mammals. Cosmopolitan. Type-species *S. cesticillus* (Molin, 1858).

Subfamily Idiogeninae Fuhrmann, 1907

Key to genera

1a. Body robust. Uterus saccular. Cirrus-sac not reaching to mid-line of proglottid .... **3.**
1b. Body delicate. Uterus inverted U-shape. Cirrus-sac large, reaching to or beyond mid-line of proglottid .... **2.**

2a. Rostellum bearing three to five irregular rows of hooks .... ***Idiogenes* Krabbe, 1868. (Figs 24.84, 24.85)**
(Syns *Ersinogenes* Spasskaya, 1961; *Paraidiogenes* Movsesyan, 1971.)

Diagnosis: Pseudoscolex may develop. Rostellum with hammer-shaped hooks. Accessory spines in annular patch proximal to rostellar hooks (easily lost). Suckers unarmed (? or armed with hooks easily lost). Proglottids distinctly craspedote. Genital pores unilateral. Testes few to several. Ovary median. Uterus inverted U-shape. Paruterine organ large; extends anteriorly to uterus. In Gruiformes. Palaearctic. Type-species *I. otidis* Krabbe, 1867.

2b. Rostellum bearing two regularly alternating rows of hooks .... ***Pseudidiogenes* Movsesyan, 1971. (Figs 24.86, 24.87)**

Diagnosis: Accessory spines in annular patch proximal to rostellar hooks (easily lost). Suckers armed (hooks easily lost). Proglottids distinctly craspedote. Genital pores unilateral. Cirrus-sac large, reaches to or beyond mid-line of proglottid.

*NB. Schmidt's (1986) placement of *Brumptiella* and *Meggittia* as synonyms of two genera 'in part' is a confusion of nomenclature and systematics. *Brumptiella* shares the same type-species as *Skrjabinia* and must, therefore, fall as an objective synonym. The type of *Meggittia, M. bolivari* López-Neyra, 1929, has unilateral genital pores; therefore, the genus is a synonym of *Paroniella. Kotlania* López-Neyra, 1929, the type of which is *Raillietina echinobothrida* (Megnin, 1880) is a synonym of *Raillietina*. The genital pores of *R. echinobothrida* are normally unilateral and only exceptionally do they alternate in short series.

24.84

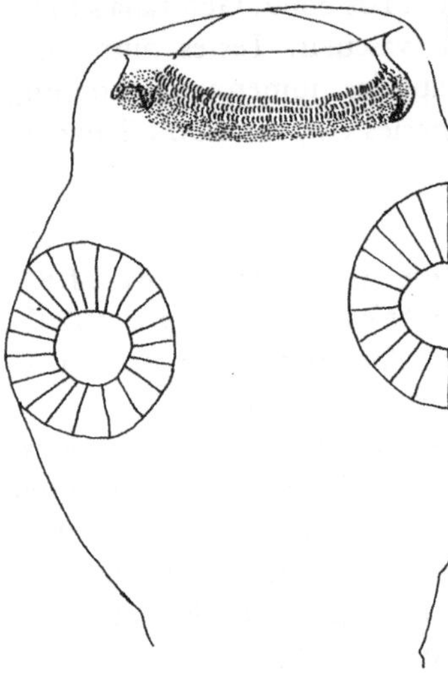

24.85

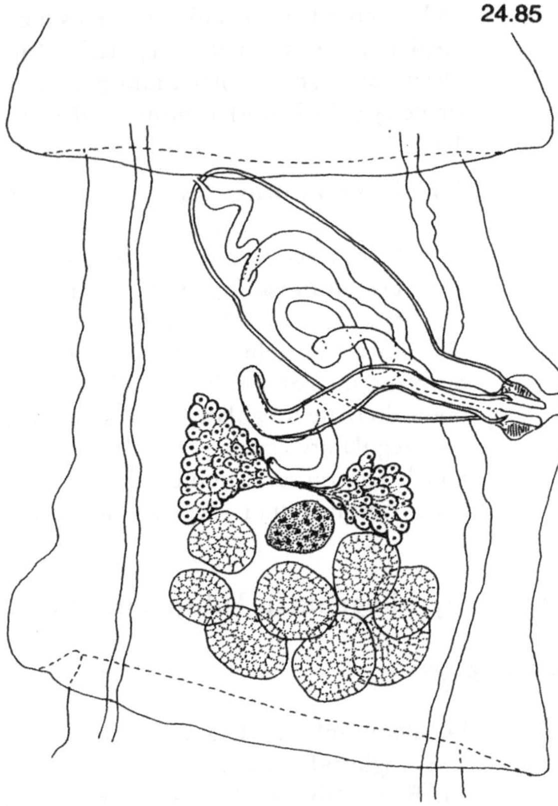

24.87

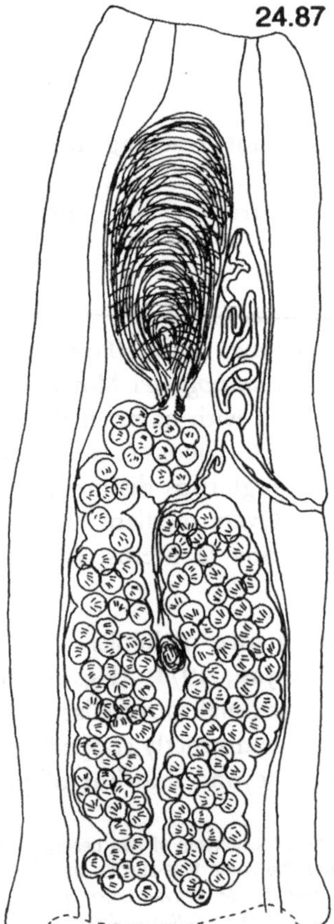

24.86

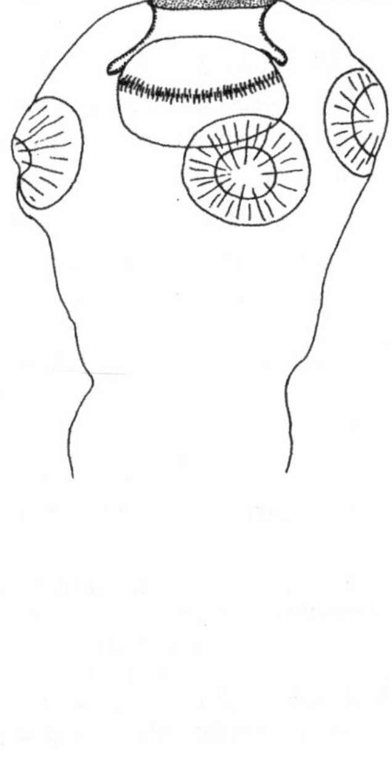

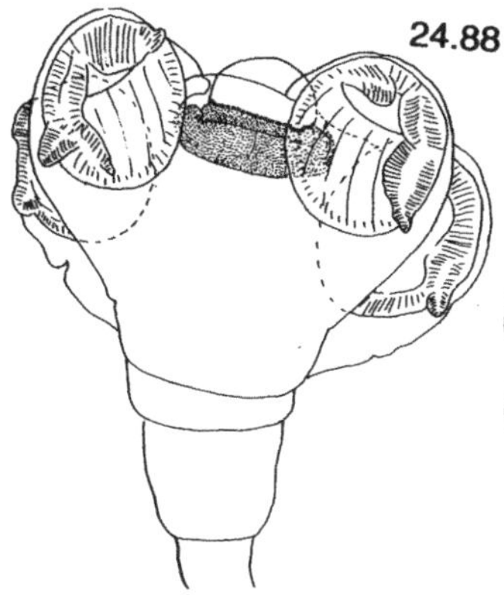

24.88

24.89

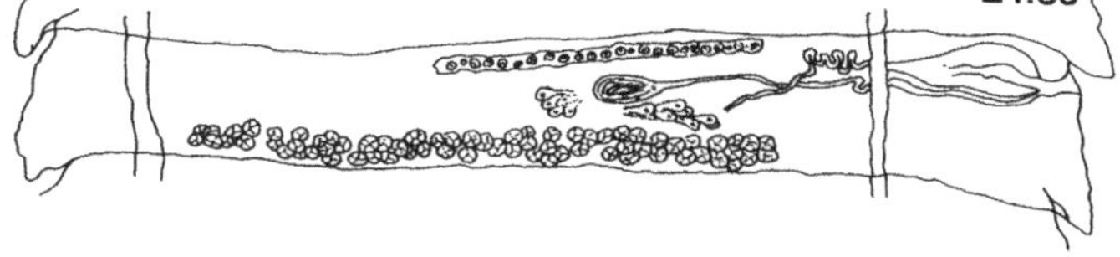

24.90

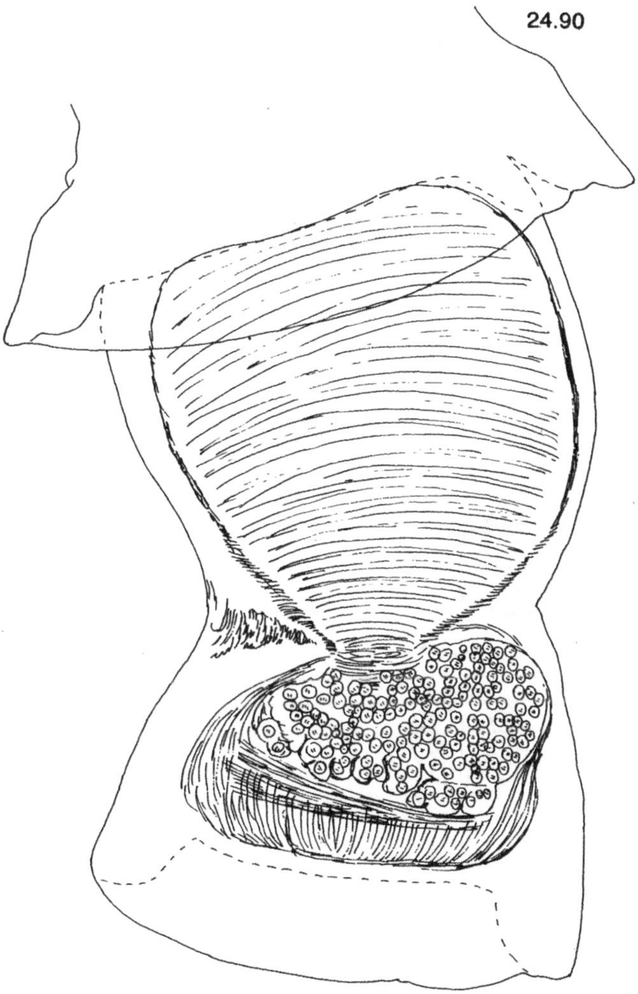

Figs 24.84, 24.85 *Idiogenes otidis* Krabbe, 1867. 24.84. Scolex. 24.85. Mature proglottid.
Figs 24.86, 24.87 *Pseudidiogenes flagellum* (Goeze, 1782). 24.86. Scolex. 24.87. Gravid proglottid.
Figs 24.88-24.90 *Otiditaenia* Beddard, 1912. 24.88. *O. macqueeni* (Woodland, 1930). Scolex. 24.89. *O. conoideis* (Bloch, 1782). Mature proglottid. 24.90. *O. conoideis*. Gravid proglottid. (Fig. 24.89 after Gomez, 1981.)

24.94

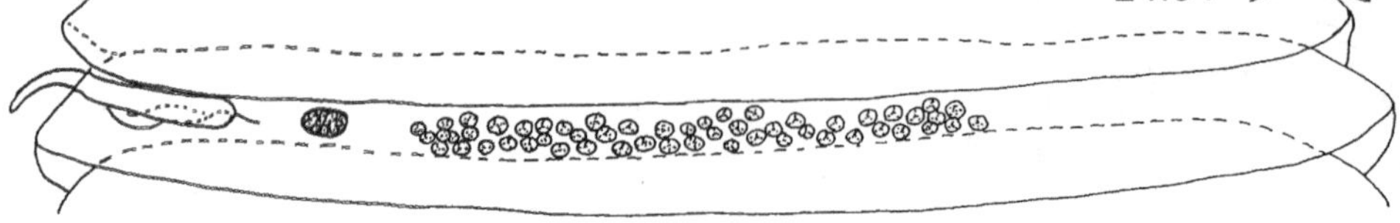

24.93

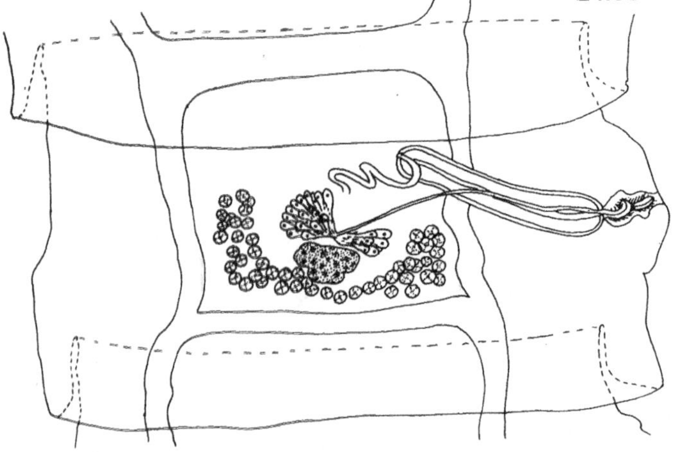

24.92

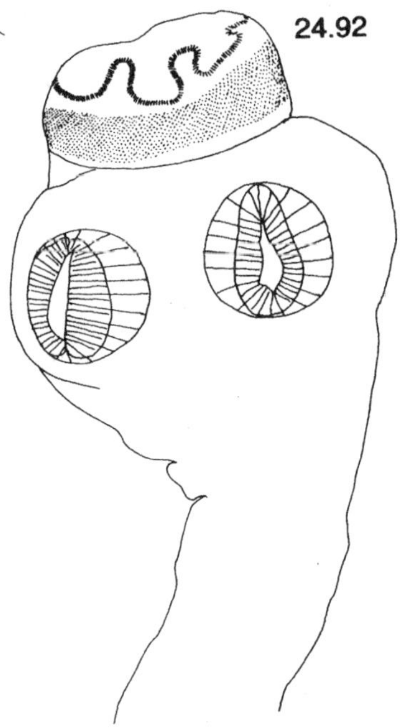

24.91

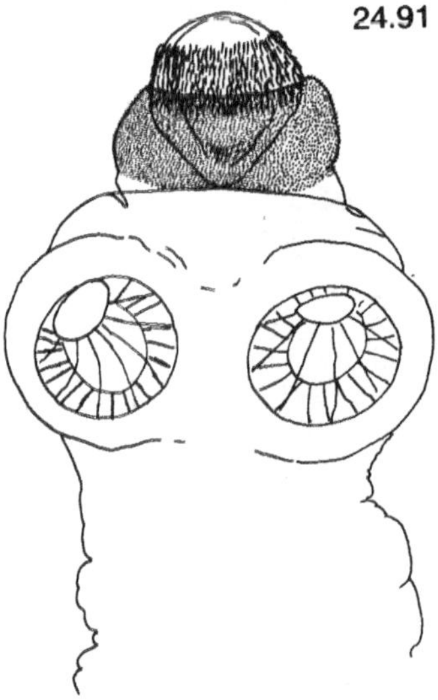

Fig. 24.91 *Sphyroncotaenia uncinata* Ransom, 1911. Scolex.
Figs 24.92, 24.93 *Chapmania tauricollis* (Chapman, 1876). 24.92. Scolex. 24.93. Mature proglottid.
Fig. 24.94 *Satyanarayana satyanarayani* Khan, 1984. (After Khan, 1984.) Mature proglottid.

Testes few to several. Ovary median. Uterus inverted U-shape. Paruterine organ large, extends anteriorly to uterus. In Gruiformes, Falconiformes, Passeriformes. Cosmopolitan. Type-species *P. flagellum* (Goeze, 1782).

3a. Suckers bearing lappets .... ***Otiditaenia* Beddard, 1912. (Figs 24.88–24.90)**
(Syns *Schistometra* Cholodkowsky, 1912; *Paraschistometra* Woodland, 1930.)

Diagnosis: Rostellum with two rows of hooks; also bears accessory spines. Suckers unarmed. Proglottids craspedote. Reproductive system single. Genital pores unilateral or alternating irregularly. Cirrus-sac small. Testes numerous. Ovary poral. Uterus divided, saccular. Bulk of paruterine organ anterior to uterus. In Gruiformes. Palaearctic, Ethiopian. Type-species *O. eupodotis* Beddard, 1912 [now considered a synonym of *O. conoideis* (Bloch, 1782)].

3b. Suckers rounded .... **4.**

4a. Two rows of small rostellar hooks .... **5.**

4b. Twelve irregular rows of small hammer-shaped hooks .... ***Sphyroncotaenia* Ransom, 1911. (Fig. 24.91)**

Diagnosis: Numerous accessory spines on rostellum proximal to hooks. Suckers rounded, unarmed. Proglottids craspedote. Osmoregulatory canals four reducing to two along strobila. Reproductive system single. Genital pores unilateral. Cirrus-sac small. Testes numerous. Ovary poral. Uterus sac-like. Paruterine organ mainly anterior to uterus. In Gruiformes. Ethiopian. Type-species *S. uncinata* Ransom, 1911.

5a. Ovary median or porally submedian. Testes numerous, poral and aporal to ovary .... ***Chapmania* Monticelli, 1893. (Figs 24.92, 24.93)**
(Syn. *Capsodavainea* Fuhrmann, 1901.)

Diagnosis: Rostellum armed with two undulating rows of small, hammer-shaped hooks; accessory spines often present. Suckers rounded, no lappets but aperture closed by two or three semicircular valves. Proglottids craspedote. Osmoregulatory canals four, may reduce to two along strobila. Reproductive system single. Genital pores alternating irregularly. Cirrus-sac large, projecting into median field. Uterus ramifying when young; deeply lobed saccular later. Paruterine organ anterior to uterus. In Gruiformes, Rheiformes, Bucerotiformes. Palaearctic, Ethiopian, Neotropical, Oriental. Type-species *C. tauricollis* (Chapman, 1876).

5b. Ovary poral. Testes numerous, entirely aporal to ovary .... ***Satyanarayana* Khan, 1984. (Fig. 24.94)**

Diagnosis: Rostellum armed with two rows of small hammer-shaped hooks. Suckers rounded, unarmed. Proglottids wider than long, craspedote. Osmoregulatory canals four. Genital organs single. Genital pores alternating irregularly. Cirrus-sac small. Uterus saccular. Paruterine organ anterior to uterus. In lungs of Ciconiiformes. India. Type-species *S. satyanarayani* Khan, 1984.

25 Family Dilepididae Railliet & Henry, 1909

F.V. BONA

Università di Torino, Dipartimento di Biologia Animale, Via Accademia Albertina 17, 10123 Torino, Italy.

Introduction

Fuhrmann's (1932) subdivision of the Dilepididae into three subfamilies, Dilepidinae (normal uterus), Paruterininae (paruterine organ) and Dipylidiinae (uterine capsules) is now obsolete. The Paruterininae and Dipylidiinae, the latter reduced to a small group of parasites of carnivorous mammals, have been raised to family rank (see Chapters 27 and 26). The term 'uterine capsules', without further specifications, no longer suits the definition of suprageneric units. Some genera have been assigned to a new family, the Metadilepididae (see Chapter 28). The remaining Dilepididae exceed one hundred genera, which do not form a monophyletic assemblage. Other new families would have to be erected to reduce the Dilepididae to a monophyletic group and, in this reduced but still large group, the potential for further divisions can be discerned. Several attempts have been made to find a suprageneric arrangement for this group. However, this requires further studies and hypotheses, taking into account the significance of new morphological characters, better knowledge of traditional characters, ranges of variation and trends in shape, anatomical homologies and revision of the generic allocation of many species.

The generic diagnoses have been updated by examining most of the type-species and some congeneric species. In a few cases where this was not possible, information has been compiled from the literature only. All drawings are original if not otherwise stated.

The genera of Dilepididae show such a high specificity for the order of their hosts that it has sometimes been considered as an important character for generic identification.

A given character does not always have the same discriminatory significance, because of its instability or wide range of variation. The shape of the hooks can be similar in an entire genus, sometimes even unique and thus becoming a diagnostic character, or can be characteristic only in a few species. The presence of a pre-atrial vaginal sphincter may be constant in one genus, occasional in another. The position of the genital pores, of the genital ducts with respect to the osmoregulatory canals, or of the vagina to the cirrus-sac can be constant or unstable in different genera. In the key, when an important character is unstable in a genus and can have two character-states, the genus may be identified starting from one or the other of them.

Some characters are more complex and more important for the systematics of the genera of the Dilepididae than has been considered in the past. Observations at high magnification with oil immersion objectives are recommended.

When examining the scolex, the structure, glandular or muscular, of the apical apparatus, the muscular sheets and bundles of the rostellum proper and the hook tip, which can be directed backward or forward when the rostellum is retracted, must be considered.

As to the uterus, it is important to define its **initial form**, indicating whether it is permanent or if it subsequently changes into a different pattern (modifying the cavity and the structure of the wall), turns into uterine capsules or disintegrates leaving the eggs free in the parenchyma. The following patterns occur: **reticulate**, either delicate, usually large-meshed with narrow tubes, or coarse, with wide irregular tubes, sometimes appearing like a cavity with spots or islets of parenchyma; **labyrinthine**, i.e. reticulate with clustered tubes and small cavities, with fairly firm walls, adhering to each other, often mamillated and moulded on the eggs; **sacciform**, more or less lobulated in outline and more or less septate internally. There are intermediate forms which are difficult to define.

The egg coats (or envelopes) are sometimes believed to be of generic significance, but more interesting is the interpretation of their various strata and membranes to establish the presence of uterine capsules. Often the outermost pellicle of the egg coats has been erroneously interpreted as a uterine capsule. Observations should be made in all stages of development. There are two egg coats. The internal coat is a syncytium with three nuclei that subdivides into two strata, the embryophore inside and, at least for a certain period of maturation, a more external stratum, often granular, called the 'external layer of the inner coat', which almost always contributes to the thickening of the embryophore (rarely, the position of the two strata of the inner envelope is inverted). The external coat is a syncytium with two larger nuclei in a polar position that may take various aspects: a thin outer membrane, more or less shrivelled around the egg in the uterus but spherical and expanded in water; a firm superficial pellicle, sometimes fairly thick, well detached from the underlying strata; a condensed layer around the inner coat, reinforcing the embryophore; or forming two polar processes varying from simple protuberances to very long and thin polar prolongations. Eventually, the remains of the three nuclei of the internal coat can be seen on the internal side of the embryophore, two at one pole and one at the other, incorporated in a thin pellicle that can detach itself by delamination from the embryophore and should not be confused with the actual membrane of the oncosphere. In the final stage of development, important changes may take place, especially in the thickening of the embryophore. In the Dilepididae, the initial 'egg capsule' (not to be confused with the 'uterine capsule'), which derives from one vitelline cell and adheres to the surface of the outer coat, is almost negligible under observation with an optical microscope, except in early stages of oncospheral development.

Uterine capsules have always been considered an important generic and suprageneric character but have caused more confusion than benefit. A uterine capsule cannot definitely be stated to exist unless the development and transformation of the egg coats and the structure of the uterine wall in its various phases can be observed.

Polyovular capsules are easily recognized, well-defined, and produced by a clear fragmentation of the uterus, the wall of which is sometimes involved in secretion processes or the production of a thin membrane round the egg packets. Monovular capsules can create problems because the isolation of the eggs can be precocious or belated, and can take place in various ways which are difficult to observe and evaluate. Furthermore, the parenchyma is differently involved in their formation. It may remain hardly modified round the pellicle of the capsule ('membranous capsules') or may adhere to the capsule with fibres and small cells, or may surround the capsule with cells that soon become vesicular (rather rare, 'parenchymatous capsules'). In a reticulate uterus, capsules form by early narrowing of the tubes between the eggs forming little cavities in which each egg remains enclosed. In a labyrinthine uterus, the uterine wall becomes moulded to the surface of the eggs, becoming mamillated and sometimes thicker (a frequent and critical case because it is difficult to ascertain whether the uterine wall actually ends up surrounding the whole egg). In both cases, the surface pellicle of the outer coat can adhere to the internal surface of the uterine capsule, forming one firm layer; however, slight detachments sometimes reveal its double structure and origin. Seldom, monovular uterine capsules derive from a sacciform uterus. When the uterine wall disappears, leaving the eggs loose in the parenchyma (sometimes producing spherical, granular cells) or in irregular lacunae without their own wall, capsules should not form.

There are some accessory criteria for recognizing the existence of monovular uterine capsules:

1. There are no capsules if the mature eggs touch each other, in clusters or in a row, or when they have long polar processes.
2. The capsules often form precociously, before completion of the oncosphere; contrary to common opinion, late ones that form in the gravid proglottid at an advanced phase of maturation are rare.
3. Spherical cavities among eggs, empty or with cell debris, well-delimited by a refractive pellicle and interpretable as abortive capsules, demonstrate the transformation in progress of the uterus into uterine capsules.
4. The uterine capsule is a layer or a pellicle outside the oncospheral outer coat and egg capsule, interposed between them and the parenchyma, to which it usually adheres.
5. When the uterus discharges eggs with laceration of the proglottid wall, there should not be any monovular capsules. In fact, monovular capsules are not a protective covering round the egg when it leaves the uterus, but organs meant for preservation of the egg within the gravid proglottid.

A cirrus with a tuft of long, bristle-like, flexible or fairly rigid spines, basally or along its entire length, is a discriminating character for a large group of genera.

This tuft should not be confused with long, fine spines on the **apex** of some cirri, armed in various ways. Often this kind of spinosity is accompanied by a **ductus masculinus** (the female orifice being more distal than the male one).

A complex atrium with specialized walls, particular muscular bundles or well-defined ductus masculinus or hermaphroditus is often characteristic of the genus. In the Dilepididae, accessory organs and structures are very rare.

The position of the genital pores in very small species with few proglottids should be checked in many specimens because an individual strobila does not always have enough proglottids to reveal the existence of an alternation.

The testicular field sometimes differs in young and adult proglottids.

True seminal vesicles do not occur in Dilepididae (except in *Proorchida*). A swelling of the internal or external vas deferens, which maintains its coils, is not considered to be a true internal or external seminal vesicle.

Principal adopted ranges. Strobila: extremely small, less than 2 mm; very small, up to 10 mm; small, from 10 to 30 mm; medium, from 30 to 80 mm; large, more than 80 mm. Hooks: few, about 10 to 14; numerous, about 16 to 30; very numerous, more than 35. Testes: only 4; few, from 5 to 13; numerous from 14 to 35; very numerous, from about 36 to 60, seldom more. Vagina: very short, meaning shorter than cirrus-sac.

Notes on genera

Ptilotolepis Spasskii, 1969. Diagnosis emended after revision of the type-species. *Emberizotaenia* is not synonymous with *Ptilotolepis* as stated by Spasskaya & Spasskii (1978), nor is *Ptilotolepis* probably synonymous with *Pseudochoanotaenia* as stated by Schmidt (1986). The last two genera are actually similar. The main difference lies in the position of the genital ducts relative to the osmoregulatory canals, and they would be further differentiated if it were confirmed that uterine capsules do not develop in *Pseudochoanotaenia*. They also have different hosts. *Rostellina* Rathore & Nama, 1984, from Falconiformes, is considered synonymous with *Ptilotolepis*. However, if its eggs really have a pyriform apparatus (which does not appear in the published figure), if the strange structure and the preovarian position of the vitellarium are confirmed and if the mature eggs scattered in the parenchyma do not have a uterine capsule similar to that of *Ptilotolepis*, this synonymy must be reconsidered. Furthermore, Falconiformes may accidentally acquire some cestodes of their prey (*Ptilotolepis* parasitizes Passeriformes). *Rostellina* is more similar to *Ptilotolepis* than to *Pseudochoanotaenia, Emberizotaenia* and *Unciunia*.

Himantaurus Spasskaya & Spasskii, 1971. The question is whether it belongs to Davaineidae or not. The type-species was first described as *Davainea minuta* Cohn, 1901. The genus bears davaineoid hooks, but also a true rostellar pouch and a rostellum similar to that of Dilepididae. The inner coat (embryophore) of the egg probably develops with five to six nuclei instead of three, thus differing from the normal pattern of Dilepididae. Its anatomy fits both the Dilepididae and Davaineidae. The lack of davaineoid spines on the suckers and of typical davaineoid uterine capsules does not prove, but increases the probability, that it does not belong to Davaineidae. Its exclusion from Davaineidae does not mean its automatic transfer to Dilepididae, where it is located with some reservation (see also *Kowalewskiella*). It is significant that three species of *Himantaurus* studied exactly answer every character enunciated in the diagnosis.

Cyclustera Fuhrmann, 1901. The proposed synonymy of *Neocyclustera* Underwood & Dronen, 1986 with *Cyclustera* is provisional. More documentation is needed.

Glossocercus Chandler, 1935. The species belonging to this genus had been ascribed to *Parvitaenia* and defined as atypical by Bona (1975). The hooks seem to form a transformation series, starting from the basic parvitaenioid pattern and ending deeply modified, with their own sclerification type and large size. Apparently, *Bancroftiella* and *Tubanguiella* resemble *Glossocercus* more than *Parvitaenia* (see notes on *Tubanguiella* and *Bancroftiella*).

Neogryporhynchus Baer & Bona, 1960. *Gryporhynchus pusillus* von Nordmann, 1832 is the name of a larval stage from freshwater fish which, according to the old drawings of von Nordmann, could be ascribed to several species, belonging to more than one genus. Therefore, this name can be used only for a collective group of larval forms of Dilepididae having a fish as second intermediate host, although it sometimes has been used for adult cestodes. *Neogryporhynchus* was proposed for such a characteristic species as *cheilancristrotus* Wedl, 1865 ascribed to *Gryporhynchus* by Ransom (1909), the larva of which most probably belongs to the *Gryporhynchus* group.

Tubanguiella Yamaguti, 1959. The type-species is insufficiently described. The presence of a single hook crown is uncertain and synonymy with *Glossocercus* Chandler, 1935 should be considered.

Eugonodaeum Beddard, 1913. Beddard's material has never been re-examined, therefore the diagnosis has been completely re-written; the only known hook was found in *Nasutaenia*, from Ciconiae, now a synonym. Probably none of the species subsequently included in the genus, from Charadriiformes or other hosts, belong to it. *Eugonodaeum* shared many characters with *Malika* and both seem to form a group different from Dilepididae.

Reticulotaenia Hoberg, 1985. The intraspecific variation of some characters exceeds the usual ranges. The diagnosis was based chiefly on Hoberg's material of *australis*, some characters of which are not to be seen in *mawsoni* Prudhoe, 1969. There are also problems about the identity of Hoberg's (1985) and Jones & William's (1967) material of *australis*.

Trichocephaloidis Sinitzin, 1896. *Diagonaliporus* is synonymized with this genus because its only differentiating character is that the pores gradually change from one side of the strobila to the other. This also occurs in some specimens of

Trichocephaloidis, mainly when fixed alive. It is a rotation of the strobila and not a shifting of pores, as proved by the spiral course of the longitudinal musculature and osmoregulatory canals.

Dilepis Weinland, 1858. Schmidt (1986) mentions *Spasskytaenia* Borgarenko, 1981 *partim* (does it mean *sensu* Borgarenko, *nec* Oshmarin ?), considering it as synonymous with *Dilepis*, and *Spasskytaenia* Oshmarin, 1956 *partim*, considering it as synonymous with *Anomotaenia*. The question can be solved only by starting from Oshmarin's genus as reported by Spasskaya & Spasskii (1978). See also the note on *Spasskytaenia* Oshmarin, 1956.

Malika Woodland, 1928. The synonymy of the type-species, *M. oedicnemus* Woodland, 1928 with *Dilepis odhneri* Fuhrmann, 1909 has still to be confirmed. As to its affinities, see *Eugonodaeum*.

Neoliga Singh, 1952. The first description of a species belonging to this genus (regularly alternate pores) was the one of *Taenia depressa* von Siebold, 1836 given by Fuhrmann (1895). Its correct name should be *Neoliga depressa* (von Siebold, 1836) n. comb. *sensu* Fuhrmann, 1895 [in accordance with Krabbe (1869) and perhaps also with Galkin (1983)] nec *T. depressa* sensu von Linstow, (1875 and 1896); nec *Anomotaenia depressa* sensu Fuhrmann (1926) and Joyeux & Baer (1936b); nec *Neoliga depressa* sensu Spasskii & Spasskaya (1959) and Spasskaya & Spasskii (1971) (all of them having irregularly alternate pores and probably belonging to *Echinotaenia*).

Hirundinicola Birova-Volosinovicova, 1969. The redescriptions of *H. chelidonariae* by Spasskaya (1959) and Spasskii & Spasskaya (1966) do not correspond to the type-species. *Anomotaenia praecox* sensu Joyeux & Timon-David, 1934 *nec* Krabbe, 1869 has to be considered synonymous.

Chitinorecta Meggitt, 1927. The herniae in the cortex of *C. agnosta* Meggitt, 1927 containing organs, mentioned by Meggitt (1927c), were not observed in normal proglottids. The double crown is sometimes difficult to discern when blade tips are on the same level and only in the handle ends an alteration is observed. In contracted strobilae, the uterus can look like a deeply septate sac. The structure of the ductus masculinus and its armature are important for distinguishing the genus from other genera having bristle-like spines on the cirrus. Probably, *agnosta* Meggitt, 1927, the type-species, is synonymous with *brevicollis* Fuhrmann, 1907.

Choanotaenia Railliet, 1896. The species *C. infundibuliformis* (Goeze, 1782) is not synonymous with *C. infundibulum* (Bloch, 1779). The species has been revalidated after examining the original type material of Goeze. Fuhrmann (1932) had already explained that *Choanotaenia*, as conceived by himself (1908b), who substituted the original type-species, *infundibulum*, with *galbulae*, represented a systematical mistake. Moreover, *galbulae* has two hook crowns and belongs to *Monopylidium*.

Spiniglans Yamaguti, 1959. In the re-examined type material of *S. microsoma* (Southwell, 1922), the uterine capsules mentioned by Southwell (1922) could not be observed due to the lack of sufficiently mature proglottids, but in the undoubted congeneric species *constricta* Molin, 1858 and *corvi* Joyeux, Baer & Martin, 1936, uterine capsules are evident.

Laritaenia Spasskaya & Spasskii, 1971. The material of *Anomotaenia hydrochelidonis* Dubinina, 1953 chosen as type-species of *Laritaenia* is probably a mixture. Dubinina's description seems to refer only to a part of the deposited material. It is what we should deduce if we consider that: (i) in Galkin's (1987b) redescription of the original material, *hydrochelidonis* is synonymized with *Anomotaenia microrhyncha*, and consequently *Laritaenia* should become synonymous with *Anomotaenia*; and (ii) species with the basic characters mentioned by Dubinina, characters that clearly distinguish *Laritaenia* from *Anomotaenia*, have been personally observed in Lariformes and Charadriiformes.

Fuhrmannolepis Spasskii & Spasskaya, 1965. Spasskii & Spasskaya (1965) erected this genus with *F. decacantha* (Fuhrmann, 1913) as type-species but without a diagnosis, which was given later by Spasskii & Poznakomkin (1966). They probably described a species different from *decacantha* Fuhrmann, 1913. The present emended diagnosis refers to its type material. The genus, as conceived by Spasskii & Spasskaya (1965) contains species belonging to other genera, including *Sacciuterina*.

Polycercus Villot, 1883. Joyeux & Baer (1939) gave the first description, which must be consulted for the identification of the adult worm corresponding to Villot's larva, under the name of *Amoebotaenia lumbrici* (Villot). See also the note on *Sacciuterina* Matevosyan, 1963.

Chimaerula n. g. The rostellar pattern is identical with that of *Stenovaria* and the anatomy very similar to that of *Himantaurus*, both from Charadriiformes. The type-species was first ascribed to *Liga* by Prudhoe (1960) and then to *Himantaurus* by Spasskii (1985).

Sacciuterina Movsesyan, 1963. Sandeman (1959) considered *Polycercus lumbrici* Villot, 1883 a synonym of *paradoxa* Rudolphi, 1802, whereas they belong to different genera. Many authors accepted his point of view, including Spasskaya & Spasskii (1970, 1977). Joyeux & Baer (1939) first described the adult of *lumbrici* (a larval stage) and gave the basic elements of the species description, while all descriptions of *paradoxa* either refer to wrongly identified material or are inadequate except Clerc's (1903) (Cohn's (1899) being inadequate too). The type material and Clerc's are lost. Baer (1962) gave a short redescription which agrees with Clerc's so that Baer's material, deposited at the Museum d'Histoire Naturelle, Geneva, has to be considered as the neotype of *paradoxa*. Both species are therefore distinct; *lumbrici* Villot is revalidated and the generic diagnosis of *Polycercus* has to be changed; *paradoxa* Rudolphi, now clearly defined, entitles us to revalidate *Sacciuterina* with an emended diagnosis, differentiating it from

Fuhrmannolepis (in which some species of *Sacciuterina* were included) and *Paricterotaenia*. Schmidt (1986) had already chosen *lumbrici* as the type-species of *Polycercus*, but did not amend the generic diagnosis.

Monosertum n. g. The type of *Taenia parina* Dujardin, 1845 is lost. The species, very little known, was sometimes wrongly identified and was said to have two crowns. It has been ascribed to *Choanotaenia, Anomotaenia* and *Paricterotaenia*. Mettrick (1958) used this last generic name and his redescription has to be considered as the only acceptable one, but his material is no longer available. A neotype has been selected from the collection of the Natural History Museum, London. *Choanofuhrmannia unicoronata* (Fuhrmann, 1908) Lopez-Neyra, 1935 (and 1944) has two crowns rather than one [Mettrick (1958) and Baer's personal communications to the author], so it becomes synonymous with *Monopylidium* and can no longer represent a group of species bearing a single crown of hooks.

Kintneria Spasskii, 1968. Kintner (1938) identified his material as *Choanotaenia passerina* (Fuhrmann, 1907) (1908 is a misprint), namely as *Monopylidium passerinum* Fuhrmann, 1907, the original, revalidated name of the species, but his description does not correspond to Fuhrmann's species, because of the single hook crown and the uterine capsules, not mentioned by Fuhrmann. Kintner's material being lost, both characters need confirmation with new material. Because of its homonymity, Kinter's species has to change its name irrespective of its change of genus; this explains the new name *Kintneria capsulata* Spasskii, 1968 (syn. *Choanotaenia passerina* Kintner, 1938 nec *Monopylidium passerinum* Fuhrmann, 1907 or *Choanotaenia passerina* (Fuhrmann, 1907) Fuhrmann, 1932). For now, we have to accept Kintner's description which justifies generic rank for the subgenus *Monopylidium (Kintneria)* Spasskii, 1968 (of which no diagnosis was supplied).

Cinclotaenia Macy, 1973. Some small European cestodes from Cinclidae may be difficult to assign to *Cinclotaenia* or *Anomotaenia*. The latter seems to be characteristic of Charadriiformes, so that the differences observed between the North American type-species and the European species of *Cinclotaenia* may merely represent an intrageneric variation that should be studied.

Megalacanthus Moghe, 1926. The species composition of the genus is now as follows: (i) *chandleri* Moghe, 1925; (ii) *macracanthoides* Fuhrmann, 1907, p. 519 (syn. *Monopylidium macracanthum* Fuhrmann, 1907, p. 530; not *Anomotaenia macracantha* Fuhrmann, 1908); (iii) *southwelli* Fuhrmann, 1932 (syn. *Monopylidium macracanthum* sensu Maplestone & Southwell, 1922, *nec* Fuhrmann, 1907); (iv) *guiarti* Shen-Tseng, 1932; (v) *Monopylidium macracanthum* sensu Shen-Tseng, 1932, *nec* Furhmann, 1907 (to be compared with *Anomotaenia macracantha* Fuhrmann, 1908 before renaming it); (vi) *lobivanelli* Burt, 1940; and, as atypical species from Brazil: (vii) *rostellata* Fuhrmann, 1908; (viii) *macracantha* Fuhrmann, 1908 (syn. *Anomotaenia macracantha* Fuhrmann, 1908; not to be mistaken for *Choanotaenia macracantha* Fuhrmann, 1908). *Monopylidium stercorarium* Baylis, 1919, re-examined, belongs to *Laritaenia*.

Spasskytaenia Oshmarin, 1956. Spasskaya & Spasskii (1978) accepted Oshmarin's (1956) genus but the validity of Oshmarin's paper, according to the *ICZN*, is doubtful. My diagnosis is based: (i) on scoleces and hooks from the Muséum d'Histoire Naturelle, Geneva, not belonging to Krabbe's type; (ii) on Dubinina's (1953) and Belopolskaya's (1977) descriptions of *platyrhyncha*, to which Spasskaya & Spasskii's (1971) may be added with some reservation, but leaving out Spasskii & Kornyushin's (1971) description; (iii) on the description of two congeneric species, namely *ochropodis* Neslobinsky, 1911 and *tringae* Spasskaya & Spasskii, 1978 (*nec* Burt, 1940). See also the note on *Dilepis* Weinland, 1858.

Ovosculpta n. g. Other synonyms of the type-species have to be added: *Choanotaenia scolopacina* Lopez-Neyra, 1944; *C. cayennensis* Lopez-Neyra, 1952; *Paricterotaenia scolopacis* (Joyeux & Baer, 1939) Sandeman, 1957; *Fuhrmannolepis scolopacina* (Lopez-Neyra, 1944) Spasskaya & Shumilo, 1971.

Spasspasskya n. g. The type-species was first described as *Anomotaenia*, then allocated to *Pseudanomotaenia* by Matevosyan (1963) and to *Monopylidium* by Spasskaya & Spasskii (1977).

Spreotaenia Spasskii, 1969. The genus was formerly ascribed to Metadilepididae.

Rallitaenia Spasskii & Spasskaya, 1975. First valid, basic redescription of the type-species by Dollfus (1934b) as *Liga gallinulae* (van Beneden, 1858). Some other redescriptions seem to refer to different species.

Liga Weinland, 1957. *Taenia punctata* Weinland, 1856 is homonymous with *T. punctata* Rudolphi, 1802 and consequently must be replaced, irrespective of its change of genus. This explains the new name *ransomi* Spasskii & Reznik, 1966. Synonyms of *ransomi* are: *L. punctata* (Weinland, 1856) Weinland, 1957; *L. brasiliensis* sensu Ransom, 1909 *nec* Parona, 1901 and Fuhrmann, 1907. Polar processes (knobs) mentioned in earlier diagnoses have never been observed in the eggs.

Dictymetra Clark, 1952. *D. numenii* Clark, 1952 is the junior, secondary homonym of *D. numenii* (Owen, 1949) (syn. *Choanotaenia numenii* Owen, 1949) and must be renamed. This explains the new name *radiaspinosa* Matevosyan, 1963 for the type-species. A pore bearing rows of radial spinules at the rostellum apex could not be observed in any of the examined species, so that this character is left out of the generic diagnosis.

Monopylidium Fuhrmann, 1899. In this genus, it is unlikely that the discontinuous film round the eggs in the parenchyma is the structure old authors considered as a uterine capsule, because it is very difficult to observe. It is more likely that the early, firm labyrinth wall of certain species, adhering to the eggs, was taken for a uterine capsule. The type-specimens of *M. musculosum* Fuhrmann, 1896 (and 1899) are a medley of histological slides. Identification can be achieved only by resorting to the first reliable redescription by Joyeux & Timon-David

(1934), confirmed by Joyeux & Baer (1955), whose specimens are deposited at the Museum d'Histoire Naturelle, Geneva. See also the notes on *Monosertum* n. g. and *Choanotaenia*. *Icterotaenia* Railliet & Henry, 1909 is synonymous with *Monopylidium*. Re-examination of the oldest extant material of the type-species, *galbulae* Gmelin, 1790 from *Oriolus galbula*, labelled by Fuhrmann and now at the Muséum d'Histoire Naturelle, Geneva, No. 106/41–42, and of material of Joyeux & Baer (1955), proves that it has two rows of rostellar hooks. Also, the development of the uterus and the anatomy is similar to that of *M. musculosum*.

Aelurotaenia Cameron, 1928. This genus is only provisionally ascribed to the Dilepididae. The rostellar apparatus is similar to that of the Metadilepididae, but the ventral uterus and the hosts do not fit with the diagnostics of this family. The hooks are similar to those of some Taeniidae.

Bancroftiella Johnston, 1911. The genus, first described in marsupials but not subsequently found in these hosts, is very similar to genera from birds. Later identifications in non-piscivorous birds are wrong and others are dubious. A comparison with *Glossocercus* and *Tubanguiella* is needed.

Kowalewskiella Baczynska, 1914. *K. longiannulata* Baczynska, 1914 is a synonym of *K. cingulifera* (Krabbe, 1868).

Ascodilepis Guildal, 1960. The type-species was ascribed to *Paradilepis* by Baer & Bona (1960).

Kotlanolepis Murai & Georgiev, 1987. The type-species, *Dilepis yorkei* Kotlan, 1923 was ascribed to *Megacirrus* by Spasskii (1965b).

Acanthocirrus Fuhrmann, 1907. *A. macrorostratus* Fuhrmann, 1907 is a synonym of the type-species.

Cuculincola n. g. The type-species, *mutabilis* Rudolphi, 1819, was ascribed to *Anomotaenia* by Fuhrmann (1907a).

Multitesticulata Meggitt, 1927. *M. aegyptica* Meggitt, 1927 is synonymous with *M. filamentosa* (Goeze, 1782) (see Baer, 1927).

A number of genera have, for various reasons, been omitted from this key.

Old larval genera

These names have no exact taxonomic meaning and can be used, at most, to indicate collective groups of genera with similar life-cycles but not necessarily systematically related. *Cryptocystis* Villot, 1882 and *Gryporhynchus* von Nordmann, 1832 (see notes on *Neogryporhynchus*) are in this category.

Nomina nuda

Choanotaenia (Cholodkovskya) Matevosyan, 1953 [synonymous neither with *Anomotaenia* as stated by Matevosyan (1963) nor with *Panuwa* (now synonymous with *Megalacanthus*) as stated by Schmidt (1986)].

Fibrocirrosa Palladwar & Kalyankar, 1980 in Passeriformes.

Genera inquirenda

Capsulana, Laiqia and *Reticulatea*, all in Passeriformes and erected by Khanum, Farooq & Jamal (1982), have been criticized by Spasskii (1989), whose comments are supported here. *Laiqia* was ascribed to the Anoplocephalidae but more resembles the Davaineidae.

Copesoma Sinitzin, 1896 in Charadriiformes and *Tetracisdicotyle* Fuhrmann, 1907 in Ciconiiformes are both old names which are now obsolete. No specimens have been recovered.

Diplochetos von Linstow, 1906, parasitic in Charadriiformes, was considered synonymous with *Anomotaenia* by Fuhrmann (1908b) and Schmidt (1986) and with *Dictymetra* by Spasskaya & Spasskii (1973). It is not known whether Fuhrmann (1908b) revised von Linstow's material in order to write 'die Beschreibung ist ganz falsch'. However, it is indisputable that the original description of *volvulus* von Linstow, 1906, the type-species, states that it has double copulatory organs and simple gonads (von Linstow, 1906). This anatomy is not characteristic of the Dilepididae but of *Diploposthe*, a parasite of Anseriformes. Possibly, there was a mistake in the host group rather than an erroneous description. In this case, *Diplochetos* cannot be considered synonymous with *Anomotaenia* or *Dictymetra* but should be regarded as a *genus inquirendum*.

Yogeshwaria Shinde, 1968 from Charadriiformes (a senior homonym of *Yogeshwaria* Chincholikar & Shinde, 1976 from Selachii) with *malabarica* Shinde, 1968 as type-species is unidentifiable. There was, however, further confusion created by Shinde's (1972) description of *Lapwingia malabarica* as a new species. The same specific name can be used in different genera, but it is surprising that the description of *malabarica* (1972) exactly corresponds to that of *malabarica* (1968), as is proved, amongst other things, by the drawings published in 1972 in Plate 1, which are the same as the 1968 ones. Either this is a simple typewriting mistake so that in the 1972 paper, instead of *Lapwingia*, we must read *Yogeshwaria* (but, if so, *malabarica* (1972) is not a new species), or we can infer that Shinde himself considered *Yogeshwaria* synonymous with *Lapwingia* Singh, 1952, having transferred to it *malabarica* (type-species of the former), without mentioning such synonymy. However, from Shinde's text (impossible to get the material), *Yogeshwaria* is likely to be different from *Lapwingia* (here placed amongst the possible synonyms of *Dictymetra*) and has to be considered as *genus inquirendum*, pending a redescription. Every attribution of species to both genera must be revised.

Johria Yamaguti, 1959 from Charadriiformes. The description of the type-species, *Eugonodaeum testifrontosa* Johri, 1934, is absolutely inadequate. Impossible to write a generic diagnosis. It is a *genus inquirendum*.

Spasskofuhrmina Palladwar & Kalyankar, 1989 from Passeriformes probably does not belong to the Anoplocephalidae, but possibly to the Davaineidae. It has some likeness to *Laiqia*; both are said to be parasites of Passeriformes, unusual hosts for Anoplocephalidae. It is poorly described and is treated as a *genus inquirendum*.

Humiditaenia Bhalya & Capoor, 1986. Interesting material from a lizard (*Hemidactylus flaviviridis*) but difficult to interpret. It could be a progamic form, aberrant and maybe accidental, of some larva intended for a carnivore. Its description is inadequate about several important characters, namely the rostellar structure, uterus, eggs, cirrus, seminal vesicles (present or absent); it is not even possible to ascribe it to a family. It is treated here as a *genus inquirendum*. According to the authors, there is some likeness to *Joyeuxiella* (Dipylidiidae), to *Echinorhynchotaenia* (Hymenolepididae) and to *Ophiovalipora* (= *Valipora*), which has in common with it only the inversion of the osmoregulatory canals. It is difficult to understand why the genus was ascribed to 'Dilepidinae'. Moreover, Dilepididae from reptiles are not known [*Ophiovalipora* must be explained by a mistaken identification of the host; see Bona (1975)].

Thaparea Johri, 1953 from Charadriiformes has a rostellum without a pouch and a large internal seminal vesicle; both characters seem to exclude it from Dilepididae. To validate it, further details are needed on the pattern of the rostellar apparatus, on the position of the vagina relative to the cirrus-sac and on the initial position, ventral or dorsal, of the uterus. The uterus actually seems sacciform, but the meaning of 'egg-capsules' mentioned in the diagnosis is not explained. It might belong to the Metadilepididae but the host, a charadriiform, advises against allocating it to this family as long as its morphology is poorly known. Because it is poorly described, it is treated as a *genus inquirendum*.

Genus incertae sedis

Dactylopsilophilus Kugi, 1990 from marsupials. Type-genus of a new family, the Dactylopsilophilidae Kugi, 1990. An interesting form with a very singular combination of characters. Unfortunately, the discussion is not exhaustive. The only characters really distinguishing the new family from the Dilepididae are the internal and external seminal vesicles, whilst the anatomy as a whole, the type of pluriovular uterine capsules as far as can be seen in Kugi's (1990) Fig. 7 and the armed suckers lead us to consider it as a davaineid with a much reduced apical apparatus, even if its very small sucker spines are not exactly like those of the Davaineidae. We lack exact data on the structure of the small apical apparatus, on the uterus and uterine capsules, on the egg coats, on the position of the genital ducts with respect to the osmoregulatory canals, as well as a drawing of the cirrus-sac and its seminal vesicle. Anyway, the genus is valid; an important element confirming it is the host, a marsupial. As we cannot yet know with what family it is most closely related, we place it in *genera incertae sedis*. Here is the original diagnosis: 'Order Dilepididea, family Dactylopsilophilidae. Scolex with an apical organ. Rostellum absent. Suckers spinose. Genital pores unilateral. Testes numerous, situated bilaterally to ovary. Internal and external seminal vesicles and seminal receptacle present.' Type-species *D. australiensis* Kugi, 1990.

Family Dilepididae Railliet & Henry, 1909

Diagnosis: Apical apparatus usually muscular or musculo-glandular, with rostellar pouch and armed rostellum*; glandular pouch, unarmed or feebly armed, less frequent; apical apparatus rarely absent. Rostellum may be differently structured. Rostellar pouch may be replaced by fibrous muscular mass, or very seldom lacking. Suckers usually unarmed, occasionally with hooklets on border or bearing very small thick spines, not of davaineoid pattern. Hooks very diversified in shape and size; one or two crowns, regular or irregular, seldom zigzagging. Complete strobilae extremely small to large. Reproductive organs usually single per proglottid†. Ovary bilobed, rarely a single mass. Cirrus usually armed. Vitellarium postovarian‡. Uterus ventral, exceptionally dorsal; initially reticulate, labyrinthine or sacciform; either persistent or changing structure during maturation, or splitting into monovular or polyovular uterine capsules, or disappearing, leaving eggs loose in parenchyma. Genital ducts dorsal to or between osmoregulatory canals, rarely ventral§. True external and internal seminal vesicles usually absent**. In birds (predominantly) and mammals. Type-genus *Dilepis* Weinland, 1858.

Key to genera

1a. Parasites of birds .... **2.**
1b. Parasites of mammals .... **105.**

2a. Huge spherical pseudoscolex cutting off scolex in adult strobilae .... ***Apokrimi* Bona, 1994. (Fig. 25.1)**
Diagnosis: Rostellar apparatus muscular, delicate. Pouch and rostellum very small. Hooks unknown. Strobila and scolex very small. Neck forms glandulo-muscular, sucker-like pseudoscolex. Genital pores irregularly alternate, far anterior. Genital ducts between osmoregulatory canals (most commonly) or dorsal, in same strobila. Vagina posterior to cirrus-sac, in same horizontal plane, thick-walled. Uterus reticulate, then labyrinthine. Atrium small; no ductus masculinus. Ovary with deep isthmus; compact. Testes numerous, postovarian. Cirrus-sac small, subspherical. Cirrus short with fine bristle-like spines, forming

* An unarmed, true rostellum was ascertained only in a species (unpublished) seemingly belonging to *Krimi*, in which the rostellum is usually armed.
† Double set only in *Eurycestus* and *Mirandula*.
‡ Only exception, *Multitesticulata*.
§ Ventral, with reservation, in *Dilepidoides* and rarely in some proglottids of *Himantaurus* and *Hunkeleria*.
** Except in *Proorchida*.

small, narrow, straight bundle when withdrawn. In Piciformes, Picidae. Neotropical (Peruvian Amazonas). Type-species *A. pseudoscolecis* Bona, 1994.

2b. No real pseudoscolex (simple widening of the neck is not pseudoscolex) .... 3.

3a. Strobila aberrant, enormously wider than long. Scolex very precociously detached. Two sets of reproductive organs per proglottid. (In young strobilae, pores may alternate regularly) .... *Eurycestus* Clark, 1954. (Fig. 25.2)

Diagnosis: Rostellar apparatus musculo-glandular, delicate. Rostellum small. Hooks in two circles, few (16). Proglottids few, acraspedote, disproportionately wider than long; pore at anterior end of proglottid; organs transversely extended. Scolex very small, early isolated amongst villi. Suckers very small with two rows of hooklets anteriorly. Longitudinal osmoregulatory canals absent. Vagina posterior to cirrus-sac, in same horizontal plane, very long. Uterus reticulate, then sacciform, deeply lobulate. Eggs small with long polar processes (in every species?). Atrium anteriorly directed. Ovary narrow. Testes numerous, post-ovarian, along posterior margin of proglottid mainly poral to vitellarium. Cirrus-sac oval or very long, narrow, bent forwards distally. Cirrus long, strong, spiny. Seminal receptacle very long, narrow, sinuous. Vas deferens very small. In Charadriiformes. Holarctic. Type-species *E. avoceti* Clark, 1954.

3b. Strobila normal. Scolex attached. One set of reproductive organs per proglottid .... 4.

4a. Internal seminal vesicle small, external large, spherical. Hooks triangular, blade claw-shaped .... *Proorchida* Fuhrmann, 1908. (Fig. 25.3)

Diagnosis: Apical apparatus unknown. Hooks in two circles. Strobila medium; velum expanded; prominent, stable papilla. Genital pores unilateral, right (type-species). Genital ducts between osmoregulatory canals. Vaginal orifice dorsal to cirrus-sac. Vagina very small, developing into seminal receptacle just outside atrium wall. Uterus sacciform. Atrium spherical; base flat, spiny. Ovary very small, bilobed, hardly lobular. Testes few, surrounding ovary and dorsal to it. Cirrus-sac delicate. Cirrus very thin, short, unarmed. Seminal receptacle very large, elongate, reaching far posterior. In Ciconiiformes, Ardeidae. Brazil. Type-species *P. lobata* Fuhrmann, 1908.

4b. No true seminal vesicles (dilated but coiled internal and external vas deferens are not true vesicles). Hooks very seldom triangular .... 5.

Fig. 25.1 *Apokrimi pseudoscolecis* Bona, 1994, type. Pseudoscoleces. [For proglottid anatomy, see *Krimi* Figs 25.189-25.192.]

Fig. 25.2 *Eurycestus avoceti* Clark, 1954. Very young specimens already without scoleces.

Fig. 25.3 *Proorchida lobata* Fuhrmann, 1908, type, mature proglottid. (From Bona, 1975.)

Fig. 25.4 *Arostellina reticulata* Neiland, 1955, type. Mature proglottid, the reticulate uterus forms exactly under the reticulate ovary.

Fig. 25.5 *Ethiopotaenia trachyphonoides* Mettrick, 1961, type. Uterus surrounded with condensed parenchyma.

Figs 25.6, 25.7 *Arctotaenia tetrabothrioides* (Lönnberg, 1890). 25.6. Scolex. 25.7. Mature proglottid.

Fig. 25.8 *Cotylorhipis furnarii* (Del Pont, 1906). Mature proglottid.

25.1

25.2

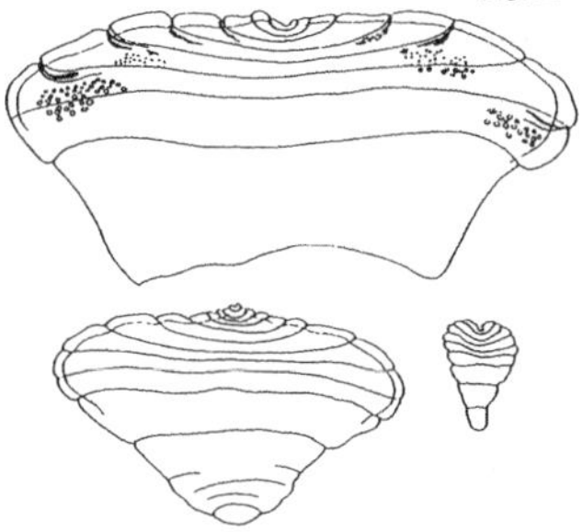

25.3

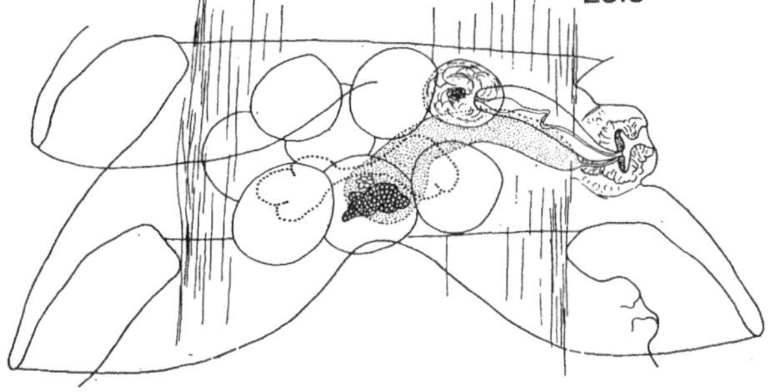

25.4

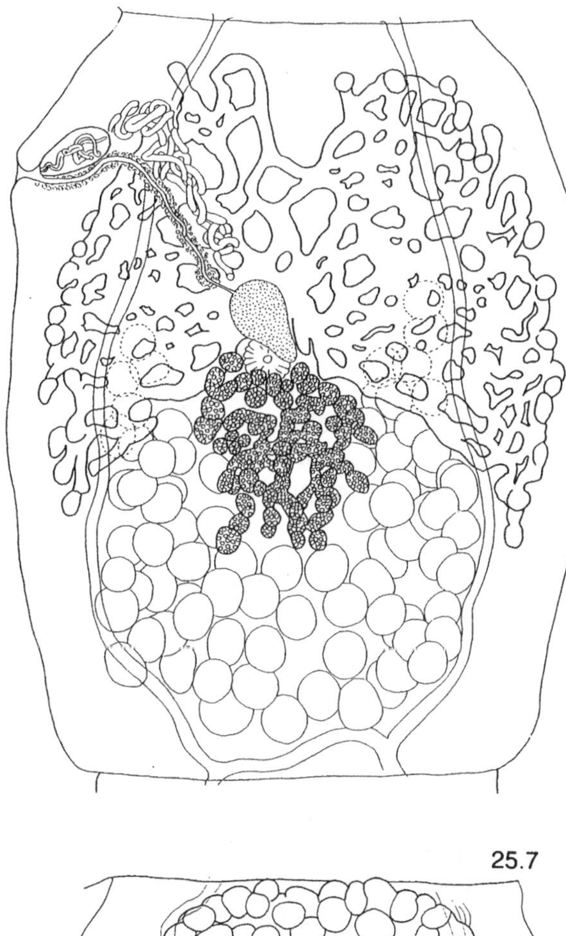

25.5

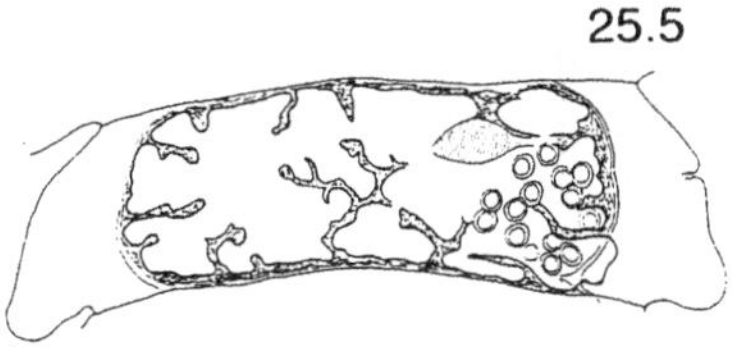

25.6

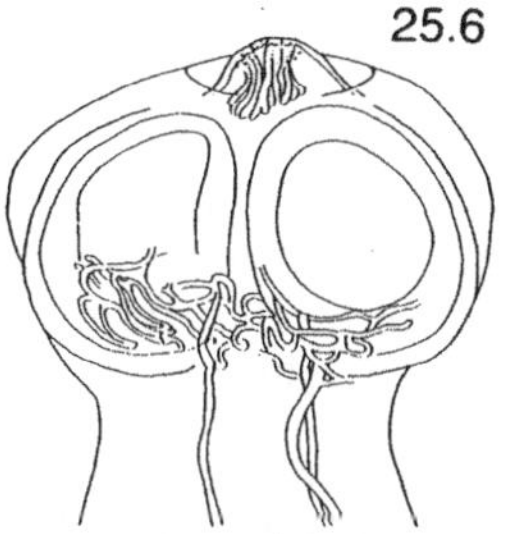

25.7

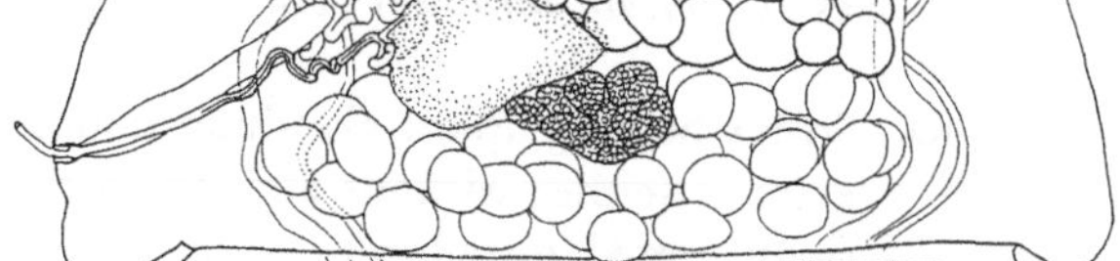

25.8

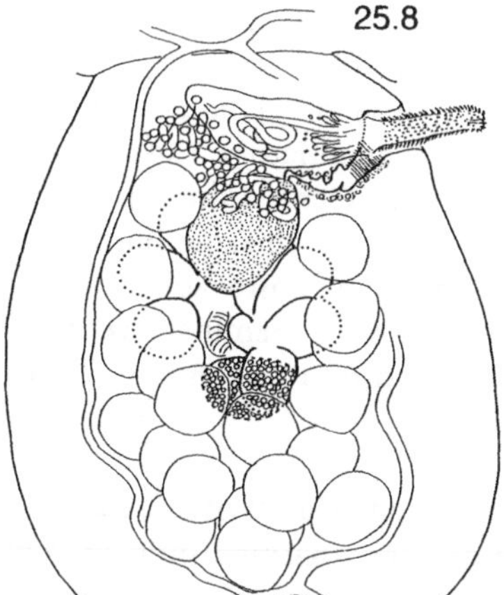

5a. Scolex without apical apparatus (may have vestigial, hardly visible structure) .... 6.

5b. Scolex with apical apparatus either glandular or muscular .... 8.

6a. Genital ducts dorsal to osmoregulatory canals .... ***Arostellina*** **Neiland, 1955. (Fig. 25.4)**

Diagnosis: Apical apparatus absent. Strobila small; proglottids acraspedote. Scolex narrower than neck; no posterior demarcation. Genital pores irregularly alternate. Vagina posterior to cirrus-sac in same horizontal plane, curving backwards. Uterus reticular, underlying ovarian reticulum, becoming labyrinthine. No uterine capsules. Atrium simple. Ovary very large, persistent reticulum, crossing osmoregulatory canals ventrally. Vitellarium reticulate; loose, small lobules. Testes very numerous, posterior to female glands. Cirrus-sac very small. Cirrus apparently unarmed. Vas deferens small, anteroporal, crossing osmoregulatory canals, following vagina backwards. In Trochiliformes. Central America. Type-species *A. reticulata* Neiland, 1955.

6b. Genital ducts between osmoregulatory canals .... 7.

7a. Pores irregularly alternate, but with long unilateral series. Ovary bilobed, lobules irregular. Uterus at first sacciform. In Piciformes, Capitonidae .... ***Ethiopotaenia*** **Mettrick, 1961. (Fig. 25.5)**

Diagnosis: Apical apparatus absent (apical sucker-like structure cannot be excluded). Strobila medium. Genital ducts between osmoregulatory canals. Vagina posterior to cirrus-sac, in same horizontal plane; straight, not very long. Uterus with large irregular lobules surrounded by compressed parenchyma. Eggs eventually loose in parenchyma, no uterine capsules. Atrium inconspicuous. Vitellarium far posterior. Testes numerous; initially in two posterolateral fields, then posterior and lateral to ovary, scarce behind vitellarium. Cirrus-sac small. Cirrus small, probably unarmed. Vas deferens small, with thick prostatic cells. In Capitonidae (frugivorous Piciformes). Africa. Type-species *E. trachyphonoides* Mettrick, 1961.

7b. Pores usually unilateral, left or right, at times irregularly alternate. Ovary not bilobed, very large, at first ramified, then with heaped lobules. Uterus labyrinthine. In Charadriiformes .... ***Arctotaenia*** **Baer, 1956. (Figs 25.6, 25.7)**

Diagnosis: Apical apparatus proper absent; small mass of glandular cells at scolex apex. Strobila medium; pore posterior. Osmoregulatory canals form reticulum behind scolex; dorsal slightly aporal to ventral. Genital ducts between osmoregulatory canals. Vagina slightly posterior to cirrus-sac, in same horizontal plane; long, narrow. Atrium small, simple. Testes numerous; postovarian, porally close to seminal receptacle and vagina. Cirrus-sac long, narrow. Seminal receptacle small. Vas deferens small, posterior to ovary. In Charadriiformes. Arctic. Type-species *A. tetrabothrioides* (Lönnberg, 1890).

8a. Scolex bearing apical glandular pouch with small anterior cavity opening to exterior (either without any structure inside or with unarmed bulb or very small armed rostellum) .... 9.

8b. Scolex bearing apical muscular rostellar apparatus, usually with rostellar pouch, seldom without; rostellar pouch with or without visible glandular component .... 18.

9a. Glandular pouch only with simple anterior cavity .... 10.
9b. Glandular pouch with anterior cavity and small unarmed bulb at base of cavity .... 14.
9c. Glandular pouch with anterior cavity and small, armed, rostellum-like structure at base of cavity .... 15.

10a. Suckers armed with hooklets .... *Cotylorhipis* Blanchard, 1909. (Fig. 25.8)

Diagnosis: Apical apparatus glandular, unarmed. Pouch spherical with small apical cavity and hardly visible spherical structure at base, simulating bulb. Strobila very small (2–6 mm). Proglottids weakly craspedote. Scolex very small. Suckers normally muscular, like flat discs; posterior rim more developed; posterior hooklets large, in one row, anterior ones in three to four rows. Genital pores irregularly alternate, far anterior. Genital ducts dorsal to, exceptionally between, osmoregulatory canals. Vagina short, funnel-shaped, sinuous with robust wall, posterior to cirrus-sac, orifice ventral. Uterus coarse-meshed reticulum then sacciform, deeply septate. Atrium simple. Ovary small, few lobules. Testes numerous, posterior, lateral and anterolateral to ovary, without surrounding it. Cirrus-sac large, in front of all other organs. Cirrus large with strong rosethorn spines. Vas deferens posterior and partly ventral to cirrus-sac. In Passeriformes, Furnariidae and Dendrocolaptidae. Neotropical. Type-species *C. furnarii* (Del Pont, 1906).

10b. Suckers unarmed .... 11.

11a. Genital ducts between osmoregulatory canals .... 12.
11b. Genital ducts dorsal to osmoregulatory canals .... 13.

12a. Cirrus armed with small spines. Atrium small, simple; no ductus masculinus .... *Pseudochoanotaenia* Burt, 1938. (Figs 25.9–25.11)

Diagnosis: Apical apparatus glandular, unarmed. Pouch with small apical cavity only. Strobila very small. Genital pores irregularly alternate. Genital ducts between osmoregulatory canals. Vagina posterior to cirrus-sac, usually dorsal, sometimes in same horizontal plane or ventral; short, pyriform; with thick wall proximally. Uterus delicate reticulum, never observed fully ripe. Eggs probably scattered in parenchyma without true uterine capsules. Ovary small, transversely elongate. Testes numerous, postovarian. Cirrus-sac long, narrow. Cirrus spiny. Vas deferens more anterior than ovary, median, with dense glandular sleeve. In Apodiformes. Sri Lanka, Africa. Type-species *P. collocaliae* Burt, 1938.

12b. Cirrus armed with long bristle-like spines. Short ductus masculinus present (vaginal orifice being more distal than male one), lined with long, thin spines continuous with cirrus bristles .... *Unciunia* Skryabin, 1914. (Figs 25.12, 25.13)

Diagnosis: Apical apparatus glandular, unarmed. Pouch with small apical cavity

25.9

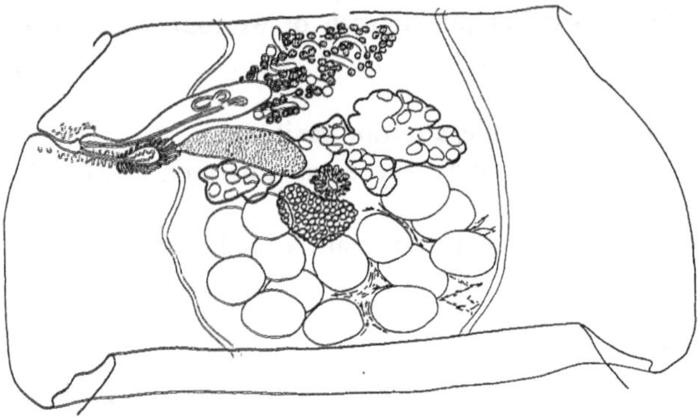

25.10

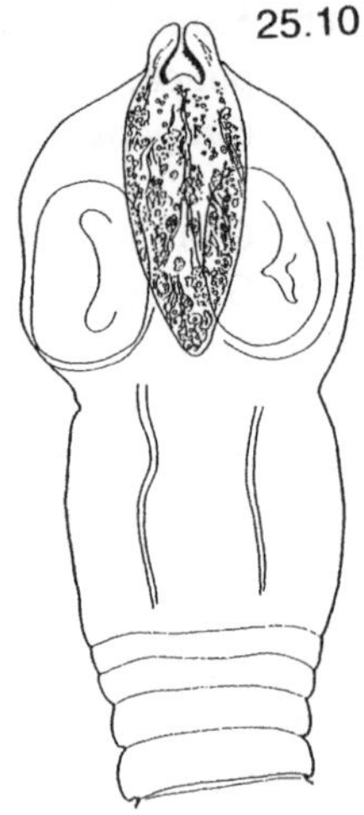

25.11

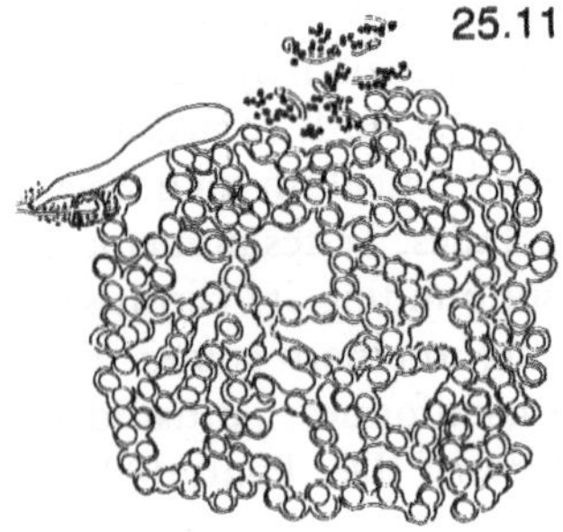

25.13

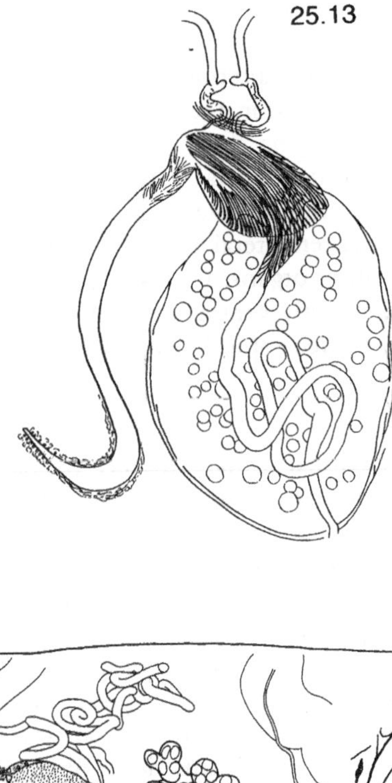

25.12

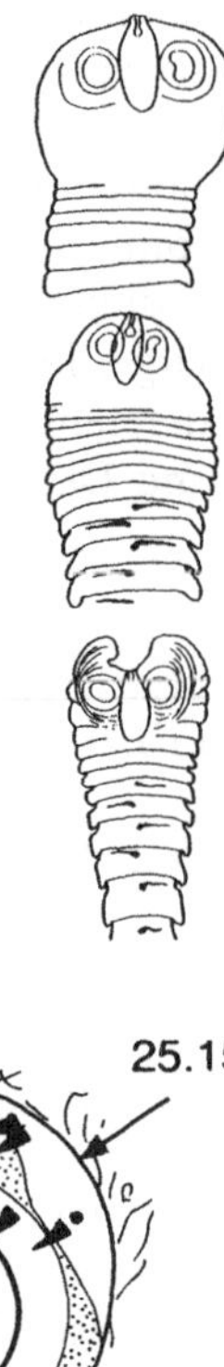

25.14

25.15

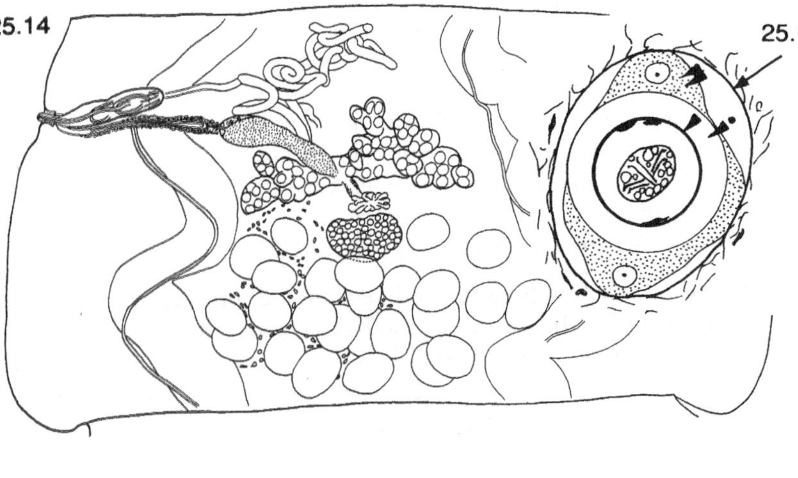

only. Strobila small. Neck often swollen. Genital pores irregularly alternate. Genital ducts between osmoregulatory canals. Vagina posterior to cirrus-sac, in same horizontal plane or slightly dorsal; short, diverging from cirrus-sac. Uterus delicate reticulum, then labyrinthine. Eggs eventually loose in parenchyma. Atrium proper small. Ovary small. Testes numerous, postovarian. Cirrus-sac small, subspherical. Cirrus large, delicate; long, dense, bristle-like spines basally, shorter, filiform ones distally; wide conical tuft engaged into ductus masculinus when cirrus withdrawn. Vas deferens farther on than ovary, nearly median. In Accipitriformes. Neotropical, South Africa, Palaearctic. Type-species *U. trichocirrosa* Skryabin, 1914.

13a. Uterus delicately reticulate, replaced by late mature monovular, membranous uterine capsules. Ovary small, transversely extended, median; few oocytes. In Passeriformes .... *Ptilotolepis* Spasskii, 1969. (Figs 25.14, 25.15)
(Syn. *Rostellina* Rathore & Nama, 1984.)

Diagnosis: Apical apparatus glandular, unarmed. Pouch with small apical cavity only. Strobila small to medium. Genital pores irregularly alternate. Genital ducts dorsal to osmoregulatory canals. Vagina posterior to cirrus-sac, orifice ventral, course approximately in same horizontal plane; slightly wider with thicker wall proximally. Atrium small. Ovary far from anterior margin of proglottid. Testes numerous, postovarian. Cirrus-sac small, elongate. Cirrus apparently unarmed. Seminal receptacle elongate, transverse, in front of poral lobe of ovary. Vas deferens more anterior than ovary, median, isolated. In Passeriformes. Australia. Type-species *P. meliphagidarum* (Johnston, 1911).

13b. Uterus sacciform, deeply lobulate. Ovary very large, bulky, close to anterior proglottid rim. In Anseriformes .... *Platyscolex* Spasskaya, 1962. (Figs 25.16, 25.17)

Diagnosis: Apical apparatus glandular, unarmed. Pouch with small apical cavity only. Strobila medium. Scolex with fleshy excrescence along posterior edge. Genital pores irregularly alternate. Genital ducts dorsal to osmoregulatory canals. Vagina posterior to cirrus-sac, probably ventral; long. Sphincter near atrium. Eggs with polar processes. Atrium wide, sucker-like, very fine 'cilia' at base. Ovary extending ventrally beyond osmoregulatory canals, aporal lobe much larger, extending farther porally in front of poral lobe. Testes very numerous, small, postovarian. Cirrus-sac elongate, narrow. Cirrus finely spinous. Vas deferens

---

Figs 25.9–25.11 *Pseudochoanotaenia collocaliae* Burt, 1938, type. 25.9. Mature proglottid. 25.10. Scolex. 25.11. Detail of immature uterus.

Figs 25.12, 25.13 *Unciunia trichocirrosa* Skryabin, 1914, type. 25.12. Scoleces with swollen necks. 25.13. Terminal genitalia. [For apical apparatus, see *Pseudochoanotaenia* Fig. 25.10 and for proglottid anatomy see *Choanotaenia* Fig. 25.185.]

Figs 25.14, 25.15 *Ptilotolepis meliphagidarum* (Johnston, 1911), type. 25.14. Mature proglottid. 25.15. Egg coats and uterine capsule (schematic); arrow: uterine capsule, double triangle: egg outer coat, single triangle: embryophore, triangle with spot: outer layer of inner coat. [For glandular apical apparatus, see *Pseudochoanotaenia*, Fig. 25.10.]

posterior to poral extension of aporal lobe of ovary. In Anseriformes. Palaearctic. Type-species *P. ciliata* (Fuhrmann, 1913).

14a. Suckers densely covered with punctiform spines. Ovary delicate not very large, lobes longitudinally elongated; narrow isthmus. Eggs with thin embryophore, cap-shaped polar masses and thin polar processes; eventually loose in parenchyma, transformed into small granular cells .... *Eburneotaenia* n. g. (Figs 25.18, 25.19)

Diagnosis: Apical apparatus glandular, unarmed. Pouch long; small bulb at base of apical cavity. Strobila small. Proglottids longer than wide. Scolex with subtegumental glands. Osmoregulatory canals form fine network behind scolex. Genital pores irregularly alternate. Genital ducts between osmoregulatory canals. Vagina posterior to cirrus-sac, in same horizontal plane; straight, fairly wide, backwardly inclined. Uterus labyrinthine. Atrium simple. Ovary symmetrical. Testes numerous, postovarian. Cirrus-sac very small. Cirrus strong, unarmed. Vas deferens more anterior than ovary, median, isolated. In Passeriformes. Africa. Type-species *E. eburnea* (Mariaux & Vaucher, 1988) n. comb. (syn. *Pseudochoanotaenia eburnea* Mariaux & Vaucher, 1988).

14b. Suckers unarmed. Ovary massive, filling anterior half of medulla. Eggs with thick embryophore, may have firm, reinforcing outer coat, eventually loose in parenchyma ... *Emberizotaenia* Spasskaya, 1970. (Figs 25.20–25.22)

Diagnosis: Apical apparatus glandular, unarmed. Pouch with bulb at base of small apical cavity. Strobila small to medium. Neck often laterally swollen. Proglottids wider than long. Scolex large, with subtegumental glands. Osmoregulatory canals form fine network behind scolex. Genital pores irregularly alternate. Genital ducts between osmoregulatory canals. Vagina posterior to cirrus-sac, ventral or dorsal; orifice in same horizontal plane; long, sinuous. Uterus labyrinthine, then sacciform, deeply septate. Atrium simple. Ovary aporal lobe

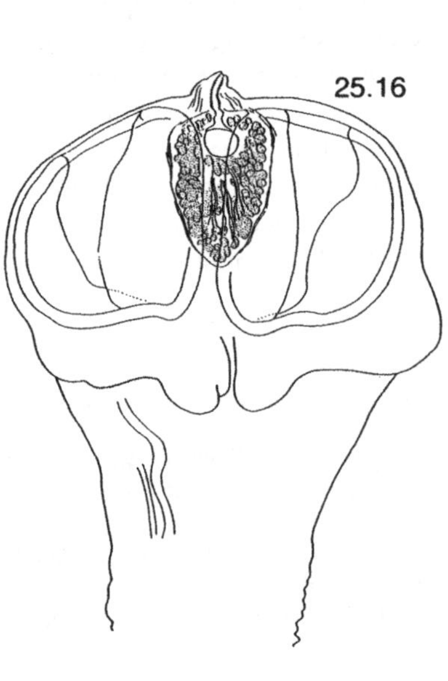

25.16

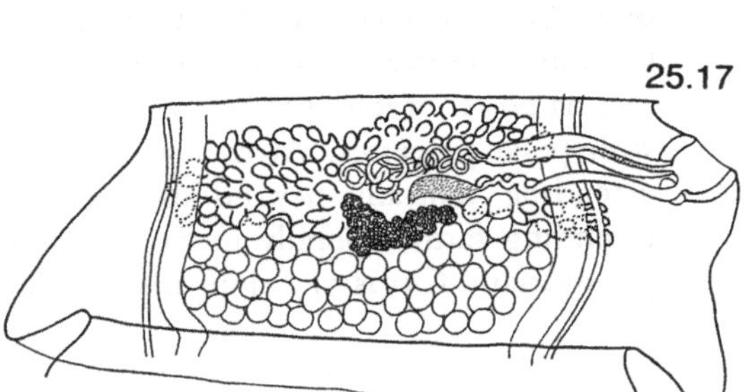

25.17

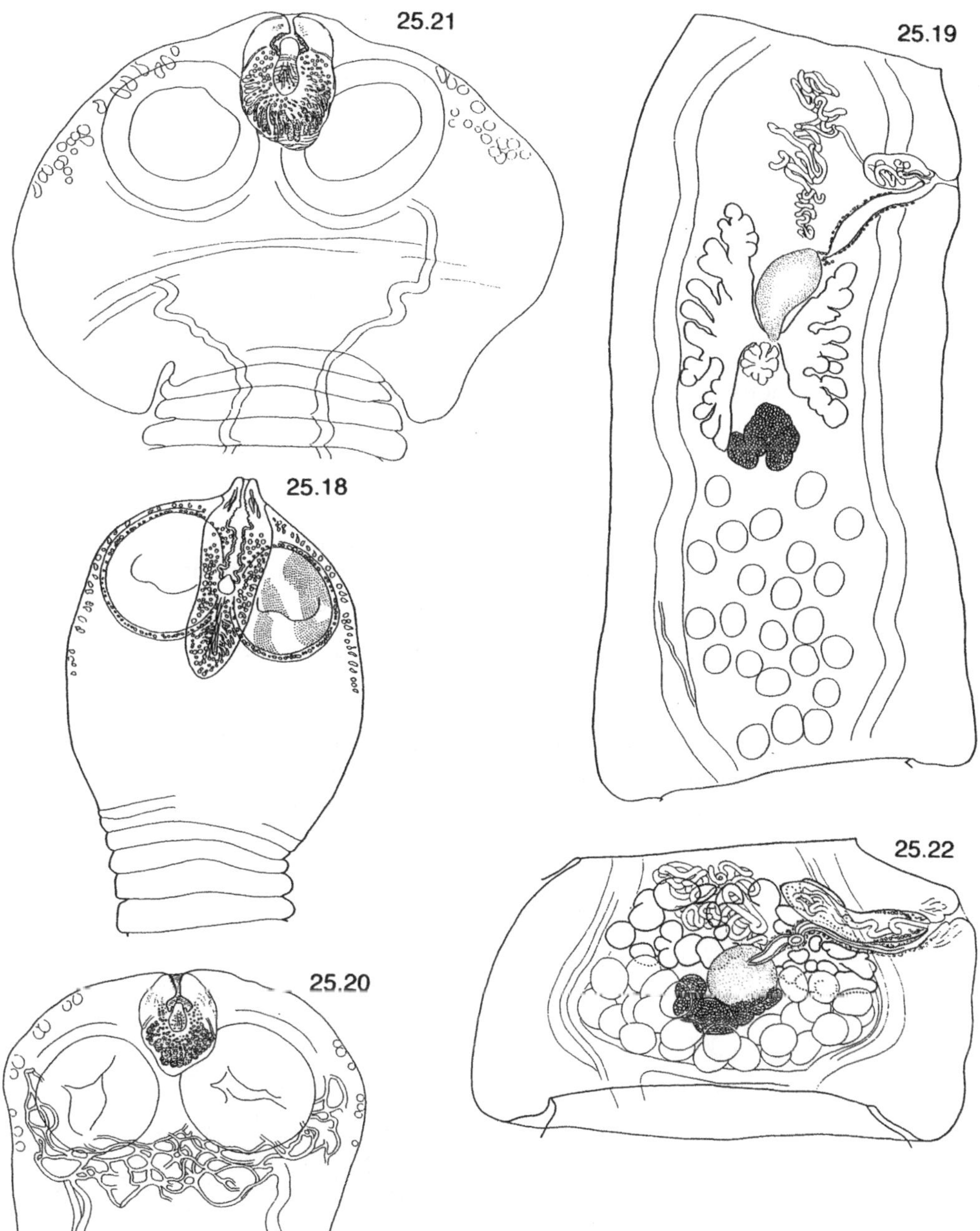

Figs 25.16, 25.17 *Platyscolex ciliata* (Fuhrmann, 1913). 25.16. Scolex, type (original). 25.17. Mature proglottid. (From Spasskaya (1962), slightly modified.)
Figs 25.18, 25.19 *Eburneotaenia eburnea* (Mariaux & Vaucher, 1988), type. 25.18. Scolex, spines partly drawn only in one sucker. 25.19. Mature proglottid.
Figs 25.20–25.22 *Emberizotaenia*. 25.20. *E. raymondi* (Gigon & Beuret, 1991) n. comb., type, scolex. 25.21. *E. chlamyderae* (Krefft, 1871) n. comb., type; scolex with enormously swollen neck. 25.22. *Emberizotaenia* sp., mature proglottid.

larger, often reaching anterior margin of proglottid. Testes very numerous, postovarian. Cirrus-sac more or less elongate. Cirrus long, finely spinous or unarmed (?). Vas deferens anterior to poral or aporal ovary lobe, overlapping it. In Passeriformes. Palaearctic, Africa, Australia, Sri Lanka. Type-species *E. reductorhyncha* (Spasskaya, 1957).

15a. Pores unilateral, right. Atrium large, strong, asymmetrical, deeper anteriorly, with prominent papilla. In Gaviiformes .... *Neovalipora* Baer, 1952. (Figs 25.23–25.26)

Diagnosis: Apical apparatus glandular, armed. Pouch small. Rostellum very small, at base of deep cavity. Hooks probably in two circles, very small. Strobila medium to large. Pore in middle of lateral margin. Scolex large. Suckers very large. Osmoregulatory canals form fine reticulum behind scolex. Genital ducts between osmoregulatory canals. Vagina posterior and ventral to cirrus-sac, straight, wide with pre-atrial sphincter. Uterus sacciform with large diverticula, then deeply septate. Ovary large, apparently not bilobed (with poral lobe extremely reduced?), with deep clavate digitations. Testes very numerous, postovarian. Cirrus-sac small, elongate. Cirrus unarmed. Vas deferens behind principal mass of ovary, dorsal to seminal receptacle, in front of poral testes. In Gaviiformes. Arctic. Type-species *N. parvispinae* (Linton, 1927).

15b. Pores irregularly alternate. Atrium simple or different in structure .... 16.

16a. Hooks in one circle. Testes few. Vagina club-shaped. (See also normal musculo-glandular rostellar apparatus, at 74a and 83a) .... *Glanduluncinata* n. g. (Figs 25.27, 25.28)

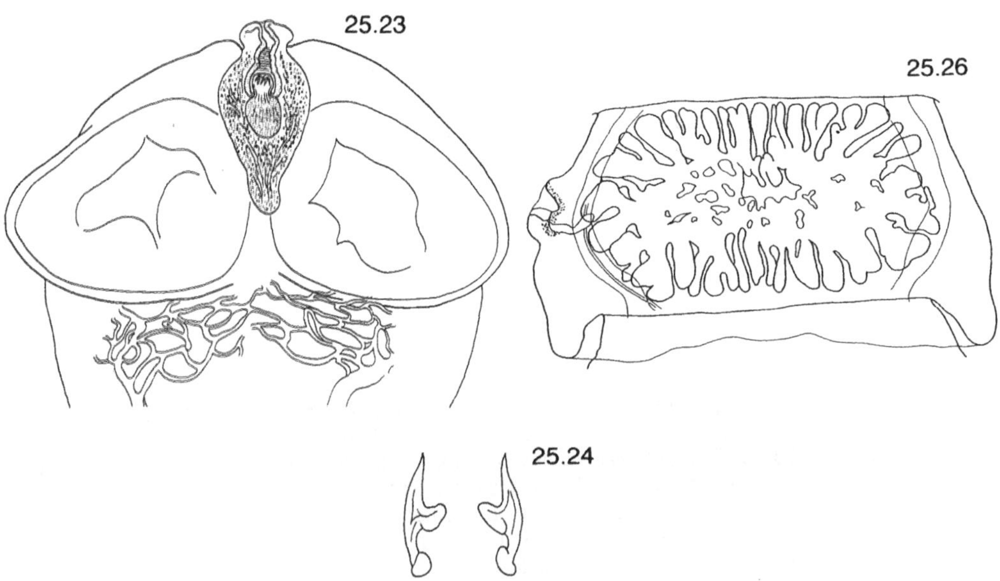

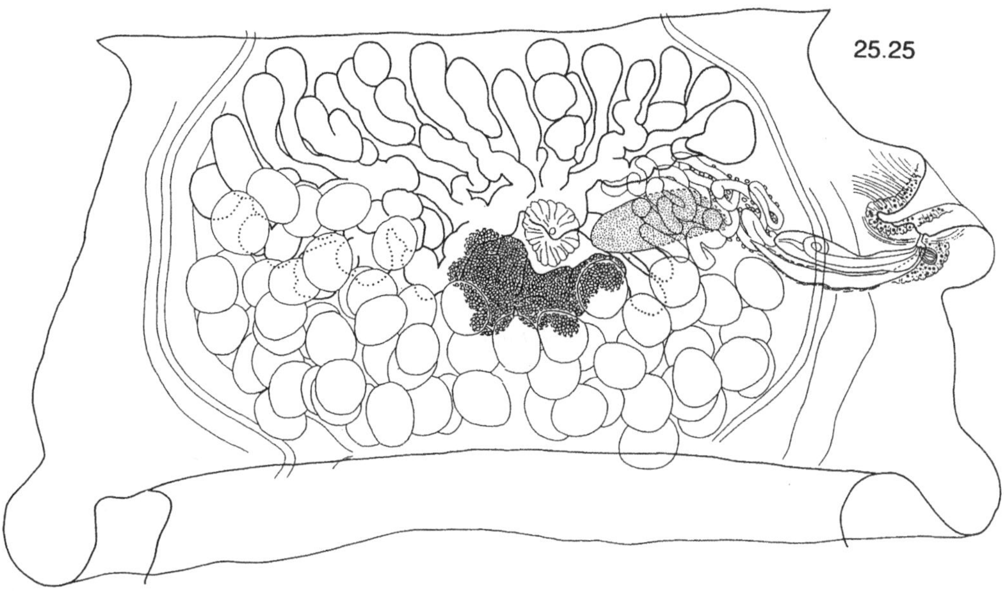

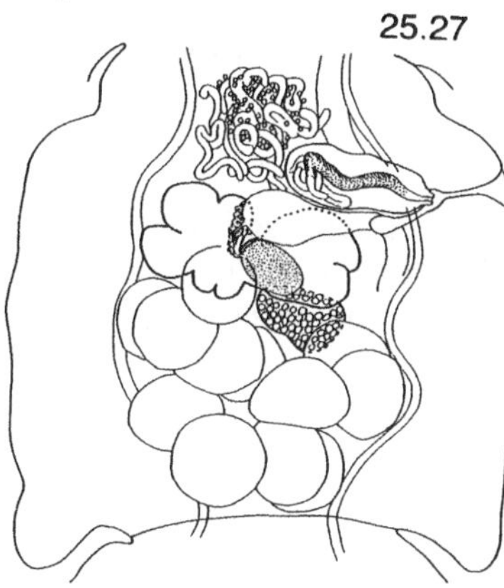

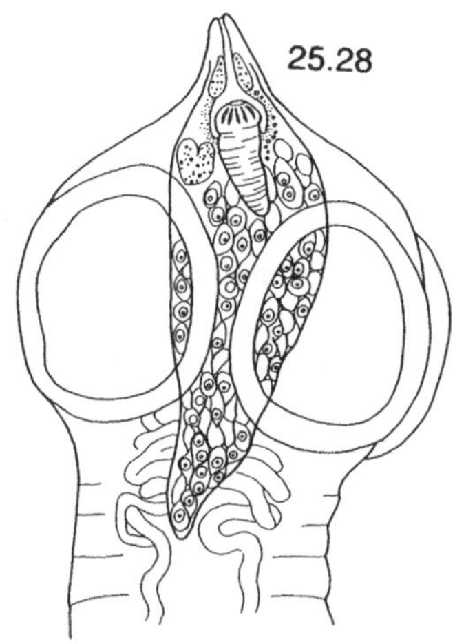

Figs 25.23-25.26 *Neovalipora parvispinae* (Linton, 1927). 25.23. Scolex. 25.24. Hooks. 25.25. Mature proglottid. 25.26. Gravid proglottid.
Figs 25.27, 25.28 *Glanduluncinata uncinata* (Fuhrmann, 1918). 25.27. Type; adult proglottid not fully mature [ovary might be 20% larger]. 25.28. Scolex, from Fuhrmann (1918), simplified.

Diagnosis: Apical apparatus glandular, armed. Pouch large with small apical cavity. Rostellum very small, normal structure. Hooks few (16), very small. Strobila very small. Suckers large. Genital pores irregularly alternate, may have long regular sequences. Genital ducts between osmoregulatory canals. Vagina posterior to cirrus-sac in same horizontal plane or dorsal; thick cellular cap at proximal end. Uterus labyrinthine, fully ripe uterus unknown. Atrium simple, deep. Ovary bilobed, compact. Testes postovarian. Cirrus-sac oval, wide. Cirrus strong with small, close spines. Vas deferens anterior, median. In Apodiformes. New Caledonia. Type-species *G. uncinata* (Fuhrmann, 1918) n. comb. (syn. *Choanotaenia uncinata* Fuhrmann, 1918).

16b. Hooks in two circles. Testes numerous. Vagina normal .... 17.

17a. Uterus reticulate, replaced by large polyovular capsules, containing packet of adhering eggs, with long polar processes. Ovary bilobed, large, reticulate, ramified. Testes postovarian, symmetrically arranged. In Passeriformes, Cinclidae. (See also musculo-glandular rostellar apparatus, diagnosis and synonyms at 91a and Figs 25.268–25.271) .... *Cinclotaenia* Macy, 1973 (North American species).

17b. Uterus sacciform. Ovary compact, probably not bilobed, small, horseshoe-shaped, markedly anteroporal. Testes mostly lateral aporal to ovary, some posterior. In Sphenisciformes. (See also musculo-glandular rostellar apparatus and diagnosis at 98b and Figs 25.287–25.289. Note: apical apparatus only apparently true glandular) .... *Parorchites* Fuhrmann, 1932.

18a. Rostellum unarmed. (Note: hooks do not develop; exceptional condition, ascertained only in as yet undescribed species, seemingly belonging to a genus where normally the rostellum is armed). In neotropical Picidae. (See diagnosis at 66b and Figs 25.188–25.193) .... *Krimi* Burt, 1944.

18b. Rostellum armed .... 19.

19a. Hooks hammer-shaped, 'davaineoid' pattern and position, small, very numerous. Muscular rostellar pouch weak. Rostellum delicate, transverse fibres not always visible, with flat or dome-shaped anterior face. In Charadriiformes .... 20.

19b. Hooks with different pattern. Rostellar pouch (seldom absent) and rostellum pronounced, usually much more muscular .... 21.

20a. Strobila small. Four testes. Pores regularly alternate. Hooks in two circles. Uterus sacciform with deep septa .... *Himantaurus* Spasskaya & Spasskii, 1971. (Figs 25.29–25.31)

Diagnosis: Rostellar apparatus muscular and probably glandular. Pouch weak. Rostellum large, delicate, conical or cylindrical; flat anteriorly, with delicate transverse fibres. Hooks numerous, davaineoid. Gravid proglottids much bigger. Scolex small, spherical. Suckers unarmed. Ventral osmoregulatory canals converge medially at posterior end of proglottid, no true transverse vessels. Genital ducts between, dorsal, or at times ventral to osmoregulatory canals in same specimen. Vagina posterior and dorsal to cirrus-sac; proximal part fusiform, dilatable, posteriorly inclined; wall very strong. Uterus ventral, sacciform with some large spots of parenchyma; often circular; thin-walled; deep septa; wall not persistent, but eggs heaped and adhering to each other by their spherical, firm outer coat. No uterine capsules. Embryophore thin (perhaps more than three cells form the inner coat). Atrium simple. Ovary small, bilobed; lobes subspherical. Cirrus-sac very large, long. Cirrus very large, strongly armed with rosethorns and apical tuft

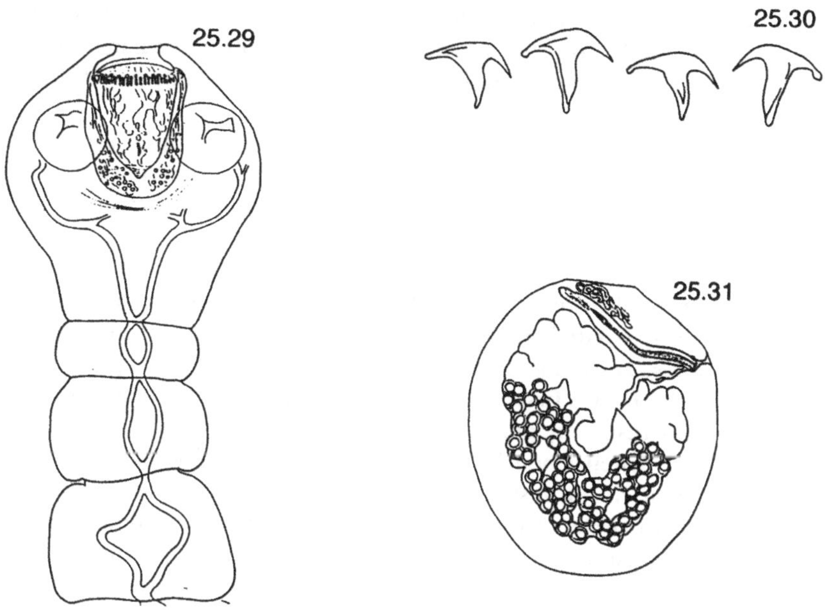

Figs 25.29-25.31 *Himantaurus* sp. 25.29. Scolex. 25.30. Hooks. 25.31. Gravid proglottid, uterus maturation half-way.

of thin, long spines. In Charadriiformes. Cosmopolitan. Type-species *H. minuta* (Cohn, 1901).

20b. Strobila medium to large. Testes numerous. Pores irregularly alternate, may form long unilateral sequences. Hooks in one (?) circle. Uterus reticulate turning into prevalently monovular uterine capsules .... *Kowalewskiella* Baczynska, 1914. (Figs 25.32–25.34)

Diagnosis: Rostellar apparatus muscular (small glandular component?). Pouch wall thin, firm. Rostellum large, delicate, dome-shaped. Hooks numerous (30–60), minute davaineoid. Proglottids much longer than wide when gravid or when testicular fields are two. Scolex small. Suckers small, unarmed. Genital ducts between osmoregulatory canals. Vagina posterior to cirrus-sac in same horizontal plane (dorsal tendency); very short, small, stiff, spiny funnel. Uterus ventral, large-meshed reticulum; thin walled with few eggs. Precocious scattered monovular uterine capsules (seldom with two to four eggs), soon reinforced by small spherical parenchymatous cells, eventually becoming fibrous. Atrium small. Ovary bilobed, small, median when testicular fields are two, otherwise poral. Testes in two groups, anterior and posterior to female glands or in continuous field, surrounding them. Cirrus-sac small, firm-walled. Cirrus uniformly spinous. Seminal receptacle decidedly poral with thin prolongation towards vagina. In Charadriiformes. Cosmopolitan. Type-species *K. cingulifera* (Krabbe, 1868).

21a. **Hook tips anteriorly directed and rostellar apex concave when rostellum retracted. Rostellar apparatus mostly highly muscular, having special patterns [onderstepoortioid (22a), cyclusteroid (25a), parvitaenioid (25b)], or normal pattern [valiporoid (25c)] .... 22.**

21b. **Hook tips directed backwards and rostellum apex usually convex when rostellum retracted. Rostellar apparatus of normal pattern, either muscular or musculo-glandular (only exception, *Tubanguiella* at 35a) .... 35.**

22a. **No rostellar pouch; delicate, loose, longitudinal and transverse fibres may surround rostellum. Rostellum very large, strong; central mass finely transversely striated, with one superficial, thick layer of longitudinally inclined fibres; apical disc like flat sucker; onderstepoortioid pattern. Hooks in one circle. Uterus first reticulate or sacciform, wall not persistent. Seminal receptacle fusiform, poral, far from vitellarium .... 23.**

22b. **True rostellar pouch present or replaced by thick mass of longitudinal, concentric fibres; rostellar structure different, apex not like flat sucker. Hooks in two circles. Uterus always sacciform, wall persistent. Seminal receptacle either subspherical, median or transverse, porally elongate; reaches vitellarium .... 25.**

23a. **Ovary not bilobed, horseshoe-shaped, bulky. Uterus reticular-labyrinthine turning into close, membranous, monovular uterine capsules. Atrium shallow, simple .... 24.**

23b. **Ovary bilobed. Uterus ventral, sacciform lobulate; later, eggs loose in parenchyma, very thick embryophore. Atrium deep, thick muscular wall .... *Burhinotaenia* Spasskii & Spasskaya, 1965. (Figs 25.35–25.37)**

Diagnosis: Rostellar apparatus onderstepoortioid (see 21a, 22a.) Hooks in one circle, very long, slender, handle shorter than blade. Strobila medium. Proglottids much wider than long. Genital pores irregularly alternate. Genital ducts between osmoregulatory canals. Vagina posterior to cirrus-sac, in same horizontal plane; sinuous. Ovary small, transverse. Testes numerous, postovarian in transversely elongated field. Cirrus-sac long, strong-walled. Vas deferens transversely extended, close-packed, anteroporal mass with prostatic sleeve. In Charadriiformes, mostly Burhinidae. Africa, Sri Lanka, Spain. Type-species *B. delachauxi* (Baer, 1925).

24a. **Testes surround ovary. Genital ducts between osmoregulatory canals. Uterus dorsal. Vagina posterior to cirrus-sac, in same horizontal plane. Osmoregulatory vessels normal. In Charadriiformes .... *Onderstepoortia* Ortlepp, 1938. (Figs 25.38–25.41)**

Diagnosis: Rostellar apparatus onderstepoortioid (see 21a, 22a). Hooks in one circle; very large, strong. Strobila large. Genital pores irregularly alternate. Vagina with pre-atrial sphincter. Uterus reticulate or labyrinthine, turning into firm, monovular, precocious, membranous uterine capsules. Atrium small, simple. Ovary not bilobed, horseshoe-shaped with thick, small lobules. Testes very numerous. Cirrus-sac small, wall strong. Cirrus thin, seemingly unarmed, may be long. Internal vas deferens long, simple loop or normally coiled; quite dilatable, in part with circular fibres. Vas deferens poral, in tight coils, with prostatic sleeve.

25.32

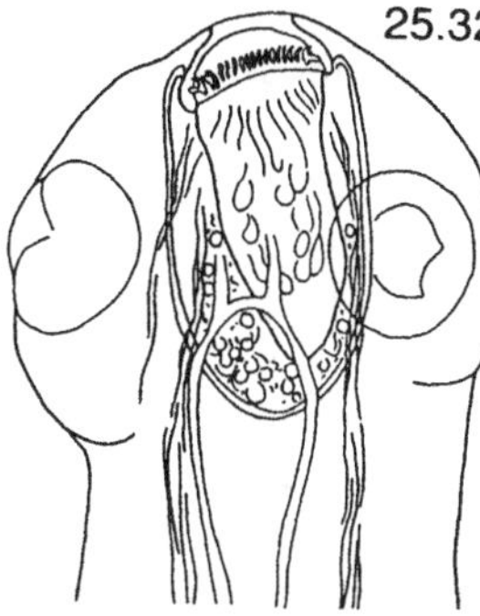

25.33

25.34

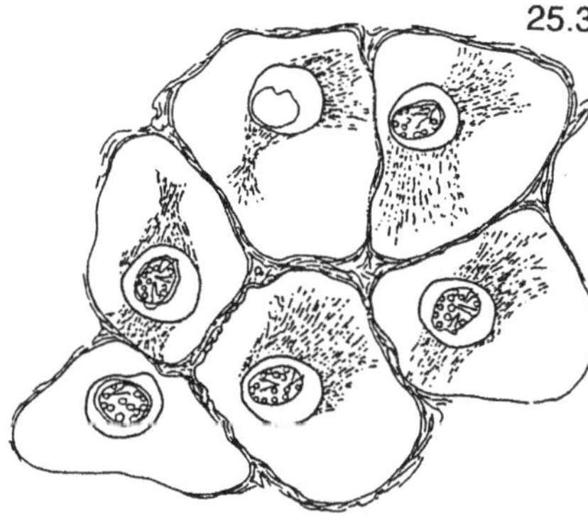

25.37

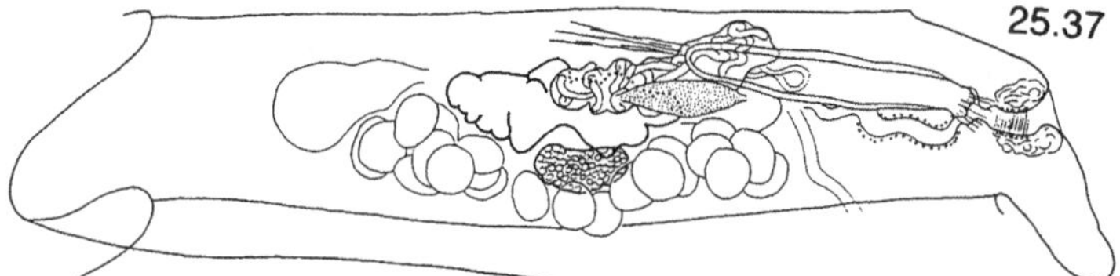

25.35

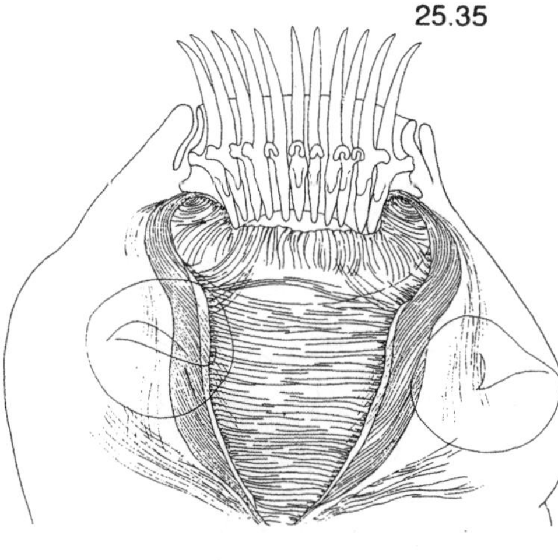

25.36

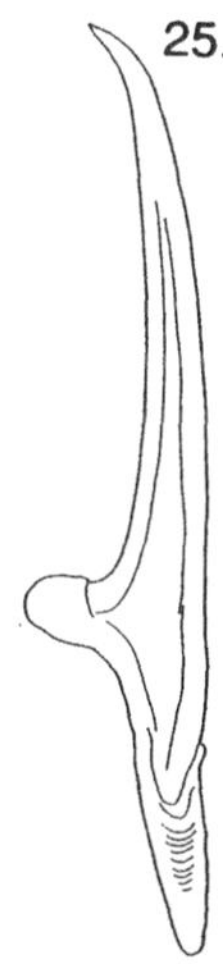

Figs 25.32-25.34 *Kowalewskiella stagnatilidis* (Burt, 1940), type. 25.32. Scolex. 25.33. Hooks. 25.34. Fully mature uterine capsules.

Figs 25.35-25.37 *Burhinotaenia delachauxi* (Baer, 1925), type. 25.35. Scolex. 25.36. Hook. 25.37. Mature proglottid, initial uterus on the left and under seminal receptacle.

In Charadriiformes. Africa, Sri Lanka, probably wider. Type-species *O. taeniaeformis* Ortlepp, 1938.

24b. Testes postovarian. Genital ducts dorsal to osmoregulatory canals, exceptionally between, in same strobila. Uterus ventral. Vagina posterior to cirrus-sac, probably ventral on right, dorsal on left. Huge transverse commissure of ventral osmoregulatory canals. In Passeriformes .... *Spreotaenia* Spasskii, 1969. (Figs 25.42–25.44)

Diagnosis: Rostellar apparatus onderstepoortioid (see 21a, 22a). Hooks in one circle; few, typical, strong. Strobila large. Proglottids longer than wide. Genital pores irregularly alternate. Vaginal wall thicker at distal end. Uterus reticulate or labyrinthine, turning into monovular, membranous, uterine capsules. Atrium small, simple. Ovary not bilobed, horseshoe-shaped with thick, small lobules. Testes very numerous. Cirrus-sac very small, globular; wall strong with sleeve of circular fibres distally during contraction. Cirrus short, wide, spinous. Internal vas deferens short, wide; one or two loops, can be mistaken for internal vesicle. Vas deferens poral; tight coils forming a long rope-like mass with prostatic sleeve, bent forwards from sac, then turning backwards. In Passeriformes. Africa. Type-species *S. abassenae* (Joyeux, Baer & Martin, 1936).

25a. Rostellar pouch strong (wall not necessarily thick), very large; cavity at times negligible when rostellum retracted. Rostellum very complex, strong, stiff, with two muscular superficial layers, at least one of them consisting of diagonal, spiral fibres; cyclusteroid pattern (see 21a). Pores either regularly alternate or unilateral, left. Vagina does not change dorsoventral position with respect to cirrus-sac in right and left pores. Scolex spherical, usually very large .... 26.

25.39

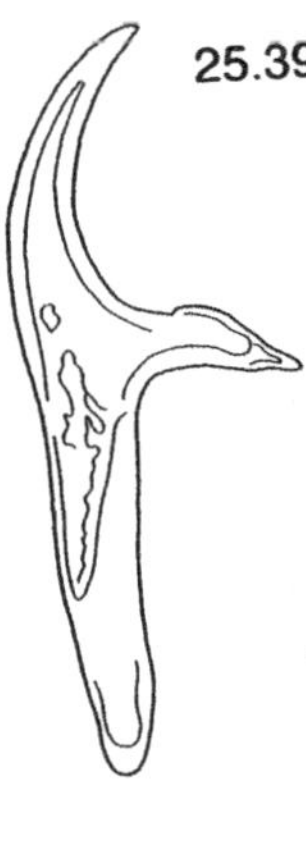

25.40

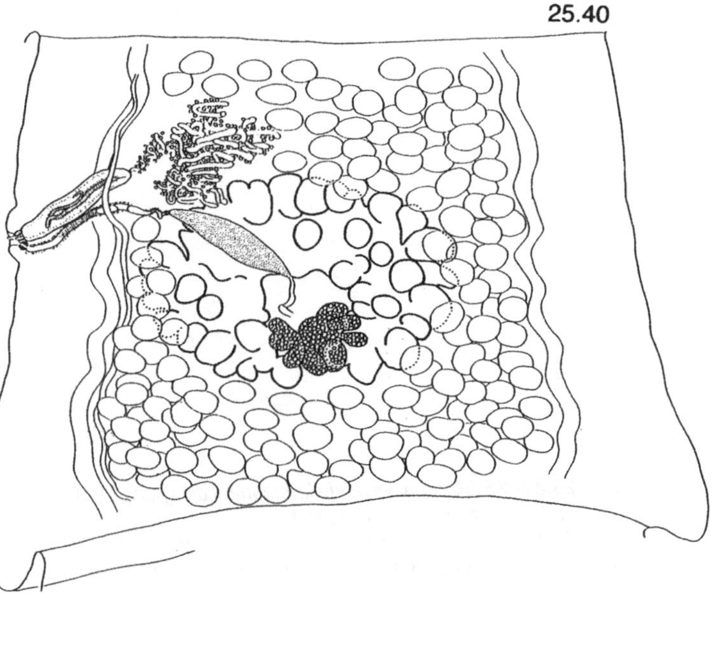

25.41

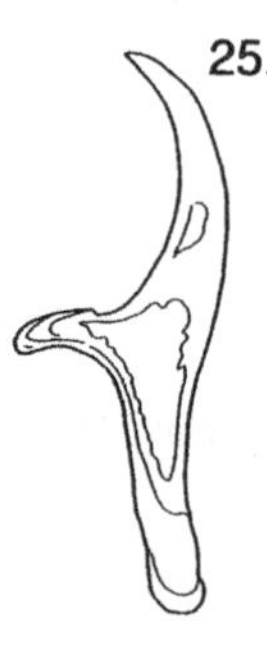

25.38

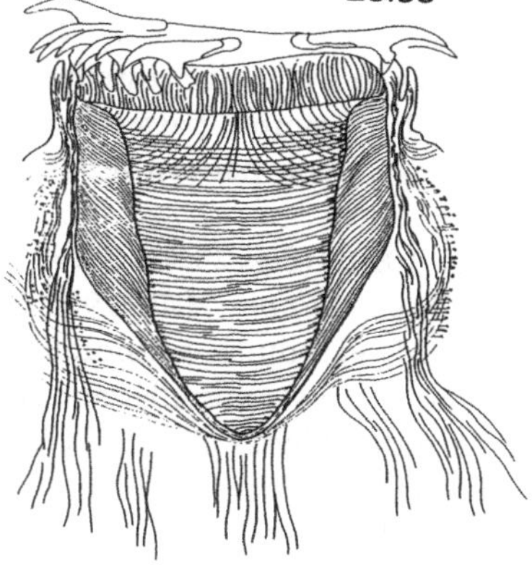

25.44

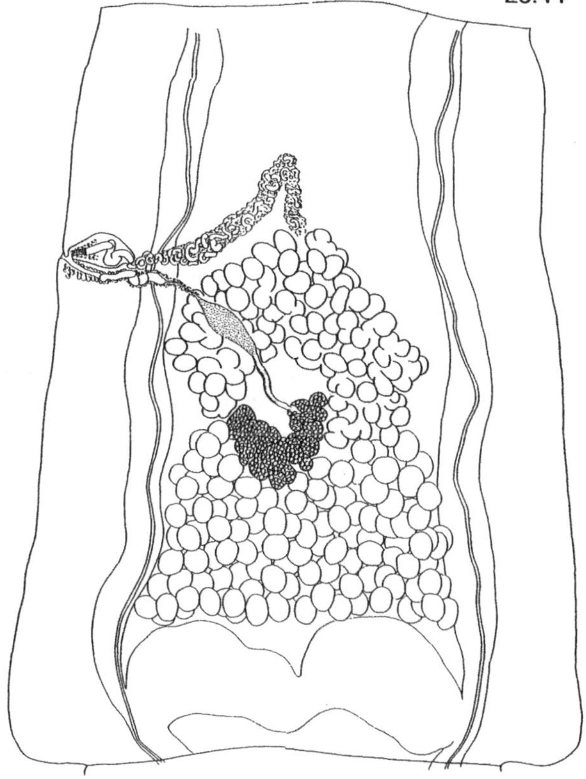

25.43

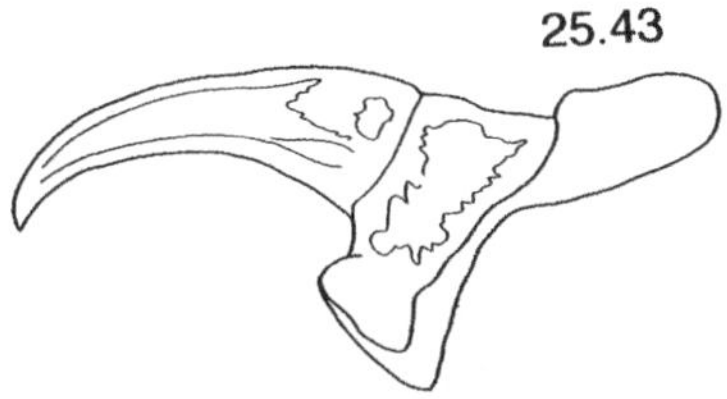

25.42

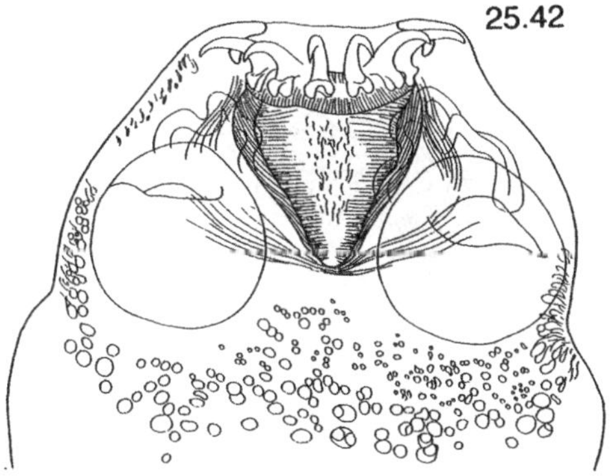

Figs 25.38-25.41 *Onderstepoortia*. 25.38-25.40. *O. taeniaeformis* Ortlepp, 1938. 25.38. rostellum and connected muscular bundles. 25.39. Hook. 25.40. Mature proglottid. 25.41. *O. tringae* (Joyeux, Baer & Martin, 1937), type, hook.

Figs 25.42-25.44 *Spreotaenia abassenae* (Joyeux, Baer & Martin, 1936), type. 25.42. Scolex. 25.43. Hook. 25.44. Mature proglottid.

25b. **Rostellar pouch replaced with dense spherical mass of longitudinal, concentric fibres, surrounding rostellum cap-wise when retracted, lying behind it, bent or transverse, when protruded. Rostellum simple, triangular, often stumpy, clearly formed by dense mass of longitudinal spiral fibres (not transverse circular), shortening and becoming compact pad when protruded; parvitaenioid pattern (see 21a). Pores either irregularly alternate or unilateral with right and left strobilae in same species. Dorsoventral position**

of vagina to cirrus-sac inverted in right and left pores. Scolex either normal or spherical, large .... **28.**

25c. Rostellar pouch and rostellum normal, delicate; pouch wall thin; rostellum simple, with hardly visible transversely striated wall; valiporoid pattern (see 21a). Pores unilateral, right. Scolex small, tapering. Hook pattern usually typical, valiporoid (see diagnosis below) .... ***Valipora* Linton, 1927. (Figs 25.45–25.48)**
(Syn. *Ophiovalipora* Hsü, 1935.)

Diagnosis: Rostellar apparatus muscular, possibly slightly glandular, valiporoid. Pouch subspherical or elongate (seldom reaching behind suckers). Rostellar apical pad, when retracted, infolds with hooks in X-shaped bundle. Hooks in two circles (20); usually small, delicate; blade of distal hooks straight, tip bent at right angle; size and shape very different in each crown. Strobila small to very small. Osmoregulatory canals aporally inverted. Genital ducts between osmoregulatory canals. Vagina ventral to cirrus-sac, posterior or along sac axis; much diversified, spiny; thick wall; may have distal sphincter. Uterus sacciform, horseshoe-shaped or transverse, lobular. Atrium large, thick-walled, at times spinous; prominent papilla; much diversified atriogenital system. Ovary bilobed. Testes few, (seldom more than 15) posterior and lateral (mostly aporal) to ovary, may surround aporal lobe. Cirrus-sac very small (seldom) to very large. Cirrus armed with various kinds of spines. In Ciconiiformes, Ardeidae. Cosmopolitan. Type-species *V. mutabilis* Linton, 1927.

25.45

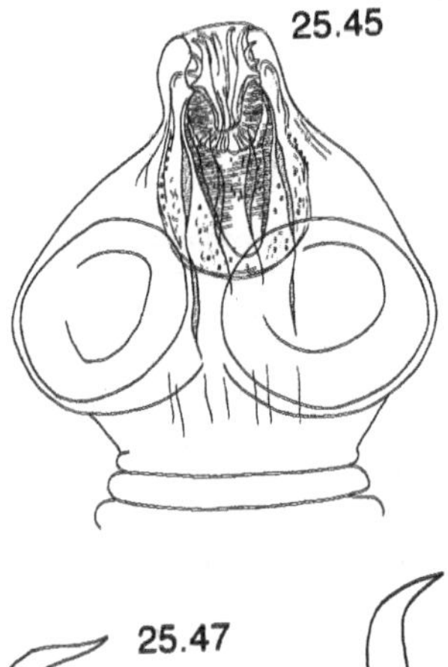

25.46

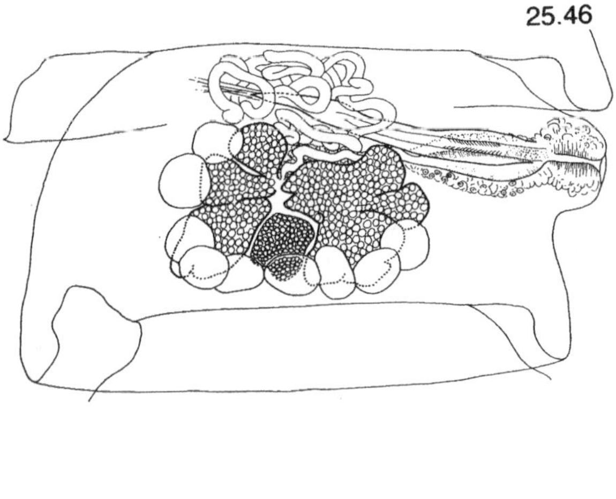

25.47

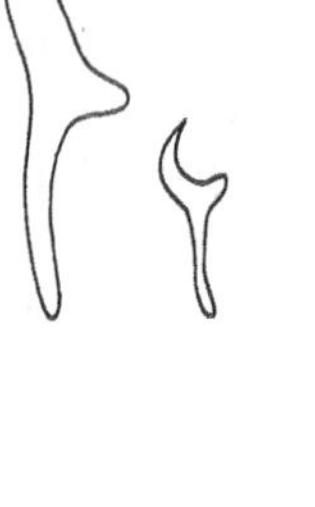

25.48

Figs 25.45–25.48 *Valipora*. 25.45. *Valipora* sp., scolex. 25.46. *V. mutabilis* Linton, 1927, type; mature proglottid. 25.47. *V. campylancristrota* (Wedl, 1855), neotype, valiporoid hooks. 25.48. *V. mutabilis*, hooks. (25.46–25.48 from Bona, 1975.)

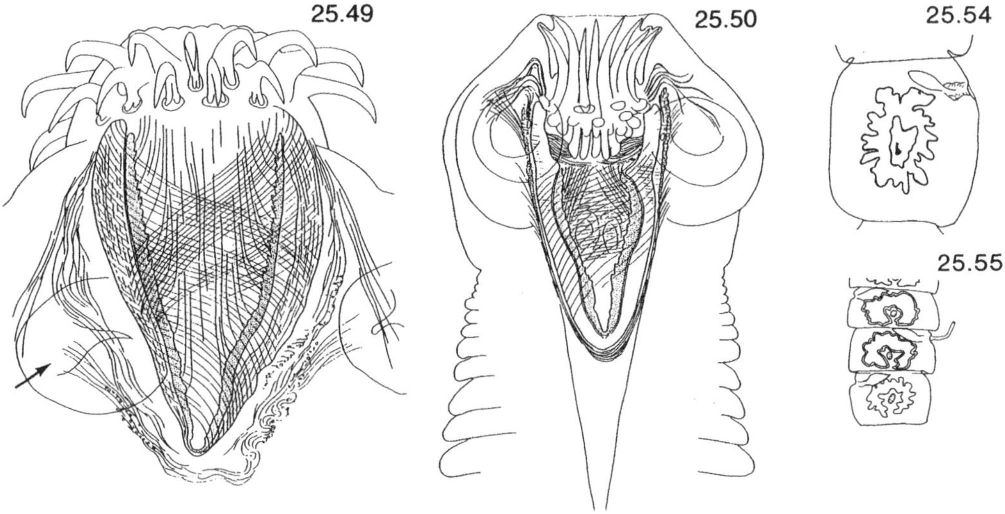

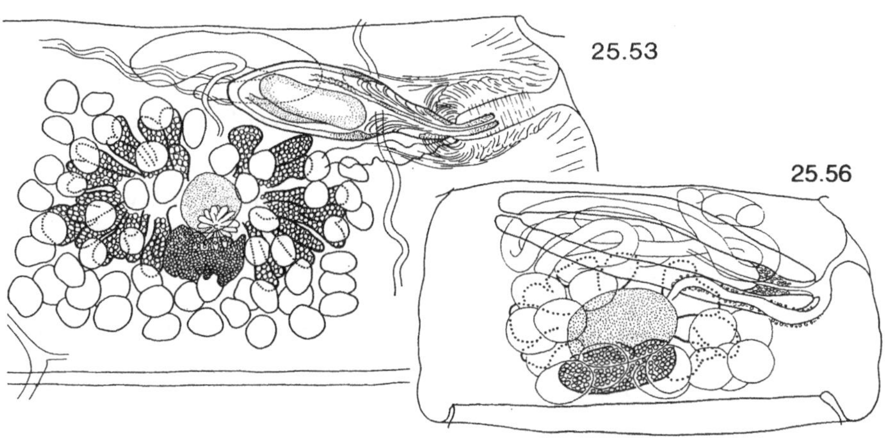

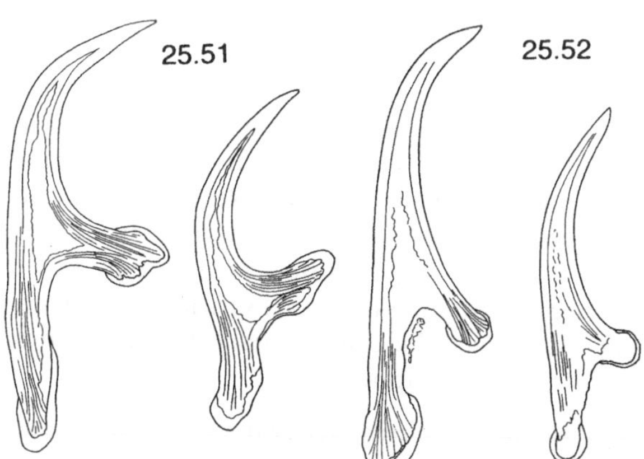

Figs 25.49–25.56 *Cyclustera*. 25.49. *C. capito* (Rudolphi, 1819); scolex, on the right side the outer layer of the rostellar stem is detached from the internal muscular layer (shown dotted); arrow, wall of rostellar pouch. 25.50. *C. ibisae* (Schmidt & Bush, 1972), type; scolex, pouch wall adheres to rostellum (spiral fibres). 25.51, 25.52. Hooks (left, distal; right, proximal): 25.51. *C. capito*. 25.52. *C. ibisae*. 25.53, 25.54. *C. capito*. 25.53. Mature proglottid. 25.54. Gravid proglottids. 25.55. *C. ibisae*, gravid proglottid. 25.56. *C. ralli* (Underwood & Dronen, 1986) n. comb., type, mature proglottid. (All from Bona (1975) except 25.49 and 25.56, original.)

26a. Pores regularly alternate (seldom two consecutive ones). Testes numerous. Ovary with deep isthmus, lobes large, deeply lobulate, fanned when extended. Uterus sacciform, deeply lobulate, either ring- or horseshoe-shaped .... *Cyclustera* Fuhrmann, 1901. (Figs 25.49–25.56)
(Syn. *Neocyclustera* Underwood & Dronen, 1986.)

Diagnosis: Rostellar apparatus cyclusteroid (see 25a). Pouch very large and long; cavity negligible when rostellum retracted. Rostellum with both superficial layers consisting of diagonal spiral fibres, variously inclined in each layer. Hooks in two circles (20–28), cyclusteroid; very large; handle and guard striated lengthwise; sometimes guard of larger hooks inclined backwards. Strobila small, thick, often narrowing backwards. Scolex very large, fleshy. Suckers small. Genital ducts between or dorsal to osmoregulatory canals in different species. Vagina posterior to cirrus-sac or along sac axis, dorsal or ventral according to species. Atrium thick-walled, either deep, sinuous or simple. Testes numerous, dorsal to and around female glands; in very small species, only on posterior portion of ovary. Cirrus-sac large, wall very thick, compact. Cirrus finely spined. Seminal receptacle spherical, central, may be very large. Internal vas deferens may be short loop with dilated portion (simulating internal vesicle) or long, narrow loop, or normally coiled. Vas deferens either with few loops with dilated portion (simulating external vesicle), or with wide coils. In Ciconiiformes, Ciconiae, perhaps Ardeae; Pelecaniformes; Anseriformes (*Mergus*); Ralliformes. South and North America, Asia, Africa. Type-species *C. capito* (Rudolphi, 1819).

26b. Pores unilateral. Testes four (seldom five or three). Ovary lobes compact, subspherical or scarcely lobulate. Uterus sacciform, forming two connected pouches, then one lobulate or ramified cavity .... 27.

27a. Genital organs without accessory structures. Vagina ventral to cirrus-sac, along its axis .... *Paradilepis* Hsü, 1935. (Figs 25.57–25.62)
(Syns *Meggittiella* Lopez-Neyra, 1942; *Skrjabinolepis* Matevosyan, 1945; *Neodilepis* Baugh & Saxena, 1974.)

Diagnosis: Rostellar apparatus cyclusteroid (see 25a). Pouch large, subspherical. Rostellum wide; superficial layer of the two-layered wall, consisting of densely packed, longitudinal, slightly spiral fibres when rostellum retracted, loses compactness and adhesion to underlying layer during protrusion. Hooks in two circles (20–36); two patterns: blade thin, much longer than handle, slightly curved, guard slightly protruding (urceus pattern); blade as long as handle, at times sharply curved at end (scolecina pattern). Strobila small to large. Proglottids acraspedote, wider than long. Genital pores unilateral, left. Genital ducts dorsal to osmoregulatory canals. Vagina short, carrot-shaped with pre-atrial sphincter. Atrium simple. Testes four, exceptionally in some species most proglottids have five, seldom only three. Cirrus-sac oval, rather large. Cirrus strong, heavily armed with rosethorn spines and long fine apical spines. No internal vesicle. Vas deferens anteroaporal. In Ciconiiformes, Ciconiae; Pelecaniformes; Accipitriformes. Cosmopolitan. Type-species *P. scolecina* (Rudolphi, 1819).

27b. Genital organs with accessory vagina and cirrus-sac. True vagina ventral and anterior to cirrus-sac ... *Ascodilepis* Guildal, 1960. (Figs 25.63–25.66)

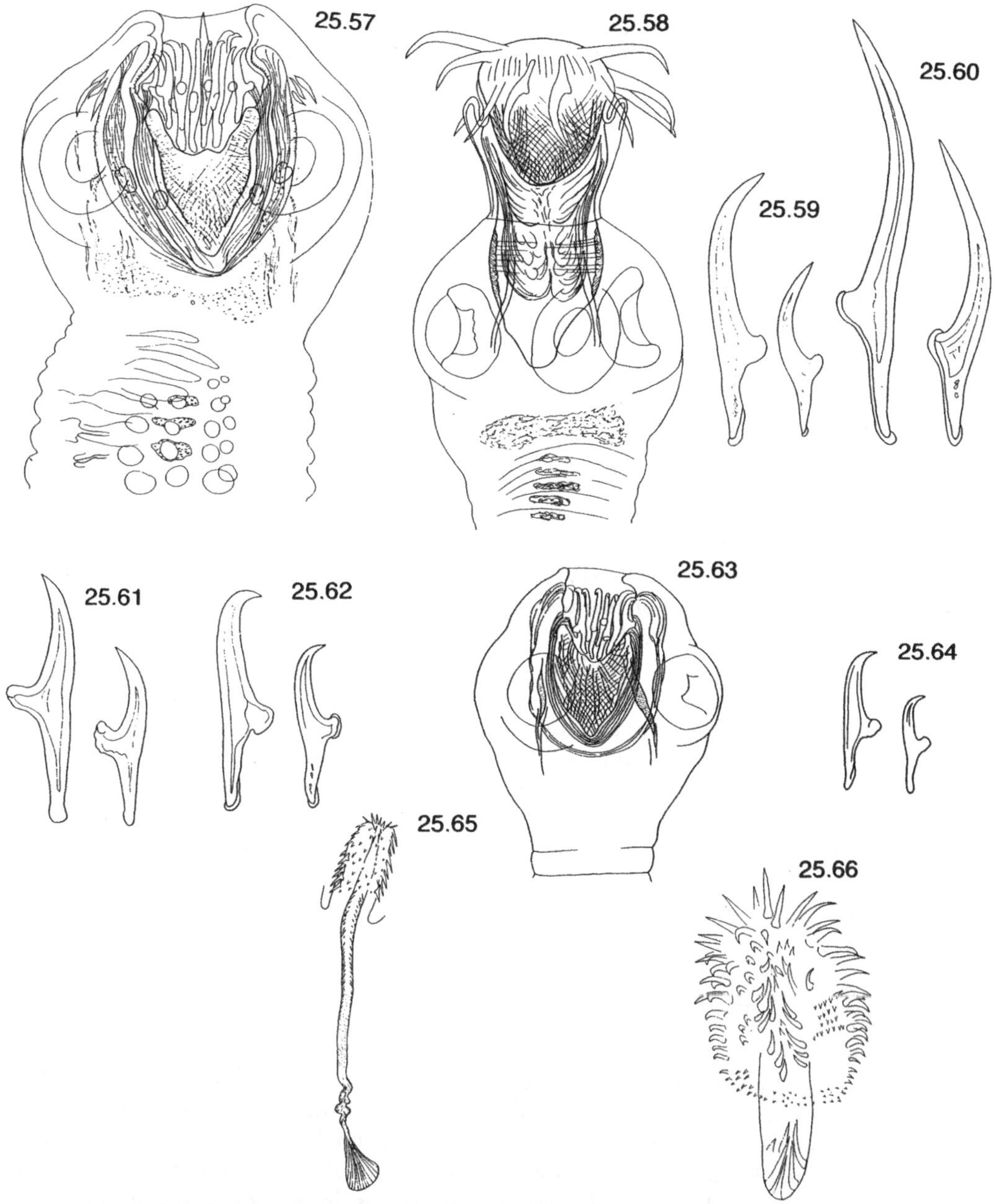

Figs 25.57-25.62 *Paradilepis*. 25.57. *P. patriciae* Baer & Bona, 1960, type; scolex, the four spherical masses apparently on the outer layer of the rostellum and on the pouch wall are contracted bundles of the longitudinal muscles from the neck penetrating the pouch. 25.58. *P. urceus* (Wedl, 1855), scolex. 25.59–25.62. Hooks (left, distal; right proximal): 25.59. *P. urceus* and 25.60. *P. kempi* Southwell, 1921, type, urceus pattern; 25.61. *P. scolecina* (Rudolphi, 1819) and 25.62, *P. patriciae*, type, scolecina pattern. (All from Bona, 1975.)

Figs 25.63-25.66 *Ascodilepis transfuga* (Krabbe, 1869), type. 25.63. Scolex. 25.64. Hooks (left, distal; right, proximal): scolecina pattern with recurved tip, similar to *Paradilepis patriciae* (see Fig. 25.62). 25.65. Cirrus. 25.66. Pseudocirrus. (All from Bona, 1975.)

Diagnosis: Rostellar apparatus cyclusteroid (see 25a) fairly delicate. Pouch proportionally large, subspherical. Hooks in two circles (20), scolecina pattern (see 27a). Strobila very small; usually acraspedote. Genital pores unilateral, left. Genital ducts dorsal to osmoregulatory canals. Vagina dilated proximally; with circular fibres but no true sphincter. Uterus sacciform, septate. Atrium simple, large. Large posterior sac (pseudovagina, dorsal, more posterior) and large pouch with huge, armed, evaginable organ (pseudocirrus, ventral) open into atrium, passing dorsally to osmoregulatory canals. Ovary small. Testes four. Cirrus narrow with delicate spines and small apical tuft of fine spines. No true internal vesicle. Vas deferens anteroaporal. In Ciconiiformes, Ciconiae. North and South America. Type-species *A. transfuga* (Krabbe, 1869).

28a. Pores irregularly alternate .... **29.**
28b. Pores unilateral .... **30.**

29a. Strobila very small to small. Hook pattern parvitaenioid: straight blade on same axis as handle, suddenly curved at tip; handle of distal hooks thickened; different in two crowns. Genital ducts between osmoregulatory canals. Few testes (6–15), postovarian, surrounding aporal lobe or forming separate group of few testes in front of it (seldom an anteroporal isolated element). Ovary bulky, irregular outline. Uterus sacciform, horseshoe-shaped, slightly lobulate .... *Parvitaenia* **Burt, 1940. (Figs 25.67–25.72)**

Diagnosis: Rostellar apparatus parvitaenioid (see 25b). Hooks in two circles (20). Width of adult proglottids 0.150–0.600 mm. Genital pores irregularly alternate. Vagina posterior to cirrus-sac, ventral on right, dorsal on left (at least vaginal orifice). Vagina varies in shape; usually spiny, may have pre-atrial sphincter. Atrium globular with thick wall. Testes initially nearly always in two distinct fields, one posterior and one less numerous, anteroaporal, tending to merge; partly dorsal to ovary. Cirrus-sac usually large, elongate; size, shape and orientation varying. Cirrus variously armed; apical tuft of thin long spines. Vas deferens large; irregular in outline; prostatic cells present. In Ciconiiformes, Ardeidae. Cosmopolitan. Type-species *P. ardeolae* Burt, 1940.

29b. Strobila medium to large, seldom small. Hook pattern glossocercoid (modified parvitaenioid, see 29a); massive sclerification in handle and guard; in distal hooks, sclerification discontinuity separates both guard and handle, or only blade, from hook body or, nearer hook tip, delimits beak in blade; guard may be posteriorly directed. Genital ducts between or dorsal to osmoregulatory canals, according to species, in both cases with intrastrobilar exceptions. Testes numerous, either subdivided into defined groups, postovarian, anteroaporal, at times into small anteroporal group too, or in a continuous field, posterior and lateral to ovary or surrounding ovary, but in front of poral lobe. Ovary transversely elongate or with fan-shaped lobes. Uterus sacciform, with numerous, deep lobules and narrow posterior fissure .... *Glossocercus* **Chandler, 1935. (Figs 25.73–25.84)**

Diagnosis: Rostellar apparatus parvitaenioid (see 25b). Hooks in two circles (20 or more) often very large. Width of adult proglottids 0.6–1.2 mm. Proglottids

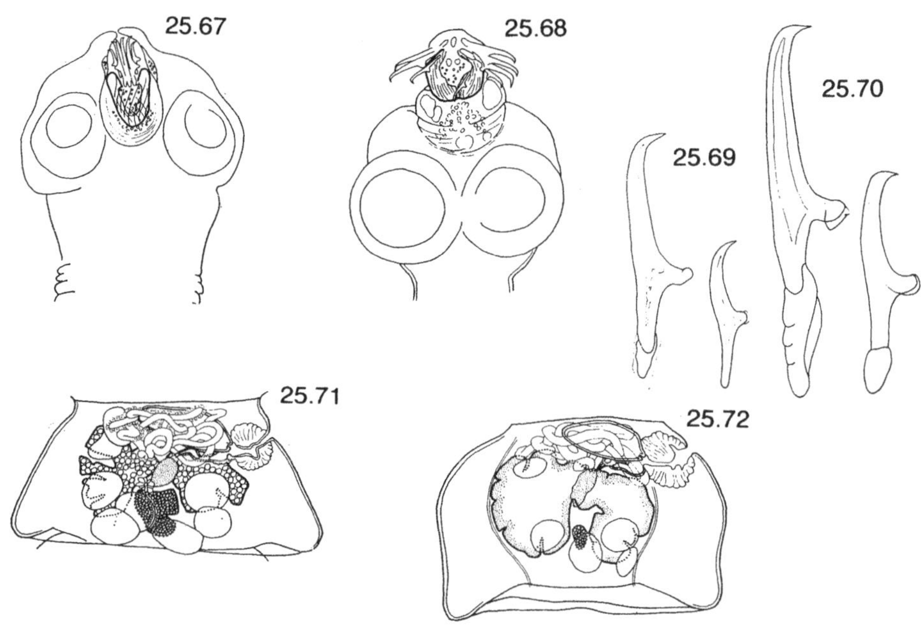

Figs 25.67-25.72 *Parvitaenia*. 25.67. *P. macropeos* (Wedl, 1855), neotype, scolex. 25.68. *P. milvi* (Singh, 1952) scolex [drawing of the type of *buckleyi* Saxena, 1970, synonymous with *milvi*]; see deformation of the rostellum during protrusion, as is usual in *Parvitaenia*. 25.69, 25.70. Hooks (left, distal; right, proximal), parvitaenioid pattern: 25.69. *P. ardeolae* Burt, 1940, type; 25.70. *P. cochleari* Coil, 1955, type. 25.71, 25.72. *P. ardeolae*, type. 25.71. Mature proglottid. 25.72. Gravid proglottid. (All from Bona, 1975.)

wider than long. Scolex large, globular. Genital pores irregularly alternate. Vagina posterior to cirrus-sac, ventral on right, dorsal on left (at least, vaginal orifice); long, sinuous, thick-walled; may have pre-atrial sphincter. Uterus with strong wall at times dorsoventrally perforate. Atrium globular, wall of varying structure, sometimes radiate. Ovary with distinct isthmus, at times perforated dorsoventrally. Cirrus-sac oval or elongate, at times club-shaped, wall very firm, even thick. Cirrus with homogeneous fine or punctiform spines, usually with apical tuft of thin long spines. Vas deferens large, dense, far anterior mass with long dense prostatic sleeve. In Ciconiiformes, Ardeidae; Pelecaniformes. South and North America, Africa, Australia, Indonesia. Type-species *G. cyprinodontis* Chandler, 1935, larval stage; *G. aurita* (Rudolphi, 1819) n. comb. (syn. *Parvitaenia aurita* (Rudolphi, 1819) Baer & Bona, 1960), adult stage.

30a. Osmoregulatory canals normal .... 31.
30b. Osmoregulatory canals aporally inverted .... 34.

31a. Hooks typically parvitaenioid (see 29a). Strobila delicate, narrow. Testes few, postovarian, lateral on aporal side and more anterior .... 32.
31b. Hooks different. Strobila usually strong, wide, fleshy. Testes numerous, surround ovary .... 33.

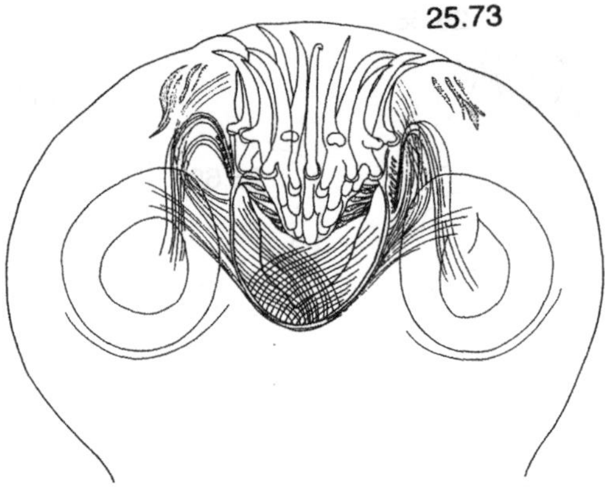
25.73

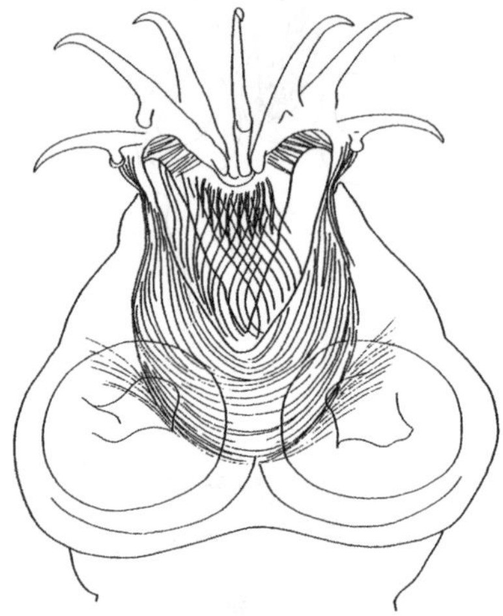
25.74

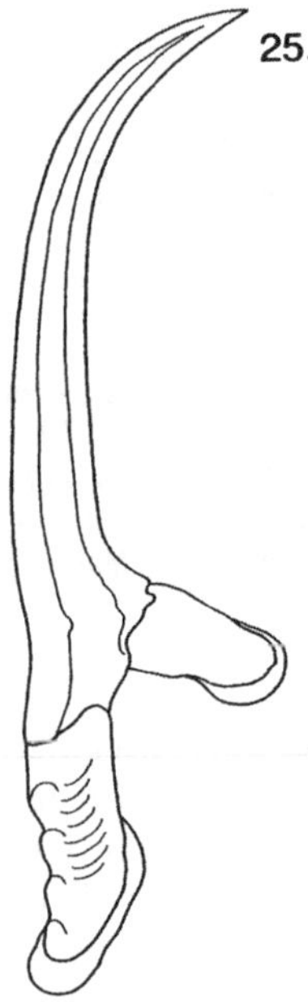
25.76

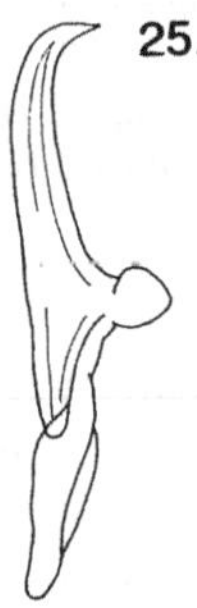
25.75

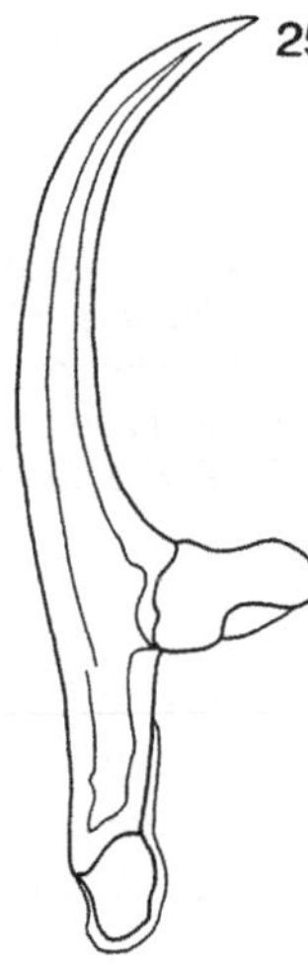
25.77

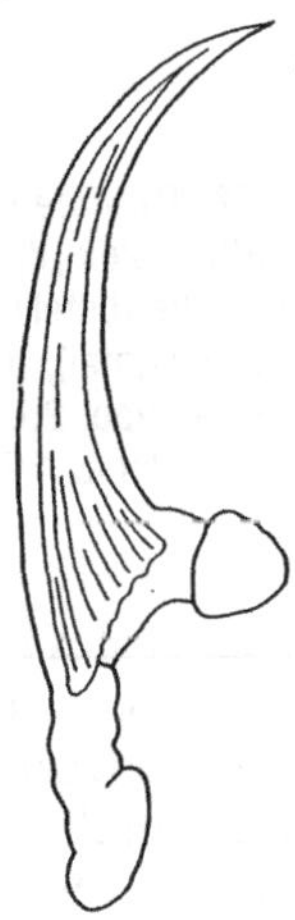
25.78

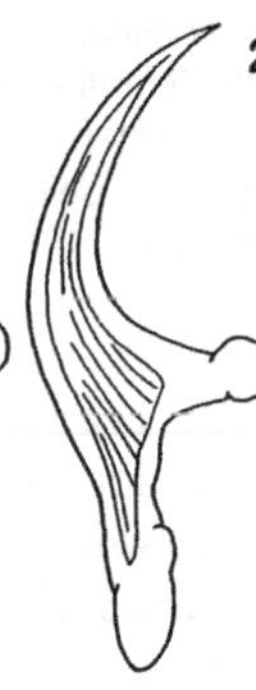
25.79

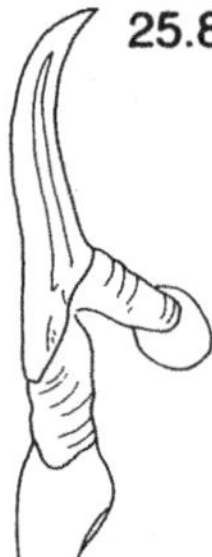
25.80

25.81

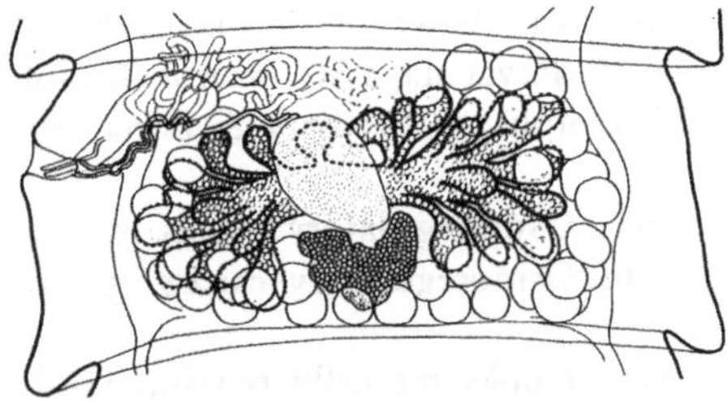
25.82

25.84 25.83

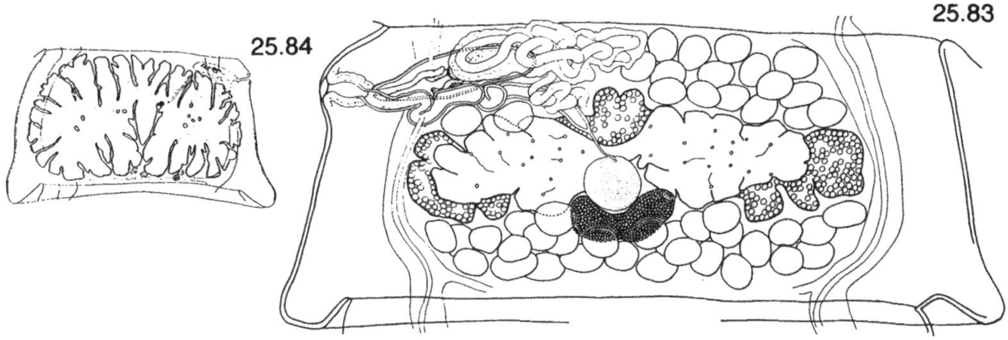

Figs 25.73–25.84 *Glossocercus.* 25.73, 25.74. *G. caribaensis* (Rysavy & Macko, 1971) n. comb., type, scoleces. 25.75–25.81. Hooks: 25.75. *G. ardeae* (Johnston, 1911) n. comb., lectotype; parvitaenioid pattern (compared with the other hooks it should be half as large), but the anatomy of the species is glossocercoid (see Bona, 1975). 25.76, 25.77. *G. aurita* (Rudolphi, 1819) n. comb. 25.78–25.79. *G. cyprinodontis* Chandler, 1935, larva. 25.80. *G. glandularis* (Fuhrmann, 1905) n. comb., type. 25.81. *G. paracyclorchida* (Baer & Bona, 1960) n. comb., type. 25.82. *G. aurita*, mature proglottid. 25.83, 25.84. *G. glandularis*, type. 25.83. Mature proglottid. 25.84. Gravid proglottid. (25.78 and 25.79 from Chandler, 1935; all others from Bona, 1975.)

32a. **Testes four. Atrium very large with adnexa: pouches, large spines, spiny furrow round sexual orifices. Cirrus short, wide, delicate, bearing very thin, long spines** .... ***Neogryporhynchus* Baer & Bona, 1960. (Figs 25.85–25.87)**
(Syn. *Gryporhynchus* auctorum, *nec* von Nordmann, 1832.)

Diagnosis: Rostellar apparatus parvitaenioid (see 25b). Spherical, fibrous mass round rostellum not very large. Hooks in two circles (20), parvitaenioid pattern (see 29a). Strobila very small to medium, narrow. Scolex globular. Genital pores unilateral, right or left in same species. Genital ducts between osmoregulatory canals. Vagina along cirrus-sac axis, ventral in right pores, dorsal in left ones; short, wide distally, thick-walled. Uterus sacciform, horseshoe-shaped. Ovary with few, large lobules, posteriorly extended. Testes four, posterior ones partly dorsal to ovary. Cirrus-sac oval or spherical, wide, thin-walled. In Ciconiiformes, Ardeidae. Holarctic, Africa, Japan. Type-species *N. cheilancristrotus* (Wedl, 1855) Baer & Bona, 1960.

32b. **Testes few, more than four. Atrium large, thick-walled but simple. Cirrus very large, strong, heavily armed** .... ***Clelandia* Johnston, 1909. (Figs 25.88, 25.89)**

Diagnosis: Rostellar apparatus parvitaenioid (see 25b). Hooks in two circles (20); parvitaenioid pattern (see 29a). Strobila very small. Genital pores unilateral, left (unilateral right in same species cannot be excluded). Genital ducts between osmoregulatory canals. Vagina posterodorsal to cirrus-sac (in left pores), very wide, winding, strong wall. Uterus sacciform, scarcely lobulate. Ovary large, bulky, posteriorly extended. Vitellarium far posterior. Testes postovarian, aporally lateral and anteriorly extended, tending to form two groups aporally, partly dorsal to

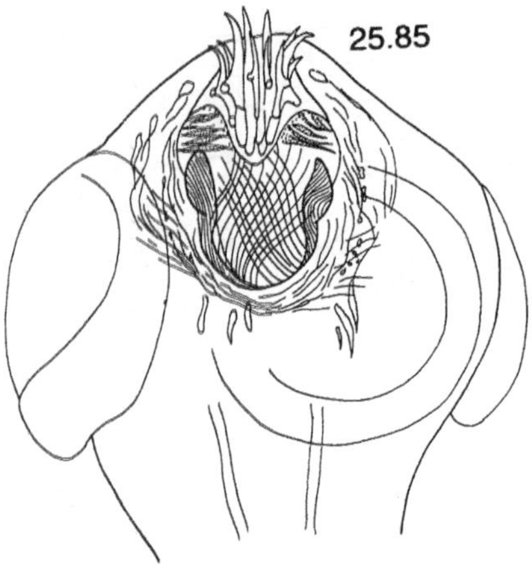

25.85

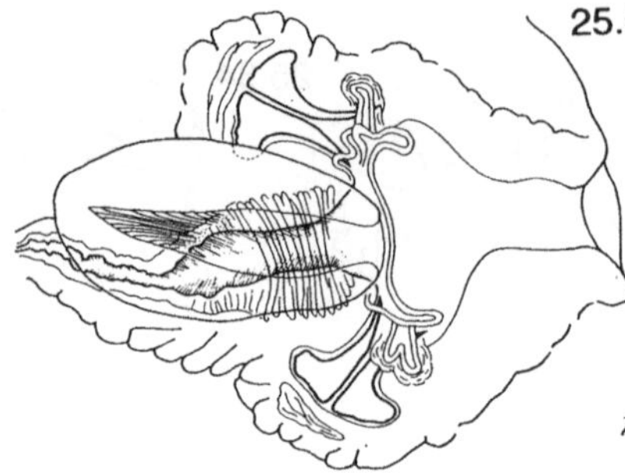

25.86

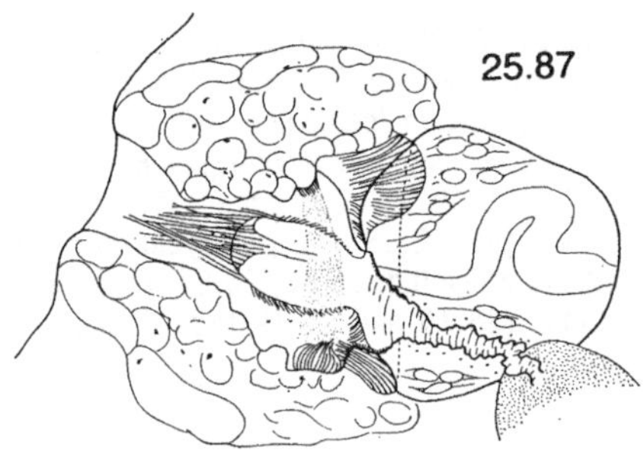

25.87

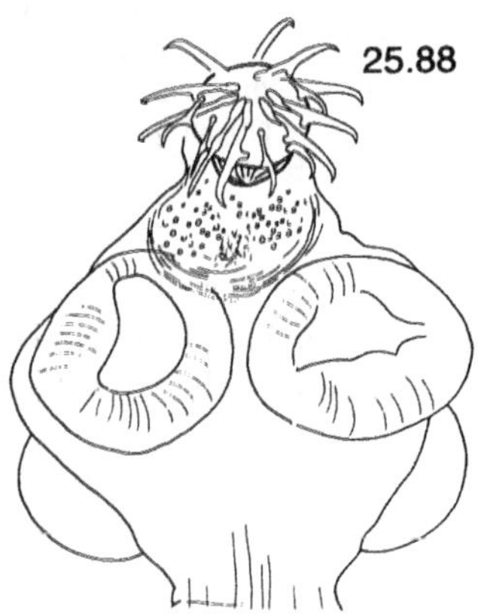

25.88

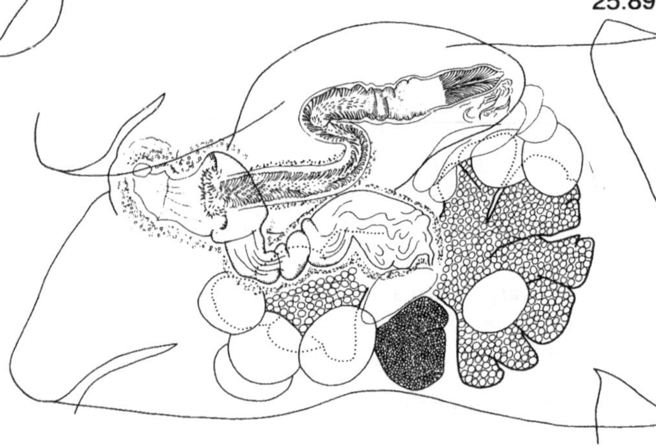

25.89

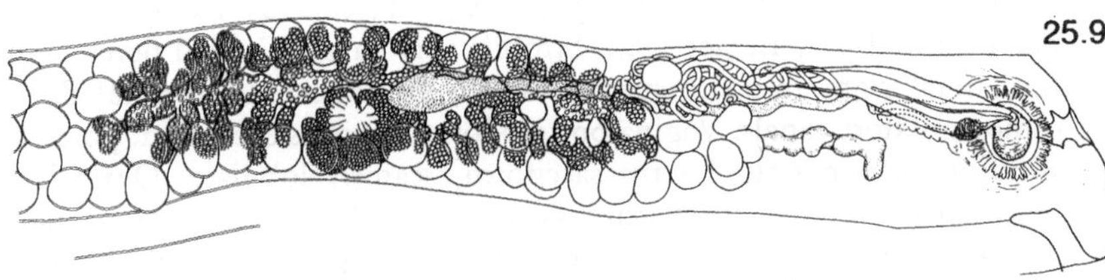

25.91

25.90

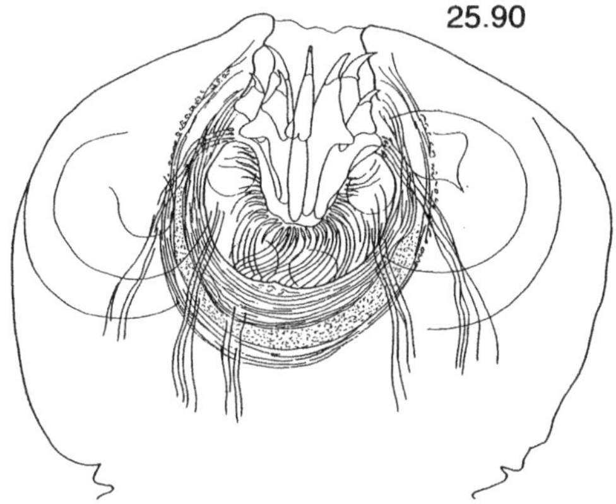

25.92

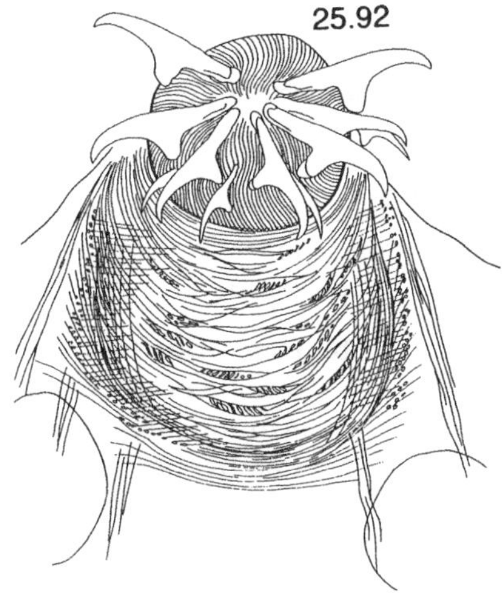

25.93

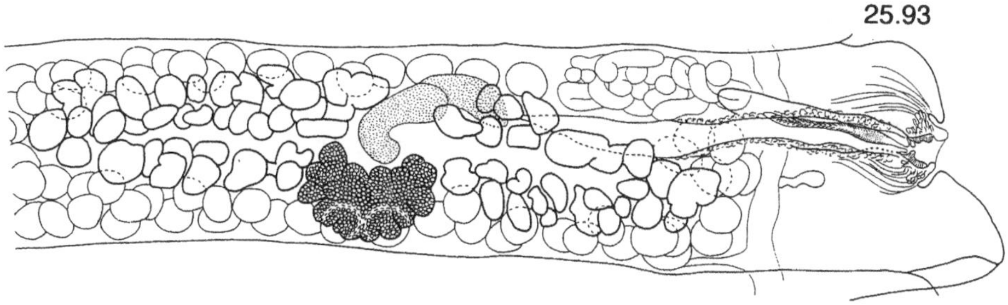

Figs 25.85-25.87 *Neogryporhynchus*. 25.85, 25.86. *N. cheilancristrotus* (Wedl, 1855). 25.85. Scolex. 25.86. Terminal genitalia. 25.87. *N. lasiopeius* Baer & Bona, 1960, type, terminal genitalia. (From Bona, 1975.)
Figs 25.88, 25.89 *Clelandia parva* Johnston, 1909, type. 25.88. Scolex. 25.89. Mature proglottid. (From Bona, 1978.)
Figs 25.90, 25.91 *Cyclorchida*. 25.90. *C. congolensis* Bona, 1975, type, scolex. 25.91. *C. o. omalancristrota* (Wedl, 1855), mature proglottid. (From Bona, 1975.)
Figs 25.92, 25.93 *Amirthalingamia macracantha* (Joyeux & Baer, 1935). 25.92. Rostellum. 25.93. Mature proglottid.

ovary. Cirrus-sac very large. Vas deferens small, anteroaporal. In Ciconiiformes, Ardeidae (probably). Australia. Type-species *C. parva* Johnston, 1909.

33a. **Hooks triangular, massive; blade beaked, distinct from hook body. Crowns regular. Atrium globular, complex, thick-walled, radiate fibres, with internal stable male papilla. Vagina very small, short, ventral to cirrus-sac if strobila with pores on right, dorsal if on left .... *Cyclorchida* Fuhrmann, 1907. (Figs 25.90, 25.91)**
(Syn. *Vogea* Johri, 1959.)

Diagnosis: Rostellar apparatus parvitaenioid (see 25b). Hooks in two circles (20). Strobila medium to large; proglottids much wider than long. Scolex large, spherical. Suckers small. Genital pores unilateral, right or left in same species.

Genital ducts between osmoregulatory canals. Vagina slightly posterior to or along cirrus-sac axis, with two pre-atrial sphincters. Uterus sacciform; with deep anterior and posterior outpockets. Ovary not clearly bilobed, wide transverse trunk, small. Testes very numerous, surround ovary. Cirrus-sac small, elongate; terminal part narrow, displaced by atrial papilla. Cirrus very delicate, with thin spines centrally and long fine ones at apex. Seminal receptacle very long, narrow, overlapping poral lobe. In Ciconiiformes, Ciconiae. Europe, Asia (India), Africa, Australia. Type-species *C. omalancristrota* (Wedl, 1855).

33b. **Hooks different in shape, striated longitudinally, of two different sizes in distal circle, arranged in bilaterally symmetrical alternating sequence. Atrium globular, sucker-like, deformable; thick wall with long outer fibres towards cirrus-sac. Vagina long and wide, dorsal to cirrus-sac if strobila with pores on right, ventral if on left .... *Amirthalingamia* Bray, 1974. (Figs 25.92, 25.93)**

Diagnosis: Rostellar apparatus parvitaenioid (see 25b). Hooks in two circles, regularly alternate but with two sizes of distal hooks in sequence s(small) – l(large) – s – s – l, resulting in bilateral symmetry of crown; very large. Strobila medium. Proglottids wider than long. Scolex large. Suckers small. Genital pores unilateral, right or left in same species. Genital ducts between osmoregulatory canals. Vagina slightly posterior to cirrus-sac or along sac axis. Uterus sacciform with deep anterior and posterior outpockets. Ovary bilobed, large with transverse trunk. Testes very numerous, surround ovary (field may become interrupted behind vitellarium and vas deferens). Cirrus-sac small, elongate. Cirrus with strong, straight spines. Seminal receptacle sinuous, on ovarian isthmus. Internal vas deferens forms swollen loops simulating internal vesicle. In Pelecaniformes. Africa. Type-species *A. macracantha* (Joyeux & Baer, 1935).

34a. **Hooks longitudinally striated, cyclusteroid pattern (see 26a). Genital ducts between osmoregulatory canals (when dorsal one has not disappeared). Testes surround ovary (may be interrupted in front of poral lobe). Uterus sacciform, deeply lobulate; eggs bear long polar processes .... *Baerbonaia* Deblock, 1966. (Figs 25.94–25.99)**

Diagnosis: Rostellar apparatus parvitaenioid (see 25b). Hooks in two circles (20); cyclusteroid pattern (see 26a), typical slight pit and prominence near blade tip. Strobila small to medium. Osmoregulatory canals aporally inverted, dorsal one at times missing. Genital pores unilateral, right. Vagina posterior to cirrus-sac, ventral, very thick-walled with pre-atrial sphincter. Atrium spherical, deep, strong. Ovary large, with isthmus. Testes numerous, some dorsal to ovary. Cirrus-sac large, elongate. Cirrus large with fine, short, dense spines and apical long, thin spines. In Ciconiiformes, Ardeidae. Africa, Madagascar, Australia. Type-species *B. baeribonae* Deblock, 1966.

34b. **Hooks different in shape. Genital ducts dorsal to osmoregulatory canals. Testes postovarian, at times also aporally lateral or in anterior group. Uterus sacciform with crowded, deep, often digitiform, ramified lobules, cavity transversely extended, or horseshoe-shaped with near extremities or circular with anterior ramified branch. (See also normal muscular rostellar apparatus**

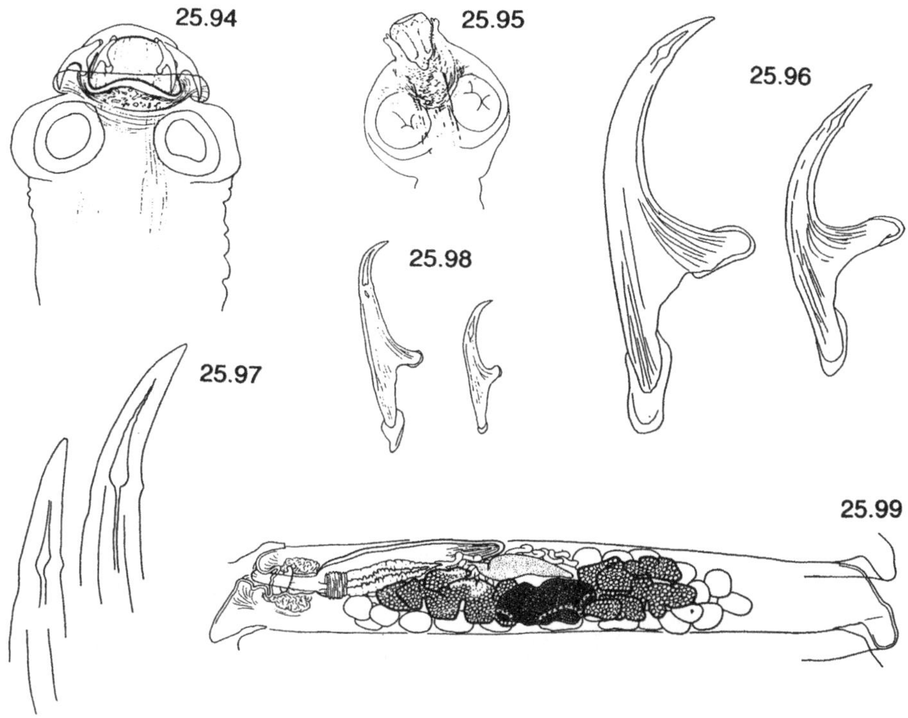

Figs 25.94–25.99 *Baerbonaia.* 25.94 and 25.99. *B. baeribonae* Deblock, 1966, type. 25.94. Scolex, see the deformation of the rostellum when protruding, common in paravitaenioid rostella. 25.99. Mature proglottid. 25.95. *B. p. parvitaeniunca* (Baer & Bona, 1960), type, scolex. 25.96–25.98. Hooks (left, distal; right, proximal): 25.96, 25.97. *B. baeribonae*; 25.98. *B. p. parvitaeniunca.* (All from Bona, 1975.)

at 49a) .... ***Dendrouterina*** **Fuhrmann, 1912. (Figs 25.100–25.106)** (Syns *Mashonalepis* Beverley-Burton, 1960; *Cyclouterina* Gulyaev & Tkachev, 1988.)

Diagnosis: Rostellar apparatus muscular. Pouch, if any, or mass of fibres round rostellum spherical. Rostellum strong, wide, probably parvitaenioid. Hooks in two circles (20); two characteristic patterns: herodiae, two crowns greatly different, at times valiporoid (see 25c); macrosphincter, two crowns very similar, at times difficult to distinguish, handle long. Strobila large. Osmoregulatory canals aporally inverted; dorsal one may be missing. Genital pores unilateral, right. Genital ducts very seldom between osmoregulatory canals (in connection with irregular appearance of dorsal canal). Vagina posterior to or along cirrus-sac axis, ventral; distal sphincter may be present. Atrium large, globular, wall strong, complex, seldom with sphincter at base round both sexual orifices; may be finely spined; often with protruding papilla. Ovary transversely extended or horseshoe-shaped with radiate external outline, with isthmus. Testes very numerous. Cirrus-sac elongate, seldom small, oval. Cirrus long with small, close spines. In Ciconiiformes, Ardeidae; Passeriformes (?). Cosmopolitan. Type-species *D. herodiae* Fuhrmann, 1912.

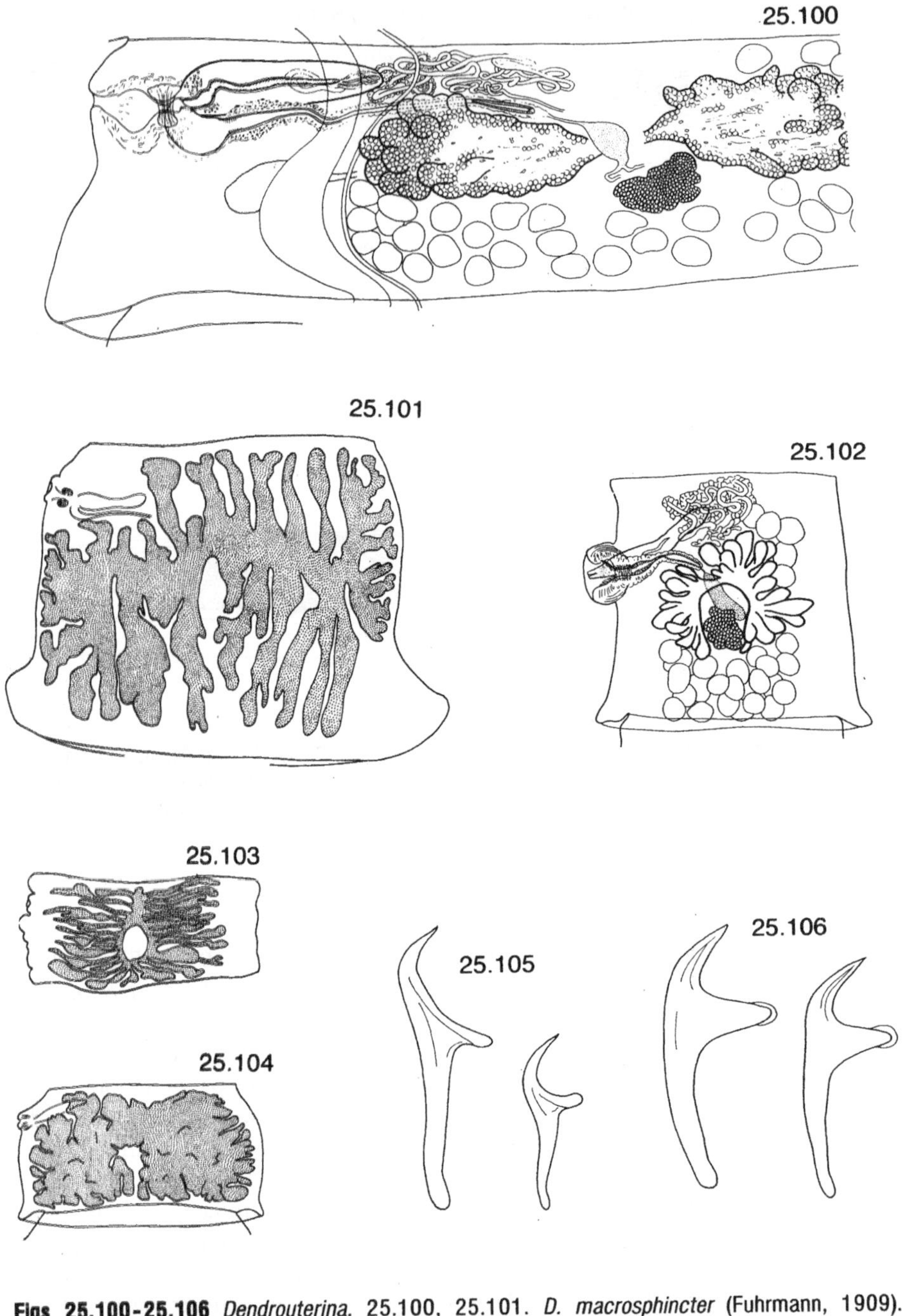

Figs 25.100–25.106 *Dendrouterina*. 25.100, 25.101. *D. macrosphincter* (Fuhrmann, 1909). 25.100. Mature proglottid, ventral view. 25.101. Gravid proglottid. 25.102. *D. papillifera* (Fuhrmann, 1908), type, mature proglottid. 25.103. *D. ixobrichi* Bona, 1975, type; uterus circular with anterior branch. 25.104. *D. pilherodiae* Mahon, 1956, type; uterus sacciform, deeply lobulate and septate. 25.105, 25.106. Hooks (left, distal; right, proximal): 25.105. *D. herodiae* Fuhrmann, 1912, herodiae pattern, at times resembling valiporoid pattern; 25.106. *D. macrosphincter*, macrosphincter pattern. (All from Bona, 1975.)

[Note: The exact structure of the rostellar apparatus is still uncertain. The pattern seems to be parvitaenioid, but a muscular pattern with backwardly directed hook tips when the rostellum is retracted cannot be excluded.]

35a. Hook tips directed backwards when rostellum retracted, but rostellar apparatus supposedly parvitaenioid (see 25b). Hooks and anatomy glossocercoid (see 29b). In Accipitriformes; Ardeidae .... *Tubanguiella* Yamaguti, 1959. (Figs 25.107–25.109)

Diagnosis: Rostellar apparatus not described. Rostellum pad-shaped when protruded. Hooks in one circle, number doubtful (10–20). Strobila large, proglottids wider than long. Genital pores irregularly alternate. Genital ducts between osmoregulatory canals. Vagina posterior to cirrus-sac (no data whether dorsal or ventral), long, sinuous. Uterus (probably ventral) sacciform, transverse, deeply lobulate. Atrium simple. Ovary with clear isthmus, large, transversely extended. Testes numerous in postovarian and anteroaporal field. Cirrus-sac elongate, far anterior. Cirrus armed. Vas deferens supposedly small, median, far anterior, dilatable near cirrus-sac (?). In Accipitriformes; Ciconiiformes, Ardeidae. Philippines, Pakistan. Type-species *T. buzzardia* (Tubangui & Masilungan, 1937).

35b. Hook tips directed backwards when rostellum retracted and rostellar apparatus muscular or musculo-glandular, consisting of true rostellar pouch and rostellum with single, thick, transversely striated wall [neither parvitaenioid (25b) nor cyclusteroid (25a) pattern] .... 36.

36a. Pores unilateral .... 37.
36b. Pores alternate regularly or irregularly .... 51.

37a. Rostellum armed but rostellar crown unknown. Ovary not bilobed, anteroporal. Aporal osmoregulatory canals inverted. Atrium bearing strong, spaced, radiate bundles. Uterus at first dorsal, transverse tube, then reticulate, very delicate, turning into membranous monovular capsules. (Note:

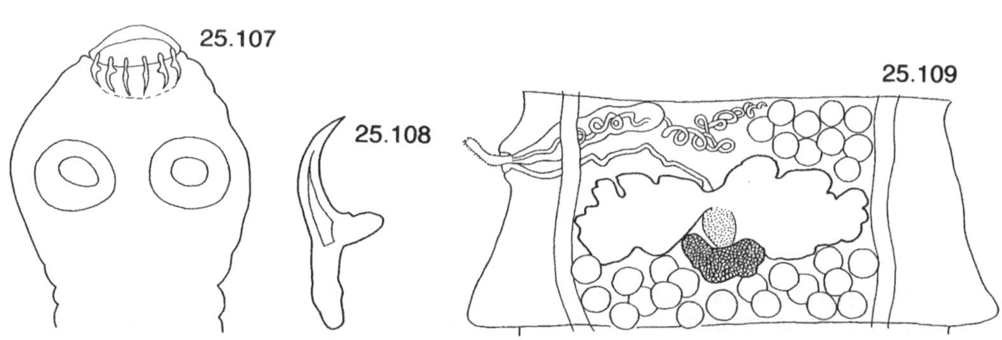

Figs 25.107–25.109 *Tubanguiella buzzardia* (Tubangui & Masilungan, 1937). 25.107. Scolex. 25.108. Hook. 25.109. Mature proglottid. (From Tubangui & Masilungan, 1937.)

unique character combination in group 36a, so that genus recognizable even without crowns) .... ***Eugonodaeum*** **Beddard, 1913. (Figs 25.110–25.112)**

(Syn. *Nasutaenia* Baer & Bona, 1960.)

Diagnosis: Rostellar apparatus insufficiently known, most probably normal muscular. Strobila medium. Proglottids much wider than long, largely overlapping, organs spaced, cortex wide. Genital pores unilateral, left, postequatorial. Genital ducts between osmoregulatory canals. Vagina posterior to cirrus-sac, mainly ventral. Atrium large, wall strong, complex with fine circular fibres distally, radiate bundles proximally; wide stable papilla present. Ovary small,

25.110

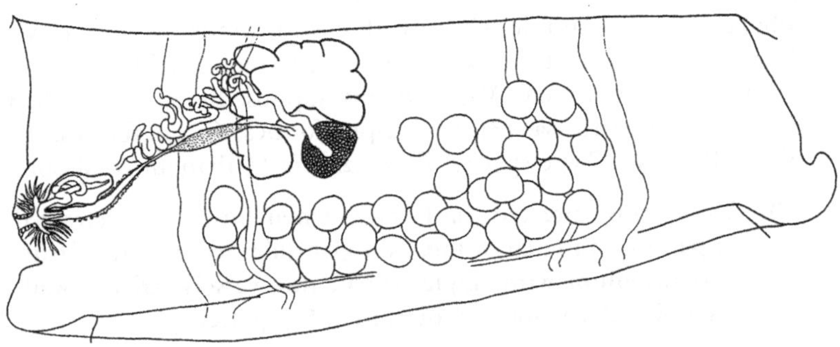

25.111

25.112

25.114

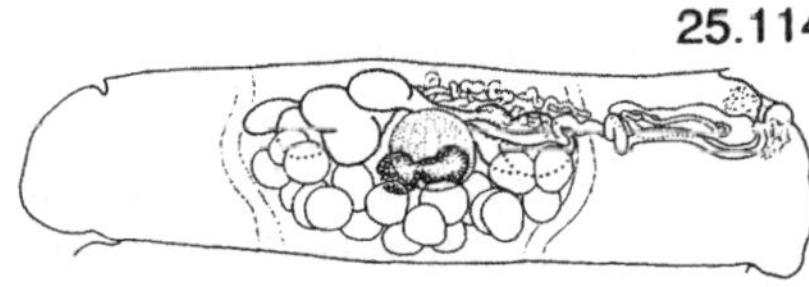

25.113

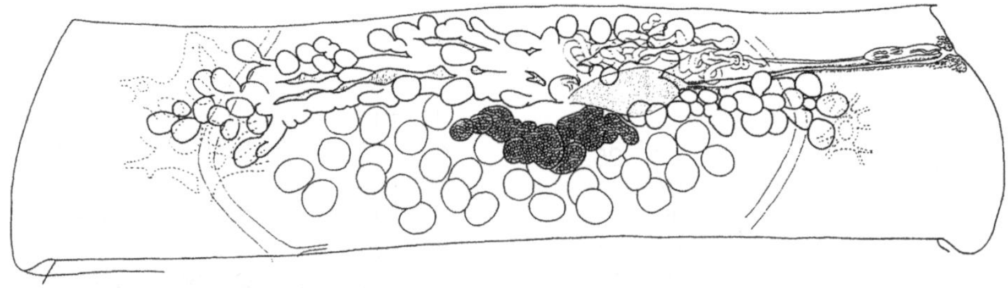

Figs 25.110-25.112 *Eugonodaeum*. 25.110, 25.111. *E. oedicnemus* Beddard, 1913, type. 25.110. Mature proglottid. 25.111. Uterine capsules, mature and abortive ones (arrow indicate the thin outer coat of the egg). 25.112. Hook of *E. nasuta* (Fuhrmann, 1908) n. comb., type. **Figs 25.113, 25.114** *Reticulotaenia*. 25.113. *R. australis* (Jones & Williams, 1967), mature proglottid, early uterus dotted, Hoberg's (1985) material. 25.114. *R. mawsoni* (Prudhoe, 1969), type, mature proglottid. (Both drawings at same scale.)

compact, horseshoe-shaped. Testes numerous, postovarian, field more extended aporally. Cirrus-sac very small. Cirrus unarmed, very delicate. Seminal receptacle poral, small, reaching or even intersecting osmoregulatory canals. No internal seminal vesicle. Vas deferens poral, small, transverse, entering cortex. In Charadriiformes; Ciconiiformes, Ciconiae. South America. Type-species *E. oedicnemus* Beddard, 1913.

37b. Hooks in one circle, regular (character combination different from 37a) .... **38.**

37c. Hooks in two circles, regular (character combination different from 37a) .... **41.**

38a. Genital ducts between osmoregulatory canals. Uterus coarsely reticulate, then labyrinthine. Hooks few, about ten, spaced. Protandric maturation, early insemination. Eggs with long polar processes. In Charadriiformes, Chionidae .... ***Reticulotaenia* Hoberg, 1985. (Figs 25.113, 25.114)**

Diagnosis: Rostellar apparatus muscular. Rostellum strong, long, wide; apical pad small. Hooks in one circle. Strobila very small to medium. Osmoregulatory canals form fine reticulum in posterior part of scolex (type-species). Genital pores unilateral, right or left in same species. Vagina posterior to cirrus-sac, orifice in same horizontal plane, course irregularly dorsal or ventral, very long; pre-atrial sphincter in type-species. Uterus early extending into cortex, wall reinforced at end. Atrium small with dense, irregular wall. Ovary either (a) first reticulate, then finely lobulate, reticulum effaced, extending beyond osmoregulatory canals, aporal lobe larger, may form anterior branch extending far porally (type-species), or (b) not reticulate, bulky, relatively large lobules, not going beyond osmoregulatory canals. Testes numerous; postovarian, at times partly dorsal, exceptionally reaching anterior margin of proglottid aporally. Cirrus-sac small, elongate, thin-walled. Cirrus with extremely small spines (at times unarmed?). In Charadriiformes, Chionidae. Antarctic. Type-species *R. australis* (Jones & Williams, 1967).

38b. Genital ducts dorsal to osmoregulatory canals, exceptionally ventral (doubtful). Uterus sacciform, persistent or not. Crown and maturation not as above; eggs without polar processes .... **39.**

39a. Strobila very small. Atrium simple. Testes few (six to eight), in transverse posterior line dorsal and lateral to female glands. Ovary very small, lobes smooth. Uterus sacciform not lobulate, wall not persistent, eggs eventually scattered in parenchyma turned into granular cells .... ***Dilepidoides* Spasskii & Spasskaya, 1954. (Figs 25.115–25.117)**

Diagnosis: Rostellar apparatus muscular. Pouch with many internal, longitudinal bundles connected to external bundles from neck. Rostellum strong, wide; flat apical disc. Hooks in one circle, large number (35). Proglottids much wider than long. Genital pores unilateral, left. Genital ducts stated as ventral to osmoregulatory canals. Vagina dorsal to cirrus-sac; poral part anterior or posterior; wide, long, sinuous, strong walled. Eggs few. Cirrus-sac very long, ventral. Cirrus extremely long with several successive fields of spines mixed with filaments and

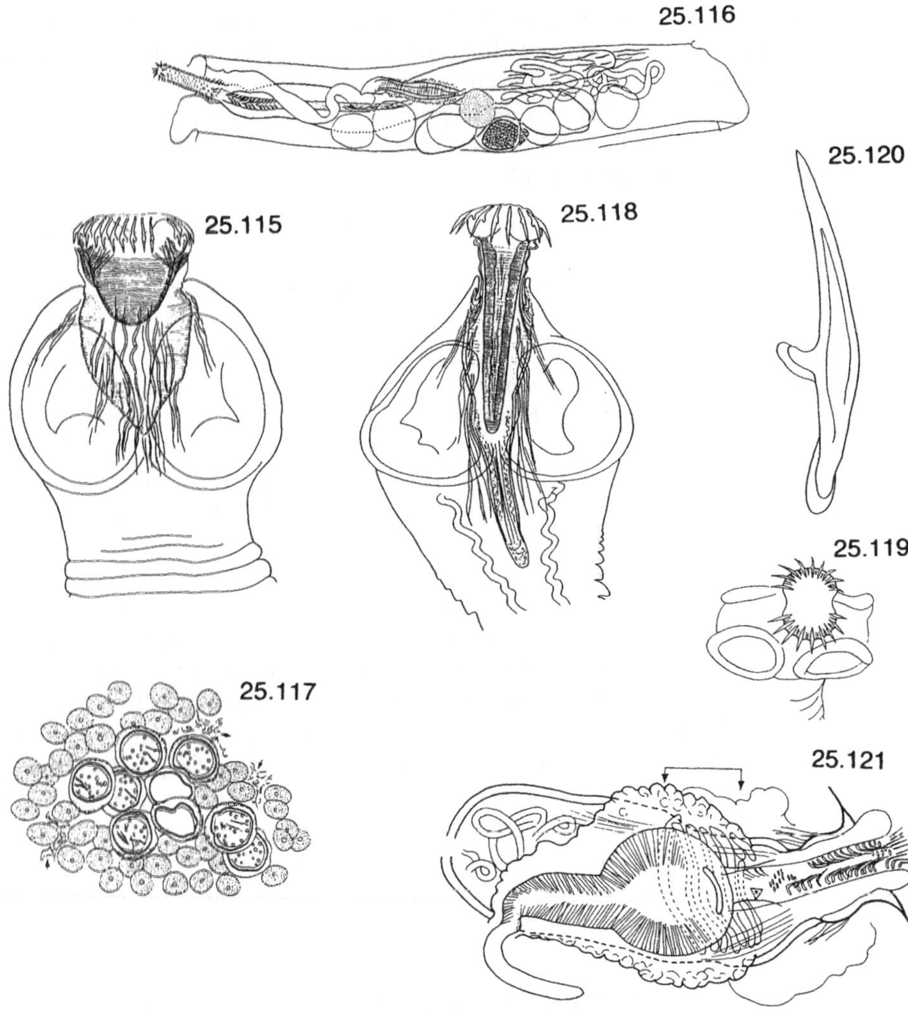

Figs 25.115–25.117 *Dilepidoides bauchei* (Joyeux, 1924), type. 25.115. Scolex. 25.116. Mature proglottid. 25.117. Embryophore, thicker line; egg outer coat outside, thin, with residues of granules.
Figs 25.118–25.121 *Trichocephaloidis*. 25.118, 25.119. *Trichocephaloidis* sp., scolex, lateral and apical views. 25.120, 25.121. *T. megalocephala* (Krabbe, 1869), Clerc's material. 25.120. Hook. 25.121. Terminal genitalia; arrows indicate the length of the horizontal layer of muscular fibres (partly dotted) between vagina and cirrus-sac, triangle with spot indicates the terminal, narrow part of the vagina.

apical tuft of straight, fine spines. Internal vas deferens short, narrow. Vas deferens anteroaporal. In Galliformes. Indochina. Type-species *D. bauchei* (Joyeux, 1924).

39b. **Strobila small to medium. Atrium wall very strong. Testes in expanded field. Ovary proportionally large, lobulate. Uterus sacciform lobulate, wall persistent** .... **40.**

40a. Pores submarginal, dorsal. Rostellar apex bilobed, hook-crown divided into dorsal and ventral semicircles. Vagina dorsal to cirrus-sac. Uterus ventral. Cirrus wide, bearing long spines .... *Trichocephaloidis* Sinitzin, 1896. (Figs 25.118–25.121)

(Syn. *Diagonaliporus* Krotov, 1951.)

Diagnosis: Rostellar apparatus muscular. Pouch long, carrot-shaped. Rostellum very long, narrow not coiled when retracted. Hooks in one row (eight or nine in each semicircle). Strobila small to medium. Pores fairly anterior, unilateral, on right. Genital ducts dorsal to osmoregulatory canals. Vagina mostly anterior to cirrus-sac; wall very thick; large distal cavity joined to atrium by short, narrow duct; long, dense 'cilia'; distally, between cirrus-sac and vagina, wide horizontal layer of muscle fibres, some of which coiled round duct towards atrium. Uterus sacciform, sometimes horseshoe-shaped. Embryophore thin, outer coat large, adhering to neighbouring eggs and to septa, at times simulating late uterine capsules. Atrium with dorsoventral axis, cavity small, wall thick. Ovary transversely extended. Testes few to numerous (7–20), posterior, in part dorsal to ovary. Cirrus-sac rather large, wall strong, distal portion bent dorsally towards pore. Cirrus with long spines, sharply curved at their base. In Charadriiformes. Palaearctic, Neotropical, probably wider. Type-species *T. megalocephala* (Krabbe, 1869).

40b. Pores marginal. Rostellum not bilobed, hooks in X-shaped bundle when retracted. Vagina ventral to cirrus-sac. Uterus dorsal. Cirrus narrow, very long, may be armed with small spines .... *Lateriporus* Fuhrmann, 1907. (Figs 25.122, 25.123)

Diagnosis: Rostellar apparatus probably only muscular. Pouch large. Rostellum large, elongate. Hooks in one circle. Strobila large. Genital pores unilateral, right (type-species). Genital ducts dorsal to osmoregulatory canals. Vagina wide, fusiform when dilated (type-species). Uterus sacciform, deeply lobulate. Atrium

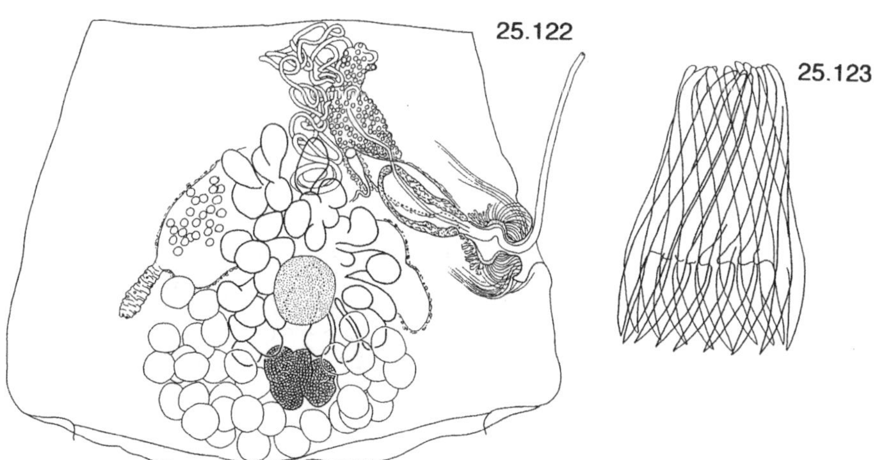

Figs 25.122, 25.123 *Lateriporus teres* (Krabbe, 1869), type. 25.122. Mature proglottid. 25.123. Hooks.

large; wall strong, sucker-like, radiate fibres bent towards pore and outer fibres towards inside of proglottid. Ovary not clearly bilobed, mainly horseshoe-shaped. Vitellarium more dorsal than ovary. Testes numerous, postovarian. Cirrus-sac (with everted cirrus) oval, wall thick. Cirrus narrow, very long. No internal seminal vesicle. Vas deferens far anterior, median. In Anseriformes, Charadriiformes (?). Palaearctic, Arctic Region. Type-species *L. teres* (Krabbe, 1869).

41a. Uterus at first reticulate, fine or coarse-meshed, or firm-walled labyrinthine, then (i) turning into uterine capsules, or (ii) remaining labyrinthine, sometimes disappearing, or (iii) becoming sacciform .... 42.

41b. Uterus at first sacciform, persistent or developing into capsules ... 47.

42a. Uterus finely reticulate or labyrinthine, developing into capsules .... 43.

42b. Uterus at first reticulate, then either persistently labyrinthine or becoming septate sac; no uterine capsules .... 45.

43a. Uterine capsules polyovular, large, with an egg packet each. Testes surround ovary. Seminal receptacle huge .... *Capsulata* Sandeman, 1959. (Figs 25.124, 25.125)

Diagnosis: Rostellar apparatus muscular. Pouch and rostellum very long. Hooks in two circles (16). Strobila medium; neck long, rigid, deeply craspedote, like a

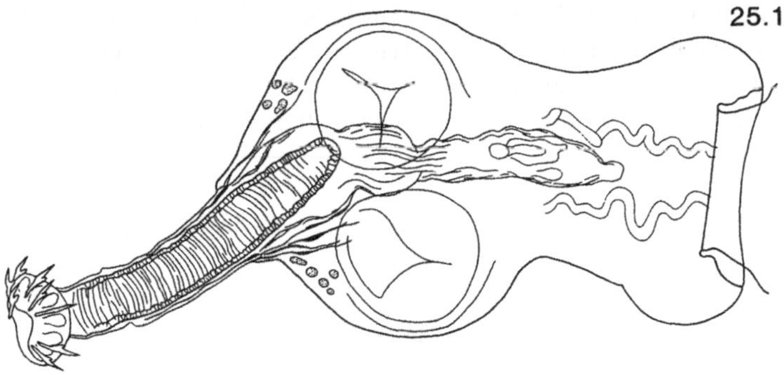

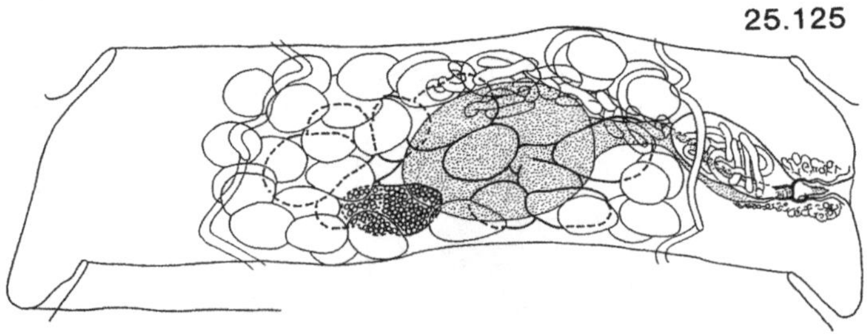

Figs 25.124, 25.125 *Capsulata edenensis* Sandeman, 1959, type. 25.124. Scolex and its proglottid-like neck. 25.125. Mature proglottid.

proglottid. Genital pores unilateral, right. Genital ducts usually between osmoregulatory canals, but may be dorsal in same strobila. Vagina ventral, slightly posterior to cirrus-sac; extremely short, wall firm, spiny, warped funnel, embedded in atrium wall. Uterus initially labyrinthine. Atrium wall thick, bulbous. Ovary large; few, large, subspherical lobules. Testes numerous, few posteriorly in median line. Cirrus-sac roundish-oval, firm-walled. Cirrus small, basal part spinescent, then abruptly filiform. Seminal receptacle prolonged as tube as far as atrium. In Charadriiformes. North Europe. Type-species *C. edenensis* Sandeman, 1959.

43b. Uterine capsules monovular. Testicular field and seminal receptacle different .... 44.

44a. Uterine capsules vesicular. Rostellar apparatus exceptionally strong; rostellum has small apical cavity with pore. Atrium wall bulbous, dense, without longitudinal, concentric fibres; inner layer thick at base. Vaginal orifice normal. Ovary small, hardly bilobed and lobulate, poral. (See also irregularly alternate pores; diagnosis and synonyms at 93a and Figs 25.272–25.276) .... *Megalacanthus* Moghe, 1926.

44b. Uterine capsules membranous. Rostellar apex normal. Atrium wall bulbous, with concentric, longitudinal, bent fibres, and inner layer thick throughout whole cavity. Vaginal orifice with sphincter-like thickening. Ovary bilobed, deeply lobulate, central. (See also irregularly alternate pores and diagnosis at 103a and Figs 25.303–25.305) .... *Birovilepis* Spasskii, 1975.

45a. Uterus persistently labyrinthine, but unusual development (see diagnosis). Eventually eggs may be linked in chains. Hooks about 20–32. Atrium not very deep, but wall strong, complex. Poral part of vagina forming sharp posterior bend before entering atrium. In Charadriiformes .... *Anomolepis* Spasskii, Yurpalova & Kornyushin, 1968. (Figs 25.126–25.130)

Diagnosis: Rostellar apparatus musculo-glandular. Pouch large. Rostellum strong; apical disc wide. Hooks in two circles, slender, blade long, guard anteriorly directed. Strobila medium. Proglottids wider or much wider than long. Osmoregulatory canals may form fine network behind suckers. Genital pores unilateral, right or left. Genital ducts between osmoregulatory canals. Vagina posterior, dorsal to or in same horizontal plane as cirrus-sac, according to species (ventral position rare). Uterus wall at first with small spindle-like cells; when septa and parenchyma degenerate, eggs lie inside transparent, elastic, stretched tubes, probably corresponding to initial reticulum, simulating very long polar processes. No uterine capsules. Embryophore strong, oval; outer coat firm with two polar caps (true polar processes doubtful). Eggs attached by polar caps. Atrium with recurved longitudinal, concentric fibres, forming ventral half-sphincter at cirrus-sac end. Ovary very large, as wide as testicular field, reaching osmoregulatory canals; aporal lobe much larger. Testes numerous, postovarian. Cirrus-sac elongate, narrow. Cirrus thin, very long. In Charadriiformes. Palaearctic, Oriental, North Africa, probably wider. Type-species *A. glareola* (Dubinina, 1953).

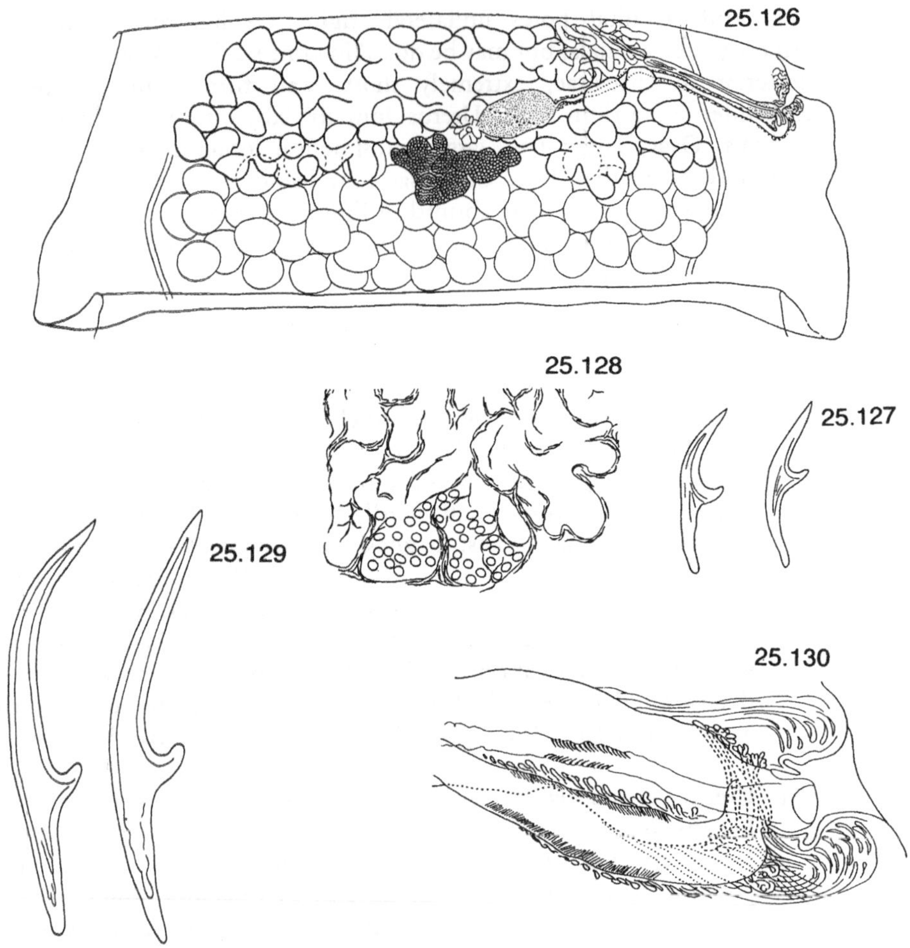

Figs 25.126–25.130 *Anomolepis*. 25.126–25.128. *Anomolepis* sp. 25.126. Mature proglottid. 25.127. Hooks (left, distal; right, proximal). 25.128. Uterus; see spindle-like cells all along the wall in early stages. 25.129, 25.130. *A. limosa* (Fuhrmann, 1907), type. 25.129. Hooks (left, distal; right, proximal). 25.130. Terminal genitalia; vagina dorsal; half-sphincter (dotted) ventrally to distal end of cirrus-sac; concentric atrial fibres connected with the distal part of the cirrus-sac.

45b. Uterus at first coarse-meshed reticulum, then sacciform. Eggs not lined in chains. Hooks 40–70. Atrium small, simple, wall may be slightly thickened. Poral part of vagina seldom posterior to cirrus-sac. Hosts different .... 46.

46a. Uterus at first coarsely reticulate or labyrinthine, often in ventral and dorsal layer, then sacciform, weakly septate. Dorsal osmoregulatory canals (at least poral side) slightly aporal to ventral ones. Usually hook blade short, stumpy, usually much shorter than handle, unduloid pattern. Rostellum strong, massive; apical pad wide, dome-shaped, indistinct. Chiefly in Passeri-

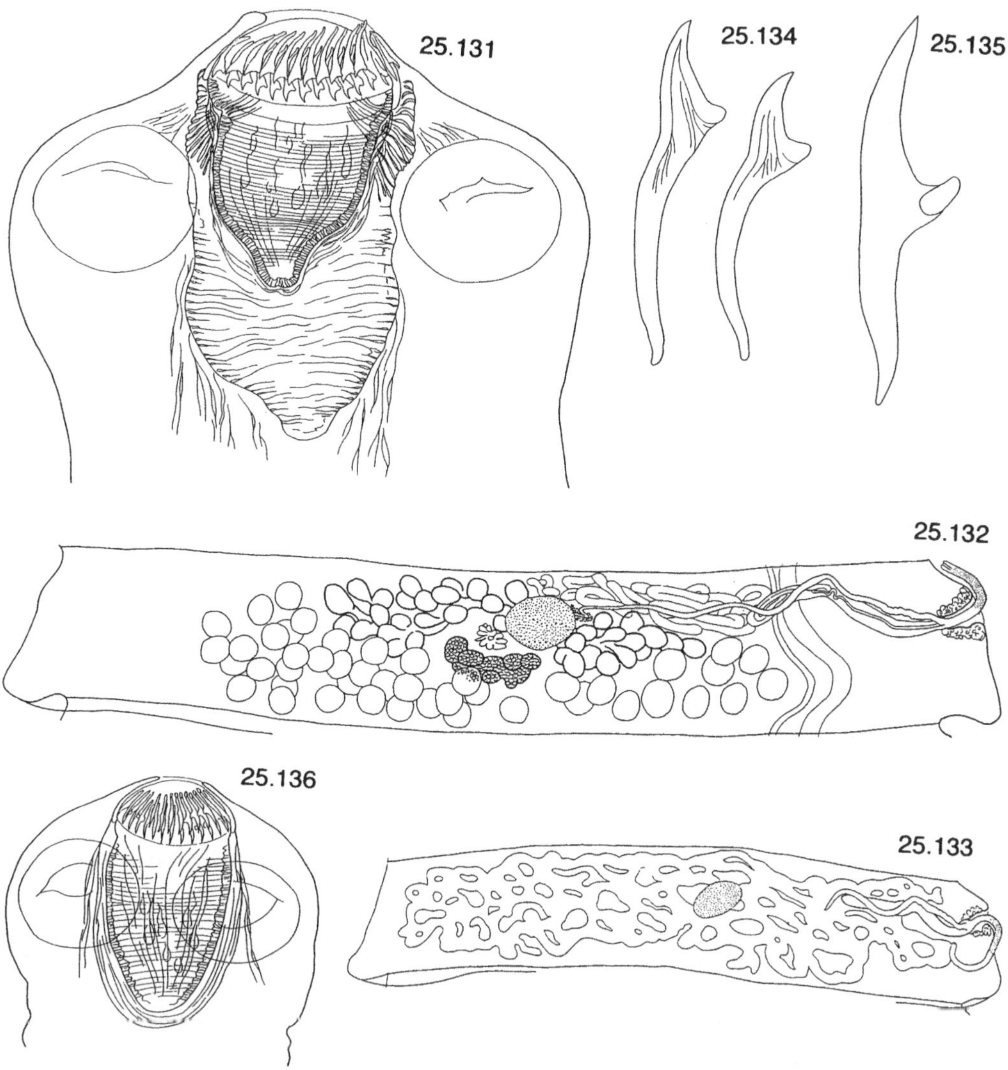

Figs 25.131–25.136 *Dilepis.* 25.131–25.134. *D. undula* (Schrank, 1788). 25.131. Scolex. 25.132. Mature proglottid. 25.133. Early uterus. 25.134. Unduloid hooks (left, distal; right, proximal). 25.135. *D. bicoronata* Fuhrmann, 1908, type, hook of different pattern. 25.136. *D. dacelonis* (Johnston, 1909) n. comb., type (syn. *Similuncinus dacelonis* Johnston, 1909), scolex.

formes .... ***Dilepis*** **Weinland, 1858. (Figs 25.131–25.136)** (Syns *Ershovilepis* Matevosyan, 1963; *Brasiliolepis* Spasskii, 1965b; *Similuncinus* Johnston, 1909; *Southwellia* Moghe, 1925.)

Diagnosis: Rostellar apparatus muscular. Pouch large, wall strong. Hooks in two circles, usually very numerous (from 50–70 to more than 100). Strobila medium to large, thick, often blunt anteriorly. Proglottids wider than long. Genital pores unilateral, right. Genital ducts usually dorsal to osmoregulatory canals, seldom between, in same strobila. Vagina dorsal to cirrus-sac, usually anterior, long. Embryophore thin, outer coat firm. Atrium small, wall globular, reduced. Ovary less extended than testicular field, usually not reaching osmoregulatory canals; aporal lobe close to anterior margin of proglottid. Testes numerous, postovarian, at times also aporally lateral. Cirrus-sac long, narrow. Cirrus long with tiny spines. Vas deferens anteroporal. In Passeriformes (chiefly Turdidae and Corvidae), more seldom Galliformes and other terrestrial birds; stomach of Alcedinidae; exceptionally Charadriiformes (?), Ciconiiformes (?) and mammals (Insectivora, Rodentia (?), Carnivora). Cosmopolitan. Type-species *D. undula* (Schrank, 1788).

46b. Uterus ventral, coarse-meshed reticulum, with islets of parenchyma, then sacciform, some deep septa. Osmoregulatory canals in normal position. Hooks shape different, blade slender, slightly or much longer than handle. Rostellum strong, stem narrow, apical pad very wide, delimited. In Galliformes, Megapodidae ... *Megacirrus* Beck, 1951. (Figs 25.137–25.141)

Diagnosis: Rostellar apparatus muscular (no data about glandular component). Pouch large, cuneiform, at times reaching first segments. Hooks in two circles (40–54), typical. Strobila medium. Genital pores unilateral, right. Genital ducts dorsal or between osmoregulatory canals in different species. Vagina dorsal, anterior (distal portion) to cirrus-sac; long, thick-walled, spiny near orifice. Atrium small, simple. Ovary relatively large, may reach osmoregulatory canals but not anterior margin of proglottid. Testes numerous, postovarian. Cirrus-sac elongate, narrow, at times very long. Cirrus long, spines small. Seminal receptacle small, spherical. Vas deferens anteromedian. In Galliformes, Megapodidae. Papua New Guinea. Type-species *M. megapodii* Beck, 1951.

47a. Uterus sacciform, dorsal, ramifying coarsely, radially and turning into both mono- and polyovular capsules (mixed). Ovary poral. Seminal receptacle very small, poral .... *Malika* Woodland, 1929. (Figs 25.142–25.144)

Diagnosis: Rostellar apparatus muscular, typical; glands inside rostellum. Pouch cavity negligible and wall poorly defined, when rostellum retracted. Rostellum robust; wall two-layered; inner layer normal, thick, with thin transverse fibres, outer layer adhering, thin, hardly visible, compact, with longitudinal fibres, thickening before reaching apical pad; apical pad deeply demarcated. Hooks in two circles. Strobila large. Proglottids largely overlapping, organs spaced, cortex wide. Genital pores unilateral, either left or right (rare pores on opposite side). Genital ducts between osmoregulatory canals. Vagina posterior to cirrus-sac, usually in same horizontal plane, straight, strong-walled, spiny. Uterine capsules firm-walled with special, loose inner structure. Atrium large, spherical, wall thick, radial fibres; atrial papilla large. Ovary small, usually not bilobed. Testes numerous, postovarian. Cirrus-sac oval, firm-walled. Cirrus large, spiny. Internal vas deferens very short; when turgid, one or two small loops persist. Vas deferens anteroporally isolated; with dense prostatic sleeve. In Charadriiformes. India, Sri Lanka, probably wider. Type-species *M. oedicnemus* Woodland, 1929.

47b. Uterus sacciform, ventral, persistent. Ovary not poral. Seminal receptacle large .... 48.

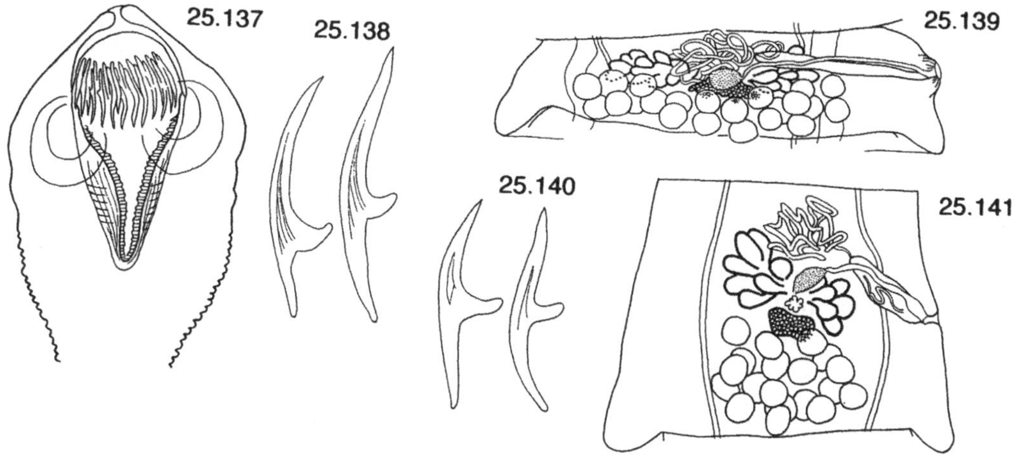

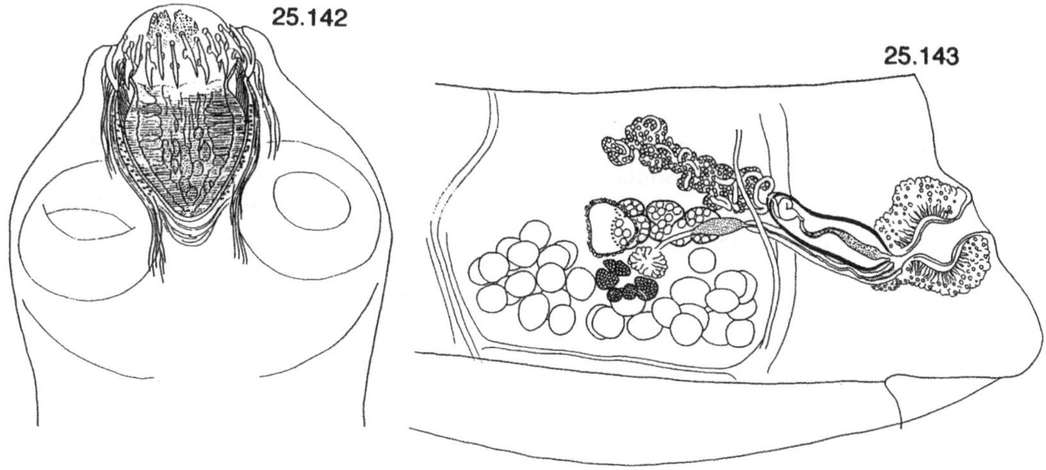

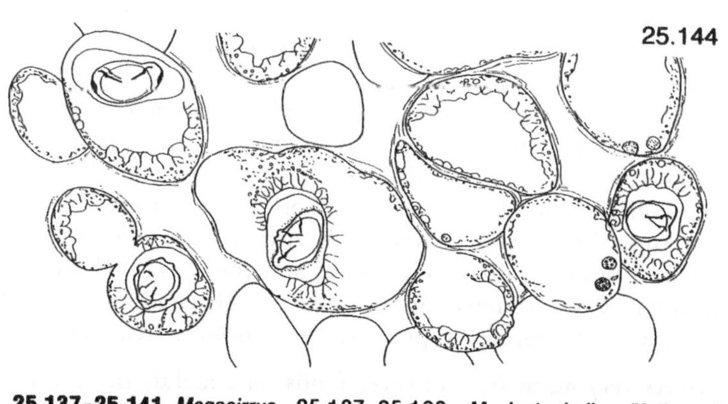

Figs 25.137–25.141 *Megacirrus*. 25.137–25.139. *M. leptophallos* (Kotlan, 1923). 25.137. Scolex. 25.138. Hooks (left, distal; right, proximal). 25.139. Mature proglottid. 25.140, 25.141. *M. horvathi* (Kotlan, 1923). 25.140. Hooks (left, distal; right, proximal). 25.141. Mature proglottid. (From Murai & Georgiev, 1987.)

Figs 25.142–25.144 *Malika himantopodis* Burt, 1940, type. 25.142. Scolex. 25.143. Mature proglottid. 25.144. Uterine capsules.

48a. Vagina dorsal to cirrus-sac. Hooks nearly 50. Ovary bilobed, small; lobules few (four), relatively large, subspherical. Testes few (four). In Galliformes, Megapodidae ....
.... ***Kotlanolepis*** **Murai & Georgiev, 1988. (Figs 25.145, 25.146)**

Diagnosis: Rostellar apparatus muscular (no data about glandular component). Pouch large, fairly narrow, reaching beyond suckers. Rostellum strong; apical pad wide. Hooks in two circles; megacirroid pattern (see 46b). Strobila small. Genital pores unilateral, right. Genital ducts dorsal to osmoregulatory canals. Vagina posterior to cirrus-sac, short, narrow, thin-walled. Uterus always sacciform. Atrium small, simple. Ovary central. Testes postovarian. Cirrus-sac long, narrow. Cirrus short, thin, unarmed (?). Seminal receptacle large, fusiform, transverse, reaching beyond osmoregulatory canals. Internal vas deferens long, swollen loop. External vas deferens with few, long, swollen loops. In Galliformes, Megapodidae. New Guinea. Type-species *K. yorkei* (Kotlan, 1923).

48b. Vagina ventral to cirrus-sac. Hooks 16–26. Ovary bilobed, large with many more than four lobules, or not bilobed (50b). Testes more than four. Different hosts .... **49.**

49a. Aporal osmoregulatory canals dorsoventrally inverted. Genital ducts dorsal. Uterus deeply lobulate and ramified; transverse, horseshoe-shaped or circular with branched anteromedian diverticulum. Ovary bilobed with numerous deep lobules. (See also parvitaenioid rostellar apparatus, diagnosis and synonyms at 34b and Figs 25.100–25.106) ....
.... ***Dendrouterina*** **Fuhrmann, 1912.**

49b. Aporal osmoregulatory canals in normal position. Genital ducts in between. Uterus simply lobulate. Ovary bilobed or not .... **50.**

50a. Testes few, postovarian. Cirrus with swollen, armed base and two long, strong spines; when withdrawn, proximally smooth with sphincter before internal vas deferens. Ovary bilobed rather large, with few, large lobules
.... ***Acanthocirrus*** **Fuhrmann, 1907. (Figs 25.147, 25.148)**

Diagnosis: Rostellar apparatus muscular. Pouch long, reaches far beyond suckers. Rostellum as long as pouch, sometimes folded once; stem narrow, apical disc wide. Hooks in two circles (about 20). Strobila medium. Genital pores unilateral, right or left in same species. Genital ducts between osmoregulatory canals. Vagina posterior, orifice ventral to cirrus-sac; short, funnel-shaped. Uterus sacciform, lobulate, superficially septate; wall thick. Atrium deep, wall thick, cavity at base round sexual orifices with circular fibres and thick setae. Cirrus-sac long, thin-walled. Internal vas deferens single long loop. In Charadriiformes (Passeriformes, very uncertain). Palaearctic. Type-species *A. retirostris* (Krabbe, 1869).

50b. Testes very numerous, in three fields, one mainly postovarian in medulla and two posterolateral in cortex. Cirrus simple, probably unarmed. Ovary not bilobed, very small, compact, central. In Upupiformes, formerly Coraciformes, Bucerotidae ....
.... ***Bucerolepis*** **Spasskii & Spasskii, 1968. (Fig. 25.149)**

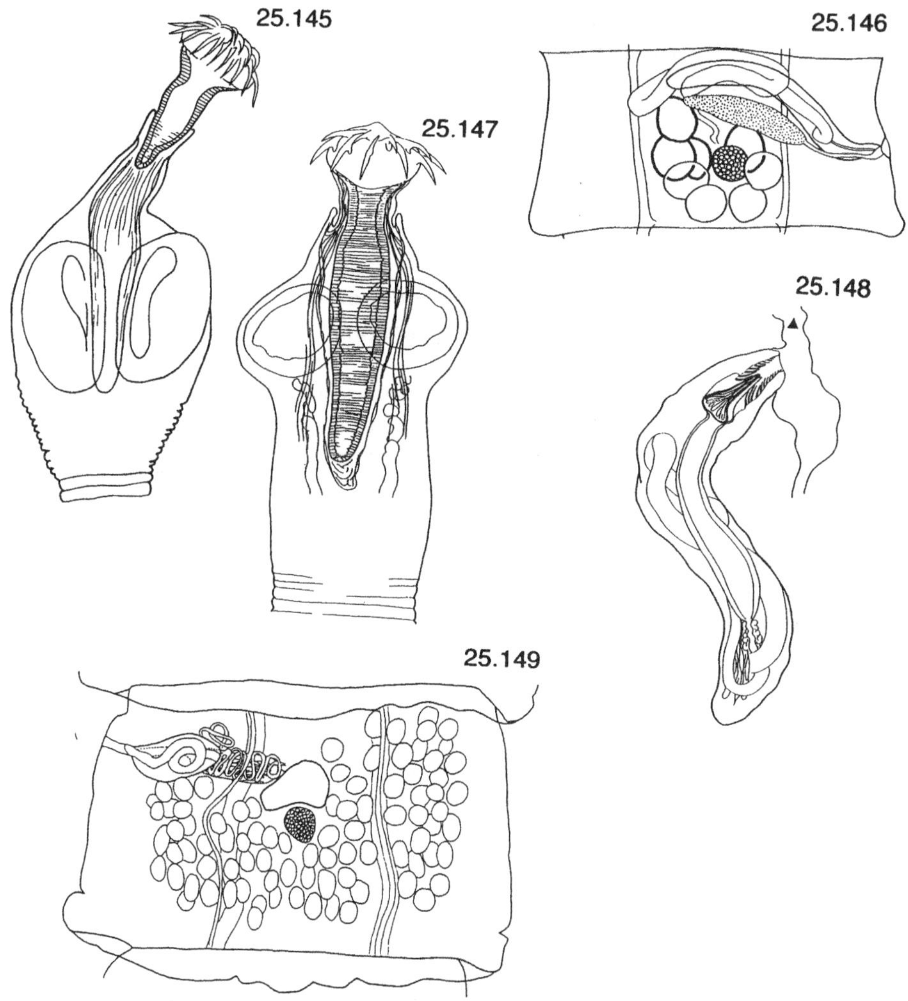

Figs 25.145, 25.146 *Kotlanolepis yorkei* (Kotlan, 1923). 25.145. Scolex. 25.146. Mature proglottid. (From Murai & Georgiev, 1987). (For hook pattern, see *Megacirrus* Fig. 25.138.)
Figs 25.147, 25.148 *Acanthocirrus retirostris* (Krabbe, 1869). 25.147. Scolex. 25.148. Terminal genitalia, probably type; triangle indicates atrial cavity.
Fig. 25.149 *Bucerolepis bicanistis* (Mahon, 1954), mature proglottid. (From Mahon, 1954.)

Diagnosis: Apical apparatus probably muscular. Pouch large, reaching first segments. Rostellum large. Hooks in two circles (24–26). Strobila probably medium; strongly protandrous. Genital pores unilateral. Genital ducts between osmoregulatory canals. Vagina ventral to cirrus-sac along sac axis. Uterus sacciform, lobulate, persistent. Atrium simple. Cirrus-sac small, subspherical to oval, anterior. Seminal receptacle underlying vas deferens. Vas deferens small, on axis of cirrus-sac. In Upupiformes (formerly Coraciformes), Bucerotidae. Africa. Type-species *B. bicanistis* (Mahon, 1954).

51a. **Dumb-bell organ (X-shaped sclerified valve) between vagina and seminal receptacle. Cirrus shape and spinosity typical in the whole group. Ductus hermaphroditus present at base of atrium, strong-walled with sphincter-like circular fibres (see 52a, 53a, 53b). In Apodiformes; uncertain Passeriformes, Hirundinidae .... 52.**

51b. **No dumb-bell organ. Cirrus shape and spinosity seldom similar. Ductus hermaphroditus very rare .... 54.**

52a. **Hooks in zigzag line. Rostellum disproportionately large with respect to scolex. Tegument of proglottids smooth .... *Pseudangularia* Burt, 1938. (Figs 25.150–25.152)**
(Syn. *Mehdiangularia* Shinde, 1969.)

Diagnosis: Rostellar apparatus muscular (and glandular?). Rostellar apical disc very wide. Hooks in three to four circles, sometimes partly only two; usually numerous (40–60); typical: guard prominent, longer than blade, oriented anteriorly. Strobila small. Suckers small. Genital pores irregularly alternate (at times with long regularly alternate series). Genital ducts dorsal to osmoregulatory canals. Vaginal orifice anterodorsal to cirrus-sac; course inclined posteriorly; usually very large; wall exceptionally thick; sclerified dumb-bell organ. Uterus sacciform, septate; tangle of small communicating cavities. Eggs may have long poral

25.150 25.151 25.152

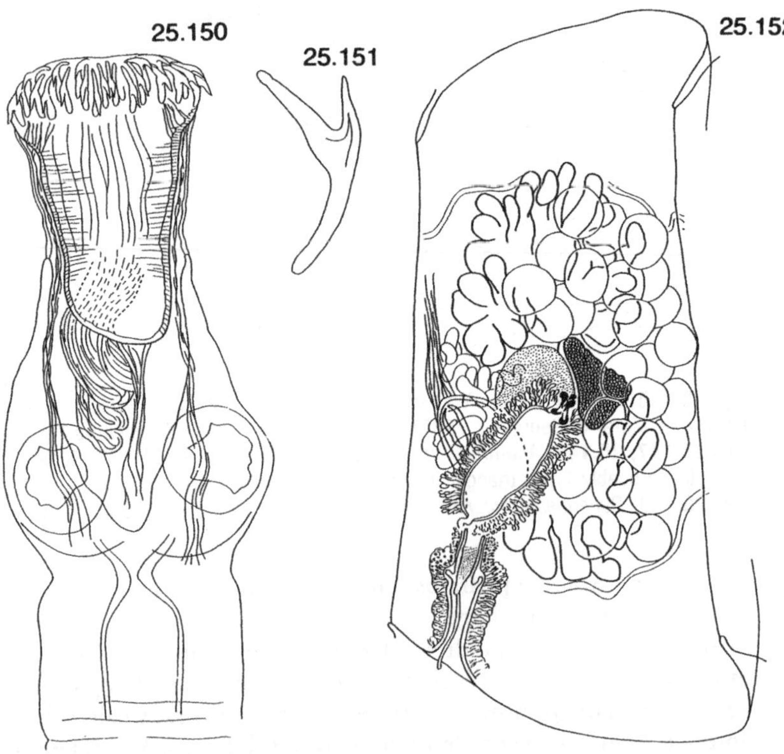

Figs 25.150-25.152 *Pseudangularia thompsoni* Burt, 1938, type. 25.150. Scolex. 25.151. Hook. 25.152. Mature proglottid. (For uterus, see *Neoliga* Fig. 25.156.)

processes. Atrium deep; wall very strong; ductus hermaphroditus strong-walled (see 51a). Ovary extended with narrow isthmus, may be laterally fanned; lobules large, clavate or digitiform, extended posteriorly. Testes numerous (12–35); posterior and dorsal to ovary. Cirrus-sac large; wall strong; muscular bulb inside at base. Cirrus long, very thin; basal part wider with tiny spines; may have second distal spinous field. Internal vas deferens swollen; two elongate loops. In Apodiformes. Sri Lanka, Morocco, Europe, probably wider. Type-species *P. thompsoni* Burt, 1938.

52b. Hooks in two circles, regular. Rostellum proportionate to scolex. Posterior part of proglottid often with visible microtriches or spines .... 53.

53a. Pores nearly always regularly alternate. Posterior part of proglottids smooth, or bearing visible microtriches or tiny spinules. Ductus hermaphroditus at base of atrium, well-developed, strong-walled (see 51a) .... *Neoliga* Singh, 1952. (Figs 25.153–25.156)
(Syns *Neoangularia* Singh, 1952; *Sureshia* Ali & Shinde, 1967.)

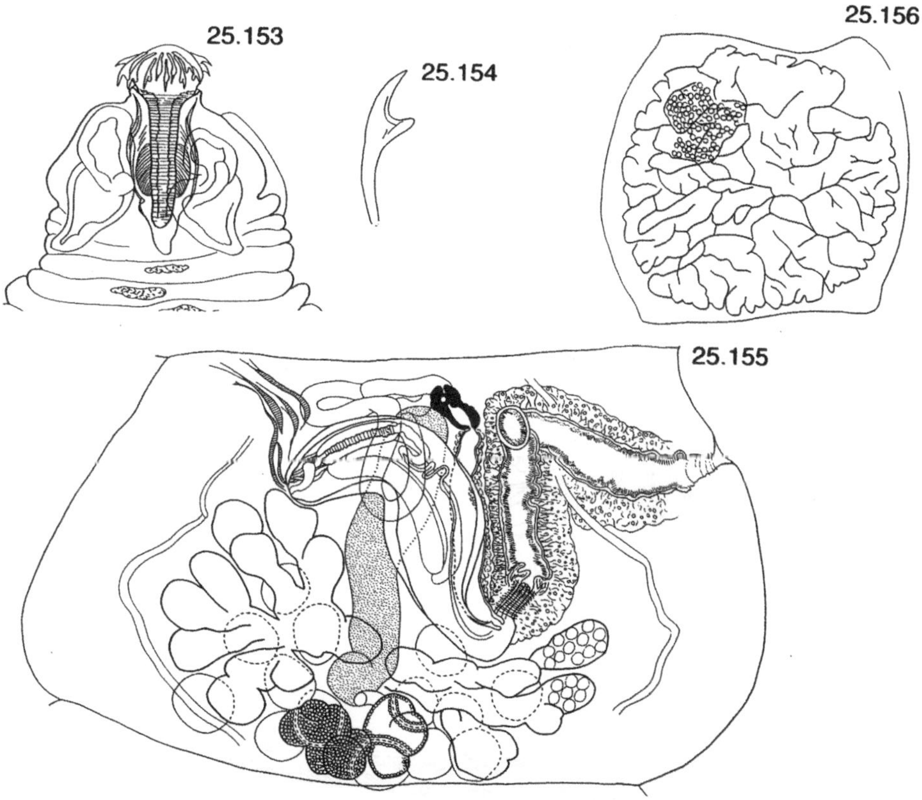

Figs 25.153-25.156 *Neoliga*. 25.153-25.155. *N. frigida* (Meggitt, 1927), type. 25.153. Scolex, two dense lenticular, muscular masses on both sides of rostellum stem. 25.154. Hook. 25.155. Mature proglottid; exceptionally deep and muscular atrium, ductus hermaphroditus with sleeve of circular fibres. 25.156. *Neoliga* sp., uterus.

Diagnosis: Rostellar apparatus muscular (and glandular?). Hooks in two circles, about 26–32. Strobila usually small. Suckers large. Genital ducts dorsal to osmoregulatory canals. Vagina anterodorsal (very seldom anteroventral) to cirrus-sac; varying size; wall strong, at times thick; sclerified dumb-bell organ. Uterus sacciform, septate; tangle of irregular communicating cavities. Egg with long polar processes. Atrium may be extremely deep; wall very strong, at times thick setae. Ovary with narrow isthmus, may be laterally fanned; lobules clavate or digitiform, extended posteriorly. Testes numerous, posterior, lateral, dorsal to ovary. Cirrus-sac large; wall strong; muscular bulb inside at base. Cirrus long, very thin; basal part wider with tiny spines; may have second distal, hardly visible, spinous field. Internal vas deferens swollen, two elongate loops. Vas deferens with few loops. In Apodiformes; uncertain in Passeriformes, Hirundinidae. India, Morocco, Europe, probably wider. Type-species *N. diplacantha* Singh, 1952.

53b. **Pores irregularly alternate. Posterior margin of proglottids armed with true spines. Ductus hermaphroditus at base of atrium, short, reduced to simple sphincter distal to sexual orifices** .... ***Echinotaenia*** **Mokhehle, 1951. (Figs 25.157–25.159)**

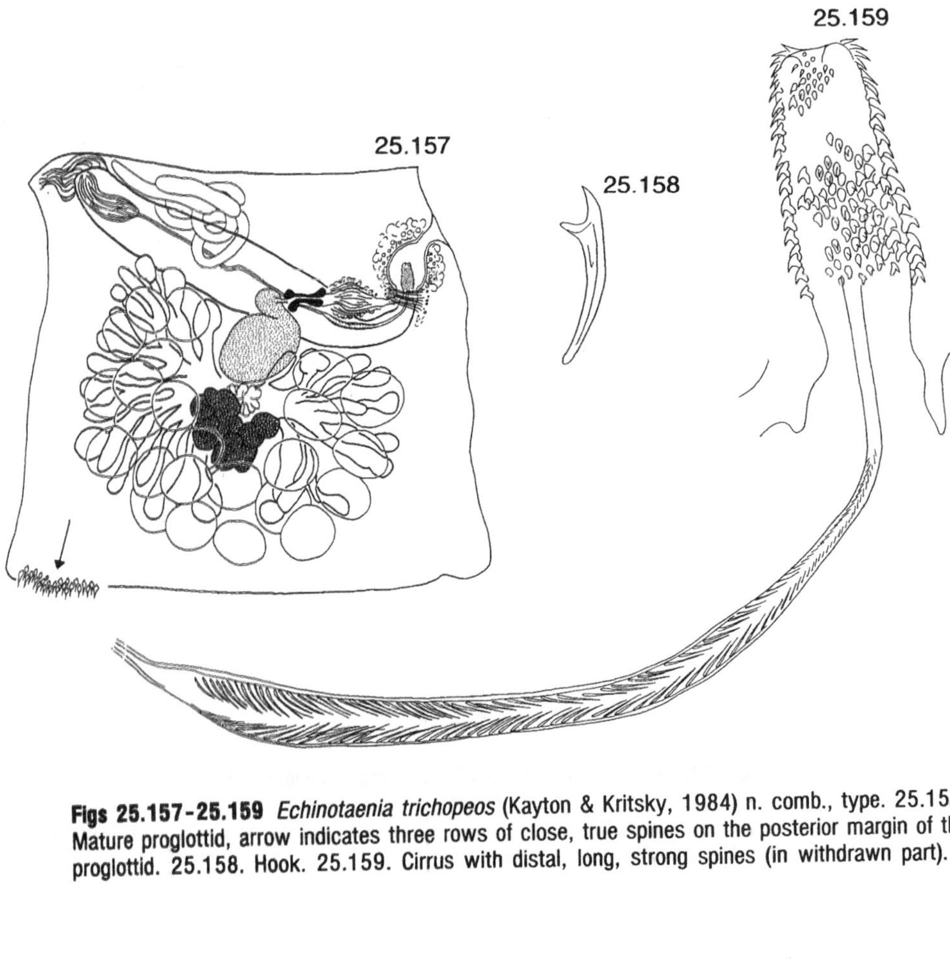

Figs 25.157-25.159 *Echinotaenia trichopeos* (Kayton & Kritsky, 1984) n. comb., type. 25.157. Mature proglottid, arrow indicates three rows of close, true spines on the posterior margin of the proglottid. 25.158. Hook. 25.159. Cirrus with distal, long, strong spines (in withdrawn part).

Diagnosis: Rostellar apparatus muscular. Hooks in two circles (about 26–32). Strobila small. Genital ducts dorsal to osmoregulatory canals. Vagina anterodorsal to cirrus-sac; small; thickness of wall variable; sclerified dumb-bell organ. Uterus sacciform, lobular, septate. Egg with long polar processes. Atrium deep, extensible, wall not particularly thick. Ovary with narrow isthmus, laterally fanned; lobules clavate or digitiform, extended posteriorly. Testes numerous, posterior, lateral and dorsal to ovary. Cirrus-sac large; wall strong; muscular bulb inside at base. Cirrus long, thin; basal part wider with tiny spines; second distal field with fine, fairly long, spines. Internal vas deferens swollen; two elongate loops. Vas deferens forms few wide loops. In Apodiformes; uncertain in Passeriformes, Hirundinidae. India, Java, Palaearctic, Nearctic. Type-species *E. lehaqasia* Mokhehle, 1951.

54a. Vagina dorsal or ventral to cirrus-sac in same strobila, according to pore side, right or left. Hooks in (i) zigzag line (more than two circles), or in two very irregular circles, (ii) in two regular circles with 1:1 or 1:2 alternation in different species. Genital ducts always dorsal to osmoregulatory canals. In Passeriformes, Hirundinidae; perhaps Apodiformes .... 55.

54b. Vagina never changing its dorsoventral position relative to cirrus-sac when pore side changes in same strobila. Hooks never in zigzag line, seldom two irregular circles. Genital ducts dorsal to or between osmoregulatory canals. In several orders of birds .... 57.

55a. Hooks in zigzag line or in very irregular circles with 1:2 alternation mixed with zigzag portions. Vagina ventral to cirrus-sac in right pores, dorsal in left ones; long, crossing ovarian lobe. Seminal receptacle small, nearly central, between ovarian lobes. Ovary large, no clear-cut lobes, not far extended posteriorly. Testes mainly postovarian, in transverse field .... 56.

55b. Hooks in two circles, regular, either with 1:1 or with 1:2 alternation in different species. Vagina dorsal to cirrus-sac in right pores, ventral in left ones; very short, funnel-like, spiny. Seminal receptacle poral, large, transversely elongate, crossing cirrus-sac before reaching vagina. Ovary very large, isthmus deep, thin, lobes fan-like extended posteriorly. Testes posterior, lateral and dorsal to ovary .... *Vitta* Burt, 1938. (Figs 25.160, 25.161)

Diagnosis: Rostellar apparatus muscular. Pouch adhering to rostellum; wall strong. Rostellum strong, stumpy, apical disc wide. Hooks 40–60. Strobila small. Scolex wide. Suckers small. Genital pores irregularly alternate, anterior. Genital ducts dorsal to osmoregulatory canals. Uterus sacciform, at times coarse reticulum, septa deep, close. Eggs with thin polar processes. Atrium of variable depth, usually large; wall thick, extending round vagina. Ovary may fill whole medulla. Testes very numerous (60–100). Cirrus-sac elongate, far anterior. Cirrus narrow, spines tiny, thick. Internal vas deferens forms muscular bulb just proximal to withdrawn cirrus. In Passeriformes, Hirundinidae; perhaps Apodiformes. Cosmopolitan. Type-species *V. magniuncinata* Burt, 1938.

56a. Pores irregularly alternate. Hooks in zigzag line. Uterus sacciform, septate, no deep outgrowths initially. Cirrus narrow, spines small, very fine.

25.160

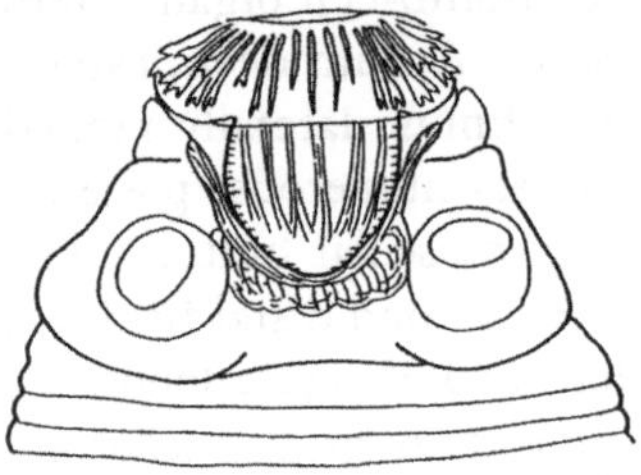

25.161

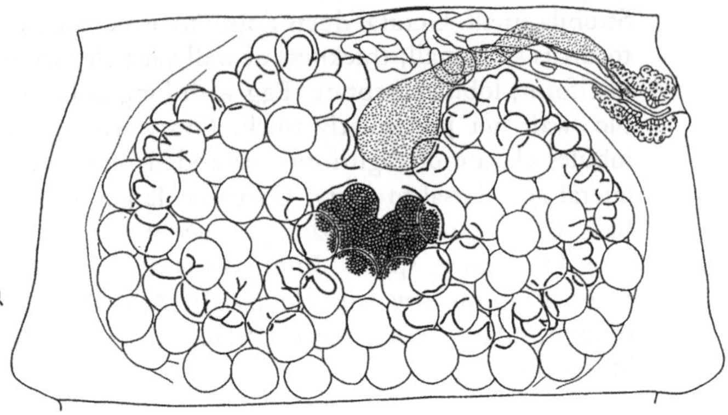

25.162

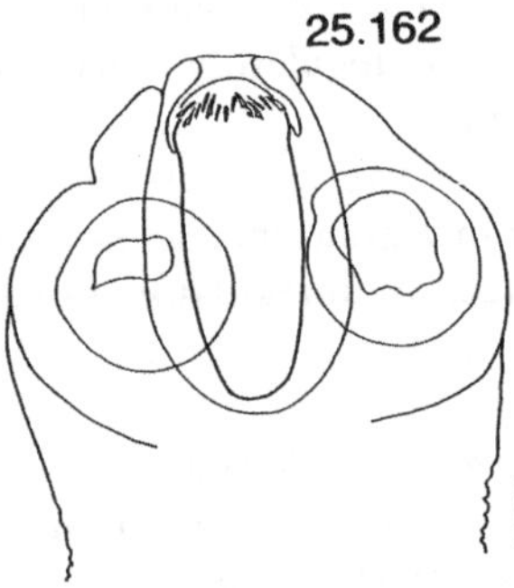

25.163

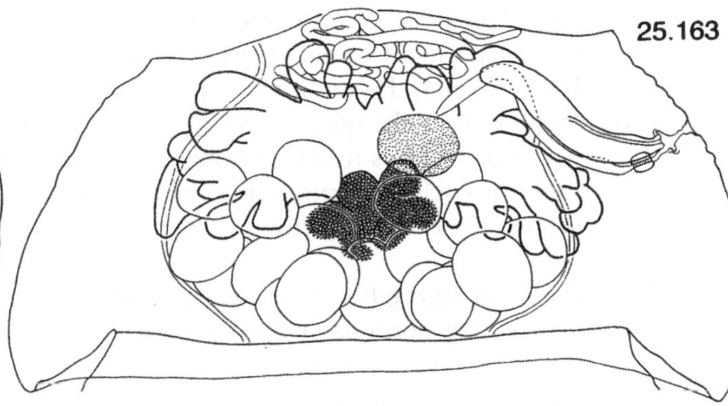

25.164

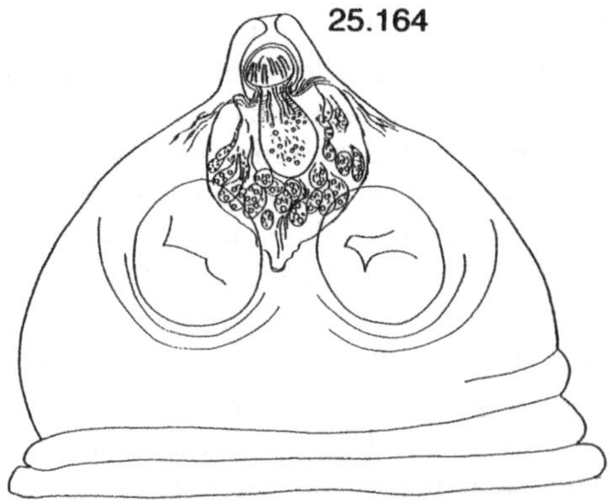

25.165

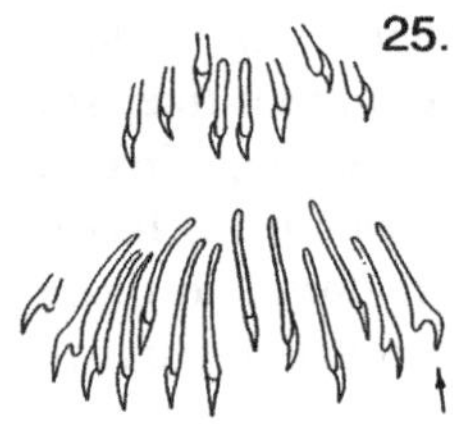

25.167

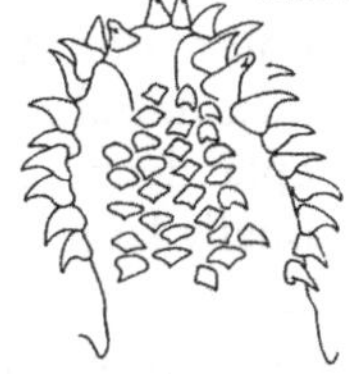

25.166

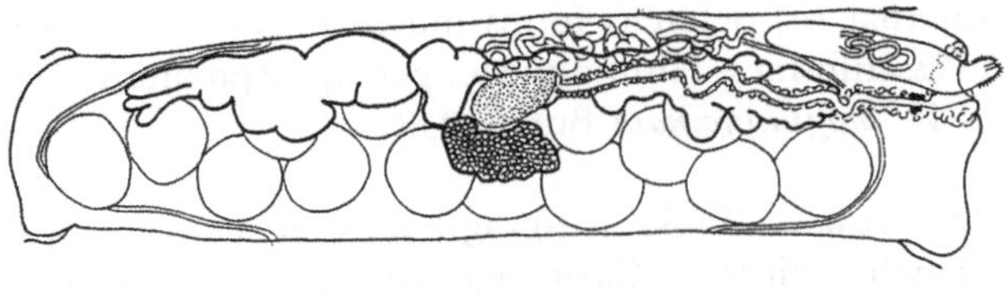

Rostellar pouch large, strong. Rostellum strong, long. Ovary bulky. Cortex fairly wide .... *Angularella* Strand, 1928. (Figs 25.162, 25.163)
(Syn. *Angularia* Clerc, 1906 preoccupied.)

Diagnosis: Rostellar apparatus musculo-glandular. Hooks in three or four circles; very numerous. Strobila small to medium. Genital ducts dorsal to osmoregulatory canals. Vagina posterior to cirrus-sac, ventral in right pores, dorsal in left ones; long, may have pre-atrial sphincter. Egg with long polar processes (types?). Atrium simple. Ovary may extend slightly beyond osmoregulatory canals and ventrally to cirrus-sac. Testes numerous, posterior and partly dorsal to ovary and vitellarium. Cirrus-sac elongate. In Passeriformes, Hirundinidae. Cosmopolitan. Type-species *A. beema* (Clerc, 1906).

56b. **Pores regularly alternate (very seldom two consecutive). Hooks disorderly, 1:2 alternation and zigzag portions mixed. Uterus sacciform, septate, initially forming longitudinally extended outgrowths. Cirrus wide, strongly spiny at base. Rostellar pouch delicate, highly glandular; rostellum small. Ovary transversely extended. Cortex narrow .... *Hirundinicola* Birova-Volosinovicova, 1969. (Figs 25.164–25.167)**
(Syn. *Vitta (Hirundinicola)* Birova-Volosinovicova, 1969.)

Diagnosis: Rostellar apparatus musculo-glandular. Rostellar apical pad clearly defined. Hooks small, thin; handle long, blade short. Strobila small (type-species). Genital ducts dorsal to osmoregulatory canals. Vagina posterior to cirrus-sac, ventral in right pores, dorsal in left ones; very long, overlying poral ovarian lobe; pre-atrial sphincter. Atrium simple. Ovary reaching osmoregulatory canals. Testes numerous in transverse narrow field, posterior and dorsal to ovary and vitellarium. Cirrus-sac small elongate, far anterior. In Passeriformes, Hirundinidae. Europe. Type-species *H. chelidonariae* (Spasskaya, 1957).

57a. **Cirrus armed with tuft or bundle of bristle-like spines (long, fine, flexible, at times curling up), extending to whole cirrus or at least to long basal part. Vaginal orifice often more distal than male one, so that ductus masculinus develops, where cirrus spines can extend. When cirrus withdrawn, tuft appears as 'carrot-shaped' brush or narrow bundle, longitudinally striated, with extremity of spines more or less engaged in ductus masculinus or even in atrium cavity. No unilateral pores. Uterus reticulate initially, never sacciform, except in *Spinilepis* (see also *Apokrimi* at 2a and *Unciunia* at 12b) .... 58.**

Figs 25.160, 25.161 *Vitta.* 25.160. *V. magniuncinata* Burt, 1938, scolex. (From Burt, 1938). 25.161. *V. undulatoides* (Fuhrmann, 1908) n. comb., type, mature proglottid.
Figs 25.162, 25.163 *Angularella.* 25.162. *A. magniuncinata* (Burt, 1938), scolex. (From Georgiev & Bray, 1988.) 25.163. *Angularella* sp., mature proglottid.
Figs 25.164-25.167 *Hirundinicola chelidonariae* (Spasskaya, 1957); (material of *praecox* Joyeux & Timon-David, 1934, *nec* Krabbe, 1869, synonymous with *chelidonariae*). 25.164. Scolex. 25.165. Hook crown showing irregularities in the hook position. 25.166. Mature proglottid, ventral view, pore left; poral end of the vagina dorsal to cirrus-sac. 25.167. Cirrus base at the beginning of evagination.

57b. Cirrus armed with spines of different pattern, or apparently unarmed; fine long spines, if present, limited to apex of organ, in cirri also bearing other kinds of spines. No ductus masculinus. Uterus initially reticulate or sacciform .... 70.

58a. Pores regularly alternate. Strobila usually small .... 59.
58b. Pores irregularly alternate. Strobila often medium or large .... 65.

59a. Hooks in one circle. In Piciformes, Picidae .... 60.
59b. Hooks in two circles .... 61.

60a. Ovary small, median. Vas deferens anteromedian. Cirrus-sac inclined anteriorly. Proglottids slightly wider than long. Testicular field symmetrically extended laterally beyond ovary. Hooks 24 .... *Ivritaenia* Singh, 1962. (Figs 25.168, 25.169)

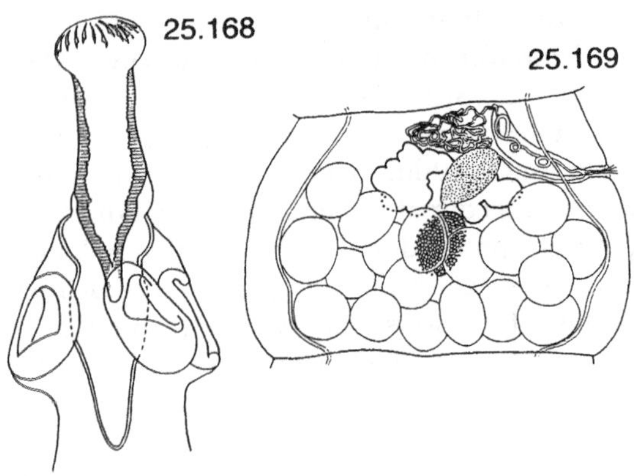

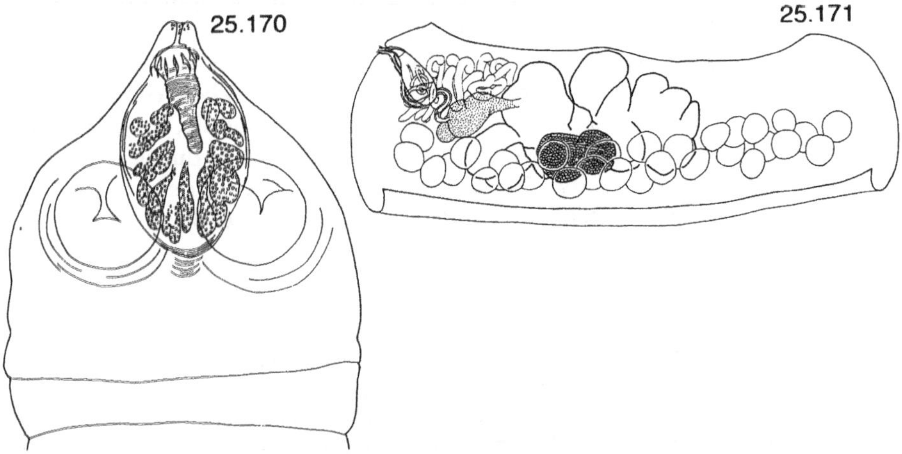

Figs 25.168, 25.169 *Ivritaenia mukteswarensis* Singh, 1962. 25.168. Scolex. 25.169. Proglottid, probably not fully mature. (From Singh, 1962.)
Figs 25.170, 25.171 *Monoliga amazonica* Bona, 1994, type. 25.170. Scolex. 25.171. Mature proglottid.

Diagnosis: Rostellar apparatus muscular and supposedly glandular. Pouch long, carrot-shaped. Rostellum long, strong; apical pad well-defined. Hooks in one circle. Strobila very small. Scolex well-demarcated posteriorly. Genital pores regularly alternate. Genital ducts dorsal to osmoregulatory canals. Vagina posterior to cirrus-sac (dorsoventral position unknown); thin, straight, transverse or anteriorly inclined. Uterus insufficiently described. Eggs eventually in clusters in parenchyma (no uterine capsules mentioned). Atrium small (apparently no ductus masculinus). Ovary with narrow isthmus, far from anterior and posterior margin of proglottid. Testes numerous, posterior and lateral to ovary. Cirrus-sac elongate. Cirrus small, armed with thin, bristle-like spines, projecting into atrium even when cirrus withdrawn. In Piciformes, Picidae. India. Type-species *I. muktes-warensis* Singh, 1962.

60b. Ovary large, principal mass in poral half of medulla. Vas deferens poral, overlapping part of cirrus-sac and vagina. Cirrus-sac often much inclined posteriorly. Proglottids much wider than long. Testicular field more extended beyond ovary on aporal side. Hooks 12–14 .... *Monoliga* Bona, 1994. (Figs 25.170, 25.171)

Diagnosis: Rostellar apparatus weakly muscular, glandular. Pouch oval, not very long. Rostellum small, stem narrow, apical pad well-defined. Hooks in one circle. Strobila very small. Proglottids rapidly widening. Scolex wide, not demarcated posteriorly. Genital pores regularly alternate, anterior. Genital ducts dorsal to osmoregulatory canals. Vagina posterior to cirrus-sac; proximal course mainly ventral; sinuous, backwardly directed. Uterus labyrinthine; eventually eggs in parenchyma. Embryophore thin, outer coat scarcely developed. Atrium simple, not spiny; ductus masculinus hardly visible. Ovary bulky, nearing anterior and posterior margins of proglottid when fully mature. Testes numerous, postovarian, in transverse field. Cirrus-sac small, oval. Cirrus short with long, bristle-like spines, projecting into atrium even when cirrus withdrawn. In Piciformes, Picidae. Neotropical. Type-species *M. amazonica* Bona, 1994.

61a. Genital ducts dorsal to osmoregulatory canals. In Piciformes, Picidae .... 62.

61b. Genital ducts between osmoregulatory canals .... 63.

62a. Atrium proper (distal to vaginal orifice) always spineless. Ductus masculinus spiny or not. Testes 14–25 (rarely only nine). Seminal receptacle transversely elongate, in front of or overlying poral ovarian lobe. Main mass of vas deferens in front of cirrus-sac .... *Liga* Weinland, 1857. (Figs 25.172, 25.173)

(Syn. *Fuhrmannia* Parona, 1901.)

Diagnosis. Rostellar apparatus muscular and substantially glandular. Pouch spherical to elongate. Rostellum of varying sizes. Hooks in two circles (16–28); seldom 1:2 alternations. Strobila small. Genital pores regularly alternate. Genital ducts dorsal to osmoregulatory canals. Vagina posterior to cirrus-sac, in same horizontal plane, usually bent along cirrus-sac. Uterus reticulate, then sacciform, tightly septate; walls may adhere to eggs simulating uterine capsules. Embryo-

25.172

25.173

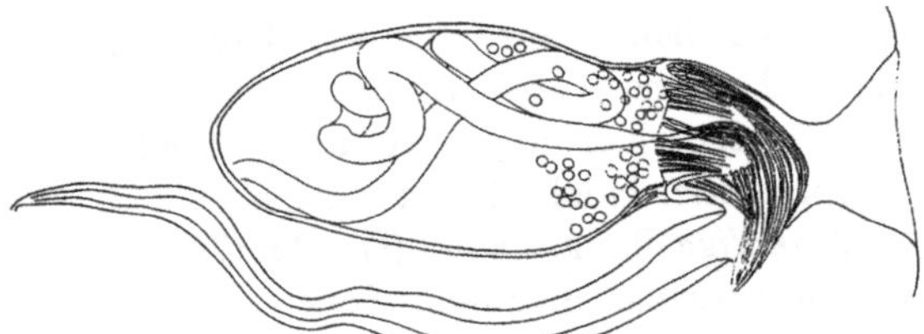

25.174

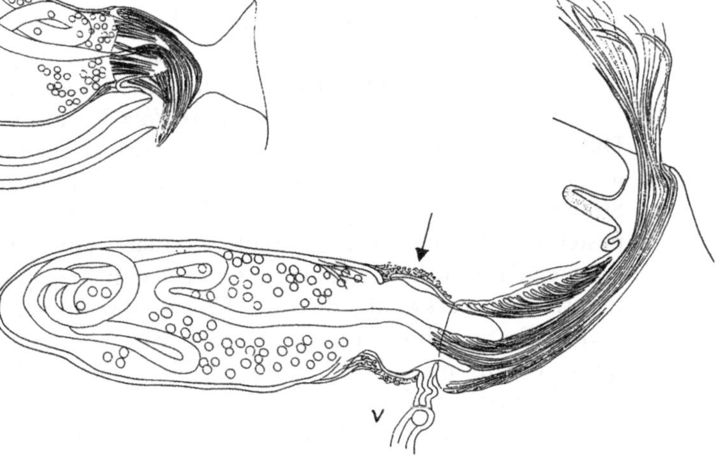

Figs 25.172, 25.173 *Liga ransomi* Spasskii & Reznik, 1966 [syn. *L. punctata* (Weinland, 1856) Weinland, 1857] neotype, Ransom's (1909) material. 25.172. Mature proglottid. 25.173. Terminal genitalia.

Fig. 25.174 *Apoliga imperialis* Bona, 1994, type; cirrus engaged in the spineless ductus masculinus (arrow) and in the spiny atrium: v. vagina.

phore strong; outer coat often shrivelled, no polar processes. Atrium simple, shallow. Ovary large, bulky. Testes posterior and lateral to ovary. Cirrus-sac subspherical or elongate, small, usually transverse. Cirrus short, with long thin bristle-like spines in conical tuft when evaginated, engaged in ductus when withdrawn. Vas deferens anteroporal, often crossing osmoregulatory canals. In Piciformes, Picidae; very seldom Cuculiformes and Trogoniformes (both from unpublished data). Nearctic, Neotropical, Palaearctic, India, Japan. Type-species *L. ransomi* Spasskii & Reznik, 1966.

62b. Atrium proper (part distal to vaginal orifice) spiny. Ductus masculinus always spineless. Testes 8–12. Seminal receptacle oval, nearly central, between ovarian lobes. Main mass of vas deferens behind cirrus-sac .... *Apoliga* Bona, 1994. (Fig. 25.174)

Diagnosis: Rostellar apparatus musculo-glandular. Pouch usually long. Rostellum long, either wide or narrow, rivet-shaped. Hooks in two circles (20). Strobila very small, gravid proglottids large, oval. Proglottids usually protandrous, with reduction in size or disappearance of testes when ovary fully ripe. Genital pores regularly alternate. Genital ducts dorsal to osmoregulatory canals. Vagina posterior to cirrus-sac, in same horizontal plane; often sinuous, diverging from cirrus-sac. Uterus labyrinthine then sacciform, tightly septate, wall may adhere to eggs simulating uterine capsules. Atrium deep, inclined posteriorly; ductus masculinus well defined; one or two small epidermal folds form on its anterior face, just proximal to pore. Ovary small. Testes posterior, sometimes also lateral and dorsal to ovary. Cirrus-sac subspherical or elongate, often inclined anteriorly. Cirrus short with very long bundle of stiff spines largely engaged in atrium when organ withdrawn. Vas deferens median. In Piciformes, Picidae. Neotropical. Type-species *A. imperialis* Bona, 1994.

63a. Cirrus-sac delicate, proportionally small, subspherical or oval elongate, thin-walled, far from poral ovarian lobe or just reaching it. Cirrus armed only with long fine bristle-like spines. Vagina regular. Ductus masculinus at times not well-defined, spiny or spineless .... 64.

63b. Cirrus-sac strong, elongate, proportionally large, usually wall firm or thick, in front of poral ovarian lobe. Cirrus armed with long bristle-like spines proximally and sometimes whip-like spines distally. Vagina divided into short, thick-walled distal part (pars copulatrix), and thin tube proximally. Ductus masculinus well-defined, strong-walled, with dense, long spines; slightly swollen on each side of base of cirrus .... *Chitinorecta* Meggitt, 1927. (Figs 25.175–25.181)

(Syn. *Chettusiana* Singh, 1960.)

Diagnosis: Rostellar apparatus musculo-glandular. Rostellum strong, long. Hooks in two circles (at times rather indistinct); handle long, blade much shorter slightly bent. Strobila small; neck short and wide; proglottids wider than long; organs packed. Scolex wide. Genital pores regularly alternate. Genital ducts between osmoregulatory canals. Vagina posterior, usually dorsal (at least orifice) to cirrus-sac or in same horizontal plane. Uterus reticulate, or labyrinthine, sometimes developing into deeply septate sac; persistent wall. Embryophore thick with adhering reduced outer coat, or thin with long polar processes. Atrium proper small, spineless. Ovary bulky; lobules large with irregular outline. Testes numerous, postovarian in transverse field. Cirrus long, with spines jutting into ductus masculinus when withdrawn. Seminal receptacle close to posterior margin of cirrus-sac. In Charadriiformes. Africa, India, Asia. Type-species *C. agnosta* Meggitt, 1927.

64a. Strobila small to large, proglottids often longer than wide. Spaced organs. Testes clearly behind ovary. Ovary with small lobules, in anterior half of medulla, extended anteriorly. Distal part of vagina straight. (See also irregularly alternating pores, diagnosis and synonyms at 68a and Figs 25.198–25.203) .... *Dictymetra* Clark, 1952.

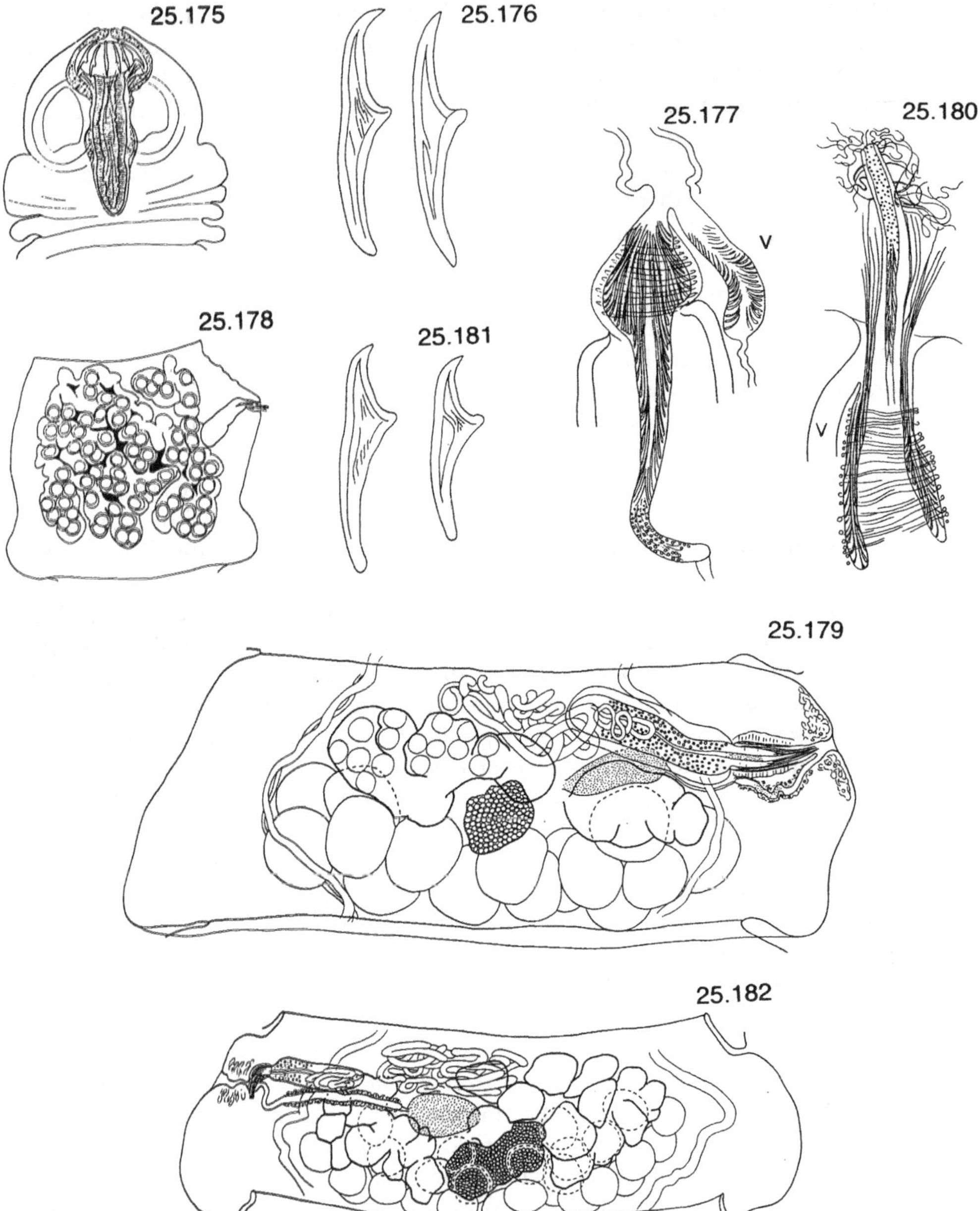

Figs 25.175–25.181 *Chitinorecta*. 25.175–25.179. *C. agnosta* Meggitt, 1927, type. 25.175. Scolex, the two crowns can be distinguished by observing the handle ends. 25.176. Hooks (left, distal; right, proximal). 25.177. Cirrus withdrawn, ductus masculinus spiny; v = vagina. 25.178. Semi-gravid proglottid. 25.179. Mature proglottid. 25.180. *Chitinorecta* sp., cirrus evaginated, whip-like spines; v: vagina. 25.181. *C. vanelli* (Fuhrmann, 1907) n. comb., type, hooks.

Fig. 25.182 *Imparmargo baileyi* Davidson, Doster & Prestwood, 1974, type, mature proglottid.

64b. Strobila small; proglottids wider than long. Organs overlap. Testes behind ovary and overlapping hind part of it. Ovary occupying large part of medulla, extended posteriorly; deep, thick lobules. Vagina bends sharply posterior before entering atrium. Protandrous maturation .... *Imparmargo* Davidson, Doster & Prestwood, 1974. (Fig. 25.182)

Diagnosis: Rostellar apparatus musculo-glandular. Hooks in two circles. Strobila small; neck short, wider than scolex. Suckers with fine microtriches. Genital pores regularly alternate. Genital ducts between osmoregulatory canals. Vagina posterior to cirrus-sac in same horizontal plane; long, wall thick. Uterus labyrinthine with large cavities, tending to become thinly septate sac. No uterine capsules. Egg probably with polar processes. Atrium simple with thin circular fibres, ductus masculinus short, both spineless. Testes numerous. Cirrus-sac small, elongate, thin-walled. Cirrus with long, bristle-like spines on whole surface; spines jutting into atrium when cirrus withdrawn. In Galliformes. Nearctic. Type-species *I. baileyi* Davidson, Doster & Prestwood, 1974.

65a. Hooks in one circle, irregularities can occur, exceptionally no hooks (see 18a). Probably only in terrestrial birds .... 66.
65b. Hooks in two circles .... 67.

66a. Ductus masculinus short but well-defined, usually lined with thin setae. Cirrus bristles forming rather loose brush in partially evaginated organs and comma-shaped bundle when withdrawn. Genital ducts between osmoregulatory canals. Strobila usually medium. Adult proglottids trapeziform, clearly craspedote. Aporal ovarian lobe larger. Eggs mostly with small polar processes. In Galliformes and other terrestrial birds .... *Choanotaenia* Railliet, 1896. (Figs 25.183–25.187)

Diagnosis: Rostellar apparatus musculo-glandular. Pouch proportionally small. Rostellum small; apical pad well-defined. Hooks in one circle, small, delicate. Scolex small. Suckers wide. Genital pores irregularly alternate. Vagina posterior to cirrus-sac in same horizontal plane, straight. Uterus reticulate, large-meshed, wall thin; then eggs loose in parenchyma, mixed with granular cells; no uterine capsules. Egg outer coat with delicate, not filamentous, polar processes which seem to recede when embryophore fully developed. Atrium simple, small. Ovary with fairly long isthmus, lobules small, rather loose. Testes numerous, post-ovarian. Cirrus-sac small, usually subspherical, thin-walled. Cirrus short, delicate, wide truncated cone; long bristle-like flexible spines protruding beyond sac orifice when withdrawn. In Galliformes, possibly Piciformes and other terrestrial birds. Cosmopolitan. Type-species *C. infundibulum* (Bloch, 1779).

66b. Ductus masculinus absent. Cirrus bristles smaller forming small, narrow, thick, straight bundle in partly evaginated and withdrawn organs. Genital ducts between and dorsal to osmoregulatory canals in same strobila. Strobila small to very small. Adult proglottids usually barrelled, hardly craspedote. Ovarian lobes nearly symmetrical. Eggs without polar processes. In Piciformes, Picidae. (See also unarmed rostellum at 18a) .... *Krimi* Burt, 1944. (Figs 25.188–25.193)

25.183

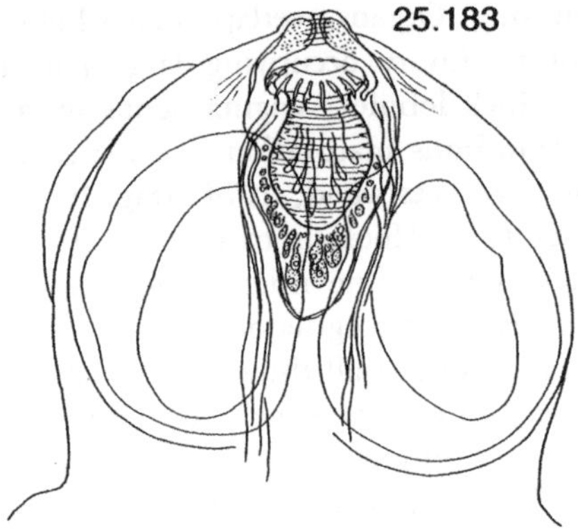

25.186

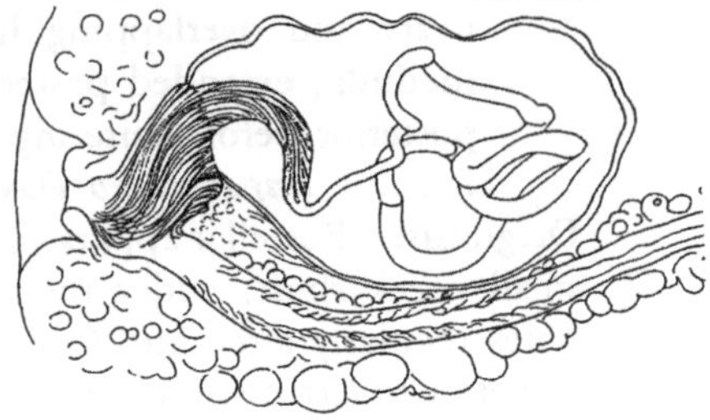

25.185

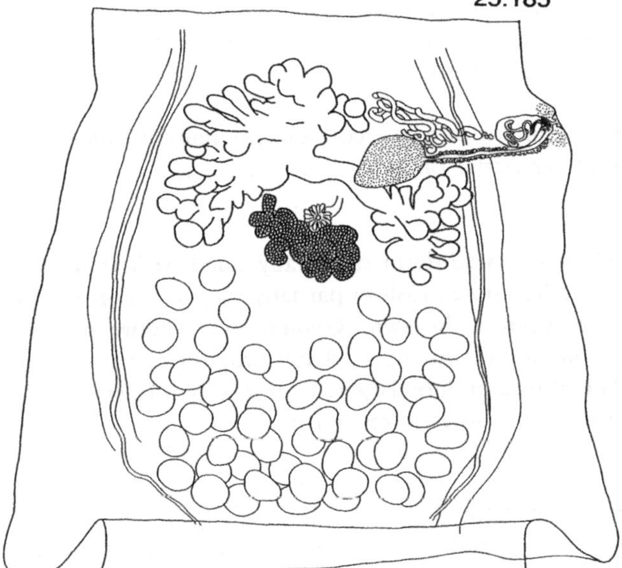

25.184

25.189

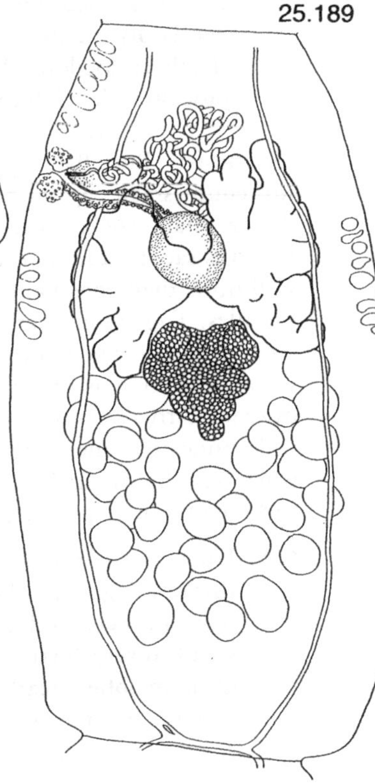

25.188

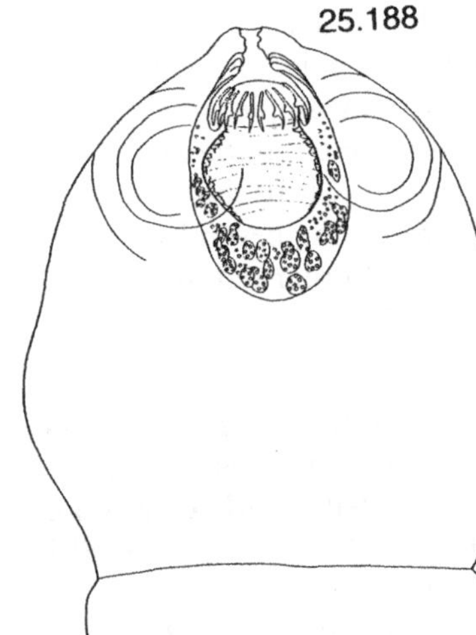

25.192

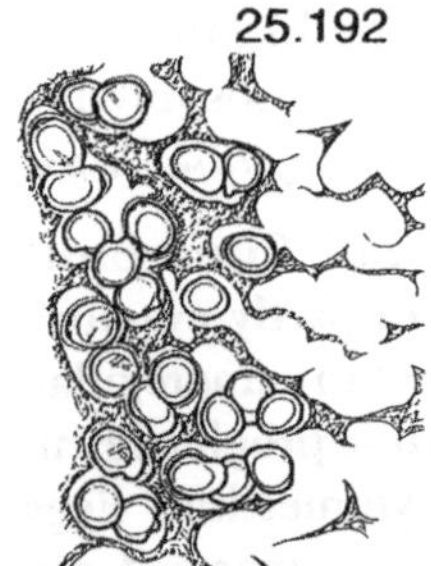

25.187

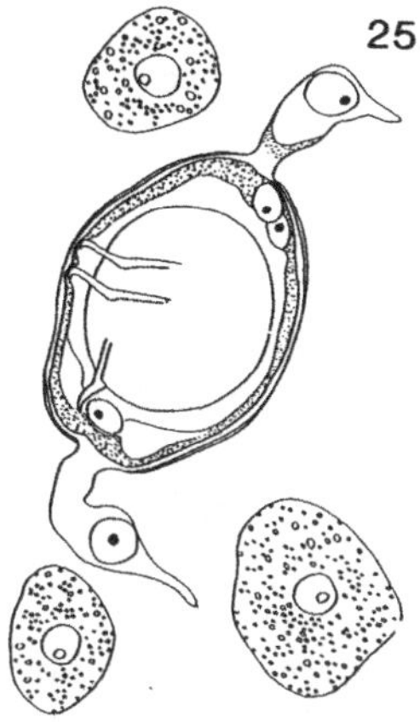

25.190

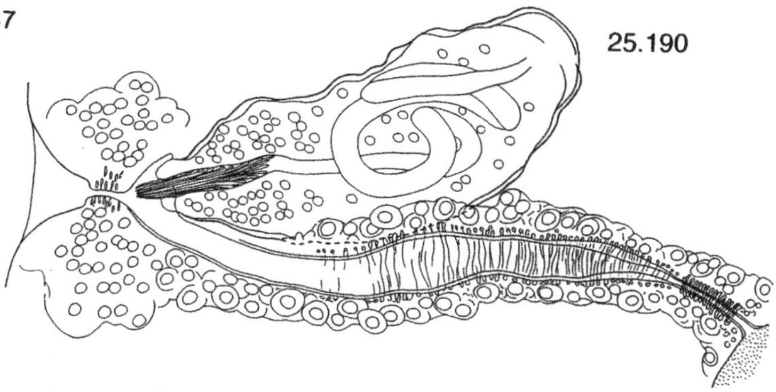

25.191

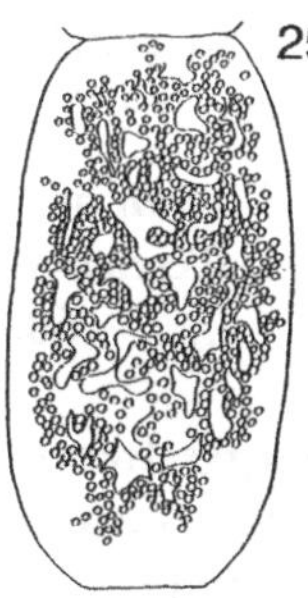

25.193

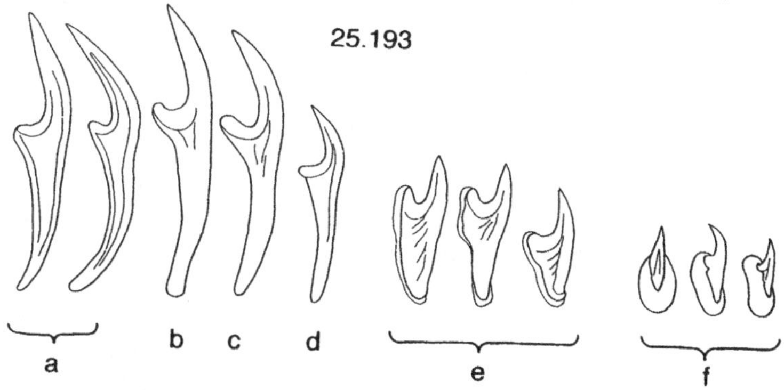

Figs 25.183–25.187 *Choanotaenia infundibuliformis* (Goeze, 1782), type. 25.183. Scolex. 25.184. Hooks (hooks show considerable intraspecific variation). 25.185. Mature proglottid. 25.186. Cirrus withdrawn, ductus masculinus short, partly spiny, chiefly toward vagina orifice. 25.187. Eggs loose in parenchyma with granular cells; wall of the uterine reticulum already disappeared; embryophore not yet formed, inner coat granular with three nuclei.

Figs 25.188–25.193 *Krimi.* 25.188–25.191. *K. chrysocolaptis* Burt, 1944, type. 25.188. Scolex. 25.189. Mature proglottid. 25.190. Cirrus withdrawn; small, straight bundle of bristle-like spines; no ductus masculinus. 25.191. Early uterus. 25.192. *Krimi* sp., details of uterus; embryophore already developed. 25.193. Hooks of several species: (a) *chrysocolaptis*, type, (b–d) *Krimi* sp. I,II,III, (e) *Krimi* sp. IV, (f) *Krimi* sp. V.

Diagnosis: Rostellar apparatus muscular, heavily glandular. Pouch proportionally large, thin-walled. Rostellum varying in size, apical pad well-defined. Hooks in one circle (exceptionally hookless), sometimes partly irregular; different patterns, sometimes very small. Gravid proglottids much larger, sometimes follow semi-adult or even juvenile ones. Scolex and neck proportionally wide. Osmoregulatory canals may form reticulum in hind part of scolex. Genital pores irregularly alternate. Vagina posterior to cirrus-sac, in same horizontal plane, straight, wall thick. Uterus reticulate or labyrinthine, thin-walled; septa mamillated. Egg outer coat adhering to strong embryophore. Atrium simple, small. Ovary with short, deep isthmus; extended lengthwise; lobules rather compact. Testes numerous, postovarian. Cirrus-sac small, oval, thin-walled, in front of ovarian lobe. Cirrus very small; fine bristle-like spines slightly protruding beyond sac orifice when withdrawn. In Piciformes, Picidae. Sri Lanka, India, Africa, Europe, South America; probably cosmopolitan. Type-species *K. chrysocolaptis* Burt, 1944.

67a. Uterus initially sacciform, persistent. Eggs with thick embryophore and two polar buttons. Expanded crown in large umbrella. Ovary very large, filling anterior half of medulla; isthmus hardly visible. Testicular field transverse. In Ralliformes .... *Spinilepis* Oshmarin, 1972. (Figs 25.194–25.197)

Diagnosis: Rostellar apparatus probably only muscular. Rostellar apical disc wide. Hooks in two circles (28); blade slender, only slightly curved, much longer than handle; guard small. Strobila small; organs crowded. Genital pores irregularly alternate. Genital ducts between osmoregulatory canals. Vagina posterior and ventral (?) to cirrus-sac. Atrium simple, shallow (ductus masculinus not described). Testes numerous, postovarian. Cirrus-sac small, wall thin. Cirrus with bundle of bristle-like spines. Vas deferens poral, crossing osmoregulatory canals. In Ralliformes. Palaearctic (Russia). Type-species *S. turnicis* Oshmarin, 1972.

67b. Uterus reticulate initially. Eggs without polar buttons. Expanded crown normal. Ovary not very large, less compact, isthmus well-defined. Testicular field more extended, often lengthwise. Not in Ralliformes .... 68.

68a. Uterus initially reticulate; then sacciform deeply septate, or eggs free in parenchyma; no uterine capsules. Egg outer coat forms resistant pellicle or polar processes, transitory or persistent. Ductus masculinus always present. Vagina rather short, seldom reaching far beyond cirrus-sac. (See also regularly alternate pores at 64a) .... *Dictymetra* Clark, 1952. (Figs 25.198–25.203)
(Syn. *Lapwingia* Singh, 1952.)

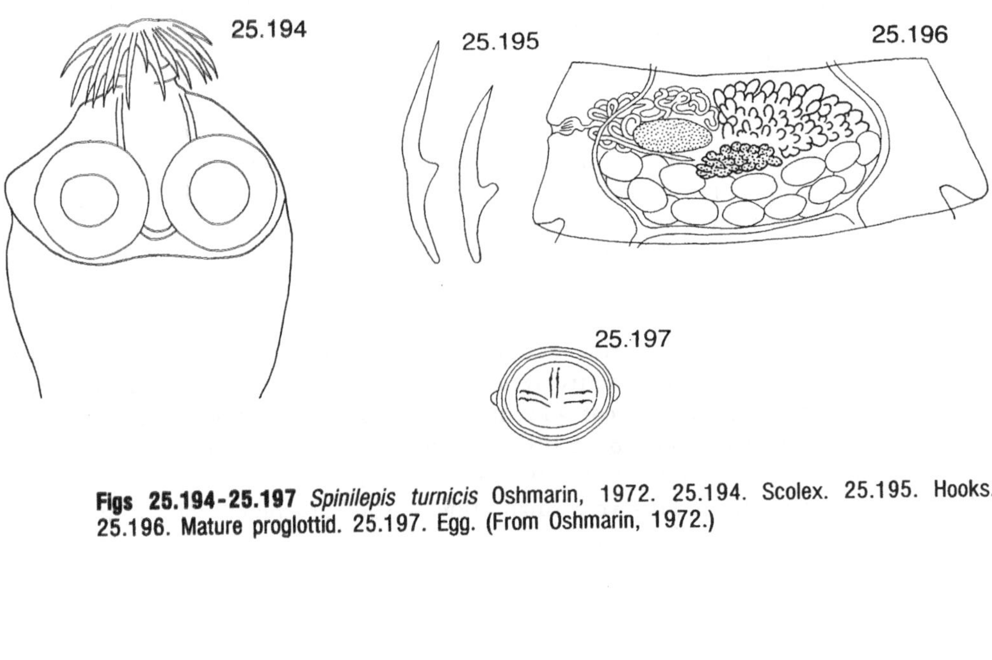

Figs 25.194-25.197 *Spinilepis turnicis* Oshmarin, 1972. 25.194. Scolex. 25.195. Hooks. 25.196. Mature proglottid. 25.197. Egg. (From Oshmarin, 1972.)

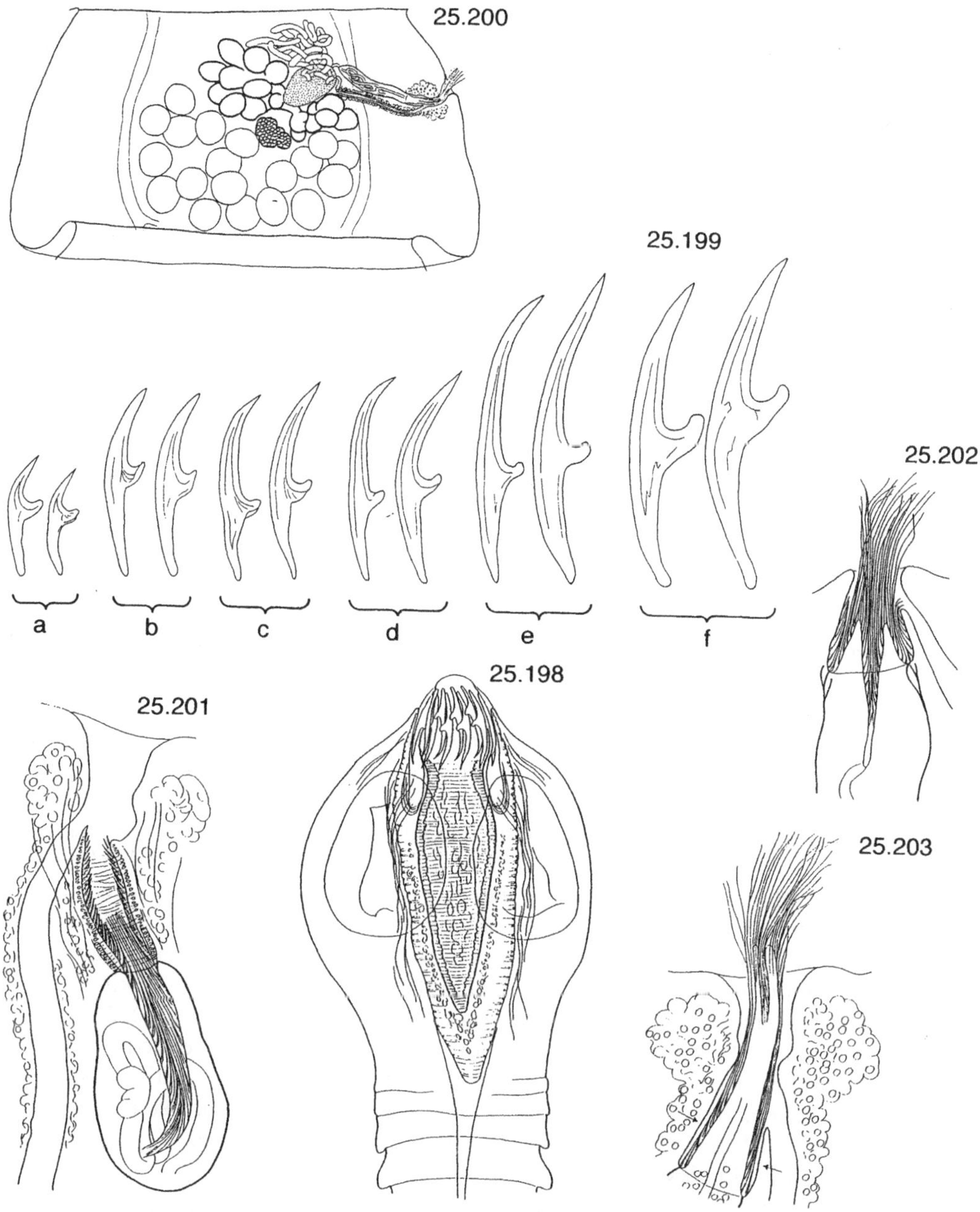

Figs 25.198–25.203 *Dictymetra*. 25.198. *D. numenii* (Owen, 1949), type, scolex. 25.199. Hooks (left, distal; right, proximal), (a,b) terrestrial birds, (c,f) Charadriiformes: (a) *D. discoidea* (van Beneden, 1868), in Ciconiae, (b) *Dictymetra* sp. in *Trogon*, (c) *D. nymphaea* (Schrank, 1790), (d) *D. radiaspinosa* Matevosyan, 1963 (syn. *D. numenii* Clark, 1952 *nec* (Owen, 1949)), type, (e) *D. numenii* (Owen, 1949), type, (f) *D. paranumenii* Clark, 1952, type. 25.200. *D. dispar* (Burt, 1940) n. comb., type, mature proglottid. 25.201. *D. numenii* (Owen, 1949), type; strong, long, spiny ductus masculinus. 25.202. *D. discoidea* (van Beneden, 1868), small, delicate, spiny ductus masculinus. 25.203. *D. nymphaea* (Schrank, 1790), small, delicate, spineless ductus masculinus (arrow).

Diagnosis: Rostellar apparatus musculo-glandular. Pouch very elongate. Rostellum long. Hooks in two circles (about 24); in typical hosts (Charadriiformes) blade long, even longer than handle, hooks arranged in twos, anterior one near posterior one; different patterns especially in other bird groups. Strobila small to large; proglottids usually longer than wide; organs spaced. Genital pores alternate irregularly; sometimes regularly in long stretches. Genital ducts between osmoregulatory canals. Vagina posterior to cirrus-sac, in same horizontal plane. Eggs often arranged in chains. Atrium small, weak; ductus masculinus rather long, annular fibres, spiny, in typical hosts (Charadriiformes); shorter, thin-walled, scarcely or not at all spinous in terrestrial birds. Ovary with narrow isthmus. Testes numerous, postovarian, usually in longitudinally elongate field. Cirrus-sac small, pyriform or elongate, thin-walled. Cirrus delicate, narrow, fairly long with fine, long, bristle-like spines throughout its length, protruding into ductus when cirrus withdrawn. In Charadriiformes; Ciconiiformes, Ciconiae; Passeriformes; Trogoniformes (from unpublished data). Holarctic, Neotropical, Africa. Type-species *D. radiaspinosa* Matevosyan, 1963.

68b. Uterus reticulate, turning into membranous uterine capsules. Egg outer coat merges with embryophore; no polar processes. No ductus masculinus. Vagina long, often reaching far beyond cirrus-sac .... 69.

69a. Hook circles irregular and difficult to distinguish; hooks small, unduloid pattern (see 46a). Uterine capsules closely packed. Vas deferens not overlapping cirrus-sac. In Passeriformes ....
.... *Spiniglans* Yamaguti, 1959. (Figs 25.204–25.209)
(Syn. *Pseudanomotaenia* Matevosyan, 1963.)

Diagnosis: Rostellar apparatus musculo-glandular. Pouch elongate, external longitudinal muscles from neck weak. Hooks in two circles (20–30). Strobila small to large; gravid proglottids longer than wide. Genital pores irregularly alternate. Genital ducts between osmoregulatory canals. Vagina posterior to cirrus-sac; orifice in same horizontal plane; varying in length, may be very long. Uterus reticulate, wall thin, then labyrinthine, turning into precocious monovular membranous uterine capsules mixed with granular cells. Atrium simple, lacking spines; no ductus masculinus. Ovary with long isthmus. Testes numerous (22–26) to very numerous (60), postovarian; field more or less transversely extended. Cirrus-sac small, oval or pyriform. Cirrus delicate, short, truncated cone, with a small tuft of very fine, flexible, bristle-like spines. Neither internal nor external seminal vesicles. In Passeriformes. India (type-species), Europe, Africa. Type-species *S. microsoma* (Southwell, 1922).

69b. Hook circles regular, well-defined; hooks large, dictymetroid pattern (see 68a). Uterine capsules scattered. Vas deferens overlapping proximal part of cirrus-sac. In Lariformes; Charadriiformes ....
.... *Laritaenia* Spasskaya & Spasskii, 1971. (Figs 25.210–25.213)

Diagnosis: Rostellar apparatus strong muscular, weakly glandular. Pouch elongate; external longitudinal muscles from neck strong. Hooks in two circles (18–20). Strobila medium, gravid proglottids much longer than wide. Genital pores irregularly alternate. Genital ducts between osmoregulatory canals. Vagina

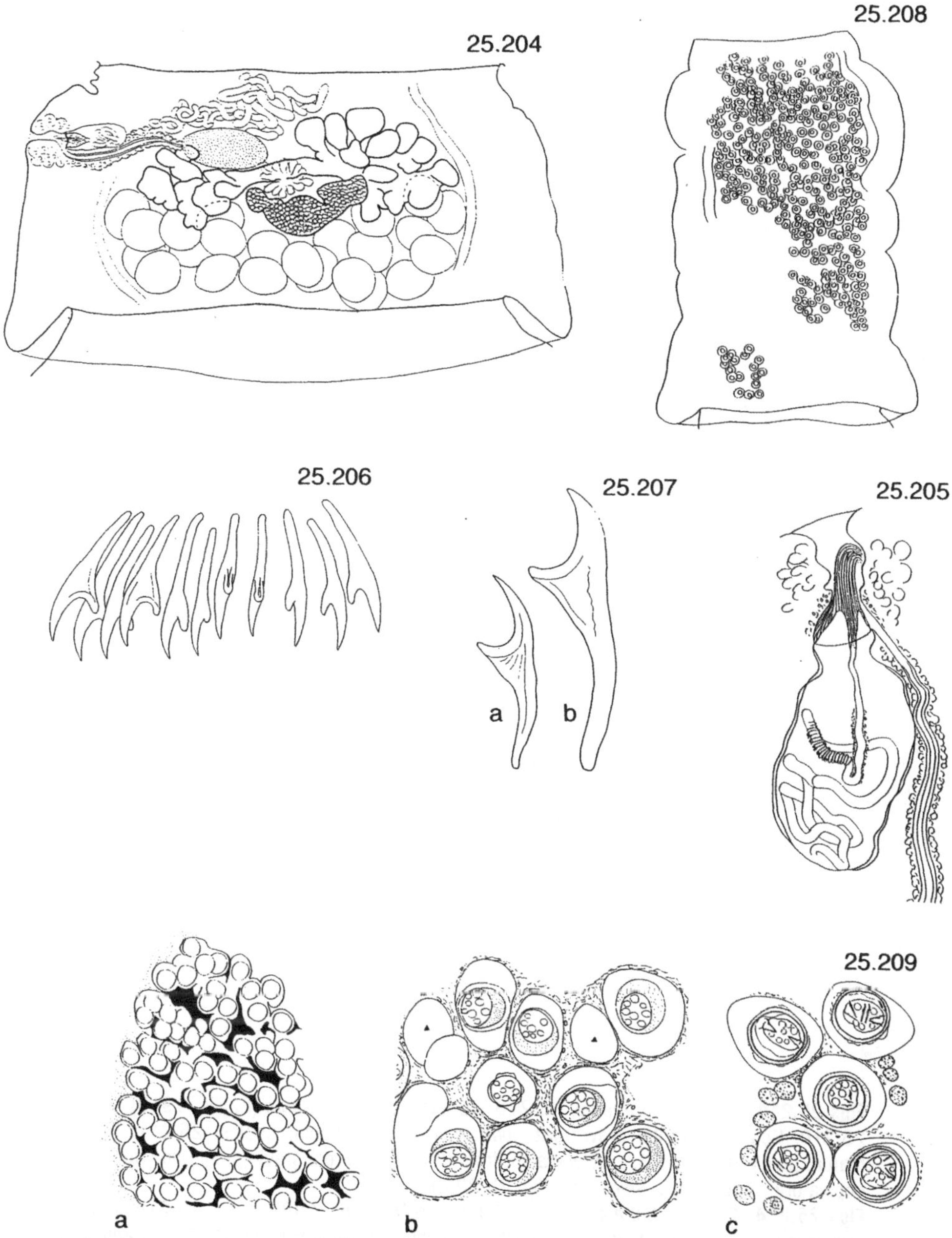

Figs 25.204-25.209 *Spiniglans*. 25.204. *S. microsoma* (Southwell, 1922), type, mature proglottid. 25.205. *S. constricta* (Molin, 1858), type, terminal genitalia; either ductus masculinus absent or very small pleat separating male and female orifices. 25.206. *S. corvi* (Joyeux, Baer & Martin, 1937), type, irregular crown. 25.207. Hooks: (a) *S. microsoma*, (b) *S. constricta*. 25.208. *S. corvi*, gravid proglottid with uterine capsules. 25.209. Three steps in the formation of uterine capsules: (a) *S. microsoma*, early uterus, (b,c) *S. corvi*, initial uterine capsules and fully mature ones; parenchyma degenerates, small granular cells among capsules.

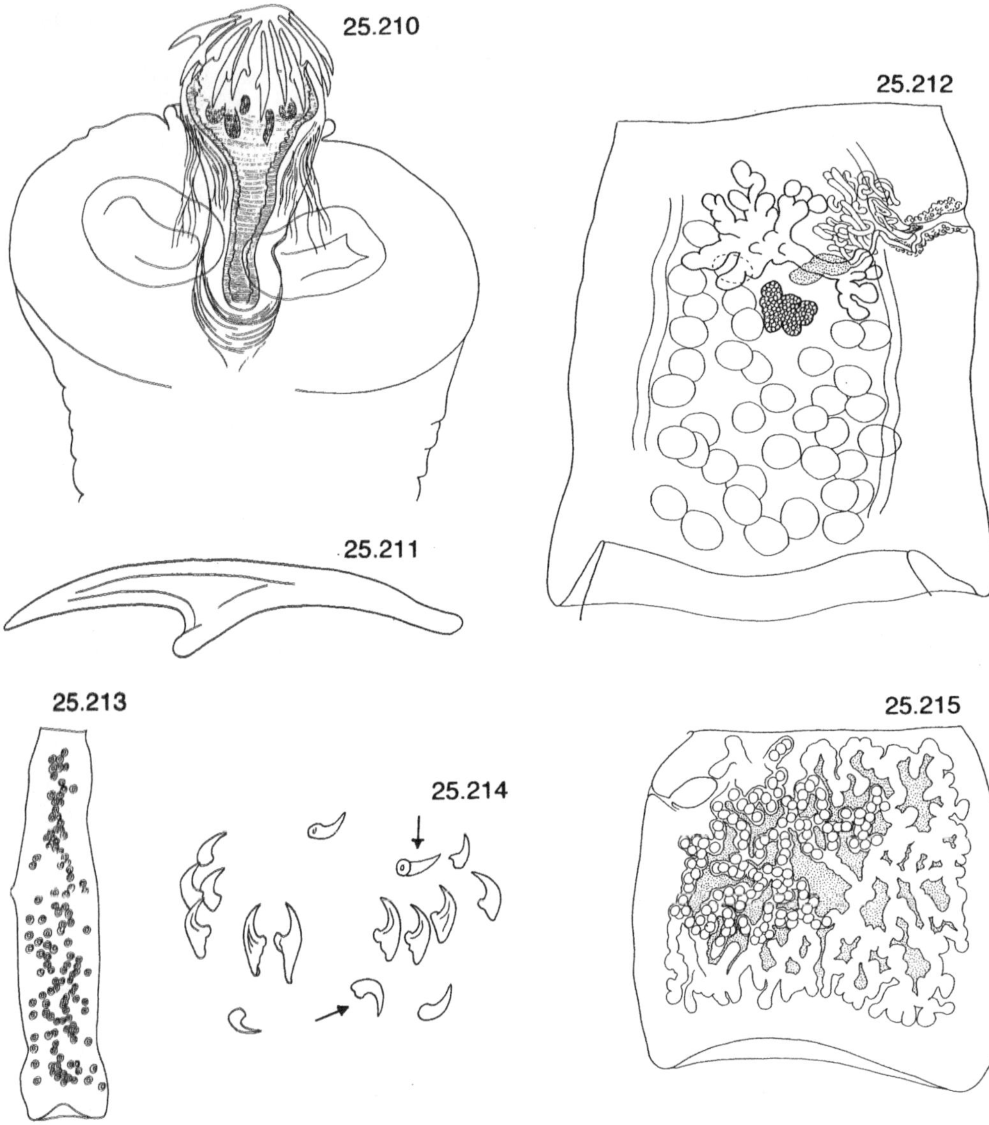

Figs 25.210–25.213 *Laritaenia*. 25.210–25.212. *L. stercorarium* (Baylis, 1919) n. comb., type. 25.210. Scolex, rostellar crown with two well-defined, regular circles. 25.211. Hook. 25.212. Mature proglottid. 25.213. *Laritaenia* sp. from Charadriiformes; gravid proglottid with capsules. (For copulatory organs, see *Spiniglans* Fig. 25.205.)

Figs 25.214, 25.215 *Bonaia africana* Mariaux & Vaucher, 1990, type. 25.214. Part of a hook crown with spine-shaped hooks in front and behind crown (arrows). 25.215. Early uterus.

posterior to cirrus-sac, in same horizontal plane. Uterus reticulate, wall thin; then precocious, monovular, membranous uterine capsules, without granular cells in parenchyma. Atrium simple, lacking spines; no ductus masculinus. Ovary with long isthmus; small if compared with testicular field. Testes numerous (35–45), postovarian, field usually longitudinally extended. Cirrus-sac small, oval. Cirrus delicate, short truncated cone, with small tuft of fine, flexible, bristle-like spines. In Lariformes; Charadriiformes. Palaearctic, Neotropical. Type-species *L. hydrochelidonis* (Dubinina, 1953).

70a. Pores regularly alternate .... 71.
70b. Pores irregularly alternate .... 81.

71a. Hooks in one circle, regular .... 72.
71b. Hooks in two circles, regular, sometimes with more or less extended irregular parts in the crown .... 76.
71c. Hooks in two circles, aberrant; crown disintegrated, partly in one, partly in two circles; hooks varying in shape in same scolex because of progressive reduction of handle. Uterus reticulate-labyrinthine. Testes in young proglottids divided into two lateral groups (very rare character in genera that follow, see *Stenovaria, Neyralla* and *Birovilepis*). In Cuculiformes .... *Bonaia* Mariaux & Vaucher, 1990. (Figs 25.214, 25.215)

Diagnosis: Rostellar apparatus musculo-glandular. Pouch small. Rostellum small; musculature delicate, normal. Hooks about 20, small, sometimes spine-shaped, also scattered in front of and behind crown proper. Strobila small, thick. Scolex wide, dome-shaped. Genital pores regularly alternate. Genital ducts dorsal to osmoregulatory canals. Vagina posterior to cirrus-sac, orifice in same horizontal plane, course mostly ventral. Uterus ventral; reticulate, then tightly labyrinthine; septa and wall firm (not observed totally gravid). Egg probably with thin polar processes. Atrium small, simple. Ovary large; very small lobules. Testes numerous, clearly postovarian. Cirrus-sac small, oval, firm-walled, far anterior. Cirrus coiled in sac; strong, long, with scarcely visible homogeneous spinules. In Cuculiformes. Africa. Type-species *B. africana* Mariaux & Vaucher, 1990.

72a. Atrium strongly muscular, at times enormous and complicated ... 73.
72b. Atrium simple, weak musculature .... 74.

73a. Atrium wall bulbous, muscular with poral sleeve of fibres. Uterus coarse-meshed reticulum, then dividing into large maturation chambers sometimes with delicate wall (similar to polyovular capsules). Testes postovarian. (See also irregularly alternate pores at 85a) .... *Fuhrmannolepis* Spasskii & Spasskaya, 1965. (Figs 25.216–25.219)
(Syn. *Taeniarhynchaena* Burt, 1983.)

Diagnosis: Rostellar apparatus muscular pattern, glands inside rostellum. Pouch long, external anterior cavity round retracted rostellum very deep. Rostellum extremely long; stem narrow, coiled inside pouch; apical disc wide, slightly bilobed. Hooks in one circle (ten), five on each rostellar lobe, small, wrench-shaped, blade much shorter than handle. Strobila small, cortex reduced. Suckers large, rim deformable. Genital pores regularly or irregularly alternating, according to species. Genital ducts between osmoregulatory canals. Vagina posterior, usually dorsal to cirrus-sac; bent posteriorly before entering atrium, with pre-atrial sphincter. Embryophore smooth. Ovary large, often extending beyond osmoregulatory canals. Testes numerous. Cirrus-sac small, elongate, narrow. Cirrus small with fine

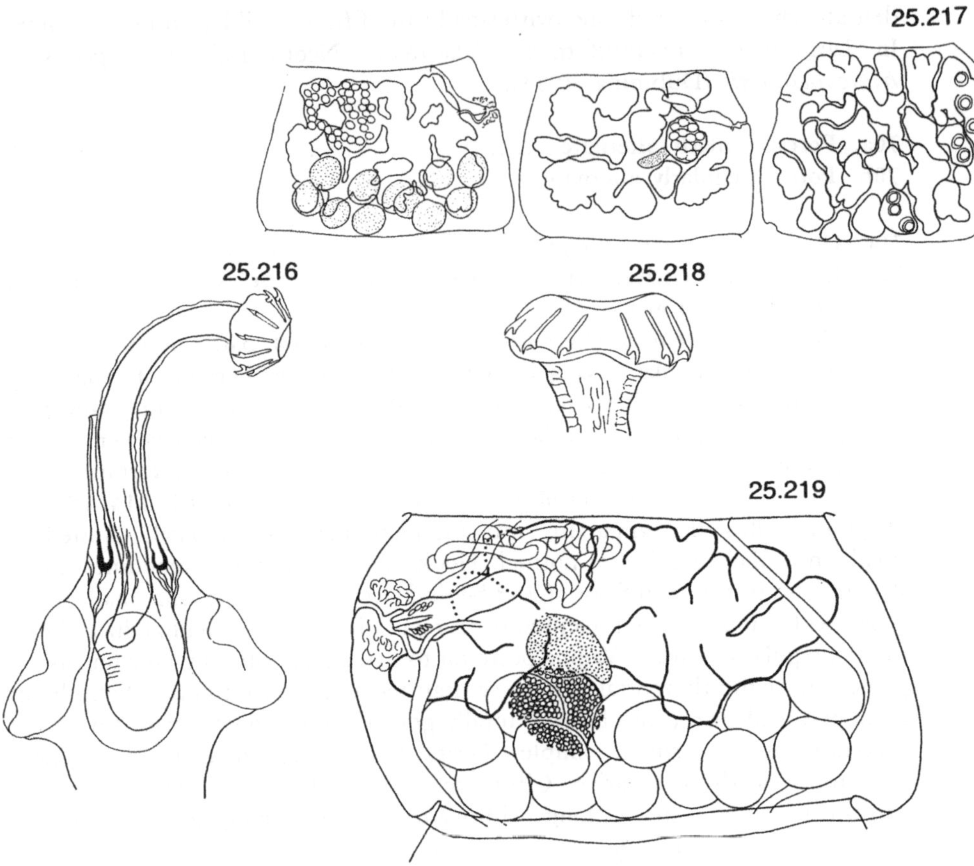

Figs 25.216–25.219 *Fuhrmannolepis*. 25.216, 25.217. *F. micropalamae* (Burt, 1983), n. comb., type. 25.216. Scolex. 25.217. Three steps of maturation of the uterus. 25.218, 25.219. *F. decacantha* (Fuhrmann, 1913), type. 25.218. Rostellum. 25.219. Mature proglottid.

spines. Vas deferens anteroporal, just cross osmoregulatory canals. In Charadriiformes. Holarctic, India. Type-species *F. decacantha* (Fuhrmann, 1913).

73b. Atrium usually enormous, sucker-like, may be extremely long, spiny; ductus hermaphroditus at times very long, spiny eversible; simpler unarmed in extremely small species. Uterus sacciform. Testes posterior and lateral to ovary seldom surrounding it, sometimes also dorsal .... *Bakererpes* Rausch, 1947. (Figs 25.220–25.223)

Diagnosis: Rostellar apparatus muscular, perhaps glands inside rostellum. Pouch very long, narrow, carrot-shaped. Rostellum long, narrow, apex often indistinct. Hooks in one circle (8–16); porosoid pattern (see 86a), changing when hooks small. Strobila extremely small (0.2–3 mm). Scolex well-defined. Proglottids few, acraspedote; all maturation stages not present at same time. Genital pores regularly alternate. Genital ducts usually between osmoregulatory canals, exceptionally dorsal in some species. Vagina posterior to cirrus-sac, prevailingly dorsal course. Uterus lobulate, wall strong. Whole atrial complex may evaginate, forming various sorts of papillae. Ovary deeply lobulate, transversely extended or rosette-shaped. Testes few to numerous, more numerous laterally. Cirrus-sac long, narrow, far anterior; sometimes deeply sunken, displaced aporally according to atrium size; wall thick. Cirrus long (few exceptions), wider proximal part with small spines followed by much thinner, hardly spiny, whip-like portion. In Caprimulgiformes. Nearctic and Neotropical. Type-species *B. fragilis* Rausch, 1947.

74a. Genital ducts between osmoregulatory canals. Uterus reticulate. Vagina thin-walled, proximal part widely dilatable, club-shaped. Rostellar apparatus with important glandular component. In Apodiformes. (See also diagnosis in apical apparatus glandular at 16a and pores irregularly alternate at 83a.) .... *Glanduluncinata* n. g.

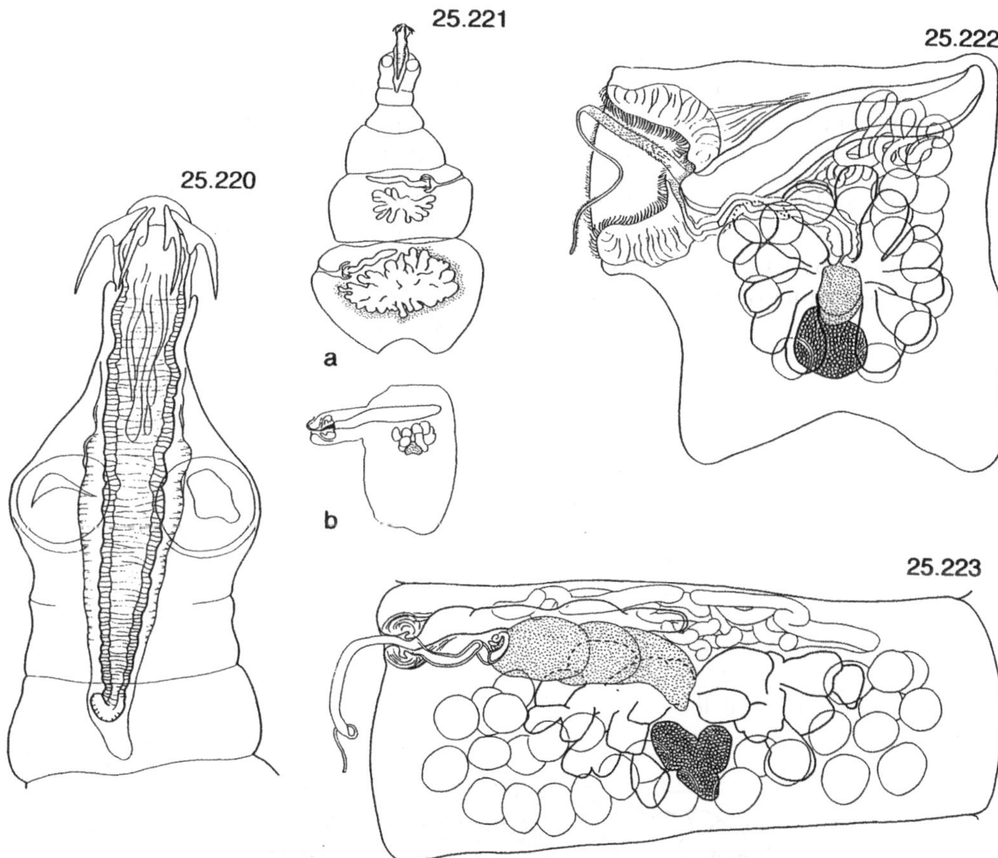

Figs 25.220–25.223 *Bakererpes* spp. 25.220, 25.221a, sp. I; scolex and strobila. 25.221b, sp. II; loose adult proglottid with huge genital papilla. 25.222, 25.223. Sp. III and sp. IV, mature proglottids.

74b. **Genital ducts dorsal to osmoregulatory canals. Uterus sacciform. Vagina firm-walled, not particularly dilated. Rostellar apparatus muscular or normal musculo-glandular .... 75.**

75a. **Scolex small compared with strobila. Vagina ventral to cirrus-sac, orifice and distal part anterior. Proglottids much wider than long. Ovary transversely elongate and narrow. Testes in row behind ovary. Hooks delicate, blade long, sharp, slightly curved towards outside, deep indentation between blade and guard, very small handle. In Galliformes .... *Amoebotaenia* Cohn, 1899. (Figs 25.224–25.226)**

Diagnosis: Rostellar apparatus muscular. Pouch with strong muscular bundles from strobila. Rostellar apex only slightly expanded. Hooks in one circle, few (10–14). Strobila small, club-shaped; no neck; cortex much reduced. Scolex small. Genital pores regularly alternate; seldom two consecutive. Genital ducts dorsal to osmoregulatory canals. Vagina long, convoluted, strong-walled, crossing cirrus-sac. Uterus sacciform, deeply lobulate, no uterine capsules. Egg outer envelope with small, delicate polar swellings. Atrium simple. Ovary lobules small, thick. Testes few, dorsal to vitellarium and part of ovary; reaching beyond osmoregulatory canals. Cirrus-sac large, elongate, thin-walled. Cirrus very long, wide, with small, thin spines. Vas deferens posterior to cirrus-sac. In Galliformes. Cosmopolitan. Type-species *A. sphenoides* (Railliet, 1892).

75b. **Scolex very large compared with strobila. Vagina dorsal and posterior to cirrus-sac. Ovary normal fairly small. Testes in more than one row behind ovary. Hooks strong, different pattern. In Charadriiformes. (See also irregularly alternate pores at 82a) .... *Polycercus* Villot, 1883. (Figs 25.227–25.229)**

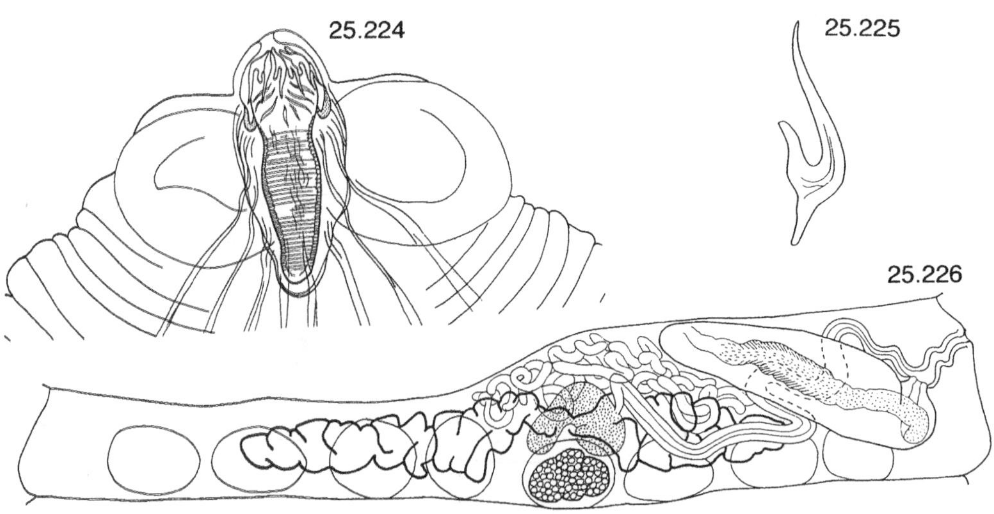

Figs 25.224–25.226 *Amoebotaenia sphenoides* (Railliet, 1892). 25.224. Scolex. 25.225. Hook. 25.226. Mature proglottid.

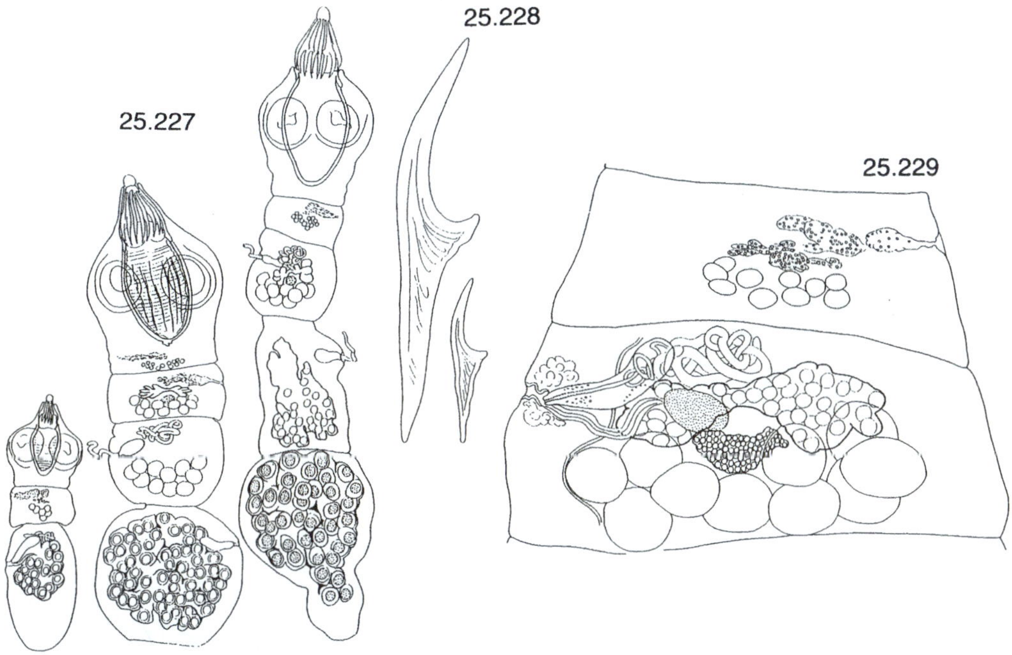

Figs 25.227-25.229 *Polycercus*. 25.227. Complete strobilae of two species; smallest, left: *Polycercus* sp. I, other two specimens: *P. lumbrici* Villot, 1883. 25.228. Hooks: large, left: *P. lumbrici*; small, right: *Polycercus* sp. I. 25.229. *Polycercus* sp. II, proglottids.

Diagnosis: Rostellar apparatus muscular. Pouch large, oval. Rostellum large, occupying whole pouch; apical pad large. Hooks in one circle, (usually 14–18); often large; blade long, slightly shorter than handle, guard small, hook base nearly straight between guard and handle (similar to porosoid pattern, see 86a). Strobila extremely small (from 0.3 to few mm); two to seven proglottids in gravid worm; neck short. Proglottids acraspedote; often all stages of maturation may not be present simultaneously. Genital pores usually regularly alternate; seldom two consecutive (proglottids being few, many specimens are needed to see irregularities). Genital ducts dorsal to osmoregulatory canals. Vagina short, slightly sinuous, lacking pre-atrial sphincter. Uterus sacciform, deeply lobulate. Eggs few, embryophore fairly large, outer coat scarcely developed. Atrium simple. Testes few (7–14). Cirrus-sac elongate, very contractile, thin-walled. Cirrus long, thin with small spines. Vas deferens anteroporal, proportionally large, may be anterior to whole ovary; partly dorsal to cirrus-sac. In Charadriiformes. Palaearctic. Type-species *P. lumbrici* Villot, 1883.

76a. Testes exceptionally numerous (250–300), also occupying cortex. Strobila large. Ovary with long, thin lobules; lobes fan-like. In Gruiformes. (See diagnosis in irregularly alternate pores at 90a and Fig. 25.267) .... *Gruitaenia* Spasskii, Borgarenko & Spasskaya, 1971.

76b. Characters different .... 77.

25.231

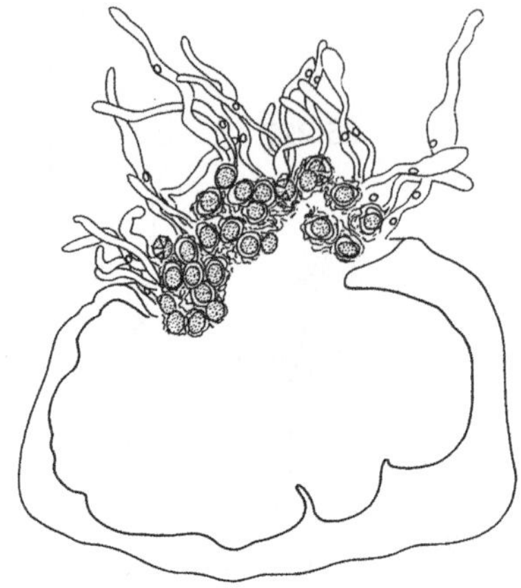

25.230

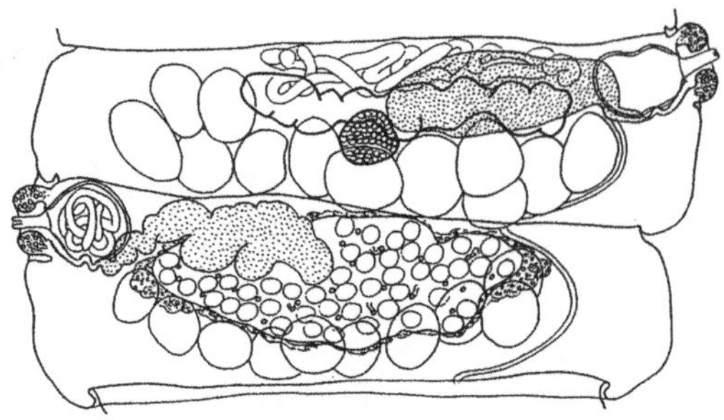

25.232

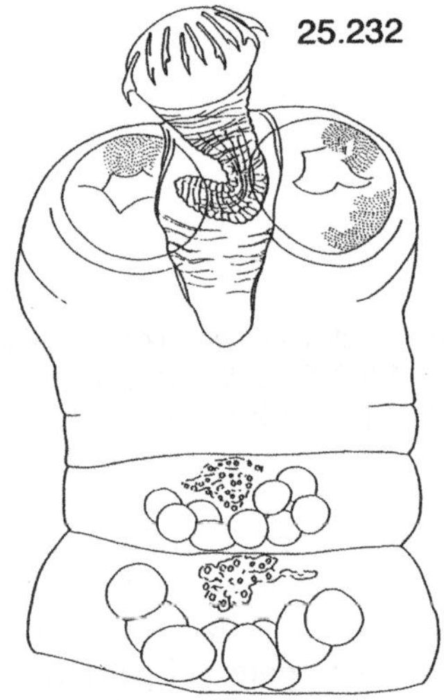

25.233

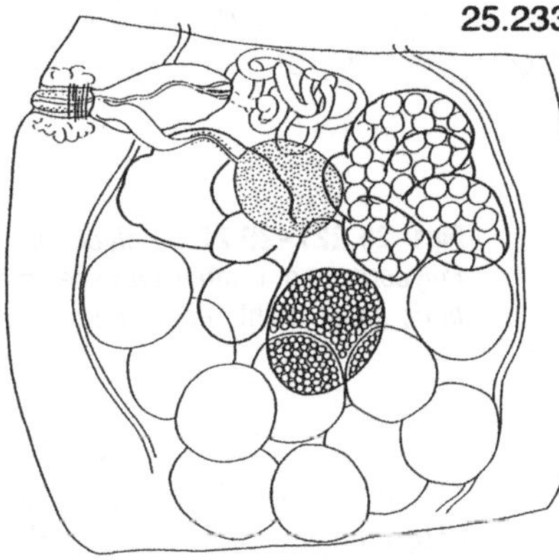

Figs 25.230, 25.231 *Paraliga celermaturus* (Deblock & Rosé, 1962), type. 25.230. Proglottids, second with uterus at first stage. 25.231. 'Oophore' (partial); prolongations surrounding egg mass not yet swollen in water; embryophore and shrivelled outer coat of eggs may be observed.
Figs 25.232, 25.233 *Rallitaenia gallinulae* (van Beneden, 1858), neotype, Dollfus' (1934b) material. 25.232. Scolex and two immature proglottids. Note the spiny suckers. 25.233. Mature proglottid.

77a. Suckers armed with hooklets on border or covered with thick rows of very small spines. Uterus sacciform. Vagina small, short .... 78.
77b. Suckers unarmed or with barely visible microtriches. Uterus reticulate (sacciform only in *Chimaerula* n. g. at 79a). Vagina usually long .... 79.

78a. Suckers armed with hooklets anteriorly. Proglottids much wider than long. Testes behind ovary, field transversely elongate, narrow. Ovary hardly bilobed, transversely elongate, narrow, compact. Atrium strong, sucker-like, with papilla. Seminal receptacle enormous, transverse, dorsal to ovary

poral half. Eggs in single uterine oophore (see diagnosis) .... *Paraliga* Belopolskaya & Kulachkova, 1973. (Figs 25.230, 25.231)

Diagnosis: Rostellar apparatus muscular, glands inside rostellum. Pouch very long, may reach second proglottid. Rostellum as long as pouch. Hooks in two circles (16); wrench-shaped (see 73a). Strobila very small (1–1.3 mm). Cortex reduced. Genital pores regularly alternate, in anterior corner of proglottid. Scolex very small. Genital ducts usually dorsal to osmoregulatory canals, sometimes between in same strobila. Vagina posterior to cirrus-sac, in same horizontal plane, tending to be dorsal; very small, short, close to cirrus-sac. Uterus sacciform, scarcely lobulate; wall firm; cavity crossed by dorsoventral strings; cells from wall form oophore, alveolar tissue where eggs get entangled, and swell in digitiform expansions round egg mass in water. Outer coat of egg forms thin crumpled membrane with polar processes, joining end to end with other polar processes, making button at junction. Testes few to numerous (10–20). Cirrus-sac small, subspherical, far anterior. Cirrus short, thin with very small stumpy spines. In Charadriiformes. Palaearctic. Type-species *P. oophorae* (Belopolskaya, 1971).

78b. **Suckers covered with very small spines. Proglottids slightly wider than long. Testicular field postovarian, more spacious. Ovary sharply bilobed, deep anterior and posterior indentations, spherical lobules. Atrium delicate, fairly deep. Seminal receptacle spherical, fairly small, on isthmus. Eggs free in uterine cavity. (See also irregularly alternate pores at 97b.) .... *Rallitaenia* Spasskii & Spasskaya, 1975. (Figs 25.232, 25.233)**

Diagnosis: Rostellar apparatus muscular. Rostellar apical pad large. Hooks in two circles. Strobila very small. Proglottids acraspedote, barrelled, cortex reduced. Genital pores regularly alternate, anterior. Genital ducts dorsal. Vagina dorsal to cirrus-sac, tending backwards; short, distally carrot-shaped (type-species). Uterus sacciform, septa thin and sparse; slightly lobular. Testes few. Cirrus-sac small, oval, thin-walled, far anterior. Cirrus fairly wide with thin, spaced needle-like spines. In Ralliformes. Palaearctic. Type-species *R. gallinulae* (van Beneden, 1858).

79a. **Uterus always clearly sacciform. Ovary small; roundish; compact lobes. Testes four. Cirrus very long, strong; spines both rosethorn and differently shaped, small apical tuft of long, fine spines .... *Chimaerula* n. g. (Figs 25.234–25.236)**

Diagnosis: Rostellar apparatus muscular. Pouch strong, very long, entering first proglottid; external anterior cavity very deep round retracted rostellum. Rostellum long, narrow; apical disc wide when protruded. Hooks in two circles very numerous (42–46) not davaineoid type. Strobila very small. Suckers large, deformable rim. Commissure of osmoregulatory canals normal. Pores regularly alternate, far anterior. Genital ducts dorsal to osmoregulatory canals. Vagina dorsal to cirrus-sac, orifice and distal part anterior; very long, convoluted; distal part wide, thick walled with sleeve of fibres; then tending backwards, crossing cirrus-sac. Uterus lobular, horseshoe-shaped (fully ripe not known). Atrium simple. Testes three postovarian, one anteroaporal. Cirrus-sac very large, long, far anterior. In Gruiformes, Aramidae; Charadriiformes. Brazil, Palaearctic. Type-

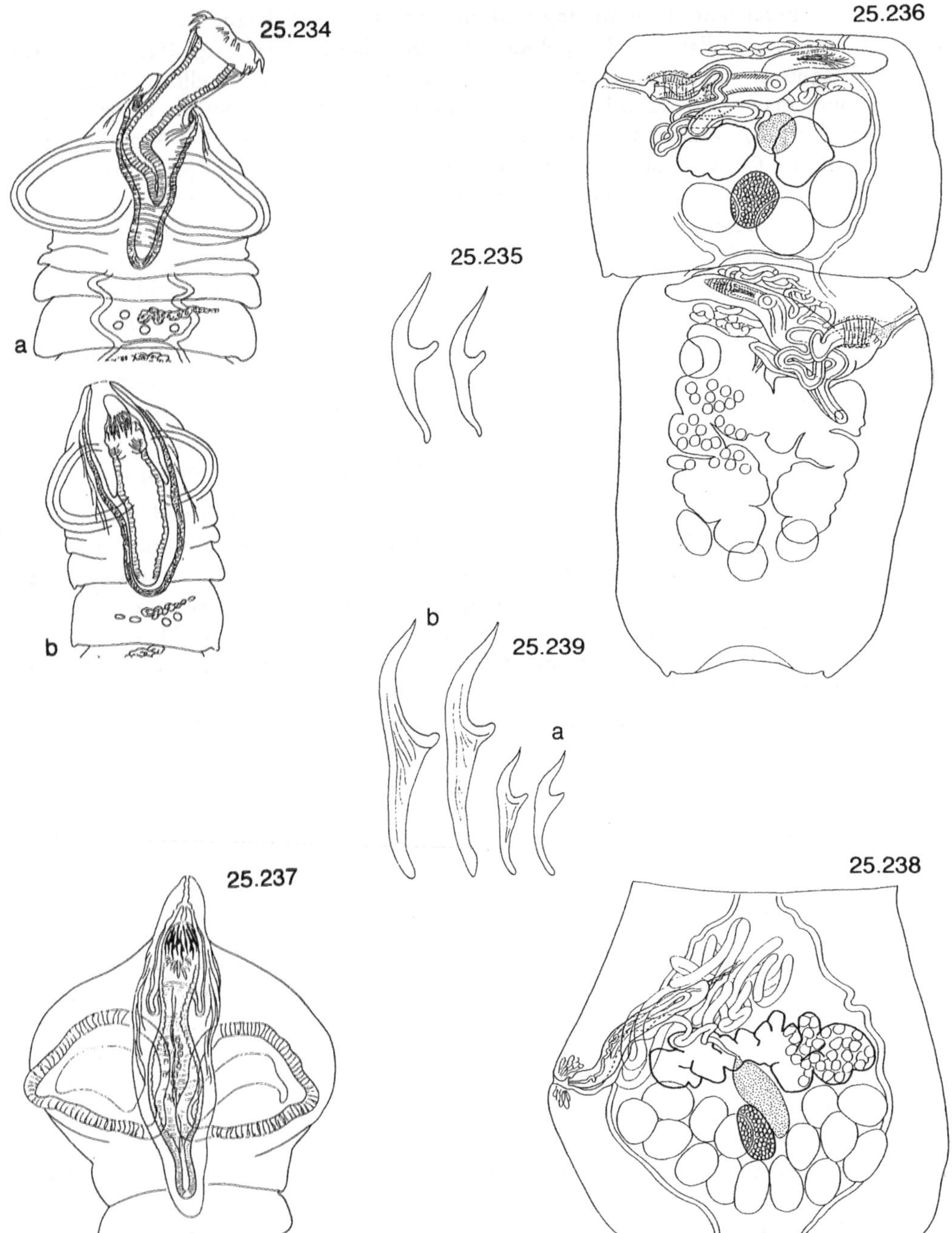

Figs 25.234-25.236 *Chimaerula woodlandi* (Prudhoe, 1960), n. comb., type. 25.234a, b. Scolex with rostellum protruding and withdrawn. 25.235. Hooks. 25.236. Proglottids.
Figs 25.237-25.239 *Stenovaria*. 25.237-25.239a. *S. falsificata* (Meggitt, 1927), type. 25.237. Scolex. 25.238. Mature proglottid. 25.239a. Hooks. 25.239b. *S. facile* (Meggitt, 1927), type, hooks.

species *C. woodlandi* (Prudhoe, 1960) n. comb. (syn. *Liga woodlandi* Prudhoe, 1960).

79b. Early uterus reticulate or labyrinthine, at times with large communicating cavities. Ovary rather small, very lobulate. More than four testes. Cirrus small, delicate, no rosethorn spines .... 80.

80a. Genital ducts dorsal to osmoregulatory canals, rarely between in some proglottids. Rostellum long and narrow, deep external anterior cavity round distal part when retracted. Ovary small, transversely extended, narrow; lobules close, irregular. Pore in middle of lateral margin. In young proglottids, testes in two lateral fields. In Charadriiformes .... *Stenovaria* Spasskii & Borgarenko, 1973. (Figs 25.237–25.239)

Diagnosis: Rostellar apparatus muscular. Pouch very long, sometimes entering first proglottid. Hooks in two circles. Strobila very small. Suckers large, deformable rim. Osmoregulatory canals converge in posterior part of proglottid; transverse commissure, at times particularly sinuous. Genital pores regularly alternate, very rarely irregular. Vagina posterior, usually ventral to cirrus-sac; long, narrow; at times pre-atrial sphincter present. Early uterus labyrinthine, later multiseptate sac with small communicating cavities containing one or several eggs. Atrium simple, very small. Testes few to numerous, postovarian. Cirrus-sac elongate, inclined anteriorly. Cirrus very small, spines extremely fine. In Charadriiformes; Falconiformes. Egypt, Russia. Type-species *S. falsificata* (Meggitt, 1927).

80b. Genital ducts mainly between osmoregulatory canals, at times dorsal. Rostellum differently shaped. Ovary small; lobules distinct, mainly subspherical. Pore anterior. Testicular field always continuous. In Passeriformes .... *Spasspasskya* n. g. (Figs 25.240–25.243)

Diagnosis: Rostellar apparatus muscular, densely glandular. Pouch wide. Hooks in two circles, sometimes irregular with some 1:2 alternations and three hooks on three different levels. Strobila small; neck wide, very short. Scolex wide, dome-shaped, hardly demarcated from strobila. Suckers small, with large microtriches, border thick. Genital pores regularly alternate, rarely two consecutive. Vagina posterior to cirrus sac, mostly dorsal, at times in same horizontal plane, rarely ventral in same strobila (unstable position); pre-atrial sphincter present. Uterus reticulate or labyrinthine, simulating uterine capsules until wall disappears (residues of septa may remain round eggs) leaving eggs in parenchyma. Eggs not numerous, embryophore thick, spherical. Atrium simple. Testes numerous, postovarian. Cirrus-sac more or less elongate, often narrow, S-shaped; far anterior. Cirrus small, spinules very tiny. In Passeriformes, Turdidae. Europe, probably wider. Type-species *S. passerum* (Joyeux & Timon-David, 1934) n. comb., syn. *Anomotaenia passerum* (Joyeux & Timon-David, 1934).

81a. Hooks in one circle, regular, seldom slightly irregular .... 82.

81b. Hooks in two circles, regular, seldom irregular (1:2 alternation, clear or mixed with small zigzag stretches) .... 89.

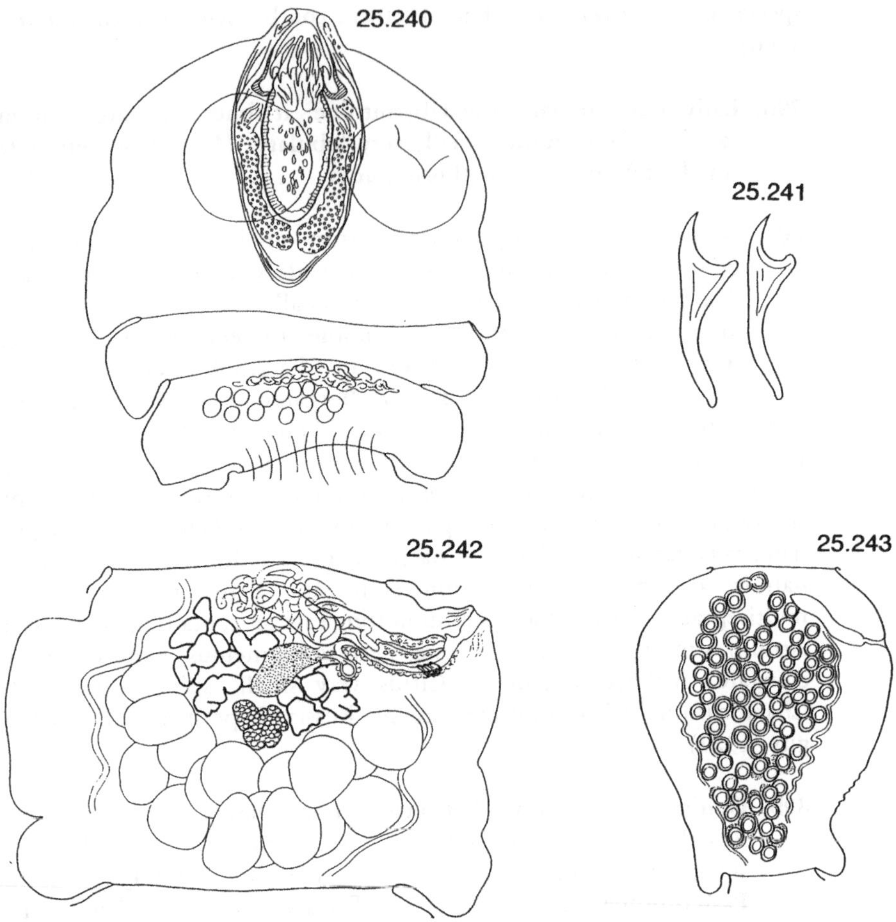

Figs 25.240–25.243 *Spasspasskya passerum* (Joyeux & Timon-David, 1934) n. comb. 25.240. Scolex. 25.241. Hooks. 25.242. Mature proglottid. 25.243. Gravid proglottid.

82a. Genital ducts dorsal. Uterus sacciform, deeply septate. Strobila extremely small. Scolex disproportionately large, wider than adult proglottids. Testes few. Eggs few. In Charadriiformes. (See diagnosis in regularly alternate pores at 75b) .... *Polycercus* Villot, 1883.

82b. Genital ducts between osmoregulatory canals. Uterus at first either labyrinthine or variously shaped reticulum, then either remaining labyrinthine or changing its structure during maturation in different ways. Strobila from small to large. Scolex proportionate. Testes numerous (except for *Glanduluncinata* n. g.). (Note: an initial real sacciform uterus is excluded in *Sacciuterina* and doubtful in *Kintneria*) .... 83.

83a. Rostellar apparatus highly glandular; rostellum very small, but normal in structure. Ovary small, lobes fairly compact, few lobules. Proximal part of vagina club-shaped. Testes few. Hooks very delicate, small. In Apodiformes. (See also regular alternate pores at 74a and diagnosis in apical apparatus glandular at 16a) .... *Glanduluncinata* n. g.

83b. Rostellar apparatus muscular or musculo-glandular; rostellum not very small, from medium to very long. Ovary clearly lobulate, lobules small and spherical or deep and irregular. Vagina not club-shaped. Testes numerous. Hooks small to very large. In other birds .... 84.

84a. Rostellar pouch and rostellar stem narrow, long or very long; glandular component of rostellar pouch scarce or absent. Uterus at first coarse-meshed reticulum or markedly labyrinthine, then (i) turning into polyovular capsules with many eggs (85a), (ii) becoming sacciform, either septate or with spots of parenchyma (86a, 87a), (iii) remaining permanently labyrinthine (87b), (iv?) disappearing, leaving eggs in parenchyma (87b). Hooks large, porosoid (see pattern at 86a) or small, wrench-shaped (see pattern at 73a). Vagina with small, pre-atrial sphincter (at times difficult to observe). In Charadriiformes; Lariformes .... 85.

84b. Rostellar pouch and rostellum not particularly long; glandular component of rostellar pouch substantial. Uterus at first finely reticulate or labyrinthine (very doubtful if initially sacciform in 88b), then (i) tubes forming transient, disappearing, cavities, eventually leaving eggs in parenchyma (88a) or (ii) tubes turn into capsules with one and few eggs (88b). Hooks small, different to above two patterns in group (i), probably wrench-shaped in group (ii). Vagina probably without pre-atrial sphincter. In Passeriformes .... 88.

85a. Rostellum narrow, extremely long, usually coiled in pouch, apical disc wide, slightly bilobed; hooks wrench-shaped (see 73a). Uterus at first coarse-meshed reticulum, then turning into maturation cavities with many eggs similar to polyovular capsules. Vagina bent sharply backwards before entering atrium. (See also diagnosis and synonyms in regularly alternate pores at 73a) .... *Fuhrmannolepis* Spasskii & Spasskaya, 1965.

85b. Rostellum narrow, long, not coiled in pouch; apical pad or disc not bilobed. No polyovular uterine capsules. Vagina usually without sharp posterior bend before entering atrium .... 86.

86a. Uterus at first coarse-and wide-meshed reticulum with irregular diverticula, then clearly sacciform with scattered spots of parenchyma. Atrium rather small, wall modest, no large papilla. Hook-shape characteristic, porosoid (see diagnosis). Eggs large, embryophore thin, smooth, outer coat forming large thin film, thickly crumpled round eggs in uterus, largely swollen in water .... *Paricterotaenia* Fuhrmann, 1932. (Figs 25.244–24.247)

Diagnosis: Rostellar apparatus musculo-glandular. Pouch and rostellum very long, narrow. Hooks in one circle, few (14) (type-species); blade long, slender, often longer than handle; guard small, forming straight line with handle. Strobila medium. Genital pores irregularly alternate. Genital ducts between osmoregulatory canals. Vagina posterior to cirrus-sac in same horizontal plane; long, straight, with S-shaped pre-atrial sphincter (type species). Ovary large, occupying

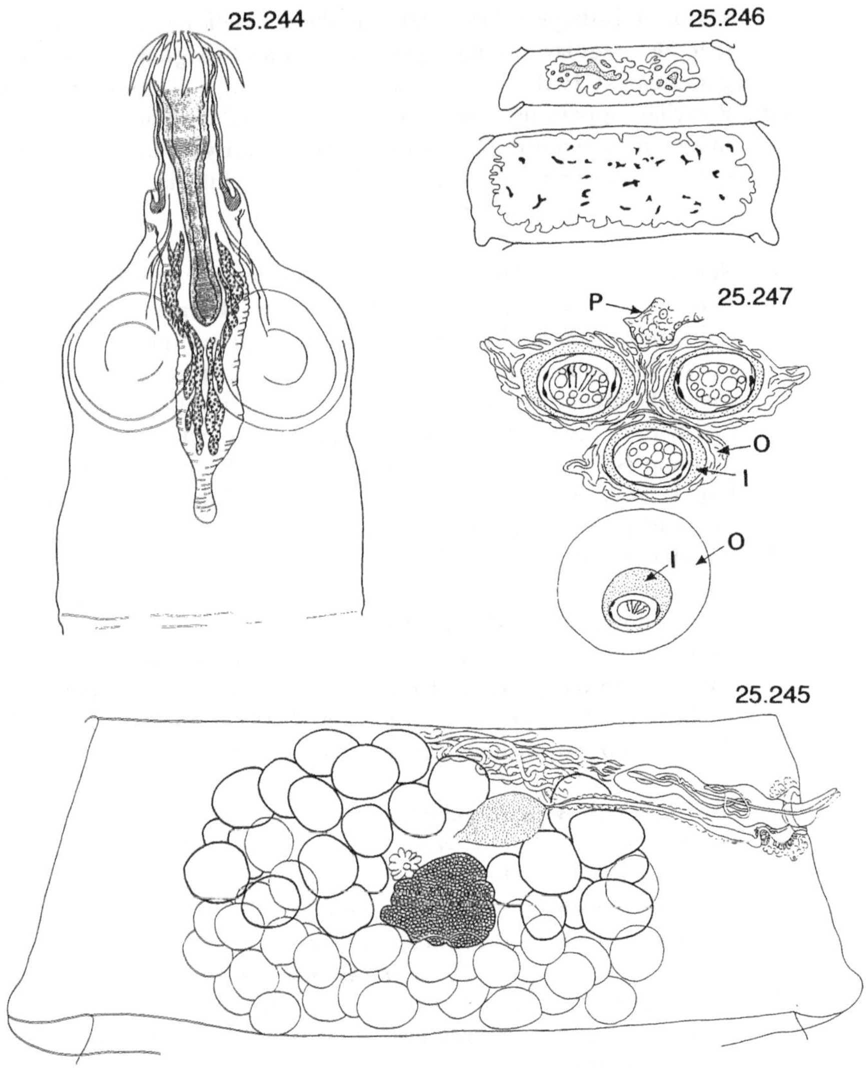

Figs 25.244–25.247 *Paricterotaenia porosa* (Rudolphi, 1810). 25.244. Scolex. 25.245. Mature proglottid. 25.246. Early and fully mature uteri. 25.247. Eggs, inside uterus and in water with swollen outer coat; I: outer layer of the inner coat, O: outer coat, P: spot of parenchyma.

anterior half of medulla, laterally surrounding vitellarium; lobules rather large, roundish (type-species). Testes numerous, posterior to ovary and vitellarium. Cirrus-sac elongate, narrow, with permanent hernia-like protrusion into atrium (type-species). Cirrus very long, with bulbous, asymmetrical base (type-species); spines small. Vas deferens anteroporal, small, thick, not overlapping cirrus-sac. In Lariformes; doubtful in other hosts. Cosmopolitan. Type-species *P. porosa* (Rudolphi, 1810).

86b. Uterus at first labyrinthine, early entering cortex, thin walled, then (i) tending to become sacciform, deeply and heavily septate, (ii) remaining

labyrinthine, (iii) tending to disappear. Atrium fairly deep, strong-walled, may develop prominent papilla. Hooks porosoid (see 86a) or different. Embryophore thick, later radiate, sculptured; outer coat coarsely granular, adhering to embryophore .... 87.

87a. Hooks few (10–14), porosoid (see 86a). Rostellum with long straight, cylindrical stem; apex delicate, indistinct, pad-shaped; when protruded, handles of hooks converge to centre of rostellar pad. Cirrus mostly small, thin, weakly spiny. Ovarian poral lobe extends forwards beyond cirrus-sac and vas deferens .... *Sacciuterina* Matevosyan, 1963. (Figs 25.248–25.252)

Diagnosis: Rostellar apparatus muscular. Pouch very long, narrow. Hooks in one circle. Strobila small, proglottid sides often deformed by prominent papilla and jutting posterior rim. Suckers large. Genital pores usually irregularly alternate, at times regularly. Genital ducts between osmoregulatory canals, rarely dorsal in some proglottids. Vagina posterior to cirrus-sac, orifice in same horizontal plane,

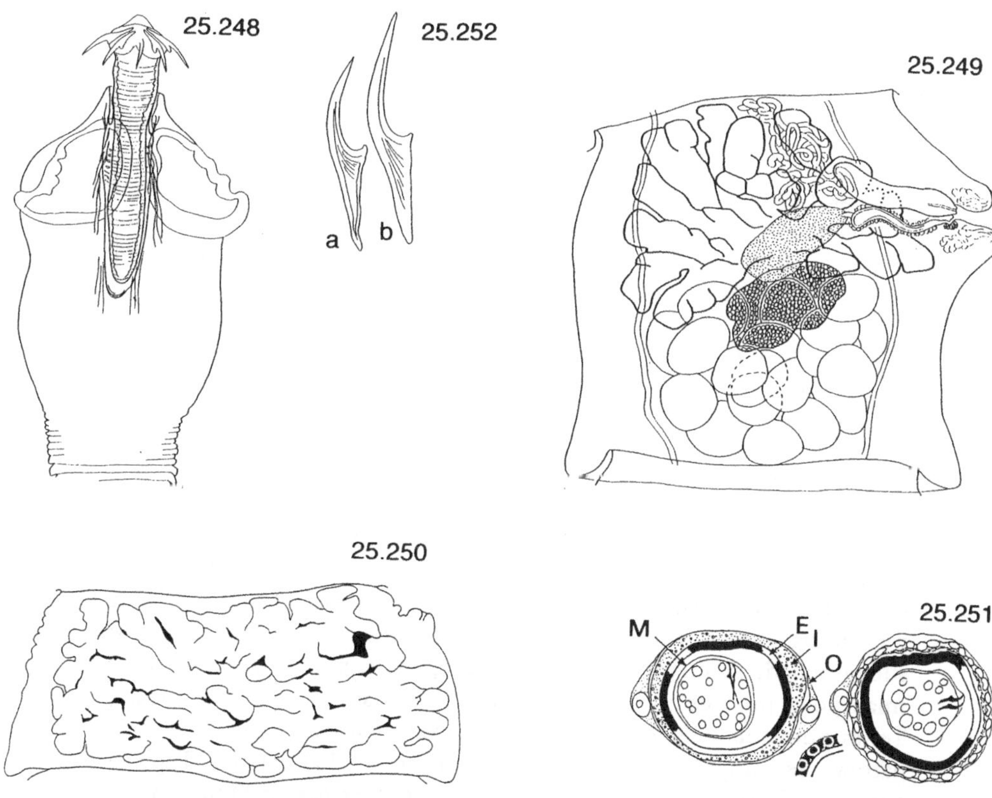

Figs 25.248-25.252 *Sacciuterina*. 25.248-25.252a. *S. paradoxa* (Rudolphi, 1802). 25.248. Scolex. 25.249. Mature proglottid. 25.250. Gravid proglottid. 25.251. Egg with late granular embryophore; M: oncospheral membrane, E: embryophore; I: outer layer of the inner coat forming large granules merging into the embryophore and giving the particular superficial structure; O: outer coat. 25.252a Hook. 25.252b. Hook of other *Sacciuterina* species.

proximal course mostly dorsal; short, with pre-atrial sphincter, at times very reduced. Uterus labyrinthine, entering cortex early, tending to become sacciform, deeply septate; no uterine capsules. Embryophore thick, becoming late radiate, sculptured; outer coat adhering to embryophore. Atrium fairly deep, wall strong, with large papilla. Ovary large, may cross osmoregulatory canals. Testes numerous, postovarian, dorsal to vitellarium. Cirrus-sac elongate, sometimes drawn into papilla. Cirrus small, thin, apparently unarmed. In Charadriiformes. Palaearctic, Neotropical, probably cosmopolitan. Type-species *S. paradoxa* (Rudolphi, 1802).

87b. Hooks numerous (18–22), wrench-shaped (see 73a). Rostellar stem often posteriorly tapering and slightly sinuous; apex disc-shaped, much wider than stem; when protruded, hooks inserted on disc rim with handle ends in a circle. Cirrus fairly wide, clearly spinous. Ovarian poral lobe behind cirrus-sac and vas deferens ..... *Ovosculpta* n. g. (Figs 25.253–25.255)

Diagnosis: Rostellar apparatus muscular. Pouch not very long. Hooks in one circle. Strobila small. Proglottid sides often deformed. Suckers large. Genital pores irregularly alternate. Genital ducts between osmoregulatory canals. Vagina posterior to cirrus-sac, orifice in same horizontal plane; course with varying position; short, bends posteriorly before entering atrium with pre-atrial sphincter. Uterus labyrinthine, entering cortex early, persistent or leaving eggs in parenchyma; no uterine capsules. Embryophore thick, becoming late radiate, sculptured; outer coat, adhering to embryophore. Atrium fairly deep, may form papilla. Ovary reaches osmoregulatory canals. Testes numerous, postovarian, dorsal to vitellarium. Cirrus-sac elongate. In Charadriiformes. Palaearctic. Type-species *O. scolopacis* (Joyeux & Baer, 1939) n. comb. (syn. *Choanotaenia cayennensis* var. *scolopacis* Joyeux & Baer, 1939).

88a. Uterus at first reticulate, then transient finely septate cavity, eventually eggs loose in parenchyma; no uterine capsules. Aporal lobe of ovary reaches anterior proglottid rim .... *Monosertum* n. g. (Figs 25.256–25.260)

Diagnosis: Rostellar apparatus musculo-highly glandular. Pouch large, oval. Rostellum small. Hooks in one circle (14–24), small; sometimes slight irregularities in alignment. Strobila small to medium; neck long, wide. Gravid proglottids longer than wide. Osmoregulatory canals form reticulum in posterior part of scolex (type-species). Genital pores irregularly alternate. Genital ducts between osmoregulatory canals. Vagina posterior to cirrus-sac in same horizontal plane; wall strong, lumen wide. Eggs with various kinds of outer coat: large with surface strengthened, simulating uterine capsule, or firm near embryophore, or with small polar processes; often in contact in short stretches. Atrium small, simple. Testes numerous (18–29), clearly postovarian. Cirrus-sac small, elongate. Cirrus narrow, seemingly unarmed. In Passeriformes. Palaearctic. Type-species *M. parinum* (Dujardin, 1845) n. comb. (syn. *Taenia parina* Dujardin, 1845).

88b. Uterus probably reticulate and not sacciform, turning into capsules with one or few eggs. Ovary far back from anterior rim of proglottid .... *Kintneria* Spasskii, 1968. (Figs 25.261–25.263)
(Syn. *Monopylidium (Kintneria)* Spasskii, 1968).

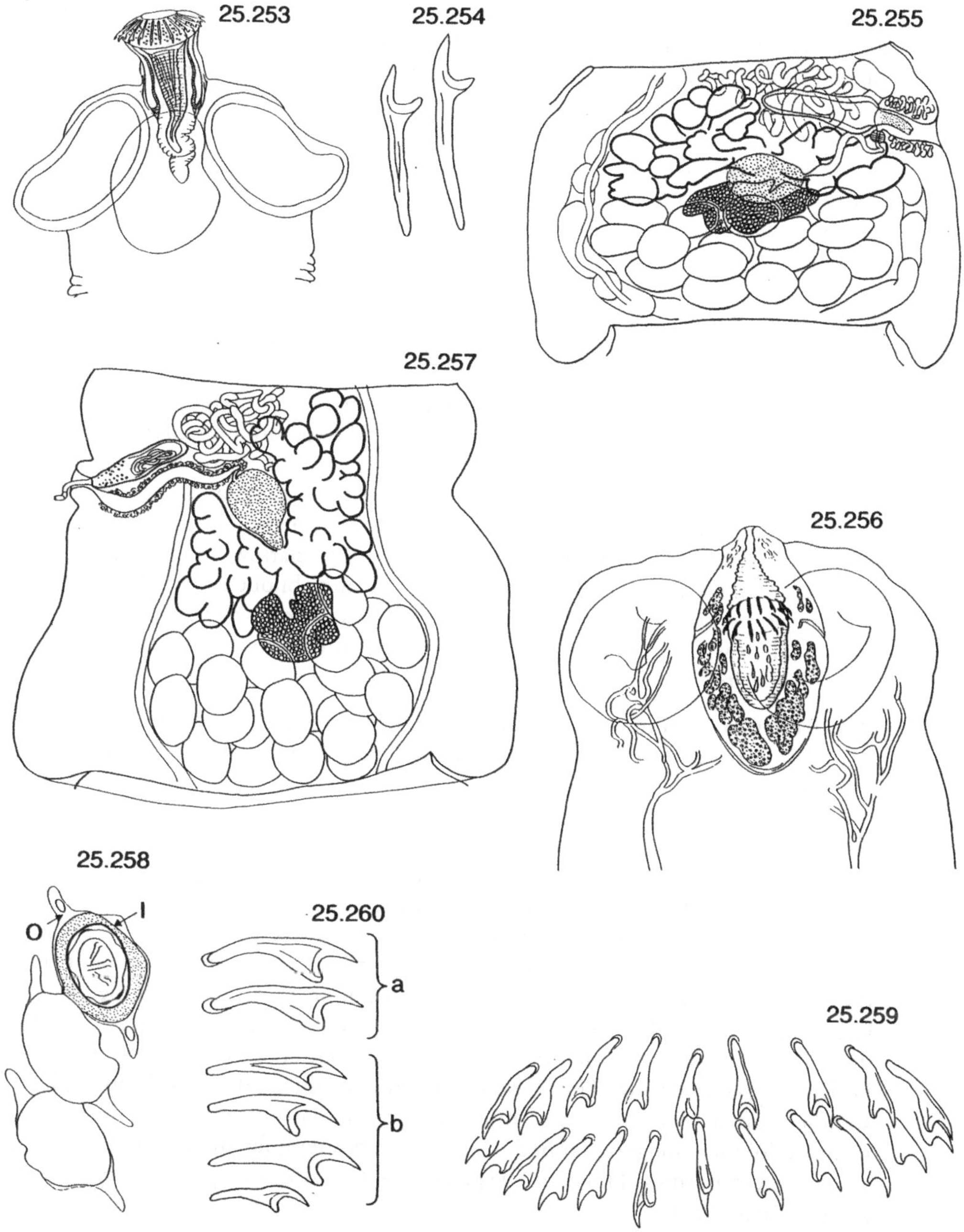

Figs 25.253-25.255 *Ovosculpta scolopacis* (Joyeux & Baer, 1939) n. comb., type. 25.253. Scolex, note the rostellar apical disc. 25.254. Hooks wrench-shaped, variation in the same species. 25.255. Mature proglottid. (For uterus and embryophore see *Sacciuterina* Figs 25.250, 25.251.)
Figs 25.256-25.260 *Monosertum parinum* (Dujardin, 1845) n. comb., neotype. 25.256. Scolex. 25.257. Mature proglottid. 25.258. Eggs; I: outer layer of the inner coat; O: outer coat. 25.259. Crown with not perfectly aligned hooks; in other species the crown may be much more regular. 25.260. Hooks: a, *parinum*; b, four other congeneric species.

Diagnosis: Rostellar apparatus musculo-highly glandular. Pouch large, oval. Rostellum deeply sunken when retracted. Hooks in one circle (15?), wrench-shaped, blade much shorter than handle. Strobila large; gravid proglottids much longer than wide. Genital pores irregularly alternate. Genital ducts between osmoregulatory canals. Vagina posterior to cirrus-sac, (horizontal plane and structure not stated). Atrium simple, shallow. Ovary with isthmus; lobules small, round, close. Testes numerous (25), postovarian. Cirrus-sac elongate. Cirrus long, narrow; short spines. In Passeriformes. Nearctic. Type-species *K. capsulata* Spasskii, 1968.

89a. Testes few in two medullary, lateral fields. Organs in posterior half of medulla. Hook guard reduced to hump. Rostellar pouch narrow, long, pyriform. In Accipitriformes .... *Neyralla* Johri, 1955. (Figs 25.264–25.266)

Diagnosis: Rostellar apparatus not described, supposedly muscular. Rostellum very narrow. Hooks in one circles (less than 20), typical. Strobila medium; no neck; proglottids craspedote. Scolex large, tapering. Genital pores irregularly alternate. Genital ducts between osmoregulatory canals. Vagina posterior to cirrus-sac, (dorsoventral position not stated); long, narrow, transverse. Uterus sacciform, finely lobulate. Atrium small. Ovary bilobed, small, lobulate. Vitellarium far posterior. Cirrus-sac elongate. Cirrus unknown. Seminal receptacle very small. Vas deferens scarcely developed(?), anteroporal. In Accipitriformes. India. Type-species *N. kotharia* Johri, 1955.

89b. Testes in continuous medullary field or in three fields, one medullary and two cortical. Organs more extended. Hooks normal. Rostellar pouch not pyriform .... 90.

90a. Testes extremely numerous (250–300), divided into three fields by osmoregulatory canals, one medullary, two cortical. Strobila very large. (See also regularly alternating pores at 76a) .... *Gruitaenia* Spasskii, Borgarenko & Spasskaya, 1971. (Fig. 25.267)

Diagnosis: Rostellum small. Hooks in two circles, very numerous (about 40). Proglottids very large, rectangular; gravid ones longer than wide. Scolex small. Genital pores irregularly alternate, with long, regularly alternate series. Genital ducts dorsal to osmoregulatory canals. Vagina posterior to cirrus-sac in same horizontal plane; narrow. Uterus reticulate-labyrinthine, persistent. Embryophore spindle-shaped. Atrium strong, muscular ring at base. Ovary large, with deep isthmus; lobules elongate, fanned. Cirrus-sac very small, far anterior. Cirrus short, spiny. In Gruiformes. Holarctic. Type-species *G. latissima* Spasskii, Borgarenko & Spasskaya, 1971.

90b. Testes much less numerous, no cortical fields .... 91.

91a. Ovary roughly reticulate, branched and lobulate. Uterus reticulate, turning into polyovular capsules. Eggs arranged in packets, with free polar processes. In Passeriformes, Cinclidae. (See also apical apparatus glandular at 17a) .... *Cinclotaenia* Macy, 1973. (Figs 25.268–25.271)

(Syn. *Borgarenkolepis* Spasskaya & Spasskii, 1977.)

25.261

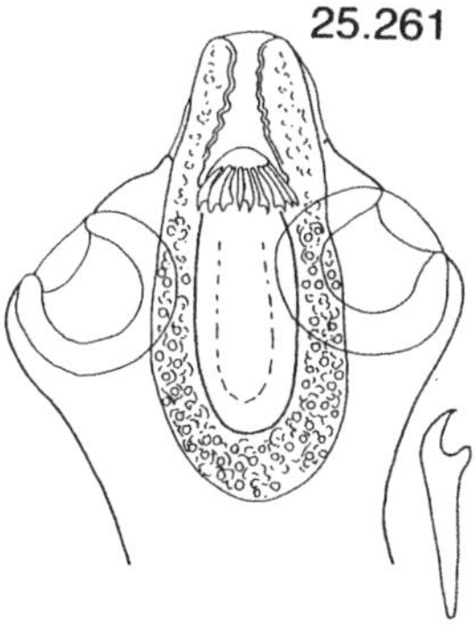

25.262

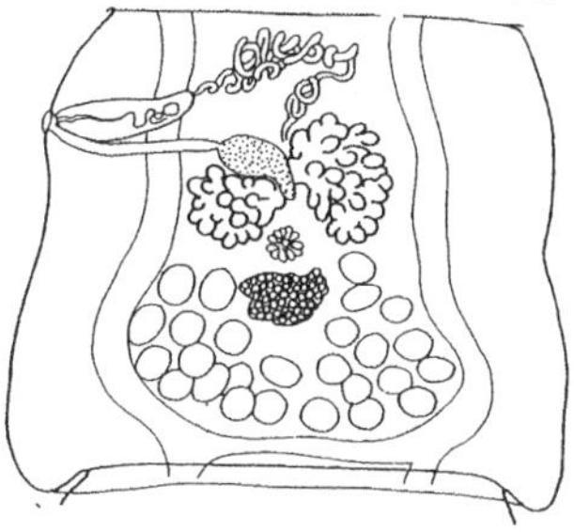

25.263

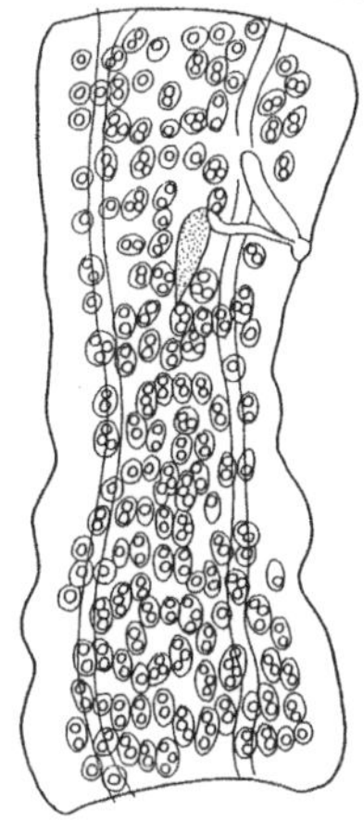

25.264

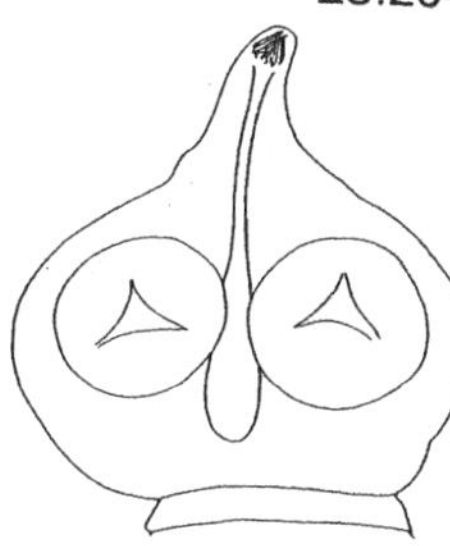

25.265

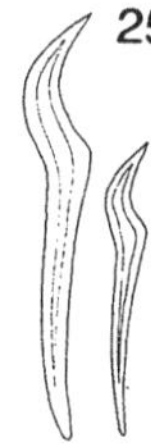

25.266

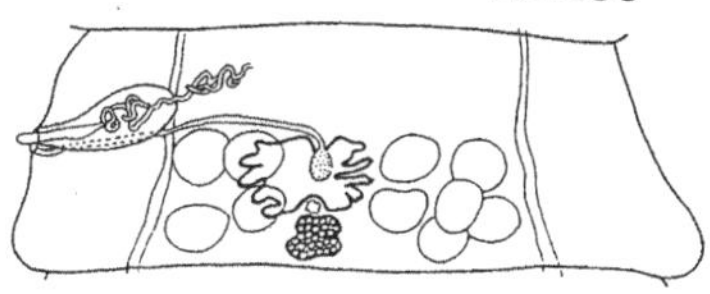

25.267

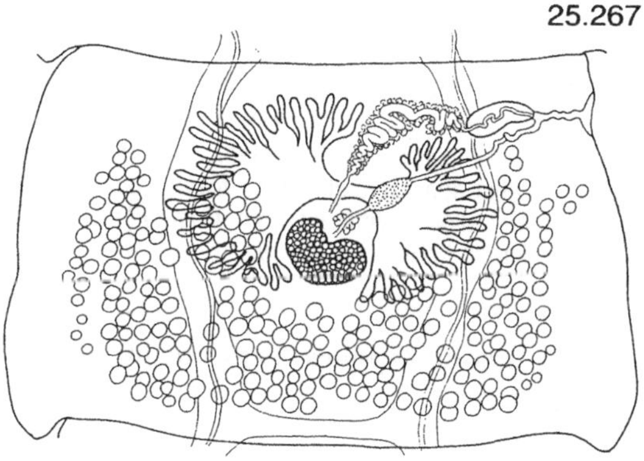

Figs 25.261-25.263 *Kintneria capsulata* Spasskii, 1968. 25.261. Scolex and hook. 25.262. Mature proglottid. 25.263. Gravid proglottid. (From Kintner, 1938.)

Figs 25.264-25.266 *Neyralla kotharia* Johri, 1955. 25.264. Scolex. 25.265. Hooks. 25.266. Mature proglottid. (From Johri, 1955.)

Fig. 25.267 *Gruitaenia gruis* Rausch & Rausch, 1985. Mature proglottid. (From Rausch & Rausch, 1985.)

Diagnosis: Rostellar apparatus muscular, highly or prevailingly glandular. Pouch large, wide; wall thin. Rostellum small, weak. Hooks in two circles, small. Strobila small. Genital pores irregularly alternate. Genital ducts dorsal to or between osmoregulatory canals in same strobila. Vagina posterior to cirrus-sac in same horizontal plane; fairly short. Atrium small, simple, wall thick. Ovary large, poral lobe at times small. Vitellarium reticulate, with small spherical lobules or compact perforate. Testes numerous, postovarian. Cirrus-sac elongate, usually small, thin-walled. Cirrus long, with very small, dense, spines. In Passeriformes, Cinclidae. Holarctic. Type-species *C. filamentosa* Macy, 1973.

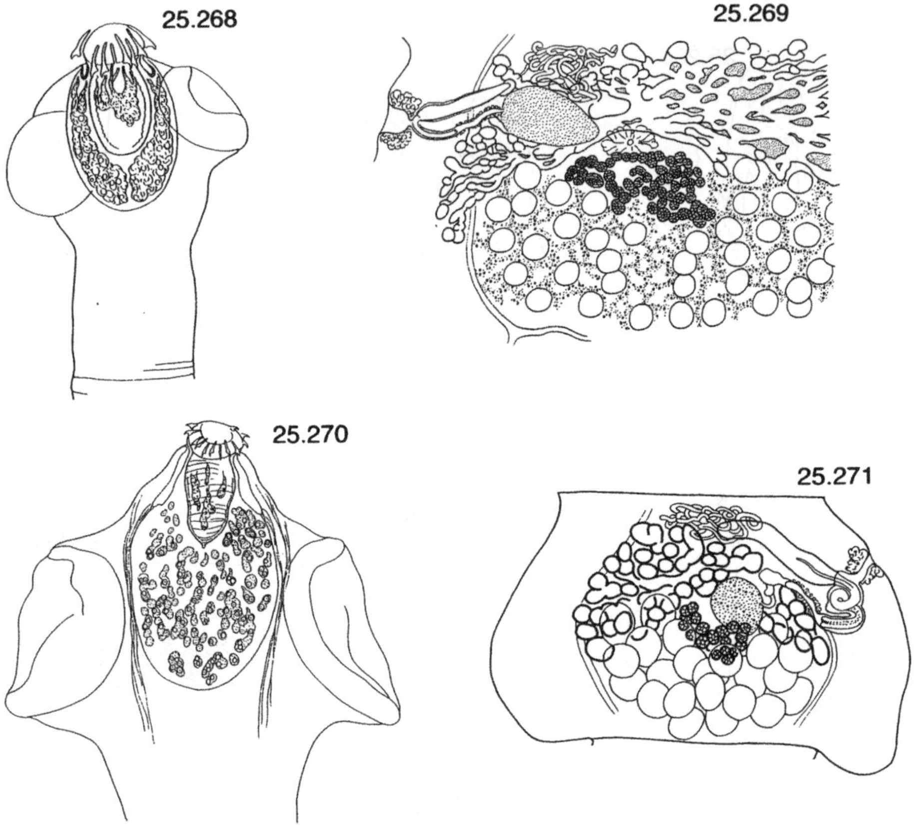

Figs 25.268–25.271 *Cinclotaenia*. 25.268, 25.269. *C. tarnogradskii* (Dinnik, 1927). 25.268. Rostellar pouch apparently slightly different from *filamentosa* pattern. 25.269. Large species with typical anatomy, similar to *filamentosa*; early uterine reticulum among testes. 25.270, 25.271. *C. dehiscens* (Krabbe, 1879) n. comb. 25.270. Rostellar apparatus similar to *filamentosa* pattern, typical of the genus. 25.271. Small species, anatomy slightly different from *filamentosa*, uterus development not completely known, apparently different from *Anomotaenia*.

91b. **Ovary not reticulate. No polyovular uterine capsules; eggs not in packets .... 92.**

92a. **Uterus reticulate, turning into characteristic monovular vesicular parenchymatous capsules. Ovary proportionately small .... 93.**

92b. **Uterus either sacciform or reticulate, not turning into vesicular uterine capsules. Ovary not particularly small (except for *Parorchites* and *Nototaenia*) .... 94.**

93a. **Rostellar apparatus heavily muscular, very large; very small cavity with pore at rostellar apex. Ovary often not bilobed in adult proglottids, compact, more or less poral. Vagina long, overlying or underlying cirrus-sac, at times just posterior. Seminal receptacle small. In Charadriiformes. (See also**

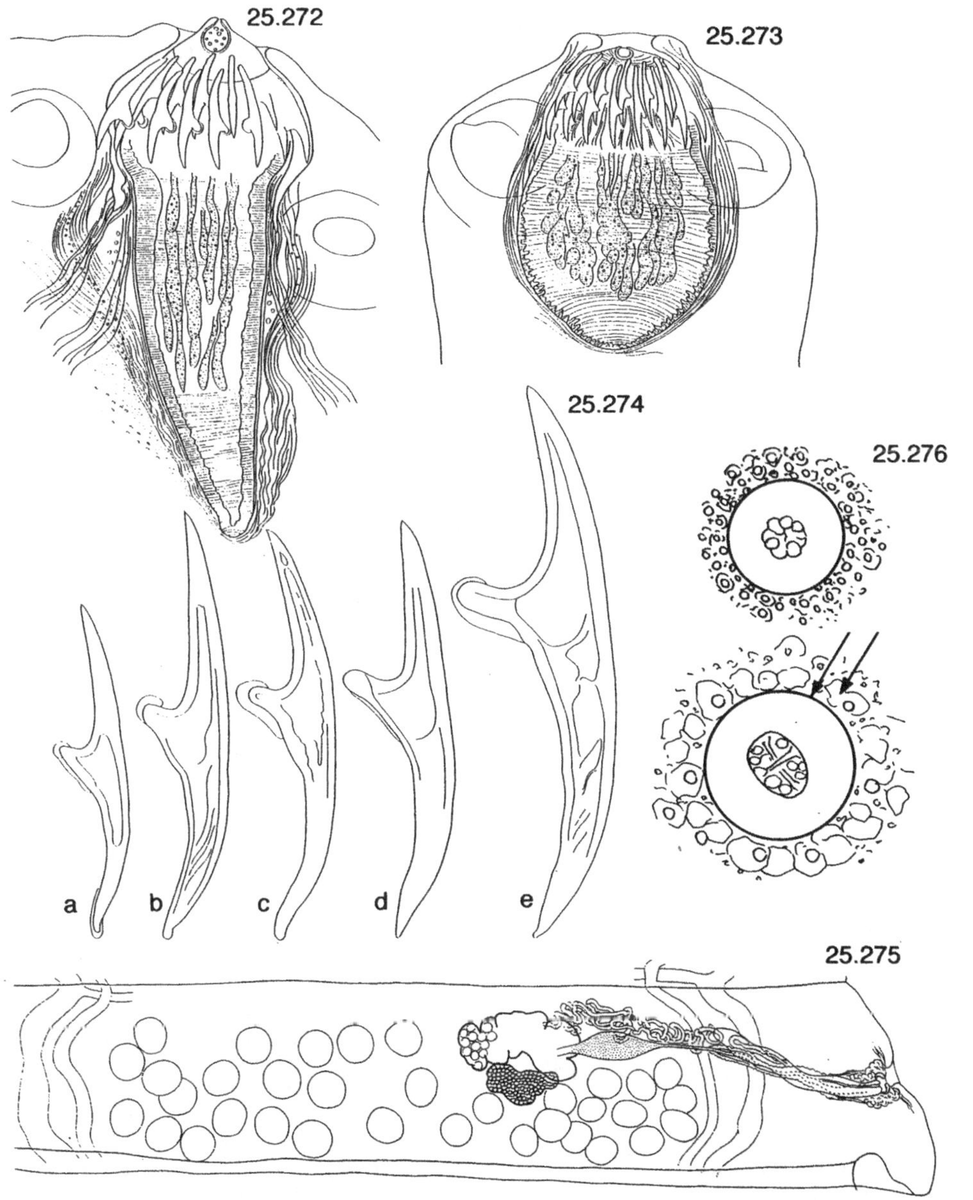

Figs 25.272–25.276 *Megalacanthus* (see note in the Introduction). All new combinations. 25.272. *M. macracanthus* (Fuhrmann, 1908), rostellum. 25.273. *M. lobivanelli* (Burt, 1940), type, rostellum. 25.274. Hooks: (a) *M. guiarti* (Shen, 1932), type; (b) *M. macracanthoides* (Fuhrmann, 1907), probably type; (c) *M. lobivanelli* (Burt, 1940), type; (d) *M. southwelli* (Fuhrmann, 1932), type; (e) *M. macracanthus* (Fuhrmann, 1908) *nec* 1907, type. 25.275. *M. rostellatus* (Fuhrmann, 1908), type, mature proglottid. 25.276. Uterine capsules, schematic; top: early capsule, before oncosphere develops; bottom: mature capsule with pellicle surrounded by vesicular cells (arrows).

unilateral pores at 44a) ....
.... *Megalacanthus* **Moghe, 1926. (Figs 25.272–25.276)**
(Syns *Monopylidium (Macracanthus)* Moghe, 1925, preoccupied; *Monopylidium (Megalacanthus)* Moghe, 1926; *Panuwa* Burt, 1940; *Parachoanotaenia* Rego, 1967, *nec* Lühe, 1910.)

Diagnosis: Rostellar pouch with thick wall of longitudinal fibres, strong outer longitudinal bundles from neck entering pouch. Rostellum very large, glands inside. Hooks in two circles (about 30), usually very large. Strobila medium, pro-

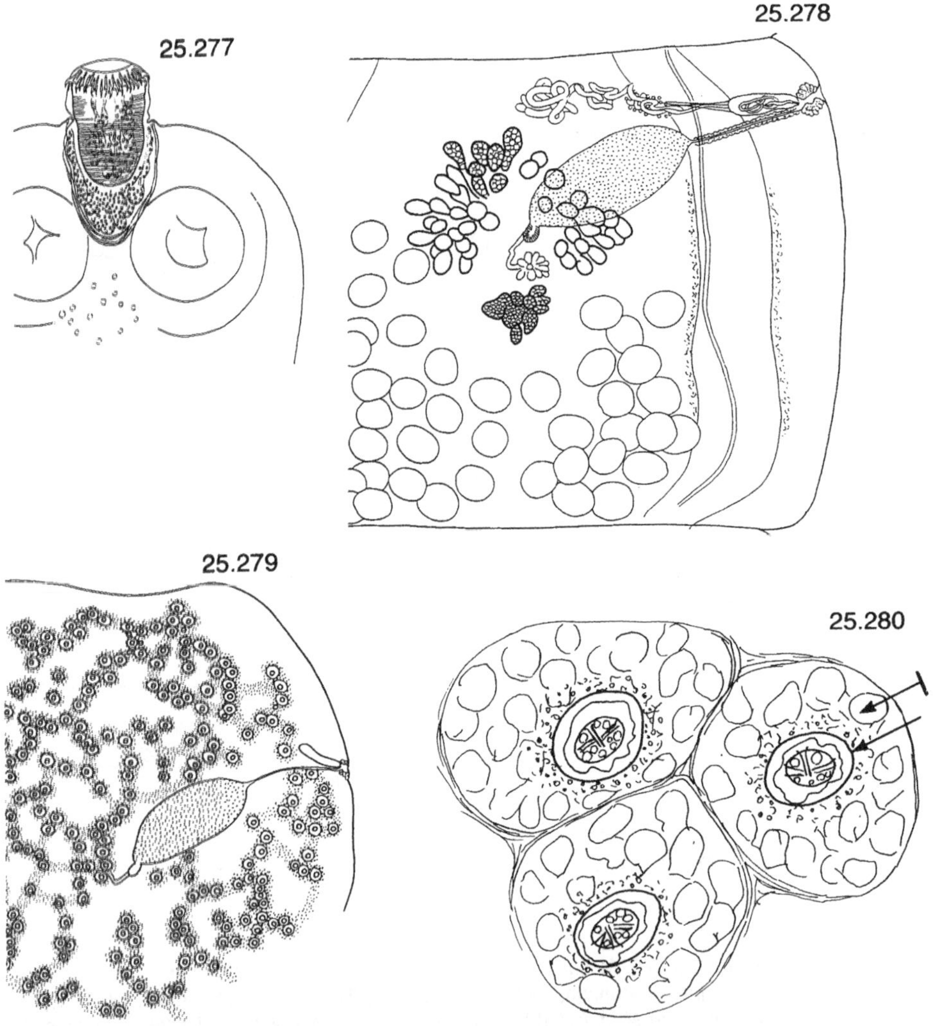

Figs 25.277–25.280 *Prochoanotaenia marchali* (Mola, 1907). 25.277. Scolex. 25.278. Mature proglottid. 25.279. Reticulate uterus, precocious uterine capsules (partial). 25.280. Mature, vesicular, parenchymatous uterine capsules; embryophore thin, undulating; the outermost membrane of the egg outer coat may (arrow), or may not, adhere to the capsule; arrow with bar: vesicular cell of the parenchyma.

glottids wider than long. Scolex very large. Suckers small. Genital pores commonly irregularly alternate, occasionally unilateral with few alternations, in different species. Genital ducts either dorsal to or prevailingly between osmoregulatory canals, in different species. Uterus reticulate, then precocious, monovular, parenchymatous, vesicular capsules. Atrium small, wall dense, globular. Vitellarium very small. Testes numerous, posterior and lateral to ovary, field nearly reaching osmoregulatory canals. Cirrus-sac long, narrow. Cirrus long, narrow; very fine, thick spines. Vas deferens small, transverse in line with cirrus-sac. In Charadriiformes. Nearly cosmopolitan. Type-species *M. chandleri* (Moghe, 1925) Moghe, 1926.

93b. Rostellar apparatus musculo-glandular, not particularly strong; rostellum normal. Ovary bilobed, small, median, with small, spaced lobules. Vagina rather short, posterior to and in same horizontal plane as cirrus-sac. Seminal receptacle large. In Ralliformes .... *Prochoanotaenia* Meggitt, 1924. (Figs 25.277–25.280)

Diagnosis: Rostellar pouch small, wall thin. Rostellum small; apex dome-shaped, not demarcated from stem. Hooks in two circles (about 40 and more). Strobila large, organs spaced. Genital pores irregularly alternate, very anterior. Genital ducts between osmoregulatory canals. Uterus reticulate, large-meshed, then forms precocious, scattered, monovular, parenchymatous, vesicular capsules; finally, eggs in small cavities lie sunken into continuous vesicular tissue. Atrium small, wall globular. Testes very numerous, postovarian. Cirrus-sac very small. Cirrus small; spines fine, close. In Ralliformes. Europe. Type-species *P. marchali* (Mola, 1907).

94a. Genital ducts dorsal to osmoregulatory canals. Uterus always sacciform, permanent (except *Alcataenia* at 95b, having at first reticulate uterus) .... 95.

94b. Genital ducts between osmoregulatory canals. Uterus at first reticulate, permanent or not .... 99.

95a. Uterus always sacciform .... 96.

95b. Uterus at first coarsely reticulate, with deep diverticula, then sacciform, much septate. Hooks in two circles, with 1:1 or 1:2 alternation, according to species. (See also genital ducts between osmoregulatory canals at 102a) .... *Alcataenia* Spasskaya, 1971. (Figs 25.281–25.283)

(Syn. *Rissotaenia* Spasskaya & Kolotilova, 1972.)

Diagnosis: Rostellar apparatus musculo-glandular. Pouch medium. Rostellum elongate, narrow. Strobila medium; proglottids largely overlapping. Genital pores irregularly alternate, may be postequatorial with cirrus-sac and distal part of vagina far posterior. Genital ducts between or dorsal to osmoregulatory canals, according to species. Vagina (at least distal part) posterior to cirrus-sac, course dorsal or ventral in same strobila; long, S-shaped, crossing cirrus-sac, or short, diverging from cirrus-sac. Atrium spherical; wall strong, sucker-like; occasionally with large papilla. Ovary bilobed, large; poral lobe very small, may be pushed backwards. Testes very numerous, postovarian. Cirrus-sac very small, elongate.

Cirrus unarmed. In Gaviiformes; Lariformes. Arctic. Type-species *A. campylacantha* (Krabbe, 1869).

96a. Suckers armed, with hooklets or very fine spines. Strobila very small .... 97.

96b. Suckers unarmed. Strobila small to medium .... 98.

97a. Sucker border with hooklets. Testes numerous, surrounding and partly overlapping ovary. Ovary small, hardly bilobed, compact, densely lobulate. In Charadriiformes, Chionidae .... *Nototaenia* Jones & Williams, 1967. (Fig. 25.284)

Diagnosis: Rostellar apparatus musculo-glandular (glands mainly in rostellum). Pouch very large and long, deeply sunken behind suckers. Rostellum very long, occupying whole pouch. Hooks in two circles. Strobila and scolex very small. Gravid proglottids much longer than wide. Pores irregularly alternating (at times difficult to ascertain because of scarcity of proglottids); very anterior. Genital ducts dorsal to osmoregulatory canals. Vagina posterior to cirrus-sac, dorsal; short, spiny. Uterus sacciform, at first posterior, septate, rectangular, elongated. Atrium deep, wall strong; large papilla. Cirrus-sac fairly large, oval, far anterior. Cirrus very large, spines thin. Seminal receptacle large, spherical, median. Vas deferens very small, tight coils. In Charadriiformes, Chionidae. Antarctic. Type-species *N. fileri* Jones & Williams, 1967.

97b. Suckers covered with very small, fine, spines. Testes few, postovarian. Ovary bilobed, deep isthmus, few large lobules. In Ralliformes. (See diagnosis in regularly alternate pores at 78b) .... *Rallitaenia* Spasskii & Spasskaya, 1975.

98a. Longitudinal osmoregulatory canals multiple, with anastomoses. Scolex large, covered with needle-like spines, embedded in deep capsule in submucosa. Ovary very large, occupying whole anterior half of medulla. In Charadriiformes .... *Rauschitaenia* Bondarenko & Tomilowskaya, 1979. (Figs 25.285, 25.286)

Diagnosis: Rostellar apparatus probably muscular (insufficiently known). Hooks in two circles. Strobila medium; neck long. Proglottids wider than long, cortex reduced. Genital pores irregularly alternate. Genital ducts dorsal to osmoregulatory canals. Vagina posterior to cirrus-sac, ventral (?); long, narrow. Uterus probably ventral; sacciform, closely lobulate. Atrium very small. Ovary apparently not bilobed, massive. Testes very numerous, postovarian. Cirrus-sac small, elongate. Cirrus finely spinulate. Vas deferens small, anteroporal, dorsal to ovary. In Charadriiformes. Palaearctic. Type-species *R. ancora* (Mamaev, 1959).

98b. Osmoregulatory canals normal. Scolex small, not spiny; neck long, swollen; both embedded in deep capsule in submucosa. Ovary decidedly anteroporal. In Sphenisciformes. (See also apical apparatus seemingly glandular at 17b) .... *Parorchites* Fuhrmann, 1932. (Figs 25.287–25.289)

Diagnosis: Rostellar apparatus muscular, highly glandular. Pouch subspherical, large; anterior cavity very large, rostellum at base, when retracted. Rostellum

25.281

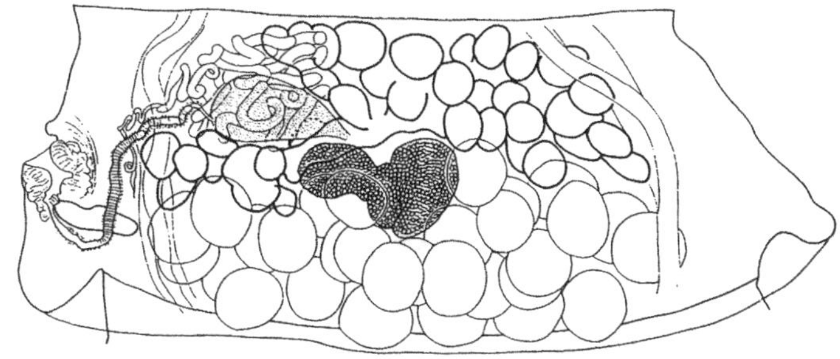

25.282

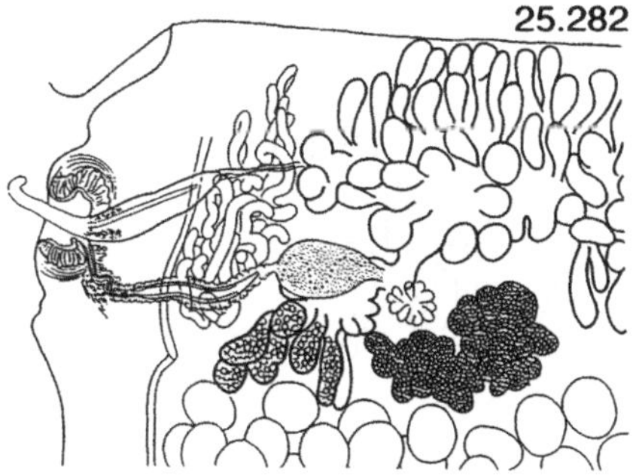

25.283

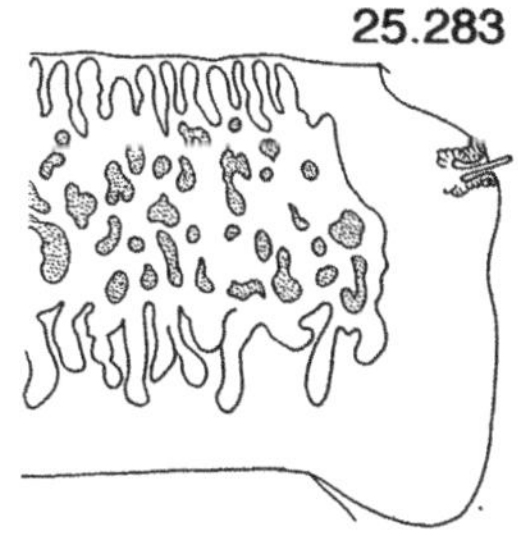

25.284

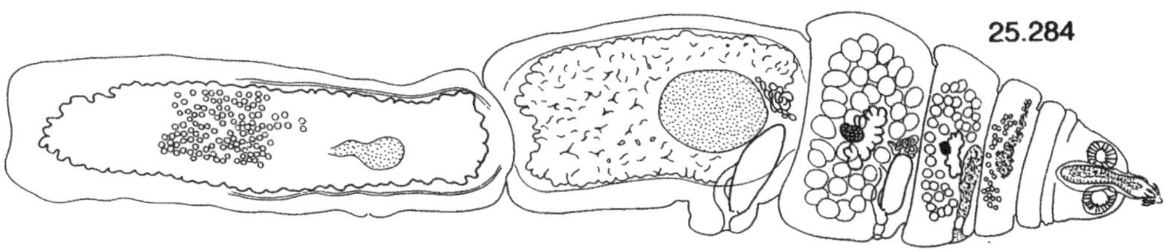

25.285

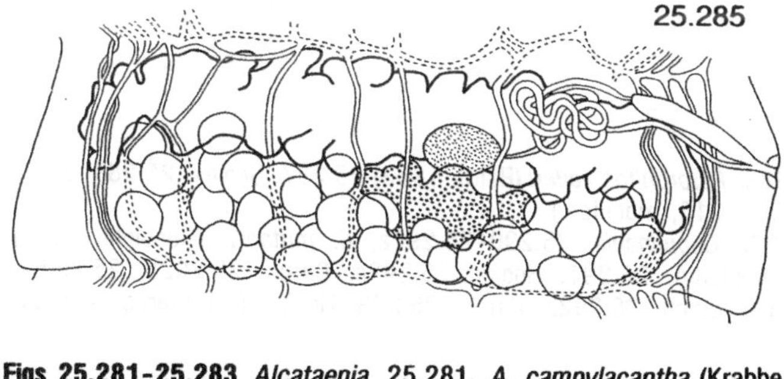

25.286

Figs 25.281–25.283 *Alcataenia*. 25.281. *A. campylacantha* (Krabbe, 1869), mature proglottid. 25.282, 25.283. *A. meinertzhageni* (Baer, 1956), type. 25.282. Part of mature proglottid. 25.283. Part of gravid proglottid.

Fig. 25.284 *Nototaenia fileri* Jones & Williams, 1967, type.

Figs 25.285–25.286 *Rauschitaenia ancora* (Mamaev, 1959). 25.285. Mature proglottid. 25.286. Hook. (From Bondarenko & Tomilowskaya, 1979.)

25.287

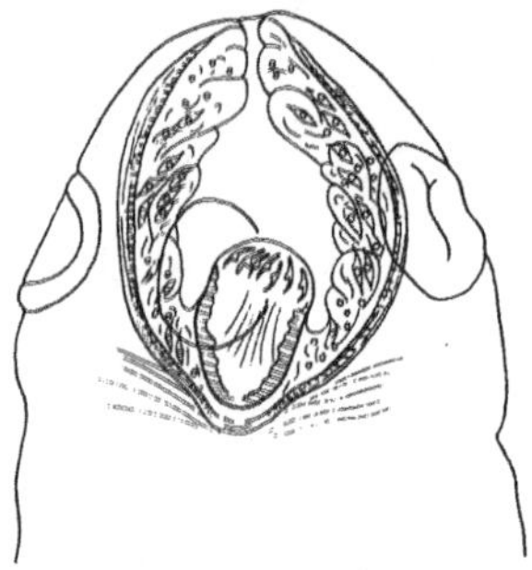

25.288

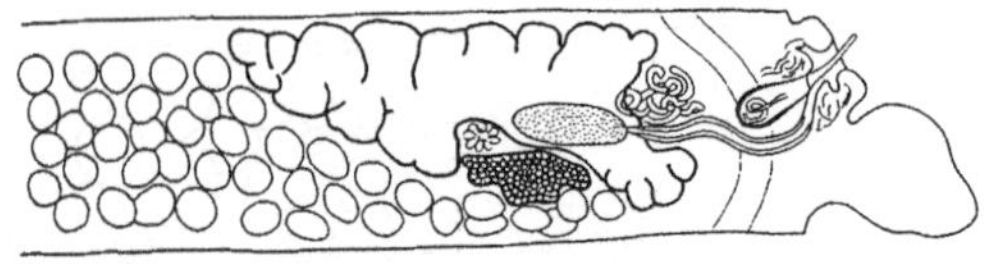

25.289

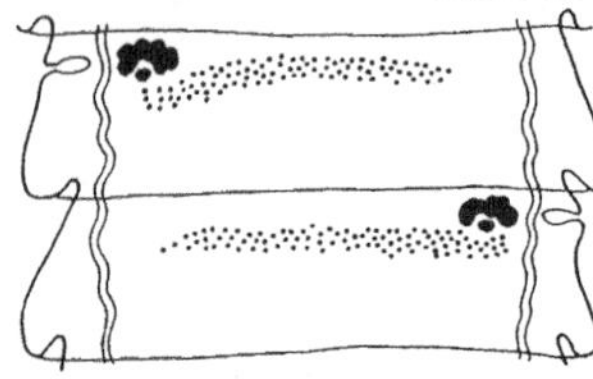

25.290

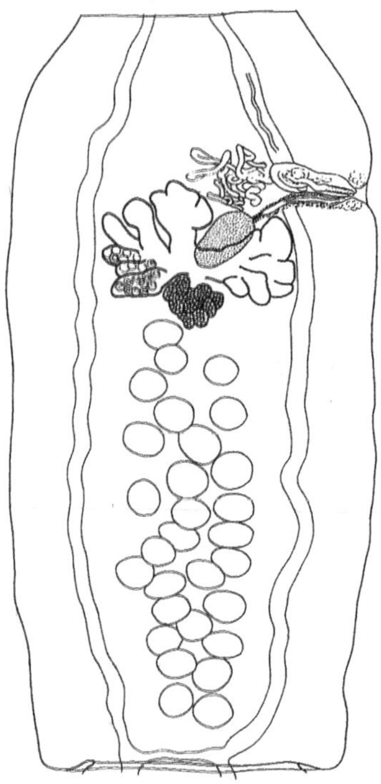

25.291

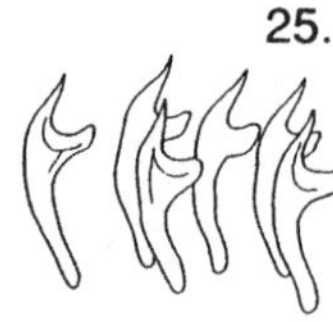

25.293

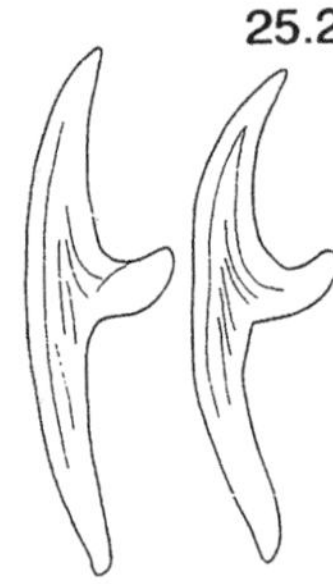

25.292

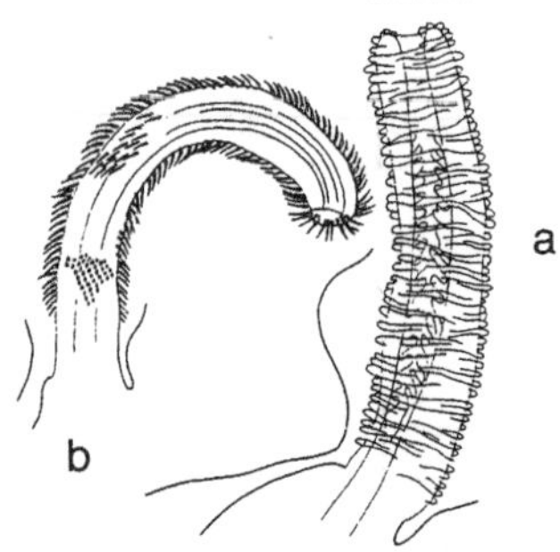

Figs 25.287-25.289 *Parorchites zederi* (Baird, 1853) 25.287. Scolex. 25.288, 25.289. Proglottids. (From Clausen, 1915, modified.)
Figs 25.290-25.293 *Cuculincola*. 25.290-25.292a, *C. mutabilis* (Rudolphi, 1819) n. comb. 25.290. Mature proglottid. 25.291. Hooks. 25.292a. Cirrus. 25.292b, 25.293. *C. acollum* (Fuhrmann, 1907) n. comb. 25.292b. Cirrus. 25.293. Hooks. (For uterus, see *Sobolevitaenia* Fig. 25.295.)

small, normal in structure. Hooks in two circles (18). Strobila large, wide. Proglottids much wider than long. Suckers small. Genital pores irregularly alternate, far anterior. Genital duct dorsal to osmoregulatory canals. Vagina posterior to cirrus-sac, ventral(?). Uterus described as initially dorsal; sacciform. Atrium forms wide papilla. Ovary weakly bilobed, small, compact, horseshoe-shaped. Testes very numerous (100) in wide transverse field, partly postovarian. Cirrus-sac very small. Cirrus small, thin (unknown if armed). In Sphenisciformes. Antarctic. Type-species *P. zederi* (Baird, 1853).

99a. Hooks in two circles, irregular, mixed 1:1 and 1:2 modules (seldom other irregularities). In Cuculiformes .... ***Cuculincola* n. g. (Figs 25.290–25.293)**

Diagnosis: Rostellar apparatus musculo-glandular. Pouch and rostellum long, narrow. Strobila small to medium. Proglottids much longer than wide; spaced organs. Scolex hardly demarcated from neck. Genital pores irregularly alternate. Genital ducts between osmoregulatory canals. Vagina posterior to cirrus-sac in same horizontal plane; circular fibres; no pre-atrial sphincter but narrow in crossing atrium wall. Uterus large-meshed reticulum, then labyrinthine; eventually, eggs loose in parenchyma. Eggs with indistinct external coat adhering to internal one and small polar processes. Atrium simple; no ductus masculinus. Ovary small with distinct isthmus. Testes numerous postovarian. Cirrus-sac small oval; wall thin. Cirrus fairly strong, long; with thin close spines or unarmed with circular superficial pleats. In Cuculiformes. Neotropical. Type-species *C. mutabilis* (Rudolphi, 1819) n. comb. (Syn. *Taenia mutabilis* Rudolphi, 1819).

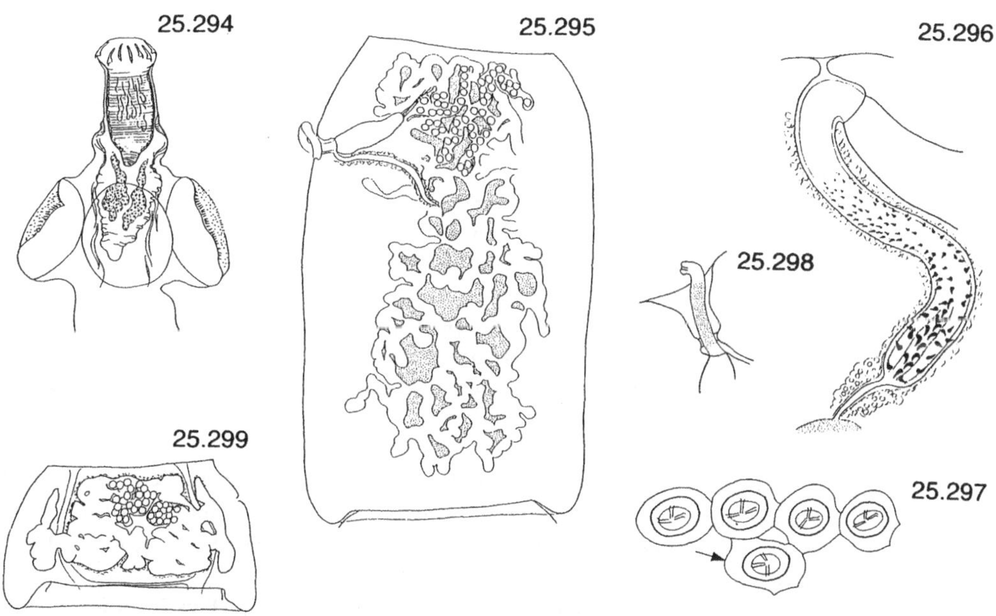

Figs 25.294–25.299 *Sobolevitaenia*. 25.294, 25.295. *S. sobolevi* Spasskaya & Makarenko, 1965. 25.294. Scolex; note spines on suckers. 25.295. Gravid proglottid. 25.296, 25.297. *Sobolevitaenia* sp. 25.296. Cirrus introduced in the vagina. 25.297. Eggs adhering to each other through their outer coat (arrow). 25.298. *S. spinosocapite* (Joyeux & Baer, 1955), type, partly evaginated cirrus. 25.299. *S. korochirei* (Voser & Vaucher, 1988) n. comb., type, gravid proglottid. (For anatomy, see *Cuculincola* Fig. 25.290.)

99b. Hooks in two circles, regular (at times either difficult to distinguish or hooks slightly disordered; see *Monopylidium*). In other orders of birds .... 100.

100a. Suckers covered with spines. In Passeriformes .... *Sobolevitaenia* Spasskaya & Makarenko, 1965. (Figs 25.294–25.299)

Diagnosis: Rostellar apparatus musculo-glandular. Pouch large. Rostellum large, wide. Hooks in two circles (about 20). Strobila small to medium; neck long; proglottids usually longer than wide. Scolex deeply demarcated from neck. Suckers large, posteriorly jutting. Genital pores irregularly alternate. Genital ducts between osmoregulatory canals. Vagina posterior to cirrus-sac, ventral, dorsal or in same horizontal plane in different species; wall strong; no pre-atrial sphincter. Uterus reticulate or thickly labyrinthine; seldom sacciform with irregular outline and spots of parenchyma; eventually, usually eggs in parenchyma. Embryophore thin; outer layer of inner coat persistent, between embryophore and delicate, thin outer coat; no polar processes; eggs form chains or not. Atrium simple, shallow; no ductus masculinus. Ovary with deep isthmus; aporal lobe may be larger. Testes numerous, postovarian (at times, partly lateral). Cirrus-sac oval, elongate, wall firm. Cirrus large; spines usually short, comma-shaped, strong, spaced; sometimes very small, fine. Vas deferens with thick prostatic cells. In Passeriformes. Palaearctic, Oriental regions; Papua New Guinea (unpublished); Neotropical; probably cosmopolitan. Type-species *S. anthusi* (Spasskaya, 1958).

100b. Suckers unarmed .... 101.

101a. Uterus at first reticulate, sometimes also lobulate, then sacciform, deeply lobulate in outline. Ovary very large, occupying all anterior half of medulla .... 102.

101b. Uterus at first reticulate or labyrinthine, then (i) turning into monovular capsules, (ii) remaining permanently labyrinthine, or disappearing, leaving eggs in parenchyma. Ovary not very large .... 103.

102a. Rostellum regular. Eggs without polar processes. Ovary very large, bilobed, extremely asymmetrical, poral lobe very small. Atrium strong, sucker-like. In Gaviiformes, Lariformes. (See also genital ducts dorsal to osmoregulatory canals, and diagnosis and synonyms at 95b) .... *Alcataenia* Spasskaya, 1971.

102b. Rostellum with narrow stem and wide apical disc or swollen pad full of glands. Eggs with very long polar processes, may form chains with processes loose. Ovary very large, apparently not bilobed. Atrium simple, shallow. In Charadriiformes .... *Spasskytaenia* Oshmarin, 1956. (Figs 25.300–25.302)

Diagnosis: Rostellar apparatus muscular and glandular chiefly inside rostellum. Pouch long, wall thin, but strong external longitudinal muscles from neck. Rostellum very long, strong. Hooks in two circles; stumpy, small; guard pronounced, thick stiffening along whole ventral part. Proglottids wider than long; medulla filled by organs. Genital pores irregularly alternate, very anterior. Genital ducts between osmoregulatory canals. Vagina posterior to cirrus-sac (dorsoventral

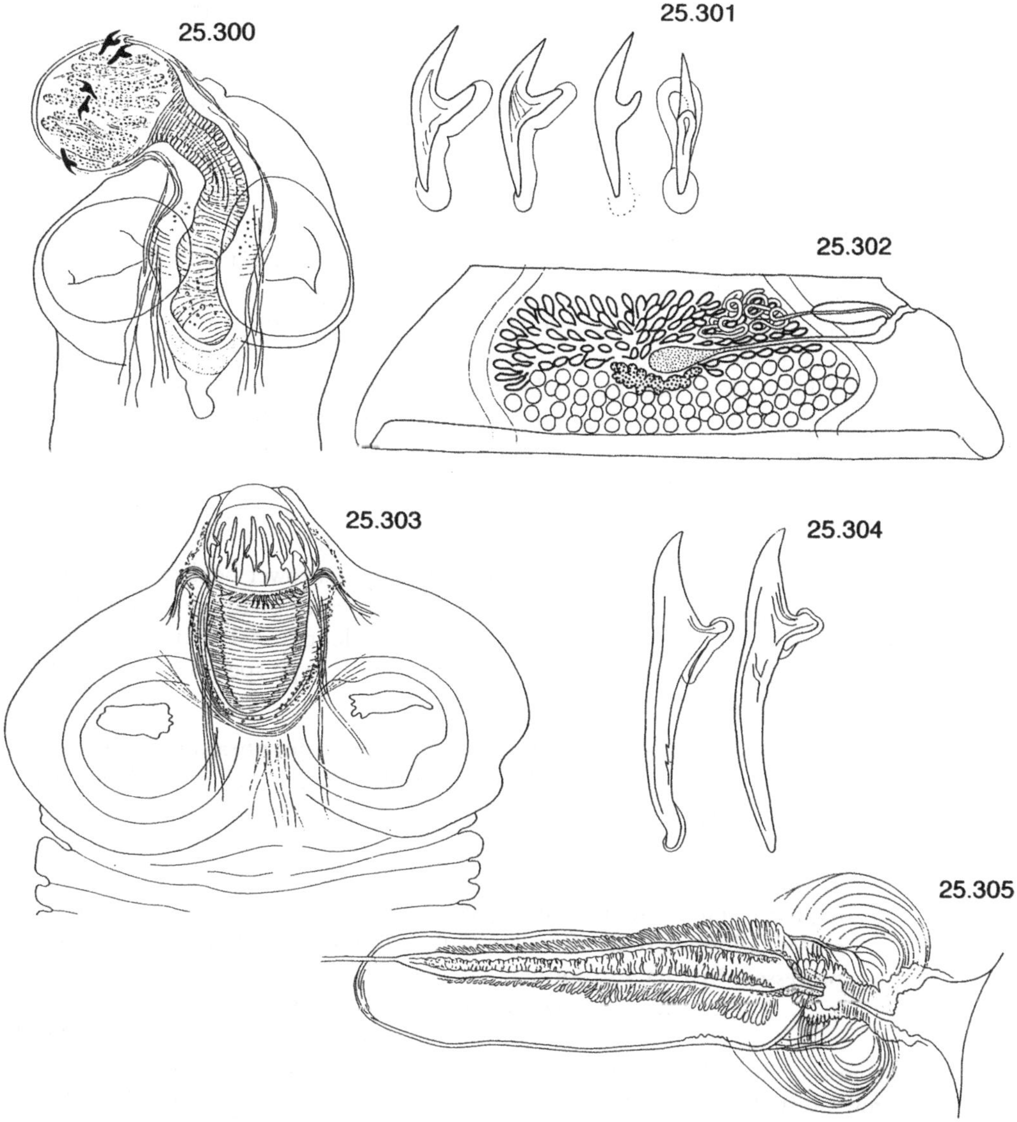

Figs 25.300–25.302 *Spasskytaenia platyrhyncha* (Krabbe, 1869). 25.300, 25.301. Doubtful whether Krabbe's type. 25.300. Scolex. 25.301. Hooks. (All original.) 25.302. Mature proglottid. (From Belopolskaya, 1977.)

Figs 25.303–25.305 *Birovilepis spasskayae* (Birova-Volosinovicova, 1967), type. 25.303. Scolex. 25.304. Hooks. 25.305. Terminal genitalia; vagina with a thick wall and a sphincter-like beak before entering atrium.

position unknown); long, straight, near posterior rim of ovary. Uterus at first reticulate, then sacciform, deeply lobular. Ovary bulky, extended along whole anterior margin of proglottid, young ovary perhaps reticulate. Testes very numerous, small, postovarian. Cirrus-sac small. Cirrus seems unarmed. Vas deferens anteroporal dorsal to ovary. In Charadriiformes. Europe, Asia. Type-species *S. platyrhyncha* (Krabbe, 1869).

103a. Uterus reticulate, then forming monovular capsules. Atrium sphincter-like, large, lined inside with thick layer; wall complex, globular, delicate with concentric fibres. Vagina divided into distal, very strong-walled, typical in structure, carrot-shaped region and long, thin duct linking it to seminal receptacle. (See also unilateral pores at 44b) .... *Birovilepis* Spasskii, 1975. (Figs 25.303–25.305)

Diagnosis: Rostellar apparatus highly muscular, typical. Pouch rather small, short, wide; wall thick with prominent concentric longitudinal fibres and feeble transverse ones; strong external bundles entering pouch. Rostellum strong, wide, normal, filling whole pouch. Hooks in two circles (22–32); unduloid (see 46a). Strobila small. Scolex large. Genital pores irregularly alternate with long unilateral series; rare strobila with unilateral pores. Genital ducts between osmoregulatory canals. Vagina posterior to cirrus-sac, or along sac axis; dorsal; orifice narrow, with sphincter-like structure. Uterine capsules precocious, membranous, firm; parenchyma gradually filled by granular, then vesicular cells. Testes numerous, postovarian, field extending laterally beyond ovary, narrower behind vitellarium, where it may split into two groups. Cirrus-sac elongate. Cirrus very thin; spines very small. Internal vas deferens only one or two loops. In Passeriformes, Corvidae, Turdidae and others. Palaearctic, Oriental. Type-species *B. spasskayae* (Birova-Volosinovicova, 1967).

103b. Uterus reticulate or labyrinthine, permanent or disappearing, leaving eggs in parenchyma. Atrium different, simple. Vagina not divided into two parts; different structure .... 104.

104a. Vagina with poral sphincter, bulbous or sleeve-like. Two hook crowns, regular, well-defined. Embryophore often oval or fusiform, outer coat seldom adhering to it, often large, pleated film or forming noticeable polar processes. In Charadriiformes .... *Anomotaenia* Cohn, 1900. (Figs 25.306–25.311)
(Syn. *Dichoanotaenia* Lopez-Neyra, 1944.)

Diagnosis: Rostellar apparatus muscular, variously glandular. Pouch fairly long, narrow. Rostellar stem relatively long, narrow. Hooks of different patterns, often very small. Strobila small to medium. Scolex small. Suckers large. Genital pores irregularly alternate. Genital ducts between osmoregulatory canals. Vagina posterior to cirrus-sac in same horizontal plane; of varying lengths, usually overlying poral lobe; wall strong but not particularly thick. Uterus reticulate thin-walled or labyrinthine, persistent with close septa adhering to eggs (at times simulating uterine capsules), or disintegrating to leave eggs in parenchyma; no uterine capsules. Atrium small; wall globular, rather thick; no ductus masculinus. Ovary varying in size, isthmus distinct. Testes numerous, postovarian. Cirrus-sac small, elongate, wall firm to fairly thick. Cirrus narrow, mostly very small; spines tiny, homogeneous, at times hardly visible; no tuft of bristle-like spines. In Charadriiformes; Lariformes. Cosmopolitan. Type-species *A. microrhyncha* (Krabbe, 1869).

25.306

25.307

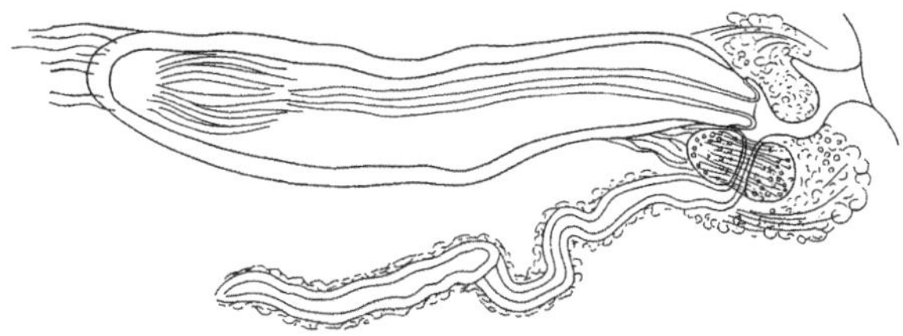

25.308

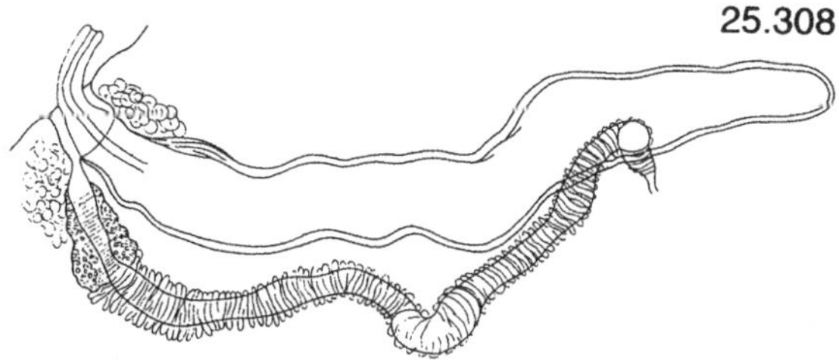

25.309

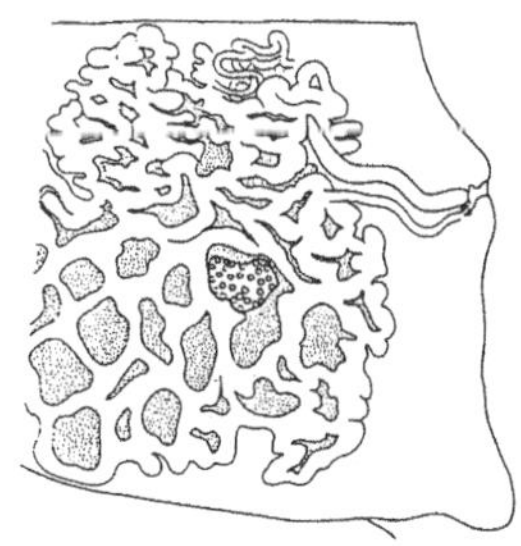

25.310

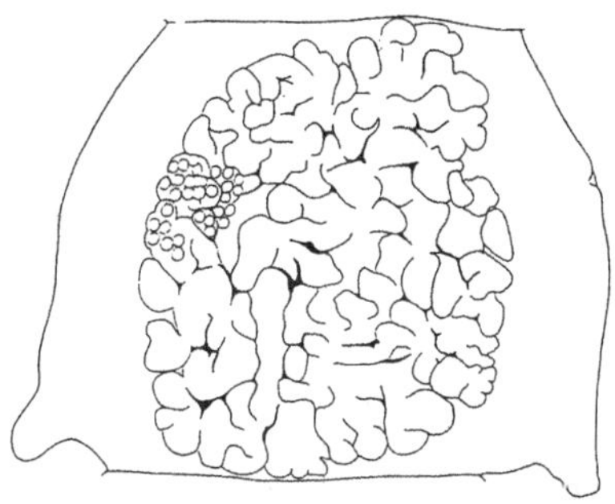

25.311

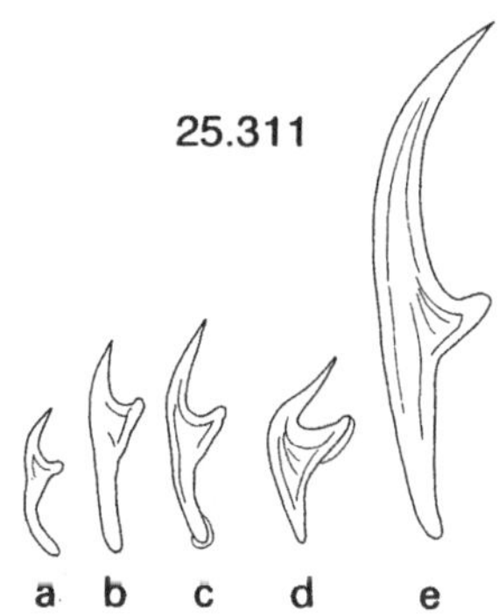

Figs 25.306–25.311 *Anomotaenia*. Several species. 25.306. Mature proglottid, 25.307. Terminal genitalia, vaginal sphincter bulbous. 25.308. Terminal genitalia, vaginal sphincter sleeve-like. 25.309. Uterus reticulate. 25.310. Uterus labyrinthine. 25.311. Hooks of five different species; first left, probably of *microrhyncha*.

104b. Vagina without poral sphincter. Two hook crowns (20–24) at times difficult to distinguish. Embryophore spherical, outer coat differently developed, often adhering to embryophore, hardly distinguishable, or pleated film; no polar processes. In Passeriformes .... *Monopylidium* Fuhrmann, 1899. (Figs 25.312–25.316)

(Syns *Icterotaenia* Railliet & Henry, 1909; *Parachoanotaenia* Lühe, 1910; *Choanofuhrmannia* Lopez-Neyra, 1935.)

Diagnosis: Rostellar apparatus musculo-glandular. Pouch relatively large; wall thin. Rostellum small, apical pad only slightly expanded. Hooks often very small,

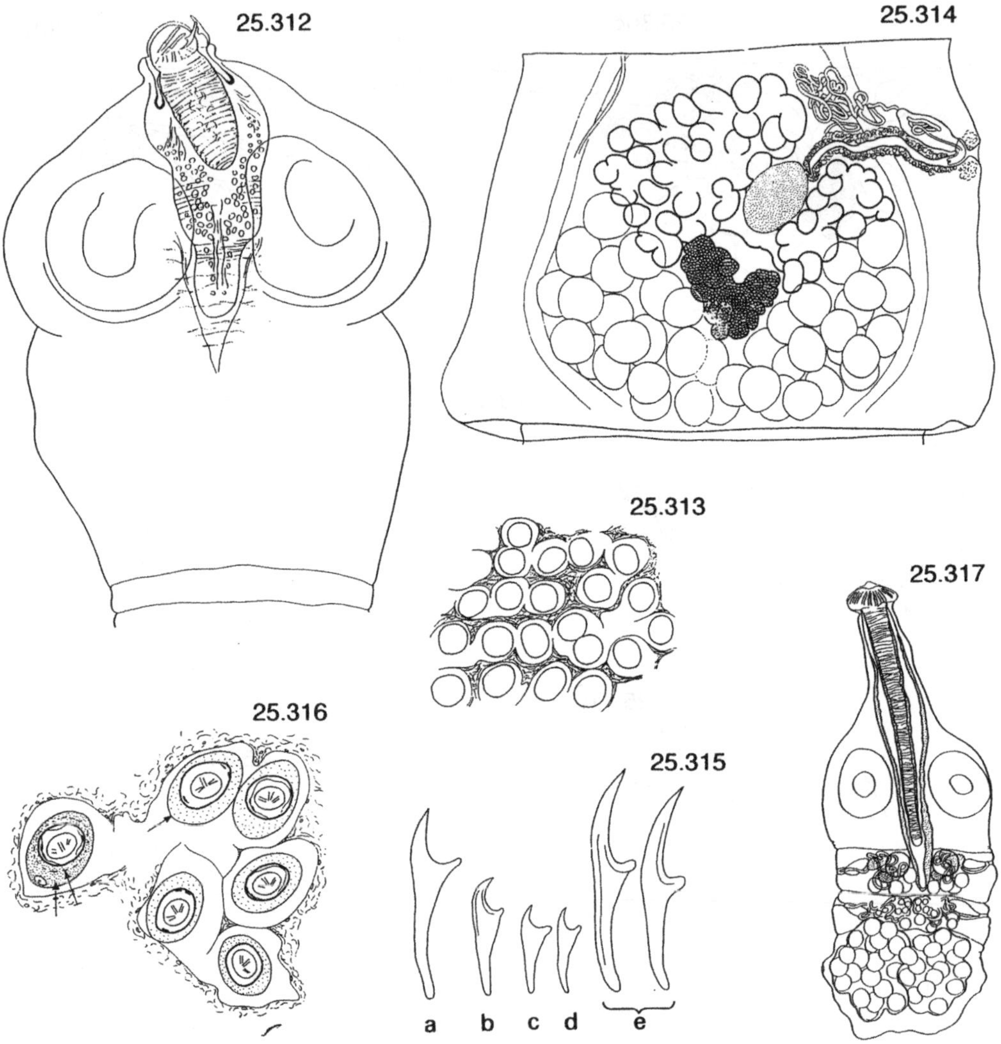

Figs 25.312-25.316 *Monopylidium*. 25.312, 25.313. *M. galbulae* (Gmelin, 1790); Joyeux & Baer's (1955) specimens. 25.312. Scolex. 25.313. Early labyrinthine uterus. 25.314. *M. musculosum* (Fuhrmann, 1896) Fuhrmann, 1899; Joyeux & Baer's (1955) specimens, mature proglottid. 25.315. Hooks: (a) *musculosum*, (b-d) three different species, (e) *galbulae*. 25.316. Uterus and eggs at an advanced stage (see diagnosis); arrow: egg outer coat; arrow with bar: residues of the outer layer of the inner coat.
Fig. 25.317 *Mirandula parva* Sandars, 1956. (From Sandars, 1956.)

delicate. Strobila small to medium. Gravid proglottids longer than wide. Scolex small. Suckers large, delicate. Genital pores irregularly alternate. Genital ducts between osmoregulatory canals; exceptionally dorsal in some proglottids. Vagina posterior to cirrus-sac in same horizontal plane or ventral; of varying length; usually between cirrus-sac and poral lobe; wall strong, thick to very thick; fairly wide lumen. Uterus either labyrinthine persistent, wall robust, mamillated, or reticulate, thin-walled, eventually disappearing leaving eggs loose in parenchyma, at times enveloped (not completely surrounded), singly or in small groups, by remains of initial tubules in which eggs get entangled; no true uterine capsules. Atrium simple, small; wall globular, delicate; no ductus masculinus. Ovary varying in size with distinct isthmus. Testes numerous, postovarian. Cirrus-sac small, oval or elongate, wall usually thin. Cirrus proportionally long, robust; spines thin, homogeneous; no tuft of bristle-like spines. In Passeriformes. Palaearctic, probably wider. Type-species *M. musculosum* (Fuhrmann, 1896).

105a. Two sets of reproductive organs per proglottid .... *Mirandula* Sandars, 1956. (Fig. 25.317)

Diagnosis: Rostellar apparatus muscular. Pouch extremely long, narrow, even dividing first proglottid. Rostellum extremely long, narrow, anterior disc well-defined. Hooks in two circles, very numerous (36). Strobila extremely small; proglottids very few, acraspedote. Scolex large. Genital ducts dorsal. Vagina probably dorsal to cirrus-sac, posterior on left side, anterior on right. Uterus ventral, sacciform; one independent cavity on each side. Eggs few, large. Ovary not bilobed, spherical. Testes few (four on both sides), postovarian. Cirrus bearing small spines. No internal seminal vesicle. Vas deferens extended, dorsal to ovary, seminal receptacle and proximal end of cirrus-sac. In marsupials. Australia. Type-species *M. parva* Sandars, 1956.

105b. Single set of reproductive organs per proglottid .... 106.

106a. Pores unilateral .... 107.
106b. Pores alternate, sometimes with long unilateral series .... 108.

107a. Rostellar apparatus of normal muscular pattern. Genital ducts usually dorsal. Vagina dorsal to cirrus-sac. Testes postovarian, at times also lateral. Atrium simple. In Carnivora, Insectivora, Rodentia(?). (See diagnosis in parasites of birds at 46a) .... *Dilepis* Weinland, 1858.
107b. Rostellar apparatus muscular without real pouch. Rostellum sucker-like, wide, concave with hook tips anteriorly directed when retracted. Genital ducts between osmoregulatory canals. Vagina ventral to cirrus-sac. Testes all around ovary. Atrium deep, very strong wall, with large sphincters. In Carnivora .... *Aelurotaenia* Cameron, 1928. (Figs 25.318–25.320)

Diagnosis: Rostellum lacks stem; surrounded by spherical cap of fibres. Hooks in two circles (about 20); stout, blade distinct from hook body. Strobila small; neck long; proglottids craspedote. Suckers small. Genital pores unilateral on left, in middle of lateral margin. Uterus ventral, sacciform; deep large lobules, thick-walled. Ovary bilobed, large, compact, extends far posteriorly. Vitellarium far posterior. Cirrus-sac long, narrow, along anterior margin of proglottid. Cirrus long, spiny. Seminal receptacle on ovarian isthmus. Vas deferens densely packed, clear-cut, transverse mass; mostly aporal, underlying cirrus-sac. In Carnivora. Malaysia (Zoological Gardens, London). Type-species *A. planicipitis* Cameron, 1928.

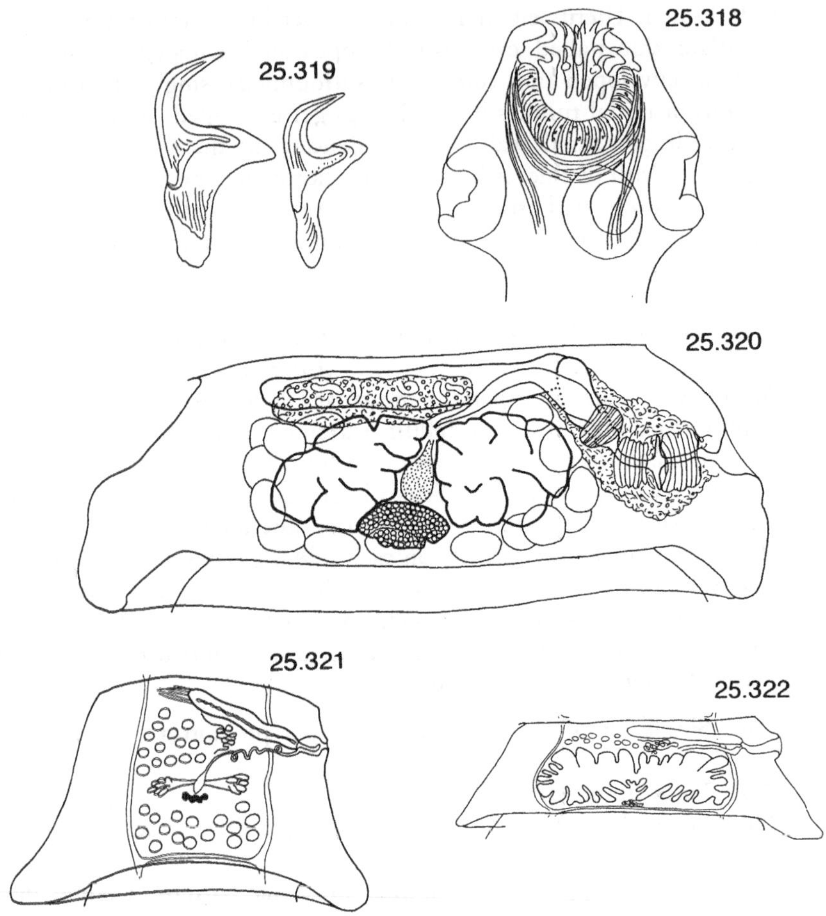

Figs 25.318–25.320 *Aelurotaenia planicipitis* Cameron, 1928, type. 25.318. Scolex. 25.319. Hooks. 25.320. Mature proglottid.
Figs 25.321, 25.322 *Bancroftiella tenuis* Johnston, 1911. 25.321. Probably semi-adult proglottid. 25.322. Gravid proglottid. (From Johnston, 1911.)

108a. Testes divided into two fields, postovarian and antero-aporal. Uterus clearly sacciform, deeply lobulate .... *Bancroftiella* Johnston, 1911. (Figs 25.321, 25.322)

Diagnosis: Rostellar apparatus probably muscular. Pouch small. Rostellum short, stumpy. Hooks in two circles. Strobila large; proglottids largely overlapping; organs spaced. Scolex small. Genital pores irregularly alternate. Genital ducts between osmoregulatory canals. Vagina posterior to cirrus-sac, ventral; long, very sinuous. Atrium rather large; thick-walled; radiating and circular fibres; large papilla may form. Uterus transversely elongate. Ovary small; clear isthmus; transversely elongate. Cirrus-sac long, narrow, strong-walled. Cirrus thin, seemingly unarmed. Vas deferens small, behind cirrus-sac. In marsupials. Australia. Type-species *B. tenuis* Johnston, 1911.

108b. Testes in continuous field, postovarian or surrounding ovary. Uterus at first reticulate or labyrinthine, then variously transformed .... 109.

109a. Ovary U-shaped, not bilobed, posterior and lateral to vitellarium (unique character in Dilepididae *sensu lato*). Testes surrounding ovary and dorsal to it. Uterus ventral, first roughly labyrinthine, then divided into large, strong-walled pouches (apparently no polyovular capsules). In Insectivora, *Talpa* .... *Multitesticulata* Meggitt, 1927. (Figs 25.323–25.326)

(Syns *Viscoia* Mola, 1929; *Rodentotaenia* Matevosyan, 1953.)

Diagnosis: Rostellar apparatus muscular. Pouch large. Hooks in two circles. Strobila medium; heaped organs; cortex reduced. Genital pores irregularly alternate, may alternate regularly in some strobilae; far anterior. Genital duct dorsal. Vagina dorsal to cirrus-sac, distal part anterior, then turning backwards (sometimes, in left pores, vaginal orifice anteroventral); short, firm-walled. Atrium simple, wall thick. Ovary very large, reaching near rear margin of proglottid;

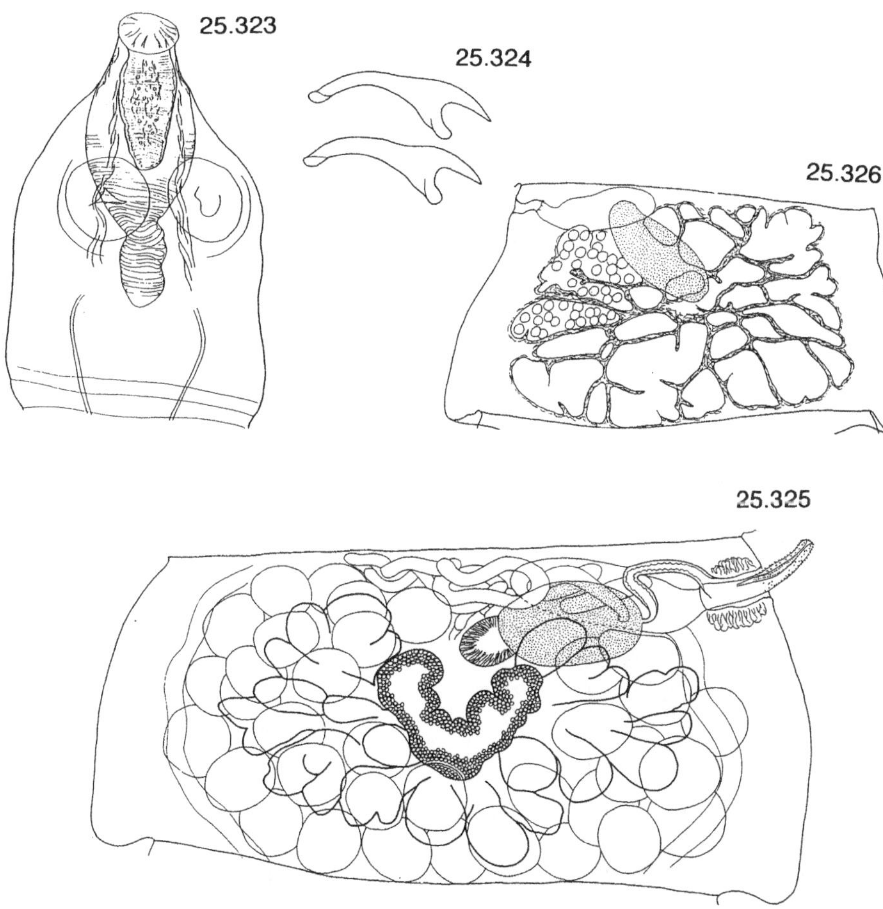

Figs 25.323-25.326 *Multitesticulata filamentosa* (Goeze, 1782). 25.323. Scolex. 25.324. Hooks. 25.325. Mature proglottid. 25.326. Gravid proglottid.

lobules large. Vitellarium anterior to ovary. Testes very numerous. Cirrus-sac near anterior margin of proglottid. Cirrus strong, spines very small. Seminal receptacle poral, far anterior, partly overlapping cirrus-sac. In Insectivora (*Talpa*). Europe. Type-species *M. filamentosa* (Goeze, 1782).

109b. Ovary normal, anterior to vitellarium. Testes postovarian. Gravid uterus not divided into large pouches .... 110.

110a. Rostellar apex large, dome-shaped when protruded, concave when retracted (not sucker-like, hook tips backwardly directed). Genital ducts between osmoregulatory canals. Uterus first reticulate in centre with lateral, transverse, long, narrow digitations. Ovary hardly bilobed .... 111.

110b. Rostellum differently shaped or absent. Genital ducts dorsal to osmoregulatory canals or unstable in position in same strobila. Uterus first reticulate or labyrinthine without lateral digitations. Ovary clearly bilobed .... 112.

25.327

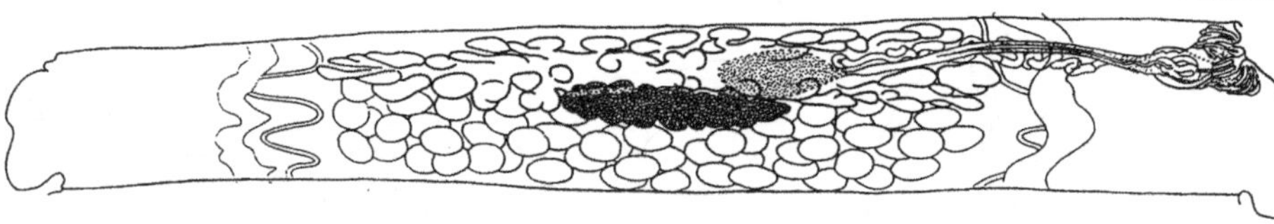

25.328

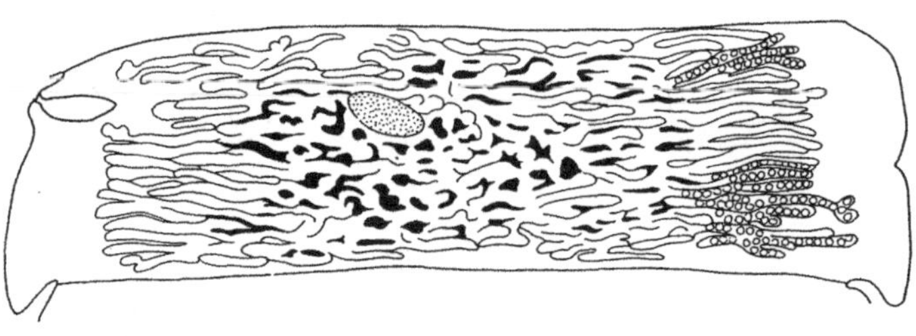

25.329

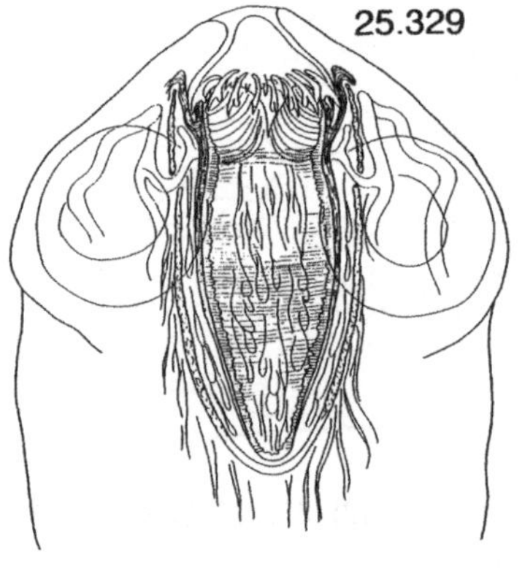

25.330

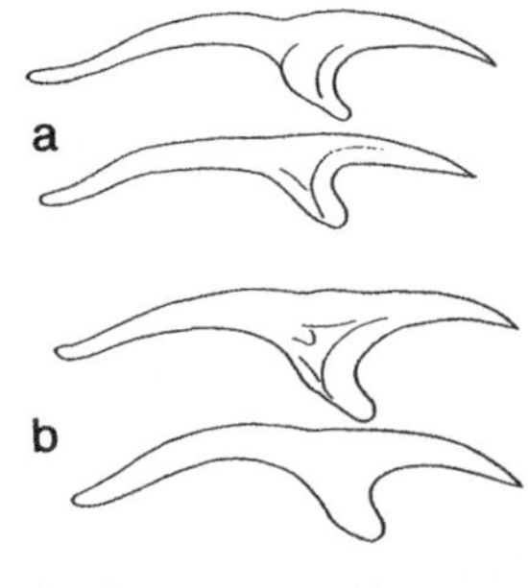

25.331

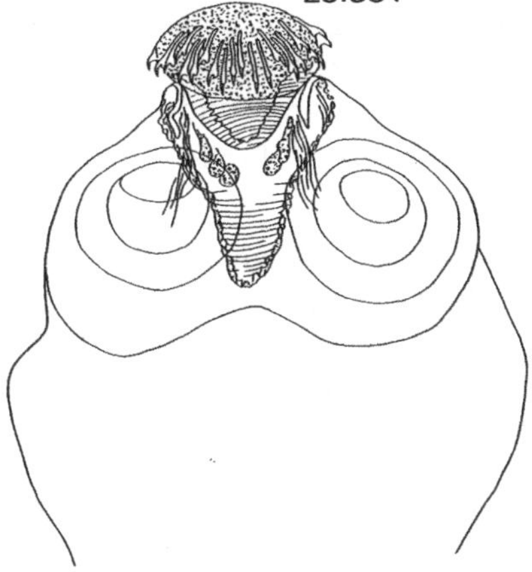

25.332

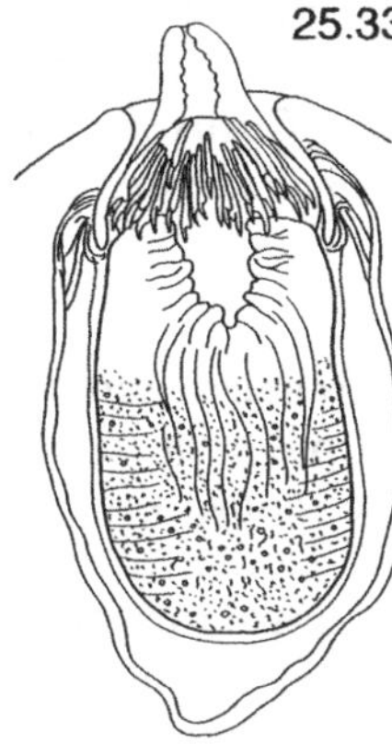

25.333

25.334

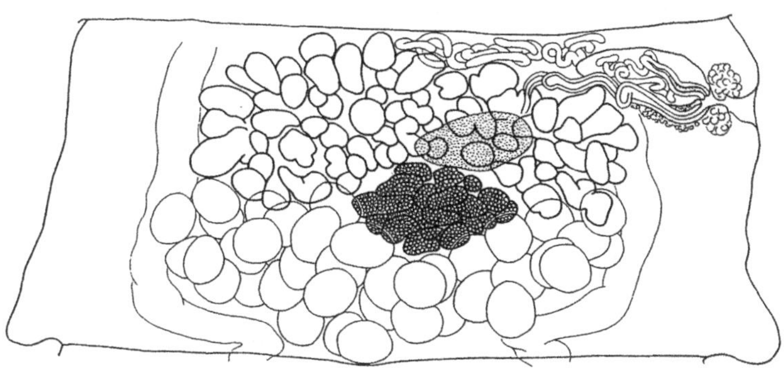

Figs 25.327–25.330 *Alproma heimi* (Quentin, 1964). 25.327, 25.328. Paratype. 25.327. Mature proglottid. 25.328. Gravid proglottid. 25.329. Scolex, material of Hunkeler (1969). 25.330. Hooks: (a) type, (b) paratype. (above, distal; below, proximal).
Figs 25.331–25.334 *Hepatocestus hepaticus* (Baer, 1932) n. comb. 25.331. Scolex. 25.332. Rostellum retracted. 25.333. Hook. 25.334. Mature proglottid. (For uterus, see *Alproma* Fig. 25.328.)

111a. Hooks in two circles, regular. Vagina anterior to cirrus-sac or along its axis. Female organs at times poral. Ovary and testes hardly reaching osmoregulatory canals. In Rodentia .... *Alproma* Spasskii, 1982. (Figs 25.327–25.330)

Diagnosis: Rostellar apparatus muscular. Pouch large, strong-walled. Rostellum large, strong; apex contracted, concave but not sucker-like and hook tips backwardly directed when retracted; wide, dome-shaped when protruded. Strobila medium; proglottids much wider than long. Genital pores alternate irregularly. Genital ducts between osmoregulatory canals. Vagina dorsal to cirrus-sac, at times ventral on left side; very long. Uterus mainly ventral, labyrinthine centrally, with long, lateral diverticula; then tubules merge and many septa persist; no uterine capsules. Embryophore thin; outer coat surface firm. Atrium simple, wall strong. Ovary hardly bilobed, transversely extended, median or poral. Testes very numerous, postovarian. Cirrus-sac very small. Cirrus partly armed; spines punctiform. Vas deferens small, transverse, in line with cirrus-sac, underlying vagina, crossing osmoregulatory vessels. In Rodentia. Africa. Type-species *A. heimi* (Quentin, 1964).

111b. Hook crown irregular, zigzagging mixed with 1:2 alternations in same crown. Vagina posterior to cirrus-sac. Female organs median. Ovary and testes largely in contact with osmoregulatory canals. In Insectivora, mainly in bile-duct .... *Hepatocestus* n. g. (Figs 25.331–25.334)

Diagnosis: Rostellar apparatus musculo-glandular. Pouch oval, strong-walled. Rostellum large, very deformable; apex contracted, concave but not sucker-like, hook tips backwardly directed when retracted; wide, dome-shaped and reduced stem when protruded. Strobila small to medium. Genital pores irregularly alternate. Genital ducts between osmoregulatory canals. Vagina dorsoventral position relative to cirrus-sac unstable; long, thick-walled. Uterus ventral, labyrinthine

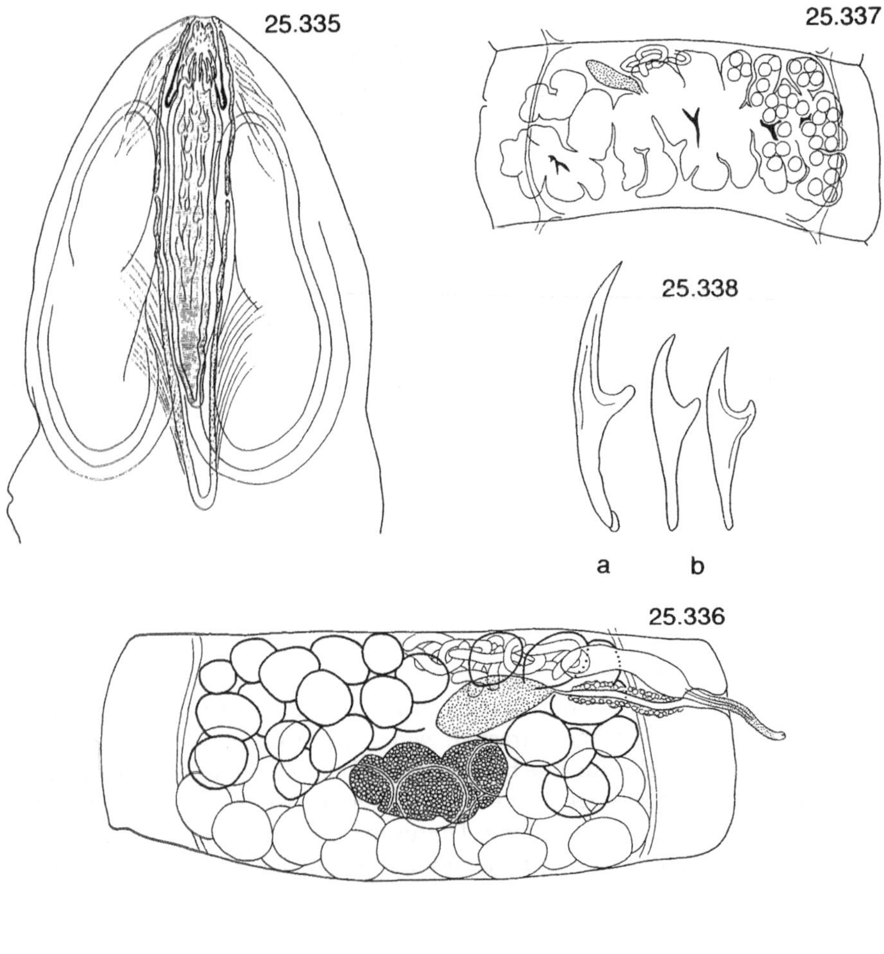

25.339

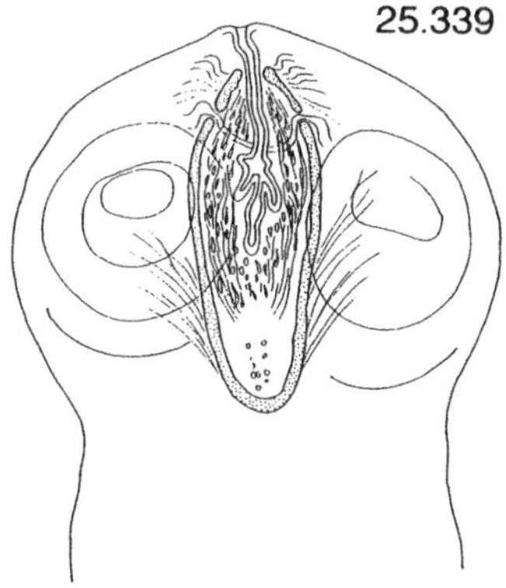

25.340

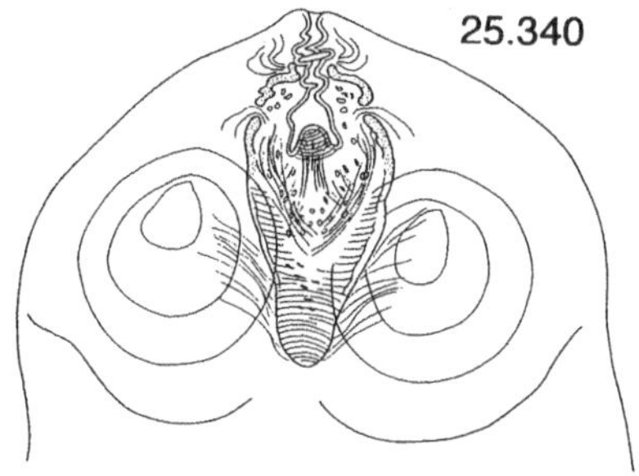

25.341

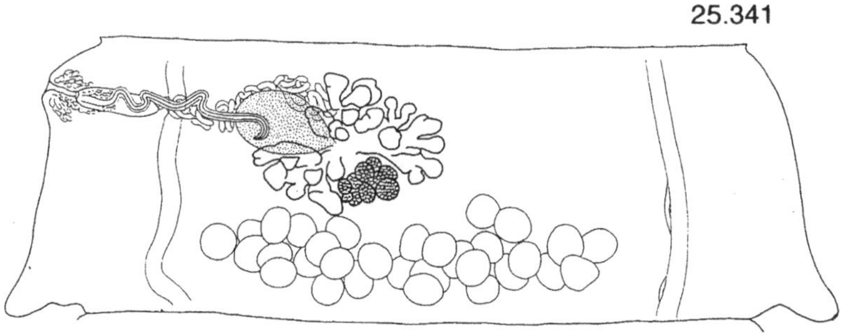

25.342

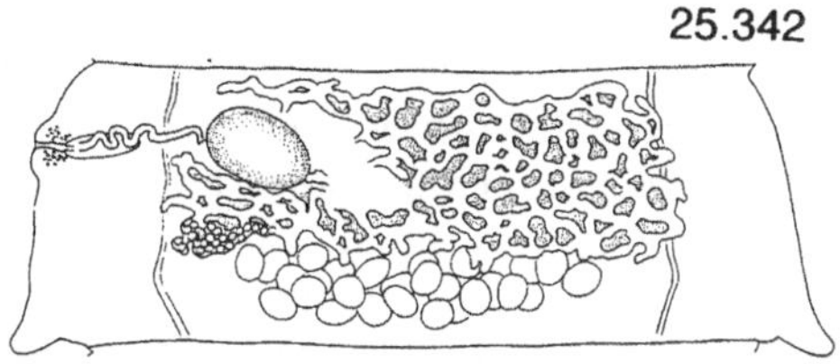

Figs 25.335-25.338 *Molluscotaenia.* 25.335. *M. crassiscolex* (von Linstow, 1890). Scolex. 25.336, 25.337. *M. soricis* (Neiland, 1953) n. comb., type. 25.336. Mature proglottid. 25.337. Gravid proglottid. 25.338. Hooks: (a) *crassiscolex*, (b) *soricis.*
Figs 25.339-25.342 *Hunkeleria steatomidis* (Hunkeler, 1972), type. 25.339, 25.340. Scoleces. 25.341. Mature proglottid. 25.342. Early reticulate uterus.

centrally, with long, lateral diverticula; then extensively labyrinthine, permanent, with many thin septa; no uterine capsules. Embryophore strong, roughly granular surface. Ovary hardly bilobed, very large; lobules small, close. Testes very numerous, postovarian. Cirrus-sac small, far anterior. Cirrus strong, finely spinulated. Vas deferens transverse, in line with cirrus-sac, far anterior. In Insectivora. Palaearctic. Type-species *H. hepaticus* (Baer, 1932) n. comb. (syn. *Monopylidium hepaticum* Baer, 1932).

112a. **Rostellum very long, narrow; apex small, rounded, not marked off from stem, armed. Scolex elongate, rounded apically. Suckers very large,**

elongate with punctiform orderly spines. Vagina short, straight, posterior to cirrus-sac. Ovary very large. In Insectivora .... *Molluscotaenia* Spasskii & Andreiko, 1971. (Figs 25.335–25.338)

Diagnosis: Rostellar apparatus typical, muscular. Pouch very long, narrow. Hooks in two circles, (16–20). Strobila small. Proglottids hardly craspedote; heaped organs; cortex reduced. Genital pores alternate irregularly in type-species (Europe), regularly in others (North America). Genital ducts dorsal. Vagina ventral to cirrus-sac in type-species, ventral and dorsal in same strobila in others; wall thick. Uterus ventral, at first coarsely reticulate, then sacciform, very deeply lobulate; eventually eggs lie in communicating lacunae, singly or in groups; uterine capsules seem not to form. Ovary bilobed, occupies whole anterior half of medulla, extends under cirrus-sac. Testes numerous, posterior and partly dorsal to vitellarium and margin of ovary. Cirrus-sac small, far anterior. Cirrus with very small, dense, spines. Vas deferens transverse, in line with cirrus-sac, far anterior. In Insectivora. Holarctic. Type-species *M. crassiscolex* (von Linstow, 1890).

112b. No rostellum; pouch with very deep, irregular, anterior channel. Scolex shape normal. Suckers spherical, thick border, unarmed. Vagina long, anterior to cirrus-sac or along its axis. Ovary small, slightly poral. In Rodentia .... *Hunkeleria* Spasskii, 1992. (Figs 25.339–25.342)

Diagnosis: Rostellar apparatus unusual, muscular and probably glandular; unarmed. Pouch strong-walled. Strobila small to medium, proglottids craspedote; organs spaced. Genital pores irregularly alternate (at times fairly long unilateral series). Genital ducts usually dorsal to, sometimes between osmoregulatory canals, very seldom ventral in the same species. Vagina dorsal to cirrus-sac, anterior (mostly left) or on sac axis (mostly right); long, sinuous. Uterus ventral, at first close reticulum, then labyrinthine; no uterine capsules. Ovary bilobed, with distinct isthmus. Testes numerous, postovarian; field narrow, extending laterally beyond ovary. Cirrus-sac very small. Cirrus small, base bearing tiny triangular spines. Vas deferens transverse, in line with cirrus-sac, underlying vagina. In Rodentia. Africa. Type-species *H. steatomidis* (Hunkeler, 1972).

26 Family Dipylidiidae Stiles, 1896

A. JONES

International Institute of Parasitology, 395A Hatfield Road, St Albans, Herts AL4 0XU, UK.

Introduction

The taxonomy of the dilepidid cestodes *sensu lato* has long been controversial. At family level and below, some authorities have recognized one family, Dilepididae, with three subfamilies Dilepidinae, Dipylidiinae and Paruterininae (e.g. Yamaguti, 1959; Schmidt, 1986). Others have given each of these subfamilies the rank of family (e.g. Matevosyan, 1953, 1963; Wardle, McLeod & Radinovsky, 1974). Revisions of the group by contributors to the present volume have resulted in the recognition of the families Dilepididae Fuhrmann, 1907, Metadilepididae Spasskii, 1959, Paruterinidae Fuhrmann, 1907 and Dipylidiidae Stiles, 1896 (see Bona, Georgiev & Kornyushin, Kornyushin & Georgiev, this volume; Mariaux, 1990, 1991a; Mariaux & Vaucher, 1989). As treated here, the Dipylidiidae corresponds to the Dipylidiinae as recognized by Witenberg (1932) in containing only the three genera *Dipylidium*, *Diplopylidium* and *Joyeuxiella*, all parasitic in carnivorous mammals. Matevosyan (1963) also included *Mirandula* Sandars, 1965 and *Diskrjabinella* Matevosyan, 1953. *Mirandula* is now in the Dilepididae (Bona, this volume) and *Diskrjabinella* is treated here as a *genus inquirendum*. The type material of *D. avicola* (Furhmann, 1906), the type-species, and of *D. columbae* (Fuhrmann, 1908) is in very poor condition and the genus may be synonymous with *Diplopylidium* (Bona, pers. comm.). Other genera placed in the Dipylidiinae by Yamaguti (1959) and Schmidt (1986) have been transferred to the family Dilepididae (Bona, this volume).

Family Dipylidiidae Stiles, 1896

Diagnosis: Scolex with protrusible rostellum armed with several rows of hooks. Hooks usually thorn-shaped, occasionally taenioid. Rostellar pouch absent. Suckers unarmed. Strobila small to medium. Proglottids numerous; mature and gravid proglottids longer than wide. Two sets of genitalia per proglottid. Genital pores bilateral. Testes numerous, occupy intervascular space, may reach anterior margin of proglottid or not. Cirrus-sac usually reaches or crosses osmoregulatory vessels. Vas deferens coiled, voluminous. Seminal vesicles absent. Ovaries lobulate, distinctly bi-alate or not. Vitellaria compact, lobed, immediately postovarian. Vaginal orifice anterior or posterior to opening of cirrus-sac. Vagina crosses cirrus-sac or not. Small seminal receptacle present at level of ovary. Uterus not persistent, replaced by thin-walled capsules containing one or several eggs. Paruterine organ absent. Adults in carnivorous mammals, larval stages in amphibians, reptiles or insects. Type-genus *Dipylidium* Leuckart, 1863.

Key to genera:

1a. Uterine capsules each contain several eggs. Vagina posterior to cirrus-sac .... *Dipylidium* Leuckart, 1863. (Figs 26.1–26.4)
(Syn. *Microtaenia* Sedgwick, 1884.)

Diagnosis: Rostellum conical, armed with several alternating rows of uniformly thorn-shaped hooks. Genital pores equatorial or postequatorial in mature and gravid proglottids. Testes fill medulla, extend to or close to anterior margin of proglottid. Cirrus-sac club-shaped or pyriform, reaches osmoregulatory vessels or not. Vagina does not cross cirrus-sac. Uterus initially reticulate, transient. Uterine capsules fill medulla, extend laterally beyond osmoregulatory canals. Adults in carnivores, larval stage in insects. Cosmopolitan. Type-species *D. caninum* (Linnaeus, 1758).

1b. Uterine capsules each contain one egg. Vagina posterior to or crossing cirrus-sac .... 2.

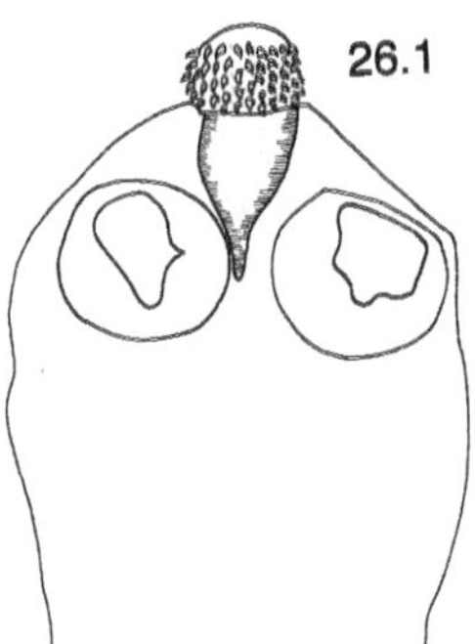
26.1

26.2

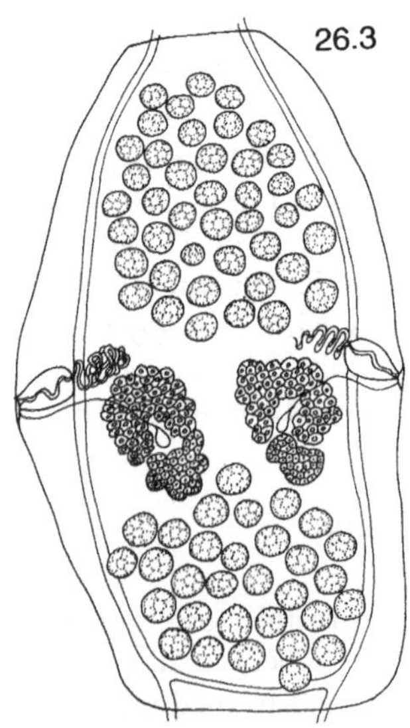
26.3

26.4

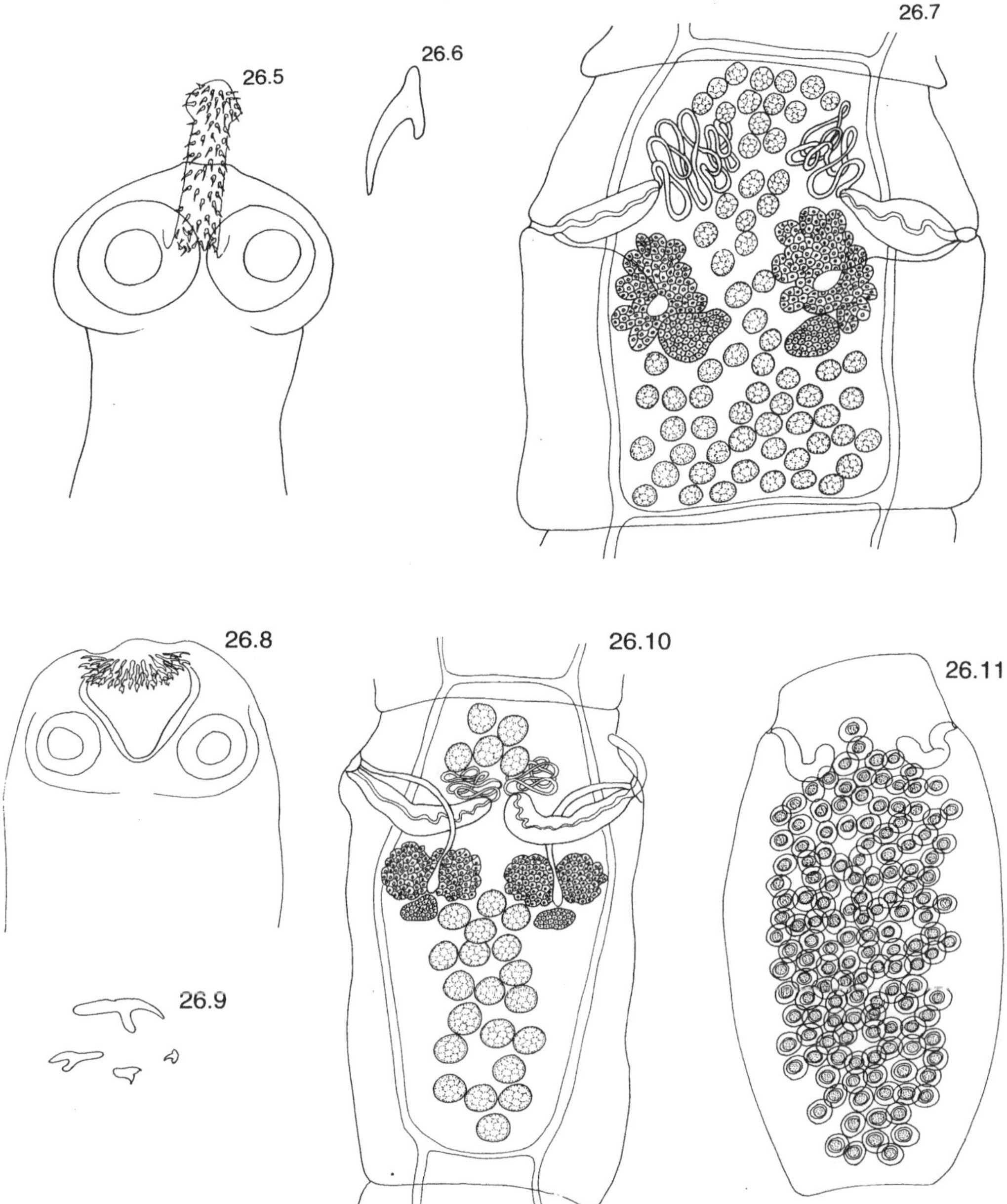

Figs 26.1–26.4 *Dipylidium caninum* (Linnaeus, 1758). 26.1. Scolex. 26.2. Rostellar hooks. 26.3. Mature proglottid. 26.4. Egg capsules.
Figs 26.5–26.7 *Joyeuxiella echinorhyncoides* (Sonsino, 1889). 26.5. Scolex. 26.6. Rostellar hooks. 26.7. Mature proglottid. (After Jones, 1983.)
Figs 26.8–26.11 *Diplopylidium* Beddard, 1913. 26.8–26.9. *D. acanthotetra* (Parona, 1887). 26.8. Scolex. 26.9. Rostellar hooks. 26.10. *D. nolleri* (Skryabin, 1924), mature proglottid. 26.11. *D. acanthotetra*, gravid proglottid.

2a. Rostellar hooks uniformly thorn-shaped. Vagina opens posteriorly to cirrus-sac .... *Joyeuxiella* Fuhrmann, 1935. (Figs 26.5–26.7)
(Syn. *Joyeuxia* López-Neyra, 1927 preoccupied.)

Diagnosis: Rostellum either conical with hooks regularly arranged in alternating rows and hook blades shorter than bases, or elongated with bulbous tip, hooks more irregular, blades longer than bases. Hooks always thorn-shaped. Genital pores pre-equatorial in mature and gravid proglottids. Testes fill space between and behind female organs, present anterior to level of vasa deferentia or not. Cirrus-sac oval, usually crosses osmoregulatory canals. Vagina does not cross cirrus-sac. Uterine capsules each contain one egg, extend laterally beyond osmoregulatory canals or not. Adults in carnivores, larval stages in reptiles. Europe, Russia, Africa, Middle East. Type-species *J. chyzeri* (Ratz, 1897), syn. of *J. pasqualei* (Diamare, 1893).

2b. Anterior rows of rostellar hooks taenioid. Vagina opens anteriorly to cirrus-sac .... *Diplopylidium* Beddard, 1913. (Figs 26.8–26.11)
(Syn. *Progynopylidium* Skryabin, 1924.)

Diagnosis: Rostellum flattened apically, armed with several alternating rows of hooks. Hooks in anterior rows have distinct handle and guard, posterior hooks thorn-shaped. Genital pores pre-equatorial in mature and gravid proglottids. Cirrus-sac elongated, proximal end flexed anteriorly, crosses osmoregulatory canals. Testes occupy most of medulla, extend close to anterior margin of proglottid. Vagina crosses cirrus-sac. Ovaries lobulate, more or less distinctly bi-alate. Uterine capsules medial to osmoregulatory canals, monovular. Adults in carnivores, larval stages in amphibians and reptiles. Europe, Africa, Asia, India, China. Type-species *D. genettae* Beddard, 1913.

27 Family Paruterinidae Fuhrmann, 1907 (*sensu lato*)

B.B. GEORGIEV[1] & V.V. KORNYUSHIN[2]

[1]*Bulgarian Academy of Sciences, Institute of Parasitology, Acad. G. Bonchev Street, Bl. 25, 1113 Sofia, Bulgaria:* [2]*Institute of Zoology, Academy of Sciences of the Ukraine, 15 Bogdan Khmelnitsky Street, 252650 Kiev–30, Ukraine.*

Introduction

The family Paruterinidae was erected as a subfamily of Dilepididae by Fuhrmann (1907b). Opinions about its rank, even during the last few decades, are quite contradictory: some authorities regard it as a subfamily within the Dilepididae (cf. Yamaguti, 1959; Joyeux & Baer, 1961; Schmidt, 1986), others recognize it as a family or as a superfamily containing two or three different families (Matevosyan, 1969; Spasskaya & Spasskii, 1971). In its present composition, this family includes practically all of the cyclophyllidean cestodes with paruterine organs which cannot be related to other families having a similar uterine apparatus. Moreover, it seems that the term 'paruterine organ', defined as a fibrous or granular appendage to the uterus that usually receives the eggs and retains them in a common capsule with protective and (or) propagative functions (after Schmidt (1986), modified), is used for structures very different in their origin, formation and morphology (Jones, 1988; personal observations particularly on the paruterinids). These organs, consequently, are a result of convergence. In the traditional concept, the family Paruterinidae is evidently a heterogeneous and polyphyletic group. Several previous attempts (e.g. Matevosyan, 1969; Spasskii, 1977, 1991) to re-arrange its system are based on a study of the literature rather than of the relevant specimens. Recently, Kornyushin (1989a) proposed the following classification (with some modifications at the generic level) of the genera traditionally considered as Paruterinidae (or Paruterinoidea):

Superfamily Davaineoidea: Family Idiogenidae: Subfamily Rhabdometrinae: *Rhabdometra*, *Metroliasthes*, *Lyruterina*, *Ascometra*, *Octopetalum*;
Superfamily Paruterinoidea: Family Paruterinidae: *Paruterina* (sensu stricto), *Cladotaenia*, *Culcitella*, *Laterotaenia*, *Matabelea*;
Superfamily Biuterinoidea: Family Biuterinidae: *Biuterina*, *Parvirostrum*, *Dictyterina*, *Spasskyterina*, *Sphaeruterina*, *Neyraia*, *Triaenorhina*, *Notopentorchis*, *Troguterina*, *Francobona*, *Orthoskrjabinia*;

Family Metadilepididae: *Metadilepis*, *Yapolepis*, *Cracticotaenia*, *Proparuterina*, *Skrjabinoporus*, *Hamatofuhrmannia*, *Pseudadelphoscolex*, *Schmidneila*; *incertae sedis*: Family Anonchotaeniidae: *Anonchotaenia*, *Mogheia*.

Although this classification gives, in our opinion, a better presentation of the relationships among the genera and groups of genera (Biuterinoidea, Biuterinidae, Metadilepididae and Anonchotaeniidae seem well-defined and homogeneous taxa), it could be regarded only as a tentative scheme. In Rhabdometrinae, there are two compact groups of genera without a proven relationship with each another. On one side, *Ascometra* and *Octopetalum* with an evident relationship to the idiogenids from bustards with an armed rostellum and muscular lappets on the suckers; on the other side, the remaining three genera with unclear connections with other cestode groups. The homogeneity of the Paruterinidae also needs further confirmation and only the genera *Laterotaenia*, *Culcitella* and *Matabelea* form a rather compact group. Taking in view these problematic points and for ease of identification, this chapter contains a common key to the genera considered traditionally as members of the Paruterinidae (*sensu lato*). Our concept of the Paruterinidae includes the following groups: Biuterinidae Meggitt, 1927; Anonchotaeniidae Matevosyan, 1965; Rhabdometrinae Matevosyan, 1965; Deltokerinae Meggitt, 1927; Orthoskrjabiniinae Matevosyan, 1966; Laterotaeniini Matevosyan, 1963.

The systematics of the paruterinid cestodes at the generic level is also very poorly worked out. There is only one large monograph, by Matevosyan (1969), which is an important source on the species described and their morphology but it has only a historical value concerning the generic concepts. The comprehensive compendium of Schmidt (1986) can be used successfully for information on the species composition of many genera: *Anonchotaenia* (incl. *Zosteropicola*), *Neyraia*, *Orthoskrjabinia* (incl. *Multiuterina* and *Anomaloporus*, pro parte), *Biuterina*, *Cladotaenia*, *Laterotaenia*, *Ascometra*, *Octopetalum*. Kornyushin's (1989a) book is useful for the identification of most genera and species occurring in Europe. There are also several recent papers dealing with *Paruterina*, *Biuterina* and *Sphaeruterina* (see Bona & Maffi, 1984a), *Anonchotaenia* (see Mariaux, 1991b; Georgiev, 1992) and *Notopentorchis* (see Georgiev & Bray, 1991).

One of the central problems in the systematics of the paruterinids concerns the species composition and concept of the type-genus, *Paruterina*. Traditionally (see Matevosyan, 1969; Schmidt, 1986), this genus includes cestodes which parasitize birds of various orders and are very different morphologically. There is also a substantial difference between species from Strigiformes with rodent intermediate hosts and other species which evidently, although their life-cycles are unknown, cannot use vertebrates as intermediates. *Paruterina* in this concept is a heterogeneous and probably polyphyletic taxon. Recently, Bona & Maffi (1984a) transferred the species from passeriform hosts to the genus *Biuterina* and divided the other species into five groups. In the present concept, following Spasskii (1977), *Paruterina* is regarded as including only species from owls (Strigiformes). Two other groups recognized by Bona & Maffi (1984a) are regarded here as distinct genera (*Triaenorhina* and *Troguterina*). *Francobona* n. g. is proposed for another group containing *F. similis* (Ransom, 1909) n. comb. and *F. similita* (Bona & Maffi, 1984) n. comb. For more information about the species of *Troguterina* and *Francobona*, see Bona & Maffi (1984b; 1985) and Bona *et al.* (1986). Additional investigations are needed to elucidate the taxonomic status of the subgenus *Paruterina* (*Unilateriporina*) Matevosyan, 1969; the species belonging to it are regarded here as *species incertae sedis*. Bona (personal communication) considers *Unilateriporina* as a probable senior synonym of *Troguterina*.

The following two genera considered usually as members of the Paruterinidae are regarded here as *incertae sedis*:

Deltokeras Meggitt, 1927 (Figs 27.1, 27.2)

Diagnosis: Scolex with sucker-like rostellum armed with double crown of numerous (about 80) triangular hooks. Proglottids craspedote, mature and postmature wider than long. Genital ducts pass between osmoregulatory canals. Genital pores unilateral. Testes in single field posteriorly, dorsally and laterally to female glands. Cirrus-sac bottle-shaped, may cross osmoregulatory canals. Cirrus unarmed. Vitellarium median, slightly lobed. Ovary wide, fan-like, lobed. Vagina thin-walled, opens dorsally to male pore. Uterus appears anteriorly and dorsally to ovary and with its further development fills anterior half of median field; saccular, lobate. Pregravid and gravid uterus not studied. Eggs and oncospheres unknown. In magpies (Passeriformes: Corvidae); South Asia. Type-species *D. ornitheios* Meggitt, 1927.

This genus evidently belongs to the Paruterinidae or the Metadilepididae. The presence of a paruterine organ is not yet ascertained (the type-specimens of the only species lack pregravid and gravid proglottids) and, therefore, its position is problematic. The redescription of the type-species on the basis of new specimens is also important from a nomenclatural point of view because it could validate Deltokerinae Meggitt, 1927 as a probable senior synonym of Metadilepididae. Almost all the other species usually considered as members of *Deltokeras* belong to *Parvirostrum* and *Dictyterina*.

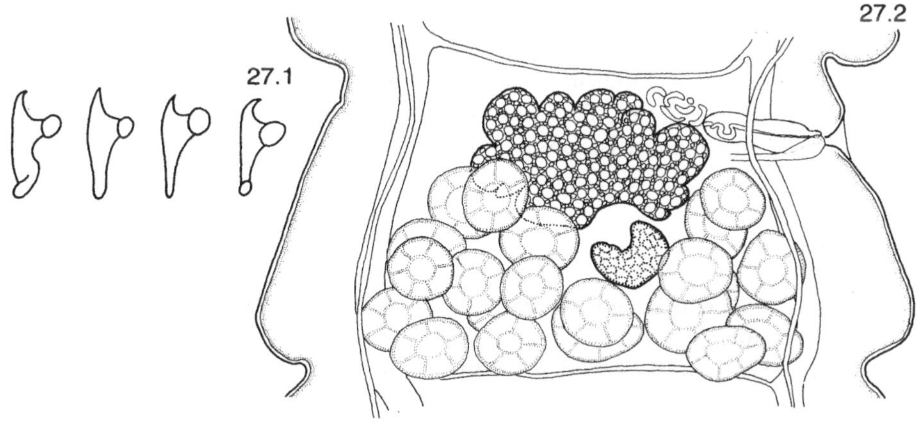

Figs 27.1, 27.2 *Deltokeras ornitheios* Meggitt, 1927. 27.1. Rostellar hooks (modified from Meggitt, 1927a). 27.2. Mature proglottid.

Dendrometra **Jordano & Diaz-Ungria, 1956 (Figs 27.3–27.5)**

Diagnosis: Scolex unarmed, with rudimentary rostellum (?). Genital pores irregularly alternating. Genital ducts pass between osmoregulatory canals. Testes encircle female glands. Cirrus-sac spherical. Vitellarium median, almost globular, compact. Ovary sac-like (?), anterior to vitellarium. Uterus initially as simple transversely arched sac; gravid uterus tree-shaped or mushroom-like, more or less lobed and with shortly pedunculated paruterine organ with round end (?). Eggs elliptical. In frigate birds (Pelecaniformes: Fregatidae); Northern coast of South America. Type-species *D. ginesi* Jordano & Diaz-Ungria, 1956.

The type material was not obtained for re-examination. The position of the testes, the shape of the cirrus-sac and genital atrium and also the systematic position of the host suggest associations with the family Tetrabothriidae.

Family Paruterinidae Fuhrmann, 1907

Diagnosis: Medium and large cestodes. Scolex with armed rostellum, sometimes without rostellum or with rudimentary unarmed rostellum. Rostellum sucker-like or discoid, without saccular sheath. Hooks usually in two rows, more or less triangular, with epiphyseal thickenings of handle and guard; sometimes hooks in more rows and spine-like (*Neyraia*). Suckers unarmed, muscular. Proglottids usually craspedote. Genital system single per proglottid. Genital ducts ventral to or between osmoregulatory canals. Vas deferens does not form seminal vesicles (exceptionally in *Lyruterina* and *Triaenorhina* internal vas deferens enlarges to form structure similar to seminal vesicle). Vitellarium and ovary present. Vitellarium usually compact or slightly lobed, median and postovarian, rarely aporal to ovary. Ovary compact, oval or two-winged or fan-like. Uterus saccular, rarely reticular (*Parvirostrum* and *Dictyterina*) or consists of median stem with lateral branches (*Cladotaenia*); usually it develops initially anteriorly and dorsally to female glands. One paruterine organ of different structure per proglottid, usually anteriorly to uterus, sometimes (when developing) anteroaporally (*Anonchotaenia*, *Mogheia*). Eggs without pyriform apparatus. Oncosphere oval, rarely vermiform (*Anonchotaenia*). In non-aquatic birds (Falconiformes, Strigiformes, Galliformes, Gruiformes, Trogoniformes, Cuculiformes, Apodiformes, Coraciiformes, Piciformes, Passeriformes), rarely (or accidentally) amphibians, aquatic birds and mammals. Cosmopolitan. Type-genus *Paruterina* Fuhrmann, 1906.

Key to genera

1a. **Scolex unarmed, without rostellum or if rostellum present it is rudimentary and without armament** .... 2.
1b. **Scolex with armed rostellum** .... 9.

2a. **Developing paruterine organ aporal or anteroaporal to uterus (Fig. 27.8)** .... 3.

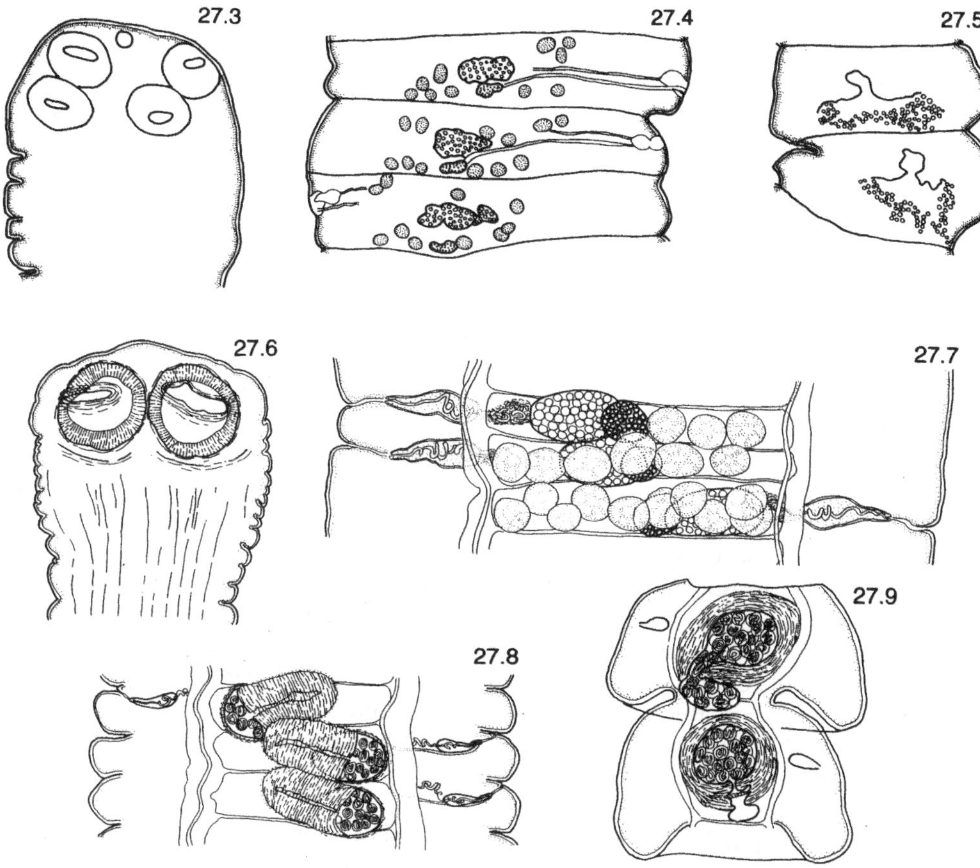

Figs 27.3–27.5 *Dendrometra ginesi* Jordano & Diaz-Ungria, 1956 (modified from Jordano & Diaz-Ungria, 1956). 27.3. Scolex. 27.4. Mature proglottids. 27.5. Gravid proglottids.
Figs 27.6–27.9 *Anonchotaenia* sp. 27.6. Scolex. 27.7. Three mature proglottids (poral testes not presented in first proglottid). 27.8. Pregravid proglottids. 27.9. Final stages of paruterine capsule.

2b. Developing paruterine organ anterior to uterus .... **4.**

3a. Testes in row dorsally to female glands (Fig. 27.7) .... ***Anonchotaenia*** **Cohn, 1900. (Figs 27.6–27.9)**
(Syns *Amerina* Fuhrmann, 1901; *Anurina* Fuhrmann, 1901; *Nematotaenoides* Ulmer & James, 1976; *Zosteropicola* Johnston, 1912.)

Diagnosis: Scolex without rostellum or rostellar hooks. Suckers muscular, deep. Mature proglottids considerably wider than long (seven to ten times or more), acraspedote; gravid proglottids with almost equal length and width, craspedote. Genital pores alternating. Genital ducts ventral to osmoregulatory canals or between them. Testes occupying all median field. Cirrus-sac may reach osmoregulatory canals. Cirrus unarmed. Vitellarium compact, round, slightly poral. Ovary compact, oval, poral. Seminal receptacle round or slightly oval, near point where osmoregulatory canals and genital ducts cross. Uterus spherical or oval. Developing paruterine organ aporal and dorsal to uterus; paruterine organ anterior to uterus in last gravid proglottids. Eggs with vermiform oncospheres. In Passeriformes (different families), sometimes (probably accidentally) in Apodiformes (Apodidae, Trochilidae), Ciconiiformes (Threskiornithidae) and amphibians. Cosmopolitan. Type-species *A. globata* (von Linstow, 1879) Cohn, 1900.

Key to subgenera

Ia. Genital ducts ventral to osmoregulatory canals .... *Anonchotaenia (Anonchotaenia)* Cohn, 1900.

Ib. Genital ducts between osmoregulatory canals .... *Anonchotaenia (Paranonchotaenia)* Mariaux, 1991.

3b. Testes in group aporally to female glands (Fig. 27.10) .... *Mogheia* Lopez-Neyra, 1944.
(Syn. *Baeria* Moghe, 1933, preoccupied.)

Diagnosis: Scolex without rostellum. Suckers deep, muscular. Proglottids wider than long, acraspedote. Genital pores irregularly alternating. Genital ducts between osmoregulatory canals. Testes aporal, including in aporal lateral field. Cirrus-sac reaches osmoregulatory canals. Vitellarium median. Ovary poral to vitellarium. Seminal receptacle present. Uterus spherical. Paruterine organ lateral to uterus. Eggs thin-shelled. Oncospheres undescribed. In babblers (Passeriformes: Muscicapidae: Timaliinae). India. Type-species *M. orbiuterina* (Moghe, 1933) Lopez-Neyra, 1944.

4a. Mature proglottids short, acraspedote. Testes in two groups laterally to female glands. Developing paruterine organ like tall cone with a wide base directed to the uterus. Young uterus narrow transversely elongate sac at posterior proglottid margin ...... *Orthoskrjabinia* Spasskii, 1947. (Figs 27.11–27.14)
(Syns *Skrjabinerina* Matevosyan, 1948; *Multiuterina* Matevosyan, 1948; *Anomaloporus* Voge & Davis, 1953.)

Diagnosis: Scolex without rostellum or with small, rudimentary, unarmed rostellum. Suckers muscular, round. Gravid proglottids craspedote. Genital pores alternating. Genital ducts ventral to osmoregulatory canals or between them.

27.10

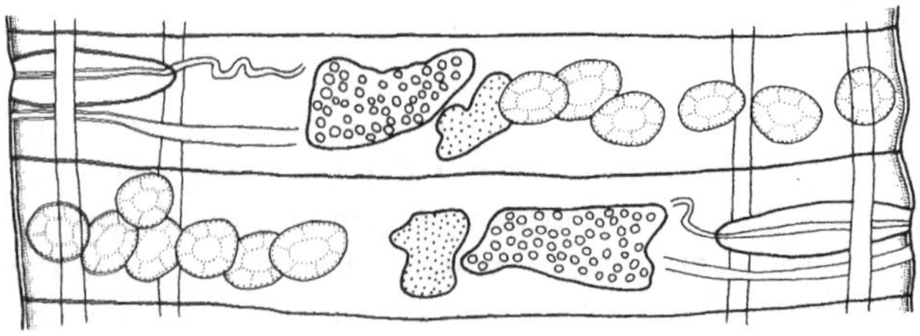

Fig. 27.10 *Mogheia orbiuterina* (Moghe, 1933). Mature proglottids (modified from Moghe, 1933).

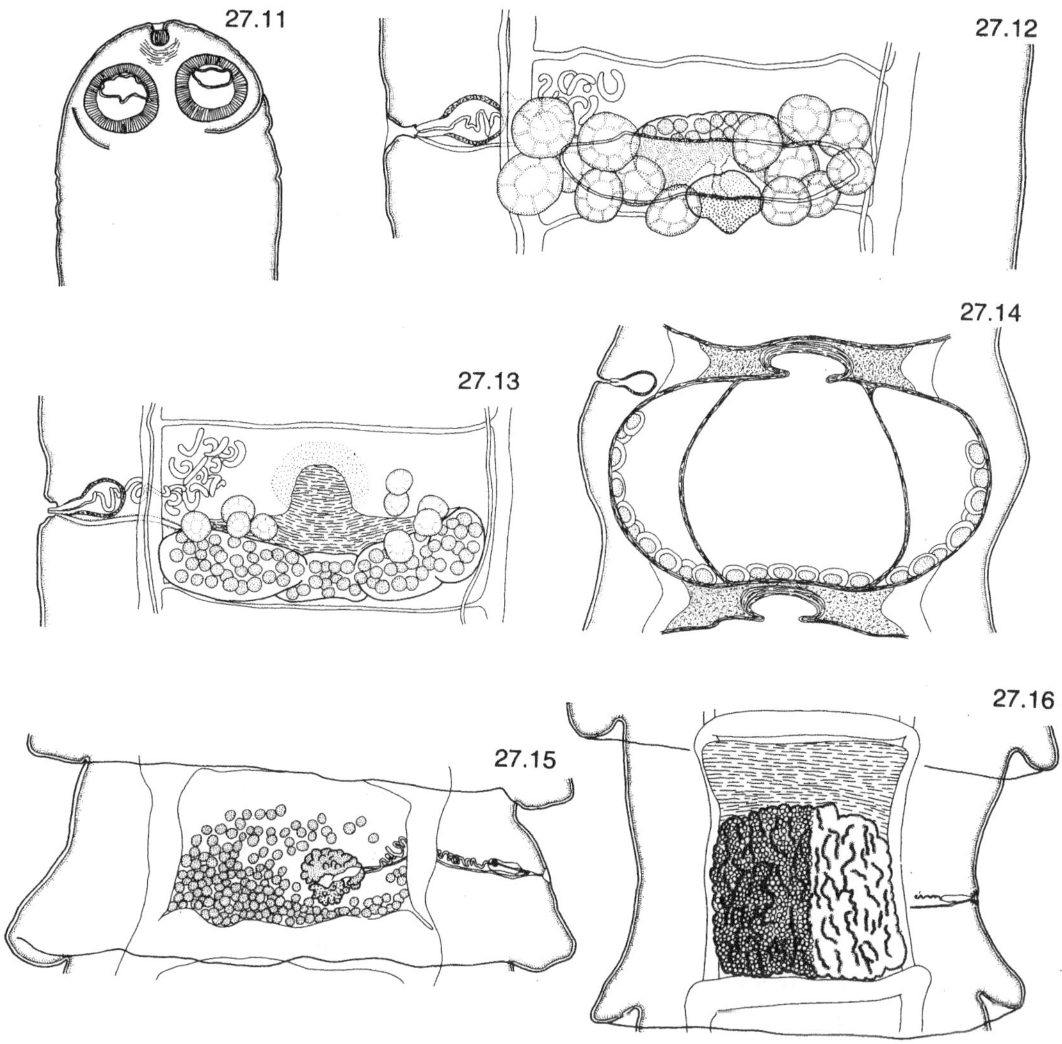

Figs 27.11–27.14 *Orthoskrjabinia conica* (Fuhrmann, 1908). 27.11. Scolex. 27.12. Mature proglottid. 27.13. Postmature proglottid. 27.14. Gravid proglottid before formation of paruterine capsule.

Figs 27.15, 27.16 *Ascometra choriotidis* Adams & Rausch, 1986. 27.15. Mature proglottid, ventral view (modified from Adams & Rausch, 1986). 27.16. Gravid proglottid.

Cirrus-sac small, pyriform, usually lateral to osmoregulatory canals, rarely crossing them. Cirrus unarmed. Vitellarium compact, postovarian. Ovary compact, transversely elongate, median. Pregravid uterus becomes multiseptate sac; in last gravid proglottids eggs move into paruterine organ which forms round capsule. Eggs and oncospheres round. In Passeriformes and Piciformes, accidentally in rodents. Holarctic, Africa, Central America. Type-species *O. bobica* (Clerc, 1903) Spasskii, 1947.

4b. Mature proglottids relatively longer, craspedote. Testes posterior to female glands or encircling them. Paruterine organ rod-like or cylindrical. Uterus otherwise .... 5.

5a. Suckers with lappets (Fig. 27.17) .... 6.
5b. Suckers without lappets .... 7.

6a. Genital ducts do not pass between osmoregulatory canals. Vitellarium and ovary situated porally. Paruterine organ (when fully developed) wider than long or almost as wide as long (Fig. 27.16) .... *Ascometra* Cholodkowsky, 1912. (Figs 27.15, 27.16)
(Syn. *Sobolevina* Spasskii, 1951.)

Diagnosis: Scolex without rostellum and hooks. Suckers muscular, each with pair of muscular lappets. Proglottids craspedote. Genital pores unilateral (in type-species) or irregularly alternating. Osmoregulatory canals in young proglottids two pairs, sometimes inverted; in mature and older proglottids dorsal canals disappear. Genital ducts ventral (? or dorsal) to osmoregulatory canals. Testes numerous, more than 100, encircle female glands and situated in several layers. Cirrus-sac not crossing osmoregulatory canals. Ovary small, slightly lobed. Seminal receptacle median, fusiform. Uterus saccular, in posterior two thirds of median field. Paruterine organ cylindrical. Eggs and oncospheres round. In bustards (Gruiformes: Otidae). Palaearctic, Africa, South Asia. Type-species *A. vestita* Cholodkowsky, 1912.

6b. Genital ducts between osmoregulatory canals. Vitellarium and ovary median. Paruterine organ considerably longer than wide when fully developed (Fig. 27.19) .... *Octopetalum* Baylis, 1914. (Figs 27.17–27.20)

Diagnosis: Scolex without rostellum and hooks. Suckers muscular, each with pair of muscular lappets. Proglottids craspedote. Genital pores irregularly alternating. Testes numerous (but usually less than 100), in one or two layers, posteriorly and laterally to female glands, sometimes (in type-species) also anteriorly to them (Fig. 27.20). Cirrus-sac from elliptical to highly elongate, may overlap or cross osmoregulatory canals. Vitellarium median, compact or slightly lobed. Ovary fan-like, lobed. Vagina opens posteriorly to male pore, thick-walled. Seminal receptacle median, oval. Uterus saccular, in posterior half of median field, with many internal septae. Paruterine organ cylindrical. In last gravid proglottids, eggs move into paruterine organ and oval capsule forms near anterior margin of proglottid. Eggs and oncospheres round. In guineafowl (Galliformes: Phasianidae: Numidinae). Africa. Type-species *O. gutterae* Baylis, 1914.

Figs 27.17-27.20 *Octopetalum* spp. 27.17-27.19. *O. numida* (Fuhrmann, 1909). 27.17. Scolex. 27.18. Mature proglottid. 27.19. Gravid proglottid. 27.20. *O. gutterae* Baylis, 1914. Mature proglottid.

Figs 27.21-27.24 *Metroliasthes lucida* Ransom, 1900. 27.21. Scolex. 27.22. Mature proglottid. 27.23. Genital ducts in pregravid proglottid. 27.24. Pregravid proglottid.

27.17

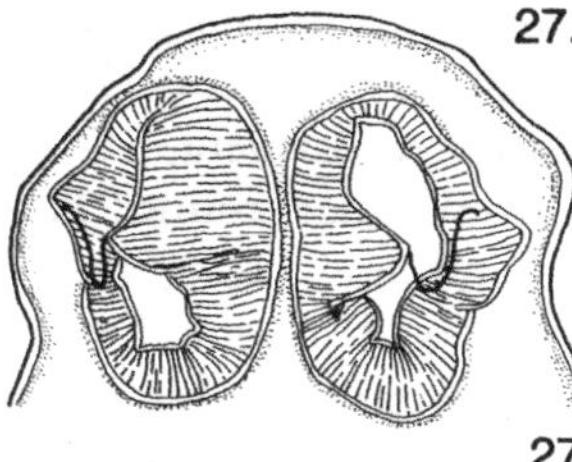

27.18

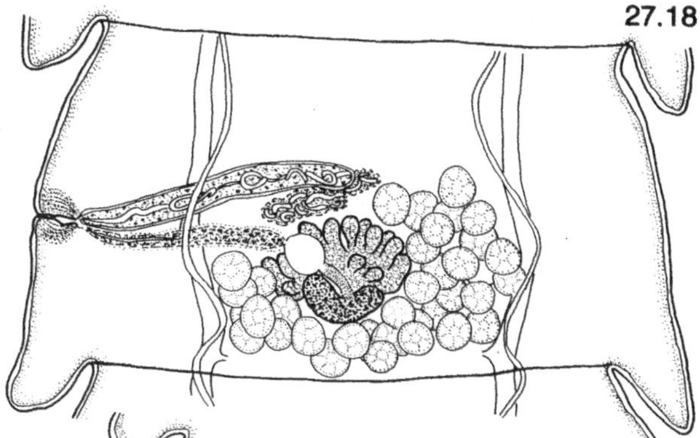

27.19

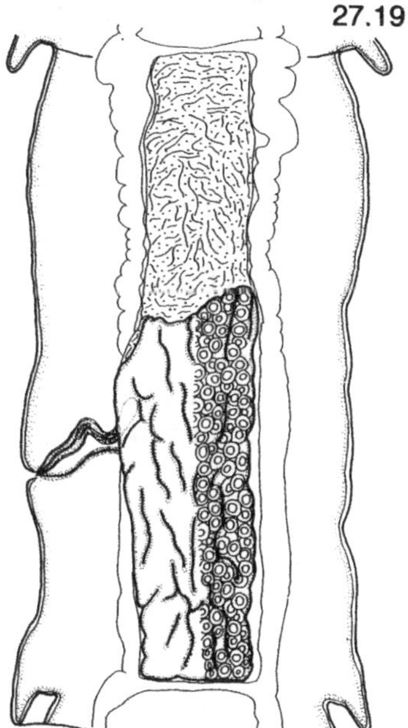

27.20

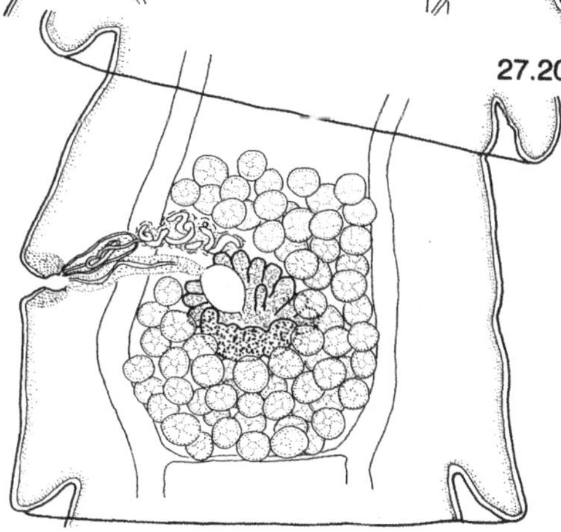

27.21

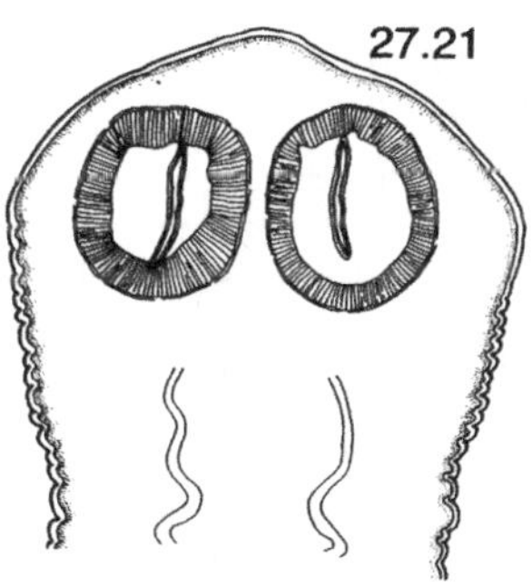

27.22

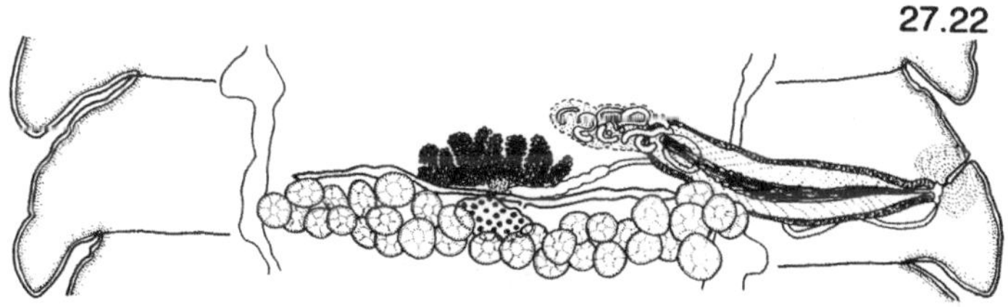

27.24

27.23

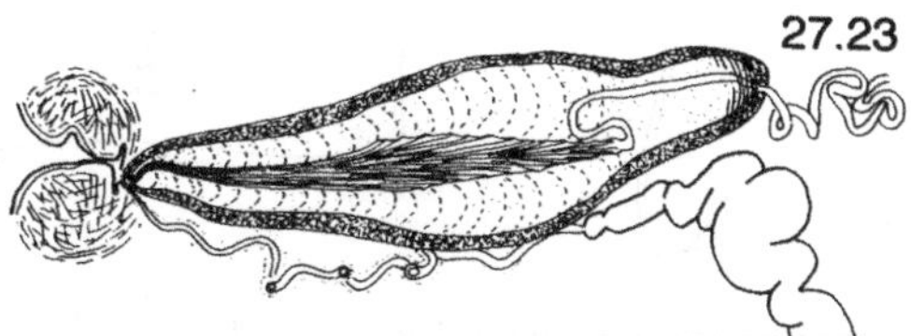

7a. **Developing uterus forming double sac. Cirrus provided with long fibres** .... *Metroliasthes* **Ransom, 1900. (Figs 27.21–27.24)**
(Syn. *Hexaparuterina* Palacios & Barroeta, 1967.)

Diagnosis: Scolex without rostellum and hooks. Suckers muscular, round. Proglottids craspedote, mature wider than long, postmature and gravid longer than wide. Genital atria deep, with highly developed muscular walls. Genital pores irregularly alternating. Genital ducts between osmoregulatory canals. Testes numerous (not less than 30), posterior to female glands and lateral to vitellarium. Cirrus-sac thick-walled, highly elongate, may cross osmoregulatory canals. Vitellarium compact, median. Ovary fan-like, lobate. Vagina opens posteriorly to male pore, with clearly distinct copulatory and conductive parts. Seminal receptacle elongate, median, directed porally. Uterus initially as transverse tube dorsal and posterior to ovary and anterior to vitellarium, later forms two sacs connected by narrow isthmus. Paruterine organ anterior to uterus, almost conical, with bifid base; in pregravid proglottids rounded body of dense parenchyma forms anteriorly to paruterine organ. In last gravid proglottids, eggs move into paruterine organ and round capsule forms in middle of proglottid. Eggs oval, thick-walled; oncospheres round. In turkeys (*Meleagris gallopavo*), several doubtful records in other galliform birds, accidentally in anuran amphibians. Cosmopolitan. Type-species *M. lucida* Ransom, 1900.

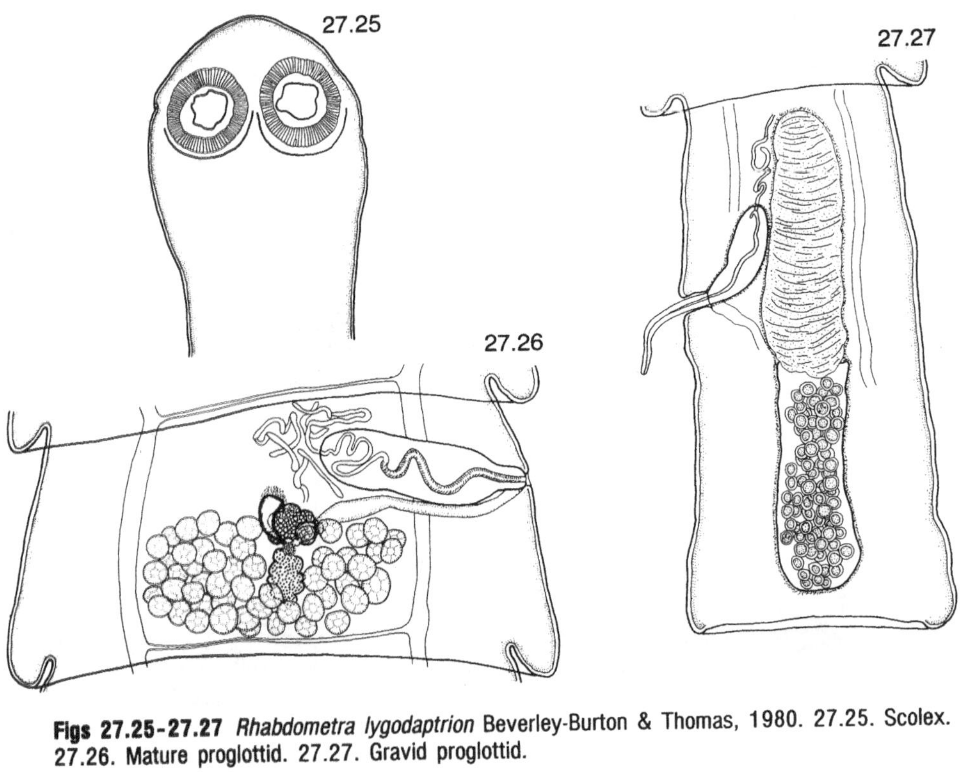

Figs 27.25–27.27 *Rhabdometra lygodaptrion* Beverley-Burton & Thomas, 1980. 27.25. Scolex. 27.26. Mature proglottid. 27.27. Gravid proglottid.

7b. Uterus forming a single sac. Cirrus unarmed or armed with spines .... 8.

8a. Developing paruterine organ straight or slightly curved, cylindrical or claviform; no pigmented tissue at its anterior end. Cirrus-sac crosses osmoregulatory canals and much of it lies in median field .... *Rhabdometra* Cholodkowsky, 1906. (Figs 27.25–27.27)

Diagnosis: Scolex without rostellum or hooks. Suckers round, deep and muscular. Proglottids craspedote, mature wider than long, gravid longer than wide. Genital pores irregularly alternating. Genital ducts between osmoregulatory canals. Testes numerous (usually more than 30) posterior and lateral to female glands. Cirrus-sac large, crossing osmoregulatory canals. Cirrus armed. Vitellarium compact or slightly lobed, median, size comparable with ovary. Ovary lobed, usually fan-like. Seminal receptacle small, oval. Vagina opens laterally to male pore. Uterus saccular. Paruterine organ anterior to uterus. Eggs and oncospheres round. In Galliformes (Phasianidae: Tetraoninae, Odontophorinae and ? Phasianinae). Holarctic, ? South Asia. Type-species *R. tomica* Cholodkowsky, 1906.

8b. Developing paruterine organ tubular, spirally curved, anterior end surrounded by dense pigmented tissue. Cirrus-sac not crossing osmoregulatory canals .... *Lyruterina* Spasskaya & Spasskii, 1971. (Figs 27.28–27.32)

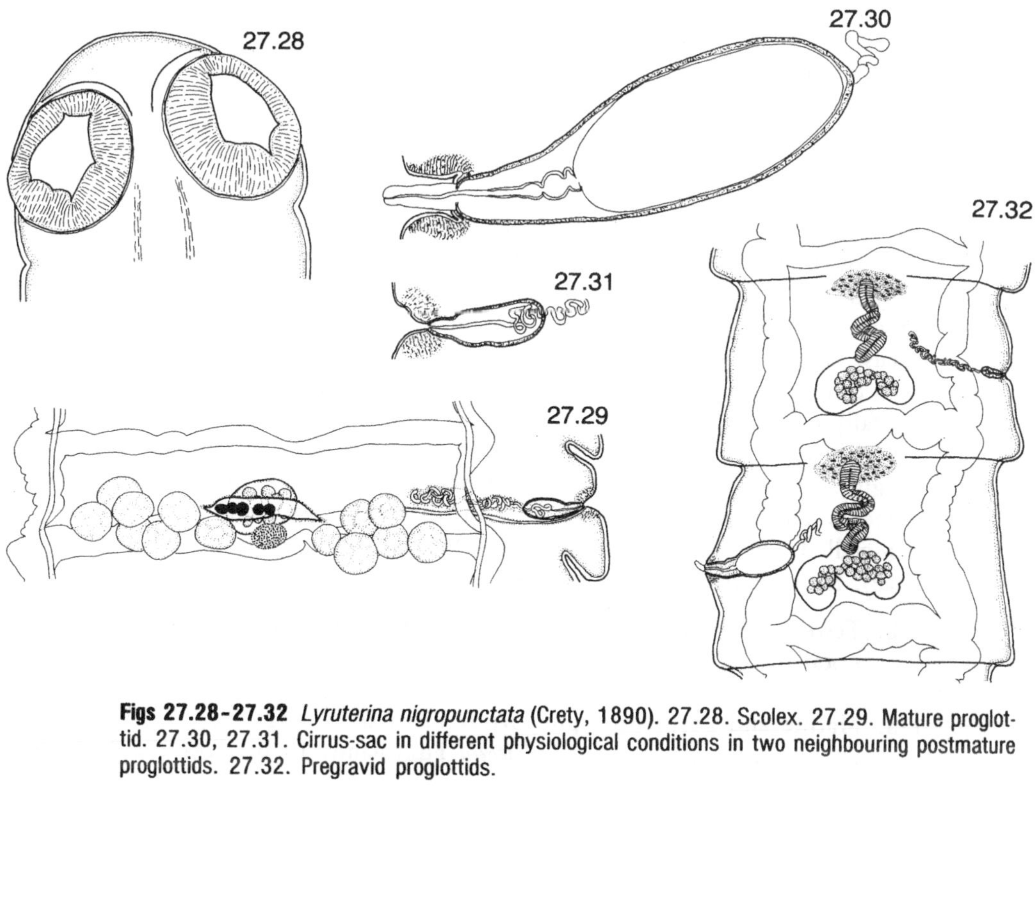

Figs 27.28–27.32 *Lyruterina nigropunctata* (Crety, 1890). 27.28. Scolex. 27.29. Mature proglottid. 27.30, 27.31. Cirrus-sac in different physiological conditions in two neighbouring postmature proglottids. 27.32. Pregravid proglottids.

Diagnosis: Scolex without hooks or rostellum (rudimentary rostellum as aggregation of dense parenchyma within scolex may be present). Suckers round, deep and muscular. Proglottids craspedote, mature wider than long, gravid can be longer than wide. Genital pores alternating. Genital ducts ventral to osmoregulatory canals (? or between them). Testes numerous (more than ten), situated posteriorly and laterally to female glands in two more or less distinct groups. Cirrus-sac small, pyriform, usually in poral lateral field; in type-species, internal vas deferens may form structure similar to seminal vesicle and then cirrus-sac overlaps osmoregulatory canals. Cirrus unarmed or armed with small spines. Vitellarium compact, globular, median or slightly poral. Ovary compact, sac-like, globular or oval, containing few oocytes. Seminal receptacle small, oval. Uterus initially dorsal to ovary as transversely elongate sac with mucronate ends; in postmature and gravid proglottids, it forms sac with irregular shape situated in posterior half of median field. Paruterine organ anterior to uterus. In last gravid proglottids, eggs move into paruterine organ and round capsule forms near anterior margin of proglottid. Eggs and oncospheres spherical. In Galliformes (Phasianidae: Phasianinae, Meleagridinae and ? Tetraoninae) and Columbiformes (Pteroclididae). Holarctic. Type-species *L. nigropunctata* (Crety, 1890) Spasskaya & Spasskii, 1971.

9a. **Paruterine organ a narrow transversely elongate body at anterior proglottis margin (Figs 27.34, 27.38, 27.45). Gravid uterus never spherical .... 10.**

9b. **Paruterine organ usually longitudinally elongate or at least as long as wide; if paruterine organ slightly wider than long then uterus more or less spherical .... 12.**

10a. **Major part of testes along the osmoregulatory canals, few central (Fig. 27.33). Female glands clearly poral .... *Laterotaenia* Fuhrmann, 1906. (Figs 27.33–27.36)**

Diagnosis: Rostellum small, armed with double crown of hooks. Proglottids craspedote. Genital pores irregularly alternating. Genital ducts between osmoregulatory canals. Cirrus-sac muscular, small, lateral to osmoregulatory canals. Cirrus with bundle of fibres. Vitellarium lobate. Ovary fan-like, reaches poral osmoregulatory canals. Seminal receptacle round. Vagina opens dorsally or ventrally to male pore, thick-walled, with sphincter at orifice. Uterus saccular, with thin walls. Paruterine organ a narrow transverse band at anterior proglottid margin, consists of slightly modified parenchyma, without fibrous or granular elements. Formation of paruterine capsule unknown. Eggs round or slightly oval, thin-walled. Oncospheres round. In New World vultures (Falconiformes: Cathartidae). South America. Type-species *L. natteri* Fuhrmann, 1906.

10b. **Testes in posterior half of proglottid. Female glands median or slightly poral .... 11.**

11a. **Mature and postmature proglottids wider than long .... *Culcitella* Fuhrmann, 1906. (Figs 27.37–27.39)**

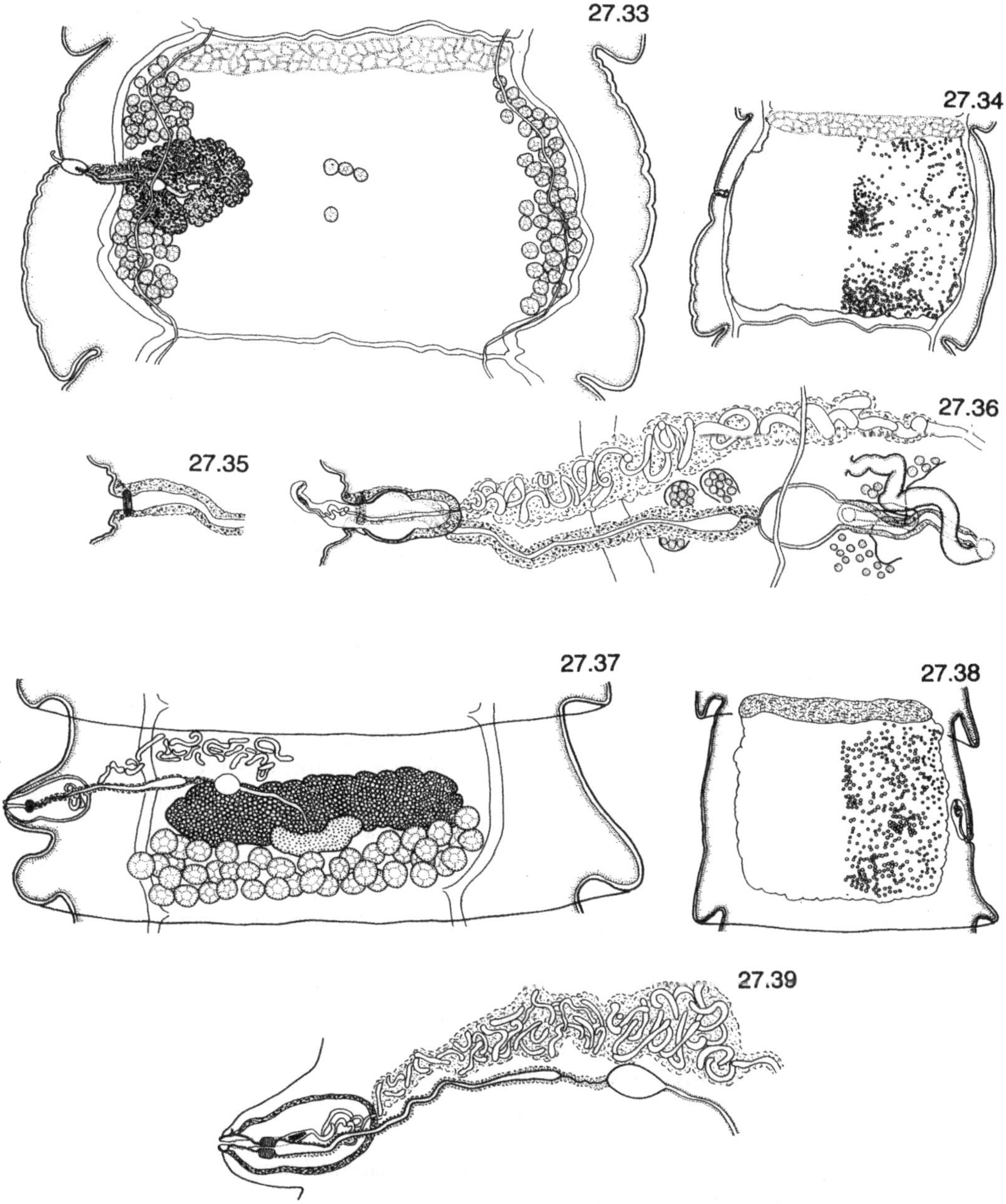

Figs 27.33–27.36 *Laterotaenia natteri* Fuhrmann, 1906. 27.33. Mature proglottid. 27.34. Gravid proglottid. 27.35. Terminal part of vagina. 27.36. Genital ducts in mature proglottid.
Figs 27.37–27.39 *Culcitella rapacicola* Fuhrmann, 1906. 27.37. Mature proglottid. 27.38. Gravid proglottid. 27.39. Genital ducts in mature proglottid.

Diagnosis: Rostellum with double crown of hooks. Hooks with handle considerably longer than blade. Genital pores unilateral (rarely irregularly alternating). Atrium may form genital papilla. Genital ducts between osmoregulatory canals. Inversion of osmoregulatory canals may occur. Testes numerous, in posterior half of median field, posterior to ovary, posterior and lateral to vitellarium. Cirrus-sac thick-walled, ovoid or pyriform, can reach osmoregulatory canals. Vitellarium median or slightly poral, reniform, compact. Ovary wide, fan-like, slightly lobed. Seminal receptacle oval or slightly elongate. Vagina thick-walled, opens dorsally to male pore, with sphincter near orifice. Uterus a transversely elongate saccular body anteriorly and dorsally to ovary; in gravid proglottids, occupies all median field. Paruterine organ transversely elongate, like narrow band at anterior proglottid margin. Formation of paruterine capsule unknown. Eggs thin-walled, oval. Oncospheres oval. In Falconiformes (Accipitridae). South America. Type-species *C. rapacicola* Fuhrmann, 1906.

11b. Mature and postmature proglottids almost equal in length and width or longer than wide .... *Matabelea* Mettrick, 1963. (Figs 27.40–27.45)

Diagnosis: Rostellum with double crown of hooks. Hooks with handle considerably longer than blade; handle and blade with epiphyseal thickenings connected with one another and forming keel. Proglottids craspedote, with highly developed velum. Genital pores unilateral. Atrium may form genital papilla. Genital ducts between osmoregulatory canals. Inversion of osmoregulatory canals may occur. Testes numerous (more than 40), posterior and lateral to female glands. Cirrus-sac elongate, may reach or overlap osmoregulatory canals. Vitellarium slightly lobed, median. Ovary wide, lobed, fan-like. Seminal receptacle globular or slightly oval, near point where osmoregulatory canals and genital ducts cross. Vagina opens dorsally to male pore, with sphincter near its orifice. Uterus initially a round or irregular body dorsal and anterior to ovary; later saccular. Developing paruterine organ like narrow band near anterior proglottid margin. Eggs and oncospheres round. In Falconiformes (Accipitridae, Falconidae). Africa. Type-species *M. aetodex* Mettrick, 1963, syn. of *M. fuhrmanni* (Southwell, 1925).

12a. Testes in two fields lateral to female glands over-reaching them and usually connected posteriorly to them by several testes (Figs 27.46, 27.48) .... 13.

12b. Testes either posterior or lateral to female glands or encircling female glands dorsally .... 14.

13a. Uterus with elongate median stem and lateral branches. Poral and aporal testes fields reach same level anteriorly ....

Figs 27.40-27.45 *Matabelea aetodex* Mettrick, 1963. 27.40. Scolex. 27.41. Anterior rostellar hook. 27.42. Posterior rostellar hook. 27.43. Mature proglottid. 27.44. Postmature proglottid. 27.45. Gravid proglottid.

Figs 27.46, 27.47 *Cladotaenia globifera* (Batsch, 1786). 27.46. Mature proglottid. 27.47. Pregravid proglottid.

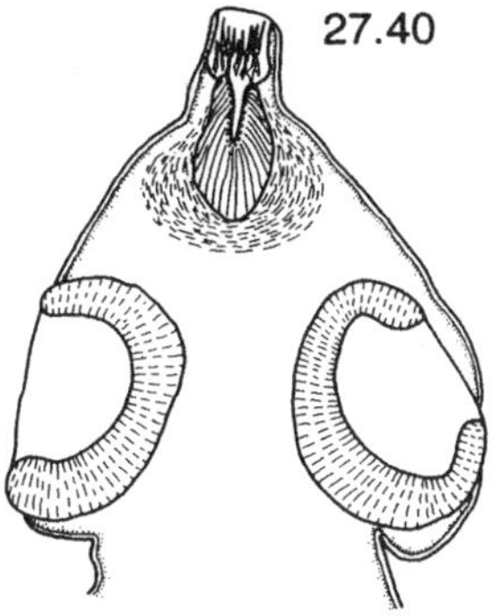

27.40

27.41

27.42

27.43

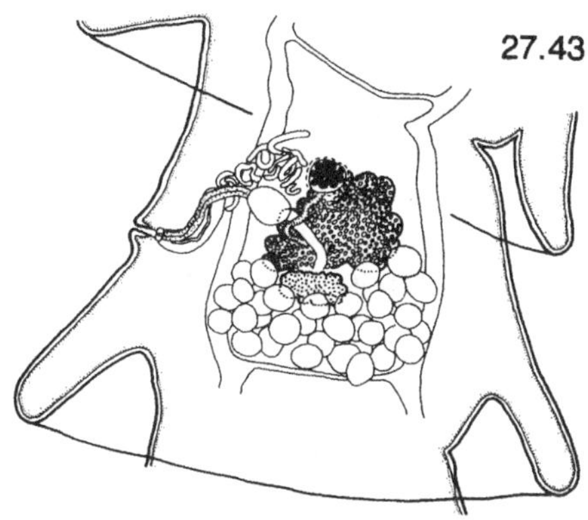

27.44

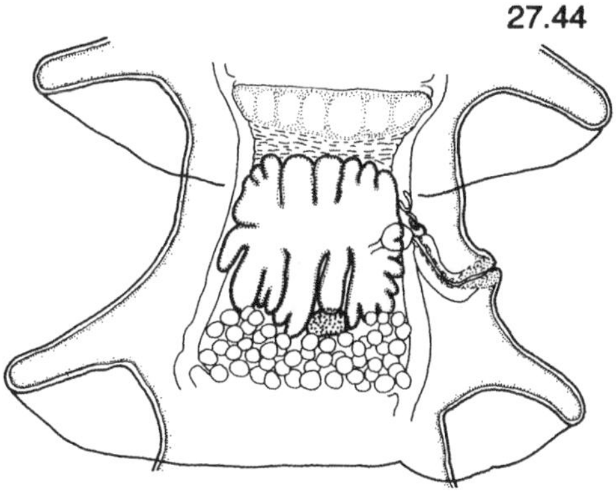

27.45

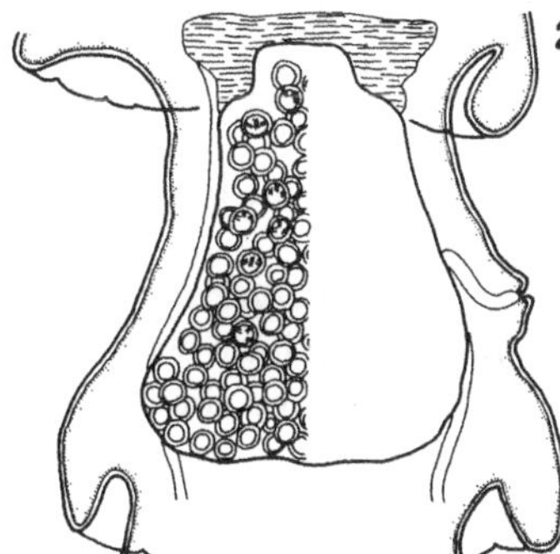

27.46

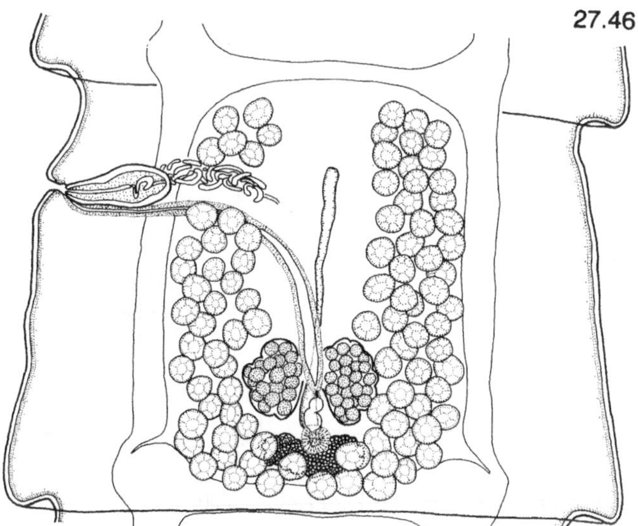

27.47

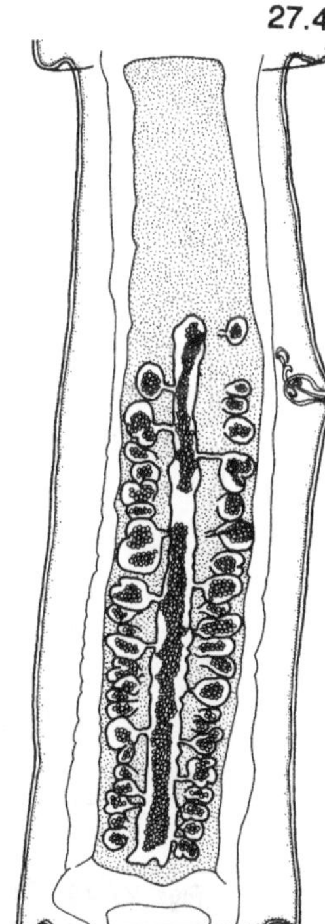

.... ***Cladotaenia*** **Cohn, 1901. (Figs 27.46, 27.47.)**
(Syn. *Paracladotaenia* Yamaguti, 1935.)

Diagnosis: Rostellum small, armed with hooks arranged in two rows. Hooks with epiphyseal thickenings on guard and handle. Proglottids craspedote, almost as long as wide (mature) or longer than wide (from postmature to gravid). Genital pores irregularly alternating. Genital ducts between osmoregulatory canals. Testes in two longitudinal lateral fields not connected posteriorly or connected by single testes situated dorsally to vitellarium. Cirrus-sac small, round or oval, not crossing osmoregulatory canals. Cirrus unarmed. Vitellarium compact, oval or with irregular shape, median, near posterior proglottid margin. Ovary two-winged, compact or slightly lobed. In pregravid and gravid proglottids, all medullary parenchyma around and anterior to uterus becomes dense and transforms into peculiar paruterine organ. Eggs and oncospheres oval. In Falconiformes, intermediate hosts rodents and insectivorous mammals. Holarctic, South Asia, Africa. Type-species *C. globifera* (Batsch, 1786) Cohn, 1901.

13b. Uterus more or less spherical. Aporal testes field extends further anteriorly than poral ....
.... ***Paruterina*** **Fuhrmann, 1906 (*sensu stricto*). (Figs 27.48, 27.49)**

Diagnosis: Rostellum with double crown of hooks of different lengths; hooks with well-developed handle and blade; handle and guard with epiphyseal

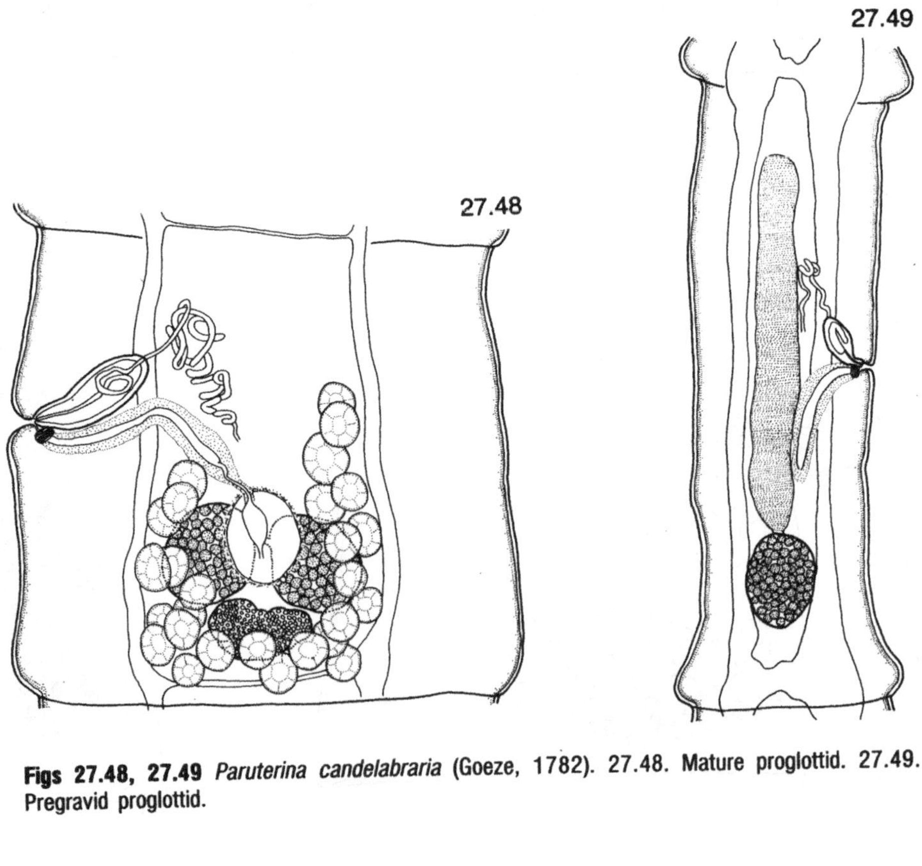

Figs 27.48, 27.49 *Paruterina candelabraria* (Goeze, 1782). 27.48. Mature proglottid. 27.49. Pregravid proglottid.

thickenings. Suckers round, muscular. Proglottids craspedote, mature proglottids may have almost equal length and width, gravid proglottids considerably longer than wide. Genital pores alternating. Genital ducts between osmoregulatory canals. Testes in two longitudinal lateral fields connected by single testes posterodorsally to vitellarium or not connected posteriorly. Cirrus-sac elongate, thick-walled, overlapping osmoregulatory canals. Vitellarium compact or slightly lobed, median, at posterior proglottid margin. Ovary two-winged, compact. Seminal receptacle small, oval, median. Vagina thick-walled, with sphincter near orifice. Uterus initially develops dorsally to ovary, then near posterior margin of proglottid when ovary has atrophied. Paruterine organ formed by densification of medullary parenchyma and fills great part of median field anteriorly to uterus; in later gravid proglottids, eggs move anteriorly and paruterine organ forms round capsule containing them. Eggs and oncospheres oval. In owls (Strigiformes), intermediate hosts rodents. Holarctic. Type-species *P. candelabraria* (Goeze, 1782) Fuhrmann, 1906.

14a. Rostellar hooks spine-like, in four rows; rows usually clearly separate, occasionally indistinct. Paruterine organ forms bicameral capsule when fully developed (Fig. 27.54) .... *Neyraia* Joyeux & Timon-David, 1934. (Figs 27.50–27.54)

(Syn. *Biuterinoides* Ortlepp, 1940.)

Diagnosis: Rostellum armed with concentric rows of numerous hooks. Hooks with large conical base and small talon-like blade; peripheral hooks larger than central. Proglottids craspedote, mature wider than long, gravid as long as wide

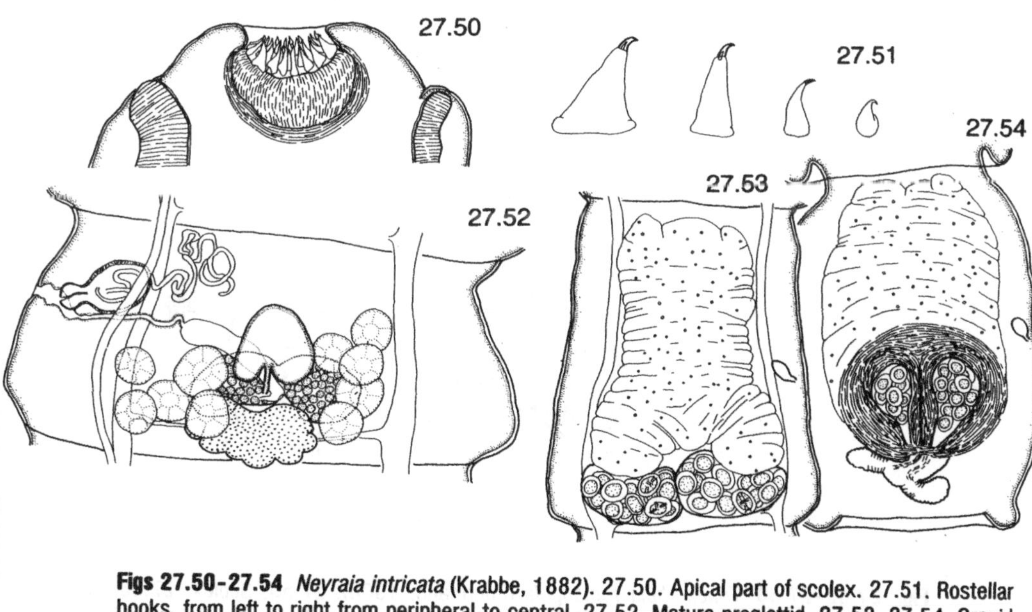

Figs 27.50–27.54 *Neyraia intricata* (Krabbe, 1882). 27.50. Apical part of scolex. 27.51. Rostellar hooks, from left to right from peripheral to central. 27.52. Mature proglottid. 27.53, 27.54. Gravid proglottids showing stages of paruterine capsule formation.

or longer than wide. Genital pores irregularly alternating. Genital ducts between osmoregulatory canals. Testes few (less than 12), in two groups laterally and dorsally to female glands. Cirrus-sac small, bottle-shaped or pyriform, may cross osmoregulatory canals. Vitellarium slightly lobed, median, at posterior proglottid margin. Ovary compact, two-winged or fan-like. Seminal receptacle elongate, directed porally. Uterus at first one sac, divides into two sacs not connected with one another in postmature proglottids. Paruterine organ anterior to uterus, fills almost all median field. Eggs and oncospheres spherical or slightly oval. In hoopoes (Coraciiformes: Upupidae and Phoeniculidae). Old World (Palaearctic, India, Africa). Type-species *N. intricata* (Krabbe, 1882) Joyeux & Timon-David, 1934.

14b. Rostellar hooks in two rows; fully developed paruterine organ forms unicameral capsule or overgrows uterus totally .... 15.

15a. Rostellar hooks with small curved triangular chitinoid blade and large irregular non-chitinized handle and guard (Figs 27.56, 27.57). Cirrus armed with small spines .... *Triaenorhina* Spasskii & Shumilo, 1965. (Figs 27.55–27.60)

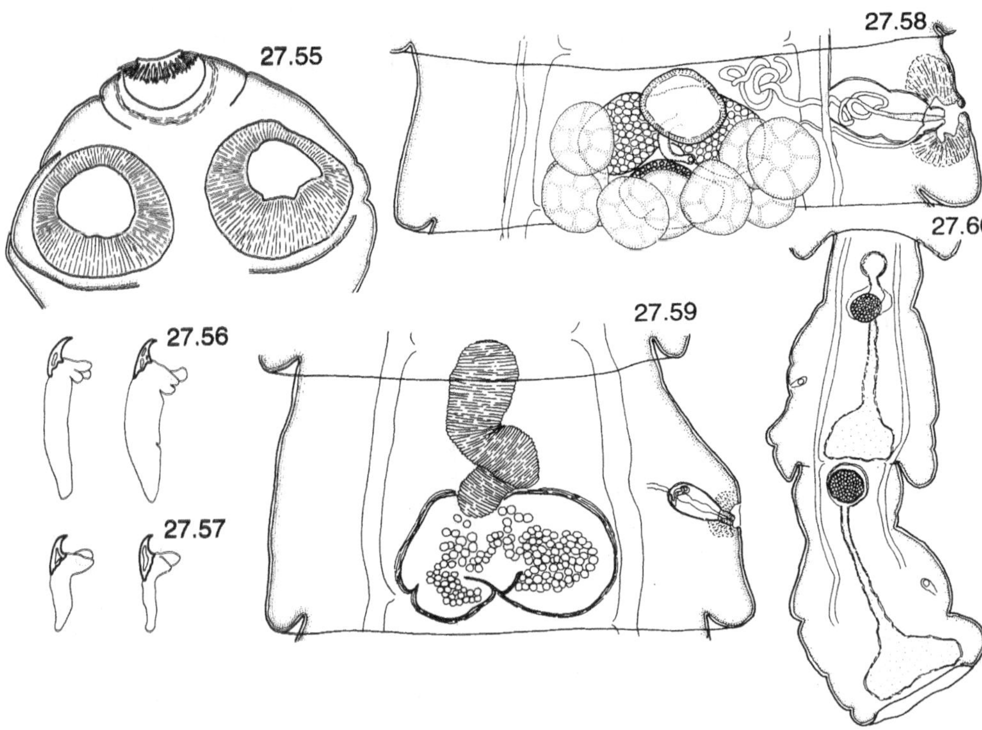

Figs 27.55-27.60 *Triaenorhina rectangula* (Fuhrmann, 1908). 27.55. Scolex. 27.56. Anterior rostellar hooks. 27.57. Posterior rostellar hooks. 27.58. Mature proglottid. 27.59. Pregravid proglottid. 27.60. Gravid proglottids.

Diagnosis: Rostellum with two regular rows of hooks; anterior hooks longer than posterior. Guard bifurcate or with more than two lobes; handle large, straight ('rectanguloid hooks'). Suckers round, muscular. Proglottids craspedote, when mature wider than long, gravid can be longer than wide. Genital pores irregularly alternating. Atrium surrounded by highly-developed circular musculature, may form genital papilla. Genital ducts between osmoregulatory canals. Testes few, dorsal and lateral to female glands. Cirrus-sac pyriform, very often constricted at middle, thin-walled, can reach or overlap osmoregulatory canals. Internal vas deferens may form structure similar to seminal vesicle. Vitellarium compact, median. Ovary reniform, compact. Seminal receptacle small, median. Uterus forming two or more deep lobed sacculations directed posteriorly. Paruterine organ cylindrical, anterior to uterus; paruterine organ forming capsule containing eggs in anterior part of last gravid proglottids. Eggs and oncospheres oval. In Coraciiformes (Coraciidae and Bucerotidae). Palaearctic, Africa, India. Type-species *T. rectangula* (Fuhrmann, 1908) Spasskii & Shumilo, 1965.
Note: Recently cestodes (innominate) with similar characteristics and probably belonging to the same genus were discovered in Neotropical Cuculiformes (see Bona & Maffi, 1987).

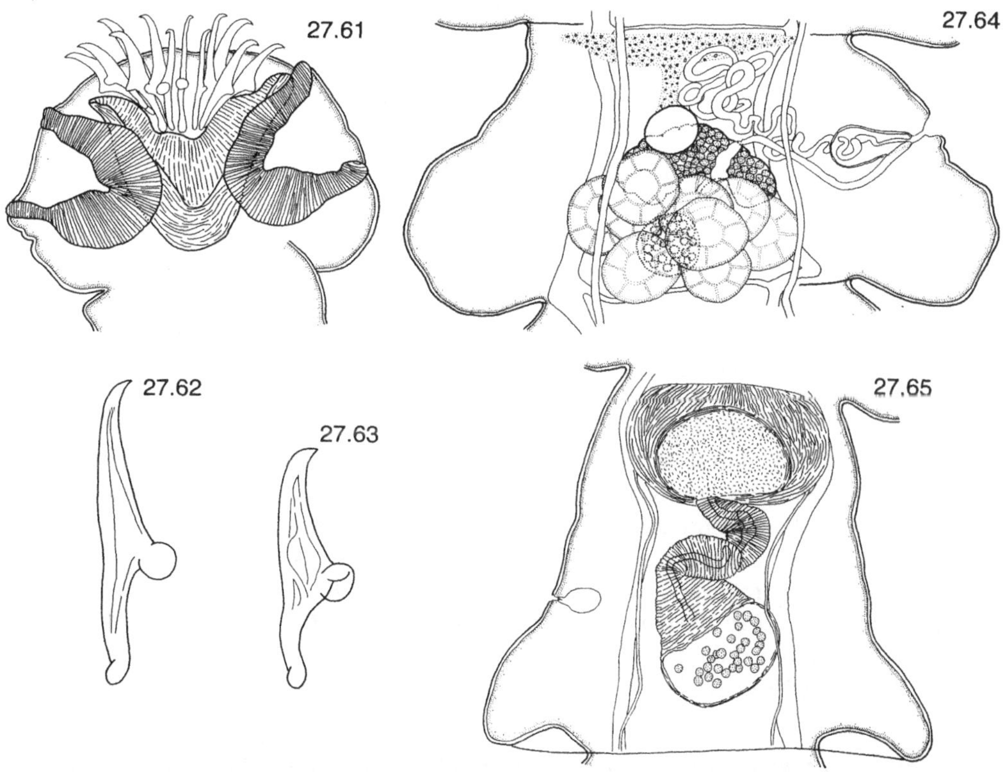

Figs 27.61–27.65 *Sphaeruterina punctata* Johnston, 1914. 27.61. Scolex. 27.62. Anterior rostellar hook. 27.63. Posterior rostellar hook. 27.64. Mature proglottid. 27.65. Pregravid proglottid.

15b. **Rostellar hooks with regular shape, more or less triangular. Cirrus unarmed** .... **16.**

16a. **Uterus reticular (Figs 27.93, 27.98)** .... **22.**
16b. **Uterus otherwise** .... **17.**

17a. **Developing uterus forms one sac** .... **18.**
17b. **Uterus forms two sacs connected anteriorly by more or less narrow canal; sometimes uterine sacs have posterior diverticula (Fig. 27.85, 27.91)** ... . **21.**

18a. **Paruterine organ consists of thick-walled, longitudinal curved tube connecting uterus with spherical body at anterior proglottid margin (Fig. 27.65)** .... ***Sphaeruterina*** **Johnston, 1914. (Figs 27.61–27.65)**

Diagnosis: Rostellum discoid, sucker-like, with double crown of hooks. Hooks with well-developed blade, handle and guard; handle and guard with epiphyseal thickenings. Proglottids craspedote, when gravid longer than wide. Genital pores irregularly alternating (typically) or unilateral. Genital ducts between osmoregulatory canals. Testes few, in one group, posterior and dorsal to female glands. Cirrus-sac small, oval or pyriform, not reaching osmoregulatory canals. Vitellarium median, compact, oval, small. Ovary two-winged, compact. Seminal receptacle median, fusiform. Vagina opens dorsally to male pore. Uterus spherical. In Passeriformes and Piciformes. Australia, Oceania, South America. Type-species *S. punctata* Johnston, 1914.

18b. **Paruterine organ otherwise** .... **19.**

19a. **Paruterine organ wider than long (Fig. 27.70). Genital ducts ventral to osmoregulatory canals** .... ***Notopentorchis*** **Burt, 1938. (Figs 27.66–27.71)**

Diagnosis: Scolex elongate, not clearly distinct from neck. Rostellum armed with double crown of triangular hooks; hooks with well-developed handle and guard with epiphyseal thickenings. Proglottids craspedote, when gravid longer than wide. Genital pores alternating. Testes not numerous (maximum known number 14), dorsal or encircling female glands. Cirrus-sac small, pyriform or slightly elongate, not crossing osmoregulatory canals. Cirrus unarmed. Vitellarium median, compact or slightly lobed. Ovary wide, compact, anterior to vitellarium. Uterus sac-shaped, initially oval or transversely elongate body anterodorsal to ovary, then becoming spherical or pyriform with short, wide paruterine organ developing anteriorly to it. In swifts (Apodiformes: Apodidae and Hemiprocnidae), one record in swallow (Passeriformes: Hirundinidae). Old World (Palaearctic, South Asia, Africa). Type-species *N. collocaliae* Burt, 1938.

Figs 27.66–27.71 *Notopentorchis* spp. 27.66–27.70. *N. iduncula* (Spasskii, 1946). 27.66. Scolex. 27.67. Anterior rostellar hook. 27.68. Posterior rostellar hook. 27.69. Mature proglottid. 27.70. Gravid proglottid. 27.71. *N. vesiculigera* (Krabbe, 1882), mature proglottid.
Figs 27.72–27.75 *Troguterina disparhamulis* (Bona, Bosco & Maffi, 1986). 27.72. Rostellar hooks. 27.73. Mature proglottid. 27.74, 2.75. Gravid proglottids.

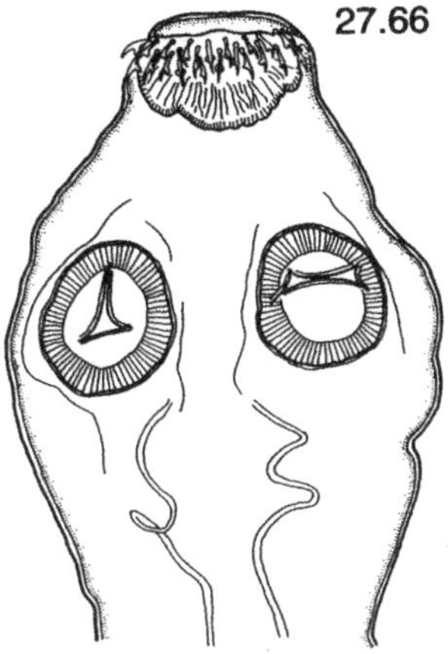
27.66

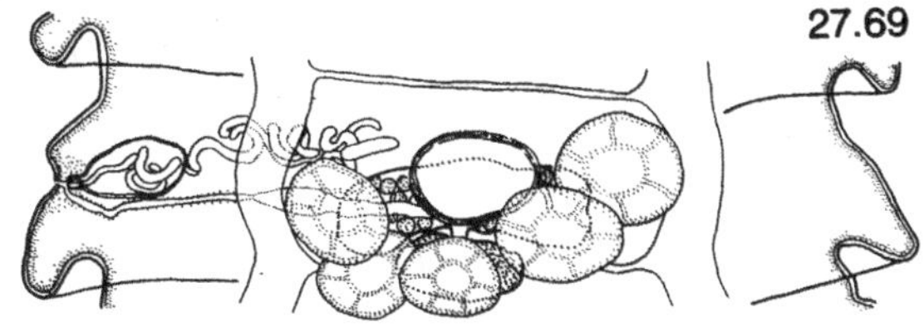
27.69

27.67

27.68

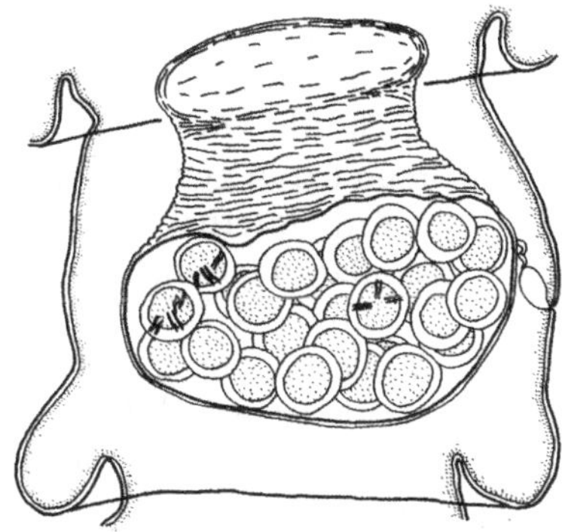
27.70

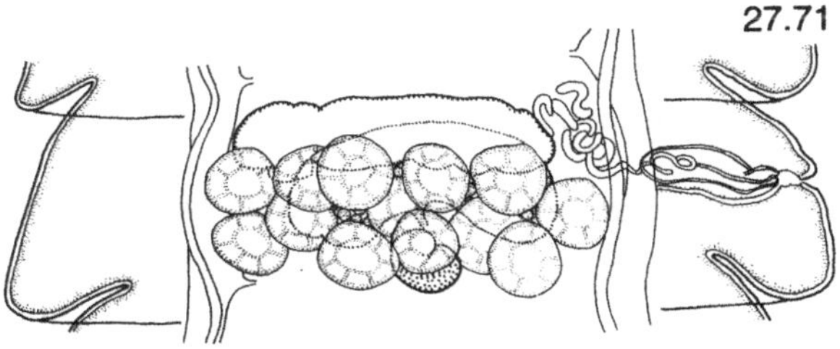
27.71

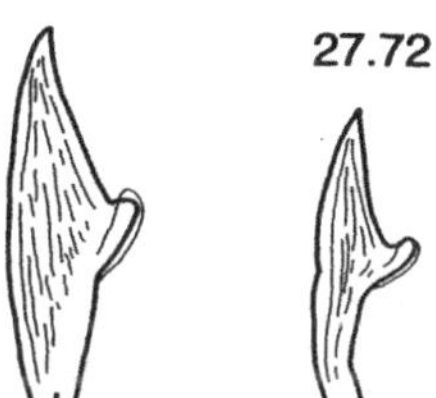
27.72

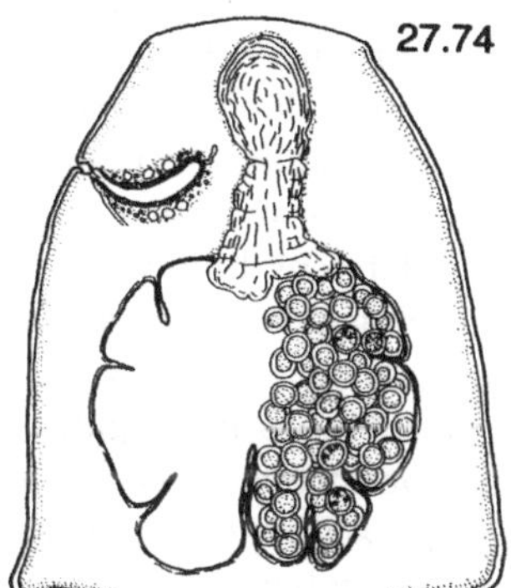
27.74

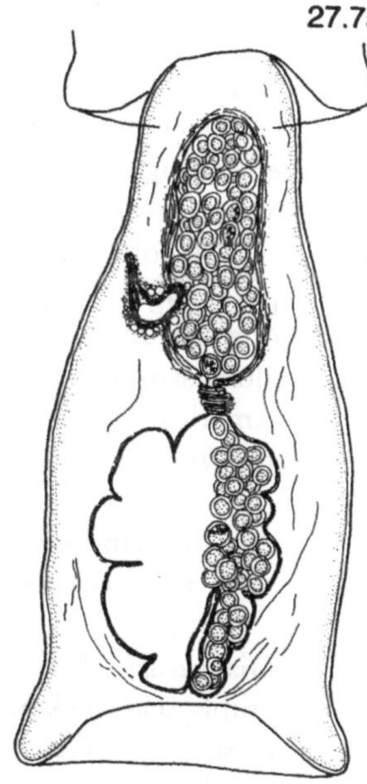
27.75

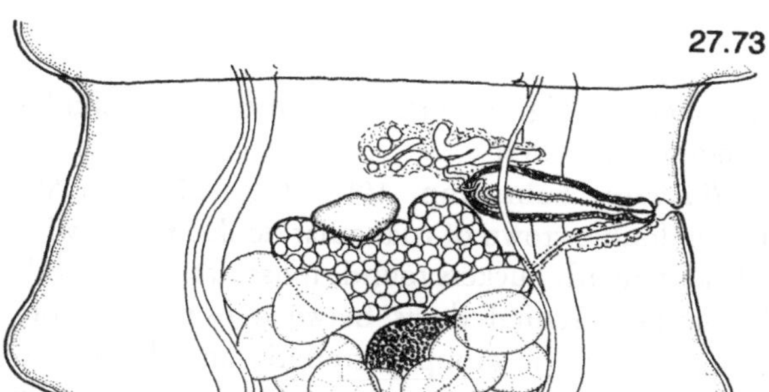
27.73

19b. Paruterine organ longer than wide (Figs 27.74, 27.79). Genital ducts between osmoregulatory canals .... 20.

20a. Genital pores unilateral. Rostellar hooks large, stylet-like. In Trogoniformes .... *Troguterina* Spasskii, 1991. (Figs 27.72–27.75)

Diagnosis: Scolex with sucker-like rostellum armed with double crown of hooks; dorsal and ventral hooks longer than lateral; handle and guard with epiphyseal thickenings. Proglottids craspedote, wider than long except pregravid and gravid which are considerably longer than wide. Genital ducts between osmoregulatory canals. Testes posterior, lateral and dorsal to ovary and vitellarium. Cirrus-sac claviform, with thick muscular walls, usually crosses osmoregulatory canals; surrounded in gravid proglottids by sleeve of large cells. External vas deferens highly coiled, internal vas deferens in postmature proglottids able to form structure similar to seminal vesicle. Vitellarium compact, median. Ovary large, compact or slightly lobed, with irregular shape or two-winged. Seminal receptacle fusiform. Vagina opens posteroventrally to male pore; copulatory and conductive parts clearly separated. Uterus lobate or horseshoe-shaped. Paruterine organ, when fully developed, longer than wide. Eggs and oncospheres oval. In Trogoniformes. South America. Type-species *T. disparhamulis* (Bona, Bosco & Maffi, 1986) Spasskii, 1991.

20b. Genital pores alternating. Rostellar hooks small, triangular. In Cuculiformes .... *Francobona* n. g. (Figs 27.76–27.79)

Diagnosis: Scolex with sucker-like rostellum armed with double crown of numerous hooks. Proglottids craspedote, wider than long except gravid (with almost equal length and width or longer than wide). Atrium consists of two parts separated by muscular ring. Genital ducts between osmoregulatory canals. Testes lateral and posterior to female glands, also sometimes dorsal to vitellarium. Cirrus-sac pyriform, with very thick muscular walls, can reach osmoregulatory canals. Cirrus unarmed. Vas deferens highly coiled; external vas deferens and encircling tissue forming elongate dense agglomerate situated obliquely. Vitellarium median, slightly lobed. Ovary with irregular shape, compact or slightly lobed. Seminal receptacle fusiform. Vagina opens posteroventrally to male pore; vaginal copulatory and conductive parts clearly distinguished. Uterus forms one sac, most often with irregular shape; uterus and paruterine organ elongate along strobilar axis. Eggs and oncospheres oval. In Cuculiformes. North and South America. Type-species *F. similis* (Ransom, 1909) n. comb. for *Rhabdometra similis* Ransom, 1909.

21a. Uterine sacs more or less spherical (Fig. 27.84); sometimes uterus similar to horseshoe with wide ends (Fig. 27.85) .... *Biuterina* Fuhrmann, 1902. (Figs 27.80–27.85)

Diagnosis: Rostellum armed with double crown of triangular hooks. Hooks with epiphyseal thickenings of handle and guard. Suckers round, muscular. Proglottids craspedote. Genital pores alternating. Genital ducts between osmoregulatory canals. Testes not numerous, situated in one group posteriorly and dorsally to ovary and dorsally and laterally to vitellarium. Cirrus-sac pyriform or oval, usually

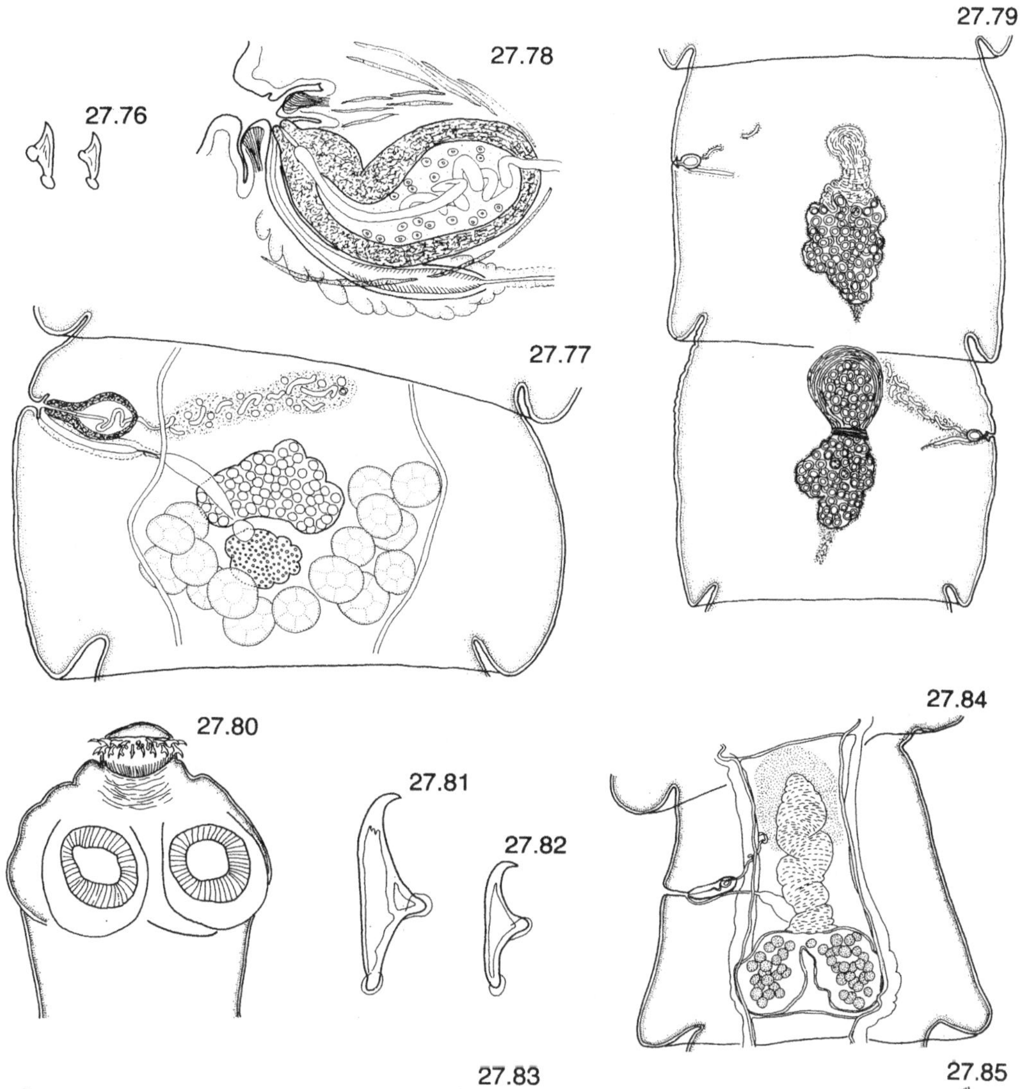

Figs 27.76–27.79 *Francobona similis* (Ransom, 1909). 27.76. Rostellar hooks. 27.77. Mature proglottid. 27.78. Atrium and terminal genital ducts. 27.79. Gravid proglottids.
Figs 27.80–27.85 *Biuterina* spp. 27.80–27.84. *B. triangula* (Krabbe, 1869). 27.80. Scolex. 27.81. Anterior rostellar hook. 27.82. Posterior rostellar hook. 27.83. Mature proglottid. 27.84. Pregravid proglottid. 27.85. *B. cordifera* Murai & Sulgostowska, 1983, gravid proglottid.

in lateral field or overlapping osmoregulatory canals. Cirrus unarmed. Vitellarium median, compact. Ovary compact or slightly lobed, usually two-winged. Seminal receptacle median, elongate, directed porally. Uterus initially small round body anterodorsal to ovary; later, two sacs connected anteriorly by narrow canal. Paruterine organ anterior to uterus, usually longitudinally elongate. Eggs and oncospheres oval or round. In Passeriformes (different families), also in Coraciiformes (Meropidae and Bucerotidae). Cosmopolitan. Type-species *B. clavulus* (Linstow, 1888) Fuhrmann, 1902.

21b. Uterine sacs each form two or three large posterior diverticula .... *Spasskyterina* Kornyushin, 1989. (Figs 27.86–27.91)

Diagnosis: Rostellum discoid, sucker-like, with regular double crown of hooks (sometimes several additional hooks may present as third incomplete row). Hooks triangular, with epiphyseal thickenings on handle and guard; anterior hooks longer, blade longer than handle; posterior hooks with handle and blade of almost equal length. Proglottids craspedote, wider than long except gravid which may be slightly longer than wide. Genital pores irregularly alternating. Genital ducts between osmoregulatory canals. Testes dorsal and lateral to vitellarium and lateral and posterior to ovary. Cirrus-sac bottle-shaped or pyriform, can overlap osmoregulatory canals. Cirrus unarmed. Vitellarium median, compact, oval, near posterior margin of proglottid. Ovary two-winged, compact. Seminal receptacle drop-like, between vitellarium and ovary. Vagina opens ventrally to male pore and passes posteriorly to cirrus-sac. Uterus initially a small round thick-walled body anterior and dorsal to ovary, transverse tube with several (usually four) posteriorly directed processes; in gravid proglottids it consists of two large sacs connected anteriorly by narrow canal. Paruterine organ wide, fills all median field anteriorly to pregravid uterus; uterus overgrown totally by paruterine organ in last gravid proglottids. In passeriform birds (Muscicapidae, Oriolidae, Pycnonotidae). Palaearctic, Africa. Type-species *S. trianguloides* Kornyushin, 1989.

22a. Gravid proglottids almost as long as wide .... *Parvirostrum* Fuhrmann, 1908. (Figs 27.92–27.93)

Diagnosis: Rostellum armed with double crown of small triangular hooks. Suckers round, muscular. Proglottids craspedote, mature wider than long. Genital pores irregularly alternating, rarely unilateral. Genital ducts pass dorsal to osmoregulatory canals or between them. Testes lateral and dorsal to female glands (in type-species in two groups lateral to vitellarium). Cirrus-sac lateral or can reach osmoregulatory canals. Vitellarium compact, median (irregular in type-species). Ovary typically two-winged, compact, sometimes fan-like and lobed. Vagina thin-walled, opens posteriorly to male pore. Uterus initially saccular, when gravid reticular. Paruterine organ develops anteriorly to uterus like wide spongiform body; in gravid proglottids uterus totally overgrown by paruterine organ. Eggs and oncospheres round. In Passeriformes. South America, ? Africa, ? New Guinea. Type-species *P. reticulatum* Fuhrmann, 1908.

22b. Gravid proglottids twice or more longer than wide .... *Dictyterina* Spasskii in Spasskaya & Spasskii, 1971. (Figs 27.94–27.98)

Diagnosis: Rostellum armed with two indistinct rows of numerous (about 40 or

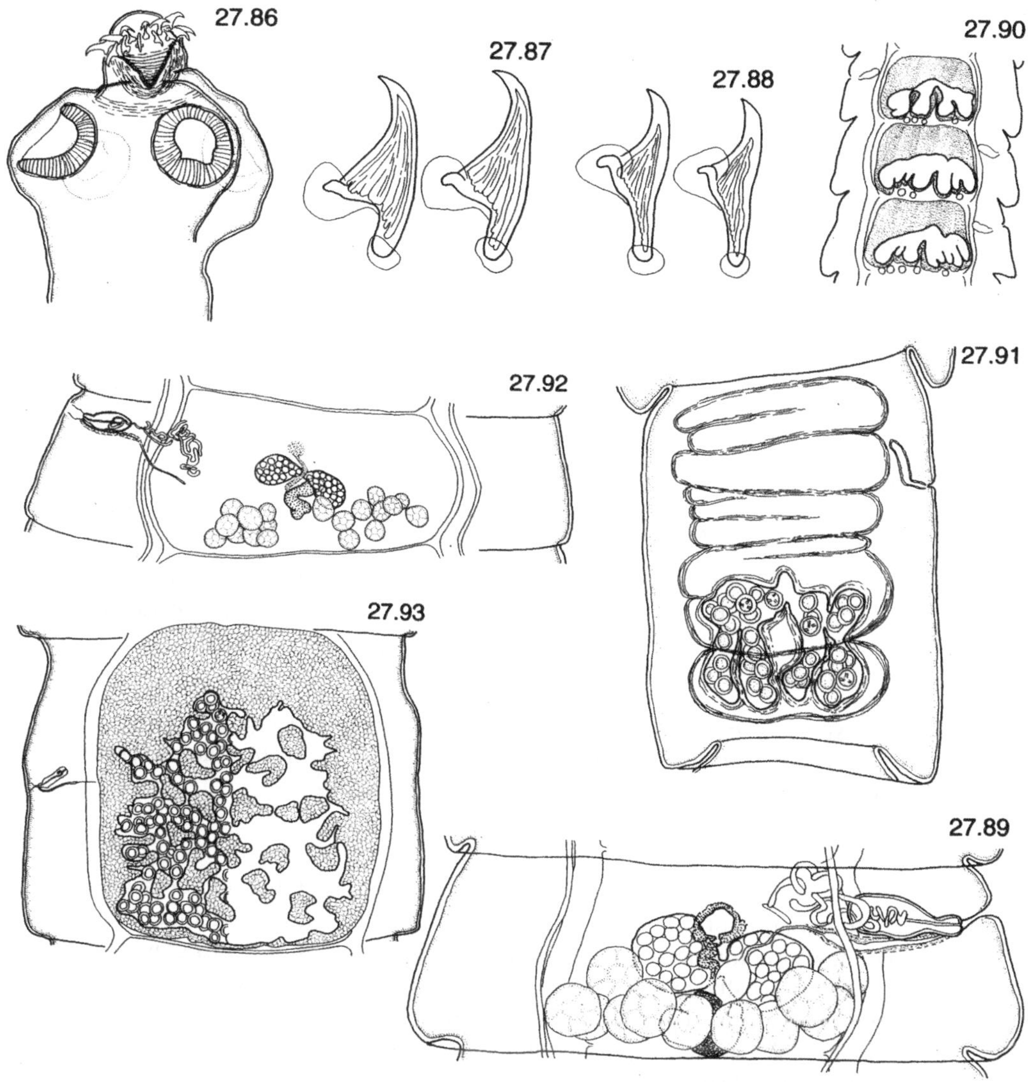

Figs 27.86–27.91 *Spasskyterina trianguloides* Kornyushin, 1989. 27.86. Scolex. 27.87. Anterior rostellar hooks. 27.88. Posterior rostellar hooks. 27.89. Mature proglottid. 27.90. Three postmature proglottids showing variability of uterine shape. 27.91. Gravid proglottid.
Figs 27.92, 27.93 *Parvirostrum reticulatum* Fuhrmann, 1908. 27.92. Mature proglottid. 27.93. Gravid proglottid.

more) small triangular hooks. Suckers round, muscular. Proglottids craspedote with small velum; mature proglottids almost as long as wide. Genital pores irregularly alternating. Genital ducts between osmoregulatory canals. Testes lateral and dorsal to vitellarium and posterior and lateral to ovary. Cirrus-sac reaches or just crosses osmoregulatory canals. Vitellarium compact, median. Ovary compact, reniform. Vagina thin-walled, opens posteriorly to male pore. Seminal

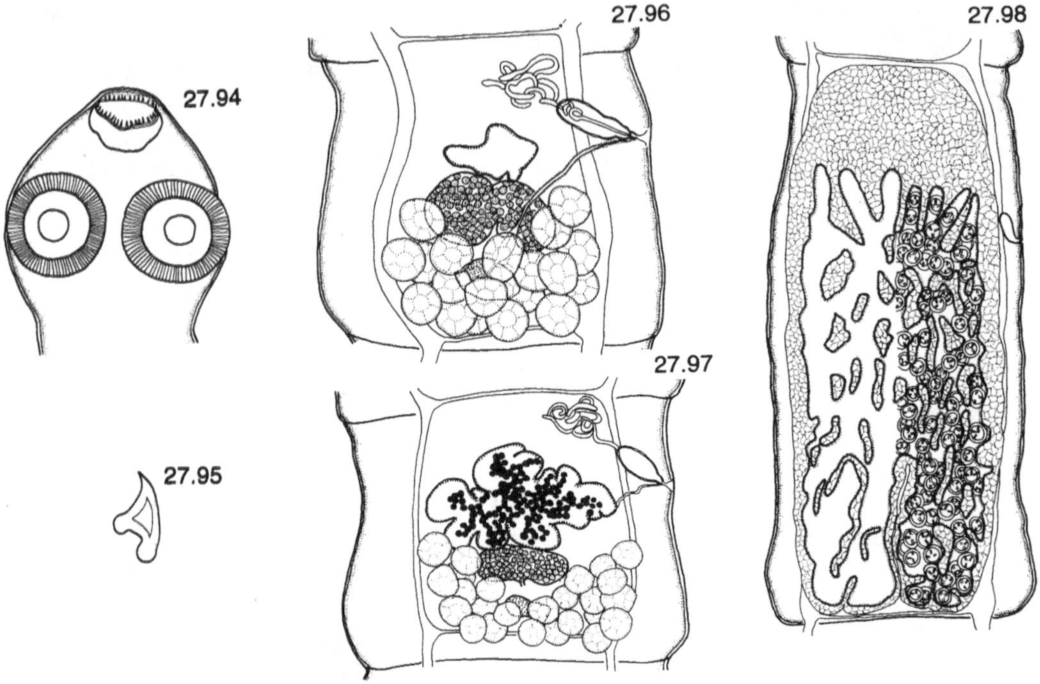

Figs 27.94-27.98 *Dictyterina cholodkowskii* (Skrjabin, 1914). 27.94. Scolex. 27.95. Rostellar hook (27.94 and 27.95 modified from Matevosyan, 1969). 27.96, 27.97. Mature proglottids with different degree of uterine development. 27.98. Gravid proglottid.

receptacle small, median. Uterus initially irregular sac anterior and dorsal to ovary; later forms anterior and posterior diverticula; when gravid reticular. Paruterine organ develops anteriorly to uterus like wide spongiform body; in gravid proglottids, uterus totally overgrown by paruterine organ, including space between uterine reticulum. Eggs and oncospheres round. In shrikes (Passeriformes: Laniidae). Palaearctic. Type-species *D. cholodkowskii* (Skryabin, 1914) Spasskii in Spasskaya & Spasskii, 1971.

28 Family Metadilepididae Spasskii, 1959

V.V. KORNYUSHIN[1] & B.B. GEORGIEV[2]

[1]*Institute of Zoology, Academy of Sciences of the Ukraine, 15 Bogdan Khmelnitsky Street, 25260 Kiev-30, Ukraine:* [2]*Bulgarian Academy of Sciences, Institute of Parasitology, Acad. G. Bonchev Street, Bl. 25, 1113 Sofia, Bulgaria.*

Introduction

The subfamily Metadilepidinae was erected by Spasskii (1959c) for three genera of Dilepididae possessing many characteristics similar to those of the family Paruterinidae. Later, Metadilepididae was recognized as a full family within the superfamily Paruterinoidea (Spasskaya & Spasskii, 1971) and eight genera were considered within it (Spasskii & Spasskaya, 1977). Its validity was accepted for a long period only by Russian-language authors (Borgarenko, 1981; Kornyushin, 1989) but was recently accepted also by Mariaux & Vaucher (1989), Mariaux (1990; 1991a) and Mariaux *et al.* (1992). The present revision confirms the validity of the Metadilepididae and the position within it of six of the genera previously referred to it. However, *Spreotaenia* corresponds more fully to Dilepididae than to Metadilepididae (Bona, personal communication). *Parvirostrum* was found to have a paruterine organ and has been transferred to Paruterinidae. In addition, two other metadilepidid genera (*Yapolepis* and *Pseudadelphoscolex*) have recently been erected (Mariaux, 1991a; Mariaux *et al.*, 1992). It can be expected that the potential number of genera is considerably larger than is currently appreciated. Since they parasitize mainly tropical birds as yet insufficiently studied, their discovery is a matter for the future.

Morphologically, the Metadilepididae are similar to the Dilepididae and Paruterinidae but they can be easily distinguished from the former by the absence of a saccular rostellar sheath and the position of the developing uterus dorsal (or anterodorsal) to the ovary, and from the latter by the absence of a paruterine organ.

Family Metadilepididae Spasskii, 1959

(Syn. Metadilepidinae Spasskii, 1959)

Diagnosis: Medium and large size cestodes. Scolex with armed rostellum, sometimes with rudimentary unarmed rostellum. Rostellum sucker-like or discoid or ovoid, without saccular sheath. Hooks in one or two rows, more or less triangular, with epiphyseal thickenings of handle and guard. Suckers unarmed, muscular. Proglottids craspedote. Genital ducts ventral to or between osmoregulatory canals, rarely dorsal (*Hamatofuhrmannia*). Vas deferens does not form seminal vesicles; rarely, (*Metadilepis* and *Yapolepis*), vas deferens in postmature and gravid proglottids widens to form structure similar to internal seminal vesicle. Genital organs single per proglottid. Vitellarium compact, postovarian. Ovary usually compact or slightly lobate, two-winged or fan-like. Uterus initially as small sac anteriorly and (or) dorsally to female glands; developed uterus saccular, in median field, without paruterine organ. Eggs without pyriform apparatus. Oncosphere oval. Parasites of non-aquatic birds (Passeriformes, Caprimulgiformes, Coraciiformes). In Old and New World tropics, rarely in temperate latitudes. Type-genus *Metadilepis* Spasskii, 1949.

Key to genera

1a. Genital pores unilateral .... **2.**
1b. Genital pores irregularly alternating .... **6.**

2a. Genital ducts ventral to osmoregulatory canals .... **3.**
2b. Genital ducts between osmoregulatory canals .... **4.**

3a. Testes in two groups lateral to female glands. Rostellar hooks in two rows, hooks in each row of almost equal length ....
.... ***Metadilepis* Spasskii, 1949. (Figs 28.1–28.6)**
(Syns *Dilepis (Metadilepis)* Spasskii, 1949; *Metaskrjabinolepis* Spasskii, 1947.)

Diagnosis: Rostellum sucker-like. Rostellar hooks with curved blade and with handle and guard with epiphyseal thickenings. Proglottids craspedote, wider than long. Genital pores unilateral. Genital ducts ventral to osmoregulatory canals. Cirrus-sac thick-walled, crossing osmoregulatory canals. External vas deferens highly coiled; internal vas deferens in gravid proglottids highly developed and similar to seminal vesicle. Cirrus armed (at least in type-species). Vitellarium median, compact, oval, near posterior margin of proglottid. Ovary slightly lobed, fan-like. Seminal receptacle small, anterior to ovary and near aporal end of cirrus sac. Vagina thick-walled, opens posteriorly and somewhat ventrally to male pore. Uterus saccular, in median field. In nightjars (Caprimulgiformes: Caprimulgidae). Holarctic, South America. Type-species *M. globacantha* (Fuhrmann, 1913) Spasskii, 1949.

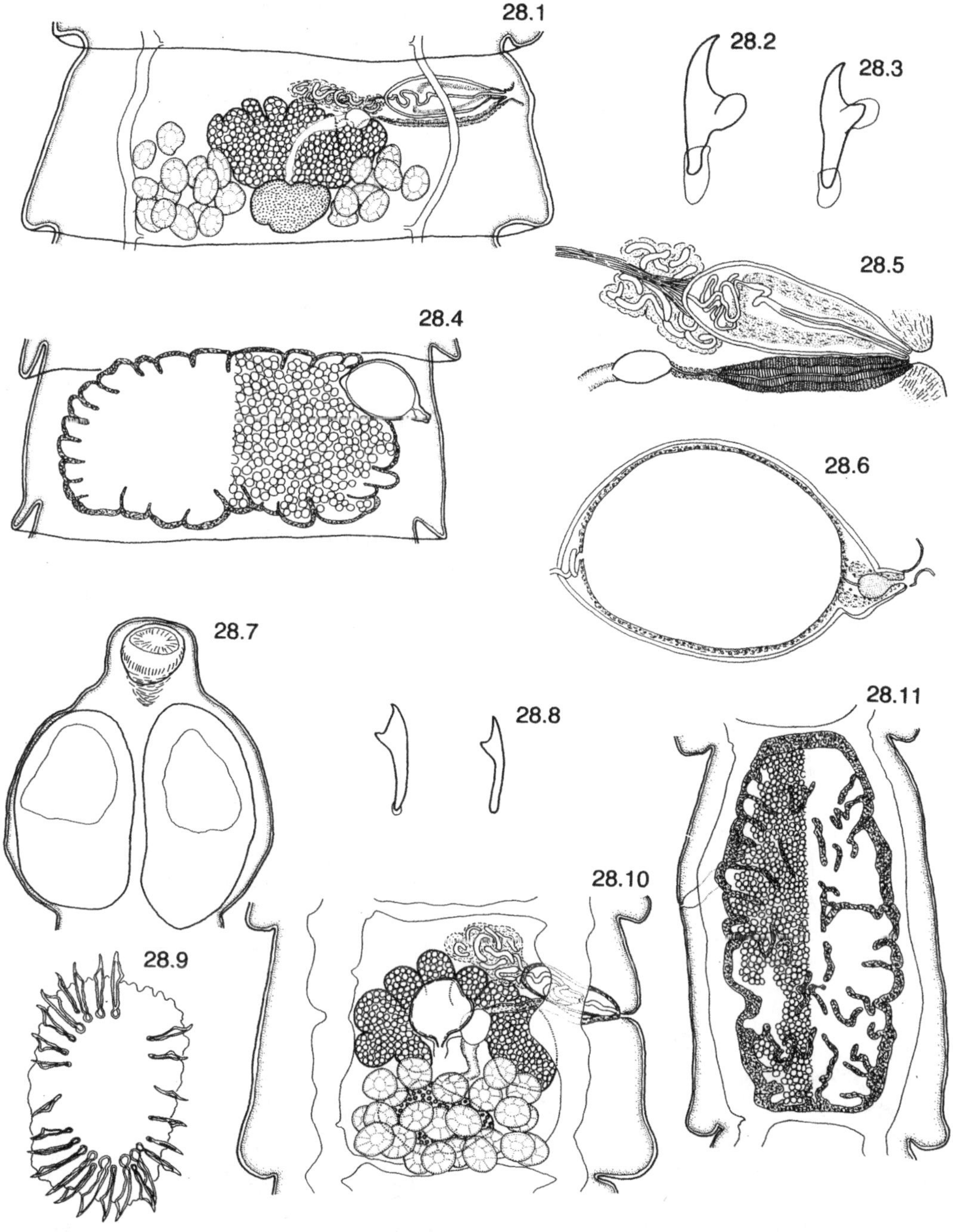

Figs 28.1-28.6 *Metadilepis globacantha* (Fuhrmann, 1913). 28.1. Mature proglottid. 28.2. Anterior rostellar hook. 28.3. Posterior rostellar hook. 28.4. Gravid proglottid. 28.5. Genital ducts in mature proglottid. 28.6. Cirrus-sac in pregravid proglottid.

Figs 28.7-28.11 *Skrjabinoporus merops* (Woodland, 1928). 28.7. Scolex. 28.8. Rostellar hooks. 28.9. Rostellar crown, apical view (modified from Spasskii & Borgarenko, 1960). 28.10. Mature proglottid. 28.11. Gravid proglottid.

3b. Testes in one group in posterior half of median field, dorsal and lateral to vitellarium. Rostellar hooks in two rows, dorsal and ventral hooks considerably larger than lateral; lateral hooks may be absent (Fig. 28.9) .... *Skrjabinoporus* Spasskii & Borgarenko, 1960. (Figs 28.7–28.11)

Diagnosis: Scolex with muscular rostellum, without rostellar sheath. Suckers elongate, with weak musculature. Proglottids craspedote, wider than long except pregravid and gravid ones. Genital pores unilateral. Genital ducts ventral to osmoregulatory canals. Testes numerous (20–30). Vas deferens coiled, covered by glandular tissue externally to cirrus-sac. Cirrus-sac elliptical or pyriform, can overlap or cross osmoregulatory canals. Cirrus unarmed. Vitellarium median, compact. Ovary more or less two-winged or fan-like, slightly lobed; anterior border more lobed than posterior. Seminal receptacle small, globular, at level of ovary. Vagina thick-walled, dorsal to cirrus-sac. Developing uterus round or oval, dorsal to ovary; gravid uterus saccular, thick-walled, in median field only. Eggs oval, with thin walls. In bee eaters (Coraciiformes: Meropidae). Old World (Southern Palaearctic, Africa). Type-species *S. merops* (Woodland, 1928) Spasskii & Borgarenko, 1960.

4a. Rostellum unarmed ...... *Yapolepis* Mariaux, 1991. (Figs 28.12–28.15)

Diagnosis: Rostellum sucker-like. Proglottids craspedote, wider than long (except part of gravid proglottids). Genital pores unilateral. Genital ducts pass between osmoregulatory canals. Testes dorsal to female glands, in one field

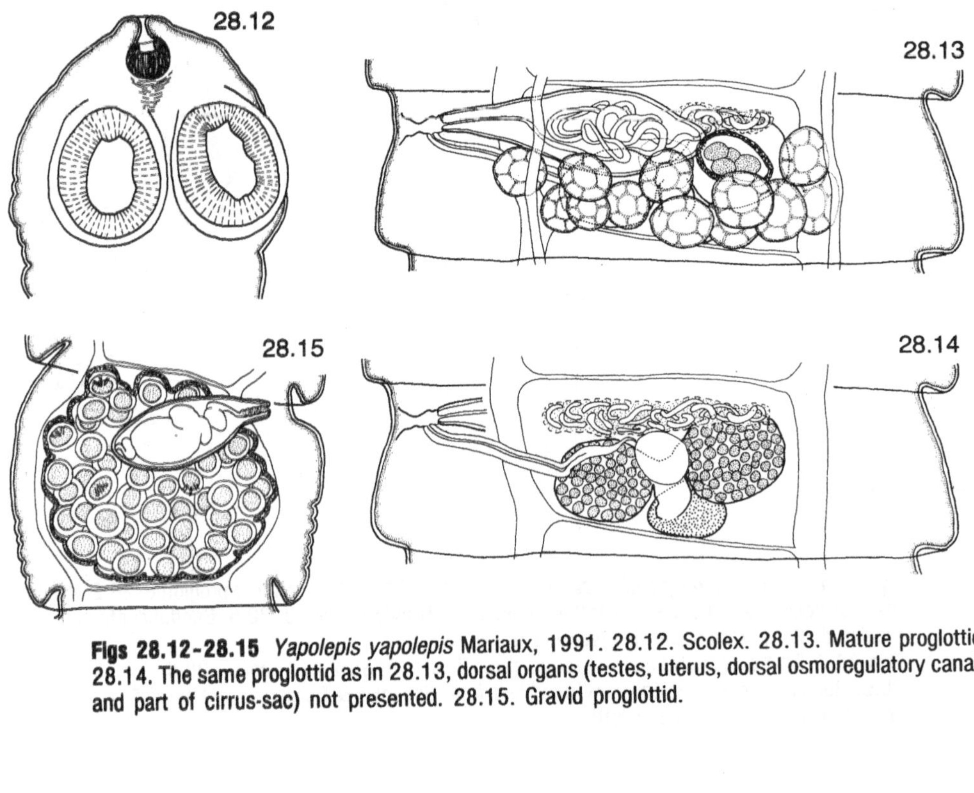

Figs 28.12–28.15 *Yapolepis yapolepis* Mariaux, 1991. 28.12. Scolex. 28.13. Mature proglottid. 28.14. The same proglottid as in 28.13, dorsal organs (testes, uterus, dorsal osmoregulatory canals and part of cirrus-sac) not presented. 28.15. Gravid proglottid.

overlapping osmoregulatory canals. Cirrus-sac pyriform, thick-walled, large, reaching middle of proglottid. External vas deferens highly coiled ventrally to cirrus-sac; internal vas deferens also coiled, in postmature proglottids can form structure similar to seminal vesicle. Cirrus armed with small spines. Vitellarium median, compact, oval, near posterior margin of proglottid. Ovary two-winged, wide, compact. Seminal receptacle globular, central. Vagina opens posteriorly to male pore, thick-walled. Young uterus an oval sac, dorsal to ovary, slightly aporal. Gravid uterus saccular, thick-walled, in median field. Eggs round. In bulbuls (Passeriformes: Pycnonotidae). Africa. Type-species *Y. yapolepis* Mariaux, 1991.

4b. Rostellum armed .... **5.**

5a. Rostellar hooks in single crown; vagina without sphincter. Half or more of cirrus-sac in median field ....

***Pseudadelphoscolex* Mariaux, Bona & Vaucher, 1992. (Figs 28.16–28.18)**

Diagnosis: Rostellum sucker-like. Hooks resemble hammer-shaped hooks of davaineids: short blade and handle and long guard; epiphyseal thickenings of handle and guard merged. Proglottids craspedote, wider than long except some gravid proglottids. Genital pores unilateral. Genital ducts pass between osmoregulatory canals. Testes in posterior two-thirds of median field, dorsally and laterally to ovary and vitellarium. Cirrus-sac thick-walled, crossing osmoregulatory canals. External vas deferens highly coiled; internal vas deferens also coiled, in postmature and gravid proglottids forming structure similar to seminal vesicle. Cirrus armed with small spines. Vitellarium median or slightly aporal, near posterior proglottid margin, slightly lobed, most often with irregular shape. Ovary median, two-winged, slightly lobed. Seminal receptacle elongate, directed

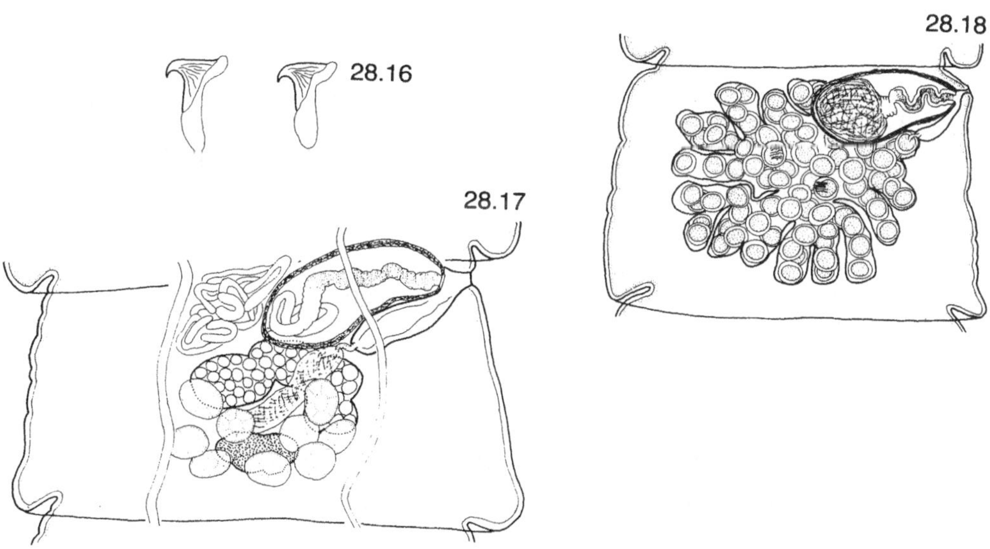

Figs 28.16–28.18 *Pseudadelphoscolex eburnensis* Mariaux, Bona & Vaucher, 1992. 28.16. Rostellar hooks. 28.17. Mature proglottid. 28.18. Gravid proglottid.

porally. Uterus initially dorsal to ovary, later thick-walled, sac-like, lobate. Eggs and oncospheres round. In paradise flycatchers (Passeriformes: Muscicapidae: Monarchinae). Africa. Type-species *P. eburnensis* Mariaux, Bona & Vaucher, 1992.

5b. Rostellar hooks in double crown. Vagina with sphincter. Cirrus-sac lateral to poral osmoregulatory canals or small proportion of it in median field ... . *Cracticotaenia* Spasskii, 1966. (Figs 28.19–28.22)

Diagnosis: Scolex with oval rostellum. Rostellar sheath lacking. Hooks almost triangular, with well-developed handle and guard with epiphyseal thickenings. Proglottids craspedote, wider than long, only gravid almost as wide as long. Genital pores unilateral. Genital ducts pass between osmoregulatory canals. Testes in one group, posterior and dorsal to female glands. Cirrus-sac elongate, with pipette-like distal extremity, may cross osmoregulatory canals. Vitellarium median, slightly lobed. Ovary lobate, fan-like or two-winged. Vagina opens posteriorly to male pore, thick-walled. Uterus in median field, thick-walled, lobate. In butcher birds (Passeriformes: Cracticidae). Australia. Type-species *C. fieldingi* (Maplestone & Southwell, 1923) Spasskii, 1966.

6a. Rostellar hooks in two rows. Uterus horseshoe-shaped .... *Proparuterina* Fuhrmann, 1911. (Figs 28.23–28.25)

Diagnosis: Scolex with sucker-like rostellum. Hooks elongate, handle longer than blade; handle with very large epiphyseal thickening; guard also with epiphyseal thickening. Proglottids craspedote. Genital pores irregularly alternating. Genital ducts pass between osmoregulatory canals. Cirrus-sac elliptical,

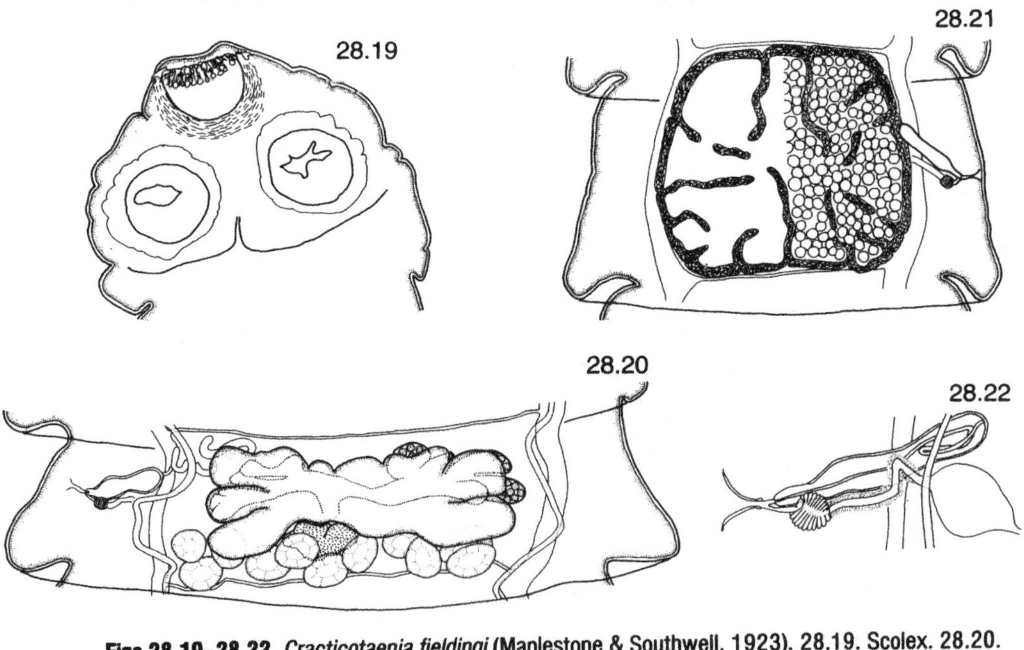

Figs 28.19-28.22 *Cracticotaenia fieldingi* (Maplestone & Southwell, 1923). 28.19. Scolex. 28.20. Mature proglottid. 28.21. Pregravid proglottid. 28.22. Genital ducts in pregravid proglottid.

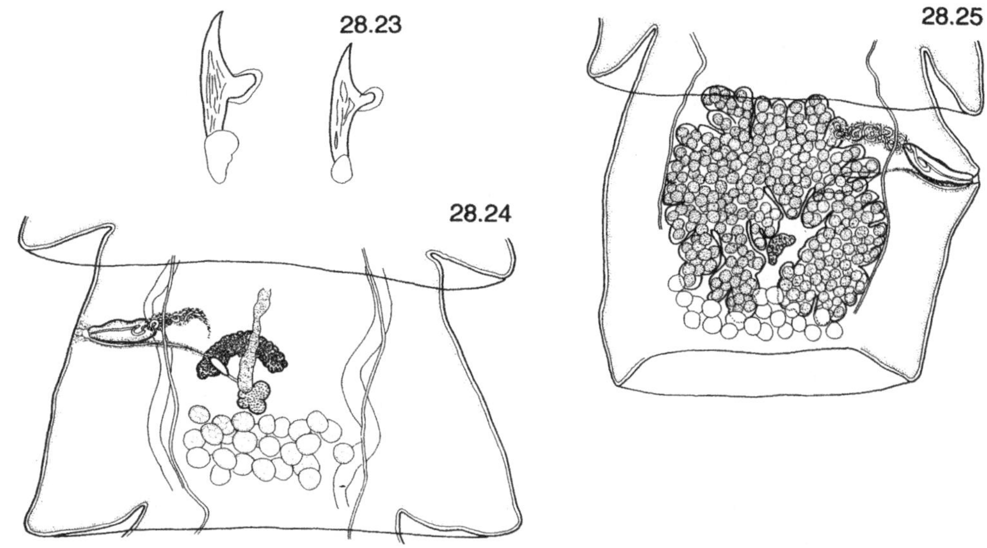

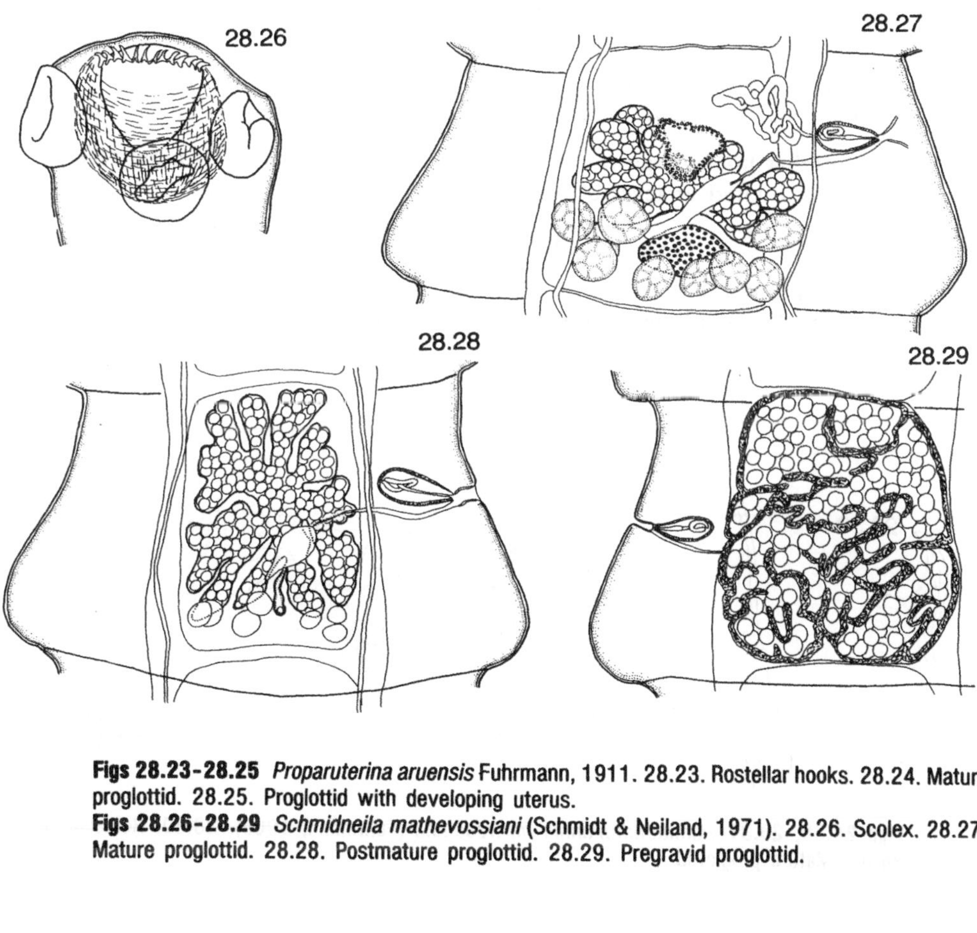

Figs 28.23–28.25 *Proparuterina aruensis* Fuhrmann, 1911. 28.23. Rostellar hooks. 28.24. Mature proglottid. 28.25. Proglottid with developing uterus.
Figs 28.26–28.29 *Schmidneila mathevossiani* (Schmidt & Neiland, 1971). 28.26. Scolex. 28.27. Mature proglottid. 28.28. Postmature proglottid. 28.29. Pregravid proglottid.

with thick muscular walls, can overlap osmoregulatory canals. External vas deferens coiled, covered by glandular cells. Testes numerous, posterior to female glands. Vitellarium median, irregular, slightly lobed. Ovary slightly lobed, usually reniform. Seminal receptacle elliptical, dorsal to poral part of ovary. Vagina opens posterolaterally to male pore; copulatory and conductive parts distinct; vaginal walls as thick cellular sleeve. Eggs and oncospheres unknown. In frogmouths (Caprimulgiformes: Podargidae). Oceania (Aru I.). Type-species *P. aruensis* Fuhrmann, 1911.

6b. Rostellar hooks in single row. Uterus otherwise .... 7.

7a. Rostellar hooks short (less than 30 μm). Eggs fusiform. Seminal receptacle elongate, situated between ovary and vitellarium .... ***Schmidneila*** **Spasskii & Spasskaya, 1973. (Figs 28.26–28.29)**

Diagnosis: Scolex with ovoid rostellum without rostellar sheath. Rostellar hooks in single crown. Proglottids craspedote, wider than long except gravid. Genital pores irregularly alternating. Genital ducts pass between osmoregulatory canals. Testes posterior and dorsal to female glands. Cirrus-sac elliptical, lateral to poral osmoregulatory canals. Cirrus unarmed. Vitellarium compact, median, near to posterior margin of proglottid. Ovary two-winged, lobed. Vagina thick-walled, opens posteriorly to male pore. Developing uterus deeply lobed; gravid uterus saccular, filling median field. Ripe eggs fusiform, oncospheres round. In gnat-wrens (Passeriformes: Muscicapidae: Polioptilinae). Central America. Type-species *S. mathevossiani* (Schmidt & Neiland, 1971) Spasskii & Spasskaya, 1973.

7b. Rostellar hooks large (more than 100 μm). Eggs round. Seminal receptacle oval, situated at level of anterior margin of ovary .... ***Hamatofuhrmannia*** **Spasskii, 1969. (Figs 28.30, 28.31)**

Diagnosis: Scolex with rostellum without sheath. Hooks in one row; blade curved; epiphyseal thickenings of handle and guard merged. Proglottids craspedote, wider than long. Genital pores irregularly alternating. Genital ducts pass

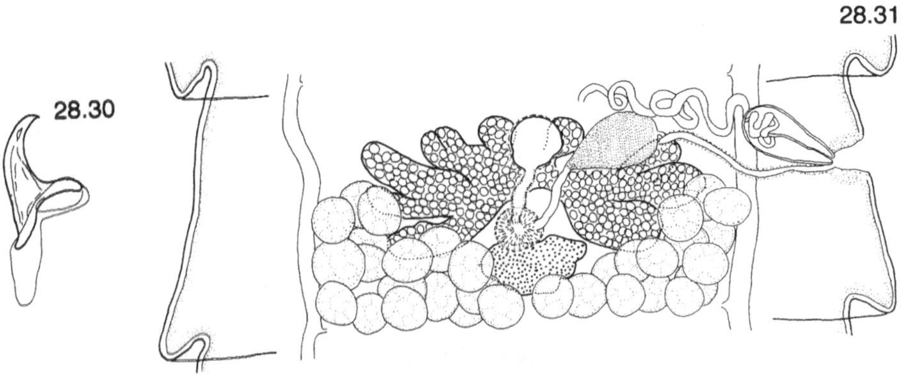

Figs 28.30, 28.31 *Hamatofuhrmannia macracantha* (Fuhrmann, 1908). 28.30. Rostellar hook. 28.31. Mature proglottid.

dorsal to osmoregulatory canals. Testes numerous, situated in posterior half of proglottid, lateral and posterior to female glands, slightly overlapping them dorsally. Cirrus-sac pyriform, thin-walled, can reach osmoregulatory canals. Vitellarium median, slightly lobate. Ovary wide, deeply lobed, more or less two-winged. Vagina thin-walled, opens posteriorly to male pore. Uterus saccular, thin-walled, filling median field. Eggs round. In antbirds (Passeriformes: Formicariidae) (?). South America. Type-species *H. macracantha* (Fuhrmann, 1908) Spasskii, 1969.

29 Family Hymenolepididae Ariola, 1899

B. CZAPLINSKI[1] & C. VAUCHER[2]

[1]*Medical School, Institute of Biostructure, Department of General Biology and Parasitology, Chalubinskiego 5, 02-004 Warsaw, Poland;* [2]*Département des Invertébrés, Muséum d'Histoire Naturelle, Case postale 6434, Route de Malagnou, CH-1211 Genève 6, Switzerland.*

Introduction

The family Hymenolepididae is the richest in number of species of all families of Cestoda with about 850 species (620 in birds and 230 in mammals). Bird hymenolepidids alone have been assigned to 142 genera and subgenera.

We have separated the parasites of birds and mammals into two different keys to facilitate the use of the keys, as there are only a few possibilities of confusion between mammal and bird parasites. There are, however, some exceptions: Prokopic (1967), on the basis of morphology, stated that *Hymenolepis muris-sylvatici* (Rudolphi, 1819) a parasite typical of wood mice (*Apodemus*) in the Palaearctic, is identical to *Variolepis crenata* (= *H. serpentulus* Schrank, 1788) parasitic in various land birds. According to the present key, this species would be classified in *Microsomacanthus* Lopez-Neyra, 1962. In our opinion, an experimental approach (life-cycle or isoenzyme electrophoresis) is necessary to solve this problem. Another case is *Hymenolepis ondatrae* Rider & Macy, 1947, discussed here with *Hymenosphenacanthoides*.

Although many genera have been synonymized, the number of genera is still high; this makes it difficult to illustrate relationships clearly and constitutes a real difficulty for identification. Too often, features useful at the specific level have been used as generic characters. Many hymenolepidids are small and delicate, easily becoming fragmented, and spurious forms based on fragments rather than entire specimens have sometimes been determined as new genera. On the other hand, many descriptions are based on single or few, incomplete, immature or not fully gravid specimens, e.g. *Allohaploparaksis sagitta* (Rosseter, 1906) with only one preserved rostellar hook and incorrectly interpreted morphological characters. Similarly, Soltys (1952) and Rybicka (1966) do not agree about the gravid uterus of *Ditestolepis diaphana* because only the first author observed fully adult specimens (see below).

Commonly, authors revising genera forget that the basic definition of a genus is given by its type-species and do not clearly refer to it in further redescriptions. As a result many generic definitions have been so modified as to accommodate almost any species (*Rodentolepis*, *Vampirolepis*). Vaucher (1992) recently revised *Vampirolepis* and noticed that this genus consistently accommodates species sharing features of the type-species regarding the shape of the rostellar hooks and the morphology of the mature and gravid segments. The genus contains species parasitic only in bats. However, all hymenolepidid tapeworms parasitic in bats cannot be assigned to *Vampirolepis* and several of them are placed in *Rodentolepis*.

The other important problem is poor knowledge of the variability of many morphological features and of their hierarchy and significance for taxonomy. The result is a lack of consistency of generic definitions and of characters considered significant at generic level, such as uterine development, scolex morphology, or arrangements of the testes. Several species have been transferred between genera without real progress towards clarity and stability of the classification: *Hymenolepis diaphana* Cholodkowsky, 1906, parasitic in *Sorex araneus* L., has been placed in *Ditestolepis*, a genus especially created for it but based on an erroneous observation of the testes. It was then placed in *Neoskrjabinolepis* because of confusion about the uterine development in the original description of the genus, but also placed in *Soricinia* because of its small size and the unarmed scolex.

The same character may be very stable in one species or group of animals and variable in another species or group. In some cases, the variation itself may be considered as a feature, for instance the variation in testis disposition in *Vampirolepis* Spasskii, 1954.

The hymenolepidids of mammals show more uniformity in their morphology than those of birds. New accurate descriptions of poorly known taxa will allow useful updates of this key.

In the present classification, we have used the following morphological features. At the subfamily level, important characters include the number of longitudinal osmoregulatory canals, the passage of the genital ducts relative to these canals, the number of crowns of rostellar hooks, the presence of a pseudoscolex, the retractability of the rostellum and the number of genital pores per proglottid.

At the generic level, important characters include the presence or absence of rostellar hooks, the hook number and shape, the presence of a pseudoscolex, the spination of the suckers, the external segmentation, the shape and distribution of the microtriches, the number of testes, the shape of the ovary and, in some cases, the development and shape of the uterus.

We offer the following comments about some controversial genera and proposed synonymies.

The subgenus *Acanthodiorchis* Spasskii, 1952, a subgenus of *Diorchis* Clerc, 1903 has not been accepted.

Acotylolepis Yamaguti, 1959 is a synonym of *Coronacanthus* Spasskii, 1954. The absence of suckers has not been confirmed (Vaucher, 1971).

The type-species of *Allohaploparaksis* Yamaguti, 1959, *A. sagitta* (Rosseter, 1906), is insufficiently described, having been found only once. The number of rostellar hooks is not known. The one small hook figured is reminiscent of the skrjabinoid type. The external seminal vesicle is interpreted as a single testis. The 'cirrus' is reminiscent of a cirrus stylet, whereas the 'cirrus sheath' is probably the cirrus itself. If the (undescribed) testes are three in number, *Allohaploparaksis* is probably a synonym of *Cladogynia* Baer, 1938.

Amazilolepis Schmidt & Dailey, 1992 is a synonym of *Hymenolepis* Weinland, 1858 because of its unarmed scolex and three testes per proglottid. The position of the vagina anterior to the cirrus-sac is not a generic feature.

The type-species of *Anserilepis* Spasskii & Tolkacheva, 1965, *A. barrowensis* (Schiller, 1952), was placed by Yamaguti (1959) and Schmidt (1986) in *Drepanidotaenia*, probably because of the antiporal position of the ovary and vitellarium which is a feature of specific, not generic significance. This synonymy is not correct, however, as *Drepanidotaenia* has eight, not ten, rostellar hooks. *Anserilepis* should be considered a synonym of *Microsomacanthus* Lopez-Neyra, 1942.

Apora Ginetsinskaya, 1944 should be considered a synonym of *Gastrotaenia* Wolffhügel, 1938. Ginetsinskaya (1944) did not know of Wolffhügel's paper (1938) and her interpretation of some morphological details was incorrect (Willers & Olsen, 1969).

Aporodiorchis Yamaguti, 1959 should be considered a synonym of *Cladogynia* Baer, 1938, because of its eight skrjabinoid hooks correlated usually with three testes and a cirrus stylet. The 'long convoluted chitinous cirrus lies in the cirrus-sac, the duct is finely spinulose' in von Linstow's original description (1906) of '*Diorchis occlusa*' is actually an armed cirrus with a long cirrus stylet. Neither the description of only two testes per proglottid or the lack of genital pores have been confirmed and are presumably, erroneous. Von Linstow (1906) did not compare his 'new species' with *Halysis liguloides* Gervais, 1847 = *Cladogynia liguloides* (Gervais, 1847) n. comb. and, therefore, did not notice the identity of these forms. In transverse section, the third testis is not visible (Spasskii, 1963).

Arhynchotaenia Saakova, 1958, preoccupied by *Arhynchotaenia* Pagenstecher, 1877, was renamed *Arhynchotaeniella* by Schmidt (1986). The type-species, *A. clausovaginata* Saakova, 1948, hitherto found only once, is close to *Hymenolepis megalops* (Nitzsch in Creplin, 1829) Ransom, 1902. The sclerotized, distal clamp of the vagina with a muscular sphincter can be recognized as a specific, not a generic character. It should be considered a synonym of *Hymenolepis* Weinland, 1858.

Armadolepis Spasskii, 1954 is a synonym of *Rodentolepis* Spasskii, 1954. The presence of minute hooks on the suckers has not been confirmed (Vaucher & Quentin, 1975).

Australiolepis Spasskii & Spasskaya, 1954 should be considered a synonym of *Hymenolepis* Weinland, 1858. The type-species, *A. southwelli* Szpotanska, 1931, is insufficiently described, but correctly classified by Szpotanska (1931) within the genus as a tritesticular form with a rudimentary unarmed rostellum. The authors of *Australiolepis* gave no differential diagnosis.

Avocettolepis Spasskii & Kornyushin, 1971 should be considered a synonym of *Cladogynia* Baer, 1938 because of the presence of eight skrjabinoid rostellar hooks, three testes and a cirrus stylet. An accessory sac is lacking. The male canal covered with minute spines is a specific rather than a generic feature. We transfer the type-species to *Cladogynia* as *C. vaginata* (Baczynska, 1914) n. comb.

Bisaccanthes Spasskii & Spasskaya, 1954, with the type-species *B. bisaccata* (Fuhrmann, 1906), was considered a synonym of *Sobolevicanthus* by Czaplinski (1956). Recently, re-examination of Fuhrmann's original material has revealed a short cirrus stylet. Its existence is difficult to detect. Thus, eight skrjabinoid rostellar hooks, a retractable rostellum, three testes and the armed accessory sac place *Bisaccanthes* as a synonym of *Sobolevicanthus* Spasskii & Spasskaya, 1954. The relatively short cirrus-sac is a specific feature.

Capiuterilepis Oschmarin, 1962, with the type-species *C. naja* (Dujardin, 1845) should be considered a synonym of *Microsomacanthus* Lopez-Neyra, 1942, because of the ten arcuatoid hooks, retractile rostellum, three testes per proglottid and lack of both accessory sac and cirrus-stylet. The oncospheres join in packets called 'egg capsules' each containing two to fifteen oncospheres. The reticular uterus is replaced by these packets in gravid proglottids, which represents a good specific but not generic criterion.

Chelacanthus Yamaguti, 1959 with the type and only species *C. parviceps* (von Linstow, 1872) is insufficiently described. The features described by von Linstow under the name *Diorchis parviceps* (von Linstow, 1872) von Linstow, 1904 combine characters of the genera *Wardium* Mayhew, 1925, *Aploparaksis* Clerc, 1903 (ten aploparaksoid hooks) and *Diorchis* Clerc, 1903 (two testes per proglottid). This form has never been found again. On the other hand, it is doubtful whether the strobila really belongs to *Diorchis* because of the entirely poral position of two testes seen in cross section. This is not known in any representative of this genus. The third testis was presumably overlooked and thus *Chelacanthus* is probably a synonym of *Wardium*, but not of *Dicranotaenia* as Spasskii (1963) and Schmidt (1986) suggested. The latter author recognized *Chelacanthus*.

Chimaerolepis Spasskii & Spasskaya, 1972 with *C. watsoni* (Prestwood & Reid, 1966) as type-species is provisionally accepted as valid. However, the presence of the accessory sac, which looks more like a delicate armament on the bottom of the genital atrium, and the unusual position of the genital ducts between the ventral and dorsal poral osmoregulatory canals both need confirmation. If the accessory sac does not exist and if the genital ducts pass dorsally to the osmoregulatory canals, this genus would be a synonym of *Wardium* Mayhew, 1925.

Cloacotaeniella Schmidt, Bauerle & Wertheim, 1988 is considered a synonym of *Hymenolepis* Weinland, 1858. The type-species, *C. tringae*, on re-examination showed: a spiny accessory sac like the armed base of the genital atrium, not a real sac; the irregular sacciform shape of the uterus; the genital ducts passing dorsally, not between, the poral osmoregulatory canals; and the absence of a genital atrium is an erroneous observation.

Colymbilepis Spasskaya, 1966 with an insufficiently described type-species, *C. multistriata* (Rudolphi, 1810) Spasskaya, 1966, should be considered a synonym of *Confluaria* because of its small, wrinkled guard and handle of the hook. Other features such as the multilobed ovary and vitellarium and the disposition of the testes are regarded as specific characters.

Craspedocotyla Jordano & Diaz-Ungria, 1960 should be considered a synonym of *Microsomacanthus* because it has ten rostellar hooks (diorchoid rather than fraternoid or cricetoid as described, but not exactly figured in the type-species, *C. margaritensis*), a retractile rostellum, three testes and lacks an accessory sac and cirrus stylet. Fig. 5 in the authors' text does not show the accessory sac.

Cucurbilepis Sadovskaya, 1965 is considered as a synonym of *Ditestolepis* because of the serial growth of the proglottids and the rudimentary rostellum.

Debrosia Spasskii, 1987 should be considered a synonym of *Wardium* Mayhew, 1925. Spasskii's diagnosis is based on the incorrect interpretation of the figure presented in the first description of *D. kwangensis* (syn. *Oligorchis kwangensis* Southwell & Lake, 1939) where the muscular sphincter of the genital atrium was determined as 'well-developed sacculus accessorius'. In addition, Spasskii (1987c) determined it as an 'internal' accessory sac, which is an error. Re-examination of the type material by Baer (1959) revealed only three testes per proglottid.

The monotypic genus *Dildotaenia* Dronen, Schmidt, Allison & Mellen, 1988 should be considered synonymous with *Cladogynia* Baer, 1938, because of its eight skrjabinoid (not nitidoid) rostellar hooks, cirrus stylet and three testes per proglottid. The position of the genital ducts between the poral osmoregulatory canals, not figured by Dronen *et al.* (1988), needs confirmation.

Dimorphocanthus Maksimova, 1989 is a synonym of *Confluaria*.

Ditestolepis Soltys, 1952 is characterized by the development of its uterus, the frequent presence of a rudimentary rostellum and by the serial growth of the strobila. The formation of a syncapsule is not itself of generic significance but represents the final state of uterine evolution, giving better protection related to the circumstances of the life-cycle. A parallel evolution occurs in genera from insectivorous mammals, e.g. *Coronacanthus*, *Triodontolepis*.

Drepanidolepis Lopez-Neyra, 1942 was erected as a subgenus of *Drepanidotaenia* Railliet, 1892 with 14 species parasitizing Anseriformes, Lariformes and Otidiformes with *D. (D.) anatina* (Krabbe, 1869) as the first species (type-species not indicated). This subgenus was only accepted by Spasskii (1963) who recognized *Drepanidolepis* with *D. anatina* (Krabbe, 1869) Spasskii, 1963 as its type species. Re-examination of the type material of '*Taenia anatina*' Krabbe, 1869 from Copenhagen revealed that it consists of scoleces of *Diorchis* and separate strobilae of *Dicranotaenia*. Consequently, *Drepanidolepis* sensu Spasskii (1963) should be regarded as a *genus inquirendum*.

Echinatrium Spasskii & Yurpalova, 1965 is a junior synonym of *Microsomacanthus*. The only character distinguishing *Echinatrium* is a deep genital atrium lined with spines but in both species of the genus, *E. skrjabini* and *E. filosomum*, the spiny base of the genital atrium looks like an armed base of the cirrus. It is not distinct morphologically.

Echinoproboscilepis Sadovskaya, 1965 is considered a *genus inquirendum*, because of its poorly documented description.

Ecrinolepis Spasskii & Karpenko, 1983 has a serial growth of the proglottids and a rudimentary rostellum. It is considered herein as a synonym of *Ditestolepis* Soltys, 1952.

Fuhrmanniella Shen, 1932 is considered a synonym of *Echinocotyle* Blanchard, 1891. The generic diagnosis repeats characters of the subfamily Hymenolepidinae emphasizing the disposition of the testes (two marginal anterior and one median posterior) but not mentioning the presence of the accessory sac. The type-species should be '*Fuhrmanniella uralensis* (Clerc, 1902)' with armed suckers and an external accessory sac, both features characteristic of *Echinocotyle*.

Geraldolepis n. g. is erected following re-examination of *Hymenolepis (Hymenolepis) deblocki* Schmidt & Neiland, 1968 which revealed a scarcely visible cirrus stylet and the doubtful existence of an unarmed single accessory sac, which resembles a muscular atrial sphincter surrounding the base of the genital atrium. The genital pores are dextral, not sinistral, as originally reported. Schmidt (1986) enlarged the diagnosis of *Retinometra* (syn. of *Cladogynia*) to include this species, but we cannot agree because representatives of *Cladogynia* have only eight skrjabinoid hooks. *Geraldolepis* is named in honour of Dr Gerald D. Schmidt for his enormous contribution to cestodology.

Globus Scopoli, 1772 is a *genus inquirendum* as it is insufficiently described and out of use; it may refer to *Echinolepis*.

Gonoscolex Saakova, 1958 is considered a synonym of *Echinocotyle*. It was considered a subgenus of *Echinocotyle* by Kornyushin (1983) because of the armament of the male genital pore which seems to be associated with the base of the cirrus and the cuticular lips closing the female pore. Both characters are of value at the specific level in our opinion.

Gopalaia Dixit, Capoor & Rengaragu, 1980 is a poorly known genus, recorded from bats in India, and is not included in this key. It is treated as a *genus inquirendum*.

Halysis Zeder, 1803 is considered a *genus inquirendum* as it was insufficiently described. Species of this genus have been referred to *Aploparaksis* (see Fuhrmann, 1908a), *Taenia* (see Yamaguti, 1959), *Dipylidium* (see Schmidt, 1986) or *Flamingolepis*.

Hunkelepis n. g. is a medium sized hymenolepidid, named in honour of Dr Pierre Hunkeler. The diagnostic characters are given in the key.

Hybridolepis Spasskii, 1959 was synonymized with *Wardium* by Mariaux (1989) when he re-examined the holotype of *Hymenolepis hughesi* Webster, 1947 and did not confirm the presence of the cirrus stylet or the accessory sac.

Hymenocoelia Capoor & Srivastava, 1966 should be considered a synonym of *Cladogynia* because of its eight skrjabinoid hooks, three testes, cirrus stylet and a vagina crossing the cirrus-sac ventrally. The authors described the cirrus as 'a penis sheath' and the stylet as a 'penis' but their figures are correct.

Hymenofimbria Skryabin, 1914 is regarded as *genus inquirendum*. It is monotypic, based on fragments found once by Skryabin (1914). The scolex belongs probably to *Wardium* or *Aploparaksis* (Figs 69 & 70 in Skryabin's paper); fragments of the strobila presented in Figs 72, 73 & 74 are reminiscent of *Dicranotaenia* and Fig. 75 is a longitudinal section of *Fimbriarioides*.

Hymenosphenacanthoides Yamaguti, 1959 is regarded as a synonym of *Cladogynia*. The insufficiently-described type-species, *H. ondatrae* (Rider & Macy, 1947), was correctly recognized by Spasskii (1963) as a junior synonym of *Weinlandia cyrtoides* Mayhew, 1925. Eight skryabinoid hooks, three testes, a cirrus stylet and no accessory sac justifies classifying this species as *Cladogynia cyrtoides* (Mayhew, 1925) n. comb., a parasite of Anseriformes, exceptionally in *Ondatra zibethica*. The existence of ten as well as eight rostellar hooks reported by Rider & Macy (1947) seems doubtful and needs confirmation.

Hymenosphenacanthus Lopez-Neyra, 1958 was a new name given for *Sphenacanthus* Lopez-Neyra, 1942 preoccupied by *Sphenacanthus* Agassiz, 1837, but replaced earlier by *Retinometra* Spasskii, 1955, a synonym of *Cladogynia*. The possibility of ten as well as eight rostellar hooks has not been confirmed.

'*Hymenolepis*' *blarinae* Rausch & Kuns, 1950, cannot be classified in the present key or allocated to a known genus. The original material is now in poor condition and a restudy of well preserved specimens is needed. This species most resembles *Lineolepis*, but differs in the anatomy of the mature proglottids and the shape of the rostellar hooks.

'*Hymenolepis*' *uncinispinosa* (Joyeux & Baer, 1930), '*Hymenolepis*' *petrodromi* (Baer, 1933) and '*Hymenolepis*' *globirostris* (Baer, 1925) all have testes more or less in a triangle in the centre of the proglottid, and few (9–14) hooks with guard and handle aligned in a straight line and a projecting curved blade. Only *S. uncinispinosa* is well known, see Hunkeler (1974), but the others are represented by poorly preserved original specimens and should be redescribed. They may belong to a new genus, which cannot be erected here without a revision of *globirostris* and *petrodromi*.

Insectivorolepis Zarnowsky, 1956 is considered herein as a synonym of *Soricinia*.

Jardugia Southwell & Hilmy, 1929 is provisionally considered a valid monotypic genus. The description of the type-species *J. paradoxa* is based on a single specimen which may be a teratological form of *Diploposthe bifaria* (Siebold in Creplin, 1946).

Jonesius Yamaguti, 1959, a monotypic genus, is based on the insufficiently-described type-species *J. ralli* (Jones, 1944), placed by Jones (1944) in *Diorchis*, which seems to be correct. Yamaguti (1959) provided no differential diagnosis, but emphasized the unarmed suckers, lack of the external seminal vesicle and the presence of numerous bundles of internal longitudinal muscles. Unarmed suckers are known in some diorchids. The external seminal vesicle was not described or figured by Jones (1944) but is not necessarily absent. The longitudinal muscles were not described, but were drawn as 18 in number. This single character cannot justify a new genus.

Joyeuxilepis Spasskii, 1947 was based on a cysticercoid, *J. biuncinata* (Joyeux & Baer, 1943) (syn. *Echinorhynchotaenia biuncinata* Joyeux & Baer, 1943 *nec* Joyeux & Gaud, 1945), placed first in the Hymenolepididae, and subsequently transferred by Spasskii & Spasskaya (1976) to Amabiliidae as *Tatria* (*Joyeuxilepis*) *decacantha* Fuhrmann, 1913, syn. *Echinorhynchotaenia biuncinata* Joyeux & Baer, 1943. [See also Chapter 23.]

Kowalewskius Yamaguti, 1959 is considered a synonym of *Microsomacanthus*. The small number of proglottids (35–40) and unconfirmed presence or absence of the internal accessory sac are not acceptable as generic characters.

Lallum Johri, 1960 with *L. magniparuterinum* as type-species is close to *Hymenolepis* especially to *H. megalops*, differing from it by the number of testes: five to seven in *L. magniparuterium* and three in *H. megalops*. The classification of the type-species within the Paruterinidae is not justified because it has features characteristic of Hymenolepididae, e.g. external seminal vesicle, few testes and unilateral genital pores. The structure determined as a paruterine organ by Johri

(1960a) seems to be a misinterpretation. We transfer *Lallum* to *Neoligorchis* Johri, 1960 as its junior synonym because of its unarmed scolex and few testes.

Linstowius Yamaguti, 1959 is based on the inadequately described *Diorchis serpentatus*, which has never been found again. Von Linstow's (1905) text does not agree with his Fig. 23 (NB the position of the vagina and vitellarium). Two of the 'testes' in this figure are the two antiporal lobes of the ovary. Fuhrmann (1908a) re-examined Linstow's material and found one testis. The ten rostellar hooks and their shape, the single testis and numerous internal longitudinal muscle bundles justify considering *Linstowius* as a junior synonym of *Aploparaksis*.

Lobatolepis Yamaguti, 1959 should be considered a junior synonym of *Wardium* because it has ten aploparaksoid rostellar hooks and three testes per proglottid. The variation in the number of hooks (8–11) originally reported by Mayhew (1925) for the type-species, *L. lobulata*, and used by Yamaguti (1959) as a generic criterion, has not been confirmed by Rysavy & Macko (1971). A knob-like enlargement of the tip of the rostellum with deep marginal lobes cannot be recognized as a character at the generic level.

Lockerrauschia Yamaguti, 1959 is monotypic. It has few affinities with other hymenolepidids with ten rostellar hooks. The genus is considered valid but should be redescribed from additional specimens.

Longirostrum Malhotra & Nanda, 1986, described from shrews in India, is insufficiently known and not included in this key. The morphology of the rostellum (narrow elongated tube), the presence of only eight hooks and the absence of the external seminal vesicle need confirmation. It is treated as a *genus inquirendum*.

Mathevolepis Spasskii, 1948, characterized by its hyperapolytic development, was originally described as lacking a rostellum, but we have found that it has a rudimentary rostellum. Spasskii (1948b) described a duct, possibly derived from the osmoregulatory system, for the release of eggs from gravid segments but this has not been confirmed.

When erecting *Matiaraensis* Dixit & Capoor, 1986, Dixit & Capoor (1986) did not mention whether all four specimens on which the description was based had rostellar hooks. Their number was described as variable (eight to ten) which is unlikely and needs confirmation. If the existence of the four accessory suckers is not a teratological character, then *Matiaraensis* would be a valid genus, close to *Wardium*. It is provisionally accepted here but is considered of doubtful validity.

Mayhewia Yamaguti, 1956 was based on at least three specimens of *M. corvi* (Mayhew, 1925) (syn. *Weinlandia corvi* Mayhew, 1925) which allegedly had eight or ten rostellar hooks. This was not confirmed by Spasskaya (1966) who found ten rostellar hooks in this species and transferred it to *Passerilepis* as *P. stylosa* (Rudolphi, 1819) with *W. corvi* as a junior synonym. Of 16 other species classified by Yamaguti (1959) in *Mayhewia*, eleven have a single crown of ten diorchoid or arcuatoid hooks, one species has eight hooks and four have 12 to 20 hooks. Variations of this nature are not acceptable in one genus. We consider *Mayhewia* to be a junior synonym of *Microsomacanthus* for species possessing ten rostellar hooks.

Spasskii (1963) transferred to Dilepididae all species of *Meggittiella* Lopez-Neyra, 1942 except *M. ficticia* (Meggitt, 1927), which remained in Hymenolepididae as *Armadoskrjabinia ficticia* (Meggitt, 1927), because other representatives of *Meggittiella* do not have external seminal vesicles.

Monorchis Clerc, 1902, preoccupied by *Monorchis* Monticelli, 1893, a trematode, is a synonym of *Aploparaksis* Clerc, 1903.

Monosaccanthes Czaplinski, 1969 is a synonym of *Anatinella*.

Myxolepis Spasskii, 1959 should be considered a ***genus inquirendum***. It is probably based on a scolex of *Microsomacanthus* or other hymenolepidid genus with ten diorchoid hooks and the strobila of *Sobolevicanthus*. Nobody has yet found an entire specimen with the characters of *Myxolepis collaris*.

Nematoparataenia Maplestone & Southwell, 1922 is a valid genus often removed from the Hymenolepididae, but which should be classified within this family because of the genital pores opening in a lateral groove, the presence of external and internal seminal vesicles and of aploparaksoid hooks. Development in the intermediate host also indicates it belongs in the Hymenolepididae. Such characters as a cylindrical body and the lack of external segmentation as in *Gastrotaenia* may result from convergence.

Neodiorchis Bilqees & Fatima, 1984 is considered valid, with a little-known type-species, *N. longicirrus*, based on four complete specimens. It differs from *Diorchis* by the shape of the ten hooks which are echinorhynchoid rather than diorchoid and by the absence of an external seminal vesicle. It is not clear if this important hymenolepidid feature was just not observed or whether it is a rare exception, characteristic of *Neodiorchis*. Another question is whether elongated gravid proglottids are usual.

Neolallum Srivastava, Pandey & Tayal, 1983 should be considered a synonym of *Neoligorchis* because of its lack of rostellum and the presence of five to eight testes per proglottid.

Neoligorchis Johri, 1960 is valid. However, the description of the type-species *N. alternatus*, based on three complete worms and a few strobilae, is insufficient and its morphology needs more detailed study, especially the unclear relationships of the genital ducts and poral longitudinal osmoregulatory canals. The text does not agree with the figure. It is very close to *Hymenolepis* but has five to six instead of three testes.

Ortleppolepis Spasskii, 1965 should be considered valid. The description of *O. multiuncinata* was based on one complete specimen, six specimens without scoleces and many fragments of this very delicate cestode. Four or five circles of rostellar hooks determined by the author on a single scolex bear witness to the inaccuracy inherent in this observation.

Oschmarinolepis Spasskii & Spasskaya, 1954 should be considered a synonym of *Microsomacanthus* because of its ten diorchoid hooks and three testes per proglottid. The position of two testes lateral to the osmoregulatory canals cannot be considered a generic character.

Otidilepis Yamaguti, 1959 was originally proposed as a subgenus of *Hispaniolepis* with *H. (O.) tetracis* (Cholodkowsky, 1906) as type-species. Neither Yamaguti (1959) nor Spasskaya (1966), who raised it to the rank of genus, justified their ideas by giving a differential diagnosis. *Otidilepis* is here considered a junior synonym of *Hispaniolepis*.

Parabiglandatrium Gvozdev & Maksimova, 1968 was erected (1968) for *P. phoenicopteri* which was later synonymized by the same authors (1985b) with *P.*

liguloides (Gervais, 1847), because of two large glands situated near the genital atrium. We consider these structures to be a double accessory sac and recognize *Parabiglandatrium* as a junior synonym of *Sobolevicanthus*.

Pararetinometra Stock & Holmes, 1982 has a type-species which does not differ from other species belonging to *Cladogynia* and should be considered as its junior synonym. Eight skryabinoid hooks, three testes per proglottid, a cirrus stylet and lack of accessory sac justify this. The armament of the posterior part of some proglottids emphasized in the diagnosis of *Pararetinometra* also occurs in other species of Hymenolepididae.

Parvibursa Yamaguti, 1959, a monotypic genus with *P. chrysolampidis* (Clapham, 1936) as type-species, should be regarded as a synonym of *Wardium* because it has ten aploparaksoid hooks and three testes per proglottid.

Passerilepis Spasskii & Spasskaya, 1954 should be considered a synonym of *Microsomacanthus* because it has ten diorchoid or arcuatoid hooks, three testes per proglottid, no accessory sac and no cirrus stylet.

Phoenicolepis Jones & Khalil, 1980, a monotypical genus with *P. nakurensis* as type-species, does not differ essentially from species belonging to *Sobolevicanthus* and should be regarded as its younger synonym. Decisive features are eight skrjabinoid hooks, three testes per proglottid, an accessory sac and numerous longitudinal muscle bundles.

Pseudobotrialepis Schaldybin, 1957, which has a huge scolex of peculiar morphology and enlarged proglottids with testes widely separated in two groups by the female organs, is considered valid. Karpenko & Gulyaev's (1990) amendment of the original spelling is unjustified according to the *International Code of Zoological Nomenclature*.

Pseudoparadilepis Brendow, 1969, is a synonym of *Skrjabinacanthus* (see Vaucher, 1971).

Rodentolepis Spasskii, 1954 differs from *Vampirolepis* mainly by the anatomy of the mature proglottid with three testes in an elongated triangle, separated into two groups by the female gonads. This resembles *Hymenolepis*. The rostellum is retractile. The genus contains many species parasitizing mainly rodents and primates. Several bat hymenolepidids are also to be placed in *Rodentolepis*. We include in this genus *R. pearsei* (Joyeux & Baer, 1930) n. comb. because the female gonads separate the testes into two groups.

Satyolepis Spasskii, 1965 is a monotypic genus with *S. lamellaris* (Yamaguti, 1956) as type-species, whose description was based on two gravid specimens. About 20 arcuatoid hooks, three testes per proglottid and inner longitudinal muscle bundles justify considering *Satyolepis* a junior synonym of *Hispaniolepis*, in spite of lacking elongate antiporal appendages of the proglottids. The latter feature may also appear in *Microsomacanthus*, e.g. *M. arcuatus*.

Schillerius Yamaguti, 1959 has *S. longiovum* (Schiller, 1953) as type-species. Schiller (1953) erroneously stated that the genital ducts pass between the dorsal and ventral poral osmoregulatory canals. This feature, exceptional in hymenolepidids, was used by Yamaguti (1959) to erect *Schillerius*, which is synonymous with *Diorchis* (see Czaplinski, 1988).

Schmelzia Yamaguti, 1959 which has *S. linderi* (Schmelz, 1941) as type-species is a synonym of *Hymenolepis* because of its rudimentary, unarmed rostellum and three testes per proglottid. It has no accessory sac or cirrus stylet.

Sinuterilepis Sadovskaya, 1965 is a synonym of *Ditestolepis* because of the serial growth of the proglottids and the presence of a rudimentary rostellum.

Skorikowia von Linstow, 1905 is a monotypic genus with *S. clausa* as type-species. It has been found once only and was recognized as a synonym of *Aploparaksis brachyphallos* (Krabbe, 1869) by Fuhrmann (1908a) who re-examined von Linstow's original material and found only one testis.

Skrjabinacanthus Spasskii & Morosov, 1959 probably has only one species parasitic in the Palaearctic (see Vaucher, 1971). It has an invaginable rostellum in which the hooks are not inserted regularly but at different levels. Furthermore, the hooks have variable dimensions. Another main feature is the testes in a triangle superimposed on the female gonads. The cirrus-sac is big and contains an armed cirrus. *Hymenolepis falculata* Rausch & Kuns, 1950 is transferred to *Skrjabinacanthus* as a new combination after re-examination of the original material.

Skrjabinoparaksis Krotow, 1949 is a valid monotypical genus close to *Diorchis*, in which the two testes are reduced to one. It also has a special kind of spines on the suckers and the male and female genitalia develop simultaneously.

Sternolepis Dixit & Capoor, 1988 is provisionally considered a valid genus, close to *Armadoskrjabinia*. The topography of the genital ducts and the poral longitudinal osmoregulatory canals was not described and needs further study. The dorsal osmoregulatory canal was not figured and the ventral one is presented by Dixit & Capoor (1988) differently in two figures.

Stylolepis Yamaguti, 1959, with *S. longistylosa* (Shen, 1932) as type-species, is based on fragments. It is synonymized with *Cladogynia* because the strobila has three testes per proglottid and a cirrus stylet. The scolex is similar to that of *Microsomacanthus*.

Vampirolepis Spasskii, 1954 is restricted (Vaucher, 1992) to species parasitic in bats. It is closely related to *Rodentolepis* and shares with it a retractile rostellum. Both genera contain many species with fraternoid hooks. *V. bihamata* Sawada & Harada, 1986, has two crowns of rostellar hooks.

Vogelepis n. g. is erected for *Hymenolepis virilis* Voge, 1950 and named in honour of Dr Marietta Voge. The differential characters are given in the key.

Weinlandia Mayhew, 1925 is a synonym of *Dicranotaenia*.

Woodlandia Yamaguti, 1959 with *W. phalacrocorax* as type-species is here considered a synonym of *Hymenolepis* because it has no rostellum, no hooks and three testes per proglottid. An unusual position of the testes, i.e. extramedullary, is a specific character but not of value at the generic level (see Czaplinski, 1990).

Family Hymenolepididae Ariola, 1899

Diagnosis: Rostellum usually present, sometimes lacking or rudimentary, armed with a single or, exceptionally, double crown of hooks (e.g. Echinorhynchotaeniinae). Suckers four; armed or not (except *Matiaraensis* with four accessory suckers all armed). Pseudoscolex sometimes present behind scolex. Proglottids broader than long, craspedote, external segmentation rarely lacking (some

Fimbriariinae, *Gastrotaenia, Parafimbriaria, Nematoparataenia*). Single, rarely double, set of reproductive organs per proglottid. One to three testes per proglottid, sometimes more, exceptionally 32 (*Polytestilepis*), most often three; in some mainly tritesticulate species the number of testes may vary from one to seven (e.g. *Hymenolepis diminuta*). External and internal seminal vesicles present. Genital pores unilateral, dextral, rarely alternating, e.g. old specimens of *H. diminuta.* Uterus usually sacciform, sometimes reticular.* Longitudinal osmoregulatory canals mostly four, but six to 11 in Fimbriariinae. Longitudinal muscle bundles mostly of one or two layers. If two layers, inner muscle bundles numerous or eight. In birds and mammals. Type-genus *Hymenolepis* Weinland, 1858.

Key to Hymenolepididae of birds (by B. Czaplinski)

Key to subfamilies

1a. Longitudinal osmoregulatory canals four .... 2.

1b. Longitudinal osmoregulatory canals 6–11. Scolex armed with single crown of ten rostellar hooks .... Fimbriariinae Wolffhügel, 1899.

Diagnosis: Genital ducts open dextrally, pass dorsally to osmoregulatory canals. Rostellum invaginable. Hooks diorchoid with blades directed mostly anteriorly when withdrawn (in *Fimbriariella* all three rami of hook are almost equal, guard bifid). Pseudoscolex well developed with slight external segmentation. Internal segmentation obscure or conspicuous. Single set of gonads with three testes per proglottid. Ovary and uterus mainly reticulate, often nonmetameric. Vitellarium compact or lobed, metameric. Oncospheres leave the strobila singly or connect with each other to form chain or packet. In birds.

2a. Genital ducts between osmoregulatory canals. Rostellum armed with double crown of hooks, 12–24. Three testes per proglottid .... Echinorhynchotaeniinae Mola, 1929.

Diagnosis: Rostellum invaginable; hooks echinorhynchoid or diorchoid, with blades directed anteriorly when withdrawn. (Rostellum of *Sternolepis* retractile.) External and internal segmentation evident. Strobila craspedote. Single set of gonads. Ovary lobate. Uterus sacciform. In birds.

2b. Genital ducts dorsal to osmoregulatory canals. Rostellum when present armed with single, exceptionally double, crown of hooks .... 3.

3a. Genital pores unilateral, mainly dextral (except *Allohymenolepis* and older specimens of *Hymenolepis diminuta*) (see Stradowski, 1993) .... Hymenolepidinae Perrier, 1897.

Diagnosis: Rostellum usually present, may be absent or rudimentary (e.g. *Hymenolepis, Neoligorchis*); when present, armed with single crown of eight, ten

* A detailed redescription of the uterus (SEM and TEM) has been given recently by Conn (1993) in *H. diminuta*, type-species of *Hymenolepis*. This author demonstrates that the labyrinthine aspect observed in light microscopy is a result of epithelial and epitheliomesenchymal folds and villi which compartmentalize the uterine lumen.

or more rostellar hooks and mostly retractile (i.e. blades directed posteriorly when rostellum withdrawn). Rostellum may be invaginable (i.e. blades directed anteriorly when withdrawn, connected mostly with aploparaksoid or similar hooks). Unfolded pseudoscolex may occur as simple enlargement of proglottids behind scolex. Proglottids craspedote or external segmentation lacking. Genital pores usually dextral, may alternate irregularly. Genital ducts usually dorsal to osmoregulatory canals. Single set of male and female genitalia per proglottid. Testes one, two or three per proglottid, exceptionally more. Ovary compact or lobate. Vitellarium compact or slightly lobate, postovarian. Uterus sacciform or reticulate. Oncospheres single with outer envelopes or connected in chains or packets. In birds and mammals.

3b. Genital pores bilateral .... Diploposthinae Poche, 1926.
Diagnosis: Rostellum retractile; armed with single crown of 10–12 arcuatoid or diorchoid hooks; blades directed posteriorly when withdrawn. Genital ducts double, dorsal to osmoregulatory canals. Single or double set of male and female gonads per proglottid. Testes few, mainly two to six, exceptionally more than 20. External and internal seminal vesicles present, double independently of single or double set of testes per segment. Cirrus-sac cylindrical, short, hardly reaching or overlapping poral excretory canals. Cirrus spinose. Ovary lobate, single or double per proglottid. Vitellarium compact or lobate, postovarian. Vagina and seminal receptacle double. Uterus single, sacciform. In birds and mammals.

Key to genera of Fimbriariinae

1a. No external segmentation of strobila. Pseudoscolex well-developed, slightly segmented .... 2.
1b. External and internal segmentation evident. Pseudoscolex poorly developed or unknown. Eight longitudinal osmoregulatory canals .... 3.

2a. Internal segmentation obscure. Six osmoregulatory canals. Pseudoscolex without genital primordia .... *Fimbriaria* Froelich, 1802. (Fig. 29.1)
(Syns *Epision* Linton, 1892; *Notobothrium* Linstow, 1905?; *Rhynchotaenia* Diesing, 1850.)
Diagnosis: Scolex small, easily lost. Rostellum invaginable with single crown of ten diorchoid hooks. Suckers unarmed. Pseudoscolex hammer-like, very well developed; strongly folded in adults. External metamerism almost absent. Only one layer of longitudinal muscle bundles. Testes ovoid, three per cirrus-sac. External seminal vesicles usually far from cirrus-sac. Cirrus-sacs small, sometimes crowded. Ovary and uterus reticulate, nonmetameric (i.e. continuous between segments). Vitellarium compact or slightly lobed, metameric. Oncospheres leave strobila singly or joined by outer envelope to form chain or packets. In Anseriformes, accidentally Galliformes. Cosmopolitan. Type-species *F. fasciolaris* (Pallas, 1781) Froelich, 1802.

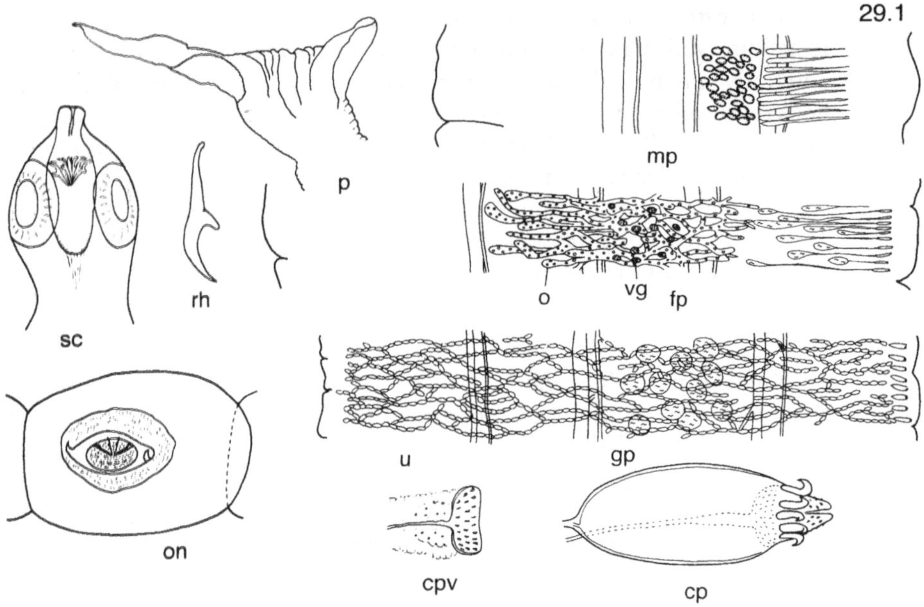

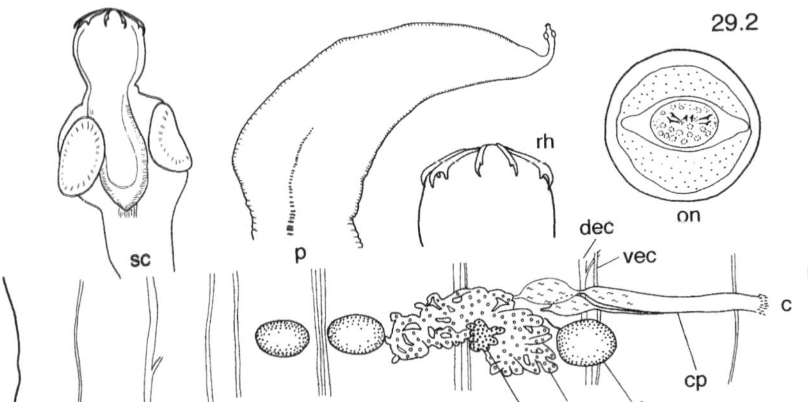

Fig. 29.1 *Fimbriaria fasciolaris* (Pallas, 1781) Froelich, 1802. (Original.)

Abbreviations

as accessory sac; asu accessory sucker; c cirrus; cp cirrus-sac; cpv copulatory part of vagina; cs cirrus stylet; csp cirrus spines; cy cysticercoid; dec dorsal excretory (osmoregulatory) canal; eas external accessory sac; esv external seminal vesicle; fp female proglottid; ga genital atrium; gp gravid proglottid; gv general view; hp hermaphroditic proglottid; ias internal accessory sac; imb internal muscle bundles; ipv intermediate part of vagina; isv internal seminal vesicle; ls longitudinal section; mp male proglottid; o ovary; on oncosphere (with envelopes); p pseudoscolex; r rostellum; rh rostellar hook; rp rostellar pouch; rs rostellar spines; s (and suc) sucker; sc scolex; sr seminal receptacle; ss sucker spine; t testis; ts transverse section; u uterus; v vagina; vec ventral excretory (osmoregulatory) canal; vg vitelline gland.

Fig. 29.2 *Fimbriarioides intermedia* (Fuhrmann, 1913) Fuhrmann, 1932. (Original.) (See Fig. 29.1 for abbreviations.)

2b. Internal segmentation conspicuous. Eight to 11 longitudinal osmoregulatory canals. Pseudoscolex containing genital primordia .... *Fimbriarioides* Fuhrmann, 1932. (Fig. 29.2)

Diagnosis: Scolex small, easily lost. Rostellum invaginable with single crown of ten diorchoid hooks. Suckers unarmed. Pseudoscolex weakly folded or flat. External segmentation lacking. One layer of numerous longitudinal muscle bundles, or outer layer rudimentary. Three smooth or lobate testes per segment. Ovary reticulate, metameric. Vitellarium lobate, metameric. Uterus reticulate, non-metameric. Oncospheres with spherical outer envelope, leave uterus singly. In Anseriformes and Charadriiformes. Europe, Asia, North America. Type-species *F. intermedia* (Fuhrmann, 1913) Fuhrmann, 1932.

3a. Pseudoscolex with genital primordia. Ten rostellar hooks. Three oval or lobate testes per proglottid .... *Fimbriariella* Wolffhügel, 1936. (Fig. 29.3)

Diagnosis: Scolex small. Suckers unarmed. Single crown of rostellar hooks with equal handle and blade, guard slightly shorter, bifid. Single layer of longitudinal muscle bundles. Ovary and vitellarium slender, tubular. Uterus originating as tubular sac forming irregular coils. In Anseriformes. North America. Type-species *F. falciformis* (Linton, 1927) Wolffhügel, 1936.

3b. Scolex and pseudoscolex unknown. Three rounded testes .... *Profimbriaria* Wolffhügel, 1936. (Fig. 29.4)

Diagnosis: Scolex and pseudoscolex unknown. Inner longitudinal muscle bundles numerous, strongly developed, outer ones rudimentary. Internal seminal vesicle absent (?). Ovary tubular. Vitellarium tubular, dorsal to ovary. Uterus unknown. In Charadriiformes. Ukraine. Type-species *P. multicanalis* (Baczynska, 1914) Wolffhügel, 1936.

Key to genera of Echinorhynchotaeniinae

1a. Rostellar hooks echinorhynchoid (handle and blade almost equal, guard a bit shorter) .... 2.

1b. Rostellar hooks diorchoid, about 12–14 .... *Sternolepis* Dixit & Capoor, 1988. (Fig. 29.5)

Diagnosis: Rostellar retractile; with double crown, hooks different in size in each crown. Suckers unarmed. Three subspherical testes in triangle in dorsal medulla, one poral and two antiporal. Ovary and vitellarium multilobate. Vitellarium postovarian. Uterus sacciform. Oncospheres oval, single (not connected). In Lariformes. India. Type-species *S. contri* Dixit & Capoor, 1988.

2a. Rostellum long, armed with numerous rosethorn-like spines below double crown of hooks. Cirrus-sac short, does not reach midline .... *Echinorhynchotaenia* Fuhrmann, 1909. (Fig. 29.6)

Diagnosis: Rostellum invaginable, double crown of echinorhynchoid hooks at apex, about 12 in each crown. Suckers unarmed. Proglottids craspedote, broader

29.3

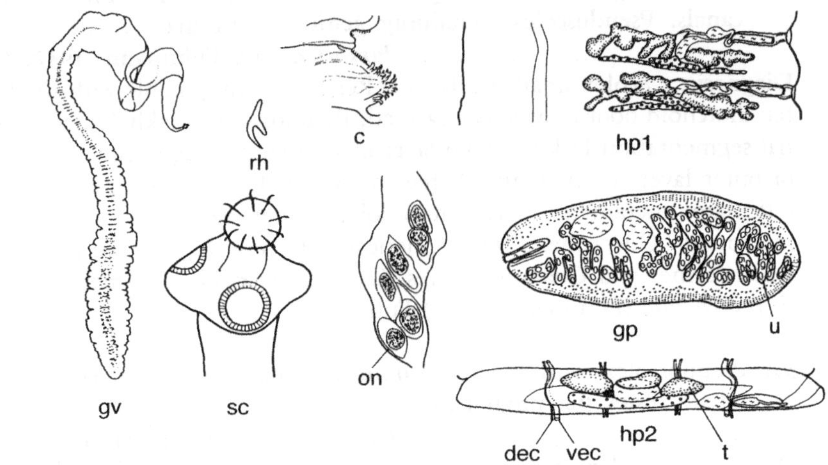

29.4

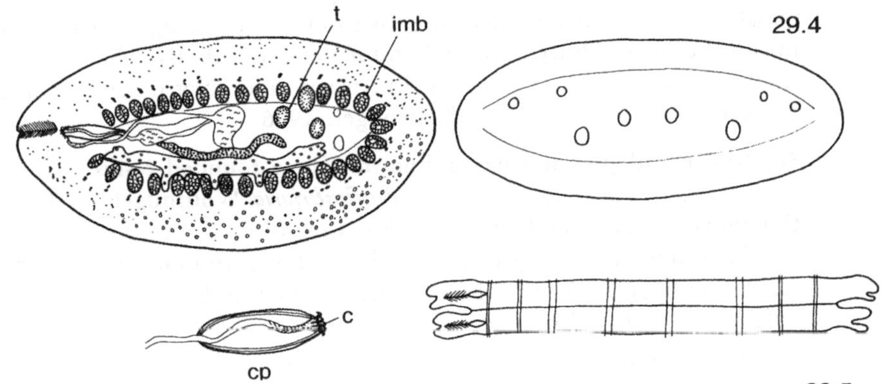

29.5

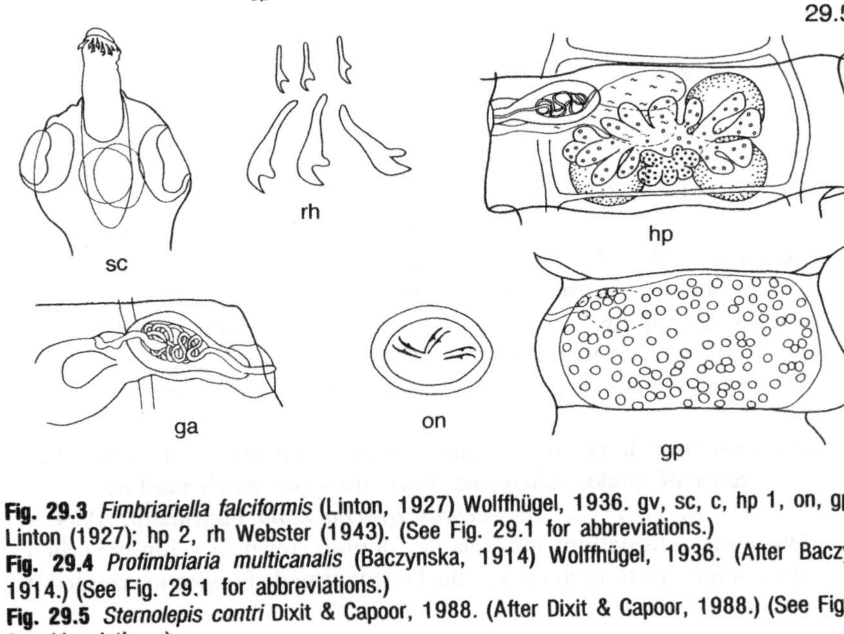

Fig. 29.3 *Fimbriariella falciformis* (Linton, 1927) Wolffhügel, 1936. gv, sc, c, hp 1, on, gp after Linton (1927); hp 2, rh Webster (1943). (See Fig. 29.1 for abbreviations.)

Fig. 29.4 *Profimbriaria multicanalis* (Baczynska, 1914) Wolffhügel, 1936. (After Baczynska, 1914.) (See Fig. 29.1 for abbreviations.)

Fig. 29.5 *Sternolepis contri* Dixit & Capoor, 1988. (After Dixit & Capoor, 1988.) (See Fig. 29.1 for abbreviations.)

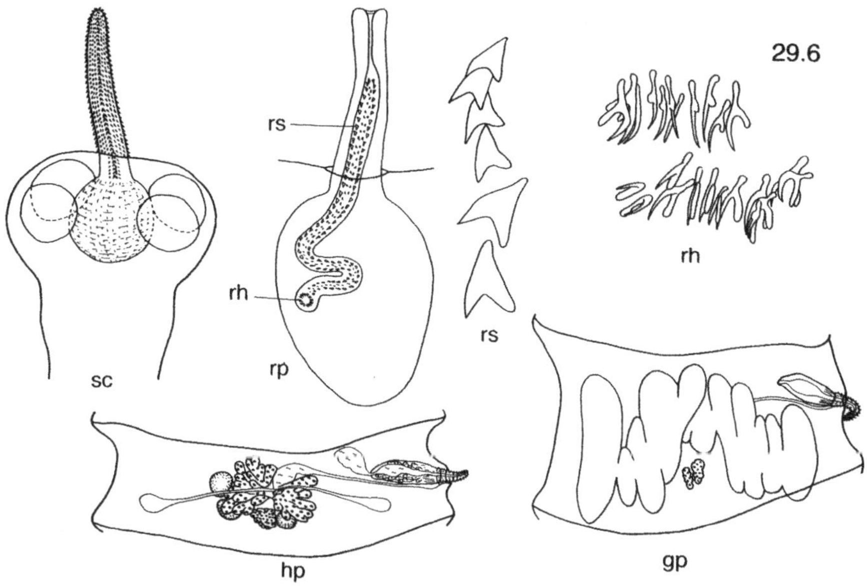

Fig. 29.6 *Echinorhynchotaenia tritesticulata* Fuhrmann, 1909. gp, hp, sc after Siddiqi (1962); rh, rp, rs after Mahon (1954). (See Fig. 29.1 for abbreviations.)

than long but last proglottid may be as long or longer than wide. Genital ducts between osmoregulatory canals. Testes three, rounded, posterior and lateral to ovary. Ovary multilobate, median. Vitellarium slightly lobate, postovarian, median. Seminal receptacle well developed. Young uterus tubular, transverse; later irregularly sacciform with wide diverticles. Oncospheres oval, numerous. In Pelecaniformes, Anseriformes. Africa, Australia, India. Type-species *E. tritesticulata* Fuhrmann, 1909.

2b. Rostellum short, unarmed below double crown of hooks. Cirrus-sac long, often crosses midline ....
.... *Armadoskrjabinia* Spasskii & Spasskaya, 1954. (Fig. 29.7)
Diagnosis: Rostellum invaginable, with double crown of 14–24 hooks of almost equal blade and handle, guard prominent but shorter than blade. Hooks different in size and shape in each crown. Suckers unarmed. Proglottids craspedote. Inner longitudinal muscle bundles numerous. Genital ducts between osmoregulatory canals. Testes three, arranged in triangle, one poral, two antiporal. Ovary multilobate. Vitellarium compact or slightly lobed, median, postovarian. Seminal receptacle well developed, median. Uterus tubular, later forming bilobed sac occupying whole proglottid. Oncospheres oval, numerous. In Pelecaniformes. Europe, Asia, Africa, Australia. Type-species *A. medici* (Stossich, 1890) Spasskii & Spasskaya, 1954.

Key to genera of Hymenolepidinae

1a. Mainly three testes per proglottid, rarely more .... **7.**

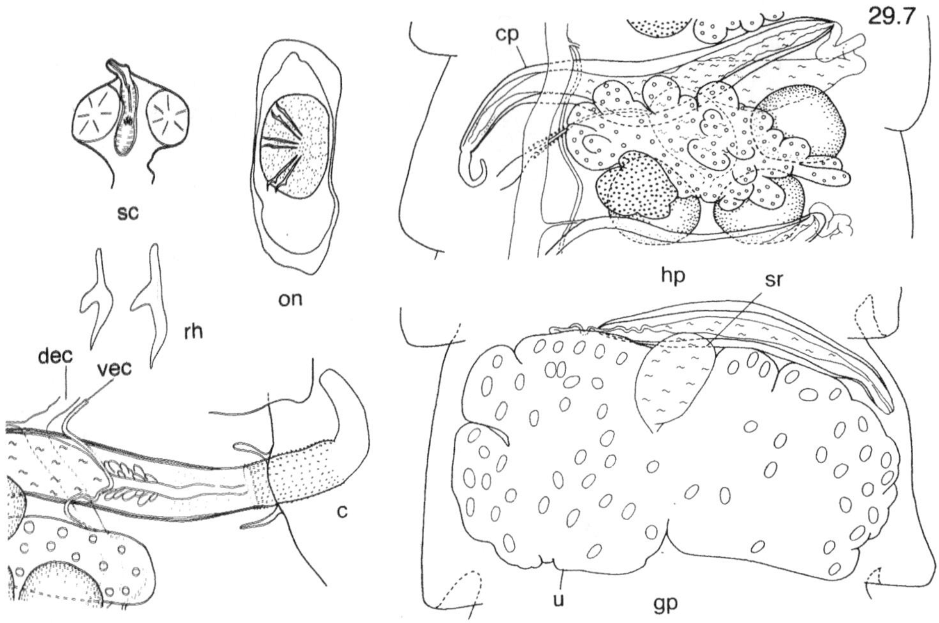

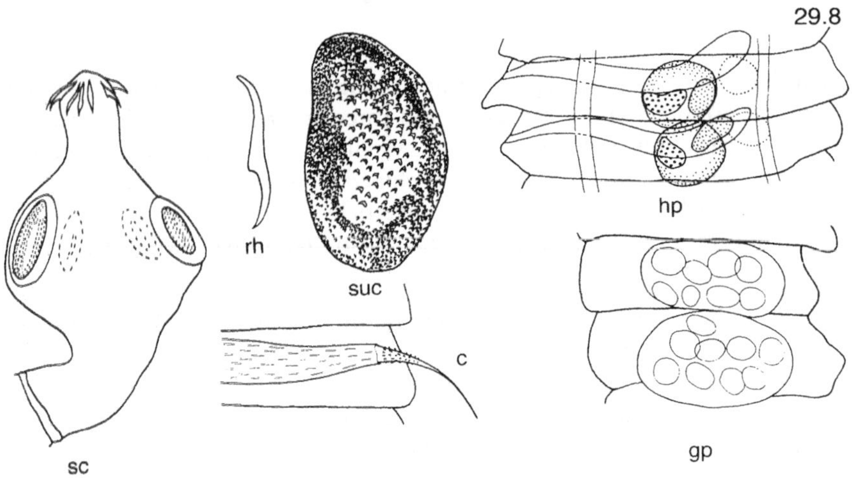

Fig. 29.7 *Armadoskrjabinia murrayensis* (Johnston & Clark, 1948) Spasskii, 1963. (Original; sc after Johnston & Clark, 1948.) (See Fig. 29.1 for abbreviations.)
Fig. 29.8 *Skrjabinoparaksis tatianae* Krotov, 1949 (gp, hp, sc after Krotov, 1949; c, rh after Maksimova, 1963; suc after Dubinina, 1953). (See Fig. 29.1 for abbreviations.)

1b. **Fewer than three testes per proglottid. Rostellum armed with single row of hooks** .... **2.**

2a. **One testis per proglottid** .... **3.**
2b. **Two testes per proglottid. Rostellum with ten hooks** .... **6.**

3a. Suckers unarmed .... **4.**
3b. Suckers armed. Rostellum with ten diorchoid hooks ....
.... ***Skrjabinoparaksis* Krotov, 1949. (Fig. 29.8)**

Diagnosis: Suckers armed with small hooklets, rosethorn in shape, larger in centre, smaller near margins of suckers. Proglottids craspedote. Testis median, spherical. Male and female genitalia develop simultaneously near neck. Cirrus-sac long, almost reaches antiporal osmoregulatory canals. Ovary compact, antiporal to compact. Vitellarium median. Uterus sacciform. Oncospheres oval, not numerous (8–26). In Anseriformes. European and Asian Russia. Type-species *S. tatianae* Krotov, 1949.

4a. Rostellum with ten hooks .... **5.**
4b. Rostellum with more than ten hooks ....
.... ***Monorcholepis* Oshmarin, 1961. (Fig. 29.9)**

Diagnosis: Rostellum bulbous, retractile with 16 to about 44 aploparaksoid hooks. Suckers unarmed. Proglottids craspedote. Internal longitudinal muscle bundles numerous. Genital ducts dorsal to osmoregulatory canals. Genital pores unilateral, dextral. Testis oval, situated antiporally. External seminal vesicle dorsal to cirrus-sac, which crosses poral osmoregulatory stems and may reach midline. Ovary trilobate, poral. Vitellarium compact, median. Seminal receptacle poral, may lie dorsal to poral osmoregulatory canals. Uterus sacciform. Oncospheres and their envelopes spherical, numerous. In Passeriformes. Europe, Asia. Type-species *M. dujardini* (Krabbe, 1869) Oshmarin, 1961.

5a. Rostellum with single crown of ten aploparaksoid hooks ....
.... ***Aploparaksis* Clerc, 1903. (Fig. 29.10)**
(Syns *Globarilepis* Bondarenko, 1966; *Haploparaksis* Neslobinsky, 1911;

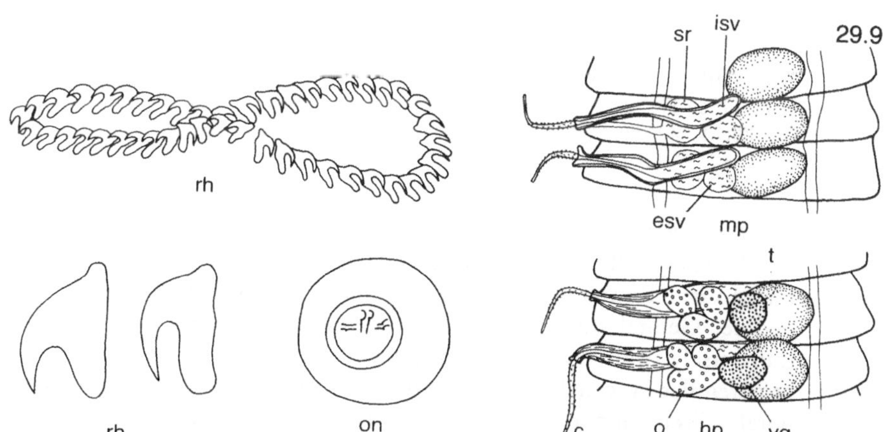

Fig. 29.9 *Monorcholepis dujardini* (Krabbe, 1869) Oshmarin, 1961 (rh after Krabbe (1869); hp, mp, on after Spasskaya (1966), modified.) (See Fig. 29.1 for abbreviations.)

29.10

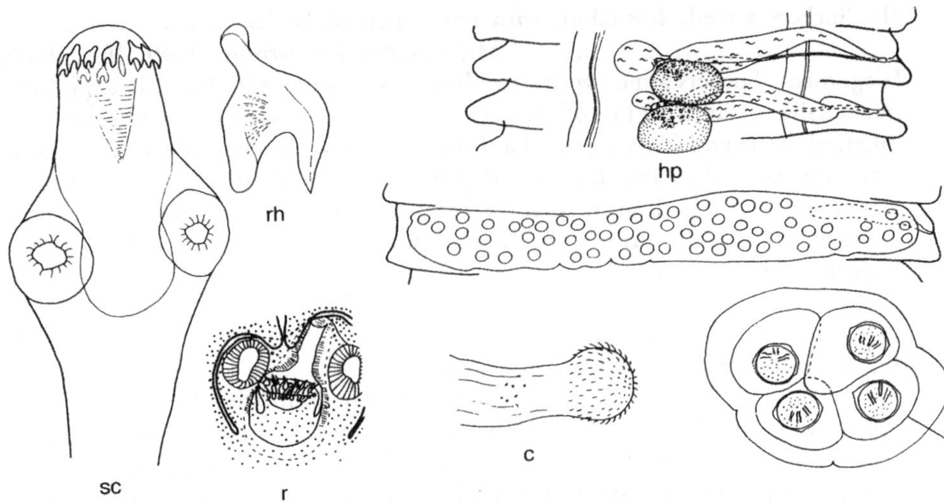

29.11

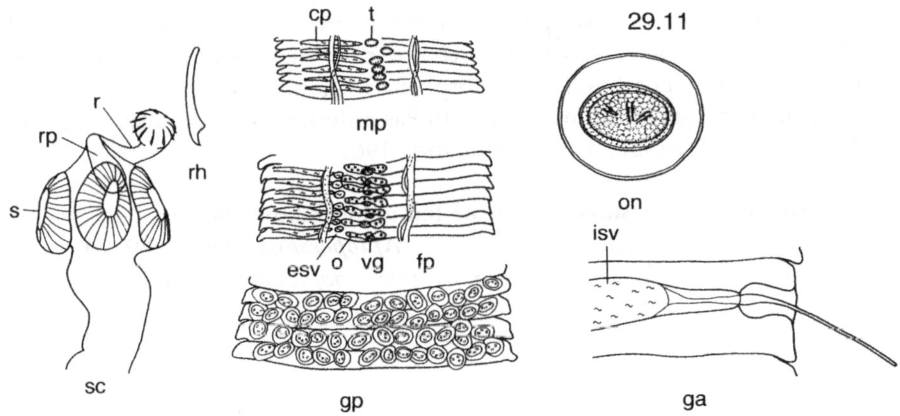

Fig. 29.10 *Aploparaksis furcigera* (Rudolphi, 1819) Fuhrmann, 1926. (Original, r after Gulyaev, 1977.) (See Fig. 29.1 for abbreviations.)
Fig. 29.11 *Monotestilepis tadornae* Gvozdev, Maksimova & Kornyushin, 1971. (After Gvozdev *et al.*, 1971.) (See Fig. 29.1 for abbreviations.)

Linstowius Yamaguti, 1959; *Monorchis* Clerc, 1902 preoccupied; *Skorikowia* Linstow, 1905; *Tanureria* Spasskii & Yurpalova, 1968.)

Diagnosis: Rostellum mostly invaginable. Suckers unarmed. Proglottids craspedote. Genital ducts dorsal to osmoregulatory canals. Testis compact, mainly transversely oval, median or antiporal. Cirrus-sac well developed, crosses poral osmoregulatory canals. Ovary compact or irregularly lobed, median or slightly poral. Vitellarium compact, rarely slightly lobed, median. Seminal receptacle situated between vitellarium and poral osmoregulatory stems. Uterus tubular when young, when gravid sacciform and occupies entire medulla. Oncospheres with thick middle envelope, with or without polar thickenings, leave the strobila singly or in pseudocapsules. In Anseriformes, Charadriiformes, Galliformes, Passeriformes, Lariformes. Cosmopolitan. Type-species *A. filum* (Goeze, 1782) Clerc, 1903.

5b. Rostellum with single crown of ten arcuatoid hooks .... ***Monotestilepis* Gvozdev, Maksimova & Kornyushin, 1971. (Fig. 29.11)**
Diagnosis: Rostellum retractile, armed. Suckers large, unarmed. Proglottids craspedote. Inner longitudinal muscle bundles eight. Genital ducts cross poral osmoregulatory canals dorsally. Testis compact, median. Cirrus-sac narrow, elongated, reaches testis. Male and female gonads develop almost simultaneously. Ovary bilobed, median. Vitellarium oval, compact, median. Uterus sacciform. Oncospheres not numerous, rounded, arranged in transverse row. In Anseriformes. Ukraine, Kazakhstan. Type-species *M. tadornae* Gvozdev, Maksimova & Kornyushin, 1971.

6a. Suckers usually armed. External and internal segmentation conspicuous. Rostellar hooks diorchoid or arcuatoid. Oncospheral envelopes oval or filamentous .... ***Diorchis* Clerc, 1903. (Fig. 29.12)**
(Syns *Diplomonorchis* Lopez-Neyra, 1944 preoccupied; *Jonesius* Yamaguti, 1959; *Nudiorchis* Matevosyan, 1941; *Schillerius* Yamaguti, 1959).
Diagnosis: Rostellum retractile. Longitudinal muscle bundles mainly eight. Genital ducts dorsal to osmoregulatory canals. Two testes per proglottid. Poral

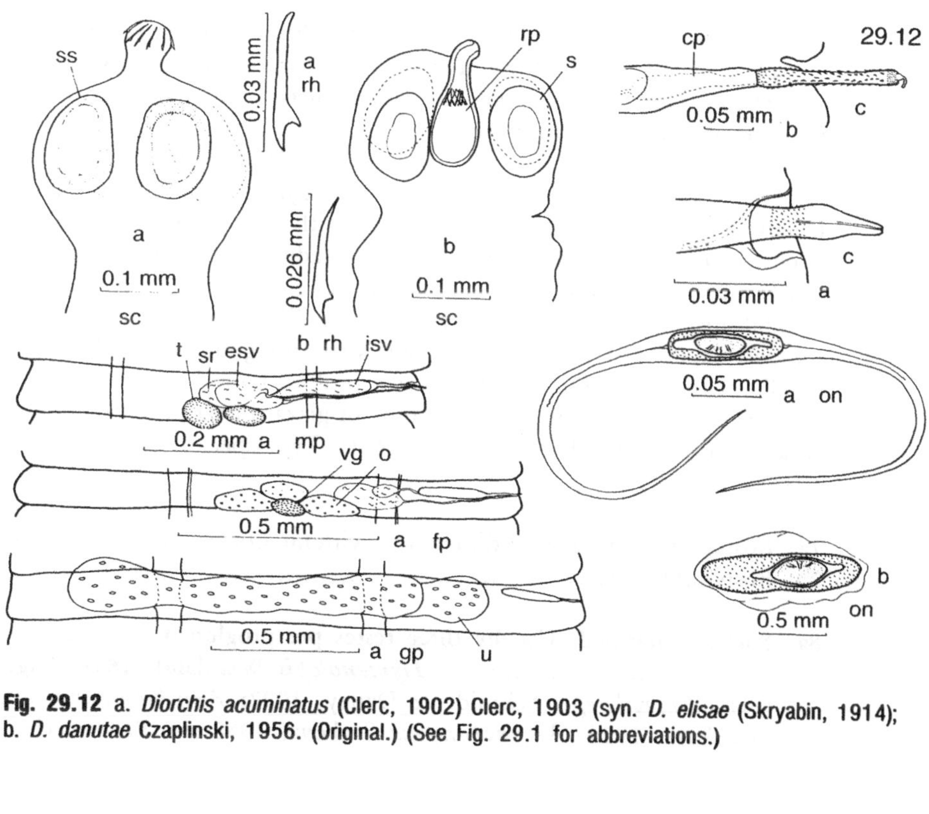

Fig. 29.12 a. *Diorchis acuminatus* (Clerc, 1902) Clerc, 1903 (syn. *D. elisae* (Skryabin, 1914); b. *D. danutae* Czaplinski, 1956. (Original.) (See Fig. 29.1 for abbreviations.)

testis may disappear, compressed by seminal receptacle (e.g. *D. stefanskii*). External and internal seminal vesicles well developed, sometimes with thin layer of circular muscles. Cirrus-sac usually crosses poral osmoregulatory canals, rarely very long reaching almost antiporal margin of proglottid. Ovary mostly trilobed, may be transversely elongated, bilobed, median, rarely submedian. Vitellarium compact, postovarian or ventral to ovary. Seminal receptacle well developed, poral to vitellarium. Distal vagina ventral or posterior to cirrus-sac, often with copulatory part of characteristic shape and size and muscular spincters. Uterus sacciform. Oncospheres mostly oval, with oval envelopes, outer envelope often filamentous, even ramified. In Anseriformes, Gruiformes, Charadriiformes, Columbiformes. Cosmopolitan. Type-species *D. acuminata* (Clerc, 1902) Clerc, 1903.

6b. Suckers unarmed. External and internal segmentation evident. Rostellar hook echinorhynchoid. Oncosphere envelopes spherical .... *Neodiorchis* Bilqees & Fatima, 1984. (Fig. 29.13)

Diagnosis: Proglottids craspedote. Gravid proglottids square or longer than wide(?). Longitudinal muscle bundles and osmoregulatory canals not described. Genital pores unilateral, submarginal. Testes rounded or transversely oval. External seminal vesicle not observed (absent ?). Cirrus-sac long, reaching midline or near antiporal margin. Internal seminal vesicle well developed. Cirrus armed (?), very long. Ovary transversely elongated, two lobes not distinct. Vitellarium compact, rounded or oval, median, ventral to ovary. Uterus not distinct; seems sacciform. Inner oncospheral envelope with polar filaments. Doubtful genus. In Anseriformes. Pakistan. Type-species *N. longicirrus* Bilqees & Fatima, 1984.

6c. Suckers armed. External segmentation lacking, internal segmentation complicated. Body cylindrical, round in transverse section with lateral groove where many radially arranged genital ducts open. Usually two testes connected with each cirrus-sac .... *Gastrotaenia* Wolffhügel, 1938. (Fig. 29.14)
(Syn. *Apora* Ginetsinskaya, 1944).

Diagnosis: Rostellum retractile, armed, with single crown of ten diorchoid hooks, guard prominent but shorter than blade. Internal and external longitudinal muscle bundles numerous. Four longitudinal osmoregulatory canals. Genital ducts dorsal to osmoregulatory canals. Cirrus-sac elongated, weak. Ovary compact, transversely oval, connected with fine-granular vitellarium situated porally. Uterus sacciform. Oncospheres oval with oval external envelopes. In Anseriformes, mainly under horny lining of gizzard near proventriculus. Europe, Asia, South and North America, Australia. Type-species *G. cygni* Wolffhügel, 1938.

7a. Rostellum lacking or rudimentary, without hooks .... 8.
7b. Rostellum armed .... 9.

8a. Suckers unarmed. Usually three testes per proglottid .... *Hymenolepis* Weinland, 1858. (Fig. 29.15)
(Syns *Amazilolepis* Schmidt & Dailey, 1992; *Amphipetrovia* Spasskii & Spasskaya, 1954; *Arhynchotaenia* Saakova, 1958 *nec* Pagenstecher, 1877;

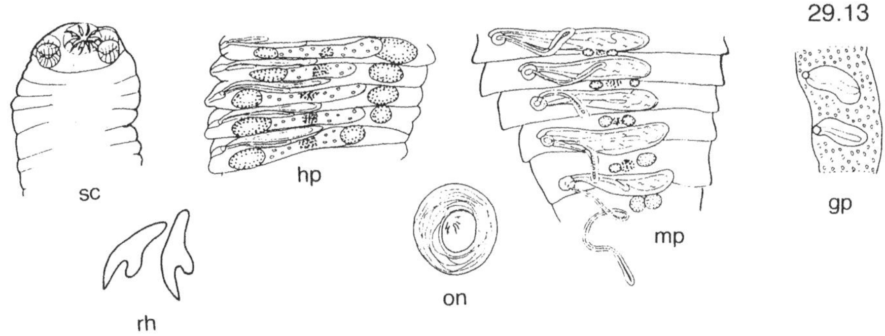

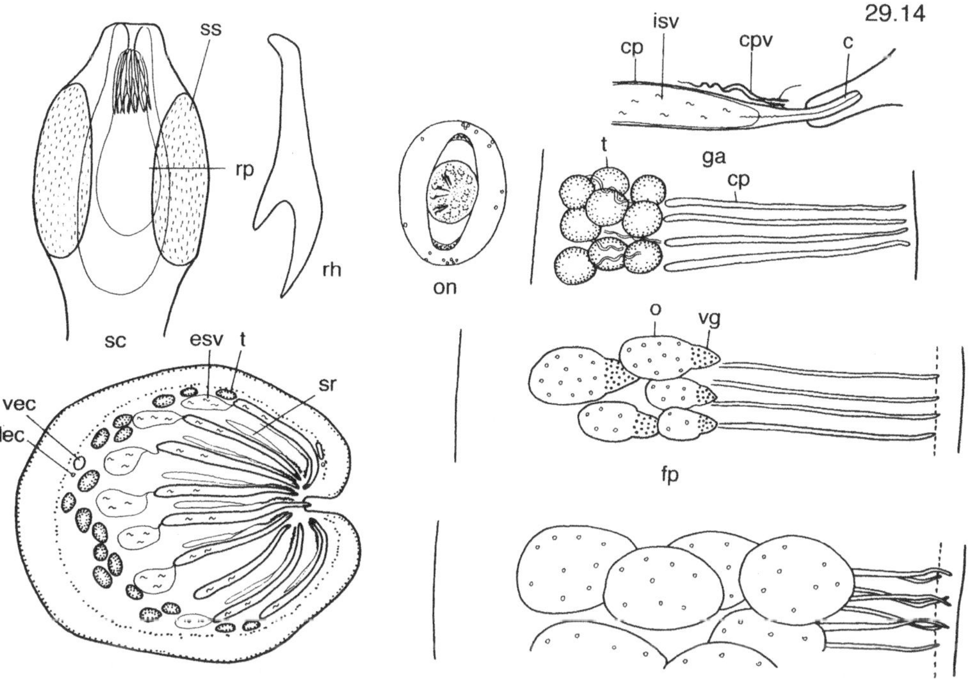

Fig. 29.13 *Neodiorchis longicirrus* Bilqees & Fatima, 1984. (After Bilqees & Fatima, 1984.) (See Fig. 29.1 for abbreviations.)

Fig. 29.14 *Gastrotaenia paracygni* Czaplinski & Ryzhikov, 1966. (All figures after Cielecka & Misiura, unpublished, except on after Wisniewski, 1971.) (See Fig. 29.1 for abbreviations.)

Arhynchotaeniella Schmidt, 1986; *Australiolepis* Spasskii & Spasskaya, 1954; *Cloacotaenia* Wolffhügel, 1938; *Cloacotaeniella* Schmidt, Bauerle & Wertheim, 1988; *Orlovilepis* Spasskii & Spasskaya, 1954; *Schmelzia* Yamaguti, 1959; *Staphylepis* Spasskii & Oshmarin, 1954; *Triorchis* Clerc, 1903 preoccupied; *Woodlandia* Yamaguti, 1959).

Diagnosis: Suckers lateral or apical. Proglottids craspedote. Inner longitudinal muscle bundles numerous. Ventral osmoregulatory canals often connected by

29.15

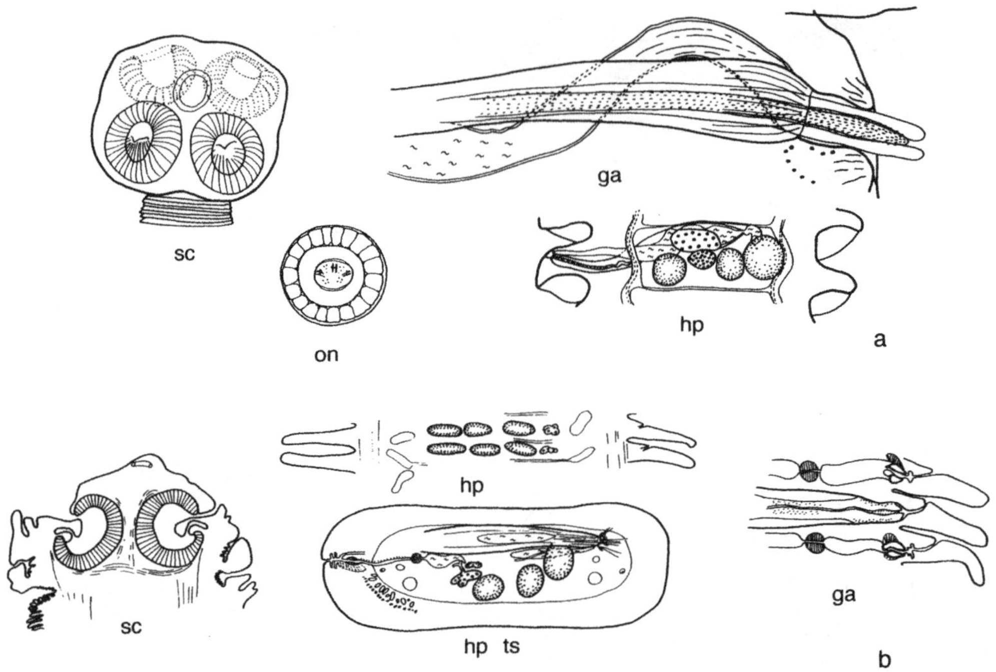

29.16

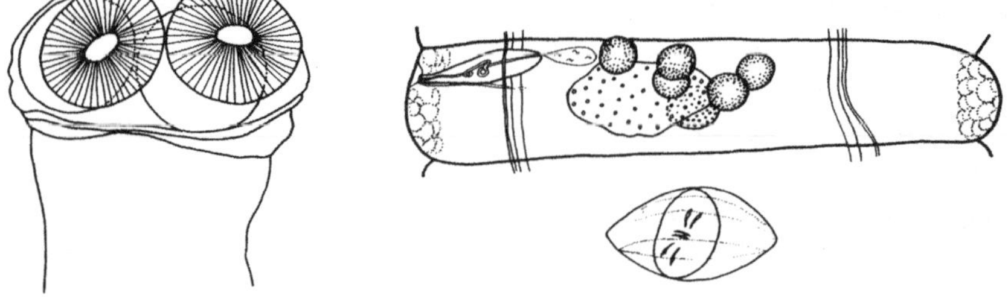

Fig. 29.15 a. *Hymenolepis megalops* (Nitzsch in Creplin, 1829) Ransom, 1902. (ga, hp, sc Czaplinski, 1956; on after Jarecka, 1958); b. *H. biaculeata* Fuhrmann, 1909. (After Fuhrmann, 1909b.) (See Fig. 29.1 for abbreviations.)
Fig. 29.16 *Neoligorchis alternatus* Johri, 1960. (After Johri, 1960b.) (See Fig. 29.1 for abbreviations.)

transverse anastomoses. Genital ducts dorsal to osmoregulatory canals. Genital pores unilateral, dextral (except in older specimens of *H. diminuta*). Three testes in transverse row or triangle, mainly in medulla. Cirrus-sac may reach midline. Ovary compact, transversely elongated or lobed, median or submedian, often pretesticular, exceptionally poral to testes. Vitellarium compact, postovarian. Seminal receptacle well developed. Vagina posterior or ventral to cirrus-sac. Uterus sacciform, rarely reticulate. Oncospheres slightly oval, envelopes mostly spherical. In birds and mammals. Cosmopolitan. Type-species *H. diminuta* (Rudolphi, 1819) Weinland, 1858.

8b. Suckers unarmed. Usually five to seven testes per proglottid .... *Neoligorchis* Johri, 1960. (Fig. 29.16)

(Syns *Lallum* Johri, 1960; *Neolallum* Srivastava, Pandey & Tayal, 1984).

Diagnosis: Scolex large. Suckers apical (forwardly directed) or lateral. Proglottids craspedote. Genital ducts dorsal to osmoregulatory canals or between them (?). Proglottids craspedote. Genital pores unilateral or irregularly alternating. External seminal vesicle well developed. Internal one not observed. Cirrus-sac may cross osmoregulatory canals. Ovary not lobed, pretesticular, or post-testicular. Vitellarium compact, postovarian. Uterus sacciform, containing numerous oncospheres. In Anseriformes and Charadriiformes. India. Type-species *N. alternatus* Johri, 1960.

8c. Suckers armed. Three testes per proglottid .... *Echinolepis* Spasskii & Spasskaya, 1954. (Fig. 29.17)

Diagnosis: Suckers with rosethorn-like to hammer-like spines. Genital ducts dorsal to osmoregulatory canals. Genital pores unilateral. External segmentation well developed, proglottids craspedote. Ventral osmoregulatory canals connected by transverse anastomoses. Inner longitudinal muscle bundles eight. Testes, one poral, two antiporal. External and internal seminal vesicles well-developed. Cirrus-sac reaches midline. Cirrus armed. Ovary bi- or trilobed, median. Vitellarium ovoid, compact, postovarian. Seminal receptacle well developed. Vagina ventral to cirrus-sac. Uterus sacciform. Oncospheres oval with subspherical envelopes. In Galliformes. Cosmopolitan. Type-species *E. carioca* (Magalhaes, 1898) Spasskii & Spasskaya, 1954.

9a. Rostellum with regular, single crown of hooks .... 10.
9b. Rostellum with irregular crown of spiniform hooks or several crowns of hammer-like hooks .... 35.

10a. Rostellum with eight hooks .... 11.
10b. Rostellum with ten hooks .... 16.
10c. Rostellum with more than ten hooks .... 28.

11a. Cirrus stylet absent. Rostellar hooks diorchoid or echinorhynchoid .... 12.
11b. Cirrus stylet present. Hooks skrjabinoid .... 14.

12a. Strobila of medium size. Hooks diorchoid, guard short .... 13.
12b. Strobila very small. Hooks diorchoid, guard long. Suckers unarmed .... *Podicipitilepis* Yamaguti, 1959. (Fig. 29.18)

Diagnosis: Small, quickly maturing forms with few craspedote proglottids. Rostellum armed with single crown of eight hooks. Longitudinal muscle bundles numerous, feeble. Genital ducts dorsal to osmoregulatory canals. Testes three, in triangle or transverse row, one poral, two antiporal. External seminal vesicle voluminous, appearing suddenly, may disappear quickly, internal vesicle small.

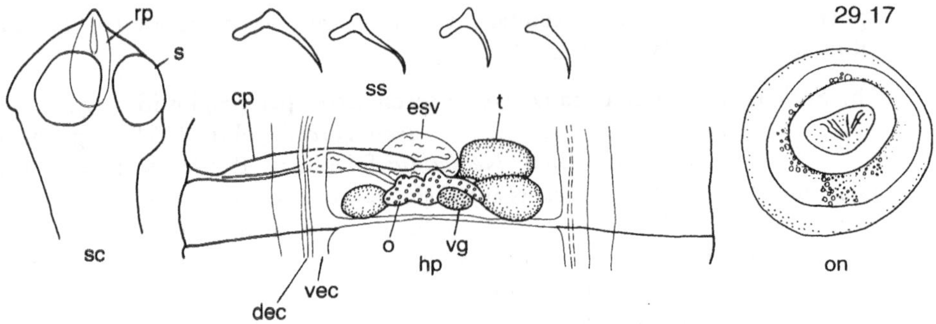

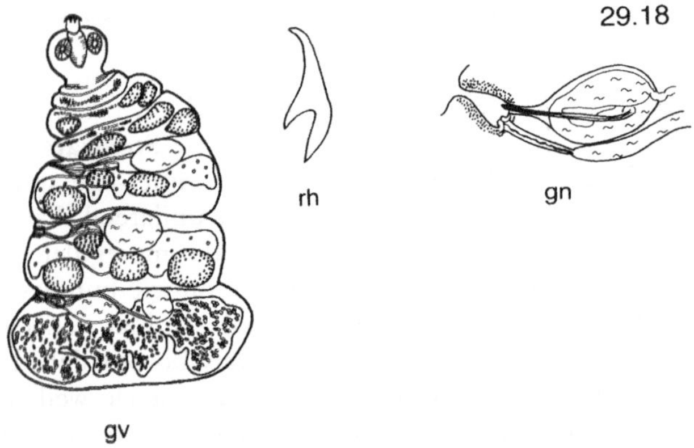

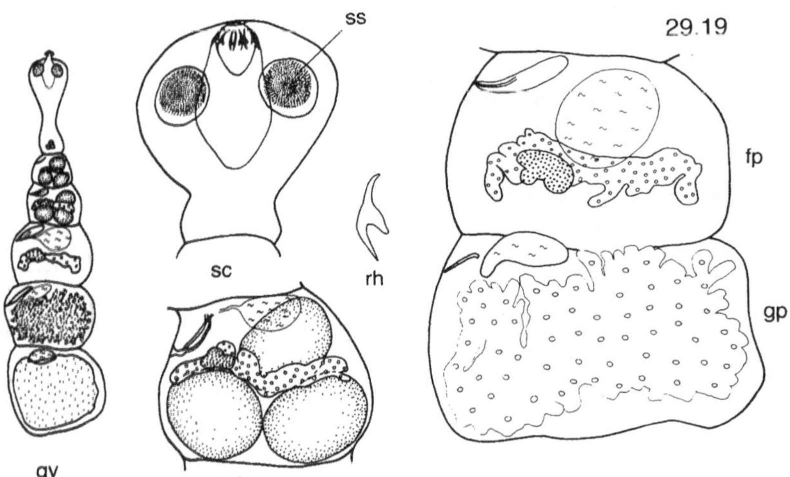

Fig. 29.17 *Echinolepis carioca* (Magalhaes, 1898) Spasskii & Spasskaya, 1954. (After Spasskii & Spasskaya, 1954.) (See Fig. 29.1 for abbreviations.)
Fig. 29.18 *Podicipitilepis laticauda* Yamaguti, 1956. (After Yamaguti, 1956b.) (See Fig. 29.1 for abbreviations.)
Fig. 29.19 *Mackoja podirufi* (Macko, 1962) Kornyushin, 1983. (After Macko, 1962b, modified.) (See Fig. 29.1 for abbreviations.)

Cirrus-sac small, bulbous. Ovary transversely elongate, reaching lateral margins of proglottid. Vitellarium compact, dorsal to ovary, submedian. Seminal receptacle voluminous in gravid proglottids. Vagina with sphincter-like thickening of copulatory part, ventral and posterior to cirrus-sac. Uterus sacciform, with posterior pouches. In Podicipediformes. Japan. Type-species *P. laticauda* Yamaguti, 1956.

12c. Strobila very small. Rostellar hooks echinorhynchoid. Suckers armed with small spines .... *Mackoja* Kornyushin, 1983. (Fig. 29.19)

Diagnosis: Rostellum with eight hooks. Proglottids few, little broader than long. Three large testes arranged in triangle. External seminal vesicle not identified, internal present. Cirrus-sac not reaching midline. External unarmed accessory sac present (?). Ovary transversely elongated, overlapping testes ventrally. Vitellarium compact, irregular in shape, submedian, near genital atrium. Seminal receptacle oval. Uterus sacciform. Oncospheres unknown. In Podicipediformes. Europe. Type-species *M. podirufi* (Macko, 1962) Kornyushin, 1983.

13a. Pseudoscolex and accessory sac present .... *Parabisaccanthes* Maksimova, 1963. (Fig. 29.20)

Diagnosis: Rostellum retractile, with eight hooks. Pseudoscolex simple, not folded, distinctly segmented, without genital primordia. Proglottids craspedote. Inner longitudinal muscle bundles numerous. Genital pores unilateral, dextral. Genital ducts dorsal to poral osmoregulatory canals. Testes three, in triangle. External and internal seminal vesicles well developed. Cirrus-sac passing beyond midline. Accessory sac double, spinose, external to cirrus-sac. Ovary trilobate,

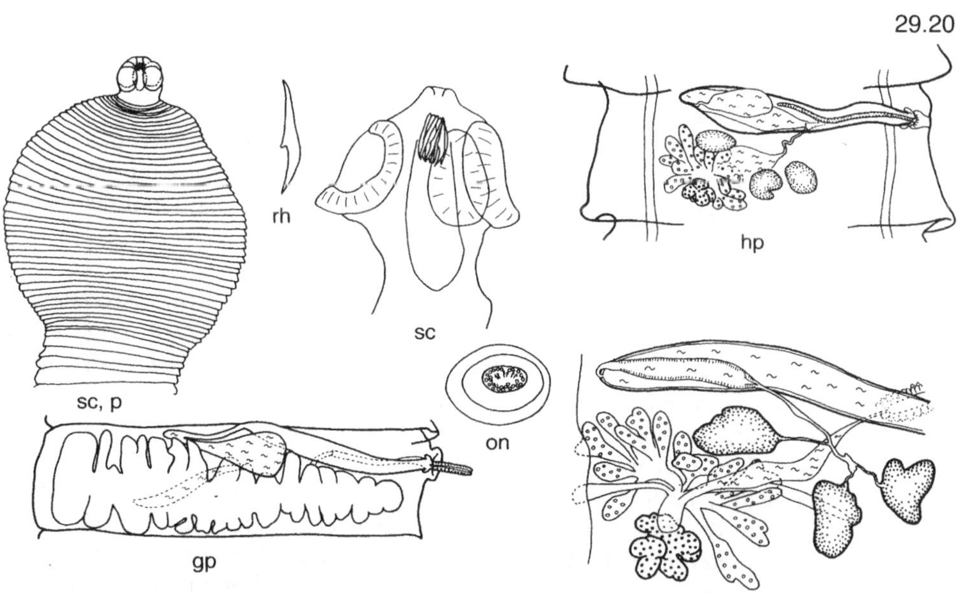

Fig. 29.20 *Parabisaccanthes philactes* (Schiller, 1951) Spasskii & Reznik, 1963. (Original.) (See Fig. 29.1 for abbreviations.)

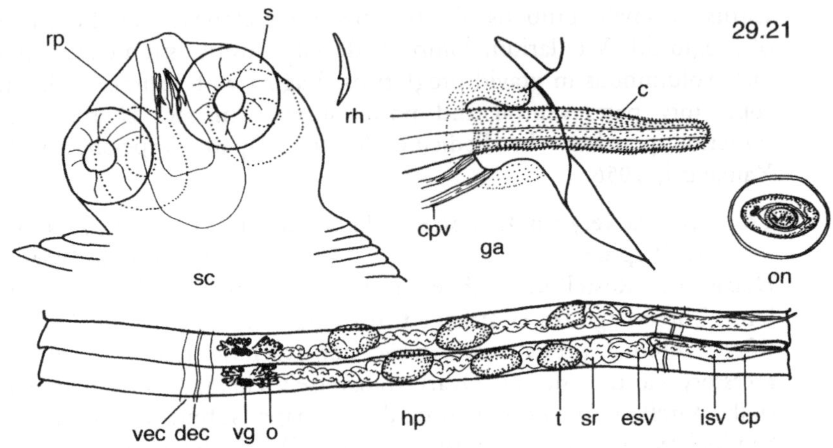

Fig. 29.21 *Drepanidotaenia lanceolata* (Bloch, 1782) Railliet, 1892. (After Czaplinski, 1956; except on after Potemkina, 1938.) (See Fig. 29.1 for abbreviations.)

when older multilobate; situated antiporally, overlapping antiporal testis ventrally. Vitellarium lobate, postovarian, ventral to ovary. Seminal receptacle well developed. Vagina ventral to cirrus-sac with distinct copulatory part surrounded by muscles and glands. Uterus sacciform with diverticles. Oncospheres oval. In Anseriformes. Europe, Asia, North America, Australia. Type-species *P. philactes* (Schiller, 1951) Spasskii & Reznik, 1963.

13b. Pseudoscolex and accessory sac absent .... ***Drepanidotaenia* Railliet, 1892. (Fig. 29.21)**

Diagnosis: Rostellum with single crown of hooks. Suckers unarmed. Proglottids short and wide. Inner longitudinal muscle bundles numerous. Genital ducts dorsal to osmoregulatory canals. Testes three, in row. Cirrus-sac crosses poral osmoregulatory canals. Ovary multilobate, near antiporal osmoregulatory canals. Vitellarium slightly lobate, postovarian, ventral to ovary. Seminal receptacle long, undulate. Uterus at first tubular, later transverse sac with many outpocketings. Oncospheres oval. In Anseriformes, exceptionally man. Cosmopolitan. Type-species *D. lanceolata* (Bloch, 1782) Railliet, 1892.

14a. Rostellar hooks skrjabinoid. Retractile cirrus stylet present. Accessory sac absent .... ***Cladogynia* Baer, 1938. (Fig. 29.22)** (Syns *Allohaploparaksis* Yamaguti, 1959 (?); *Aporodiorchis* Yamaguti, 1959; *Avocettolepis* Spasskii & Kornyushin, 1971; *Dildotaenia* Dronen, Schmidt, Allison & Mellen, 1988; *Flamingolepis* Spasskii & Spasskaya, 1952; *Hymenocoelia* Capoor & Srivastava, 1966; *Hymenosphenacanthoides* Yamaguti, 1959; *Hymenosphenacanthus* Lopez-Neyra, 1958; *Octacanthus* Spasskii & Spasskaya, 1954; *Pararetinometra* Stock & Holmes, 1981; *Retinometra* Spasskii, 1955; *Sphenacanthus* Lopez-Neyra, 1942.)

Diagnosis: Rostellum retractile, with single crown of eight hooks. In some species several rows of spines below crown of hooks. Suckers unarmed. Simple,

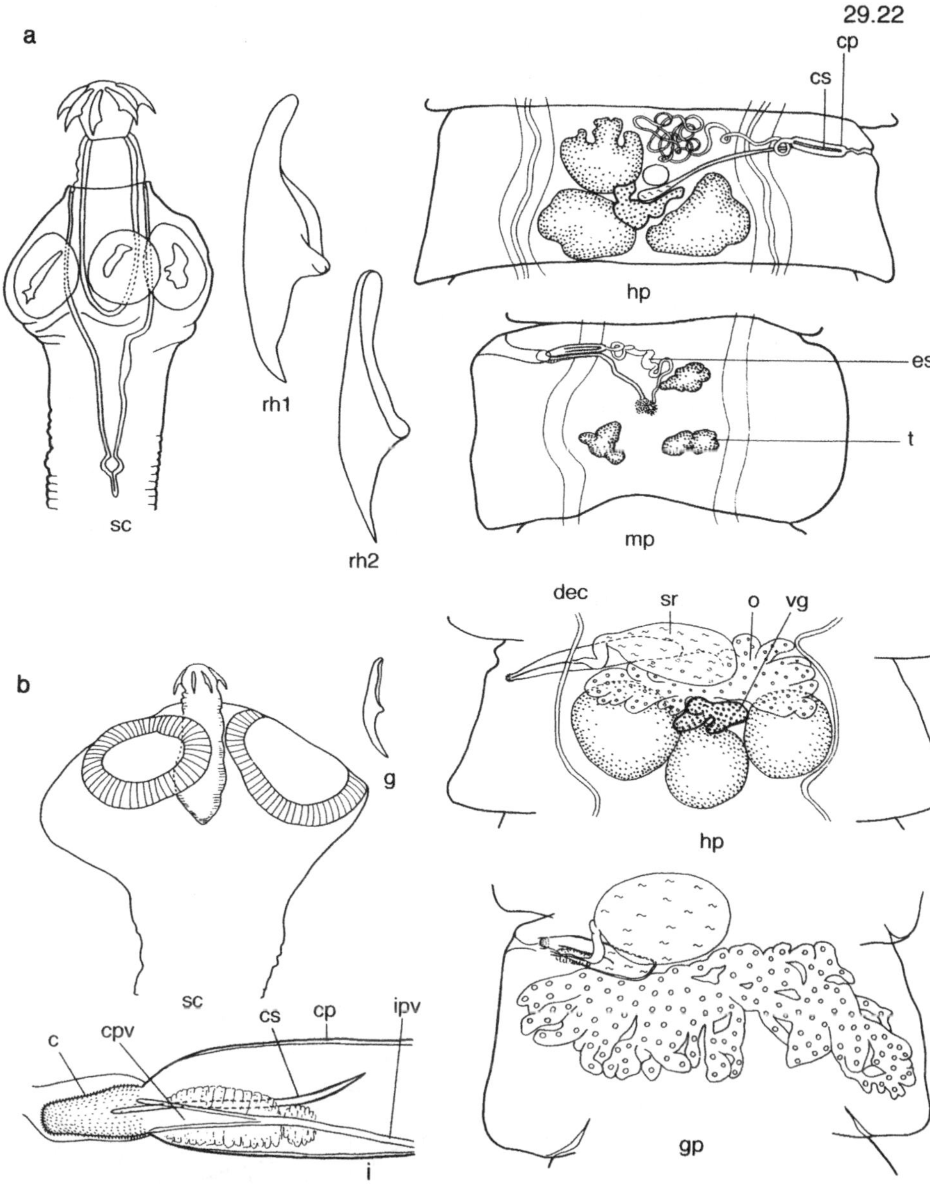

Fig. 29.22 a. *Cladogynia phoeniconaiadis* (Hudson, 1934) Baer, 1938 (sc, hp, rh 1 after Jones & Khalil, 1980; mp, rh 2 after Maksimova, 1989); b. *Cladogynia giranensis* (Sugimoto, 1934) n. comb. (syn. *Retinometra giranensis*). (Original.) (See Fig. 29.1 for abbreviations.)

unfolded pseudoscolex without genital primordia may occur. Proglottids distinctly craspedote; posterior surface sometimes covered with conical microtriches of different size. Longitudinal muscle bundles divided into two layers or not. Inner longitudinal muscle bundles numerous. Genital ducts pass dorsal to osmoregulatory canals. Testes three, arranged in row or triangle, smooth, rounded or oval, or slightly irregularly lobed, situated in medulla, exceptionally overlapping osmoregulatory canals. Cirrus-sac weak, short or long, exceptionally very long, coiled. Ovary usually strongly lobed, median, ventral to testes. Vitellarium compact or branched, postovarian, median. Seminal receptacle well developed. Vagina may cross cirrus-sac from dorsal to ventral side. Distal vagina often forms copulatory part with specialized muscles. Uterus reticular, crossing osmoregulatory canals. Oncospheres oval, envelopes rounded or oval. In Anseriformes, Phoenicopteriformes, Charadriiformes and Podicipediformes. Cosmopolitan. Type-species *C. phoeniconaiadis* (Hudson, 1934) Baer, 1937.

14b. Rostellar hooks skrjabinoid. Cirrus stylet and external accessory sac single or double .... 15.

15a. Three testes per proglottid .... *Sobolevicanthus* Spasskii & Spasskaya, 1954. (Fig. 29.23)
(Syns *Bisaccanthes* Spasskii & Spasskaya, 1954; *Parabiglandatrium* Gvozdev & Maksimova, 1968; *Phoenicolepis* Jones & Khalil, 1980.)

Diagnosis: Rostellum retractile, with single crown of eight hooks. Suckers unarmed. Simple, unfolded pseudoscolex without genital primordia may occur. Proglottids distinctly craspedote, posterior surface covered with conical micro-

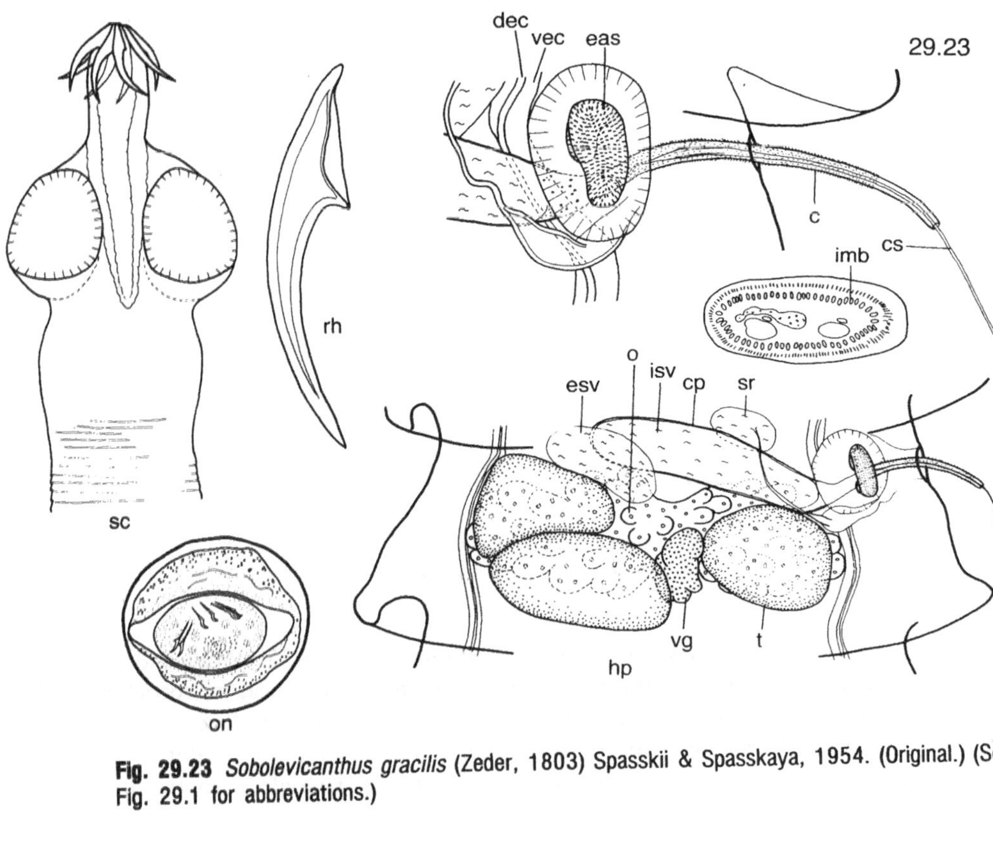

Fig. 29.23 *Sobolevicanthus gracilis* (Zeder, 1803) Spasskii & Spasskaya, 1954. (Original.) (See Fig. 29.1 for abbreviations.)

triches of varying size. Inner longitudinal muscle bundles numerous. Genital ducts dorsal to poral longitudinal osmoregulatory canals. Testes oval or rounded, smooth or slightly lobed, arranged mostly in triangle or rarely in transverse row. Internal seminal vesicle exceptionally absent. Cirrus-sac weak, usually passing midline, may be coiled, longer than proglottid is wide (e.g. *S. krabbeella*). Accessory sac external to cirrus-sac, usually provided with own muscles, glands and spines, near male genital pore, may be eversible. Ovary usually strongly lobed, median, ventral to testes. Vitellarium compact or slightly lobed, median, postovarian, dorsal to ovary. Seminal receptacle well developed. Vagina crosses cirrus-sac from dorsal to ventral side, rarely not. Distal vagina often forms copulatory part with specialized muscles. Uterus saccular or irregularly reticular. Oncospheres oval. Outer oncospheral envelope mostly rounded or spindle-shaped without filaments. In Anseriformes and Phoenicopteriformes. Cosmopolitan. Type-species *S. gracilis* (Zeder, 1803) Spasskii & Spasskaya, 1954.

15b. More than three testes per proglottid (about 32). Accessory sac single. Cirrus with stylet. Scolex unknown (probably with eight skrjabinoid hooks because of high correlation of these characters) ....
.... *Polytestilepis* Oshmarin, 1960. (Fig. 29.24)

Diagnosis: Scolex unknown. Proglottids craspedote. Genital ducts dorsal to poral osmoregulatory canals. Cirrus-sac weak, almost reaches midline. Cirrus not described exactly. External accessory sac muscular and spiny. Ovary multilobate, when mature occupies almost whole medulla. Vitellarium compact, slightly lobed

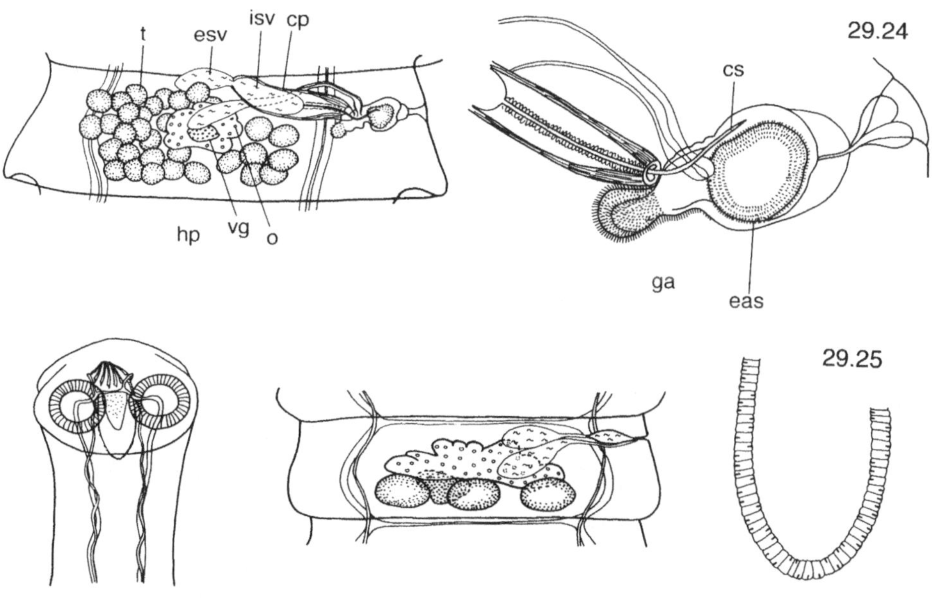

Fig. 29.24 *Polytestilepis chitinocloacis* Oshmarin, 1960. (After Spasskaya, 1966, modified.) (See Fig. 29.1 for abbreviations.)
Fig. 29.25 *Allohymenolepis mitudori* Yamaguti, 1956. (After Yamaguti, 1956a.)

posteriorly. Seminal receptacle well developed. Vagina crosses cirrus-sac from ventral to dorsal side and opens anteriorly to male pore. Genital atrium armed with chitinized spines. In Anseriformes. Russian Far East. Type-species *P. chitinocloacis* Oshmarin, 1960.

16a. Genital pores irregularly alternating (exceptional in Hymenolepidinae) .... *Allohymenolepis* Yamaguti, 1956. (Fig. 29.25)

Diagnosis: Rostellum armed with single crown of ten arcuatoid hooks. Suckers unarmed. Proglottids craspedote. Inner longitudinal muscle bundles numerous. Genital ducts dorsal to longitudinal osmoregulatory canals. Testes three, postovarian. Cirrus-sac small, not reaching median line. Ovary transversely elongated, slightly lobed anteriorly. Vitellarium compact, postovarian. Seminal receptacle large. Uterus sacciform. Oncospheres oval with three envelopes. In Passeriformes. Celebes. Type-species *A. mitudori* Yamaguti, 1956.

16b. Genital pores unilateral, dextral .... **17.**

17a. Accessory sac present .... **18.**
17b. No accessory sac .... **22.**

18a. Accessory sac internal to cirrus-sac. Hooks diorchoid. Suckers armed or not .... *Anatinella* Spasskii & Spasskaya, 1954. (Fig. 29.26)
(Syn. *Monosaccanthes* Czaplinski, 1967.)

Diagnosis: Rostellum retractile, armed with ten hooks. Pseudoscolex sometimes present simple, not folded. Proglottids craspedote. Longitudinal muscle bundles numerous or eight. Genital ducts dorsal to osmoregulatory canals. Usually three testes, oval or rounded, smooth or slightly lobed, in triangle or transverse row.

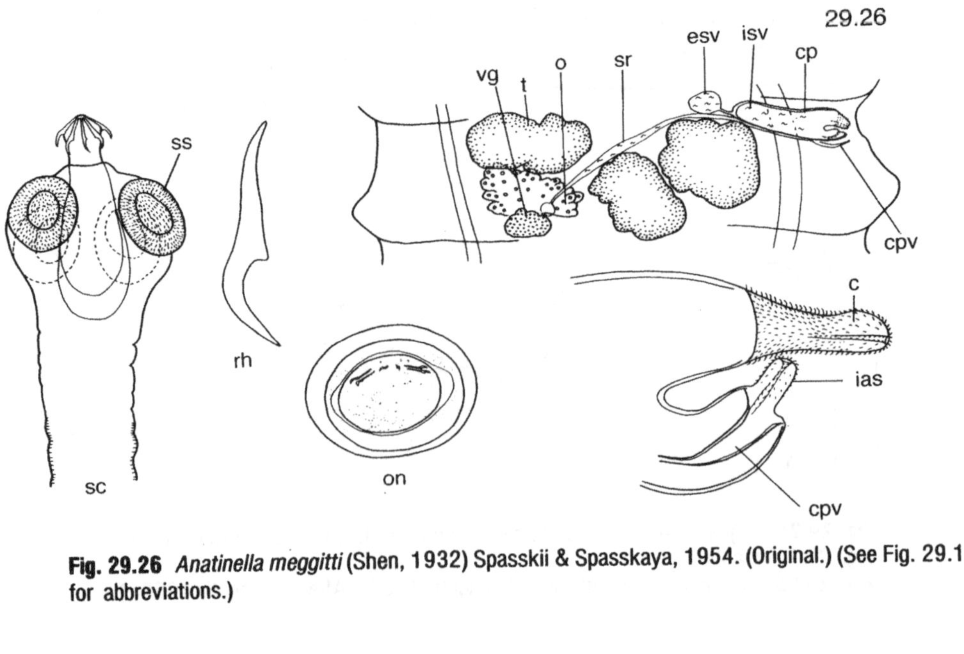

Fig. 29.26 *Anatinella meggitti* (Shen, 1932) Spasskii & Spasskaya, 1954. (Original.) (See Fig. 29.1 for abbreviations.)

29.27

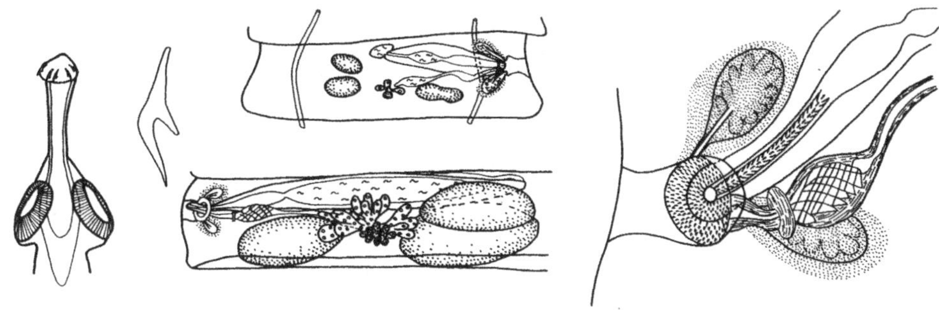

Fig. 29.27 *Biglandatrium biglandatrium* Spasskaya, 1961. (After Spasskaya, 1961a.)

Cirrus-sac small or reaching midline, may be thick-walled. Accessory sac spiny, invaginable, may look like second cirrus. Ovary lobate, mostly submedian, antiporal. Vitellarium postovarian, mainly submedian, compact or slightly lobed. Seminal receptacle well developed. Uterus irregularly sacciform. Oncospheres oval with oval or spherical outer envelope. In Anseriformes. Cosmopolitan except South America. Type-species *A. meggitti* (Shen, 1932) Spasskii & Spasskaya, 1954.

18b. Accessory sac external .... 19.

19a. Accessory sac double. Rostellar hooks echinorhynchoid .... *Biglandatrium* Spasskaya, 1961. (Fig. 29.27)
Diagnosis: Rostellum retractile, with ten hooks. Suckers unarmed. Proglottids craspedote. Genital ducts dorsal to osmoregulatory canals. Testes three, oval, in triangle, one poral. Cirrus-sac reaching midline. Cirrus and base of genital atrium armed. Accessory sac glandular, unarmed, external to cirrus-sac. Ovary lobate, median. Vitellarium lobate, postovarian. Seminal receptacle well developed. In Gaviiformes. Tuva. Type-species *B. biglandatrium* Spasskaya, 1961.

19b. Accessory sac single .... 20.

20a. Cirrus without stylet .... 21.
20b. Cirrus with stylet .... *Geraldolepis* n. g. (Fig. 29.28)
Diagnosis: Rostellum retractile, with ten hooks. Proglottids craspedote. Genital ducts dorsal to osmoregulatory canals. Three oval or rounded testes in nearly straight line. External seminal vesicle rounded. Internal seminal vesicle filling most of cirrus-sac. Cirrus-sac cylindrical, transverse or curved, reaching or crossing median line. Cirrus small, armed. Cirrus stylet protrusible. External accessory sac single, thick walled, unarmed. Ovary posterior, slightly poral, compact, transversely elongated. Vitellarium compact, posteromedial to ovary. Seminal receptacle spindle-shaped. Vagina ventral to cirrus-sac. Copulatory part of vagina spindle-like with little spines inside, surrounded by muscles and glands. Uterus sacciform. Oncospheres unknown. In Charadriiformes. Alaska. Type-species *G.*

29.28

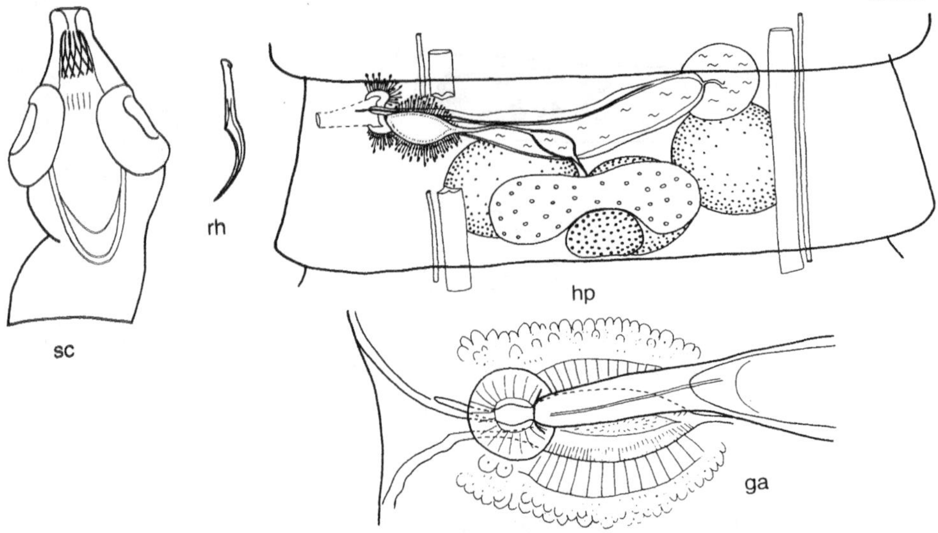

29.29

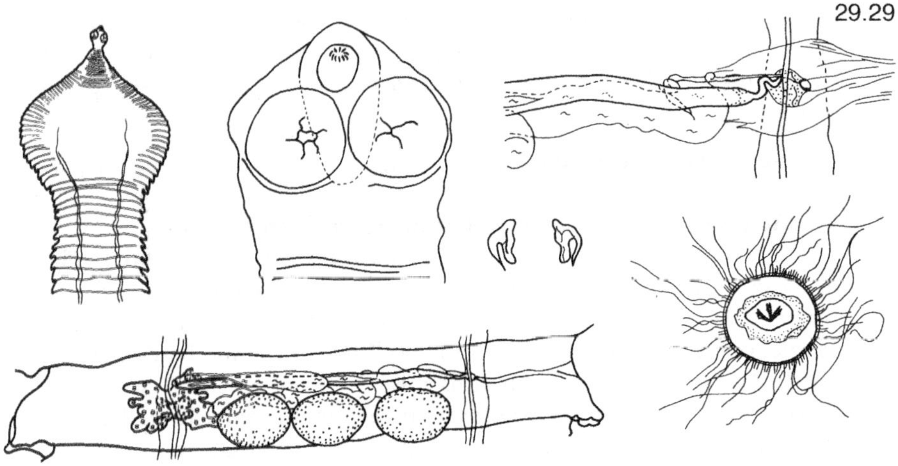

29.30

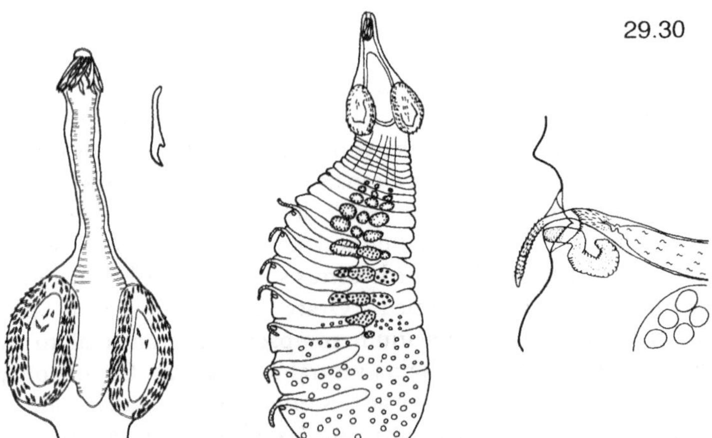

deblocki (Schmidt & Neiland, 1968) n. comb. (for *Hymenolepis deblocki* Schmidt & Neiland, 1968).

21a. **Rostellar hooks aploparaksoid. Suckers unarmed. Presence of accessory sac needs confirmation .... *Chimaerolepis* Spasskii & Spasskaya, 1972. (Fig. 29.29)**

Diagnosis: Rostellum retractile ? with ten hooks. Pseudoscolex simple, unfolded. Proglottids craspedote. Inner longitudinal muscle bundles numerous. Testes three, oval, arranged in line. Cirrus-sac weak, reaching level of ovary. Accessory sac spiny (? or armed base of genital atrium). Ovary multilobate, antiporal, overlaps antiporal osmoregulatory canals. Vitellarium multilobate, antiporal, ventral to ovary. Seminal receptacle very long, undulate. Uterus tubular, then sacciform. Oncospheres oval, external envelope spherical with hair-like projections. In Galliformes. USA. Type-species *C. watsoni* (Prestwood & Reid, 1966) Spasskii & Spasskaya, 1972.

21b. **Rostellar hooks diorchoid, nitidoid or arcuatoid. Suckers armed with rosethorn spines. Accessory sac definitely present .... *Echinocotyle* Blanchard, 1891. (Fig. 29.30)**
(Syns *Echinocotyloides* Kornyushin, 1983; *Gonoscolex* Saakova, 1958; *Larocotyloides* Kornyushin, 1983; *Mariicotyle* Kornyushin, 1983.)

Diagnosis: Rostellum retractile, armed with ten hooks. Proglottids craspedote, not numerous (up to 50) or exceeding hundred. Longitudinal muscle bundles eight or numerous. Genital ducts dorsal to osmoregulatory canals. Testes three, in row or triangle. Cirrus-sac short or long. Cirrus armed or not. Accessory sac external to cirrus-sac, opens into genital atrium, invaginable, armed with spines, muscular and often glandular. Ovary usually compact, transversely elongated, exceptionally with finger-like lobes, submedian or median. Vitellarium compact usually dorsal to ovary, submedian or median. Seminal receptacle present. Uterus sacciform, exceptionally reticulate (*E. verschureni*). Oncospheres oval, outer envelopes spherical or oval. In Anseriformes, Charadriiformes, Lariformes, Passeriformes; exceptionally Rodentia. Europe, Asia, Africa, North America. Type-species *E. rosseteri* Blanchard, 1891.

21c. **Rostellar hooks nitidoid. Suckers unarmed. Cirrus stylet absent .... *Debloria* Spasskii, 1975. (Fig. 29.31)**

Diagnosis: Rostellum with ten hooks. Proglottids craspedote. Inner longitudinal muscle bundles eight. Genital ducts dorsal to osmoregulatory canals. Testes three in straight line or triangle. Cirrus-sac crosses midline. Accessory sac muscular, with

---

Fig. 29.28 *Geraldolepis deblocki* (Schmidt & Neiland, 1968) n. comb (syn. *Retinometra deblocki* (Schmidt & Neiland, 1968) (ga original; others after Schmidt & Neiland, 1968.) (See Fig. 29.1 for abbreviations.)

Fig. 29.29 *Chimaerolepis watsoni* (Prestwood & Reid, 1966) Spasskii & Spasskaya, 1972. (After Prestwood & Reid, 1966.)

Fig. 29.30 *Echinocotyle rosseteri* Blanchard, 1891. (After Czaplinski, 1956.)

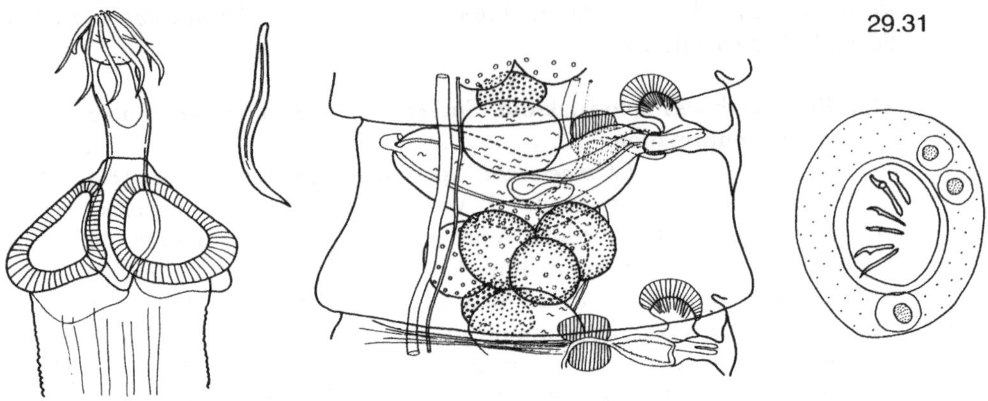

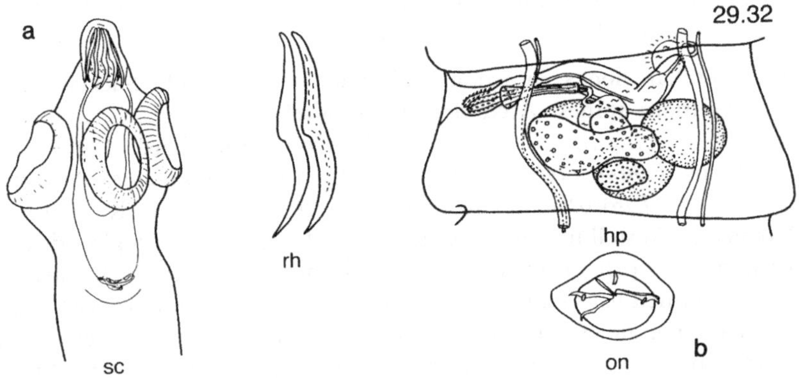

Fig. 29.31 *Debloria capetownensis* (Deblock & Rosé, 1962) Spasskii, 1975. (After Deblock & Rosé, 1962.)
Fig. 29.32 a. *Nadejdolepis belopolskaiae* (Deblock & Rosé, 1962) Spasskaya, 1966. b. Oncosphere of *Nadejdolepis patersoni* (Deblock & Rosé, 1962) n. comb. (After Deblock & Rosé, 1962.) (See Fig. 29.1 for abbreviations.)

hair-like spines inside, external and anterior to distal cirrus-sac. Ovary trilobate. Vitellarium postovarian, compact. Seminal receptacle ovoid. Vagina ventral to cirrus-sac, copulatory part with sphincter. Uterus sacciform. Oncospheres and outer envelopes oval. Embryonic hooks unequal in shape. In Charadriiformes. South Africa. Type-species *D. capetownensis* (Deblock & Rosé, 1962) Spasskii, 1975.

22a. Hooks nitidoid ....
.... ***Nadejdolepis*** **Spasskii & Spasskaya, 1954. (Fig. 29.32)**

Diagnosis: Rostellum retractile. Suckers unarmed. Proglottids craspedote. Inner longitudinal muscle bundles numerous, more than eight. Genital ducts dorsal to osmoregulatory canals. Testes three, in transverse row or triangle. Cirrus-sac crossing median line or not. Cirrus armed. Accessory sac and cirrus stylet lacking. Ovary compact, lobate. Vitellarium compact, postovarian. Seminal receptacle well developed. Vagina ventral to cirrus-sac, may form distal copulatory part. Uterus sacciform. Oncospheres and outer envelopes oval. Oncospheral hooks differ in shape. In Charadriiformes, Passeriformes, Anseriformes. Europe, Asia, North America. Type-species *N. nitidulans* (Krabbe, 1882) Spasskii & Spasskaya, 1954.

22b. Hooks not nitidoid .... **23.**

23a. Hooks aploparaksoid or echinorhynchoid .... **24.**

23b. Hooks diorchoid, ten ....
.... ***Microsomacanthus* Lopez-Neyra, 1942. (Fig. 29.33)**
(Syns *Abortilepis* Yamaguti, 1959; *Anserilepis* Spasskii & Tolkacheva, 1965; *Anterolepis* Spasskii, 1967; *Capiuterilepis* Oshmarin, 1962; *Craspedocotyla* Jordano & Diaz-Ungria, 1960; *Dicranolepis* Lopez-Neyra, 1942; *Dubininolepis* Spasskii & Spasskaya, 1954; *Echinatrium* Spasskii & Yurpalova, 1965; *Hispaniolepidoides* Yamaguti, 1959; *Kowalewskius* Yamaguti, 1959;

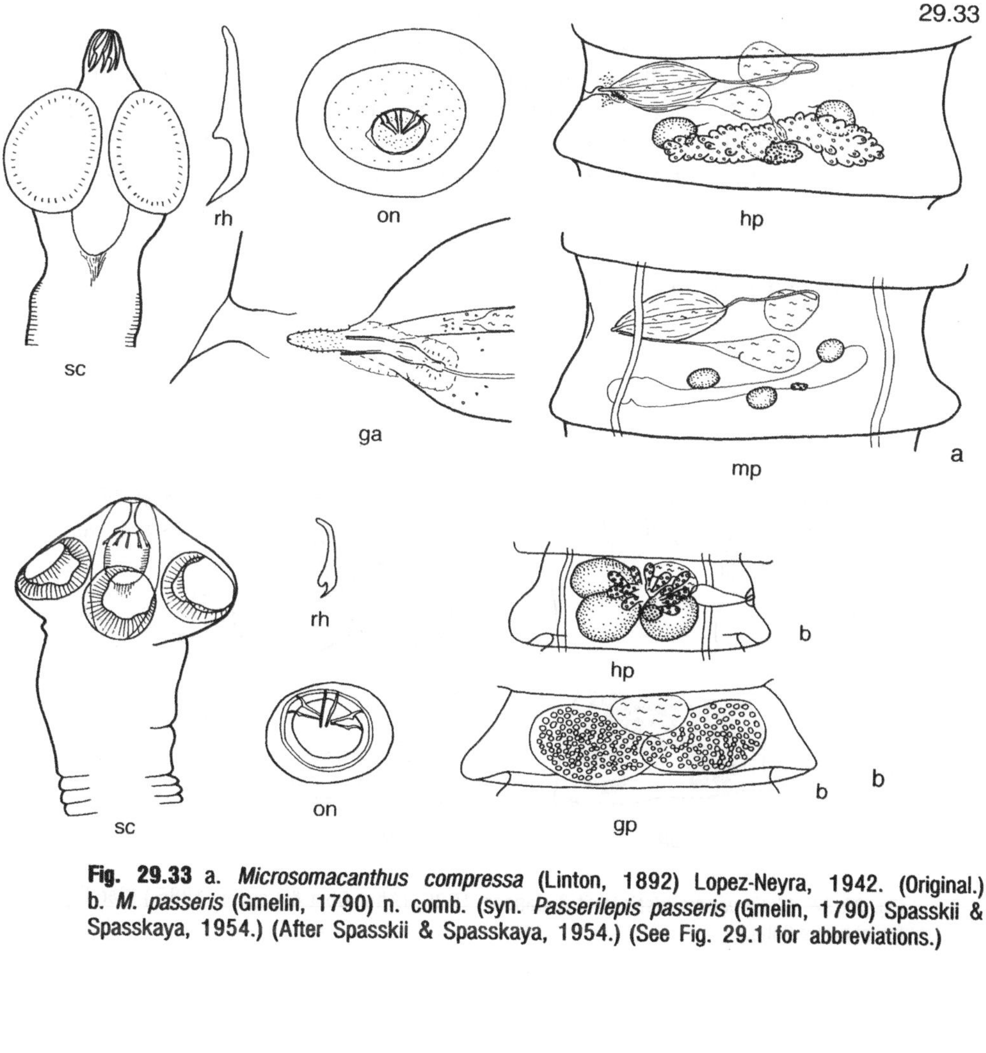

Fig. 29.33 a. *Microsomacanthus compressa* (Linton, 1892) Lopez-Neyra, 1942. (Original.) b. *M. passeris* (Gmelin, 1790) n. comb. (syn. *Passerilepis passeris* (Gmelin, 1790) Spasskii & Spasskaya, 1954.) (After Spasskii & Spasskaya, 1954.) (See Fig. 29.1 for abbreviations.)

Laricanthus Spasskii, 1963; *Mayhewia* Yamaguti, 1956; *Orientolepis* Spasskii & Yurpalova, 1964; *Oshmarinolepis* Spasskii & Spasskaya, 1954; *Passerilepis* Spasskii & Spasskaya, 1954; *Tschertkovilepis* Spasskii & Spasskaya, 1954; *Vigissotaenia* Matevosyan, 1968.)

Diagnosis: Rostellum retractile. Suckers unarmed. Proglottids craspedote. Internal longitudinal muscle bundles eight or numerous. Genital ducts dorsal to longitudinal osmoregulatory canals. Testes three, in transverse line or triangle. Cirrus-sac may be strongly muscular, reaching beyond midline or not. Cirrus usually armed. Ovary compact, transversely elongated bi-, tri- or multilobate, median or submedian. Vitellarium usually compact or slightly lobed, postovarian. Uterus sacciform. Oncosphere hooks may be variable in shape. In Anseriformes, Charadriiformes, Passeriformes, Lariformes, rarely Ciconiiformes. Cosmopolitan. Type-species *M. microsoma* (Creplin, 1829) Lopez-Neyra, 1942.

24a. Rostellar hooks ten (eight?), aploparaksoid. Four accessory apical suckers. All eight suckers armed .... ***Matiaraensis* Dixit & Capoor, 1986. (Fig. 29.34)**

Diagnosis: Rostellum retractile with single crown of hooks. Proglottids craspedote. Two layers of longitudinal muscles. Genital ducts dorsal to osmoregulatory canals. Testes three, in triangle. Cirrus-sac reaching near midline. Cirrus unarmed. Ovary bi- or trilobed, median. Vitellarium compact, postovarian, median. Seminal receptacle present. Uterus sacciform, bilobed or inverted U-shaped, with numerous oval oncospheres. Outer oncospheral envelope oval. In Passeriformes (*Acridotheres tristis*). India. Type-species *M. tristis* Dixit & Capoor, 1986.

24b. No accessory apical suckers .... **25.**

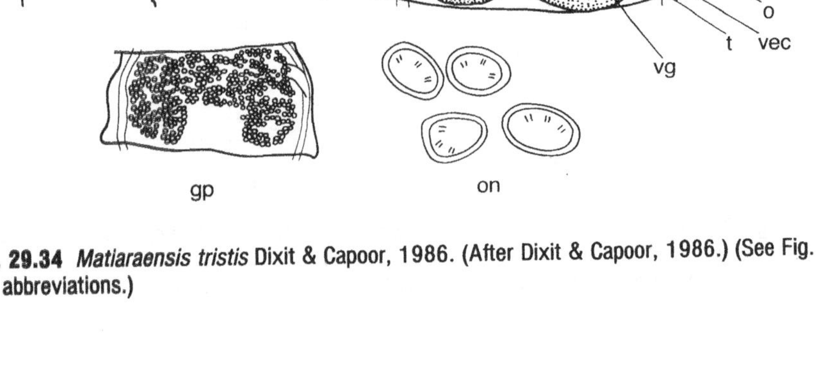

Fig. 29.34 *Matiaraensis tristis* Dixit & Capoor, 1986. (After Dixit & Capoor, 1986.) (See Fig. 29.1 for abbreviations.)

25a. External segmentation lacking ....
.... ***Parafimbriaria* Voge & Read, 1954. (Fig. 29.35)**

Diagnosis: Rostellum invaginable with single crown of ten hooks. Hooks close to aploparaksoid but with longer handle, shorter than blade. Suckers unarmed. Longitudinal muscle bundles numerous. Genital ducts dorsal to osmoregulatory canals. Ventral longitudinal osmoregulatory canals with transverse and oblique anastomoses. Testes three, irregularly spherical, in antiporal transverse row. Cirrus-sac short, not reaching midline. Cirrus armed. Ovary lobate, poral to testes. Vitellarium compact, postovarian, dorsal to ovary. Seminal receptacle elongate. Vagina ventral to cirrus-sac. Uterus sacciform, with irregular outline, metameric with possible anastomoses between proglottids. Oncospheres and outer envelopes ovoid or spherical. In Podicipediformes. Europe, Asia, North America. Type-species *P. websteri* Voge & Read, 1954.

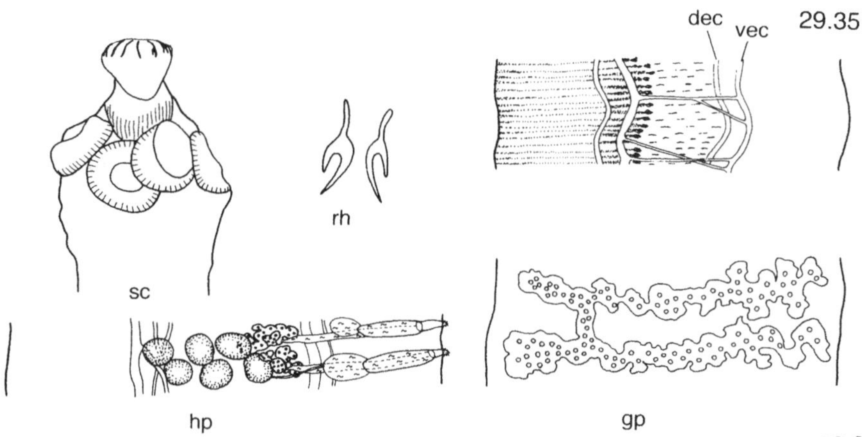

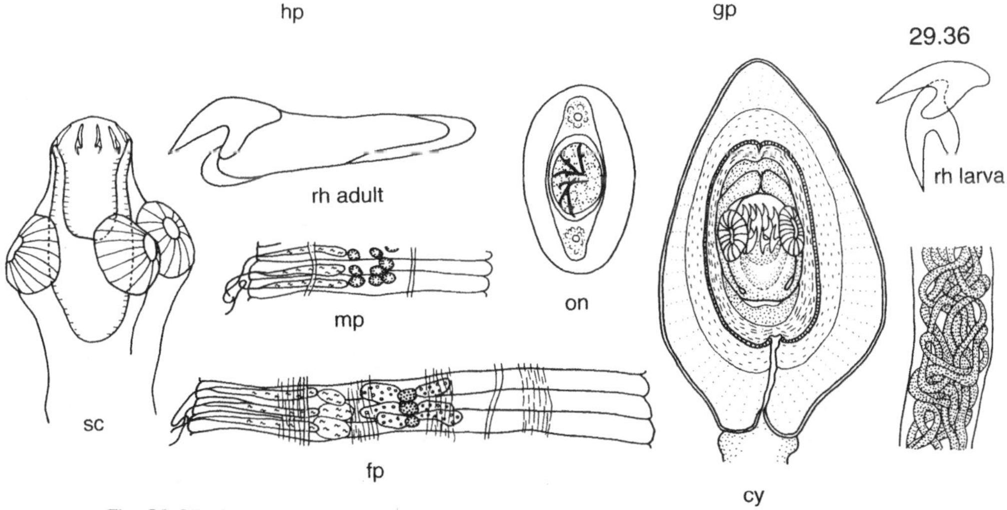

Fig. 29.35 *Parafimbriaria websteri* Voge & Read, 1954. (After Voge & Read, 1954.) (See Fig. 29.1 for abbreviations.)

Fig. 29.36 *Confluaria podicipina* (Szymanski, 1905) Spasskaya, 1966. (After Maksimova, 1989.) (See Fig. 29.1 for abbreviations.)

25b. External segmentation distinct .... **26.**

26a. Rostellar hooks ten, aploparaksoid in cysticercoids, in adults growing a special appendage on handle and guard .... ***Confluaria* Ablasov, 1953. (Fig. 29.36)**

(Syns *Colymbilepis* Spasskaya, 1966; *Dimorphocanthus* Maksimova, 1989.)

Diagnosis: Rostellum invaginable, armed with single crown of hooks. Suckers four, unarmed. Proglottids craspedote. Eight longitudinal, inner muscle bundles. Genital ducts dorsal to osmoregulatory canals. Three testes per proglottid. Cirrus-sac reaching midline or not. Cirrus armed. Accessory sac and cirrus stylet lacking. Ovary lobate, median or submedian. Vitellarium compact, oval, postovarian, median or submedian. Uterus sacciform. Oncospheres and outer envelopes oval. In Podicipediformes, rarely Anseriformes, Charadriiformes. Europe, Asia, Africa, North America. Type-species *C. spasskii* Ablasov, 1953.

26b. Rostellar hooks without special appendage in adults .... **27.**

27a. Rostellar hooks aploparaksoid or echinorhynchoid, ten. Testes three .... ***Wardium* Mayhew, 1925. (Fig. 29.37)**

(Syns *Chelacanthus* Yamaguti, 1959; *Debrosia* Spasskii, 1987; *Decacanthus* Yamaguti, 1959; *Glareolepis* Spasskii, 1967; *Hybridolepis* Spasskii, 1959; *Limnolepis* Spasskii & Spasskaya, 1954; *Lobatolepis* Yamaguti, 1959; *Parvibursa* Yamaguti, 1959; *Troglodytilepis* Yamaguti, 1956; *Variolepis* Spasskii & Spasskaya, 1954.)

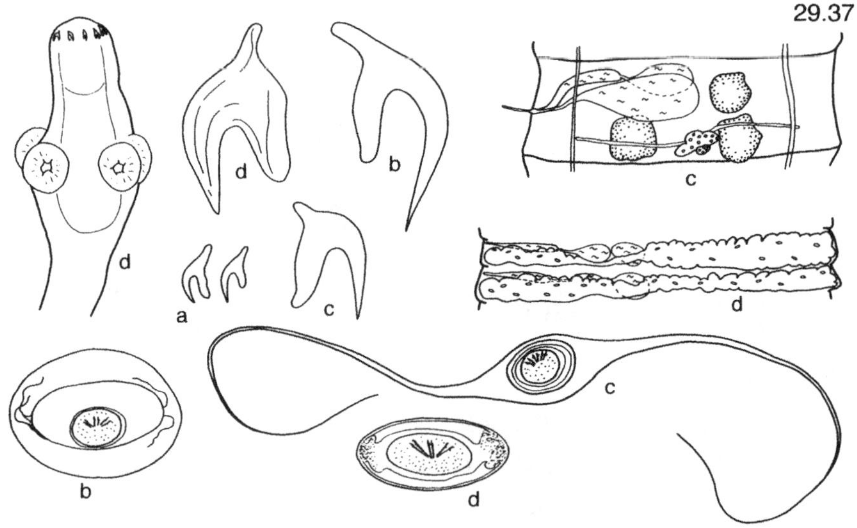

Fig. 29.37 a. *Wardium fusum* (Krabbe, 1869) Spasskii, 1961 (after Krabbe, 1869); b. *Wardium pseudofusum* Skryabin & Matevosyan, 1942 (after Macko, 1962a); c. *Wardium filamentoovatum* Macko, 1962 (after Macko, 1962a); d. *Wardium aequabile* (Rudolphi, 1810) (original). (See Fig. 29.1 for abbreviations.)

Diagnosis: Rostellum retractile or invaginable with single crown hooks. Suckers unarmed. Proglottids craspedote. Internal longitudinal muscle bundles numerous. Genital ducts dorsal to osmoregulatory canals. Three testes per proglottid, in row or triangle. Cirrus armed. Accessory sac and cirrus stylet lacking. Ovary lobate, median. Vitellarium compact or slightly lobate, postovarian. Seminal receptacle well developed. Uterus sacciform. Oncospheres oval, outer envelope oval or filamentous. In Charadriiformes, Lariformes, Anseriformes, Podicipediformes, Passeriformes. Europe, Asia, North and South America, Africa. Type-species *W. fusum* (Krabbe, 1869) Spasskii, 1961.

27b. Ten aploparaksoid hooks. Three to 11 testes per proglottid .... ***Mackolepis* Spasskii, 1962. (Fig. 29.38)**

Diagnosis: Rostellum with single crown of hooks. Inner and outer longitudinal muscle bundles numerous. Ovary bilobed. Vitellarium compact, oval. Uterus sacciform. Oncospheres oval. In Charadriiformes. Europe. Type-species *M. paucitesticulata* (Fuhrmann, 1913) Spasskii, 1962.

28a. Hooks arcuatoid .... **29.**
28b. Hooks not arcuatoid .... **30.**

29a. Three testes per proglottid .... ***Hispaniolepis* Lopez-Neyra, 1942. (Fig. 29.39)**
(Syn. *Satyolepis* Spasskii, 1965.)

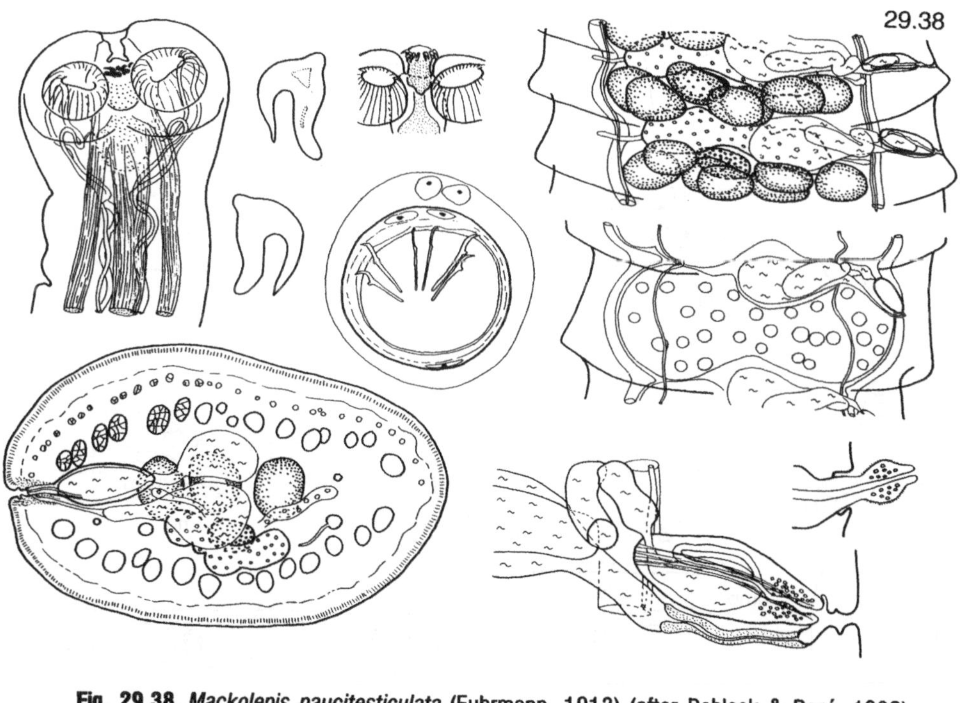

Fig. 29.38 *Mackolepis paucitesticulata* (Fuhrmann, 1913) (after Deblock & Rosé, 1962).

Diagnosis: Rostellum retractile with 14–20 hooks. Suckers unarmed. Proglottids craspedote, may have antiporal elongate appendage becoming longer as proglottids mature. Inner longitudinal muscle bundles eight or numerous. Genital ducts dorsal to osmoregulatory canals. Testes in transverse row or slight triangle, one or two antiporal. Cirrus-sac reaching median line or not, may be muscular. Cirrus armed or not, without stylet. Ovary compact or slightly lobed, median or submedian. Vitellarium compact or slightly lobed, postovarian. Seminal receptacle well developed. Uterus sacciform. Oncospheres and outer envelopes oval. In Galliformes, Gruiformes and Passeriformes. Europe, Asia, Africa, North America. Type-species *H. villosa* (Bloch, 1782) Lopez-Neyra, 1942.

29b. Typically four testes per proglottid .... ***Oligorchis*** **Fuhrmann, 1906. (Fig. 29.40)**

Diagnosis: Rostellum retractile with single crown of 14–16 arcuatoid hooks. Suckers unarmed. Proglottids craspedote. Inner longitudinal muscle bundles numerous. Genital ducts dorsal to osmoregulatory canals. Testes in more or less transverse row. Cirrus-sac short, not reaching poral osmoregulatory canal. Ovary deeply lobed, median. Vitellarium small, median. Seminal receptacle present. Uterus sacciform. Oncospheres and outer envelope subspherical. In

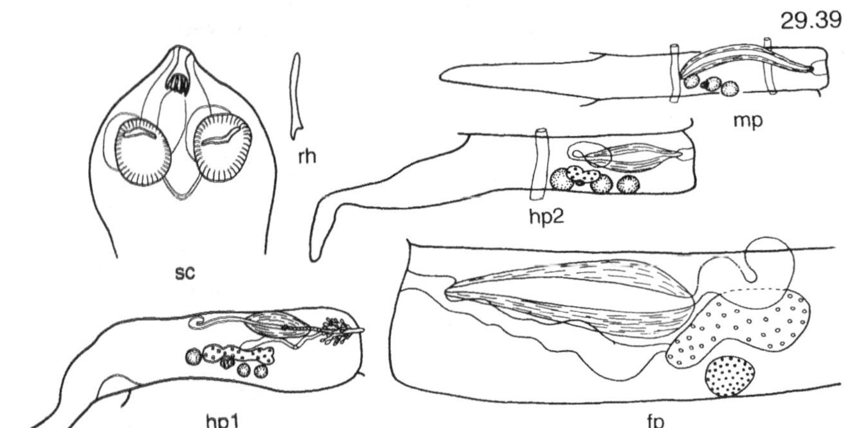

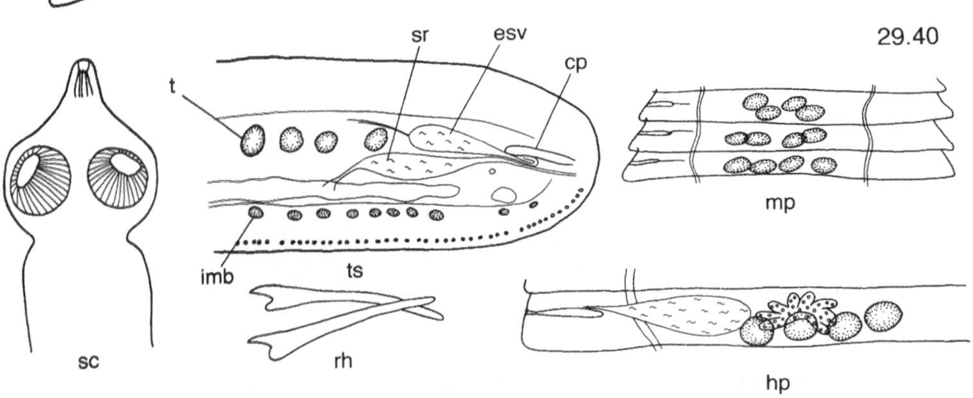

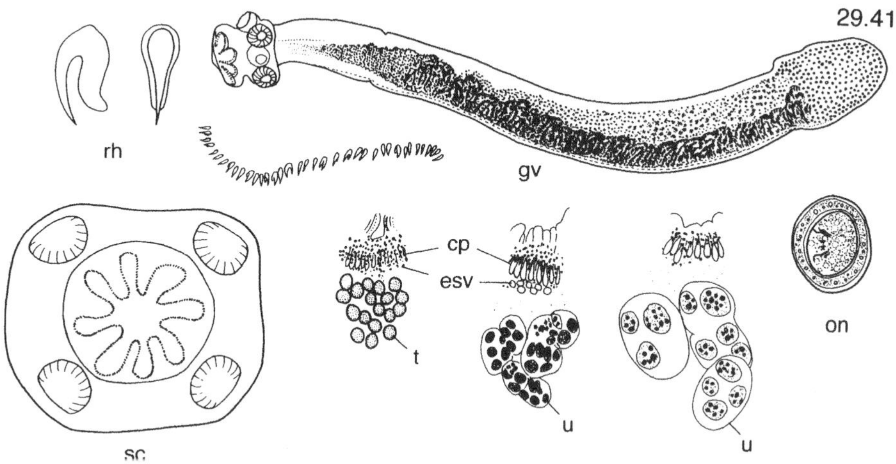

Fig. 29.39 *Hispaniolepis villosa* (Bloch, 1782) Lopez-Neyra, 1942. (hp 1 after Lopez-Neyra, 1942; rh, sc after Skryabin & Matevosyan, 1945; fp, hp 2, mp after Spasskaya, 1962). (See Fig. 29.1 for abbreviations.)

Fig. 29.40 *Oligorchis strangulatus* Fuhrmann, 1906. (After Fuhrmann, 1906a.) (See Fig. 29.1 for abbreviations.)

Fig. 29.41 *Nematoparataenia southwelli* Fuhrmann, 1934 (on, rh after Wisniewski, 1971, others after Cielecka & Misiura, unpublished.) (See Fig. 29.1 for abbreviations.)

Accipitriformes. Central and South America. Type-species *O. strangulatus* Fuhrmann, 1906.

30a. About 600–1000 aploparaksoid hooks arranged in single undulating row with eight lobes on top of large rostellum. No external segmentation, internal segmentation very complicated. Male and female genital ducts open in lateral groove. Strobila round in cross section .... *Nematoparataenia* Maplestone & Southwell, 1922. (Fig. 29.41)

Diagnosis: Scolex little wider than rostellum with unarmed suckers, directed anterolaterally. Strobila with lateral longitudinal groove. Protandry evident. Many small, rounded testes connected with external and internal seminal vesicles, latter enclosed in cylindrical, thin-walled cirrus-sac. Ovary follicular, mixed with vitelline cells in mid-region of body. Seminal receptacle elongated. Vagina parallel to cirrus-sac opens near male pore. Uterus little known, seems to be sac-like structure in posterior enlarged part of body. Oncosphere and outer envelope oval. In Anseriformes. Europe, Asia, Australia. Type-species *N. paradoxa* Maplestone & Southwell, 1922.

30b. Up to 50 aploparaksoid, coronuloid or cricetoid hooks .... 31.

31a. Single row of about 50 cricetoid rostellar hooks .... *Paradicranotaenia* Lopez-Neyra, 1943. (Fig. 29.42)

Diagnosis: Suckers unarmed. Pseudoscolex not distinct, simple enlargement behind scolex, with genital primordia. Proglottids craspedote, crowded. Inner

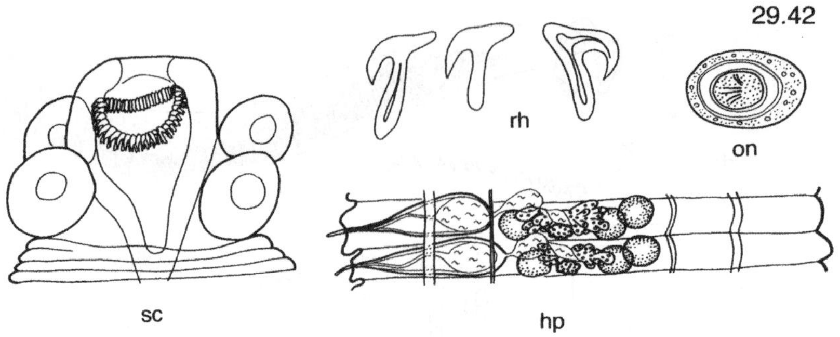

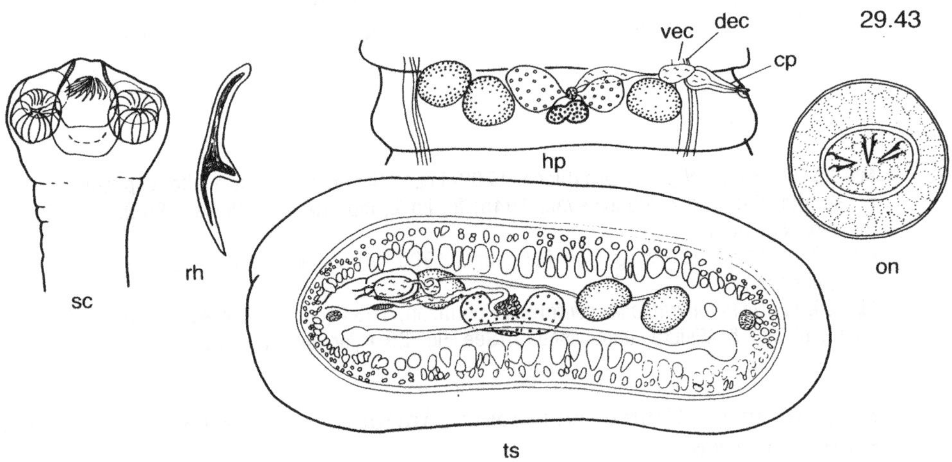

Fig. 29.42 *Paradicranotaenia anormalis* Lopez-Neyra, 1943. (After Lopez-Neyra, 1943b, modified.) (See Fig. 29.1 for abbreviations.)
Fig. 29.43 *Hamatolepis teresoides* (Fuhrmann, 1906) Spasskii, 1962. (Original, except ts after Czaplinski & Vaucher, 1986.) (See Fig. 29.1 for abbreviations.)

longitudinal muscle bundles eight. Genital ducts dorsal to poral osmoregulatory canals. Testes three, one poral, two antiporal, in transverse row or triangle. Cirrus-sac not reaching median line. Ovary multilobate, median. Vitellarium slightly lobate, postovarian, ventral to ovary. Seminal receptacle pyriform. Vagina posterior to cirrus-sac. Uterus sacciform. Oncosphere and outer envelope oval. In Columbiformes and Galliformes. Europe. Type-species *P. anormalis* Lopez-Neyra, 1943.

31b. Fewer than 50 diorchoid or coronuloid hooks .... **32.**

32a. Hooks diorchoid .... **33.**
32b. Hooks coronuloid .... **34.**

33a. Three testes per proglottid ....
.... ***Hamatolepis* Spasskii, 1962. (Fig. 29.43)**

Diagnosis: Rostellum retractile with single crown of usually 16 diorchoid hooks.

Suckers unarmed. Proglottids craspedote. Inner longitudinal muscle bundles numerous. Genital ducts dorsal to osmoregulatory canals. Cirrus-sac not reaching midline. Cirrus unarmed. Ovary compact, bilobed, median. Vitellarium compact, postovarian, median. Seminal receptacle present. Uterus sacciform. Oncosphere oval, outer envelope spherical. In Anseriformes. Europe, Asia, Australia. Type-species *H. teresoides* (Fuhrmann, 1906) Spasskii, 1962.

33b. Ten to twenty testes per proglottid .... *Fuhrmanacanthus* Spasskii, 1966. (Fig. 29.44)

Diagnosis: Rostellum retractile with single circle of usually 16 diorchoid hooks. Suckers unarmed. Proglottids craspedote. Inner longitudinal muscle bundles numerous. Genital ducts dorsal to osmoregulatory canals. Ventral osmoregulatory stems have transverse anastomoses. Cirrus-sac short, not reaching midline. Cirrus armed. No accessory sac. Ovary compact, bilobed, median. Vitellarium compact or slightly lobed. Seminal receptacle well developed. Uterus sacciform. Oncospheres and outer shell oval. In Anseriformes. South America. Type-species *F. propeteres* (Fuhrmann, 1907) Spasskii, 1966.

34a. Usually 16 coronuloid hooks (exceptionally 14 or 18). Three testes per proglottid. Armed base of genital atrium protrusible .... *Wardoides* Spasskii, 1963. (Fig. 29.45)

Diagnosis: Rostellum invaginable, with single crown of hooks. Suckers unarmed. Pseudoscolex simple, unfolded. Proglottids craspedote. Genital ducts dorsal to osmoregulatory canals. Inner longitudinal muscle bundles numerous. Testes transversely oval, usually smooth, arranged in transverse row or slight triangle, overlapping ovary anteriorly. Cirrus-sac may reach antiporal osmoregulatory canals. Cirrus armed. Ovary deeply multilobate, poral, may overlap the antiporal osmoregulatory canals dorsally. Vitellarium irregularly lobed, overlaps the antiporal testis. Seminal receptacle spindle-like, oblique. Vagina with distal copulatory part ventral or posterior to distal cirrus-sac. Uterus sacciform reticulate when gravid. Oncosphere and outer envelope oval, latter with hair-like

29.44

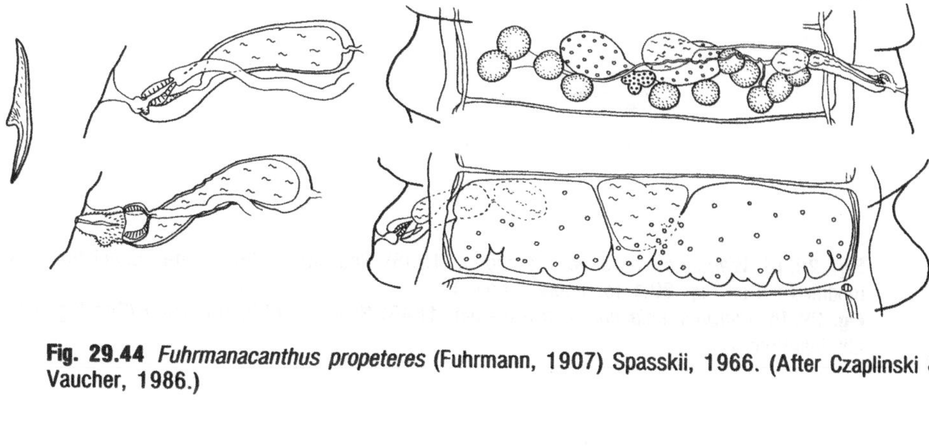

Fig. 29.44 *Fuhrmanacanthus propeteres* (Fuhrmann, 1907) Spasskii, 1966. (After Czaplinski & Vaucher, 1986.)

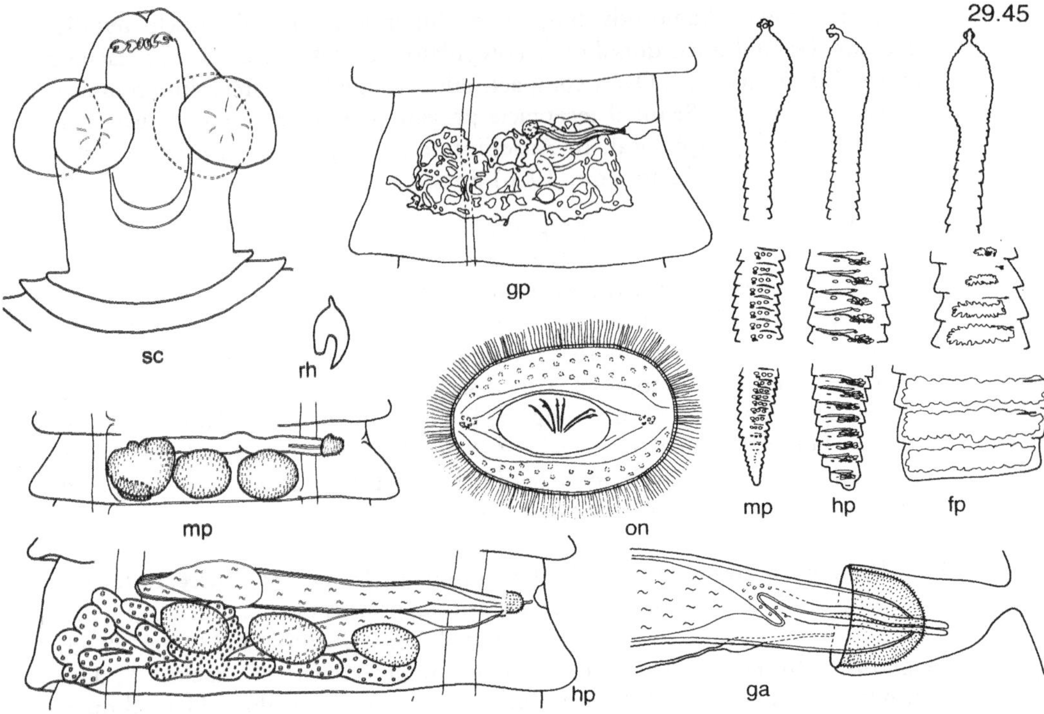

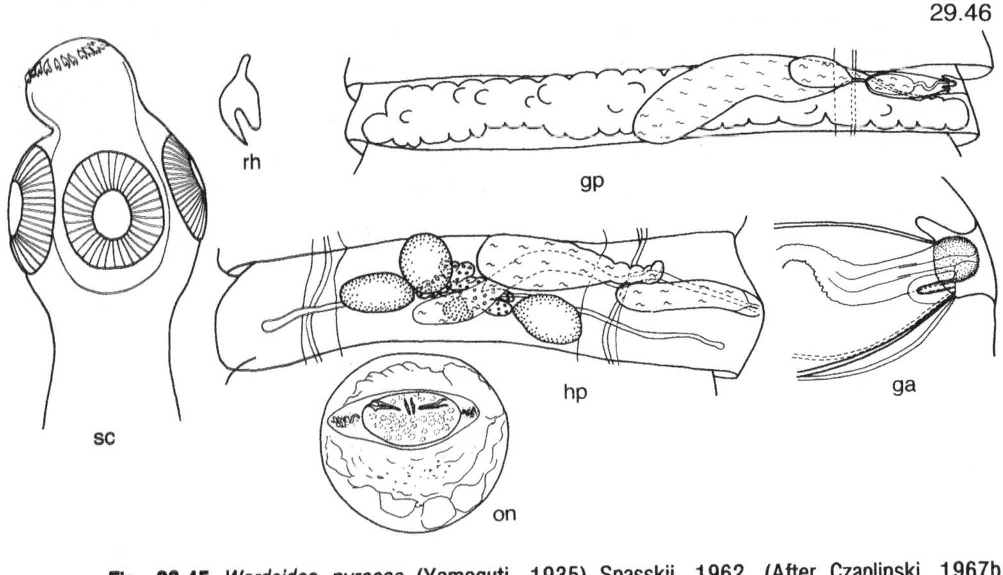

Fig. 29.45 *Wardoides nyrocae* (Yamaguti, 1935) Spasskii, 1962. (After Czaplinski, 1967b, modified.) (See Fig. 29.1 for abbreviations.)

Fig. 29.46 *Dicranotaenia coronula* (Dujardin, 1845) Railliet, 1892. (Original.) (See Fig. 29.1 for abbreviations.)

projections. In Anseriformes, mainly *Cygnus*. Europe, Asia. Type-species *W. nyrocae* (Yamaguti, 1935) Spasskii, 1962.

34b. About 20–30 coronuloid hooks. Three testes per proglottid. Internal accessory sac present .... *Dicranotaenia* Railliet, 1892. (Fig. 29.46)
(Syn. *Weinlandia* Mayhew, 1925.)

Diagnosis: Rostellum invaginable, rarely retractile, with single crown of hooks. Suckers unarmed. Proglottids craspedote. Inner longitudinal muscle bundles numerous. Genital ducts dorsal to osmoregulatory canals. Testes smooth or slightly lobed, in triangle or transverse row, one testis usually poral. Cirrus-sac usually short, not reaching median line. Protrusible, armed accessory sac present, internal to cirrus-sac, may be fused with base of armed, invaginable cirrus. Ovary multilobate, median or submedian, overlaps middle testis ventrally or also other two testes. Vitellarium slightly lobate, postovarian. Seminal receptacle well developed. Uterus sacciform. Oncospheres oval, outer envelope spherical. In Anseriformes, exceptionally *Gallus gallus* and *Fulica atra*. Cosmopolitan except Australia and South America. Type-species *D. coronula* (Dujardin, 1845) Railliet, 1892.

35a. Irregular crown of 16–34 small spiniform rostellar hooks .... *Thaumasiolepis* Mariaux & Vaucher, 1989. (Fig. 29.47)

Diagnosis: Rostellum invaginable, armed with single irregular crown of very small hooks, variable in number. Suckers four, unarmed. Internal segmentation

29.47

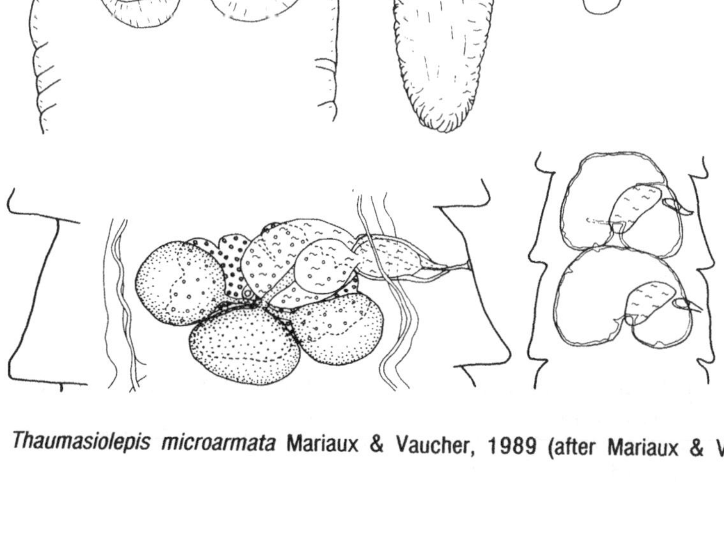

Fig. 29.47 *Thaumasiolepis microarmata* Mariaux & Vaucher, 1989 (after Mariaux & Vaucher, 1989).

evident. Proglottids craspedote. Genital ducts pass between dorsal and ventral osmoregulatory canals. Three testes per proglottid. Cirrus-sac crosses osmoregulatory canals. Ovary lobate, median. Vitellarium compact, postovarian. Seminal receptacle well developed. Vagina ventral to cirrus-sac. Uterus sacciform. Oncospheres small, oval, with oval envelopes. In Piciformes. Ivory Coast. Type-species *T. microarmata* Mariaux & Vaucher, 1989.

35b. Four or five rows of numerous (about 70) hammer-like rostellar hooks .... *Ortleppolepis* Spasskii, 1965. (Fig. 29.48)

Diagnosis: Rostellum rounded. Suckers unarmed. Neck long. Proglottids craspedote with long sinistral appendage. Genital ducts dorsal to osmoregulatory canals. Ventral, wider pair of longitudinal osmoregulatory canals irregularly con-

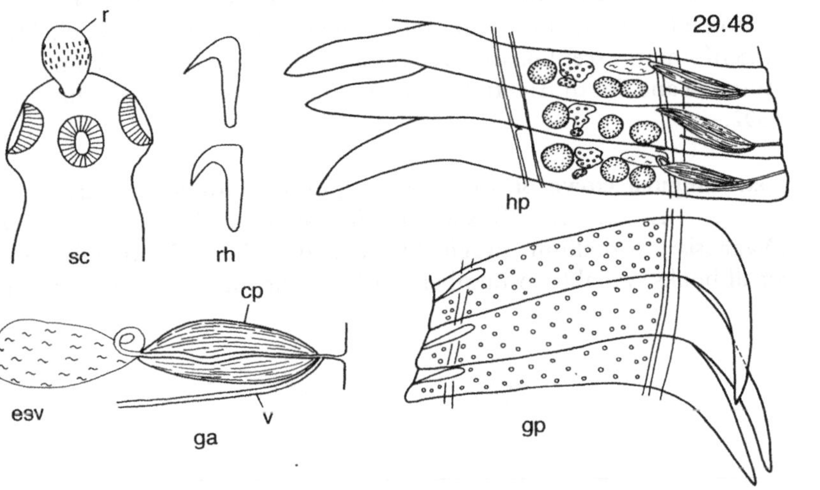

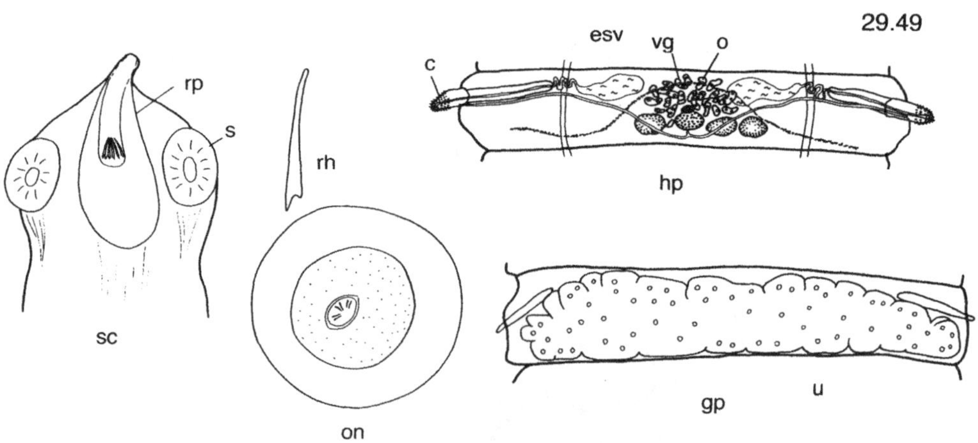

Fig. 29.48 *Ortleppolepis multiuncinata* (Ortlepp, 1963) Spasskii, 1965. (After Ortlepp, 1963, modified.) (See Fig. 29.1 for abbreviations.)

Fig. 29.49 *Diploposthe laevis* (Bloch, 1782) Jacobi, 1896. (gp, on, rh, sc original; hp after Sulgostowska, 1977.) (See Fig. 29.1 for abbreviations.)

nected by transverse canals. Testes three, round, in transverse row, two poral and one antiporal. External seminal vesicle well developed, internal one not observed (absent ?). Cirrus-sac oval to almost tubular, overlaps poral osmoregulatory canals. Ovary irregularly lobate, submedian. Vitellarium slightly oval, postovarian. Seminal receptacle oval. Vagina weakly developed, opens ventral to cirrus-sac. Uterus sacciform. Oncospheres numerous, large, rounded. In Galliformes. Africa. Type-species *O. multiuncinata* (Ortlepp, 1963) Spasskii, 1965.

Key to genera of Diploposthinae

1a. Female gonads single per proglottid .... **2.**
1b. Female gonads double per proglottid .... **3.**

2a. Male and female gonads single, in median field. Double set of male and female genital ducts. Ten arcuatoid rostellar hooks .... ***Diploposthe* Jacobi, 1896. (Fig. 29.49)**
Diagnosis: Rostellum retractile with single crown of hooks. Suckers unarmed. Proglottids craspedote, numerous. Inner longitudinal muscle bundles numerous (14–18). Genital pores bilateral. Genital ducts dorsal to osmoregulatory canals. Single set of 2–20, usually three or 6–13 testes. Vas deferens unpaired. External and internal seminal vesicles double. Cirrus-sac double, not reaching midline. Cirrus armed. Ovary single, bilobed with many finger-like lobules. Vitellarium single, compact or slightly lobed, postovarian. Ovary and vitellarium pretesticular. Uterus initially tubular, later sacciform with diverticula. Oncosphere and external envelope rounded. Tendency to dioecism in *D. bifaria*. In Anseriformes. Europe, Asia, Africa, Australia, North America. Type-species *D. laevis* (Bloch, 1782) Jacobi, 1896.

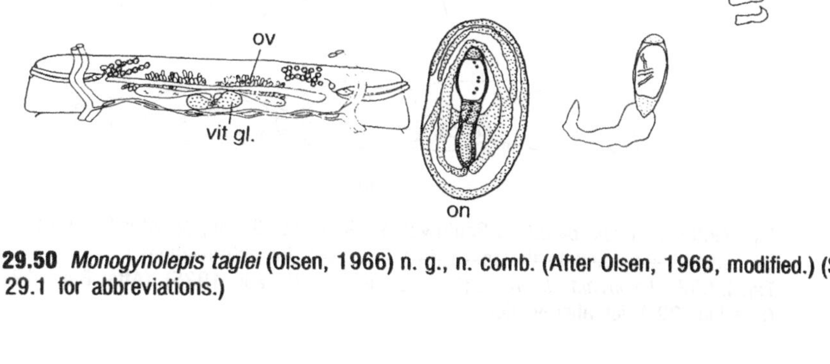

Fig. 29.50 *Monogynolepis taglei* (Olsen, 1966) n. g., n. comb. (After Olsen, 1966, modified.) (See Fig. 29.1 for abbreviations.)

2b. Male genitalia double with two groups of 25–36 testes each per proglottid in submedian fields, pre-ovarian. Twelve diorchoid rostellar hooks .... *Monogynolepis* n. g. (Fig. 29.50)

Diagnosis: Rostellum retractile. Suckers unarmed. Proglottids craspedote. Inner longitudinal muscle bundles numerous. Both dorsal and ventral longitudinal osmoregulatory canals with transverse connections that anastomose with each other. Cirrus-sac short, overlaps osmoregulatory canals ventrally (?). Cirrus spinose. Ovary single, bilobed, elongated transversely with many finger-like lobules. Vitellarium oval to reniform, single, postovarian. Seminal receptacle elongated, double. Uterus single, tubular in mature proglottids, sacciform when gravid. Oncospheres oval, inner envelope with tail, outer envelope oval. In

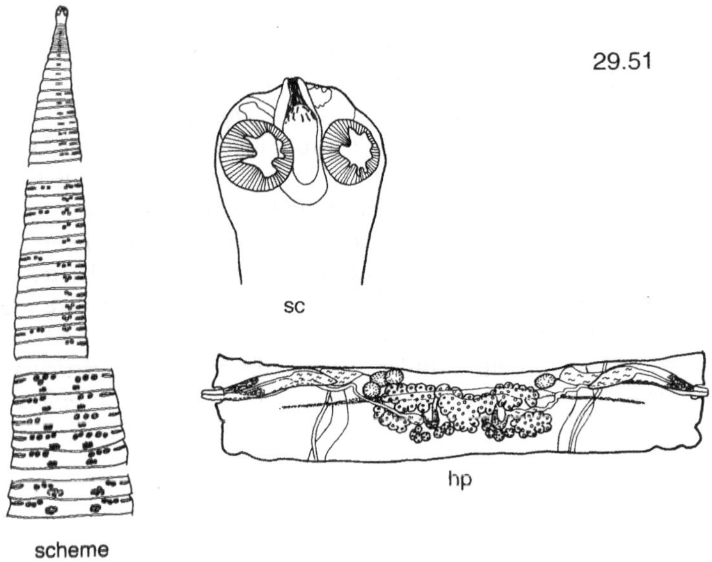

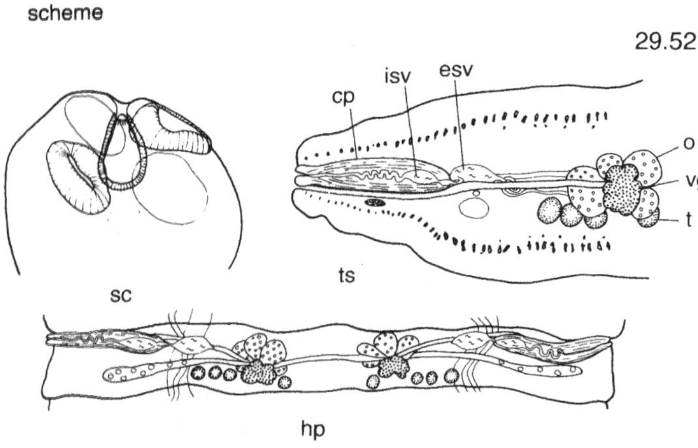

Fig. 29.51 *Jardugia paradoxa* Southwell & Hilmy, 1929. (hp, sc after Southwell & Hilmy, 1929; gv scheme according to the text.) (See Fig. 29.1 for abbreviations.)
Fig. 29.52 *Diplogynia oligorchis* (Maplestone, 1922) Baer, 1925). (After Baer, 1925a, modified.) (See Fig. 29.1 for abbreviations.)

Lagidium peruanum (Mammalia, Chinchillidae). South America, Chile. Type-species *M. taglei* (Olsen, 1966) n. g., n. comb. (syn. *Diplophallus taglei* Olsen, 1966).

3a. **Male and female genitalia double in mature proglottids, two to six testes in two submedian fields, pre-ovarian, young proglottids contain single set of male and female organs. About ten arcuatoid rostellar hooks .... *Jardugia* Southwell & Hilmy, 1929. (Fig. 29.51)**

Diagnosis: Rostellum retractile with single crown of hooks. Suckers unarmed. Proglottids craspedote. Three layers (?) of longitudinal muscles. Inner longitudinal muscle bundles numerous. Genital pores bi- or unilateral. Genital ducts dorsal to osmoregulatory canals. Single set of male genitalia in most young proglottids. Cirrus-sac irregularly alternating. In mature proglottids male and female genitalia become double. Testes two to six per proglottid in two groups anterior to ovary. External and internal seminal vesicles present. Cirrus-sac small, reaches or overlaps osmoregulatory canals. Cirrus spinose. Duplication of female organs irregular at first. In mature proglottids ovary lobate, double. Vitellarium deeply lobate, double, postovarian. Seminal receptacle weak. Vagina double. Uterus single, tubular when young, then sacciform. Oncosphere unknown. In Ardeiformes. Nigeria. Type-species *J. paradoxa* Southwell & Hilmy, 1929.

3b. **Double set of male and female genital organs. Three to five testes on each side of proglottid, postovarian .... *Diplogynia* Baer, 1925. (Fig. 29.52)**

Diagnosis: Rostellum retractile, with single crown of small hooks, shape (arcuatoid?) and exact number (ten?) unknown. Suckers unarmed. Proglottids craspedote. Longitudinal muscle bundles in two layers, inner ones numerous. Male and female genitalia double. Genital ducts dorsal to both osmoregulatory canals. Cirrus-sac cylindrical, hardly reaches poral osmoregulatory canals. Ovary lobed, submedian, double. Vitellarium slightly lobed, postovarian. Seminal receptacle elongated, double. Vagina ventral to cirrus-sac. Uterus single, initially tubular, then gravid sacciform. Developed oncospheres unknown. In Anseriformes, Ciconiiformes. Australia, North America, Java. Type-species *D. oligorchis* (Maplestone, 1922) Baer, 1925.

Key to Hymenolepididae of mammals* (by C. Vaucher)

1a. **Three testes per proglottid (only abnormal proglottids have more or less) .... 10.**

1b. **One, two or more than three testes per proglottid (when more than three the number is variable) .... 2.**

2a. **One or two testes per proglottid .... 3.**

2b. **More than three testes per proglottid .... 4.**

3a. **One testis per proglottid. Strobila very small as in *Urocystis* (Fig. 29.53). Rostellum has numerous minute hooklets. Parasites of Insectivora (shrews) from the Nearctic region .... *Protogynella* Jones, 1943. (Fig. 29.54)**

* See also *Monogynolepis*, p. 644.

Diagnosis: Strobila with few proglottids, much broader than long. Scolex with retractile armed rostellum. Rostellar hooks of nearly identical length. One testis. Ovary transverse. Vitelline gland round, asymmetrical, in aporal half of proglottid. Cirrus-sac long, reaching middle of proglottid. Cirrus armed with minute spines. Uterus sacciform, containing only few relatively big eggs. Type-species *P. blarinae* Jones, 1943.

3b. Two testes per proglottid. Rostellum with numerous minute hooks. Parasites of Insectivora (shrews) from the Nearctic region .... *Pseudodiorchis* Skryabin & Matevosyan, 1948. (Fig. 29.55)

Diagnosis: Strobila with numerous proglottids and progressive maturation. Scolex with retractile, armed rostellum. Rostellar hook number nearly constant. Hooks of nearly identical length. Two testes, in posterior, median part of proglottid, partially overlapping female gonads. Ovary trilobed or multilobed. Vitelline gland round and asymmetrical. Cirrus-sac reaching middle of proglottid. Cirrus armed with minute spines. Uterus at first bilobed, entirely filling gravid proglottids. Type-species *P. reynoldsi* (Jones, 1944).

4a. Scolex with armed rostellum. Hooks of characteristic shape (rose-thorn). Ten or more testes. Parasitic in Carnivora and rodents in the Ethiopian region .... *Pseudandrya* Fuhrmann, 1943. (Figs 29.56, 29.57)

Diagnosis: Strobila with numerous proglottids and progressive maturation. Number of rostellar hooks nearly constant. Hooks of nearly identical length. Ten or more testes, most aporal, few ventral and poral. Ovary multilobed. Vitelline gland lobed, median. Cirrus-sac short, not reaching middle of proglottid. Cirrus unarmed. Uterus strongly labyrinthine, finally filling whole proglottid, basic structure remaining in gravid proglottids. Type-species *P. monardi* Furhmann, 1943.

4b. Scolex without hooks. Rostellum either absent or present but unarmed .... 5.

5a. Testes 25–27 in two groups, one poral and one aporal. Parasite of Suiformes and rodents, in Sri Lanka, China and Japan .... *Pseudanoplocephala* Baylis, 1927. (Fig. 29.58)

Diagnosis: Strobila with numerous proglottids and progressive maturation. Scolex without rostellum. 25–27 testes in two groups, separated by female glands. Ovary with numerous finger-like lobes. Vitelline gland somewhat lobate, median. Cirrus-sac short, not reaching middle of proglottid. Uterus an elongate pouch with numerous diverticula penetrating cortex. Type-species *P. crawfordi* Baylis, 1927.

5b. Fewer than 25 testes .... 6.

6a. Testes in two groups, one poral and one aporal, generally forming two single rows in posterior part of proglottid .... 7.
6b. Testes surrounding female gonads .... 9.

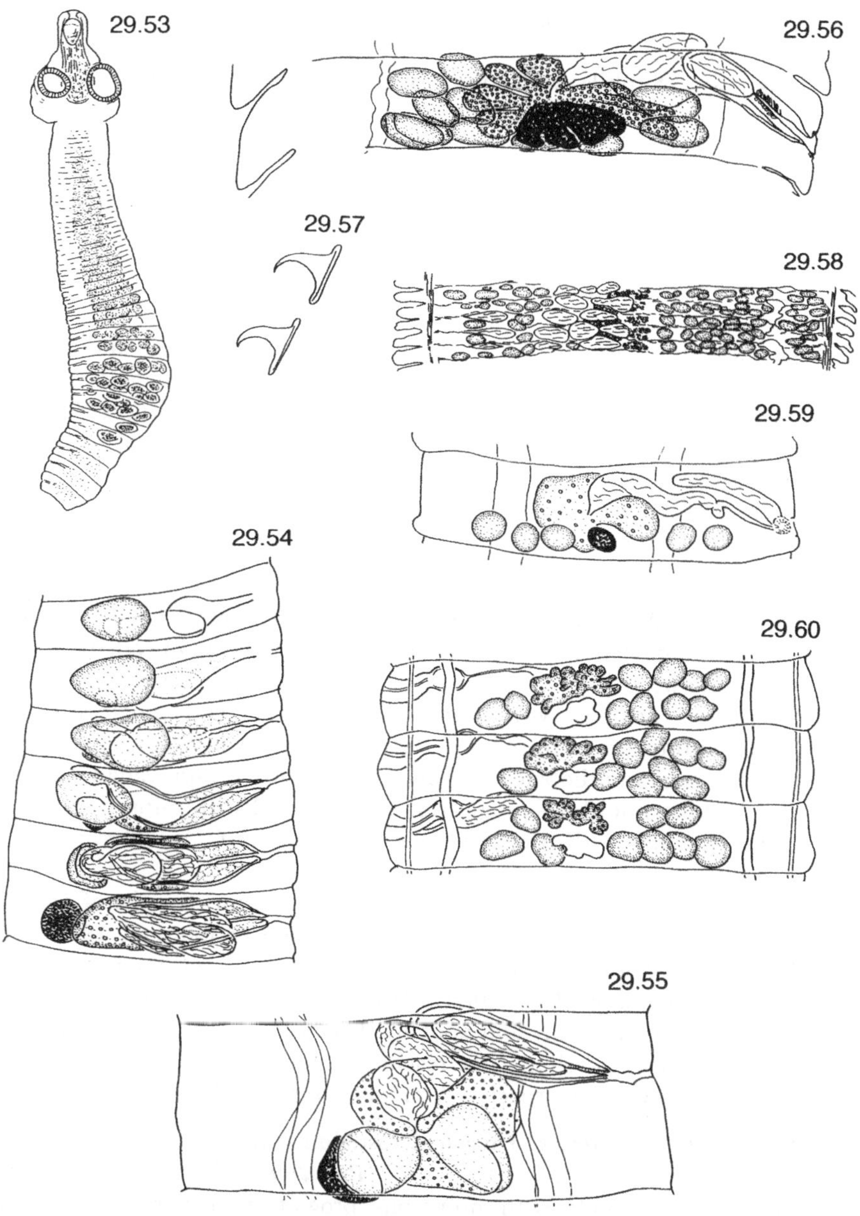

Figs 29.53–29.60 29.53. Strobila of *Urocystis prolifer* (Villot, 1880) (original). 29.54. Part of strobila of *Protogynella blarinae* Jones, 1943 (after Vaucher & Durette-Desset, 1978). 29.55. Proglottid of *Pseudodiorchis reynoldsi* (Jones, 1944) (after Vaucher & Durette-Desset, 1978). 29.56. Proglottid of *Pseudandrya monardi* Fuhrmann, 1943 (after Fuhrmann, 1943). 29.57. Two rostellar hooks of *Pseudandrya monardi* (after Fuhrmann, 1943). 29.58. Part of strobila of *Pseudanoplocephala crawfordi* Baylis, 1927, frontal section (after Baylis, 1927). 29.59. Proglottid of *Pentorchis arkteios* Meggitt, 1927 (after Meggitt, 1927a). 29.60. Proglottids of *Hymenandrya thomomysi* Smith, 1954 (after Smith, 1954).

7a. Five testes. Parasite of Carnivora (*Ursus*) in Burma .... *Pentorchis* Meggitt, 1927. (Fig. 29.59)

Diagnosis: Strobila with numerous proglottids and progressive maturation. Scolex with unarmed (?) rostellum. Testes five, in two groups (two poral), more or less separated by female glands. Ovary elongated, median. Vitelline gland entire. Cirrus-sac short, reaching osmoregulatory vessels. No external seminal vesicle. Genital pore surrounded by muscular sphincter. Uterus pouch-like, slightly lobed, with few minute incomplete septa. Type-species *P. arkteios* Meggitt, 1927.

7b. Seven testes or more .... 8.

8a. Rostellum absent; 7–15 testes separated into poral and aporal groups. Embryophores with filiform processes. Parasite of rodents (*Thomomys*) in USA .... *Hymenandrya* Smith, 1954. (Fig. 29.60)

Diagnosis: Strobila with numerous proglottids and progressive maturation. Scolex without rostellum. Testes 7–15, in two groups separated by female glands. Ovary multilobate. Vitelline gland ovoid or slightly lobate. Cirrus-sac short, not reaching middle of proglottid. Uterus reticulate, network still visible in gravid proglottids. Type-species *H. thomomysi* Smith, 1954.

8b. Unarmed rostellum present, 9–12 testes forming poral and aporal rows. Embryophores without filiform processes. Parasite of rodents (*Leopoldamys*) in Borneo .... *Chitinolepis* Baylis, 1926. (Fig. 29.61)

Diagnosis: Strobila with numerous proglottids and progressive maturation. Scolex with unarmed rostellum. Testes 9–12, in two groups, separated by female glands. Ovary multilobate. Vitelline gland compact, postovarian. Cirrus-sac short, not reaching middle of proglottid. Cirrus armed with minute spines. Uterus an elongate pouch with irregular walls. Type-species *C. mjoebergi* Baylis, 1926.

9a. Four to seven testes. Unarmed rostellum present, parasite of rodents (*Tatera*) in India .... *Paraoligorchis* Wason & Johnson, 1977. (Fig. 29.62)

Diagnosis: Strobila with numerous proglottids and progressive maturation. Scolex with unarmed rostellum. Testes four to seven, separated, surrounding female gonads ventrally and aporally. Ovary multilobate. Vitelline gland lobed, median. Cirrus-sac short, not reaching middle of proglottid. Cirrus unarmed. Uterus saccular. Type-species *P. taterae* Wason & Johnson, 1977.

9b. Eight to 12 testes. Unarmed rostellum present. Parasite of Chiroptera in India .... *Pseudoligorchis* Johri, 1934. (Fig. 29.63)

Diagnosis: Strobila with numerous proglottids and progressive maturation. Genital ducts pass between osmoregulatory canals. Testes 8–12, surrounding female gonads ventrally and aporally. Ovary somewhat lobate. Vitelline gland lobed, median. Cirrus-sac short, not reaching middle of proglottid. Uterus an irregularly lobed sac. Type-species *P. magnireceptaculata* Johri, 1934.

10a. Scolex with an armed rostellum, the hooks may be very small ... 11.

10b. Scolex without hooks: either rostellum absent or unarmed rostellum present .... 27.

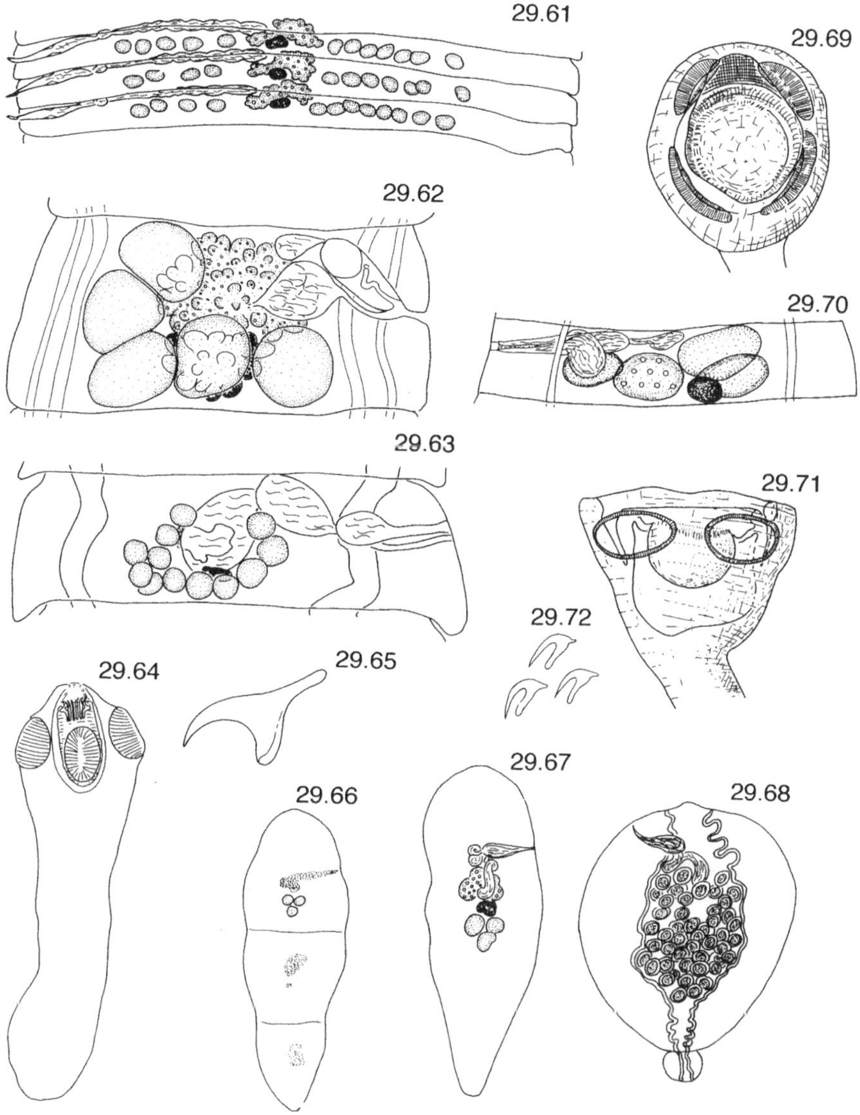

Figs 29.61-29.72 29.61. Proglottids of *Chitinolepis mjoebergi* Baylis, 1926 (after Smith, 1954). 29.62. Proglottid of *Paraoligorchis taterae* Wason & Johnson, 1977 (after Wason & Johnson, 1977). 29.63. Proglottid of *Pseudoligorchis magnireceptaculata* Johri, 1934 (after Johri, 1934). 29.64. Scolex and undifferentiated strobila of *Pseudhymenolepis redonica* Joyeux & Baer, 1935 (original). 29.65. Rostellar hook of *Pseudhymenolepis redonica* Joyeux & Baer, 1935 (original). 29.66-29.68. Proglottids of *Pseudhymenolepis redonica* at different stages of maturation (after Joyeux & Baer, 1936c). 29.69, 29.71. Scolex of *Coronacanthus integrus* (Hamann, 1891) (after Vaucher, 1971). 29.70. Proglottid of *Coronacanthus integrus* (after Vaucher, 1971). 29.72. Three rostellar hooks of *Coronacanthus integrus* (after Vaucher, 1971).

11a. Strobila hyperapolytic, parasite of white-toothed shrews in the Old World .... *Pseudhymenolepis* Joyeux & Baer, 1935. (Figs 29.64–29.68)

Diagnosis: Hyperapolytic strobila. Scolex with invaginable, armed rostellum. Rostellar hook number nearly constant; hooks of nearly identical length. Testes three in triangle behind female gonads. Ovary transverse, more or less lobed. Vitelline gland somewhat lobed. Cirrus-sac reaching middle of proglottid. Cirrus smooth. Uterus breaking down into uniovular capsules. Type-species *P. redonica* Joyeux & Baer, 1935.

11b. Strobila with typical development and maturation .... **12.**

12a. Scolex with apical depression, inside which suckers open. Rostellum armed with small hooks of typical shape. Ovary and vitelline gland round. Parasite of water-shrews in Palaearctic region .... *Coronacanthus* Spasskii, 1954. (Figs 29.69–29.72) (Syn. *Acotylolepis* Yamaguti, 1959).

Diagnosis: Strobila with numerous proglottids and progressive maturation. Scolex with armed rostellum. Suckers open in large depression. Rostellar hooks very small and numerous, of nearly identical length. Testes three, in a triangle, overlapping female gonads. Ovary rounded, entire. Vitelline gland rounded, entire, aporal. Cirrus-sac long, reaching middle of proglottid. Cirrus armed with spines or hair-like structures. Uterus sac-like, ovoid, generally forming thick envelope in gravid proglottid. Envelope containing eggs may be released in intestine. Type-species *C. integrus* (Hamann, 1891) (syn. *C. polyacantha* (Baer, 1931)).

12b. Scolex not as above .... **13.**

13a. Scolex with huge rostellum, bearing numerous hooks near its base. Suckers behind rostellum. Proglottids enlarged, testes in one irregular row, one poral and two aporal, more or less overlapping female gonads. Parasite of white-toothed shrews from Palaearctic and Ethiopian regions .... *Hilmylepis* Skryabin & Matevosyan, 1942. (Figs 29.73, 29.74)

Diagnosis: Strobila normal, with numerous proglottids and progressive maturation. Scolex with much enlarged rostellum. Number of rostellar hooks nearly constant; hooks of nearly identical length. Ovary trilobed. Vitelline gland lobed, median. Cirrus-sac short, not reaching middle of proglottid. Cirrus without spines. Uterus transverse, more or less lobate, completely filling gravid proglottids and extending laterally beyond excretory vessels. Type-species *H. nagatyi* (Hilmy, 1936).

13b. Rostellum not as above .... **14.**

14a. Very short gravid strobila, about 1–2 mm long with few proglottids. Numerous minute hooklets less than 10 μm long. Parasite of shrews in the Palaearctic region .... *Urocystis* Villot, 1880. (Fig. 29.53)

Diagnosis: Proglottids much broader than long. Scolex with retractile, armed rostellum. Testes three, in transverse row, more or less overlapping female gonads.

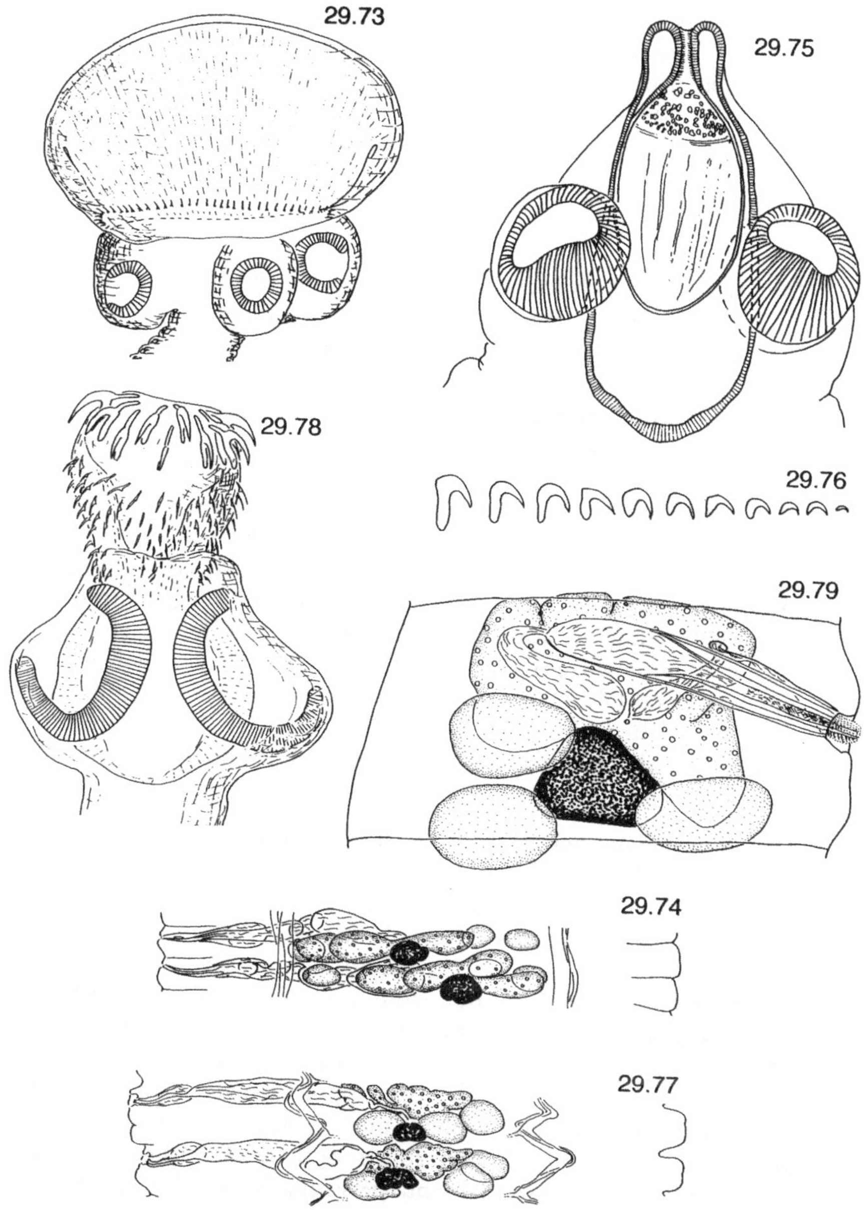

Figs 29.73–29.79 29.73. Scolex of *Hilmylepis nagatyi* (Hilmy, 1936) (original). 29.74. Mature proglottid of *Hilmylepis nagatyi* (Hilmy, 1936) (after Hunkeler, 1974). 29.75. Scolex of *Lophurolepis petteri* (Quentin, 1964) (after Quentin, 1964). 29.76. Some rostellar hooks of *Lophurolepis petteri* (after Hunkeler, 1974). 29.77. Mature proglottid of *Lophurolepis petteri* (Quentin, 1964) (after Quentin, 1964). 29.78. Scolex of *Vigisolepis spinulosa* (Cholodkowsky, 1906) (original). 29.79. Mature proglottid of *Vigisolepis spinulosa* (after Vaucher, 1971).

Ovary transverse, more or less lobate. Vitelline gland lobed in aporal half of proglottid. Cirrus-sac not reaching middle of proglottid. Cirrus armed with spines or hair-like structures. Uterus sacciform, containing few eggs. Type-species *U. prolifer* Villot, 1880.

14b. Strobila longer when gravid; sometimes short, with small number of proglottids: if so, fewer than 30 hooks measuring more than 10 μm .... 15.

15a. Rostellum with irregular circle of hooks, which may be interpreted as two or more circles. Hooks in same circle variable in length and may be accompanied by triangular rows of small spines. Or rostellum with many hooklets on its surface and disposed without order .... 16.

15b. Rostellum with regular circle of hooks, without additional spines. Hooks of nearly identical length .... 19.

16a. Rostellum conical, with 120–500 hooks not arranged in circles. Short cirrus-sac, not crossing osmoregulatory vessels. Parasitic in African rodents ... . *Lophurolepis* Spasskii, 1973. (Figs 29.75–29.77)

Diagnosis: Strobila with numerous proglottids and progressive maturation. Scolex with retractile, armed rostellum. Hooks of different lengths. Testes three, separated into two groups by female gonads. Ovary trilobed or multilobed. Vitelline gland lobed, median. Cirrus-sac short, not reaching middle of proglottid. Cirrus armed with spines or hair-like structures. Uterus lobate, filling gravid proglottid and extending laterally beyond osmoregulatory vessels. Type-species *L. petteri* (Quentin, 1964).

16b. Morphology not as above. Parasitic in shrews .... 17.

17a. Rostellum with additional spines behind hooks. Spines form triangular areas. Cirrus-sac large. Testes in triangle. Parasitic in shrews in Palaearctic region .... *Vigisolepis* Matevosyan, 1945. (Figs 29.78, 29.79)

Diagnosis: Strobila with numerous proglottids and progressive maturation. Scolex with invaginable rostellum, armed with hooks and additional vertical rows of spines. Number of rostellar hooks nearly constant; hooks of different lengths. Testes three, forming triangle, overlapping female gonads. Ovary trilobed or multilobed. Vitelline gland lobed, median. Cirrus-sac long, reaching middle of proglottid. Cirrus armed with spines or hair-like structures. Uterus sac-like, slightly lobed with horizontal septa persisting in gravid proglottids. Type-species *V. spinulosa* (Cholodkowsky, 1906).

17b. Rostellum without additional spines behind hooks. Marked variation in size of rostellar hooks in same circle .... 18.

18a. Ovary and vitelline gland lobate. Rostellar hooks with obvious handle, blade and guard. Parasitic in shrews in Holarctic region .... *Skrjabinacanthus* Spasskii & Morosov, 1959. (Figs 29.80, 29.81.)
(Syn. *Pseudoparadilepis* Brendow, 1969.)

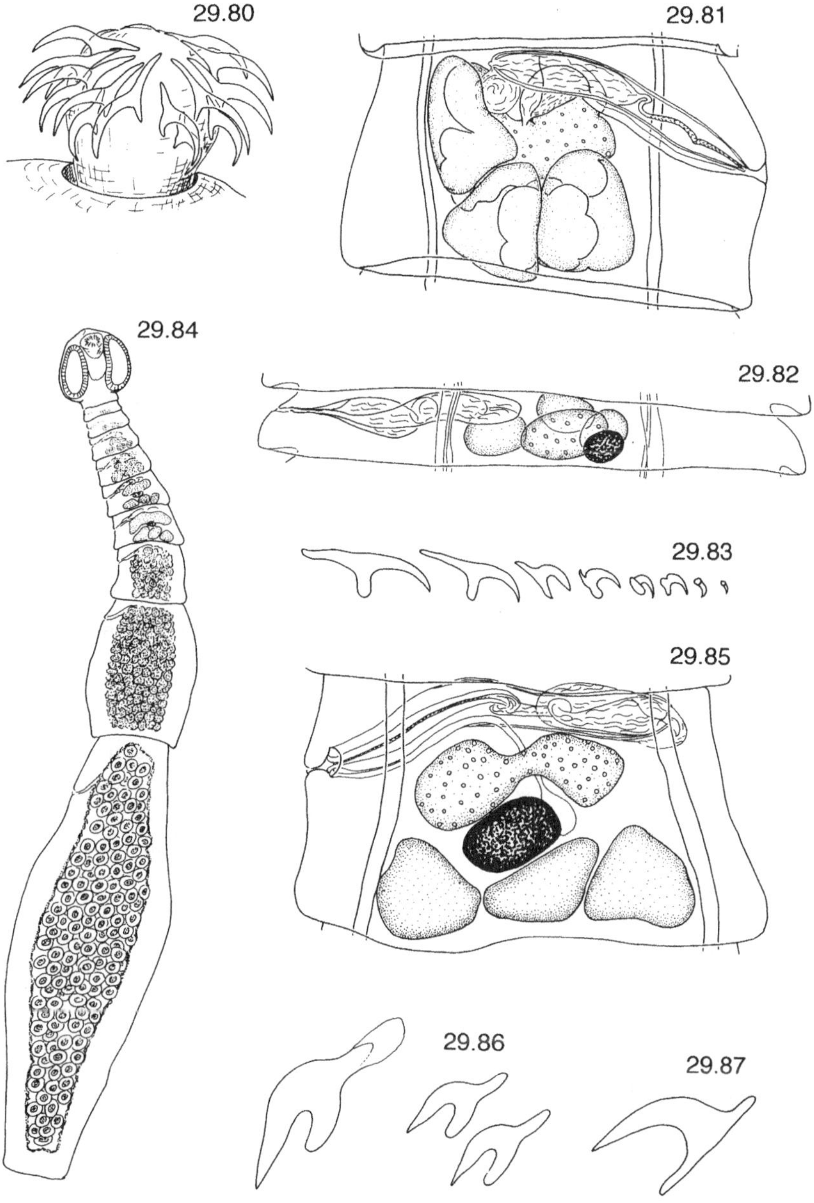

Figs 29.80–29.87 29.80. Evaginated rostellum of *Skrjabinacanthus jacutensis* Spasskii & Morosov, 1959) (after Vaucher, 1971). Fig. 29.81. Mature proglottid of *Skrjabinacanthus jacutensis* (after Vaucher, 1971). 29.82. Mature proglottid of *Hunkelepis multihami* (Hunkeler, 1972) (after Hunkeler, 1974). 29.83. Some rostellar hooks of *Hunkelepis multihami* (after Hunkeler, 1974). 29.84. Gravid specimen of *Staphylocystoides stefanskii* Zarnowski, 1954 (original). 29.85. Mature proglottid of *Staphylocystoides stefanskii* (after Vaucher, 1971). 29.86. Shape of rostellar hooks of members of *Neoskrjabinolepis*: left, *N. singularis* (Cholodkowsky, 1912); right, *N. schaldybini* Spasskii, 1947 (original). 29.87. Rostellar hook of '*Hymenolepis*' *petrodromi* Baer, 1933 (after Baer, 1933).

Diagnosis: Strobila with numerous proglottids and progressive maturation. Scolex with invaginable armed rostellum. Number of rostellar hooks nearly constant; hooks of different lengths. Testes three, in triangle, overlapping female gonads. Ovary trilobed or multilobed. Vitelline gland lobed, median. Cirrus-sac long, reaching middle of proglottid. Cirrus armed with spines or hair-like structures. Uterus sac-like, moderately lobate. Type-species *S. diplocoronatus* Spasskii & Morosov, 1959.

18b. Ovary and vitelline gland rounded. Smaller rostellar hooks reduced to spine-like structures. Parasite of shrews in the Ethiopian region .... *Hunkelepis* n. g. (Figs 29.82, 29.83)

Diagnosis: Strobila with progressive and regular maturation. Scolex with invaginable armed rostellum. Rostellar hooks very variable in length, at various levels on rostellum, not forming regular crown. Number of hooks nearly constant. Testes three, in triangle, with median testis more anterior than poral and aporal. Ovary entire, oval, median. Vitelline gland entire, behind ovary or slightly aporal. Uterus lobate, small with few eggs. Comparatively big oncospheral hooklets. Type-species *H. multihami* (Hunkeler, 1972) n. comb. (syn. *Hymenolepis multihami* Hunkeler, 1972).

19a. Strobila short with few proglottids. Rostellum invaginable; 8–18 hooks. Cirrus-sac long, usually crosses middle of proglottid. Testes in transverse line behind female gonads. Parasite of shrews in the Holarctic region .... *Staphylocystoides* Yamaguti, 1959. (Figs 29.84, 29.85)
(Syn. *Zarnowskiella* Spasskii & Andreiko, 1970.)

Diagnosis: Strobila shortened, with few proglottids, usually only one or two gravid segments. Rostellar hooks of nearly identical length. Testes three, linear. Ovary rounded, entire. Ovary transverse, more or less lobate. Vitelline gland round, entire. Vitelline gland median or submedian. Cirrus-sac long, extending beyond middle of proglottid. Cirrus armed with spines or hair-like structures. Gravid uterus sacciform, filling last, much elongated proglottid. Type-species *S. sphenomorphus* (Locker & Rausch, 1952).

19b. Strobila and anatomy not as above .... 20.

20a. Ten* rostellar hooks of typical shape (Fig. 29.86) .... 21.
20b. Rostellar hooks of other shape†; if shape more or less similar, more than ten hooks .... 23.

21a. Strobila with numerous proglottids, enlarged when mature and more elongated when gravid. Testes in transverse line. Cirrus-sac not reaching

* Ten hooks of 23 μm with small blade and small guard, parasite of rodents in the Palaearctic region: '*Hymenolepis*' *muris-sylvatici* (Rudolphi, 1819). Not classified in the present section; see Introduction. Ten hooks of 33 μm of triangular shape; parasite of *Blarina brevicauda* in USA: '*Hymenolepis*' *blarinae* Rausch & Kuns, 1950. Not classified in the present section.
† Nine to 14 hooks like Fig. 29.87. Hook length 11-13 μm: '*Hymenolepis*' *petrodromi* Baer, 1933, from *Petrodromus dactylus*, Africa; hook length 24 μm: '*Hymenolepis*' '*globirostris* Baer, 1925 from a rodent, Africa and Asia; hook length 37-42 μm: '*Hymenolepis*' *uncinispinosa* Baer, 1930 from various rodents, Africa. Not classified in the present section.

middle of proglottid. Parasitic in shrews of Palaearctic region .... *Neoskrjabinolepis* Spasskii, 1947. (Fig. 29.88)

Diagnosis: Strobila with progressive maturation. Scolex with invaginable armed rostellum. Rostellar hooks ten; hooks of nearly identical length. Testes three, linear, more or less overlapping female gonads. Ovary transverse, more or less lobate. Vitelline gland rounded, entire, in aporal half of proglottid. Cirrus-sac short, not reaching middle of proglottid. Cirrus armed with spines or hair-like structures. Uterus sac-like, filling gravid proglottids. Type-species *N. schaldybini* Spasskii, 1947.

21b. Short strobila, testes in triangle or in oblique line. Cirrus crosses middle of proglottid. Parasitic in shrews in Nearctic region .... 22.

22a. Cirrus-sac crosses middle of proglottid. Cirrus armed with few huge spines. Vagina without particular structures ...... *Vogelepis* n. g. (Fig. 29.89)

Diagnosis: Small strobila with few proglottids. Scolex with invaginable, armed rostellum. Rostellar hooks ten, of nearly identical length. Testes in triangle, more or less overlapping female gonads. Cirrus large. Ovary transverse, moderately lobed. Vitelline gland compact. Uterus sac-like with few eggs. Parasitic in shrews in Nearctic region. Type-species *V. virilis* (Voge, 1955) n. comb. (syn. *Hymenolepis virilis* Voge, 1955).

22b. Cirrus-sac comparatively enormous, reaching aporal margin of proglottid. Vagina with terminal, bulb-like, heavily spined section and additional structures supported by rigid spicule-like structures. Parasitic in shrews in Nearctic region .... *Lockerrauschia* Yamaguti, 1959. (Figs 29.90, 29.91)

Diagnosis: Strobila with numerous proglottids and progressive maturation. Scolex with armed rostellum. Rostellar hooks ten, of nearly identical length. Testes in transverse row, more or less overlapping female gonads. Ovary rounded, entire. Cirrus armed with spines or hair-like structures. Terminal genital ducts with accessory structures. Uterus sac-like. Type-species *L. intricata* (Locker & Rausch, 1952).

23a. Rostellar hooks with bifurcated guard. Testes in triangle or irregular line in middle of proglottid. Ovary and vitelline gland rounded, entire. Gravid uterus usually a rounded sac. Parasitic in Insectivora in Palaearctic region (water shrews *Neomys*) .... *Triodontolepis* Yamaguti, 1959. (Figs 29.92, 29.93)

Diagnosis: Strobila with numerous proglottids and progressive maturation. Scolex with invaginable armed rostellum. Rostellar hooks of nearly identical length and nearly constant number. Testes three, separated into two groups by female gonads. Vitelline gland median. Cirrus-sac not reaching middle of proglottid. Cirrus armed with spines or hair-like structures. Uterus sac-like, ovoid, generally forming thick envelope in gravid proglottid. Envelope containing eggs may be released in intestine. Type-species *T. tridontophora* (Soltys, 1954).

23b. Rostellar hooks with guard entire or rarely bifurcated: in latter case, anatomy of mature and gravid proglottids not as above .... 24.

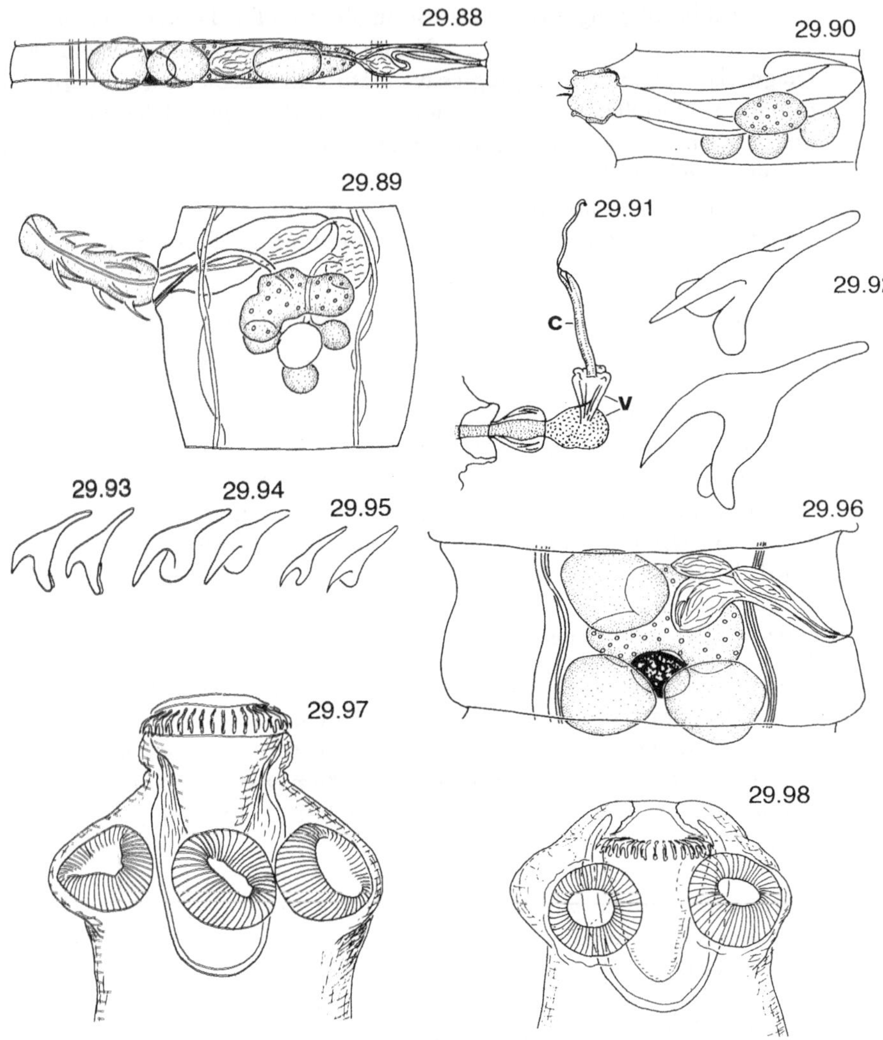

Figs 29.88–29.98 29.88. Mature proglottid of *Neoskrjabinolepis singularis* (Cholodkowsky, 1912) (after Vaucher, 1971). 29.89. Mature proglottid of *Vogelepis virilis* (Voge, 1955) (after Voge, 1955). 29.90. Mature proglottid of *Lockerrauschia intricata* (Locker & Rausch, 1952) (after Locker & Rausch, 1952). 29.91. Semidiagrammatic drawing of cirrus (c) and vagina (v) of *Lockerrauschia intricata* (after Locker & Rausch, 1952). 29.92, 29.93. Rostellar hooks of members of *Triodontolepis*: 29.92. *T. bifurcus* (Hamann, 1891); 29.93. *T. hamanni* (Mrazek, 1891) (after Vaucher, 1971). 29.94, 29.95. Rostellar hooks of members of *Staphylocystis*: 29.94. *S. scalaris* (Dujardin, 1845); 29.95. *S. pistillum* Dujardin, 1845) (after Vaucher, 1971). 29.96. Mature proglottid of *Staphylocystis dodecacantha* (Baer, 1925) (after Vaucher, 1971). 29.97, 29.98. Scolex of *Vampirolepis temmincki* (Vaucher, 1986) showing the retractable rostellum (original).

24a. Rostellar hooks generally with rather short guard and somewhat flattened, pointed blade, handle variable, often short. Testes in triangle in middle of proglottids, overlapping female organs. If short, strobila not enlarged posteriorly and with more than one or two gravid proglottids (compare with *Staphylocystoides*). Rostellum retractile. Parasitic in Insectivora, mainly white-toothed shrews; Old World .... *Staphylocystis* Villot, 1877. (Figs 29.94–29.96)

(Syn. *Crocidolepis* Spasskii & Karpenko, 1992.)

Diagnosis: Strobila with numerous proglottids and progressive maturation. Scolex with armed rostellum. Number of rostellar hooks nearly constant. Hooks of nearly identical length. Testes three, in triangle, overlapping female gonads. Ovary trilobate or multilobate. Vitelline gland rarely rounded, mostly lobate, median. Cirrus-sac short, not reaching middle of proglottid. Cirrus smooth, without spines. Uterus sac-like, more or less horseshoe-shaped at first, then sac-like and lobate, filling gravid proglottids and usually extending laterally to margins. Type-species *S. pistillum* (Dujardin, 1843).

24b. Hooks and testes disposition not as above .... 25.

25a. Ten hooks. Strobila long, thin. Gravid proglottids usually much longer than wide. Ten rostellar hooks of typical shape. Rostellum invaginable. Testes in transverse line in posterior half of proglottid. Vitellarium usually slightly aporal. Parasitic in Insectivora (Holarctic shrews) .... *Lineolepis* Spasskii, 1959. (Figs 29.99–29.102)

Diagnosis: Strobila with numerous proglottids and progressive maturation. Scolex with invaginable armed rostellum. Rostellar hooks ten, of nearly identical length. Testes overlap female gonads. Ovary transverse, trilobate or multilobate. Vitelline gland rounded, entire. Cirrus-sac long, reaching middle of proglottid. Cirrus armed with spines or hair-like structures. Uterus bilobed and horseshoe-shaped at first filling gravid proglottids, extending laterally to margins. Type-species *L. parva* (Rausch & Kuns, 1950).

25b. More than ten hooks .... 26.

26a. Twelve to more than 50 fraternoid rostellar hooks arranged in two crowns in one case (*V. bihamata* Sawada & Harada, 1986). Testes arranged in transverse line in posterior part of proglottid, sometimes aporal testes slightly diagonal, not separated into two groups by female organs. Uterus clearly two-winged in pregravid proglottids. Parasitic in Chiroptera. Cosmopolitan .... *Vampirolepis* Spasskii, 1954. (Figs 29.97, 29.98, 29.104–29.106)

Diagnosis: Strobila with numerous proglottids and progressive maturation. Scolex with retractile, armed rostellum. Number of rostellar hooks nearly constant; hooks of nearly identical length. Ovary trilobate or multilobate. Vitelline gland lobate, median. Cirrus-sac short, not reaching middle of proglottid. Cirrus smooth or armed with minute spines. Uterus initially with two lateral wings, finally filling whole proglottid. Type-species *V. skrjabinariana* (Skarbilovitsch, 1946).

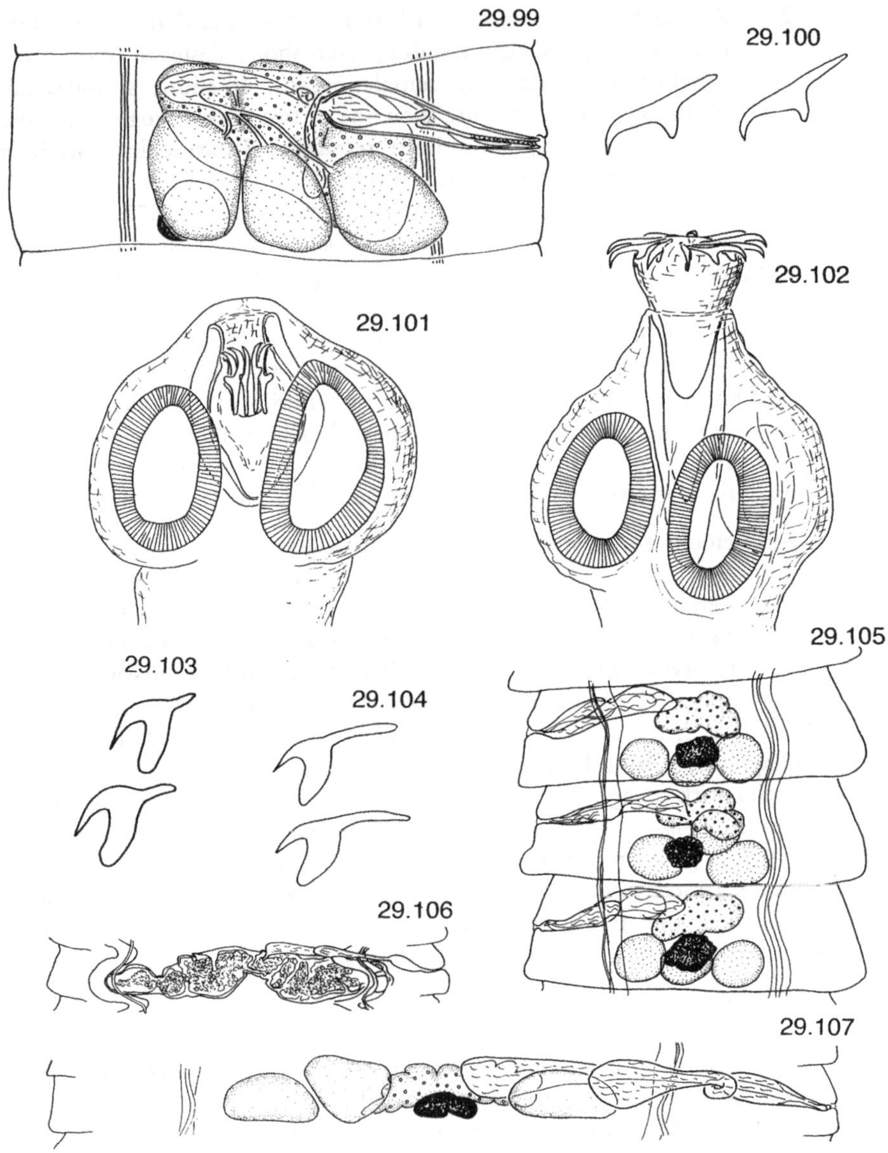

Figs 29.99-29.107 29.99. Mature proglottid of *Lineolepis scutigera* (Dujardin, 1845) (after Vaucher, 1971). 29.100. Two rostellar hooks of *Lineolepis scutigera* (after Vaucher, 1971). 29.101, 29.102. Scolex of *Lineolepis scutigera* showing the invaginable rostellum (original). 29.103. Rostellar hooks of *Rodentolepis straminea* (Goeze, 1782) (original). 29.104. Two rostellar hooks of *Vampirolepis decipiens* (Diesing, 1850) (after Vaucher, 1986a). 29.105. Mature proglottids of *Vampirolepis decipiens* (after Vaucher, 1986a). 29.106. Uterus of *Vampirolepis mazanensis* (Vaucher, 1986) (after Vaucher, 1986b). 29.107. Mature proglottid of *Rodentolepis straminea* (Goeze, 1782) (original).

26b. **Several to numerous hooks of various shape, often fraternoid. Testes arranged in transverse line or in elongated triangle, separated into two groups by female gonads. Parasitic in various mammals (rodents, marsupials, chiropterans, primates) .... *Rodentolepis* Spasskii, 1954. (Figs 29.103, 29.107)**
(Syn. *Diplacanthus* Weinland, 1858, preoccupied.)

Diagnosis: Strobila with numerous proglottids and progressive maturation. Scolex with retractile, armed rostellum. Number of rostellar hooks nearly constant; hooks of nearly identical length. Ovary trilobate or multilobate. Vitelline gland lobate, median. Cirrus-sac short, not reaching middle of proglottid. Cirrus smooth, or armed with small spines. Uterus deeply lobate, labyrinthine, filling gravid proglottids and extending laterally beyond osmoregulatory vessels. Type-species *R. straminea* (Goeze, 1782).

27a. **Strobila hyperapolytic** .... **28.**
27b. **Strobila with classical development and maturation** .... **29.**

28a. **Scolex without rostellum. Aporal margins of proglottids with lateral expansions. Testes in line. Parasitic in Lagomorpha in Palaearctic region .... *Gvosdevilepis* Spasskii, 1953. (Figs 29.108, 29.109)**

Diagnosis: Hyperapolytic strobila. Testes in a line in poral half of proglottid and poral to female gonads. Cirrus-sac short, not reaching middle of proglottid. Vitelline gland elongated. Uterus sac-like, transverse. Type-species *G. fragmentata* (Gvozdev, 1948).

28b. **Scolex with unarmed rostellum. Strobila not as above but with only two to three proglottids. Testes in triangle. Parasitic in Insectivora (shrews) in Palaearctic region .... *Mathevolepis* Spasskii, 1948. (Fig. 29.110)**

Diagnosis: Strobila very short (two to four proglottids), hyperapolytic. Scolex with unarmed, rudimentary rostellum. Testes in triangle. Cirrus-sac reaching middle of proglottid. Ovary irregularly lobate. Vitelline gland postovarian, somewhat lobate. Uterus sac-like, elongate. Type-species *M. petrotschenkoi* Spasskii, 1948.

29a. **Gravid strobila short or very short, with few proglottids. No rostellum. Few eggs in gravid uterus. Parasitic in shrews in Palaearctic region .... *Soricinia* Spasskii & Spasskaya, 1954. (Fig. 29.111)**
(Syn. *Insectivorolepis* Zarnowsky, 1956.)

Diagnosis: Testes three in posterior part of proglottid, in triangle. Cirrus-sac short, not reaching middle of proglottid. Ovary somewhat lobate. Vitelline gland compact. Uterus sac-like. Type-species *S. soricis* Baer, 1927 (syn. *H. minuta* Baer, 1925).

29b. **Strobila not as above. Rostellum unarmed or absent** .... **30.**

30a. **Strobila divided into sections in which proglottids present same maturation state. Usually with unarmed rostellum. Testes in triangle overlapping female gonads. Gravid proglottids with thick uterine wall. Uterus of several gravid proglottids may join together to form syncapsule. Parasitic in shrews**

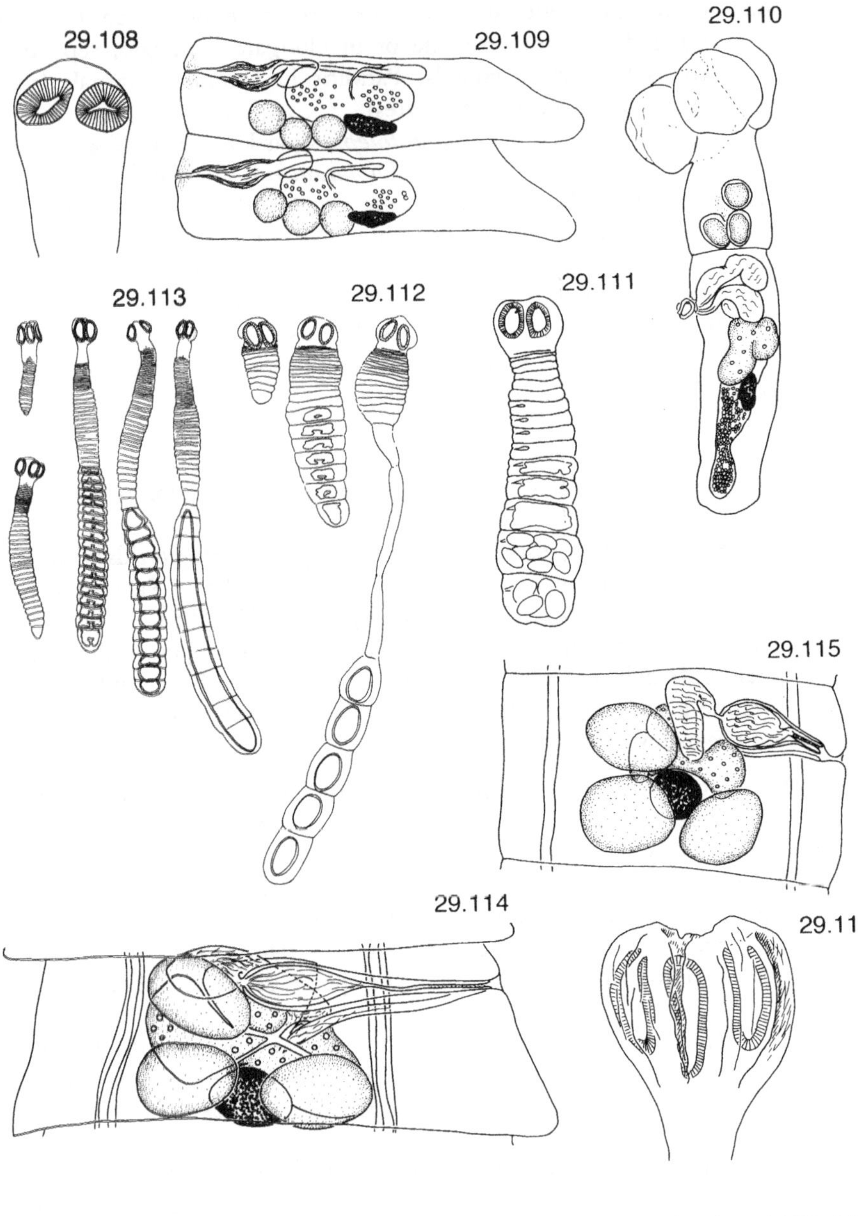

Figs 29.108–29.116 29.108. Scolex of *Gvosdevilepis fragmentata* (Gvozdev, 1948) (after Spasskii, 1954). 29.109. Mature proglottids of *Gvosdevilepis fragmentata* (after Spasskii, 1954). 29.110. Pregravid specimen of *Mathevolepis petrotschenkoi* Spasskii, 1948 (after Spasskii, 1954). 29.111. Gravid specimen of *Soricinia infirma* (Zarnowski, 1955) (after Vaucher, 1971). 29.112, 29.113. Growth of strobila and schematic evolution of the uterus in members of *Ditestolepis*. 29.112. *D. tripartita* (Zarnowski, 1955); 29.113. *D. diaphana* (Cholodkowsky, 1906) (after Vaucher, 1971). 29.114. Mature proglottid of *Ditestolepis diaphana* (after Vaucher, 1971). 29.115. Mature proglottid of *Cryptocotylepis anthocephalus* (van Gundy, 1935) (after Vaucher & Durette-Desset, 1978). 29.116. Scolex of *Cryptocotylepis anthocephalus* (after Vaucher & Durette-Desset, 1978).

in Holarctic region ... ***Ditestolepis*** **Soltys, 1952. (Figs 29.112–29.114)** (Syns *Cucurbilepis* Sadovskaya, 1965; *Ecrinolepis* Spasskii & Karpenko, 1983.)

Diagnosis: Strobila with serial growth of proglottids. Scolex usually with unarmed, rudimentary rostellum. Cirrus-sac small or medium-sized, may reach middle of proglottid. Cirrus armed with minute spines. Ovary elongated transversely, bi- or trilobed. Vitelline gland ovoid, median. Uterus sacciform, commonly horseshoe-shaped at first, with thick wall when gravid; eventually forms syncapsule by confluence of uteri of several proglottids. Type-species *D. diaphana* (Cholodkowsky, 1906).

30b. Strobila presenting progressive maturation of proglottids .... 31.

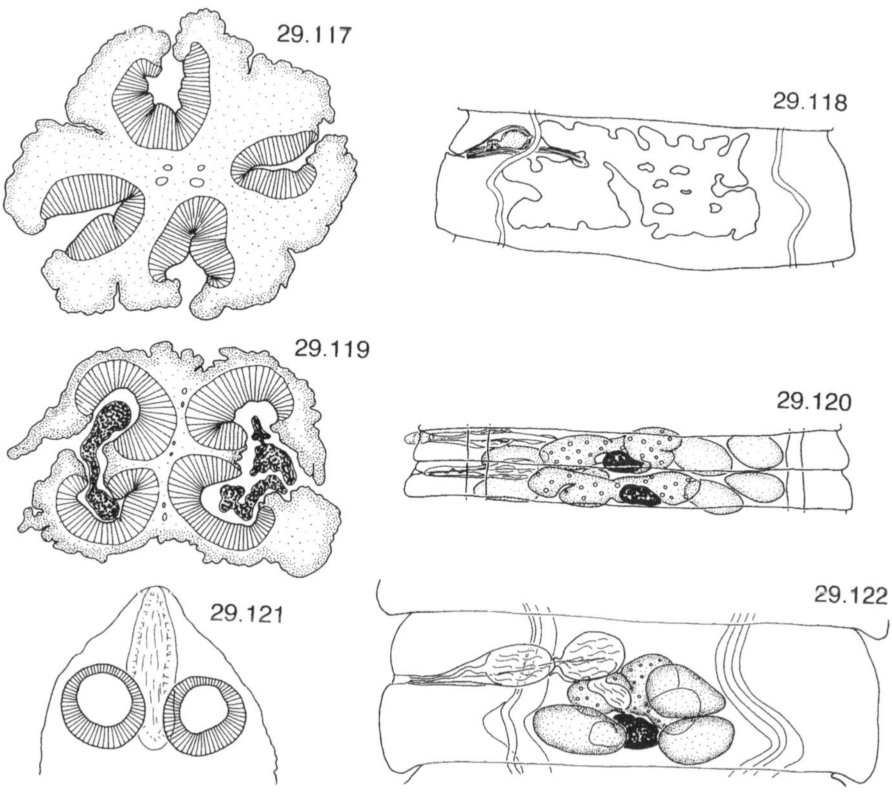

Figs 29.117-29.122 29.117. Transverse section of a scolex of *Cryptocotylepis anthocephalus* showing the four suckers embedded in the scolex tissue (original). 29.118. Pregravid proglottid of *Cryptocotylepis anthocephalus* with two-winged uterus (after Vaucher & Durette-Desset, 1978). 29.119. Transverse section of a scolex of *Pseudobotrialepis globosoides* (Soltys, 1954) showing the paired suckers opening in two lateral folds. Fragments of host's intestinal villi remain in the lateral cavities (original). 29.120. Mature proglottid of *Pseudobotrialepis globosoides* (after Vaucher, 1971). 29.121. Scolex of *Milina grisea* van Beneden, 1873 (original). 29.122. Mature proglottid of *Milina grisea* (original).

31a. Suckers deeply embedded in scolex tissues, opening in folds or even cavities of scolex. No rostellum. Parasitic in shrews .... 32.

31b. Suckers not as above, opening on surface of scolex. Rostellum unarmed, rudimentary (Fig. 29.121). Parasitic in various mammals .... 33.

32a. Testes in triangle in middle of proglottid, superimposed on female organs. Uterus labyrinthine in pregravid proglottids. Suckers opening in deep folds

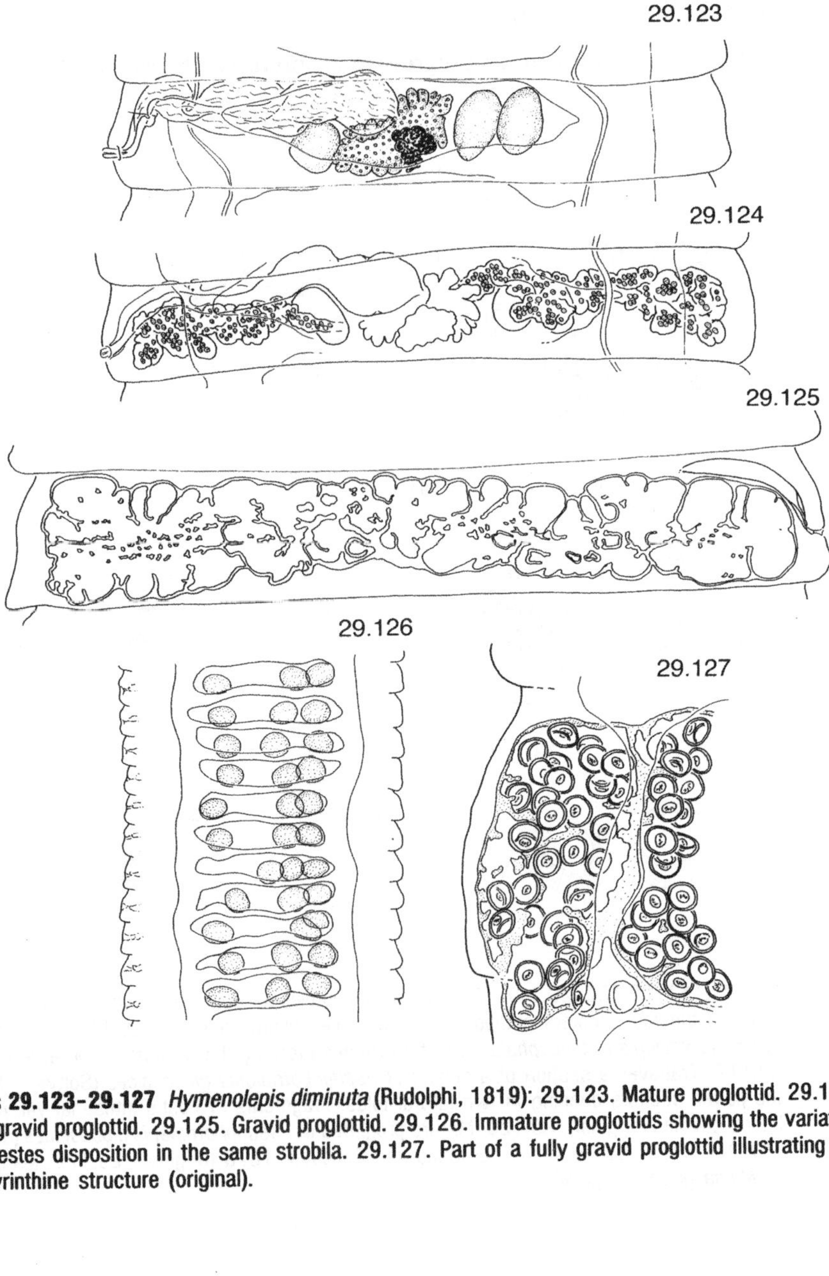

Figs 29.123–29.127 *Hymenolepis diminuta* (Rudolphi, 1819): 29.123. Mature proglottid. 29.124. Pregravid proglottid. 29.125. Gravid proglottid. 29.126. Immature proglottids showing the variation of testes disposition in the same strobila. 29.127. Part of a fully gravid proglottid illustrating the labyrinthine structure (original).

of scolex. Parasitic in *Blarina brevicauda* in Nearctic region ....
... *Cryptocotylepis* Skryabin & Matevosyan, 1948. (Figs 29.115–29.118)

Diagnosis: Strobila with numerous proglottids and progressive maturation. Scolex without rostellum. Suckers opening in deep folds of scolex. Ovary trilobate. Vitelline gland compact, postovarian. Cirrus-sac short, not reaching middle of proglottid. Cirrus armed with minute spines. Uterus beginning as irregular and horseshoe-shaped network, finally filling proglottids, with parts of uterine wall remaining in gravid proglottids. Type-species *C. anthocephalus* (van Gundy, 1944).

32b. Testes in straight line or in enlarged triangle and divided into two groups by female organs. Suckers in pairs, opening into two common lateral cavities of scolex. Parasitic in shrews in Palaearctic region ....
.... *Pseudobotrialepis* Schaldybin, 1957. (Figs 29.119, 29.120)

Diagnosis: Strobila with numerous proglottids and progressive maturation. Scolex without rostellum. Suckers opening in pairs into two lateral folds of scolex. Ovary trilobate. Vitelline gland ovoid, median. Cirrus-sac short, not reaching middle of proglottid. Cirrus armed with spines. Uterus sac-like, transverse, deeply lobate or labyrinthine, with septa persisting in gravid proglottid. Type-species *P. globosoides* (Soltys, 1954).

33a. Testes in triangle in middle of proglottid, overlapping female organs. Parasitic in Chiroptera in Palaearctic region ....
.... *Milina* van Beneden, 1873. (Figs 29.121, 29.122)
(Syn. *Myotolepis* Spasskii, 1954.)

Diagnosis: Strobila with numerous proglottids and progressive maturation. Scolex with unarmed elongate rostellum. Testes three, not separated into two groups by female gonads. Ovary somewhat lobate. Vitelline gland ovoid, postovarian. Cirrus-sac short, not reaching middle of proglottid. Cirrus smooth. Uterus beginning as irregular horseshoe-shaped network. Type-species *M. grisea* van Beneden, 1873.

33b. Testes arranged in transverse line or elongated triangle, separated into two groups by female gonads. Uterus transverse, labyrinthine. Parasites mainly of rodents. Cosmopolitan ....
.... *Hymenolepis* Weinland, 1858. (Figs 29.123–29.127)
(Syn. *Lepidotrias* Weinland, 1858.)

Diagnosis: Strobila with numerous proglottids and progressive maturation. Scolex with unarmed rostellum. Testes three, separated into two groups by female gonads. Ovary trilobed or multilobed. Vitelline gland lobed, median. Cirrus-sac short, not reaching middle of proglottid. Cirrus smooth or armed with minute spines. Uterus an irregular network, labyrinthine, finally filling proglottids, with parts of uterine walls remaining in gravid proglottids. Type-species *H. diminuta* (Rudolphi, 1819).

30 Family Taeniidae Ludwig, 1886

R.L. RAUSCH

University of Washington, Department of Comparative Medicine, T-142 Health Sciences, SB-42, Seattle, Washington 98195, USA.

Introduction

Because of their medical or veterinary importance, some species of the family Taeniidae have been intensively studied; consequently, the taeniids are the best known cestodes. As conceived here, their systematic arrangement in the order Cyclophyllidea is as follows: suborder Taeniata Skryabin & Shul'ts, 1937 (the subordinal category in the Cyclophyllidea is not universally accepted); family Taeniidae Ludwig, 1886; subfamily Taeniinae Stiles, 1896, with the single genus *Taenia* Linnaeus, 1758; and the subfamily Echinococcinae Abuladze, 1960, with the single genus *Echinococcus* Rudolphi, 1801. Cestodes in the Taeniinae are characterized by having a ribbon-like strobila consisting of many proglottids and a larval stage (metacestode) that is a cysticercus or a modification of a cysticercus. Those in the Echinococcinae are very small, not ribbon-like, with not more than seven proglottids and a larval stage in which protoscoleces are produced in brood-capsules. The Echinococcinae are further distinguished by the presence of a laminated membrane in the larval stage. The strobilar stage of taeniids occurs in the alimentary canal of mammals. Mammals also serve as the intermediate host.

In the discrimination of genera in the Taeniidae, disagreement still exists because of conceptual differences concerning the relative significance of taxonomic characters. On the basis of superficial morphological similarities (convergence) and the involvement of two mammalian hosts in their cycles, cestodes of genera in other families have been placed until recently in the Taeniidae (all in the subfamily Taeniinae): *Cladotaenia* Cohn, 1901; *Paracladotaenia* Yamaguti, 1935; *Anoplotaenia* Beddard, 1911; *Dasyurotaenia* Beddard, 1912; and apparently *Insinuarotaenia* Spasskii, 1948.

The tendency to retain these genera in the Taeniidae is difficult to understand since, in most cases, grounds for removing them have been well established. Schmelz (1941) placed *Paracladotaenia* in synonymy with *Cladotaenia*, on the basis that the type material represented *Cladotaenia* sp. from which the rostellar hooks had been lost. Fuhrmann & Baer (1943) transferred *Cladotaenia* to the Dilepididae Fuhrmann, 1907, confirmed by Freeman (1959, 1973). Yamaguti (1959) listed *Paracladotaenia* as a synonym of *Cladotaenia* but established a separate subgenus for it in the Taeniidae. Without explanation, Joyeux & Baer (1961) again included *Cladotaenia* in the Taeniidae and provisionally accepted *Paracladotaenia* as valid. Abuladze (1964) retained both genera in the Taeniidae. Schmidt (1986) appropriately replaced *Cladotaenia* (syn. *Paracladotaenia*) in the Dilepididae.

The exclusion of *Anoplotaenia* and *Dasyurotaenia* can be justified on both zoogeographical and morphological grounds (Rausch, 1985). Thin-shelled eggs (without the typical taeniid embryophore) in cestodes of both genera, the complex genital atrium in *Anoplotaenia*, unilateral genital pores, the form of the rostellar hooks, and the penetration of the intestinal wall by *Dasyurotaenia* make these cestodes incompatible with the Taeniidae. Moreover, the larval stage of *Anoplotaenia* resembles a linstowiine cysticercoid (Beveridge *et al.*, 1975). In contrast, Spasskii (1990) has proposed a new subfamily, Anoplotaeniinae in the Taeniidae for *Anoplotaenia*.

The systematic status of *Insinuarotaenia* is uncertain. The type-species, *I. schikobalovi* (= *I. schikhobalovae*) Spasskii, 1948, was described from a single, immature specimen from a badger, *Meles meles*. Its scolex is unlike that of any known taeniid in that it lacks a rostellum and is adapted for deep penetration of the intestinal wall (Spasskii, 1948a). Without knowledge of the structure of the egg and characteristics of the larval stage, the genus cannot be assigned to family. A second species, *I. spasskii* Andreiko & Iun' Lian', 1963, was evidently based on specimens of *Taenia mustelae* Gmelin, 1780, from which the rostellar hooks had been lost (Rausch, 1985). Schmidt (1986) retained this genus in the Taeniidae.

Excluding the aforementioned, the number of nominal genera in the Taeniidae, subfamily Taeniinae, has ranged from three (Schmidt, 1986) to six (Abuladze, 1964). More recently, *Fimbriotaenia* Kornyushin & Sharpilo, 1986 has been established for *Taenia martis* (Zeder, 1803) (see Kornyushin & Sharpilo, 1986). Two of the nominal genera have been distinguished by means of characters associated with the scolex, the remainder by organization of the larval stage. In agreement with Verster (1969), I consider the following to be synonyms of *Taenia* Linnaeus, 1758: *Multiceps* Goeze, 1782; *Taeniarhynchus* Weinland, 1858; *Hydatigera* Lamarck, 1816; *Tetratirotaenia* Abuladze, 1964; *Monordotaenia* Little, 1967; and *Fossor* Honess, 1937. To those I add *Fimbriotaenia* Kornyushin & Sharpilo, 1986. The grounds for synonymizing those genera are presented briefly below.

Multiceps. The form of the handle of the large hooks ('wavy' or sinuous) and the curving of the vagina near the longitudinal osmoregulatory canals are of no significance at generic level (see also Verster, 1969). The organization of the larval stage, a coenurus, is the major criterion by which *Multiceps* has been distinguished. 'Typical' coenuri, such as those of *Taenia multiceps* Leske, 1780 and *T. serialis* (Gervais, 1847), are polycephalic cysticerci. When *Multiceps* was established, larval stages were known for few species of taeniids, so it is not remarkable that undue emphasis was placed on what then appeared to be striking differences in organization. Subsequently, with the description of many additional species, a wide range of forms derived from the simple cysticercus became evident. Diverse types of polycephalic larvae have been placed in *Multiceps* (cf. Bondareva, 1963), including for example the long-armed types of *Taenia twitchelli* Schwartz, 1924 and *T. endothoracicus* Kirschenblatt, 1948, that occur in the pleural cavities of the intermediate host. The form of the larval *T. mustelae* differs with species of host. It is typically monocephalic in small arvicoline rodents, but is usually polycephalic (coenurus-type) in the muskrat, *Ondatra zibethicus*, as well as in some other rodents. Freeman (1956) found that either type could occur in hosts of a single species, and both were sometimes present in individual animals. If *Multiceps* were not placed in synonymy with *Taenia*, it would be necessary to establish additional genera for other species whose larval stages are polycephalic but cannot be designated coenuri.

Taeniarhynchus. The lack of a rostellum is the only character which distinguishes *Taeniarhynchus* from *Taenia*. Šlais & Machnická (1976) and Ždárská (1976) found that the rostellum and rostellar hooks begin development in the early-stage cysticercus of *Taenia saginata* (Goeze, 1782) and then regress, which detracts significantly from the value of the absence of the rostellum as a generic character in this case. Schmidt (1986) retained *Taeniarhynchus* to contain *Taenia saginata*. Another taeniid, *T. punica* Cholodkovsky, 1908, described from a dog in Tunis, also lacks a rostellum (Cholodkovsky, 1908, fig. 2), and is distinguished also in having the testes in two separate lateral fields.

Hydatigera. This genus was characterized by lack of a neck in the strobilar stage and a larval stage consisting of a long pseudostrobila derived from a simple, persistent bladder. As remarked by Verster (1969), the presence or absence of a neck may depend on methods of fixation. Species assigned to *Hydatigera* exhibit much morphological diversity in the larval stage, and none closely resembles that of the generic type, *Taenia taeniaeformis* (Batsch, 1786). Abuladze (1964) included *T. hyperborea* von Linstow, 1905 in *Hydatigera* without knowledge of its larval stage [*T. hyperborea* was synonymized with *T. crassiceps* (Zeder, 1800) by Rausch (1959)], along with *T. rileyi* Loewen, 1929 and *T. krepkogorski* (Schulz & Landa, 1934). Abuladze placed *T. parva* Baer, 1926, of which the larval stage is a polycephalic strobilocysticercus, in *Taenia*. *T. selousi* Mettrick, 1962, which has a similar larval stage, was not considered by Abuladze. Murai *et al.* (1989) illustrated polycephalic larval stages of the strobilocysticercus type. Esch & Self (1965) preceded Verster (1969) in the conclusion that *Hydatigera* is synonymous with *Taenia*.

Monordotaenia. Honess (1937) established *Fossor* (preoccupied, = *Monordotaenia* Little, 1967) for a cestode from the Nearctic badger, *Taxidea taxus*, because it had a single row of rostellar hooks. The cestode had been described already from the same host as *Taenia taxidiensis* Skinker, 1935. The larval stage, a typical cysticercus, occurs in ground squirrels, *Citellus* spp. (Pederson & Leiby, 1969). All of the other taxa assigned to *Monordotaenia* are of uncertain identity, since the loss of one row of rostellar hooks (usually the large hooks) is not unusual in taeniids that are not removed and fixed promptly after the death of the host. When convinced that a single row of hooks is the normal condition, taxonomists usually have made comparisons only with other nominal species characterized in part by that feature, without considering the possibility of a loss of hooks. Abuladze (1964) transferred *Taenia monostephanos* von Linstow, 1905 to *Fossor* (= *Monordotaenia*), being aware that von Linstow had originally believed the specimens to represent *T. laticollis* Rudolphi, 1819, from which a row of hooks was missing. Von Linstow's first impression was probably correct, since subsequent surveys of helminths of the lynx, *Felis lynx*, have not revealed any cestodes identifiable as *T. monostephanos*. *Monordotaenia alopexi* Obushenkov, 1983 was based on specimens of *Taenia polyacantha* Leuckart, 1856, which lacked one row of hooks (Rausch & Fay, 1988b). The description of *M. leonina* He & Liu, 1981, from a captive lion, *Panthera leo* (see He & Liu, 1981), corresponds closely to that of *Taenia gonyamai* Ortlepp, 1938, from the same host in South Africa. The status of *M. honessi* Hendrickson, Grieve & Kingston, 1975, from a dog, remains to be determined. For all cestodes placed in *Taenia*, the arrangement and characteristics of the rostellar hooks can be established with certainty by study of the larval stage. The examination of early-age cysticerci of *T. taxidiensis* is needed to determine whether two rows of hooks begin development, with subsequent regression of one row.

Tetratirotaenia. The genus was established by Abuladze (1964) for *Taenia polyacantha*, which has a modified cysticercus closely resembling that of *T. martis* (Zeder, 1803) (see Rausch & Fay, 1988b). Whereas the process of asexual multiplication in *T. polyacantha* may be unique among taeniids, the form of the larval stage does not justify removal of the species to a monotypic genus (Rausch & Fay, 1988a).

Fimbriotaenia. This genus would be synonymous with *Tetratirotaenia* if the latter were valid, since it also was distinguished by minor morphological features of the larval stage of another cestode, *Taenia martis*. Kornyushin & Sharpilo (1986) proposed to include in *Fimbriotaenia* such diverse species (in the larval stage) as *Taenia twitchelli, T. mustelae* and *T. brachyacantha* Baer & Fain, 1951.

The subfamily Echinococcinae Abuladze, 1964 is now usually considered to be monotypic, but some taxonomists still recognize *Alveococcus* Abuladze, 1960, established for *Echinococcus multilocularis* Leuckart, 1863. According to Abuladze (1960), characters that distinguish *Alveococcus* include the form of the gravid uterus, the location of the genital pore on the proglottidal margin, and mammalian hosts of species different from those of *E. granulosus*. Use of such characters, valid for distinguishing species, as generic criteria would require that the two other recognized species of *Echinococcus* be placed in monotypic genera. *Alveococcus* is here considered synonymous with *Echinococcus*.

Except for the two species of *Taenia* that occur in the strobilar stage in man, all taeniids inhabit the alimentary canal of fissiped carnivores. The respective larval stages occur in herbivorous mammals. A predator-prey relationship exists between the final and intermediate host of each species of taeniid, but cycles of different taeniids may involve the same hosts (in northern Alaska, for example, the wolf, *Canis lupus*, and deer, *Rangifer tarandus* or *Alces alces*, serve as hosts for *Taenia hydatigena* Pallas, 1766, *T. ovis* (Cobbold, 1869), and *Echinococcus granulosus*). Taeniids are indigenous to all continents except Australia. The family appears to be little represented in South America, where canids and felids of early North American origin have undergone a high degree of radiation. The present cosmopolitan distribution of several species of taeniids is attributable to the worldwide dispersal of domestic animals, particularly since colonization by Europeans began in the early 16th century (Rausch, 1967).

Family Taeniidae Ludwig, 1886

Diagnosis: Strobila ribbon-like, with many proglottids, or small, with few proglottids. Rostellum usually well developed and bearing two rows of hooks of characteristic form; rostellum rarely with one row of hooks, or absent. If in two rows, rostellar hooks of anterior row usually larger and alternating with those of second row. Single set of reproductive organs in each proglottid. Genital pores marginal, alternating irregularly. Testes numerous. Ovary bilobed, median, posterior to majority of testes. Vitelline gland compact, posterior to ovary. Uterus typically with numerous lateral branches. Eggs with thick-walled embryophore, composed of truncated blocks. Larval stage a cysticercus or modification thereof; or a cystic or multicystic structure producing protoscoleces in brood-capsules by means of asexual proliferation. Strobilar and larval stages in mammals. Type-genus *Taenia* Linnaeus, 1758.

Key to subfamilies and genera

1a. **Strobila ribbon-like, with many proglottids. Scolex usually with two rows of rostellar hooks of characteristic shape. Gravid uterus with numerous lateral branches, often with secondary branching. Larval stage a cysticercus or modification thereof; scolex invaginated within, or associated with a bladder .... Taeniinae Stiles, 1896.**

Diagnosis: Immature and mature proglottids wider than long, with relative length increasing posteriad. Rostellum usually with two rows of hooks of typical shape; hooks of anterior row larger, alternating with those of second row. Rostellum rarely with one row of hooks, or absent. Genital pores irregularly alternating. Female genital organs situated posteriorly. Ovary bilobed, median. Vitelline gland simple, posterior to ovary. Testes abundant, mostly anterior and lateral to female organs. Uterus arising as median, longitudinal tube. Gravid uterus with lateral branches, often secondarily branched. Eggs spherical, abundant.

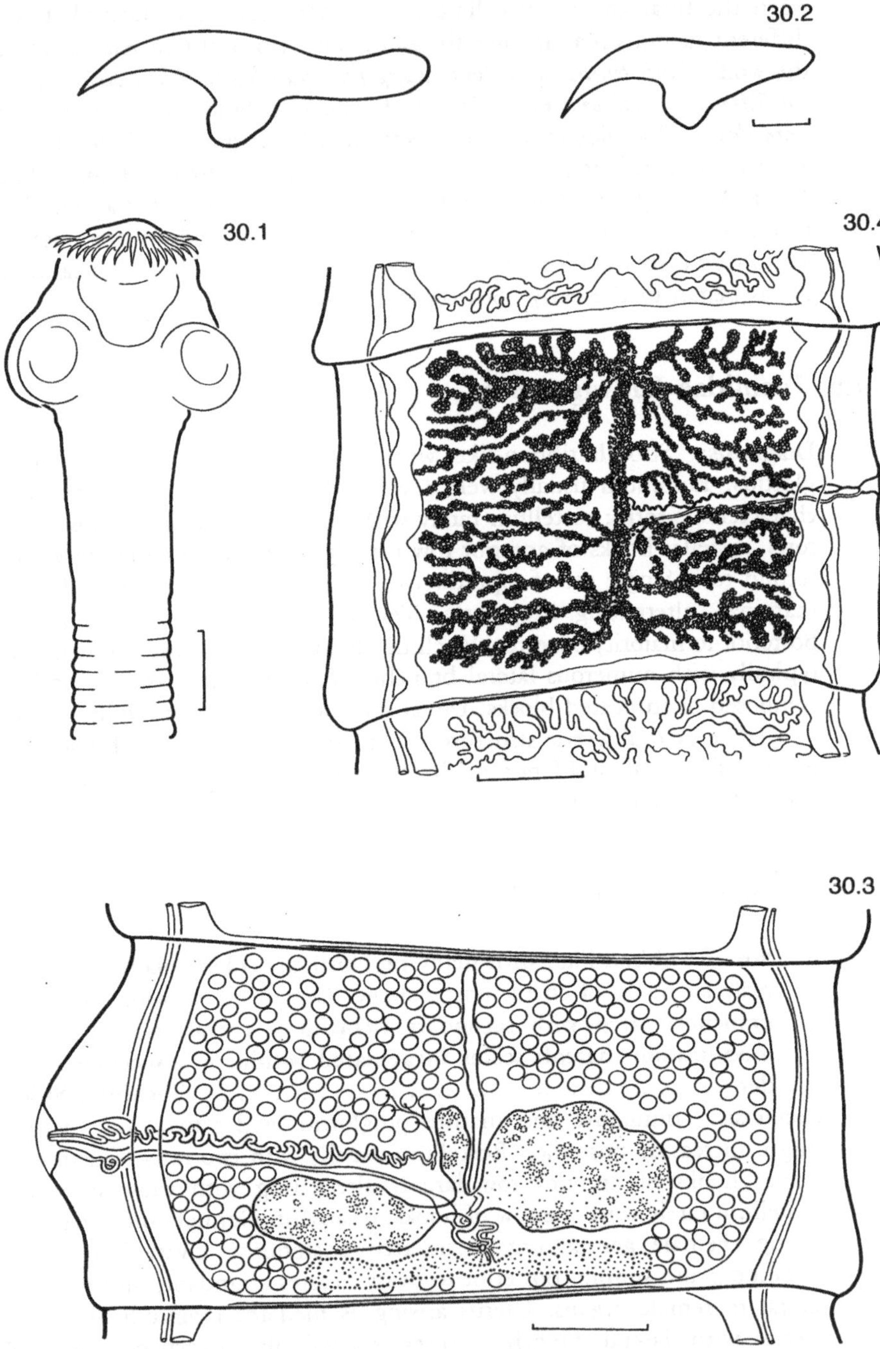

Figs 30.1–30.4 30.1. *Taenia macrocystis* (Diesing, 1850). Scolex. Scale 500 μm. 30.2. *Taenia solium* Linnaeus, 1758. Rostellar hooks. Scale 25 μm. 30.3. *Taenia solium*. Mature proglottid. Scale 500 μm. 30.4. *Taenia rileyi* Loewen, 1929. Gravid proglottid. Scale 1 mm.

Genus *Taenia* Linnaeus, 1758 (Figs. 30.1–30.4)

(Syns. *Haerucula* Pallas, 1760; *Taeniola* Pallas, 1760; *Tenia* Scopoli, 1777; *Hydatigena* Goeze, 1782; *Megocephalos* Goeze, 1782; *Multiceps* Goeze, 1782; *Pseudoechinorhynchus* Goeze, 1782; *Finna* Werner, 1786; *Vesicaria* Müller, 1787; *Vesicaria* Schrank, 1788; *Haeruca* Gmelin, 1790; *Hydatula* Abildgaard, 1790; *Hydatis* Blumenbach, 1797; *Alyselminthus* Zeder, 1800; *Cysticercus* Zeder, 1800; *Polycephalus* Zeder, 1800; *Hygroma* Schrank, 1802; *Halysis* Zeder, 1803; *Cisticercus* Rudolphi, 1805; *Coenurus* Rudolphi, 1808; *Finna* Brera, 1809; *Physchiosoma* Brera, 1809; *Goeziana* Rudolphi, 1810; *Hydatigera* Lamarck, 1816; *Pentastoma* Virey, 1823; *Fischiosoma* della Chiaje, 1825; *Trachelocampylus* Frédault, 1847; *Arhynchotaenia* Diesing, 1850; *Cysticercus* Goldberg, 1855; *Halisis* Goldberg, 1855; *Hidatula* Goldberg, 1855; *Acanthotrias* Weinland, 1858; *Taeniarhynchus* Weinland, 1858; *Cystotaenia* Leuckart, 1863; *Neotenia* Sodero, 1886; *Neotaenia* Braun, 1894; *Reditaenia* Sambon, 1924; *Fossor* Honess, 1937; *Tetratirotaenia* Abuladze, 1964; *Monordotaenia* Little, 1967; *Fimbriotaenia* Kornyushin & Sharpilo, 1986.)

Diagnosis: With the characters of the subfamily. Type-species *T. solium* Linnaeus, 1758.

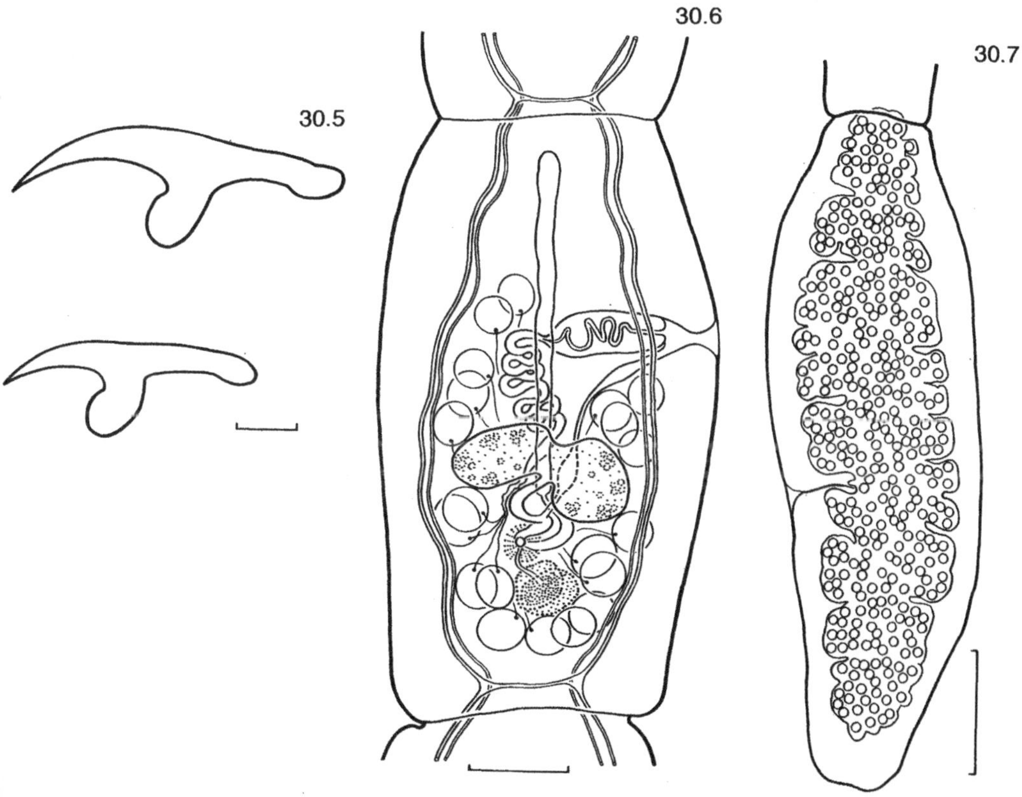

Figs 30.5–30.7 30.5. *Echinococcus multilocularis* Leuckart, 1863. Rostellar hooks. Scale 5 μm. 30.6. *Echinococcus multilocularis*. Mature proglottid. Scale 100 μm. 30.7. *Echinococcus granulosus* (Batsch, 1786). Gravid proglottid. Scale 500 μm.

1b. Strobila up to 12 mm in length, with three to seven proglottids. Scolex with two rows of rostellar hooks of characteristic shape. Gravid uterus with slightly developed lateral branches, or lateral branches lacking; uterus fully gravid only in terminal proglottid. Larval stage a single cyst, or multicystic; protoscoleces produced within brood-capsules. Laminated membrane present .... Echinococcinae Abuladze, 1960.

Diagnosis: Mature proglottids usually longer than wide; gravid proglottids elongate. Genital pores irregularly alternating. Female genital organs situated posteriorly. Ovary usually bilobed, median. Vitelline gland simple, posterior to ovary. Testes relatively few, mostly anterior and lateral to female organs. Uterus arising as median, longitudinal tube. Gravid uterus with or without lateral branches. Eggs spherical.

Genus *Echinococcus* Rudolphi, 1801 (Figs 30.5–30.7)

(Syns *Hydatis* Goeze, 1782; *Acephalocystis* Laennec, 1804; *Echinococcifer* Bremser, 1819; *Liococcus* Bremser, 1819; *Splanchnococcus* Bremser, 1819; ?*Astoma* Goodsir, 1844; ?*Diskostoma* Goodsir, 1844; ?*Sphaeridion* Goodsir, 1844; *Echinokokkus* Buhl, 1856; *Echinococcifer* Weinland, 1858; ?*Discostoma* Braun, 1894; *Alveococcus* Abuladze, 1960.)

Diagnosis: With the characters of the subfamily. Type-species *E. granulosus* (Batsch, 1786).

31 Lists of Genera

A. JONES[1], R.A. BRAY[2] & L.F. KHALIL[1]

[1]*International Institute of Parasitology, 395A Hatfield Road, St Albans, Herts AL4 0XU, UK:* [2]*Department of Zoology, The Natural History Museum, Cromwell Road, London SW7 5BD, UK.*

Any generic names that do not appear in the Keys as valid or as synonyms, or which are new, are listed in this chapter. Frequently, they will be the subject of comments in the text, which can be located by reference to the index. Where authors have concluded that the taxa should be considered *genera inquirenda, genera incertae sedis*, etc., the names have been listed under the appropriate heading. Also presented are names of genera which have been omitted inadvertently or which are derived from the early, often medical, literature and may not have been used in a Linnaean sense.

Genera inquirenda

Adelobothrium Shipley, 1900
Aocobothrium Mola, 1927
Aphanobothrium Seurat, 1906
Armandia Guiart, 1927
Barbusa Capoor & Srivastava, 1975
Bilocularia Obersteiner, 1913
Biporophyllaeus Subramanian, 1939
Blanchardella Moniez, 1895
Bothriotaenia Railliet, 1892
Capooria Malhotra, 1985
Capsulana Khanum, Farooq & Jamal, 1982
Cephalobothrium Shipley & Hornell, 1906
Chelacanthus Yamaguti, 1959
Circulobothrium Shinde & Chincholikar, 1975
Cleberia Rego, 1967
Coelodela Shipley, 1901
Copesoma Sinitzin, 1896

Corynesoma Leuckart in Diesing, 1835
Cotylogenes Lühe, 1902
Cyatocotyle Mola, 1908
Diagonobothrium Shipley & Hornell, 1906
Diplochetos Linstow, 1906
Diporus Diesing, 1863
Diskrjabinella Matevosyan, 1953
Dittocephalus Parona, 1887
Doublesetina Srivastava & Srivastav, 1989
Drepanidolepis Lopez-Neyra, 1942
Echinoproboscilepis Sadovskaya, 1954
Flapocephalus Deshmukh, 1979
Globus Scopoli, 1772
Gopalaia Dixit, Capoor & Rengaragu, 1980
Halysis Zeder, 1803 [Ch.29; syn. of *Taenia* in Ch.30]
Hoaleshwaria Shinde & Chincholikar, 1975
Humiditaenia Bhalya & Capoor, 1986
Hymenofimbria Skryabin, 1914
Kowsalyabothrium Muralidhar, Shinde & Jadhav, 1989
Kystocephalus Shipley & Hornell, 1906
Laiqia Khanum, Farooq & Jamal, 1982
Leuckartia Moniez, 1880 preoccupied
Longirostrum Malhotra & Nanda, 1986
Manaosia Woodland, 1935
Merocestus Cohn, 1902
Moravekia Sahay & Sahay, 1976
Myxolepis Spasskii, 1959
Neoaporina Saxena & Baugh, 1973
Neolitobothrium Bilqees, 1983
Neolytocestus Sahay & Sahay, 1976
Oschmarenia Spasskii, 1951
Phanobothrium Mola, 1907
Platybothrium Hornell, 1923
Polipobothrium Mola, 1908
Pseudobothrium Guiart, 1935
Pseudogilquinia Bilqees & Khatoon, 1982
Ptychobothrioides Yamaguti, 1959
Reticulatea Khanum, Farooq & Jamal, 1982
Rhoptrobothrium Shipley & Hornell, 1906
Sciadocephalus Diesing, 1850
Shindeiobothrium Jadhav, Shinde & Deshmukh, 1981
Shindeobothrium Shinde & Chincholikar, 1975
Spasskofuhrmina Palladwar & Kalyankar, 1989
Spongiobothrium Linton, 1889
Tetracisdicotyle Fuhrmann, 1907
Thaparea Johri, 1953
Tiarabothrium Shipley & Hornell, 1906

Trygonicephalum Shinde & Jadhav, 1984
Unilateriporina Matevosyan, 1969
Yogeshwaria Chincholikar & Shinde, 1976
Yogeshwaria Shinde, 1968

Genera incertae sedis

Acanthorhynchus Diesing, 1850
Anoplotaenia Beddard, 1911
Anthemobothrium Shipley & Hornell, 1906
Balanophorus Braun, 1894
Bombicirhynchus Pintner, 1931
Bothriorhynchus Jeude, 1829
Calycobothrium Southwell, 1911
Clujia Guiart, 1935
Dactylopsilophilus Kugi, 1990
Dasyurotaenia Beddard, 1912
Daveneolepis Spasskii, 1979
Deltokeras Meggitt, 1927
Dendrometra Jordano & Diaz-Ungria, 1956
Eniochobothrium Shipley & Hornell, 1906
Eulacistorhynchus Subhapradha, 1957
Francolina Capoor, Sawada, Bhalya & Rastogi, 1987
Guptaia Malhotra, 1985
Hornellobothrium Shipley & Hornell, 1906
Insinuarotaenia Spasskii, 1948
Johria Yamaguti, 1959
Maccallumiella Yamaguti, 1959
Mastacembellophyllaeus Shinde & Chincholikar, 1977
Microbothriorhynchus Yamaguti, 1952
Multicapsiferina Fuhrmann, 1922 (syns *Zschokkea* Fuhrmann, 1902 and *Zschokkeella* Ransom, 1909, both preoccupied).
Myliobatibothrium Shinde & Mohekar[*sic*], 1983
Pelichnibothrium Monticelli, 1889
Pillersium Southwell, 1927
Pintneriella Yamaguti, 1934
Rhopalothylax Guiart, 1935
Rhynchobothrium Rudolphi, 1819
Rhynchobothrius Rudolphi, 1819
Shindeia Gaikwad & Shinde, 1985
Staurobothrium Shipley & Hornell, 1905
Tetrabothriorhynchus Diesing, 1850
Tetrarhynchus Rudolphi, 1809
Vaullegeardia Guiart, 1927
Wageneria Monticelli, 1892

Of the genera listed above, *Anoplotaenia* Beddard, 1911 and *Dasyurotaenia* Beddard, 1912 are both well characterized and of undoubted validity. Readers are referred to the redescriptions by Sandars (1957) and Beveridge (1984), respectively.

Nomina nuda

Amphicephalus van Beneden, 1871
Arostilepis Mas Coma, 1982
Biacetabuloides Mackiewicz, 1960
Cestoideum Rudolphi, 1819 (syn. *Orygmatoscolex* Diesing, 1854)
Cholodkovskya Matevosyan, 1953
Corderolepis Mas Coma, 1982
Cylindrophorus Diesing, 1863
Fibrocirrosa Palladwar & Kalyankar, 1980
Hunterium Mackiewicz, 1960
Palaia Shipley, 1900
Pleudobothrium van Beneden, 1871
Priapacanthus Dollfus, 1975

Larval and collective names

Cephalocotyleum Diesing, 1850
Cercocystis Villot, 1882
Cestoscolex van Beneden, 1870
Cisticercoideum Parona, 1901
Cryptocystis Villot, 1882
Cysticercoides 'Leuckart'
Dithyridium Rudolphi, 1919
Dubium Rudolphi, 1910
Gryporhynchus von Nordmann, 1832
Gymnoscolex Diesing, 1854
Ilisha Southwell & Prashad, 1918, preoccupied
Monocercus Villot, 1882
Onchoscolex Diesing, 1853
Piestocystis Diesing, 1850
Plerocercoides Neumann, 1892
Plerocercus Braun, 1883
Pseudoscicus Polonio, 1860
Scolex Müller, 1787
Scolexia Rafinesque, 1815
Slossia Meggitt, 1931
Sparganum Diesing, 1854
Steganobothrium Diesing, 1854
Stenotenia Gervais, 1870
Strobilocercus Sambon, 1924

Tetrastoma Forbes & Goodsir, 1841
Tetrathyridium Rudolphi, 1819
Trichoda Müller, 1773
Vermis

New genera

Bulbultaenia Beveridge
Chimaerula Bona
Cuculincola Bona
Dollfusiella Campbell & Beveridge
Eburneotaenia Bona
Francobona Georgiev & Kornyushin
Geraldolepis Czaplinski in Czaplinski & Vaucher
Glanduluncinata Bona
Hepatocestus Bona
Hunkelepis Vaucher in Czaplinski & Vaucher
Monogynolepis Czaplinski in Czaplinski & Vaucher
Monosertum Bona
Ovosculpta Bona
Spasspasskya Bona
Vogelepis Vaucher in Czaplinski & Vaucher

The following genera came to our notice too late to be included in the keys:

Paraorygmatobothrium Ruhnke, 1994 (Tetraphyllidea, Phyllobothriidae)

Paucicapsula Kugi, 1993 (Cyclophyllidea, Anoplocephalidae).

Glossary

accessory seminal vesicle a saclike diverticulum adjoining the external seminal vesicle at its junction with the cirrus-sac but not penetrated by the vas deferens; also, 'cirromotionblase' of Zerny.

accessory sucker a sucker, often small, located on the anterior margin of a bothridium (*Phyllobothrium*), or on a muscular apical pad (members of family Onchobothriidae).

acini small sac-like lobules composing a compound gland; as of the vitelline gland.

acraspedote adjoining anterior and posterior borders of proglottids do not overlap; or scolex does not overlap its junction with the strobila.

adnexa structure adjoining or attached to a larger organ and having a related function.

alternate arrangement of principal rows of hooks in successive half-spiral turns to the right and left; as seen on the internal face of the tentacle of heteroacanths and poeciloacanths.

anapolysis gravid proglottids are not shed from the strobila and the pygidium is always the last proglottid in the strobila.

anapolytic lacking apolysis; see anapolysis.

anterior in the direction of the apex of the scolex or the tips of the extended tentacles.

antibothridial face the side (half-turn) of the tentacle opposite the nearest bothridium.

apical organ structure, probably glandular, localized in anterior part of scolex in some roteocephalid species.

aploparaksoid hook guard and blade well developed, almost of same length, handle rudimentary.

apolysis release of a terminal proglottid that is gravid, which then degenerates and frees its eggs (Pintner, 1913).

appendix a posterior extension of the growth zone from the scolex of a trypanorhynch plerocercus.

arcuatoid hook handle much longer than blade, weakly developed, small blade and guard almost equal in length.

armature the hooks or spines covering the surface of a structure; tentacle armature of trypanorhynchs is categorized into three basic types designated homeoacanthous, heteroacanthous and poeciloacanthous.

armed possessing spines or hooks; as an armed cirrus or armed tentacles.

atrium the cavity receiving the male and female ducts and opening at the genital pore.

atypical heteroacanthous heteroacanthous armature in which there are more rows of hooks on the one face (usually external face) than on the opposite face (internal face).

auricles fleshy to muscular ear-like appendages usually projecting laterally (less often medially or posteriad) from the anterior margins of the bothridia among the tetrabothriids (excluding *Priapocephalus*).

axis a straight line about which a geometrical object may be conceived to rotate.

basal armature hooks at the tentacle base that differ in form, number and arrangement from those of the metabasal armature; basal armature often on a tentacular swelling.

biacetabulate with two acetabula or true suckers on scolex.

bilateral symmetry the simplest plane isometry in which there is reflection in a line, i.e. mirror images; the figure looks the same on both sides of a line, except that the two sides are mirror images of each other.

biloculate suckers septate suckers, forming two loculi as in the proteocephalids, *Peltidocotyle*, *Zygobothrium*.

blade see hook blade

blastocyst the bladder-like structure into which the trypanorhynch larva is withdrawn.

bothridial face the side (half-turn) of a tentacle adjacent to the nearest bothridium.

bothridium (pl.-ia) in trypanorhynchs one of the dorsoventrally paired subapical outgrowths of the scolex, ear-like or elongated in shape, either sessile or pedicellate, whose musculature is contiguous with the general musculature, and generally very mobile but rarely capable of a firm grip because of its shallow cavity and lack of suckers or subcompartments.

bothrioloculodiscate with bothria, accessory loculi and terminal flattened knob at apex of scolex.

bothrium dorsal and ventral longitudinal grooves in the pseudophyllidean scolex.

bulbate shaped like a bulb, expanded; type of scolex.

bulboacuminate bulb or knob-like with pointed apex; type of scolex.

bulboloculate bulb or knob-like with pair of loculi; type of scolex.

cambrensoid hook handle longer than blade, guard directed backwards and shorter than blade.

chainette a file or files of alternating hooks on the external face of the tentacle, separated and distinct from the principal rows, in which each hook is aligned with a principal row. Three types have been described: 1) **single chainette**, comprised of a single file of similar, alternating hooks (*Lacistorhynchus tenuis*); 2) **double chainette**, consisting of a double file of closely spaced hook pairs (*Dasyrhynchus variouncinatus*); 3) **compound chainette**, combining hooks of several forms or numbers (*Chimaerarhynchus* and *Hornelliella*).

chainette elements hooks that comprise the chainette.

choanocampanulate shaped like a bell, with terminal funnel; type of scolex.

circumcortical forming a complete circle within the cortical zone.

circum-medullary forming a complete circle within the boundary of the medullary zone.

cirromotionblase (Zerny, 1912) an accessory seminal vesicle once believed to be involved with eversion of the cirrus.

cirrus the male copulatory organ.

cirrus-pouch see cirrus-sac.

cirrus-sac the bladder-like enclosure of the male terminal genitalia.

compound chainette a chainette composed of hooks of more than one type or number.

convergent the condition in which hooks 1(1') are closely approximate and tend to form a V-shaped intersection of the opposing rows.

cordiform having a conventional heart-shape; or reverse cordiform, pointed anteriorly and indented or notched posteriorly. As in reverse cordiform bothridia (viz. *Grillotia erinaceus*).

coronulate shaped like a head with a crown on it; dome-like with basal frill; type of scolex.

coronuloid hook guard and blade well developed, almost equal in length, each longer than handle.

cortex region or zone surrounding the inner longitudinal muscle layer.

cortical referring to location in the cortex.

craspedote posterior border of a proglottid overlaps the anterior border of adjoining proglottid; scolex overlaps at its junction with the strobila often forming a velum.

crescentic crescent shape of lateral vitelline clusters as viewed in body cross-section.

cricetoid hook close to coronuloid but guard longer than blade.

cuneicrispitate wedge-like with thin edge curled; type of scolex.

cuneifimbriate wedge-like with thin edge having fringe or frill; type of scolex.

cuneiform like a wedge; type of scolex.

cuneiloculate wedge-shaped, with median and accessory loculi; type of scolex.

cyclusteroid apical apparatus (see Figs 25.49, 25.50)

dictymetroid hook with long blade and handle and shorter guard (see Fig. 25.199).

digitiform shaped like a finger; type of scolex.

diorchoid hook handle longer than well-developed blade, guard shorter than blade.

distal anatomically located far from the point of origin or attachment.

divergent the condition in which hooks 1(1') are widely spaced apart from a point that would form the V-shaped intersection of the opposing rows.

dorsal in trypanorhynchs, the surface opposite that paralleled by the preovarian portion of the vagina. (Trypanorhyncha.)
refers to the part of the parenchyma generally occupied by the testes. (Proteocephalidea.)

double chainette two files of hooks abut forming the chainette.

ductus yamagutii term coined by Bandoni & Brooks (1987a) for a duct of unknown structure and function which emerges from a glandular mass adjacent to the seminal receptacle of the amphilinid genus *Gigantolina* and extends anteriorly alongside the terminal limb of the uterus to open at the anterior extremity of the body close to the uterine pore.

echinorhynchoid hook handle and blade almost equal, guard slightly shorter.

egg a fertilized ovum surrounded by several membranes forming a shell or capsule.

enigmatical prebulbar organ see prebulbar organ.

euapolysis type of release of proglottids in which the mature proglottid leaves the strobila when the uterus has begun to fill with eggs. When detached it continues to live in the intestine of the host and grow until fully gravid, may develop a pseudo-holdfast and change shape (Pintner, 1913).

external face that half of the tentacle surface facing the nearest lateral margin of the scolex; described by dividing a cross section of the tentacle dorsoventrally in the midsagittal plane (12 o'clock to 6 o'clock); the surface on which the principal hook rows end in heteroacanths and poeciloacanths.

external seminal vesicle sac-like dilation of the vas deferens found outside the cirrus-sac that serves as a reservoir for sperm.

extravascular referring to location in the area lateral to the longitudinal osmoregulatory canals.

falciform hook shape that has a scythe-like blade, slender, tapering and gradually curved.

fertilization duct portion of the oviduct where the sperm duct from the seminal receptacle or vagina enters.

file a longitudinal row of hooks in the tentacle armature. Tentacles usually taper distally and the number of files may decrease distally or their number remain constant if hooks decrease in size toward the tip of the tentacle. Determine constancy of file number by counting the number of hooks per principal row.

fixomegabothriate immovable, with huge bothria; type of scolex.

flabellate fan-shaped; type of scolex.

foliiform leaf-shaped.

follicle a small, globular subdivision of a compound vitelline gland.

fossette an eversible sensory pit on the bothridial margins; see *Otobothrium* and *Paranybelinia*.

fraternoid hook handle longer than guard and blade, guard as long or longer than blade, thicker than blade.

genital atrium see atrium.

genital papilla a papilla of varying complexity, projecting from the wall of the genital atrium and on which is located the aperture(s) of the male or male and female genital ducts among some tetrabothriids.

genital pore gonopore, the external opening for the genitalia.

glide reflection symmetry pattern symmetry composed of a glide (translation) by a certain distance along some line, accompanied by a reflection (mirror-image) along that line.

glossocercoid hook pattern see Chapter 25, couplet 29b.

gravid containing eggs, as a gravid or ripe proglottid.

hammer-like hook handle stick-like, no guard; blade sharp, directed backwards.

hastate a shape resembling a spearhead; type of scolex.

heel see hook heel.

helical see Chapter 7.

hermaphroditic duct a common duct formed by the union of the vagina and male sperm duct.

hermaphroditic sac sac in which the male and female genital terminalia have united to form an hermaphroditic duct; also, the hermaphroditic vesicle described by Yamaguti (1954a), illustrated by Campbell & Beveridge (1987).

hermaphroditic vesicle a common vesicle, within the hermaphroditic sac of *Hornelliella*, joined by the male sperm duct and vagina, that exits via a hermaphroditic duct into the genital atrium (Yamaguti, 1954a; Campbell & Beveridge, 1987).

heteroacanthous hook pattern symmetry consisting of bilaterally distinct half-spiral rows instead of quincunxes, always demonstrating glide reflection symmetry; the number of hook rows may be constant or increased from internal to external surfaces of the tentacle; never with chainette; also, heteracanthous.

heteromorphous referring to hooks of different forms.

homeoacanthous hook pattern symmetry in which all hooks are arranged in quincunxes.

homeomorphous referring to hooks of the same form.

hook-base region of the hooks affixed to the tentacle. (Trypanorhyncha.)

hook blade the elongated portion projecting from the surface, usually tapered to a sharp point.

hook file a linear arrangement of adjacent hooks that parallels the longitudinal axis of the tentacle.

hook forms winged, single winged vs. two-winged, doubly winged or V-shaped, hastate, deltoid, spiniform, bent spiniform, falciform, uncinate, sagittate, serpe à bec.

hook heel extension at either end of the hook base beyond its junction with the blade.

hook row a transverse or oblique linear arrangement of adjacent hooks oriented across the longitudinal axis of the tentacle.

hook talon see heel.

hooklets very small hooks, commonly interpolated between principal rows.

hyperapolysis release of a proglottid precociously before it contains any eggs in the uterus to complete its development while free in the intestine of the host.

inner seminal receptacle an ovoid to elongate dilation of the proximal vagina situated posterior to the ovary in the region of the Mehlis' gland among some tetrabothriids.

insertion as previously used in the literature, insertion is the point at which the tentacle retractor muscle is attached within the bulb; however, in muscle anatomy the insertion of a muscle is the moveable end of a muscle attachment in contradistinction to the muscle origin which does not move. In trypanorhynchs the true insertion of the retractor muscle is in the distal end of the tentacle and the muscle origin within the bulb or proximal end of the tentacle sheath. Contraction of the retractor muscle causes invagination of the tip of the tentacle, turning the armature to the inside, and subsequent withdrawal of the tentacle into the tentacle sheath within the scolex.

intercalary see interpolated.

internal face that side of the tentacle facing away from the nearest lateral margin of the scolex; side opposite the external face of the tentacle.

internal seminal vesicle saccate enlargement of the male duct within the cirrus sac.

interpolated said of hooks or row of hooks located between the principal rows of hooks; also, intercalary hooks or rows. Best observed from the bothridial/antibothridial faces or external tentacle face.

intervascular located medial to the longitudinal osmoregulatory canals.

isometry see plane isometry.

lappets folds of tissue found in the scolex, metascolex or proglottids in some genera of proteocephalids.

loculomonobothriate with single terminal introvert or funnel and lateral loculi; type of scolex.

loculopapillate with a nipple and loculi; type of scolex.

loculotholate shaped like a dome, with loculi; type of scolex.

loculotruncate appearing as if cut off, with loculi; type of scolex.

loculus shallow cavity on the scolex that can appear or disappear, depending on scolex mobility (Caryophyllidea).

longitudinal lengthwise, pertaining to the greatest length.

longitudinal muscle bundles the principal muscular system in proglottids; it divides the parenchyma into central or medullary parenchyma and the external or cortical region. These muscles are weakly developed or practically nonexistent in some species.

medial pertaining to the middle or median plane that divides a bilaterally symmetrical object into right and left halves.

medulla the region or zone internal to the inner layer of longitudinal muscle bundles; the region which contains the osmoregulatory ducts, nerve cords and most of the reproductive organs.

medullary referring to location in the medulla.

Mehlis' gland glands surrounding the ootype where egg shells are formed; shell gland.

metabasal armature hook rows distal to the basal region of the tentacle; usually of consistent pattern all the way to the tip.

metascolex the posterior portion of a divided scolex. Consists of folds of tissue generally encircling or hiding the suckers. It either originates from the base of the scolex, sometimes forming a sort of 'collar', or from folds of tissue between the suckers.

monobothriate with single terminal introvert or funnel; type of scolex.

monopleuroid mature stage of cestodes lacking a cercomer, internal or external proglottidization and with a single set of reproductive organs.

montanate shaped like a mountain with peak; type of scolex.

multiple chainettes more than one chainette per tentacle.

neck an unsegmented region between the scolex and the first distinct proglottids of the strobila. A neck is not present in all species of tapeworms.

nitiduloid hook handle and blade almost equal in length, guard rudimentary.

oncotaxy classification based upon hook pattern symmetry.

onderstepoortioid apical apparatus (Dilepididae); see Fig. 25.38.

oophore (Dilepididae) formed from uterus; eggs entangled in alveolar tissue, swells in water; digitiform expansion.

ootype an expansion of the oviduct surrounded by Mehlis' gland where egg shells are formed.

origin the beginning; see proximal.

osmoregulatory canal also osmoregulatory duct. Main canals of the osmoregulatory system, mainly used in water balance. They are mainly longitudinal, often with transverse anastomoses. They may be reticular.

ova unfertilized egg cells produced by the ovary.

ovarian isthmus the transverse hollow connection between the ovarian lobes through which ova travel to reach the oviduct.

ovary the female gonad, produces ova.

ovicapt sphincter controlling entry of ova from ovary into the oviduct; also, oocapt.

oviduct duct controlling passage of ova from the ovary to the ootype.

parenchyma the storage tissue filling the body and surrounding the internal organs.

parenchymal muscle thin layer of muscle internal to the subtegumental longitudinal muscle bundles; separates interior of proglottid into cortical and medullary zones.

pars antica scolecis scolex and adjoining germinal region.

pars bothridialis portion of the scolex with bothridia.

pars bulbosa scolex region containing the bulbs.

pars metabothridialis scolecis region combining the pedunculus scolecis and pars proliferans.

pars postbulbosa region of the scolex posterior to the bulbs.

pars postica portion posterior to the scolex and germinal region; the blastocyst.

pars proliferans the growth zone immediately posterior to the scolex.

pars vaginalis the scolex region containing the tentacle sheaths.

paruterine capsules saccate, egg-bearing, thin-walled capsules that develop from fibrous regions at the apex of the paruterine organs (in Nematotaeniidae).

paruterine organ conical, apparently fibrous structures, lying paired or singly in proximity to uterus in pre-gravid proglottids (in Nematotaeniidae).

parvitaenioid apical apparatus (Dilepididae) (see Figs 25.67, 25.68).

patelliform having the shape of a shallow, almost square plate. As in patelliform bothridia (viz. *Grillotia*).

pedicel a stalk-like support; pterobothriids have pedicellated bothridia (viz. *Pterobothrium*).

peduncle stalk-like portion of the scolex from which several pedicels may arise.

pedunculated possessing a peduncle or pedicel.

pedunculus scolecis elongated portion of the scolex to which the bothridia are attached.

piercing organ a specialized rostellum of a muscular glandular nature, found in species of subfamily Acanthotaeniinae (Proteocephalidae).

Pintner's cells muscle cells disposed in tiers in the scolex peduncle between the bothridia and the bulbs, their fibres radiate transversely giving an internal pseudosegmented appearance to the pars vaginalis; see Eutetrarhynchidae.

plane isometry geometric measure of rigid motion where the original figure and its image are congruent. There are four types of rigid motions in a plant: reflection (also called bilateral symmetry), rotation, translation and glide reflection. Every rigid motion of an infinite pattern corresponds with one of these four symmetries. This allows a simple classification of all patterns, according to which of the four fundamental rigid motions leaves a pattern invariant.

plerocercoid third stage cestode larva characterized by a solid body; postlarvae have this appearance but are considered to have lost their blastocyst.

plerocercus third stage larva of a trypanorhynch which possesses a blastocyst.

poeciloacanthous also pecilacanthous. (See Chapter 7.)

polyzoic condition where body is formed of two or more proglottids.

porosoid hooks (Dilepididae) with long blade and handle, short guard (see Fig. 25.252).

posterior direction away from the scolex end.

prebulbar organ nucleated enlargement at the junction of the tentacle and its bulb characteristic of some trypanorhynchs; also enigmatic prebulbar organ, enigmatic organ.

principal row a row of large hooks traversing a half-turn around a tentacle.

proboscis in trypanorhynchs, one of the tentacles.

proglottid the area surrounding each set of reproductive organs or genitalium; also proglottis; not synonymous with segment, see proglottization (Wardle & McLeod, 1952).

proglottization metameric repetition of the reproductive organs, characteristic of all polyzoic tapeworms; not synonymous with segmentation (i.e. in the Order Spathebothriidea proglottization is not attended by segmentation, also true of some members of the Order Pseudophyllidea).

protandry in monoecious worms, maturation of the male system precedes that of the female system.

proximal nearest the central part of the body, the origin or point of attachment.

punctiform having the form of a point, puncture or dot.

pygidium the first proglottid formed and last or oldest on the strobila; it contains the excretory pore.

pyriform pear-shaped.

quincunx an arrangement of five objects with one at each corner of a rectangle and one at the centre. In homeoacanthous armatures the hooks of any three adjacent files are arranged in quincunxes.

rank a near horizontal row of hooks or arrangement of hooks, closely spaced, that traverse the circumference of the tentacle; ranks form ascending spirals in homeoacanths but separate, alternate, ascending half-spirals in heteroacanths and poeciloacanths.

rebated hook shape in which the point is blunted.

rectanguloid hooks with a small, curved, triangular chitinous blade and a large irregularly-shaped non-chitinized handle and guard (Paruterinidae: *Triaenorhina*).

recurved hook shape in which there is a strong downward curvature, said of the blade or its tip.

retractor muscle long, thin muscle causing invagination of the tentacles into their sheaths.

rhyncheal apparatus apparatus for eversion and retraction of the tentacles; includes bulbs, retractor muscles, tentacles and their sheaths.

rosethorn-shaped hook shape with broad base, broad blade of relatively short length and gradual curvature, heel extensions often prominent.

rostellar pouch muscular or membraneous sac around embedded part of rostellum; usually clearly distinct from rostellum.

rostellum muscular organ of various shapes embedded at apex of scolex, invaginable or not, often bearing spines.

rotational symmetry plane symmetry in which a figure has a centre about which it can be rotated by a certain angle without changing its appearance. Rotation of a square or quincunx by 90° about its centre leaves it unchanged. Rotational pattern symmetry is only infinite in the homeoacanthous (homeomorphous) armature.

rugomonobothriate with single terminal introvert or funnel and longitudinal furrows or rugae; type of scolex.

rugomontanate with wrinkles, longitudinal furrows or rugae on scolex shaped like a mountain.

sagittate hook shape that resembles the head of an arrow; especially, triangular with a deeply indented base.

satellite hook a principal hook displaced in its alignment with remaining hooks of its respective row, often in tandem (viz. *Lacistorhynchus tenuis*).

scolex term proposed by van Beneden for the anterior end of a tapeworm often modified for attachment or locomotion; also the head, or holdfast organ of same. The scolex of trypanorhynchs is complex and usually possesses four armed tentacles projecting from its apex.

secondary grooves wrinkles or furrows found on the surface of the proglottids.

segment the unit of subdivision of the strobila containing one or more proglottids and defined by transverse grooves or partitions (Wardle & McLeod, 1952).

segmentation the separation of adjoining proglottids into segments by superficial transverse grooves, membranous partitions or by internal zones in which longitudinal muscle fibres are lacking.

seminal receptacle a dilation of the vagina that temporarily stores sperm.

seminal vesicle a dilation of the vas deferens that stores sperm; may be located external to the cirrus-sac, inside it, or both.

sensory fossette a pit on the bothridial margin that is eversible and presumed of sensory function.

serpe à bec S-shaped hook in which the tip is curved downward resembling the bill of a raptor; a hook with an undulating curvature, serpentine.

sessile said of a bothridium that is stalkless; attached directly at the base.

shell gland the gland which fertilized ova enter and are enclosed by a tanned 'shell' and emerge as 'eggs' to be stored in the uterus.

simple chainette a chainette composed of a single hook file.

single chainette see chainette.

skryabinoid hook handle and blade well developed, one shorter or longer than the other; guard rudimentary; ventral concavity between proximal part of handle and guard.

spatulate shaped like a spatula, flat, expanded and round apex.

spiniform hook shape which is needle-like, a long slender blade, gradually tapering to a thin point (Trypanorhyncha).
handle and guard together form a spherical structure; blade much shorter (Hymenolepididae).

strobila the region of a tapeworm posterior to the scolex; the body of a tapeworm.

talon heel-like extension of the base of a hook, may be anterior or posterior.

tentacle elongated, flexible, protrusions; the four tentacles of trypanorhynchs are hyaline, retractable, hollow cylinders armed with hooks.

tentacle bulb the hollow, muscular, oval structure from which the tentacle sheaths arise, usually contains a retractor muscle.

tentacle sheath the tube within the scolex into which the armed tentacle can be withdrawn.

testes male reproductive organs, produce spermatozoa.

tholate dome-like; type of scolex.

translation symmetry plane symmetry in which a slide in a certain direction by a certain distance does not change the appearance of the figure.

typical heteroacanthous heteroacanthous armature in which the number of hook rows is the same on all sides of the tentacle; principal rows meet on the external face to form inverted V or overlap.

unarmed without hooks or spines.

uncinate claw-like hook shape.

unduloid hooks with short blade and guard, long handle (see Fig. 25.134).

uniloculate suckers the most common type of sucker; generally of acetabular type.

uterine diverticulum a digit-like extension of the uterus, may be formed prior to the entry of eggs; often develop as multiple saccate extensions in a gravid uterus.

uterine duct a narrow duct leading from the ootype to the uterus.

uterine glands dark-staining cells projecting from the uterine surface into the parenchyma.

uterine pore opening from the uterus to the external surface through which eggs escape, may be preformed.

uterus the tubular organ receiving eggs from the uterine duct or shell gland.

vagina the tube through which sperm travel to reach the oviduct; part of the female reproductive system extending from the genital pore to the oviduct.

vaginal pore entrance to the vagina.

vaginal seminal receptacle an elongate dilation of the distal vagina involved in the storage of sperm following copulation, usually at the level of the poral osmoregulatory canals among tetrabothriids.

valiporoid apical apparatus see Fig. 25.45.

vas deferens the single ducts formed by union of the vasa efferentia and entering the cirrus-sac.

vas efferens the small ducts connecting the testes to the vas deferens.

velum thin overlapping membrane-like portion of a bothridia.

ventral the surface paralleled by the vagina in trypanorhynchs; by the uterus in tetraphyllideans; the surface with the uterine pore.
the part of the parenchyma generally occupied by the uterus.

vitelline follicles acini that produce the cells and substances necessary for formation of 'eggshells'.

vitelline reservoir a sac-like dilation at the junction of the main vitelline ducts in which vitelline cells accumulate prior to their entry into the oviduct.

winged see hook forms.

Bibliography

Abuladze, K.I. (1960) [Taeniate taxonomy.] In: *Sbornik Nauchnaya konferentsiya, pozvyashchennaya 40 -letiyu Moskovskoi veterinarnoi akademii. Materialy sektsii infektsionnykh i invazionnykh boleznei*, pp. 66–70. [In Russian.]

Abuladze, K.I. (1964) [*Principles of Cestodology. Vol. IV. Taeniata of Animals and Man and Diseases Caused by Them.*] [Ed. K.I. Skryabin.] Moscow: Izdatel'stvo 'Nauka', 530 pp. [In Russian.] [English translation, 1970, Israel Program for Scientific Translations, iv + 549 pp.]

Adams, A.M. & Rausch, R.L. (1986) Two new species of *Ascometra* Kholodkovskii, 1912 (Cestoda: Paruterinidae) from the kori bustard, *Choriotis kori* (Burchell), in Kenya. *Journal of Parasitology* 72, 101–106.

Agassiz, L. (1846a) *Nomenclatoris Zoologicus index universalis. Continens nomina systematica classium, ordinum, familiarum et generum animalium omnium.* Soloduri, Jent et Gassmann (1846), pp. 115, 368.

Agassiz, L. (1846b) *Nomenclatoris Zoologicus. Continens nomina systematica classium, ordinum, familiarum et generum animalium omnium.* Soloduri, Jent et Gassmann (1842–1846). 'Nomina systematica generum Neuropterorum', p. 8; 'Nomina systematica generum Orthopterorum', p. 13.

Ahern, W.B. & Schmidt, G.D. (1976) Parasitic helminths of the American avocet *Recurvirostra americana*: four new species of the families Hymenolepididae and Acoleidae (Cestoda: Cyclophyllidea). *Parasitology* 73, 381–398.

Akhmerov, A.K. (1941) *Amurotaenia perccotti* n. gen., n. sp – Vertreter einer neuen Cestoden-Ordnung. *Doklady Akademii Nauk SSSR* 30, 191–192.

Akhmerov, A.K. (1960) [Tapeworms of fishes of the River Amur.] *Trudy Gel'mintologicheskoi Laboratorii* 10, 15–31. [In Russian.]

Akhmerov, A.K. (1969) [A new cestode, *Postgangesia orientale* gen. et sp. n., and a new subfamily Postgangesiinae (Cestoda: Proteocephalidae) from silurids of the river Amur.] *Trudy Gel'mintologicheskoi Laboratorii* 20, 3–7. [In Russian.]

Albers, D., Campbell, P.J., Crowe, D., Schuster, S. & Thompson, M. (1988) On size and shape. Part IV. In: Steen, L.A. (ed.), *For All Practical Purposes. Introduction to Contemporary Mathematics*. New York, W.H. Freeman Co., 450 pp.

Alexander, C.G. (1963) Tetraphyllidean and diphyllidean cestodes of New Zealand selachians. *Transactions of the Royal Society of New Zealand* 3, 117–142.

Andersen, K.I. & Lysfjord, S. (1982) The functional morphology of the scolex of two *Tetrabothrius* Rudolphi 1819 species (Cestoda; Tetrabothriidae) from penguins. *Zeitschrift für Parasitenkunde* 67, 299–307.

Ariola, V. (1899) Il gen. *Scyphocephalus* Rigg. e proposta di nuova classificazione dei cestodi. Genova, Tipografia di Angelo Ciminago, 12 pp.

Ariola, V. (1902) Richerche anatomo-zoologiche sui cestodi parassiti del *Centrolophus pompilus* C.V. *Atti della R. Università di Genova* 17, 117–170.

Artyukh, E.S. (1966) [*Principles of Cestodology. Vol. VI. Davaineata, Cestodes of Wild and Domestic Animals.*] Moscow, USSR: Akademiya Nauka, 511 pp. [In Russian.]

Baczynska, H. (1914) Études anatomiques et histologiques sur quelques nouvelles espèces de cestodes d'oiseaux. *Bulletin de la Société Neuchâteloise des Sciences Naturelles* 40, 185–240.

Baer, J.G. (1925a) Quelques cestodes d'oiseaux nouveaux et peu connus. *Bulletin de la Société Neuchâteloise des Sciences Naturelles* 49, 138–154.

Baer, J.G. (1925b) Some Cestoda described by Beddard, 1911–1920. *Annals of Tropical Medicine and Parasitology* 19, 1–22.

Baer, J.G. (1925c) Contributions to the helminth fauna of South Africa. Thesis. Université de Neuchâtel, 79 pp.

Baer, J.G. (1927) Monographie des cestodes de la famille des Anoplocephalides. *Bulletin Biologique de la France et Belgique* 10, 1–241 (suppl.).

Baer, J.G. (1933) Contribution à l'étude de la faune helminthologique africaine. *Revue Suisse de Zoologie* 40, 31–84.

Baer, J.G. (1937) Un genre de cestode d'oiseaux. *Bulletin de la Société Neuchâteloise des Sciences Naturelles* 62, 149–156.

Baer, J.G. (1940) Some avian tapeworms from Antigua. *Parasitology* 32, 174–197.

Baer, J.G. (1948) Contribution à l'étude des cestodes de sélaciens. I–IV. *Bulletin de la Société Neuchâteloise des Sciences Naturelles* 71, 63–122.

Baer, J.G. (1954a) Revision taxonomique et étude biologique des cestodes de la famille des Tetrabothriidae parasites d'oiseaux de haute mer et de mammiféres marins. *Memoires de l'Université Neuchâtel*, Serie In-Quarto 1, 4–122.

Baer, J.G. (1954b) The tapeworm genus *Wyominia* Scott, 1941. *Proceedings of the Helminthological Society of Washington* 21, 48–52.

Baer, J.G. (1955a) Revision critique de la sous-famille Idiogeninae Fuhrmann 1907 (Cestodes: Davaineidae) et étude analytique de la distribution des espèces. *Revue Suisse de Zoologie* 62 (Fascicule supplémentaire), 1–51.

Baer, J.G. (1955b) Incidence de la spécificité parasitaire sur la taxonomie. Problèmes d'évolution chez les cestodes cyclophyllidiens. *Bulletin de la Société Zoologique de France* 30, 275–287.

Baer, J.G. (1956) Parasitic helminths collected in West Greenland. *Meddelelser om Gronland udgivne af Kommiss. for Unders. Gronland. Bd.* 124, 55 pp.

Baer, J.G. (1959) Helminthes parasites. *Exploration des Parcs Nationaux du Congo Belge. Mission J.G. Baer – W. Gerber (1958)* 1, 3–163.

Baer, J.G. (1962) Cestoda. *The Zoology of Iceland* 2, 63 pp.

Baer, J.G. & Bona, F. (1960) Révision des cestodes Dilepididae Fuhrm., 1907 des Ardéiformes. Note préliminaire. *Bollettino dell'Instituto e Museo di Zoologia della Università di Torino* 6, 1–53.

Baer, J.G. & Euzet, L. (1955) *Prosobothrium armigerum* Cohn, 1902 (Cestoda). Historique, synonymie, description et position systématique. *Recueil des Travaux des Laboratoires de Botanique, Géologie et Zoologie de la Faculté des Sciences de l'Université de Montpellier. Zoologie* 1, 44–55.

Baer, J.G. & Euzet, L. (1962) Revision critique des cestodes tétraphyllides décrits par T. Southwell (1re partie). *Bulletin de la Societé Neuchâteloise des Sciences Naturelles* 85, 143–172.

Baer, J.G. & Fain, A. (1955) Les cestodes de pangolins. *Bulletin de la Société Neuchâteloise des Sciences Naturelles* 78, 37–52.

Bandoni, S.M. & Brooks, D.R. (1987a) Revision and phylogenetic analysis of the Amphilinidea Poche, 1922 (Platyhelminthes: Cercomeria: Cercomeromorphae). *Canadian Journal of Zoology* 65, 1110–1128.

Bandoni, S.M. & Brooks, D.R. (1987b) Revision and phylogenetic analysis of the Gyrocotylidae Poche, 1926 (Platyhelminthes: Cercomeria: Cercomeromorphae). *Canadian Journal of Zoology* 65, 2369–2389.

Bates, R.M. (1990) A checklist of the Trypanorhyncha (Platyhelminthes: Cestoda) of the World (1935–1985). *National Museum of Wales, Zoological Series*, No. 1, 218 pp.

Baverstock, P.R., Adams, M., Bailey, P. & Swincer, D. (1985a) Isozyme electrophoresis and species identification. In: *Proceedings of the Fourth Australian Applied Entomological Research Conference, Adelaide 24–28 September 1984. Pest control: recent advances and future prospects. [Edited by Bailey, P. & Swincer, D.], 1984*. Adelaide, Australia: South Australian Government Printer, 428–431.

Baverstock, P.R., Adams, M. & Beveridge, I. (1985b) Biochemical differentiation in bile duct cestodes and their marsupial hosts. *Molecular Biology and Evolution* 2, 321–337.

Baverstock, P.R., Fielke, R., Johnson, A.M., Bray, R.A. & Beveridge, I. (1991) Conflicting phylogenetic hypotheses for the parasitic platyhelminths tested by partial sequencing of 18S ribosomal RNA. *International Journal for Parasitology* 21, 329–339.

Baylis, H.A. (1926) Some tetrabothriid cestodes from whales of the genus *Balaenoptera*. *Journal of the Linnaean Society of London (Zoology)* 36, 161–172.

Baylis, H.A. (1927) On two adult cestodes from wild swine. *Annals and Magazine of Natural History* 9, 417–425.

Baylis, H.A. (1928) Some parasitic worms, mainly from fishes, from Lake Tanganyika. *Annals and Magazine of Natural History* 1, 552–562.

Baylis, H.A. (1935) Note on the cestode *Moniezia (Fuhrmannella) transvaalensis* (Baer, 1925). *Annals and Magazine of Natural History*, Series 10, 15, 673–675.

Baylis, H.A. (1947) Some roundworms and flatworms from the West Indies and Surinam. II. Cestodes. *Journal of the Linnaean Society of London (Zoology)* 41, 406–414.

Bazitov, A.A. (1976) [The status of caryophyllideans in the system of Platyhelminthes.] *Zoologicheskii Zhurnal* 55, 1779–1787. [In Russian.]

Beaver, P.C. (1989) *Mesocestoides corti*: mouse type host, uncharacteristic or questionable? *Journal of Parasitology* 75, 815.

Beaver, P.C. & Simer, P.H. (1940) A restudy of the three existing species of the cestode genus *Marsipometra* Cooper (Amphicotylidae) from the spoonbill, *Polyodon spathula* (Wal.). *Transactions of the American Microscopical Society* 59, 167–182.

Beddard, F.E. (1912) Contributions to the anatomy and systematic arrangement of the Cestoidea. III. On a new genus of tapeworms (*Otiditaenia*) from the bustard (*Eupodotis kori*). *Proceedings of the Zoological Society of London*, Part 1, 194–221.

Beddard, F.E. (1913) Contributions to the anatomy and systematic arrangement of the Cestoidea. VIII. On six species of tapeworms from reptiles belonging to the genus *Ichthyotaenia* (s.l.). *Proceedings of the Zoological Society of London*, Part 1, 4–36.

Beddard, F.E. (1914) Contributions to the anatomy and systematic arrangement of the Cestoidea. XV. On a new species and genus of the family Acoleidae. *Proceedings of the Zoological Society of London*, Part 4, 1039–1055.

Belopolskaya, M.M. (1977) [Cestodes of waders of the genus *Anomotaenia* Cohn, 1900 and *Dichoanotaenia* Lopez-Neyra, 1944 (Cestoda: Cyclophyllidea).] *Vestnik Leningradskogo Gosudarstvennogo Universiteta (Ser. Biol)* 21, 31–44. [In Russian.]

Beneden, P.J. van (1850) Recherches sur la faune littorale de Belgique (Cestoides). *Mémoires de l'Academie Royale des Sciences de Belgique* 25, 1–200.

Beneden, P.J. van (1858) *Mémoires sur les vers intestinaux*. Supplément aux Comptes Rendus de l'Academie des Sciences, Paris, 1–376.

Berland, B., Bristow, G.A. & Grahl-Nielsen, O. (1989) *Gyrocotyle* species – fatty acid profiles as taxonomic characters. *Information*, Abo Akademi 20, 79.

Beumer, J.P., Ashburner, L.D., Burbury, M.E., Jette, E. & Latham, D.J. (1982) A checklist of the parasites of fishes from Australia and its adjacent Antarctic Territories. *Commonwealth Institute of Parasitology Technical Communication No. 48*, Farnham Royal: Commonwealth Agricultural Bureaux, 99 pp.

Beveridge, I. (1975) On two new davaineid cestodes from Australian marsupials. *Journal of Helminthology* 49, 129–136.

Beveridge, I. (1976) A taxonomic revision of the Anoplocephalidae (Cestoda: Cyclophyllidea) of Australian marsupials. *Australian Journal of Zoology*, Supplementary Series No. 44, 1–110.

Beveridge, I. (1977) On two new linstowiid cestodes from Australian dasyurid marsupials. *Journal of Helminthology* 51, 31–40.

Beveridge, I. (1978) A taxonomic revision of the genera *Cittotaenia* Riehm, 1881, *Ctenotaenia* Railliet, 1893, *Mosgovoyia* Spasskii, 1951 and *Pseudocittotaenia* Tenora, 1976. (Cestoda: Anoplocephalidae). *Mémoires du Muséum National d'Histoire Naturelle Paris, Nouvelle Série, Série A, Zoologie* 107, 1–64, figs 1–102.

Beveridge, I. (1980) *Echidnotaenia tachyglossi* (Johnston) gen. et comb. nov. (Anoplocephalata: Linstowiidae) from the monotreme *Tachyglossus aculeatus* Shaw in Australia. *Journal of Helminthology* 54, 129–134.

Beveridge, I. (1983) The genus *Linstowia* Zschokke, 1899 (Cestoda: Anoplocephalidae) in Australian mammals with the description of a new species, *L. macrouri*. *Systematic Parasitology* 5, 291–304.

Beveridge, I. (1984) *Dasyurotaenia robusta* Beddard, 1912, and *D. dasyuri* sp. nov., from carnivorous Australian marsupials. *Transactions of the Royal Society of South Australia* 108, 185–195.

Beveridge, I. (1985) The genus *Bertiella* (Cestoda: Anoplocephalidae) from Australasian mammals: new species, new records and redescriptions. *Systematic Parasitology* 7, 241–289.

Beveridge, I. (1990a) Taxonomic revision of Australian Eutetrarhynchidae Guiart (Cestoda: Trypanorhyncha). *Invertebrate Taxonomy* 4, 785–845.

Beveridge, I. (1990b) A revision of the family Gilquiniidae Dollfus (Cestoda: Trypanorhyncha) from elasmobranch fishes. *Australian Journal of Zoology* 37, 481–520.

Beveridge, I. & Campbell, R.A. (1987) *Trimacracanthus* gen. nov. (Cestoda: Trypanorhyncha: Eutetrarhynchidae), with redescriptions of *T. aetobatidis* (Robinson, 1959) comb. nov. and *T. binuncus* (Linton, 1909) comb. nov. *Transactions of the Royal Society of South Australia* 111, 163–171.

Beveridge, I. & Campbell, R.A. (1988a) *Cetorhinicola* n. g., *Shirleyrhynchus* n. g. and *Stragulorhynchus* n. g., three new genera of trypanorhynch cestodes from elasmobranchs in Australian waters. *Systematic Parasitology* 12, 47–60.

Beveridge, I. & Campbell, R.A. (1988b) A review of the Tetrarhynchobothriidae Dollfus, 1969 (Cestoda: Trypanorhyncha) with descriptions of two new genera, *Didymorhynchus* and *Zygorhynchus*. *Systematic Parasitology* 12, 3–29.

Beveridge, I. & Campbell, R.A. (1989) *Chimaerarhynchus* n. g. and *Patellobothrium* n. g., two new genera of trypanorhynch cestodes with unique poeciloacanthous armatures, and a reorganisation of the poeciloacanthous trypanorhynch families. *Systematic Parasitology* 14, 209–225.

Beveridge, I. & Campbell, R.A. (1992) Redescription of *Halysiorhynchus macrocephalus* (Cestoda:

Trypanorhyncha) a genus newly recorded from the Australasian region. *Systematic Parasitology* 22, 151–157.

Beveridge, I. & Campbell, R.A. (1993) A revision of *Dasyrhynchus* Pintner (Cestoda: Trypanorhyncha), parasitic in elasmobranch and teleost fishes. *Systematic Parasitology* 24, 129–157.

Beveridge, I. & Campbell, R.A. (1994) Redescription of *Diesingium lomentaceum* (Diesing, 1850) (Cestoda: Trypanorhyncha). *Systematic Parasitology* 27, 149–157.

Beveridge, I. & Sakanari, J.A. (1987) *Lacistorhynchus dollfusi* sp. nov. (Cestoda: Trypanorhyncha) in elasmobranch fishes from Australian and North American coastal waters. *Transactions of the Royal Society of South Australia* 111, 147–154.

Beveridge, I. & Smith, K. (1988) Ultrastructure of the rhyncheal system of *Trimacracanthus aetobatidis* (Cestoda: Trypanorhyncha). *International Journal for Parasitology* 18, 623–632.

Beveridge, I., Rickard, M.D., Gregory, G.G. & Mundy, B.L. (1975) Studies on *Anoplotaenia dasyuri* Beddard, 1911 (Cestoda: Taeniidae), a parasite of the Tasmanian devil: observations on the egg and metacestode. *International Journal for Parasitology* 5, 257–267.

Bilqees, F.M. & Fatima, H. (1984) *Neodiorchis longicirrus*, new genus et new species (Cestoda: Hymenolepididae: Hymenolepidinae) from duck in Karachi. *Pakistan Journal of Zoology* 16, 125–127.

Bilqees, F.M. & Khatoon, A. (1980) A new trypanorhynchid family Pseudogilquiniidae including a new genus and species *Pseudogilquinia karachiense*. *Philippine Journal of Science* 19, 89–92.

Bilqees, F.M. & Kurshid, N. (1985) *Pseudogilquinia magna* n. sp. (Cestoda: Trypanorhyncha: Pseudogilquiniidae) from the fish *Pseudosciaena diacanthus* of Karachi coast. *Pakistan Journal of Agricultural Science* 22, 101–107.

Binder, N. (1971) Beiträge zur Morphologie, Invasionsdynamik und Entwicklung der Helminthen der Amsel (*Turdus merula* L.). *Zoologische Beiträge* 17, 83–150.

Bjodtvedt, G. & Hendricks, G.M. (1982) *Mesogyna hepatica* (Cestoda) in kit foxes. *Canine Practice* 9, 17–18; 20–23.

Blair, D. (1993) The phylogenetic position of the Aspidobothrea within the parasitic flatworms inferred from ribosomal RNA sequence data. *International Journal for Parasitology* 23, 169–178.

Blanchard, R. (1891) Notices helminthologiques (2). Sur les teniades à ventouses armées. *Mémoires de la Société Zoologique de France* 4, 420–489.

Bloch, M. (1782) *Abhandlung von der Erzeugung der Eingeweidewuermer und den Mitteln wider dieselben*. Berlin, 54 pp.

Bona, F.V. (1975) Etude critique et taxonomique des Dilepididae Fuhrm., 1907 (Cestoda) parasites de Ciconiiformes. Considérations sur la specificité et la spéciation. *Monitore Zoologico Italiano N.S. Monografia* 1, 750 pp.

Bona, F. (1978) The genus *Clelandia* Johnston, 1909 and its affinities with *Parvitaenia* and *Neogryporhynchus* (Cestoda, Dilepididae). *Annales de Parasitologie Humaine et Comparée* 53, 163–180.

Bona, F.V. (1994) Three new species and genera of Dilepididae (Cestoda, Cyclophyllidea) from Neotropical Picidae (Aves). *Revue Suisse de Zoologie* 101, 195–213.

Bona, F.V. & Maffi, A. (1984a) La estructura del genero *Paruterina* Fuhrm., 1906 y consideraciones sobre los generos *Biuterina* Fuhrm. 1902, *Proparuterina* Fuhrm., 1911 y *Sphaeruterina* Johnston, 1914 (Cestoda, Paruterinidae). *Bolletino del Museo Regionale di Scienze Naturali* 2, 411–444.

Bona, F.V. & Maffi, A. (1984b) *Paruterina similita* n. sp. (Cestoda, Paruterinidae) en Cuculidae (Aves) del Norte de Argentina. *Bolletino del Museo Regionale di Scienze Naturali* 2, 181–198.

Bona, F.V. & Maffi, A. (1985) Redescription y variabilidad de *Paruterina similis* (Ransom, 1909)

(Cestoda, Paruterinidae), recolectada en un nuevo hospedador, *Coccyzus melacoryphus* (Cuculiformes) en el Norte de Argentina. *Rivista di Parassitologia* 46, 115–139.

Bona, F.V. & Maffi, A. (1987) Los Paruterinidae (Cestoda) con ganchos de patron 'rectanguloide'. Parte I. Observaziones sobre los ganchos rectanguloides y revision de la literatura. *Bolletino del Museo Regionale di Scienze Naturali* 5, 455–489.

Bona, F.V., Bosco, M.C. & Maffi, A. (1986) Tre nuove specie del genere *Paruterina* (Cestoda, Paruterinidae) in Trogoniformi (Aves) neotropicali. *Bolletino del Museo Regionale de Scienze Naturali* 4, 1–61.

Bondarenko, S.K. (1966) [New trematodes and cestodes from Charadriiformes in northern Asia.] *Materialy k nauchnoi konferentsii VOG* 3, 50–60. [In Russian.]

Bondarenko, S.K. & Tomilovskaya, N.S. (1979) [New dilepidid species, *Rauschitaenia* n. g., and the life-cycle of *R. ancora* (Mamaev, 1959) n. comb., a parasite of snipes.] In: Sonin, M.D. (ed.), *Ekologiya i morfologiya gel'mintov pozvonochnykh chukotki.* Moscow, USSR: 'Nauka', pp. 29–37. [In Russian.]

Bondareva, V.I. (1963) [*Coenurus Infections of Domestic and Wild Animals. (The Eradication of Cerebral Coenuriasis in the USSR)*.] Alma-Ata: Izdatel'stvo Akademii Nauk Kazakhskoi SSR, 356 pp. [In Russian.]

Borgarenko, L.F. (1981) [*Helminths of Birds of Tadzhikistan. Book 1. Cestodes.*) Dushanbe: Izdatel'stvo 'Donish', 327 pp. [In Russian.]

Borgarenko, L.F. & Gulyaev, V.D. (1990) [On the morphology of the type species of the genus *Joyeuxilepis* (Cestoda, Schistotaeniinae)] *Parazitologiya* 24, 350–353. [In Russian.]

Borgarenko, L.F. & Gulyaev, V.D. (1991) [Two new species of amabiliid cestodes from grebes in Tadzhikistan.] *Izvestiya Sibirskogo Otdeleniya Akademii Nauk SSSR. Sibirskii Biologicheskii Zhurnal*, No. 1, 44–48. [In Russian.]

Bosc, L.A.G. (1811) Sur des nouveaux genres des vers. *Bulletin des Sciences par la Société Philomatique, Paris* 2, 384–385.

Braun, M. (1894–1900) In: Bronn, H.G. (ed.), *Klassen und Ordnungen des Thierreichs, Band IV, Vermes, Abt.1, Cestodes*, pp. 927–1731. Leipzig.

Bristow, G.A. & Berland, B. (1988) A preliminary electrophoretic investigation of the gyrocotylid parasites of *Chimaera monstrosa* L. *Sarsia* 73, 75–77.

Brooks, D.R. (1977) Six new species of tetraphyllidean cestodes, including a new genus, from a marine stingray *Himantura schmardae* (Werner, 1904) from Colombia. *Proceedings of the Helminthological Society of Washington* 44, 51–59.

Brooks, D.R. (1978) Evolutionary history of the cestode order Proteocephalidea. *Systematic Zoology* 27, 312–323.

Brooks, D.R. (1989) A summary of the database pertaining to the phylogeny of the major groups of parasitic platyhelminths, with a revised classification. *Canadian Journal of Zoology* 67, 714–720.

Brooks, D.R. & McLennan, D.A. (1993) *Parascript. Parasites and the Language of Evolution.* Smithsonian Institution Press, Washington & London, 429 pp.

Brooks, D.R. & Rasmussen, G. (1984) Proteocephalidean cestodes from Venezuelan siluriform fishes with a revised classification of Monticelliidae. *Proceedings of the Biological Society of Washington* 97, 748–760.

Brooks, D.R. & Thorson, T.B. (1976) The tetraphyllidean cestodes from the freshwater stingray *Potamotrygon magdalenae* Dumeril, 1852 (Chondrichthyes, Potamotrygonidae) from Colombia. *Journal of Parasitology* 62, 943–947.

Brooks, D.R., Hoberg, E.P. & Weekes, P.J. (1991) Preliminary phylogenetic systematic analysis of the major lineages of the Eucestoda (Platyhelminthes: Cercomeria). *Proceedings of the Biological Society of Washington* 104, 651–668.

Brooks, D.R., O'Grady, R.T. & Glen, D.R. (1985) The phylogeny of the Cercomeria Brooks, 1982 (Platyhelminthes). *Proceedings of the Helminthological Society of Washington* 52, 1–20.

Burt, D.R.R. (1938) New avian cestodes of the subfamily Dilepidinae from the eastern swallow (*Hirundo rustica gutturalis*) with descriptions of *Vitta magniuncinata* gen. et sp. nov. *Ceylon Journal of Science* 21, 21–30.

Burt, D.R.R. (1939a) On the cestode family Acoleidae, with a description of a new dioecious species, *Infula burhini* gen. et sp. nov. *Ceylon Journal of Science* 21, 195–208.

Burt, D.R.R. (1939b) New cestodes of the genus *Paronia*. *Ceylon Journal of Science* 21, 209–218.

Burt, D.R.R. (1973) On a new anoplocephalid cestode *Biporouterina psittaculae* gen. et sp. nov. from Layard's paroquet *Psittacula calthropae* (Layard, 1849). *Zoological Journal of the Linnean Society* 53, 81–86.

Burt, D.R.R. & Burt, M.D.B. (1984) On *Davainea minuta* Cohn, 1901 (Cestoda: Cyclophyllidea) and its variation in North American Recurvirostridae (Aves). *Journal of Natural History* 18, 15–23.

Burt, M.D.B. (1962) A contribution to the knowledge of the cestode genus *Ophryocotyle* Friis, 1870. *Journal of the Linnean Society of London, Zoology* 44, 645–668.

Butler, S.A. (1987a) The taxonomic history of the family Lecanicephalidae Braun, 1900, a little known group of marine cestodes. *Systematic Parasitology* 10, 105–115.

Butler, S.A. (1987b) Taxonomy of some tetraphyllidean cestodes from elasmobranch fishes. *Australian Journal of Zoology* 35, 343–371.

Byrd, E.E. & Ward, J.W. (1943) Observations on the segmental anatomy of the tapeworm, *Mesocestoides variabilis* Mueller, 1928, from the opossum. *Journal of Parasitology* 29, 217–226.

Cable, R.M. & Michaelis, M.B. (1967) *Plicatobothrium cypseluri* n. gen., n. sp. (Cestoda: Pseudophyllidea) from the Caribbean flying fish, *Cypselurus bahiensis* (Ranzani, 1842). *Proceedings of the Helminthological Society of Washington* 34, 15–18.

Cable, R.M. & Myers, R.M.E. (1956) A dioecious species of *Gyrocoelia* (Cestoda: Acoleidae) from the naped plover. *Journal of Parasitology* 42, 510–515.

Caira, J.N. (1987) Scolex structural homologies and the systematic position of the genus *Spiniloculus* (Cestoda: Tetraphyllidea). *Proceedings of the Helminthological Society of Washington* 54, 115–117.

Caira, J.N. (1990) The tapeworm *Spiniloculus mavensis* (Tetraphyllidea: Onchobothriidae) from the brownbanded bambooshark in Australia. *Australian Journal of Zoology* 37, 705–710.

Caira, J.N. (1992) Verification of multiple species of *Pedibothrium* in the Atlantic nurse shark with comments on the Australasian members of the genus. *Journal of Parasitology* 78, 289–308.

Calentine, R. (1962) *Archigetes iowensis* sp. n. (Cestoda: Caryophyllaeidae) from *Cyprinus carpio* L. and *Limnodrilus hoffmeisteri* Claparède. *Journal of Parasitology* 48, 513–524.

Calentine, R. & DeLong, B.C. (1966) *Archigetes sieboldi* (Cestoda: Caryophyllaeidae) in North America. *Journal of Parasitology* 52, 428–431.

Calentine, R. & Mackiewicz, J.S. (1966) *Monobothrium ulmeri* n. sp. (Cestoda: Caryophyllaeidae) from North American Catostomidae. *Transactions of the American Microscopical Society* 85, 516–520.

Calentine, R. & Ulmer, M. (1961) *Khawia iowensis* n. sp. (Cestoda: Caryophyllaeidae) from *Cyprinus carpio* L. in Iowa. *Journal of Parasitology* 47, 795–805.

Campbell, R.A. (1977) A new family of pseudophyllidean cestodes from the deep-sea teleost *Acanthochaenus lutkenii* Gill, 1884. *Journal of Parasitology* 63, 301–305.

Campbell, R.A. (1979) Two new genera of pseudophyllidean cestodes from deep-sea fishes. *Proceedings of the Helminthological Society of Washington* 46, 74–78.

Campbell, R.A. & Beveridge, I. (1987) *Hornelliella macropora* (Shipley & Hornell, 1906) comb. n. (Cestoda: Trypanorhyncha) from Australian elasmobranch fishes and a re-assessment of the family Hornelliellidae. *Transactions of the Royal Society of South Australia* 111, 195–200.

Campbell, R.A. & Beveridge, I. (1988) *Mustelicola antarcticus* sp. nov. (Cestoda: Trypanorhyncha) from Australian elasmobranchs and a re-assessment of the family Mustelicolidae Dollfus, 1969. *Transactions of the Royal Society of South Australia* 112, 153–161.

Campbell, R.A. & Beveridge, I. (1990) *Pseudochristianella* (Cestoda: Trypanorhyncha), a new genus for *Tentacularia minuta* (Van Beneden, 1849) *sensu* Southwell, 1929 and its relationships with the family Eutetrarhynchidae. *Transactions of the Royal Society of South Australia* 114, 219–222.

Campbell, R.A. & Beveridge, I. (1993) New species of *Grillotia* and *Pseudogrillotia* (Cestoda: Trypanorhyncha) from Australian sharks, and definition of the family Grillotiidae Dollfus, 1969. *Transactions of the Royal Society of South Australia* 117, 37–46.

Campbell, R.A. & Williams, A.D. (1984) *Tylocephalum* Linton, 1890 (Cestoda: Lecanicephalidea) from the cownose ray, *Rhinoptera bonasus* (Mitchill, 1815) with a discussion of its validity and systematic relationships. *Proceedings of the Helminthological Society of Washington* 51, 121–134.

Campbell, R.A., Correia, S.J. & Haedrich, R.L. (1982) A new monogenean and cestode from the deep-sea fish, *Macrourus berglax* Lacépède, 1802, from the Flemish Cap off Newfoundland. *Proceedings of the Helminthological Society of Washington* 49, 169–175.

Capoor, V.N. & Srivastava, V.C. (1966) On a new dioecious cestode *Hymenocoelia chauhani* n. g., n. sp. from the Indian pigeon *Columba livia* (Gmelin) with diagnosis of the new subfamily Hymenocoelinae and the key to the family Dioecocestidae Southwell, 1930. *Journal of the Zoological Society of India* 16, 99–104.

Capoor, V.N. & Srivastava, V.C. (1967) On the synonymity [*sic*] of the genus *Columbia* Srivastava & Capoor, 1966 [*sic*] with *Killigrewia* Meggitt, 1927 and on a new combination *Killigrewia allahabadi* n. comb. and with its redescription. *Journal of the Zoological Society of India* 17, 123–124.

Capoor, V.N. & Srivastava, V.C. (1975) On a new cestode, *Barbusa passeri* n. g., n. sp., (Cestoda: Davaineidae) from *Passer domesticus* from India along with a provisional definition of a new tribe Barbuseini n. tribe. *Proceedings of the National Academy of India* 45(B), II, 101–104.

Carter, W.J. (1943) *Proteocephalus sandgroundii*, a new tetraphyllidean cestode from an east Indian monitor lizard. *Transactions of the American Microscopical Society* 62, 301–305.

Carvajal, J. & Campbell, R.A. (1975) A revision of some trypanorhynchs from the western North Atlantic described by Edwin Linton. *Journal of Parasitology* 61, 1016–1022.

Carvajal, J., Campbell, R.A. & Cornford, E.M. (1976) Some trypanorhynch cestodes from Hawaiian fishes with descriptions of four new species. *Journal of Parasitology* 62, 70–77.

Chambrier, A. de (1987) *Vaucheriella bicheti* n. gen, n. sp. (Cestoda: Monticelliidae: Zygobothriinae) parasite de *Tropidophis* cf. *taczanowskyi* (Steindachner, 1880) (Serpentes: Trophidophinae) des Andes Equatoriens. *Revue Suisse de Zoologie* 94, 829–840.

Chambrier, A. de (1989) Revision du genre *Crepidobothrium* Monticelli, 1900 (Cestoda: Proteocephalidae) parasite d'ophidiens neotropicaux. *Revue Suisse de Zoologie* 96, 345–380.

Chambrier, A. de & Vaucher, C. (1992) *Nomimoscolex touzeti* n. sp. (Cestoda). A parasite of *Ceratophrys cornuta* (L.): first record of a Monticellidae in an amphibian host. *Memorias do Instituto Oswaldo Cruz*, 87 (Supplement) 61–67.

Chandler, A.C. (1935) Parasites of fishes in Galveston Bay. *Proceedings of the United States National Museum* 83, 123–157.

Chandler, A.C. (1942a) Some cestodes from Florida sharks. *Proceedings of the United States National Museum* 92, 25–31.

Chandler, A.C. (1942b) Helminths of tree squirrels in southeast Texas. *Journal of Parasitology* 28, 135–140.

Chandler, A.C. (1942c) The helminths of raccoons in East Texas. *Journal of Parasitology* 28, 255–268.

Chandler, A.C. (1946) Observations on the anatomy of *Mesocestoides*. *Journal of Parasitology* 32, 242–246.

Chandler, A.C. (1947) The anatomy of *Mesocestoides* – corrections. *Journal of Parasitology* 33, 444.

Chandler, A.C. (1948) New species of the genus *Schistotaenia* with a key to the known species. *Transactions of the American Microscopical Society* 67, 169–176.

Chen, C.L. (1973) [*An Illustrated Guide to Fish Disease and Causative Pathogenic Fauna and Flora in Hupei Province*.] Beijing, China: Science Press, 456 pp. [In Chinese.]

Chertkova, A.N. & Kosupko, G.A. (1977) [Revision of the cestode suborder Mesocestoidata Skryabin, 1940.] *Trudy Vsesoyuznogo Instituta Gel'mintologii im. K.I. Skryabina* 23, 141–153. [In Russian.]

Chertkova, A.N. & Kosupko, G.A. (1978) [The suborder Mesocestoidata Skryabin, 1940.] In: *Principles of Cestodology. Vol. 9. Tetrabothriata and Mesocestoidata – Cestodes of Birds and Mammals*. [Ed. K.M. Ryzhikov.] Moscow: Nauka, pp. 118–229. [In Russian.]

Chin, T.G. (1938) A new species of cestode of the family Anoplocephalidae (Cestoda) from tapirs. *Lingnan Science Journal* 17, 605–607.

Chincholikar, L.N. & Shinde, G.B. (1976) On a new cestode *Yogeshwaria nagabushani* (Cestoda: appendix to Lecanicephalidea, Genera Inseartae [sic] Sedis) gen. et sp. nov. from a marine fish at Ratnagiri, India. *Marathwada University Journal of Science (Natural Sciences)* 15, 273–276.

Chincholikar, L.N. & Shinde, G.B. (1977) A new species of *Circumonchobothrium* Shinde, 1968 (Cestoda, Pseudophyllidea Carus, 1863) from a freshwater fish in India. *Marathwada University Journal of Science* 16, 183–185.

Cholodkovsky, N. (1908) Über eine neue Tänie des Hundes. *Zoologischer Anzeiger* 33, 418–420.

Cholodkovsky, N.A. (1912) [*Descriptive Catalogue of the Collection of Parasitic Worms of the Zoological Cabinet of the Imperial Military Medical Academy. Part I. Cestodes (Cyclophyllidea)*.] St. Petersburg: Knigopechatnya Shmidt', 98 pp. [In Russian.]

Chomicz, L. & Czubaj, A. (1991) Transmission electron micrograph studies of developing oncospheral envelopes of *Fimbriaria fasciolaris* (Hymenolepididae). *Parasitology Research* 77, 503–508.

Clarke, M.R. (1962) *Multiductus physeteris* gen. et sp. nov. – a new diphyllobothriid cestode from a sperm whale. *Journal of Helminthology* 36, 1–10.

Clausen, E. (1915) Recherches anatomiques et histologiques sur quelques cestodes d'oiseaux. Thesis, Neuchâtel, 111 pp.

Clerc, W. (1902) Contribution à l'étude de la faune helmintologique de l'Oural. Communication préliminaire 2. *Zoologischer Anzeiger*, Leipzig 25 (681), 658–664.

Clerc, W. (1903) Contribution à l'étude de la faune helmintologique de l'oural. *Revue Suisse de Zoologie* 2, 241–368.

Cohn, L. (1899) Zur Systematik der Vogeltänien. Vorläufige Mitteilung. *Zentralblatt für Bakteriologie, Parasitenkunde, Infecktionskrankheiten und Hygiene (Abteilung 1)* 25, 415–422.

Cohn, L. (1902a) Helminthologische Mittheilungen. *Archiv für Naturgeschichte* 1, 47–68.

Cohn, L. (1902b) Zur Kenntnis des Genus *Wageneria* Monticelli und anderen Cestoden. *Zentralblatt für Bakteriologie und Parasitenkunde* 33, 53–60.

Coil, W.H. (1955) *Infula macrophallus* sp. nov., dioecious cestode parasitic in the black-necked stilt, *Himantopus mexicanus*. *Journal of Parasitology* 41, 291–294.

Coil, W.H. (1963) The genera *Gyrocoelia* Fuhrmann, 1899 and *Infula* Burt, 1939 with observations on the histochemistry of the egg membranes. *Proceedings of the Helminthological Society of Washington* 30, 111–117.

Coil, W.H. (1966) Studies on some dioecious tapeworms. [Abstract.] *Proceedings of the First International Congress of Parasitology*, Rome, Sept. 21–26, 1964. Vol. 1, pp. 31–32.

Coil, W.H. (1968) Observations on the embryonic development of the dioecious tapeworm *Infula macrophallus* with emphasis on the histochemistry of the egg membrane. *Zeitschrift für Parasitenkunde* 30, 301–317.

Coil, W.H. (1970a) Studies on the biology of the tapeworm *Dioecocestus acotylus* with emphasis on the oogenotop. *Zeitschrift für Parasitenkunde* 33, 314–328.

Coil, W.H. (1970b) Studies on the biology of the tapeworm *Shipleya inermis* Fuhrmann, 1908. *Zeitschrift für Parasitenkunde* 35, 40–54.

Colin, J.A., Williams, H.H. & Halvorsen, O. (1986) One or three gyrocotylideans (Platyhelminthes) in *Chimaera monstrosa* (Holocephali)? *Journal of Parasitology* 72, 10–21.

Conn, D.B. (1985) Fine structure of the embryonic envelopes of *Oochoristica anolis* (Cestoda: Linstowiidae). *Zeitschrift für Parasitenkunde* 71, 639–648.

Conn, D.B. (1993) Ultrastructure of the gravid uterus of *Hymenolepis diminuta* (Platyhelminthes: Cestoda). *Journal of Parasitology* 79, 583–590.

Cooper, A.R. (1914a) On the systematic position of *Haplobothrium globuliforme* Cooper. *Transactions of the Royal Society of Canada, Series III* 8, 1–5.

Cooper, A.R. (1914b) A new cestode from *Amia calva* L. *Transactions of the Royal Canadian Institute* 10, 81–119.

Cooper, A.R. (1917) A morphological study of bothriocephalid cestodes from fishes. *Journal of Parasitology* 4, 33–39.

Cooper, A.R. (1918) North American pseudophyllidean cestodes from fishes. *Illinois Biological Monographs* 4, 1–243.

Cordero, E.H. (1946) *Ophiotaenia cohospes* n. sp., de la tortuga fluvial *Hydromedusa tectifera* Cope, una larva plerocercoide en el parenquima dei *Temnocephala brevicornis* Mont., y su probable metamorfosis. *Comunicaciones Zoológicas del Museo de Historia Natural de Montevideo* 2, 12 pp.

Creplin, F.C.H. (1829) Novae observationes de Entozois. *Berolini*, 134 pp.

Cuvier, G. (1817) *Le Regne Animal Distribue d'Apres son Organization*. 4 vols. Paris.

Czaplinski, B. (1956) Hymenolepididae Fuhrmann, 1907 (Cestoda) parasites of some domestic and wild Anseriformes in Poland. *Acta Parasitologica Polonica* 4, 175–373.

Czaplinski, B. (1967a) Genus *Monosaccanthes* g. n. (Cestoda: Hymenolepididae) redescription of *M. tenuirostris* (Rud., 1819 pp) comb. n. (syn. *Hymenolepis tritesticulata* Fuhrmann, 1907) and *M. kazachstanica* (Maksimova, 1963) comb. n. *Acta Parasitologica Polonica* 14, 327–350.

Czaplinski, B. (1967b) The anatomy of *Wardoides nyrocae* (Yamaguti, 1935) Spassky, 1962 (Cestoda: Hymenolepididae) from *Cygnus olor* (Gmelin). *Acta Parasitologica Polonica* 15, 113–121.

Czaplinski, B. (1988) Redescription of *Diorchis longiovum* Schiller, 1953 with remarks on the genus *Schillerius* Yamaguti, 1959 (Cestoda: Hymenolepididae). *Acta Parasitologica Polonica* 33, 115–122.

Czaplinski, B. (1990) Redescription of *Retinometra guadeloupensis* (Graber et Euzéby, 1976) comb. n. and remarks on the taxonomic value of some morphological features of this species. *Acta Parasitologica Polonica* 35, 297–303.

Czaplinski, B. & Ryzhikov, K.M. (1966) *Gastrotaenia paracygni* sp. n. (Hymenolepididae) a new cestode of *Cygnus olor* and *C. cygnus*. *Acta Parasitologica Polonica* 14, 113–119.

Czaplinski, B. & Szelenbaum-Cielecka, D. (1986) Morphological and biological studies on *Diorchis danutae* Czaplinski, 1956 and *D. asiatica* Spassky, 1963 (Cestoda, Hymenolepididae). *Acta Parasitologica Polonica* 30, 127–142.

Czaplinski, B. & Vaucher, C. (1986) Helminthes parasites de Paraguay VII. Redescription de *Fuhrmanacanthus propeteres* (Fuhrmann, 1907) Spassky, 1966 (syn. *Lateriporus biuterinus* Fuhrmann, 1908) (Cestoda: Hymenolepididae) et remarques sur les genres *Fuhrmanacanthus* Spassky, 1966 et *Hamatolepis* Spassky, 1962. *Systematic Parasitology* 8, 83–90.

Dailey, M.D. (1969) *Litobothrium alopias* and *L. coniformis*, two new cestodes representing a new order from elasmobranch fishes. *Proceedings of the Helminthological Society of Washington* 36, 218–224.

Dailey, M.D. & Overstreet, R.M. (1973) *Cathetocephalus thatcheri* gen. et sp. n. (Tetraphyllidea: Cathetocephalidae fam. n.) from the bull shark: a species demonstrating multistrobilization. *Journal of Parasitology* 59, 469–473.

Dailey, M. & Vogelbein, W. (1982) Mixodigmatidae, a new family of cestode (Trypanorhyncha) from a deep-sea, planktivorous shark. *Journal of Parasitology* 68, 145–149.

Dailey, M.D. & Vogelbein, W. (1990) *Clistobothrium carcharodoni* gen. et sp. n. (Cestoda: Tetraphyllidea) from the spiral valve of the great white shark (*Carcharodon carcharias*). *Journal of the Helminthological Society of Washington* 57, 108–112.

Davies, T.I. & Rees, G. (1947) *Andrepigynotaenia haematopodis* n. g., n. sp., a new protogynous tapeworm from the oyster catcher *Haematopus ostralegus occidentalis* Neumann. *Parasitology* 38, 93–100.

Deblock, S. & Diaouré, A. (1962) Quelle est la valeur du genre *Thysanotaenia* Beddard, 1911. (A propos d'une redescription de *T. lemuris* Beddard, Anoplocephalidae de Madagascar). *Annales de Parasitologie Humaine et Comparée* 37, 73–82.

Deblock, S. & Rosé, F. (1962) Les *Hymenolepis* (sensu lato) de Charadriiformes (à propos de 23 descriptions), *Annales de Parasitologie Humaine et Comparée* 37, 767–847.

Deblock, S., Rosé, F. & Broussart, J. (1962) *Ophiotaenia ventosaloculata* n. sp., parasite d'*Ithocyphus miniatus* (Ophidien, Colubridae) à Nossi-Be. *Archives de l'Institut Pasteur de Madagascar* 31, 15–60.

Della Santa, E. (1956) Revision du genre *Oochoristica* Lühe (Cestodes). *Revue Suisse de Zoologie* 63, 1–113.

Delyamure, S.L. (1955) [*The Helminth Fauna of Marine Mammals in the Light of their Ecology and Phylogeny.*] Moscow: Izdatel'stvo Akademii Nauk, SSSR, 517 pp. [In Russian.]

Demshin, N.I. (1985) [Postembryonal development of the cestode *Nippotaenia mogurndae* (Nippotaeniidea, Nippotaeniidae).] *Parazitologiya* 19, 39–43. [In Russian.]

Deshmukh, R.A. (1979) On a new cestode *Flapocephalus trygonis* gen. et sp. nov. (Cestoda: Lecanicephalidae) from *Trygon sephen* from west coast of India. *Rivista di Parassitologia* 40, 261–265.

Deshmukh, R.A. (1980) On a new cestode *Spinocephalum rhinobati* gen. et sp. nov. (Cestoda: Lecanicephalidae) from a marine fish from west coast of India. *Rivista di Parassitologia* 41, 27–32.

Deshmukh, R.A. & Shinde, G.B. (1980) *Spinibiloculus ratnagiriensis* gen. n., sp. n. (Cestoda: Onchobothriidae) from a marine fish *Ginglymostoma concolor* of the west coast of India. *Acta Parasitologica Polonica* 27, 431–435.

Diesing, K.M. (1835) *Tropisurus* und *Thysanosoma*, zwei neue Gattungen von Binnenwürmern aus Brasiliens. *Medicinische Jahrbucher des K.K. Österreichischen Staates, Wien* 16, 83–116.

Diesing, K.M. (1850) *System Helminthum. Vol. I.*, xvi + 680 pp. Vindobonae.

Diesing, K.M. (1854) Über eine naturgemässe Vertheilung der Cephalocotyleen. *Sitzungsberichte der Kaiserlichen Akademie der Wissenschaften*, Wien, XIII, pp. 556–616.

Diesing, K.M. (1855) Sechzehn Gattungen von Binnenwürmern und ihre Arte. *Denkschr. der Kaiserlichen. Akademie der Wissenschaften Wien Math. Naturwiss. Classe* IX, Abt.I., 171–185.

Diesing, K.M. (1863) Revision der Cephalocotyleen. Abteilung 1. Paramecocotyleen. *Sitzungsberichten der Akademie der Wissenschaften in Wien Mathematik-Naturwissenschaften Klasse* 48, 200–345.

Dixit, G.R. & Capoor, V.N. (1986) An interesting new hymenolepidid cestode *Matiaraensis tristis* n. g., n. sp. (Cestoda: Hymenolepididae) from *Acridotherus* [sic] *tristis* L. from India. *Indian Journal of Helminthology* 38, 83–87.

Dixit, G.R. & Capoor, V.N. (1988) A new cestode *Sternolepis contri* n. g., n. sp. (Cestoda:

Hymenolepididae Railliet et Henry, 1909) from *Sternus contra* (Linn.) from Allahabad, India. *Indian Journal of Helminthology* 40, 130–134.

Dollfus, R.P. (1929) Sur les tétrarhynques. I. Definition des genres. *Bulletin de la Société Zoologique de France* 54, 308–342.

Dollfus, R.P. (1930) Sur les tétrarhynques (2me contribution. *Dibothriorhynchus, Sphyriocephalus, Tentacularia, Nybelinia*). *Mémoires de la Société Zoologique de France* 29, 139–216.

Dollfus, R.P. (1932) Identification d'un cestode de la collection du laboratoire de parasitologie de la Faculté de Médicine de Paris. *Bulletin de la Société de Zoologie de France* 57, 246–258.

Dollfus, R.P. (1934a) Sur un cestode pseudophyllide parasite de poisson d'ornement. *Bulletin de la Société Zoologique de France* 59, 476–490.

Dollfus, R.P. (1934b) Sur le *Taenia gallinulae* van Beneden, 1858. *Annales de Parasitologie Humaine et Comparée* 12, 267–272.

Dollfus, R.P. (1935) Sur les tétrarhynques (notes preliminaires). *Bulletin de la Société Zoologique de France* 60, 353–357.

Dollfus, R.P. (1940) Parasites du germon. *Annales de l'Institut Océanographique* 20, 276–283.

Dollfus, R.P. (1942) Etudes critiques sur les tétrarhynques du Muséum de Paris. *Archives du Muséum National d'Histoire Naturelle, Paris*, 6 ser. 19, 1–466.

Dollfus, R.P. (1946) Notes diverses sur les tétrarhynques. *Mémoires du Museum National d'Histoire Naturelle, Paris* 22, 179–220.

Dollfus, R.P. (1953) Miscellanea helminthologica maroccana. X. *Catenotaenia chabaudi* n. sp. de *Xerus (Atlantoxerus) getulus* (Linné, 1785). *Archives de l'Institut Pasteur de Maroc* 4, 533–540.

Dollfus, R.P. (1954) Miscellanea helminthologica maroccana. XVIII. Quelques cestodes du groupe *Oochoristica* auctorum récoltés au Maroc, avec une liste des cestodes des hérissons (Erinaceidae) et une liste des sauriens et ophidiens (exclus. Amerique et Australie) ou ont été trouvé des *Oochoristica*. *Archives de l'Institut Pasteur du Maroc* 4, 654–714.

Dollfus, R.P. (1957) Miscellanea helminthologica maroccana. XIX. Nouvelles récoltes d'*Oochorista* chez des sauriens du Maroc. *Archives de l'Institut Pasteur du Maroc* 5, 272–299.

Dollfus, R.P. (1960) Sur une collection de tétrarhynques homeacanthes de la famille des Tentaculariidae, récoltés principalement dans la region de Dakar. *Bulletin de l'Institut Francais d'Afrique Noire, Ser. A., Sciences naturelles* 22, 788–852.

Dollfus, R.P. (1963) Mission Yves-J. Golvan et Jean-A. Rioux en Iran. Cestodes d'oiseaux. I. Cestode d'Accipitriformes. *Annales de Parasitologie Humaine et Comparée* 38, 23–27.

Dollfus, R.P. (1969a) De quelques cestodes tétrarhynques (hétéracanthes et pécilacanthes) récoltés chez des poissons de la Méditerranée. *Vie et Milieu* 20, 491–542.

Dollfus, R.P. (1969b) Quelques espèces de cestodes tétrarhynques de la côte Atlantique des Etats Unis, dont l'une n'était pas connue à l'état adulte. *Journal of the Fisheries Research Board of Canada* 26, 1037–1061.

Dollfus, R.P. (1970) D'un cestode ptychobothrien parasite de cyprinide en Iran. Mission C.N.R.S. (Théodore Monod), février 1969. *Bulletin du Muséum National d'Histoire Naturelle, 2e Série, Year 1969* 41, 1517–1521.

Douglas, L.T. (1958) The taxonomy of nematotaeniid cestodes. *Journal of Parasitology* 44, 261–273.

Dronen, N.O., Schmidt, G.D., Allison, B.R. & Mellen, J.W. (1988) Some parasitic helminths from the American oystercatcher, *Haematopus palliatus* Temminck, from the Texas Gulf Coast and the common pied oystercatcher, *H. ostralegus* Linnaeus, from New Zealand, including *Dildotaenia latovarium* n. gen. and n. sp. (Cestoda: Hymenolepididae). *Journal of Parasitology* 74, 864–867.

Dubinina, M.N. (1953) [Cestodes of birds nesting in western Siberia.] *Parazitologicheskii Sbornik* 15, 117–233. [In Russian.]

Dubinina, M.N. (1960) [On the morphology of the Amphilinidae (Cestodaria) in relation to its position amongst the flatworms.] *Doklady Akademii Nauk Soyuza Sovetskikh Sotsialisticheskikh Respublik* 135, 501–504. [In Russian.]

Dubinina, M.N. (1962) Class Cestoidea Rud., 1808 Tapeworms. In: Bykhovskaya-Pavlovskaya, I.E. *et al.* (eds) *Key to Parasites of Freshwater Fish of the USSR.* [English translation, 1964. Israel Program for Scientific Translations, Jerusalem, pp. 445–510.]

Dubinina, M.N. (1966) [*Tapeworms (Cestoda, Ligulidae) of the Fauna of the USSR.*] Akademii Nauk SSSR, Moscow and Leningrad, 262 pp. [In Russian. English translation 1980. Amerind Publishing co., New Delhi.]

Dubinina, M.N. (1971) [Cestodes from the fishes of the River Amur basin.] *Parasitologicheskii Sbornik* 25, 77–119. [In Russian.]

Dubinina, M.N. (1982a) [Parasitic worms of the class Amphilinida.] *Trudy Zoologicheskogo Instituta, Leningrad* 100, 143 pp. [In Russian.]

Dubinina, M.N. (1982b) [On the synonymy of species of the genus *Bothriocephalus* (Cestoda: Bothriocephalidae), parasites of Cyprinidae of the USSR.] *Parazitologiya* 16, 41–45. [In Russian.]

Dubinina, M.N. (1987) [Class tapeworms – Cestoda Rudolphi, 1808]. In: Bauer, O.N. (ed.) [*Keys to the Parasites of Freshwater Fish Fauna of the USSR. Vol. 3. Parasitic Metazoa (second part).*] Leningrad: 'Nauka', pp. 5–76. [In Russian.]

Dujardin, F. (1845) *Histoire Naturelle des Helminthes ou Vers Intestinaux.* Librairie encyclopedique de Roret, Paris, 654 pp.

Eckert, J., von Brand, T. & Voge, M. (1969) Asexual multiplication of *Mesocestoides corti* (Cestoda) in the intestine of dogs and skunks. *Journal of Parasitology* 55, 241–249.

Ehlers, U. (1985) *Das Phylogenetische System der Plathelminthes.* Stuttgart, New York: Gustav Fischer, 317 pp.

Ehlers, U. (1986) Comments on a phylogenetic system of the Platyhelminthes. *Hydrobiologia* 132, 1–12.

Esch, G.W. & Self, J.T. (1965) A critical study of the taxonomy of *Taenia pisiformis* Bloch, 1780; *Multiceps multiceps* (Leske, 1780); and *Hydatigera taeniaeformis* Batsch, 1786. *Journal of Parasitology* 51, 932–937.

Essex, H.E. (1928) The structure and development of *Corallobothrium*. *Illinois Biological Monographs* 11, 7–64.

Essex, H.E. (1929) The life-cycle of *Haplobothrium globuliforme* Cooper, 1914. *Science* 69, 677–678.

Etges, F.J. (1991) The proliferative tetrathyridium of *Mesocestoides vogae* sp. n. (Cestoda). *Journal of the Helminthological Society of Washington* 58, 181–185.

Euzet, L. (1952) Sur deux cestodes tetraphyllides. *Bulletin de la Société Neuchâteloise des Sciences Naturelles* 75, 169–178.

Euzet, L. (1953) Suggestions pour une nouvelle classification des cestodes tetraphyllides. *Congrès International de Zoologie Copenhague.*

Euzet, L. (1955) Quelques cestodes de *Myliobatis aquila* L. *Travaux des Laboratoires de Botanique, Geologie et Zoologie de la faculté des Sciences de l'Université de Montpellier (ser. Zool.)* 1, 18–27.

Euzet, L. (1959) Recherches sur les cestodes tétraphyllides des sélaciens des côtes de France. Thèses présentées à la Faculté des Sciences de Montpellier (1956), 263 pp.

Euzet, L. & Combes, C. (1965) Contribution à l'étude de *Tetragonocephalum uarnak* (Shipley et Hornell, 1906). *Bulletin de la Société Neuchâteloise des Sciences Naturelles* 88, 101–122.

Euzet, L. & Radujkovic, B.M. (1989) *Kotorella pronosoma* (Stossich, 1901) n. gen., n. comb., type des Kotorellidae, nouvelle famille de Trypanorhyncha (Cestoda), parasite intestinal de *Dasyatis pastinaca* (L., 1758). *Annales de Parasitologie Humaine et Comparée* 64, 420–425.

Faust, E.C., Campbell, H.E. & Kellogg, C.R. (1929) Morphological and biological studies on the species of *Diphyllobothrium* in China. *American Journal of Hygiene* 9, 560–583.

Feigenbaum, D.L. (1975) Parasites of the commercial shrimp *Penaeus vannamei* Boone and *Penaeus brasiliensis* Latreille. *Bulletin of Marine Science* 25, 491–514.

Feliu, C., Mas-Coma, S. & Gallego, J. (1984) Contribucion al conocimiento de la helmintofauna de micromamiferos ibéricos. VIII. Nuevos datos sobre parasitos de *Apodemus sylvaticus* Linnaeus, 1758 (Rodentia: Muridae). *Revista Ibérica de Parasitología* 44, 109–128.

Fischthal, J.H. (1954) *Bialovarium nocomis* Fischthal, 1953 (Cestoda: Caryophyllaeidae) from the hornyhead chub, *Nocomis biguttatus* (Kirtland). *Proceedings of the Helminthological Society of Washington* 21, 117–119.

Fotedar, D.N. & Chishti, M.Z. (1976) *Pseudoschistotaenia* n. gen. (Amabiliidae Fuhrmann, 1908: Cestoda) from *Podiceps ruficollis capensis* in Kashmir. [Abstract.] *Proceedings of the Indian Science Congress* 63, 219.

Fotedar, D.N. & Chishti, M.Z. (1977a) On a new species of the genus *Pseudoschistotaenia* Fotedar and Chishti, 1976 from a bird *Podiceps reficollis*[sic] *capensis* in Kashmir. *All-India Symposium on Helminthology, Srinagar, Kashmir, India*, 8–11 August, 1977. [Abstract.] p. 37.

Fotedar, D.N. & Chishti, M.Z. (1977b) Studies on helminth parasites of some passeriform birds of Kashmir. *Abstracts of the 1st National Congress of Parasitology, Baroda, 24–26 February, 1977*, pp. 47–48.

Fotedar, D.N. & Chishti, M.Z. (1980) *Pseudoschistotaenia indica* gen. et sp. nov. (Amabiliidae Fuhrmann, 1908: Cestoda) from *Podiceps ruficollis capensis* in Kashmir. *Rivista di Parassitologia* 41, 39–43.

Freeman, R.S. (1949) Notes on the morphology and life cycle of the genus *Monoecocestus* Beddard, 1914 (Cestoda: Anoplocephalidae) from the porcupine. *Journal of Parasitology* 35, 605–612.

Freeman, R.S. (1952) The biology and life history of *Monoecocestus* Beddard, 1914 (Cestoda: Anoplocephalidae) from the porcupine. *Journal of Parasitology* 38, 111–129.

Freeman, R.S. (1956) Life history studies on *Taenia mustelae* Gmelin, 1890 and the taxonomy of certain taenioid cestodes from Mustelidae. *Canadian Journal of Zoology* 34, 219–242.

Freeman, R.S. (1957) Life cycle and morphology of *Paruterina rauschi* n. sp. and *P. candelabraria* (Goeze, 1782) (Cestoda) from owls, and the significance of plerocercoids in the order Cyclophyllidea. *Canadian Journal of Zoology* 35, 349–370.

Freeman, R.S. (1959) On the taxonomy of the genus *Cladotaenia*, the life histories of *C. globifera* (Batsch, 1786) and *C. circi* Yamaguti, 1935, and a note on distinguishing between the plerocercoids of the genera *Paruterina* and *Cladotaenia*. *Canadian Journal of Zoology* 37, 317–340.

Freeman, R.S. (1973) Ontogeny of cestodes and its bearing on their phylogeny and systematics. *Advances in Parasitology* 11, 481–557.

Freze, V.I. (1965) [*Principles of Cestodology. Vol. 5. Proteocephalata, Cestodes of Fishes, Amphibians and Reptiles.*] Izdatel'stvo Nauk (Ed. K.I. Skryabin), Moscow (Israel Program of Scientific Translation), 597 pp.

Fritsch, G. (1886) Die Parasiten der Zitterweises. *Sitzungsberichte Preussischen Akademie Wissenschaftliche* 6, 99–108.

Froelich, J.A. (1802) Beiträge zur Naturgeschichte der Ein geweidewürmer. *Der Naturforscher, Halle* 29, 5–96.

Fuhrmann, O. (1895) Beitrag zur Kenntnis der Vogeltaenian. I. *Revue Suisse de Zoologie* 2, 433–458.

Fuhrmann, O. (1899a) Zwei eigentümliche Vogeltänien (Vorläufige Mitteilung) (Mitteilungen über Vogeltänien. 2). *Zentralblatt für Bakteriologie, Parasitenkunde und Infektionskrankheiten. I. Abteilung Originale* 26, 618–622, 748.

Fuhrmann, O. (1899b) Deux singuliers taenias d'oiseaux: *Gyrocoelia perversus* n. g. n. sp. et *Acoleus armatus* n. g. n. sp. *Revue Suisse de Zoologie* 7, 341–451.
Fuhrmann, O. (1899c) *Taenia musculosa* Fuhrm. und *T. crateriformis* Goeze. (*Monopylidium* nov. gen.). *Zentralblatt für Bakteriologie, Parasitenkunde und Infektionskrankheiten. I. Abteilung Originale* 26, 622–627.
Fuhrmann, O. (1900a) Neue eigenthümliche Vögeltaenie. *Zoologischer Anzeiger* 23, 48–51.
Fuhrmann, O. (1900b) Zur Kenntnis de Acoleinae. *Zentralblatt für Bakteriologie, Parasitenkunde und Infektionskrankheiten. I. Abteilung Originale* 28, 363–376.
Fuhrmann, O. (1902a) Sur deux nouveaux genres de cestodes d'oiseaux. *Zoologischer Anzeiger* 25, 357–360.
Fuhrmann, O. (1902b) Die Anoplocephaliden der Vögel. *Zentralblatt für Bakteriologie, Parasitenkunde und Infektionskrankheiten. I. Abteilung Originale* 32, 122–147.
Fuhrmann, O. (1906a) Die Taenien der Raubvögel, *Zentralblatt für Bakteriologie, Parasitenkunde und Infektionskrankheiten. I. Abteilung Originale* 41, 212–221.
Fuhrmann, O. (1906b) Die Hymenolepisarten der Vögel. *Zentralblatt für Bakteriologie, Parasitenkunde und Infektionskrankheiten. I. Abteilung Originale* 41, 440–452.
Fuhrmann, O. (1907a) Bekannte und neue Arten und Genera von Vögeltaenien. *Zentralblatt für Bakteriologie, Parasitenkunde und Infektionskrankheiten. I. Abteilung Originale* 45, 516–536.
Fuhrmann, O. (1907b) Die Systematik der Ordnung der Cyclophyllidea. *Zoologischer Anzeiger* 32, 289–297.
Fuhrmann, O. (1908a) Die Cestoden der Vögel. *Zoologische Jahrbucher, Supplement 10* 1, 232 pp.
Fuhrmann, O. (1908b) Nouveaux taenias d'oiseaux. *Revue Suisse de Zoologie* 16, 27–73.
Fuhrmann, O. (1909a) Cestoden. In: *Wissenschaftliche Ergebnisse der Schwedischen Expedition nach dem Kilimandjaro, dem Meru und den umgebenden Masai Steppen Deutsch, Ost-Afrikas, 1905–1906* 3 (22). Vermes, No. 2, 11–22.
Fuhrmann, O. (1909b) Die Cestoden der Vögel des weissen Nils. *Results of the Swedish Zoological Expedition to Egypt and the White Nile, 1901* 27, 1–55.
Fuhrmann, O. (1909c) Neue Davaineiden. *Zentralblatt für Bakteriologie, Parasitenkunde und Infektionskrankheiten. I. Abteilung Originale* 49, 94–124.
Fuhrmann, O. (1911) Vogelcestoden der Aru-Inseln. *Abhandlungen der Senckenbergischen Naturforschenden Gesellschaft* 34, 251–266.
Fuhrmann, O. (1913) Nordische Vögelcestoden aus dem Museum von Göteborg. *Meddelanden fran Göteborgs Musei Zoologiska, Afdelning 1* 1–41.
Fuhrmann, O. (1918) Cestodes d'oiseaux de la Nouvelle-Calédonie et des Iles Loyalty. *Nova Caledonia (Sarasin et Roux), A. Zoologie* 2, 399–449.
Fuhrmann, O. (1920a) Considérations générales sur les Davainea. *Festschrift für Zschokke, Bale* No. 27, 19 pp.
Fuhrmann, O. (1920b) Die Cestoden der Deutschen Südpolar Expedition, 1901–1903. *Deutsche Südpolar-Expedition, 1901–1903. Vol. 16, Zoologie* 8, pp. 469–524.
Fuhrmann, O. (1926) *Cestodes, Catalogue des Invertébrés de la Suisse, 17.* Musée d'Histoire Naturelle de Génève, 150 pp.
Fuhrmann, O. (1931) Dritte Klasse des Cladus Plathelminthes. Cestoidea. In: Kukenthal, W., *Handbuch der Zoologie.* Kukenthal und Krumbach, Berlin und Leipzig, 1930–1031, Vol. 2, 141–416.
Fuhrmann, O. (1932) Les ténias des oiseaux. *Mémoires de l'Université de Neuchâtel* 8, 381 pp.
Fuhrmann, O. (1934a) Un cestode aberrant. *Bulletin de la Société Neuchâteloise des Sciences Naturelles* 58, 107–120.
Fuhrmann, O. (1934b) Vier Diesing'sche Typen (Cestoda). *Revue Suisse de Zoologie* 41, 545–564.
Fuhrmann, O. (1936a) *Gynandrotaenia stammeri* n. g., n. sp. *Revue Suisse de Zoologie* 43, 517–518.

Fuhrmann, O. (1936b) Un singulier ténia d'oiseaux, *Gynandrotaenia stammeri* n. g., n. sp. *Annales de Parasitologie Humaine et Comparée* 14, 261–271.

Fuhrmann, O. (1943) Cestodes d'Angola. *Revue Suisse de Zoologie* 50, 449–471.

Fuhrmann, O. & Baer, J.G. (1943) Cestodes. *Bulletin de la Société Neuchâteloise des Sciences Naturelles* 68, 113–140.

Galkin, A.K. (1983) [Cestodes of *Apus apus* L. from the Kurish Split.] *Trudy Zoologicheskogo Instituta, Leningrad* 121, 57–67. [In Russian.]

Galkin, A.K. (1987a) [Origin of non-cyclophyllidean cestodes, parasites of gulls.] *Trudy Zoologicheskogo Instituta Akademiya Nauk SSSR (Leningrad)* 161, 3–23. [In Russian.]

Galkin, A.K. (1987b) [Synonymy of *Anomotaenia microrhyncha* (Krabbe, 1869) (Cestoidea, Dilepididae).] *Parazitologiya* 21, 669–672. [In Russian.]

Genov, T. & Georgiev, B.B. (1988) Review of the species of the genus *Anoplocephaloides* Baer, 1923 emend. Rausch, 1976 (Cestoda: Anoplocephalidae) parasitizing rodents in Bulgaria, with an analysis of the taxonomic structure of the genus. *Parasitologia Hungarica* 21, 31–52.

Genov, T., Murai, E., Georgiev, B.B. & Harris, E.A. (1990) The erection of *Leporidotaenia* n. g. (Cestoda: Anoplocephalidae) for *Anoplocephaloides* spp. parasitising Leporidae (Lagomorpha). *Systematic Parasitology* 16, 107–126.

Georgiev, B.B. (1992) Invalidation of the genus *Zosteropicola* Johnston, 1912 (Cestoda: Paruterinidae: Anonchotaeniinae). *Systematic Parasitology* 21, 39–44.

Georgiev, B.B. & Bray, R.A. (1988) Cestodes of the genera *Angularella* Strand, 1928 and *Vitta* Burt, 1938 (Dilepididae) from the collection of the British Museum (Natural History). *Systematic Parasitology* 12, 61–75.

Georgiev, B.B. & Bray, R.A. (1991) *Notopentorchis cyathiformis* (Frölich, 1791) n. comb. and *N. iduncula* (Spassky, 1946) (Cestoda: Paruterinidae) from Palaearctic swifts (Aves: Apodiformes), with a review of the genus *Notopentorchis* Burt, 1938. *Systematic Parasitology* 20, 121–133.

Gibson, D.I., Bray, R.A. & Powell, C.B. (1987) Aspects of the life history and origins of *Nesolecithus africanus* (Cestoda: Amphilinidea). *Journal of Natural History* 21, 785–794.

Ginetsinskaya, T.A. (1944) Yavlenie neotenii u Cestodes. *Zoologicheskii Zhurnal* 23, 35–42.

Gómez, M.P.I. (1981) *Idiogenes otidis* Krabbe, 1868; parasito intestinal de *Otis tarda* L. *Revista Ibérica de Parasitología* 41, 475–484.

Graber, M., Blanc, P. & Delavenay, R. (1980) Hélminthes des animaux sauvages d'Ethiopie. I. Mammifères. *Revue d'Élevage et Médecin Vétérinaire des Pays Tropicaux* 33, 143–158.

Gubanov, N.M. & Fedorov, K.P. (1970) [Fauna of helminths of mouselike rodents of Yakutiya.] In: Cherepanov, A.I. (ed.) *Fauna Sibiri*. Novosibirsk: Nauka, pp. 18–47. [In Russian.]

Guiart, J. (1927) Classification des tétrarhynques. *Association Francais pour l'Avancement de la Science, 50ème session*, Lyon (1926), 397–401.

Guiart, J. (1931) Considérations historiques sur la nomenclature et sur la classification des tétrarhynques. *Bulletin de l'Institut Oceanographique, Monaco*, No. 575, 1–27.

Gulyaev, V.D. (1977) [Larvogenesis of diplocyst of *Aploparaksis furcigera* (Rud., 1819) Fuhrmann, 1926 (Cestoda, Hymenolepididae).] *Parazitologia* 11, 17–23. [In Russian.]

Gulyaev, V.D. (1989) [New morpho-ecological types of cysticercoids in cestodes of the subfamily Schistotaeniinae Johri, 1959.] In: Fedorov, K.P. (ed.) *Ekologiya gel'mintov pozvonochnykh Sibiri. Sbornik nauchnykh trudov*. Novosibirsk, USSR: Nauka, Sibirskoe Otdelenie, 199–213. [In Russian.]

Gulyaev, V.D. & Tolkacheva, L.M. (1987) [A new cestode genus, *Ryjikovilepis* gen. nov., from grebes and redescription of *R. dubininae* (Ryzhikov & Tolkacheva, 1981) comb. nov.] *Novye i Maloizvestnye Vidy Fauny Sibiri*, No. 19, 80–88. [In Russian.]

Gulyaev, V.D., Korotaeva, V.D. & Kurochkin, Y.V. (1989) [*Paratelemerus seriolella* gen. et sp. n and *P. psenopsis* sp. n. new pseudophyllidean cestodes from the perciform fishes of the

Australian Shelf.] *Izvestiya Sibirskogo Otdeleniya akademii Nauk SSSR*, No. 2, 86–91. [In Russian.]

Gupta, S.P. (1961) Caryophyllaeids (Cestoda) from freshwater fishes of India. *Proceedings of the Helminthological Society of Washington* 28, 38–50.

Gupta, V. & Parmar, S. (1988) *Echinobothrium deeghai* sp. n. from a marine fish *Trygon sephen* of West Bengal. *Proceedings of Parasitology* 6, 78–81.

Gupta, V. & Singh, S (1992) On a new cestode *Pseudogephyrolina guptai* gen. et sp. nov. from a freshwater fish *Rita rita* from Lucknow. *Indian Journal of Helminthology* 43, 96–99.

Gvozdev, E.V. (1961a) [*Diuterinotaenia spasskyi* n. g., n. sp. from Ochotonidae in the Kazakh SSR.] *Helminthologia* 3, 139–142. [In Russian.]

Gvozdev, E.V. (1961b) [The helminth fauna of Pterocletiformes in Kazakhstan.] *Conference on the Natural Focal Occurrence of Diseases and Problems in Parasitology of Kazakhstan and Central Asian Republics (4th), Alma-Ata*, 1959, Part 3, pp. 309–316. [In Russian.]

Gvozdev, E.V. & Maksimova, A.P. (1968) [New cestode species (Cestoda: Hymenolepididae) from the flamingo (*Phoenicopterus roseus* Pall.).] *Kazakhskaya SSR, Izvestiya AN Kaz. SSR, Seriya biologicheskaya*, No. 5, 30–37. [In Russian.]

Gvozdev, E.V. & Maksimova, A.P. (1985a) [Helminths of the biocoenosis of the saline Lake Tengiz.] *Gel'minty zhivotnykh v ekosystemikh Kazakhstana* [ed. Gvozdev, E.V.] Alma-Ata, USSR; 'Nauka' Kazakhskoi SSR, pp. 3–12. [In Russian.]

Gvozdev, E.V. & Maksimova, A.P. (1985b) [On the systematic position of hymenolepidids (Cestoda) of the flamingo *Phoenicopterus roseus*.] *Trudy-Gel'mintologicheskoi Laboratorii Issledovaniya po Morfologii, Taksonomii i Biologii Gel'mintov Ptits* 33, 36–43. [In Russian.]

Gvozdev, E.V., Maksimova, A.P. & Kornyushin, V.V. (1971) *Monotestilepis tadornae* n. g., n. sp., a new cestode (Hymenolepididae) from *Tadorna tadorna* L. *Trudy Instituta Zoologii, Alma-Ata* 31, 47–50. [In Russian.]

Hart, J.F. (1936) Cestodes from fishes of Puget Sound. III. Phyllobothrioidea. *Transactions of the American Microscopical Society* 55, 488–496.

Hart, J.F. & Guberlet, J.E. (1936) Cestoda from fishes of Puget Sound. I. Spathebothrioidea, a new superfamily. *Transactions of the American Microscopical Society* 55, 199–207.

Hayden, B.P. & Campbell, R.A. (1981) *Zyxibothrium* (Tetraphyllidea, Phyllobothriidae) a new genus of cestodes from skates with suggestions for diagnoses, classification and revision of the Rhinebothriinae Euzet, 1953. *Journal of Parasitology* 67, 262–267.

He, Cheng-de & Liu, Shi-xiu (1981) A new species of tapeworm from an African lion. *Acta Zootaxonomica Sinica* 6, 244–246.

Heinz, M.L. & Dailey, M. (1974) The Trypanorhyncha (Cestoda) of elasmobranch fishes from southern California and Northern Mexico. *Proceedings of the Helminthological Society of Washington* 41, 161–169.

Hess, E. (1972) Contribution à la biologie larvaire de *Mesocestoides corti* Hoeppli, 1925 (Cestoda, Cyclophyllidea). Note préliminaire. *Revue Suisse de Zoologie* 79, 1031–1037.

Hildreth, M.B. & Lumsden, R.D. (1985) Description of *Otobothrium insigne* plerocercus (Cestoda: Trypanorhyncha) and its incidence in catfish from the Gulf coast of Louisiana. *Proceedings of the Helminthological Society of Washington* 52, 44–50.

Hine, P.M. (1977) New species of *Nippotaenia* and *Amurotaenia* (Cestoda: Nippotaeniidae) from New Zealand freshwater fishes. *Journal of the Royal Society of New Zealand* 7, 143–155.

Hoberg, E.P. (1985) *Reticulotaenia* n. gen. for *Lateriporus australis* Jones and Williams, 1967 and *Lateriporus mawsoni* Prudhoe, 1969 (Cestoda: Dilepididae), from sheathbills, *Chionis* spp., in Antarctica, with a consideration of infraspecific variation and speciation. *Journal of Parasitology* 71, 319–326.

Hoberg, E.P. (1987a) Recognition of larvae of the Tetrabothriidae (Eucestoda): implications for the origins of tapeworms in marine homeotherms. *Canadian Journal of Zoology* 65, 997–1000.

Hoberg, E.P. (1987b) *Tetrabothrius shinni* sp. nov. (Eucestoda) from *Phalacrocorax atriceps bransfieldensis* (Pelecaniformes) in Antarctica with comments on morphological variation, host-parasite biogeography and evolution. *Canadian Journal of Zoology* 65, 2969–2975.

Hoberg, E.P. (1989) Phylogenetic relationships among genera of the Tetrabothriidae (Eucestoda). *Journal of Parasitology* 75, 617–626.

Hoberg, E.P. (1990) *Trigonocotyle sexitesticulae* sp. n. (Eucestoda: Tetrabothriidae) a parasite of pygmy killer whales (*Feresa attenuata*). *Canadian Journal of Zoology* 68, 1835–1838.

Hoberg, E.P. (1992) Congruent and synchronic patterns in biogeography and speciation among seabirds, pinnipeds and cestodes. *Journal of Parasitology* 78, 601–615.

Hoberg, E.P. & Adams, A.M. (1992) Phylogeny, historical biogeography, and ecology of *Anophryocephalus* spp. (Eucestoda: Tetrabothriidae) among pinnipeds of the Holarctic during the late Tertiary and Pleistocene. *Canadian Journal of Zoology* 70, 703–719.

Hoberg, E.P., Adams, A.M. & Rausch, R.L. (1991) Revision of the genus *Anophryocephalus* Baylis, 1922 from pinnipeds in the Holarctic with descriptions of *Anophryocephalus nunivakensis* sp. n. and *A. eumetopii* sp. n. (Tetrabothriidae) and evaluation of records from the Phocidae. *Canadian Journal of Zoology* 69, 1653–1668.

Hoeppli, R.J.C. (1925) *Mesocestoides corti*, a new species of cestode from the mouse. *Journal of Parasitology* 12, 91–96.

Hörchner, F. (1962) Zur Helminthenfauna des Nandu (*Rhea americana*). *Zeitschrift für Parasitenkunde* 21, 181–186.

Hoffman, G.L. (1967) *Parasites of North American Freshwater Fishes*. University of California Press, Berkeley, California 486 pp.

Honess, R.F. (1937) Un nouveau cestode: *Fossor angertrudae* n. g., n. sp. du blaireau d'Amerique *Taxidea taxus taxus* (Schreber, 1788). *Annales de Parasitologie Humaine et Comparée* 15, 363–366.

Hornell, J. (1912) New cestodes from Indian fishes. *Records of the Indian Museum* 7, 198–204, plates IX–X.

Hudson, J.R. (1934) Notes on some avian cestodes. *Annals and Magazine of Natural History* 14, 314–318.

Hungerbühler, M. (1910) Studien an *Gyrocotyle* un Cestoden. *Denkschriften Medicin-Naturwissenschaftlich Gesellschaft zu Jena* 16, 496–521.

Hunkeler, P. (1972) Les cestodes parasites des petits mammifères (rongeurs et insectivores) de Côte-d'Ivoire et de Haute-Volta. (Note préliminaire.) *Bulletin de la Société Neuchâteloise des Sciences Naturelles* 95, 121–132.

Hunkeler, P. (1974) Les cestodes parasites des petits mammifères (rongeurs et insectivores) de Côte-d'Ivoire et de Haute-Volta. *Revue Suisse de Zoologie* 80, 809–930.

Hunter, G.W., III (1927) Notes on the Caryophyllaeidae of North America. *Journal of Parasitology* 14, 16–26, plates I–II.

Hunter, G.W., III (1930) Studies on the Caryophyllaeidae of North America. *Illinois Biological Monographs* 11, (1927), 186 pp.

Hyman, L.H. (1951) *The Invertebrates: Platyhelminthes and Rhynchocoela. Vol. II.* McGraw-Hill Book Company, New York, Toronto, London, 550 pp.

Inamdar, N.B. (1942) A new species of avian cestode from India. *Journal of the University of Bombay, Section B, Biological Sciences (New Series)* 11, 77–81.

Jacobi, A. (1896) *Diploposthe*, eine neue Gattung von Vögeltaenien. *Zoologischer Anzeiger* 19, 268–269.

Janicki, C. (1902) Ueber zwei neue Arten des Genus *Davainea* aus celebensischen Säugern. *Archives de Parasitologie* 6, 257–292.

Janicki, C. (1928) Cestoden s.str. aus Fischen und Amphibien. In: Jägerskiöld, L.A. *Result of the Swedish Zoological Expedition to Egypt and the White Nile, 1901*, Part 5, Uppsala, 1–58.

Janicki, C. (1930) Über die jüngsten Zustände von *Amphilina foliacea* in der Fischleibeshöle, sowie Generalles zur Auffassung des Genus *Amphilina* G. Wagen. *Zoologischer Anzeiger* 90, 190–205.

Janiszewska, J. (1954) Caryophyllaeidae europejeskie ze szczególnym uwzglednieniem Polski.

Travaux de la Société des Sciences et des Lettres de Wroclaw, Seria B, No. 66, 73 pp.

Jarecka, L. (1958) Plankton crustaceans in the life cycle of tapeworms occurring at Druzno Lake. *Acta Parasitologica Polonica* 6, 65–109.

Jensen, L.A. (1976) *Parabothriocephalus sagitticeps* (Sleggs, 1927) comb. n. (Cestoda: Parabothriocephalidae) from *Sebastes paucispinis* of southern and central California. *Journal of Parasitology* 62, 560–562.

Jensen, L.A. & Heckmann, R.A. (1977) *Anantrum histocephalum* sp. n. (Cestoda: Bothriocephalidae) from *Synodus lucioceps* (Synodontidae) of southern California. *Journal of Parasitology* 63, 471–472.

Johnston, T.H. (1911) The entozoa of Monotremata and Australian Marsupialia. No. ii. *Proceedings of the Linnean Society of New South Wales* 36, 47–57.

Johnston, T.H. & Clarke, H.G. (1948) Cestodes from Australian birds. I. Pelicans. *Transactions of the Royal Society of South Australia* 72, 77–82.

Johri, G.N. (1959) Descriptions of two amabilid cestodes from the little grebe, *Podiceps ruficollis*, with remarks on the family Amabiliidae Braun, 1900. *Parasitology* 49, 454–461.

Johri, G.N. (1960a) A new paruterinid cestode, *Lallum magniparuterina* gen. et sp. nov. from the intestine of a common teal, *Nettion crecca* Linn. *Parasitology* 50, 269–272.

Johri, G.N. (1960b) Studies on some cestode parasites. IV. On four new species including a new genus belonging to the family Hymenolepididae. *Proceedings of the National Academy of Sciences, India* 30B, 192–202.

Johri, L.N. (1934) Report on a collection of cestodes from Lucknow (V.P., India). *Records of the Indian Museum* 36, 153–177.

Johri, L.N. (1955) On a new genus and species, *Neyralla kotharia* of the subfamily Dipylidiinae Stiles, 1896 (Cestoda) from Delhi State, India. *Revista Ibérica de Parasitología*, Tomo Extraordinario, 165–122.

Jones, A. (1980) *Proteocephalus pentastoma* (Klaptocz, 1906) and *Polyonchobothrium polypteri* (Leydig, 1853) from species of *Polypterus* Geoffroy, 1802 in the Sudan. *Journal of Helminthology* 54, 25–38.

Jones, A. (1982) A redescription of *Triuterina anoplocephaloides* (Fuhrmann, 1902) (Cestoda: Anoplocephalidae) from African parrots. *Systematic Parasitology* 4, 253–255.

Jones, A. (1983) A revision of the cestode genus *Joyeuxiella* Fuhrmann, 1935 (Dilepididae: Dipylidiinae). *Systematic Parasitology* 5, 203–213.

Jones, A. & Khalil, L.F. (1980) The helminth parasites of the lesser flamingo, *Phoeniconaias minor* (Geoffroy), from Lake Nakuru, Kenya, including a new cestode, *Phoenicolepis nakurensis* n. g., n. sp. *Systematic Parasitology* 2, 61–76.

Jones, A., Khalil, L.F. & Bray, R.A. (1992) *Pluviantaenia kassalensis* n. g., n. sp. (Davaineidae), a new cestode from the Egyptian plover *Pluvianus aegyptius* (L.) in the Sudan. *Systematic Parasitology* 22, 205–213.

Jones, A.W. (1944) *Diorchis reynoldsi* n. sp., a hymenolepidid cestode from the shrew. *Transactions of the American Microscopical Society* 63, 46–49.

Jones, A.W. & Mackiewicz, J. (1969) Naturally occurring triploidy and parthenogenesis in *Atractolytocestus huronensis* Anthony (Cestoidea: Caryophyllidea) from *Cyprinus carpio* L. in North America. *Journal of Parasitology* 55, 1105–1118.

Jones, H.I. (1988) Notes on the parasites in penguins (Spheniscidae) and petrels (Procellariidae) in the Antarctic and Sub-antarctic. *Journal of Wildlife Diseases* 24, 166–167.

Jones, M.K. (1987a) *Nematotaenoides ranae* (Cyclophyllidea, Nematotaeniidae) transferred to the genus *Anonchotaenia* (Paruterininae). *Proceedings of the Helminthological Society of Washington* 54, 158–160.

Jones, M.K. (1987b) A taxonomic revision of the Nematotaeniidae Lühe, 1910 (Cestoda: Nematotaeniidae). *Systematic Parasitology* 10, 165–245.

Jones, M.K. (1988) Formation of the paruterine capsules and embryonic envelopes in *Cylindrotaenia hickmani* (Jones, 1985) (Cestoda: Nematotaeniidae). *Australian Journal of Zoology* 36, 545–563.

Jones, N.V. & Williams, I.C. (1967) The cestode parasites of the sheathbill, *Chionis alba* (Gmelin), from Signy Island, South Orkney Islands. *Journal of Helminthology* 41, 151–160.

Jordano, D. & Diaz-Ungria, C. (1956) Cestodes de Venezuela II. *Dendrometra ginesi* (nov. gen. nov. sp.) (Cestoda: Dilepididae) nueva tenia parasita de la tijereta (*Fregata magnificens*). *Novedades Cientificas. Contribuciones Ocasionales del Museo de Historia Natural La Salle, Serie Zoologica* 18, 1–4.

Jordano, D. & Diaz-Ungria, C. (1960) *Craspedocotyla margaritensis* nov. gen. nov. sp. (Cestoda: Hymenolepididae) nueva tenia parasita del turpial (*Icterus icterus*) (Aves: Passeres). *Memorias de la Sociedad de Ciencias Naturales La Salle* 20, 198–210.

Joyeux, C. (1945) Cestodes du Brésil. I. *Arkiv för Zoologi* 37, 1–4.

Joyeux, C. & Baer, J.G. (1934) Sur quelques cestodes de France. *Archives du Muséum National d'Histoire Naturelle*, Paris, series 6, 11, 157–171.

Joyeux, C. & Baer, J.G. (1936a) Notices helminthologiques. *Bulletin de la Société Zoologique de France* 60, 482–501.

Joyeux, C. & Baer, J.G. (1936b) Cestodes. *Faune de France* 30, 1–613.

Joyeux, C. & Baer, J.G. (1936c) Quelques helminthes nouveaux ou peu connus de la musaraigne *Crocidura russula. Revue Suisse de Zoologie* 43, 25–50.

Joyeux, C. & Baer, J.G. (1939) Sur quelques cestodes des Charadriiformes. *Bulletin de la Société Zoologique de France* 64, 171–186.

Joyeux, C. & Baer, J.G. (1943) Sur quelques helminthes du Maroc. Note préliminaire. *Bulletin de la Société de Pathologie Exotique* 36, 86–88.

Joyeux, C. & Baer, J.G. (1951) Le genre *Gyrocotyle* Diesing, 1850 (Cestodaria). *Revue Suisse de Zoologie* 8, 371–381.

Joyeux, C. & Baer, J.G. (1955) Cestodes d'oiseaux récoltés dans le centre de la France. *Bulletin de la Société Zoologique de France* 80, 174–196.

Joyeux, C. & Baer, J.G. (1961) Classe des cestodes. In: Grassé, P.P. (ed.) *Traité de Zoologie. Anatomie, Systématique, Biologie, Vol. 4, Fasc. 1. Platyhelminthes, Mésozaires, Acanthocéphales, Némertiens*. Masson et Cie, Paris, pp. 347–560.

Joyeux, C. & Timon-David, J. (1934) Sur quelques cestodes d'oiseaux. *Annales du Musée d'Histoire Naturelle de Marseille* 26 (mem.II), 1–26.

Joyeux, C., Gendre, E. & Baer, J.G. (1928) Recherches sur les hélminthes de l'Afrique occidentale française. *Collection de la Societé de Pathologie Exotique, Monographie II* 120 pp.

Kamegai, S., Ichihara, A., Nonobe, H. & Machida, M. (1967) The 6th and 7th records of human infection with *Mesocestoides lineatus* (Cestoda) in Japan. *Research Bulletin of the Meguro Parasitological Museum* 1, 1–7.

Karpenko, S.W. & Gulyaev, V.D. (1990) [Redescription of the shrew cestode *Pseudobothrialepis mathevossianae* Schaldybin, 1957 (Hymenolepididae).] *Izvestiya Sibirskogo Otdeleniya Akademii Nauk SSSR, Ser. Biol. Nauk* 3, 68–71. [In Russian.]

Kayton, R.J. & Kritsky, D.C. (1984) *Sureshia trychopea* sp. n. (Cestoda: Dilepididae) from white-throated swift, *Aeronautes s. saxatilis*, (Aves: Apodiformes) with note on dilepidine cestodes having a vaginal sclerite. *Proceedings of the Helminthological Society of Washington* 51, 305–308.

Kennedy, C.R. (1965) Taxonomic studies on *Archigetes* Leuckart, 1878 (Cestoda: Caryophyllaeidae). *Parasitology* 55, 439–451.

Khalil, L.F. (1960) On a new genus *Sandonella* for *Proteocephalus sandoni* Lynsdale, 1960 (Proteocephalidae) and the erection of a new subfamily Sandonellinae. *Journal of Helminthology* 34, 47–54.

Khalil, L.F. (1971) *Ichthybothrium ichthybori* gen. et sp. nov. (Cestoda: Pseudophyllidea) from the

African freshwater fish *Ichthyborus besse* (Joannis, 1835). *Journal of Helminthology* 45, 371–379.

Khalil, L.F. & Abdul-Salam, J. (1989) *Macrobothridium rhynchobati* n. g., n. sp. from the elasmobranch *Rhynchobatus granulatus*, representing a new family of diphyllidean cestodes, the Macrobothridiidae. *Systematic Parasitology* 13, 103–109.

Khan, M.S.B. (1984) *Satyanarayana satyanarayani* n. g. n. sp. from the lungs of stork, *Ardea cinerea. Proceedings of the Indian Academy of Parasitology* 5, 59–62.

Khanum, Z. (1968) A new species of hymenolepid cestode *Oligorchis funambuli* from the intestine of the common striped squirrel *Funambulus pennanti argentescens* Wroughton from Lower Sind. *University Studies. University of Karachi* 5, 80–85.

Khanum, Z., Farooq, M. & Jamal, N. (1982) Three new cestode genera and new species from birds of Karachi, Pakistan. *Pakistan Journal of Zoology* 14, 11–19.

Kintner, K.E. (1938) Notes on the cestode parasites of English sparrows in Indiana. *Parasitology* 30, 347–357.

Kornyushin, V.V. (1983) [Analysis of the structure of the genus *Echinocotyle* (Cestoda, Hymenolepididae).] *Parazitologiya* 17, 268–271. [In Russian.]

Kornyushin, V.V. (1989) [Monogenea and Cestoda. Vol. 3. Davaineoidea, Biuterinoidea, Paruterinoidea.] *Fauny Ukraini* 33, 252 pp. [In Russian.]

Kornyushin, V.V. & Sharpilo, L.D. (1986) [A new taeniid genus (Cestoda, Taeniidae) parasitic in Mustelidae.] *Vestnik Zoologii*, No. 3, 10–16. [In Russian.]

Korpaczewska, W. & Sulgostowska, T. (1974) Revision of the genus *Tatria* Kow., 1904 (Cestoda, Amabiliidae), including description of *Tatria iunii* sp. n. *Acta Parasitologica Polonica* 22, 67–91.

Kotlán, A. (1921) Uj-Guinea Madádar-cestodák. I. Papagály-cestodák. *Annales Musei Nationalis Hungarici* 18, 1–27.

Krabbe, H. (1868) Trappens Baendelorme. *Videnskabelige Meddelelser Naturhistoriska Forening i Kjöbenhavn* (1867), 122–126.

Krabbe, H. (1869) Bidrag til Kundskab om Fuglenes Baendelor me. Bianco Lunos Bogtrykkeri ved F.S. Muhle. Kjobenhavn, 251–368.

Krotov, A.I. (1949) [The hymenolepidid fauna of Anseriformes of the USSR.] *Trudy Gelmintologicheskoi Laboratorii* 2, 99–110. [In Russian.]

Kugi, G. (1990) *Dactylopsilophilus australensis* n. g., n. sp., a representative of a new family Dactylopsilophilidae, from a common striped possum, *Dactylopsila trivirgata. Japanese Journal of Parasitology* 39, 581–584.

Kugi, G. (1993) Three new cestodes from reptiles collected on Amami-oshima Island, Japan. *Japanese Journal of Parasitology* 42, 415–421.

Kulakovskaya, O.P. (1961) [Materials on the fauna of Caryophyllaeidae (Cestoda, Pseudophyllidea) of the Soviet Union.] *Parazitologicheskii Sbornik* 20, 339–355. [In Russian.]

Kulakovskaya, O.P. (1962) [A new genus and species of tapeworm, *Breviscolex orientalis* (Caryophyllaeidae, Cestodes), from fish in the Amur Basin.] *Doklady Akademii Nauk SSSR* 143, 1001–1004. [English translation in National Science Foundation translation of same journal, 1962, 143, 386–388.]

Kulakovskaya, O.P. & Demshin, N.I. (1978) [Origin and phylogenetic relationships of caryophyllideans.] In: *Problemy Gidro-Parazitologii*. Markevich, A.P. (ed.). Kiev: Naukova Dumka, pp. 95–104. [In Russian.]

Kuperman, B.I. (1973) [*Tapeworms of the Genus* Triaenophorus, *Parasites of Fishes*.] Leningrad: Izdatel'stvo Nauka, 208 pp. [In Russian, English translation (1981), New Delhi: Amerind Publishing Co. Pvt. Ltd., 222 pp.]

Larsch, J.E. (1941) *Corallobothrium parva* n. sp., a cestode from the common bullhead, *Ameiurus nebulosus. Journal of Parasitology* 27, 221–227.

La Rue, G.R. (1914) A revision of the cestode family Proteocephalidae. *Illinois Biological Monographs* 1, 350 pp.

Li, Min-min (1964) A new genus and three new species of cestodes (Caryophyllaeidae) from *Cyprinus carpio* in China. *Acta Zootaxonomica Sinica* 1, 355–363, 3 pl.

Linnaeus, C. (1758) *Systema Naturae per Regna Tria Naturae, Secundum Classes, Ordines, Genera, Species, cum Characteribus, Differentiis, Synonymis, Locis*. Editio decima, reformata. Vol. 1. Holmiae, 823 pp.

Linstow, O.F.B. von (1875) Beobachtungen an neuen und bekannten Helminthen. *Archiv für Naturgeschichte* 41, 183–207.

Linstow, O.F.B. von (1896) Helminthologische Mittheilungen. *Archiv für Mikroskopische Anatomie* 48, 375–397.

Linstow, O.F.B. von (1905) Helminthen der russischen Polarexpedition 1900–1903. *Mémoires de l'Academie Imperiale des Sciences de St. Petersbourg* Ser. 8, 18, 1–17.

Linstow, O.F.B. von (1906) Helminths from the collection of the Colombo Museum. *Ceylon Journal of Science* 3, 163–188.

Linton, E. (1890) Notes on Entozoa of marine fishes of New England. II. *Annual Report, U.S. Commission of Fish and Fisheries (1887), Washington, D.C.* 15, 719–899.

Linton, E. (1892) Notes on avian Entozoa. *Proceedings of the United States National Museum* 15, 87–113.

Linton, E. (1927) Notes on cestode parasites of birds. *Proceedings of the United States National Museum* 70, 1–75.

Llewellyn, J. (1986) Phylogenetic inference from platyhelminth life-cycle stages. In: Howell, M.J. (ed.) *Parasitology – quo vadit?* [Proceedings of the Sixth International Congress of Parasitology, Brisbane.] Canberra: Australian Academy of Sciences, 281–289.

Locker, B. & Rausch, R.L. (1952) Some cestodes from Oregon shrews, with descriptions of four new species of *Hymenolepis* Weinland, 1858. *Journal of the Washington Academy of Sciences* 42, 26–31.

Loos-Frank, B. (1980) *Mesocestoides leptothylacus* n. sp. und das nomenklatorische Problem in der Gattung *Mesocestoides* Vaillant, 1863 (Cestoda: Mesocestoididae). *Tropenmedizin und Parasitologie* 31, 2–14.

López-Neyra, C.R. (1929) Consideraciones sobre el género *Davainea* (s.l.) y descripción de dos especies nuevas. *Boletin de la Real Sociedad Española de Historia Natural* 29, 345–359.

López-Neyra, C.R. (1931) Revisión del género *Davainea*. *Memorias de la Academia de Ciencias Exactas, Físicas y Naturales de Madrid* 1, 1–177.

López-Neyra, C.R. (1942) División del género *Hymenolepis* Weinland (s.l.) en otros mas naturales. *Revista Ibérica de Parasitología* 2, 46–93; 113–256.

López-Neyra, C.R. (1943a) Reflexiones sobre la importancia de la evolucion uterina en la sistematica de los ciclofilididos. *Revista Ibérica de Parasitología* 3, 30–58.

López-Neyra, C.R. (1943b) *Paradicranotaenia anormalis* n. g. n. sp. y consideraciones sobre los Nematoparataeniidae. *Revista Ibérica de Parasitología* 3, 229–256.

López-Neyra, C.R. (1944) Compendio de helmintología ibérica. Parte II. *Revista Ibérica de Parasitologi!a* 4, 75–96; 136–198; 209–342; 403–492.

López-Neyra, C.R. (1952) *Gyrocoelia albaredai* n. sp. Relaciones con Tetrabothriidae y Dilepididae. *Revista Ibérica de Parasitología* 12, 319–344.

López-Neyra, C.R. (1954) Anoplocephalidae. *Revista Ibérica de Parasitología* 14, 13–130, 225–290, 303–396.

López-Neyra, C.R. (1955) Anoplocephalidae. *Revista Ibérica de Parasitología* 15, 33–84.

López-Neyra, C.R. (1958) *Hymenosphenacanthus* nomen novum par *Sphenacanthus* López-Neyra, 1942 (Cestoda, Hymenolepididae) nec Agassiz, 1837 (pez fósil). *Revista Ibérica de Parasitología* 18, 315.

Lühe, M.F.L. (1902) Revision meines Bothriocephalidensystemes. *Zentralblatt für Bakteriologie, Parasitenkunde, Infektionskrankheiten und Hygiene*, Abt. Orig. 31, 318–331.

Lühe, M.F.L. (1910) Parasitische Plattwürmer. 2. Cestodes. *Süsswasserfauna Deutschlands* 16, iv + 153 pp.

McCrae, R.C. (1961) Studies on the Caryophyllaeidae (Cestoda) of the white sucker, *Catostomus commersoni* (Lacépède) in northern Colorado. *Dissertation Abstracts* 21, 2835–2836.

Mackiewicz, J.S. (1965) *Isoglaridacris bulbocirrus* gen. et sp. n. (Cestoidea: Caryophyllaeidae) from *Catostomus commersoni* in North America. *Journal of Parasitology* 51, 377–381.

Mackiewicz, J.S. (1968) Two new caryophyllaeid cestodes from the spotted sucker, *Minytrema melanops* (Raf.) (Catostomidae). *Journal of Parasitology* 54, 808–813.

Mackiewicz, J.S. (1969) *Penarchigetes oklensis* gen. et sp. n. and *Biacetabulum carpiodi* sp. n. (Cestoidea: Caryophyllaeidae) from catostomid fish in North America. *Proceedings of the Helminthological Society of Washington* 36, 119–126.

Mackiewicz, J.S. (1972) Parasitological review. Caryophyllidea (Cestoidea): a review. *Experimental Parasitology* 31, 417–512.

Mackiewicz, J.S. (1974) *Isoglaridacris calentinei* n. sp. (Cestoidea: Caryophyllidea) from catostomid fish in Western United States. *Transactions of the American Microscopical Society* 93, 143–147.

Mackiewicz, J.S. (1976) *Glaridacris vogei* n. sp. (Cestoidea: Caryophyllidea) from catostomid fishes in Western North America. *Transactions of the American Microscopical Society* 95, 92–97.

Mackiewicz, J.S. (1981a) Caryophyllidea (Cestoidea): evolution and classification. *Advances in Parasitology* 19, 139–206.

Mackiewicz, J.S. (1981b) Synoptic review of the Caryophyllidea (Cestoidea) of India, Pakistan and Nepal. *Himalayan Journal of Science* 1, 1–14.

Mackiewicz, J.S. (1982) Caryophyllidea (Cestoidea): perspectives. *Parasitology* 84, 397–417.

Mackiewicz, J.S. & Blair, D. (1980) *Caryoaustralus* gen. n. and *Tholophyllaeus* gen. n. (Lytocestidae) and other caryophyllid cestodes from *Tandanus* spp. (Siluriformes) in Australia. *Proceedings of the Helminthological Society of Washington* 47, 168–178.

Mackiewicz, J.S. & Deutsch, W.G. (1976) *Rowardleus* and *Janiszewskella*, new caryophyllid genera (Cestoidea: Caryophyllidea) from *Carpiodes cyprinus* (Catostomidae) in Eastern North America. *Proceedings of the Helminthological Society of Washington* 43, 9–17.

Mackiewicz, J.S. & McCrae, R.C. (1962) *Hunterella nodulosa* gen. n., sp. n. (Cestoidea: Caryophyllaeidae) from *Catostomus commersoni* (Lacépède) (Pisces: Castostomidae). *Journal of Parasitology* 48, 798–806.

MacKinnon, B.M. & Burt, M.D.B. (1985a) Ultrastructure of spermatogenesis and the mature spermatozoon of *Haplobothrium globuliforme* Cooper, 1914 (Cestoda: Haplobothrioidea). *Canadian Journal of Zoology* 63, 1478–1487.

MacKinnon, B.M. & Burt, M.D.B. (1985b) The comparative ultrastructure of the plerocercoid and adult primary scolex of *Haplobothrium globuliforme* Cooper, 1914 (Cestoda: Haplobothrioidea). *Canadian Journal of Zoology* 63, 1488–1496.

MacKinnon, B.M. & Burt, M.D.B. (1985c) Histological and ultrastructural observations on the secondary scolex and strobila of *Haplobothrium globuliforme* (Cestoda: Haplobothrioidea). *Canadian Journal of Zoology* 63, 1995–2000.

MacKinnon, B.M., Jarecka, L. & Burt, M.D.B. (1985) Ultrastructure of the tegument and penetration glands of developing procercoids of *Haplobothrium globuliforme* Cooper, 1914 (Cestoda: Haplobothrioidea). *Canadian Journal of Zoology* 63, 1470–1477.

Macko, J. (1962a) Neue Art eines Bandwurms *Wardium filamento-ovatum* sp. n. (Hymenolepididae) aus dem Wirt *Larus*. *Biologia* 17, 355–364.

Macko, J. (1962b) Neue Art eines Bandwurms aus der Gattung *Echinocotyle* Blanchard, 1891 (Hymenolepididae). *Biologia* 17, 503–507.

Macko, J. (1990) On the variability and taxonomy of dicranotaeniiids (Hymenolepididae, Cestoda) of Anseres in Eastern Slovakia. II. On the most important taxonomic characters of *Dicranotaenia coronula* (Dujardin, 1845). *Helminthologia* 27, 55–65.

Magalhaes, P.S. (1898) Notes d'helminthologie brésilienne. 8. Deux nouveaux ténias de la poule domestique. *Archives de Parasitologie, Paris* 5, 442–451.

Mahon, J. (1954) Contributions to the helminth fauna of tropical Africa. Tapeworms from the Belgian Congo. *Annales du Musée Royal du Congo Belge, C-Zoologie, Serie 5* 1, 137–264.

Maksimova, A.P. (1963) [New species of cestodes from swans in the Kazakh SSR.] *Trudy Instituta Zoologii Akademii Nauk Kazakhskoi SSR* 19, 126–132. [In Russian.]

Maksimova, A.P. (1989) [*Hymenolepidid Cestodes of the Waterfowl of Kazakhstan.*] Alma-Ata, Nauka, 223 pp. [In Russian.]

Mamaev, Yu.L. (1971) [Larval helminths in crustaceans of Primorskii Krai.] In: Mamaev, Yu.L. (ed.) [*Parasites of Animals and Plants of the Soviet Far-East.*] Vladivostok: Dal'nevostochnoe Knizhnoe, pp. 120–132. [In Russian.]

Maplestone, P.A. (1922) Notes on Australian cestodes. III. *Cotugnia oligorchis* n. sp. *Annals of Tropical Medicine and Parasitology* 16, 55–60.

Maplestone, P.A. & Southwell, T. (1922) Notes on Australian cestodes. V. Three cestodes from the black swan. *Annals of Tropical Medicine and Parasitology* 16, 189–198.

Mariaux, J. (1989) Cestodes d'oiseaux de Côte d'Ivoire. III. Sur quelques parasites de Charadriiformes. *Revue Suisse de Zoologie* 96, 541–549.

Mariaux, J. (1990) Metadilepididae: a family going West! *Bulletin de la Société Française de Parasitologie* 8 (Supplement no. 1), 222.

Mariaux, J. (1991a) Cestodes of birds from the Ivory Coast. *Yapolepis yapolepis* n. g., n. sp., a new metadilepidid (Cyclophyllidea: Paruterinoidea) parasite of the icterine greenbul (Aves: Pycnonotidae). *Systematic Parasitology* 18, 187–191.

Mariaux, J. (1991b) Cestodes of birds from the Ivory Coast. Species of the genus *Anonchotaenia* Cohn, 1900. *Systematic Parasitology* 20, 109–120.

Mariaux, J. & Vaucher, C. (1989) Cestodes d'oiseaux de Côte d'Ivoire. II. Parasites de Coraciiformes et Piciformes. *Systematic Parasitology* 14, 117–133.

Mariaux, J., Bona, F.V. & Vaucher, C. (1992) A new genus of Metadilepididae (Cestoda: Cyclophyllidea) parasitic in *Terpsiphone rufiventer* (Aves: Muscicapidae) from the Ivory Coast. *Journal of Parasitology* 78, 309–313.

Markowski, S. (1952) The cestodes of seals from the Antarctic. *Bulletin of the British Museum (Natural History), Zoology* 1, 123–150.

Mateo, E. & Bullock, W.L. (1960) *Neobothriocephalus aspinosus* gen. et sp. n. (Pseudophyllidea: Parabothriocephalidae), from the Peruvian marine fish, *Neptomenus crassus. Journal of Parasitology* 52, 1070–1073.

Matevosyan, E.M. (1953) [Re-organisation of the classification of dilepidid cestodes.] *Papers on helminthology presented to Academician K.I. Skryabin on his 75th birthday*. Moscow: Izdatel'stvo Akademii Nauk SSSR, pp. 392–397. [In Russian.]

Matevosyan, E.M. (1963) [*Principles of Cestodology. Vol. III. Dilepidoidea, Tapeworms of Domestic and Wild Animals.*] Moscow, USSR: Izdatel'stvo Akademii Nauk SSSR, 687 pp. [In Russian.]

Matevosyan, E.M. (1968) [The analysis of some hymenolepidids (Cestoda: Hymenolepididae).] *Trudy Vsesoyuznogo Instituta Gelmintologii* 14, 229–232. [In Russian.]

Matevosyan, E.M. (1969) [*Principles of Cestodology. Vol. VII. Paruterinoidea – Tapeworms of Domestic and Wild Birds.*] Moscow: Izdatel'stvo 'Nauka', 304 pp. [In Russian.]

Matevosyan, E.M. & Movsesyan, S.O. (1966) [*Pentocoronaria rusannae* n. g., n. sp. (Cestoda: Davaineidae) from *Turtur turtur.*] *Materialy Nauchnoi Konferentsii Obshchestva Gelmintologii* Year 1966, Part 3, 170–175. [In Russian.]

Matevosyan, E.M. & Movsesyan, S.O. (1970) [A revision of the genera *Ascometra* and *Octopetalum*

(Cestoda: Paruterinoidea).] *Trudy Vsesoyuznogo Instituta Gel'mintologii imeni K.I. Skryabina* 16, 137–146. [In Russian.]

Mayes, M.A., Brooks, D.R. & Thorson, T.B. (1981) Two new tetraphyllidean cestodes from *Potamotrygon circularis* Garman (Chondrichthyes: Potamotrygonidae) in the Itacui river, Brazil. *Proceedings of the Helminthological Society of Washington* 48, 38–42.

Mayhew, R.L. (1925) Studies on the avian species of the cestode family Hymenolepididae. *Illinois Biological Monographs* 10, 1–125.

Meggitt, F.J. (1915) A new species of tapeworm from a parakeet, *Brotogerys typica*. *Parasitology* 8, 42–55.

Meggitt, F.J. (1924a) *The Cestodes of Mammals*. London [no publisher], 282 pp.

Meggitt, F.J. (1924b) On the life-history of a reptilian tapeworm (*Sparganum reptans*). *Annals of Tropical Medicine and Parasitology* 18, 195–204.

Meggitt, F.J. (1927a) On cestodes collected in Burma. *Parasitology* 19, 141–153.

Meggitt, F.J. (1927b) Report on a collection of Cestoda, mainly from Egypt. Part 1. Families Anoplocephalidae, Davaineidae. *Parasitology* 19, 314–327.

Meggitt, F.J. (1927c) Report on a collection of Cestoda, mainly from Egypt. Part II. Cyclophyllidea: Family Hymenolepididae. *Parasitology* 19, 420–450.

Meggitt, F.J. (1927d) Remarks on the cestode families Monticelliidae and Ichthyotaeniidae. *Annals of Tropical Medicine and Parasitology* 21, 69–87.

Meggitt, F.J. (1930) Report on a collection of Cestoda, mainly from Egypt. Part IV. Conclusion. *Parasitology* 22, 338–345.

Meinkoth, N.A. (1947) Notes on the life cycle and taxonomic position of *Haplobothrium globuliforme* Cooper, a tapeworm of *Amia calva* L. *Transactions of the American Microscopical Society* 66, 256–261.

Mendes, M.V. (1944) Sôbre Cestoda de teleósteos marinhos. *Boletim da Faculdade de Filosofia, Ciências e Letras. Universidade de São Paulo* 43 (Zoologia 8), 173–188.

Meszaros, F. & Murai, E. (1979) Contribution to the knowledge of helminths of rodents in Roumania. *Parasitologia Hungarica* 12, 53–70.

Mettrick, D.F. (1958) Helminth parasites of Hertfordshire birds. II. Cestoda. *Journal of Helminthology* 32, 159–164.

Mettrick, D.F. (1962) A re-description of the female, and a description of the male form of the dioecious cestode, *Gyrocoelia kiewietti* Ortlepp, 1937. *Journal of Helminthology* 36, 149–156.

Moghe, M.A. (1933) Four new species from avian cestodes from India. *Parasitology* 25, 333–341.

Mola, P. (1903) Su di un Cestodi del *Carcharodon rondeletii* Muller et Henle. *Archivio Zoologico Italiano, Napoli* 1, 345–346.

Mola, P. (1906) Di alcune specia poco studiate o mal note di Cestodi. *Annuario del Museo Zoologico della Università di Napoli* 2, 1–12.

Mola, P. (1908) Due nuove forme di Tetraphyllidae. *Bolletino della Societa Adriatica di Sciencia Naturale* 24, 1–16.

Mola, P. (1929) Descriptio platodorum sine exstis. *Zoologischer Anzeiger* 86, 101–113.

Monticelli, F.S. (1889) Notes on some Entozoa in the collection of the British Museum. *Proceedings of the Zoological Society of London* 321–325.

Motomura, I. (1927) On *Caryophyllaeus gotoi* n. sp., a new monozoic cestode from Korea. *Science Reports of the Tohoku Imperial University, 4th Series, Biology* (Sendai) 3, 51–53.

Movsesyan, S.O. (1966) [Revision of *Raillietina* Fuhrmann, 1920 (Cestoda: Davaineidae).] *Tematicheskii Sbornik Rabot po Gel'mintologii* [*Trudy VIGIS*.] 12, 5–10. [In Russian.]

Movsesyan, S.O. (1969) [Revision of the genus *Cotugnia* Diamare, 1893 (Cestoidea: Davaineidae).] *Trudy Vsesoyuznogo Instituta Gel'mintologii imeni K.I. Skryabina* 15, 195–217. [In Russian.]

Movsesyan, S.O. (1970) [Reorganization of the cestode system of the suborder Davaineata Skryabin,

1940.] *Trudy Vsesoyuznogo Instituta Gel'mintologii imeni K.I. Skryabina* 16, 171–175. [In Russian.]

Movsesyan, S.O. (1971) [Revision of the genus *Idiogenes* Krabbe, 1867 (Cestoidea: Idiogenidae Mola, 1929).] In: Yarnikh, A.M. (ed.) [*Collected papers on helminthology presented to Academician K.I. Skryabin on his 90th Anniversary*.] Moscow: Izdatel'stvo 'Kolos', 227–247. [In Russian.]

Movsesyan, S.O. (1977) [*Cestodes of the Fauna of the USSR and Adjacent Territories (Davaineata)*.] Moscow: Izdatel'stvo 'Nauka', 272 pp. [In Russian.]

Movsesyan, S.O. (1987) [On the validity of generic taxons of tapeworms of the suborder Davaineata Skryabin, 1940.] *Biologicheskii Zhurnal Armenii* 40, 707–717. [In Russian.]

Movsesyan, S.O. (1989) New zoological system of cestodes of the suborder Davaineata Skrjabin, 1940. *Helminthologia* 26, 117–127.

Mudaliar, S.V. (1939) *Cotugnia brotogerys* Meggitt, 1915 (from *Gallus domesticus*, Hosur cattle farm, Madras). *Indian Journal of Veterinary Science and Animal Husbandry* 9, 333.

Murai, É. & Georgiev, B.B. (1987) Redescription of three dilepidid species from New Guinean *Megapodius freycinet* (Aves, Galliformes) erected by Sandor Kotlan, with the description of *Kotlanolepis* gen. n. (Cestoda). *Parasitologia Hungarica* 20, 65–79.

Murai, É., Tenora, F. & Stanek, M. (1989) A typical strobilocercus (Cestoda: Taeniidae) – a parasite in experimental stocks of *Ondatra zibethicus* (Rodentia). *Miscellanea Zoologica Hungarica* 5, 21–27.

Murav'eva, S.I. (1975) [Morphological and anatomical bases of the systematics of tetrabothriids.] *Problemy Parazitologii* 2, 49. [In Russian.]

Murav'eva, S.I. & Popov, V.N. (1976) [Taxonomic position and some data on the ecology of *Anophryocephalus skrjabini* (Cestoda, Tetrabothriidae), a parasite of pinnipeds.] *Zoologicheskii Zhurnal* 55, 1247–1250. [In Russian.]

Neave, S.A. (1939) Nomenclator Zoologicus. A list of the names of genera. *Zoological Society of London* 1, p. 953.

Neave, S.A. (1940) Nomenclator Zoologicus. A list of the names of genera. *Zoological Society of London* 4, pp. 458, 459.

Neiland, K.A. (1952) Helminths of northwestern mammals. II. *Oligorchis nonarmatus* n. sp. (Cestoda, Hymenolepididae) from the yellow-bellied squirrel. *Journal of Parasitology* 38, 341–345.

Neslobinsky, N. (1911) Zur Kenntnis der Vögeltaenien Mittelrusslands. *Zentralblatt für Bakteriologie und Parasitenkunde, Originale* 57, 436–442.

Nock, A.M. & Caira, J.N. (1988) *Disculiceps galapagoensis* n. sp. (Lecanicephalidea: Disculicepitidae) from the shark, *Carcharhinus longimanus* with comments on *D. pileatus*. *Journal of Parasitology* 74, 153–158.

Nybelin, O. (1922) Anatomisch-systematische Studien über Pseudophyllidien. *Göteborgs Kgl. Vetenskaps.-Akad. Handl.* 26, 1–228.

Nybelin, O. (1940) Sur le *Tetrarhynchus minutus* P.J. van Beneden. *Göteborgs Kungl. Vetenskapsakad-oh Vitterhetssamhälles Handlingar, 5 Följden, Ser. B* 6, 1–20.

Nybelin, O. (1942) Zur Helminthfauna der Süsswasserfische Schwedens. II. Der Cestoden der Welses. *Göteborgs Kungl. Vetenskapsakad-och Vitterhetssamhälles Handlingar, Ser. B* 1, 1–24.

Odening, K. (1982) Cestoden aus Flugvögeln der Südshetlands (Antarktis) und der Falklandinseln (Malwinen). *Angewandte Parasitologie* 23, 202–223.

Olsen, O.W. (1966) *Diplophallus taglei* n. sp. (Cestoda, Cyclophyllidea) from the viccacha *Lagidium peruanum* Mayer, 1832 (Chinchillidae) from the Chilean Andes. *Proceedings of the Helminthological Society of Washington* 33, 49–53.

Ortlepp, R.J. (1963) Observations on cestode parasites of guinea-fowl from Southern Africa. *Onderstepoort Journal of Veterinary Research* 30, 95–118.

Oshmarin, P.G. (1956) [Parasitic worms of mammals and birds. (Fauna, experimental ecology and geographical characteristics.).] Unpublished dissertation.
Oshmarin, P.G. (1960) [*Polytestilepis chitinocloacis* n. g., n. sp., from ducks.] *Soobshcheniya Dalnevostochnogo filiala im. V.L. Komarova Sibirskogo Otdeleniya Akademii Nauk SSR* No. 12, 133–136. [In Russian.]
Oshmarin, P.G. (1961) Paraziticheskie chervi mlekopitayushchikh i ptic Primorskogo kraya. *Avtoreferat doktorskoy dissertacii biologicheskikh nauk. Moskva, VIGIS.*
Oshmarin, P.G. (1972) [A new species and genus of cestode from *Turnix tanki.*] *Parazitologiya* 6, 557–561. [In Russian.]
Overstreet, R.M. (1968) Parasites of the inshore lizardfish *Synodus foetens* from south Florida including a new genus of Cestoda. *Bulletin of Marine Science* 18, 444–470.
Pagenstecher, H. (1877) Zur Naturgeschichte der Cestoden. *Zeitschrift für Wissenschaftliche Zoologie* 30, 171–193.
Palacios, N.M. & Barroeta, L.F. (1967) Cestodos de vertebrados. XI. *Revista Ibérica de Parasitología* 27, 43–62.
Palladwar, T.D. & Kalyankar, S.D. (1989) On two new tapeworms from Indian vertebrates. *Indian Journal of Helminthology* 41, Supplement: Fotedar Commemoration Volume, 191–196.
Pallas, P.S. (1781) Bemerkungen über die Bandwürmer in Menschen und Thieren, Neue Nordische Beytrage zur physik und geograph. Erd-und Voelkerbeschreibung. *Naturgeschichte und Oekonomie* I, 39–112.
Parra, B.E. (1953) *Perutaenia threlkeldi*, n. g., n. sp. (Cestoda: Anoplocephalidae) from *Lagidium peruanum. Journal of Parasitology* 39, 252–255.
Paspalewa, A. & Woidowa, S.M. (1969) *Raillietina macracanthos* sp. n. (Davaineidae) vom Specht *Picus viridis* (L.). *Izvestiya na Zoologicheskiya Institut s Musei* 30, 133–139.
Pavanelli, G.C. & Machado dos Santos, M.H. (1991) *Cangatiella arandasi* gen. n., sp. n. (Cestoda – Proteocephalidae), parasito de *Parauchenipterus galeatus* (Siluriformes – Auchenipteridae) do Rio Paraná, pr. *Revista Brasileira de Biologia* 7, 535–539.
Pavanelli, G.C. & Rego, A.A. (1991) Cestoídes proteocefalídeos de *Sorubim lima* (Schneider, 1801) (Pisces – Pimelodidae) do rio Paraná e reservatório de Itaipu. *Revista Brasileira de Biologia* 51, 7–12.
Pederson, E.D. & Leiby, P.D. (1969) Studies on the biology of *Monordotaenia taxidiensis*, a taeniid cestode of the badger. *Journal of Parasitology* 55, 759–765.
Perrenoud, W. (1931) Recherches anatomiques et histologiques sur quelques cestodes de sélaciens. *Revue Suisse de Zoologie* 38, 469–555.
Perrier, E. (1897) *Traité de Zoologie*, Fasc. 4, 1345–2136. Masson et Cie, editeurs, Paris.
Pintner, T. (1913) Vorarbeiten zu einer Monographie der Tetrarhynchoideen. *Sitzungsberichte der Osterreichische Akademie der Wissenschaften. Mathematik-Naturwissenschaftliche Klasse* 122, 171–253.
Pintner, T. (1928) Helminthologische Mitteilungen. I. *Zoologischer Anzeiger* 76, 318–322.
Pintner, T. (1931a) Wenigbekanntes und Unbekanntes von Rüsselbandwürmen II. *Sitzungsberichte der Osterreichische Akademie der Wissenschaften. Mathematik-Naturwissenschaftliche Klasse*, Abteilung 1, 140, 777–820.
Pintner, T. (1931b) Über fortgesetzte Tetrarhynchenuntersuchungen, II. *Anzeiger der (Kaiserlichen) Akademie der Wissenschaften. Mathematische Naturwissenschaftliche Klasse, Wien* 68, 107–117.
Poche, F. (1922) Zur Kenntnis der Amphilinidea. *Zoologischer Anzeiger* 54, 276–287.
Poche, F. (1924) Die Entstehung der Rüssel der Tetrarhynchiden. *Zoologischer Anzeiger* 59, 100–104.
Poche, F. (1926) Das System der Platodaria. *Archiv für Naturgeschichte* 91, 241–459.
Potemkina, V.A. (1938) [The diagnosis of hymenolepidiasis in geese and the biology of the pathogen.] *Trudy Vsesoyuznogo Instituta Gelmintologii* 3, 97–126. [In Russian.]

Pozdniakov, S.E. (1983) [*Pachybothrium dogieli* sp. n., gen. n. (Cestoda, Amphicotylidae, Pachybothriinae subfam. nov.), a parasite of *Mola ramsayi* (Gigl.) from the south Pacific.] *Parazitologiya* 17, 482–487. [In Russian.]

Prestwood, A.K. & Reid, W.M. (1966) *Drepanidotaenia watsoni* sp. n. (Cestoda, Hymenolepididae) from the wild turkey of Arkansas. *Journal of Parasitology* 52, 432–436.

Prokopic, J. (1967) *Hymenolepis muris-sylvatici* (Rudolphi, 1819) = *Variolepis crenata* (Goeze, 1782). *Folia Parasitologica* 14, 365–369.

Protasova, E.N. (1977) [*Principles of Cestodology. Edited by K.M. Ryzhikov. Volume VIII. Bothriocephalata – Cestodes of Fish.*] Moscow, USSR: Izdatel'stvo 'Nauka', 298 pp. [In Russian.]

Protasova, E.N. & Mordvinova, T.N. (1986) [A ptychobothriid cestode, *Tetrapapillocephalus magnus* n. g., n. sp. (Pseudophyllidea, Ptychobothriidae) from marine fish.] *Parazitologiya* 20, 313–316. [In Russian.]

Protasova, E.N. & Parukhin, A.M. (1986) [New genera and species of cestodes (Pseudophyllidea, Amphicotylidae) from marine fishes.] *Parazitologiya* 20, 278–287. [In Russian.]

Prudhoe, S. (1960) On two new parasitic worms from Brazil. In: *Libro Homenaje al Doctor Eduardo Caballero y Caballero* 415–421.

Prudhoe, S. (1969) Cestodes from fish, birds and whales. *British Australian and New Zealand Antarctic Research Expedition, 1924–1931*, Ser. B, 8, 171–193.

Purvis, G.B. (1932) Cestodes from domestic animals in Malaya, with descriptions of new species. *Veterinary Record* 12, 1407–1409.

Purvis, G.B. (1940) Corrections. *Veterinary Record* 52, 757.

Quentin, J.C. (1964) Cestodes de rongeurs de République Centrafricaine. *Cahiers de La Maboké*, 2, 117–140.

Quentin, J.C. (1967) *Atriotaenia (Ershovia) baltazardi* n. sp. (Cestoda, Linstowiidae) parasite d'un rongeur du Brésil; *Galea spixii* (Wagner). *Bulletin du Muséum National d'Histoire Naturelle, Paris* 39, 595–602.

Quentin, J.C. (1971) Cestodes *Skrjabinotaenia* de rongeurs muridés et dendromuridés de Centrafrique. Hypothése sur l'evolution des cestodes Catenotaeniinae. *Cahiers de La Maboké* 5, 57–79.

Railliet, A. (1892) Notices parasitologiques: *Taenia tenuirostris* Rud. chez l'oie domestique: remarques sur la classification des cestodes parasites des oiseaux. *Bulletin de la Société Zoologique de France* 17, 110–117.

Ramadevi, P. (1969) *Echinobothrium reesae* (Cestoda: Diphyllidea) from the sting rays of Waltair Coast. *Annales de Parasitologie Humaine et Comparée* 44, 231–239.

Ransom, B.H. (1909) The taenioid cestodes of North American birds. *Smithsonian Institute, U.S. National Museum Bulletin* 69, 1–141.

Rao, K.H. (1954) A new bothriocephalid parasite (Cestoda) from the gut of the fish *Saurida tumbil* (Bloch). *Current Science* 23, 333–334.

Rao, K.H. (1960) Studies on *Penetrocephalus ganapatii*, a new genus (Cestoda: Pseudophyllidea) from the marine teleost *Saurida tumbil* (Bloch). *Parasitology* 50, 155–163.

Rathore, G.S. & Nama, H.S. (1984) On a new cestode from peregrin [*sic*] falcon. *Indian Journal of Helminthology(n.s.)* 1, 60–63.

Rausch, R.L. (1959) Studies on the helminth fauna of Alaska. XXXV. On the identity of certain cestodes (Taeniidae) from foxes. *Proceedings of the Helminthological Society of Washington* 26, 125–131.

Rausch, R.L. (1967) On the ecology and distribution of *Echinococcus* spp. (Cestoda: Taeniidae), and characteristics of their development in the intermediate host. *Annales de Parasitologie Humaine et Comparée* 42, 19–63.

Rausch, R.L. (1970) Studies on the helminth fauna of Alaska. XLV. *Schistotaenia srivastavai* n. sp. (Cestoda: Amabiliidae) from the red-necked grebe, *Podiceps grisegena* (Boddaert). In:

Singh, K.S. & Tandan, B.K. (eds), *H.D. Srivastava Commemoration Volume*, Indian Veterinary Research Institute, Izatnagar, Uttar Pradesh, pp. 109–115.

Rausch, R.L. (1972) Cestodes in mammals: the zoogeography of some host-parasite assemblages. *Mémoires du Muséum National d'Histoire Naturelle, Nouvelle série, serie A, Zoologie* 132, 179–183.

Rausch, R.L. (1976) The genera *Paranoplocephala* Luehe, 1910 and *Anoplocephaloides* Baer, 1923. *Annales de Parasitologie Humaine et Comparée* 51, 513–562.

Rausch, R.L. (1981) Morphological and biological characteristics of *Taenia rileyi* Loewen, 1929 (Cestoda: Taeniidae). *Canadian Journal of Zoology* 59, 653–666.

Rausch, R.L. (1982) Cestodes in mammals: the zoogeography of some host-parasite assemblages. *Mémoires du Muséum National d'Histoire Naturelle, Nouvelle série, Série A, Zoologie* 132, 179–183.

Rausch, R.L. (1985) Parasitology: retrospect and prospect. *Journal of Parasitology* 71, 139–151.

Rausch, R.L. & Fay, F.H. (1988a) Postoncospheral development and cycle of *Taenia polyacantha* Leuckart, 1856 (Cestoda: Taeniidae). First part. *Annales de Parasitologie Humaine et Comparée* 63, 263–277.

Rausch, R.L. & Fay, F.H. (1988b) Postoncospheral development and cycle of *Taenia polyacantha* Leuckart, 1856 (Cestoda: Taeniidae). Second part. *Annales de Parasitologie Humaine et Comparée* 63, 334–348.

Rausch, R.L. & Hilliard, D.K. (1970) Studies on the helminth fauna of Alaska. XLIX. The occurrence of *Diphyllobothrium latum* (Linnaeus, 1758) (Cestoda: Diphyllobothriidae) in Alaska, with notes on other species. *Canadian Journal of Zoology* 48, 1201–1219.

Rausch, R.L. & Margolis, (1969) *Plicobothrium globicephalae* gen. et sp. nov. (Cestoda: Diphyllobothriidae) from the pilot whale, *Globicephala melaena* Traill, in Newfoundland waters. *Canadian Journal of Zoology* 47, 745–750.

Rausch, R.L. & Maser, C. (1977) *Monoecocestus thomasi* sp. n. (Cestoda: Anoplocephalidae) from the northern flying squirrel, *Glaucomys sabrinus* (Shaw) in Oregon. *Journal of Parasitology* 63, 793–799.

Rausch, R.L. & Rausch, V.R. (1985) *Gruitaenia gruis* sp. n. (Cestoda, Dilepididae) from the sandhill crane, *Grus canadensis* (L.), in Alaska. *Helminthologia* 22, 153–159.

Rausch, R.L. & Rausch, V.R. (1990) Reproductive anatomy and gametogenesis in *Shipleya inermis* (Cestoda: Dioecocestidae). *Annales de Parasitologie Humaine et Comparée* 65, 229–237.

Rees, G. (1956) The scolex of *Tetrabothrius affinis* (Lönnberg), a cestode from *Balaenoptera musculus* L., the blue whale. *Parasitology* 46, 425–442.

Rees, G. (1959) *Ditrachybothridium macrocephalum* gen. nov., sp. nov., a cestode from some elasmobranch fishes. *Parasitology* 49, 191–209.

Rees, G. (1969) Cestodes from Bermuda fishes and an account of *Acompsocephalum tortum* (Linton, 1905) gen. nov. from the lizard fish *Synodus intermedius* (Agassiz). *Parasitology* 59, 519–548.

Rego, A.A. (1964) *Lentiella machadosi* g. n., sp. n. e *Raillietina (R.) trinitatae* (Cameron and Reesal, 1951) parasitos do roedor (Cestoda, Cyclophyllidea). *Revista Brasileira de Biologia* 24, 211–220.

Rego, A.A. (1967a) Um novo cestódeo parasito de *Agouti paca* L. (Anoplocephalidae, Linstowiinae). *Atas da Sociedade de Biologia do Rio de Janeiro* 11, 79–80.

Rego, A.A. (1967b) Sôbre alguns cestódeos parasitos de roedores do Brasil (Cestoda, Cyclophyllidea). *Memórias do Instituto Oswaldo Cruz* 65, 1–18.

Rego, A.A. (1968) Sobre tres cestodes de aves Charadriiformes. *Memórias do Instituto Oswaldo Cruz* 66, 107–115.

Rego, A.A. (1984a) Proteocefalídeos (Cestoda) de *Phractocephalus hemilopterus*, peixe da Amazonia. *Memórias do Instituto Oswaldo Cruz* 79, 257–261.

Rego, A.A. (1984b) Proteocephalidea from Amazonian freshwater fishes: new systematic arrangement for the species described by Woodland as *Anthobothrium* (Tetraphyllidea). *Acta Amazonica* 14, 86–94.

Rego, A.A. (1987) Cestóides proteocefalídeos do Brasil. Reorganização taxonomica. *Revista Brasileira de Biologia* 47, 203–212.

Rego, A.A. (1989) Cestóides proteocefalídeos de 'cachara' *Pseudoplatystoma fasciatus* (L.) (Pisces, Pimelodidae) de Mato Grosso. *Memórias do Instituto Oswaldo Cruz* 84 Suppl. 4, 455–461.

Rego, A.A. (1990) Cestóides proteocefalídeos parasitas de pintado, *Pseudoplatystoma corruscans* (Agassiz) (Pisces, Pimelodidae). *Ciência e Cultura* 42, 997–1002.

Rego, A.A. (1991) *Nomimoscolex piraeeba* Woodland, 1934 (Cestoda, Proteocephalidea) parasite from South American freshwater fish. Synonyms. Considerations on Endorchiinae and *Endorchis*. *Memorias do Instituto Oswaldo Cruz* 86, 229–232.

Rego, A.A. & Pavanelli, G.C. (1985) *Jauella glandicephalus* gen. n., sp. n. e *Megathylacus brooksi* sp. n., cestóides proteocefalídeos patogênicos para o jau, *Paulicea luetkeni*, peixe pimelodideo. *Revista Brasileira de Biologia* 45, 643–652.

Rego, A.A. & Pavanelli, G.C. (1987) Cestóides proteocefalideos do jau, *Paulicea luetkeni*, peixe pimelodídeo do Brasil. *Revista Brasileira de Biologia* 47, 357–361.

Rego, A.A. & Pavanelli, G.C. (1990) Novas especies de cestoides proteocefalideos parasitas de peixes nao siluriformes. *Revista Brasileira de Biologia* 50, 91–101.

Reid, W.M., Ackert, J.E. & Case, A.A. (1938) Studies on the life history and biology of the fowl tapeworm *Raillietina cesticillus* (Molin). *Transactions of the American Microscopical Society* 57, 65–76.

Renaud, F. & Gabrion, C. (1988) Speciation of Cestoda. Evidence for two sibling species in the complex *Bothrimonus nylandicus* (Schneider, 1902) (Cestoda: Cyathocephalidea). *Parasitology* 97, 139–147.

Richmond, C. & Caira, J.N. (1991) Morphological investigations into *Floriceps minacanthus* (Trypanorhyncha: Dasyrhynchidae) with analysis of the systematic utility of scolex microtriches. *Systematic Parasitology* 19, 25–32.

Ride, W.D.L. *et al.* (1985) *International Code of Zoological Nomenclature, Third Edition. International Trust for Zoological Nomenclature*. British Museum (Natural History), London, 338 pp.

Rider, C.L. & Macy, R.W. (1947) Preliminary survey of the helminth parasites of muskrats in Northwestern Oregon, with description of *Hymenolepis ondatrae* n. sp. *Transactions of the American Microscopical Society* 66, 176–181.

Riggenbach, E. (1895) Beiträge zur Kenntnis der Tänien der Süsswasserfische. Vorlaufige Mitteilung. *Zentralblatt für Bakteriologie, Parasitenkunde und Infektionskrankheiten. I. Abteilung Originale* 18, 609–613.

Riser, N.W. (1955) Studies on cestode parasites of sharks and skates. *Journal of the Tennessee Academy of Sciences* 30, 265–311.

Robinson, E.S. (1959) Some new cestodes from New Zealand marine fishes. *Transactions of the Royal Society of New Zealand* 86, 381–392.

Roitman, V.A. (1965) [*Fissurobothrium unicum* n. g., n. sp. (Amphicotylidae, Marsipometrinae), a new pseudophyllidean from fish in the Amur basin.] *Trudy Gel'mintologicheskoi Laboratorii* 15, 127–131. [In Russian.]

Rosseter, T.B. (1906) On a new tapeworm *Drepanidotaenia sagitta*. *Journal of the Queckett Microscopical Club* 2, 275–278.

Ruhnke, T.R. (1994) *Paraorygmatobothrium barberi* n. g., n. sp. (Cestoda: Tetraphyllidea), with amended descriptions of two species transferred into the genus. *Systematic Parasitology* 28, 65–79.

Rybicka, K. (1966) Embryogenesis in cestodes. *Advances in Parasitology* 4, 107–186.

Rysavy, B. (1966) Nuevas especies de céstodos (Cestoda: Cyclophyllidea) de aves para Cuba. *Poeyana, Serie A*, 19, 1–22.

Rysavy, B. & Macko, J.K. (1973). Bird cestodes of Cuba. I. Cestodes of birds of the orders Podicipediformes, Pelecaniformes and Ciconiiformes. *Anales del Instituto Biologia, Universidad Nacional Autonoma, Mexico*, Seria Zoologia 42, 1–28.

Ryzhikov, K.M. & Tolkacheva, L.M. (1975) [Taxonomic review of the Amabiliidae (Cestoda, Cyclophyllidea.] *Zoologicheskii Zhurnal* 54, 498–502. [In Russian.]

Ryzhikov, K.M. & Tolkacheva, L.M. (1981) [*Principles of Cestodology. Vol. X. Acoleata – Cestodes of Birds*.] Moscow, USSR: Akademiya Nauka, 215 pp. [In Russian.]

Ryzhikov, K.M., Rysavy, B., Khokhlova, I.G., Tolkacheva, L.M. & Kornyushin, V.V. (1985). *Helminths of Fish-eating Birds of the Palaearctic Region. II. Cestoda and Acanthocephales*. Academia Publishing House, Prague, Czechoslovakia, 412 pp.

Saakova, E.O. (1958) [Two new cestode genera of the family Hymenolepididae from birds of the Danube Delta.] [*Papers on helminthology presented to Acad. K.I. Skryabin on his 80th birthday*.], pp. 310–314. [In Russian.]

Sahay, S.N. & Sahay, U. (1976a) On *Neolytocestus vitellodiscontinutus* gen. et sp. nov. from *Claris* [sic] *batrachus* a fresh water fish in Chotanagpur. *First National Convention on Indian Helminthology, Bhubaneswar, India, Abstracts* pp. 53–54.

Sahay, U. & Sahay, S.N. (1976b) On a new caryophyllaeid cestode *Moravekia chotanagpurensis* gen. et sp. nov. from a fresh water fish *Clarius* [sic] *batrachus* in Chotanagpur. *First National Convention on Indian Helminthology, Bhubaneswar, India, Abstracts* pp. 54–55.

Sakanari, J. & Moser, M. (1989). Complete life cycle of the elasmobranch cestode, *Lacistorhynchus dollfusi* Beveridge and Sakanari, 1987 (Trypanorhyncha). *Journal of Parasitology* 75, 806–808.

Sandars, D.F. (1956) *Mirandula parva* gen. et sp. nov. (Cestoda, Dilepididae) from the long-nosed bandicoot (*Perameles nasuta* Geoff.). *Journal of Helminthology* 30, 183–188.

Sandars, D.F. (1957) Redescription of some cestodes from marsupials. I. Taeniidae. *Annals of Tropical Medicine and Parasitology* 51, 317–329.

Sandeman, I.M. (1959) A contribution to the revision of the dilepid tapeworms from Charadrii-formes. Preliminary note. *Zoologischer Anzeiger* 163, 278–288.

Satpute, L.R. & Agarwal, S.M. (1974) 'Diverticulosis' of the fish duodenum infested with cestodes. *Indian Journal of Experimental Biology* 12, 373–375.

Satpute, L.R. & Agarwal, S.M. (1980) *Introvertus raipurensis* gen. nov. sp. nov. a fish cestode (Cestoda: Caryophyllidea: Lytocestidae) from Raipur, India. *Proceedings of the Indian Academy of Parasitology* 1, 17–19.

Saxena, S.K. (1969) Studies on the cestodes of the common Indian vulture, *Neophron percnopterus* (Linn.). V. On the morphology and systematic position of *Neophronia percnopteri* (Singh, 1952) n. comb. with notes on other species of the genus *Neophronia*. *Zoologischer Anzeiger* 183, 455–462.

Saxena, S.K. & Baugh, S.C. (1973) The taxonomy of the cestode genus *Aporina* with a discussion on the systematic positions of the species assigned to it. *Angewandte Parasitologie* 14, 236–245.

Sawada, I. & Tongchai, P. (1966) *Paranoplocephala indicata* n. sp. (Cestoda: Anoplocephalidae) from the Malayan tapir, *Tapirus indicus*. *Japanese Journal of Zoology* 15, 125–128.

Schell, S.C. (1955) *Schistotaenia colymba* n. sp. from the horned grebe (*Colymbus auritus* L.). *Transactions of the American Microscopical Society*, 74, 347–350.

Schiller, E.L. (1951) Studies on the helminth fauna of Alaska. VI. The parasites of the emperor goose (*Philactes canagica* L.) with the description of *Hymenolepis philactes*. *Journal of Parasitology* 37, 217–220.

Schiller, E.L. (1953) Studies on the helminth fauna of Alaska. XIV. Some cestode parasites of the Aleutian teal (*Anas crecca* L.) with the description of *Diorchis longiovum* n. sp. *Proceedings of the Helminthological Society of Washington* 20, 7–12.

Schmelz, M.O. (1941) Quelques cestodes nouveaux d'oiseaux d'Asie. *Revue Suisse de Zoologie* 48, 143–199.

Schmidt, G.D. (1969) *Dioecotaenia cancellata* (Linton, 1890) gen et comb. n. a dioecious cestode (Tetraphyllidea) from the cownosed ray, *Rhinoptera bonasus* (Mitchill) in Chesapeake Bay, with a proposal of a new family, Dioecotaeniidae. *Journal of Parasitology* 55, 271–275.

Schmidt, G.D. (1970) *How to Know the Tapeworms*. Dubuque, Iowa: W.C. Brown Co. Publishers, 266 pp.

Schmidt, G.D. (1972) Cyclophyllidean cestodes of Australian birds, with three new species. *Journal of Parasitology* 58, 1085–1094.

Schmidt, G.D. (1986) *CRC Handbook of Tapeworm Identification*. Boca Raton, Florida, USA: CRC Press, Inc., 675 pp.

Schmidt, G.D., Bauerle, S.R. & Wertheim, G. (1988) *Cloacotaeniella tringae* n. gen., n. sp. (Cestoidea: Hymenolepididae) from a redshank *Tringa totanus* (Charadriiformes) in Israel. *Journal of Parasitology* 74, 868–869.

Schmidt, G.D. & Canaris, A.G. (1992) Tapeworms of the families Progynotaeniidae Fuhrmann, 1936 and Dioecocestidae Southwell, 1930 from 8 shore-birds of South Africa. *Systematic Parasitology* 23, 37–42.

Schmidt, G.D. & Neiland, K.A. (1968) *Hymenolepis (Hym.) deblocki* sp. n. and records of other helminths from charadriiform birds. *Canadian Journal of Zoology* 46, 1037–1040.

Schulze, F.E., Kukenthal, W. & Heider, K. (1928) Nomenclator animalium generum et subgenerum. *Preussischen Akademie der Wissenschaften zu Berlin*, 1929, 2, p. 908. [Issued separately, 1928].

Schulze, F.E., Kukenthal, W. & Heider, K. (1936) Nomenclator animalium generum et subgenerum. *Preussischen Akademie der Wissenschaften zu Berlin*, 1936–1954, 5, p. 908. [Issued separately, 1936].

Schumacher, G. (1914). Cestoden aus *Centrolophus pompilus* (L.). *Zoologische Jahrbücher* 36, 149–198.

Scott, J.W. (1941) A new genus and species of tapeworm from the bighorn sheep. *Anatomical Record* 81, 65–66.

Semenova, N.S. (1972) [*Eranuides mathevossianae* n. g., n. sp. (Anoplocephalidae), a new cestode from reindeer on Taimyr Peninsula.] *Trudy Vsesoyuznogo Instituta Gelmintologii im. K.I. Skryabina* 19, 171–175. [In Russian.]

Seurat, G. (1906) Sur un cestode parasite des huitres perlières déterminant la production des perles fines aux Iles Gambier. *Comptes Rendus de l'Académie des Sciences, France* 141, 801–803.

Shen, T. (1932) Etude sur les cestodes d'oiseaux de Chine. *Annales de Parasitologie* 5, 105–128.

Shinde, G.B. (1968a) On *Circumoncobothrium ophiocephali* n. gen. n. sp. from a fresh water fish (*Ophiocephalus leucopunctatus*) in India. *Rivista di Parassitologia* 29, 111–114.

Shinde, G.B. (1968b) On a new genus *Yogeshwaria malabarica* from a lapwing in India. *Rivista di Parassitologia* 29, 257–260.

Shinde, G.B. (1972) New avian cestodes of the genus *Lapwingia* Singh, 1952 in India. *Marathwada University Journal of Science* 11, Sci.4 (Biol.Sci.), 21–29.

Shinde, G.B. (1977) On a new species of *Circumoncobothrium* Shinde, 1968 (Cestoda: Pseudophyllidea, Carus, 1963) from a fresh water fish in India. *Marathwada University Journal of Science* 16, 129–132.

Shinde, G.B. & Chincholikar, L.N. (1975) On *Hoaleshwaria marathwadensis* (Cestoda: Phyllobothriidae Braun, 1900) n. gen., n. sp. from marine fish at Ratnagiri, India. *Marathwada University Journal of Sciences (Natural Sciences)* 14, 359–361.

Shinde, G.B. & Chincholikar, L.N. (1977) A new species of *Circumoncobothrium* Shinde, 1968 (Cestoda: Pseudophyllidea Carus, 1863) from a fresh water fish in India. *Marathwada University Journal of Sciences* 16, 177–179.

Shinde, G.B. & Jadhav, B.V. (1976) New species of the genus *Circumoncobothrium* Shinde, 1968 (Cestoda: Pseudophyllidea Carus, 1863) from a fresh water fish, from Maharashtra, India. *Marathwada University Journal of Sciences* 15, 269–272.

Shinde, G.B. & Jadhav, B.V. (1984) *Trygonicephalum ratnagiriensis* gen. nov., sp. nov. (Cestoda: Lecanicephalidae) from a marine fish. *Rivista di Parassitologia* 45, 403–405.

Shinde, G.B. & Mohekar, A.D. (1983) *Myliobatibothrium alii* gen. and sp. n. (Cestoda: Tetraphyllidea) from ray *Myliobatis nieuhofii* (Müller et Henle). *Rivista di Parassitologia* 44, 247–251.

Shinde, G.B., Deshmukh, R.A. & Jadhav, B.V. (1981) On a new cestode *Echeneibothrium smitii* sp. n., from marine fish. *Rivista di Parassitologia* 42, 267–270.

Shipley, A.E. (1900) A description of the entozoa collected by Dr Willey during his sojourn in the western Pacific. *Willey Zoological Results, Part V*, Cambridge, pp. 531–536; 545–547.

Shipley, A.E. & Hornell, J. (1906) Cestode and nematode parasites from marine fishes of Ceylon. *Report to the Government of Ceylon on the Pearl Oyster Fisheries of the Gulf of Manaar* Part V, 43–96.

Siddiqi, A.H. (1960) On a new unisexual cestode, *Neodioecocestus cablei* n. g., n. sp., from the little grebe, *Podiceps ruficollis*. *Zeitschrift für Parasitenkunde* 20, 381–384.

Siddiqi, A.H. (1962) On *Echinorhynchotaenia tritesticulata* Fuhrmann, 1909 (Cestoda: Dilepididae) with remarks on its relationship. *Zeitschrift für Parasitenkunde* 21, 207–211.

Singh, J.P. & Baugh, S.C. (1984) Embryos and embryonic envelopes in eggs of two cyclophyllidean cestodes. *Angewandte Parasitologie* 25, 12–16.

Singh, K.P. (1960) Some avian cestodes from India. I. Species belonging to families Davaineidae and Biuterinidae. *Indian Journal of Helminthology* 11, 1–24.

Singh, K.S. (1962) Parasitological survey of Kumaun region. Part XV. *Ivritaenia mukteswarensis* n. g., n. sp. (Cestoda: Dipylidiinae: Dilepididae) from a woodpecker. *Indian Journal of Helminthology* 14, 127–132.

Skarbilovich, T.S. (1946) [On the helminth fauna of bats in the USSR.] *Skryabin Commemoration Volume*, Akademiya Nauk SSSR, pp. 235–244. [In Russian.]

Skryabin, K.I. (1914) Vögelcestoden aus Russisch Turkestan. *Zoologische Jahrbücher, Abteilung Systematik, Geographie und Biologie* 37, 411–492.

Skryabin, K.I. (1926) Zur Kenntnis der Systematik der Anoplocephaliden. *Trudy Gosudarstvennogo Instituta Eksperimental'noi Veterinarii* 3, 58–64.

Skryabin, K.I. & Matevosyan, E.M. (1942) [Corrections to errors and controversies in the taxonomy of the cestodes of the family Hymenolepididae.] *Doklady Akademii Nauk SSSR* No. 36, 188–191. [In Russian.]

Skryabin, K.I. & Matevosyan, E.M. (1945) [*Tapeworms – Hymenolepidids of Domestic and Game Birds*.] Selkhozgiz, Moscow, 488 pp. [In Russian.]

Skryabin, K.I. & Matevosyan, E.M. (1948) [Hymenolepididae from mammals.] *Trudy Gel'mintologicheskii Laboratorii* 1, 15–92. [In Russian.]

Slais, J. & Machnická, B. (1976) Appearance of a temporary hook anlage in the early development of *Taenia saginata* Goeze, 1782. *Zoologischer Anzeiger* 196, 85–92.

Smith, C.F. (1954) Four new species of cestodes of rodents from the High Plains, Central and Southern Rockies and notes on *Catenotaenia dendritica*. *Journal of Parasitology* 40, 245–254.

Soldatova, A.P. (1944) A contribution to the study of the development cycle in the cestode *Mesocestoides lineatus* (Goeze, 1782), parasitic in carnivorous mammals. *Comptes Rendus (Doklady) de l'Académie des Sciences de l'URSS* 45, 310–312.

Sołtys, A. (1952) [The helminths of common shrew (*Sorex araneus* L.) of the Białowieza National Park (Poland).] *Annales Universitatis Maria Curie Skłodowska*, 6, 165–209. [In Polish.]

Sołtys, A. (1954) Helmintofauna ryjówkowatych (Soricidae) Białowieskiego Parku Narodowego. *Acta Parasitologica Polonica* 1, 353–402.

Southwell, T. (1911) Description of nine new species of cestode parasites, including two new genera from marine fishes of Ceylon. *Ceylon Marine Biological Reports* 1, 216–225, 230.

Southwell, T. (1922) Cestodes from Indian birds with a note on *Ligula intestinalis*. *Annals of Tropical Medicine and Parasitology* 16, 355–382.

Southwell, T. (1925) A monograph on the Tetraphyllidea with notes on related cestodes. *Memoirs of the Liverpool School of Tropical Medicine (New Series)* No. 2, 363 pp.

Southwell, T. (1926) Cestodes in the collection of the Liverpool School of Tropical Medicine. *Annals of Tropical Medicine and Parasitology* 20, 221–228.

Southwell, T. (1927) On a collection of cestodes from marine fishes of Ceylon. *Annals of Tropical Medicine and Parasitology* 21, 351–373.

Southwell, T. (1929a) On the classification of the Cestoda. *Ceylon Journal of Science* 15, 49–72.

Southwell, T. (1929b) A monograph on cestodes of the order Trypanorhyncha from Ceylon and India, Part I. *Ceylon Journal of Science* 15, 159–312.

Southwell, T. (1930a) *The Fauna of British India including Ceylon and Burma. Cestoda.* Vol. 1. Taylor & Francis, London, xxxi + 391 pp.

Southwell, T. (1930b) *The Fauna of British India, including Ceylon and Burma. Cestoda.* Vol. 2. Taylor & Francis, London, ix + 262 pp.

Southwell, T. & Hilmy, J.S. (1929) *Jardugia paradoxa*, a new genus and species of cestode with some notes on the families Acoleidae and Diploposthidae. *Annals of Tropical Medicine and Parasitology* 23, 397–406.

Spasskaya, L.P. (1954) [Helminth fauna of birds from the Barabinsk Steppe. (From material collected by the 257th Soviet Helminthological Expedition).] [Abstract of thesis.] *Trudy Gel'mintologicheskoi Laboratorii* 7, 390–392. [In Russian.]

Spasskaya, L.P. (1959) [Cestodes of birds from the autonomous Territory of Tuva, USSR. I. Dilepididae.] *Acta Veterinaria Academiae Scientiarum Hungaricae* 9, 77–100. [In Russian.]

Spasskaya, L.P. (1961a) [A new genus of hymenolepidids, *Biglandatrium* n. g., (Cestoda: Hymenolepididae.] *Helminthologia* 3, 340–345. [In Russian.]

Spasskaya, L.P. (1961b) [Species of the family Idiogenidae (Cestoda) from the Tuva region.] *Acta Veterinaria Academiae Scientiarum Hungaricae* 11, 423–440.

Spasskaya, L.P. (1962) [Cestodes of birds in Tuva. V. Species from galliforms and some raptor birds.] *Acta Veterinaria Academiae Scientiarum Hungaricae* 12, 423–444. [In Russian.]

Spasskaya, L.P. (1964) [Cestodes of birds in Tuva. VII. Hymenolepididae.] *Acta Veterinaria Academiae Scientiarum Hungaricae* 14, 13–34. [In Russian.]

Spasskaya, L.P. (1966) [*Cestodes of Birds of the USSR. Hymenolepididae.*] Moscow: Izdatel'stvo 'Nauka', 698 pp. [In Russian.]

Spasskaya, L.P. & Spasskii, A.A. (1970) [Species composition of the genus *Polycercus* (Cestoda: Dilepididae).] *Parazity Zhivotnykh i Rastenii* 5, 38–44. [In Russian.]

Spasskaya, L.P. & Spasskii, A.A. (1971) [*Cestodes of Birds in Tuva.*] Kishinev: Izdatel'stvo 'Shtiintsa'. 252 pp. [In Russian.]

Spasskaya, L.P. & Spasskii, A.A. (1973) [*Dictymetra belopolskajae* sp. n. (Cestoda: Dilepididae) – a new cestode of birds.] *Izvestiya Akademii Nauk Moldavskoi SSR* No. 4, 63–68. [In Russian.]

Spasskaya, L.P. & Spasskii, A.A. (1977) [*Cestodes of Birds in the USSR. Dilepididae of Terrestrial Birds.*] Izdatel'stvo Nauka, Moscow, 299 pp. [In Russian.]

Spasskaya, L.P. & Spasskii, A.A. (1978) [*Cestodes of Birds in the USSR. Dilepididae of Limnophilous Birds.*] Izdatel'stvo Nauka, Moscow, 314 pp. [In Russian.]

Spasskaya, L.P., Spasskii, A.A. & Borgarenko, L.F. (1971). [*Diporotaenia colymbi* n. g., n. sp – a new genus and species of amabiliid cestodes from grebes.] *Izvestiya Akademii Nauk Moldavskoi SSR, Seriya Biologicheskikh i Khimicheskikh Nauk* No. 6, 49–53. [In Russian.]

Spasskii, A.A. (1947) [The position of the genus *Echinorhynchotaenia* Fuhrmann, 1909 in the Cestoda.] *Doklady Akademii Nauk SSSR* 58, 513–515. [In Russian.]

Spasskii, A.A. (1948a) [Change of function of the attachment apparatus in the cestode *Insinuaro-*

taenia schikhobalovi gen. et sp. nov.] *Doklady Akademii Nauk SSSR* 59, 825–827. [In Russian.]

Spasskii, A.A. (1948b) [*Mathevolepis petrotschenkoi* nov. gen. nov. sp. – a new cestode with a uterine duct for the development of eggs.] *Doklady Akademii Nauk SSSR* 59, 1513–1515. [In Russian.]

Spasskii, A.A. (1951) *Essentials of Cestodology. Vol. 1. Anoplocephalata.* Moscow; Akademiya Nauk SSSR. [English translation, Israel Program for Scientific Translations, Jerusalem, 1961, 783 pp.]

Spasskii, A.A. (1954) [Classification of Hymenolepididae from mammals.] *Trudy Gel'mintologicheskoi Laboratorii* 7, 120–167. [In Russian.]

Spasskii, A.A. (1955) [On the independent formation of the character of a net-like uterus in representatives of different groups of hymenolepidids (Cestoda).] *Zoologicheskii Zhurnal* 34, 1012–1018. [In Russian.]

Spasskii, A.A. (1958) [Short analysis of the classification of cestodes.] *Ceskoslovenskă Parasitologie* 5, 163–171. [In Russian.]

Spasskii, A.A. (1959a) [A more precise definition of the types of relative positions of the genitalia in hymenolepidids.] *Zoologicheskii Zhurnal* 38, 31–37. [In Russian.]

Spasskii, A.A. (1959b) [The polyphyletic origin of the hymenolepidid genus *Oligorchis* Fuhrmann.] *Trudy Gel'mintologicheskoi Laboratorii* 19, 298–310. [In Russian.]

Spasskii, A.A. (1959c) [On the phylogenetic relations of the subfamily Metadilepidinae nov. subfam. (Cestoda: Cyclophyllidea).] *Helminthologia* 1, 155–158. [In Russian.]

Spasskii, A.A. (1961) Breve revisione di Hymenolepididae. Parte seconda. *Parassitologia* 3, 179–198.

Spasskii, A.A. (1962a) Genera composition of the family Hymenolepididae. Parasitic worms and aquatic conditions. Symposium Prague 29.X.–3.XI. Praha, Part I,II, 159–198.

Spasskii, A.A. (1962b) [Secondary polymerization of gonads in hymenolepidids.] *Doklady Akademii Nauk SSSR* 142, 734–736. [In Russian.]

Spasskii, A.A. (1963) [*Principles of Cestodology. Vol. II. Hymenolepididae – Parasites of Wild and Domestic Birds. Part I.*] Moscow: Izdatel'stvo 'Nauka', 418 pp. [In Russian.]

Spasskii, A.A. (1965a) [Two new hymenolepidid genera from birds: *Ortleppolepis* n. g. and *Satyolepis* n. g. (Cestoda, Cyclophyllidea).] *Trudy Gelmintologicheskoi Laboratorii* 15, 145–150. [In Russian.]

Spasskii, A.A. (1965b) [Revision of the genus *Dilepis* (Cestoda, Cyclophyllidea).] *Parazity Zhivotnykh e Rastenii* 1, 65–83. [In Russian.]

Spasskii, A.A. (1966) [Phylogenetic analysis of cestodes of the genus *Lateriporus* (Cyclophyllidea).] *Parazity Zhivotnykh i Rastenii*, No. 2, 50–63. [In Russian.]

Spasskii, A.A. (1967) [Two new genera of hymenolepidids of birds.] *Izvestiya Akademii Nauk Moldavskoi SSR* 1, 11–16. [In Russian.]

Spasskii, A.A. (1969a) [A comparative ecological and morphological analysis of cestodes of the genus *Choanotaenia*.] In: Spasskii, A.A. (ed.), [*Parasites of Vertebrates*.] Kishinev: Izdatel'stvo 'Kartya Moldovenyaski', pp. 3–30. [In Russian.]

Spasskii, A.A. (1969b) [*Quentinia* n. g. (Anoplocephalidae), a new cestode genus from mammals.] [*All-Union Conference (7th) on the natural focal occurrence of diseases and general problems in animal parasitology. Alma-Ata, 1969. General helminthology*.] pp. 100–103. [In Russian.]

Spasskii, A.A. (1973) [New genera of cyclophyllidean cestodes.] *Parazity Zhivotnykh i Rastenii* No. 9, 38–48. [In Russian.]

Spasskii, A.A. (1975) [The species composition of the genera *Oligorchis* and *Wardium* and remarks on the systematics of hymenolepidids from Charadriiformes.] *Parazity Zhivotnykh i Rastenii*, No. 11, 3–26. [In Russian.]

Spasskii, A.A. (1977) [Identity of the genera *Hexaparuterina* and *Metroliasthes* (Cestoda, Cyclophyllidea) and remarks on the systematics of the Paruterinidae.] *Izvestiya Akademii*

Nauk Moldavskoi SSR. Biologicheskikh i Khimicheskikh Nauk No. 5, 65–70. [In Russian.]

Spasskii, A.A. (1978) [Determination of species series in the cyclophyllidean cestodes.] *Izvestiya Akademii Nauk Moldavskoi SSR (ser. Biol. i Khim)*, No. 5, 72–77. [In Russian.]

Spasskii, A.A. (1979) [On the alien taxa in the family Davaineidae Braun, 1900 (Cestoda, Cyclophyllidea).] *Izvestiya Akademii Nauk Moldavskoi SSR, Biologicheskikh i Khimicheskikh Nauk*, No. 1, 67–70. [In Russian.]

Spasskii, A.A. (1985) [The species composition and life-cycle of cestodes of the genus *Himantaurus* (Cestoda, Dilepididae).] *Vestnik Zoologii*, No. 2, 7–9. [In Russian.]

Spasskii, A.A. (1987a) [The recent genera of Linstowiidae.] In: Spasskii, A.A. (ed.) [*Parasites and Parasitocoenoses of Animals and Plants of the Dnestrovsk-Prutsk Interfluvium*], AN Moldavsk SSR, Shtiintsa: Kishinev, pp. 3–27. [In Russian.]

Spasskii, A.A. (1987b) [A new linstowiid genus (Cestoda: Cyclophyllidea).] *Izvestiya Akademii Nauk Moldavskoi SSR, Biologicheskikh i Khimicheskikh Nauk* No. 5, 36–37. [In Russian.]

Spasskii, A.A. (1987c) [A new genus of hymenolepidid (Cestoda, Cyclophyllidea).] *Izvestiya Akademii Nauk Moldavskoi SSR. Biologicheskikh i Khimicheskikh Nauk* No. 1, 41–44. [In Russian.]

Spasskii, A.A. (1988a) [A new anoplocephalid genus *Beveridgia*, gen. n. Cestoda, Cyclophyllidea]. *Izvestiya Akademii Nauk Moldavskoi SSR, Biologicheskikh i Khimicheskikh Nauk* 1988, No. 4, 68–69. [In Russian.]

Spasskii, A.A. (1988b) [A new skrjabinochorid genus (Cestoda: Cyclophyllidea]. *Izvestiya Akademii Nauk Moldavskoi SSR, Biologicheskikh i Khimicheskikh Nauk* No. 5, 42–44. [In Russian.]

Spasskii, A.A. (1989) [Systematic position of five genera of higher cestodes of warm-blooded vertebrates in South-East Asia.] *Izvestiya Akademii Nauk Moldavskoi SSR Biologicheskikh i Khimicheskikh Nauk* No. 2, 41–46. [In Russian.]

Spasskii, A.A. (1990) [New subfamily of taeniid tapeworms (Cestoda: Cyclophyllidea).] *Izvestiya Akademii Nauk Moldavskoi SSR. Seriya Biologicheskie i Khimicheskie Nauk*, No. 1, p. 73. [In Russian.]

Spasskii, A.A. (1991) [A brief essay on the systematics of the Paruterinidae.] *Izvestiya Akademii Nauk SSR Moldavskoi SSR, Biologicheskikh i Khimicheskikh Nauki* No. 2, 43–52. [In Russian.]

Spasskii, A.A. (1992) [Classification of cestodes.] *Selskokhozyaistvennaya Biologiya* 6, 107–114. [In Russian.]

Spasskii, A.A. & Borgarenko, L.F. (1960) [*Skrjabinoporus* gen. nov. (Cestoda: Cyclophyllidea).] *Izvestiya Akademii Nauk Tadzhikskoi SSR. Otdelenie Sel'skokhozyaistvenykh Khimicheskikh Nauk* No. 2, 113–118. [In Russian.]

Spasskii, A.A. & Kornyushin, V.V. (1971) [Morphological evolution of gonads in hymenolepidid and dilepidid cestodes.] *Doklady Akademii Nauk SSSR* 198, 1232–1234. [In Russian.]

Spasskii, A.A. & Kornyushin, V.V. (1977) [Revision of classification of Ophryocotylidae (Cestoda, Davaineoidea).] *Vestnik Zoologii* No. 5, 34–42. [In Russian.]

Spasskii, A.A. & Oshmarin, P.G. (1954) [A new genus of hymenolepidid – *Staphylepis* n. g. from domestic and wild fowls.] *Trudy Gorkovskogo Gosudarstvennogo Pedagogicheskogo Instituta im. Gorkovo* 7, 181–184. [In Russian.]

Spasskii, A.A. & Poznakomkin, S.D. (1966) [Genus *Fuhrmannolepis* (Cestoda: Dilepididae).] *Parazity Zhivotnykh i Rastenii* 2, 87–92. [In Russian.]

Spasskii, A.A. & Reznik, V.N. (1963) [A revision of the genus *Drepanidotaenia* (Cestoda, Hymenolepididae).] In: *Parazity Zhivotnykh i Rastenii Moldavii* 84–90. [In Russian.]

Spasskii, A.A. & Spasskaya, L.P. (1954) [Construction of a classification of hymenolepidids parasitic in birds.] *Trudy Gelmintologicheskoi Laboratorii* 7, 55–119. [In Russian.]

Spasskii, A.A. & Spasskaya, L.P. (1959) Cestodes of *Riparia riparia* and *Hirundo rustica* in phylogenetically distant, but oecologically related birds. *Helminthologia* 1, 85–98.

Spasskii, A.A. & Spasskaya, L.P. (1960) [Cestodes of passeriform birds parasitic in rodents.] *Trudy Gelmintologicheskoi Laboratorii* 10, 212–216. [In Russian.]

Spasskii, A.A. & Spasskaya, L.P. (1965) [Revision of the genus *Paricterotaenia* (Cestoda, Dilepididae).] *Parazity Zhivotnykh i Rastenii* 1, 84–103. [In Russian.]

Spasskii, A.A. & Spasskaya, L.P. (1966) [Morphological and ecological analysis of the genus *Amoebotaenia* (Cestoda: Dilepididae).] [In: Parasites of Animals and Plants.] *Kartya Moldovenyaske* 2, 75–86. [In Russian.]

Spasskii, A.A. & Spasskaya, L.P. (1968) [Insufficient morphological criteria of the genus *Himantocestus* (Gyrocoeliinae: Diploposthidae).] *Helminthologia* 8–9, 531–536. [In Russian.]

Spasskii, A.A. & Spasskaya, L.P. (1972) [The genera *Monorcholepis* Oshmarin, 1961 and *Chimaerolepis* n. g., and the subfamily Aploparaksinae (Cestoda: Hymenolepididae).] *Parazity Zhivotnykh i Rastenii* No. 8, 69–75. [In Russian.]

Spasskii, A.A. & Spasskaya, L.P. (1976) [The systematics of amabiliids and davaineids (Cestoda: Amabiliidae, Davaineidae).] *Parazity Teplokrovnykh Zhivotnykh Moldavii*. Kishinev, USSR: Izdatel'stvo 'Shtiintsa', 3–30. [In Russian.]

Spasskii, A.A. & Spasskaya, L.P. (1977) [Brief results of the phylogenetic analysis of two dilepidid tapeworm tribes, Dilepidini and Laterotaeniini.] In: Spasskii, A.A., Shumilo, R.P. & Kharshun, A.I. (eds) *Ekto-i Endoparazity Zhivotnykh Moldavii*. Kishinev: Izdatel'stvo 'Shtiintsa', 3–30. [In Russian.]

Spasskii, A.A. & Tolkacheva, L.M. (1965) [*Anserilepis* n. g. (Cyclophyllidea, Hymenolepididae). A new cestode genus from anseriform birds.] *Trudy Gelmintologicheskoi Laboratorii* 15, 151–155. [In Russian.]

Spasskii, A.A. & Yurpalova, N.M. (1964) [*Orientolepis* n. g. (Cestoda: Hymenolepididae) from domestic chicken.] *Trudy Gelmintologicheskoi Laboratorii* 14, 197–200. [In Russian.]

Spasskii, A.A. & Yurpalova, N.M. (1965) [*Echinatrium* n. g. a new hymenolepidid genus from Anseriformes in Chukotka.] *Parazity Zhivotnykh i Rastenii* No. 1, 104–112. [In Russian.]

Spasskii, A.A. & Yurpalova, N.M. (1968) [Cestodes of the genus *Nadejdolepis* (Hymenolepididae) from Charadriiformes of Chukotka.] *Parazitologiya* 2, 249–257. [In Russian.]

Specht, D. & Voge, M. (1965) Asexual multiplication of *Mesocestoides* tetrathyridia in laboratory animals. *Journal of Parasitology* 51, 268–272.

Srivastav [*sic*], A.K. & Capoor, V.N. (1982) A new cestode, *Ophryocotylus dinopii* gen. n. et sp. n. (Cyclophyllidea: Davaineidae) from *Dinopium benghalense* (L.). *Helminthologia* 19, 129–134.

Srivastava, A.K. & Capoor, V.N. (1977) On a new cestode, *Ophryocotylus dinopilium* n. g., n. sp., (family Davaineidae Fuhrmann, 1907) from the goldenbacked woodpeckers, *Dinopium benghalense* from Amethi, Sultanpur (U.P.). *Annual Number. National Academy of Sciences, India*, 47th Annual Session, 191.

Srivastava, B.K. & Srivastav [sic], A.K. (1989) First record of a new tapeworm, *Doublesetina fotedari* n. g. from domestic fowl, *Gallus gallus* (Linnaeus). *Uttar Pradesh Journal of Zoology* 9, 25–28.

Stiles, C.W. & Orleman, M. (1926) La nomenclature des genres de cestodes *Raillietina, Ransomia* et *Johnstonia. Annales de Parasitologie Humaine et Comparée* 4, 65–67.

Stock, T.M. & Holmes, J.C. (1981) *Pararetinometra lateralacantha* gen. et sp. nov. (Cestoda: Hymenolepididae) in grebes (Podicipedidae) from Alberta. *Canadian Journal of Zoology* 59, 2319–2321.

Stradowski, M. (1989) Correlations between testes position and genital pore placement in *Hymenolepis diminuta. Acta Parasitologica Polonica* 34, 145–153.

Stradowski, M. (1993) Variability of the position of organs in *Hymenolepis diminuta* tapeworms at different ages. *Acta Parasitologica* 38, 23–28.

Stunkard, H.W. (1961) *Cycloskrjabinia taborensis* (Loewen, 1934), a cestode from the red bat, *Lasiurus borealis* (Mueller, 1776), and a review of the family Anoplocephalidae. *Journal of Parasitology* 47, 847–855.

Stunkard, H.W. (1969) *Panceriella* nom. n., for *Pancerina* Fuhrmann, 1899, preoccupied by *Pancerina* Chun, 1879, and systematic relations of the genus. *Journal of Parasitology* 55, 1162–1168.

Subhapradha, C.K. (1951) On the genus *Polypocephalus* Braun, 1878 (Cestoda), together with descriptions of six new species from Madras. *Proceedings of the Zoological Society of London* 121, 205–235.

Subhapradha, C.K. (1955) Two new bothriocephalids from the marine fish *Saurida tumbil* (Bloch). *Proceedings of the Indian Academy of Sciences* 41B, 20–30.

Subhapradha, C.K. (1957) Cestode parasites of fishes of Madras Coast. *Indian Journal of Helminthology* 7, 41–132.

Sugimoto, M. (1934) Morphological studies on the avian cestodes from Formosa. *Department of Agriculture Government Research Institute, Taihoku, Formosa, Japan, Report No. 64* 52 pp.

Sulgostowska, T. (1977) Redescription of the species *Diploposthe laevis* (Bloch, 1782) and *D. bifaria* (Siebold in Creplin, 1846) and revision of the genus (Cestoda: Hymenolepididae). *Acta Parasitologica Polonica* 24, 231–248.

Szidat, L. (1937) Über einige neue Caryophyllaeiden aus ostpreussischen Fischen. *Zeitschrift für Parasitenkunde* 9, 771–786.

Szpotanska, I. (1931) Note sur une espèce du genre *Liga* Weinland. *Annales Musei Zoologici Polonici* 9, 237–246.

Szymanski, M. (1905) Przyczynek do helmintologii. Rozprawy Wydziału matematyczno-przyrodniczego Akademii Umiejetnosci w Krakowie, 44, Serya B, 1–4.

Tadros, G. (1966) On the classification of the family Bothriocephalidae Blanchard, 1849 (cestode). *Journal of Veterinary Science of the United Arab Republic* 3, 39–43.

Tadros, G. (1968) A re-description of *Polyoncobothrium clarias* (Woodland, 1925) Meggitt, 1930 (Bothriocephalidae: cestode) with a brief review of the genus *Polyoncobothrium* Diesing, 1854 and the identity of the genera *Tetracampos* Dollfus, 1935, and *Oncobothriocephalus* Yamaguti, 1959. *Journal of Veterinary Science of the United Arab Republic* 5, 53–84.

Temirova, S.I. & Skryabin, A.S. (1978) [Suborder Tetrabothriata (Ariola, 1899) Skryabin, 1940.] In: [*Principles of Cestodology. Vol IX. Tetrabothriata and Mesocestoidata, Cestodes of Birds and Mammals.*] Moscow, USSR: Izdatel'stvo Akademii Nauk SSR, pp. 7–117. [In Russian.]

Tenora, F. (1976) Tapeworms of the family Anoplocephalidae Cholodkowsky, 1902. Evolutionary implications. *Acta Scientiarum Academiae Scientiarum Bohemoslovacae Brno, Nova Series* 10 (5), 1–37.

Tenora, F. & Mas-Coma, S. (1978) Records of *Gallegoides arfaai* (Mobedi and Ghadirian, 1977) n. comb. (Cestoda: Anoplocephalidae) in *Apodemus sylvaticus* L. from Western Europe. Proposition of *Gallegoides* nov. gen. *Saügetierkundliche Mitteilungen* 26, 222–226.

Tenora, F., Mas-Coma, S., Murai, E. & Feliu, C. (1980) The system of the suborder Catenotaeniata Spassky, 1963. *Parasitologia Hungarica* 13, 39–57.

Tenora, F., Murai, E. & Vaucher, C. (1984) On Anoplocephalidae (Cestoda), parasitising Rodentia and Lagomorpha in Europe. *Parasitologia Hungarica* 17, 51–57.

Tenora, F., Murai, E. & Vaucher, C. (1986) On *Andrya* Railliet, 1893 and *Paranoplocephala* Luehe, 1910 (Cestoda, Monieziinae). *Parasitologia Hungarica* 19, 43–75.

Thenius, E. (1972) *Grundzüge der Verbreitungs geschichte der Säugetiere*. Jena: Gustav Fischer Verlag, 345 pp.

Thenius, E. (1979) *Die Evolution der Säugetiere. Eine Übersicht über Ergebnisse und Probleme.* Stuttgart & New York: Gustav Fisher Verlag, 295 pp.

Thomas, L.J. (1930) Notes on the life history of *Haplobothrium globuliforme* Cooper, a tapeworm of *Amia calva* L. *Journal of Parasitology* 16, 140–145.

Thomas, L.P. (1983) Fine structure of the tentacles and associated microanatomy of *Haplobothrium globuliforme* (Cestoda: Pseudophyllidea). *Journal of Parasitology* 69, 719–730.

Thompson, D.W. (1942) *On Growth and Form.* Vol. II. Cambridge University Press, Cambridge, 1116 pp.

Thornely, L.R. (1904) Report on the Hydroida collected by Professor Herdman, at Ceylon, in 1902. *Ceylon Pearl Oyster Fisheries and Marine Biology*, Part II, Supplementary Reports VII, p. 108, 109, 112.

Tolkacheva, L.M. (1979) [*Echinoshipleya semipalmati* n. g., n. sp., a new dioecious cestode from Charadriiformes.] *Trudy Gel'mintologicheskoi Laboratorii* 29, 149–154. [In Russian.]

Toth, L.M. (1990) A critical revision of *Oncomegas* Dollfus, 1929 (Cestoda: Trypanorhyncha), with addition of two new species, and comments on its classification within the Eutetrarhynchidae Guiart, 1927. Master's thesis, Southeastern Massachusetts University, N. Dartmouth, Massachusetts, 89 pp.

Toth, L.M., Campbell, R.A. & Schmidt, G D. (1992). A revision of *Oncomegas* Dollfus, 1929 (Cestoda: Trypanorhyncha), the description of two new species and comments on its classification. *Systematic Parasitology* 22, 167–187.

Tubangui, M.A. (1938) Pseudophyllidean cestodes occurring in the Philippines. *Livro Jubilar do Professor Lauro Travassos, 1938*, Brasil, Rio de Janeiro, pp. 489–494 + 4 pl.

Tubangui, M.A. & Masiluñgan, V.A. (1937) Tapeworm parasites of Philippine birds. *Philippine Journal of Science* 62, 409–438.

Ukoli, F.M.A. (1965) Three cestodes from the families Diploposthidae Poche, 1926, Dioecocestidae Southwell, 1930 and Progynotaeniidae Fuhrmann, 1933 found in the black-winged stilt, *Himantopus himantopus* (Linn., 1758) in Ghana. *Journal of Helminthology* 39, 383–398.

Ukoli, F.M.A. (1972) Occurrence, morphology and systematics of caryophyllaeid cestodes of the genus *Wenyonia* Woodland, 1923 from fishes in River Niger, Nigeria. *Journal of the West African Science Association* 17, 49–67.

Ulmer, M.J. & James, H.A. (1976) *Nematotaenoides ranae* gen. et sp. n. (Cyclophyllidea: Nematotaeniidae), from the leopard frog (*Rana pipiens*) in Iowa. *Proceedings of the Helminthological Society of Washington* 43, 185–191.

Van der Land, J. & Dienske, H. (1968) Two new species of *Gyrocotyle* (Monogenea) from chimaerids (Holocephali). *Zoologische Mededelingen* 43, 97–105.

Vaucher, C. (1971) Les cestodes parasites des Soricidae d'Europe. Etude anatomique, révision taxonomique et biologie. *Revue Suisse de Zoologie* 78, 1–113.

Vaucher, C. (1986a) Helminthes parasites du Paraguay XI: Hymenolepididae (Cestoda) parasites de chiroptères Molossidae, avec description de deux espèces nouvelles. *Revue Suisse de Zoologie* 92, 393–407.

Vaucher, C. (1986b) Cestodes parasites de chiroptères en Amérique du Sud II: *Hymenolepis mazanensis* n. sp., chez *Saccopteryx bilineata* (Temm.) et *Rhynchonycteris naso* (Wied. Neuwied) en Amazonie péruvienne. *Revue Suisse de Zoologie* 93, 817–821.

Vaucher, C. (1992) Revision of the genus *Vampirolepis* Spasskii, 1954 (Cestoda: Hymenolepididae). *Memórias Instituto Oswaldo Cruz* 87, Suppl. 1, 299–304.

Vaucher, C. & Durette-Desset, M.C. (1978) Nouvelles données sur les helminthes parasites de la musaraigne *Blarina brevicauda* (Say). *Revue Suisse de Zoologie* 85, 361–378.

Vaucher, C. & Quentin, J.C. (1975) Présence du cysticercoide de *Hymenolepis myoxi* (Rud., 1819) chez la puce du lërot et redescription du ver adulte. *Bulletin de la Société Neuchâteloise de Sciences Naturelles* 98, 27–34.

Verma, S.C. (1928) Some cestodes from Indian fishes including four new species of Tetraphyllidea and revised keys to the genera *Acanthobothrium* and *Gangesia*. *Allahabad University Studies* 4, 119–176.

Verster, A. (1969) A taxonomic revision of the genus *Taenia* Linnaeus, 1758 s. str. *Onderstepoort Journal of Veterinary Research* 36, 3–58.

Voge, M. (1952) *Mesogyna hepatica* n. g., n. sp., (Cestoda: Cyclophyllidea) from the kitfox, *Vulpes macrotis*. *Transactions of the American Microscopical Society* 71, 350–354.

Voge, M. (1955) *Hymenolepis virilis* n. sp., a cestode from the shrew *Sorex trowbridgei* in California. *Journal of Parasitology* 41, 270–272.

Voge, M. (1969) Systematics of cestodes – present and future, In: Schmidt, G.D. (ed.) *Problems in Systematics of Parasites*. University Park Press, Baltimore, pp. 49–72.

Voge, M. & Read, C.P. (1953) *Diplophallus andinus* n. sp. and *Monoecocestus rheiphilus* n. sp., avian cestodes from the high Andes. *Journal of Parasitology* 39, 558–567.

Voge, M. & Read, C.P. (1954) A description of *Parafimbriaria websteri* n. g. n. sp., a cestode from grebes, and notes on three species of *Hymenolepis*. *Journal of Parasitology* 40, 564–570.

Wagener, G.R. (1854) Die Entwicklung der Cestoden. *Nova Acta Acad. Nat. Curios* 24, 21–91.

Wang, P.Q. (1984) [Notes on some cestodes of fishes in Fujian Province, with a list of fish cestodes recorded in China.] *Wuyi Science Journal* 4, 71–83. [In Chinese, English summary.]

Wardle, R.A. & McLeod, J.A. (1952) *The Zoology of Tapeworms*. University of Minnesota Press, Minneapolis, USA, xxiv + 780 pp.

Wardle, R.A., McLeod, J.A. & Radinovsky, S. (1974) *Advances in the Zoology of Tapeworms, 1950–1970*. University of Minnesota Press, Minneapolis, USA, xvi + 275 pp.

Wason, A. & Johnson, S. (1977) A new genus of hymenolepid cestodes from the Indian gerbil, *Tatera indica*. *Journal of Helminthology* 51, 309–312

Webster, J.D. (1943) A revision of the Fimbriariinae (Cestoda, Hymenolepididae). *Transactions of the American Microscopical Society* 62, 390–397.

Webster, J.D. (1947) Studies on the genus *Hymenolepis*, with descriptions of three new species. *Journal of Parasitology* 33, 99–106.

Weekes, P.J. (1981) *Pulluterina nestoris*, an anoplocephalid cestode parasite of the kea, *Nestor notabilis*. *New Zealand Journal of Zoology* 8, 387–390.

Weekes, P.J. (1986) Growth and development of *Amurotaenia decidua* Hine, 1977 in a copepod. In: Howell, K.J. (ed.) *Parasitology – Quo Vadit? Handbook* (Sixth International Congress of Parasitology, Brisbane), Canberra: Australian Academy of Science, p. 255.

Weinland, D.F. (1858) Human cestoides. An essay on the tapeworms of man giving a full account of the nature, organization, and embryonic development; the pathological symptoms they produce, and the remedies which have proved successful in modern practice, to which is added an appendix, containing a catalogue of all species of helminthes hitherto found in man. Cambridge, x + 93 pp.

Wertheim, G., Schmidt, G.D. & Greenberg, Z. (1986) *Witenbergitaenia sinaica* gen. n., sp. n. (Anoplocephalidae) and other cestodes from small mammals in Israel and in the Sinai Peninsula. *Bulletin du Muséum National d'Histoire Naturelle, Paris, 4 Sér.*, 8, 543–550.

Willers, W.B. & Olsen, O.W. (1969) The morphology of *Gastrotaenia cygni* Wolffhügel, 1938 (Cestoda: Aporidea) with a redescription of the genus. *Journal of Parasitology* 55, 1004–1011.

Williams, A.D. & Campbell, R.A. (1978) *Duplicibothrium minutum* gen. et sp. n. (Cestoda: Tetraphyllidea) from the cownose ray *Rhinoptera bonasus* (Mitchill, 1815). *Journal of Parasitology* 64, 835–837.

Williams, A.D. & Campbell, R.A. (1980) *Echinobothrium bonasum* sp. n., a new cestode from the cownose ray, *Rhinoptera bonasus* (Mitchill, 1815) in the western North Atlantic. *Journal of Parasitology* 66, 1036–1038.

Williams, D.D. (1977) A key to caryophyllaeid cestodes of Wisconsin. *Iowa State Journal of Research* 51, 471–477.

Williams, D.D. (1978) A key to caryophyllidean cestodes of Iowa fishes. *Iowa State Journal of Research* 52, 401–409.

Williams, D.D. & Sutherland, D.R. (1981). *Khawia sinensis* (Caryophyllidea: Lytocestidae) from *Cyprinus carpio* in North America. *Proceedings of the Helminthological Society of Washington* 48, 253–255.

Williams, E.H., Jr (1978) *Diefluvium unipapillatum* n. g., n. sp. (Cestoda: Caryophyllaeidae) from the river redhorse *Moxostoma carinatum* (Cope) from the southeastern United States. *Transactions of the American Microscopical Society* 97, 601–605.

Williams, H.H. (1964) Some new and little known cestodes from Australian elasmobranchs with a brief discussion on their possible use in problems of host taxonomy. *Parasitology* 54, 737–748.

Williams, H.H. (1966a) The ecology, functional morphology and taxonomy of *Echeneibothrium* Beneden, 1849 (Cestoda: Tetraphyllidea), a revision of the genus and comments on *Discobothrium* Beneden, 1870, *Pseudanthobothrium* Baer, 1956 and *Phormobothrium* Alexander, 1963. *Parasitology* 56, 227–284.

Williams, H.H. (1966b) The genus *Acanthobothrium* Beneden, 1849 (Cestoda, Tetraphyllidea). *Nytt Magasin för Zoologi* 17, 1–56.

Williams, H.H. & Bray, R.A. (1984) *Chimaerocestos prudhoei* gen. et sp. nov. representing a new family of tetraphyllideans and the first record of strobilate tapeworms from a holocephalan. *Parasitology* 88, 105–116.

Williams, H.H., Colin, J.A. & Halvorsen, O. (1987) Biology of gyrocotylideans with emphasis on reproduction, population ecology and phylogeny. *Parasitology* 95, 173–207.

Williams, I.C. (1960) The anatomy of the scolex of *Andrepigynotaenia haematopodis* Davies & Rees, 1947, a protogynous cestode from the oyster-catcher, *Haematopus ostralegus occidentalis* Neumann, and a further record of *Progynotaenia odhneri* Nybelin, 1914 (Cestoda: Cyclophyllidea) in the British Isles. *Annals and Magazine of Natural History* Series 13, 3, 353–376.

Wisniewski, R.J. (1971) Studies on the development of *Nematoparataenia southwelli* Fuhrmann, 1934 and *Gastrotaenia paracygni* Czaplinski et Ryzhikov, 1966 (Cestoda: Hymenolepididae) in the intermediate hosts. *Acta Parasitologica Polonica* 19, 49–61.

Witenberg, G. (1932) On the cestode subfamily Dipylidiinae Stiles. *Zeitschrift für Parasitenkunde* 4, 542–584.

Wolffhügel, K. (1899) Beitrag zur Kenntnis der Anatomie einiger Vögelcestoden. *Zoologischer Anzeiger* 22, 217–223.

Wolffhügel, K. (1936) Fimbriariinae (Cestoda). *Zeitschrift für Infektionskrankheiten und Hygiene der Haustiere* 49, 257–291.

Wolffhügel, K. (1938) Nematoparataeniidae. Scolex und Verdauung. *Zeitschrift für Infektionskrankheiten und Hygiene der Haustiere* 53, 9–42.

Woodland, W.N.F. (1925) On some remarkable new *Monticellia*-like and other cestodes from Sudanese siluroids. *Quarterly Journal of Microscopical Science* 69, 703–729.

Woodland, W.N.F. (1927) A revised classification of the tetraphyllidean Cestoda, with descriptions of some Phyllobothriidae from Plymouth. *Proceedings of the Zoological Society of London*, Part 3, 519–548.

Woodland, W.N.F. (1928) On some new avian cestodes from the Sudan. *Parasitology* 20, 305–314, plates XVII, XVIII.

Woodland, W.N.F. (1930) On the genus *Polycephalus* Braun, 1878, (Cestoda). *Proceedings of the Zoological Society of London* Year 1930, (2), 347–354.

Woodland, W.N.F. (1933a) On the anatomy of some fish cestodes described by Diesing from the Amazon. *Quarterly Journal of Microscopical Science* 76, 176–214.

Woodland, W.N.F. (1933b) On a new subfamily of proteocephalid cestodes – the Othinoscolecinae – from the Amazon siluroid fish *Platystomatichthys sturio* (Kner). *Parasitology* 25, 401–500.

Woodland, W.N.F. (1934a) On six new cestodes from Amazon fishes. *Proceedings of the Zoological Society of London*, Part 1, 33–34.

Woodland, W.N.F. (1934b) On the Amphilaphorchidinae, a new subfamily of proteocephalid cestodes, and *Myzophorus admonticellia* gen. n., sp. n., parasitic in *Pirinampus* spp. from the Amazon. *Parasitology* 26, 141–149.

Woodland, W.N.F. (1934c) On some remarkable new cestodes from the Amazon siluroid fish *Brachyplatystoma filamentosum* (Licht.). *Parasitology* 26, 268–277.

Woodland, W.N.F. (1935a) Additional cestodes from the Amazon siluroids pirarará, dorad and sudobim. *Proceedings of the Zoological Society of London*, 851–864.

Woodland, W.N.F. (1935b) Some more remarkable cestodes from Amazon siluroid fish. *Parasitology* 27, 207–225.

Woodland, W.N.F. (1937) Some cestodes from Sierra Leone II. A new caryophyllaeid, *Marsypocephalus*, and *Polyonchobothrium*. *Proceedings of the Zoological Society of London*, 1937, Ser. B, 189–197.

Xylander, W. (1986) Zur Biologie und Ultrastruktur der Gyrocotylida und Amphilinida sowie ihre Stellung im phylogenetischen System der Plathelminthes. Thesis: Göttingen, 304 pp.

Yamaguti, S. (1934) Studies on the helminth fauna of Japan. Part 4. Cestodes of fishes. *Japanese Journal of Zoology* 6, 1–112.

Yamaguti, S. (1935) Studies on the helminth fauna of Japan. Part 6. Cestodes of birds. I. *Japanese Journal of Zoology* 6, 183–232.

Yamaguti, S. (1939) Studies on the helminth fauna of Japan. Part 28. *Nippotaenia chaenogobii*, a new cestode representing a new order from freshwater fishes. *Japanese Journal of Zoology* 8, 287–289.

Yamaguti, S. (1942) Studies on the helminth fauna of Japan. Part 42. Cestodes of mammals. II. Published by the Author, 18 pp.

Yamaguti, S. (1951) Early stages of postembryonic development of *Nippotaenia chaenogobii* Yamaguti, 1939 (Cestoda). *Arbeiten aus der Medizinischen Fakultät Okayama* 7, 335–337.

Yamaguti, S. (1952) Studies on the helminth fauna of Japan. Part 49. Cestodes of fishes, II. *Acta Medicinae Okayama* 8, 1–76.

Yamaguti, S. (1954a) Parasitic worms mainly from Celebes. Part 6. Cestodes of fishes. *Acta Medicinae Okayama* 8, 353–374.

Yamaguti, S. (1954b) Parasitic worms, mainly from Celebes. Part 7. Cestodes of reptiles. *Acta Medicinae Okayama* 8, 375–385.

Yamaguti, S. (1956a) Parasitic worms mainly from Celebes. Part 11. Cestodes of birds. Published by the Author, 41 pp.

Yamaguti, S. (1956b) Studies on the helminth fauna of Japan. Part 50. Cestodes of birds, III. Published by the Author, 23 pp.

Yamaguti, S. (1959) *Systema Helminthum. Vol. II. The Cestodes of Vertebrates*. New York & London: Interscience Publishers, Inc., vii + 860 pp.

Yamaguti, S. (1968) Cestode parasites of Hawaiian fishes. *Pacific Science* 22, 21–36.

Yeh, L.S. (1957) A collection of helminths from the great bustard, *Otis tarda* from Spain, with a description of a new species of *Oxyspirura* (Nematoda). *Proceedings of the Zoological Society of London* 128, 279–286.

Zarnowski, E. (1955) Parasitic worms of forest micromammalians (Rodentia and Insectivora) of the environment of Pulawy (district Lublin). *Acta Parasitologica Polonica* 3, 279–368.

Zdárská, Z. (1976) Morphological and histochemical demonstration of hooks in the rostellar region of *Cysticercus bovis*. *Folia Parasitologica* 23, 131–134.

Zeder, I.G.H. (1803) *Anleitung zur Naturgeschichte der Eingeweidewuermer*. Bamberg, x + 432 pp.

Zerny, H. (1912) Beitrag zur Kenntnis des anatomischen Baues von *Rhynchobothrius tetrabothrius* v. Ben. *Arbeiten aus den Zoologischen Instituten zu Wein* XIX, 297–316.

Zmeev, G. Ya. (1936) [Trematodes and cestodes of fishes of the Amur.] *Parazitologicheskii Sbornik* 6, 405–436. [In Russian.]

Index

Page numbers in bold type are those on which the diagnosis of the taxon appears.

Printed and bound by CPI Group (UK) Ltd, Croydon, CR0 4YY

16/09/2026

14963433-0006